ENDOTHELIUM AND CARDIOVASCULAR DISEASES
Vascular Biology and Clinical Syndromes

ENDOTHELIUM AND CARDIOVASCULAR DISEASES

Vascular Biology and Clinical Syndromes

Edited by

PROTÁSIO L. DA LUZ

PETER LIBBY

ANTONIO C. P. CHAGAS

FRANCISCO R. M. LAURINDO

ELSEVIER

ACADEMIC PRESS
An imprint of Elsevier

Academic Press is an imprint of Elsevier
125 London Wall, London EC2Y 5AS, United Kingdom
525 B Street, Suite 1800, San Diego, CA 92101-4495, United States
50 Hampshire Street, 5th Floor, Cambridge, MA 02139, United States
The Boulevard, Langford Lane, Kidlington, Oxford OX5 1GB, United Kingdom

Notices
Knowledge and best practice in this field are constantly changing. As new research and experience broaden our understanding,
changes in research methods, professional practices, or medical treatment may become necessary.

Practitioners and researchers must always rely on their own experience and knowledge in evaluating and using any
information, methods, compounds, or experiments described herein. In using such information or methods they should be mindful
of their own safety and the safety of others, including parties for whom they have a professional responsibility.

To the fullest extent of the law, neither the Publisher nor the authors, contributors, or editors, assume any liability for any
injury and/or damage to persons or property as a matter of products liability, negligence or otherwise, or from any use or
operation of any methods, products, instructions, or ideas contained in the material herein.

Library of Congress Cataloging-in-Publication Data
A catalog record for this book is available from the Library of Congress

British Library Cataloguing-in-Publication Data
A catalogue record for this book is available from the British Library

ISBN 978-0-12-812348-5

For information on all Academic Press publications visit our
website at https://www.elsevier.com/books-and-journals

Working together
to grow libraries in
developing countries

www.elsevier.com • www.bookaid.org

Publisher: Mica Haley
Acquisition Editor: Stacy Masucci
Editorial Project Manager: Tracy Tufaga
Production Project Manager: Anusha Sambamoorthy
Cover Designer: Christian Bilbow

Typeset by SPi Global, India

About the Editors

Protásio Lemos da Luz
- Senior Full Professor of Cardiology, Faculty of Medicine, Universidade de São Paulo (FMUSP);
- Senior Researcher of Cardiology, Instituto do Coração (InCor), Faculty of Medicine, Universidade de São Paulo (FMUSP) (current);

Peter Libby
- Mallinckrodt Professor of Medicine, Harvard Medical School. Brigham and Women's Hospital, Harvard Medical School, Boston, MA, United States.

Antonio Carlos Palandri Chagas
- Full Professor, Head of Cardiology, ABC School of Medicine;
- Associate Professor, Faculty of Medicine, Universidade de São Paulo (USP).

Francisco Rafael Martins Laurindo
- Associate Professor of Cardiology, Faculty of Medicine, Universidade de São Paulo (FMUSP);
- Director of Vascular Biology Laboratory, Instituto do Coração (InCor), Hospital das Clínicas, Faculty of Medicine, Universidade de São Paulo.

About the Contributors

Alexander R. Moise   Department of Pharmacology and Toxicology, School of Pharmacy, University of Kansas, Lawrence, KS, United States.

Alexandre Abizaid   Director of Invasive Cardiology Service, Instituto Dante Pazzanese de Cardiologia (IDPC), São Paulo, SP, Brazil. Associate Professor, Universidade de São Paulo (USP), São Paulo, SP, Brazil. Medical Doctor, Escola Paulista de Medicina, Universidade Federal de São Paulo (EPM/Unifesp), São Paulo, SP, Brazil.

Alfredo Inácio Fiorelli   Collaborator Professor, School of Medicine, Universidade de São Paulo (FMUSP). Associate Professor, FMUSP. Director of Cardiorespiratory Perfusion Unit, Instituto do Coração (InCor), FMUSP.

Aline Alexandra Iannoni de Moraes   Cardiologist and Professor of Medicine, Pontifícia Universidade Católica do Paraná (PUC-PR), Brazil.

Allan Robson Kluser Sales   PhD in progress in Cardiopneumology, School of Medicine, Universidade de São Paulo (FMUSP).
Doctor of Cardiovascular Sciences, Universidade Federal Fluminense (UFF) and Collaborating Researcher of Cardiovascular Rehabilitation and Exercise Physiology Unit, Instituto do Coração (InCor), FMUSP.

Amit Nussbacher   Attending Physician at Hospital Israelita Albert Einstein. Doctor of Cardiology, School of Medicine, Universidade de São Paulo (FMUSP).

Ana Cristina Andrade   Assistent Physician at Instituto do Coração (InCor) Clinical Division, Hospital das Clínicas, School of Medicine, Universidade de São Paulo (HC-FMUSP). Doctor of Cardiology, Universidade de São Paulo (USP), São Paulo, SP, Brazil. Collaborating Professor, Department of Cardiopneumology, InCor, HC-FMUSP.

Ana Maria Lottenberg   Nutritionist of the Endocrinology Subject, School of Medicine, Universidade de São Paulo (FMUSP). PhD in Nutrition, School of Pharmaceutical Sciences, Universidade de São Paulo (FCF-USP).

Ana P. Azambuja   National Biosciences Laboratory, Centro Nacional de Pesquisa em Energia e Materiais, Brazil.

Ana Paula Dantas   Researcher Doctor of Institut d'Investigacions Biomèdiques August Pi i Sunyer (IDIBAPS), Barcelona, Spain.

Ângela M.S. Costa   National Biosciences Laboratory, Centro Nacional de Pesquisa em Energia e Materiais, Brazil.

Antonio Augusto Lopes   Associate Professor of Clinical Unit of Pediatric Cardiology and Adult Congenital Cardiopathies, Instituto do Coração (InCor), Hospital das Clínicas, School of Medicine, Universidade de São Paulo (HC-FMUSP).

Antonio Carlos Palandri Chagas   Full Professor, Head of Cardiology, ABC School of Medicine. Associate Professor, Faculty of Medicine, Universidade de São Paulo (USP).

Antonio Casela Filho   MS in Cardiology Medical Clinic, Ribeirão Preto School of Medicine, Universidade de São Paulo (FMRP-USP). PhD and Post-doctor in Cardiology, Instituto do Coração (InCor), Hospital das Clínicas, School of Medicine, Universidade de São Paulo (HC-FMUSP). Fellow of American College of Cardiology.

Breno Bernardes de Souza   Division of Cardiology, Department of Medicine, Brigham and Women's Hospital, Harvard Medical School, Boston, MA, United States.

C. Noel Bairey Merz   Barbra Streisand Women's Heart Center, Cedars Sinai Heart Institute, Cedars-Sinai Medical Center, Los Angeles, CA, United States.

Carlos Eduardo Negrão   Full Professor, Department of Biodynamics of the Human Body Movement, School of Physical Education and Sport, Universidade de São Paulo (USP) with subsidiary link to Department of Cardiopneumology, School of Medicine, USP. PhD by University of Wisconsin-Madison, Madison, WI, United States and Director of the Cardiovascular Rehabilitation and Exercise Physiology Unit, Instituto do Coração (InCor), Hospital das Clínicas, School of Medicine, Universidade de São Paulo (HC-FMUSP).

Carlos J. Rocha Oliveira   PhD in Molecular Biology. Full Professor at Universidade Anhembi Morumbi, School of Health Sciences, São Paulo, SP, Brazil.

Cinthya Echem MS in Pharmacology by Universidade de São Paulo (USP), São Paulo, SP, Brazil. PhD in progress of the Department of Pharmacology, Institute of Biomedical Sciences, USP.

Clara Nobrega MS in progress in Sciences at School of Physical Education and Sport, Universidade de São Paulo (EEFE-USP). Member of the Biochemistry and Exercise Molecular Biology Laboratory, EEFE-USP. Member of Sociedade Brasileira de Fisiologia (SBF).

Cristina Pires Camargo Post-PhD in progress at the School of Medicine, Universidade de São Paulo (FMUSP).

Daniel Chamie Interventional Cardiologist of the Invasive Cardiology Service, Instituto Dante Pazzanese de Cardiologia (IDPC), São Paulo, SP, Brazil. Director of the Optical Coherence Tomography Lab, Cardiovascular Research Center, São Paulo, SP, Brazil.

Daniel Umpierre PhD in Cardiovascular Sciences, Universidade Federal do Rio Grande do Sul (UFRGS). Professor of the Graduate Program in Cardiology and Cardiovascular Sciences, School of Medicine, UFRGS.

Denise C. Fernandes Vascular Biology Laboratory, Instituto do Coração (InCor), Hospital das Clínicas, School of Medicine, Universidade de São Paulo (HC-FMUSP).

Desidério Favarato PhD in Medicine at Universidade de São Paulo (USP), São Paulo, SP, Brazil. Assistant Physician, Atherosclerosis Unit, InCor, USP.

Edilamar Menezes de Oliveira Associate Professor, Department of Human Body Movement Biodynamics of the School of Physical Education and Sport, Universidade de São Paulo (EEFE-USP). Post-PhD at Stem Cells Lab of Keck Graduate Institute of Applied Life Science, Claremont, CA, United States. PhD in Biochemistry and Molecular Biology at Universidade Federal do Rio Grande do Sul (UFRGS). Instituto do Coração (InCor), Hospital das Clínicas, School of Medicine, Universidade de São Paulo (HC-FMUSP). Coordinator, Exercise Biochemistry and Molecular Biology Lab, EEFE-USP.

Elaine Guadelupe Rodrigues Biochemical Pharmacist, School of Pharmaceutical Sciences of Universidade de São Paulo (USP), São Paulo, SP, Brazil. MS and PhD in Sciences at Microbiology and Immunology Program of Universidade de São Paulo (Unifesp), São Paulo, SP, Brazil. New York University, Department of Parasitology, New York, NY, United States. Assistant Professor, Department of Microbiology, Immunology and Parasitology, Universidade Federal de São Paulo (EPM/Unifesp), São Paulo, SP, Brazil. Head of Cancer Immunobiology Lab, Department of Microbiology, Immunology and Parasitology of EPM/Unifesp, São Paulo, SP, Brazil.

Elaine Marques Hojaij Psychologist, Instituto do Coração (InCor), Hospital das Clínicas, School of Medicine, Universidade de São Paulo (HC-FMUSP).

Eliana Hiromi Akamine PhD, Professor of the Department of Pharmacology, Institute of Biomedical Sciences, Universidade de São Paulo (USP), São Paulo, SP, Brazil.

Eliete Bouskela Laboratory of Clinical and Experimental Researches in Vascular Biology (BioVasc), Biomedical Center, Universidade do Estado do Rio Janeiro (UERJ), Brazil.

Elisa Alberton Haas Post-graduate student in Cardiology, InCor, School of Medicine, Universidade de São Paulo (FMUSP).

Emilio Ros Lipid Clinic, Endocrinology and Nutrition Service, Institut d'Investigacions Biomèdiques August Pi Sunyer (IDIBAPS), Hospital Clinic, Barcelona and CIBEROBN, ISCIII, Spain.

Erich Vinicius de Paula Hematologist, PhD in Medical Physiopathology, Universidade Estadual de Campinas (Unicamp), Brazil. Professor of Hematology and Hemotherapy of Unicamp School of Medical Sciences.

Erik Svensjö Instituto de Biofísica Carlos Chagas Filho, Universidade Federal do Rio de Janeiro (UFRJ).

Erika Jones Barbra Streisand Women's Heart Center, Cedars Sinai Heart Institute, Cedars-Sinai Medical Center, Los Angeles, CA, United States.

Fabiano Vanderlinde Physician Collaborator of Behavioral Changes in Elderly of the Geriatrics Service, Hospital das Clínicas, School of Medicine, Universidade de São Paulo (HC-FMUSP).

Fabiola Zakia Mónica BPharm, PhD, Universidade Estadual de Campinas (Unicamp), Brazil.

Fernanda Fatureto Borges Universidade Federal da Grande Dourados (UFGD). PhD in progress in Sciences at School of Medicine, Universidade de São Paulo (FMUSP).

Fernanda Marciano Consolim-Colombo Associate Professor of Cardiology at School of Medicine, Universidade de São Paulo (FMUSP). Assistant Physician of the Hypertension Unit of Instituto do Coração (InCor), Hospital das Clínicas, FMUSP. Coordinator of Medicine Graduate Program at Universidade Nove de Julho (Uninove).

Fernanda Roberta Roque MS and PhD in Sciences at the School of Physical Education and Sport, Universidade de São Paulo (EEFE-USP). PhD in progress at Exercise Biochemistry and Molecular Biology Lab, EEFE-USP. Member of Sociedade Brasileira de Fisiologia (SBF) and Sociedade de Cardiologia do Estado de São Paulo.

Fernando Bacal   Associate Professor of Cardiology at the School of Medicine, Universidade de São Paulo (FMUSP). Director, Clinical Cardiac Transplantation Unit, Instituto do Coração (InCor).

Francisco Antonio Helfenstein Fonseca   Associate Professor, Head of the Department of Lipids, Atherosclerosis and Vascular Biology of the Subject Cardiology, São Paulo School of Medicine, Universidade Federal de São Paulo (EPM/Unifesp), São Paulo, SP, Brazil.

Francisco R.M. Laurindo   Associate Professor of Cardiology. Director of Vascular Biology Laboratory, Instituto do Coração (InCor), Hospital das Clínicas, Faculty of Medicine, Universidade de São Paulo.

Geraldo Lorenzi-Filho   Director, Sleep Laboratory, Instituto do Coração (InCor). Associate Professor of the School of Medicine, Universidade de São Paulo (FMUSP). President of Associação Brasileira do Sono (ABS).

Gilbert Alexandre Sigal   Endocrinologist. MS and PhD in Endocrinology at Hospital das Clínicas, School of Medicine, Universidade de São Paulo (HC-FMUSP). Assistant Physician, Laboratory of Metabolism and Lipids, Instituto do Coração (InCor), HC-FMUSP.

Gilberto De Nucci   MD, BPharm, PhD, Universidade Estadual de Campinas (Unicamp), Brazil.

Guillermo Garcia-Cardeña   Center for Excellence in Vascular Biology, Division of Cardiovascular Medicine, Department of Medicine and Division of Vascular Research, Department of Pathology, Brigham and Women's Hospital, Harvard Medical School, Boston, MA, United States.

Gustavo Henrique Oliveira de Paula   Department of Pharmacology, Ribeirão Preto School of Medicine, Universidade de São Paulo (FMRP-USP).

Haniel Alves Araújo   Division of Cardiology, Brigham and Women's Hospital, Harvard, Medical School, Boston, MA, United States.

Heitor Moreno Júnior   Cardiologist, Full Professor of Medicine. Universidade Estadual de Campinas (Unicamp), Brazil. Associate Professor, School of Medical Sciences.

Helena Coutinho Franco de Oliveira   Full Professor, Physiology, Department of Structural and Functional Biology, Institute of Biology, Universidade Estadual de Campinas (Unicamp), Brazil.

Hozana A. Castillo   National Laboratory of Biosciences, Centro Nacional de Pesquisa em Energia e Materiais, Brazil.

Hugo P. Monteiro   Full Professor, Department of Biochemistry and Director of the Center for Cellular and Molecular Therapy, São Paulo School of Medicine, Universidade Federal de São Paulo (EPM/Unifesp), São Paulo, SP, Brazil.

Iáscara Wozniak de Campos   Heart Transplant Clinic Unit, Instituto do Coração (InCor), Hospital das Clínicas, School of Medicine, Universidade de São Paulo (HC-FMUSP).

Ivan Aprahamian   Coordinator, Ambulatory of Behavioral Changes in Elderly, Geriatrics Service, Hospital das Clínicas, FMUSP. Associate Professor and Head of Department of Medical Clinic, Faculdade de Medicina de Jundiaí (FMJ). Fellow, American College of Physicians.

J. Eduardo Sousa   Director, Center for Intervention in Structural Heart Diseases, Instituto Dante Pazzanese de Cardiologia (IDPC), São Paulo, SP, Brazil. Head of the Hemodynamics Laboratory, Hospital do Coração (HCor), Associação do Sanatório Sírio, São Paulo, SP, Brazil. Associate Professor of Cardiology at São Paulo School of Medicine, Universidade Federal de São Paulo (EPM/Unifesp), São Paulo, SP, Brazil.

J. Ribamar Costa, Jr.   Head of the Medical Section of Coronary Intervention, Instituto Dante Pazzanese de Cardiologia (IDPC), São Paulo, SP, Brazil. Interventional Cardiologist, Hospital do Coração (HCor), Associação do Sanatório Sírio, São Paulo, SP, Brazil.

Janet Wei   Barbra Streisand Women's Heart Center, Cedars Sinai Heart Institute, Cedars-Sinai Medical Center, Los Angeles, CA, United States.

Jenna Maughan   Barbra Streisand Women's Heart Center, Cedars Sinai Heart Institute, Cedars-Sinai Medical Center, Los Angeles, CA, United States. University of California, Berkeley, CA, United States.

Jordi Merino   Vascular Medicine and Metabolism Unit, Research Unit on Lipids and Atherosclerosis, Hospital Sant Joan, Universitat Rovira i Virgili, IISPV, Reus and CIBERDEM, Instituto de Salud Carlos III (ISCIII), Spain.

José Eduardo Krieger   Professor of Genetics and Molecular Cardiology, School of Medicine, Universide de São Paulo (USP). Director, Laboratory of Genetics and Molecular Cardiology, Instituto do Coração (InCor), Hospital das Clínicas, School of Medicine, Universidade de São Paulo (HC-FMUSP).

José Eduardo Tanus dos Santos   Department of Pharmacology, Ribeirão Preto School of Medicine, Universidade de São Paulo (FMRP-USP).

Jose Jayme Galvão de Lima   Associate Professor, Instituto do Coração (InCor), Hospital das Clínicas, School of Medicine, Universidade de São Paulo (FMUSP).

José Leudo Xavier Júnior   Heart Failure Specialist/ Clinical Unit of Cardiac Transplantation, Instituto do Coração (InCor), School of Medicine, Universidade de São Paulo (FMUSP).

José Rocha Faria Neto   Full Professor of Cardiology, Pontifícia Universidade Católica do Paraná (PUC-PR), Brazil. Director, Post-graduate program in Cardiology.

José Xavier-Neto   National Laboratory of Biosciences, Centro Nacional de Pesquisa em Energia e Materiais, Brazil.

Joyce M. Annichino-Bizzacchi   Full Professor of Hematology, Universidade Estadual de Campinas (Unicamp), Brazil.

Juan Carlos Yugar-Toledo   Full Professor of Hematology, Universidade Estadual de Campinas (Unicamp), Brazil. Cardiologist Doctor. PhD Professor of Pharmacology, School of Medical Sciences, Unicamp, Brazil. Graduate Program Professor of Health Sciences, São José do Rio Preto School of Medicine (Famerp).

Juliana Carvalho Tavares   Department of Physiology and Biophysics, Institute of Biological Sciences, Universidade Federal de Minas Gerais (UFMG).

Julio Flavio Marchini   MD, PhD. Hemodynamics and Interventional Cardiology Service, Instituto do Coração (InCor), Hospital das Clínicas, School of Medicine, Universidade de São Paulo (HC-FMUSP).

Keyla Yukari Katayama   MS in progress, Graduate Program, Universidade Nove de Julho (Uninove). BS in Physical Education at the Universidade Estadual do Centro-oeste (Unicentro).

Leonardo Y. Tanaka   Laboratory of Vascular Biology, Instituto do Coração (InCor), School of Medicine, Universidade de São Paulo (FMUSP).

Lucas Giglio Colli   PhD in progress, Department of Pharmacology, Institute of Biomedical Sciences, Universidade de São Paulo (USP), São Paulo, SP, Brazil.

Luciano F. Drager   Associate Professor, School of Medicine, Universidade de São Paulo (FMUSP). Assistant Physician of Hypertension Unit, Instituto do Coração, Hospital das Clínicas, School of Medicine, Universidade de São Paulo.

Luís Henrique Wolff Gowdak   Assistant Physician of the Laboratory of Genetics and Molecular Cardiology and Clinical Unit of Chronic Coronary Artery Disease, Instituto do Coração (InCor), Hospital das Clínicas, School of Medicine, University of São Paulo (HC-FMUSP). PhD in Cardiology at FMUSP. Fellow of European Society of Cardiology.

Luiz Aparecido Bortolotto   Director, Hypertension Unit of Instituto do Coração (InCor), Hospital das Clínicas, School of Medicine, Universidade de São Paulo (HC-FMUSP). Associate Professor, Department of Cardiology, FMUSP. Coordinator of Arterial Hypertension Nucleus, Hospital Alemão Oswaldo Cruz.

Marcel Liberman   Clinical Cardiologist at InCor/HC-FMUSP. Cardiologist, CTI-A and Researcher of IIEP, Hospital Albert Einstein, São Paulo.

Marcelo Nicolás Muscará   Department of Pharmacology, Institute of Biomedical Sciences, Universidade de São Paulo (USP), São Paulo, SP, Brazil.

Marcelo Nishiyama   Post-graduate student in Cardiology, InCor, School of Medicine, Universidade de São Paulo (FMUSP). Cardiologist, Graphic Methods Sector, Hospital Israelita Albert Einstein.

Marcelo Zugaib   Full Professor of the Department of Obstetrics and Gynecology, School of Medicine, Universidade de São Paulo (FMUSP).

Marco Aurélio Lumertz Saffi   Center for Excellence in Vascular Biology, Division of Cardiovascular Medicine, Department of Medicine and Division of Vascular Research, Department of Pathology, Brigham and Women's Hospital, Harvard Medical School, Boston, MA, United States.

Maria Cristina O. Izar   Associate Professor of Cardiology, Department of Medicine of São Paulo School of Medicine, Universidade Federal de São Paulo (EPM/Unifesp), São Paulo, SP, Brazil. Coordinator, Laboratory of Molecular Biology of the Subject Cardiology, EPM/Unifesp, São Paulo, SP, Brazil.

Maria Helena Catelli de Carvalho   Full Professor, Department of Pharmacology, Institute of Biomedical Sciences, Universidade de São Paulo (USP), São Paulo, SP, Brazil.

Maria Janieire de Nazaré Nunes Alves   Assistant Physician, Cardiovascular Rehabilitation and Exercise Physiology Unit, Instituto do Coração (InCor), Hospital das Clínicas, School of Medicine, Universidade de São Paulo (HC-FMUSP). Collaborating Professor, Department of Cardiology, FMUSP. PhD in Sciences from FMUSP.

Maria Silvia Ferrari Lavrador   Nutritionist from Universidade Federal de Alfenas (Unifal), Minas Gerais, MG. MS PhD in progress from the School of Medicine, Universidade de São Paulo (FMUSP) at the Laboratory of Lipids (LIM-10).

Maria Theresa O. M. Albuquerque   PhD in progress in Science, Department of Translational Medicine, Universidade Federal de São Paulo (Unifesp). MS in Science, Department of Medical Clinic, Unifesp. Assistant Professor of Faculdade Santa Marcelina, São Paulo.

Mariana Meira Clavé   Biomedical from Universidade Luterana do Brasil (ULBRA), Canos, Rio Grande do Sul. PhD in progress in Cardiology, Graduate Program in Cardiology, Instituto do Coração (InCor), Hospital das Clínicas, School of Medicine, Universidade de São Paulo (HC-FMUSP).

Marina Beltrami Moreira Center for Excellence in Vascular Biology, Division of Cardiovascular Medicine, Department of Medicine and Division of Vascular Research, Department of Pathology, Brigham and Women's Hospital, Harvard Medical School, Boston, MA, United States.

Marina Maria Biella Assistant Physician, Geriatrics Service of Hospital das Clínicas, School of Medicine, Universidade de São Paulo (HC-FMUSP).

Mario J. A. Saad Full Professor, Department of Clinical Medicine, School of Medicine, Universidade Estadual de Campinas (FCM-Unicamp), Brazil.

Marli F. Curcio PhD in Biochemistry, Department of Biochemistry and Molecular Biology, São Paulo School of Medicine, Universidade Federal de São Paulo (EPM/Unifesp), São Paulo, SP, Brazil. Associate Researcher (PhD) from the Department of Infectious and Parasitic Diseases (DIPA), EPM/Unifesp, São Paulo, SP, Brazil.

Mauricio Wajngarten Associate Professor of Cardiology, School of Medicine, Universidade de São Paulo (FMUSP).

Mayra Luciana Gagliani Psychologist, Instituto do Coração (InCor), Hospital das Clínicas, School of Medicine, Universidade de São Paulo (HC-FMUSP).

Michael Andrades PhD in Biological Sciences, Universidade Federal do Rio Grande do Sul (UFRGS). Professor, Graduation Program in Cardiology and Cardiovascular Sciences from the School of Medicine, UFRGS.

Milessa Silva Afonso Nutritionist, Universidade Estadual Paulista "Júlio de Mesquita Filho" (UNESP). MS in Food Science from the School of Pharmaceutical Sciences, USP (FCF-USP). PhD in Science from the Program of Endocrinology, School of Medicine, Universidade de São Paulo (FMUSP).

Murilo Carvalho Laboratory of Ichthyology, Department of Zoology, Institute of Biosciences, Universidade de São Paulo. National Laboratory of Biosciences, Centro Nacional de Pesquisa em Energia e Materiais, Brazil.

Nadine Clausell Full Professor, School of Medicine, Universidade Federal do Rio Grande do Sul (UFRGS).

Noedir Antonio Groppo Stolf Senior Professor of Cardiovascular Surgery, School of Medicine, Universidade de São Paulo (FMUSP).

Otavio Berwanger Cardiologist. Director of the Research Institute, Hospital do Coração (HCor), Associação do Sanatório Sírio, São Paulo, SP, Brazil. Professor Post-Graduate Course of Cardiology, Instituto do Coração (InCor), Hospital das Clínicas, School of Medicine, Universidade de São Paulo (FMUSP).

Paulo Ferreira Leite PhD in Sciences, School of Medicine, Universidade de São Paulo (FMUSP). Assistant Phisician of Chronic Coronary Diseases Unit, Instituto do Coração (InCor), FMUSP.

Paulo Magno Martins Dourado PhD in Cardiology at Universidade de São Paulo (USP), São Paulo, SP, Brazil. Fellow of American College of Cardiology, American Heart Association, American Society of Echocardiography and European Society of Cardiology. Researcher Assistant, Laboratory of Vascular Biology, Instituto do Coração (InCor), Hospital das Clínicas, School of Medicine, Universidade de São Paulo.

Paulo Roberto B. Evora PhD in General Surgery, Ribeirão Preto School of Medicine, Universidade de São Paulo (FMRP/USP). Full Professor, Department of Surgery and Anatomy, FMRP/USP.

Pedro A. Lemos MD, PhD, Hemodynamic and Interventional Cardiology Service, Instituto do Coração (InCor), Hospital das Clínicas, School of Medicine, Universidade de São Paulo (HC-FMUSP). Associate Professor of Cardiology.

Peter Libby Mallinckrodt Professor of Medicine, Harvard Medical School. Brigham and Women's Hospital, Harvard Medical School, Boston, MA, United States.

Prediman K. Shah MD, MACC at the Division of Cardiology and Oppenheimer Atherosclerosis Research Center, Cedars Sinai Heart Institute, and UCLA School of Medicine, Los Angeles, CA, United States.

Protásio Lemos da Luz Senior Full Professor of Cardiology. Senior Researcher of Cardiology, Instituto do Coração (InCor), Faculty of Medicine, Universidade de São Paulo (FMUSP).

Puja Mehta Barbra Streisand Women's Heart Center, Cedars Sinai Heart Institute, Cedars-Sinai Medical Center, Los Angeles, CA, United States.

Raul Cavalcante Maranhão Full Professor of Clinical Biochemistry, School of Pharmaceutical Sciences, Universidade de São Paulo. Director of the Laboratory of Metabolism and Lipids, Instituto do Coração (InCor), Hospital das Clínicas, School of Medicine, Universidade de São Paulo (FMUSP).

Riccardo Lacchini Department of Psychiatric Nursing and Human Sciences, Ribeirão Preto School of Nursing, Universidade de São Paulo (EERP-USP).

Richard Kones Cardiometabolic Research Institute, Houston, TX, United States.

Rita C. Tostes Full Professor, Department of Pharmacology, FMRP/USP.

Roberta Eller Borges   PhD in Cell and Molecular Biology at Universidade Federal de São Paulo (Unifesp).

Roberta Marcondes Machado   PhD in Sciences, Endocrinology Program, School of Medicine, Universidade de São Paulo (FMUSP). Specialist in Nutrition in Non-Communicable Chronic Diseases, Institute of Education and Research, Hospital Israelita Albert Einstein (IEP-HIAE). MS in Nutrition at Universidade Anhembi Morumbi.

Roberto Rocha C. V. Giraldez   Associate Professor, Universidade de São Paulo (USP), São Paulo, SP, Brazil. Professor, School of Medicine, Universidade de São Paulo (FMUSP). Assistant Physician, Acute Coronary Unit of Instituto do Coração (InCor), Hospital das Clínicas, School of Medicine, Universidade de São Paulo (HC-FMUSP).

Robson Augusto Souza dos Santos   PhD in Physiology at Universidade de São Paulo (USP), São Paulo, SP, Brazil. Full Professor of the Department of Physiology and Biophysics, Universidade Federal de Minas Gerais (UFMG).

Rodrigo Modolo   PhD Professor of Pharmacology, School of Medical Sciences, Universidade Estadual de Campinas (Unicamp), Brazil. Interventional Cardiologist, Laboratory of Cardiac Catheterization, Hospital das Clínicas, Unicamp, Brazil.

Rolf Gemperli   Full Professor, Plastic Surgery, Department of Surgery, Universidade de São Paulo (USP), São Paulo, SP, Brazil.

Rossana Pulcineli Vieira Francisco   Associate Professor, Professor of the School of Medicine, Universidade de São Paulo (FMUSP).

Salvador Moncada   Director of Cancer Sciences, University of Manchester, United Kingdom.

Santiago A. Tobar   PhD in Biosciences at Universidade do Estado do Rio de Janeiro (UERJ). Post-PhD in progress from the Graduate Program in Cardiology and Cardiovascular Sciences.

Soubhi Kahhale   Associate Professor, School of Medicine, Universidade de São Paulo (FMUSP).

Stefany B. A. Cau   PhD and Post-PhD in Pharmacology at Ribeirão Preto School of Medicine, Universidade de São Paulo (FMRP/USP). Assistant Professor, Department of Pharmacology, Instituto de Ciências Biológicas (ICB), Universidade Federal de Minas Gerais (UFMG).

Thaís L. S. Araujo   Post-PhD in Laboratory of Vascular Biology, Instituto do Coração (InCor), Hospital das Clínicas, School of Medicine, Universidade de São Paulo.

Tiago Fernandes   MS and PhD in Sciences at the School of Physical Education and Sports, Universidade de São Paulo (EEFE-USP). Post-PhD in progress at Exercise Biochemistry and Molecular Biology Laboratory, EEFE-USP. Member of Sociedade Brasileira de Fisiologia (SBF) and Sociedade de Cardiologia do Estado de São Paulo (SOCESP).

Tiago Januário da Costa   MS in Pharmacology from Universidade de São Paulo (USP), São Paulo, SP, Brazil. PhD in progress from the Department of Pharmacology, Institute of Biomedical Sciences, USP.

Valéria Costa-Hong   PhD in Sciences, School of Medicine, Universidade de São Paulo (FMUSP). Researcher of the Hypertension Unit, Instituto do Coração (InCor), Hospital das Clínicas, School of Medicine, Universidade de São Paulo (HC-FMUSP).

Vinicius Esteves   MD. Hemodynamics and Interventional Cardiology Service, Instituto do Coração (InCor), Hospital das Clínicas, School of Medicine, Universidade de São Paulo (HC-FMUSP).

Viviane Zorzanelli Rocha Giraldez   Physician of the Dyslipidemia and Prevention of Atherosclerosis Unit, Instituto do Coração (InCor), Hospital das Clínicas, School of Medicine, Universidade de São Paulo (HC-FMUSP). Doctor in Sciences from FMUSP. Post-PhD Fellow at Brigham and Women's Hospital, Harvard Medical School, Boston, MA, United States.

Wagner Luiz Batista   Assistant Professor of Microbiology, Department of Pharmaceutical Sciences, Universidade Federal de São Paulo (Unifesp/Campus Diadema). PhD in Sciences of the Subject Microbiology and Immunology from São Paulo School of Medicine, Universidade Federal de São Paulo (EPM/Unifesp), São Paulo, SP, Brazil.

Walkyria Oliveira Sampaio   PhD in Physiology, Universidade Federal de Minas Gerais (UFMG). Assistant Professor at Universidade de Itaúna (UIT).

Dedication

To our wives, Rosália, Beryl Benacerraf, Ieda and Jusete, and our children, Salvador and Raphael da Luz, Oliver and Brigitte Libby, Rafael and Lucas Laurindo, João Paulo, Luis Fernando, and Laura Beatriz Chagas.

To Professors Luiz V. Decourt, Fulvio Pileggi, Eduardo Moacyr Krieger and Eugene Braunwald, our mentors and examples of humanism, competence and scientific spirit.

Acknowledgments

We are deeply grateful to the national and international collaborators whose contributions allowed the composition of this work.

We thank Instituto do Coração (InCor) of the Faculty of Medicine, Universidade de São Paulo (FMUSP) and Fundação Zerbini for the work environment and research incentive that characterize our academic environment.

We thank Banco Bradesco for supporting the research work conducted in our group.

The work of our group was financed by FAPESP (Fundação de Amparo à Pesquisa do Estado de São Paulo), CNPq (Conselho Nacional de Desenvolvimento Científico e Tecnológico), and FINEP (Financiadora de Estudos e Projetos).

The collaborations of many post-PhD, undergraduate and graduate students were essential to InCor's research projects reported herein.

We also thank the excellent technical and administrative support of numerous individuals who contributed to our work.

Our special thanks to Dr. Michelle Pereira—Research Coordinator—for her competent organizational work.

Preface

The discovery that the endothelium is a master regulator of vascular function and structure was a profound scientific revolution, among the most transformative ones ever to occur in biology and medicine. Such a major role of the monolayer of cells interfacing the blood vessel wall and the circulation was simultaneously perplexing and challenging, while at the same time filling important gaps in physiology and pathology. The identification of nitric oxide as a major endothelium-derived vasorelaxing factor was a landmark achievement in this regard, as it was accompanied by a steady flow of novel and relevant information connecting endothelium-derived mediators to most vascular processes and to systemic integrated physiological programs. As is often the case with genuine scientific revolutions, such a flow of novel concepts has maintained itself at high levels and continues at a surprisingly high pace. Over the recent years, an enormous amount of information has been systematically integrated into knowledge related to genetic, molecular, biochemical, and physiological pathways. Most importantly, such pathways provided important mechanistic avenues to understand disease processes and paved theway to a number of related pharmacological interventions, while integrating with medical advances related to invasive vascular procedures. With these advances, endothelial function became deeply embedded and essentially reinvented vascular biology and pathophysiology.

Such a massive knowledge evolution created another challenge: the need to get acquainted with this complex array of information. This challenge belongs to research scientists, clinicians, healthcare professionals, and ultimately to everyone willing to become familiar with basic as well as advanced scientific concepts in the area and how they integrate with disease pathophysiology and therapeutics. As a result, there has been a clear need for a text that updates and summarizes both basic and translational knowledge of vascular biology as it relates to endothelial pathobiology. This need was the basic motivation for the present textbook, which follows the previous one published in 2003. The fundamental motivation and overall architecture of this textbook in its second edition remain the same as in the first one. Such similarities involve the overall interplay between basic science and clinical implications, providing a bridge between them. Essentials of vascular structure, molecular biology, pharmacology, and physiology are presented side by side with the roles of endothelial function and dysfunction in a variety of clinical syndromes related to the cardiovascular system. The success of the first book indicated that it fulfilled its aim to provide not only a syllabus for investigators but also a basic text for clinicians wishing to familiarize themselves with the fundamentals of endothelial biology. The overall tone of providing a mechanistic basis underlying medical applications was a clear feature of this book. In parallel with these similarities, this second book comes up with profound differences, which reflect a number of advances. The first obvious difference is the size of the book, which is an evident remark of the amount of knowledge evolution throughout the period separating the two editions. The second difference is the scope of basic science chapters, which has grown substantially in diversity and sophistication, reflecting the significant advances in molecular biology, genetics, cell signaling, systems biology, and stem cell biology over the last decade. The third advance has been the significant growth of the number and scope of chapters discussing the role of endothelial biology in clinical syndromes. The latter reflects the substantial amount of integration of endothelial biology to cardiovascular medicine and also to many interrelated disciplines, which allowed established or emerging clinical applications.

The book is divided, for didactic purposes, into two parts: vascular structure and function, and endothelial dysfunction and clinical syndromes. The first part is essentially related to basic science concepts, discussing vascular structure, signal transduction mechanisms, and molecular pharmacology and physiology. Special attention is given to established and novel paracrine mediators and how they relate not only to vascular function but also to integrated (patho) physiological events. Also, reflecting the advance in genetics and stem cell biology, there are chapters dedicated to these discussions. The second part is dedicated to the discussion of several clinical syndromes that essentially translate into the role of endothelial dysfunction in cardiovascular medicine, while also introducing in a subtle way the foundations of what might be classified as the emerging discipline of vascular medicine. There is an extensive discussion of the role of lipids and how they integrate to vascular (dys)function, in parallel with the roles of diet and exercise as interventions to counteract vascular dysfunction.

Discussions about kidney dysfunction integrate well with the role of endothelium in systemic hypertension, while the discussion on pulmonary hypertension indicates the importance of endothelial dysfunction for organ-specific diseases. This is particularly important with respect to the discussion of endothelial dysfunction in heart failure.

While the two book parts (basic and clinical) might seem to indicate distinct reader targets, this is clearly not the case; their integration is evident as one of the strongest parts of this book. This becomes clear as we approach the chapters dealing with inflammation and immunology of atherosclerosis, in which basic concepts are closely tied to the clinical implications and the avenues of multidisciplinary research opening ahead of us. These concepts interplay with discussions of mechanistic and pathophysiological aspects of pharmacological, surgical, and interventional therapeutic procedures. Particularly interesting and opportune is the discussion of a number of novel state-of-the-art topics such as cognitive dysfunction, gender issues, microcirculation, and even cancer, as they relate to endothelial biology. Chapters addressing wide perspectives in endothelial biology and clinical implications will facilitate the approach of those who wish to start with a general overview of the problem and those who are less familiar with the basic sciences.

The book is headed by an authoritative editorial board, and the list of authors is marked by categorized specialists and a number of prominent international collaborators. Their experience is commensurate with the fact that the book is much more than just an abridged catalogue of information; rather it is an organic body of discussions that provide a dynamic contextualized knowledge in vascular biology and its several related medical areas. Those willing to get acquainted with or to review basic and clinical applied concepts in the area will find a thoughtful and valuable text, which will be accessible to a wide readership spanning basic science investigators, clinical investigators, and also practicing clinicians and healthcare professionals. The leitmotif of this book—the integration between basic science and clinical implications—provides the foundations for a much-needed yet accessible translational science.

I am pleased and honored to be invited to write this preface and warmly welcome all readers approaching this book with the aim of immersing themselves in this source of genuinely translational work.

Valentin Fuster

Director, Mount Sinai Heart Center; Physician in Chief, Mount Sinai Hospital, NY, United States

Presentation

References to cells that cover the capillaries were already being made by Rudolf Virchow in 1860. In 1865, the Swiss anatomist Wilhelm His introduced the term "endothelium." He thought that the primary function of the endothelium was to be a barrier, and added, "I have no reason to attribute any secretory function to the endothelium." Evidently, the scenario changed, especially with the advent of new techniques such as electron microscopy and the culture of tissues. But it was the discovery of the endothelium's vasculature modulating function—and its specific agent, nitric oxide—that won Robert Furchgott, Louis Ignarro, and Ferid Murad the Nobel Prize in Medicine in 1988. Salvador Moncada has also made essential contributions in this saga.

A few years ago, we published the book *Endotélio e Doenças Cardiovasculares* (Endothelium and Cardiovascular Diseases). Since then, several discoveries have occurred, in both the basic and clinical area; these discoveries led naturally to the elaboration of this new book.

In the area of basic research, the molecular mechanisms that permeate the multiple functions of endothelial substances have been better clarified. Thus, the intimacy of the functioning of substances, such as nitric oxide, endothelins, renin-angiotensin system, cytokines, reactive oxygen and nitrogen species, elements of the coagulation and fibrinolysis system, among others, was unveiled. Subunits of enzyme systems, such as NAPH oxidase, have been better studied, as well as other molecules that modulate them. With this, fundamental concepts emerged about the role of inflammation, oxidative stress, intracellular signaling, and cellular subunits such as mitochondria, caveolae, and endoplasmic reticulum. The role of shear stress as an essential motor force for the functioning of the endothelium has been emphasized, just as the redox signaling pathways have become better known. The participation of genes in the endothelium—and its epigenetic control—deserved great attention. It can thus be said that the basic knowledge about the structure and functions of the endothelium has progressed extraordinarily. Fifteen chapters are devoted to these basic topics.

Some consequences stem from this sequence of events: first, the increase of physiological and pathophysiological knowledge; second, the expansion of endothelium participation in clinical syndromes; third, the search for new

therapeutic targets; and finally, there was the impact on medical education.

Pathophysiological knowledge based on cellular and molecular biology has illuminated and expanded our understanding of the mechanisms and clinical courses of various diseases, such as atherosclerosis, diabetes, arterial hypertension, eclampsia, and cognitive disorders of aging, among others. This has a direct impact on diagnostic research and therapeutic behavior.

However, the expansion of endothelium participation in clinical syndromes is notorious. In addition to the situations mentioned, we highlight angina with nonobstructive coronary arteries, sleep disorders, emotional stress, aging and cognitive function, cardiovascular health and lifestyle, and diet and exercise. In view of this, we added 20 new chapters to the new edition, completing a total of 50 chapters.

Due to advances in pathophysiology at the cellular and molecular level, new therapeutic targets are being considered. As an example, the possibility of developing a vaccine for atherosclerosis based on LDL particle epitopes and the possibility of using an artificial LDL that could carry antiatherosclerotic drugs are evaluated.

No less significant is the impact of new knowledge on medical education. It is no longer possible to comprehend the human cardiovascular system without a deep understanding of the endothelium functions. As widely demonstrated in several chapters, changes in endothelial functions constitute basic mechanisms of clinical syndromes, such as acute coronary syndrome and myocardial infarction. By extension, it is clear that medical schools should train workers in laboratories in biology and genetics techniques to attend to these new demands, as well as provide staff with training appropriate to this new reality. It also transposes from the texts presented the new fundamental characteristic of modern science—translational medicine, in which a constant interaction between basic science and clinical application is fundamental.

The participation of foreign authors—Peter Libby, Pk Shah, Noel Bairey Merz, Emilio Ross, Salvador Moncada, and Valentin Fuster—is of special significance; their contributions enrich the book, given its recognized scientific qualification. The participation of national authors is also highly relevant, with all supplying original contributions from their own laboratories.

Therefore, the spirit of the book is to associate an in-depth review of basic concepts about the endothelium with applications of such knowledge to the clinical area. With these characteristics, we hope that this book will serve undergraduate and graduate students, multilevel researchers, and care physicians.

We are deeply grateful to all collaborators; without their participation, this book would not exist.

The Editors

Contents

II

ENDOTHELIAL DYSFUNCTION AND CLINICAL SYNDROMES

SECTION III

METHODS OF INVESTIGATION

SECTION IV

AGING AND COGNITIVE FUNCTIONS

SECTION IX
HEART FAILURE

SECTION X
PERCUTANEOUS CORONARY INTERVENTIONS AND CARDIAC SURGERY

CONTENTS

SECTION XIII
TREATMENT OF ENDOTHELIAL DYSFUNCTION

STRUCTURE AND FUNCTIONS

STRUCTURE AND DEVELOPMENT

1

The Vascular Endothelium

Salvador Moncada

INTRODUCTION

The vascular endothelium, the innermost layer of blood and lymphatic vessels, covers the whole surface of the vascular system and provides the interface between circulating blood or lymph and the vessel wall. It is one cell thick and, following its discovery in the 19th century, it was long believed to be an inert layer that merely facilitated the circulation of fluids around the body. It was Florey who, while describing his early pioneering work on the ultrastructure of the vascular endothelial cell, predicted that important discoveries could be made when pursuing the study of these cells, despite the fact that the endothelium had until then been considered to be just a kind of cellophane wrapping [1]. He was right.

In the last 40 years, research on the vascular endothelium has been very productive and its results have contributed greatly to our understanding of the normal functioning of the vasculature. This has provided important clues for unravelling the mystery of cardiovascular disease, its origin, development, and complications, and its prevention or treatment [2–9]. The endothelium is now considered to be an organ with significant physiological roles rather than an inert surface. Fig. 1.1 shows the number of publications on this subject over the last four decades as well as some key discoveries that have directly contributed to the developing interest in the vascular endothelium. These include the discovery of the vasodilator prostacyclin, that of the endothelium-derived relaxing factor and its identification as nitric oxide (NO). The author has described elsewhere his own contribution to this research [10].

The purpose of this brief review is to revisit some areas in which research is generating significant new information or where translation into clinical medicine is taking place. These include: the significance of the balance between prostacyclin and thromboxane A_2 for vascular homeostasis, the potential use of aspirin and related compounds in the prevention of cancer, endothelial dysfunction and the use of prostacyclin as a drug for the management of pulmonary hypertension.

The unexpected finding of the vasodilator prostacyclin while we were looking for the vasoconstrictor thromboxane A in the vascular wall [10] revealed that two compounds with opposing biological functions, derived from the same precursor (arachidonic acid), are synthesized by cyclooxygenase enzymes in the platelets and the vascular wall. This led us to the hypothesis that a balance exists between the generation of these two compounds (thromboxane A_2 from the platelets and prostacyclin from the vessel wall), and that this is not only important for the understanding of the homeostasis of platelet-vessel wall interactions, but also for the understanding of disease. A closely related question concerns the net effect achieved in the vasculature following treatment with aspirin and aspirin-like drugs, which have the ability to inhibit the synthesis of both prostacyclin and thromboxane A_2. This has proven to be an enduring question, the answer to which has only been becoming clear in the last few years [2].

The unique action of aspirin was unravelled in the late 1970s when it was demonstrated that the platelet cyclooxygenase, unlike the vessel wall enzyme, is exquisitely sensitive to aspirin. The acetylation by aspirin of a serine residue at the active site of the enzyme was shown to be irreversible and lasts for the duration of the life of the platelets [11–13], which are unable to synthesize new proteins. This, together with the demonstration that a small dose of aspirin is more effective than a large dose in increasing cutaneous bleeding time in humans [14], led to the understanding of the now well-recognized protective effect of low doses of aspirin against vascular disease. This protective effect has been demonstrated in a large number of clinical trials in different cardiovascular conditions [15,16]. The work on aspirin lent support to the hypothesis of the significance of the balance between prostacyclin and thromboxane A, since a low dose of aspirin achieves a shift in the balance in favor of prostacyclin by selectively inhibiting generation of thromboxane A_2.

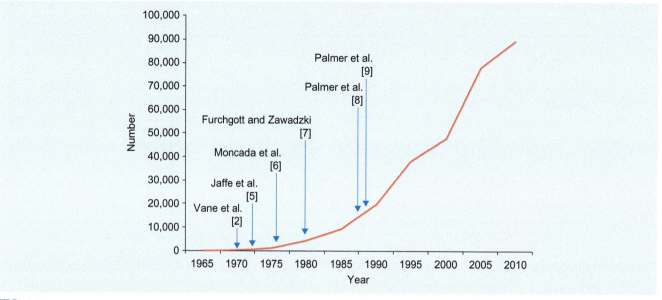

FIG. 1.1   Some key publications in the field of vascular endothelium research and number of publications in the field overtime [2–9].

Further support for the hypothesis came from an unexpected source. In the 1990s, it was discovered that cyclooxygenase exists in two forms: one constitutive (called COX1) which generates prostaglandins for physiological functions, and a second, inducible form (called COX2), which is expressed during pathological conditions and generates prostaglandins involved in inflammation [17]. Each enzyme is encoded by a different gene and their molecular structure is sufficiently different to warrant the pursuit of selective inhibitors of the COX2 enzyme. It was believed that these types of compounds would possess anti-inflammatory activity without the side effects (particularly gastric side effects) that bedevil the classical non-steroidal anti-inflammatory drugs (NSAIDs). In the event, such compounds were synthesized and the objective of achieving similar anti-inflammatory activity to the traditional aspirin-like drugs with reduced gastric side effects was achieved [18,19]. However, during the development of these compounds, a potential problem was identified [20–22], which was later confirmed in patients, namely that COX2 inhibitors increase the risk of cardiovascular events [18,19]. Studies indicated that this serious side effect was due to inhibition of the generation of prostacyclin in the vasculature, leading to an increase in blood pressure and thus to a pro-thrombotic state. Over the last few years, animal experiments [23,24] and clinical studies have produced overwhelming evidence in support of this suggestion, confirming that this is not a side effect related to any specific molecule but is associated with the pharmacological action of the whole class of compounds and is dependent on the strength and duration of inhibition of the synthesis of prostacyclin [19]. Fittingly, a concomitant inhibition of COX1 with low-dose aspirin protects against

this side effect through inhibition of the generation of thromboxane A_2 [23,25].

Two problems remain to be clarified fully. The first is whether the generation of prostacyclin in the vasculature is due to an inducible COX2 resulting from a subliminal inflammatory condition of the vasculature, or is due to a constitutive enzyme. There is a body of evidence in favor of the latter [19]. However, recent evidence indicates that it may be a mixture of the two enzymes [26,27], a fact that would be in agreement with the early observation that the concentration of 6-oxo PGF1α, the stable end product of the metabolism of prostacyclin, is elevated in patients with atherosclerosis [28]. If that is correct, then COX2 overexpression would be part of an inflammatory condition and thus a defensive mechanism.

The second problem relates to the question of whether classical NSAIDs also carry the risk of cardiovascular side effects. This remains a highly controversial issue which may be resolved in further clinical trials. However, it is reasonable to assume that the cardiovascular risk of these drugs will be associated with the degree and duration of inhibition of COX2, and that the ratio between COX1 and COX2 will be determinant in their relative tendency to cause this side effect [29,30]. Indeed, the use of diclofenac (which has a ratio of COX1:COX2 inhibition similar to the COX2 inhibitor Celebrex) is associated with increased cardiovascular risk, while the use of Naproxen (which is a more selective COX1 inhibitor) is not [31]. The data on ibuprofen, which is also a more selective COX1 inhibitor, remains controversial [32]. In summary, the concept of the balance between prostacyclin and thromboxane A_2 in the homeostasis of the vascular system has been validated. Its relevance in health and disease is now well understood and is guiding further development of therapies. Recently, however, a genetic variant of the gene responsible for the encoding of COX2 (PTGS2), associated with lower COX2 activity, has been identified in humans. The relationship between this condition and cardiovascular risk has so far proven to be controversial [33–36]. One of the reasons for this may be that this genetic variant, although associated with a decrease in excretion of 6-oxo PGF1α, seems also to be associated with decreased concentrations of thromboxane A_2; this complicates interpretation of the results using the prostacyclin/thromboxane A_2 balance hypothesis.

One of the most exciting discoveries in the use of NSAIDs has been the finding that these compounds prevent the development of different forms of cancer. This effect, which was identified some years ago in large prospective clinical trials [37,38], was later attributed to the inhibition of prostaglandin synthesis, specifically that of prostaglandin E2 (PGE2), generated by a COX2 enzyme induced by inflammation associated with premalignant lesions [39]. This prostaglandin was believed to be responsible, at least in part, for the neo-plastic transformation through its activation of pro-survival pathways [40–43]. This led to the testing of COX2 inhibitors in the chemoprevention of colorectal cancers, in which a protective action was demonstrated. These trials were, however, marred by concerns related to the potential cardiovascular side-effects of these drugs, which hampered their full evaluation [44]. Studies which demonstrated that the enzyme converting PGH2 to PGE2, the so-called microsomal prostaglandin E synthase-1 (mPGE-1), is overexpressed in inflammation and couples with COX2 to enhance PGE2 generation. This has led more recently to the suggestion that selective inhibitors of this enzyme may be an important therapeutic target that will result in selective inhibition of the pathological PGE2, allowing PGH2 to be converted into the physiologically active prostaglandin, prostacyclin [45,46]. Overexpression of MPGE-1 has been shown in different forms of cancers and its presence is significantly correlated with a worse prognosis, at least in colorectal cancers [47,48]. Although animal studies in which deletion of this enzyme has been carried out show controversial results in relation to cancer [49,50], the development and early in vitro testing of selective inhibitors of this enzyme is proceeding [51,52], and clinical trials are likely to clarify before long the viability of this hypothesis.

The origin of the inflammatory reaction in premalignant lesions has been linked to platelet activation. Evidence for this originally came from the long-term follow-up clinical trials mentioned above in which the efficacy of aspirin as an antithrombotic agent was investigated [53–55]. It was noticed that ingestion of aspirin, even at the low doses used to protect against arterial thrombosis, reduced the incidence of mortality due to cancer, particularly of those of the gastrointestinal tract.

These results pointed to the platelets as a culprit [55]—a suspicion that has been strengthened by several observations including the fact that aggregating platelets can produce inflammation and induction of COX2 [56,57], and that the doses of aspirin that are protective do not reach plasma concentrations sufficiently high to inhibit COX2 and are therefore likely to be inhibiting the platelet COX1 [52,58]. If these results are correct, they point towards a key role of platelet activation not only in atherosclerosis and thrombosis, where their role is now fully accepted, but also in the process of neoplastic transformation. Both effects take place via a two-step process which involves the activation of COX1 and other pathways in the platelets, followed by the induction of COX2 in a number of cells participating in the development of the atherosclerotic plaque or the tumor.

As far as prevention or antineoplastic therapy is concerned, low-dose aspirin therefore emerges as a particularly attractive option for antithrombotic and antitumor therapy. It is clearly superior to the more selective COX2, inhibitors which possess cardiovascular side effects, and also superior to the classical NSAIDs, none of which shares aspirin's unique selectivity of inhibition of the platelet COX-1 enzyme.

Although the idea of a dysfunctional vascular endothelium was mooted many years ago [59], it has become one of the most studied areas of vascular biology only in the last 20 years. Indeed, early detection of "endothelial dysfunction" is proving to be predictive of cardiovascular disease and may indicate ways of preventing its development. Endothelial dysfunction occurs in a number of conditions including hypertension, diabetes (types 1 and 2), coronary artery disease, and chronic renal failure [60]. It has been equated with a decrease in generation of NO by the vascular endothelium and it is likely that this may indeed be its major pathophysiological cause. However, more recently, other changes have been identified which indicate that, besides a decrease in availability of NO, endothelial dysfunction also comprises an increase in vasoconstrictor, pro-inflammatory and prothrombotic parameters [60].

The decrease in activity of NO has been attributed largely to a decrease in its availability resulting from the interaction with oxygen-derived species, mainly superoxide anion [61,62], which may be generated by a number of enzymes including NADPH oxidase, xanthine oxidase, uncoupling of NO synthase or from the mitochondrial oxidative phosphorylation chain [63–65].

More recently, it has been suggested that increases in the concentration of asymmetric dimethylarginine (ADMA) may be involved in endothelial dysfunction. This compound was discovered some years ago to be an endogenous inhibitor of the NO synthase and shown to be increased in patients with renal insufficiency [66]. Since then, evidence in favor of its role in endothelial dysfunction and in cardiovascular disease has been mounting. Indeed, an increase in plasma concentration of ADMA is associated with hypercholesterolemia [67], and with increased cardiovascular risk factors in patients with renal failure [68]. Furthermore it is predictive of acute coronary events [69], overall mortality of patients with chronic renal failure [70], and mortality in critically ill patients [71]. Two independent pieces of evidence have added support to the suggestion that ADMA plays a role in vascular disease. First, it has been shown that in some forms of vascular pathology the intracellular concentration of ADMA is elevated three- to ninefold over physiological concentrations; these concentrations, unlike physiological concentrations, are sufficient to inhibit NO synthase, indicating that endogenous inhibitors of NO synthesis are critical factors in vascular dysfunction following injury [72]. Secondly, a genetic mutation has been identified in the enzyme dimethylarginine dimethylaminohydrolase (DDAH, the enzyme responsible for the metabolism of ADMA) in some individuals with a susceptibility to preeclampsia [73]. In summary, although a great deal of evidence has been generated supporting the concept of endothelial dysfunction, much work is still required to clarify fully the physiopathological mechanisms involved in this early manifestation of vascular disease. It will be important to establish whether, and to what extent, early intervention has a significant effect on the development of vascular disease.

Although the powerful vasodilator and antiplatelet effect of prostacyclin suggested early on its potential use in clinical conditions associated with thrombosis and vasoconstriction [74], its main clinical use at present is in the management of primary pulmonary hypertension [75,76], where it has been shown to improve symptoms, induce remodeling of the pulmonary vasculature, and reduce mortality. The difficulties related to its intravenous usage as an unstable compound, requiring continuous administration, led to the development of different formulations of prostacyclin or its analogs for intravenous, subcutaneous, and inhaled administration [75,76]. In addition to the use of these compounds, two different approaches have also proven to be useful in the management of primary pulmonary hypertension. These are the use of endothelin receptor antagonists and inhibitors of the enzyme 5-phosphodiesterase to boost the effect of endogenous NO on its receptor, the soluble guanylyl cyclase. These compounds, used alone or in different combinations and schedules, have revolutionized the treatment of this complex and fatal disease to the point that the long-term treatment with orally active compounds is now being investigated. Prostacyclin and nitric oxide receptor agonists are at present the subject of long-term clinical trials [76–80]. Furthermore, the proliferative nature of the disease, at least in part associated with the release of platelet-derived growth factors,

has led to the development and use of different kinase inhibitors [79].

In summary, research on the vascular endothelium and in closely related areas continues to generate a great deal of interest. As this work matures, translational developments into medicine are becoming prominent, and clear clinical benefits are being demonstrated. Almost half a century after Florey, it is still likely that the endothelial cell has many secrets yet to be uncovered and that, when this occurs, further avenues for the prevention and treatment of disease will be identified.

References

[1] Florey L. The endothelial cell. Br Med J 1966;2:487–90.
[2] Vane JR. Inhibition of prostaglandin synthesis as a mechanism of action for aspirin-like drugs. Nat New Biol 1971;231:232–5.
[3] Smith JB, Willis AL. Aspirin selectively inhibits prostaglandin production in human platelets. Nat New Biol 1971;231:235–7.
[4] Ferreira SH, Moncada S, Vane JR. Indomethacin and aspirin abolish prostaglandin release from the spleen. Nat New Biol 1971;231:237–9.
[5] Jaffe EA, Nachman RL, Becker CG, et al. Culture of human endothelial cells derived from umbilical veins. identification by morphologic and immunologic criteria. J Clin Invest 1973;52:2745–56.
[6] Moncada S, Gryglewski R, Bunting S, et al. An enzyme isolated from arteries transforms prostaglandin endoperoxides to an unstable substance that inhibits platelet aggregation. Nature 1976;263:663–5.
[7] Furchgott RF, Zawadzki JV. The obligatory role of endothelial cells in the relaxation of arterial smooth muscle by acetylcholine. Nature 1980;288:373–6.
[8] Palmer RMJ, Ferrige AG, Moncada S. Nitric oxide release accounts for the biological activity of endothelium-derived relaxing factor. Nature 1987;327:524–6.
[9] Palmer RMJ, Ashton DS, Moncada S. Vascular endothelial cells synthesize nitric oxide from L arginine. Nature 1988;333:664–6.
[10] Moncada S. Adventures in vascular biology: a tale of two mediators. Philos Trans R Soc Lond B Biol Sci 2006;361:735–59.
[11] Roth GJ, Majerus PW. The mechanism of the effect of aspirin on human platelets. i. acetylation of a particulate fraction protein. J Clin Invest 1975;56:624–32.
[12] Burch JW, Baenziger NL, Stanford N, et al. Sensitivity of fatty acid cyclooxygenase from human aorta to acetylation by aspirin. Proc Natl Acad Sci U S A 1978;75:5181–4.
[13] Patrono C, Baigent C, Hirsh J, et al. Antiplatelet drugs: American College of Chest Physicians evidence-based clinical practice guidelines (8th edition). Chest 2008;133(6 Suppl):199S–233S.
[14] O'Grady J, Moncada S. Aspirin: a paradoxical effect on bleeding-time. Lancet 1978;2:780.
[15] Oates JA, FitzGerald GA, Branch RA, et al. Clinical implications of prostaglandin and thromboxane a2 formation (1). N Engl J Med 1988;319:689–98.
[16] Baigent C, Blackwell L, Collins R, et al. Aspirin in the primary and secondary prevention of vascular disease: collaborative meta-analysis of individual participant data from randomised trials. Lancet 2009;373:1849–60.
[17] Masferrer JL, Zweifel BS, Seibert K, et al. Selective regulation of cellular cyclooxygenase by dexamethasone and endotoxin in mice. J Clin Invest 1990;86:1375–9.
[18] Hinz B, Renner B, Brune K. Drug insight: cyclo-oxygenase-2 inhibitors—a critical appraisal. Nat Clin Pract Rheumatol 2007;3:552–60.

[19] Grosser T, Yu Y, Fitzgerald GA. Emotion recollected in tranquility: lessons learned from the cOX-2 saga. Annu Rev Med 2010;61:17–33.

[20] Mcadam BF, Catella-Lawson F, Mardini IA, et al. Systemic biosynthesis of prostacyclin by cyclooxygenase (cOX)-2: the human pharmacology of a selective inhibitor of cOX-2. Proc Natl Acad Sci U S A 1999;96:272–7.

[21] Catella-Lawson F, Mcadam B, Morrison BW, et al. Effects of specific inhibition of cyclooxygenase-2 on sodium balance, hemodynamics, and vasoactive eicosanoids. J Pharmacol Exp Ther 1999;289:735–41.

[22] Mn Muscará, Vergnolle N, Lovren F, et al. Selective cyclooxygenase-2 inhibition with celecoxib elevates blood pressure and promotes leukocyte adherence. Br J Pharmacol 2000;129:1423–30.

[23] Cheng Y, Wang M, Yu Y, et al. Cyclooxygenases, microsomal prostaglandin E synthase-1, and cardiovascular function. J Clin Invest 2006;116:1391–9.

[24] Yu Y, Ricciotti E, Grosser T, et al. The translational therapeutics of prostaglandin inhibition in atherothrombosis. J Thromb Haemost 2009;(Suppl 1):222–6.

[25] Farkouh ME, Kirshner H, Harrington RA, et al. Comparison of lumiracoxib with naproxen and ibuprofen in the therapeutic arthritis research and Gastrointestinal Event trial (tarGEt), cardiovascular outcomes: randomised controlled trial. Lancet 2004; 364:675–84.

[26] Bishop-Bailey D, Mitchell JA, Warner TD. COX-2 in cardiovascular disease. Arterioscler Thromb Vasc Biol 2006;26:956–8.

[27] Caughey GE, Celeland LG, Penglis PS, et al. Roles of cyclooxygenase (COX)-1 and COX-2 in prostanoid production by human endothelial cells: selective up-regulation of prostacyclin synthesis by COX-2. J Immunol 2001;167:2831–8.

[28] FitzGerald GA, Smith B, Pedersen AK, et al. Increased prostacyclin biosynthesis in patients with severe atherosclerosis and platelet activation. N Engl J Med 1984;310:1065–8.

[29] White WB. Cardiovascular effects of the cyclooxygenase inhibitors. Hypertension 2007;49:408–18.

[30] Patrono C, Baigent C. Low-dose aspirin, coxibs, and other NSAIDs: a clinical mosaic emerges. Mol Interv 2009;9:31–9.

[31] FitzGerald GA, Patrono C. The coxibs, selective inhibitors of cyclooxygenase-2. N Engl J Med 2001;345:433–42.

[32] Bhala N, Emberson J, Mrhi A, et al. Vascular and upper gastrointestinal effects of non-steroidal anti-inflammatory drugs: meta-analyses of individual participant data from randomised trials. Lancet 2013;382:769–79.

[33] Papafili A, Hill MR, Brull DJ, et al. Common promoter variant in cyclooxygenase-2 represses gene expression: evidence of role in acute-phase inflammatory response. Arterioscler Thromb Vasc Biol 2002;22:1631–6.

[34] Cepollone F, Toniato E, Martinotti S, et al. A polymorphism in the cyclooxygenase 2 gene as an inherited protective factor against myo-cardial infarction and stroke. JAMA 2004;291:2221–8.

[35] Lee CR, North KE, Bray MS, et al. Cyclooxygenase polymorphisms and risk of cardiovascular events: the atherosclerosis risk in communities (ARIC) study. Clin Pharmacol Ther 2008;83:52–60.

[36] Ross S, Eikelboom J, Anand SS, et al. Association of cyclooxygenase-2 genetic variant with cardiovascular disease. Eur Heart J 2014;35:2242–8.

[37] Thun MJ, Namboodiri MM, Heath Jr. CW. Aspirin use and reduced risk of fatal colon cancer. N Engl J Med 1991;325:1593–6.

[38] Thun MJ, Namboodiri MM, Calle EE, et al. Aspirin use and risk of fatal cancer. Cancer Res 1993;53:1322–7.

[39] Sahin IH, Hassan MM, Garrett CR. Impact of non-steroidal anti-inflammatory drugs on gastrointestinal cancers: current state-of-the science. Cancer Lett 2014;345:249–57.

[40] Plescia OJ, Smith AH, Grinwich K. Subversion of immune system by tumor cells and role of prostaglandins. Proc Natl Acad Sci U S A 1975;72:1848–51.

[41] Ben-av P, Crofford LJ, Wilder RL, et al. Induction of vascular endothelial growth factor expression in synovial fibroblasts by prostaglandin E and interleukin-1: a potential mechanism for inflammatory angiogenesis. FEBS Lett 1995;372:83–7.

[42] Sheng H, Shao J, Washington MK, et al. Prostaglandin E2 increases growth and motility of colorectal carcinoma cells. J Biol Chem 2001;276:18075–81.

[43] Wang D, Dubois RN. Eicosanoids and cancer. Nat Rev Cancer 2010;10:181–93.

[44] Rodriguez LAG, Cea-Soriano L, Tacconelli S, et al. Coxibs: pharmacology, toxicity and efficacy in cancer clinical trials. Recent Results Cancer Res 2013;191:67–93.

[45] Samuelsson B, Morgenstern R, Jakobsson PJ. Membrane prostaglandin E synthase-1: a novel therapeutic target. Pharmacol Rev 2007;59:207–24.

[46] Wang M, FitzGerald GA. Cardiovascular biology of microsomal prostaglandin E synthase-1. Trends Cardiovasc Med 2010;20:189–95.

[47] Nakanishi M, Gokhale V, Meuillet EJ, et al. mPGES-1 as a target for cancer suppression: a comprehensive invited review "Phospholipase a2 and lipid mediators" Biochimie 2010;92:660–4.

[48] Chang HH, Meuillet EJ. Identification and development of mPGES-1 inhibitors: where we are at? Future Med Chem 2011; 3:1909–34.

[49] Nakanishi M, Montrose DC, Clarck P, et al. Genetic deletion of mPGES-1 suppresses intestinal tumorigenesis. Cancer Res 2008; 68:3251–9.

[50] Elander N, Ungerbäck J, Olsson H, et al. Genetic deletion of mPGES-1 accelerates intestinal tumorigenesis in aPc(Min/+) mice. Biochem Biophys Res Commun 2008;372:249–53.

[51] Leclerc P, Idborg H, Spahiu L, et al. Characterization of a human and murine mPGES-1 inhibitor and comparison to mPGES-1 genetic deletion in mouse models of inflammation. Prostaglandins Other Lipid Mediat 2013;107:26–34.

[52] GuillemLlobat P, Dovizio M, Alberti S, et al. Platelets, cyclooxygenases, and colon cancer. Semin Oncol 2014;41:385–96.

[53] Patrono C, Patrignani P, Garcia Rodriguez LA. Cyclooxygenase-selective inhibition of prostanoid formation: transducing biochemical selectivity into clinical read-outs. J Clin Invest 2001; 108:7–13.

[54] Rothwell PM, Fowkes FG, Belch JF, et al. Effect of daily aspirin on long-term risk of death due to cancer: analysis of individual patient data from randomised trials. Lancet 2011;377:31–41.

[55] Dovizio M, Tacconelli S, Sostres C, et al. Mechanistic and pharmacological issues of aspirin as an anticancer agent. Pharmaceuticals (Basel) 2012;5:1346–71.

[56] Barry OP, Kazanietz MG, Praticò D, et al. Arachidonic acid in platelet microparticles up-regulates cyclooxygenase-2-dependent prostaglandin formation via a protein kinase c/mitogen-activated protein kinase-dependent pathway. J Biol Chem 1999;274:7545–56.

[57] Lindemann S, Tolley ND, Dixon DA, et al. Activated platelets mediate inflammatory signaling by regulated interleukin 1 beta synthesis. J Cell Biol 2001;154:485–90.

[58] Dovizio M, Bruno A, Tacconelli S, et al. Mode of action of aspirin as a chemopreventive agent. Recent Results Cancer Res 2013;191: 39–65.

[59] Stemerman MB. Vascular injury: platelets and smooth muscle cell response. Philos Trans R Soc Lond B Biol Sci 1981;294:217–24.

[60] Endemann DH, Schiffrin EL, editors. Endothelial dysfunction. 2004/07/31 ed. J Am Soc Nephrol 2004;15:1983–92.

[61] Gryglewski RJ, Palmer RM, Moncada S. Superoxide anion is involved in the breakdown of endothelium-derived vascular relaxing factor. Nature 1986;320:454–6.

[62] Moncada S, Palmer RM, Higgs EA. Nitric oxide: physiology, pathophysiology, and pharmacology. Pharmacol Rev 1991;43:109–42.

[63] Hamilton CA, Brosnan MJ, Al-Benna S, et al. Nad(P)H oxidase inhibition improves endothelial function in rat and human blood vessels. Hypertension 2002;40:755–62.

[64] Landmesser U, Spiekermann S, Dikalov S, et al. Vascular oxidative stress and endothelial dysfunction in patients with chronic heart failure: role of xanthine-oxidase and extracellular superoxide dismutase. Circulation 2002;106:3073–8.

[65] Moncada S. Mitochondria as pharmacological targets. Br J Pharmacol 2010;160:217–9.

[66] Vallance P, Leone A, Calver A, et al. Accumulation of an endogenous inhibitor of nitric oxide synthesis in chronic renal failure. Lancet 1992;339:572–5.

[67] Boger RH, Bode-Böger SM, Szuba A, et al. Asymmetric dimethylarginine (ADMA): a novel risk factor for endothelial dysfunction: its role in hypercholesterolemia. Circulation 1998;98:1842–7.

[68] Zoccali C, Benedetto FA, Maas R, et al. CrEEd investigators. Asymmetric dimethylarginine, c-reactive protein, and carotid intima-media thickness in end-stage renal disease. J Am Soc Nephrol 2002;13:490–6.

[69] Valkonen VP, Päivä H, Salonen JT, et al. Risk of acute coronary events and serum concentration of asymmetrical dimethylarginine. Lancet 2001;358:2127–8.

[70] Zocalli C, Bode-Böger S, Mallamaci F, et al. Plasma concentration of asymmetrical dimethylarginine and mortality in patients with end-stage renal disease: a prospective study. Lancet 2001;358:2113–2117.

[71] Nijveldt RJ, Teerlink T, Van Der Hoven B, et al. Asymmetrical dimethylarginine (ADMA) in critically ill patients: high plasma ADMA concentration is an independent risk factor of ICU mortality. Clin Nutr 2003;22:23–30.

[72] Cardounel AJ, Cui H, Samouilov A, et al. Evidence for the pathophysiological role of endogenous methylarginines in regulation of endothelial nO production and vascularfunction. J Biol Chem 2007;282:879–87.

[73] Akbar F, Heinonen S, Pirskanen M, et al. Haplotypic association of ddaH1 with susceptibility to pre-eclampsia. Mol Hum Reprod 2005;11:73–7.

[74] Moncada S, Vane JR. Prostacyclin in perspective, in: Prostacyclin. New York: Raven Press; 1979 p. 5–16.

[75] Safdar Z. Treatment of pulmonary arterial hypertension: the role of prostacyclin and prostaglandin analogs. Respir Med 2011;105:818–827.

[76] Duarte JD, Hanson RL, Machado RF. Pharma-cologic treatments for pulmonary hypertension: exploring pharmacogenomics. Futur Cardiol 2013;9:335–49.

[77] Stamm JA, Risbano MG, Mathier MA. Overview of current therapeutic approaches for pulmonary hypertension. Pulm Circ 2011;1:138–59.

[78] Galie N, Ghofrani AH. New horizons in pulmonary arterial hypertension therapies. Eur Respir Rev 2013;22:503–14.

[79] Morrell NW, Archer SL, Defelice A, et al. Anticipated classes of new medications and molecular targets for pulmonary arterial hypertension. Pulm Circ 2013;3:226–44.

[80] Seferian A, Simonneau G. Therapies for pulmonary arterial hypertension: where are we today, where do we go tomorrow? Eur Respir Rev 2013;22:217–26.

2

Development of the Coronary System: Perspectives for Cell Therapy From Precursor Differentiation

Alexander R. Moise, Ângela M.S. Costa, Murilo Carvalho, Ana P. Azambuja, José Xavier-Neto, and Hozana A. Castillo

In the initial stages of heart development, all nutrition of myocardial tissue occurs by diffusion, since at this stage the heart is a cardiac tube consisting of only two cell layers: one internal endocardial tube and one external myocardial layer. However, as myocardial thickness increases, diffusion becomes insufficient. Evidence suggests that at this stage cardiac hypoxia behaves as a molecular trigger to initiate development of coronary circulation, that is, a vascular network is formed in the heart and ensures that the myocardium is irrigated appropriately for its function and further development [1]. In this chapter we discuss the cellular processes and signaling pathways involved in the development of coronary circulation, as well as the possible role of coronary circulation precursor cells in cardiac and vascular regeneration. We use a comparative approach, since the studies in different animal models provide data on the molecular bases of coronary and cardiac regeneration processes. These data provide important clues that may lead to the development of new therapeutic strategies to treat coronary artery disease (CAD) and induce myocardium revascularization after a heart attack.

THE EVOLUTIONARY ORIGIN OF CORONARY CIRCULATION

The evolutionary emergence of the coronary vascular system has been the subject of discussion in the literature. Hypotheses, currently less accepted, suggest several origins along the evolution of vertebrates (i.e., homoplasy). Others argue that there was a single origin, shared by all descendants (i.e., homology), although several independent losses have occurred during evolution.

The most basal strains of vertebrates (fish with no jaws, agnatha) have no coronary vascular system. In these animals, the heart has a spongy myocardium (trabecular), which allows venous blood of the lumen to reach the epicardial layer, conveying nutrients (O_2 and CO_2) to the tissue. A coronary system is present in all cartilaginous fish—elasmobranches (sharks and stingrays) [2–4] and chimeras [5]—as well as in more basal bony fishes [6] and tetrapods [5,7], which supports the hypothesis of single origin in jawed vertebrates. The advent of a coronary circulation was associated with the emergence of a compact myocardium, increased body size, and exploration of hypoxic environments. In jawed fish, the ventricle often displays an additional muscular type, compact, with its own circulation around and external to the spongy myocardium. Consequently, a major fraction of the ventricle possesses an blood supply proceeding directly from the gills via hypobranchial artery or dorsal aorta. On the other hand, reptiles, birds and mammals have a complete coronary system, with arterial and venous segments and a dense network of coronary vessels that supply oxygen to the elevated metabolic demands of the thick myocardial wall.

As mentioned above, several independent evolutionary events of loss of coronary vasculature occurred in vertebrate evolution. Most teleost fishes have no coronary circulation [8], and those that do have it, are larger predator fish, or inhabit environments with little oxygen. The same is valid for amphibians, which in this case, display only vestigial vessels on the surface of the outflow tract that do not penetrate the myocardial wall [9].

Although a more detailed discussion regarding the evolutionary origin of the coronary vasculature is out of the scope of this revision, it is important to observe that a comparative perspective allows clearer comprehension of the system, providing insights to normal and pathological developmental processes. Therefore, we shall discuss the establishment of coronary vasculature components in different organisms, in amniotes (birds and mammals) and teleost fish (*Danio rerio*).

ARCHITECTURE OF THE ADULT CORONARY SYSTEM

In the adult human heart, coronary arteries branch out of the aortic root and then first route along the outer layer of the heart (epicardium) and occasionally tunnel through the myocardium to reach the cardiac apex. A capillary network irrigates the heart and the venous blood returns to via the subepicardial coronary veins, which drain into the coronary sinus; though smaller veins, such as the right cardiac vein, drain directly into the right atrium. The mechanism guiding the timing and connection of the coronary plexus to the aorta is complex and poorly understood, despite being associated with congenital coronary anomalies [10].

In addition to congenital anomalies, there is also a great deal of variation in the anatomy, position and branching pattern of the coronary system among individuals [11]. Some of these variants can affect blood supply to the heart, leading to increased risk of heart disease in affected individuals. Therefore, a better understanding of the genetic and environmental factors that give rise to the disease-associated variations in the coronary vasculature is warranted.

Coronary Artery Disease (CAD) is the most common cause of mortality in the developed world, and one for which effective treatments are limited in options and for which recovery is often incomplete. Unmitigated CAD leads to cardiac ischemia, angina and arrhythmia and can eventually result in myocardial infarction, or heart failure. The pathological basis of CAD is often a dysfunctional coronary endothelium as a result of arteriosclerosis, which is most often caused by a buildup of oxidized fats and cholesterol in the coronary vascular wall (atherosclerosis). In addition to preventive therapies and lifestyle modification, more invasive procedures such as, angioplasty, stenting and bypass procedures are required in the management of CAD. Though the causes of CAD are derived from exposure of the endothelium to proatherogenic factors during postnatal life, a better understanding of the development of the coronary vasculature during embryonic life could reveal unexplored approaches in the treatment of CAD by harnessing the regenerative potential of the vascular wall [12].

DEVELOPMENT OF THE CORONARY VASCULATURE

The development of the coronary vasculature is unique in that the coronary plexus develops before being connected to the systemic circulation via the aortic root [13]. This is well exemplified in mouse where the coronary vascular system starts to develop at embryonic day 10.5 (E10.5), the plexus is nearly complete by E14.5, and then it becomes connected to the aorta at E15.5. The architecture of the coronary vessel is, however, a typical vascular wall. The three main cell types of the coronary vasculature are represented in the endothelial cells (EC) lining the lumen, which are supported by mural cells composed of pericytes and smooth muscle cells (SMC).

There appear to be multiple developmental sources of coronary EC and the exact origin of various coronary vascular EC populations is still being disputed. Based on a survey of current literature, the origin of coronary EC forming the primitive coronary plexus comprises three potential sources: the epicardium, the sinus venosus and the endocardium. In the more conventional view, the coronary EC are derived from epicardial endothelial precursors which migrate into the myocardium differentiate and assemble into tubules via the process of vasculogenesis. The more recent models based on data from genetic lineage-tracing studies propose that the coronary plexus is formed via sprouting angiogenesis from either the sinus models or the endocardium. Each model will be discussed briefly in the following sections.

As the coronary plexus matures, extensive remodeling occurs driven mainly by intussusceptive angiogenesis, and mural cells are recruited to stabilize the coronary vascular wall. Molecular mechanisms that guide coronary maturation are still unknown. Recently, it has been shown that chemokine CXCL12, which is expressed in the epicardium, acts in a paracrine way to activate its CXCR4 receptor, expressed in coronary endothelial cells, to influence vascular coronary plexus maturation, with CXCL12 mutants showing excess immature capillary chains and arterial maturation deficiency [14]. To allow for complete vascularization of the myocardium, the vessels of the coronary plexus also undergo anastomosis. Finally, the coronary plexus is connected to the aorta by growing towards the aorta and forming coronary stems and by creating ostia openings in the aortic wall [13, 15]. Further vascular remodeling of the coronary plexus results in adaptations which affect the size of the lumen and the mechanoelastic properties of the wall. These adaptive changes occur in response to the needs to accommodate different rates of blood flow and shear stress in of arterial and venous vasculature [16]. The arterial-venous remodeling allows for considerable plasticity; this plasticity forms the basis of the coronary artery bypass graft surgery by allowing the reprogramming of the grafted vein into an artery.

CORONARY FORMATION FROM EPICARDIUM-DERIVED PRECURSORS PROCEEDING FROM THE EPICARDIUM

The Proepicardium

Proepicardium (PE) is a trasient embryonic structure that emerges as a mesothelial protuberance of the transverse septum from the venous sinus, next to the embryonic liver, at embryonic stage E9.5 in mouse embryos and at embryonic stage E2.5 in bird embryos (Fig. 2.1A) [20]. PE cells display an adhesive interaction with myocardial cells and, through this process, migrate over the embryonic heart while they proliferate, forming the epicardium. There are two different contact mechanisms between PE cells and the myocardium. One is predominant in fish embryos and some mammals, and consists of cell aggregates formed by PE villus structures, which detach from the PE and passively cross pericardial space to reach myocardial surfaces at different points, from which they migrate to the myocardium wall [21,22]. In bird and amphibian embryos, the contact mechanism between PE cells and the myocardium is different. In this mechanism, an extracellular matrix bridge is formed between the PE and the developing heart. This bridge fosters connections of PE villus with the myocardium in a single point, from which the PE cells migrate to cover the cardiac surface (Fig. 2.1B) [21,23]. Some authors state that this extracellular matrix bridge acts not only as physical connection between the PE and the heart, but also as a signaling center that controls the transference of cells from the PE to the heart, since it is rich in proteoglycans, and these molecules may function as storage of growth factors, which are potential paracrine signaling centers [23].

The PE role in coronary development has been primarily shown in the beginning of the 1930s, by experiments in which the contact between the PE and the heart was prevented by introducing a piece of egg shell membrane between the structures. Prevention of direct contact between PE and the heart delayed formation of the epicardium, promoted development of hypoplastic ventricular myocardium, and caused coronary underdevelopment [20]. The first studies using vital dye markers in avian hearts, together with transplants and chicken-quail chimeras showed the PE as in

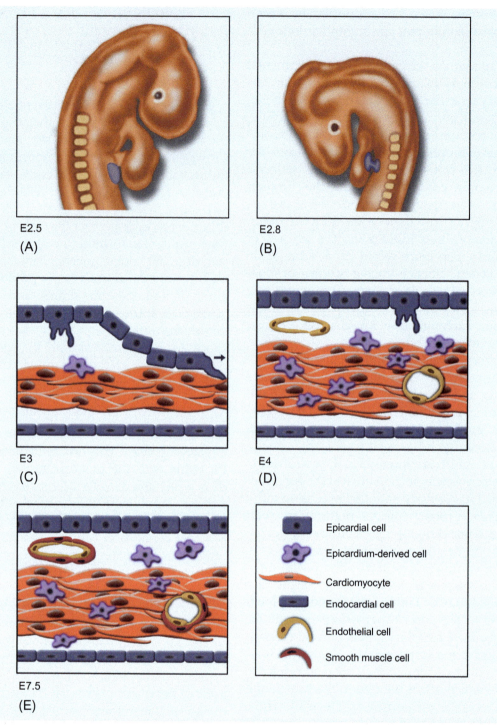

FIG. 2.1   Model of coronary formation from proepicardium-derived cells. (A) The traditional model for the development of coronary vasculature assumes that it is formed from proepicardium-derived cells, highlighted in *blue* in the scheme of a avian embryo in embryonic stage E2.5. (B) In avian embryos, PE cells adhere to the heart through of an extracellular matrix bridge (shown in *blue*), at embryonic stage E2.8. (C) Beginning of epicardium formation occurs at embryonic stage E3. Even before complete formation of the epicardium, the first mesenchymal cells derived from the PE are already observed in the subepicardium [17]. (D) As the epithelial to mesenchymal transition of the epicardium cells proceeds, the first endothelial tubes are observed at embryonic stage E4, in the atrioventricular region [18]. (E) The first SMC (smooth muscle cells) appear at embryonic stage E7.5, composing the vessel wall [18,19]. Modified from Azambuja AP. Mecanismos embrionários de diferenciação de precursores coronários: princípios para aplicação em terapia celular. In: Cell and developmental biology. São Paulo: University of São Paulo; 2009. p. 127.

important source of cells in the formation of coronary vasculature [24]. Later on, it was shown that PE contains precursors of the epicardium, endothelial cells, and smooth muscle cells that form the coronary arteries, besides precursors of interstitial fibroblasts [24–28].

Although the plasticity and contribution of cells derived from PE have been extensively studied in recent years, factors and processes that lead to the formation of this transitory structure have been poorly explored. A piece of information that may provide clues regarding the origin of the PE is the identification of PE and PE-derived molecular markers from the initial stages of formation. In the last few years, some transcription factors, structural molecules, and enzymes were proposed as PE/epicardium markers, namely: transcription factors WT1 (Wilms Tumor-1), T-box 18 (TBX18) and Epicardin (also known as Capsulin, TCF21, and POD); structural protein cytokeratin and Aldehyde Dehydrogenase 1 Family Member A2 (ALDH1A2), which is the main enzyme for retinoic acid synthesis during embryonic development [29–33]. These molecules are not exclusively expressed in PE, but in co-expression they have been used as molecular markers of PE/epicardium cells.

Another fundamental piece of information to understand the PE origin is identification of the progenitor cells of this transitory structure. Zhou et al. [34] demonstrated that PE cells descend from a population of precursors that express genes Nkx2.5 and Isl1, makers of cardiac lineages. However, from the beginning of its formation, PE cells do not express Nkx2.5 or Isl1, which indicates that precursor cells Nkx2.5+/Isl1+ are in an early position in the hierarchy of cells that originate PE. The authors also show that expression of Nkx2.5 in PE precursor cells is fundamental for the development of this structure [34]. Interestingly, the contribution of cardiac lineage cells as precursors of PE is enhanced by the ability to differentiate PE cells in cardiomyocytes in vitro [35]. The contribution of PE-derived cells to cardiac development goes beyond coronary formation, since it also originates cells that will participate in the formation of septa and valves. As mentioned, PE-derived cells also have the potential to differentiate into cardiomyocytes and it has already been shown that these cells contribute substantially to myocytes in the interventricular septum, atrial and ventricular walls during mammalian cardiac development [36,37]. In addition, signaling for thickening of the myocardial wall and for formation of the heart's conduction system also comes from PE-derived cells [17,38–44]. These data demonstrate the plasticity and different roles of PE-derived cells in cardiac development.

THE ROLE OF THE EPICARDIUM IN CORONARIOGENESIS GENESIS

The epicardium is a fundamental tissue in the process of coronary circulation formation. Its origins date back to the embryonic period after formation of the primordial cardiac tube, when proepicardium cells delaminate, migrate to the myocardium, and form a layer of cells that cover the heart and the epicardium (Fig. 2.1C).

In early epicardium development, cells from this mono-layer adhere directly to myocardial surface. Next, the space between the epicardium in migration and the myocardium emerges and is filled by the extracellular matrix, called the subepicardial space [45]. Initially, this space is formed in the atrioventricular, conoventricular, and interventricular junctions, and then along the ventricles and in the ventral portion of the atrium [22]. The subepicardial space contains elements of the extracellular matrix, such as collagen type IV, fibronectin, and laminin [46]. Since these extracellular matrix elements accumulate discontinuously, there is higher concentration in the atrioventricular channel and interventricular septum regions, which makes these territories micro environments appropriate for hematopoiesis and vasculogenesis [46]. Promotion of hematopoiesis in these locations is probably due to accumulation of growth factors produced by the myocardium, such as vascular endothelial growth factor (VEGF), beta transforming growth factor (TGFβ), fibroblast growth factors (FGFs), and bone morphogenetic proteins (BMPs) [45].

The subepicardial space is populated by mesenchymal cells that detach from the epicardium when undergoing a process known as epithelial-mesenchymal transition (EMT) (Fig. 2.1C) [21,25,45]. After passing through EMT, epicardial cells, now with mesenchymal characteristics, acquire migration capacity and invade the myocardium (Fig. 2.1D). This myocardial migration process starts at stages before complete formation of the epicardium, which occurs at stage E4.5 in avian embryos, and at stage E3 it is possible to observe the presence of migratory cells derived from the epicardium invading the subepicardial space and the myocardium, respectively [17,33,47]. The population of cells that fill the subepicardial space includes endothelial precursors (hemangioblasts), which contribute to formation of blood islets and start the coronary plexus vasculogenesis process (Fig. 2.1D). In addition to contribution of endothelial cells, the proepicardium/epicardium is the source of subepicardial precursors of vascular smooth muscle cells (SMC) and fibroblasts (Fig. 2.1E).

These precursors follow the same path of angioblasts and form mural cells that stabilize the coronary vessel wall. However, in order to form the blood vessels, endothelial cells must form the endothelial tubules before the smooth muscle cells separate and migrate. Interestingly, initial studies with quail chickens/chimeras showed that the population of migratory cells derived from the epicardium consists of two sub-populations: the first is responsible for the invasion of endothelial precursors, while the second is negative for endothelial markers [42]. This suggests that the second sub-population of migratory cells would give rise to smooth muscle cells and cardiac fibroblasts. In fact, further studies showed that, during coronariogenesis, endothelial cells separate earlier and organize themselves in endothelial tubes, forming the first coronary vessel cell layer [48]. It is only after this endothelial distinction that smooth muscle cell precursors are attracted to these endothelial tubes and complete their differentiation (Fig. 2.1D and E).

The epicardium plays a critical role in the formation of the coronary vasculature, not just as a source of precursors of the cells that make up the coronary vascular wall, but also as the site of production of trophic factors required for myocardial growth and coronary vessel formation. The epicardium also secretes factors that directly affect the epicardial EMT, the differentiation of endothelial and SMC precursors, as well as the connection of the coronary plexus to the aorta. In addition, the epicardium secretes factors that promote myocardial growth and thus indirectly impact the coronary vascular plexus by increasing the thickness of the ventricular wall. Such growth promoting factors include retinoic acid, IGF2, Shh, Wnt, TGFβ and numerous FGF family members [41,49, 50]. Recently, it has been shown that the thin compact ventricular layer, resulting from removal, or blocking of the signaling

proceeding from the epicardium, is secondary on formation of small cardiomyocytes and immature sarcomeres, and this type of phenotype is caused by reduction of signaling by TGFβ and FGF, emanating from the epicardium [51].In turn, the epicardium is also responsive to signals from the myocardium via VEGF, and angiopoietins, as well as from the liver, via erythropoietin, which allows for modulation of the epicardial secretion of trophic factors [52, 53]. Finally, migrating epicardially-derived vascular progenitors respond to signals and cues from their myocardial environments to migrate and differentiate along the correct path. Therefore, the epicardium contributes to the coronary plexus both as a source of precursor cells and as signaling center that coordinates myocardial growth with the processes of assembly of the endothelial tubules and stabilization of the coronary vessels by mural cells. The primary importance of the epicardium in coronariogenesis is underscored by the prevalence of coronary defects in mouse strains in which either the formation, or function of the epicardium has been altered.

RETINOIC ACID AND VEGF SYNCHRONIZE THE DIFFERENTIATION OF CORONARY SMOOTH MUSCLE CELLS TO ENDOTHELIAL PLEXUS FORMATION

Coronary morphogenesis has received strong attention in recent years, as it is a model to understand vascular development, and also because discovery of the cellular and molecular processes involved in coronary morphogenesis provides clues regarding the potential development of new regenerative therapies. Coronary vessels are formed by subepicardial endothelial and primary intramyocardial plexus. Morphogenesis of the subepicardial/intramyocardial vessels constitutes one of the most intriguing characteristics of coronary development. For vessels to be formed correctly, there must be a delay between endothelial cell and smooth muscular cell differentiation. Such a delay is dramatic, and can be easily observed in the context of avian development. At embryonic day 4 endothelial cells are already observed in the subepicardium and within the myocardium, where they already organize themselves to form the first vascular tubes. In contrast, the first smooth muscle cells are only found at embryonic stage E7.5, when the endothelial bed is already connected to the aorta, directional blood flow is established, and coronary stretching begins. Curiously, the differentiation of PE cells, in culture, into smooth muscular cells, is extraordinarily efficient, and occurs concomitantly with the differentiation of endothelial cells. This suggests that there are embryonic/cardiac mechanisms that *prevent premature differentiation* of smooth muscular cells before endothelial support is formed.

The time interval between the differentiation of precursors derived from the epicardium into endothelial and smooth muscular cells is critical to ensure extensive remodeling of the endothelial tubes, before a complex vascular plexus is formed by recruitment of smooth muscle cells. The first model for the mechanisms that control epicardium-derived cell differentiation was proposed by Azambuja et al. [54]. Using adenoviral vectors for super-expression of ALDH1A2 and VEGF and in vivo inhibition of retinoic acid (RA) synthesis, the authors showed that the RA signaling pathways, together with VEGF, act in autocrine and paracrine fashion to inhibit expression of smooth muscle markers, such as: CRP2, GATA-6, SRF, and SM22A, and therefore, delay smooth muscle differentiation. In this model, epicardial cells produce retinoic acid, which diffuses, creating an environment rich in RA in the subepicardial space and in the epicardial face of the myocardium. RA is responsible for maintaining cells derived from the epicardium in an undifferentiated state, by activating a positive feedback loop of ALDH1A2 and WT1, and also by preventing the differentiation of these cells into smooth muscles. In fact, RA signaling regulates not only WT1 expression, but also expression by the POD1 transcription factor. In chicken embryos, inhibition of RA signaling decreases POD1 expression, which leads to early differentiation of smooth muscle cells [55]. In addition, knockout mice for POD1 show an increase in smooth muscle cell differentiation [55]. These data confirm the role of RA in preventing smooth muscle differentiation from the epicardium in smooth muscle cells. According to the model proposed by Azambuja et al. [54], once in the myocardium, the cells derived from the epicardium come into contact with the VEGF signaling that induces differentiation into endothelial cells, and the consequent formation of an endothelial vascular plexus (Fig. 2.2). At this point,

VEGF also acts as an additional mechanism to inhibit the differentiation of smooth muscle cells. As maturation of endothelial cells occurs, their production of RA decreases and so does the diffusion of such morphogen. The decrease or absence of RA signaling creates myocardial environment that is permissive to differentiation of epicardium-derived cells into smooth muscle. Hence, these coronary precursors respond to factors released by the endothelial cells, such as platelet-derived growth factor (PDGF), and terminally differentiate into smooth muscle cell (Fig. 2.2). The model proposed by Azambuja et al. [54] suggests new approaches to guide differentiation of coronary precursors towards specific fates, which involves RA, VEGF, and PDGF signaling pathways. These data also suggest that regenerative and/or revascularization therapies may be improved, considering the discovery of molecular mechanisms that coordinate the several steps of coronary development and that involve both internal and external coronary vessel morphogenesis.

The contribution of proepicardial-derived cells to the formation of coronary EC has been questioned by genetic tracing studies using epicardial markers such as WT1 and Tbx18 [36, 37, 56]. These studies show that few endothelial cells are labeled by using Wt1 or Tbx18-driven Cre lines. Nevertheless, the (pro)epicardial-precursor vasculogenesis model is supported by the frequent observation of hemangioblasts and blood islands in the subepicardium. It remains to established whether endothelial cells are, in fact, derived from the small population of proepicardial precursors that do not express the epicardial markers WT1 and Tbx18 [44].

NEW ACTORS IN CORONARY VASCULATURE FORMATION: CONTRIBUTIONS OF THE VENOUS SINUS AND ENDOCARDIUM

Recent studies have shown that coronary vessels can also be generated through the process of sprouting angiogenesis, by extending the vasculature already present in the sinus venosus [56-58]. The sinus venosus, as a chamber, is a transient developmental feature. Later, the sinus venosus either involutes, or becomes incorporated into the atrium. [19].

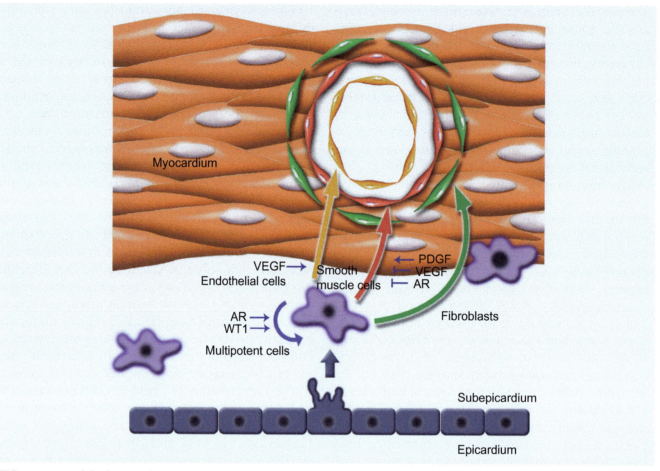

FIG. 2.2 Model of control mechanisms for the differentiation of epicardium-derived cells into coronary vessels. The epicardium is a source of retinoic acid (RA) signaling, which is involved in maintaining the epicardium-derived cells in an undifferentiated state, within the subepicardial space, by activating a positive feedback loop of retinol dehydrogenase enzyme 1A2 (ALDH1A2) and Wilms Tumor-1 (WT1). The RA signals derived from the epicardium also act together with the vascular endothelial growth factor (VEGF) to delay the differentiation of epicardium-derived cells into smooth muscle cells. This delay is essential because it allows construction and remodeling of the coronary endothelial network before the final vessels are stabilized by the aggregation of smooth muscle cells, which are differentiated by inactivation of VEGF and RA signaling and by stimulus of the platelet-derived growth factor (PDGF). Modified from Azambuja AP. Mecanismos embrionários de diferenciação de precursores coronários: princípios para aplicação em terapia celular. In: Cell and developmental biology. São Paulo: University of São Paulo; 2009. p. 127.

Genetic lineage tracing studies show that the process of sprouting angiogenesis of the sinus venosus can contribute to both coronary arteries and veins. In this model, subepicardial venous cells derived from the sinus venosus dedifferentiate and form a progenitor endothelial cell population that can give rise to arterial, capillary and venous fates [56, 58]. This conclusion, however, contradicts other tracing studies, which indicate that coronary arteries and veins are derived from distinct developmental lineages [48, 59]. However, it is possible that some coronary arteries and veins can be derived from common origins, while others from can be generated from distinct lineages.

The source of the endothelial cells precursors giving rise to the vasculature is found in the ventricular septum. Although, it is easier to envision how the sinus venosus, or the epicardium, can contribute to the formation of subepicardial vessels, it is more difficult to imagine the same endothelial precursors giving rise to the vasculature present in the ventricular septum, or deep in the endocardium. Not surprisingly, lineage tracing studies show that neither the epicardium, nor the vessels sprouting from the sinus venosus give rise to vasculature deep into the myocardium and within the septum wall. Rather, this second coronary vascular bed was recently shown to be generated from the angiogenesis of the endocardial endothelium [59, 60]. The mechanism through which the endocardium contributes to the deep coronary vascular bed is not known, although it has been proposed that endocardial-derived coronary vessels form by endocardial-to-vascular endothelial cells lineage conversion. The lineage conversion was proposed to be driven by the coalescence of the trabecular layer during the process of compaction of the ventricular wall [60]. An interesting concept was revealed by lineage tracing studies using *Fatty acid binding protein 4* (FABP4)-driven Cre. According to this study, the endocardium continues to be a source of new coronary vasculature after birth [61]. This observation could be explained by the fact that the trabecular compaction continues in the first few days after birth. Thus, the process of coronary angiogenesis from the endocardium could, in theory, allow for the expansion of the coronary vascular bed much later in life than other mechanisms that rely on vasculogenesis via subepicardial precursors, or via sprouting from the sinus venosus.

DEVELOPMENT OF CORONARY CIRCULATION IN ZEBRAFISH: A MODEL TO STUDY CARDIAC REGENERATION

Zebrafish (*Danio rerio*), a teleost fish, has coronary vessels on the heart surface and on the ventricular compact layer wall [62]. However, in contrast to amniotes, in which the vasculature develops at the embryonic stage, in zebrafish the coronary vessels appear only at the late juvenile stage (1–2 months after fertilization), with the emergence of endothelial cells in the atrioventricular region. Later, endothelial cells migrate towards the *arterial bulb* (*AB*) and establish vascular connections with the branchial arches, allowing oxygen supply directly to the heart through the coronary artery [63].

The absence of coronary plexus formation during the entire larval stage until the end of the juvenile stage, may be rationalized considering the small size of the zebrafish, which displays a proportionally smaller ventricular layer, composed by a relatively thin external compact layer, when compared to the thick myocardial layer of amniotes. However, as the heart develops and increases in size, coronary capillaries invade the ventricular region and connect to the subtrabecular layer next to the compact layer, which is highly vascularized in adult animals.

Although a few aspects of coronary vascular system formation vary among different species, recent studies show that the fundamental mechanisms and molecular programs involved in the construction of these vessels in mammals and birds are extraordinarily conserved in zebrafish. For example, recent studies show the importance of the epicardium for myocardium vascularization. Similar to other vertebrates, in zebrafish the epicardium forms early during embryogenesis (48 h after fertilization) from the proepicardium, a transient structure formed by spherical agglomerations of cells located close to the sinus venosus. During zebrafish embryonic development the epicardium plays a crucial role in the growth and maturation of the ventricular myocardium compact layer, contributing to formation of cardiomyocytes.

Fate mapping studies show that cells expressing the *tcf21/epicardin* gene specify the epicardium in zebrafish and originate coronary vasculature elements, such as smooth muscle cells that surround the bulbus arteriosus, as well as fibroblasts located between the cortical and trabecular regions of the myocardium [64]. Interestingly, this study displays no evidence for direct epicardial contribution for the formation of endothelial cells, considering the co-location of *tcf21* positive cells (epicardial) and cells expressing the *fli1* gene (marker of endothelial cells).

However, a better understanding of the role of the epicardium in coronary vessel formation in zebrafish comes from important observations from cardiac regeneration studies in adults, in which the epicardium is fundamental for the neovascularization of the injured region and for complete regeneration after partial amputation of the ventricle [65–67]. During this process, coronary blood vessels are formed de novo in the amputated region, involving quick activation and proliferation of epicardial cells by reactivating the expression of genes *Tbx18* and *ALDH1A2*, as well as paralogs *Wt1a* and *Wt1b* (orthologs of gene *WT1*), all of which are important genes in the specification of the epicardium and development of vascular coronary system in amniotes. This quick molecular reactivation of epicardial cells is followed by a EMT process controlled by multiple signaling pathways, including FGFs [66] and PDGF [68].

Therefore, the epicardium contributes to the formation of coronary vessels in zebrafish and displays quick and robust response during ventricular regeneration, in a process highly reminiscent of the role of the epicardium during cardiac development in amniotes, during which the epicardial cells start proliferating again, undergo EMT, invade the subepicardial and myocardial regions, and contribute to the assembly of new coronary vessels.

As in amniotes, the epicardial origins of coronary endothelium is also controversial in zebrafish. Recent studies [63] showed, by means of elegant Cre recombination experiments guided by endothelial promoter and clonal analysis that endothelial cells derive from endocardial cells of the atrioventricular (AV) region angiogenesis. Cxcr4/Cxcl12 seems to be a key regulator of this process. Mutant animals for the gene *Cxcr4a* neither develop, nor regenerate coronary vessels. Cxcr4/Cxcl12 signaling is also involved in coronary artery development in mammals [69]. However, it is important to emphasize that these findings do not exclude the possibility of different cell lineages contributing to formation of coronary endothelium, which represents a most likely scenario, since in the clonal analysis made by Harrison et al. [63] some coronary vessels do not seem to derive from endocardial cells.

Considering that epicardial cells also express gene Cxcl12a after injury [70,71] and that inhibition of Cxcr4 receptor prevents normal regeneration of the myocardium [71], it is probable that cells derived from the endocardium and from the epicardium contribute to formation of different coronary vasculature components in zebrafish, reconciling the apparently controversial views regarding the origin of the coronary system.

Future studies are necessary to better understand coronariogenesis, and genetic studies in the zebrafish will contribute not only to find more details regarding molecular mechanisms involved during coronary development, but also for development of therapeutic strategies in cardiac regeneration and repair.

EXPLOITING CORONARY DEVELOPMENTAL PATHWAYS TO REGENERATE THE INJURED HEART

One of the main reasons for gaining a good understanding of coronary vascular development is that it could provide a potential tool to unlock the process of vasculogenesis and to regenerate the injured, or diseased vasculature [12]. A promising cell-based therapeutic approach in heart repair involves exploiting the well-known heart regenerative functions of the epicardium. Though, the epicardium becomes quiescent after the birth, it can become reactivated in response to myocardial injury, and was shown to play a critical role in mediating myocardial regeneration and repair [37, 72, 73]. In addition to regenerating myocyte repair, epicardial reactivation also contributes to the coronary angiogenic program allowing for remodeling, and repair and, potentially, for the development of collateral irrigation to the injured myocardium.

In one example of epicardial-mediated repair, a study by Smart et al. showed the potential of thymosin β4 to prime epicardially-derived progenitors in differentiating along the endothelial lineage [74]. This was followed by a second study from the same group that showed epicardial cells can also regenerate cardiomyocytes in response to thymosin β4 [75]. The mechanism for epicardial repair in the studies by Smart et al. was proposed to involve a cell-autonomous mechanism, but a contribution by the epicardium to heart repair via paracrine mechanisms was not ruled out. In fact, subsequent studies by other groups suggest that thymosin β4-mediated heart repair may involve mechanisms independent of epicardial differentiation to myocyte and endothelial precursors [76, 77]. Therefore, there is tremendous potential for engaging in multipronged approaches towards heart regeneration and repair. However, a thorough understanding of the basic biology of coronary vascular development and vascular differentiation program is needed before the implementation of such therapies.

References

[1] Yue X, Tomanek RJ. Stimulation of coronary vasculogenesis/angiogenesis by hypoxia in cultured embryonic hearts. Dev Dyn 1999;216:28–36.

[2] Tota B, Cimini V, Salvatore G, et al. Comparative study of the arterial and lacunary systems of the ventricular myocardium of elasmobranch and teleost fishes. Am J Anat 1983;167:15–32.

[3] Tota B. Myoarchitecture and vascularization of the elasmobranch heart ventricle. J Exp Zool 1989;252:122–35.

[4] Grimes AC, Kirby ML. The outflow tract of the heart in fishes: anatomy, genes and evolution. J Fish Biol 2009;74:983–1036.

[5] Durán AC, López-Unzu MA, Rodríguez C, et al. Structure and vascularization of the ventricular myocardium in Holocephali: their evolutionary significance. J Anat 2015;226:501–10.

[6] McKenzie DJ, Brauner CJ, Farrell AP. Primitive fishes. Fish physiology. Amsterdam: Elsevier Academic Press; 2007. p. 562.

[7] Farrell AP, Farrell ND, Jourdan H, et al. A perspective on the evolution of the coronary circulation in fishes and the transition to terrestrial life. In: Sedmera D, Wang T, editors. Ontogeny and phylogeny of the vertebrate heart. New York: Springer; 2012. p. 75–102.

[8] Icardo JM. The teleost heart: a morphological approach. In: Sedmera D, Wang T, editors. Ontogeny and phylogeny of the vertebrate heart. New York: Springer; 2012. p. 35–53.

[9] Sedmera D, Wang T. Ontogeny and phylogeny of the vertebrate heart. New York: Springer; 2012. p. 231.

[10] Dyer L, Pi X, Patterson C. Connecting the coronaries: how the coronary plexus develops and is functionalized. Dev Biol 2014;395:111–9.

[11] Loukas M, Groat C, Khangura R, et al. The normal and abnormal anatomy of the coronary arteries. Clin Anat 2009;22:114–28.

[12] Michelis KC, Boehm M, Kovacic JC. New vessel formation in the context of cardiomyocyte regeneration—the role and importance of an adequate perfusing vasculature. Stem Cell Res 2014;13:666–82.

[13] Bogers AJ, Gittenberger-de Groot AC, Poelmann RE, et al. Development of the origin of the coronary arteries, a matter of ingrowth or outgrowth? Anat Embryol (Berl) 1989;180:437–41.

[14] Cavallero S, Shen H, Yi C, et al. CXCL12 signaling is essential for maturation of the ventricular coronary endothelial plexus and establishment of functional coronary circulation. Dev Cell 2015;33:469–77.

[15] Tian X, Hu T, He L, et al. Peritruncal coronary endothelial cells contribute to proximal coronary artery stems and their aortic orifices in the mouse heart. PLoS ONE 2013;8:e80857.

[16] Baeyens N, Nicoli S, Coon BG, et al. Vascular remodeling is governed by a VEGFR3-dependent fluid shear stress set point. Elife 2015;4:e04645.

[17] Lie-Venema H, Eralp I, Maas S, et al. Myocardial heterogeneity in permissiveness for epicardium-derived cells and endothelial precursor cells along the developing heart tube at the onset of coronary vascularization. Anat Rec A: Discov Mol Cell Evol Biol 2005;282:120–9.

[18] Vrancken Peeters MP, Gittenberger-de Groot AC, Mentink MM, et al. The development of the coronary vessels and their differentiation into arteries and veins in the embryonic quail heart. Dev Dyn 1997;208:338–48.

[19] Vrancken Peeters MP, Gittenberger-de Groot AC, Mentink MM, et al. Differences in development of coronary arteries and veins. Cardiovasc Res 1997;36:101–10.

[20] Manner J. Experimental study on the formation of the epicardium in chick embryos. Anat Embryol (Berl) 1993;187:281–9.

[21] Manner J, Pérez-Pomares JM, Macías D, et al. The origin, formation and developmental significance of the epicardium: a review. Cells Tissues Organs 2001;169:89–103.

[22] Perez-Pomares JM, Macías D, García-Garrido L, et al. Contribution of the primitive epicardium to the subepicardial mesenchyme in hamster and chick embryos. Dev Dyn 1997;210:96–105.

[23] Nahirney PC, Mikawa T, Fischman DA. Evidence for an extracellular matrix bridge guiding proepicardial cell migration to the myocardium of chick embryos. Dev Dyn 2003;227:511–23.

[24] Mikawa T, Gourdie RG. Pericardial mesoderm generates a population of coronary smooth muscle cells migrating into the heart along with ingrowth of the epicardial organ. Dev Biol 1996;174:221–232.

[25] Dettman RW, Denetclaw Jr. W, Ordahl CP, et al. Common epicardial origin of coronary vascular smooth muscle, perivascular fibroblasts, and intermyocardial fibroblasts in the avian heart. Dev Biol 1998;193:169–81.

[26] Mikawa T, Fischman DA. Retroviral analysis of cardiac morphogenesis: discontinuous formation of coronary vessels. Proc Natl Acad Sci U S A 1992;89:9504–8.

[27] Perez-Pomares JM, Macías D, García-Garrido L, et al. The origin of the subepicardial mesenchyme in the avian embryo: an immunohistochemical and quail-chick chimera study. Dev Biol 1998;200:57–68.

[28] Vrancken Peeters MP, Gittenberger-de Groot AC, Mentink MM, et al. Smooth muscle cells and fibroblasts of the coronary arteries derive from epithelial-mesenchymal transformation of the epicardium. Anat Embryol (Berl) 1999;199:367–78.

[29] Carmona R, González-Iriarte M, Pérez-Pomares JM, et al. Localization of the Wilm's tumour protein WT1 in avian embryos. Cell Tissue Res 2001;303:173–86.

[30] Kraus F, Haenig B, Kispert A. Cloning and expression analysis of the mouse T-box gene Tbx18. Mech Dev 2001;100:83–6.

[31] Robb L, Mifsud L, Hartley L, et al. Epicardin: a novel basic helix-loop-helix transcription factor gene expressed in epicardium, branchial arch myoblasts, and mesenchyme of developing lung, gut, kidney, and gonads. Dev Dyn 1998;213:105–13.

[32] Viragh S, Gittenberger-de Groot AC, Poelmann RE, et al. Early development of quail heart epicardium and associated vascular and glandular structures. Anat Embryol (Berl) 1993;188:381–93.

[33] Xavier-Neto J, Shapiro MD, Houghton L, et al. Sequential programs of retinoic acid synthesis in the myocardial and epicardial layers of the developing avian heart. Dev Biol 2000;219:129–41.

[34] Zhou B, von Gise A, Ma Q, et al. Nkx2-5- and Isl1-expressing cardiac progenitors contribute to proepicardium. Biochem Biophys Res Commun 2008;375:450–3.

[35] Kruithof BP, van Wijk B, Kruithof-de Julio M, et al. BMP and FGF regulate the differentiation of multipotential pericardial mesoderm into the myocardial or epicardial lineage. Dev Biol 2006;295:507–22.

[36] Cai CL, Martin JC, Sun Y, et al. A myocardial lineage derives from Tbx18 epicardial cells. Nature 2008;454:104–8.

[37] Zhou B, Ma Q, Rajagopal S, et al. Epicardial progenitors contribute to the cardiomyocyte lineage in the developing heart. Nature 2008;454:109–13.

[38] Eralp I, Lie-Venema H, Bax NA, et al. Epicardium-derived cells are important for correct development of the Purkinje fibers in the avian heart. Anat Rec A: Discov Mol Cell Evol Biol 2006;288:1272–80.

[39] Manner J, Schlueter J, Brand T. Experimental analyses of the function of the proepicardium using a new microsurgical procedure to induce loss-of-proepicardial-function in chick embryos. Dev Dyn 2005;233:1454–63.

[40] Chen T, Chang TC, Kang JO, et al. Epicardial induction of fetal cardiomyocyte proliferation via a retinoic acid-inducible trophic factor. Dev Biol 2002;250:198–207.

[41] Stuckmann I, Evans S, Lassar AB. Erythropoietin and retinoic acid, secreted from the epicardium, are required for cardiac myocyte proliferation. Dev Biol 2003;255:334–49.

[42] Gittenberger-de Groot AC, Vrancken Peeters MP, Mentink MM, et al. Epicardium-derived cells contribute a novel population to

the myocardial wall and the atrioventricular cushions. Circ Res 1998;82:1043–52.

[43] Wilting J, Buttler K, Schulte I, et al. The proepicardium delivers hemangioblasts but not lymphangioblasts to the developing heart. Dev Biol 2007;305:451–9.

[44] Katz TC, Singh MK, Degenhardt K, et al. Distinct compartments of the proepicardial organ give rise to coronary vascular endothelial cells. Dev Cell 2012;22:639–50.

[45] Wessels A, Perez-Pomares JM. The epicardium and epicardially derived cells (EPDCs) as cardiac stem cells. Anat Rec A: Discov Mol Cell Evol Biol 2004;276:43–57.

[46] Kalman F, Viragh S, Modis L. Cell surface glycoconjugates and the extracellular matrix of the developing mouse embryo epicardium. Anat Embryol (Berl) 1995;191(5):451–64.

[47] Perez-Pomares JM, Phelps A, Sedmerova M, et al. Experimental studies on the spatiotemporal expression of WT1 and RALDH2 in the embryonic avian heart: a model for the regulation of myocardial and valvuloseptal development by epicardially derived cells (EPDCs). Dev Biol 2002;247:307–26.

[48] Lavine KJ, Long F, Choi K, et al. Hedgehog signaling to distinct cell types differentially regulates coronary artery and vein development. Development 2008;135:3161–71.

[49] Lavine KJ, Ornitz DM. Fibroblast growth factors and Hedgehogs: at the heart of the epicardial signaling center. Trends Genet 2008;24:33–40.

[50] Lavine KJ, Yu K, White AC, et al. Endocardial and epicardial derived FGF signals regulate myocardial proliferation and differentiation in vivo. Dev Cell 2005;8:85–95.

[51] Takahashi M, Yamagishi T, Narematsu M, et al. Epicardium is required for sarcomeric maturation and cardiomyocyte growth in the ventricular compact layer mediated by transforming growth factor beta and fibroblast growth factor before the onset of coronary circulation. Congenit Anom (Kyoto) 2014;54:162–71.

[52] Brade T, Kumar S, Cunnigham TJ, et al. Retinoic acid stimulates myocardial expansion by induction of hepatic erythropoietin which activates epicardial Igf2. Development 2011;138:139–148.

[53] Shen H, Cavallero S, Estrada KD, et al. Extracardiac control of embryonic cardiomyocyte proliferation and ventricular wall expansion. Cardiovasc Res 2015;105:271–8.

[54] Azambuja AP, Portillo-Sánchez V, Rodrigues MV, et al. Retinoic acid and VEGF delay smooth muscle relative to endothelial differentiation to coordinate inner and outer coronary vessel wall morphogenesis. Circ Res 2010;107:204–16.

[55] Braitsch CM, Combs MD, Quaggin SE, et al. Pod1/Tcf21 is regulated by retinoic acid signaling and inhibits differentiation of epicardium-derived cells into smooth muscle in the developing heart. Dev Biol 2012;368:345–57.

[56] Red-Horse K, Ueno H, Weissman IL, et al. Coronary arteries form by developmental reprogramming of venous cells. Nature 2010;464:549–53.

[57] Chen HI, Sharma B, Akerberg BN, et al. The sinus venosus contributes to coronary vasculature through VEGFC-stimulated angiogenesis. Development 2014;141:4500–12.

[58] Tian X, Hu T, Zhang H, et al. Subepicardial endothelial cells invade the embryonic ventricle wall to form coronary arteries. Cell Res 2013;23:1075–90.

[59] Wu B, Zhang Z, Lui W, et al. Endocardial cells form the coronary arteries by angiogenesis through myocardial-endocardial VEGF signaling. Cell 2012;151:1083–96.

[60] Tian X, Hu T, Zhang H, et al. Vessel formation. De novo formation of a distinct coronary vascular population in neonatal heart. Science 2014;345:90–4.

[61] He L, Tian X, Zhang H, et al. Fabp4-CreER lineage tracing reveals two distinctive coronary vascular populations. J Cell Mol Med 2014;18:2152–6.

[62] Norman H, Yost HJ, Edward BC. Cardiac morphology and blood pressure in the adult zebrafish. Anatom Rec 2001;264:1–12.

[63] Harrison MR, Bussmann J, Huang Y, et al. Chemokine-guided angiogenesis directs coronary vasculature formation in zebrafish. Dev Cell 2015;33:442–54.

[64] Kikuchi K, Gupta V, Wang J, et al. tcf21 + epicardial cells adopt non-myocardial fates during zebrafish heart development and regeneration. Development 2011;138:2895–902.

[65] Masters M, Riley PR. The epicardium signals the way towards heart regeneration. Stem Cell Res 2014;13:683–92.

[66] Lepilina A, Coon AN, Kikuchi K, et al. A dynamic epicardial injury response supports progenitor cell activity during zebrafish heart regeneration. Cell 2006;127:607–19.

[67] Kikuchi K, Holdway JE, Major RJ, et al. Retinoic acid production by endocardium and epicardium is an injury response essential for zebrafish heart regeneration. Dev Cell 2011;20:397–404.

[68] Jieun K, Wu Q, Zhang Y, et al. PDGF signaling is required for epicardial function and blood vessel formation in regenerating zebrafish hearts. Proc Natl Acad Sci 2010;107:17206–10.

[69] Ivins S, Chappell J, Vernay B, et al. The CXCL12/CXCR4 axis plays a critical role in coronary artery development. Dev Cell 2015;33:455–68.

[70] Gonzalez-Rosa JM, Peralta M, Mercader N. Pan-epicardial lineage tracing reveals that epicardium derived cells give rise to myofibroblasts and perivascular cells during zebrafish heart regeneration. Dev Biol 2012;370:173–86.

[71] Itou J, Oishi I, Kawakami H, et al. Migration of cardiomyocytes is essential for heart regeneration in zebrafish. Development 2012;139:4133–42.

[72] Zangi L, Lui KO, von Gise A, et al. Modified mRNA directs the fate of heart progenitor cells and induces vascular regeneration after myocardial infarction. Nat Biotechnol 2013;31:898–907.

[73] Wang J, Cao J, Dickson AL, et al. Epicardial regeneration is guided by cardiac outflow tract and Hedgehog signalling. Nature 2015;522:226–30.

[74] Smart N, Risebro CA, Melville AA, et al. Thymosin beta4 induces adult epicardial progenitor mobilization and neovascularization. Nature 2007;445:177–82.

[75] Smart N, Bollini S, Dubé KN, et al. Myocardial regeneration: expanding the repertoire of thymosin beta4 in the ischemic heart. Ann N Y Acad Sci 2012;1269:92–101.

[76] Zhou B, Honor LB, Ma Q, et al. Thymosin beta 4 treatment after myocardial infarction does not reprogram epicardial cells into cardiomyocytes. J Mol Cell Cardiol 2012;52:43–7.

[77] Kispert A. No muscle for a damaged heart: thymosin beta 4 treatment after myocardial infarction does not induce myocardial differentiation of epicardial cells. J Mol Cell Cardiol 2012;52:10–2.

Further Reading

[1] Azambuja AP. Mecanismos embrionários de diferenciação de precursores coronários: princípios para aplicação em terapia celular. In: Cell and developmental biology. São Paulo: University of São Paulo; 2009. p. 127.

3

Signal Transduction Pathways in Endothelial Cells: Implications for Angiogenesis

Hugo P. Monteiro, Maria Theresa O.M. Albuquerque,
Carlos J. Rocha Oliveira, and Marli F. Curcio

KINASE PROTEINS AND PHOSPHATASE PROTEINS

Initially protein kinases were thought to be involved only with the regulation of cellular metabolism. However, the easiness with which phosphate groups can be added or removed from the amino acids that constitute the proteins, altering therefore their location, enzymatic activity, or association with other proteins, makes protein reversible phosphorylation a very effective way for cells to respond to environmental changes. Therefore, cell adhesion, distinction, proliferation, transformation, or death are processes reversibly controlled by protein phosphorylation and require not only a protein kinase (PK), but also a phosphatase protein (PP) as well. Target proteins are phosphorylated in specific amino acids by one of more PKs, and these phosphate groups are removed by specific PPs, maintaining therefore system homeostasis [1]

PKs, more commonly found in eukaryotic cells, are those that catalyze phosphorylation; that is, the transference of a phosphate group in position 7 of the ATP molecule for serin, threonine, and/or tyrosine residues. In general terms, there are two basic consequences resulting from serin/threonine phosphorylation into tyrosine. Phosphoserine/phosphothreonine change the conformation of enzymes and/or its connection to substrates, modifying their activities [1]. On the other hand, phosphorylation into tyrosine performs a fundamental role in the formation of protein-protein complexes in cellular signaling pathways [2].

The last 10 years have accumulated evidence of growing participation of other posttransformation changes, of redox nature, such as S-nitrosylation of cysteine residues and nitration of tyrosine residues that follow tyrosine phosphorylation and/or threonine residues, resulting in gain or loss of functions of the proteins that have gone through these changes. This set of posttranslational changes has been frequently detected in endothelial cell signaling processes [3–6].

REGULATION OF CELL SIGNALING PATHWAYS DEPENDING ON PROTEIN TYROSINE PHOSPHORYLATION RESIDUES—TYROSINE/PPs

Due to the central importance of protein tyrosine kinases (PTKs) in cell signaling processes, their activity must be finely controlled, either by specific PPs—protein tyrosine phosphatase (PTP)—or by other PKs (tyrosine and/or serin/threonine), by self-regulation mechanisms [7], or due to intracellular redox status changes [8]. Given the complexity

☆ The scientific production of the Cellular Signaling Laboratory of the Center for Cellular and Molecular Therapy at Unifesp, mentioned herein, was obtained through works by students and colleagues, and by the indispensable financial aid of the following development institutions: Fapesp, CNPq, and Capes.

and variety of external signals received, it is possible for control mechanisms to operate simultaneously.

The PTPs are responsible for dephosphorylation of phosphorylated tyrosine residues in proteins that constitute signaling pathways, mainly by this type of posttranslation modification (PTM). PTPs are located in two cellular compartments: cytoplasm and plasma membrane. Cytoplasmic PTPs present a single catalytic domain and several amino and carboxy terminal extensions, which probably have regulatory functions [9]. Cytoplasmic PTPs can connect to specific sites belonging to intracellular domains of the proteins involved in the signaling process. The catalytic domain of PTP1B, the first isolated intracellular PTP (human placenta), is located in the amino terminal region and does not present structural homology with serin/threonine PPs. Some cytoplasmic PTPs were classified as double specificity phosphatase, because they dephosphorylated phosphorylated tyrosine and/or serine/threonine residues as exemplified by the phosphatase called VH-1, isolated from the vaccinia [10], virus, which acts dephosphorylating threonine and tyrosine residues from the substrate protein. From this same group of phosphatase, the MAPK phosphatase, MKP1/2, stands out, and its activity is well characterized as inhibitor of ERK1/2 MAP kinase activity in converging endothelial cells confluence [11].

The PTPs of the receptor type have one extracellular domain and one or two intracellular catalytic domains. The PTP extracellular domains of the receptor type are similar to cellular adhesion molecules N-CAM, Ng-CAM, and tenascin [12], with repetitions of immunoglobulin and fibronectin type III domains.

As a representative of this group of PTPs, CD45 is an enzyme present in all hematopoietic cells (except red blood cells and its precursors), and is involved with the immune response of T and B cells [9, 10]. Expression of a receptor PTP, RPTPα, found in several cell types, is associated to dephosphorylation, activation of Src kinase and other members of this family of protein tyrosine kinases [13] (Fig. 3.1).

The activity of receptor PTPs has been associated to cellular growth inhibition by means of cell-cell contact. It has been proposed that receptor PTPs may act as tumor suppressors, and their inactivation could lead to uncontrolled proliferation with consequential cell transformation [12]. All PTPs, without exception, have at least 230 amino acids in the catalytic domain, which presents a highly conserved sequence of amino acids in its catalytic site [9, 12]. Site-directed oxidation or mutation of an essential cysteine residue, present in this sequence, makes any PTP catalytically inactive [14]. Kinetic and chemical studies regarding the action of such PTPs showed that the cysteine residue present in the catalytic site is

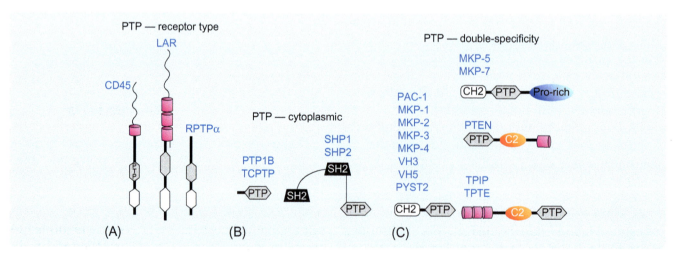

FIG. 3.1   Schematic representation of the PTPs family. The double-specificity PTPs are represented by VH3, VH5, MKPs, PAC-1, PYST2, TPIP, TPTE, and PTEN. The cytoplasmic PTPs are represented by PTP1B, SHP1 SHP2, and TCTP. Receptor type PTPs are represented by CD45, LAR, and RPTPa. The conserved catalytic domains are represented by small half-moon structures (see text for abbreviation).

essential for the enzymes to remove the phosphate group from the substrate phosphoproteins. These studies suggest the probable catalysis mechanism of the dephosphorylation reaction used by this enzyme, as illustrated in Fig. 3.2. The nucleophilic cysteine attacks the phosphate associated to the tyrosine residue of the target protein, releasing this dephosphorylated protein and forming an enzyme-substrate intermediate of thiophosphate nature. In the second stage of the reaction, this intermediate suffers slow hydrolysis, regenerating the active enzyme and releasing inorganic phosphate.

The PTPs act as intracellular redox sensors in signaling processes that have tyrosine amino acid phosphorylation as main mediator. Excessive levels of phosphorylation, resulting from chronic oxidative stress conditions, which promote oxidation of the cysteine residue essential for PTP activity, may contribute for the characteristic uncontrolled proliferation profile of tumor cells [15]. Meanwhile, maintaining the cysteine amino acid essential for PTP activity in its reduced state may result in inhibition of cell proliferation stimulated by growth factors that bind to PTK receptors [16].

Several PTPs were identified as being encoded by tumor suppressor genes. One of these phosphatases, PTEN, is an important component of cell signaling pathway mediated by the protein kinase phosphatidyl-inositol-3-kinase (PI3K). Mutations in coding genes for components of this path, in particular mutations of the coding gene PTEN, have been associated with several types of tumors [9].

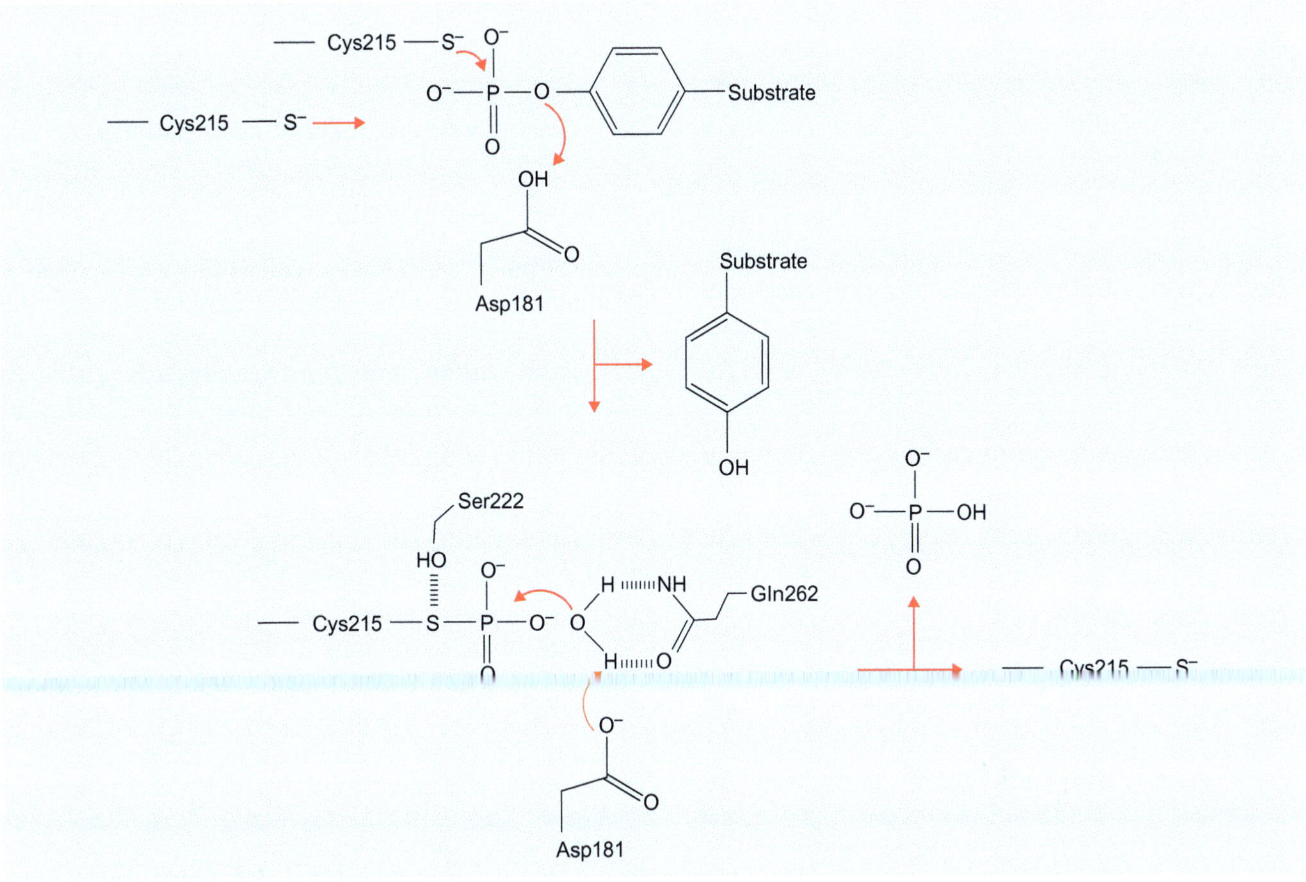

FIG. 3.2   The dephosphorylation catalysis mechanism of tyrosine amino acid by PTPs is a process that occurs in two stages. On the first stage, there is a nucleophilic attack on the phosphate group linked to the tyrosine amino acid by the cysteine residue sulfur atom (in the form of the thiolate anion), essential for the catalysis of the dephosphorylation reaction. This stage is linked to the protonation of the tyrosyl group of the substrate by the aspartic acid conserved residue. The second stage involves the hydrolysis of the intermediate, enzyme-phosphate, mediated by a conserved glutamine residue (see text for abbreviations).

PROTEIN TYROSINE KINASES—RECEPTOR PROTEIN TYROSINE KINASES

The key importance of the PKs, especially PTKs, in cellular signaling processes, can be demonstrated by examining the structures and functions of receptors for a variety of ligands that include hormones, growth and cellular distinction factors, and cytokines (e.g., epidermal growth factor receptor (EGF), platelet-derived growth factor (PDGF), vascular endothelial growth factor (VEGF), insulin receptor, and basic fibroblast growth factor (bFGF)). The specific insulin EGF and PDGF receptors are the most studied prototypes of three subfamilies of receptors that have tyrosine kinase activity.

Basically, the PTK receptors are structured in three domains:

- one extracellular domain, of variable structure that essentially recognizes the binder;
- one hydrophobic transmembrane domain, consisting of a single α-helix that crosses the membrane; and
- one cytoplasmic domain consisting of one catalytic domain with tyrosine kinase activity and one regulatory domain. The catalytic domains of PTK receptors have a relatively conserved structure with homologies between catalytic domains for the several PTK receptors, ranging between 32% and 95% [3]. In addition to the catalytic domain, the cytoplasmic domain also has a regulatory domain that contains tyrosine residues that are phosphorylated after association with the receptor binder.

Based on structural characteristics of the extracellular domains, the PTK receptors may be classified in two large groups:

- Receptors of group I, exemplified by EGF, PDGF, VEGF, and FGF receptors, which constitute a single polypeptide chain and are monomeric in the absence of specific ligand. The extracellular domain also presents two repeated sequences, rich in cysteine (EGF receptors and other subfamily members), or sequences equivalent to immunoglobulins, with cysteine residues between these sequences (PDGF, VEGF, and FGF receptors and other subfamily members).
- Receptors of group II, exemplified by the insulin receptor and insulin-like growth factor receptor type 1, which are dimmers with two polypeptide

chains, associated by disulfide bridges that originate a $\alpha^2 \beta^2$ heterotetrameric structure.

The association between the ligand and the extracellular domain of the PTK receptor promotes the dimerization of monomeric receptors or a rearrangement of the quaternary structure of heterotetrameric receptors, resulting in autophosphorylation of specific tyrosine residues, present in the cytoplasmic domain [17].

It has been demonstrated that most of PTK receptors have from one to three tyrosine residues in a catalytic domain region known as kinase activation *loop* or *A-loop* [18]. The autophosphorylation of tyrosine residues located in the loop results in deep region shape alterations, facilitating access of ATP and proteins-substrates to their linking sites. Several studies evidenced the importance of phosphorylation of such tyrosine residues in the catalytic activity and consequent biological function of PTK receptors [18] (Fig. 3.3).

PROTEIN TYROSINE KINASES—CYTOPLASMIC PROTEIN TYROSINE KINASES

Cytoplasmic PTKs, as well as those with receptor functions, have a catalytic domain; however, unlike the latter, cytoplasmic PTKs do not have extracellular or transmembrane domains, presenting modules to interact with lipids and other proteins that are mediators of signal transduction processes. Examples of these cytoplasmic PTKs, are Src kinase subfamilies and focal adhesion kinase proteins [19]. These and other proteins (e.g., GTPase activity activating protein and phospholipase Cgamma, etc.) [17], were characterized with two domains: one consisting of approximately 10 amino acids (SH2) and the other with 50–75 amino acids (SH3), with specific functions of interacting with tyrosine phosphorylated proteins or with proline-rich sequence, respectively. These domains, SH2 (Src Homology Type 2) and SH3 (Src Homology Type 3) were so called because they were originally characterized as part of the protein sequence product of proto-oncogene Src [2]. The regulation of cytoplasmic PTK activity frequently involves the interaction between SH2 and SH3 domains. In the case of the Src kinase subfamily, which consists of 11 members (Blk, Lyn, Brk, Fgr, Fyn, Hck, Lck, Src, Srm, and Yes) with high level of structural homology regarding the

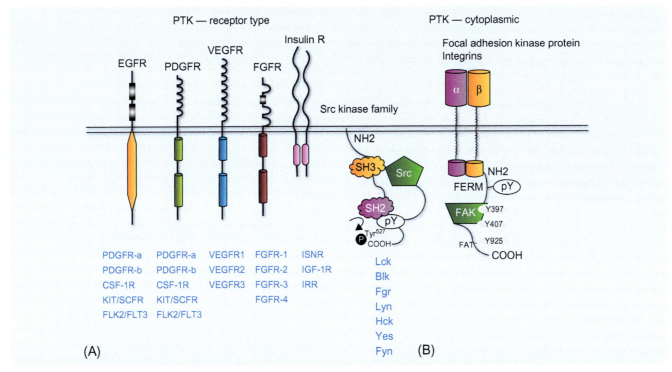

FIG. 3.3 Schematic representation of kinase tyrosine protein prototypes of the receptor type of groups I and II. The name of the prototype receptor is indicated above its respective schematic representation. *EGFR*, epidermal growth factor receptor; *insulin R*, insulin receptor; *PDGFR*, platelet-derived growth factor receptor; *VEGFR*, vascular endothelial growth factor receptor; *FGFR*, fibroblast growth factor receptor. Schematic representation of the prototypes of PTK cytoplasmic FAK and Src-kinase, which are prototypes of their respective families. Highlight: SH2, SH3, and Src-kinase domains, integrin link domains (FERM), and FAK Focal adhesion (FAT) (see text for abbreviations).

regulator domain, the activity is regulated basically in the same fashion [19, 20]. The Src, Fyn, and Yes proteins are expressed in all cellular types [21].

When cells have no stimuli from growth factors or stress conditions, the Src kinase are found in an inactive shape. In this inactive or "closed" shape, the SH2 domain of the enzyme interacts with the tyrosine phosphorylated residue in 527 (p-Tyr527), located in the C-terminal position of its kinase domain. At the same time, domains SH2 and SH3 interact with each other and with the kinase domain, preventing phosphorylation of tyrosine residue 416 (Tyr416) located in the A-loop of the kinase domain [7]. The restrictions imposed by the interactions between different domains shall be cleared when the phosphorylated tyrosine residues belonging to other proteins effectively compete for the SH2 domain of the Src kinase [22] (Fig. 3.3). Alternatively, the Src may be activated by interacting with proteins that have sequences rich in proline, which shall bind to its SH3 domain, clearing the restrictions caused by intramolecular interactions with the SH2

domain and, consequently, the enzyme kinase domain. These domains and their importance in cellular signaling processes shall be detailed appropriately in the next section.

The Src kinase family participates of the several signal transduction pathways, including those that result in cell division and/or activation, by oxidative stress and cytoskeleton rearrangements [19]. The Src signaling activity in endothelial cells is associated to increased permeability of such cells. A rupture between the intercellular junctions of endothelial cells allows increased passage of macromolecules through the endothelium, event that contributes to establish a proangiogenic micro environment [23]. Transformation of cells due to excessive Src kinase expression results in increased tyrosine phosphorylation of proteins associated to focal adhesion, causing loss of adhesion and changes in cell morphology [19]. In particular, this transformation results in phosphorylation of tyrosine residues from another cytoplasmic PTK, the focal adhesion protein tyrosine kinase (FAK) [24].

The structure of FAK is different from other cytoplasmic PTK family proteins, consisting essentially of a central catalytic domain (kinase) flanked by very wide amino and carboxy terminal domains [25]. The amino terminal contains the association region of FAK with cytoplasmic domain of integrin subunit β (FERM). The carboxy terminal domain has the specific amino acid sequences (FABD sequences), which shall connect FAK to the focal adhesion sites [26] (Fig. 3.3). After appropriate stimulus, FAK associated with the integrins and located in a focal adhesion region suffers autophosphorylation in amino acid tyrosine 397 (Tyr397) located between the amino terminal and catalytic domains of the FAK. This tyrosine residue, now phosphorylated, shall be the site in which the FAK associates with the Src kinase SH2 domains. FAK may be phosphorylated in other tyrosine residues, especially residue Tyr925, which some experimental evidence suggests is an interaction site with adapter proteins such as Grb2 or, even, with Pl3K regulator subunit [25]. These proteins consist essentially of SH2 and SH3 domains. In addition to the stimulus provided by the integrins, tyrosine phosphorylation of FAK may be induced by ligands that stimulate receptors coupled to G proteins or those that activate PTK receptors [25]. FAK is part of a subfamily of cytoplasmic kinase tyrosine constituted by the FAK itself and enzyme PYK2/CAKb [27]. Later works showed that PYK2/CAKb kinase is involved in cell response to stress [28].

SIGNAL TRANSMISSION—SH2 AND SH3 DOMAINS

Autophosphorylation in PTK tyrosine residues of the receptor or cytoplasmic type plays two essential roles in biological signal transmission. One is regulating the enzymatic activity of these enzymes, and the other is creating binding sites with high affinity to other signaling molecules.

In general, there are three basic signal transduction mechanisms from interactions between activated PTK and the signaling molecules that contain SH2/SH3 domains (Fig. 3.4).

In the first, tyrosine phosphorylation plays an essential role, and the first evidence of occurrence of such process were obtained with studies of the interactions between the EGF and PDGF receptors with the enzyme phospholipase Cg [29]. The

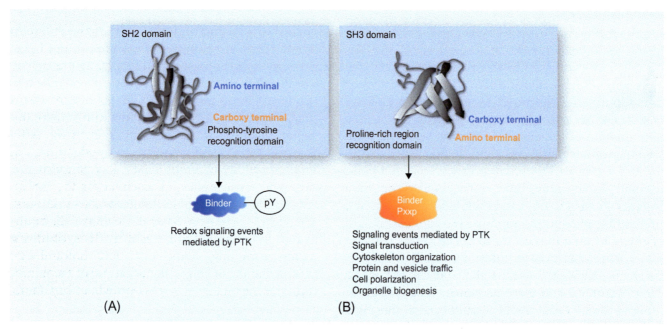

FIG. 3.4 The SH2 and SH3 domains recognize, respectively, phosphorylated tyrosine residues and proline-rich amino acid sequences. These domains act in the regulation of: signaling events mediated by PTKs, redox events, and other biological events (see text for abbreviations).

association of EGF or PDGF growth factors and their specific receptors induces phosphorylation of tyrosine residues present in the cytoplasmic domain of such receptors, creating specific binding sites for the PLCg SH2 domains. By associating to the receptor, the PLCg is phosphorylated into tyrosine and activated, catalyzing the production of diacylglycerol (DAG) e inositol trisphosphate (IP3), two second messengers essential for the activation of protein kinase C, and releasing Ca^{2+} ions from their intracellular stocks, respectively [30].

In the second mechanism, tyrosine phosphorylation also performs an essential role, which could be exemplified by the association of PDGF receptors, or tyrosine phosphorylated interleukin-3 with PI3K. Here, the activation of PI3K recruited by the phosphorylated receptor involves conformational changes. PI3K is a heterodimeric protein, consisting of a regulating subunit of 85 kDa (p85) and a catalytic subunit of 110 kDa (p110). The SH2 domain, present in p85, associates with the phosphotyrosine residue of the PTK receptor. This association causes a conformational change in p85 that shall be transmitted to the catalytic subunit, making the heterodimer, now activated, migrate to the plasma membrane and promote phosphatidylinositol phosphorylation. In this new location, PI3K can interact with the Sos protein that promotes GDP/GTP exchange in Ras. Besides that, PI3K shall participate in the activation of ribosomal S6 kinase, Akt/protein kinase B, and transcription factor NF-κB [31].

A third way to conduct the signal is from changes in the intracellular location of the signaling pathway constituents. In this case, Grb2, by means of its SH3 domains, interacts with a proline-rich region of the Sos protein, which has the function of promoting GDP/GTP exchange in Ras. The Grb2/Sos complex binds to the phosphorylated tyrosine residue of the PTK, by means of domain SH2 of protein Grb2, moving to the plasma membrane. In this new location, the Grb2/Sos complex is phosphorylated into tyrosine and binds to the SHC adapter protein. This multiprotein complex activates Ras, promoting exchange of GDP for GTP in the proto-oncogene. Ras associated to the GTP starts a new signaling cascade with activation of PKs in the cytoplasm, culminating with the activation of transformation factors in the nucleus [31]. This is the Ras-ERK1/2 MAP kinase signaling module.

CONFORMATIONAL CHANGES AND POSTTRANSLATIONAL MODIFICATIONS (PTM) IN RAS

The signal sent by receptors of cytoplasmic PTKs is almost invariably conducted by Ras, a 21-kDa oncoprotein with GTPase activity which has three isoforms: N-Ras, H-Ras, and K-Ras. Ras acts as regulator of cellular growth and other functions in all eukaryotic cells. As a GTPase, the Ras biological activity is regulated by the release of GDP and association with GTP. Two groups of enzymes that present antagonistic activity control the Ras-GDP/Ras-GTP cycle. Guanine (GEFs), RasGRF1/2, and Sos nucleotide exchange factors promote the formation of Ras-GTP, the active form of Ras. In the opposite direction, GTPase activity stimulating proteins (GAPs), p120 GAP e NF1, accelerate the hydrolytic activity of GTP, intrinsic of Ras, promoting the generation of the inactive form of Ras associated to GDP.

The GDP-GTP exchanges do not imply Ras PTM. However, new findings have evidenced a decisive role for Ras PTM regarding the distribution of the GTPase in the different intracellular compartments [32]. Some of these PTMs are constitutive, and occur immediately after the translation, and others are conditional. Ras farnesylation is a constitutive and irreversible Ras PTM, catalyzed by the farnesyltransferase enzyme, which changes the cysteine residue of the CAAX sequence. Subsequently, proteolytic cleavage takes place with the removal of the AAX group and, finally, the cysteine residue suffers a methyl esterification process. These three changes promote remodeling of the carboxy terminal domain of the Ras proteins of a hydrophilic region to a hydrophobic region, making them capable of entering cellular membranes and collaborating for the signaling activity. Among constitutive PTMs, Ras palmitoylation, which is irreversible and essential for the traffic of Ras isoforms between endomembranes and plasma membranes, deserves special mention [32]. In addition to irreversible and reversible PTMs, conditional PTMs, including phosphorylation, ubiquitination, glycosylation, and S-nitrosylation, are of great importance for the signaling processes mediated by Ras [32].

RAS INTRACELLULAR COMPARTMENTALIZATION

Farnesylation and palmitoylation of Ras will determine its location in different intracellular compartments. The newly synthesized Ras protein is present, transiently, in the endoplasmic reticulum and Golgi apparatus. In addition, two Ras isoforms, N-Ras and H-Ras, are expressed permanently and abundantly in the Golgi apparatus [33, 34].

When receptors in the cell surface are activated, the Src tyrosine kinase protein is recruited to a region in the plasma membrane next to the intracellular domain of the receptor that received the signal. Simultaneously, PLC-γ1 is recruited by the receptor and phosphorylated by Src. As mentioned, the activation of PLC-γ1 results in the production of important second messengers, DAG and IP3, and the action of the latter on the specific receptors of the endoplasmic reticulum causes the release of Ca^{2+} ions of such stocks. DAG e Ca^{2+} lead to translocation of a cytoplasmic protein, Ras-GRP1 (member of the GEFs family), to the Golgi apparatus, activating the Ras associated to this organelle. This compartmentalized signaling pathway coexists with the "classic" Ras activation pathway in the endoplasmic membrane, which is independent from the

intracellular levels of Ca^{2+}. Elevated intracellular levels of Ca^{2+} also induce activation of CAPRI (member of the GAPs family), which possibly inhibit the activation of Ras present in the plasma membrane (Fig. 3.5) [35].

RAS SIGNALING MODULE—ERK1/2 MAP KINASE

Ras is a central mediator of the flow of information generated from activated receptor PTKs, promoting a cascade of reactions catalyzed by PKs. Among these PKs, the serin/threonine PK, Raf (c-Rafl, A-Raf, and B-Raf) isoforms have been studied the most. It is already well established that Raf is the main effector of Ras in cellular signaling pathways. Ras interacts with Raf by means of two distinct regions located in the Raf amino terminal domain, promoting its displacement from the cytoplasm to the membrane, and facilitating its activation. In

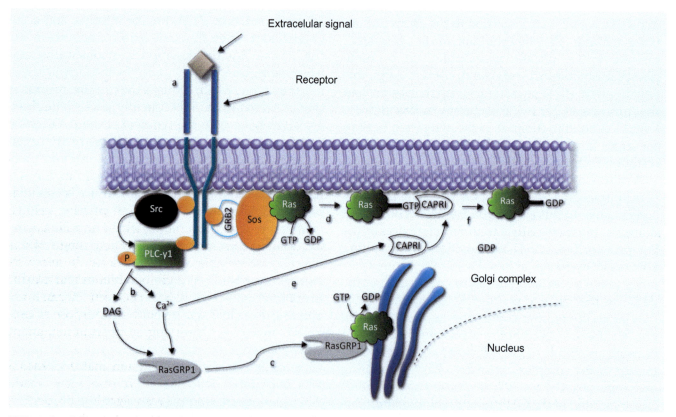

FIG. 3.5 Schematic representation of the Ras compartmentalized signaling pathway describing the events that lead to Ras signaling from the plasma membrane in connection with the signaling events of this protein in the Golgi apparatus, as suggested by Bivona et al. [35] (see text for abbreviations).

addition to Ras, phospholipids, PKs, and other proteins also promote Raf activation, evidencing the formation of an apparatus involving several proteins that would take part in this process. After activation, Raf shall promote MEK phosphorylation, which is a PK that phosphorylates other proteins in the threonine and tyrosine amino acids, with the specific function of activating MAP kinase ERK1/ERK2, phosphorylating these PKs in threonine and tyrosine residues [31, 36]. In turn, ERK1/ERK2 shall phosphorylate other proteins in the cytoplasm and migrate to the nucleus, stimulating the activity of transformation factors [37]. The Raf kinase are enzymes able to convert the signals sent through tyrosine phosphorylation in signals carried by serine/threonine phosphorylation means, connecting PTK receptors or cytoplasmic PTKs to the transcription factors. However, other evidence suggests that the signaling events that occur after Ras involve more than Raf activation. In addition to the cascade reactions of Raf-dependent phosphorylation that culminate with the activation of ERK1/ERK2, two other MAP kinase, Jun NH2-amino terminal kinase (JNK) and p38 MAP kinase, are activated by Ras regardless of Raf. [31]. Therefore, in addition to Raf, other Ras effectors were characterized, including p120 Ras-GAP, MEKK, and PI3K [36]. The action of this set of effectors expands the Ras action spectrum, placing signaling pathways started by growth factors, hormones, cytokines, and stress conditions, including oxidative and nitrosative stress, under the control of this protein [8, 37–39].

In the signaling cascade, right below Ras, a group of PKs performs an essential function in the transformation of signals generated originally by a growth factor, cytokine, or active redox species [37]. These kinases that we described are organized in a signaling module consisting basically of three prototype enzymes (Fig. 3.6A):

(1) serin/threonine PKs, commonly known as MAP3Ks (e.g., Raf-1, MEKK);
(2) MAPKKs, which are double-specificity PKs that catalyze the phosphorylation of MAP kinase (MAPKs) into threonine and tyrosine amino acids (e.g., MEK, SEK); and
(3) MAPKs, which are serine/threonine PKs that migrate to the nucleus, phosphorylating and activating transformation factors (e.g., ERK1/ERK2, JNK, and p38 MAPK).

CELLULAR SIGNALING IN THE ENDOTHELIUM: ANGIOGENESIS AND PARTICIPATION OF NITRIC OXIDE (NO)

Formation of new capillary vessels from pre-existing vascular structures is of fundamental importance to maintain vascular integrity in repair of damaged tissue, healing of injuries, and formation of vessels in respect to ischemia events. Angiogenesis is a complex process, orchestrated by cytokines and growth factors. It begins with stimulation of endothelial cells in quiescence, evolving to a state of quick proliferation and migration, ending with redifferentiation of activated endothelial cells that compose the vascular tubes.

Angiogenesis regulation is carried out by several growth factors, such as EGF, VEGF, and FGF. Among these factors, VEGF was characterized as one of the most potent, and may regulate angiogenic processes of pathological or physiological order. The VEGF performs a fundamental role in the angiogenesis process, by binding to its receptor (VEGFR), especially when the expression levels of this factor rise after hypoxia stimulus. Hypoxia stimulates vascular expansion, and the signaling mediated by hypoxia transcription factors is fundamental in the positive regulation of several genes related to angiogenesis [40].

The biological responses to VEGF are mediated by the activation of its PTK, VEGFR-1, VEGFR-2, and VEGFR-3 receptors. VEGFR-2 and VEGFR-3 participate of the activation of the Ras-MAP kinase signaling module, while the VEGFR-1 receptor preferably activates the PI3K/Akt signaling module [41, 42]. Besides the VEGFRs, whose activity is essential for the development and growth of new vessels, another class of PTK receptors, Tie-1 and Tie-2, performs an important role in the final stage of the angiogenesis process, more specifically in remodeling new vessels [43]. The use of inhibitors for the different isoforms of VEGFR has been increasingly frequent in clinical studies, with the purpose of testing strategies to prevent angiogenesis in malignant diseases [44].

Pioneering studies have shown that endothelial cells stimulated with VEGF release nitric oxide (NO), gaseous free radical with signaling properties, and this release was demonstrated by the neutralizing action of specific monoclonal antibodies for the VEGFR-1 receptor [45].

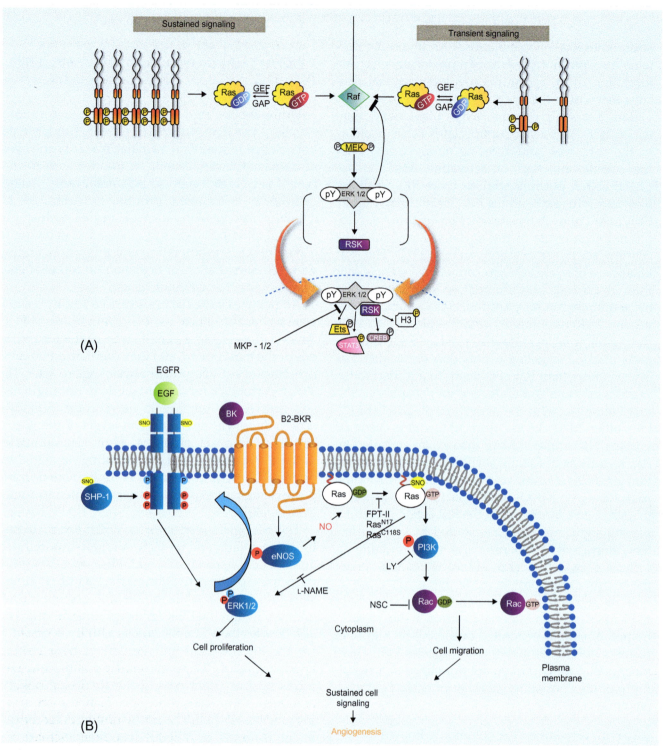

FIG. 3.6 (A) After binding growth factors to the respective receptors with Ras activation, the Raf (MAP3K) is recruited by the plasma membrane and are activated. Activated Raf phosphorylates and activates the MEK1/2 (MAP2K) protein kinase, which, in turn, phosphorylates the ERK1/2 MAP (MAPK) protein kinase. The activated ERK1/2 MAP kinase phosphorylates other proteins located on the inner face of the plasma membrane and cytoplasm. The ERK1/2 MAP kinase activated migrates to the nucleus, where it shall phosphorylate transcription factors and transcription regulators. High concentrations of growth factors shall mediate sustained signaling processes. Low concentrations of these factors shall mediate transient signaling processes. (B) Cell signaling pathways, triggered by BK and mediated by NO in endothelial cells. According to this model, three different pathways are integrated: activation of the EGF receptor, activation of the B2 receptor by peptide BK, and activation of low molecular weight Ras protein G, and their interactions with the Rac1 protein. The integration between these three pathways results in cell proliferation and migration, culminating in the angiogenesis process (see text for abbreviations).

NO is produced enzymatically from L-arginine and O_2 by NO synthase enzymes (NOS), isolated in endothelial cells, vascular smooth muscle cells, macrophages, neuronal cells, fibroblasts, platelets, and in cells of different types of tumors. There are three main NOS isoforms, with well-characterized properties: the two constitutive isoforms, which depend on calcium/calmodulin, and were isolated initially in endothelial and neuronal cells, and the inducible isoform, present in macrophages, vascular smooth muscle cells, fibroblasts, and tumor cells [46]. Besides the enzymatic production, the NO may also be produced in physiological conditions from the oxidation of compounds known as NO donors [47]. As mentioned, due to its chemical characteristics, the NO is able to stimulate signaling processes through reactions with sulfhydryl groups, with reactive oxygen species, or transition metals. In physiological conditions, the NO reacts with O_2, transition metals, and sulfhydryls, generating nitrogen oxides, peroxynitrite, metal-NO adducts, and S-nitrosothiols [48].

Angiogenesis begins with vasodilation, a process mediated by NO that, when reacting with an iron-heme group of the soluble form the guanylyl cyclase enzyme, stimulates the production of the second cGMP messenger and/or acts by modifying cysteine residues of the proteins that participate of several cell signaling pathways. This modification was defined as S-nitrosylation, a PTM that turns NO into a free radical that participates in a wide spectrum of cellular signaling processes [39], in particular signaling processes that occur in endothelial cells.

The first works published by Ziche et al. [49] established the basis for understanding the role of NO as mediator of endothelial cell proliferation. In these studies, the authors showed that the NO donors of different classes, sodium nitroprusside (SNP), glyceryl trinitrate, and isosorbide dinitrate promoted capillary endothelial cell growth and mobilization in growing concentrations. In addition, vasoactive substances such as substance P and VEGF had their ability to induce endothelial proliferation associated to the production of NO by such cells [49, 50]. Increase of intracellular cGMP levels and stimulation of ERK1/2 MAP kinase were somehow related to the NO proproliferation activity [51].

In studies conducted by our research group, with fibroblasts of HER14 mouse incubated with NO, SNP, and S-nitroso-N-acetylpenicillamine (SNAP) donors, in the absence or presence of EGF, we detected an increased tyrosine phosphorylation in a group of 126, 56, 43, and 40 kDa proteins [52]. Src kinase was immunoprecipitated from HER14 cells treated with SNP or EGF, and we observed that they were identical to protein 56 kDa, with the tyrosine phosphorylation stimulated by NO [53]. Another cytoplasmic PTK that apparently has activity regulated by the intracellular redox state is the FAK kinase. Protein 126 kDa, whose tyrosine phosphorylation levels were stimulated after HER14 mice fibroblast incubation with NO donor, were identical to FAK [53]. Pioneer studies published by Lander et al. [54] described a new NO signaling mechanism independent from cGMP. Lander et al. [54] showed that NO would signal through the activation of protein G (which connects to guanine nucleotides). In vitro studies, using recombinant Ras, revealed that NO promoted pronounced conformational change of the protein associated to a GDP release stimulus and consequent association with GTP [38]. The Ras conformational change, resulted in its activation, depended on S-nitrosylation of cysteine residue 118 (Cys118), located in the guanine nucleotide binding domain in Ras [55].

After demonstrating Ras regulation by NO, the investigation was directed towards the possibility of MAPKs also being prone to the same type of regulation. Therefore, studies conducted by our research group and others showed the participation of mechanisms of redox nature in activity regulation of PKs belonging to the three MAPK subfamilies: ERK1/ERK2, JNK, and p38 MAPKs [56].

In subsequent studies, we focused on unraveling this relation by describing a cellular signaling pathway triggered by NO donors in endothelial cells of rabbit aorta, which uses elements of the dependent and independent pathway of guanylyl cyclase activation [57]. Initially, we observed that low concentrations of the NO donors SNP and SNAP raised the intracellular levels of cGMP. Both donors and the cGMP stable analog, 8BrcGMP, stimulated tyrosine phosphorylation, and also promoted activity of Ras and other elements that compose the Ras-MAP kinase expression signaling cascade. These findings lead us to propose the expression of a guanine nucleotide-exchange protein (CNRasGEF) in the endothelium, with activity positively modulated by intracellular cGMP/cAMP levels, and with previously documented expression in neural cells [58]. In cells stimulated with NO and 8Br-cGMP donors, we observed transactivation of the receptor EGF (EGFR) mediated by ERK1/2 MAP kinase [57].

We have also demonstrated in rabbit aortic endothelial cells (RAEC) that the activation of signaling module Ras-ERK1/2 MAP kinase, by means of low concentrations of NO SNAP donor, depended essentially on S-nitrosylation of Ras cysteine residue 118 (Cys118), resulting in the activation of GTPase and other elements of the downstream cascade signaling related to Ras. Activation of ERK1/2 MAP kinase in this process resulted in its migration to the nucleus and activation of the Elk1 transcription factor by such kinase. The activation of Elk 1 led to stimulation of cyclins synthesis and activity of cyclin-dependent kinases (CDKs) with progression through the cellular cycle and culminating in endothelial cell proliferation [59].

In subsequent studies, we demonstrated that different NO generation sources, such as s-nitrosothiol, s-nitrosoglutathione (GSNO) and production of NO by human umbilical vein endothelial cell (HUVEC) stimulation with Bradykinin peptide (BK) mediated Ras activation in different cellular compartments. Human cervical tumor HeLa cells stimulated with low concentrations of GSNO presented a Ras activation profile that involved two cellular compartments, the plasma membrane and the Golgi apparatus. In the experimental conditions of this model, we observed a quick and transient Ras activation in the plasma membrane followed by sustained Ras activation in the Golgi apparatus. Ras activation in the Golgi apparatus, mediated by GSNO in these cells, depended on S-nitrosylation and Src cytoplasmic PTK activation. On the other hand, Ras activation by endogenous NO sources in HUVEC cells, stimulated with peptide BK, resulted in a process that occurs in a single compartment, the plasma membrane of these cells. Unlike what was observed when the agonist was GSNO, Ras activation resulting from the action of NO generated by cell stimulation with peptide BK was not dependent on the activity of Src kinase [60].

In two other recent studies [61, 62], we evidenced the central role played by NO as positive regulator of the signaling pathways associated with endothelial cell proliferation and migration that culminated in angiogenesis processes. In these studies, the NO was generated from stimulation of RAEC and HUVEC endothelial cells with peptide BK. This stimulus promoted the activation of the NOS endothelial isoform (eNOS) and stimulation of tyrosine phosphorylation and cysteines-nitrosylation of EGFR. In addition to the nitrosylation of EGFR,

NO promoted S-nitrosylation of PTP SHP-1 and GTPase Ras. Ras activation by S-nitrosylation from endogenous NO sources also promoted activation of ERK1/2 MAP kinase, corroborating observations we made when we incubated these cells with low concentration of s-nitrosothiol SNAP.59 The ERK1/2 MAP kinase, on the other hand, when activated, phosphorylated specific threonine residues of the EGFR cytoplasmic domain, which prevented immediate internalization of this receptor.

BK stimulated endothelial cells proliferation. The use of L-NAME, nonspecific NOS inhibitor, and PD98059, specific inhibitor of MEK kinase protein, inhibited the proliferation of HUVEC endothelial cells stimulated by BK. The BK stimulus also induced the expression of VEGF. Still in the same study, we showed that VEGF expression depends on the activation of EGFR, BK receptor B2, and Ras. We also confirmed that VEGF expression depends on endogenous NO production.

Our research group is implementing a new experimental approach, in which we investigated the relation between the response of HUVEC endothelial cells to the BK stimulation, resulting in an in vitro angiogenesis process and the expression of different microRNAs. These noncoding RNAs of approximately 18–22 nucleotides, which regulate gene expression by suppressing translation or degradation of the equivalent messenger RNA. Initially, by means of in silico prediction, nine proangiogenic miRs (miR-27b, miR-92a, miR-126, miR-130a, miR-132, miR-210, miR296-3p, miR-296-5p, and miR-378) and eight antiangiogenic miRs (miR-15a, miR-15b, miR-16, miR-20a, miR-20b, miR-21, miR-221, and miR-222) were identified, which regulated the expression of proteins that participate in cell signaling pathways mediated by VEGF, EGFR, and eNOS [61]. By means of analysis using appropriate techniques, we identified several microRNAs in absence of stimulus. In these conditions, we identified miR-16-5p, miR-125b-5p, miR-126-3p, and miR-222-3p. However, in the presence of BK stimulation, we identified the following microRNAs: miR-15b-5p, miR-20a-5p, miR-21-5p, and miR-125a-5p. MiR-375 is currently expressed when we compared the following experimental situations: nonstimulated cells, cells stimulated with BK, cells stimulated with BK and preincubated with specific eNOS inhibitor, L-N5-(1-Iminoethyl) ornithine. On the other hand, microRNAs: miR-15a-5p, miR-15b-5p, miR-20b-5p, miR-130a-3p, and miR-210-3p, are under expressed in cells not stimulated by BK. The levels of expression of microRNA miR-33a-5p are also reduced when HUVEC cells are stimulated by BK. These new findings allowed us to describe a cellular signaling pathway regulation process mediated by NO, which leads to angiogenesis performed by specific microRNAs [63].

Another essential element in the angiogenesis process is cell migration stimulated by NO. Several independent observations describe the role of NO as positive regulator of alpha5beta3 integrin and extracellular matrix metalloproteinase expression levels, MT1-MMP, MMP-9, and MMP-13, allowing the invasion of endothelial cells and formation of new vessels [64–67]. New observations corroborate and expand these observations when we proceeded with RAEC endothelial cell stimulation with BK during shorter periods than those determined to follow up cell proliferation. By proceeding this way, we observed the occurrence of migration of these cells depending on the endogenous production of NO. The NO indirectly promoted GTPase Racl activation by means of Ras S-nitrosylation and activation of PI3K [62].

Together, these observations allowed us to attribute to the activation of signaling pathways initiated by EGFR/BK-B2R receptors and/or by GTPase Ras to NO, an essential role in angiogenesis. These findings are summarized in Fig. 3.6B.

References

[1] Hunter T. Protein kinases and phosphatases: the Yin and Yang of protein phosphorylation and signaling. Cell 1995;80:225–36.

[2] Schlessinger J. SH2/SH3 signaling proteins. Curr Opin Genet Dev 1994;4:25–30.

[3] Monteiro HP, Arai RJ, Travassos LR. Protein tyrosine phosphorylation and protein tyrosine nitration in redox signaling. Antioxid Redox Signal 2008;10:843–89.

[4] Thibeault S, Rautureau Y, Oubaha M, et al. S-nitrosylation of β-catenin by eNOS-derived NO promotes VEGF-induced endothelial cell permeability. Mol Cell 2010;39:468–76.

[5] Curcio MF, Batista WL, Linares E, et al. Regulatory effects of nitric oxide on Src kinase, FAK, p130Cas, and receptor protein tyrosine phosphatase alpha: a role for the cellular redox environment. Antioxid Redox Signal 2010;13:109–25.

[6] Feng X, Sun T, Bei X, et al. S-nitrosylation of ERK inhibits ERK phosphorylation and induces apoptosis. Sci Rep 2013;3:1814.

[7] Hubbard SR, Mohammadi M, Schlessinger J. Autoregulatory mechanisms in protein tyrosine kinases. J Biol Chem 1998;273: 11987–90.

[8] Monteiro HP, Stern A. Redox modulation of tyrosine phosphorylation-dependent signal transduction pathways. Free Radic Biol Med 1996;21:323–33.

[9] Tonks N. Protein tyrosine phosphatases: from genes, to function, to disease. Nat Rev Mol Cell Biol 2006;7:833–46.

[10] Sun H, Tonks NK. The coordinated action of protein tyrosine phosphatases and kinases in cell signaling. Trends Biochem Sci 1994;19:480–5.

[11] Viñals F, Pouyssegur J. Confluence of vascular endothelial cells induces cell cycle exit by inhibiting p42/44 mitogen-activated protein kinase activity. Mol Cell Biol 1999;19:2763–72.

[12] Brady-Kalnay SM, Tonks NK. Protein tyrosine phosphatases as adhesion receptors. Curr Opin Cell Biol 1995;7:650–7.

[13] Su J, Muranjan M, Sap J. Receptor protein tyrosine phosphatase a activates Src-family kinase and controls integrin-mediated responses in fibroblasts. Curr Biol 1999;9:505–11.

[14] Monteiro HP, Ivaschenko Y, Fischer R, et al. Inhibition of protein tyrosine phosphatase activity by diamide is reversed by epidermal growth in fibroblasts. FEBS Lett 1991;295:146–8.

[15] Hussain SP, Hofseth LJ, Harris CC. Radical causes of cancer. Nat Rev Cancer 2003;3:276–85.

[16] Sundaresan M, Yu ZX, Ferrans VJ, et al. Requirement for generation of HO for platelet-derived growth factor signal transduction. Science 1995;270:296–9.

[17] Schlessinger J. Cell signaling by receptor tyrosine kinases. Cell 2000;103:211–25.

[18] Mohammadi M, Schlessinger J, Hubbard S. Structure of the FGF receptor tyrosine kinase domain reveals a novel autoinhibitory mechanism. Cell 1996;86:577–87.

[19] Abram CL, Courtneidge SA. Src family tyrosine kinases and growth factor signaling. Exp Cell Res 2000;254:1–13.

[20] Manning GDB, Whyte RM, Hunter T, et al. The protein kinase complement of the human genome. Science 2002;298:1912–34.

[21] Thomas SM, Brugge JS. Cellular functions regulated by Src family kinase. A comprehensive review of the physiology and substrates of Src and Src-family kinases. Annu Rev Cell Dev Biol 1997;13:513–609.

[22] Cobb BS, Schaller MD, Lee TH, et al. Stable association of pp60src and p59fyn with the focal adhesion-associated protein tyrosine kinase, pp125FAK. Mol Cell Biol 1994;14:147–55.

[23] Eliceiri BP, Paul R, Schwartzberg PL, et al. Selective requirement for Src kinases during VEGF-induced angiogenesis and vascular permeability. Mol Cell 1999;4:915–24.

[24] Schaller MD, Hildebrand JD, Shannon JD, et al. Autophosphorylation of the focal adhesion kinase pp125FAK, directs SH2-dependent binding of pp60src. Mol Cell Biol 1994;14:1680–8.

[25] Schaller MD. The focal adhesion kinase. J Endocrinol 1996;150:1–7.

[26] Blume-Jensen P, Hunter T. Oncogenic kinase signaling. Nature 2001;411:355–65.

[27] Sasaki H, Nagura K, Ishino M, et al. Cloning and characterization of cell adhesion kinase beta, a novel protein-tyrosine kinase of the focal adhesion kinase subfamily. J Biol Chem 1995;270:21206–19.

[28] Dikic I, Tokiwa G, Lev S, et al. A role for PYK2 and Src in linking G-protein coupled receptors with MAP kinase activation. Nature 1996;000:01F 00.

[29] Margolis B, Rhee SG, Felder S, et al. EGF induces tyrosine phosphorylation of phospholipase C-II: a potential mechanism for EGF receptor signaling. Cell 1989;57:1101–7.

[30] Nishizuka Y. Intracellular signaling by hydrolysis of phospholipids and activation of protein kinase C. Science 1992;258:607–14.

[31] Vojtek AB, Der CJ. Increasing complexity of the Ras signaling pathway. J Biol Chem 1998;273(32):19925–8.

[32] Ahearn IM, Haigis K, Bar-Sagi D, et al. Regulating the regulator: post-translational modifications of Ras. Nat Rev Mol Cell Biol 2012;1:39–51.

[33] Apolloni A, Prior IA, Lindsay M, et al. H-ras but not K-ras traffics to the plasma membrane through the exocytic pathway. Mol Cell Biol 2000;20(7):2475–87.

[34] Choy E, Chiu VK, Silletti J, et al. Endomembrane trafficking of Ras: the CAAX motif targets proteins to the ER and Golgi. Cell 1999; 98:69–80.

[35] Bivona TG, Pérez De Castro I, Ahearn IM, et al. Phospholipase Cgamma activates Ras on the Golgi apparatus by means of Ras-GRP1. Nature 2003;424:694–8.

[36] Kolch W. Meaningful relationships: the regulation of the Ras/Raf, MEK/ERK pathway by protein interactions. Biochem J 2000;351: 289–305.

[37] Garrington TP, Johnson CL. Organization and regulation of mitogen-activated protein kinase signaling pathways. Curr Opin Cell Biol 1999;11:211–8.

[38] Lander HM, Ogiste JS, Pearce SFA, et al. Nitric oxide-stimulated guanine nucleotide exchange on p21 Ras. J Biol Chem 1995;270: 7017–20.

[39] Hess DT, Matsumoto A, Kim SO, et al. Protein S-nitrosylation: purview and parameters. Nat Rev Mol Cell Biol 2005;6:150–66.

[40] Carmeliet P. Angiogenesis in health and disease. Nat Med 2003;9:653–60.

[41] Koch S, Tugues S, Li X, et al. Signal transduction by vascular endothelial growth factor receptors. Biochem J 2011;437:169–83.

[42] Tvorogov D, Anisimov A, Zheng W, et al. Effective suppression of vascular network formation by combination of antibodies blocking VEGFR ligand binding and receptor dimerization. Cancer Cell 2010;18:630–40.

[43] Augustin HG. Angiogenesis in the female reproductive system. EXS 2005;94:35–52.

[44] Jeltsch M, Leppanen VM, Saharinen P, et al. Receptor tyrosine kinase-mediated angiogenesis. Cold Spring Harb Perspect Biol 2013;5:a009183.

[45] Bussolati B, Dunk C, Grohman M, et al. Vascular endothelial growth factor recptor-1 modulates vascular endothelial growth factor mediated angiogenesis via nitric oxide. Am J Pathol 2001;159: 993–1008.

[46] Fleming I, Busse R. Signal transduction of eNOS activation. Cardiovasc Res 1999;43:532–41.

[47] Cruetter DY, Cruetter CA, Barry BK, et al. Activation of coronary arterial guanylate cyclase by nitric oxide, nitroprusside, and nitrosoguanidine—inhibition by calcium, lanthanum, and other cations, enhancement by thiols. Biochem Pharmacol 1980;29:2943–50.

[48] Stamler JS, Lamas S, Fang FC. Nitrosylation: the protoypic redox-based signaling mechanism. Cell 2001;106:675–83.

[49] Ziche M, Morbidelli L, Masini E, et al. Nitric oxide mediates angiogenesis in vivo and endothelial cell growth and migration in vitro promoted by substance P. J Clin Invest 1994;94:2036–44.

[50] Bussolati B, Dunk C, Grohman M, et al. Vascular endothelial growth factor receptor-1 modulates vascular endothelial growth factor-mediated angiogenesis via nitric oxide. Am J Pathol 2001;159:993–1008.

[51] Parenti A, Morbidelli L, Cui XL, et al. Nitric oxide is an upstream signal of vascular endothelial growth factor-induced extracellular-signal-regulated kinase 1/2 activation in postcapillary endothelium. J Biol Chem 1998;273:4220–6.

[52] Peranovich TMS, da Silva AM, Fries DM, et al. Nitric oxide stimulates tyrosine phosphorylation in murine fibroblasts in the absence and presence of epidermal growth factor. Biochem J 1995;305: 613–9.

[53] Monteiro HP, Gruia-Gray I, Peranovich TMS, et al. Nitric oxide stimulates tyrosine phosphorylation of focal adhesion kinase, Src kinase and mitogen-activated protein kinases in murine fibroblasts. Free Radic Biol Med 2000;28:174–82.

[54] Lander HM, Sehaipal P, Levine DM, et al. Activation of human peripheral blood mononuclear cells by nitric oxide-generating compounds. J Immunol 1993;150:1509–16.

[55] Lander HM, Milbank AJ, Tauras JM, et al. Redox regulation of cell signaling. Nature 1996;381:380–1.

[56] Allen RG, Tresini M. Oxidative stress and gene regulation. Free Radic Biol Med 2000;28:463–99.

[57] Rocha Oliveira CL. Óxido nítrico e cGMP ativam a via de Ras-MAP quinase estimulando fosforilação de resíduos de tirosina em proteínas intracelulares e proliferação em céulas endoteliais de aorta de coelho [Dissertação de Mestrado]. Departamento de Bioquímica, Disciplina Biologia Molecular, Escola Paulista de Medicina, Universidade Federal de São Paulo; 2001.

[58] Pham N, Cheglakov CA, Koch CL, et al. The guanine nucleotide exchange factor CNrasGEF activates Ras in response to cAMP and cGMP. Curr Biol 2000;10:555–8.

[59] Rocha Oliveira CJ, Curcio MF, Moraes MS, et al. The low molecular weight s-nitrosothiol, S-nitroso-N-acetylpenicillamine, promotes cell cycle progression in rabbit aortic endothelial cells. Nitric Oxide 2008;18:241–55.

[60] Batista WL, Ogata FT, Curcio MF, et al. S-nitrosoglutathione and endothelial nitric oxide synthase-derived nitric oxide regulate compartmentalized Ras S-nitrosylation and stimulate cell proliferation. Antioxid Redox Signal 2013;18:221–38.

[61] Moraes MS, Costa PE, Batista WL, et al. Endothelium-derived nitric oxide (NO) activates NO-epidermal growth factor receptor mediated signaling pathway in bradykinin-stimulated angiogenesis. Arch Biochem Biophys 2014;558:14–27.

[62] Borges RE, Batista WL, Costa PE, et al. Ras, Rac1, and phosphatidylinositol-3kinase (PI3K) signaling in nitric oxide-induced endothelial cell migration. Nitric Oxide 2015;47:40–51.

[63] Medeiros Albuquerque MTO, Carvalho CV, Rosa H, et al. Prediction of MicroRNAs associated with the free radical NO during angiogenesis. "Annals of the 23rd congress of the international union for biochemistry and molecular biology and 44th annual meeting of the Brazilian society for biochemistry and molecular biology." Brazilian Society for Biochemistry and Molecular Biology (SBBq); 2015.

[64] Lee PC, Kibbe MR, Schuchert MJ, et al. Nitric oxide induces angiogenesis and up regulates alpha(v)beta(3) integrin expression on endothelial cells. Microvasc Res 2000;60:269–80.

[65] López Rivera E, Lizarbe TR, Martínez-Moreno M, et al. Matrix metalloproteinase 13 mediates nitric oxide activation of endothelial cell migration. Proc Natl Acad Sci U S A 2005;102:3685–90.

[66] Genís L, Gonzalo P, Tutor AS, et al. Functional interplay between endothelial nitric oxide synthase and membrane type 1 matrix metalloproteinase in migrating endothelial cells. Blood 2007;110: 2916–23.

[67] Lee CZ, Xue Z, Hao Q, et al. Nitric oxide in vascular endothelial growth factor-induced focal angiogenesis and matrix metal-loproteinase-9 activity in the mouse brain. Stroke 2009;40:2879–81.

4

Endothelial Barrier: Factors That Regulate Its Permeability

Erik Svensjö and Eliete Bouskela

1 INTRODUCTION

The endothelium plays an important role in vascular function maintenance. Endothelial barrier dysfunction leads to increased permeability and vascular leakage is associated to several pathological conditions, such as edema and sepsis. Therefore, a very active research field today is discovery/development of drugs that may improve endothelial function/dysfunction. In the last decades, several diseases were characterized as having an inflammatory component due to new discoveries regarding changes in endothelial cell functions induced by them. Atherosclerosis, asthma, and diabetes are today considered inflammatory diseases, and even a clinical condition such as obesity may be considered an inflammatory disease of low degree with changes in endothelial function that precede the development of more severe consequences such as diabetes or hypertension [1].

The vascular endothelium is the mono-layer of cells that covers the cardiovascular system responsible for the barrier between tissue and blood. Therefore, the endothelium is strategically positioned to serve as a communication sentinel between the information proceeding from the interstice to the immunological system cells (e.g., leukocytes). Therefore, the endothelium plays a critical role in directing traffic patterns, as well as leukocyte activation states. The endothelium can also be injured by inappropriate immunological/inflammatory responses that transform it into a dysfunctional barrier, characteristic of several inflammatory diseases such as sepsis, ischemia/reperfusion (I/R), anaphylaxis, atherosclerosis, and arthritis.

In recent years, the continuous and massive publication of studies and reviews regarding the function of the endothelial barrier illustrates the need to understand the endothelial cell (EC) function in different diseases—diabetes, atherosclerosis, asthma, periodontitis—as well as tropical diseases, Chagas disease, leishmaniasis, and dengue fever. All of these examples show that breakage of the endothelial barrier, which allows leakage of plasma, may be the initial event for progressive disease development.

The physiological balance between the volume of blood circulating and volume of extravascular liquid is maintained by the healthy endothelium that covers the entire cardiovascular system. The exchange of solutes and liquids in normal conditions has been the object of many studies and reviews, among which it is worth mentioning Michel [2], Michel and Curry [3], and more recently the review by Levick and Michel [4] regarding the importance of the glycocalyx for the exchange of solutes that takes place in microcirculation. The endothelial glycocalyx is formed by a semipermeable thin layer, with thickness of 400–500 nm, consisting of proteoglycans and glycoproteins bound to the endothelial cell membrane [5]. The mechanical properties of this layer limit the access of plasma components to the membrane of endothelial cells. The barrier properties of this endothelial surface layer are deduced from the penetration speeds of markers, such as fluorescein-labeled dextran, or FITC-dextran, and by the mechanics of red blood cells and leukocytes in capillary vessels [6]. The glycocalyx in glomerular vessels is essential for normal filtration of protein by the kidneys [7]. Metformin, drug used in diabetes treatment, administered to diabetic mice, for 2 weeks, improved the barrier function of the glycocalyx, measured by the duration of macromolecule retention in circulation [8].

An overview of the microvascular permeability or tissue edema is beyond the scope of this chapter, since these topics have already been widely reviewed in detail [3,9,10]. This chapter intends to show the most important aspects of regulation of endothelial permeability in some clinical conditions in which inflammation is an important component, and indicate the progress of the knowledge of regulatory mechanisms of adherens junctions between endothelial cells in the postcapillary region of the microcirculation, which have advanced significantly in recent years, and to inform about references wherein detailed information can be found.

The key for the inflammatory process is injury or destruction of the endothelial barrier initiated by active contraction of the EC cytoskeleton. Majno and Palade [11] and Majno et al. [12] showed in their classic studies with electron microscopy that, in vascular leakage induced by histamine and serotonin, colloidal particles passed through wide spaces (gaps) between endothelial cells in postcapillary venules, and suggested that the separation between endothelial cells originated in endothelial cell contraction, subsequently confirmed in studies by Majno et al. [13,14]. Simionescu et al. [15] showed that endothelial cells in postcapillary venules were more separated than in other parts of the microcirculation, confirming the initial observation of Majno and Palade [11] that vascular leakage occurs between endothelial cells in postcapillary venules. In a study using electron microscopy with histamine bound to ferritin, small ferritin aggregates were used to detect histamine binding sites in the diaphragm microcirculation of mice. The histamine-ferritin conjugate maintained histamine capacity of opening endothelial junctions in venules, and therefore allowed the demonstration that there were more histamine-binding sites in venular endothelial

cells than in other parts of the microcirculation [16]. A detailed review of the morphological and physiological aspects of the microvascular endothelium, including several photographs of the ultrastructure of these cells, was published by Palade et al. [17].

In a study with intravital microscopy of the hamster cheek pouch, as shown in Fig. 4.1, it was confirmed with electron microscopy of the same tissue that the locations of endothelial cells separated in the fixed tissue coincided with the FITC-dextran leakage sites in postcapillary venules observed a few minutes before its fixation [18]. Pharmacological evidence of endothelial contraction as an explanation for increased leakage of plasma was demonstrated by the inhibition of macromolecular leakage induced by bradykinin [19] and histamine [20], by means of β_2-adrenergic agonists. Later, Erlansson et al. [21], Persson et al. [22], Svensjö and Grega [23], and Svensjö et al. [24] confirmed the permeability-reducing effect of terbutaline (β_2-adrenergic agonist), and Baluk and McDonald [25], using formoterol (a long-acting β_2-adrenergic agonist), emphasized the relaxation or anticontraction action of these drugs. The issue regarding gaps between endothelial cells in inflammation was reviewed by McDonald et al. [26], who examined the evidence regarding the formation of such gaps, and if they could be differentiated from the transendothelial holes or other potential leakage pathways. The results, using five different methods to observe the formation of the gaps, showed that they are rare or absent in normal conditions, and appear with the beginning of leakage of plasma induced by neurogenic inflammation, histamine, or substance P. Morphology changes of the gaps are complex and accompanied by cellular finger-like processes which seem to anchor adjacent endothelial cells to each other and also participate in the gap closure process. Holes through endothelial cells have a frequency lower than 1% as compared to the gaps [26].

The hypothesis of endothelial cell contraction as an explanation for increased permeability was, and still is, attractive to many investigators, and gained indirect support from pharmacological studies in which drugs that may relax vascular smooth musculature (phosphodiesterase, adenyl cyclase activators, and β_2-adrenergic agonists) were tested for their potency to inhibit plasma leakage induced with bradykinin and histamine. Inhibition was

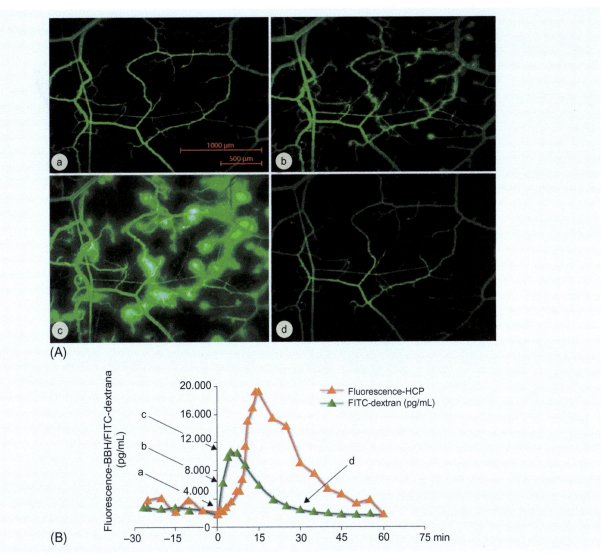

FIG. 4.1   (A) Four images obtained with a digital camera of a rectangular area of 5 mm² from a hamster cheek pouch (HCP) at points indicated in (B), before (a), at 1 min (b), 4 min (c), and 30 min, and (d) after topical application of histamine. (B) Fluorescence in a rectangular area of 5 mm² from Hamster cheek pouch (HCP) (A) and the concentrations of FITC-dextran in the superfusion fluid of the same HCP (pg/mL of superfusion fluid). Histamine (10 mM) was applied topically for 2 min causing increased HCP fluorescence and FITC-dextran concentration in the fluid that comes out of the preparation. The letters and arrows indicate the fluorescence measured before (a), at 1 min (b), at 4 min (c), and at 30 min (d) after histamine application. Peak values for both curves differ by approximately 10 min. Linear regression analysis showed significant correlation between fluorescence of an area of 5 mm² and concentrations of FITC-dextran ($n = 12$, $r = 0.973$, and $P = 0.00001$). Adapted from Svensjö E, Saraiva EM, Bozza MT, et al. Salivary gland homogenates of Lutzomyia longipalpis and its vasodilatoy peptide maxadilan cause plasma leakage via PAC1 receptor activation. J Vasc Res 2009;46:435–46.

shown with aminophylline, milrinone, rolipram, and terbutaline [20,24] (Fig. 4.2).

Historically, studies regarding vascular permeability can be separated in two categories: those conducted on noninflamed endothelial cell layers, reviewed by Michel [2], Michel and Curry [3], Levick and Michel [4]; and other studies, in which many stimuli were used to interfere with the

contractile intracellular mechanism located in endothelial cell junctions, for example, histamine, bradykinin, prostaglandin, leukotrienes, and mediators formed by activated leukocytes, such as TNF-α and reactive oxygen species (ROS). However, there are exceptions to this division, as shown by Curry et al. [27], and Adamson et al. [28,29], who demonstrated that the concentration of sphingosine-1-phosphate

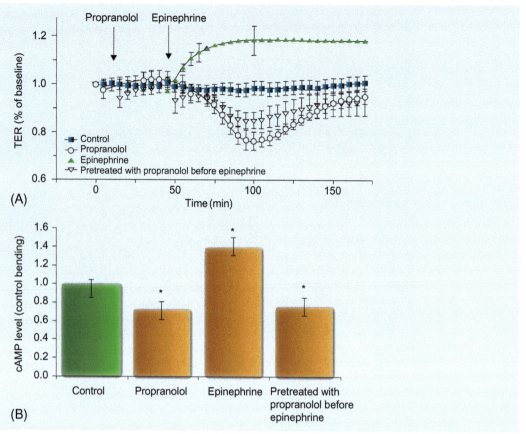

FIG. 4.2 Effects of beta-adrenergic receptors on transendothelial electrical resistance (TER) and cAMP levels in microvascular endothelial cell cultures (A). TER was measured after addition of propranolol, epinephrine, or propranolol followed by epinephrine, and started to decrease 60 min after adding propranolol, while epinephrine increased TER quickly. Pretreatment with propranolol blocked TER increase due to epinephrine. Results are expressed with average ± S.E. of six independent experiments. (B) cAMP levels measured in microvascular endothelial cell culture. Propranolol decreased cAMP levels after 60 min when compared to the nontreated control, while epinephrine increased cAMP levels after incubation for 60 min. Effects of epinephrine were blocked with pretreatment with propranolol. Results represent an average ± S.E. of five independent experiments. *$P < 0.05$. Adapted from Spindler V, Waschke J. Beta-adrenergic stimulation contributes to maintenance of endothelial barrier functions under baseline conditions. Microcirculation 2011;18:118–27.

(S1P) during perfusion of mesenteric vessels with albumin determines the quantity of albumin that exits the perfused venule through the venular gaps. It was suggested that red blood cells may donate S1P to albumin, which stabilizes the endothelial barrier by interacting with the contractile machinery of adherens type junctions.

Endothelial junctions in postcapillary venules have been accepted as the most important locations of the cardiovascular system to regulate permeability to macromolecules (plasma proteins), in inflammation, and also as locations for leukocyte migration in pioneer works by Majno and Palade [11], Simionescu et al. [15], Heltianu et al. [16], confirmed by McDonald et al. [26], and others. Several studies concerning endothelial permeability have

focused on the micro anatomy of endothelial junctions of the adherens and tight types, as well as signaling mechanisms in adjacent endothelial cells, as illustrated in several reviews [9,30–33].

VASCULAR PERMEABILITY REGULATION BY OPENING AND CLOSING ADHERENS TYPE JUNCTIONS—PRINCIPAL MECHANISMS

Endothelial Barrier Maintenance: The Role of cAMP and β₂-Adrenergic Agonists

Intraarterial infusion of histamine into the forelimbs of dogs causes significant efflux of protein and formation of edema (swollen paw) due to increased microvascular permeability to plasma proteins [34]. On the other hand, systemic infusions of intravenous histamine or in the left ventricle, in doses equal or higher than those that produced massive efflux or proteins and formation of edema during intraarterial infusion, did not cause paw edema [35,36]. However, if a beta-blocker, such as propranolol, is infused before intravenous infusion of histamine, the response to histamine is dramatically different, since treatment with propranolol increase the filtration of liquids and efflux of proteins, indicating that histamine promotes endogenous release of catecholamines that antagonize the direct action of histamine on microcirculation by stimulating beta-adrenergic receptors [37]. The antagonism of histamine and bradykinin of the efflux of proteins by catecholamines is completely independent from blood flow variations, in microvascular pressure and in the perfused area [38,39], and reflects a direct action of catecholamines in the microcirculation that counterbalances the endothelial effects produced by histamine or bradykinin.

The use of intravital microscopy and fluorescent markers allowed demonstration that histamine and bradykinin increased the number of venular gaps (leakage) by separating endothelial cells in small venules, through which the macromolecules pass quickly into the extravascular space [19,40].

Inflammatory mediators increase vascular permeability basically by forming intracellular gaps between endothelial cells of postcapillary venules. In these conditions, cell-cell contact, such as tight and adherens junctions open to allow the passage of the fluid by means of a paracellular pathway [41]. This study illustrates the existence of physiological mechanisms to protect the endothelium, probably elevation of cyclic adenosine monophosphate (cAMP), stimulated by epinephrine, which stabilizes the endothelial barrier, as shown by Spindler and Waschke [42]. They studied propranolol, specific β_2-adrenergic antagonist, in measurements of in vivo hydraulic conductivity (L_p) in postcapillary venules of rats and compared the results with changes in cellular cAMP levels, transendothelial electrical resistance (TER), and VE-cadherin in vascular endothelium in monolayers of dermis microcirculation, in vitro. It was noted that superfusion of endothelial cells in culture with propranolol increased transendothelial permeability and superfusion with epinephrine

decreased it, as illustrated in Fig. 4.2 by Spindler and Waschke [42].

Results from in vitro and in vivo experiments, using pharmacological tools, were reinforced by simultaneous observations of morphological changes in VE-cadherin and distribution of actin filaments in endothelial cell layers. The results suggest that epinephrine is secreted by endothelial cells and is involved with regulation of endothelial barrier properties by means of a mechanism that depends on cAMP in paracrine or autocrine fashion. This local stimulation may also be influenced by epinephrine present in systemic circulation (Fig. 4.3) [42]. Furthermore, the important role of cAMP in Rac1 activation was demonstrated in an investigation that combined functional and ultrastructural in vivo and in vitro of the microvascular endothelium, which show that endothelial barrier improvement via Rac1 activation only occurred in microvascular endothelium and not in the macrovascular endothelium [41]. Other studies conducted by Waschke's group, and also other studies that reviewed this point in details, suggest that cAMP activates a small GTPase known as Rac1, which goes in the same direction of the S1P secreted by activated erythrocytes and platelets [43]. The main conclusion of this work is that the cAMP/Rac1 axis is critical for endothelial barrier regulation, and as stimulation of cAMP/Rac1 signaling is effective to block increased permeability in respect to most inflammatory mediators, this pathway must be considered a therapeutic target [43].

Stabilization of the Endothelial Barrier With S1P

The discovery of the S1P capacity to stabilize endothelial barrier began with the clinical observation that a reduced number of circulating platelets accelerated capillary leakage and tissue edema, which was confirmed in animals that suffered platelet depletion [44,45]. To examine the specific role of platelets, rat hearts were perfused with platelet-enriched or platelet-depleted plasma in order to investigate microvascular permeability, using intravital microscopy with fluorescence and FITC marked with albumin, showing that permeability was higher in hearts perfused with platelet-depleted plasma when compared to hearts perfused with platelet-enriched plasma [46].

Under the hypothesis that S1P released by circulating platelets would improve the barrier function, Garcia et al. [47] demonstrated, for the first time, that

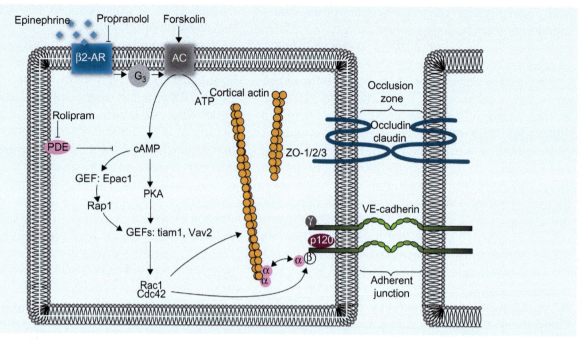

FIG. 4.3 Endothelial barrier regulation scheme by beta-adrenergic receptors. Local or circulating epinephrine activated β-adrenergic receptors in endothelial cells that promote the activation of adenyl cyclase by stimulation of proteins G, increasing the production of cAMP from the ATP. cAMP increases Rac1 activity via protein kinase A (PKA) or promotes exchange activated directly by the factor (Epac) of the guanine nucleotide. cAMP, at least partially with Rac1, is able to modulate the properties of the endothelial barrier by different mechanisms, such as increased intercellular junction molecule adhesion, combining junctional complexes with actin cytoskeleton or strengthening cortical actin cytoskeleton. Adapted from Spindler V, Waschke J. Beta-adrenergic stimulation contributes to maintenance of endothelial barrier functions under baseline conditions. Microcirculation 2011;18:118–27.

S1P neutralizes the effect of plasma leakage induced by thrombin, and this effect was measured by the contraction of the cytoskeleton of endothelial cells or reduction of transendothelial electrical resistance. After extensive studies, Garcia et al. [47] concluded that S1P can be added to the list of key endothelial growth factors that regulate vascular function and homeostasis, and also to the select list of agonists that promote vascular bed integrity (Fig. 4.4).

S1P is a natural bioactive sphingolipid that acts in the extracellular compartment by means of the S1P receptor coupled to protein G, and also in the intracellular compartment in several targets. It has been shown that S1P is a potent angiogenic factor that increases integrity of lung endothelial cells and inhibits vascular permeability and alveolar filling in preclinical animal models of acute pulmonary injury. Besides that, S1P and S1P analogs such as 2-amino-2-(2-[4-octylphenyl]ethyl)-1,3-propanediol (FTY720), FTY720 phosphate and FTY720 phosphonate offer therapeutic potential in murine models of pulmonary injury. A recent review summarizes the roles of S1P, its analogs, enzymes for its metabolization, and its

physiopathological receptors and Fig. 4.4 illustrates the signaling cascade related to stabilization of the endothelial barrier [48] and how S1P acts to preserve the endothelial barrier.

As mentioned before, Curry et al. [27] and Adamson et al. [28,29] showed the importance of S1P in regulating vascular permeability in the absence of inflammation. In subsequent works, they studied the role of albumin as endothelial barrier stabilizer, showing that the component really responsible for solution stabilization was not albumin, but S1P dissolved and transported by albumin [29] (Figs. 4.5 and 4.6). Using a selective S1P inhibitor (W-146), they showed that, in the presence of W-146 + S1P, vascular permeability, measured by hydraulic conductivity, increased to the same proportion as perfusion, without the presence of S1P, and concluded that the effect of albumin (defined as protection of normal permeability) depends on the action of albumin as enhancer of S1P release and transportation by red blood cells that normally provide a significant quantity of S1P to the endothelium [29].

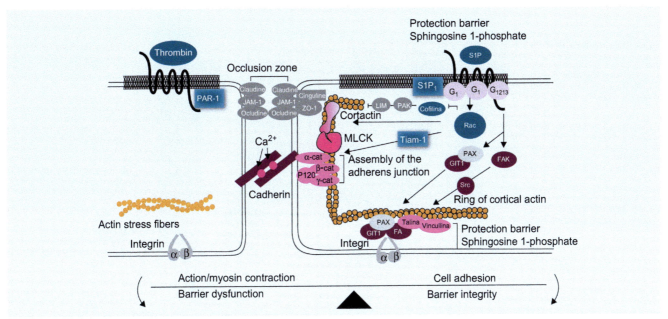

FIG. 4.4 Endothelial barrier regulation by S1P. S1P bound to protein G coupled to active S1P1 and Rac1 induces a series of signaling cascades, including reorganization of the cytoskeleton, set of adherens and tight junctions, and formation of adhesions that act together to increase endothelial barrier function. However, breakage of activated receptor by protease (PAR-1) by thrombin induces stress in the formation of actin fibers and breaks the set of adherens and tight junctions together with focal proteins. *LIM*, LIM kinase; *PAK*, p21-kinase-2 activated; *ZO*, occludin zone; *JAM*, junctional adhesion molecule; *PXN*, paxillin; *GIT*, protein G coupled to the kinase receptor that interacts with the protein; *FAK*, focal adhesion kinase; *MLCK*, myosin light-chain kinase; *cat*, catenin; *Src*, oncogene of the homologous rous sarcoma cell; *Rac1*, related to Ras C3 of botulinum toxin 1. Adapted from Natarajan V, Dudek SM, Jacobson JR, et al. Sphingosine-1-phosphate, FTY720, and sphingosine-1-phosphate receptors in the pathobiology of acute lung injury. Am J Respir Cell Mol Biol 2013;49:6–17.

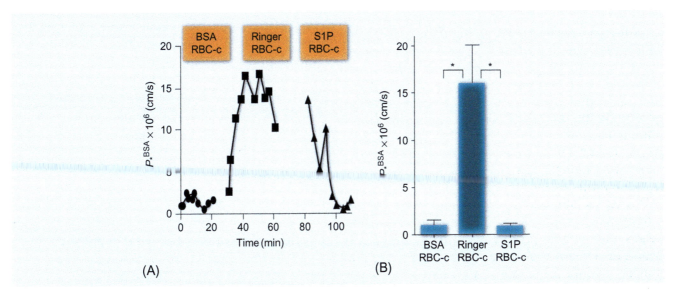

(A) (B)

FIG. 4.5 (A) Representative data showing that Ringer solution conditioned with red blood cells does not maintain the normal permeability to albumin contained in BSA solution (10 mg/mL) of conditioning to red blood cells (RBCc; $n = 9$). In another experiment group, the vessels were also perfused with S1P phosphate (1000 nM) in Ringer solution that restored the permeability of solutes to normal values ($n = 6$). Solutions were conditioned (20–22°C) with red blood cells in concentrations typically used as flow markers in microperfusion (hematocrit [Hct]: 1.3% Hct). The BSA marker (Alexa fluor 555, concentration: 0.5–1 mg/mL) alone did not maintain low permeability. (B) Vessel data such as item (A). $P = 0.05$ (by means of Kruskal-Wallis test with Dunn as posttest). Adapted from Adamson RH, Clark JF, Radeva M, et al. Albumin modulates S1P delivery from red blood cells in perfused microvessels: mechanism of the protein effect. Am J Physiol Heart Circ Physiol 2014;306(7):H1011–7 [Erratum in: Am J Physiol Heart Circ Physiol 2014;307:H120].

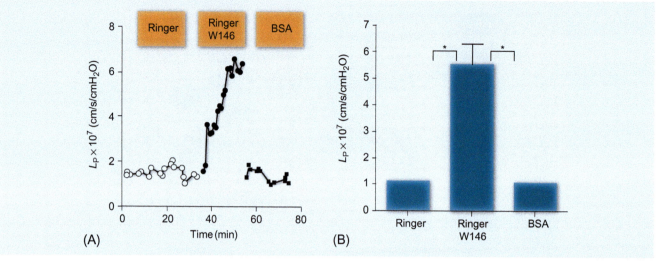

FIG. 4.6   (A) Measurement of hydraulic conductivity (L_p) to test the "protein effect" by comparing the L_p measured with bovine serum albumin (BSA) and the one measured with the Ringer solution, without protein. The experiment shows that both are similar, but adding W-146 (S1P antagonist) to the perfusion solution increased L_p, demonstrating that red blood cells are able to donate S1P to the endothelium, in the absence of BSA. The perfusion solution contained red blood cells as flow markers (hematocrit 1.3%). (B) Summary data of experiments conducted according to the conditions shown in item (A) ($n = 12$) *$P < 0.05$ according to the Friedman test with Dunn posttest. Adapted from Adamson RH, Clark JF, Radeva M, et al. Albumin modulates S1P delivery from red blood cells in perfused microvessels: mechanism of the protein effect. Am J Physiol Heart Circ Physiol 2014;306(7):H1011–7 [Erratum in: Am J Physiol Heart Circ Physiol 2014;307:H120].

VE-Cadherin Phosphorylation and Its Relation to Endothelial Permeability

Adherens junctions play an important role in vascular permeability control. These structures are located in cell-cell contact, serve as mediators of cell adhesion, and transfer intracellular signals. Adhesion is mediated by cadherins that interact homophylically in the *trans* version and form lateral interactions in the *cis*version. VE-cadherin (also known as CDH5 and CD144) is the largest component of endothelial adherens type junctions, and is specific for these cells. Endothelial cells of different types of vessels, such as lymphatic, arteries, and veins, have different compositions and junction organizations. Vascular permeability increases with changes in expression and function of adherens type junction components [49].

Several studies suggest that *Tyrosine phosphorylation of VE-cadherin* is required to open endothelial junctions [50–52]. It seems that phosphorylation induced by the vascular endothelial growth factor (VEGF) is highly selective and three molecules of tyrosine in the VE-cadherin chain are important for increased permeability: Tyr658, Tyr685, and Tyr731. A recent study showed that Tyr731 phosphorylation only caused leukocyte migration, while Tyr685 phosphorylation only caused plasma leakage, showing a distinct regulation of VE-cadherin related to the opening of junctions in vivo [53]. In an editorial comment, Sidibé and Imhof [54] proposed Fig. 4.7 to illustrate the consequences of highly selective phosphorylation in a simplified fashion. These studies suggest that the main leukocyte migration pathway is paracellular, but the leukocytes may also pass through endothelial cells [55,56].

ENDOTHELIAL BARRIER CHANGES INDUCED BY INFLAMMATION IN OBESITY

There are evidences, both clinical and experimental, that obesity induces a low-level inflammatory state. Activation products (adipokines) of an increased pool of adipocytes probably represent a connection between obesity and inflammation. The largest tissue target of the proinflammatory actions

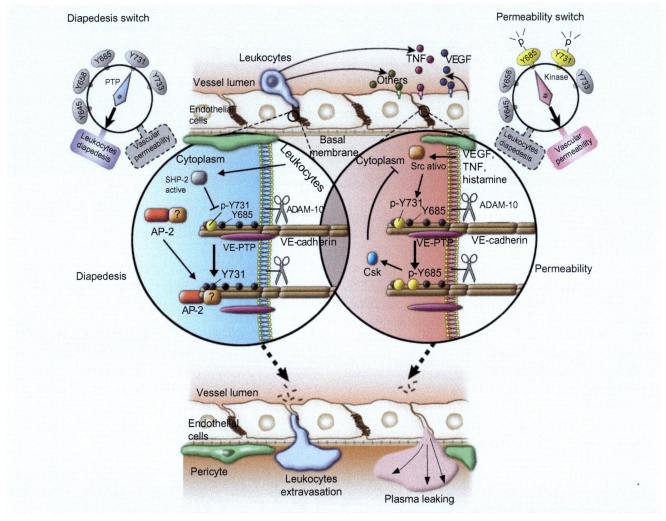

FIG. 4.7 In vivo control mechanisms of leakage of leukocyte and vascular permeability by VE-cadherin phosphorylation. A change in VE-cadherin phosphorylation by leukocyte or other soluble factors is responsible for the final result. Leukocytes induce activation of SHP-2, which leads to dephosphorylation of VE-cadherin Tyr731 (p-Y731) phosphorylated residue that associates to the AP-2 complex with consequent VE-cadherin endocytosis. On the other hand, factors that increase permeability, such as VEGF, TNF-α, or histamine, induce phosphorylation, mediated by Src, of VE-cadherin into Tyr 685 (p-Y685) residue. Breakage of VE-cadherin dependent on metalloproteinase ADAM-10 is induced by two processes, but is related only to permeability and transmigration, pending demonstration in mutant VE-cadherin studies, with altered proteolytic binding site. Adapted from Sidibé A, Imhof BA. VE-cadherin phosphorylation decides: vascular permeability or diapedesis. Nat Immunol 2014;15:215–7.

of the adipokines released locally and systematically is the microcirculation, which shows similar dysfunction in arterioles, capillary vessels, and venules. Microvascular dysfunction induced by obesity contributes to the initiation and propagation of the inflammatory response. The consequence of the systemic inflammatory state, associated to obesity, predisposes tissues to higher inflammatory/injury responses in case of additional inflammatory stimulus, which can explain the role of obesity as risk factor for several diseases [57].

EFFECTS OF ANTIOXIDANTS IN INCREASING VASCULAR PERMEABILITY INDUCED BY I/R

Experimental ischemia of the hamster cheek pouch, followed by reperfusion, may cause immediate microcirculation changes observed in intravital microscopy. Initially, there is adhesion and accumulation of leukocytes in postcapillary venules, and in <2 min, there is visible and measurable FITC-dextran macromolecule leakage, which reaches the maximum value 5–7 min after the beginning of reperfusion [58]. Reversible increase in leakage of plasma induced by I/R is due to the oxidative effects of ROS, since it can be prevented by endovenous administration of a recombinant extracellular superoxide dismutase (EC-SOD), or by CuZn-SOD before 30-min ischemia, suggesting that the ROS are formed when the oxygenated blood enters the ischemic area [59]. The effects of I/R can also be prevented by antioxidants of less selective action, such as flavonoids [60–65]. The flavonoid hesperidin methylchalcone also has antipermeability effects in diabetic hamsters [66]. Polyunsaturated fatty acids (n-3 PUFA), such as eicosapentaenoic acid (EPA, 20:5n-3) and the docosahexaenoic acid (DHA, 22:6n-3), found in fish oil, have several antiinflammatory properties, and its beneficial potential for I/R injury was studied in the hamster cheek pouch showing that these polyunsaturated fatty acids prevented the increase of leukocyte adhesion and reduced leakage of plasma by 70% [67]. Iloprost, analog of prostacyclin, administered intravenously before ischemia, also inhibited leukocyte adhesion and accumulation and leakage of plasma almost completely during reperfusion, suggesting that the antiadhesion effect of iloprost reduced leukocyte adhesion and subsequent ROS activation and release [68]. The same drug, iloprost, prevented alterations observed in the initial stage of endotoxemia by *Escherichia coli* lipopolysaccharide (LPS) [69].

IMPORTANCE OF PLASMA EXTRAVASATION AT THE CONTACT SITE WITH TISSUE, PARASITE, AND BACTERIA

Several experimental studies showed that infection caused by parasites such as *Leishmania major* and *Trypanosoma cruzi*, as well as some bacteria (*Porphyromonas gingivalis*) is followed by alterations in vascular permeability at the contact site with the tissue and parasite/bacteria [70–75]. Several pharmacological agents, such as B_2R-antagonist (HOE-140), mepyramine (antihistamine), endothelin antagonists, stabilizer of mast cells (cromoglycate), and reparixin (CCR2-antagonist), showed the capacity to inhibit leukocyte accumulation and leakage of plasma [72–78]. Studies conducted to explore the hypothesis that extravascular plasma, formed by action of *T. cruzi* trypomastigotes

infection or by histamine, may be used as a substrate for extravascular or tissue proteases (tissue kallikrein) to generate bradykinin by cleavage of kininogen, showed that the presence of a small quantity of plasma around postcapillary venules significantly increased the response to *T. cruzi* trypomastigotes and to a synthetic activator of the kallikrein-kinin cascade. Pharmacological interventions suggested that the activation of mast cells (MC) resulted in the generation of bradykinin via the kallikrein-kinin cascade [79].

CONCLUSIONS

During the last decades, it has become clear that inflammatory mediators induce interendothelial gaps, an essential mechanism to regulate paracellular permeability and leukocyte transendothelial migration. In many vascular beds of the systemic circulation, gaps are observed mainly in postcapillary venules, and endothelial cytoskeleton and VE-cadherin in adherens junctions suffer dynamic and reversible reorganization to allow opening and closing these gaps. There are several physiological and pharmacological possibilities to stabilize the endothelial barrier and counterbalance leakage of plasma: (1) maintenance of cAMP levels in endothelial cells with phosphodiesterase inhibitors and β_2-adrenergic agonists (epinephrine, terbutaline, salmeterol, formoterol); (2) Bradykinin1R- and bradykinin2R-antagonists (HOE-140); (3) S1P and agonists related to S1P; (4) inhibitors of Rho-kinase (Fasudil [approved for clinical use], Y-27632); (5) tyrosine kinase inhibitors that may counterbalance the effects of VEGF; (6) glucocorticoids; and (7) antioxidants (flavonoids, resveratrol, etc.). Some inflammatory mediators (including I/R) may induce preconditioning of endothelial cells, a condition that reduces leakage of plasma and represents a physiological mechanism to preserve the endothelial barrier.

References

[1] Kraemer-Aguiar LG, Miranda ML, Bottino DA, et al. Increment of body mass index is positively correlated with worsening of endothelium-dependent and independent changes in forearm blood flow. Front Physiol 2015;6:223.
[2] Michel CC. Transport of macromolecules through microvascular walls. Cardiovasc Res 1996;32:644–53.
[3] Michel CC, Curry FE. Microvascular permeability. Physiol Rev 1999;79:703–61.

[4] Levick JR, Michel CC. Microvascular fluid exchange and the revised Starling principle. Cardiovasc Res 2010;87:198–210.

[5] Vink H, Duling BR. Identification of distinct luminal domains for macromolecules, erythrocytes, and leukocytes within mammalian capillaries. Circ Res 1996;79:581–9.

[6] Curry FE, Adamson RH. Endothelial glycocalyx: permeability barrier and mechanosensor. Ann Biomed Eng 2012;40(4):828–39.

[7] Fridén V, Oveland E, Tenstad O, et al. The glomerular endothelial cell coat is essential for glomerular filtration. Kidney Int 2011;79 (12):1322–30.

[8] Eskens BJ, Zuurbier CJ, van Haare J, et al. Effects of two weeks of metformin treatment on whole-body glycocalyx barrier properties in db/db mice. Cardiovasc Diabetol 2013;12:175.

[9] Mehta D, Malik AB. Signaling mechanisms regulating endothelial permeability. Physiol Rev 2006;86:279–367.

[10] Sukriti S, Tauseef M, Yazbeck P, et al. Mechanisms regulating endothelial permeability. Pulm Circ 2014;4:535–51.

[11] Majno G, Palade GE. Studies on inflammation. 1. The effect of histamine and serotonin on vascular permeability: an electron microscopic study. J Biophys Biochem Cytol 1961;11:571–605.

[12] Majno G, Palade GE, Schoefl GI. Studies on inflammation. II. The site of action of histamine and serotonin along the vascular tree: a topographic study. J Biophys Biochem Cytol 1961;11:607–26.

[13] Majno G, Gilmore V, Leventhal M. On the mechanism of vascular leakage caused by histamine type mediators. A microscopic study in vivo. Circ Res 1967;21:833–47.

[14] Majno G, Shea SM, Leventhal M. Endothelial contraction induced by histamine-type mediators: an electron microscopic study. J Cell Biol 1969;42:647–72.

[15] Simionescu N, Simionescu M, Palade GE. Open junctions in the endothelium of the postcapillary venules of the diaphragm. J Cell Biol 1978;79(1):27–44.

[16] Heltianu C, Simionescu M, Simionescu N. Histamine receptors of the microvascular endothelium revealed in situ with a histamine-ferritin conjugate: characteristic high-affinity binding sites in venules. J Cell Biol 1982;93(2):357–64.

[17] Palade GE, Simionescu M, Simionescu N. Structural aspects of the permeability of the microvascular endothelium. Acta Physiol Scand Suppl 1979;463:11–32.

[18] Hultström D, Svensjö E. Intravital and electron microscopic study of bradykinin-induced vascular permeability changes using FITC-dextran as a tracer. J Pathol 1979;129:125–33.

[19] Svensjö E, Persson CG, Rutili G. Inhibition of bradykinin induced macromolecular leakage from post-capillary venules by a beta2 adrenoreceptor stimulant, terbutaline. Acta Physiol Scand 1977;101:504–6.

[20] O'Donnell SR, Persson CG. Beta-adrenoceptor mediated inhibition by terbutaline of histamine effects on vascular permeability. Br J Pharmacol 1978;62:321–4.

[21] Erlansson M, Persson NH, Svensjö E, et al. Macromolecular permeability increase following incomplete ischemia in the hamster cheek pouch and its inhibition by terbutaline. Int J Microcirc Clin Exp 1987;6:265–71.

[22] Persson NH, Erlansson M, Bergqvist D, et al. Terbutaline and budesonide as inhibitors of postischaemic permeability increase. Acta Physiol Scand 1987;129:517–24.

[23] Svensjö E, Grega GJ. Evidence for endothelial cell-mediated regulation of macromolecular permeability by postcapillary venules. Fed Proc 1986;45:89–95.

[24] Svensjö E, Andersson KE, Bouskela E, et al. Effects of two vasodilatory phosphodiesterase inhibitors on bradykinin-induced permeability increase in the hamster cheek pouch. Agents Actions 1993;39:35–41.

[25] Baluk P, McDonald DM. The beta 2-adrenergic receptor agonist formoterol reduces microvascular leakage by inhibiting endothelial gap formation. Am J Physiol 1994;266:L461–8.

[26] McDonald DM, Thurston G, Baluk P. Endothelial gaps as sites for plasma leakage in inflammation. Microcirculation 1999;6:7–22.

[27] Curry FE, Clark JF, Adamson RH. Erythrocyte-derived sphingosine-1-phosphate stabilizes basal hydraulic conductivity and solute permeability in rat microvessels. Am J Physiol Heart Circ Physiol 2012;303:H825–34.

[28] Adamson RH, Sarai RK, Altangerel A, et al. Sphingosine-1-phosphate modulation of basal permeability and acute inflammatory responses in rat venular microvessels. Cardiovasc Res 2010;88:344–51.

[29] Adamson RH, Clark JF, Radeva M, et al. Albumin modulates S1P delivery from red blood cells in perfused microvessels: mechanism of the protein effect. Am J Physiol Heart Circ Physiol 2014;306(7): H1011–7 [Erratum in: Am J Physiol Heart Circ Physiol 2014;307: H120].

[30] Komarova YA, Mehta D, Malik AB. Dual regulation of endothelial junctional permeability. Sci STKE 2007;2007(412):re8.

[31] Komarova Y, Malik AB. Regulation of endothelial permeability via paracellular and transcellular transport pathways. Annu Rev Physiol 2010;72:463–93.

[32] Ochoa CD, Stevens T. Studies on the cell biology of interendothelial cell gaps. Am J Physiol Lung Cell Mol Physiol 2012;302: L275–86.

[33] Trani M, Dejana E. New insights in the control of vascular permeability: vascular endothelial-cadherin and other players. Curr Opin Hematol 2015;22:267–72.

[34] Grega GJ, Kline RL, Dobbins DE, et al. Mechanisms of edema formation by histamine administered locally into canine forelimbs. Am J Physiol 1972;223:1165–71.

[35] Daugherty Jr RM, Scott JB, Emerson Jr TE, et al. Comparison of IV and IA infusion of vasoactive agents on dog forelimb blood flow. Am J Physiol 1968;214:611–9.

[36] Marciniak DL, Dobbins DE, Maciejko JJ, et al. Effects of systemically infused histamine on transvascular fluid and protein transfer. Am J Physiol 1977;233:H148–53.

[37] Grega GJ, Marciniak DL, Jandhyala BS, et al. Effects of intravenously infused histamine on canine forelimb transvascular protein efflux following adrenergic receptor blockade. Circ Res 1980;47:584–91.

[38] Marciniak DL, Dobbins DE, Maciejko JJ, et al. Antagonism of histamine edema formation by catecholamines. Am J Physiol 1978;234:H180–5.

[39] Rippe B, Grega GJ. Effects of isoprenaline and cooling on histamine induced changes of capillary permeability in the rat hindquarter vascular bed. Acta Physiol Scand 1978;103:252–62.

[40] Svensjö E, Arfors KE, Raymond RM, et al. Morphological and physiological correlation of bradykinin-induced macromolecular efflux. Am J Physiol 1979;236:H600–6.

[41] Spindler V, Peter D, Harms GS, et al. Ultrastructural analysis reveals cAMP-dependent enhancement of microvascular endothelial barrier functions via Rac1-mediated reorganization of intercellular junctions. Am J Pathol 2011;178:2424–36.

[42] Spindler V, Waschke J. Beta-adrenergic stimulation contributes to maintenance of endothelial barrier functions under baseline conditions. Microcirculation 2011;18:118–27.

[43] Schlegel N, Waschke J. cAMP with other signaling cues converges on Rac1 to stabilize the endothelial barrier—a signaling pathway compromised in inflammation. Cell Tissue Res 2014;355:587–96.

[44] Shepro D, Welles SL, Hechtman HB. Vasoactive agonists prevent erythrocyte extravasation in thrombocytopenic hamsters. Thromb Res 1984;35:421–30.

[45] Lo SK, Burhop KE, Kaplan JE, et al. Role of platelets in maintenance of pulmonary vascular permeability to protein. Am J Physiol 1988;254:H763–71.

[46] McDonagh PF. Platelets reduce coronary microvascular permeability to macromolecules. Am J Physiol 1986;251:H581–7.

I. STRUCTURE AND FUNCTIONS

[47] Garcia JG, Liu F, Verin AD, et al. Sphingosine 1-phosphate promotes endothelial cell barrier integrity by Edg-dependent cytoskeletal rearrangement. J Clin Invest 2001;108:689–701.

[48] Natarajan V, Dudek SM, Jacobson JR, et al. Sphingosine-1-phosphate, FTY720, and sphingosine-1-phosphate receptors in the pathobiology of acute lung injury. Am J Respir Cell Mol Biol 2013;49:6–17.

[49] Dejana E, Orsenigo F. Endothelial adherens junctions at a glance. J Cell Sci 2013;126:2545–9.

[50] Esser S, Lampugnani MG, Corada M, et al. Vascular endothelial growth factor induces VE-cadherin tyrosine phosphorylation in endothelial cells. J Cell Sci 1998;111:1853–65.

[51] Wallez Y, Cand F, Cruzalegui F, et al. Src kinase phosphorylates vascular endothelial-cadherin in response to vascular endothelial growth factor: identification of tyrosine 685 as the unique target site. Oncogene 2007;26:1067–77.

[52] Orsenigo F, Giampietro C, Ferrari A, et al. Phosphorylation of VE-cadherin is modulated by haemodynamic forces and contributes to the regulation of vascular permeability in vivo. Nat Commun 2012;3:1208.

[53] Wessel F, Winderlich M, Holm M, et al. Leukocyte extravasation and vascular permeability are each controlled in vivo by different tyrosine residues of VE-cadherin. Nat Immunol 2014;15:223–30.

[54] Sidibé A, Imhof BA. VE-cadherin phosphorylation decides: vascular permeability or diapedesis. Nat Immunol 2014;15:215–7.

[55] Carman CV, Sage PT, Sciuto TE, et al. Transcellular diapedesis is initiated by invasive podosomes. Immunity 2007;26:784–97.

[56] Carman CV. Mechanisms for transcellular diapedesis: probing and pathfinding by invadosome-like protrusions. J Cell Sci 2009;122:3025–35.

[57] Vacharajani V, Granger DN. Adipose tissue: a motor for the inflammation associated with obesity. IUBMB Life 2009;61:424–30.

[58] Persson NH, Erlansson M, Svensjö E, et al. The hamster cheek pouch—an experimental model to study postischemic macromolecular permeability. Int J Microcirc Clin Exp 1985;4(3):257–63.

[59] Erlansson M, Bergqvist D, Marklund SL, et al. Superoxide dismutase as an inhibitor of postischemic microvascular permeability increase in the hamster. Free Radic Biol Med 1990;9:59–65.

[60] Bouskela E, Cyrino FZ, Marcelon G. Inhibitory effect of the Ruscus extract and of the flavonoid hesperidine methylchalcone on increased microvascular permeability induced by various agents in the hamster cheek pouch. J Cardiovasc Pharmacol 1993;22:225–30.

[61] Bouskela E, Cyrino FZ, Marcelon G. Possible mechanisms for the inhibitory effect of Ruscus extract on increased microvascular permeability induced by histamine in hamster cheek pouch. J Cardiovasc Pharmacol 1994;24:281–5.

[62] Bouskela E, Donyo KA, Verbeuren TJ. Effects of Daflon 500 mg on increased microvascular permeability in normal hamsters. Int J Microcirc Clin Exp 1995;15(Suppl. 1):22–6.

[63] Bouskela E, Svensjö E, Cyrino FZ, et al. Oxidant-induced increase in vascular permeability is inhibited by oral administration of S-5682 (Daflon 500 mg) and alpha-tocopherol. Int J Microcirc Clin Exp 1997;17(Suppl. 1):18–20.

[64] Bouskela E, Donyo KA. Effects of oral administration of purified micronized flavonoid fraction on increased microvascular permeability induced by various agents and on ischemia/reperfusion in diabetic hamsters. Int J Microcirc Clin Exp 1995;15:293–300.

[65] Bouskela E, Donyo KA. Effects of oral administration of purified micronized flavonoid fraction on increased microvascular permeability induced by various agents and on ischemia/reperfusion in the hamster cheek pouch. Angiology 1997;48:391–9.

[66] Svensjö E, Bouskela E, Cyrino FZ, et al. Antipermeability effects of Cyclo 3 Fort in hamsters with moderate diabetes. Clin Hemorheol Microcirc 1997;17:385–8.

[67] Md dS, Conde CM, Laflôr CM, et al. n-3 PUFA induce microvascular protective changes during ischemia/reperfusion. Lipids 2015; 50(1):23–37.

[68] Erlansson M, Bergqvist D, Persson NH, et al. Modification of postischemic increase of leukocyte adhesion and vascular permeability in the hamster by Iloprost. Prostaglandins 1991;41(2):157–68.

[69] Bouskela E, Rubanyi GM. Effects of iloprost, a stable prostacyclin analog, and its combination with NW-nitro-L-arginine on early events following lipopolysaccharide injection: observations in the hamster cheek pouch microcirculation. Int J Microcirc Clin Exp 1995;15:170–80.

[70] Todorov AG, Andrade D, Pesquero JB, et al. Trypanosoma cruzi induces edematogenic responses in mice and invades cardiomyocytes and endothelial cells in vitro by activating distinct kinin receptor (B1/B2) subtypes. FASEB J 2003;17:73–5.

[71] Svensjö E, Batista PR, Brodskyn CI, et al. Interplay between parasite cysteine proteases and the host kinin system modulates microvascular leakage and macrophage infection by promastigotes of the Leishmania donovani complex. Microbes Infect 2006;8(1):206–20.

[72] Monteiro AC, Schmitz V, Svensjo E, et al. Cooperative activation of TLR2 and bradykinin B2 receptor is required for induction of type 1 immunity in a mouse model of subcutaneous infection by Trypanosoma cruzi. J Immunol 2006;177:6325–35.

[73] Monteiro AC, Scovino A, Raposo S, et al. Kinin danger signals proteolytically released by gingipain induce Fimbriae-specific IFN gamma- and IL-17-producing T cells in mice infected intramucosally with Porphyromonas gingivalis. J Immunol 2009;183:3700–11.

[74] Andrade D, Serra R, Svensjö E, et al. Trypanosoma cruzi invades host cells through the activation of endothelin and bradykinin receptors: a converging pathway leading to chagasic vasculopathy. Br J Pharmacol 2012;165:1333–47.

[75] Schmitz V, Almeida LN, Svensjö E, et al. C5a and bradykinin receptor cross-talk regulates innate and adaptive immunity in Trypanosoma cruzi infection. J Immunol 2014;193:3613–23.

[76] Svensjö E, Saraiva EM, Amendola RS, et al. Maxadilan, the Lutzomyia longipalpis vasodilator, drives plasma leakage via PAC1-CXCR1/2-pathway. Microvasc Res 2012;83:185–93.

[77] Svensjö E, Nogueira de Almeida L, Vellasco L, et al. Ecotin-like ISP of L. major promastigotes fine-tunes macrophage phagocytosis by limiting the pericellular release of bradykinin from surface-bound kininogens: a survival strategy based on the silencing of proinflammatory G-protein coupled kinin B2 and B1 receptors. Mediators Inflamm 2014;2014:143450.

[78] Scharfstein J, Andrade D, Svensjö E, et al. The kallikrein-kinin system in experimental Chagas disease: a paradigm to investigate the impact of inflammatory edema on GPCR-mediated pathways of host cell invasion by Trypanosoma cruzi. Front Immunol 2013;3:396.

[79] Nascimento CR, Andrade D, Carvalho-Pinto CE, et al. Mast cell coupling to the kallikrein-kinin system fuels intracardiac parasitism and worsens heart pathology in experimental Chagas disease. Front Immunol 2017;8:840. https://doi.org/10.3389/fimmu.2017.00840.

Further Reading

Svensjö E, Saraiva EM, Bozza MT, et al. Salivary gland homogenates of Lutzomyia ongipalpis and its vasodilatoy peptide maxadilan cause plasma leakage via PAC1 receptor activation. J Vasc Res 2009;46:435–46.

CHAPTER

5

Vascular Growth Factors, Progenitor Cells, and Angiogenesis

Luís Henrique Wolff Gowdak and José Eduardo Krieger

Blood vessels will respond to the body's requirements and grow in length an diameter as and when required to do so
John Hunter (1728–93)

In: "A treatise on the blood, inflammation and gun-shot wounds," London, John Richardson (ed.), 1794 (posthumous publication).

HISTORICAL ASPECTS

Current knowledge regarding the complex processes that involve vascular growth, as well as factors involved in its regulation, began with a series of experiments conducted in the late 1960s, which culminated with two publications [1,2] authored by the man who would, later on, be deservedly recognized as the "father" of modern angiogenesis, Judah Folkman (1933–2008). At the time, working at Harvard Medical School, in the Department of Surgery and Pediatrics in Boston, Folkman and his team conducted experiments on artificial perfusion systems and solid tumors, aiming, among other aspects of tumoral biology, to determine their growth factors.

Since the mid-1940s, with works by Algire and Chalkley [3], it was already known that tumoral cells have the capacity of continuously inducing development of new capillary endothelial cells in vivo. For that matter, Greene [4] had noted that small tumors, when implanted for more than 1 year in the anterior chamber of the eyeball of a guinea pig, did not experience appreciable growth,

probably due to the lack of local vascularization; on the other hand, when these same tumors are removed and re-implanted in the muscular tissue of rabbits, a situation in which tumor vascularization is possible, the tumors grow amazingly. Similarly, Folkman et al. [5] also demonstrated that tumors implanted experimentally in different organs and maintained by an artificial perfusion system did not grow more than 3–4 mm in diameter. These observations on the differences between tumor growth in artificial avascular systems and vascularized biological tissues led Folkman to formulate the hypothesis that there would be strict interdependency between tumoral growth and vascularization; this relationship, he imagined, would obligatorily involve previously unknown diffusible substances, released by tumoral cells and capable of stimulating capillary endothelial cell growth from a distance. Driven by this idea, and after successive animal experiments with several tumoral cell strains, in 1971 Folkman and his colleagues were finally able to isolate from Walker tumor ascitic cell extracts, for the first time, a substance able to stimulate capillary growth in vivo, which they called tumor angiogenic factor (TAF) [1].

After Folkman's discovery, his attention turned quickly to the possibility of clinical application of this new knowledge about tumor growth to treat cancer by blocking the growth of new blood vessels that nourish the tumor and allow its expansion. Thus, a new pioneer and innovative concept emerged almost simultaneously with the discovery of the first substance with angiogenic capacity to be applied in solid tumor therapy: angiogenesis [6].

While Folkman and others tried to antagonize the angiogenic effects of this new substance, the scientific community lingered to consider other possibilities, diametrically opposed, namely to use this recently discovered angiogenic factor to revascularize ischemic tissue. Thus it was only in 1977, in a letter with no more than three paragraphs sent to the editors of *The Lancet*, that two Russian scientists, Svet-Moldavsky and Chimishkyan, from the Moscow Cancer Research Center Virology laboratory, contradicting current tendencies, suggested that "The Folkman tumor angiogenic factor could be used to induce vascularization in ischemic and infarcted tissue (especially in myocardial infarction)" [7].

Years later, Kumar et al. [8] isolated, in myocardial tissue obtained by necropsy from patients who had suffered from acute infarction, a potent angiogenic factor similar to TAF, which they called human myocardial infarct angiogenic factor (MIAF). The authors suggested that this factor could modulate the opening of preexisting vessels or even lead to the growth of new collateral vessels in infarcted tissue, and therefore could have some clinical value in treating myocardial ischemic syndromes.

The possibility of therapeutic application of vascular growth factors gained impetus when Vallee et al. [9–11] isolated and characterized biochemically a tumor angiogenic factor, determined its sequence of amino acids, and cloned the gene responsible for its production. After this detailed characterization, Vallee's group proposed the name "angiogenin" for this new angiogenic factor. Immediately, editorials were published [12,13] emphasizing the importance of the new discovery, including the possibility of clinical application in diverse situations such as wound healing, heart attacks, and cerebrovascular accidents. Finally, in 1993, Höckel et al. proposed the term "therapeutic angiogenesis," defined as follows in the original publication:

> We suggest the term therapeutic angiogenesis for interventions with the purpose of inducing or stimulating local growth of blood vessels as therapeutic principle for clinical conditions characterized by hipovascularity [14].

This would start an era of great advancements in the recently emerged field of therapeutic angiogenesis, all resulting from the integration of vascular physiology and molecular biology, culminating with its clinical application in the mid-1990s.

VASCULOGENESIS, ANGIOGENESIS, AND ATHEROGENESIS

A vessel may be formed by different processes [15–17]. In the beginning of embryonic development, vessels are formed from precursor mesodermal cells, angioblasts, in a process called vasculogenesis (Fig. 5.1). This dynamic process, involving cell-cell and cell-extracellular matrix interactions, occurs in extra- and intra-embryonic tissue.

The angioblasts (a) are already committed as artery or vein precursors from the lateral mesoderm; arterial precursors migrate towards the mid line (b).

In response to secretion of vascular growth factors, such as VEGF, in this path, angioblasts can align in cords, forming a network (c).

The angioblasts then coalesce in larger vessels, such as the dorsal aorta (d), even if the precise moment of initiation of vessel lumen formation is still unknown.

Fig. 5.1 shows vasculogenesis in a vertebrate embryo.

In turn, angiogenesis is the term that describes budding vascular growth, cell division, migration, and organization of endothelial cells from the preexisting vasculature [16]. Although it has fundamental importance for embryonic vascular development, angiogenesis is not a common phenomenon in adults. It occurs physiologically in the female reproductive system, during the menstrual cycle, and in the placenta during pregnancy [19]. In addition it also takes part in reparative processes such as wound and bone healing [20]. However, vascular neoformation is also closely associated with the appearance, development, and progression of several diseases such as diabetic retinopathy [21], rheumatoid arthritis [22], psoriasis [23], and, as already mentioned, solid tumor growth [24].

Basically there are two types of angiogenesis: by budding and intussusceptive angiogenesis. The budding vascular growth process implies a series of events that occur in sequence, orchestrated in time and space, involving endothelial cells and pericytes, in addition to the extracellular matrix itself [16] (Fig. 5.2).

After activation of endothelial cells and pericytes, morphological changes take place in these cells; activated endothelial cells produce proteases (e.g., collagenases and tissue plasminogen activator), responsible for basal membrane degradation. Chemotactic and mitogenic factors produced by several

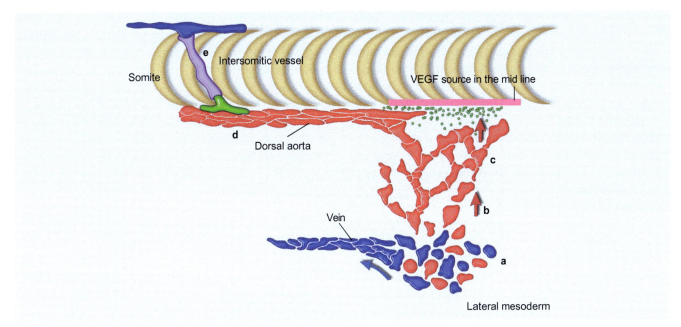

FIG. 5.1 Vertebrate embryo vasculogenesis. *VEGF*, vascular endothelial growth factor. (a) Artery *(red)* or vein *(blue)* precursors; (b) Artery precursors migrate to the mid line in response to VEGF; (c) On the path, they form a plexus; (d) Angioblasts gather forming large vessels; (e) Intersomitic vessels are organized sequentially from endothelial cells. Modified from Hogan BL, Kolodziej PA. Organogenesis: molecular mechanisms of tubulogenesis. Nat Rev Genet 2002;3:513–23.

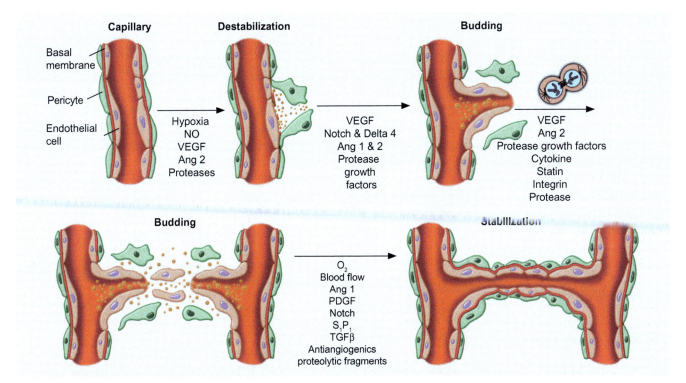

FIG. 5.2 Angiogenesis by budding [18]. *VEGF*, vascular endothelial growth factor; *PDGF*, platelet derived growth factor; *TGFβ*, transforming growth factor beta (see text and Table 5.2). Modified from Carmeliet P. Mechanisms of angiogenesis and arteriogenesis. Nat Med 2000;6(3):389–95.

cell types indicate signaling for endothelial budding, migration, and proliferation [25]. The end of endothelial cell differentiation is associated with reestablishment of cell quiescence phenotype characterized be ceased cell proliferation and migration, redefinition of cell-cell contact, and formation of vessel lumen. Subsequently, pericytes migrate towards the external surface of the recently formed vascular structure; a new basal membrane is synthesized and deposited, which marks the final stages of vessel maturation process; and the beginning of blood flow inside the newly formed vessel establishes its functional integration with the collateral system.

Intussusceptive angiogenesis is also called vascular growth by division, because the vessel wall extends towards the lumen, dividing a single vessel in two (Fig. 5.3). This type of vascular growth is considered quick and efficient when compared to budding angiogenesis, because it initially requires only reorganization of existing endothelial cells, and is not based on immediate endothelial proliferation or migration. The most remarkable characteristic of this type of vascular growth is the formation of the transluminal pillar (Fig. 5.3, arrows). The pillar formation process begins with migration of opposite endothelial walls inside a vessel, followed by rearrangement of interendothelial junctions and invasion of pericytes and myofibroblasts, consequently leading to vessel division [26].

Finally, another vascular growth process in adult tissue, atherogenesis, was defined as in situ growth of collateral muscle arteries (arterioles that interconnect two main arteries) in response to arterial occlusion [27]. Several stages are currently recognized during atherogenesis: an initial stage of endothelial activation; a proliferation stage, characterized by maximum mitotic activity of endothelial cells,

smooth muscle cells, and fibroblasts; a synthetic stage, in which internal elastic lamina is broken, facilitating migration of smooth muscle cells to the subendothelial space; and, finally, a maturation stage, defined by acquisition of vasomotor and viscoelastic properties by the newly formed vessel. In adult organisms, low arterial flow sets both vascular growth mechanisms described: angiogenesis and atherogenesis.

GROWTH FACTORS AND ANGIOGENESIS MODULATION

The events described above, involved with vascular growth, are the result of a balance between the action of pro-angiogenic and antiangiogenic factors. Table 5.1 lists the main factors involved with vascular growth regulation.

These cytokines are released by several cell types in response to multiple angiogenic stimuli, such as tissue hypoxia and mechanical forces, like shear stress. The low tissue O_2 tension, the most potent angiogenic stimulus known, sets a coordinated angiogenic and atherogenic response by inducing expression of VEGF, VEGFR-1, VEGFR-2, neuropilin-1, neuropilin-2, Ang 2, NO synthase, TGFβ-1, PDGF-BB, endothelin-1, IL-8, IGF-1, Tie-2, cyclooxygenase-2, among others [28]. The existence of a transcriptional complex consisting of hypoxia inducible factors (HIF)—HIF-1α, HIF-1β and HIF-2α—leads to overexpression of many genes involved with angiogenesis, in the presence of hypoxia [29]. In addition to the increase in angiogenic cytokine expression, hypoxia also promotes increased local expression of cell receptors for these growth factors, therefore explaining the vascular growth in ischemia sites [30].

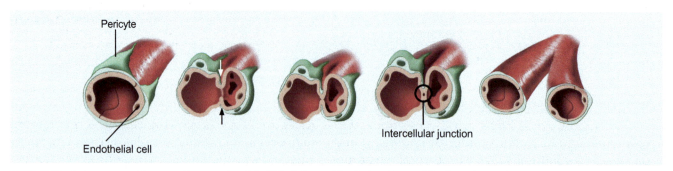

FIG. 5.3 Intussusceptive angiogenesis. Modified from De Spiegelaere W, Casteleyn C, Van den Broeck W, et al. Intussusceptive angiogenesis: a biologically relevant form of angiogenesis. J Vasc Res 2012;49:390–404.

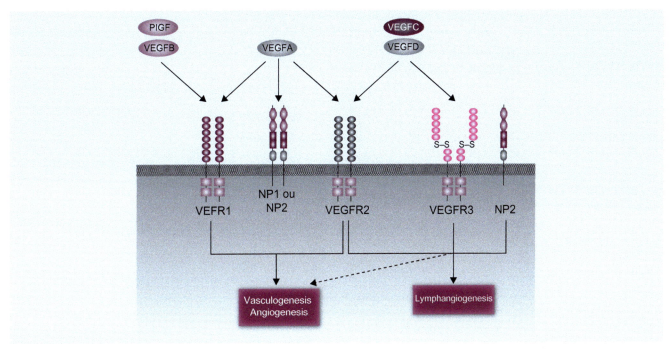

FIG. 5.4 Multiple interactions between VEGF family members and their receptors. Modified from Ellis LM, Hicklin DJ. VEGF-targeted therapy: mechanisms of anti-tumour activity. Nat Rev Cancer. 2008;8:579–91.

TABLE 5.1 Main Factors Involved With Vascular Growth Regulation

Process	Stimulators	Inhibitors
Vasculogenesis	VEGF, GM-CSF, bFGF, IGF-1	Unknown
Angiogenesis	VEGF-A, VEGF-B, VEGF-C, VEGF-D, PlGF, VEGFR-1, VEGFR-2, VEGFR-3, Ang1, Ang2, Tie2, FGF, PDGF, IGF-1, HGF, TNF-α, TGFb-1, αvb3, α5b3, PA, MMP, PECAM, VE-cad, NO, CXC, HIF-1α, COX-2, IL-8	TSP-1, TSP-2, endostatin, angiostatin, vasostatin, PF4, IFNg, IFNβ, IL-12, IL-4, Id1, Id3, TFPI, VEGFI, TIMP, PEX, IP-10
Atherogenesis	MCP-1, GM-CSF, bFGF, FGF-4, FGFR-1, PDGF-B, VCAM, TNF-α, PA, MMP, selectin, TGFb-1	PAI-1, TIMP

Ang, angiopoietin; *bFGF*, basic fibroblast growth factor; *COX*, cyclooxygenase; *CXC*, CXC family chemokines; *FGF*, fibroblast growth factor; *GM-CSF*, granulocyte macrophage colony stimulating factor; *HGF*, hepatocyte growth factor; *HIF*, hypoxia inducible factor; *IFN*, interferon; *IGF*, insulin-like growth factor; *IL*, interleukin; *IP*, interferon gamma induced protein; *MCP*, monocyte chemotactic protein; *MMP*, matrix metalloproteinase; *NO*, nitric oxide; *PA*, plasminogen activator; *PAI*, plasminogen activator inhibitor; *PDGF*, platelet derived growth factor; *PECAM*, platelet endothelium cell adhesion molecule; *PEX*, metalloproteinase 2 fragment; *PlGF*, placental growth factor; *TFPI*, tissue factor pathway inhibitor; *TGF*, transformation growth factor; *TIMP*, tissue inhibitor metalloproteinase; *TNF*, tumoral necrosis factor; *TSP*, thrombospondin; *VCAM*, vascular endothelium adhesion molecule; *VE-cad*, vascular endothelial cadherin; *VEGF*, vascular endothelial growth factor; *VEGFR*, vascular endothelial growth factor receptor.

There is also considerable evidence that shear stress caused by blood flow affects development of collateral circulation both under physiological and pathological conditions [31], by means of the participation of transcription factors (c-Fos and Egr-1), enzymes (ACE and NO synthase), growth factors (PDGF-A and B, TGF-β), and signaling molecules (integrins and adhesion molecules). It is noteworthy that, even if specific factors play a central role in vascular growth and knock-out experimental models of their respective genes have led to embryonic lethality (such as VEGF), the complexity of the angiogenic

TABLE 5.2   Main Functions of Selected Factor Functions Involved With Vascular Growth Regulation

Factor	Function
ACTIVATORS	
VEGF	Stimulation of angiogenesis; increased permeability
VEGF-C	Stimulation of lymphangiogenesis
PlGF	Involved in pathological angiogenesis
Angiopoietin-1	Vessel stabilization by reinforcing the interaction between smooth muscle cells and endothelial cells
PDGF-BB	Recruitment of smooth muscle cells
TGF-β1	Vessel stabilization by stimulating extracellular matrix production
FGF	Stimulation of angiogenesis and atherogenesis
HGF	Stimulation of angiogenesis
MCP-1	Stimulation of atherogenesis
Ephrins	Regulation of artery and vein differentiation
Activators of plasminogen and matrix metalloproteinase	Involved in cell migration and matrix remodeling; release of FGF and VEGF; activation of TGF-β1
NO synthase and cyclooxygenase	Stimulation of angiogenesis and vasodilation
INHIBITORS	
Angiopoietin-2	Destabilization of vessels before budding; induces vessel regression in the absence of angiogenic stimulation
Angiostatin	Inhibition of endothelial migration and cell survival
Endostatin	Inhibition of endothelial migration and cell survival
Vasostatin	Inhibition of vascular growth
Interferon and interleukins	Inhibition of endothelial migration

process implies participation of several of these factors in crucial moments for establishment of physiologically functional vessels integrated to the native vascular network. Table 5.2 lists some selected factors (activators and inhibitors) and their respective roles in vascular growth.

Adding to the complexity of the vascular growth process, there is the observation that some growth factors interact with more than one type of receptor, such as some members of the VEGF family (Fig. 5.4).

VASCULAR ENDOTHELIAL GROWTH FACTOR (VEGF)

Among the pro-angiogenic factors, VEGF forms a family of polypeptides secreted with a binding site to its receptor, similarly to the platelet-derived growth factor [32]. Since its discovery in 1983 [33], and subsequent gene cloning in 1989 [34], VEGF emerged as the most important regulator of blood vessel formation in physiological and pathological conditions; it is essential for vascular development in the embryo, and is a key mediator in cancer neo-vascularization [35]. The VEGF family consists of five members: VEGF-A, VEGF-B, VEGF-C, VEGF-D, and the placental growth factor (PlGF). There are also other isoforms of VEGF-A, according to the different lengths transcribed (in number of amino acids, excluding signaling peptides): 121, 145, 165, 183, 189, and 206 amino acids. The VEGF-A$_{121}$ and VEGF-A165 factors are the main isoforms in mammals.

The main receptors involved in the initiation of signal transduction cascade in response to VEGF include a family of three strongly related tyrosine kinase receptors, currently called VEGFR-1 (previously known as Flt-1), VEGFR-2 (previously known as KDR or Flk-1), and VEGFR-3 (previously known as Flt-3).

Among the VEGF gene expression regulating mechanisms, O_2 is one of the most important mediators, both in vitro and in vivo. Expression of VGEF RNAm is quick and reversibly induced by exposure to low partial pressure of O_2 in a variety of normal or transformed cell cultures. In addition to the transcriptional activation of VEGF in response to hypoxia, the increase in VEGF RNAm stability has been viewed as the posttranscriptional element responsible for VEGF superexpression in these conditions [36].

Accumulated knowledge regarding VEGF family indicated that the secretion of this potent and fundamental vascular growth regulator must be rigorously controlled in quantity, time, and space, in order to avoid uncontrolled blood vessel growth. An elegant demonstration of VEGF importance for normal vasculature development comes from the observation that suppression of even one single VEGF allele in mice leads to embryonic lethality due to severe vascular anomalies, providing, perhaps, the only example of embryonic lethality due to the "half-dose" effect [37].

Angiopoietins

Despite the indispensable role that the members of the VEGF family play in vascular formation, they must work together with other factors. Angiopoietins seem to be one of the most important partners of VEGF. They were discovered as binders of Tie receptors, a family of tyrosine kinase receptors expressed selectively inside the vascular endothelium in the same fashion as VEGF receptors.

Angiopoietins add up to the members of the VEGF family as growth factors known for being highly specific for vascular endothelium. The angiopoietins family includes an agonist molecule (angiopoietin-1), as well as an antagonist (angiopoietin-2), both acting through the receptor Tie-2. Recently, molecular biology studies led to the identification of two new family members, angiopoietin-3 (mice) and angiopoietin-4 (human) [38]. Apparently, angiopoietin-3 acts as an antagonist, while angiopoietin-4 works as a vascular growth agonist.

Ang-1 is predominantly expressed in perivascular cells, such as pericytes, vascular smooth muscle cells, fibroblasts, and tumor cells [39]. Hypoxia, VEGF-A, and PDGF-B increase Ang-1 expression in pericytes and smooth muscle cells [40]. After Ang-1 binds to the Tie-2 receptor, several intracellular signaling pathways are activated, leading to NO synthesis and higher survival of endothelial cells due to decreased apoptosis.

On the other hand, Ang-2 is expressed in endothelial cells, and its production is regulated by several growth factors (including VEGF-A) and physiopathological conditions (such as tissue hypoxia) [41]. While almost absent in quiescent vasculature, Ang-2 expression is quickly increased after angiogenic activation of the endothelium. Release of Ang-2 by endothelial cells allows autocrine endothelium destabilization, and this destabilizing effect may be antagonized by Ang-1. In addition, it has been shown that Ang-2 induces blood vessel regression in the absence of VEGF, but promotes angiogenesis in its presence [42]. Therefore, the proportion between Ang-1 and Ang-2 is essential for Tie-2 receptor signaling balance and vascular homeostasis regulation in response to angiogenic stimuli.

Ephrins

The Eph receptors are a unique family of tyrosine kinase receptors that play a critical role in the embryonic stage, in vascular development, and in neo-vascularization in adult individuals. Coupling between Eph receptors and its binders (ephrins) mediates critical angiogenesis phenomena, including cell-cell contact, cellular adhesion to the extracellular matrix, and cell migration [43]. Recent evidence from in vitro angiogenesis assays and analysis of mice that have deficiency of one or more members of the ephrin family established the role of ephrin signaling in budding angiogenesis and blood vessel remodeling during vascular development. In addition, elevated expression of Eph receptors and ephrin binders is present in several types of cancer, suggesting also a critical role in tumor angiogenesis and tumor growth. In the developing cardiovascular system, ephrin receptors and its binders seem to control angiogenic remodeling of blood and lymphatic vessels, and perform essential roles in endothelial cells, as well as pericytes and vascular smooth muscle cells.

Unlike many angiogenic factors, such as VEGF of bFGF, ephrins do not directly promote cell proliferation or transformation. However, subexpression of EphA1 in fibroblasts and EphA1 in normal breast epithelial cells promoted the growth of cell colonies and formation of tumor in mice, indicating the capacity of the EphA2 receptor to transform normal cells [44].

Several members of the ephrin receptor family and their binders are expressed in the cardiovascular system, but the receptor EphB4 and its binder ephrin-B2 have attracted the greatest interest. Efrin-B2 is predominantly expressed in arterial endothelial cells, while EphB4 is more specific to veins [45].

PROGENITOR CELLS AND ANGIOGENESIS

The importance of the endothelium in cardiovascular homeostasis has long been known. In fact, endothelial dysfunction has been involved in several clinical conditions, including, but not limited to, atherosclerosis, thrombosis, and arterial hypertension. Likewise, the balance between endothelial injury and recovery seems to be related to the occurrence of cardiovascular events. Therefore, and considering the observation that mature endothelial cells have limited regeneration capacity, the interest in the role of progenitor endothelial cells in maintaining endothelial integrity and in postnatal neo-vascularization has grown [46].

Other than endothelial progenitor cells, several other cell types are implied in vascular growth, and some have already been tested in clinical cell therapy studies for ischemic tissue [47]. We emphasize pro-angiogenic cells derived from bone marrow, resident vascular stem cells, mesenchymal stem cells, and embryonic stem cells or induced pluripotent stem cells (Fig. 5.5).

The initial landmark of the advances in this field of knowledge came from the identification, in the late 1990s, by Asahara et al. [48] of endothelial progenitor cells, opening the possibility of therapeutic application mediated by cell therapy in ischemic syndrome treatment, something that would only

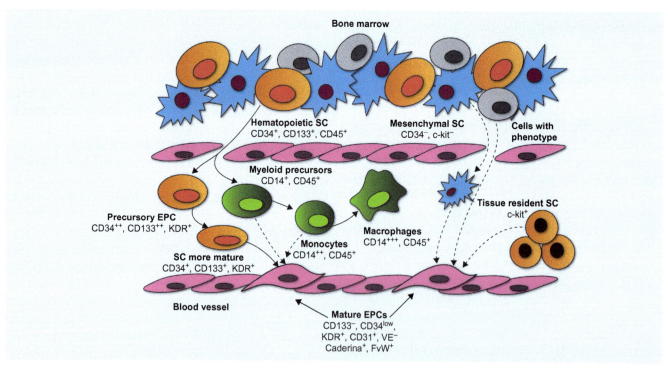

FIG. 5.5   Potential endothelial progenitor cell origins and differentiation pathways [46]. *SC*, stem cells; *EPC*, endothelial progenitor cells. Modified from Shantsila E, Watson T, Lip GYH. Endothelial progenitor cells in cardiovascular disorders. J Am Coll Cardiol 2007;49:741–52.

occur a decade later. These cells are implicated in vascular growth by migration, implantation, and differentiation in vascular structures, found, for example, during fetal development; in angiogenesis, vascular growth is due to endothelial cell migration, proliferation, and budding from the preexisting vasculature [48]. Even if there is still some controversy regarding the function of endothelial progenitor cells, it is acknowledged that these cells participate in the vascular growth and endothelial reparation processes.

Several experimental studies have shown that pro-angiogenic cells derived from bone marrow and peripheral blood are essential for recovery of ischemic tissue functions [49–51]. It is interesting to note that vascular risk factors, such as oxidative stress, may alter the function of pro-angiogenic cells, resulting in impairment of functional recovery of ischemic tissues [52]. Furthermore, knowing that cardiovascular risk factors contribute for the atherosclerosis process by causing abnormalities in endothelial function, whose integrity is a fundamental condition for atherogenesis, it is reasonable to suppose the association between risk factors, regenerative capacity of endothelial cells, and progression of atherosclerosis. Hill et al. [53] studied the number of endothelial progenitor cell colonies obtained in peripheral blood samples of 45 men with an average age of 50 ± 2 years, with different levels of cardiovascular risk, but in the absence of clinically manifested or past cardiovascular disease. Independent and endothelium-dependent vasodilatation was determined by high-resolution ultrasonography of the brachial artery. Fig. 5.6A shows the strong association between cardiovascular risk (evaluated by the Framingham risk score) and the number of

endothelial progenitor cell colony units. On the other hand, Fig. 5.6B illustrates the relation between the number of endothelial progenitor cells and endothelial function.

The investigators concluded that, in healthy men, the levels and functionality of endothelial progenitor cells may be a surrogate marker of endothelial function and cumulative cardiovascular risk, and consequently, predisposes the individual to cardiovascular disease progression. More than its physiopathological importance, the closer relationship between endothelial progenitor cells and cardiovascular diseases was documented in another clinical study with 519 patients with angiographically confirmed coronary disease [54]. In this study, endothelial progenitor cell levels at baseline were inversely associated to cardiovascular event risks, including cardiovascular death, acute myocardial infarction, and the need of revascularization in the first year of follow-up. Specifically, patients with higher circulating levels of endothelial progenitor cells had a reduction of 69% in the relative risk of cardiovascular death compared with patients with lower levels.

After understanding the importance of endothelial progenitor cells in the atherosclerosis physiopathological process, and its direct relation with cardiovascular risk factor and events, an experimental model of endothelial injury mediated by a balloon catheter documented that simvastatin promoted quick re-endothelization of the injured vessel by mobilizing and incorporating endothelial progenitor cells derived from bone marrow (Fig. 5.7A), leading to a lower level of neointimal thickening (Fig. 5.7B) [55], without changing lipid levels.

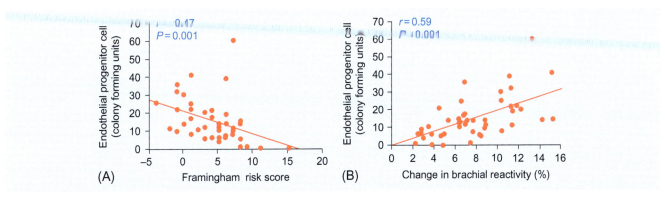

FIG. 5.6   Association between the Framingham risk score (A) and endothelial function (B) with the number of endothelial progenitor cell colony forming units. Adapted from Hill JM, Zalos G, Halcox JP, et al. Circulating endothelial progenitor cells, vascular function, and cardiovascular risk. N Engl J Med 2003;348:593–600.

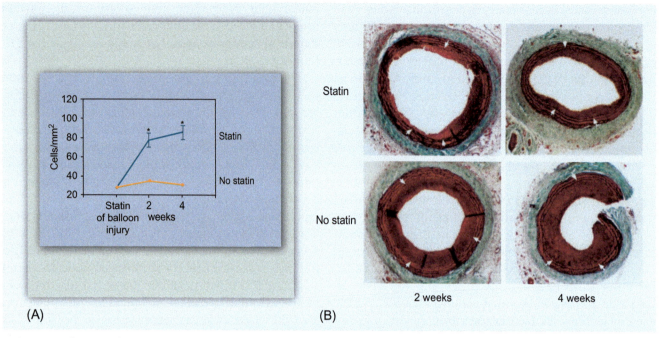

FIG. 5.7 The use of simvastatin promoted increased number of endothelial progenitor cells derived from bone marrow (A), and lower neointimal thickening (B) in an endothelial injury model mediated by balloon catheter. Adapted from Walter DH, Rittig K, Bahlmann FH, et al. Statin therapy accelerates reendothelialization: a novel effect involving mobilization and incorporation of bone marrow-derived endothelial progenitor cells. Circulation 2002;105:3017–24.

The functional modulation of endothelial progenitor cells was already documented with other drugs, including atorvastatin [56], ramipril [57], perindopril [58], and rosiglitazone [59]. Note that not only pharmacological interventions affect circulating angiogenic cell mobilization: dietetic interventions such as consumption of food rich in flavonoids [60] or resveratrol [61,62] and physical activity [63] may increase functionality of circulating angiogenic cells.

THERAPEUTIC APPLICATION OF STEM CELLS AND PROGENITOR CELLS

It has been shown that the three tunicas of blood vessels (intima, media, and adventitia) contain resident progenitor cells, including endothelial progenitor cells and stromal mesenchymal cells [64]. In vascular disease physiopathology, stem cells and progenitor cells may participate in vascular reparation and neo-intimal injury formation processes. In the adventitial layer, a large number of progenitor cells expressing Sca-1 were identified, which may contribute to both endothelial regeneration and accumulation of smooth muscle cells in neo-intimal injuries [65]. These few examples illustrate the complexity of vascular stem cells, which can be directly related to reparation or disease development processes.

Therefore, with advances in cellular biology techniques and the emergence of techniques to characterize, isolate, and purify several cell types implied in vascular growth, their therapeutic potential was tested in peripheral or myocardial ischemia animal models to subsequently reach the clinical arena [66].

One of the first successful examples of vascular progenitor cell application comes from a study in which systemic injection of these cells increased vessel formation in animals subjected to coronary artery ligation, with functional recovery compared to control-group animals [67]. In addition to the probable direct effect on vascular growth due to cell differentiation, the release of angiogenic cytokines (for example, VEGF or HGF) may strongly contribute to vascular growth induction as observed in peripheral and myocardial ischemia models [68].

Stimulated by the success accumulated in the last decade, with the transplant of hematopoietic progenitor cells derived from adult bone marrow, application of this cell type began in patients with coronary artery disease and heart failure. Assmus et al. [69] transplanted, by intracoronary infusion, progenitor cells derived from bone marrow ($n=9$) or from peripheral blood ($n=11$) in patients with acute myocardial infarction, within 4.9 ± 1.5 days after the acute event. During the following 4 months, treated patients presented increased LV ejection fraction, better regional motility in the infarct zone, decreased final systolic volume, and increased coronary flow reserve in the artery related to acute myocardial infarction (AMI), results sustained during a 1-year follow-up with a larger number of patients ($n=59$) [70]. A late 5-year follow-up confirmed the initial results obtained 1 year after the procedure, which were: sustained increase of LV ejection fraction from 46% to 57%, and decreased infarct area evaluated by cardiac magnetic resonance [71]. On the other hand, intracoronary infusion of bone marrow derived mononuclear cells in patients with left ventricle dysfunction 2–3 weeks post-AMI did not promote increased global or regional function of LV in a 6-month follow-up [72]. These conflicting results show the complexity of clinical assays in cell therapy, not only regarding the best cell type, route of administration, and cell concentration, but also to the characteristics of the micro-environment at the moment of cell transplant.

The use of adult bone marrow derived cells to treat severe heart ischemic diseases, associated with heart failure, was proposed by Perin et al. in a study conducted in 14 patients [73]. Patients were subjected to transendocardial injection guided by electromechanical mapping in feasible but ischemic areas (hibernating myocardium). The authors showed that, at a 4-month follow-up, functional class improved, perfusion defects evaluated by SPECT were reduced significantly, and the ejection fraction increased significantly from 20% to 29%.

Stamm et al. [74] proposed the use of intramyocardial injection of bone marrow-derived stem cells with the potential to induce angiogenesis, combined with surgical myocardial revascularization in six post-AMI patients. During surgery, $\sim 1.5 \times 10^6$ cells were injected on the edge of the infarct zone. After 3- to 9-month follow-ups, they documented increased global motility (in four out of six patients) and perfusion in the infarct area (in five out of six patients). Gowdak et al. [75] adopted a similar strategy to treat patients with severe and diffuse coronary artery disease, who were refractory to clinical treatment, and not suitable to complete surgical revascularization due to the extent of the disease. Stem cells and autologous hematopoietic progenitor cells were injected in these patients during surgery, in myocardial areas previously identified as viable and ischemic. Analysis of myocardial perfusion in injected and nonrevascularized segments indicated ischemia reversion and contractile improvement. Obviously, due to the open type of study without control group, it was not possible to exclude the contribution of graft performed at distance for the improvement observed in injected segments; the completion of a large national multicenter study will answer this unsettled question [76].

More recently, a new therapeutic strategy based on the use of bone marrow-derived lymphomono nuclear cells was proposed for patients with refractory angina who were not candidates to conventional myocardial revascularization. The combination of transmyocardial laser revascularization and intramyocardial cell injection was conducted in patients with refractory angina, with a subsequent increment in physical exercise tolerance (improvement in functional capacity) and decreased ischemic burden of the left ventricle, during the first 6 months of follow-up [77].

The RENEW [78] study, currently in progress, shall test the efficacy and safety of intramyocardial injection of CE34+ autologous cells in patients with refractory angina on top of optimal medical therapy who are not candidates to revascularization procedures. Another recently started study, IMPACT-CABG [79], shall test the safety and efficacy of intramyocardial injection of autologous CD133$^+$ cells in patients subjected to myocardial revascularization surgery.

Although most probably by means of additional mechanisms, beyond its angiogenic potential, one of the last cell types explored in treatment of patients with ischemic cardiomyopathy was the result of the identification of resident cardiac stem cells with myocardial regeneration potential [80]. Some preclinical studies demonstrated the potential benefit of these cells in treating postinfarction ventricular dysfunction [81,82]. In SCIPIO [83], a clinical study, resident heart stem cells were obtained from the right atrial appendage during myocardial revascularization surgery. Once isolated, the cells were expanded and infused by means of an intracoronary pathway ~ 4 months after surgery. Heart function evaluated by magnetic resonance showed a significant increase of left ventricular ejection fraction in the treated group, from 27.5% (baseline) to 35.1% and 41.2%, respectively, 4 and 12 months after cell infusion, in addition to significant reduction of the infarction area.

CONCLUSIONS

The observation made by Hunter more than 200 years ago regarding vascular growth in response to body needs may be naive in face of the complex mechanisms involved in angiogenesis. In fact, some of the interactions between growth factors and their receptors and between pro- and antiangiogenic factors remain only partially understood. Furthermore, recognition that several clinical conditions as diverse as rheumatoid arthritis, psoriasis, diabetic retinopathy, and cancer have common aspects related to pathological angiogenesis, created a new investigation front to be explored. The possibility of therapeutic strategies founded on blocking vascular growth or antiangiogenic therapy was envisaged, and some have already been implemented in clinical practice [84]. On the other hand, for clinical conditions marked by chronic ischemia, such as peripheral arterial disease and heart ischemic disease, the therapeutic proposal is diametrically opposed—that is, the objective is to stimulated vascular growth or therapeutic angiogenesis [85,86]. In the last strategy, due to the large number of factors involved, regulated accurately in time, space, and quantities secreted, it is possible to anticipate the difficulty inherent to this attempt to recapitulate a physiological process, fundamental to maintain organ and tissue vitality.

However, it is important to keep in mind that the real contribution of cellular therapy to cardiology shall be defined by randomized double-blind clinical trials, controlled by placebo, and with sufficiently large number of patients followed-up for a long period of time. Several questions remain unanswered or were only partially answered, regarding the elements that regulate the complex processes involved in transdifferentiation of several cell types and the paracrine effects, responsible for myocyte regeneration and/or vascular growth induction [87]. The influence of the micro-environment in which the cells are injected may determine not only their survival, but also their final destination (phenotype) [88]. The best access to implant cells in the heart shall be defined, but it must consider the moment in the natural history of the disease (acute event versus chronic disease), in isolated or associated procedures [89].

The therapeutic contribution of bone marrow derived progenitor cells in the treatment of patients with ischemic cardiomyopathy was recently reviewed by meta-analysis involving 48 randomized studies and including 2602 patients [90]. In comparison to usual therapy, cellular therapy promoted increase of left ventricular ejection fraction by ~3%, reduction of little more than 2% in infarct size, and ~6 mL in the final systolic volume. Although numerically modest, these differences translated into a lower risk of clinical outcomes, including all-cause mortality, recurring myocardium infarct, or ventricular arrhythmia.

Eventually, advances in molecular and cellular biology may enable not only deeper comprehension of the vascular growth process, but also allow precise therapeutic modulation with antiangiogenic or pro-angiogenic purposes.

References

[1] Folkman J, Merler E, Abernathy C, et al. Isolation of a tumor factor responsible for angiogenesis. J Exp Med 1971;133(2):275–88.
[2] Folkman J. Tumor angiogenesis: therapeutic implications. N Engl J Med 1971;285(21):1182–6.
[3] Algire GH, Chalkley HW. Vascular reactions of normal and malignant tissues in vivo. I. Vascular reactions of mice to wounds and to normal and neoplastic implants. J Natl Cancer Inst 1945;6:73–85.
[4] Greene HS. Heterologous transplantation of mammalian tumors. J Exp Med 1941;73(4):461–74.
[5] Folkman J, Cole P, Zimmerman S. Tumor behavior in isolated perfused organs. Ann Surg 1966;164(3):491–502.
[6] Folkman J. Anti-angiogenesis: new concept for therapy of solid tumors. Ann Surg 1972;175:409–16.
[7] Svet-Moldavsky GJ, Chimishkyan KL. Tumour angiogenesis factor for revascularisation in ischaemia and myocardial infarction. Lancet 1977;1:913.
[8] Kumar S, West D, Shahabuddin S, et al. Angiogenesis factor from human myocardial infarcts. Lancet 1983;2:364–8.
[9] Fett JW, Strydom DJ, Lobb RR, et al. Isolation and characterization of angiogenin, an angiogenic protein from human carcinoma cells. Biochemistry 1985;24:5480–6.
[10] Strydom DJ, Fett JW, Lobb RR, et al. Amino acid sequence of human tumor derived angiogenin. Biochemistry 1985;24:5486–94.
[11] Kurachi K, Davie EW, Strydom DJ, et al. Sequence of the cDNA and gene for angiogenin, a human angiogenesis factor. Biochemistry 1985;24:5494–9.
[12] Angiogenin. Lancet 1985;2:1340–1.
[13] Liotta LA. Isolation of a protein that stimulates blood vessel growth. Nature 1985;318:14.
[14] Höckel M, Schlenger K, Doctrow S, et al. Therapeutic angiogenesis. Arch Surg 1993;128:423–9.
[15] Risau W. Mechanisms of angiogenesis. Nature 1997;386:671–4.
[16] Carmeliet P. Mechanisms of angiogenesis and arteriogenesis. Nat Med 2000;6:389–95.
[17] Semenza GL. Vasculogenesis, angiogenesis, and arteriogenesis: mechanisms of blood vessel formation and remodeling. J Cell Biochem 2007;102:840–7.
[18] Hogan BL, Kolodziej PA. Organogenesis: molecular mechanisms of tubulogenesis. Nat Rev Genet 2002;3:513–23.

[19] Okada H, Tsuzuki T, Shindoh H, et al. Regulation of decidualization and angiogenesis in the human endometrium: mini review. J Obstet Gynaecol Res 2014;40:1180–7.

[20] Stegen S, van Gastel N, Carmeliet G. Bringing new life to damaged bone: the importance of angiogenesis in bone repair and regeneration. Bone 2015;70:19–27.

[21] Behl T, Kotwani A. Exploring the various aspects of the pathological role of vascular endothelial growth factor (VEGF) in diabetic retinopathy. Pharmacol Res 2015;99:137–48.

[22] Azizi G, Boghozian R, Mirshafiey A. The potential role of angiogenic factors in rheumatoid arthritis. Int J Rheum Dis 2014;17:369–83.

[23] Chua RA, Arbiser JL. The role of angiogenesis in the pathogenesis of psoriasis. Autoimmunity 2009;42:574–9.

[24] Lee SH, Jeong D, Han YS, et al. Pivotal role of vascular endothelial growth factor pathway in tumor angiogenesis. Ann Surg Treat Res 2015;89:1–8.

[25] Clapp C, Thebault S, Jeziorski MC, et al. Peptide hormone regulation of angiogenesis. Physiol Rev 2009;89:1177–215.

[26] De Spiegelaere W, Casteleyn C, Van den Broeck W, et al. Intussusceptive angiogenesis: a biologically relevant form of angiogenesis. J Vasc Res 2012;49:390–404.

[27] Cai W, Schaper W. Mechanisms of arteriogenesis. Acta Biochim Biophys Sin Shanghai 2008;40:681–92.

[28] Conway EM, Collen D, Carmeliet P. Molecular mechanisms of blood vessel growth. Cardiovasc Res 2001;49:507–21.

[29] Semenza GL. Hypoxia-inducible factor 1 and cardiovascular disease. Annu Rev Physiol 2014;76:39–56.

[30] Semenza GL. Regulation of tissue perfusion in mammals by hypoxia-inducible factor 1. Exp Physiol 2007;92:988–91.

[31] Wragg JW, Durant S, McGettrick HM, et al. Shear stress regulated gene expression and angiogenesis in vascular endothelium. Microcirculation 2014;21:290–300.

[32] Holmes DI, Zachary I. The vascular endothelial growth factor (VEGF) family: angiogenic factors in health and disease. Genome Biol 2005;6:209.

[33] Senger DR, Galli SJ, Dvorak AM, et al. Tumor cells secrete a vascular permeability factor that promotes accumulation of ascites fluid. Science 1983;219:983–5.

[34] Leung DW, Cachianes G, Kuang WJ, et al. Vascular endothelial growth factor is a secreted angiogenic mitogen. Science 1989;246:1306–9.

[35] Ferrara N, Gerber HP, LeCouter J. The biology of VEGF and its receptors. Nat Med 2003;9:669–76.

[36] KumarVB Binu S, Soumya SJ, et al. Regulation of vascular endothelial growth factor by metabolic context of the cell. Glycoconj J 2014;31:427–34.

[37] Carmeliet P, Ferreira V, Breier G, et al. Abnormal blood vessel development and lethality in embryos lacking a single VEGF allele. Nature 1996;380:435–9.

[38] Valenzuela DM, Griffiths JA, Rojas J, et al. Angiopoietins 3 and 4: diverging gene counterparts in mice and humans. Proc Natl Acad Sci U S A 1999;96:1904–9.

[39] Fagiani E, Christofori G. Angiopoietins in angiogenesis. Cancer Lett 2013;328:18–26.

[40] Park YS, Kim NH, Jo I. Hypoxia and vascular endothelial growth factor acutely up-regulate angiopoietin-1 and Tie2 mRNA in bovine retinal pericytes. Microvasc Res 2003;65:125–31.

[41] Oh H, Takagi H, Suzuma K, et al. Hypoxia and vascular endothelial growth factor selectively up-regulate angiopoietin-2 in bovine microvascular endothelial cells. J Biol Chem 1999;274:15732–9.

[42] Hanahan D. Signaling vascular morphogenesis and maintenance. Science 1997;277:48–50.

[43] Cheng N, Brantley DM, Chen J. The ephrins and Eph receptors in angiogenesis. Cytokine Growth Factor Rev 2002;13:75–85.

[44] Zelinski DP, Zantek ND, Stewart JC, et al. EphA2 overexpression causes tumorigenesis of mammary epithelial cells. Cancer Res 2001;61:2301–6.

[45] Kuijper S, Turner CJ, Adams RH. Regulation of angiogenesis by Eph-ephrin interactions. Trends Cardiovasc Med 2007;17:145–51.

[46] Shantsila E, Watson T, Lip GYH. Endothelial progenitor cells in cardiovascular disorders. J Am Coll Cardiol 2007;49:741–52.

[47] Zhang L, Xu Q. Stem/progenitor cells in vascular regeneration. Arterioscler Thromb Vasc Biol 2014;34:1114–9.

[48] Asahara T, Murohara T, Sullivan A, et al. Isolation of putative progenitor endothelial cells for angiogenesis. Science 1997;275:964–7.

[49] Rafii S, Lyden D. Therapeutic stem and progenitor cell transplantation for organ vascularization and regeneration. Nat Med 2003;9:702–12.

[50] Assmus B, Honold J, Schächinger V, et al. Transcoronary transplantation of progenitor cells after myocardial infarction. N Engl J Med 2006;355:1222–32.

[51] Kawamoto A, Gwon HC, Iwaguro H, et al. Therapeutic potential of ex vivo expanded endothelial progenitor cells for myocardial ischemia. Circulation 2001;103:634–7.

[52] Bae ON, Wang JM, Baek SH, et al. Oxidative stress-mediated thrombospondin-2 upregulation impairs bone marrow-derived angiogenic cell function in diabetes mellitus. Arterioscler Thromb Vasc Biol 2013;33:1920–7.

[53] Hill JM, Zalos G, Halcox JP, et al. Circulating endothelial progenitor cells, vascular function, and cardiovascular risk. N Engl J Med 2003;348:593–600.

[54] Werner N, Kosiol S, Schiegl T, et al. Circulating endothelial progenitor cells and cardiovascular outcomes. N Engl J Med 2005;353:999–1007.

[55] Walter DH, Rittig K, Bahlmann FH, et al. Statin therapy accelerates reendothelialization: a novel effect involving mobilization and incorporation of bone marrow-derived endothelial progenitor cells. Circulation 2002;105:3017–24.

[56] Vasa M, Fichtlscherer S, Adler K, et al. Increase in circulating endothelial progenitor cells by statin therapy in patients with stable coronary artery disease. Circulation 2001;103:2885–90.

[57] Min TQ, Zhu CJ, Xiang WX, et al. Improvement in endothelial progenitor cells from peripheral blood by ramipril therapy in patients with stable coronary artery disease. Cardiovasc Drugs Ther 2004;18:203–9.

[58] Ferrari R, Guardigli G, Ceconi C. Secondary prevention of CAD with ACE inhibitors: a struggle between life and death of the endothelium. Cardiovasc Drugs Ther 2010;24:331–9.

[59] Pistrosch F, Herbrig K, Oelschlaegel U, et al. PPARgamma-agonist rosiglitazone increases number and migratory activity of cultured endothelial progenitor cells. Atherosclerosis 2005;183:163–7.

[60] Heiss C, Jahn S, Taylor M, et al. Improvement of endothelial function with dietary flavanols is associated with mobilization of circulating angiogenic cells in patients with coronary artery disease. J Am Coll Cardiol 2010;56:218–24.

[61] Hamed S, Alshiek J, Aharon A, et al. Red wine consumption improves in vitro migration of endothelial progenitor cells in young, healthy individuals. Am J Clin Nutr 2010;92:161–9.

[62] Huang PH, Chen YH, Tsai HY, et al. Intake of red wine increases the number and functional capacity of circulating endothelial progenitor cells by enhancing nitric oxide bioavailability. Arterioscler Thromb Vasc Biol 2010;30:869–77.

[63] Kaźmierski M, Wojakowski W, Michalewska-Włudarczyk A, et al. Exercise-induced mobilisation of endothelial progenitor cells in patients with premature coronary heart disease. Kardiol Pol 2015;73:411–8.

[64] Torsney E, Xu Q. Resident vascular progenitor cells. J Mol Cell Cardiol 2011;50:304–11.

[65] Hu Y, Zhang Z, Torsney E, et al. Abundant progenitor cells in the adventitia contribute to atherosclerosis of vein grafts in ApoE— deficient mice. J Clin Invest 2004;113:1258–65.

[66] O'Neill CL, O'Doherty MT, Wilson SE, et al. Therapeutic revascularisation of ischaemic tissue: the opportunities and challenges for therapy using vascular stem/progenitor cells. Stem Cell Res Ther 2012;3:31.

[67] Asahara T, Kalka C, Isner JM. Stem cell therapy and gene transfer for regeneration. Gene Ther 2000;7:451–7.

[68] Ziebart T, Yoon CH, Trepels T, et al. Sustained persistence of transplanted proangiogenic cells contributes to neovascularization and cardiac function after ischemia. Circ Res 2008;103:1327–34.

[69] Assmus B, Schächinger V, Teupe C, et al. Transplantation of progenitor cells and regeneration enhancement in acute myocardial infarction (TOPCARE-AMI). Circulation 2002;106:3009–17.

[70] Schächinger V, Assmus B, Britten MB, et al. Transplantation of progenitor cells and regeneration enhancement in acute myocardial infarction: final one-year results of the TOPCARE-AMI trial. J Am Coll Cardiol 2004;44:1690–9.

[71] Leistner DM, Fischer-Rasokat U, Honold J, et al. Transplantation of progenitor cells and regeneration enhancement in acute myocardial infarction (TOPCARE-AMI): final 5-year results suggest long-term safety and efficacy. Clin Res Cardiol 2011;100:925–34.

[72] Traverse JH, Henry TD, Ellis SG, et al. Effect of intracoronary delivery of autologous bone marrow mononuclear cells 2 to 3 weeks following acute myocardial infarction on left ventricular function: the late TIME randomized trial. JAMA 2011;306:2110–9.

[73] Perin EC, Dohmann HF, Borojevic R, et al. Transendocardial, autologous bone marrow cell transplantation for severe, chronic ischemic heart failure. Circulation 2003;107:2294–302.

[74] Stamm C, Westphal B, Kleine HD, et al. Autologous bone-marrow stem-cell transplantation for myocardial regeneration. Lancet 2003;361:45–6.

[75] Gowdak LH, Schettert IT, Rochitte CE, et al. Early increase in myocardial perfusion after stem cell therapy in patients undergoing incomplete coronary artery bypass surgery. J Cardiovasc Transl Res 2011;4:106–13.

[76] Tura BR, Martino HF, Gowdak LH, et al. Multicenter randomized trial of cell therapy in cardiopathies—MiHeart study. Trials 2007;8:2.

[77] Gowdak LH, Schettert IT, Rochitte CE, et al. Transmyocardial laser revascularization plus cell therapy for refractory angina. Int J Cardiol 2008;127:295–7.

[78] Povsic TJ, Junge C, Nada A, et al. A phase 3, randomized, double-blinded, active-controlled, unblinded standard of care study assessing the efficacy and safety of intramyocardial autologous CD34+ cell administration in patients with refractory angina: design of the RENEW study. Am Heart J 2013;165:854–61.

[79] Forcillo J, Stevens LM, Mansour S, et al. Implantation of CD133+ stem cells in patients undergoing coronary bypass surgery: IMPACT-CABG pilot trial. Can J Cardiol 2013;29:441–7.

[80] Beltrami AP, Barlucchi L, Torella D, et al. Adult cardiac stem cells are multipotent and support myocardial regeneration. Cell 2003;114:763–76.

[81] Dawn B, Stein AB, Urbanek K, et al. Cardiac stem cells delivered intravascularly traverse the vessel barrier, regenerate infarcted myocardium, and improve cardiac function. Proc Natl Acad Sci U S A 2005;102:3766–71.

[82] Tang XL, Rokosh G, Sanganalmath SK, et al. Intracoronary administration of cardiac progenitor cells alleviates left ventricular dysfunction in rats with a 30-day-old infarction. Circulation 2010;121:293–305.

[83] Bolli R, Chugh AR, D'Amario D, et al. Cardiac stem cells in patients with ischaemic cardiomyopathy (SCIPIO): initial results of a randomised phase 1 trial. Lancet 2011;378:1847–57.

[84] Jain RK. Antiangiogenesis strategies revisited: from starving tumors to alleviating hypoxia. Cancer Cell 2014;26:605–22.

[85] Gowdak LH, Poliakova L, Wang X, et al. Adenovirus-mediated VEGF(121) gene transfer stimulates angiogenesis in normoperfused skeletal muscle and preserves tissue perfusion after induction of ischemia. Circulation 2000;102:565–71.

[86] Hinkel R, Trenkwalder T, Kupatt C. Gene therapy for ischemic heart disease. Expert Opin Biol Ther 2011;11:723–37.

[87] Jadczyk T, Faulkner A, Madeddu P. Stem cell therapy for cardiovascular disease: the demise of alchemy and rise of pharmacology. Br J Pharmacol 2013;169:247–68.

[88] Maher KO, Xu C. Marching towards regenerative cardiac therapy with human pluripotent stem cells. Discov Med 2013;15:349–56.

[89] Mummery CL, Davis RP, Krieger JE. Challenges in using stem cells for cardiac repair. Sci Transl Med 2010;2. 27 ps17.

[90] Afzal MR, Samanta A, Shah ZI, et al. Adult bone marrow cell therapy for ischemic heart disease: evidence and insights from randomized controlled trials. Circ Res 2015;117:558–75.

Further Reading

[1] Ellis LM, Hicklin DJ. VEGF-targeted therapy: mechanisms of antitumour activity. Nat Rev Cancer 2008;8:579–91.

6

Characteristics of the Endothelium in Both Sexes

*Tiago Januário da Costa[a], Cinthya Echem[a], Lucas Giglio Colli[a],
Eliana Hiromi Akamine, Ana Paula Dantas,
and Maria Helena Catelli de Carvalho*

INTRODUCTION

Interest in studying sex as a biological and decisive variable of several diseases has grown in the last few years, mainly after the publication of *Exploring the Biological Contributions to Human Health: Does Sex Matter?* [1].

Several works published in scientific literature refer to sexual differences, inherent to a biological context, classifying them without distinguishing sex and gender. Currently, it is still possible to find overlapping between these terms in literature, without the correct distinction between them.

The term "sex" refers to the biological and physiological condition of the individual, human and animal, categorized as male and female. Indicators of this condition are sex chromosomes, gonads, internal reproductive organs, and external genitalia. The term "gender" refers to attributes that involve economic, social, and cultural aspects associated with the conception of being a man or a woman [1,2].

In sex chromosomes, genes are expressed differently according to the sex of the organism, since females have XX chromosomes while males have XY chromosomes. There are 1100 genes in chromosome X, and most of them are not expressed in chromosome Y. In addition, cellular mosaic created by the inactivation of chromosome X provides a biological advantage for females, that is, genes expressed in chromosome X are randomized between alleles proceeding from the father and the mother, while in males these genes are exclusive from the maternal X chromosome [3]. It is also important to consider the different meiotic processes and gene imprinting [1,4,5].

For decades, most of the research involving cardiovascular disease and the endothelium used males as experimental models, and extrapolated the results to females. In addition, many studies conducted in cell cultures do not specify the sex of the cell strain used. The gene imprint due partially to the presence of both "X" chromosomes in females and the "Y" chromosome in males may influence several biochemical and molecular pathways differently, which shall be decisive for cellular physiology [6,7].

Behavioral, anatomic, physiological, cellular, and molecular differences between males and females are common characteristics observed in several vertebrate species. Some of the differences are already evident on birth, mainly due to the inherent influence of sex chromosomes and, to a lesser extent, to fetal exposure to gonadal sex hormones. Currently, it is known that may differences observed may appear with sexual maturity, since the concentration of gonadal sex hormones are different since the intrauterine period and last the entire life [8–10].

Cardiovascular diseases are the largest cause of morbimortality in developed countries in both

---

[a] These authors have equal participation in this work.

sexes [11]. Women of childbearing age have lower risk of developing cardiovascular diseases than men in the same age group; however, after menopause, in which estrogen plasmatic concentration reduces, the risk of cardiovascular diseases becomes similar in both sexes [11–14]. There is scientific consensus that men and women respond differently to risk factors, regarding the development and severity of cardiovascular diseases [15].

Most studies evaluating the differences associated with sex in the cardiovascular system focus especially on the endothelial function, since vascular endothelium is an important tissue to regulate cardiovascular homeostasis and may present differences between males and females [16,17]. Cardiovascular diseases such as arterial hypertension and coronary diseases are more severe in young men, as well as in several animal experiment models, than women/females of the same species and age group [11,18,19]. Sexual differences in the progression and severity of cardiovascular diseases may be attributed, largely, to differentiated regulation of the endothelial function, which may depend on part on the male and female gene imprints, and largely on the hormonal regulation of cardiovascular functional regulating molecules, such as some receptors, α-adrenergics and bradykinin B2, as well as endothelial nitric oxide synthase enzyme (eNOS) [20].

It was demonstrated that acetylcholine (ACh), an endothelium-dependent vasodilator, binds to the muscarinic receptors and promotes higher vasodilatation of aorta rings isolated from female than in males, in both normotensive rat strains [21] and SHR (spontaneously hypertensive rats) strains [22,23] (Fig. 6.1). These animals also displayed differences associated with sex in endothelial regulation of vasoconstrictor agents, such as angiotensin-II, endothelin-1, and noradrenaline [24]. The reduced plasmatic concentration of estrogen promoted by the surgical removal of the ovaries (ovariectomy) reduces ACh vasodilation in comparison to SHR females with ovaries and hormonal treatment with 17β-estradiol or conjugated equine estrogens was effective in restoring the reduced ACh vasodilation in ovariectomized SHR females [25,26] (Fig. 6.2).

In 1996, Taddei et al. studied sexual differences in endothelial function associated with aging. By measuring forearm blood flow change after administration of ACh (endothelium-dependent vasodilator) or sodium nitroprusside (Nitric Oxide donor) in normotensive and essential hypertension men and women, they observed constant and age-related maximum ACh vasodilation decline in normotensive and hypertensive men. In contrast, women (normotensive and hypertensive) showed only a slight reduction in ACh vasodilation per year, until middle age

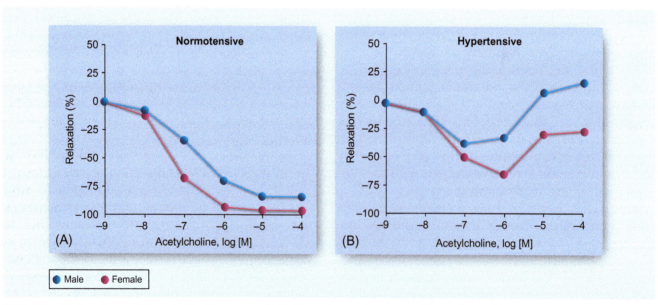

FIG. 6.1 Relaxation with acetylcholine, an endothelium-dependent vasodilator, is smaller in aorta rings, with endothelium, isolated from normotensive (Wistar) and hypertensive (SHR) male rats when compared to normotensive (Wistar) and hypertensive (SHR) females. Adapted from Kauser K, Rubanyi GM. Gender difference in endothelial dysfunction in the aorta of spontaneously hypertensive rats. Hypertension 1995;25:517–23.

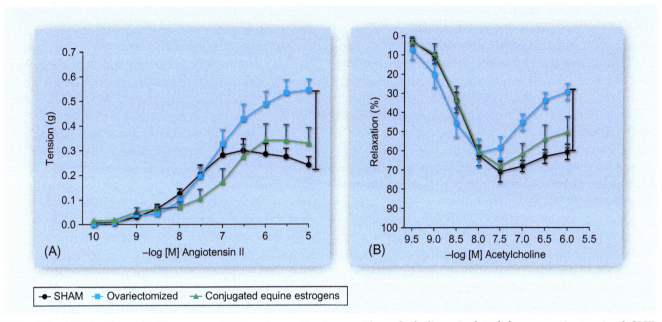

FIG. 6.2 (A) Angiostesin-II vasoconstriction in aortic rings with endothelium isolated from ovariectomized SHR females is increased when compared to aortic rings isolated from control SHR females (SHAM-operated), and those of ovariectomized females treated with conjugated equine estrogens. (B) Relaxation with acetylcholine, an endothelium-dependent vasodilator, is reduced in aortic rings isolated from ovariectomized SHR females when compared to those of control SHR females (SHAM-operated), and those of ovariectomized females treated with conjugated equine estrogens. Adapted from Costa TJ, Ceravolo GS, dos Santos RA, et al. Association of testosterone with estrogen abolishes the beneficial effects of estrogen treatment by increasing ROS generation in aorta endothelial cells. Am J Physiol Heart Circ Physiol 2015;308:H723–32.

(~50 years old). After that, the decline in endothelium-dependent vasodilator response accelerated and became more accentuated in comparison to men [27]. With that, they also evaluated the influence of postmenopause and, consequently, estrogen deficiency in endothelium-dependent vasodilation [27].

Arterial pressure values of women of childbearing age are lower than those of men of the same age group, and this difference has been attributed mainly to female gonadal sex hormones. Frequently, arterial pressure decreases during pregnancy, since in this period the concentration of circulating estrogen and progesterone are high [28–30] and in postmenopause, when the plasmatic concentration of estrogen is reduced, arterial pressure increases [31].

ESTROGENS, PROGESTERONE, AND TESTOSTERONE—MECHANISMS OF VASCULAR ACTION

The endothelial function of both sexes may suffer the influence of two important variables: sex chromosomes and sex hormones. In the prenatal period,

the impact of sex chromosomes and genetic regulation contribute greatly to determining sexual differences. Female human umbilical vein endothelial cells (FHUVECs) have higher expression of eNOS messenger ribonucleic acid (mRNA) than male human umbilical vein endothelial cells (MHUVECs) [32], although in this cell type the protein expression of estrogen (ERα, ERβ e GPER) and androgen receptors (AR) do not differ between sexes [32].

After birth and during the entire life, sex-dependent characteristics are also determined by chronobiology of gonadal sex hormones. Therefore, influences of chromosomes and sex hormones may act together or in parallel to define the sex-dependent phenotype. During their lives, men and women are exposed to different concentrations of gonadal sex hormones. The difference includes variation between sexes and variations that are intrinsic to females for example, during the menstrual period and hormone decline after menopause.

The main sex hormones in men and women (estrogen, progesterone, and testosterone) act on specific receptors in target cells to promote multiple actions on nonsexual tissues, including the cardiovascular system [33–38]. Steroid receptors were identified on the plasmatic membrane, cytosol, and nucleus of target cells [39]. Vascular endothelium expresses all subtypes of estrogen receptors (ER), as well as AR and progesterone receptors (PR) [22,40,41].

As gonadal sex hormones are liposoluble, they penetrate the cell and cross the plasmatic membrane by passive diffusion, and when they connect to specific receptors, forming the hormone-receptor complex, they promote genomic and nongenomic effects [34–38]. The nongenomic effects occur regardless of gene transcription and protein synthesis. These are considered quick effects, which occur in a matter of seconds or minutes, after the formation of the hormone-receptor complex [42] and involve the activation of kinase, ion channels present in the membrane, and production of nitric oxide (NO) [43,44].

On the other hand, genomic effects are delayed in response to the nuclear translocation of the hormone-receptor complex, positively or negatively regulating the gene and/or protein expression of some target genes [37] (Fig. 6.3).

The classic actions of the estrogens (17β-estradiol, estrone, and estriol) occur through three receptors, two of which are nuclear, called alpha (ERα/ERS1) [45] and beta (ERβ/ERS2) [46], and one membrane G-protein coupled receptor (GPR30) [47] (currently called GPER). The three types of receptors [48–50], as well as aromatase enzyme to metabolize androgens into estrogens [51], are expressed in both in smooth muscles and vascular endothelium of males and females.

The estrogen, coupled to its receptors associated with the membrane, ERα, or coupled to protein G, GPER, activates PI3K (phosphatidylinositol 3-kinase) and MAPK (mitogen activated kinase-like protein) signaling, contributing with NO generation in the vascular endothelium [52]. Although less studied, the ERβ located in the plasmatic membrane may activate MAPK, ERK and Src (extracellular signal-regulated kinase) pathways, and therefore modulate phosphorylation of several proteins involved in cell migration and proliferation processes [53].

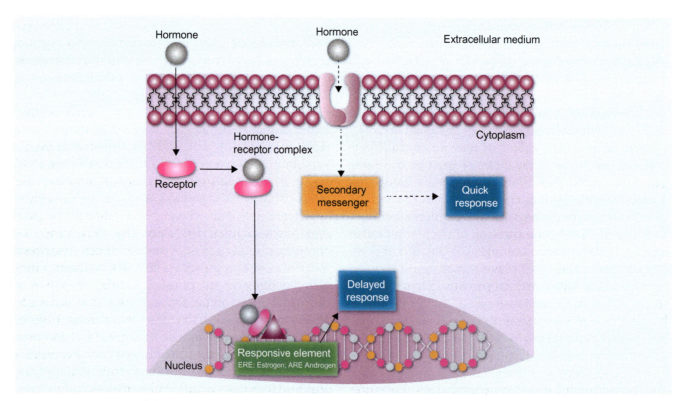

FIG. 6.3   Overview of the genomic and nongenomic action mechanisms of sex hormones. *ERE*, estrogen responsive element; *ARE*, androgen responsive element.

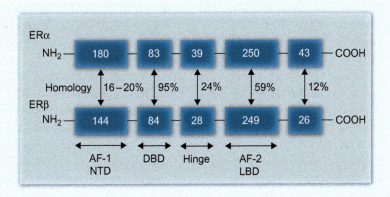

FIG. 6.4 Functional domains and homology of the estrogen receptor subtypes: ERα and ERβ. *NTD*, (NH₂)-terminal domain; *DBD*, DNA binding domain; *LBD*, ligand binding domain (see text for abbreviations).

The nuclear ERs consist of four functional domains:

(1) (NH₂)-terminal domain (NTD, N-terminal domain);
(2) DNA binding domain (DBD);
(3) Hinge domain; and
(4) Ligand binding domain (LBD), located in the carboxy terminal portion (COOH) of the receptor (Fig. 6.4).

The NTD contains an autonomous transcriptional activation region called AF-1 (activation function 1), which regulates transcription specifically in each gene and each cell. This regulation domain is considerably different in both ERs (only 16%–20% homology) and, in some cases, in ERβ, the AF1 may be significantly modified or absent [54]. The DBD is next to the NTD, highly conserved between ERα and ERβ (95% homology). The DBD is a domain structured as two zinc fingers able to recognize specific deoxyribonucleic acid (DNA) sequences in target genes (called estrogen responsive element, or ERE). The Hinge domain is essential for receptor dimerization, and is also a rotation point (therefore a "hinge"), which is fundamental for the receptor to acquire several conformations required to link to DNA. Regardless of the high-DBD homology in both receptor subtypes, suggesting that they bind to the DNA in similar fashion, the low homology between the NTD domains and Hinge indicates that both receptors may move differently and modulate different forms of gene transcription.

The LDB domain is in the carboxy terminal portion of the receptor, which contains the hormone biding site responsible for most of the functions activated by the agonist, such as dimerization of the receptor and translocation to the nucleus, besides attracting gene transcription co-regulator molecules, by means of autonomous activation region, called AF-2 (activation function 2) [55,56]. Differently from the great homology between the two DBD receptors, the LBD ERα and ERβ domains show lower homology (~59%), suggesting that the affinity between the agonists/antagonists for the receptor subtypes may differ significantly.

Estrogens influence vascular reactivity by direct effects on endothelial cells [57]. Vascular protection of estrogens in females is mediated partially via ERα [58,59], while in males this effect has more participation of ERβ [60,61]. In fact, mesenteric arterioles of ERβ knockout males have increased response to phenylephrine [62] and higher pressure levels than ERβ knockout females [62,63].

Due to the alternative splicing process, several forms of ERα have been described, although only a few has shown physiopathological relevance: ERαΔ3 which loses exon 3, which codifies part of the DBD and consequently its transcriptional capacity; and ERα36 and ERα46, variants with lower molecular weight (36 and 46 kDa, respectively) than the original ERα (with 66 kDa) [64]. For the ERβ, until the moment, at least four isoforms have been described: ERβ2, ERβ3, ERβ4, and ERβ5 [55]. In most cases, alternative splicing receptors lose their function or part of it and may act as negative domains that is able to inhibit the action of native ERs by forming a dimer with these receptors and inhibiting their action [64]. It is believed that the ratio between native receptors and their alternative splicing may change the response to estrogen and lead to tissue dysfunction.

Although this theory has been confirmed in several types of gynecological cancer, participation of the ER variants in the vascular endothelium still needs to be clarified. On the other hand, studies have described that isoform ERα46 may facilitate its connection to the plasmatic membrane, and therefore improve endothelial function by quick activation of eNOS [65]. However, increased ERα46 expression over ERα66 (native) in cytosol significantly changes the genomic effects induced by estrogen on the vascular wall [66–68]. Up to now, ERs have been the sex hormone receptors with the best structural and functional characterization, although other nuclear receptors are also expressed in the vascular wall and have endothelial function modulation effects.

Progesterone is a natural steroid hormone, produced by gonads, by adrenal cortex, and by placenta. There are several progesterone derivatives, such as: medroxyprogesterone, norgesterone, and acetate, which have similar activities. The PR are called A (PR-A) and B (PR-B). Although they are codified by a single gene, the gene that codifies the PRs uses separate promoters and different translation starting points to produce two isoforms, which are practically identical, except for an additional group of amino acids in the N-terminal portion of PR-B [69]. Although PR-A and PR-B share several structural domains, the transcriptional activity is distinct and measures their own genes with physiological response and effect with little overlap [70]. Both PRs were identified in smooth muscle and vascular endothelium of humans, mice, rats, rabbits, and primates [71]. PR-B is expressed equally in the aorta of men and women, while PR-A has higher expression in females [72]. The role of progesterone the endothelium is relevant, but is not as well characterized as the effects of estrogens. These actions have been generally associated with regulation of angiogenesis process in tumors [73], although the isolated effects of progesterone also have been associated with decreased arterial pressure and antiinflammatory potential [40,74–76]. It was demonstrated that acute administration of progesterone induced quick vasodilation (nongenomic pathway) in the coronary artery of ovariectomized female rhesus macaque (*Macaca Mulatta*) [77]. In endothelial cell cultures, the administration of progesterone increases, by means of genomic pathway, eNOS activity, and production of NO [78]. For nongenomic signaling, the membrane progesterone receptors (mPR) are strong candidates. The mPRs are receptors with seven transmembrane domains coupled to protein G, and have five subtypes: mPRα, mPRβ, mPRγ, mPRδ and mPRε. The mPRs are expressed both in human umbilical vein endothelial cells (HUVECs) and vascular smooth muscle cells.

Treatment with both progesterone and the specific mPRα agonist increases NO production quickly and reduces the concentration of cAMP, suggesting that the receptor is coupled to an inhibitory protein G. On the other hand, treatment with specific PR agonist does not cause the same effect on NO production [30]. The influence of these mPRs on the vascular wall of both sexes still requires clarification.

Despite direct vascular effects, it is believed that progesterone may antagonize estrogen's effects. Administration of progesterone in an ovariectomized mouse treated with 17β-estradiol reduced the antioxidant effects of estrogen, leading to increased nicotinamide adenine dinucleotide phosphate (NADPH) oxidase activity and reduced level of mRNA antioxidant enzymes as manganese-dependent superoxide dismutase (MnSOD) and extracellular superoxide dismutase (SOD) [79]. In vascular endothelium of females, progesterone inhibits arginine transportation through cationic amino acid transporter 1, impairing eNOS activity [80].

Testosterone is the main natural androgen produced in men and women, being responsible for male sexual characteristics, libido, and increase of bone and muscle mass in both sexes [81]. Testosterone exercises its actions by interacting with the target receptor, one of them being the cytosolic receptor belonging to the family of steroid hormone nuclear receptors, and the other located in the plasmatic membrane [82]. In the cardiovascular system, these receptors are expressed in smooth muscle cells and in vascular endothelium [41,82,83].

Although they are associated with male characteristics, the gene that codifies the testosterone receptor is located in the X chromosome, is codified by eight exons, and one of its products is a protein with molecular weight of ~110 kDa [84]. Similar to the ERs, the AR have a LBD and the AF-1 and AF-2 domains that recognize androgen response elements (AREs) in DNA [84–86]. When activated by its agonist, the AR is translocated to the nucleus and binds, in its dimerized form, to the target gene AREs, activating or repressing the expression of such genes. Changes in AR sequence consist mainly of highly polymorphic trinucleotidic repeats (CAG) in exon 1 and the number of repeats is inversely correlated to the transcriptional activity of the androgen target genes [87]. In men, the number of CAG repeats is not correlated to total or free testosterone serum concentration, but few CAG repeats entail low levels of high-density lipoproteins (HDL) and reduced vasodilation mediated by brachial arterial flows, therefore increasing the risk of developing cardiovascular diseases [88]. Also similarly to the ERs, testosterone may exercise nongenomic actions by means of activation of an AR located in the plasmatic membrane [82].

The role of androgens on the cardiovascular system is still controversial. Studies have shown both beneficial and harmful effects of these hormones [89–91]. For example, in men, low concentration of testosterone is associated with higher body mass index, higher waist circumference, diabetes, hypertension, low HDL, and risk of developing coronary arterial diseases [90–92], while in women in postmenopause, high concentration of testosterone is associated within sulin resistance, metabolic syndrome, and cardiovascular diseases [93].

Androgens exercise specific effects on each sex regarding functions regulated by endothelial cells, including angiogenesis and interaction between monocytes and the endothelium via AR. MHUVECs exposed to dihydrotestosterone (DHT) increased the gene expression of vascular cell adhesion molecule-1 (VCAM-1), effect abolished when the AR receptor antagonist, hidroxyflutamide, is used. However, when the HUVECs are from female donors, the phenomenon was not observed [94]. It has been shown in studies developed in vivo and in vitro that endogenous androgens are required for angiogenesis in males, but not in females [95].

Testosterone may exercise part of the effects on the vascular endothelium by means of metabolization of estrogen by the aromatase enzyme present in the endothelial cell. In fact, administration of testosterone in HUVECs decreased VCAM-1 gene and protein expression due to the conversion in estrogen [96]. However, testosterone may have a direct effect on the vascular endothelium, since in rat aorta endothelial cells, the hormone increases the production of NO, which was abolished in the presence of the androgen receptor antagonist (flutamine), but not with the aromatase inhibitor (anastrol) [97].

ACTION OF SEX HORMONES ON ENDOTHELIUM-DERIVED RELAXATION FACTORS AND ENDOTHELIUM-DERIVED CONTRACTING FACTORS

Nitric Oxide (NO)

The NO molecule is able to promote vascular relaxation, induce angiogenesis, and inhibit vascular smooth muscle cell proliferation, leukocyte adhesion, platelet aggregation, and thrombosis, among other functions. It is formed from the transformation of L-arginine into L-citrulline by a family of enzymes called nitric oxide synthases (NOS), present in several tissues. Mammals have three NOS isoforms, of which two are constitutive isoforms, the endothelial NOS (eNOS/NOS3) and the neuronal NOS (nNOS/NOS1), and one inducible isoform, inducible NOS (iNOS/NOS2), produced in response to inflammatory stimuli [98]. The increase in NO production or bioavailability via gonadal sex hormones, mainly estrogen, may involve several mechanisms, such as increased protein expression of eNOS [99], reduced generation of reactive oxygen species (ROS), such as superoxide anion [100,101], increased intracellular calcium ($[Ca^{2+}]$) in endothelial cells [102], activation of the PI3K pathway [103], decreased asymmetric dimethylarginine (ADMA) an eNOS endogenous inhibitor, and increased concentration of L-arginine (Fig. 6.5) [104].

In several studies, the NO released by the vascular endothelium of females was higher than in males, probably due to the higher expression/activity of eNOS observed in females [105,106]. In fact, aortic rings isolated from SHR females showed higher ACh vasodilation and higher phenylephrine vasoconstriction after incubation with NOS inhibitor, L-NAME (N^G-nitro-L-arginine methyl ester) when compared to SHR males [107].

Progesterone and testosterone also may increase NO production, positively modulating expression and eNOS activity in Wistar female aortas [78,108]. In Wistar female aortas, progesterone increases NO production, positively regulating eNOS activity [97,98]. In Wistar normotensive female endothelial cell culture, acute treatment with testosterone increases NO production via RA activation [97].

The role of estrogen as potent NO stimulator becomes evident when reduced endogenous levels of estrogen, due to ovariectomy, decreased the expression of eNOS in female aortas of normotensive Sprague-Dawley [109] and hypertensive SHR [25]. Hormonal treatment with conjugated equine estrogens in ovariectomized SHR females restored the mRNA expression of eNOS, and consequently improved endothelial function [25].

Although the effects of estrogen on NO production are well characterized in females, biological effects are less known and more controversial in males. In males, it has already been shown that acute and chronic administration of estrogen improves

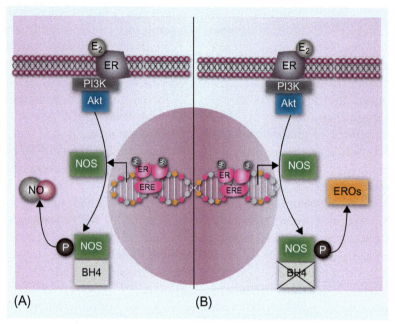

FIG. 6.5   Representation of estrogen genomic and nongenomic signaling pathway in NO production regulation in young (A) and old (B) females. *ER*, estrogen receptor; E_2, estrogen; *NOS*, nitric oxide synthase; *BH4*, tetrahydrobiopterin; *ROS*, reactive oxygen species; *PI3K*, phosphatidylinositol-3-Kinase; *ERE*, estrogen responsive element. Adapted from Murphy E. Estrogen signaling and cardiovascular disease. Circ Res 2011;109:687–96.

endothelial function and increase release of NO in carotids and aorta of Sprague-Dawley and SHR, respectively [110,111]. On the other hand, the estrogen may increase vascular damage in males by activating iNOS [112]. The opposite and controversial effects of estrogen in males and females may result from the differential expression of ER subtypes in both sexes [52, 113].

Prostaglandin (PG)

In addition to NO, endothelial cells produce and release prostacyclin (PGI_2), a vasodilator prostaglandin produced by conversion of arachidonic acid by cyclooxygenase (COX). Although PGI_2 is the main prostanoid produced in endothelial cells, the balance between production of vasodilator prostanoid and vasoconstrictor prostanoid, such as thromboxane A_2 (TXA_2), is extremely important to regulate vascular tonus [114]. Therefore, sexual differences in endothelium-dependent relaxation of macro and microvessels of rats and mouse may be partially explained by the imbalance in the production of prostanoids derived from COX-1 or COX-2 [115–117].

Deletion of the PGI2 receptor in C57B16 females reduced cardiovascular protection of ovariectomized females treated with estrogen [46], suggesting

that PGI2 is one of the important mediators in vascular protection of females. Other studies have demonstrated that estrogen was able to modulate vascular function by decreasing the production of vasoconstricting prostanoids [26].

Effects of estrogen on gene and protein expression of COX-1 and COX-2 on the endothelium are still controversial. On the one hand, estrogen increases mRNA expression and COX-2 protein expression on women uterine circulation endothelium; [118] on the other hand, it reduced its expression on female mice dermis microcirculation and vena cava [119], showing that regulation of COX expression by estrogen may be specific for each vascular bed.

In Sprague-Dawley males, surgical removal of testicles (orchiectomy) increased COX-2 protein expression in the aorta, and induced unbalanced production of vasodilator and vasoconstrictor prostanoids, with predominant production of vasoconstrictor prostanoids [120]. Despite that fact, chronic administration of testosterone in males orchiectomized Fisher-344 strain increased vascular tonus of cerebral arteries, by increasing TXA_2 [121], showing that the physiological levels of sex gonadal hormones are important for the control of the vascular tone, since they may influence the production of various vasoactive agents.

ENDOTHELIUM-DERIVED HYPERPOLARIZING FACTORS (EDHFS)

The EDHFs are vascular relaxation mediators with an important role in tonus control. Studies have proposed that the contribution of EDHFs for vascular relaxation increases in vascular beds and physiopathological conditions in which there is reduced participation of NO [122]. The release of EDHFs may be modulated by binding the agonist to specific receptors and by the shear stress on the blood vessel walls. The EDHFs may act on resistance or conductance arteries, but the vasodilator effect is more pronounced in resistance arteries of humans and experimental models [123,124].

Estrogen can also regulate vascular relaxation mediated by EDHFs, being one of the possible mechanisms by means of which hormonal treatment exercises a protective effect on the cardiovascular system [125]. Studies have shown reduced hyperpolarization mediated by EDHF in mesenteric bed arterioles of Wistar females submitted to ovariectomy, which was reversed after treatment with 17β-estradiol [125].

The differences associated with sex in the contribution of EDHFs in resistance artery relaxation have been described in several vascular beds, although opposite effects have been observed [126,127]. In resistance arteries of the mesenteric bed and other peripheral beds, the release of EDHFs is higher in females than in males [126]. On the other hand, in cerebral circulation, the contribution of EDHFs for vascular relaxation is smaller in females than in males [126]. Despite the lower contribution of EDHF in females, the infarcted area after induced ischemia in Wistar and stroke-prone SHR (animals prone to cerebrovascular accident) is greater in males than in females. These results suggest greater postischemia cerebral protection in females, as it is independent from EDHF release [128–130].

Although the nature of EDHFs is unknown, the sexual difference in vasodilation mediated by hyperpolarization may be correlated to the soluble epoxide hydrolase (EHs) enzyme, responsible for metabolizing epoxyeicosatrienoic acids, important vasodilator and candidate to one of the EDHFs [126].

Female C57Bl6 mice, subjected to ovariectomies, showed increased protein expression of EHs, which was reverted with estrogen treatment. Studies in vivo have shown that female mice knock-out to EHs showed infarcted areas similar to males of the same species [131]. It is known that even a temporary reduction of estrogen can affect the vascular reactivity of mesenteric circulation microvessels, mediated by EDHF, because female C57B16 mice in the diestro phase in which the plasmatic concentration of estrogen is reduced, showed diminished response mediated by EDHF [125]. In Sprague-Dawley males, treatment with 17β-estradiol increased endothelium-dependent relaxation, via EDHFs, in isolated aorta rings [132]. Therefore, estrogen may promote vascular relaxation in different vascular beds of males and females via nonidentified EDHFs.

REACTIVE OXYGEN SPECIES (ROS)

ROS play an important role in endothelial function, either directly as vasodilator agents (H_2O_2) and vasoconstrictor agents ($ONOO^-$), or indirectly by reducing NO bioavailability. In the latter case, superoxide anion (O_2^-) quickly reacts with the NO, promoting its inactivation and decreasing its beneficial effects on the vascular wall [133–135]. ROS production and the bioavailability of NO are important factors to determine endothelial dysfunction [133–135].

It has been demonstrated in experiments developed in SHR mesenteric arterioles studied by intravital microscopy [136] and aortic rings isolates from SHR [137] that the ROS production was lower in females than that of males and, therefore, blood vessels of females tend to respond less strongly to vasoconstriction agonists. The prooxidant environment also is less accentuated in women, as shown by the lower plasmatic concentration of malondialdehyde and thiobarbituric acid reactive substances (TBARS), human oxidative stress markers [138,139]. Also, reduced plasmatic concentration of estrogen, induced by ovariectomy, in SHR [136,137] and Sprague-Dawley [140] females increased the concentration of superoxide anion in mesenteric arterioles [136] and aortas [137,140]. Hormonal treatment with 17β-estradiol [136,140] or conjugated equine estrogens [137] reduced the concentration of ROS in the aorta and mesenteric arterioles in the female experimental models described above [136].

The effect of the estrogens (estradiol, estrone, and estriol) in reducing ROS may be related to the phenolic structure of these hormones, since, regardless of the interaction with its receptors, they may remove

ROS. However, this effect was only observed in concentrations 1000 times higher than the physiological concentrations [141,142], while the estrogen antioxidant effects are observed in females that showed physiological estrogen levels [101].

Actions of sex hormones regarding ROS regulation have been associated with modulation of NADPH oxidase expression/activity, which requires recruitment of cytosolic subunits (p40^{phox}, p47^{phox}, and p67^{phox}), and association with membrane subunits (gp91^{phox} and p22^{phox}). Changes in expression or phosphorylation of these subunits induce higher or lower enzyme activity, and consequently ROS production. Seven subunits equivalent to gp91^{phox} have been described, called Nox, and four of them were described in the vasculature (Nox1, Nox2, Nox4, and Nox5) [143].

In 2004, Dantas et al. [136] showed within in vivo studies on SHR animals that there was lower ROS production in mesenteric arterioles of hypertensive females than males. The sexual difference in ROS generation in SHR vasculature was followed by higher protein expression in subnuits gp91^{phox}, p22^{phox}, p47^{phox}, and p67^{phox} of NADPH oxidase in males. In addition, the aging process, a physiological condition associated with ROS increase, showed higher increase and more anticipated generation of ROS in the male mouse aorta when compared to the female mouse. At 7 months of age (middle age in mice), males presented higher ROS production than young males (3 months). In females, this difference becomes evident only at 12 months of age (aging in mice). In these females, treatment with apocynin, NADPH oxidase inhibitor, reduced the ROS generation [144].

Male mouse knockout for Nox2 presented reduced angiostesin-II vasoconstriction in the middle cerebral artery in comparison to native mouse. Reduced vasoconstriction described above was higher in knockout males than in females [145]. In addition, ovariectomized SHR females showed increased production of ROS, associated with positive regulation of mRNA of gp91^{phox}, p22^{phox} and protein of p47^{phox}, subunits of NADPH oxidase. Treatment with conjugated equine estrogens [25,137] or 17β-estradiol reduced ROS generation and mRNA expression of subunits gp91^{phox}, p22^{phox} of NADPH oxidase [25], and protein of p47^{phox} [137].

The effects of testosterone on ROS generation seem oppose the effects of estrogen. Ovariectomized SHR females treated with testosterone alone [19] or with testosterone associates with equine conjugated estreogens [25] presented an increase in the generation of vascular ROS, mainly due to the increase in the active (phosphorylation) of the p47^{phox} subunit of NADPH oxidase [25].

ROS generation may also be related to mitochondria and its modulation by estrogens [146]. Reduction in ROS generation in situation such as reoxygenation, followed by anoxia, occurred in a more pronounced way in cardiomyocyte mitochondria in isolated culture from females than males [146]. Generation of mitochondrial ROS can be modulated by estrogens, since the ERα and ERβ receptors are present in the mitochondrial membrane [147,148]. The cells of the aerobic organisms have developed a complex system of antioxidant enzymes that maintain the control of the production of ROS, avoiding the cellular damage. By the action of SOD, O_2^- is transformed into hydrogen peroxide (H_2O_2). The H_2O_2 by catalase is convert to water (H_2O) and oxygen (O_2). Thus, the regulation of the expression or activity of antioxidant enzymes, influenced directly by gonadal sex hormones, may contribute to the circulating redox state [133–135].

It has been demonstrated that the aorta of ovariectomized SHR females showed reduced protein expression of SOD and catalase, which was corrected by treatment with conjugated estrogens [137]. Bellanti et al. [149] analyzed redox balance in mononuclear cells of the peripheral blood of premenopause women with bilateral ovariectomy (surgical menopause). This was evaluated 30 days after the surgery, without treatment with estrogen, and 30 days after treatment with estrogen. After surgery, increased oxidative stress was observed due to reduced expression of mRNA for SOD and glutathione peroxidase, and recovered after estrogen treatment. The expression of mRNA of catalase and glutathione transferase was not modified in any of these conditions. The authors concluded that menopause is associated with significant changes in antioxidant enzyme gene expression that, in turn, change the circulating redox state.

Females and males knockout for MnSOD were infused with nonhypertensive doses of Ang II, and the authors observed endothelial dysfunction associated to increased ROS in basilar artery rings more pronounced in males than in females [46]. In fact, they observed more SOD activity in vascular smooth muscle cell culture of females than in males [150].

It is important to emphasize that most of the works described above clearly showed the participation of antioxidant enzymes in the endothelial function of experimental animals and women in postmenopause; however, few results compare sexual differences. The sexual difference was found in vascular smooth muscle cells, in the brain, and in the liver, since in the latter tissues catalase and glutathione peroxidase are more expressed in females than in male rats [151,152].

RENIN-ANGIOTENSIN SYSTEM (RAS)

The RAS is an important hormone complex that regulates arterial pressure, salts, and bodily fluids. Angiotensin II (Ang II), main RAS vasoconstrictor peptide, acting on AT1 (AT1R) receptors, contributes to increase vasoconstriction and ROS generation [153].

Aortic rings isolated from SHR males were more responsive to Ang II when compared to SHR females in physiological estrus, although AT1R antagonism reduces vasoconstriction in both sexes [23]. Similar results were demonstrated in animals in aging process. Aorta rings of male CD-1 mice with 12 months of age were more responsive to Ang II than females of the same age [154].

The ovariectomy procedure in SHR females increased vasoconstriction to Ang II, equating with SHR males. This response was corrected after treatment of the females with 17β-estradiol [136], suggesting that estrogen may reduce the response to Ang II. The sexual difference regarding response to Ang II is partially related to the capacity of estrogen in reducing AT1R mRNA expression [155] and increase AT2R after vascular injury [156]. In fact, it was demonstrated that AT1R expression in aortas was higher in males that in SHR females. However, AT2R expression was higher in females than in SHR males [23]. Activation of AT2R in blood vessels is associated with vasodilation by increased production of NO, by means of eNOS, FHDE, and the B2 receptor of bradykinin [157].

Okomura et al. [156] demonstrated that, after vascular occlusion and induction of inflammatory process, young female mouse presented higher AT2R expression in femoral artery than males, but this difference was smaller in old females with reduced estrogen levels. The sexual difference in AT2R expression in the cardiovascular system is due to

the X chromosome, as in the iliac arter of male knockout mouse for the Y chromosome, the vasoconstricting response to Ang II became the same as that of females [157].

Estrogen also may act on the angiotensin converting enzyme (ACE), responsible for converting angiotensin I into Ang II. In large vessels, such as the aorta, estrogen reduced the expression of mRNA of ACE and the plasmatic content of this enzyme, reducing local and systemic production of Ang II and the deleterious effects of this peptide on the endothelium [158]. The lower plasmatic content of ACE contributes to increasedbioavailability of Ang 1-7 [1,8,32,60,74,128,159] and bradykinin, which has vascular actions that are opposite from Ang II, therefore maximizing the beneficial effects of estrogen [160]. In fact, Sullivan et al. [161] demonstrated that the production of renal Ang 1-7 [1,8,32,60,74,128,159] was higher in females that in SHR males.

ENDOTHELINS (ETS)

In 1985, Hickey et al. [162] described a vasoconstrictor polypeptide derived from the endothelium that regulates vascular muscle contractility. Later, this potent vasoconstrictor peptide of 21 amino acids was isolated and called endothelin-1 (ET-1) [163]. Currently, three different endogenous isoforms of the 21 amino acids peptide (ET-1, ET-2, and ET-3) and three of 31 amino acids (ET-1^{1-31}, ET-2^{1-31} and ET-3^{1-31}) were identified [164–166]. There are two main subtypes of endothelin receptors: ETA and ETB, that belongs to the super family of receptors coupled to protein G and expressed in the endothelium, smooth vascular muscle, and mesangial cells [166].

ET-1 is responsible for promoting potent vasoconstriction, cellular growth, and inflammation, other than stimulating ROS generation, deposition of collagen in tissues, and expression of adhesion molecules in endothelial cells [165]. ET-1 binds mainly with the ETA receptor, but the existence of this subtype of endothelin receptor in the endothelium is still controversial. However, in human aortic endothelial cells, it was demonstrated that the presence of ETA e ETB receptors and peptide ET-1 is predominantly nuclear (including its envelope) [167]. ETB is expressed in vascular endothelial cells and its activation by ET-1 releases NO and prostaglandin [168].

The endothelin system, i.e., ET-1 and endothelin receptors, contribute to sexual differences present in cardiovascular diseases and arterial hypertension [169]. Women have a lower quantity of endothelin receptors in the saphenous vein, in the ration ET_A to ET_B when compared to men [170]. The expression of vascular mRNA of ET_B is increased in rats with DOCA-salt hypertension (uninefrectomized rats treated chronically with deoxycorticosterone acetate and sodium chloride) when compared to females [171]. The vasoconstriction induced by ET-1 is two times higher in samples of saphenous vein of men submitted to bypass surgery then in women submitted to the same procedure [170].

The sex hormones have a modulatory action on endothelin plasmatic concentration. In both hypertensive and nonhypertensive patients, the plasmatic concentration of ET-1 is higher in men than in women [165,169,172]. The concentration of ET-1 in plasma changes according to the stage of the menstrual cycle, suggesting a modulatory action of estrogens on endothelins. During the menstrual period, in which the concentration of circulating estrogen is lower, the concentration of plasmatic ET-1 is higher than in the follicular and luteal phases [108]. During the gestational period, concentration of ET-1 decreases. In female transgenders, treatment with estradiol and progestational substance cyproterone acetate decreased the plasmatic concentration of ET-1, while in male transgenders treated with testosterone the concentration of ET-1 increased [108].

ESTROGEN HORMONE TREATMENT— CLINICAL STUDIES

Observational studies suggested that estrogen treatment reduced the risk of mortality due to cardiovascular complications in postmenopause women in 30%–50% [173–175]. The study conducted with the nurses of Framingham Hospital (Nurse Health Study) was considered one of the largest observational studies regarding cardiovascular system of hormonal therapy. In this study, more than 48,000 women were followed for a period of 10 years. After adjustments by age and cardiovascular risk factors, the authors concluded that women who used estrogen therapy presented lower risk of developing acute coronary diseases or death due to cardiovascular diseases [173,176].

These results stimulated smaller clinical studies—randomized and double-blind—to evaluate the effects of treatment with estradiol in endothelium-dependent vasodilation mediated by brachial artery flow in postmenopause women. One of these studies showed that the hormone has endothelium-dependent vasodilator effects [176]. Not only with chronic treatment with 17β-estradiol, but acute treatment as well, have an effect on vascular tissue of women. Besides that Gilligan et al. demonstrated that intra-arterial infusion of physiological doses of estradiol in postmenopause women, with or without diagnosed coronary atherosclerosis, promoted endothelium-dependent vasodilation. The same was observed in normotensive women that were submitted to ovariectomy procedures (surgical menopause) and treated with transdermal estradiol [177,178]. However, large randomized, double-blind, and placebo-controlled clinical tests raised doubts regarding the beneficial effects of hormone therapy with conjugated equine estrogens on the cardiovascular system.

The purpose of the HERS (Heart and Estrogen/Progestin Replacement Study) was to evaluate the role of estrogen in secondary prevention of cardiovascular diseases. Started in the 1990s, the study followed a group of 2763 women with average age of 66.7 years, who received conjugated equine estrogen (0.625 mg) associated with medroxyprogesterone acetate (2.5 mg) for a period of 4.1 years. In this study it was observed that hormone therapy increased coronary and venous thromboembolism events in the first year of follow-up after acute myocardial infarction [179]. In the second segment of HERS, HERS II, which was expanded by 2.7 years and started to evaluate the role of estrogen in primary prevention of cardiovascular diseases, no difference was observed in respect to acute nonfatal myocardial infarction, death by coronary diseases, or other cardiovascular events, except for nonfatal ventricular arrhythmia, which was higher in the group treated with estrogen [180].

Afterwards, the Women's Health Initiative (WHI), the largest study regarding the effects of hormone therapy on women's health (more than 15,000 women in postmenopause), evaluated the effects of therapy with conjugated equine estrogen associated with medroxyprogesterone in primary prevention of cardiovascular diseases. The study demonstrated that hormone therapy may result in increased risk of cardiovascular events such as myocardial infarction and cerebrovascular accident [181,182].

Several questions emerged from these randomized studies (HERS and WHI). One of them is that administration of conjugated equine estrogens, concomitantly with progestogens, may influence the impact of the hormone on the vascular endothelium [182].

Progesterone is commonly administered with estrogen to reduce the risk of developing endometrial cancer; however, little is known regarding its effects on the cardiovascular system. Sorensen et al. [182a] demonstrated, in a randomized study, that administration of estrogen together with norethisterone—a progestogen, in women in postmenopause—did not improve dilation mediated by brachial artery flow, which is reduced in menopause. On the other hand, McCrohon et al. [183] demonstrated that medroxyprogesterone did not interfere in vasodilation promoted by estrogen in the brachial artery of women in postmenopause [183]. The therapeutic branch of the WHI study, which analyzed hormone therapy with estrogen alone, did not show differences regarding cardiovascular results; furthermore, it was interrupted in advance due to increased risk of breast cancer [184].

Another important point to be considered is the age in which estrogen therapy begins, since in the studies described the groups of women were in postmenopause, on average, for 10 years. This could lead to erroneous interpretations, because little is currently known about the relationship between the vascular effects of estrogens and the changes resulting from the vascular aging process [182]. In this regard, emerged the timing hypothesis or therapeutic opportunity window, a hypothesis created by WHI researchers proposing that the potential benefits mediated by estrogen to prevent cardiovascular diseases only appear when the hormone therapy is initiated before the deleterious effects of aging or before vascular dysfunction subclinical is present in the vascular wall [185,186].

In fact, the ELITE (Early versus Late Intervention Trial with Estradiol) clinical study, which evaluated 673 women with less than 6 and more than 10 years in postmenopause, observed that treatment with 17β-estradiol reduced the progression of atherosclerosis, measured in the carotid by ultrasound, in women who were in postmenopause for less than 6 years [187].

In men, plasmatic concentration of testosterone also has favorable direct and indirect effects on the cardiovascular system. The replacement of testosterone in elderly men with heart failure is associated with improved physical activity capacity, muscular strength, and glucose metabolism [188]. In men with angina pectoris, acute administration of testosterone promotes vasodilation in coronaries, by means of the potassium channel [189]. In contrast, transsexual women who received chronic administration of testosterone showed reduced vasodilation response induced by nitrates, increased concentration of triglicerides (TG), total cholesterol, low density lipoproteins (LDL), and apolipoprotein-B, as well as reduced levels of HDL [190].

Given these reports, the conclusion is that several aspects regarding the action of gonadal sex hormones on the cardiovascular system are controversial, and still are not completely clarified. It is important to emphasize that the endothelial effects induced by the action of gonadal sex hormones and sexual differences depend on the vascular bed, the animal model used, the plasmatic concentration of hormones, and the association between the different hormones.

NEW EXPERIMENTAL APPROACHES

Sex differences in several cardiovascular diseases is a well-established fact and regardless of the significant importance of gonadal sex hormones on regulating the mechanisms involved in these differences, it is still unknown if the expression of sexually dimorphous genes or if the sexual differences, that are intrinsic to the cells itself, may influence sexual dimorphism in cardiovascular physiopathology [191].

Currently, in this context, besides being widely used to study endothelial physiology and pathology regarding the cardiovascular system, HUVECs have also been used to study sexual differences present in cardiovascular diseases [32].

Curiously, when primary cultures of HUVECs obtained from male (MHUVECs) and female (FHUVECs) donors were studied independently, several sexual differences were noted; for example, MHUVECs synthesize higher concentrations of prostacyclins and prostaglandins E2 than FHUVECs when stimulated with thrombin, a molecule with important role in platelet aggregation [192]. In HUVECs stimulated with DHT, it was shown that in MHUVECs this androgen had proinflammatory and proatherogenic action, increased migration of endothelial cells, proliferation, tubulogenesis, and production of endothelial vascular growth factor, while there were no similar changes in FHUVECs, suggesting that androgens can regulate vascular changes differently in both sexes [94–96,193].

Other studies showed that FHUVECs presented higher cell migration and proliferation rates, higher expression of genes associated with metabolism, stress, and immune response, as well as higher expression of eNOS when compared to MHUVECs. In addition, after these cells were submitted to shear stress, the cells proceeding from FHUVECs showed higher expression of eNOS, SOD-1, and HO-1 (Heme Oxygenase 1) genes, as well as lower expression of the ET-1 gene when compared to MHUVECs. On the other hand, MHUVEC cells showed higher production of hydrogen peroxide, higher expression of beclin-1, and LC3-II/LC3-I ratio (microtubule-associated protein 1 light chain 3), indicating higher autophagic activity in these cells when compared to FHUVECs [32,191].

It is important to emphasize that no differences were found regarding the expression of estrogen and AR (ER-α, ERβ, GPER, and AR) and aromatase 5α-redutase 1 and 5α-redutase 2 enzymes (convert testosterone in DHT) in HUVECs proceeding from both sexes, as well as no differences in Akt (protein kinase B) and mTOR (mammalian target of rapamycin) expression, a pathway involved in several cellular processes, including proliferation, growth, and survival [32,159]. Furthermore, images obtained using an inverted microscope also did not show differences in size, shape, and morphology of cells proceeding from MHUVECs and FHUVECs. On the other hand, images obtained using a transmission electronic microscope revealed that MHUVECs and FHUVECs showed different ultrastructural patterns. In this context, cells proceeding from MHUVECs showed, for example, pinocytic vesicles distributed uniformly in the cell membrane, several autophagic vacuoles, and absence of lipid vacuoles, while cells proceeding from FHUVECs showed, for example, pinocytic vesicles distributed eccentrically in the cell membrane, several lysosomes, and lipid vacuoles, which could justify the sexual dimorphism found in several cell responses [32].

Sexual differences observed in endothelial cells can contribute to improve the comprehension of endothelial functions on cardiovascular diseases. Together, the findings described herein emphasize the importance of the use of male and female cells in cellular culture experiments, which should be applied not only to endothelial cells, but also to other types of cells, organs, and tissues, as well as using both sexes in studies to understand the mechanisms involved in the development and progression of cardiovascular diseases.

CLINICAL PERSPECTIVE

Regardless of evidence on sexual differences in cardiovascular system regulation, effective treatment of cardiovascular diseases in women in pre or postmenopause is a difficult issue in medicine, mainly due to the lack of understanding of the mechanisms involved in the initial stage of cardiovascular diseases, symptoms, and menopause process. Cardiovascular diseases are the main cause of death of women in postmenopause, and therefore should receive high priority among women health issues. However, there is still an alarming gap regarding the knowledge and understanding of the effects of estrogen on the cardiovascular system and general awareness of the medical and scientific societies regarding how to treat cardiovascular diseases in women. Medical training is still dominated by studies conducted in men, and therefore the clinical trends and instructions on drug and procedure recommendations are focused on this data. The lack of crucial information and the differences in the data available regarding women cardiovascular system regulation many times lead to inappropriate diagnosis and treatment. Therefore, women are still treated in the same way as men, regardless of the notable sexual differences in cardiovascular function. Increased awareness regarding risk factors and specific treatment for cardiovascular diseases in women is required to allow early diagnosis and more effective treatment.

CONCLUSIONS

Considering the importance of sex for the differences in cardiovascular morbimortality, it is important to have different diagnosis and therapeutic strategies for men and women. In basic research, the use of both sexes, male and female, in experimental and cellular models, is decisive to characterize the vascular effects of gonadal sex hormones.

References

[1] Wizemann TM, Pardue ML. Exploring the biological contributions to human health: does sex matter? J Women's Health Gend Based Med 2001;10:433–9.

[2] Arnold AP. Promoting the understanding of sex differences to enhance equity and excellence in biomedical science. Biol Sex Differ 2010;1:1.

[3] Deng X, Berletch JB, Nguyen DK, et al. X chromosome regulation: diverse patterns in development, tissues and disease. Nat Rev Genet 2014;15:367–78.

[4] Itoh Y, Arnold AP. Are females more variable than males in gene expression? Meta-analysis of microarray datasets. Biol Sex Differ 2015;6:18.

[5] Liu H, Lamm MS, Rutherford K, et al. Large-scale transcriptome sequencing reveals novel expression patterns for key sex-related genes in a sex-changing fish. Biol Sex Differ 2015;6:26.

[6] Chandra R, Federici S, Haskó G, et al. Female X-chromosome mosaicism for gp91phox expression diversifies leukocyte responses during endotoxemia. Crit Care Med 2010;38:2003–10.

[7] Shah K, McCormack CE, Bradbury NA. Do you know the sex of your cells? Am J Physiol Cell Physiol 2014;306:C3–C18.

[8] Alonso LC, Rosenfield RL. Oestrogens and puberty. Best Pract Res Clin Endocrinol Metab 2002;16:13–30.

[9] Korstanje R, Li R, Howard T, et al. Influence of sex and diet on quantitative trait loci for HDL cholesterol levels in an SM/J byNZB/BlNJ intercross population. J Lipid Res 2004;45:881–8.

[10] Ober C, Loisel DA, Gilad Y. Sex-specific genetic architecture of human disease. Nat Rev Genet 2008;9:911–22.

[11] Mozaffarian D, Benjamin EJ, Go AS, et al. Heart disease and stroke statistics—2015 update: a report from the American Heart Association. Circulation 2015;131:e29–e322.

[12] Bairey Merz CN, Shaw LJ, Reis SE, et al. Insights from the NHLBI-sponsored Women's Ischemia Syndrome Evaluation (WISE) Study: part II: gender differences in presentation, diagnosis, and outcome with regard to gender-based pathophysiology of atherosclerosis and macrovascular and microvascular coronary disease. J Am Coll Cardiol 2006;47:S21–9.

[13] Messerli FH, Garavaglia GE, Schmieder RE, et al. Disparate cardiovascular findings in men and women with essential hypertension. Ann Intern Med 1987;107:158–61.

[14] Shaw LJ, Bairey Merz CN, Pepine CJ, et al. Insights from the NHLBI-sponsored Women's Ischemia Syndrome Evaluation (WISE) Study: part I: gender differences in traditional and novel risk factors, symptom evaluation, and gender-optimized diagnostic strategies. J Am Coll Cardiol 2006;47:S4–S20.

[15] Mosca L, Barrett-Connor E, Wenger NK. Sex/gender differences in cardiovascular disease prevention: what a difference a decade makes. Circulation 2011;124:2145–54.

[16] Denton K, Baylis C. Physiological and molecular mechanisms governing sexual dimorphism of kidney, cardiac, and vascular function. Am J Physiol Regul Integr Comp Physiol 2007;292:R697–9.

[17] Orshal JM, Khalil RA. Gender, sex hormones, and vascular tone. Am J Physiol Regul Integr Comp Physiol 2004;286:R233–49.

[18] Go AS, Mozaffarian D, Roger VL, et al. Executive summary: heart disease and stroke statistics—2014 update: a report from the American Heart Association. Circulation 2014;129:399–410.

[19] Reckelhoff JF, Zhang H, Srivastava K. Gender differences in development of hypertension in spontaneously hypertensive rats: role of the renin-angiotensin system. Hypertension 2000;35:480–3.

[20] Nunes RA, Barroso LP, Pereira AC, et al. Gender-related associations of genetic polymorphisms of α-adrenergic receptors, endothelial nitric oxide synthase and bradykinin B2 receptor with treadmill exercise test responses. Open Heart 2014;1:e000132.

[21] Nigro D, Fortes ZB, Scivoletto R, et al. Simultaneous release of endothelium-derived relaxing and contracting factors induced by noradrenaline in normotensive rats. Gen Pharmacol 1990;21:443–6.

[22] Kauser K, Rubanyi GM. Gender difference in endothelial dysfunction in the aorta of spontaneously hypertensive rats. Hypertension 1995;25:517–23.

[23] Silva-Antonialli MM, Fortes ZB, Carvalho MH, et al. Sexual dimorphism in the response of thoracic aorta from SHRs to losartan. Gen Pharmacol 2000;34:329–35.

[24] Fortes ZB, Nigro D, Scivoletto R, et al. Influence of sex on the reactivity to endothelin-1 and noradrenaline in spontaneously hypertensive rats. Clin Exp Hypertens A 1991;13:807–16.

[25] Costa TJ, Ceravolo GS, dos Santos RA, et al. Association of testosterone with estrogen abolishes the beneficial effects of estrogen treatment by increasing ROS generation in aorta endothelial cells. Am J Physiol Heart Circ Physiol 2015;308:H723–32.

[26] Dantas AP, Scivoletto R, Fortes ZB, et al. Influence of female sex hormones on endothelium-derived vasoconstrictor prostanoid generation in microvessels of spontaneously hypertensive rats. Hypertension 1999;34:914–9.

[27] Taddei S, Virdis A, Ghiadoni L, et al. Menopause is associated with endothelial dysfunction in women. Hypertension 1996;28:576–82.

[28] Wilson M, Morganti AA, Zervoudakis I, et al. Blood pressure, the renin-aldosterone system and sex steroids throughout normal pregnancy. Am J Med 1980;68:97–104.

[29] Wiinberg N, Høegholm A, Christensen HR, et al. 24-h ambulatory blood pressure in 352 normal Danish subjects, related to age and gender. Am J Hypertens 1995;8:978–86.

[30] Pang Y, Dong J, Thomas P. Progesterone increases nitric oxide synthesis in human vascular endothelial cells through activation of membrane progesterone receptor-α. Am J Physiol Endocrinol Metab 2015;308:E899–911.

[31] Cannoletta M, Cagnacci A. Modification of blood pressure in postmenopausal women: role of hormone replacement therapy. Int J Womens Health 2014;6:745–57.

[32] Addis R, Campesi I, Fois M, et al. Human umbilical endothelial cells (HUVECs) have a sex: characterisation of the phenotype of male and female cells. Biol Sex Differ 2014;5:18.

[33] dos Santos RL, da Silva FB, Ribeiro RF, et al. Sex hormones in the cardiovascular system. Horm Mol Biol Clin Invest 2014;18:89–103.

[34] Franconi F, Campesi I, Occhioni S, et al. Sex and gender in adverse drug events, addiction, and placebo. Handb Exp Pharmacol 2012;214:107–26.

[35] Mendelsohn ME, Karas RH. Molecular and cellular basis of cardiovascular gender differences. Science 2005;308:1583–7.

[36] Mendelsohn ME, Karas RH. The protective effects of estrogen on the cardiovascular system. N Engl J Med 1999;340:1801–11.

[37] Paech K, Webb P, Kuiper GG, et al. Differential ligand activation of estrogen receptors ERalpha and ERbeta at AP1 sites. Science 1997;277:1508–10.

[38] Spoletini I, Vitale C, Malorni W, et al. Sex differences in drug effects: interaction with sex hormones in adult life. Handb Exp Pharmacol 2012;214:91–105.

[39] Morrill GA, Kostellow AB, Gupta RK. Transmembrane helices in "classical" nuclear reproductive steroid receptors: a perspective. Nucl Recept Signal 2015;13:e003.

[40] Goddard LM, Murphy TJ, Org T, et al. Progesterone receptor in the vascular endothelium triggers physiological uterine permeability preimplantation. Cell 2014;156:549–62.

[41] Torres-Estay V, Carreño DV, San Francisco IF, et al. Androgen receptor in human endothelial cells. J Endocrinol 2015;224:R131–7.

[42] Farhat MY, Abi-Younes S, Ramwell PW. Non-genomic effects of estrogen and the vessel wall. Biochem Pharmacol 1996;51:571–6.

[43] Hammes SR, Levin ER. Extranuclear steroid receptors: nature and actions. Endocr Rev 2007;28:726–41.

[44] Simoncini T, Mannella P, Fornari L, et al. Genomic and non-genomic effects of estrogens on endothelial cells. Steroids 2004;69:537–42.

[45] Kuiper GG, Enmark E, Pelto-Huikko M, et al. Cloning of a novel receptor expressed in rat prostate and ovary. Proc Natl Acad Sci U S A 1996;93:5925–30.

[46] Chrissobolis S, Zhang Z, Kinzenbaw DA, et al. Receptor activity-modifying protein-1 augments cerebrovascular responses to calcitonin gene-related peptide and inhibits angiotensin II-induced vascular dysfunction. Stroke 2010;41:2329–34.

[47] Revankar CM, Cimino DF, Sklar LA, et al. A transmembrane intracellular estrogen receptor mediates rapid cell signaling. Science 2005;307:1625–30.

[48] Colburn P, Buonassisi V. Estrogen-binding sites in endothelial cell cultures. Science 1978;201:817–9.

[49] Orimo A, Inoue S, Ikegami A, et al. Vascular smooth muscle cells as target for estrogen. Biochem Biophys Res Commun 1993;195:730–6.

[50] Takada Y, Kato C, Kondo S, et al. Cloning of cDNAs encoding G protein-coupled receptor expressed in human endothelial cells exposed to fluid shear stress. Biochem Biophys Res Commun 1997;240:737–41.

[51] Villablanca AC, Jayachandran M, Banka C. Atherosclerosis and sex hormones: current concepts. Clin Sci (Lond) 2010;119:493–513.

[52] Murphy E. Estrogen signaling and cardiovascular disease. Circ Res 2011;109:687–96.

[53] Levin ER. Plasma membrane estrogen receptors. Trends Endocrinol Metab 2009;20:477–82.

[54] Giguère V, Tremblay A, Tremblay GB. Estrogen receptor beta: re-evaluation of estrogen and antiestrogen signaling. Steroids 1998;63:335–9.

[55] Jia M, Dahlman-Wright K, Gustafsson J. Estrogen receptor alpha and beta in health and disease. Best Pract Res Clin Endocrinol Metab 2015;29:557–68.

[56] Kumar R, Zakharov MN, Khan SH, et al. The dynamic structure of the estrogen receptor. J Amino Acids 2011;2011:812540.

[57] Khalil RA. Estrogen, vascular estrogen receptor and hormone therapy in postmenopausal vascular disease. Biochem Pharmacol 2013;86:1627–42.

[58] Douglas G, Cruz MN, Poston L, et al. Functional characterization and sex differences in small mesenteric arteries of the estrogen receptor-beta knockout mouse. Am J Physiol Regul Integr Comp Physiol 2008;294:R112–20.

[59] Kublickiene K, Svedas E, Landgren BM, et al. Small artery endothelial dysfunction in postmenopausal women: in vitro function, morphology, and modification by estrogen and selective estrogen receptor modulators. J Clin Endocrinol Metab 2005;90:6113–22.

[60] Aavik E, du Toit D, Myburgh E, et al. Estrogen receptor beta dominates in baboon carotid after endothelial denudation injury. Mol Cell Endocrinol 2001;182:91–8.

[61] Lindner V, Kim SK, Karas RH, et al. Increased expression of estrogen receptor-beta mRNA in male blood vessels after vascular injury. Circ Res 1998;83:224–9.

[62] Luksha L, Poston L, Gustafsson JA, et al. Gender-specific alteration of adrenergic responses in small femoral arteries from estrogen receptor-beta knockout mice. Hypertension 2005;46:1163–8.

[63] Zhu Y, Bian Z, Lu P, et al. Abnormal vascular function and hypertension in mice deficient in estrogen receptor beta. Science 2002;295:505–8.

[64] Herynk MH, Fuqua SA. Estrogen receptor mutations in human disease. Endocr Rev 2004;25:869–98.

[65] Li L, Haynes MP, Bender JR. Plasma membrane localization and function of the estrogen receptor alpha variant (ER46) in human endothelial cells. Proc Natl Acad Sci U S A 2003;100:4807–12.

[66] Figtree GA, McDonald D, Watkins H, et al. Truncated estrogen receptor alpha 46-kDa isoform in human endothelial cells: relationship to acute activation of nitric oxide synthase. Circulation 2003;107:120–6.

[67] Novella S, Dantas AP, Segarra G, et al. Aging enhances contraction to thromboxane A2 in aorta from female senescence-accelerated mice. Age (Dordr) 2013;35:117–28.

[68] Novella S, Heras M, Hermenegildo C, et al. Effects of estrogen on vascular inflammation: a matter of timing. Arterioscler Thromb Vasc Biol 2012;32:2035–42.

[69] Hagan CR, Faivre EJ, Lange CA. Scaffolding actions of membrane-associated progesterone receptors. Steroids 2009;74:568–72.

[70] Scarpin KM, Graham JD, Mote PA, et al. Progesterone action in human tissues: regulation by progesterone receptor (PR) isoform expression, nuclear positioning and coregulator expression. Nucl Recept Signal 2009;7:e009.

[71] Goletiani NV, Keith DR, Gorsky SJ. Progesterone: review of safety for clinical studies. Exp Clin Psychopharmacol 2007;15:427–44.

[72] Nakamura Y, Suzuki T, Inoue T, et al. Progesterone receptor subtypes in vascular smooth muscle cells of human aorta. Endocr J 2005;52:245–52.

[73] Simoncini T, Mannella P, Fornari L, et al. In vitro effects of progesterone and progestins on vascular cells. Steroids 2003;68:831–6.

[74] Aksoy AN, Toker A, Celık M, et al. The effect of progesterone on systemic inflammation and oxidative stress in the rat model of sepsis. Indian J Pharm 2014;46:622–6.

[75] Goddard LM, Ton AN, Org T, et al. Selective suppression of endothelial cytokine production by progesterone receptor. Vasc Pharmacol 2013;59:36–43.

[76] Kristiansson P, Wang JX. Reproductive hormones and blood pressure during pregnancy. Hum Reprod 2001;4:13–7.

[77] Minshall RD, Pavcnik D, Browne DL, et al. Nongenomic vasodilator action of progesterone on primate coronary arteries. J Appl Physiol (1985) 2002;92:701–8.

[78] Selles J, Polini N, Alvarez C, et al. Progesterone and 17 beta-estradiol acutely stimulate nitric oxide synthase activity in rat aorta and inhibit platelet aggregation. Life Sci 2001;69:815–27.

[79] Wassmann S, Bäumer AT, Strehlow K, et al. Endothelial dysfunction and oxidative stress during estrogen deficiency in spontaneously hypertensive rats. Circulation 2001;103:435–41.

[80] Bentur OS, Schwartz D, Chernichovski T, et al. Estradiol augments while progesterone inhibits arginine transport in human endothelial cells through modulation of cationic amino acid transporter-1. Am J Physiol Regul Integr Comp Physiol 2015;309:R421–7.

[81] Matsumoto T, Sakari M, Okada M, et al. The androgen receptor in health and disease. Annu Rev Physiol 2013;75:201–24.

[82] Liu PY, Death AK, Handelsman DJ. Androgens and cardiovascular disease. Endocr Rev 2003;24:313–40.

[83] Tostes RC, Carneiro FS, Carvalho MH, et al. Reactive oxygen species: players in the cardiovascular effects of testosterone. Am J Physiol Regul Integr Comp Physiol 2016;310:R1–R14.

[84] Lubahn DB, Joseph DR, Sullivan PM, et al. Cloning of human androgen receptor complementary DNA and localization to the Xchromosome. Science 1988;240:327–30.

[85] Callewaert L, Christiaens V, Haelens A, et al. Implications of a polyglutamine tract in the function of the human androgen receptor. Biochem Biophys Res Commun 2003;306:46–52.

[86] Tan MH, Li J, Xu HE, et al. Androgen receptor: structure, role in prostate cancer and drug discovery. Acta Pharmacol Sin 2015;36:3–23.

[87] Zitzmann M, Nieschlag E. The CAG repeat polymorphism within the androgen receptor gene and maleness. Int J Androl 2003;26:76–83.

[88] Zitzmann M, Brune M, Kornmann B, et al. The CAG repeat polymorphism in the AR gene affects high density lipoprotein cholesterol and arterial vasoreactivity. J Clin Endocrinol Metab 2001;86:4867–73.

[89] Herring MJ, Oskui PM, Hale SL, et al. Testosterone and the cardiovascular system: a comprehensive review of the basic science literature. J Am Heart Assoc 2013;2:e000271.

[90] Oskui PM, French WJ, Herring MJ, et al. Testosterone and the cardiovascular system: a comprehensive review of the clinical literature. J Am Heart Assoc 2013;2:e000272.

[91] Ruige JB, Ouwens DM, Kaufman JM. Beneficial and adverse effects of testosterone on the cardiovascular system in men. J Clin Endocrinol Metab 2013;98:4300–10.

[92] Srinath R, Hill Golden S, Carson KA, et al. Endogenous testosterone and its relationship to preclinical and clinical measures of cardiovascular disease in the atherosclerosis risk in communities study. J Clin Endocrinol Metab 2015;100:1602–8.

[93] Patel SM, Ratcliffe SJ, Reilly MP, et al. Higher serum testosterone concentration in older women is associated with insulin resistance, metabolic syndrome, and cardiovascular disease. J Clin Endocrinol Metab 2009;94:4776–84.

[94] Death AK, McGrath KC, Sader MA, et al. Dihydrotestosterone promotes vascular cell adhesion molecule-1 expression in male human endothelial cells via a nuclear factor-kappaB-dependent pathway. Endocrinology 2004;145:1889–97.

[95] Sieveking DP, Lim P, Chow RW, et al. A sex-specific role for androgens in angiogenesis. J Exp Med 2010;207:345–52.

[96] Mukherjee TK, Dinh H, Chaudhuri G, et al. Testosterone attenuates expression of vascular cell adhesion molecule-1 by conversion to estradiol by aromatase in endothelial cells: implications in atherosclerosis. Proc Natl Acad Sci U S A 2002;99:4055–60.

[97] Campelo AE, Cutini PH, Massheimer VL. Cellular actions of testosterone in vascular cells: mechanism independent of aromatization to estradiol. Steroids 2012;77:1033–40.

[98] Bielli A, Scioli MG, Mazzaglia D, et al. Antioxidants and vascular health. Life Sci 2015;143:209–16.

[99] Lamas AZ, Caliman IF, Dalpiaz PL, et al. Comparative effects of estrogen, raloxifene and tamoxifen on endothelial dysfunction, inflammatory markers and oxidative stress in ovariectomized rats. Life Sci 2015;124:101–9.

[100] Borgo MV, Claudio ER, Silva FB, et al. Hormonal therapy with estradiol and drospirenone improves endothelium-dependent vasodilation in the coronary bed of ovariectomized spontaneously hypertensive rats. Braz J Med Biol Res 2016;49(1)e4655.

[101] Dantas AP, Tostes RC, Fortes ZB, et al. In vivo evidence for antioxidant potential of estrogen in microvessels of female spontaneously hypertensive rats. Hypertension 2002;39:405–11.

[102] Thor D, Uchizono JA, Lin-Cereghino GP, et al. The effect of 17 beta-estradiol on intracellular calcium homeostasis in human endothelial cells. Eur J Pharmacol 2010;630:92–9.

[103] Chen W, Cui Y, Zheng S, et al. 2-methoxyestradiol induces vasodilation by stimulating NO release via PPARg/PI3K/Akt pathway. PLoS One 2015;10.e0118902.

[104] Valtonen P, Punnonen K, Saarelainen H, et al. ADMA concentration changes across the menstrual cycle and during oral contraceptive use: the Cardiovascular Risk in Young Finns Study. Eur J Endocrinol 2010;162:259–65.

[105] Kleinert H, Wallerath T, Euchenhofer C, et al. Estrogens increase transcription of the human endothelial NO synthase gene: analysis of the transcription factors involved. Hypertension 1998;31:582–8.

[106] Knot HJ, Lounsbury KM, Brayden JE, et al. Gender differences in coronary artery diameter reflect changes in both endothelial Ca2+ and ecNOS activity. Am J Physiol 1999;276:H961–9.

[107] Loria AS, Brinson KN, Fox BM, et al. Sex-specific alterations in NOS regulation of vascular function in aorta and mesenteric arteries from spontaneously hypertensive rats compared to Wistar Kyoto rats. Physiol Rep 2014;2(8).

[108] Cutini PH, Campelo AE, Massheimer VL. Differential regulation of endothelium behavior by progesterone and medroxyprogesterone acetate. J Endocrinol 2014;220:179–93.

[109] Yung LM, Wong WT, Tian XY, et al. Inhibition of renin-angiotensin system reverses endothelial dysfunction and oxidative stress in estrogen deficient rats. PLoS One 2011;6:e17437.

[110] Sobey CG, Weiler JM, Boujaoude M, et al. Effect of short-term phytoestrogen treatment in male rats on nitric oxide-mediated responses of carotid and cerebral arteries: comparison with 17beta-estradiol. J Pharmacol Exp Ther 2004;310:135–40.

[111] Yen CH, Lau YT. 17beta-Oestradiol enhances aortic endothelium function and smooth muscle contraction in male spontaneously hypertensive rats. Clin Sci (Lond) 2004;106:541–6.

[112] Francisco YA, Dantas AP, Carvalho MH, et al. Estrogen enhances vasoconstrictive remodeling after injury in male rabbits. Braz J Med Biol Res 2005;38:1325–9.

[113] Tsutsumi S, Zhang X, Takata K, et al. Differential regulation of the inducible nitric oxide synthase gene by estrogen receptors 1 and 2. J Endocrinol 2008;199:267–73.

[114] Kang KT. Endothelium-derived relaxing factors of small resistance arteries in hypertension. Toxicol Res 2014;30:141–8.

[115] Duckles SP, Krause DN. Cerebrovascular effects of oestrogen: multiplicity of action. Clin Exp Pharmacol Physiol 2007;34:801–8.

[116] Geary GG, Krause DN, Duckles SP. Estrogen reduces mouse cerebral artery tone through endothelial NOS- and cyclooxygenase-dependent mechanisms. Am J Physiol Heart Circ Physiol 2000;279:H511–9.

[117] Graham DA, Rush JW. Cyclooxygenase and thromboxane/prostaglandin receptor contribute to aortic endothelium-dependent dysfunction in aging female spontaneously hypertensive rats. J Appl Physiol (1985) 2009;107:1059–67.

[118] Tamura M, Deb S, Sebastian S, et al. Estrogen up-regulates cyclooxygenase-2 via estrogen receptor in human uterine microvascular endothelial cells. Fertil Steril 2004;81:1351–6.

[119] Hertrampf T, Schmidt S, Laudenbach-Leschowsky U, et al. Tissue-specific modulation of cyclooxygenase-2 (Cox-2) expression in the uterus and the v. cava by estrogens and phytoestrogens. Mol Cell Endocrinol 2005;243:51–7.

[120] Martorell A, Blanco-Rivero J, Aras-López R, et al. Orchidectomy increases the formation of prostanoids and modulates their role in the acetylcholine-induced relaxation in the rat aorta. Cardiovasc Res 2008;77:590–9.

[121] Gonzales RJ, Ghaffari AA, Duckles SP, et al. Testosterone treatment increases thromboxane function in rat cerebral arteries. Am J Physiol Heart Circ Physiol 2005;289:H578–85.

[122] Edwards G, Félétou M, Weston AH. Endothelium-derived hyperpolarising factors and associated pathways: a synopsis. Pflugers Arch 2010;459:863–79.

[123] Félétou M, Vanhoutte PM. Endothelium-derived hyperpolarizing factor: where are we now? Arterioscler Thromb Vasc Biol 2006;26:1215–25.

[124] Urakami-Harasawa L, Shimokawa H, Nakashima M, et al. Importance of endothelium-derived hyperpolarizing factor in human arteries. J Clin Invest 1997;100:2793–9.

[125] Liu MY, Hattori Y, Fukao M, et al. Alterations in EDHF-mediated hyperpolarization and relaxation in mesenteric arteries of female rats in long-term deficiency of oestrogen and during oestrus cycle. Br J Pharmacol 2001;132:1035–46.

[126] Davis CM, Siler DA, Alkayed NJ. Endothelium-derived hyperpolarizing factor in the brain: influence of sex, vessel size and disease state. Women's Health (Lond Engl) 2011;7:293–303.

[127] Villar IC, Hobbs AJ, Ahluwalia A. Sex differences in vascular function: implication of endothelium-derived hyperpolarizing factor. J Endocrinol 2008;197:447–62.

[128] Alkayed NJ, Harukuni I, Kimes AS, et al. Gender-linked brain injury in experimental stroke. Stroke 1998;29:159–65 [discussion 166].

[129] Haast RA, Gustafson DR, Kiliaan AJ. Sex differences in stroke. J Cereb Blood Flow Metab 2012;32:2100–7.

[130] Liu M, Dziennis S, Hurn PD, et al. Mechanisms of gender-linked-ischemic brain injury. Restor Neurol Neurosci 2009;27(3):163–79.

[131] Zhang W, Iliff JJ, Campbell CJ, et al. Role of soluble epoxide hydrolase in the sex-specific vascular response to cerebral ischemia. J Cereb Blood Flow Metab 2009;29:1475–81.

[132] Woodman OL, Boujaoude M. Chronic treatment of male rats with daidzein and 17 beta-oestradiol induces the contribution of EDHF to endothelium-dependent relaxation. Br J Pharmacol 2004;141:322–8.

[133] Azevedo LC, Pedro MA, Souza LC, et al. Oxidative stress as a signaling mechanism of the vascular response to injury: the redox hypothesis of restenosis. Cardiovasc Res 2000;47:436–45.

[134] Li H, Horke S, Forstermann U. Oxidative stress in vascular disease and its pharmacological prevention. Trends Pharmacol Sci 2013;34:313–9.

[135] Li H, Horke S, Forstermann U. Vascular oxidative stress, nitric oxide and atherosclerosis. Atherosclerosis 2014;237:208–19.

[136] Dantas AP, Franco Mdo C, Silva-Antonialli MM, et al. Gender differences in superoxide generation in microvessels of hypertensive rats: role of NAD(P)H-oxidase. Cardiovasc Res 2004;61:22–9.

[137] Ceravolo GS, Filgueira FP, Costa TJ, et al. Conjugated equine estrogen treatment corrected the exacerbated aorta oxidative stress in ovariectomized spontaneously hypertensive rats. Steroids 2013;78:341–6.

[138] Ide T, Tsutsui H, Ohashi N, et al. Greater oxidative stress in healthy young men compared with premenopausal women. Arterioscler Thromb Vasc Biol 2002;22:438–42.

[139] Powers RW, Majors AK, Lykins DL, et al. Plasma homocysteine and malondialdehyde are correlated in an age- and gender-specific manner. Metabolism 2002;51:1433–8.

[140] Florian M, Freiman A, Magder S. Treatment with 17-beta-estradiol reduces superoxide production in aorta of ovariectomized rats. Steroids 2004;69:779–87.

[141] Ceravolo GS, Tostes R, Fortes Z, et al. Efeitos do estrógeno no sistema cardiovascular. Hipertensão 2007;10:124–30.

[142] Dubey RK, Gillespie DG, Imthurn B, et al. Phytoestrogens inhibit growth and MAP kinase activity in human aortic smooth muscle cells. Hypertension 1999;33:177–82.

[143] Lassègue B, Griendling KK. NADPH oxidases: functions and pathologies in the vasculature. Arterioscler Thromb Vasc Biol 2010;30:653–61.

[144] Miller AA, De Silva TM, Judkins CP, et al. Augmented superoxide production by Nox2-containing NADPH oxidase causes cerebral artery dysfunction during hypercholesterolemia. Stroke 2010;41:784–9.

[145] Doughan AK, Harrison DG, Dikalov SI. Molecular mechanisms of angiotensin II-mediated mitochondrial dysfunction: linking mitochondrial oxidative damage and vascular endothelial dysfunction. Circ Res 2008;102:488–96.

[146] Lagranha CJ, Deschamps A, Aponte A, et al. Sex differences in the phosphorylation of mitochondrial proteins result in reduced production of reactive oxygen species and cardioprotection in females. Circ Res 2010;106:1681–91.

[147] Razmara A, Sunday L, Stirone C, et al. Mitochondrial effects of estrogen are mediated by estrogen receptor alpha in brain endothelial cells. J Pharmacol Exp Ther 2008;325:782–90.

[148] Yager JD, Chen JQ. Mitochondrial estrogen receptors—new insights into specific functions. Trends Endocrinol Metab 2007;18:89–91.

[149] Bellanti F, Matteo M, Rollo T, et al. Sex hormones modulate circulating antioxidant enzymes: impact of estrogen therapy. Redox Biol 2013;1:340–6.

[150] Morales RC, Bahnson ES, Havelka GE, et al. Sex-based differential regulation of oxidative stress in the vasculature by nitric oxide. Redox Biol 2015;4:226–33.

[151] Capel ID, Smallwood AE. Sex differences in the glutathione peroxidase activity of various tissues of the rat. Res Commun Chem Pathol Pharmacol 1983;40:367–78.

[152] Pajović SB, Saicić ZS. Modulation of antioxidant enzyme activities by sexual steroid hormones. Physiol Res 2008;57:801–11.

[153] Touyz RM. Reactive oxygen species in vascular biology: role in arterial hypertension. Expert Rev Cardiovasc Ther 2003;1:91–106.

[154] Garabito M, Costa G, Jimenez-Altayo F, et al. Sex-associated differences in oxidative stress and renin-angiotensin system contribute to a differential regulation of vascular aging. Cardiovasc Res 2014;103(suppl. 1):S137–8.

[155] Nickenig G, Bäumer AT, Grohè C, et al. Estrogen modulates AT1 receptor gene expression in vitro and in vivo. Circulation 1998;97:2197–201.

[156] Okumura M, Iwai M, Nakaoka H, et al. Possible involvement of AT2 receptor dysfunction in age-related gender difference in vascular remodeling. J Am Soc Hypertens 2011;5:76–84.

[157] Pessôa BS, Slump DE, Ibrahimi K, et al. Angiotensin II type 2 receptor- and acetylcholine-mediated relaxation: essential contri- bution of female sex hormones and chromosomes. Hypertension 2015;66:396–402.

[158] Gallagher PE, Li P, Lenhart JR, et al. Estrogen regulation of angiotensin-converting enzyme mRNA. Hypertension 1999;33:323–8.

[159] Annibalini G, Agostini D, Calcabrini C, et al. Effects of sex hormones on inflammatory response in male and female vascular endothelial cells. J Endocrinol Investig 2014;37:861–9.

[160] Brosnihan KB, Senanayake PS, Li P, et al. Bi-directional actions of estrogen on the renin-angiotensin system. Braz J Med Biol Res 1999;32:373–81.

[161] Sullivan JC, Bhatia K, Yamamoto T, et al. Angiotensin (1–7) receptor antagonism equalizes angiotensin II-induced hypertension in male and female spontaneously hypertensive rats. Hypertension 2010;56:658–66.

[162] Hickey KA, Rubanyi G, Paul RJ, et al. Characterization of a coronary vasoconstrictor produced by cultured endothelial cells. Am J Physiol 1985;248:C550–6.

[163] Yanagisawa M, Kurihara H, Kimura S, et al. A novel potent vasoconstrictor peptide produced by vascular endothelial cells. Nature 1988;332:411–5.

[164] Kishi F, Minami K, Okishima N, et al. Novel 31-amino-acid-length endothelins cause constriction of vascular smooth muscle. Biochem Biophys Res Commun 1998;248:387–90.

[165] Tostes RC, Fortes ZB, Callera GE, et al. Endothelin, sex and hypertension. Clin Sci (Lond) 2008;114:85–97.

[166] Tostes RC, Muscará MN. Endothelin receptor antagonists: another potential alternative for cardiovascular diseases. Curr Drug Targets Cardiovasc Haematol Disord 2005;5:287–301.

[167] Avedanian L, Riopel J, Bkaily G, et al. ETA receptors are present in human aortic vascular endothelial cells and modulate intracellular calcium. Can J Physiol Pharmacol 2010;88:817–29.

[168] Schiffrin EL. Vascular endothelin in hypertension. Vasc Pharmacol 2005;43:19–29.

[169] Kitada K, Ohkita M, Matsumura Y. Pathological importance of the endothelin-1/ET(B) receptor system on vascular diseases. Cardiol Res Pract 2012;2012:731970.

[170] Ergul A, Shoemaker K, Puett D, et al. Gender differences in the expression of endothelin receptors in human saphenous veins in vitro. J Pharmacol Exp Ther 1998;285:511–7.

[171] David FL, Montezano AC, Rebouças NA, et al. Gender differences in vascular expression of endothelin and ET(A)/ET(B) receptors, but not in calcium handling mechanisms, in deoxycorticosterone acetate-salt hypertension. Braz J Med Biol Res 2002;35:1061–8.

[172] Miyauchi T, Yanagisawa M, Iida K, et al. Age- and sex-related variation of plasma endothelin-1 concentration in normal and hypertensive subjects. Am Heart J 1992;123:1092–3.

[173] Stampfer MJ, Colditz GA. Estrogen replacement therapy and coronary heart disease: a quantitative assessment of the epidemiologic evidence. Prev Med 1991;8:47–63.

[174] Bush TL. Evidence for primary and secondary prevention of coronary artery disease in women taking oestrogen replacement therapy. Eur Heart J 1996;17(Suppl. D):9–14.

[175] Limacher MC. Hormones and heart disease: what we thought, what we have learned, what we still need to know. Trans Am Clin Climatol Assoc 2002;113:31–40 [discussion 40-31].

[176] Lieberman EH, Gerhard MD, Uehata A, et al. Estrogen improves endothelium-dependent, flow-mediated vasodilation in postmenopausal women. Ann Intern Med 1994;121:936–41.

[177] Gilligan DM, Sack MN, Guetta V, et al. Effect of antioxidant vitamins on low density lipoprotein oxidation and impaired endothelium-dependent vasodilation in patients with hypercholesterolemia. J Am Coll Cardiol 1994;24:1611–7.

[178] Pinto S, Virdis A, Ghiadoni L, et al. Endogenous estrogen and acetylcholine-induced vasodilation in normotensive women. Hypertension 1997;29:268–73.

[179] Hulley S, Grady D, Bush T, et al. Randomized trial of estrogen plus progestin for secondary prevention of coronary heart disease in postmenopausal women. Heart and Estrogen/progestin Replacement Study (HERS) Research Group. JAMA 1998;280:605–13.

[180] Grady D, Herrington D, Bittner V, et al. Cardiovascular disease outcomes during 6.8 years of hormone therapy: Heart and Estrogen/progestin Replacement Study follow-up (HERS II). JAMA 2002;288:49–57.

[181] Howard BV, Rossouw JE. Estrogens and cardiovascular disease risk revisited: the Women's Health Initiative. Curr Opin Lipidol 2013;24:493–9.

[182] Virdis A, Taddei S. Endothelial aging and gender. Maturitas 2012;71:326–30.

[182a] Sorensen KE, Dorup I, Hermann AP, Mosekilde L. Combined hormone replacement therapy does not protect women against the age-related decline in endothelium-dependent vasomotor functio, Circulation 1998;97:1234–8. https://doi.org/10.1161/01.CIR.97.13.1234.

[183] McCrohon JA, Adams MR, McCredie RJ, et al. Hormone replacement therapy is associated with improved arterial physiology in healthy post-menopausal women. Clin Endocrinol 1996;45: 435–41.

[184] Rossouw JE, Anderson GL, Prentice RL, et al. Risks and benefits of estrogen plus progestin in healthy postmenopausal women: principal results from the Women's Health Initiative randomized controlled trial. JAMA 2002;288:321–33.

[185] Manson JE. The 'timing hypothesis' for estrogen therapy in menopausal symptom management. Women's Health (Lond Engl) 2015;11:437–40.

[186] Harman SM. Estrogen replacement in menopausal women: recent and current prospective studies, the WHI and the KEEPS. Gend Med 2006;3:254–69.

[187] Hodis HN, Mack WJ. Estrogen therapy and coronary-artery calcification. N Engl J Med 2007;357:1252–3 [author reply 1254].

[188] Caminiti G, Volterrani M, Iellamo F, et al. Effect of long-acting testosterone treatment on functional exercise capacity, skeletal muscle performance, insulin resistance, and baroreflex sensitivity in elderly patients with chronic heart failure a double-blind, placebo-controlled, randomized study. J Am Coll Cardiol 2009;54:919–27.

[189] Wu SZ, Weng XZ. Therapeutic effects of an androgenic preparation on myocardial ischemia and cardiac function in 62 elderly male coronary heart disease patients. Chin Med J 1993;106:415–8.

[190] McCredie RJ, McCrohon JA, Turner L, et al. Vascular reactivity is impaired in genetic females taking high-dose androgens. J Am Coll Cardiol 1998;32:1331–5.

[191] Lorenz M, Koschate J, Kaufmann K, et al. Does cellular sex matter? Dimorphic transcriptional differences between female and male endothelial cells. Atherosclerosis 2015;240:61–72.

[192] Batres RO, Dupont J. Gender differences in prostacyclin and prostaglandin E2 synthesis by human endothelial cells. Prostaglandins Leukot Med 1986;22:159–71.

[193] Egan KM, Lawson JA, Fries S, et al. COX-2-derived prostacyclin confers atheroprotection on female mice. Science 2004;306:1954–7.

ENDOCRINE FUNCTIONS AND METABOLIC INTERACTIONS

7

Hemodynamic Forces in the Endothelium: From Mechanotransduction to Implications on Development of Atherosclerosis

Denise C. Fernandes, Thaís L.S. Araujo, Francisco R.M. Laurindo, and Leonardo Y. Tanaka

THE ENDOTHELIUM

Introduction

Cells are constantly exposed to physical forces, which generate mechanical cues according to each micro-environment. Forces acting on biological systems include gravity, adhesion, pressure, turgescence, and shear stress. These forces are transduced into biochemical signals, composing signaling networks activated by structural elements (mechanotransduction), which culminate in physiological responses of adaptation in response to the physical environment. Signaling networks depending on mechanical forces regulate several cellular processes, including growth, differentiation, migration, angiogenesis, and apoptosis [1,2].

In mammals, the major mechanical stimuli affecting vascular cells are forces associated with blood flow or blood pressure: tangential forces (shear stress) or stretching forces (pulsatile distention) (Fig. 7.1). Both stretching and hydrostatic pressure act on all vessel wall cells, but have only received attention in the literature in recent years. On the other hand, shear stress, affecting almost exclusively the blood/endothelial interface, targets the endothelium, and has been studied for more than 25 years [1]. Those three forces cause adaptive pathophysiological responses according to their magnitude, and each force promotes different cell responses, depending on the context, for example, the type of angiogenesis of endothelial cells submitted to in vitro shear stress or stretching [3].

In this chapter, we will address the pathway of mechanotransduction occurring in the endothelium submitted to shear stress. This response includes foursequential stages: activation of mechanosensors, transduction of the mechanical stimuli into biochemical signals, intracellular signal integration, which finally determine changes in cell structure, metabolism and/or gene expression. These changes allow the endothelium to respond and adapt to specific micro-environment changes. Mechanotransduction by stretching and implications of shear stress on the development of atherosclerosis pathologies will also be discussed briefly.

Endothelium Shear Stress

Shear stress is defined as the frictional force generated by blood flow in the endothelium, that is, the force that the blood flow exerts on the vessel wall, expressed in force-area unit (typically dynes/cm^2). As the blood flows without turbulence or admixture, more specifically without convective mass transference [4], flow is designated as laminar and occurs predominantly in straight artery regions. In arterial system bifurcations or curves, the flow may show turbulence and/or random movements, and is classified as oscillatory or turbulent (Table 7.1).

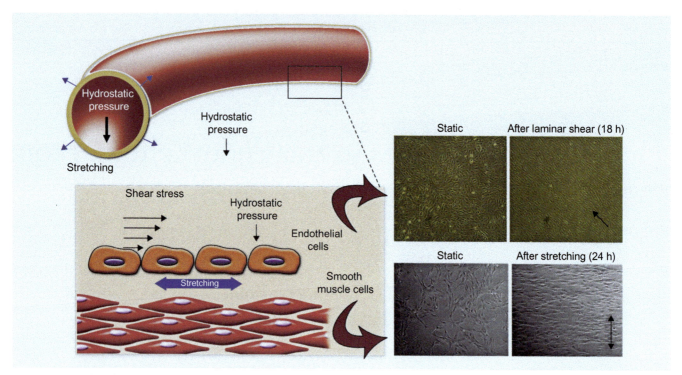

FIG. 7.1   Main forces associated to vessel blood flow. Vascular cells are constantly exposed to hydrostatic pressure (represented by blood pressure), stretching (caused by blood flow pulsatility), and, in the case of the endothelium, also to shear stress (caused by the tangential frictional force of the blood flow on the endothelium). Both *shear stress* and stretching change cell morphology, and endothelial and smooth muscle cells tend to be oriented according to the flow direction (shown in the picture as *arrows*). The cell alignment pictures refer to Human Umbilical Vein Endothelial Cells (HUVECs) after 18 h of exposure to laminar shear stress in a cone and plate equipment, and rat (A7r5) aortic smooth muscle cells after 24 h of stretching (frequency of 1 Hz).

TABLE 7.1   Shear Stress Patterns, Location, and Magnitude in the Vascular System

Arterial system			
Location		**Pattern**	**Shear stress** **10–70 dynes/cm^2**
Great arteries	Straight regions	Laminar	15–30 dynes/cm^2
	Bifurcations/curves	Turbulent/oscillatory	$\pm$4 dynes/cm^2
Stenotic arteries		Turbulent/oscillatory	30–40 dynes/cm^2
Small arteries		Laminar	>12 dynes/cm^2
Venous system			1–6 dynes/cm^2

Laminar flow: no turbulence, characterized by parallel layers that do not mix during the movement; Turbulent/oscillatory flow: flow with irregular and chaotic movement, with a certain degree of randomness.
Adapted from Nigro P, Abe J, Berk BC. Flow shear stress and atherosclerosis: a matter of site specificity. Antioxid Redox Signal 2011;15:1405–14.

Changes in shear stress determine instantaneous vasomotor changes which are regulated on a beat-to-beat basis in order to maintain constant shear stress [5] and optimize the conductance artery flow distributive function. Therefore, increased blood flow in these vessels induces vasodilation and, conversely, reduced flow determines vasoconstriction. These vasomotor changes are strongly dependent on endothelial function integrity, in particular the production of nitric oxide (NO) by endothelial nitric oxide synthase (eNOS) enzyme. This accords to the concept that the endothelium is the primary sensor of shear stress changes [6,7], and that shear stress is the main eNOS [7,8] tonic activation physiological mechanism (see "Physiological responses" below). Persistent changes in shear stress patterns, in particular, oscillatory flow, has been associated to vessel regions with higher rates of lipoprotein oxidation [9], increased endothelial apoptosis [10], and propensity to develop atheroma plaques [11]. In general, one can say that high laminar shear stress is atheroprotective, while low shear stress or oscillatory shear stress tends to enhance atherogenicity [2]. In fact, persistent high laminar shear stress reduces neointimal cell proliferation [12,13] and may also determine regression of an already installed neointimal layer [14] (see Atherosclerosis pathology below).

SHEAR STRESS SENSORS

Mechanotransduction is a subcellular signaling process that involves sensor and integrative pathways composing the cellular response to mechanical forces. In the case of shear stress mechanotransduction in endothelial cells, several proteins and microdomains at the plasma membrane have been proposed as mechanosensors. The conformational structure of protein mechanosensors is directly or indirectly altered by shear stress, depending on the location within the endothelial cell and in this new conformation, can activate intracellular signaling pathways. In addition, plasma membraned microdomains have their fluidity altered through the mechanical stimulus, promoting spatial rearrangement of several proteins, with consequent activation of signaling pathways, such domains include caveolae (invaginations found in endothelial cell apical regions) and lipid rafts (regions with enhanced rigidity due to higher cholesterol concentrations).

The endothelial cell mechanosensors that act in response to shear stress are ion channels, receptors for Vascular Endothelial Growth Factor type 2 (VEGFR2 or Flk-1), adhesion molecules such as Platelet Endothelial Cell Adhesion Molecules (PECAM-1), G-protein-coupled receptors (GP-CRs), and trimeric G-proteins [15] (Table 7.2). There is no universal sensor among these mechanosensors (Fig. 7.2), but multimeric complexes of several biomolecules that must act together to transduce mechanical signals into biochemical signals.

Transduction of forces by the cytoskeleton has been proposed on the basis of the tensegrity model,

in which a series of compression-resistant structures is surrounded by tension elements, creating an internal tension that promotes support to the structure. When force is applied to the structure as a whole, the cytoskeleton elements are rearranged without losingtension. Therefore, cytoskeleton rearrangement may allow the local activation of signaling molecules in response to shear stress. However, it is still unclear whether the cytoskeleton acts directly as a mechanosensor or as a structure mediating the spatial regulation of intracellular signaling events [16]. The cytoskeleton can transmit the tension force from the apical endothelial cell region to adhesion regions, such as focal adhesions (regions that bind to the extracellular matrix) and cell–cell junctions. Such focal adhesions incorporate one of the main mechanosensors: integrins. Integrins are examples of adhesion proteins: heterodimeric transmembrane glycoproteins that bind to the extracellular matrix. When cells are submitted to shear stress, distinct integrins are activated through conversion from low to high affinity binding state to substrates that include, for example, extracellular matrix proteins such as fibronectin, vitronectin, collagen, and laminin [17]. The conformation change that promotes the activation of integrins such as αvβ3, αIIbβ3, and α2β1 involves changes in their redox state by means of reduction or isomerization of intramolecular disulfide bonds. Once activated, integrins convert external stimuli into intracellular responses consisting of signaling pathways governing cytoskeleton organization and cell motility.

PECAM-1 adhesion molecule is another well studied mechanosensor that mediates the activation of ERK in response to mechanical tension. In fact, PECAM-1 is the mechanosensor component of a protein complex recently called "mechanosome," [18] located in cell–cell junctions, which also contains the adaptive protein VE-cadherin and VEGFR2. After mechanical stimulus, PECAM-1 activates VEGFR2, leading to activation of kinases from Src and Akt families, as well as PI3 kinase, which in turn activate integrins. Such PECAM-1 responses are initial and transitory after laminar shear, and sustained in oscillatory shear.

In addition to the classic mechanosensors (Table 7.2), it is important to mention that cathepsin L is a mechanosensitive matrix protease regulated by laminar shear. Unidirectional laminar shear of 15 dynes/cm^2 for 24 h leads to decreased proteolytic capacity of the endothelial cell matrix, due to down regulation of cathepsin L activity [19].

TABLE 7.2　Characteristics of some endothelial cell mechanosensors

Mechanosensor	Description	Effect as sensor	How the mechanical stimulus is mechano-transduced into biochemical signal
Ion channels	Potassium channel (K^+)	Changes in membrane fluidity due to mechanical stimulus change channel conformation, generally causing enhanced opening	Causes local hyperpolarization of the plasma membrane
	Calcium channels (Ca^{2+})		Increase of intracellular Ca^{2+}, which binds to calmodulin and increases its affinity for eNOS and, consequently, production of NO
G Protein Coupled Receptors (GPCR)	Transmembrane receptors that, once activated, activate cytoplasmic proteins, such as G protein	Due to the mechanical force, there are conformational changes	The receptors are activated independently from the agonist, for example, angiotensin II (AT1R) receptor, beta-2 adrenergic (B2) receptor
Caveolae	Lipid-rich invaginations in the lumenal region of the plasma membrane	Via redistributed/increased number of caveolae in the plasma membrane	Possibly serve as structure for mechanosensitive elements close to effector biomolecules (e.g., kinase)
Cytoskeleton	Structure composed mainly by microtubules, microfilaments, and intermediate filaments	Due to deformation in the cell surface structure, tension may be transmitted to other parts of the structure	Still unclear if acts directly as mechanosensor it or as structure mediating the spatial regulation intracellular signaling events
Adhesion proteins	Integrins: transmembrane proteins that bind cytoskeletal proteins to extracellular matrix components	The cytoskeleton tension force is redistributed by integrins to the proteins connected to them	Activation of kinases such as c-Src and Focal Adhesion Kinase (FAK)
	PECAM-1: Adhesion molecule expressed at the endothelial surface, found in cell–cell junctions	Formation of the mechanosensor complex with Fyn kinase, VEGF receptor (VEGFR2), and VE-cadherin	The tension force of cell–cell junctions changes PECAM-1 conformation, which activates Fyn, which in turn promotes ligand-independent activated VEGFR2 activation

From Chatterjee S, Fujiwara K, Pérez NG, et al. Mechanosignaling in the vasculature: emerging concepts in sensing, transduction and physiological responses. Am J Physiol Heart Circ Physiol 2015;308:H1451–62.

PHYSIOLOGICAL RESPONSES OF ENDOTHELIAL CELL UNDER SHEAR STRESS

The opposite physiological effects of endothelial laminar and oscillatory shear are described in Fig. 7.3.

EXPERIMENTAL HEMODYNAMIC FORCE STUDY METHODS

Most signaling pathways described herein were discovered based on reductionist models in cell culture (endothelial or smooth muscle cells), isolated from vessels (of human or animal origin). These

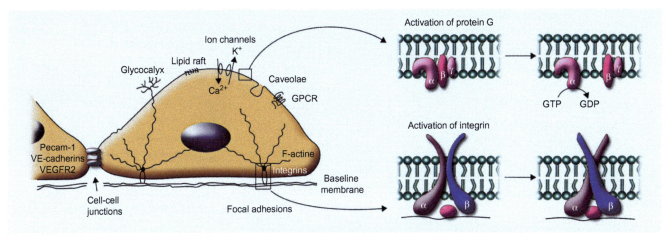

FIG. 7.2 Endothelial cell mechanosensors. Location of mechanosensors such as cytoskeleton, integrins, cell-cell junctions, caveolae, lipid rafts, cell surface glycocalyx, G protein-coupled Receptors (GPCR), and ion channels. While mechanosensors in the apical region (lumenal) are activated directly by shear stress (such as G proteins), the cytoskeleton (represented by actin fibers, F-actin) is responsible for transmitting forces to the mechanosensors at the basal region of endothelial cells (such as integrins). G protein activation occurs due to local changes in plasma membrane fluidity, therefore directly due to shear stress and independent from an agonist, causing hydrolysis of GTP into GDP. On the other hand, the structure of mechanosensitive integrins is changed from inactive to active when submitted to shear stress, possibly due to transmission of the mechanical force to the cytoskeleton. In active conformation, integrins have higher affinity for cognate proteins in the extracellular matrix.

hemodynamic force studies generally compare cells without mechanical stimulation (static) with those submitted to mechanical tension, different patterns of shear stress (laminar versus oscillatory), or even different types of mechanical stress (shear versus stretch).

Shear stress: the unidirectional movement of the culture medium over the mono-layer of endothelial cells mimics the laminar blood flow in arteries. Generally, laminar shear is simulated using parallel plate or cone-and-plate equipment. The latter can also be used to mimic oscillatory flow, if motorized, so that it turns 180° to each side at a controlled frequency. Some parallel plate chambers can be designed with bifurcations, so that in the same experiment they simulate regions with different stimuli, such as laminar and turbulent/oscillatory shear.

Stretching: to submit the cells to stretching, they are generally cultured in silicone membranes, which are stretched mechanically, indirectly inducing the stimulation that occurs in vivo (Fig. 7.4).

Exposure of endothelial cells to laminar shear translates into atheroprotective responses such as inhibition of thrombosis, inhibition of platelet adhesion and monocyte recruitment, inhibition of endothelial cell apoptosis [20], and reduced proliferation of neointimal cells [12,13]. In contrast, oscillatory shear induces pro-inflammatory and pro-atherogenic responses in the endothelium, such as thrombosis, leukocyte adhesion, and endothelial cell apoptosis [20]. Alignment among stress fibers in the direction of flow is an important characteristic of regions resistant to atherosclerosis in vivo, that is, those submitted to laminar shear stress (Fig. 7.1). Chronic laminar shear in endothelial cells has a similar effect to that observed in vivo, being characterized by lower expression of inflammation markers [21].

The time course of responses induced by shear stress were studied in isolated endothelial cells (check box Experimental hemodynamic force study models—how to study shear in vitro). Some seconds after shear stress, the cells show activation of ion channels associated with the plasma membrane, such as K^+ and Ca^{2+} channels, activation of G proteins, phosphorylation of PECAM-1 and production of oxidants. In a few minutes, several intracellular signaling pathways are activated, including calcium-dependent eNOS phosphorylation and its activation, with NO release [22], activation of

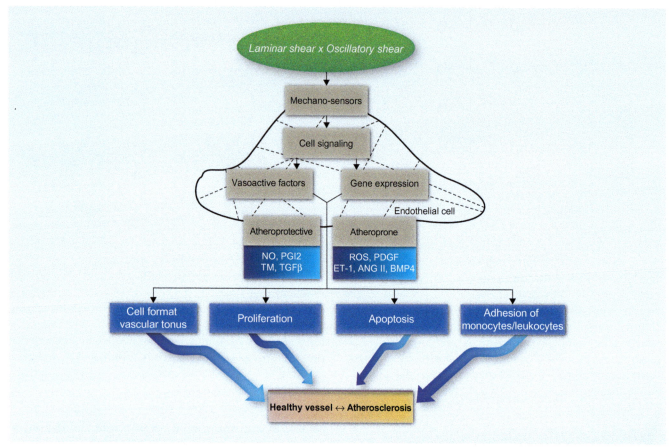

FIG. 7.3 Different effects of laminar and oscillatory shear on cell function and atherosclerosis. The *dotted lines* represent the endothelial cell cytoskeleton. Laminar and oscillatory shear forces are recognized in endothelial cells by mechanosensors and the mechanosignals initiate signaling cascades that regulate the production of vasoactive factors. While laminar shear stimulates the production of atheroprotective factors, oscillatory shear stimulates the production of atherogenic factors, and the balance between these factors determines the vessel tendency to stay healthy or to develop atherosclerotic plaques. *PGI2*, prostacyclin; *TM*, thrombomodulin; *TGFb*, Transforming Growth Factor beta; *PDGF*, Platelet-Derived Growth Factor; *ET-1*, Endothelin-1; *BMP4*, Bone Morphogenetic Protein 4. Adapted from Jo H, Song H, Mowbray A. Role of NADPH oxidases in disturbed flow- and BMP4-induced inflammation and atherosclerosis. Antioxid Redox Signal 2006;8:1609–19.

phosphatidylinositol-3-kinase (PI3K), and signaling mediated by integrins [16].

From minutes to hours, several signaling pathways mediated by Rho GTPase family proteins are activated, as well as the kinases c-Src, Focal Adhesion Kinase (FAK), Protein-Kinase C (PKC), and c-Jun N-terminal kinase (JNK). Several shear stress-responsive transcription factors are activated, such NF-κB, AP-1, and erythroid nuclear factor 2 type 2, Nrf2. Full response of adaptation to laminar shear occurs within 24–48 h, when there is increased transcription of several genes related to intracellular signaling, cytoskeleton, extracellular matrix, metabolism, and angiogenesis [23–25].

Regulation of gene expression by endothelial shear stress is, in part, mediated by sequences called "Shear Stress-Responsive Elements" (SSREs), when present in the promoter gene region. However, SSRE-independent genes are also transcribed in cells submitted to shear stress [26,27].

Among the transcription factors activated during shear stress, KLF2 is specifically induced by laminar shear in endothelial cells. KLF2 induces the expression of anticoagulant and anti-inflammatory proteins, especially eNOS and thrombomodulin, and reduces the expression of pro-inflammatory and anti-fibrinolysis genes, as it inhibits the NF-κB and AP-1 transcription factors. The Nrf2 transcription

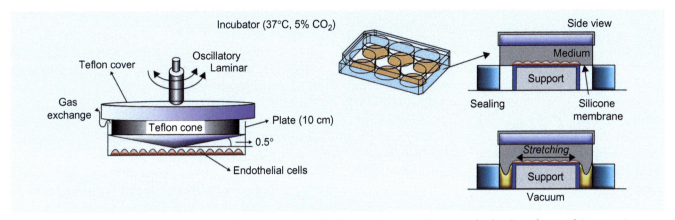

FIG. 7.4 Schematic representation of shear stress simulation equipment (cone-and-plate) and stretching equipment used in cell culture studies. In the cone-and-plate system, cells are exposed to laminar or oscillatory flow due to rotation of the Teflon cone (bearing an angle of 0.5 degree from the center to the edges of the cone) over the culture medium. In stretching systems (such Flexercell systems), the membrane is stretched (by application of vacuum) together with the cells adhered to its surface. Adapted from http://www.flexcellint.com/BioFlex.htm.

factor, which increases the expression of several antioxidant genes, is also induced by laminar shear. The combined action of KLF2 and Nrf2 corresponds to approximately 70% of gene expression response induced by laminar shear [28].

On the other hand, NF-κB and AP-1 transcription factors are induced in endothelial cells submitted to oscillatory shear stress. These factors increase the expression of pro-inflammatory and pro-atherogenic genes, such as intercellular adhesion molecule-1 (ICAM-1), E-selectin, Platelet-Derived Growth Factor (PDGF), Interleukin-1 (IL-1a), Bone Morphogenetic Protein-4 (BMP4), Monocyte Chemoattractant Protein-1 (MCP-1), and the vasoconstrictor Endothelin-1 (ET-1). Therefore, while laminar shear stress activated KLF2 and Nrf2, which inhibit inflammation, oscillatory shear stress preferentially induces NF-κB and AP-1, which promote inflammation [2,29].

Shear stress is one of the most potent activators of endothelial NO synthase (eNOS) enzyme [30]. eNOS is a dimeric enzyme with complex activation, regulated by several co-factors, including calmodulin, and several phosphorylation sites (see Chapter 8). The activation of eNOS by laminar shear stress occurs mainly by phosphorylation of the serine in position 1179 (into human eNOS) by Akt kinase, which increases NO production at low

calcium concentrations by increasing the transference of electrons from NADPH to L-arginine substrate and decreasing calmodulin sensitivity to calcium [31,32]. Shear stress also increases eNOS activity by promoting higher physical interaction between eNOS and the chaperone Hsp90, which increases NO production possibly due to altered conformation of eNOS and by promoting eNOS phosphorylation in serine 116 (which inhibits its interaction with caveolin-1, a protein that associates with eNOS in the membrane and decreases its activity) [33]. Another important feature of the atheroprotective role of sustained laminar shear is the increased stability of eNOS [34] messenger RNA and of its protein expression, ultimately increasing NO bioavailability.

Another remarkable aspect of laminar shear stress is reorganization of the cytoskeleton to polarize the endothelial cell towards the flow direction, which results in alignment observed after sustained laminar shear stress (Fig. 7.1).[1] In part, these changes in cell alignment follow the activation of integrins (Fig. 7.3). Through pathways involving kinase activation and phosphorylation of small GTPase effectors (Ras and Rap1) activating Raf-MEK-ERK [35] pathways and RhoGTPases. One such RhoGTPase is RhoA, which, if suppressed, leads to inhibition of alignment and shear-induced stress fiber

[1] Endothelial cells derived from the aorta and most vessels align in the flow direction, but this behavior depends on cell type. For example, endothelial cells derived from heart valves tend to align perpendicularly to flow.

formation [35]. Remodeling of actin cytoskeleton and endothelial cell alignment occur after 6 h of laminar shear stress, while the typical increase of stress fibers occurs after 24 h. In addition, alignment of actin filaments in the flow direction depends on activation of the mechanosensor complex PECAM-1/VE-cadherin/VEGFR2 (Table 7.2). In vivo, mice knockout for PECAM-1 show defective organization of actin fibers (F-actin) and undergo NF-κB activation [36].

REDOX PROCESSES IN SHEAR STRESS-INDUCED SIGNAL TRANSDUCTION IN THE ENDOTHELIUM

An early response to sudden increased shear stress is the production of superoxide free radical, described initially by our group in assays with isolated rabbit vessels perfused with saline solution. Electron Paramagnetic Resonance Spectroscopy (ESR) allowed identification of superoxide radical anion production in vessels perfuse with spin traps or through formation of ascorbyl radical in the plasma of these animals (indicative the production of free radicals during shear stress) [37]. Such superoxide production required endothelial integrity. In fact, studies in endothelial cell cultures showed that NADPH oxidase is activated both by laminar and oscillatory shear stress [6].

NADPH oxidase is the main source of superoxide for redox signaling purposes in vascular cells [38] both in physiological situations and in several diseases, including those associated with atherosclerosis, hypercholesterolemia, diabetes, and hypertension [39,40]. The multiprotein complex NADPH oxidase is formed by one catalytic nox subunit (among 5 distinct types) and p22^{phox} anchored in cell membranes. Enzyme activation is triggered by recruitment of regulatory cytoplasmic subunits to the membrane for the final assembly of the complex and production of superoxide. The main types of NADPH oxidase expressed in endothelial cells are NOX1 and NOX4; NOX5 is also expressed in mammals except mice and rats. While laminar shear stress transiently generates superoxide production oscillatory shear stress produces higher levels of such reactive species in sustained fashion, in a way depending on p47^{phox} NADPH oxidase subunit [41]. Sustained superoxide production during oscillatory shear stress increases the expression of NADPH oxidase subunits, such as p22^{phox}, NOX2, NOX4 and p47^{phox} in cell culture [2]. In fact, in vivo carotid arteries with abnormal high flow show chronic increase in NADPH oxidase-dependent superoxide production [42].

In addition to upregulation of NADPH oxidase, a few hours of mechanical stimulation usually lead to increased expression of antioxidant enzymes such as Superoxide Dismutase (Cu-ZnSOD, and MnSOD, isoforms) which convert superoxide to H_2O_2, Glutathione Peroxidase (GPx), Peroxiredoxin1 (Prx1)—both involved in H_2O_2 removal—and Thioredoxin 1, which is important for GPx and Prx recycling. There is also increased eNOS expression. Finally, sustained laminar shear stress increases the activity of the enzyme Glucose-6-phosphate dehydrogenase (G6PD), which by increasing the formation of NADPH also contributes to maintain higher levels of glutathione in its reduced form (GSH) (see Chapter 8). Therefore, during sustained laminar shear stress, there is shift towards a more reducing environment, with higher NO availability [2].

MECHANICAL RESPONSE TO CYCLIC STRETCH

Models that mimic in vitro the stretching stress suffered by endothelial and vascular smooth muscle cells during each cardiac cycle (Fig. 7.1) allowed the demonstration of their important influence on associated physiological mechanisms and different pathological processes [43]. Perpendicular repositioning of the cells with respect to the stretching direction is observed and has been used to understand how exogenous forces affect dynamic processes that control the cytoskeleton. In this direction, several mechanisms related to distinct hierarchical levels of signaling have been demonstrated, such as integrin activation, cytoskeleton rearrangement and focal adhesion component reorganization [44,45]. For example, stretching activates JNK kinase, and consequently, the AP-1 transcription factor, which increases the expression of several inflammatory genes, in a way similar to oscillatory shear stress, as commented before. However, the activation of JNK depends of the direction of the stretching force with respect to actin fibers; if stretching is perpendicular to the actin fibers, JNK is not activated [46].

During smooth muscle cell stretching, there is NADPH oxidase activation, mainly NOX4, important to activate the cofilin, a protein that increases the depolymerization speed of actin fibers—therefore, allowing cytoskeleton rearrangement and cell alignment [47]. Finally, paracrine signaling mediated by vascular cells has also been investigated, with respect to smooth muscle cell proliferation mechanisms regulated by endothelial cells [44]. However, there is still a substantial investigative avenue in order to better understand how stretching, as well as other hemodynamic forces, interact and converge during mechano-adaptive processes in the cardiovascular system.

HEMODYNAMIC FORCES AND ATHEROSCLEROSIS PATHOPHYSIOLOGY

Hemodynamic forces influence several aspects atherosclerosis, including atheroma formation, definition of plaque phenotype, and evolution into acute rupture events [48]. Atherosclerosis is a local disease, with development of plaques in preferential regions of the vascular system, such as artery bifurcations or curve regions. In particular, these regions coincide with shear stress patterns to which these vessels are submitted. Regions with lower or oscillatory shear stress, normally bifurcations or curves, are preferred locations for atherogenesis. In contrast,

regions with moderate/physiological shear stress are naturally more resistant to atherosclerosis (Table 7.1 and Fig. 7.5). The mechanisms regulated by shear stress that support endothelial function maintenance have been widely investigated in in vitro models, and were described before. In particular, low shear stress affects endothelial integrity due to: (a) reduced production of NO, with concomitant increase of reactive species, as well as other vasoactive molecules such as endothelin-1; (b) induction of apoptosis and change in endothelial cell phenotypes, which assume a polygonal pattern; (c) accumulation of LDL cholesterol and its modification by ROS. Low shear stress significantly contributes to the activation of inflammatory response via NF-κB, which regulates the expression of adhesion molecules and cytokines. Therefore, monocytes are recruited in order to differentiate into macrophages, and then evolve into foam cells, performing an important role in atherosclerosis progression. Also depending on NF-κB, decreased shear stress results in increased expression and activity of metalloproteinases that act in plaque formation, as well as other vascular architecture changes [48].

Progression of the atherosclerotic plaque itself further changes the blood flow pattern, promoting other alterations [49]. Initially, there is a positive (or expansive) remodeling compensation response,

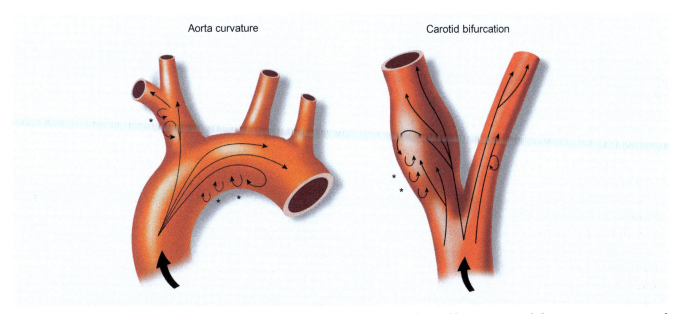

FIG. 7.5 Aorta and carotid flow pattern. The blood flow direction is indicated by *arrows*, and the aorta curvature and carotid bifurcation regions have flow turbulence (marked with *red asterisks*) and low shear stress. These are the regions more prone to development of atherosclerotic plaques. Adapted from White CR, Frangos JA. The shear stress of it all: the cell membrane and mechanochemical transduction. Philos Trans R Soc Lond B Biol Sci 2007;362:1459–67.

characterized by increased vessel area and preservation of lumen caliber. This process, known as the Glagov phenomenon, is able to accommodate an increase in plaque mass of approximately 40% stenosis, and has also been described in other vascular diseases such as hypertension and response to angioplasty. However, positive remodeling may sustain low shear stress, intensifying the alterations mentioned above and consequently promoting atheroma evolution into a more unstable or vulnerable phenotype, which has a thin fibrous cap layer, showing a necrotic center rich in macrophages and with low collagen content.

Finally, vascular architecture heterogeneity promoted by plaque evolution influences the way that biomechanical forces affect plaque rupture. When vascular wall stress, determined by blood pressure, vessel geometry and plaque composition, exceeds the force borne by the fibrous layer, such layer ruptures and may promote, through the formation of a mural thrombus, acute clinical events. It is believed that increased shear stress in unstable plaque regions, with high levels of stenosis, contributes to rupture or erosion [49]. However, additional studies are required to understand how plaque characteristics influences its rupture, since only 5% of plaques with vulnerable phenotype were associated with rupture [50,51].

CONCLUSIONS

Although there is already extensive characterization of signaling pathways during mechanotransduction of endothelial cells submitted to shear stress, there are several aspects that require additional studies, mainly regarding the initial events of signal transduction. A profound understanding, at several levels of complexity, namely molecular, biochemical, and physiological, related to vascular cell regulation by different mechanical stimuli, such as shear stress, stretching, and hydrostatic pressure, is fundamental to interpret the detailed mechanisms associated to inflammation, atherosclerosis, and aging. Based on these mechanisms, one can anticipate future vascular therapeutic interventions able to mimic or reproduce, in part, mechanotransduction protective pathways or even possible biomarkers based on these mechanisms.

Analysis of cell responses to mechanical stimuli in different regions of the aorta (with different flow patterns) has also been conducted in vivo. Experimental atherosclerosis models, for example, are used to analyze protein expression, RNA expression (messenger, microRNAs, long non-coding RNA), location of proteins, etc. [52]. Some of these strategies are used to simulate oscillatory shear stress in a more controlled fashion, such as partial carotid obstruction in mice [53]. The great advantage of using mice models is the possibility to use genetically modified mice (transgenic or knock-out). It is also possible to analyze gene expression separately in the endothelium versus smooth muscle or, in addition, performing physical separation of both cells types with more refined protocols. Several mechanisms tested and explored in detail in reductionist cell culture models derive from exploratory studies that generate a large amounts of data and hypothesis [52].

References

[1] Hahn C, Schwartz MA. Mechanotransduction in vascular physiology and atherogenesis. Nat Rev Mol Cell Biol 2009;10:53–62.
[2] Nigro P, Abe J, Berk BC. Flow shear stress and atherosclerosis: a matter of site specificity. Antioxid Redox Signal 2011;15:1405–14.
[3] Shiu YT, Weiss JA, Hoying JB, et al. The role of mechanical forces in angiogenesis. Crit Rev Biomed Eng 2005;33:431–50.
[4] Jacobs CR, Huang H, Kwon RY. Introduction to cell mechanics and mechanobiology. London: Garland Science; 2013. p. 94.
[5] Kamiya A, Togawa T. Adaptive regulation of wall shear stress to flow change in the canine carotid artery. Am J Physiol 1980;239:H14–21.
[6] De Keulenaer GW, Chappell DC, Ishizaka N, et al. Oscillatory and steady laminar shear stress differentially affect human endothelial redox state: role of a superoxide-producing NADH oxidase. Circ Res 1998;82:1094–101.
[7] Lehoux S, Tedgui A. Cellular mechanics and gene expression in blood vessels. J Biomech 2003;36:631–43.
[8] Hendrickson RJ, Cappadona C, Yankah EN, et al. Sustained pulsatile flow regulates endothelial nitric oxide synthase and cyclooxygenase expression in co-cultured vascular endothelial and smooth muscle cells. J Mol Cell Cardiol 1999;31:619–29.
[9] Hwang J, Ing MH, Salazar A, et al. Pulsatile versus oscillatory shear stress regulates NADPH oxidase subunit expression: implication for native LDL oxidation. Circ Res 2003;93:1225–32.
[10] Davies PF, Remuzzi A, Gordon EJ, et al. Turbulent fluid shear stress induces vascular endothelial cell turnover in vitro. Proc Natl Acad Sci U S A 1986;83:2114–7.
[11] Pedrigi RM, Poulsen CB, Mehta VV, et al. Inducing persistent flow disturbances accelerates atherogenesis and promotes thin cap fibroatheroma development in D374Y-PCSK9 hypercholesterolemic minipigs. Circulation 2015;132:1003–12.
[12] Wentzel JJ, Krams R, Schuurbiers JC, et al. Relationship between neointimal thickness and shear stress after Wallstent implantation in human coronary arteries. Circulation 2001;103:1740–5.
[13] Kohler TR, Kirkman TR, Kraiss LW, et al. Increased blood flow inhibits neointimal hyperplasia in endothelialized vascular grafts. Circ Res 1991;69:1557–65.

[14] Mattsson EJ, Kohler TR, Vergel SM, et al. Increased blood flow induces regression of intimal hyperplasia. Arterioscler Thromb Vasc Biol 1997;17:2245–9.

[15] Li YS, Haga JH, Chien S. Molecular basis of the effects of shear stress on vascular endothelial cells. J Biomech 2005;38:1949–71.

[16] Collins C, Tzima E. Hemodynamic forces in endothelial dysfunction and vascular aging. Exp Gerontol 2011;46:185–8.

[17] Tzima E, Irani-Tehrani M, Kiosses WB, et al. A mechanosensory complex that mediates the endothelial cell response to fluid shear stress. Nature 2005;437:426–31.

[18] Chatterjee S, Fujiwara K, Pérez NG, et al. Mechanosignaling in the vasculature: emerging concepts in sensing, transduction and physiological responses. Am J Physiol Heart Circ Physiol 2015;308:H1451–62.

[19] Platt MO, Ankeny RF, Jo H. Laminar shear stress inhibits cathepsin L activity in endothelial cells. Arterioscler Thromb Vasc Biol 2006;26:1784–90.

[20] Traub O, Berk BC. Laminar shear stress: mechanisms by which endothelial cells transduce an atheroprotective force. Arterioscler Thromb Vasc Biol 1998;18:677–85.

[21] Boon RA, Leyen TA, Fontijn RD, et al. KLF2-induced actin shear fibers control both alignment to flow and JNK signaling in vascular endothelium. Blood 2010;115:2533–42.

[22] Kumagai R, Lu X, Kassab GS. Role of glycocalyx in flow-induced production of nitric oxide and reactive oxygen species. Free Radic Biol Med 2009;47:600–7.

[23] Cullen JP, Sayeed S, Sawai RS, et al. Pulsatile flow-induced angiogenesis: role of G(i) subunits. Arterioscler Thromb Vasc Biol 2002;22:1610–6.

[24] Abumiya T, Sasaguri T, Taba Y, et al. Shear stress induces expression of vascular endothelial growth factor receptor Flk-1/KDR through the CT-rich Sp1 binding site. Arterioscler Thromb Vasc Biol 2002;22:907–13.

[25] Nagel T, Resnick N, Dewey Jr CF, et al. Vascular endothelial cells respond to spatial gradients in fluid shear stress by enhanced activation of transcription factors. Arterioscler Thromb Vasc Biol 1999;19:1825–34.

[26] Resnick N, Yahav H, Khachigian LM, et al. Endothelial gene regulation by laminar shear stress. Adv Exp Med Biol 1997;430:155–64.

[27] Tzima E. Role of small GTPases in endothelial cytoskeletal dynamics and the shear stress response. Circ Res 2006;98:176–85.

[28] Fledderus JO, Boon RA, Volger OL, et al. KLF2 primes the antioxidant transcription factor Nrf2 for activation in endothelial cells. Arterioscler Thromb Vasc Biol 2008;28:1339–46.

[29] Ji H, Song H, Mowbray A. Role of NADPH oxidases in disturbed flow- and BMP4-induced inflammation and atherosclerosis. Antioxid Redox Signal 2006;8:1609–19.

[30] Malek AM, Jiang L, Lee I, et al. Induction of nitric oxide synthase mRNA by shear stress requires intracellular calcium and G-protein signals and is modulated by PI 3 kinase. Biochem Biophys Res Commun 1999;254:231–42.

[31] Dimmeler S, Fleming I, Fisslthaler B, et al. Activation of nitric oxide synthase in endothelial cells by Akt-dependent phosphorylation. Nature 1999;399:601–5.

[32] McCabe TJ, Fulton D, Roman LJ, et al. Enhanced electron flux and reduced calmodulin dissociation may explain "calcium-independent" eNOS activation by phosphorylation. J Biol Chem 2000;275:6123–8.

[33] Qian J, Fulton D. Post-translational regulation of endothelial nitric oxide synthase in vascular endothelium. Front Physiol 2013;4:347.

[34] Weber M, Hagedorn CH, Harrison DG, et al. Laminar shear stress and 3' polyadenylation of eNOS mRNA. Circ Res 2005;96:1161–8.

[35] Shyy JY, Chien S. Role of integrins in endothelial mechanosensing of shear stress. Circ Res 2002;91:769–75.

[36] Tzima E, Irani-Tehrani M, Kiosses WB, et al. A mechanosensory complex that mediates the endothelial cell response to fluid shear stress. Nature 2005;437:426–31.

[37] Laurindo FR, Pedro Mde A, Barbeiro HV, et al. Vascular free radical release. Ex vivo and in vivo evidence for a flow-dependent endothelial mechanism. Circ Res 1994;74:700–9.

[38] Clempus RE, Griendling KK. Reactive oxygen species signaling in vascular smooth muscle cells. Cardiovasc Res 2006;71:216–25.

[39] Guzik TJ, West NE, Black E, et al. Vascular superoxide production by NAD(P)H oxidase: association with endothelial dysfunction and clinical risk factors. Circ Res 2000;86:E85–90.

[40] Rajagopalan S, Kurz S, Munzel T, et al. Angiotensin II mediated hypertension in the rat increases vascular superoxide production via membrane NADH/NADPH oxidase activation. Contribution to alterations of vasomotor tone. J Clin Invest 1996;97:1916–23.

[41] Hwang J, Saha A, Boo YC, et al. Oscillatory shear stress stimulates endothelial production of O2 from p47^{phox}-dependent NAD(P)H oxidases, leading to monocyte adhesion. J Biol Chem 2003;278:47291–8.

[42] Castier Y, Brandes RP, Leseche G, et al. p47phox-dependent NADPH oxidase regulates flow-induced vascular remodeling. Circ Res 2005;97:533–40.

[43] Lu D, Kassab GS. Role of shear stress and stretch in vascular mechanobiology. J R Soc Interface 2011;8:1379–85.

[44] Lehoux S, Castier Y, Tedgui A. Molecular mechanisms of the vascular responses to haemodynamic forces. J Intern Med 2006;259:381–92.

[45] Halka AT, Turner NJ, Carter A, et al. The effects of stretch on vascular smooth muscle cell phenotype in vitro. Cardiovasc Pathol 2008;17:98–102.

[46] Kaunas R, Usami S, Chien S. Regulation of stretch-induced JNK activation by stress fiber orientation. Cell Signal 2006;18:1924–31.

[47] Montenegro MF, Valdivia A, Smolensky A, et al. Nox4-dependent activation of cofilin mediates VSMC reorientation in response to cyclic stretching. Free Radic Biol Med 2015;85:288–94.

[48] Wentzel JJ, Chatzizisis YS, Gijsen FJ, et al. Endothelial shear stress in the evolution of coronary atherosclerotic plaque and vascular remodelling: current understanding and remaining questions. Cardiovasc Res 2012;96:234–43.

[49] Kwak BR, Bäck M, Bochaton-Piallat ML, et al. Biomechanical factors in atherosclerosis: mechanisms and clinical implications. Eur Heart J 2014;35:3013–20. 3020a–3020d.

[50] Tanaka LY, Araújo HA, Hironaka GK, et al. Peri/epicellular protein disulfide isomerase sustains vascular lumen caliber through an anticonstrictive remodeling effect. Hypertension 2016;67:613–22.

[51] Stone GW, Maehara A, Lansky AJ, et al. A prospective natural-history study of coronary atherosclerosis. N Engl J Med 2011;364:226–35.

[52] Davies PF, Civelek M, Fang Y, et al. The atherosusceptible endothelium: endothelial phenotypes in complex haemodynamic shear stress regions in vivo. Cardiovasc Res 2013;99:315–27.

[53] Nam D, Ni CW, Rezvan A, et al. Partial carotid ligation is a model of acutely induced disturbed flow, leading to rapid endothelial dysfunction and atherosclerosis. Am J Physiol Heart Circ Physiol 2009;297:H1535–43.

Further Reading

[1] White CR, Frangos JA. The shear stress of it all: the cell membrane and mechanochemical transduction. Philos Trans R Soc Lond B Biol Sci 2007;362:1459–67.

8

Endothelium-Dependent Vasodilation: Nitric Oxide and Other Mediators

Francisco R.M. Laurindo, Marcel Liberman, Denise C. Fernandes, and Paulo Ferreira Leite

ENDOTHELIUM-DEPENDENT VASODILATORS

Vasodilation has become the archetype of endothelial cell function. The concept that the endothelium controls vascular tone in a paracrine fashion (i.e., by secreting diffusible soluble mediators able to act on physically contiguous cells, in this case smooth muscle) was extremely innovative and relevant to vascular physiology. Therefore, the study of endothelium-dependent vasodilation mechanisms has commanded the investigations in the field of endothelial function. Even today, the term endothelial dysfunction is frequently assumed to mean loss of endothelium-dependent vasodilation, regardless of the several other functions of these cells [1]. Similarly, the identification of nitric oxide (NO) as the main endothelium-derived relaxing factor [2] was such an important landmark of these studies that endothelium-dependent vasodilation is usually associated with NO, although there are several other mediators capable of such an effect (e.g., endothelium-derived hyperpolarizing factors, EDHFs). In this chapter, we shall discuss the biologically relevant factors that have been more thoroughly studied. The identification of NO as a paracrine vasodilator opened another innovative chapter in vascular physiology: endogenous gaseous mediators, recently joined by carbon monoxide (CO) and hydrogen sulfide (H_2S), are also briefly discussed herein.

NITRIC OXIDE

Although NO has several biological effects, it is important to mention that most studies regarding this mediator are still centered on its vasodilator effect. The NO molecule is a free radical (as it has an unpaired electron in its last layer) in the gaseous state [2,3]. NO is freely diffusible and permeable to cell membranes. This nature gives NO the ability to react with other free radicals or with molecular oxygen (which is a diradical) [3]. Among the products of NO oxidation, nitrite (NO_2^-) and nitrate (NO_3^-) are the main NO physiological metabolites in aqueous medium, and can be measured as an index of NO production in a given biological system [4,5].

NO Reactions: Determinant of Its Biological Effects

A peculiar characteristic of biological redox systems, which distinguishes them from nonredox signaling mechanisms, is their strong dependency on chemical reactivity of the involved for the final biological effect. In the case of NO, its intermediates chemical reactivity is one of the factors responsible for multiple biological effects [6]. Another important factor is the activity and physiology of NO synthases, discussed in subsequent sections.

The effects of NO can be classified as protective, regulatory, and deleterious (Fig. 8.1) [6]. In general, the protective and regulatory effects are mediated

97

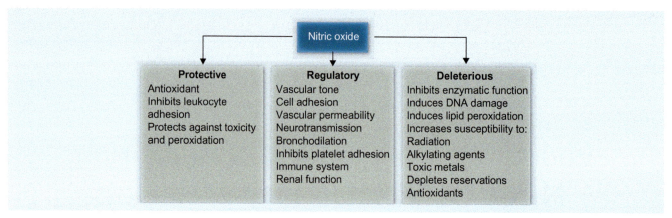

FIG. 8.1   Nitric oxide biological effects and reactivity. Adapted from Wink DA, Mitchell JB. Chemical biology of nitric oxide: insights into regulatory, cytotoxic, and cytoprotective mechanisms of nitric oxide. Free Radic Biol Med 1998;25:434–56.

by direct coordination reactions between NO and metals, mainly with the heme group of guanylyl cyclase, or radical combinations with lipid peroxides [7]. Reactions of NO with thiol groups (—SH present in peptides or proteins) possibly explain several regulatory effects of NO [6,7]. The deleterious effects are generally associated with excessive production of NO and its reaction with superoxide radical, generating reactive intermediates [5,6]. This simplification is didactic, and certainly several exceptions may be found as the complex mechanisms of NO reaction are unraveled in greater depth.

At this point, it is useful to define the meaning of some terms that are frequently confused.

- *Nitrosylation*: in a strict sense, the term nitrosylation must be employed to designate a coordination reaction between NO and metals, for example, hemoglobin or guanylyl cyclase heme group, or iron-sulfur core of the aconitase enzyme. However, frequently on the field of signal transduction, the term nitrosylation is employed with a wider meaning, designating any modification of a protein by the NO group, associated with either heme or thiols, regardless of the chemical process involved [7].
- *Nitrosation*: it is the reaction of the donation of a nitrosonium ion (NO^+) to a nucleophilic substrate, exemplified by the nitrosation of thiols (however, not all nitrosation of thiols involves NO^+ chemistry) [7].
- *Nitration*: it is the addition of NO_2 groups to a substrate, generally secondary to reactions involving nitrogen dioxide radical ($^\bullet NO_2$), which, in turn, is an intermediate of secondary

peroxynitrite reactions. The main nitration substrate consists of residues of tyrosine or tryptophan in proteins [7].

Guanylyl Cyclase: An Important Redox Sensor Activated by NO

The vascular effects of NO include: vasodilation, inhibition of smooth muscle cell proliferation, inhibition of platelet activation, and tonic inhibition of leukocyte adhesion to the vascular endothelium [1]. All these effects have as an important intracellular transducing mechanism—although by no means the only one—the activation of the soluble fraction of guanylyl cyclase and elevation of cyclic guanosine monophosphate (cGMP) levels [7,8].

Guanylyl cyclase is a hemeprotein present in membranes (i.e., activated, e.g., by atrial natriuretic factor) or in cytosol, composing the soluble fraction of the enzyme sGC, which is the NO receptor [8]. sGC catalyzes the conversion of guanosine-5'-triphosphate (GTP) into cGMP and pyrophosphate [8]. sGC is activated by the conformational change of the heme group induced upon NO binding to the iron atom [9,10]. Another physiological activator of sGC is CO, produced endogenously by heme oxygenase enzymes [9]. However, the preference of sGC for CO is approximately six orders of magnitude smaller than for NO, making the latter the main physiological activator [9]. In fact, regardless of the several physiological effects of endogenous CO, which are superficially similar to those of NO, the physiological roles of endogenous CO in diseases are relatively less evident; on the other hand, the therapeutic potential of several pharmacological CO donors is centered on such functional effects [9]. Some molecules, such as methylene blue, LY83583, and ODQ, may interact with the sGC heme group and inhibit the enzyme. The mechanisms whereby cGMP induces vasodilation are more complex than previously anticipated. Cyclic cGMP activates a specific kinase designated GK (also called PKG, cGMP-dependent protein kinase) [8]. Kinases are proteins that phosphorylate a given protein substrate, inducing its activation or inhibition. There are several substrates phosphorylated by GKs, including, among other proteins: VASP (vasodilator-stimulated phosphoprotein, which has cell adhesion signaling functions via integrins), phosphodiesterase V, phospholamban (increases the reuptake of calcium by smooth muscle cell reticulum), some ion channels, and heat shock protein 20 [8,10].

The phosphorylation of these substrates, especially VASP, is a useful marker of sGC activation in biological systems. Once cGMP rises intracellularly, its effects are autoregulated in the cell environment by at least two mechanisms: (a) effect of phosphatases, which have an unclear role in the case of NO; and (b) phosphodiesterases, which degrade cGMP, some in specific fashion, and others in nonspecific fashion (i.e., also degrade other nucleotides such as cAMP) [10].

Reaction Between NO and Lipid Peroxides: The NO Antioxidant Effect

The addition reaction between NO and lipid peroxides is one of the most kinetically favorable ones, with a high rate constant ($\sim 10^{10}$ M^{-1} s^{-1}) [1]. Therefore, NO is able to interrupt the propagation of enzymatic and nonenzymatic lipid peroxidation, and in this sense it is one of the most effective antioxidants [11]. The by-products of this reaction are nitrosated lipoperoxides LONOs and LOONOs, which constitute not only target molecules whose measurement reflects the endogenous production of NO, but also NO donors of potential pharmacotherapeutic importance [11]. In fact, nitrolipids are agents with significant therapeutic potential against atherosclerosis and thrombosis [12]. NO may form adducts with some peroxides formed enzymatically in the metabolism of arachidonic acid by cyclooxygenase or lipoxygenase. Due to the high rate of this reaction, the activity of these enzymes

represents a functional NO scavenger system, which may modulate its physiological effects. For example, peroxides formed by cyclooxygenase-1 (PGHS-1) of platelets may accelerate the catalytic consumption of NO of endothelial origin and favor platelet activation [6,11].

Thiol Nitrosation: A New and Important Signaling Effector Mediated by NO

Some studies suggest that nitrosation (or nitrosylation, mentioned above) of thiols (SH groups, generally of the amino acid cysteine) plays a key role in posttranslational modification of proteins associated with NO-dependent cell signaling [6]. In fact, nitrosation is analogous to phosphorylation, the classic posttranslational modification, in the sense that both are reversible and specific, allowing the cell to change protein function in response to environmental signals.

The nitrosation reaction consists of: RSH (protein with thiol group) + NO ⇌ RSNO (protein with nitrosothiol, also called nitrosoprotein). The same reaction may occur with low-molecular-weight-thiols, for example, glutathione [6]. In this case, as well as with certain proteins such as albumin, nitrosation represents a storage mechanism of NO, which can be released remotely or locally from this stock.

Some proteins can also be nitrosated by transnitrosation reactions, which consist of direct transference of thiol groups from another nitrosated protein or low-molecular-weight-nitrosothiols. Some examples of cell protein targets modulated by nitrosation include calcium channels, calcium-activated potassium channels, NMDA receptors in the brain, thioredoxin, and especially the protein *Ras*, which can control cell proliferation.

NO-Superoxide Interaction: The Chemistry of Peroxynitrite and Related Species

One of the main and fastest reactions involving NO occurs with superoxide radicals. This reaction occurs at a rate constant of 10^{10} M^{-1} s^{-1}; the favorable kinetics of this reaction renders its occurrence limited by diffusion capacity of the reactants in each specific biological system [13,14]. The NO-superoxide reaction deprives the system from NO bioactivity and is considered one of the main regulatory mechanisms of its action on aerobic biological systems. It is debatable, however, to what extent the NO-superoxide interaction is involved in the physiological control of NO action or primarily represents a mechanism of pathological phenomena. This discussion is in the scope of the controversial biology of one of the main by-products of this reaction, the peroxynitrite anion, according to the following equation:

$$NO^\bullet + O_2^{-\bullet} \rightarrow ONOO^- \text{ (peroxynitrite)}$$
$$\rightarrow \text{reactive radical and nonradical intermediates}$$

The complex chemical pathways that involve the reactivity and decay of peroxynitrite were revised by Radi et al. [14] and Augusto et al. [5]. There are several factors that determine the decomposition pathways of this intermediate in vivo, including the chemical nature, reaction constant, reactant concentration, compartmentalization, specific removal or separation, and target-protein biophysical properties. These factors behave as specific variables for each physiological situation, and even for each particular microenvironment.

In biologically relevant systems, only a small fraction of peroxynitrite undergoes spontaneous decomposition, since, in most cases, peroxynitrite is directly consumed by reactions with a variety of molecules such as glutathione, cysteine, ascorbate, and bicarbonate [5,14], as well as thiol-proteins.

Therefore, it is likely that pathways other than decay to nitrate are preferential routes of peroxynitrite decomposition. Data from our laboratory, after arterial injury in rabbits, support this notion. While NO, a highly diffusible gaseous free radical and permeable to membranes, has a biological half-life around seconds, the decay of the $O_2^{\bullet-}$ radical is around milliseconds, and it is able to cross membranes only through anion channels. Thus, the highest probability of peroxynitrite formation is close to the source of superoxide radical generation [15,16]. Once formed, peroxynitrite may diffuse within ~5–20 μm of the formation site [14].

Peroxynitrite may exert various harmful effects on cells, including oxidative stress secondary to the generation of nitrogen dioxide and hydroxyl radicals.

Thus, peroxynitrite represents, above all, an oxidative stress effector. In particular, however, peroxynitrite or closely related intermediates are associated with the nitration of tyrosine residues in various proteins [14,15]. The situation in which the generation of such intermediates leads to protein nitration is called nitrative stress (sometimes also

referred to as nitrosative stress, although in reality nitrative stress is in fact a subtype of oxidative stress). Tyrosine nitration may lead to the loss or modification of protein function and thus appears to represent an important route of pathological protein modification of underinflammatory conditions associated with increased expression/activity of NOSs, particularly iNOS, conditions in which there are high flows of NO [7,15,17]. Some examples of proteins which are nitrated under pathological conditions include, among many, superoxide dismutase, albumin, and actin. However, although a potential marker of peroxynitrite formation, it is now well accepted that several other pathways can form nitrotyrosine, although peroxynitrite seems to be its most likely source in vivo, in situations of simultaneous equimolar flows of the two radicals.

Another pathway that may also contribute to tyrosine nitration involves the myeloperoxidase enzyme, which is secreted by neutrophil phagocytes in inflammatory conditions and may generate tyrosyl radicals that can react with NO, leading to the formation of nitrotyrosine in the presence of additional oxidants such as H_2O_2 [18]. In the presence of H_2O_2, myeloperoxidase may cause nitration by oxidation of NO_2^- to $^\bullet NO_2$ [15,16].

Finally, myeloperoxidase catalyzes the formation of hypochlorous acid (HOCl), which reacts nonenzymatically with NO_2, forming the potent nitrating agent nitrile chloride [16]. The generation of $^\bullet NO_2$ formed by autoxidation of $NO^\bullet$ (i.e., reaction between NO and oxygen) could influence nitration in biological systems, particularly considering that the reaction of NO with O_2 is accelerated approximately 300 times in the lipid phase of biological membranes [19]. Nitrotyrosine formation is not the only and probably not even the major pathway of peroxynitrite reaction, which is further suggested by the simultaneous oxidation of other residues, such as cysteine, methionine, and tryptophan, concomitant with tyrosine [15,16]. Altogether, these considerations indicate that the interpretation of tyrosine nitration indexes must be careful and take into account several possibilities dictated by the chemical reactivity of the intermediates coming from the NO-superoxide interaction.

Nitrate and Nitrite: NO Storage Forms

For some time, it was thought that nitrite (NO_2^-) and nitrate (NO_3^-), which are products of NO oxidation by molecular oxygen or hemeproteins, were inactive forms of NO. In fact, the vasorelaxing activity of nitrite, for example, is very low in water medium. However, in the presence of hemoglobin, which has nitrite reductase activity, nitrite is reduced to NO and shows substantial vasodilator activity [4]. Many evidences indicate that this mechanism is physiologically and clinically relevant. For example, endogenous nitrite generates vasodilation and inhibits platelet aggregation in the presence of the reductase activity of intraerythrocytic hemoglobin [4,20]. In addition, plasma nitrite can be a remote preconditioning mediator. In this case, NO derived from eNOS stimulation by shear forces circulates as nitrite and can be reduced to NO by cardiomyocyte myoglobin, promoting cardioprotection [21] due to better mitochondrial function. Cytoglobin, an intracellular protein similar to hemoglobin, may have a similar effect of reducing nitrite to NO [18]. These data indicate the physiological and potential pharmacological relevance of nitrite.

Nitric Oxide Synthesis Mechanisms

The integration of redox signaling pathways involves not only the reactive intermediates themselves, but also the enzymatic sources that produce them. This concept is entirely applicable to NO, which is produced by enzymes called nitric oxide synthases (NOS), which have different isoforms. Understanding the regulatory mechanisms and NOS activity profiles is essential to assess the possible pathophysiological role of NO, in particular the multiplicity and some apparently contradictory aspects of its effects. The existing contrasts among the different isoforms of NOS strongly affect the biological function of NO and its pathophysiological implications.

NOS has three well-characterized isoforms, encoded by different genes, described in Table 8.1: the isoform originally identified in the brain (neuronal NOS or nNOS), macrophages (inducible NOS or iNOS), and endothelial cells (eNOS) [22,23].

Studies with cloning by homology have not identified, and therefore make it unlikely, that there are other NOS with molecular similarity to these isoforms. These isoforms share 50%–60% homology in their amino acid sequence, determined by its genes, called NOS1, NOS2, and NOS3, respectively [22,23]. Each side of the NOS molecule has two functionally complementary portions. The carboxy-terminal portion contains a reductase domain (i.e., which abstracts electrons from NADPH) equivalent to cytochrome P450,

TABLE 8.1 Isoforms of Mammalian NO Synthases (NOS)

Characteristic	nNOS (NOS I)	eNOS (NOS III)	iNOS (NOS II)
Molecular weight (kDa)	160	135	125–30
Inducibility	Constitutive/ inducible	Constitutive	Inducible
Binding to calmodulin (M)	$\sim 3.0 \times 10^9$	$\sim 3.0 \times 10^9$	$> 3.0 \times 10^9$
Posttranslational modification	Specific phosphorylation sites present	Myristoylation, palmitoylation, sites Specific phosphorylation	Specific phosphorylation sites present
$O_2^{\bullet-}$ formation sources	Heme and reductase domain	Mainly heme domain	Mainly reductase domain
Protein-protein interaction	PSD-95, caveolin 3	Caveolin 1, HSP90, bradykinin receptor	
Major physiological function	Neuronal transmission	Vasodilation	Hot defenses, cytotoxicity
Role in diseases	Stroke	Endothelial dysfunction, hypercholesterolemia, hypertension, vascular remodeling	Sepsis, inflammation, autoimmune disease

HSP, heat shock protein; *AVC*, stroke.

while the amino-terminal portion includes the oxidase domain (i.e., which abstracts an electron from the L-arginine substrate), which has heme binding sites for the tetrahydrobiopterin (BH4) cofactor and L-arginine substrate. Both portions are united by a calcium-calmodulin binding site essential for the functional coupling of electron transfer from flavin groups to heme [22,23]. Therefore, NO production occurs after five steps involving electron transfer, in the following order: NADPH—FAD—FMN—calmodulin—heme/oxygen—L-arginine/NO. The transference of electrons to heme iron induces its activation, with consequent binding to molecular oxygen, and this complex catalyzes the oxidation of nitrogen of L-arginine guanidine terminal, resulting in NO synthesis and L-citrulline by-product, as shown in Fig. 8.1 [22,24,25]. One of the most relevant structural aspects of all NO synthases, shown clearly in recent years, is their dimerization (Fig. 8.1). This process is important to optimize the enzyme electron flow, using the reductase domain of one pair's component and the oxidase domain of the other pair's components, allowing appropriate NO synthesis. Interventions that disturb dimerization cause NOS uncoupling (see below), and reduce the production of bioactive NO [24,25].

Mechanisms Regulating eNOS Activation Component

The mechanisms of eNOS activation revealed to be quite elaborate, maybe reflecting the complexity of physiological control of distinct vascular beds [2,5]. The most well-known and classic eNOS activation mechanism is increased cytosolic concentration of calcium. This binds to calmodulin, which executes electron transport, apparently the only example known in which calmodulin exerts such an electron transport role (Fig. 8.2). Such cytosolic calcium dependent-pathways are responsible for eNOS activation after exposure to acetylcholine and, in good part, to bradykinin [22]. On the other hand, the tonic or phasic activation of eNOS in response to blood flow changes, the main physiological NO release mechanism, occurs in the absence of cytosolic calcium concentration changes. In this case, eNOS is activated by a mechanism involving phosphorylation of serine amino acid in position 1177. The enzyme responsible for this phosphorylation is Akt kinase (or protein kinase B) [26,27], which in turn is phosphorylated by PI-3 lipid kinase (phosphatidylinositol 3-kinase). This phosphorylation significantly increases eNOS sensitivity to baseline levels of calcium/calmodulin, leading to an apparent independence from cytosolic calcium (which is not an absolute truth, but instead an independence from changes in this cation levels) [22,23]. eNOS phosphorylation via Akt has been documented in several other conditions, for example, production of NO in the initial stages of portal hypertension, penile erection, and, in particular, migration of endothelial cells in VEGF-stimulated angiogenesis (vascular endothelial growth factor). The latter effect

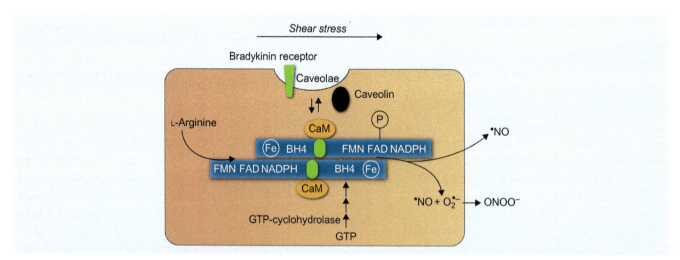

FIG. 8.2   Structure of endothelial NOS in its active form (dimer) The FMN, FAD, and NADPH components form the reductase domain, while the BH4 and heme iron binding sites form the oxidase domain. *BH4*, tetrahydrobiopterin; *CaM*, calmodulin; *FMN*, flavin mononucleotide; *FAD*, flavin adenine dinucleotide; *GTP*, guanosine trisphosphate. Modified from Andrew PJ, Mayer B. Enzymatic function of nitric oxide synthases. Cardiovasc Res 1999;43:521–31.

grants Akt with a central role in angiogenesis, considering, moreover, that this kinase is an important endothelial cell survival pathway [26,27]. In fact, phosphorylation via Akt is also responsible for certain antiapoptotic effects of eNOS, such as in ischemia/reperfusion [27]. A third eNOS activation mechanism, apparently involved in vascular cell proliferation and maturation, is the sphingolipid pathway. Sphingosine-1-phosphate activates eNOS at physiological concentrations, possibly by interacting with Akt kinase [22,23].

The eNOS activation pathways herein described occur in the context of a complex intracellular traffic of this protein and its protein-protein interactions. The baseline location of nonactivated eNOS is still uncertain, and may include the Golgi and post-Golgi apparatus, as well as an association with cytoskeletal proteins [2]. In particular, the entire eNOS activation process is apparently located in caveolae [22,23]. These are invaginations of the plasma membrane rich in sphingolipids and cholesterol, and poor in phospholipids, which form and disappear dynamically depending on the caveolin protein, and may occupy up to 30% of the surface of capillary endothelial cells. Caveolae perform the function of an integrative cell signaling microenvironment, congregating several receptors, kinases, and enzymes [23]. The traffic of eNOS to caveolae depends on the enzymatic addition of fatty acid residues. Enrichment of eNOS with a myristate molecule, added during transduction, and two palmitate molecules, added after transduction, anchor eNOS in caveolae, in apparent physical interaction with the scaffolding domain of caveolin-1, which inhibits eNOS [2,5]. This eNOS inhibiting conformation is attenuated and reversed by elevating the concentration of calcium/calmodulin and/or by Akt mediated phosphorylation. Therefore, the combination between loss of interaction with caveolin-1 and direct effect of calmodulin/calcium induces eNOS enzymatic activity. Mice with caveolin-1 gene deletion show increased activity of eNOS, which stays free at the endothelial cell cytoplasm [28]. Similarly, mice with specific caveolin mutations are able to form caveolae, but show increased NO bioactivity, another evidence for the direct inhibitory effect of caveolin on eNOS [29].

Due to the specific lipid composition of caveolae, it has been proposed that the known association between hypercholesterolemia and endothelium-dependent vasorelaxation dysfunction may involve physiological alterations in these structures. Exposure of cells to cholesterol per se increases the number of caveolae, as well as eNOS expression, its association with caveolae and NO production [22,30]. On the other hand, exposure to oxidative stress or oxidized lipoproteins reduces the number of caveolae, as well as their eNOS association, leading to lower NO production. Although complex, the relative importance of these mechanisms in the context of endothelial dysfunction is more and more evident [22,29,30].

iNOS and nNOS Regulatory Mechanisms

In comparison with eNOS, mechanisms that regulate iNOS and nNOS activation, in particular in the cardiovascular system, are less evident. As mentioned, iNOS activation does not depend on elevated concentrations of intracellular calcium due to the high binding affinity between the enzyme and calmodulin. Once induced, the enzyme seems to be located in the cytosol and is continuously activated, producing a high flow of NO according to the specific cofactors discussed below [23].

The intracellular location of nNOS is different from other isoforms and seems to be the endoplasmic reticulum or unidentified specific organelles; there is an isolated observation of association between nNOS and caveolae. The activation of the protein, similarly to the other constitutive eNOS isoform, is closely connected to cytosolic calcium elevation [22,23].

A particularly interesting mechanism influencing NOS activity is the allosteric inhibition by NO itself, able to bind to the enzyme heme group and inhibit electron transport [31]. Inhibition due to NO seems particularly important for nNOS, at a lower level for eNOS, and is characteristically absent in iNOS [7,31].

Control of NOS Gene Expression

Although the eNOS and nNOS isoforms are defined as constitutive, differently from the canonically inducible isoform, the expression of both may be induced by several factors. In vessels, under nonstimulated basal condition, eNOS expression is strong in the vascular endothelium and insignificant in other layers [22,32]. There is a minor nNOS expression in the adventitial layer, and a small or absent expression of iNOS in medial smooth muscle cells, while iNOS is present in eventual adventitial phagocytes. Shear stress, estrogens, angiogenic factors such as VEGF and endothelial regeneration, are associated with increased eNOS RNA and/or protein expression [22,33]. Experimental observations from our laboratory indicate that nNOS expression is induced in the vascular system during the vascular injury repair process [17], similarly to the process observed by other groups in the initial stages of atherosclerosis [32]. The pathways governing such expression are not entirely clear.

iNOS expression can be induced in practically any cell type present in the cardiovascular system by cytokines (such as tumor necrosis factor alpha and interleukin-1), other inflammatory stimuli, bacterial lipopolysaccharide, and oxidative stress (e.g., oxidized lipoproteins) [34]. This induction is strongly dependent on the activation of transcription factor NF-κB.

Mitochondrial NO Synthase

The inhibitory action of NO on mitochondrial respiration has been documented in ex vivo models. NO inhibits mitochondrial respiration by binding to the heme group of cytochrome oxidase, and possibly by nitrosation of complex-1 dehydrogenase thiols [35]. The presence of NO synthase in mitochondria has also been documented [35], although its functional role and genetic characterization are controversial. In other words, regardless of the physical existence of a protein able to produce NO, the in vivo physiological role of NO in mitochondrial respiration is not entirely clear, as well as the importance of NO synthase in mitochondria. It is also unclear if this NO synthase represents a specific particular modification of one of the known NOS isoforms [35].

Cardiovascular Effects of the Different NOS Isoforms: Studies in Genetically Modified Models

Studies in genetically modified models represent the most powerful tool and most important reference to understand the global physiological effects of a protein. Studies with mice in which the genes of specific NO isoforms were modified or knock-in mice, in which a permanent mutation of the enzyme is produced, clearly indicate an association between eNOS and endothelial dysfunction. eNOS knockout mice (i.e., without the functioning gene) are hypertensive, show endothelial dysfunction,

and have exacerbated response to vascular injury [36,37]. In response to hyperlipemic diets, genetic deletion of eNOS exacerbates atherosclerosis, indicating an eNOS protective role [36,37]. Intriguingly, induced eNOS gene overexpression exacerbates atherosclerosis, showing a double face of NO derived from eNOS [38]. Possibly, the excess enzyme may generate cofactor insufficiency, and consequently its functional uncoupling, with oxidant production (discussed ahead). Different eNOS mutations, for example, in residues associated with phosphorylation activation, produce intermediate endothelial dysfunction levels, overall in line with the effects of total deletion of eNOS [36]. Mice knockout for nNOS gene have less cell injury in response to cerebral ischemia, but on the other hand develop higher extension of diet-induced atherosclerosis, emphasizing that this isoform also shows cardiovascular effects [37]. Mice knockout for the iNOS gene show lower hypotension in response to septic shock. Since knockout of a specific isoform of NO synthase can promote compensatory effects by other isoforms, a knockout model was developed for all three isoforms simultaneously: eNOS, nNOS, and iNOS. This triple knockout unequivocally confirms that NO exerts cardiovascular protection, since its phenotype involves exacerbated cardiovascular manifestations in comparison with knockout of a single isoform, including atherosclerosis, dyslipidemia, and myocardial infarction [39].

Cofactors of NO Synthase: NO Synthase Uncoupling

Several cofactors are required for appropriate electron transfer during enzyme-mediated NO synthesis, including availability of heme groups, reduced thiols (SH groups), flavin mononucleotides and dinucleotides, and especially the folic acid derivative BH4, besides the substrate L-arginine [1,2,22,23]. Of these, the two last cofactors have been particularly studied and may significantly affect enzyme function, as discussed below. In addition to these, the connection between NOS and heat shock protein 90 (hsp90) has been recently studied as a facilitator of NO synthesis [22,40].

The dependence on the described cofactors occurs for all NOS isoforms. However, it is particularly important in situations of high enzyme activity output, such as normal iNOS activity or in situations of

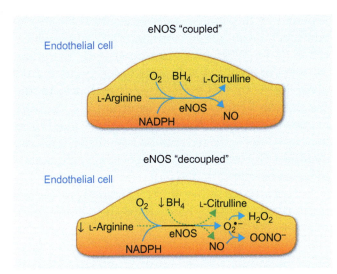

FIG. 8.3 Schematic representation of the "uncoupling" of eNOS. Suboptimal concentrations of L-arginine and/or tetrahydrobiopterin (BH4) can lead to incomplete transfer of electrons and leakage of such electrons into oxygen (the "universal" electron acceptor of the cell), leading to superoxide radical generation. $O_2^{\bullet-}$, superoxide radical; $ONOO^-$, peronitrite; H_2O_2, hydrogen peroxide. Modified from Hamilton SJ, Watts GF. Endothelial dysfunction in diabetes: pathogenesis, significance and treatment. Rev Diabet Stud 2013;10:133–56.

overstimulated or overexpressed eNOS or nNOS. In these situations, relative deficiency of cofactors induces the NOS uncoupling phenomenon [41,42].

NOS uncoupling is the situation in which the transfer of electrons from NOS is not completed appropriately. These electrons "leak" during enzyme activity and are captured by molecular oxygen, which is by excellence the acceptor of electrons in aerobic organisms, generating superoxide radical (i.e., $O_2 + electron \rightarrow O_2^{\bullet-}$) (Fig. 8.3). In other words, uncoupled NOS works as an NADPH oxidase, because the balance of its activity finally transfers an electron from NADPH to molecular oxygen [41]. Therefore, uncoupled NOS not only deprives the cell environment of NO synthesis, but also produces the main NO antagonist, superoxide radical, promoting an ideal situation to generate the toxic peroxynitrite intermediate [41,42]. NOS uncoupling is an important (although not the only one) mechanism involved in the so-called "NO double face," which indicates the possibility that NO has both beneficial and toxic effects on biological systems. In case of toxic effects, it is important to notice that, in general, the toxicity effector is not NO, but peroxynitrite and/or superoxide generated by the enzyme. eNOS uncoupling seems to constitute an important endothelial dysfunction mechanism, particularly in the initial stages of atherosclerosis, *diabetes mellitus*, arterial hypertension, and nitrate tolerance [23,42]. For example, native or oxidized LDL is capable of inducing eNOS uncoupling [42]. As commented in the previous section, a paradoxical effect studied recently is the increased susceptibility to atherosclerosis in mice with eNOS overexpression [38]. This effect seems to be explained by the disproportion between eNOS protein expression and activity of enzyme cofactors (which are not increased in parallel), possibly leading to eNOS uncoupling. In other words, simply increasing eNOS expression is not sufficient to ensure that the final enzyme product is bioactive NO. Therefore, it is interesting to understand in better detail the pathophysiology of some NOS cofactors.

L-Arginine is a normal constituent amino acid of proteins and has, in its free form, which is the NOS substrate, concentrations in the order of 100–1000 mM, well above the K_m of NOS, which is around 5 mM (K_m is the concentration of substrate that promotes 50% of the maximum activity of an enzyme). Despite this excess of substrate, the enzymatic activity is strongly influenced by the exogenous administration of L-arginine. A classic example of this phenomenon is the reversal by L-arginine of the competitive antagonism of NOS exerted by N-methyl-L-arginine (L-NMMA) or nitro-L-arginine (L-NA). The mechanisms of this apparent paradox are not currently known, but may involve specific processes associated with the subcompartimentalization or uptake of L-arginine [42,43]. Some conditions may provide a deficiency in the availability of L-arginine and, therefore, a greater chance of NOS uncoupling: (a) competition with other amino acids, for example, glutamine; (b) inhibition of intracellular L-arginine, which may occur, for example, by oxidized LDL; (c) changes in activity of enzymes that synthesize L-arginine (L-arginine synthase), still insufficiently investigated; and (d) increased activity of enzymes that degrade L-arginine, such as arginase, which in excess may deplete the specific L-arginine pool. A peculiar situation is the reduction in NOS function caused by changes in L-arginine induced by pathological conditions [43]. Asymmetric dimethylarginine (ADMA) is the best-known example and may lead to reduced eNOS function not only by constituting a false enzyme substrate, but also by inducing superoxide radical production [26]. ADMA is formed enzymatically by L-arginine methyltransferases, which use S-adenosylmethionine as a methyl group donor. ADMA formation is favored in oxidative stress situations. Increased ADMA concentrations have been described in apparent correlation with endothelial dysfunction in patients under chronic hemodialysis, aging, *diabetes mellitus*, and atherosclerosis risk factors [44,45]. Oxidized LDL promotes lower removal of plasma ADMA [44]. Finally, the significant protective effects of acute or chronic exogenous administration of L-arginine in experimental or clinical atherosclerosis studies [23,43] suggest that deficiency of this substrate (by means of mechanisms described above) is a pathogenic factor of endothelial dysfunction in this condition.

BH4 is an essential cofactor for the catalytic activity of the three NOS isoforms, and is a decisive factor for NO production in physiological and pathological environments [42,45]. This cofactor is a remote derivative of the folic acid synthesized by GTP cyclohydrolase I, and has profound effects on NOS structure, including the ability to shift heme iron to a higher spin, favoring L-arginine binding, and stabilizing the enzyme active dimeric form. The proximity between BH4 and heme, as well as flavins when the enzyme is in its active form (dimer), is possibly important in electron transfer favored by BH4 [45].

The folic acid cycle and GTP cyclohydrolase I activity are closely related to NOS activity. Gene transfer of this enzyme increases the concentration of BH4 and production of NO in endothelial cells [45]. Exogenous administration of BH4 or its metabolic precursors minimizes endothelial dysfunction in experimental or clinical hypercholesterolemia/atherosclerosis studies, indicating the importance of this cofactor [42]. Experimental studies also show that the lower NO synthesis in diabetic mice is correlated with BH4 deficiency [46]. BH4 deficiency may be induced by higher oxidative stress (e.g., oxidized LDL), particularly via the effects of peroxynitrite (the main by-product of the reaction between NO and superoxide radical) [47]. Peroxynitrite oxidizes BH4 into its inactive by-product dihydrobiopterin (BH2), so that the BH2/BH4 ratio is a relevant indicator of bioactive NO production by eNOS. Certain antioxidants, particularly vitamin C, are able to stabilize BH4 and increase NO synthesis in endothelial cells, regardless of the interaction with superoxide. In summary, BH4 activity is an important mechanism able to influence NO synthesis by eNOS (and other NOS isoforms) and, therefore, the level of endothelial dysfunction.

hsp90 is an eNOS cofactor that binds to the enzyme and increases NO synthesis significantly in endothelial cells. Recent studies show that blocking hsp90 not only reduces NO synthesis, but also increases NO uncoupling and, consequently, superoxide radical production [40].

Recent studies show other factors that can induce uncoupling of NO synthase. One of them is glutathionylation, that is, formation of a mixed NO synthase and glutathione disulfide, induced by oxidants [46]. Glutathionylation reduces enzyme activity and may be a relevant functional enzyme regulatory mechanism. Another important cofactor may be NADP(H). During postischemic dysfunction, the recovery of endothelial function is not complete after BH4 replacement alone, suggesting additional deficiencies of other cofactors. NADP(H) depletion due to NADPase activity of the CD38 molecule contributes to NOS uncoupling after reperfusion [47].

Polymorphisms and Genetic Variations of eNOS and Atherosclerosis

Some studies have investigated the relationship between genetic polymorphisms of eNOS and occurrence of cardiovascular disease, more specifically atherosclerosis. The association between the replacement of glutamate residue 298 by aspartate in exon 7 and coronary arterial disease was investigated in the population of the CHAOS (Cambridge Heart Association Study) study [48]. In two different series, there was a higher prevalence of angiographic disease in patients homozygous for variant Asp298 [30,48]. In another study, the presence of allele 4 of eNOS was associated with higher prevalence of coronary arterial disease in nonsmokers [30,48]. Although the predictive value of single gene polymorphisms is limited in clinical conditions and must be seen with caution, these data may help composing future predictive indexes based on multiple genes.

Nonenzymatic Generation of NO

In physiological pH, NO is irreversibly and spontaneously oxidized into nitrite and nitrate. However, nitrite reduction may occur in acid pH, generating NO by means of a nonenzymatic pathway [49]. The possible situations in which this process may occur are found in the stomach and cells submitted to prolonged ischemia [49]. The importance of nonenzymatic generation of NO is still controversial.

EDHFs

Many physiological preparations show that even with prostacyclin and NO inhibition, a variable level of endothelium-dependent vasodilation remains in response to agonists such as acetylcholine, bradykinin, substance P, or hemodynamic alterations such as reduced arterial perfusion pressure [50]. The same phenomenon was recently observed in mice with gene silencing of the three NO synthase isoforms [39].

This phenomenon occurs at variable levels, depending on the size and type of vascular bed, species, and relative concentration of other vasodilators. This vasodilation is followed by membrane hyperpolarization (i.e., higher transmembrane potential negativity) in endothelial cells and, subsequently, in smooth muscle cells. This hyperpolarization involves cellular efflux of potassium by means of calcium-activated potassium channels (K_{Ca}) and rectifying potassium channels (K_{IR}). Hyperpolarization of smooth muscle cells reduces the open probability of voltage-dependent calcium channels, leading to decreased cytosolic concentration of calcium and consequent relaxation. Therefore, endothelium-dependent relaxation in vessels treated with cyclooxygenase and NO synthase inhibitors, inhibitable by K_{Ca} channel inhibitors or excessive extracellular potassium, is attributable to one or more factors called EDHFs. Much has been discussed if in fact EDHF would be a soluble mediator released in paracrine fashion by the endothelium. Together, there is considerable support for the idea of an autonomous chemical mediator, independent of NO and prostacyclin, although its identity is not fully understood and it is likely that several distinct factors may play the role of EDHF in specific circumstances. The most likely identities of EDHF are: arachidonic acid metabolites derived from cytochrome P450 [51], hydrogen peroxide [52,53], potassium ions [54], or gap junction [55] communications between smooth muscle and endothelial cells, in addition to H_2S, discussed in a separate section.

Epoximetabolites of Arachidonic Acid

Evidence that the cytochrome P450 enzyme system, which has a broad spectrum of effects, participates in the genesis of EDHF is based on the fact that its inhibitors suppress vasodilation dependent on this mechanism [51]. Cytochrome P450 centered-epoxygenases metabolize arachidonic acid, generating metabolites such as 8,9-, 14,15-, 5,6- or 11,12-epoxyeicosatrienoic acids EETs and their corresponding dihydroxy-EET derivatives. In human endothelial cells, the isoform 2C of cytochrome P450 fulfills criteria to be an EDHF synthase. EETs released by bradykinin-stimulated endothelial cells have a hyperpolarizing effect on smooth muscle cells and kinetic characteristics compatible with those previously suggested for EDHF. EETs promote coronary artery vasorelaxation concomitant to higher probability of K_{Ca} channel opening and reduced cytosolic calcium. The passage of EETs to smooth muscle cells seems to occur predominantly via gap junctions, which are opened due to stimulation from EET itself [55]. It is possible, therefore, that EETs work as important secondary intracellular messengers to initiate and transmit the hyperpolarization spread from endothelial cells. Also, EDHF synthase (cytochrome P450 2C9) has the effect of generating reactive oxygen species [51] in coronary artery endothelial cells. These oxygen species induce the activation of transcription factor NF-κB and increased expression of leukocyte adhesion molecules.

Hydrogen Peroxide

In mice genetically deficient in endothelial NO synthase, acetylcholine-dependent relaxation is inhibited by catalase [52], an enzyme that removes hydrogen peroxide (H_2O_2, or oxygenated water). In addition to effecting vasodilation, H_2O_2 induces hyperpolarization of smooth muscle cells via activation of K_{Ca} channels and has physicochemical characteristics compatible with those expected for EDHF. H_2O_2 is permeable to cell membranes, and is therefore able to diffuse from the endothelium. In addition, H_2O_2 plays a physiological role in controlling the autoregulation of small coronary arteries in response to both acetylcholine and gradual reductions in perfusion pressure [53]. The mechanisms by which hydrogen peroxide induces vasodilation include the formation of a disulfide bond in protein kinase G (GK), capable of activating the enzyme in a way independent of guanylyl cyclase. Mice with a single cysteine mutation of this disulfide have exacerbated vasoconstriction and are hypertensive [56]. This is one of the most important examples of redox signaling and potentially protective effects of an oxidant species.

Potassium Ions and Gap Junctions

The activation of K_{Ca} channels in endothelial cells, for example, by acetylcholine, promotes efflux of potassium into the myoendothelial space. In turn, the potassium ions activate Na-K ATPase of smooth muscle cells and potassium rectifying channels (K_{IR}) of endothelial cells, and relax the smooth muscle by means of inhibitory mechanisms similar to the combination of ouabain and barium ions (inhibitor of the K_{IR} channel). In vessels without endothelium, relaxation due to potassium is blocked by ouabain, but not by barium ions, indicating a synergistic action between endothelial cell hyperpolarization (via K_{IR}) and Na-K ATPase of smooth muscle cells [50,55]. Certain effects of EDHF, in contrast to NO, are also inhibited by ouabain. Therefore, it has been proposed that potassium could be an EDHF potentially able to transit between endothelial cells and smooth muscle cells by means of gap junctions. The importance of these cell junctions in intermediating relaxation via EDHF has been recognized [55]. Just like EDHF, the prevalence of gap junctions is higher in arterioles and does not exist in large-sized arteries. The existence of electric coupling between endothelial cells and smooth muscle cells supports the importance of these communications. The hyperpolarizing response is not conducted in the absence of endothelium, and is abolished in mice deficient for connexin-40 (key protein of the gap junction). Therefore, gap junctions could directly mediate electric coupling between the endothelium and smooth muscle cells or, alternatively, conduct chemical mediators [55].

Given these considerations, it is apparent that only one type of EDHF is not able to respond for all the characteristics of the hyperpolarization phenomenon in different conditions or vascular beds. In addition, the different EDHFs share several elements in common and have complementary activities, as illustrated in Fig. 8.4.

Gap junctions, for example, could participate in the propagation of each of these mediators. The

action of superoxide dismutase may generate hydrogen peroxide from superoxide generated by cytochrome P450 or other sources. H_2O_2, via protein C kinase, stimulates the activation of phospholipase A_2, producing, therefore, arachidonic acid, which participates directly in opening K_{Ca} channels, and is the substrate of cytochrome P450 that generates EETs.

Pathological Dysfunction of Endothelium-Dependent Hyperpolarization

Endothelial dysfunction found in several vascular diseases has been described and studied in connection with the NO-NO synthase system. Growing attention has been given to EDHF dysfunction in these circumstances. In *diabetes mellitus*, several studies show loss or reduced EDHF activity as a component of vascular tone control [50]. In mice with insulin resistance, EETs lose their vascular relaxation and hyperpolarization capacity. Inhibitors of cytochrome P450 lose their inhibitory effect of relaxation to acetylcholine in diabetic mice. In diabetic mice, certain studies show more deficiency of hyperpolarizing responses than NO-dependent responses. Hypertensive animals also have deficient EDHF-dependent relaxation [50], which can be improved by treatment with angiotensin converting enzyme inhibitors. In hypertensive patients, calcium channel inhibitors can improve the lower forearm flow response to acetylcholine infusion. EDHF also takes part in the vascular component of estrogen effects.

H₂S: A New EDRF and EDHF

H_2S, which for a long time was considered only a toxic (and stinky) gas present in the atmosphere and metabolized by bacteria, is the most recent gas molecule described as the second messenger of signaling pathways [57], in addition to NO and CO. Low concentrations of H_2S regulate vascular tone and blood pressure by mechanisms not yet elucidated, as discussed below.

H_2S can be produced in eukaryotic cells by three different enzymes that use the amino acid L-cysteine as substrate: cystathionine γ-lyase (CSE), cystathionine β-synthase (CBS), and mercaptopyruvate S-transferase (MST). Another source of H_2S in mammals are gut microbes [58]. In the endothelium, there is evidence to date of H_2S production mainly

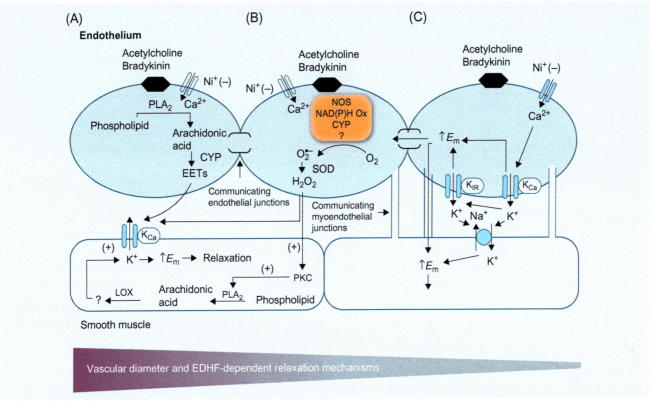

FIG. 8.4 Proposed hyperpolarization and relaxation mechanisms by: (A) Epoxyeicosatrienoic acids (EETs); (B) hydrogen peroxide; or (C) potassium ion. Endothelial stimulation by acetylcholine, bradykinin, or increased flow induces release of EDHFs. The diffusion and action on smooth muscle cells leads to membrane hyperpolarization (increased E_m). Electric propagation also occurs by means of endothelial and myoendothelial gap junctions. The contribution of the several types of EDHFs changes according to vessel size. *CYP*, cytochrome P450; K_{Ca}, calcium-activated potassium channel; K_{IR}, potassium rectifying channel; *LOX*, lipoxygenase; *NAD(P)H Ox*, NAD(P)H oxidase; *NOS*, nitric oxide synthase; *PKC*, protein kinase C; PLA_2, phospholipase A_2; *SOD*, superoxide dismutase. Adapted from Campbell WB, Gauthier KM. What is new in endothelium-derived hyperpolarizing factors? Curr Opin Nephrol Hypertens 2002;11:177–83.

by the enzyme CSE, known to be inducible, similarly to iNOS. The H_2S vasodilator action was initially described with the use of H_2S donors and/or pharmacological inhibitors of CSE and later confirmed with studies with knockout mice for this enzyme (CSE-KO), which show compromised production of H_2S and develop aging-associated hypertension [57].

Deficient production of H_2S has also been implicated in early development of atherosclerosis. CSE-KO mice submitted to proatherogenic diet (Paigen diet), presented, as compared to control mice, early atherosclerotic injury in aorta curvature, with increased intimal layer, higher cholesterol and LDL plasma levels, and increased expression of ICAM-1 and VCAM-1 adhesion molecules in smooth muscle cells [59]. In addition, they show disrupted plasma redox homeostasis (increased levels

of malondialdehyde, decreased GSH concentration, and lower activity of antioxidant enzymes such as glutathione peroxidase, glutathione reductase, and superoxide dismutase), besides increased levels of oxidized LDL in the endothelial subspace. Some of these parameters were reversed by treatment with NaHS, an exogenous H_2S donor [59].

Regardless of the protective effects of H_2S in vivo, the description of H_2S mechanisms of action are still scarce. One of the mechanisms by which H_2S acts as a signaling molecule is by modifying cysteine residues in target proteins, a process called "sulfhydration" or persulfhydration [60]. H_2S is a weak acid in solution ($pK_a = 7$) in equilibrium with its conjugated base (HS^-) and does not react directly with reduced thiols (RSH). Among the possible H_2S action mechanisms in biological systems, the most plausible is the reaction between the conjugated base (HS^-)

and disulfides (RSSR) or sulfenic acid (RSOH), which produces persulfides (RSSH or RS_nH), the product of "sulfhydration" [61].

There is also a discussion in the area that some of the effects of H_2S could be mediated by polysulfides, formed from H_2S oxidation, found in H_2S solutions [62].

Sulfhydration generally increases the reactivity of the modified cysteine, unlike the nitrosation process (formation of RSNO), which generally decreases the reactivity of the target cysteine. Sulfhydration appears to be a highly prevalent modification; for example, 25% of the enzyme glyceraldehyde 3-phosphate dehydrogenase (GAPDH) is sulfhydrated in liver cell lysates. However, in some cases, sulfhydration precedes nitrosation in the same cysteine; the p65 subunit of NF-κB is initially sulfhydrated within the first 2 hours of treatment with TNF-α and then is nitrosated [63].

Vascular tone is regulated by H_2S dependently or independently from the endothelium, by means of sulfhydration of the ATP-dependent potassium channel ATP (K_{ATP}). The K_{ATP} channel is modified in cysteine 43, inhibiting its association to ATP, and binding with phosphatidylinositol 4,5-bisphosphate (PIP2), which maintains it in open format, favoring vasodilation [60,64]. This H_2S can be produced by the smooth muscle cells themselves or by an endothelial source. In addition, the deletion of enzyme CSE specifically in the endothelium of mice decreased, in addition to H_2S production, endothelium-dependent relaxation induced by acetylcholine [57]. These observations suggest that H_2S is, similarly to NO, an endothelium-dependent relaxation factor (EDRF) [65].

Besides EDRF, H_2S has also been considered an EDHF, because its endothelium-dependent vasorelaxation effects are more prominent in resistance arteries than in conductance arteries, and require hyperpolarization of the plasma membrane, endothelial cells, and smooth muscle cells [66]. The H_2S produced in the endothelium activates the potassium-dependent calcium channels (SK_{Ca} and IK_{Ca}), causing endothelial hyperpolarization by electric coupling via myoendothelial gap junctions and/or increased potassium efflux. Calcium channel blockers activated by potassium (K_{Ca}) inhibit hyperpolarization of both cells.

Another form by which H_2S can affect vascular tone is via NO regulation. Treatment of mice with Na_2S (donor of H_2S) increases eNOS expression [67], and in cultured cells it was observed that H_2S increases eNOS activity by inducing posttranslational modifications, such as sulfhydration, phosphorylation, and dimerization [68–70]. In addition, CSE-KO mice showed dysfunctional low levels of NO and dysfunctional eNOS [71]. On the other hand, NO can inhibit CSE activity, via nitrosation, but also increases CSE expression. These opposite effects may reflect fine regulation of H_2S production, depending on NO concentration.

Deficiency in CSE expression/activity or low endogenous levels of H_2S have been discussed as potential risk factors for the development of atherosclerosis. In fact, hypertensive patients (degrees 2 and 3) and diabetics with hypertension have lower H_2S plasma levels than healthy individuals [66,72]. However, the methods for obtaining reliable measurements of H_2S in biological samples are still under development, and there is not yet a consensus about the physiological levels of H_2S in the blood. From a therapeutic standpoint, a good strategy should be the modulation of pathways that produce endogenous H_2S, but there are still many open related issues, for example, how CSE is activated in endothelial cells. Furthermore, the development of specific inhibitors/activators of CSE (and other endogenous potential sources of H_2S in the endothelium), controlled donors of H_2S, as well as specific methods to monitor circulating H_2S levels, can be essential tools for future antiatherosclerosis therapeutic use.

Interrelation Between Endothelium-Dependent Vasodilation Mediators

NO is able to inhibit EDHF synthase activity by binding to the heme group. Therefore, the endothelial dysfunction states in which NO bioactivity is reduced lead to higher activation of cytochrome P450 and increased EDHF-dependent vasodilator mechanism. With inhibited NO production or in eNOS genetically deficient mice, EDHF assumes higher importance as paracrine vasodilation mechanism. EDHF may maintain myocardial perfusion upon loss of NO in cardiovascular diseases. These data indicate a functional connection between relaxing and hyperpolarizing factors from the endothelium. However, the physiological role of each system is different, according to the species and vascular size. In particular, in conductance arteries (e.g., aorta, epicardial arteries), NO is hierarchically the most important mediator of endothelium-dependent vasodilation. On the other hand, in small arteries and arterioles < 100 mm, (resistance vessels), the predominant endothelium-mediated vasodilation mechanism is EDHF. EDHF activity is important as an in vivo coronary auto regulatory mechanism, in cooperation with NO and adenosine.

Recent data suggest interesting mechanisms responsible for hierarchical effects of EDRFs (NO) versus EDHF(s) related to a structure that, although known for a long time, had not been studied in detail: the myoendothelial junction [73]. These structures consist of an interface between specific portions of the endothelial cell and the smooth muscle cell, with a relatively fixed space and reduced cellular intermembranes, constituting a true synapse. Myoendothelial junctions are much more abundant in arterioles of resistance than conductance arteries. The endothelial cell in this region shows a redistribution and accumulation of rough endoplasmic reticulum and ribosomes, as well as enrichment in eNOS [73]. In this region, there is also an intraendothelial accumulation of the alpha chain of hemoglobin, bound to Fe^{3+}, and it is associated with cytochrome b5R3, which promotes reduction of Fe^{3+} to Fe^{2+}, capable of sequestering and inactivating NO. Silencing of cytochrome b5R3 promotes increased NO bioavailability. This mechanism may thus help clarify why resistance arterioles have a lower relative contribution of EDRF (NO) vs. other EDHFs [73].

CONCLUSIONS

Paracrine vasodilation mediated by the endothelium is a primordial function of these cells, which integrates with its other multiple functions. These vasodilation mechanisms represent an orchestrated system, which executes first-order regulation of conductance artery vasomotor tone, as well as resistance vessel flow modulation, and distributive adjustments between different organ layers. These effects play essential physiological roles in the genesis and adaptation of vascular pathological conditions, constituting, in these cases, primary therapeutic intervention targets.

A considerable part of the NO biology complexity can be attributed to its different chemical reactivity pathways. Their understanding generated important implications, some of them of therapeutic interest, such as, for example, vasodilation and cell protection effects of nitrite in the presence of heme reductase. In parallel, NO biology depends strongly on the different NOS isoforms. Although the exact role of the cofactors, such as BH4 and L-arginine and their regulation in diseases, is not completely understood, it may generate relevant clinical implications. In the same fashion, understanding the role of superoxide radical produced by NOS uncoupling in cell signaling processes must provide relevant advances in pathophysiology and treatment of several vascular conditions. Finally, growing evidence shows that the structural differences between NOS isoforms allow these enzymes to play distinct roles in different compartments. These structural differences must be explored regarding the possibility of gene therapy that selectively modify a specific isoform. Recent advances regarding an improved understanding of vascular relaxation mediators associated with hyperpolarization must significantly increase the scope of endothelial functions and provide global understanding of these processes, helping clarify how NO integrates with hydrogen peroxide, arachidonic acid derivatives, and other EDHFs, including new gaseous mediators such as CO and H_2S.

Acknowledgments

Studies from our laboratory described in the present chapter were supported by grants from Fapesp, CNPq, Finep/Pronex e Fundação EJ Zerbini. FRML and DCF are members of Cepid de Processos Redox em Biomedicina ("Redoxoma," Fapesp nº 13/07937-8) and Instituto Nacional de Ciência e Tecnologia de Processos Redox em Biomedicina (Fapesp/CNPq).

References

[1] Harrison DG. Cellular and molecular mechanisms of endothelial cell dysfunction. J Clin Invest 1997;100:2153–7.

[2] Moncada S, Higgs EA. Nitric oxide and the vascular endothelium. Handb Exp Pharmacol 2006;176(Pt 1):213–54.

[3] Kojda G, Harrison D. Interactions between NO and reactive oxygen species: pathophysiological importance in atherosclerosis, hypertension, diabetes and heart failure. Cardiovasc Res 1999;43:562–71.

[4] Kim-Shapiro DB, Gladwin MT. Mechanisms of nitrite bioactivation. Nitric Oxide 2014;38:58–68.

[5] Augusto O, Bonini MG, Amanso AM, et al. Nitrogen dioxide and carbonate radical anion: two emerging radicals in biology. Free Radic Biol Med 2002;32:841–59.

[6] Wink DA, Mitchell JB. Chemical biology of nitric oxide: insights into regulatory, cytotoxic, and cytoprotective mechanisms of nitric oxide. Free Radic Biol Med 1998;25:434–56.

[7] Espey MG, Miranda KM, Thomas DD, et al. A chemical perspective on the interplay between NO, reactive oxygen species, and reactive nitrogen oxide species. Ann N Y Acad Sci 2002;962:195–206.

[8] Lucas KA, Pitari GM, Kazerounian S, et al. Guanylyl cyclases and signaling by cyclic GMP. Pharmacol Rev 2000;52:375–414.

[9] Andreadou I, Iliodromitis EK, Rassaf T, et al. The role of gasotransmitters NO, H₂S and CO in myocardial ischaemia/reperfusion injury and cardioprotection by preconditioning, postconditioning and remote conditioning. Br J Pharmacol 2015;172:1587–606.

[10] Koesling D, Friebe A. Soluble guanylyl cyclase: structure and regulation. Rev Physiol Biochem Pharmacol 1999;135:41–65.

[11] Freeman BA, O'Donnell VB. Interactions between nitric oxide and lipid oxidation pathways: implications for vascular disease. Circ Res 2001;88:12–21.

[12] Trostchansky A, Bonilla L, González-Perilli L, et al. Nitro-fatty acids: formation, redox signaling, and therapeutic potential. Antioxid Redox Signal 2013;19:1257–65.

[13] Fridovich I. Fundamental aspects of reactive oxygen species, or what's the matter with oxygen? Ann N Y Acad Sci 1999;893:13–8.

[14] Radi R, Peluffo G, Alvarez MN, et al. Unraveling peroxynitrite formation in biological systems. Free Radic Biol Med 2001;30:463–88.

[15] Ischiropoulos H. Biological tyrosine nitration: a pathophysiological function of nitric oxide and reactive oxygen species. Arch Biochem Biophys 1998;356:1–11.

[16] Brennan ML, Wu W, Fu X, et al. A tale of two controversies: defining both the role of peroxidases in nitrotyrosine formation in vivo using eosinophil peroxidase and myeloperoxidase-deficient mice, and the nature of peroxidase-generated reactive nitrogen species. J Biol Chem 2002;277:17415–27.

[17] Leite PF, Danilovic A, Moriel P, et al. Sustained decrease in superoxide dismutase activity underlies constrictive remodeling after balloon injury in rabbits. Arterioscler Thromb Vasc Biol 2003;23:2197–202.

[18] Halligan KE, Jourd'heuil FL, Jourd'heuil D. Cytoglobin is expressed in the vasculature and regulates cell respiration and proliferation via nitric oxide dioxygenation. J Biol Chem 2009;284: 8539–47.

[19] Liu X, Miller MJ, Joshi MS, et al. Accelerated reaction of nitric oxide with O₂ within the hydrophobic interior of biological membranes. Proc Natl Acad Sci U S A 1998;95:2175–9.

[20] Apostoli GL, Solomon A, Smallwood MJ, et al. Role of inorganic nitrate and nitrite in driving nitric oxide-cGMP-mediated inhibition of platelet aggregation in vitro and in vivo. J Thromb Haemost 2014;12:1880–9.

[21] Rassaf T, Totzeck M, Hendgen-Cotta UB, et al. Circulating nitrite contributes to cardioprotection by remote ischemic preconditioning. Circ Res 2014;114:1601–10.

[22] Stuehr D, Pou S, Rosen GM. Oxygen reduction by nitric-oxide synthases. J Biol Chem 2001;276:1433–6.

[23] Michel T. Nitric oxide synthases: which, where, how, and why. J Clin Invest 1997;100:2146–52.

[24] Stuehr DJ, Tejero J, Haque MM. Structural and mechanistic aspects of flavoproteins: electron transfer through the nitric oxide synthase flavoprotein domain. FEBS J 2009;276:3959–74.

[25] Campbell MG, Smith BC, Potter CS, et al. Molecular architecture of mammalian nitric oxide synthases. Proc Natl Acad Sci U S A 2014;111:E3614–23.

[26] Fulton D, Gratton J-P, McCabe TJ, et al. Regulation of endothelium derived nitric oxide production by the protein kinase Akt. Nature 1999;399:597–601.

[27] Dimmeler S, Fleming I, Fisslthaler B, et al. Activation of nitric oxide synthase in endothelial cells by Akt-dependent phosphorylation. Nature 1999;399:601–5.

[28] Drab M, Verkade P, Elger M, et al. Loss of caveolae, vascular dysfunction, and pulmonary defects in caveolin-1 gene-disrupted mice. Science 2001;293:2449–52.

[29] Kraehling JR, Hao Z, Lee MY, et al. Uncoupling caveolae from intracellular signaling in vivo. Circ Res 2016;118(1):48–55.

[30] Goligorsky MS, Brodsky S, Chen J. Relationship between caveolae and eNOS: everything in proximity and the proximity of everything. Am J Physiol Renal Physiol 2002;283:F1–F10.

[31] Griscavage JM, Fukuto JM, Komori Y, et al. Nitric oxide inhibits neuronal nitric oxide synthase by interacting with the heme prosthetic group. J Biol Chem 1994;269:21644–9.

[32] Wilcox JN, Subramanian RR, Sundell CL, et al. Expression of multiple isoforms of nitric oxide synthase in normal and atherosclerotic vessels. Arterioscler Thromb Vasc Biol 1997;17:2479–88.

[33] Chavakis E, Dimmeler S. Regulation of endothelial cell survival and apoptosis during angiogenesis. Arterioscler Thromb Vasc Biol 2002;22:887–93.

[34] Kibbe M, Billiar T, Tzeng E. Inducible nitric oxide synthase and vascular injury. Cardiovasc Res 1999;43:650–7.

[35] Giulivi C. Characterization and function of mitochondrial nitric oxide synthase. Free Radic Biol Med 2003;34:397–405.

[36] Liu VW, Huang PL. Cardiovascular roles of nitric oxide: a review of insights from nitric oxide synthase gene disrupted mice. Cardiovasc Res 2008;77:19–29.

[37] Huang PL. Mouse models of nitric oxide synthase deficiency. J Am Soc Nephrol 2000;11(Suppl. 16):S120–3.

[38] Ozaki M, Kawashima S, Yamashita T, et al. Overexpression of endothelial nitric oxide synthase accelerates atherosclerotic lesion formation in apoE-deficient mice. J Clin Invest 2002;110:331–40.

[39] Tsutsui M, Tanimoto A, Tamura M, et al. Significance of nitric oxide synthases: lessons from triple nitric oxide synthases null mice. J Pharmacol Sci 2015;127:42–52.

[40] Ou J, Ou Z, Ackerman AW, et al. Inhibition of heat shock protein 90 (hsp90) in proliferating endothelial cells uncouples endothelial nitric oxide synthase activity. Free Radic Biol Med 2003;15:269–76.

[41] Vasquez-Vivar J, Kalyanaraman B, Martasek P, et al. Superoxide generation by endothelial nitric oxide synthase: the influence of cofactors. Proc Natl Acad Sci U S A 1998;95:9220–5.

[42] Cai H, Harrison DG. Endothelial dysfunction in cardiovascular diseases: the role of oxidant stress. Circ Res 2000;87:840–4.

[43] Cooke JP. Does ADMA cause endothelial dysfunction? Arterioscler Thromb Vasc Biol 2000;20:2032–7.

[44] Böger RH, Sydow K, Borlak J, et al. LDL cholesterol upregulates synthesis of asymmetrical dimethylarginine in human endothelial cells. Circ Res 2000;87:99–105.

[45] Thomazella MC, Góes MF, Andrade CR, et al. Effects of high adherence to Mediterranean or low-fat diets in medicated secondary prevention patients. Am J Cardiol 2011;108:1523–9.

[46] Chen CA, Wang TY, Varadharaj S, et al. S-glutathionylation uncouples eNOS and regulates its cellular and vascular function. Nature 2010;468:1115–8.

[47] Reyes LA, Boslett J, Varadharaj S, et al. Depletion of NADP(H) due to CD38 activation triggers endothelial dysfunction in the postischemic heart. Proc Natl Acad Sci U S A 2015;112:11648–53.

[48] Hingorani AD, Liang CF, Fatibene J, et al. A common variant of the endothelial nitric oxide synthase (glu298 → Asp) is a major risk factor for coronary artery disease in the UK. Circulation 1999;100: 1515–20.

[49] Zweier JL, Wang P, Samouilov A, et al. Enzyme-independent formation of nitric oxide in biological tissues. Nat Med 1995;8:804–9.

[50] Busse R, Edwards G, Feletou M, et al. EDHF: bringing the concepts together. Trends Pharmacol Sci 2002;23:374–80.

[51] Fleming I. Cytochrome P450 2C is an EDHF synthase in coronary arteries. Trends Cardiovasc Med 2000;10:166–70.

[52] Matoba T, Shimokawa H, Nakashima M, et al. Hydrogen peroxide is an endothelium-derived hyperpolarizing factor in mice. J Clin Invest 2000;106:1521–30.

[53] Yada T, Shimokawa H, Hiramatsu O, et al. Hydrogen peroxide, an endogenous endothelium-derived hyperpolarizing factor, plays an important role in coronary autoregulation in vivo. Circulation 2003;107:1040–5.

[54] Edwards G, Dora KA, Gardener MJ, et al. K+ is an endothelium-derived hyperpolarizing factor in rat arteries. Nature 1998;396: 269–72.

[55] Popp R, Brandes RP, Ott G, et al. Dynamic modulation of interendothelial gap junctional communication by 11,12-epoxyeicosatrienoic acid. Circ Res 2002;19(90):800–6.

[56] Burgoyne JR, Madhani M, Cuello F, et al. Cysteine redox sensor in PKGIa enables oxidant-induced activation. Science 2007;317: 1393–7.

[57] Yang G, Wu L, Jiang B, et al. H$_2$S as a physiologic vasorelaxant: hypertension in mice with deletion of cystathionine gamma-lyase. Science 2008;322(5901):587–90.

[58] Carbonero F, Benefiel AC, Alizadeh-Ghamsari AH, et al. Microbial pathways in colonic sulfur metabolism and links with health and disease. Front Physiol 2012;3:448.

[59] Mani S, Li H, Untereiner A, et al. Decreased endogenous production of hydrogen sulfide accelerates atherosclerosis. Circulation 2013;127(25):2523–34.

[60] Mustafa AK, Gadalla MM, Sen N, et al. H$_2$S signals through protein S-sulfhydration. Sci Signal 2009;2(96):ra72.

[61] Cuevasanta E, Lange M, Bonanata J, et al. Reaction of hydrogen sulfide with disulfide and sulfenic acid to form the strongly nucleophilic persulfide. J Biol Chem 2015;290(45):26866–80.

[62] Kimura Y, Mikami Y, Osumi K, et al. Polysulfides are possible H$_2$S-derived signaling molecules in rat brain. FASEB J 2013;27 (6):2451–7.

[63] Sen N, Paul BD, Gadalla MM, et al. Hydrogen sulfide-linked sulfhydration of NF-κB mediates its antiapoptotic actions. Mol Cell 2012;45(1):13–24.

[64] Mustafa AK, Sikka G, Gazi SK, et al. Hydrogen sulfide as endothelium-derived hyperpolarizing factor sulfhydrates potassium channels. Circ Res 2011;109(11):1259–68.

[65] Wang R. Hydrogen sulfide: a new EDRF. Kidney Int 2009;76 (7):700–4.

[66] Wang R, Szabo C, Ichinose F, et al. The role of H$_2$S bioavailability in endothelial dysfunction. Trends Pharmacol Sci 2015;36 (9):568–78.

[67] Kram L, Grambow E, Mueller-Graf F, et al. The anti-thrombotic effect of hydrogen sulfide is partly mediated by an upregulation of nitric oxide synthases. Thromb Res 2013;132(2):e112–7.

[68] Altaany Z, Ju Y, Yang G, et al. The coordination of S-sulfhydration, S-nitrosylation, and phosphorylation of endothelial nitric oxide synthase by hydrogen sulfide. Sci Signal 2014;7(342):ra87.

[69] Kida M, Sugiyama T, Yoshimoto T, et al. Hydrogen sulfide increases nitric oxide production with calcium-dependent activation of endothelial nitric oxide synthase in endothelial cells. Eur J Pharm Sci 2013;48(1–2):211–5.

[70] Coletta C, Papapetropoulos A, Erdelyi K, et al. Hydrogen sulfide and nitric oxide are mutually dependent in the regulation of angiogenesis and endothelium-dependent vasorelaxation. Proc Natl Acad Sci U S A 2012;109(23):9161–6.

[71] King AL, Polhemus DJ, Bhushan S, et al. Hydrogen sulfide cytoprotective signaling is endothelial nitric oxide synthase-nitric oxide dependent. Proc Natl Acad Sci U S A 2014;111(8):3182–7.

[72] Whiteman M, Gooding KM, Whatmore JL, et al. Adiposity is a major determinant of plasma levels of the novel vasodilator hydrogen sulphide. Diabetologia 2010;53(8):1722–6.

[73] Straub AC, Zeigler AC, Isakson BE. The myoendothelial junction: connections that deliver the message. Physiology (Bethesda) 2014;29:242–9.

Further Reading

[1] Andrew PJ, Mayer B. Enzymatic function of nitric oxide synthases. Cardiovasc Res 1999;43:521–31.

[2] Hamilton SJ, Watts GF. Endothelial dysfunction in diabetes: pathogenesis, significance and treatment. Rev Diabet Stud 2013;10:133–56.

[3] Campbell WB, Gauthier KM. What is new in endothelium-derived hyperpolarizing factors? Curr Opin Nephrol Hypertens 2002;11: 177–83.

CHAPTER

9

Vasoconstrictor Substances Produced by the Endothelium

Stefany B.A. Cau, Paulo Roberto B. Evora, and Rita C. Tostes

INTRODUCTION

The last three decades have witnessed the discovery of endothelium-derived relaxing factors (EDRF) such as prostacyclin, nitric oxide (NO), and EDHF—endothelium-derived hyperpolarizing factors, represented by arachidonic acid metabolites (EET) derived from cytochrome P450, hydrogen peroxide, potassium ions (K^+) and communications via gap junctions. The mechanisms by which EDRF trigger their effects [(1) activation of guanylyl cyclase (GC)—cyclic GMP; (2) activation of the adenylyl cyclase (AC)—cyclic AMP pathway; and (3) activation of mechanisms that control the membrane potential of vascular smooth muscle cells (VSMC) producing hyperpolarization (such as opening K+ channels and activating the Na+/K+ pump)] were also unraveled (see Chapter 8).

It also became evident that the endothelium produces endothelium-dependent vasoconstrictor substances or contracting factors (EDCF) [1,2]. Initially, EDCF were identified as products of cyclooxygenase (COX), the so called prostanoids, since inhibitors of this enzyme blocked the endothelium-dependent vascular contractions [3]. Subsequently, it was demonstrated that endothelial cells produce other EDCF, including endothelin-1 (ET-1), angiotensin II (Ang II), reactive oxygen species (ROS), as superoxide anion ($\bullet O_2^-$), and uridine adenosine tetraphosphate [uridine adenosine tetraphosphate (Up4A)].

Increased concentration of intracellular calcium ($[Ca^{2+}]$) in endothelial cells is the initial event that leads to the release of EDRF and EDCF.[1] This conclusion is based on the following facts:

- Activation of cell membrane receptors by agonists that release EDRF and EDCF, such as acetylcholine, ADP, and ATP, increases intracellular $[Ca^{2+}]$.
- Reduced extracellular concentration of Ca^{2+} decreases endothelium-dependent relaxation and contractions.
- Ca^{2+} ionophores, such as A23187, induce endothelium-dependent relaxation and contractions.

In electrically nonexcitable cells, such as endothelial cells, the increased $[Ca^{2+}]$ induced by agonists is generally biphasic. The first stage shows a transient increase of $[Ca^{2+}]$ due to the release of Ca^{2+} from the endoplasmic reticulum (ER). For example, activation of plasma membrane receptors coupled to Phospholipase-C (PLC) promotes the formation of 1,4,5-inositol trisphosphate (IP$_3$), which activates IP$_3$ receptors in the ER and therefore promotes release of Ca^{2+} in the cytosol. The Ca^{2+} released is re-uptaken in the ER or taken to the extracellular medium by means of transporters located in the

---

[1] In the initial proposal by Prof. Paul Vanhoutte, EDCFs were classified as: EDCF1—cyclooxygenase pathway-dependent vasoconstrictor factors (COX); EDCF2—family of endothelins; EDCF3—vasoconstrictors released during hypoxia; and EDCF4—oxygen-free radicals independent from the COX pathway.

plasma membrane, which makes the increase of $[Ca^{2+}]$ transitory.

The second stage shows sustained $[Ca^{2+}]$ increase, only in the presence of extracellular Ca^{2+}, by influx of Ca^{2+} through plasma membrane channels. Since the second stage occurs after the release of the intracellular stores of Ca^{2+}, it is called Ca^{2+} store-operated calcium entry (SOCE) and the channels permeable to Ca^{2+} in the plasma membrane are called store-operated channels (SOC).

The most recent hypothesis to explain the SOCE activation mechanism suggests that the protein STIM (stromal interaction molecule), which is located in the ER membrane, oligomerizes and moves to micro regions of contact between the ER and the cell plasma membrane, which have other proteins that control Ca^{2+} influx, such as ORAI or TRPC (Ca^{2+} ion nonselective channels, transient receptor potential cation channels). Therefore, the interaction between STIM and ORAI/TRPC proteins promotes SOCE stimulation and increased $[Ca^{2+}]$ in endothelial cells, as illustrated in Fig. 9.1. Subsequently, the mechanisms that lead to EDCF production, its main actions and participation in endothelial dysfunction associated with vascular complications in cardiovascular and metabolic diseases will be discussed.

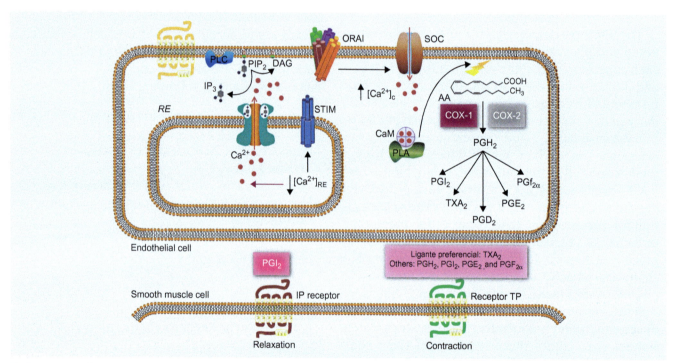

FIG. 9.1   Calcium signaling in endothelial cells increases the release of EDCF derived from arachidonic acid. The increased concentration of calcium in cytosol, $[Ca^{2+}]$, may occur via activation of a G protein-coupled receptor, which activates phospholipase C (PLC), cleaving PIP_2 in diacylglycerol (DAG) and 1,4,5—inositol trisphosphate (IP_3). IP_3 binds to its receptors (RIP_3), which are channels for Ca^{2+}, present in the Endoplasmic Reticulum (ER) membrane, increasing $[Ca^{2+}]$ even more. The depletion of Ca^{2+} stores activates the STIM protein present in the reticulum membrane, which associates to the ORAI plasma membrane protein that regulates Ca^{2+} Store-Operated Channels (SOC). Ca^{2+} binds to calmodulin (CaM), and the Ca^{2+}/CaM complex activates cleavage of arachidonic acid membrane lipids by activating Phospholipase-A (PLA). The arachidonic acid is converted by cyclooxygenase-1, -2 (COX-1, COX-2) in prostanoid derivatives (prostaglandins—PG and Thromboxanes—Tx). The COX products diffuse to smooth muscle cells, where they can interact with their respective receptors. The connection to the TP receptor is the one that has been more extensively related to excessive vasoconstriction in diseases such as diabetes and arterial hypertension. The IP receptor for PGI_2 mediates vasodilation induced by PGI_2 in physiological conditions (see text for abbreviations).

ENDOTHELIAL DYSFUNCTION

Endothelial dysfunction is characterized by alterations in endothelium regulating functions, resulting in imbalanced production of relaxing and contracting factors, pro-coagulant and anticoagulant mediators or growth inhibiting and promoting substances. Physiopathological mechanisms that lead to endothelial dysfunction or reduced EDRF/increased EDCF include risk factors for cardiovascular diseases such as hypercholesterolemia, tobacco use, insulin resistance, hyperglycemia, arterial hypertension, hyperhomocysteinemia, or a combination of these factors. Reduced production and/or bioavailability of NO is considered the central mechanism responsible for endothelial dysfunction. Clinically, endothelial dysfunction syndrome can be evidenced by local or general vasospasm, or thrombosis, and is associated with the development of atherosclerosis and restenosis, and vascular complications associated with several cardiovascular and metabolic diseases.

Most texts regarding endothelial dysfunction focus on the altered release of endothelial vasoactive substances, without considering vasoplegic conditions caused by increased release of vasorelaxing factors. These texts, as well as the specific bibliographic research, failed to present an endothelial dysfunction classification. Considering that this classification would make it easier to integrate etiological and physiopathological concepts and, consequently, considerations regarding the therapeutic perspectives of this pathological condition of growing importance, this chapter proposes a classification for endothelial dysfunction [4,5], aiming to stimulate discussion and a consensual definition (Table 9.1). Based on this classification, it is possible to discuss, in general, the diagnostic possibilities and therapeutic perspectives of endothelial dysfunction. This approach evidences the transition from laboratory research to clinical practice.

Considering this classification, EDCF would be more clearly related to a phenotypic, vasotonic, and partially reversible endothelial dysfunction.

Endothelial dysfunction is present in several cardiovascular and metabolic diseases, such as hypertension, chronic heart failure, peripheral arterial disease, atherosclerosis, diabetes, obesity, septic shock, and chronic renal failure. In atherosclerosis, endothelial dysfunction contributes to the onset and evolution of thrombotic, proinflammatory, and proliferative events. Endothelial dysfunction

TABLE 9.1 Proposed Classification of Endothelial Dysfunction

Endothelial dysfunction classification
I. ETIOLOGICAL CLASSIFICATION
(a) *Primary or "genotypic" endothelial dysfunction*: shown in children with homozygous Homocystinuria, and in normotensive patients with a family history of essential arterial hypertension
(b) *Secondary or "phenotypic" endothelial dysfunction*: present in all cardiovascular diseases (atherosclerosis, coronary artery diseases, arterial hypertension, diabetes, and others)
II. FUNCTIONAL CLASSIFICATION
(a) *"Vasotonic" endothelial dysfunction*: present in cardiovascular diseases, implying risk of vasospasm and thrombosis
(b) *"Vasoplegic" endothelial dysfunction*: present in distributive shock status (sepsis, anaphylactic shock, anaphylactoid reactions, and vasoplegy related to extracorporeal circulation) due to the action of cytokines that stimulate increased pathological release of endothelial relaxing factors, especially NO
III. PROGRESSIVE OR PROGNOSTIC CLASSIFICATION
(a) *"Reversible" endothelial dysfunction*: most probable occurrence in the initial stages of "vasoplegic" dysfunctions. "Vasotonic" dysfunctions associated with cardiovascular diseases are hardly completely reverted
(b) *"Irreversible" endothelial dysfunction*: present in advanced states of cardiovascular diseases and sepsis

is also evident in aging and may be caused by drugs such as the immunosuppressive agent cyclosporin A and by toxic substances in the environment. Increased release or action of EDCF contributes significantly to the endothelial dysfunction associated with these diseases/conditions, as will be discussed below.

ENDOTHELIUM-DERIVED CONTRACTING FACTORS

Products of Cyclooxygenase

All eicosanoids and prostanoids, including arachidonic acid epoxy metabolites (EETs) and thromboxane A_2 (TxA_2), are formed after increases of $[Ca^{2+}]$, due to the metabolization of arachidonic acid or other lipid substrates, such as eicosapentaenoic acid, docosahexaenoic acid, and linoleic acid [6]. For example, arachidonic acid is metabolized by one of the two isoforms of prostaglandin H_2 (PGH_2) synthase [more commonly known as cyclooxygenases, cyclooxygenase-1 (COX-1) and cyclooxygenase-2 (COX-2)] to form the intermediate PGH_2, which is then metabolized by several subtypes of prostaglandin synthase and isomerase, originating a variety of vasoactive metabolites or prostanoids [6]. More specifically, TxA is formed after PGH_2 by a TxA synthase, member of the superfamily epoxygenase cytochrome P450 (CYP) (CYP5 in human beings). Eicosanoid and prostanoid receptors, such as the TP receptor for TxA_2, expressed both in VSMCs and endothelial cells, mediate the paracrine and autocrine actions of COX metabolites formed in endothelial cells (Fig. 9.1).

In spontaneously hypertensive rats (SHR), as well as in other experimental models of arterial hypertension, EDCF production involves an increase in endothelial intracellular $[Ca^{2+}]$, ROS production, predominant activation of COX-1 and, to a lesser extent, of COX-2 [7]. The diffusion of EDCF to the VSMC and the subsequent stimulation of TP receptors promotes endothelium-dependent contractions, more intense in hypertensive animals than observed in normotensive and control-animals. All prostanoids can bind to TP receptors, but with different affinities. Depending on the experimental model of arterial hypertension, TxA_2, PGH_2, $PGF_{2\alpha}$, PGE_2, and (paradoxically) PGI_2 act as EDCF. [8] The explanation for this ambiguous role of PGI_2 on vascular tone is a function of its relative affinity for the IP versus TP receptors, and its relative production. The physiological production of PGI_2 and its interaction with IP receptors in VSMC are associated with vasodilation. However, in the arteries of hypertensive or elderly individuals, the levels of PGI_2 synthase and PGI_2 are increased, leading to the activation of the TP receptor and to endothelium-dependent vasoconstriction.

In humans, increased production of EDCF derived from COX is a characteristic of the blood vessels of elderly individuals and patients with essential hypertension [9], causing earlier onset and acceleration of endothelial dysfunction. Since in most cases activation of TP receptors has been the common downstream effector [10], selective antagonists of such receptors could prevent endothelial dysfunction and are of therapeutic interest.

Endothelins

In 1988, the Yanagisawa group identified endothelin-1 (ET-1) as the first endothelium-derived contracting factor [11]. ET-1 is released by endothelial cells as a biologically inactive peptide (Big ET-1 or pro-ET-1), which by the action of endothelin converting enzymes (ECE) and chymase, it is converted into its active form, a 21 amino acid peptide. There are three identified ET isoforms, ET-1, ET-2, and ET-3, and several stimuli that increase the expression of prepropeptide ET-1 in endothelial cells: thrombin, adrenaline, angiotensin II, bradykinin, hypoxia, high- and low-density lipoproteins, insulin, ischemia, shear stress, and growth factors [12]. ET-1 is a potent vasoconstrictor, both in large vessels and in microcirculation, with intramyocardial vessels being very sensitive to its actions. ET-1 also promotes vasodilation, which depends on the presence of the endothelium, both in isolated arteries and in vivo.

The effects of ET-1 are mediated by two sub-types of receptors: ET_A and ET_B receptors [11,13]. ET_A receptors are expressed in VSMCs and heart cells, and sub-type ET_B is expressed in endothelial cells, renal cells, and also in VSMCs [12,13]. In vessels where the endothelium is functionally intact, ET-1 stimulates the production of NO and PGI_2, which then negatively modulate the vasoconstrictor actions and reduce the synthesis of ET-1 itself. In endothelial dysfunctional states, such as atherosclerosis, the vasoconstrictor, proliferative, and proinflammatory actions of ET-1 are maximized without NO opposition.

In humans, in most cardiovascular diseases, such as myocardial infarction, cardiogenic shock, cerebral vasospasm after subarachnoid hemorrhage, unstable angina, coronary artery disease, heart failure, and essential hypertension, circulating ET-1 levels are increased, leading ET-1 to be considered a likely mediator of endothelial dysfunction or excessive vasoconstriction associated with them [13,14].

The pulmonary circulation is highly susceptible to ET-1. Elevated ET-1 levels are associated with pulmonary arterial hypertension and circulating levels of ET-1 are considered a marker of disease severity and prognosis. In animal models of pulmonary arterial hypertension, nonselective antagonists of ET_A and ET_B receptors (bosentan) and selective antagonists of the ET_A receptor (sitaxsentana, tardentana, TBC-3711) are effective in reducing pulmonary and vascular resistance and inhibiting remodeling.

Both types of antagonists are used in humans. Bosentan, the first antagonist of ET_A and ET_B receptors to enter clinical application (in 1993), was approved for the treatment of pulmonary arterial hypertension (PAH) in 2001 based on two clinical trials ("Study 351" with 32 patients in class III, with idiopathic PAH or associated with systemic sclerosis, and the important BREATHE-1 study, which included 150 patients with idiopathic PAH, 47 with systemic pulmonary hypertension associated with multiple sclerosis, and 16 with systemic lupus erythematosus associated with PAH) [15]. ET_A antagonists, such as ambrisentan and sitaxsentan, have also been approved for use in patients with Class II, III, and IV pulmonary hypertension in the United States, Canada, and European countries. Due to side effects, such as irreversible hepatic failure, sitaxsentan was withdrawn from global markets in 2010. This does not apply to ambrisentan, since in 2011 the FDA issued a notice on withdrawal of warning of liver damage from the leaflet of that drug. In Brazil, in 2013, the National Commission for the Incorporation of Technologies in SUS (Conitec) approved the incorporation of ambrisentan and bosentan for the treatment of PAH as the second treatment line.

Angiotensin II

In the last decade, the identification of all constituents forming angiotensin II (Ang II)—that is, angiotensinogen, renin, and angiotensin converting enzyme (ACE)—in the different layers of blood vessels, lead to the proposition that increased vascular tone associated to cardiovascular diseases would be related to higher tissue formation of Ang II, independently from the circulating angiotensinogen and renin [16]. Although the importance of tissue Ang II formation remains open, it is admitted that large quantities of ACE—responsible for the formation of Ang II from Angiotensin I (Ang I)—in endothelial cells represent a form to regulate the local concentration of Ang II.

The main effects of Ang II are mediated by receptors of sub-type AT_1, which are dominant in adults and widely distributed along the vasculature and its endothelium (Fig. 9.2). Ang II triggers a series of actions to minimize fluid and renal sodium loss, and to maintain the volume of extracellular liquid and arterial pressure [17]. Besides vasoconstriction,

decreased blood flow, and increased vascular resistance, Ang II has pro-inflammatory, pro-oxidative, proliferative, and pro-fibrotic properties. It also stimulates the release of aldosterone, increases tubular sodium reabsorption rate, stimulates thirst, and increases sympathetic nerve activity [17].

Receptors of the AT_2 sub-type trigger actions normally opposed to the actions mediated by AT_1 receptors, but have low expression in adults and are more prominent in fetal and newborn kidneys. These are expressed in endothelial cells, in which they couple to increase NO production (Fig. 9.2).

Antagonists of the AT_1 receptor, ACE inhibitors, and renin inhibitors are frequently used clinically to block the effects of endogenous Ang II in several conditions, such as arterial hypertension, congestive heart failure, and diabetic nephropathy. In conditions associated with increased Ang II activity, the inhibition of the renin-angiotensin system increases blood flow, reduces peripheral vascular resistance, and reduces arterial pressure.

At the cellular level, Ang II increases cytosolic Ca^{2+}, by means of the voltage-dependent Ca^{2+} channels and mobilization of Ca^{2+} from intracellular stores. It is important to emphasize that while in endothelial cells, the increased Ca^{2+} couples to the formation of EDRF and EDCF, in VSMC the increase of $[Ca^{2+}]$ by means of voltage-dependent Ca^{2+} channels induces cell contraction and subsequent vasoconstriction [18]. Consequently, some vasoconstrictor responses of Ang II are inhibited by blockers of the Ca^{2+} channel. Among the downstream proteins regulated by the Ca^{2+}/calmodulin complex, there is myosin light-chain kinase (MLCK), which phosphorylates myosin and energizes it for contraction (Fig. 9.2). Also important for the contracting process is the activation of RhoA, a member of the small GTPase family, which participates in vasoconstriction by activating Rho-kinase. The Rho-kinase, in turn, inactivates the myosin light-chain phosphatase (MLCP), preventing the phosphate removal of myosin and favoring the contraction process.

Signaling by Ang II via AT_1 receptor in VSMCs is complex and involves phosphorylation of several tyrosine-kinases, including c-Src, kinase of the Janusfamily (JAK), focal adhesion kinases (FAK), and others [19]. These signaling pathways contribute to vascular injury associated with clinical conditions, such as chronic renal diseases, atherosclerosis, and arterial hypertension [19].

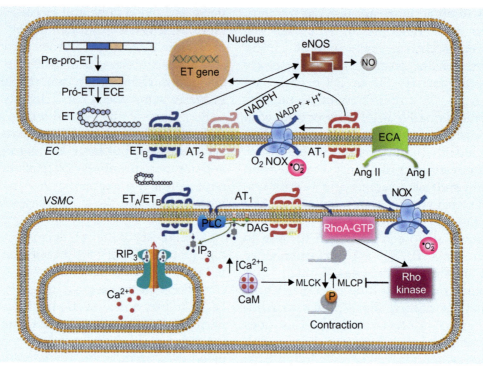

FIG. 9.2 Endothelin-1 and angiotensin-II act as EDCF. Endothelin-1 (ET-1) is generated in endothelial cells as pre-pro-endothelin, and is cleaved into pro-endothelin (or big-endothelin). Due to the action of the Endothelin Converting Enzyme (ECE), the pro-endothelin is cleaved into the active peptide ET-1. ET-1 can bind to ET_A receptors (present in Vascular Smooth Muscle Cells—VSMCs—and Endothelial Cells—ECs) and ET_B receptors (present in ECs). Locally, the Angiotensin Converting Enzyme (ACE), present on the surface of endothelial cells, which converts circulating Angiotensin I (Ang I) into Ang II, mediates the synthesis of another vasoconstrictor peptide, Angiotensin II (Ang II). Ang II may bind to AT_1 receptors (present in VSMCs and ECs) and AT_2 receptors (present in ECs). In endothelial cells, signaling by ET-1 and Ang II via ET_B and AT_2 receptors, respectively, is associated with the production of NO, the main EDRF. However, in smooth muscle cells, ET-1 and Ang II, by means of $ET_{A/B}$ and AT_1 receptors, activate cell-signaling pathways related to contraction, such as increased cytosolic calcium due to the activity of phospholipase C (PLC), activating the Myosin Light-Chain Kinase (MLCK) and Rho-kinase, which phosphorylates and inactivates the Myosin Light-Chain Phosphatase (MLCP). The increased activity of MLCK and decreased activity of MLCP lead to vasoconstriction. Besides these pathways, Ang II may also induce the formation of Reactive Oxygen Species by means of NADPH-oxidase (NOX) (see text for abbreviations).

In endothelial dysfunction, Ang II binding to AT_1 receptors present in the endothelium induces transcription of ET-1, which contributes to vasoconstriction and increased ROS production. The main enzyme involved in the production of ROS induced by Ang II is nicotinamide adenine dinucleotide phosphate (NADPH)-oxidase, or NOX [19]. Other enzymes associated with ROS production that may be activated by Ang II binding to the AT_1 receptor, both in endothelial cells and VSMCs, include xanthine oxidase [20] and uncoupled endothelial nitric oxide synthase (eNOS) [21] (apparently, the latter is exclusively present in endothelial cells).

These mechanisms will be discussed later, as well as the proposition that ROS constitute EDCF.

Reactive Oxygen Species

Reactive Oxygen Species (ROS) are highly reactive molecules that contain oxygen with one or more unpaired electrons in its outer orbital. The chemical instability of ROS is responsible for the reaction of these molecules with molecular targets, controlling several cellular events. However, in pathological conditions, there is increased production of ROS, while antioxidant defenses are reduced, commonly called oxidative stress. In endothelial dysfunction, oxidative stress is a fundamental mechanism of the coagulability, inflammation, and proliferation phenomena, as well as increased vascular tone.

The proposition that ROS act as EDCF in the vasculature comes from several experimental observations: (1) intact vessels and endothelial cells in vitro submitted to mechanical forces, which mimic shear stress, generate, besides NO, $\bullet O_2^-$ and hydrogen peroxide (H_2O_2); (2) in some vascular beds (rat renal artery and dog basilar artery), vasoconstriction induced by increasing doses of acetylcholine is inhibited by antioxidants; [22,23] (3) antioxidant agents restore vasodilation in arteries with endothelium proceeding from hypertensive and diabetic animals [24].

Several enzymatic sources, from several cell types, are responsible for ROS production, as illustrated in Fig. 9.3. In endothelial cells, the following

stand out: NOX, uncoupled eNOS, xanthine oxidase, and COX.

Differently from the latter, in which ROS are synthesized as by-products, the NOX enzymes are membrane NADPH-oxidase with the single function of producing $\bullet O_2^-$, by reducing one molecular oxygen electron using NAD(P)H as an electron donor. eNOS, the enzyme that synthesizes NO by oxidizing amino acid L-arginine, in specific situations, such as the absence of substrate or depletion of the tetrahydrobiopterin (BH_4) cofactor, transfers one electron to the molecular oxygen, forming $\bullet O_2^-$. Xanthine oxidase is an enzyme ubiquitously distributed in the organism, responsible by the catabolism of purines, producing uric acid, $\bullet O_2^-$, and H_2O_2.

The mechanisms through which ROS induce vasoconstriction are multiple, and will be divided into those that cause rapid changes in the machinery

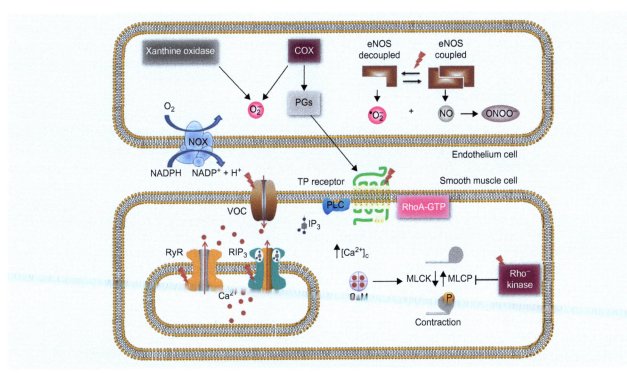

FIG. 9.3 Reactive Oxygen Species (ROS) derived from the endothelium induce vasoconstriction. These ROS can be formed in endothelial cells, mainly by the activity of NADPH-oxidase (NOX), xanthine oxidase, COX, and uncoupled Endothelial Nitric Oxide Synthase (eNOS). Superoxide anion ($\bullet O_2^-$) is a by-product of the activity of xanthine oxidase and COX; however, the single product of NOX. Instead of producing NO, the main EDRF, uncoupled eNOS forms $\bullet O_2^-$, contributing to vascular oxidative stress. $\bullet O_2^-$ acts as EDCF by means of different mechanism, such as: consumption of NO, forming peroxynitrite (ONOO$^-$) and favoring eNOS uncoupling, maximization of prostanoid receptor (TP receptor) signaling, increased activity of Rho-kinase, increased concentration of cytosol calcium ([Ca^{2+}]), modulation of Ca^{2+} release from intracellular stores by ryanodine receptors (RyR) and IP$_3$ receptors (RIP$_3$), and influx of Ca^{2+} through voltage-sensitive Ca^{2+} channels (VOC). The targets with which $\bullet O_2$ reacts are emphasized with the symbol ⚡ (see text for abbreviations).

of relaxation/contraction and in those that cause alteration of the arterial wall structure, contributing to the sustained increase of the peripheral vascular resistance.

Regarding the quick mechanisms, ROS reduce the capacity of the endothelium in promoting relaxation via NO, amplify the effects of other EDCFs, such as COX derivatives, increase cytosolic concentrations of Ca^{2+} in VSMCs, activate contraction regulating kinases, such as PKC, Rho-kinase, and mitogen-activated protein kinase (MAPK). It is important to observe that these mechanisms become more important in pathological processes, followed by oxidative stress. For example, it has been shown that in the arteries of hypertensive rats, the $\bullet O_2^-$ formed from NOX modifies BH_4, favoring uncoupling of eNOS, which becomes an important source of $\bullet O_2^-$ [25]. Besides that, it is known that $\bullet O_2^-$ can per se consume NO, producing peroxynitrite ($ONOO^-$), another ROS that is potentially harmful to vascular function. Similarly, in SHR aortas, the activation of COX by acetylcholine results in production of vasoconstrictor prostanoids and ROS. The ROS activate COX directly in endothelial cells and smooth muscles. This positive feedback mechanism leads to excessive generation of prostanoids, which activate hyper-responsive TP receptors and induces smooth muscle contraction [26]. In SHR carotids, the endothelium-dependent contracting response, which is mediated by prostaglandins and ROS derived from COX, is reversed by Rho kinase inhibition, and not by PKC inhibition [27], corroborating the in vitro findings that ROS directly activates Rho-kinase. Elegant in vitro demonstrations that ROS oxidizes redox-sensitive residues in ion channels that control input (voltage-sensitive Ca^{2+} channels) and release of Ca^{2+} stores (IP_3 receptor and ryanodine receptor) suggest a possible direct mechanism by which ROS induce vasoconstriction [28]. However, its in vivo participation has not been convincingly demonstrated.

In addition, ROS interacts with various signaling pathways involved in the proliferative responses of VSMC. For example, ROS activates MAPK, such as the kinases-1/2 regulated by extracellular signal (ERK1/2), N-terminal C-Jun kinase (JNK), and p38; activate receptors with tyrosine kinase domain, such as epidermal growth factor receptor (EGFR) and platelet-derived growth factor receptor-β (PDGFR-β). These molecular events culminate in the remodeling of arteries and arterioles, which

makes them more rigid and reduces the vascular lumen, respectively. Both in animal models and in patients with hypertension and diabetes, these mechanisms help to explain the sustained increase in peripheral vascular resistance, contributing to elevated blood pressure or target organ damage [29].

Adenosine UridineTetraphosphate (Up4A)

Uridine adenosine tetraphosphate (Up4A), a dinucleotide containing purine and pyrimidine, was identified by Jankowski et al. in 2005 as a novel and potent EDCF. [30] Up4A is released by endothelial cells in response to various stimuli, such as A23187 [30]. Although the molecular mechanisms that leads to the production/release of Up4A are still unknown, the activation of the receptor for vascular endothelial growth factor 2 (VEGFR2) leads to the production of Up4A [31].

Up4A modulates the tone of various arteries, inducing relaxation in the aorta of rats and swine coronary artery, and contraction in the rat pulmonary artery, aorta and renal arteries [32]. Up4A triggers its effects via activation of purinergic receptors (vasoconstriction via P2X1, P2Y2, and P2Y4 receptors, vasodilation via P2Y1 and P2Y2 receptors on endothelial cells, which induces NO release) [32]. In addition to vascular contractions, Up4A induces vascular calcification, activation of inflammatory mediators, proliferation and migration of VSMC [33,34], and generation of ROS via activation of NADPH oxidase. Up4A-induced vasoconstriction involves Ca^{2+}-dependent mechanisms (Ca^{2+} influx and Ca^{2+} release from intracellular stores) and Ca^{2+}-independent mechanisms, such as activation of kinases (ERK1/2). The proliferation of VSMC induced by Up4A involves PI3K/Akt and ERK1/2 signaling pathways, while pro-inflammatory responses are associated with ROS generation and activation of ERK1/2 and p38 MAPK.

Since these events are involved in vascular dysfunction associated with diabetes and hypertension [35,36], it is speculated that the regulation of Up4A-activated signaling in VSMC may be a potential therapeutic target.

In this sense, the circulating levels of Up4A are higher in young hypertensive patients [37], and in patients with chronic kidney disease [33] compared to healthy individuals. Up4A-induced contraction is altered in arteries (basilar, femoral, renal, and mesenteric) of experimental models of arterial hypertension—DOCA-sal rats [38,39] and type 2 diabetes—Goto-Kakizaki rats [40]. In renal arteries of hypertensive animals, increased contraction to Up4A is associated with increased ERK1/2 activity in, not alteration of P2Y receptors [30,38]. In renal arteries of diabetic animals, the increase in contractility at Up4A is associated with the activation of COX and TP receptors [40].

CONCLUSIONS

The concepts presented herein and in other chapters emphasize that the endothelium is a crucial regulator of vascular physiology, which in healthy conditions produces several substances with potent vasodilator, anti-thrombotic, and anti-atherosclerotic properties. Consequently, the presence of endothelial dysfunction is associated with increased risk of cardiovascular events. In the presence of pathological conditions, the EDCF, including ET-1, Ang II, COX-, and ROS-derived prostanoids, contribute actively to endothelial dysfunction observed, for example, in patients with cardiovascular and metabolic diseases.

Regardless of the efforts in the development of therapeutic strategies that improve endothelial function, only a few drugs have shown long-term clinical benefits. Currently, ACE inhibitors, AT_1 receptor antagonists, statins, and thiazolidinediones are the only medications that effectively change endothelial function. These agents are well known for their pleiotropic actions (not related to their main action mechanism). For example, statins that promote inhibition of the enzyme 3-hydroxy-3-methyl-glutaryl-CoA reductase (HMG-CoA reductase), preventing the formation of mevalonate and synthesis of cholesterol, also have secondary effects—that do not depend on the reduction of cholesterol levels, i.e., pleiotropic effects, such as modulation of the neuro-humoral and innate immunity systems, inhibition of oxidative stress, etc. These drugs also positively interfere in the treatment of underlying cardiovascular risk factors, and directly affect the vascular endothelium, improving its function.

EDCF derived from COX and ROS were identified as responsible for compromising endothelium-dependent vasodilation in patients with essential hypertension and diabetes. The production of COX-dependent EDCF is also a characteristic of the aging process, and essential hypertension seems to anticipate the phenomenon. The production of COX-derived EDCF is a characteristic of blood vessels of elderly patients with essential hypertension, causing the early onset and accelerating endothelial dysfunction. In this regard, COX inhibitors and antioxidant agents represent potential therapeutic strategies to reduce endothelial dysfunction. However, chronic use of COX inhibitors, in particular COX-2 inhibitors, is associated with important adverse cardiovascular effects, which include increased risk of myocardial infarction, cerebrovascular accident, heart failure, renal failure, and arterial hypertension.

Since in most cases the activation of TP receptors mediates the effects of COX-derived EDCF, selective antagonists of these receptors may prevent endothelial dysfunction and would be of therapeutic interest to treat cardiovascular diseases. Initial studies with TP receptor antagonists have shown improved endothelial function in patients with coronary arterial disease and patients with atherosclerosis and high cardiovascular risk [41]. However, there is a long way to go until it is demonstrated that these inhibitors are appropriate in respect to possible adverse effects in the gastrointestinal and renal systems.

Regarding the use of antioxidant agents, such as vitamins A, C, and E, coenzyme Q, beta-carotene, polyphenols, and flavonoids, to reduce endothelial dysfunction, the data from clinical trials are inconsistent and inconclusive, since most of the large clinical trials did not show beneficial cardiovascular effects from antioxidants [42]. Regarding the use of inhibitors of ROS-generating enzymes, such as xanthine oxidase and NOX, the first clinical trials have shown improved vascular function in patients with arterial hypertension, chronic renal diseases, and pulmonary hypertension. However, several questions related to these inhibitors, such as level of selectivity, oral availability, and action mechanisms, must still be clarified. Efficacy and safety studies in human beings are also indispensable before making any conclusions regarding the clinical utility of these inhibitors.

These considerations clearly illustrate that the mechanisms that regulate the balance between EDRF and EDCF, and the processes that transform the endothelium from a protection organ to a source of vasoconstrictor, pro-aggregation, and pro-mitogenic mediators are still a great challenge. Better comprehension of these processes and clinical investigation of potential therapeutic agents will ensure new approaches to treat endothelial dysfunction.

References

[1] De Mey JG, Vanhoutte PM. Heterogeneous behavior of the canine arterial and venous wall. Importance of the endothelium. Circ Res 1982;51(4):439–47.

[2] Hickey KA, Rubanyi G, Paul RJ, et al. Characterization of a coronary vasoconstrictor produced by cultured endothelial cells. Am J Physiol 1985;248(5 Pt 1):C550–6.

[3] Miller VM, Vanhoutte PM. Endothelium-dependent contractions to arachidonic acid are mediated by products of cyclooxygenase. Am J Physiol 1985;248(4 Pt 2):H432–7.

[4] Evora PRB. An open discussion about endothelial dysfunction: is it timely to propose a classification? Int J Cardiol 2000;73(3):289–92.

[5] Evora PRB, Baldo CF, Celotto AC, et al. Endothelium dysfunction classiflcation: why is it still an open discussion? Int J Cardiol 2009;137(2):175–6.

[6] Wong MSK, Vanhoutte PM. COX-mediated endothelium-dependent contractions: from the past to recent discoveries. Acta Pharmacol Sin 2010;31(9):1095–102.

[7] Gluais P, Paysant J, Badier-Commander C, et al. In SHR aorta, calcium ionophore A-23187 releases prostacyclin and thromboxane A2 as endothelium-derived contracting factors. Am J Physiol Heart Circ Physiol 2006;291(5):H2255–64.

[8] Feletou M, Huang Y, Vanhoutte PM. Vasoconstrictor prostanoids. Pflug Arch Eur J Phys 2010;459(6):941–50.

[9] Taddei S, Virdis A, Ghiadoni L, et al. Vitamin C improves endothelium-dependent vasodilation by restoring nitric oxide activity in essential hypertension. Circulation 1998;97(22):2222–9.

[10] Feletou M, Vanhoutte PM, Verbeuren TJ. The thromboxane/endoperoxide receptor (TP): the common villain. J Cardiovasc Pharm 2010;55(4):317–32.

[11] Yanagisawa M, Kurihara H, Kimura S, et al. A novel potent vasoconstrictor peptide produced by vascular endothelial cells. Nature 1988;332(6163):411–5.

[12] Barton M, Yanagisawa M. Endothelin: 20 years from discovery to therapy. Can J Physiol Pharmacol 2008;86(8):485–98.

[13] Sandoval YH, Atef ME, Levesque LO, et al. Endothelin-1 signaling in vascular physiology and pathophysiology. Curr Vasc Pharmacol 2014;12(2):202–14.

[14] Taddei S, Virdis A, Ghiadoni L, et al. Role of endothelin in the control of peripheral vascular tone in human hypertension. Heart Fail Rev 2001;6(4):277–85.

[15] Channick RN, Simonneau G, Sitbon O, et al. Effects of the dual endothclin-receptor antagonist bosentan in patients with pulmonary hypertension: a randomised placebo-controlled study. Lancet 2001;358(9288):1119–23.

[16] Cockcroft JR, O'Kane KP, Webb DJ. Tissue angiotensin generation and regulation of vascular tone. Pharmacol Ther 1995;65 (2):193–213.

[17] Navar LG. Physiology: hemodynamics, endothelial function, renin-angiotensin-aldosterone system, sympathetic nervous system. J Am Soc Hypertens 2014;8(7):519–24.

[18] Cheyou ER, Bouallegue A, Srivastava AK. Ca^{2+}/calmodulin-dependent protein kinase-II in vasoactive peptide- induced responses and vascular biology. Curr Vasc Pharmacol 2014;12 (2):249–57.

[19] Montezano AC, Nguyen Dinh Cat A, Rios FJ, et al. Angiotensin II and vascular injury. Curr Hypertens Rep 2014;16(6):431.

[20] Landmesser U, Spiekermann S, Preuss C, et al. Angiotensin II induces endothelial xanthine oxidase activation: role for endothelial dysfunction in patients with coronary disease. Arterioscler Thromb Vasc Biol 2007;27(4):943–8.

[21] Chalupsky K, Cai H. Endothelial dihydrofolatereductase: critical for nitric oxide bioavailability and role in angiotensin II uncoupling of endothelial nitric oxide synthase. Proc Natl Acad Sci U S A 2005;102(25):9056–61.

[22] Gao YJ, Lee RM. Hydrogen peroxide is an endothelium-dependent contracting factor in rat renal artery. Br J Pharmacol 2005;146 (8):1061–8.

[23] Katusic ZS, Vanhoutte PM. Superoxide anion is an endothelium-derived contracting factor. Am J Physiol 1989;257(1 Pt 2):H33–7.

[24] Cai H, Harrison DG. Endothelial dysfunction in cardiovascular diseases: the role of oxidant stress. Circ Res 2000;87(10):840–4.

[25] Landmesser U, Dikalov S, Price SR, et al. Oxidation of tetrahydrobiopterin leads to uncoupling of endothelial cell nitric oxide synthase in hypertension. J Clin Invest 2003;111(8):1201–9.

[26] Tang EH, Vanhoutte PM. Prostanoids and reactive oxygen species: team players in endothelium-dependent contractions. Pharmacol Ther 2009;122(2):140–9.

[27] Denniss SG, Jeffery AJ, Rush JW. RhoA-Rho kinase signaling mediates endothelium- and endoperoxide-dependent contractile activities characteristic of hypertensive vascular dysfunction. Am J Physiol Heart Circ Physiol 2010;298(5):H1391–405.

[28] Song MY, Makino A, Yuan JX. Role of reactive oxygen species and redox in regulating the function of transient receptor potential channels. Antioxid Redox Signal 2011;15(6):1549–65.

[29] Staiculescu MC, Foote C, Meininger GA, et al. The role of reactive oxygen species in microvascular remodeling. Int J Mol Sci 2014;15 (12):23792–835.

[30] Jankowski V, Tolle M, Vanholder R, et al. Uridine adenosine tetraphosphate: a novel endothelium-derived vasoconstrictive factor. Nat Med 2005;11(2):223–7.

[31] Jankowski V, Schulz A, Kretschmer A, et al. The enzymatic activity of the VEGFR2 receptor for the biosynthesis of dinucleoside polyphosphates. J Mol Med 2013;91(9):1095–107.

[32] Matsumoto T, Tostes RC, Webb RC. The role of uridine adenosine tetraphosphate in the vascular system. Adv Pharmacol Sci 2011;2011:435132.

[33] Schuchardt M, Tolle M, Prufer J, et al. Uridine adenosine tetraphosphate activation of the purinergic receptor P2Y enhances in vitro vascular calcification. Kidney Int 2012;81(3):256–65.

[34] Wiedon A, Tolle M, Bastine J, et al. Uridine adenosine tetraphosphate (Up4A) is a strong inductor of smooth muscle cell mi-gration via activation of the P2Y2 receptor and cross-communication to the PDGF receptor. Biochem Biophys Res Commun 2012;417(3):1035–40.

[35] Chen NX, Moe SM. Vascular calcification: pathophysiology and risk factors. Curr Hypertens Rep 2012;14(3):228–37.

[36] Touyz RM. BrionesAM. Reactive oxygen species and vascular biology: implications in human hypertension. Hypertens Res 2011;34 (1):5–14.

[37] Jankowski V, Meyer AA, Schlattmann P, et al. Increased uridine adenosine tetraphosphate concentrations in plasma of juvenile hypertensives. Arterioscler Thromb Vasc Biol 2007;27 (8):1776–81.

[38] Matsumoto T, Tostes RC, Webb RC. Uridine adenosine tetraphosphate-induced contraction is increased in renal but not pulmonary arteries from DOCA-salt hypertensive rats. Am J Physiol Heart Circ Physiol 2011;301(2):H409–17.

[39] Matsumoto T, Tostes RC, Webb RC. Alterations in vasoconstrictor responses to the endothelium-derived contracting factor uridine adenosine tetraphosphate are region specific in DOCA-salt hypertensive rats. Pharmacol Res 2012;65(1):81–90.

[40] Matsumoto T, Watanabe S, Kawamura R, et al. Enhanced uridine adenosine tetraphosphate-induced contraction in renal artery from type 2 diabetic Goto-Kakizaki rats due to activated cyclooxygenase/thromboxane receptor axis. Pflugers Archiv 2014;466(2): 331–42.

[41] Lesault PF, Boyer L, Pelle G, et al. Daily administration of the TP receptor antagonist terutroban improved endothelial function in high-cardiovascular-risk patients with atherosclerosis. Br J Clin Pharmacol 2011;71(6):844–51.

[42] Montezano AC, Touyz RM. Reactive oxygen species, vascular Noxs, and hypertension: focus on translational and clinical research. Antioxid Redox Signal 2014;20(1):164–82.

Further Reading

[1] Ruhle B, Trebak M. Emerging roles for native Orai Ca^{2+} channels in cardiovascular disease. Curr Top Membr 2013;71:209–35.

10

Redox Cellular Signaling Pathways in Endothelial Dysfunction and Vascular Disease

Francisco R.M. Laurindo

A peculiar aspect of the vascular system is the fact, not so evident in other systems, that the physiology of the organ directly mirrors cellular and subcellular physiology. In this regard, the key role of redox regulation in vascular cell biology translates into the evident importance of these processes on the global physiology of this system and consequently on vascular pathophysiology.

In perspective, the elaboration of a regulation paradigm of the vascular function and structure focused on redox processes became one of the major advances on vascular physiology over the last decades. Investigations supporting this paradigm have been triggered by the identification of the gaseous free radical NO* (nitric oxide) as an important vasorelaxant factor secreted by the endothelium. The idea that chemically simple mediators can behave in a biologically intelligent manner, being able to transmit specific autocrine and paracrine signals, has been expanded to other redox species, especially reactive species derived from oxygen (ROS, from "reactive oxygen species").

Some ROS interact directly with NO and govern their effects. In several cell types, particularly vessels, there are evidences for the involvement of ROS in signaling growth, proliferation, differentiation, senescence, and apoptosis. Therefore, endothelial dysfunction, and ultimately vascular dysfunction, under several perspectives, is a redox signaling dysfunction.

These redox paradigms have been essential to understand the pathophysiology of atherosclerosis, diabetic and hypertensive vasculopathy and postangioplasty restenosis. Specifically, the oxidative theory of atherogenesis has extended the notion about the pathophysiological role of vascular free radicals.

As a consequence of all these advances, the concept of antioxidant therapy has gained space and several studies have been conducted in this field. Nevertheless, the results have been essentially negative or questionable, bringing to light a complex scenario that indicates the need to obtain in-depth knowledge about the subcellular and molecular mechanisms of redox processes in order to turn into reality the translational potential of the area.

The main goal of this chapter is to review the biochemical basis and the physiological and pathological mechanisms of redox modulation of vascular function in connection with disease mechanisms. A key focus is to approach the link between experimental studies and their potential translation into relevant clinical advances.

WHAT IS A FREE RADICAL?

The concept of redox homeostasis, i.e., the dynamic equilibrium of electron transfer reactions, is intrinsically linked with the notion of free radicals. A free radical is a mediator capable of independent existence that has an unpaired electron on the outer layer [1]. As a consequence, it can donate this electron—reducing activity—or take an electron from another substance to stabilize itself—oxidative activity. Several chemical elements can form free radicals, but chemical mechanisms determine the special propensity of oxygen to form these radicals.

The main free radicals of biological relevance formed from molecular oxygen are the superoxide radical (O_2) and the hydroxyl radical (OH). The term "reactive oxygen species" (ROS) is used to name the free radicals and related nonradical mediators (e.g., hydrogen peroxide, H_2O_2) that together participate on electron transfer reactions. Hydrogen peroxide can be formed: (a) by spontaneous (low rate constant) or enzymatic (via superoxide dismutase, steady speed high) dismutation of superoxide radical; (b) by the bi-electronic reduction of oxygen by certain enzymes such as NADPH oxidases (e.g., from phagocytes or vessels). Although the use of the term reactive oxygen species has become widely known, its use has been often incorrect in the biological scenario. Each chemical species encompassed by this term has specific chemical reactivity, diffusional properties and distinctive physical characteristics, making it wrong to assume that ROS are a generic entity having a unified biological meaning. Thus,

it is always important to define and characterize which chemical species are involved in a given biological phenomenon [2,3]. Other chemical elements can generate free radicals of biological importance, discussed in Table 10.1.

The reactivity of a free radical is the main determinant of its effects and most importantly, of its toxicity. Such reactivity results from both thermodynamic and kinetic factors [4]. Thermodynamic factors can be roughly compared to a "spontaneous tendency" for the action to occur and depend on the reduction potential of the components of the redox pair. For example, the hydroxyl radical (OH) has a high oxidative character—i.e., the most negative reduction potential amongst mediators able to be formed in vivo. Kinetic factors (i.e., the "speed" by which the reaction occurs) in addition to reduction potentials, depend on accessibility, diffusion capacity relative concentration of reagents, on the velocity of reagent removal and on ambient conditions such as pH and temperature. Thus the toxicity and reactivity of a radical can be widely

TABLE 10.1   Free Radicals and Related Species of Biological Importance

Central elements	Radical	Properties
Oxygen	Superoxide (O_2^-)	Likely signal mediator in vessels. Relatively little reactive, crosses cellular membranes via anion channels. Quickly reacts wit NO, reducing its bioactivity
	Hydrogen peroxide	A relatively moderate oxidant has several characteristics of a second-messenger of signals and can originate other oxidants. Uncharged, it can diffuse and permeate membranes
	Hydroxyl (OH)	The most powerful oxidant potentially produced in biological conditions. Has an uncertain role in pathophysiology
	Singlet oxygen	It is not a radical, but the product of energetic excitation of an electron in oxygen outer layer, e.g., by ionizing radiation. Increasing evidences for possible biological effects
Nitrogen	Nitric oxide	Gaseous free radical, it is the main relaxing factor derived from the endothelium. Relatively little reactive per se. Lipid-soluble and permeable to membranes
	Peroxynitrite ($ONOO^-$)	Gives origin to powerful oxidants. Vasodilator properties whose mechanism is still uncertain
	Nitrogen dioxide (NO_2)	A powerful and reactive mediator, able to nitrate protein residues [3]
Carbon	Methyl	Possible involvement in changes in DNA
	Carbonate	Peroxynitrite by-product. Powerful oxidant. Biologic effects have become evident [3]
Sulfur	Thyil (RS)	Mediator of oxidation and thiolinterconversion. Probable biological effects, not clearly demonstrated yet

variable according to the location and environment in which it is produced [2–4].

Not all radicals are oxidant and not all of them are highly reactive. In fact, some of those mediators can have a sufficiently long average lifespan, combined with membrane permeability, to be able to perform roles as signaling mediators [5–7]. For example, superoxide radical, in general, a slightly reactive reduction agent, has a higher importance for vascular regulation than hydroxyl radical, a powerful and reactive oxidizing agent.

It must be noted that free radical reactivity is a relative concept under the thermodynamic standpoint, i.e., a species tends to be oxidizing in relation to another that has a more negative reduction potential, but it can be reductive in relation to another redox pair with a more positive potential [4]. Besides that, external factors such as pH are also important. The superoxide radical, for example, can act as oxidant and exhibits increased reactivity in acidic environments: in this case, the radical, which at pH 7.4 is an anion ($O_2^{\bullet-}$), is transformed into a non-charged radical, the hydroperoxyl radical (ROO) [1].

BIOCHEMICAL CHARACTERISTICS OF REDOX PROCESSES: RADICAL AND NONRADICAL PATHWAYS

The biochemistry of redox processes comprises radical and nonradical pathways [7], respectively, indicating if these processes involve transfer of one or two electrons. Some key features of the reaction pathways involving free radicals are as follows:

(a) The possibility of forming toxic products from less reactive radicals. For example, the interaction between two less reactive radicals, superoxide radical and nitric oxide, can generate peroxynitrite ($ONOO^-$), a mediator whose decomposition gives rise to powerful oxidants, including nitrogen dioxide (NO_2) radical. This is an important route of potential free radical toxicity in biological systems.

(b) The possibility of chain reactions. For example, the nonenzymatic peroxidation of lipids; such as those from cellular membrane and organelles-one of most well-known toxic effects of oxidizing radicals—is a chain reaction that can lead to changes in permeability and loss of membrane structure. Peroxidation is the introduction of a peroxyl radical (ROO) in a substrate. An example of enzymatic peroxidation is the arachidonic acid oxidation by cyclo- or lipoxygenase; another example, a beneficial action in this case, is the removal of the hydrogen peroxide and other organic peroxides by glutathione peroxidases, as in the reaction: $GSH + H_2O_2 \rightarrow GSSG + H_2O$ [1].

(c) Another important aspect of radical processes is the interaction with metal ions. The reaction between hydrogen peroxide and superoxide radical leading to hydroxyl radical is spontaneously slow, in addition to being thermodynamically unfavorable, a reaction known as the Haber-Weiss reaction. This reaction is, however, accelerated in the presence of catalytic metal complexes, especially iron and cooper on their reduced states (i.e., Fe^{2+} and Cu^+). One of the agents that can reduce Fe^{3+} to Fe^{2+} and make it available for such catalysis is superoxide radical itself. The production of hydroxyl radicals from hydrogen peroxide and reduced metals is known as the Fenton reaction. The combination of Fenton/Haber-Weiss reactions is a widely studied mechanism of potential formation of hydroxyl radicals in vivo. In this context, one of the main antioxidant mechanisms of the extracellular environment is to keep iron in a state unable to catalyze the Fenton reaction. It is also important to emphasize, though, that in vivo, the occurrence of Fenton reaction should not be accepted without criticisms and there are few clear evidences in fact supporting the importance of this mechanism [4]. Moreover, this reaction is not the sole potentially toxic mechanism of transition metals; as they can also directly induce lipid peroxidation (see below). Furthermore, superoxide radical can release iron linked to ferritin or to proteins with an iron-sulfur nucleus (4FE-4S, e.g., aconitase).

Nonradical pathways of redox processes involve species capable of perform 2-electron oxidations, thus not forming an intermediate free radical. These routes can involve, particularly, hydrogen peroxide itself, but also include lipid hydroperoxides, aldehydes, quinones, peroxynitrite, and disulfides [7]. These pathways converge to regulatory targets in thiol-proteins and are controlled by glutathione (GSH), thioredoxin and cysteine. Nonradical routes have been increasingly implicated in redox signaling and can have a high quantitative importance, although these are not completely clarified yet. It is interesting to note that these routes are probably not sensitive to antioxidants specifically aimed to target radical routes, e.g., antioxidant vitamins.

ANTIOXIDANTS: THE CLASSICAL CHEMICAL CONCEPT

The term "antioxidant" classically designates compounds that, when present in low concentrations in relation to oxidizable substrates, significantly slow down or prevent the oxidation of such substrates [1]. Broadly, there are the preventive oxidants, whose archetypes are intracellular enzymes such as superoxide dismutase (which removes superoxide radical), catalase and glutathione peroxidase (both removing hydrogen peroxide). Recently, the peroxiredoxin family was identified as the main route to detoxify and at the same time exert signaling (see the discussion below) involving hydrogen peroxide. Peroxiredoxins are present at high intracellular concentrations and have a high rate constant of reaction with hydrogen peroxide, being regenerated by reductive cystolic agents [6,8]. On the other hand, in the extracellular environment, the majority of antioxidant power is determined by small repairing molecules (i.e., which cancel the radical after it has been produced), whose main representatives are vitamins C (ascorbic acid, which is soluble in water) and E (alpha-tocopherol, which concentrates itself in the membrane lipids), urate, in addition to thiol-protein groups (especially those from albumin).

In addition to a high rate constant of reaction with the radical to be antagonized, one of the main desirable features of an oxidant is to be accessible to the site of production said oxidant. For example, the exogenous administration of unmodified form of superoxide dismutase is not always efficient, because this protein is not rapidly permeable to intact membranes and has restricted access to the cell. Another example is LDL oxidation, which is preferably inhibited by lipophilic compounds linked to the particle, e.g., vitamin E or probucol.

The notion that good antioxidants should be powerful reductive agents is incorrect since powerful reductive agents can potentially generate superoxide radical from oxygen [4]. Moreover, excess cellular reductive agents can promote a condition called reductive stress, which will be discussed in latter sections. Adequate antioxidants present a standard potential of reduction close to zero, i.e., neither very oxidative nor very reductive. The by-product of the activity of those compounds is thus a stable and innocuous radical, e.g., ascorbyl or alpha-tocopherol. However, at greater concentrations, even these radicals can perform toxic effects, explaining why, in certain circumstances, antioxidants can have a prooxidative effect. Another characteristic is the possible synergism or interaction between antioxidants. The most well-known example is the role of ascorbic acid in the action of vitamin E; ascorbic acid, besides its antioxidant power, acts by repairing alpha-tocopheroxyl radical and thus, is able to regenerate the active vitamin E [4]. The negative interaction, i.e., cancelation of one antioxidant effect by another, is also possible and has been demonstrated in some examples.

OXIDATIVE STRESS AND CELLULAR REDOX SIGNALING: ADVANCING THE CONCEPT

The involvement of ROS with the pathophysiology of diseases, including those from vascular origin, is linked to the concept of oxidative stress. The classic concept of oxidative stress was formulated in the 1980s (revised in Ref. [9]), in the context of the notion of ROS as harmful mediators toward biomolecules. In this context, oxidative stress has been defined as a disequilibrium between the production of oxidant mediators and the cellular antioxidant capacity, leading to oxidative damage to of several cellular components, such as lipids, proteins, and sugars. The concept of oxidant stress emerged from a solid biochemical base to constitute a metaphor with a great power of contextualization and communication of science in the field. However, several basic concepts have evolved and modified the premises that used to limit the mechanistic force of such metaphor, though its efficiency to communicate and contextualize the science of the field is still valid [9]. In fact, the evolution of the concept of oxidative stress has been one of the main advances of the last several years regarding redox science. Some basic points of this evolution are as follows:

(1) The inefficacy of clinical studies with classic antioxidants in cardiovascular diseases, cancer,

metabolic disease, and others indicate that redox cellular dysfunction is more complex than a simple oxidant/antioxidant disequilibrium.

(2) The study of cellular models, genetically modified animals and improvement in oxidant detection methods has conclusively demonstrated that the production of oxidants is a physiological event that does not occur "by accident."

(3) In line with this concept, several enzymatic ROS producers have been characterized at a molecular level, especially the NADPH oxidase family, whose specific function is dedicated to such production.

(4) The discovery of several proteins containing thiol groups (—SH or sulfhydryl) associated with the amino acid cysteine, whose specific function involves reversible redox changes in its structure (true "redox receptors" aimed at physiological transduction of cellular signals).

(5) In this context, new intermediate states of thiol oxidation have been characterized, suggesting an unanticipated biochemical specificity [6].

(6) Other redox reaction routes have been characterized, involving physiological phenomena.

(7) The existence of several molecular routes activated by oxidants (Nrf2 and FOXO) able to induce antioxidant protective cellular signals has been characterized.

(8) Recently, the existence of proteins that have the ability to act as redox sensors, e.g., peroxiredoxins, has been discovered [6,7,9].

Together, these advances led to the concept of redox signaling as well as to the notion of oxidative stress as, first and foremost, a disruption of redox signaling, not necessarily involving biomolecular damage, even though the latter can be associated.

WHAT IS REDOX SIGNALING?

Redox signaling is the transduction of signals coding for cellular processes in which the integrative elements are electron transfer reactions involving free radicals or related species, redox-active metals (e.g., iron, copper, etc.) or reductive equivalents. A typical reductive equivalent is the hydrogen atom donated by reductive substrates such as NADPH, reduced glutathione (GSH) or thiol-proteins (RSH). A primary attribute of redox signaling is its strict dependence on kinetics and thermodynamics of electron transfer. At the same time, biological factors such as the nature of the enzymatic sources of free radicals, their cellular subcompartmentalization and the interaction with other proteins are crucial determinants of effector redox signals. The distinction between signaling and toxic redox processes is not always obvious, and some of these characteristics are listed in Table 10.2.

The specificity of redox signaling is based on two pillars. First, it is based on the formation of specific chemical species, as mentioned previously, for example, the intermediate states of thiol oxidation. Second, the specificity is given especially through to the existence of effector target proteins, reversely modulated by redox-sensitive mechanisms, e.g., thiol-disulfide chemistry or redox-active metals. Examples of such target proteins include several kinases, phosphatases, transcription factors, receptors, adherence molecules,

TABLE 10.2   Free Radicals and Oxidative Stress: Signaling vs. Toxic Effects

	Signaling	Toxic
Quantity of radical produced	Pico/nanomolar	Nano/micromolar
Location	Restricted	Widespread
State of antioxidant defenses	Usually accessible and efficient, often induced by oxidative stress itself	Typically inaccessible or inefficient; can be induced if there is sub lethal stress
Affected cellular target	In general, proteins that are specifically controlled by redox mechanisms (e.g., thiol, metals)	Same, including several cellular components not usually controlled by redox mechanisms
Organization	Modular	Modularity loss [9]

TABLE 10.3   Some Typical Examples of Signal Transduction Mediated by Redox Processes

Critical role of the intracellular hydrogen peroxide in the effects of PDGF (platelet-derived growth factor)

Redox modulation of tyrosine-phosphatases involved in growth inhibition via cellular contact

Superoxide production in fibroblasts ras-transformed

Activation of p38 MAP kinase by angiotensin II mediated by hydrogen peroxide

Angiotensin II vasopressor effects mediated by superoxide

Inhibition of proliferation or induction of apoptosis upon catalase overexpression in smooth muscle cells

ROS effects in Akt kinase activation by angiotensin II

ER stress signaling in vascular cells

Induction of cellular senescence in vascular hypertrophy

Activation of inflammatory cell adhesion molecules

and proteases, which are critical determinants of pro-liferation, survival or apoptosis [5–7,9]. Some of the countless prototypical examples of redox signaling in vessels are listed in Table 10.3, highlighting that protective as well as harmful phenomena can be mediated by ROS. Such processes constitute the basis for the proposed involvement of ROS and redox processes in general on atherosclerosis and other vascular diseases.

From an evolutive perspective, it makes sense to consider that aerobic organisms have adapted in order to control the toxicity of ROS while at the same time developing mechanisms to use their powerful biological effects as second messengers for autocrine or paracrine regulation of cellular processes.

A CONCEPTUAL REDEFINITION OF ANTIOXIDANTS

Considering that the definition of oxidative stress has exhibited a significant evolution and that ROS can be considered second subcellular messengers, it becomes necessary to revisit the classic definition of antioxidants. In general, oxidative stress can be prevented or repaired by interventions that block metabolic pathways that produce ROS (e.g., mitochondria or NADPH oxidases) or that emulate or multiply the natural physiological defense mechanism. The hierarchy of such mechanisms varies according to the environment; in the cytoplasm the main defense mechanisms are enzymatic, while in plasma, small molecules account for the majority of antioxidant power. In the vascular interstitium, high concentrations of extracellular SOD isoenzyme suggest a

possible defensive role [10]. On the other hand, evolution of the redox signaling concept carries the notion that any intervention capable to rebalance or restore such pathways can be considered an "antioxidant." Thus, the conceptual broadness of antioxidants must be considerably expanded [9].

In addition, an emerging concept is of *Redox Hormesis*, which consists of induction of endogenous antioxidant defenses in face of a nonlethal oxidant challenge [11,12]. This seems to be the mechanism of action of several natural products such as flavonoid antioxidants. Their in vivo action is not directly antioxidant; they act instead via a hormetic mechanism. This is the case of resveratrol, lipoic acid, sulphoraphane and several other related products [11,12]. Physical training is by several angles a redox hormesis pathway and it was reported that administration of large dose of antioxidants immediately before a physical training session prevented the conditioning effects of such training [13]. Thus, the consumption of flavonoids or physical exercise activate, respectively by direct routes or by an oxidant challenge, protective antioxidant molecular pathways, promoting a final antioxidant effect. These protective pathways include the transcription factors Nrf2 and FOXO, which bind to promotor sequences in genes that code for antioxidant proteins.

ENZYMATIC SUPEROXIDE RADICAL SOURCES IN VESSELS

Enzymatic ROS generation is at the same time a requirement and a corollary of the redox signaling concept, since enzymatic pathways can provide fine mechanisms of catalytic activity regulation, thus governing their production tightly. At the same time, enzymes allow a spectrum of posttranslational changes and protein-protein interactions, capable of locating such activity in specific subcellular compartments, e.g., caveolae, vesicles, lamellipodia, or focal adhesions. Finally, enzymes allow controlled interaction with enzymatic antioxidants regarding protein-protein interactions, adding a new level of control to ROS production. Under the translational standpoint, enzymatic ROS sources are relevant therapeutic targets for development of new antioxidant therapies.

Known ROS enzymatic sources include the NADPH oxidase complex, mitochondria, uncoupled NO synthases, xanthine oxidase, cytochrome P450, and cyclooxygenase, among others (Fig. 10.1). Of those, the NADPH oxidase will be commented due to its redox signaling importance, while the role of the NO synthases uncoupling will be discussed in the chapter dealing with endothelium-dependent vasodilators.

Vascular NADPH Oxidase

One of the most important advances of redox cell biology has been the discovery that vascular cells,

and ultimately, all cell types, present the expression and activity of NADPH oxidase enzymatic complexes from the NOX family analogous to those initially identified in phagocytes [14–16]. In the latter ones, the NADPH oxidase function is bactericidal, while in other cells, the multiple functions are basically of cellular signal transduction. The NADPH oxidase complexes are the only specific sources dedicated to ROS production (e.g., ROS are not produced as a side effect of the activity, but as a product of catalysis) and have the capacity of fine and regulated signaling of processes located in specific compartments, e.g., cellular migration poles [14]. The regulation of the activity and expression of this complex is an important theme in pathophysiology of redox processes and it is involved in several vascular diseases. The NADPH oxidase complex has a modular structure with several subunits, of which one is a transmembrane catalytic subunit (Nox) exhibit binding sites for flavin and haeme, while several other subunits perform regulatory roles, and their combination varies according to the Nox isoform [14] (Table 10.4).

The catalytic transmembrane subunit transfers one electron, donated by the NADPH substrate (cytosolic) via flavin and haeme groups, to molecular oxygen, generating superoxide radical or, in the

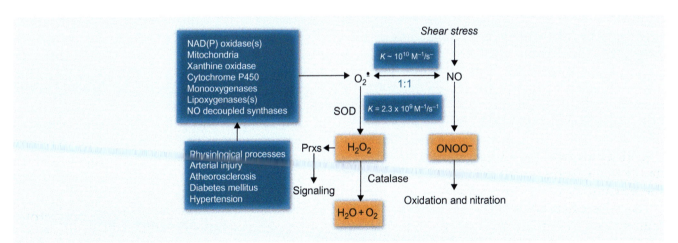

FIG. 10.1 Scheme illustrating the overall panorama of redox signaling pathways and oxidative stress in vessels. Physiologic processes and several signals that occur in disease pathophysiology trigger the activation of several ROS enzymatic sources, among which NADPH oxidase and mitochondria have been the most studied and the key ones in quantitative terms. Superoxide radical (O_2) quickly reacts with endothelial NO (whose main physiological stimulus is the laminar shear stress) in an extremely rapid reaction that overruns the superoxide scavenger superoxide dismutase (SOD). This reaction promotes the removal of bioactive NO via oxidative stress and it is the main pathway by which oxidative stress originates vascular and endothelial dysfunction. Hydrogen peroxide (H_2O_2) formed by spontaneous superoxide dismutation or as a product of SOD can be removed by catalase or used as second messenger for cellular signaling. In this case, an important pathway is the reaction with peroxidases, e.g., from the peroxiredoxin group (Prxs), which act as H_2O_2 sensors and can transfer oxidant equivalents to other target proteins.

TABLE 10.4 Vascular NADPH Oxidase

	Catalytic subunits			Associated regulatory subunit	Main cellular effect
	End	**VSMC**	**Fib**		
Nox1	+	+	0/+	p22 phox Noxo 1 Noxa 1 p47phox (?)	Proliferation Migration
Nox2	++	++	+++	p22 phox p47 phox p67 phox p40phox (?)	Proliferation Migration Inflammation
Nox4	+++	+++	+	p22 phox	Differentiation Apoptosis
Nox5	+	+	(?)	p22 phox	Ca^{2+}-dependent responses

End, endothelial cell; *VSMC*, vascular smooth muscle cell; *Fib*, fibroblast.
Quantification of + to +++ represents semiquantitative estimates collected from several literature references. (?), unclear.

case of Nox4, hydrogen peroxide. The Nox family has also two Duox subfamily members (dual oxidases), relatively less expressed in vessels. In all instances, the production of reactive species occurs at the membrane side opposite to the cystolic one, which can be the extracellular compartment or the lumen of a vesicle or phagosome [14] (Fig. 10.2). In other words, there can be Nox-related intracellular ROS production ROS production, but in compartments topologically related to the extracellular environment. The existence of these various subunits, which determine the activity of the enzyme only when arranged like a puzzle, reflects the complex regulation of an enzyme complex that can be potentially lethal to the cell. Table 10.4 summarizes the main catalytic and regulatory subunits present in vascular cells as well as their associated functions. Fig. 10.2 illustrates the arrangement of the subunits in the resting state as well as following activation of the vascular or phagocytic oxidase.

Therefore, first-order factors involved in regulating Nox regulation are the identity and structural characteristics of catalytic subunit present in each cell type, which vary according to physiologic or pathologic conditions. The second factor is the traffic and phosphorylation, (or other posttranslational modifications) of regulatory subunits listed in Table 10.4. Moreover, regulatory interactions with several proteins that do not belong to the usual NOX-regulatory group have been described. Overall, all such regulatory proteins connect the oxidase activation to specific physiologic programs and/or

target complex activation to cellular microdomains in a temporal and spatial manner.

Several studies indicate that NOX enzymatic complexes are the main producers of ROS with cellular signaling function on the vascular wall, both in endothelial cells and smooth muscle cells. In addition, the adventitial layer presents a large activity of this enzyme and is an important site of superoxide production in the vascular wall, maybe with the function of inactivating endothelium-derived NO. There is usually low/moderate basal vascular NADPH oxidase activity which accounts for nanomolar levels of reactive mediators [14–16]. However, expression and activity of vascular NADPH oxidase can be stimulated by several factors such as angiotensin II, bradykinin, thrombin, platelet-derived growth factor (PDGF), cytokines such as TNF-alpha, ceramides and mechanical stretching [14–16]. Angiotensin II is a particularly important agonist: about 70% of the vasoconstrictor effects of a sustained angio can be attributed to superoxide radical production. Vascular NADPH oxidase activation is an essential element in transducing of cell proliferation signals triggered by AT-1 angiotensin receptor activation [14]. Several angiotensin effects on kinases associated with cell proliferation and in the synthesis of proteins are antagonized by the NADPH oxidase inhibition. Thus, inhibitors of angiotensin converting enzyme or AT-1 receptor antagonists arguably behave as antioxidant interventions in vessels. Nox1 and Nox4 are expressed according to the specific smooth muscle cell

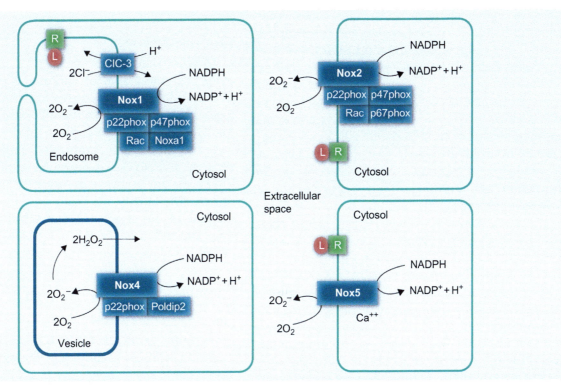

FIG. 10.2 Diagram showing the structure of several isoforms of the NADPH oxidase complex from vascular cells. The catalytic subunits (Nox1, 2, 4, or 5) are the central nucleus of electron transfer in the complex, containing the heme-transmembrane group and the binding site to flavin (in the cytosolic side). In the case of Nox 1, 2, and 4, they are heterodymerized with the transmembrane regulatory subunit p22phox, which stabilizes the complex. The one-electron reduction of O_2 generates superoxide, using NADPH as reductive equivalent source. Rac is low molecular weight G protein which has a regulatory function, while p67phox and p47phox are regulatory cytoplasmic subunits that migrate to the oxidase complex at the membrane when it is activated, following their phosphorylation. In the case of Nox1, p47phox, and p67phox analogs, known as Noxo1 and Noxa1, perform this role, even though p47phox is also considered important. The production of superoxide always occurs at the side opposite to the cytosolic. Modified from Lassègue B, San Martín A, Griendling KK. NADPH oxidases: functions and pathologies in the vasculature. Arterioscler Thromb Vasc Biol 2010; 30:6653–61.

phenotype: if in a proliferation phase there is a predominant Nox1 expression while under quiescence, there is a predominant Nox4 expression.

NADPH oxidase complexes from the vascular system or other nonphagocytic systems significantly differ from the phagocytic complex, which has an eminently bactericidal function. The vascular enzyme, compatible with its role in cellular physiology, presents lower and more sustained production of hydrogen peroxide or superoxide. Additionally, the leukocyte enzyme has a "burst-like" activity, quickly producing high concentrations levels of ROS, whilst the vascular enzyme generates more stable flows. Furthermore, the leukocyte enzyme complex is essentially activated by the arrangement of subunits already stocked in the cells, while the vascular NADPH oxidase complex requires additional synthesis transcriptional activity of the

subunits [14–17]. Nox4 and Nox2, in the carotid-body cell or in the pulmonary vessels, can work as oxygen sensors [15,16]. Analogs of gp91phox have the function of Fe^{+2} carriers in prokaryotes and plays roles in the synthesis of thyroid hormones. From an evolutive perspective, the leukocyte enzyme represents the specialization of a more ancestral enzyme in several species. In conclusion, the term "vascular NADPH oxidase" actually comprises an uneven mix of structurally different and variable (according to the cell type) enzymatic complexes.

Mitochondrial ROS Production

During oxidative phosphorylation processes, essential to the energetic cellular efficiency, mitochondria consume oxygen to produce ATP, but in this process they can also produce ROS. Earlier estimates that such ROS production could correspond up to 1%–2% of cellular oxygen consumption are probably exaggerated; the most plausible values in physiological situations are 0.1%–0.3% and can significantly increase in conditions such as mitochondrial and metabolic dysfunction [18]. However, mitochondria are perhaps, in quantitative terms, still the main source of ROS in most cells, even though this may not been the case in vascular cells. The partial reduction of molecular oxygen, leading to the production of superoxide and subsequently of hydrogen peroxide, can especially occur at the level to complexes I and 3 of the electron carrier chain [18,19]. Mitochondria express a specific SOD isoform in the matrix MnSOD (or SOD2), and in the intermembrane space there is the CuZnSOD (SOD1) isoform. Moreover, there are specific isoforms of peroxiredoxins and thioredoxin 2. Thanks to the high concentration of these antioxidants, mitochondria tend to maintain their oxidant concentrations at safe levels. In this context, mitochondria are the most evident example of the compartmentalization of subcellular redox processes, since a great amount of the produced reactive species is contained inside the organelle itself under normal conditions. Mitochondrial-derived ROS are not top candidates to exert fine regulation of cellular signaling process, since extra-mitochondrial ROS leakage is usually considered poorly compartmentalized and specific [9]. This idea has been recently changed and ROS from mitochondrial origin have been described as mediators of several signaling processes, e.g., response to hypoxia and several cellular and systemic metabolic control pathways [19], which affect processes such as gene transcription, proliferation, differentiation, mechano-transduction and many others. Particularly, it is well accepted that hydrogen peroxide of mitochondrial origin produced increases in shear stress has vasorelaxant and hyperpolarizing activity in arterioles [20]. Mitochondrial ROS are important mediators of hunger and satiety in the central nervous system [19]. Specifically, redox processes associated with mitochondria are important regulators of cellular apoptosis. Mitochondria have also a significant role in redox regulation of the physiologic response to physical exercise. The regulation of oxygen consumption by the mitochondria is highly influenced by the mitochondrial concentration of nitric oxide [18]. Mitochondria significantly converge with Nox NAPDH oxidase through of bidirectional regulatory pathways.

Xanthine Oxidase and Oxygenases

Xanthine Oxidase

Xanthine oxidase normally produces ROS during oxidative conversion of xantina to uric acid. The importance of this enzyme as ROS producer in biological systems has been initially proposed in ischemia re-oxygenation situations [1]. In humans, this role has not been uniformly confirmed, partly because the levels of xanthine oxidase are very low in certain tissues, e.g., myocardial cells. In atherosclerosis, there are evidences implicating this enzyme in the genesis of the endothelial dysfunction. The endothelial cell has detectable levels of this enzyme, which can be induced by certain cytokines. Moreover circulating xanthine oxidase attaches to the basement arterial wall membrane [21]. In smokers, xanthine oxidase inhibition with allopurinol corrects the endothelial dysfunction [21]. Other nonredox by-products of xanthine oxidase can eventually contribute to vascular pathology [22].

Cyclo and Lipoxygenases

The role of cyclooxygenase as a superoxide radical producer has been suggested from initial studies in brain circulation. Other studies ascribed to certain lipoxygenases an analogous role regarding the genesis of lipoprotein oxidation [23]. The vasoconstrictor effect of cyclic endoperoxides (PGG$_2$ and PGH$_2$) has been attributed to the production of superoxide. Lipoxygenases are markers of atherosclerosis susceptibility in mice. However, clinical studies have not confirmed the robust importance of these enzymes in the production of vascular free radicals in atherosclerosis.

Cytochrome P450

The potential role of monooxygenases centered on cytochrome P450 for the production of ROS has been postulated several years ago. However, only recently experimental data have consistently demonstrated the roles of these enzymes in the production of ROS by vascular cells. The relative importance of cytochrome P450 in the total production of ROS in physiological and pathological situations is still unclear, but it is interesting to mention that this enzymatic pathway is associated in the endothelium to the production of a family of epoxides with a smooth muscle hyperpolarizing activity, reminiscent of the so-called EDHF (endothelium derived hyperpolarizing factor).

ENDOPLASMIC RETICULUM AND OXIDATIVE STRESS

The endoplasmic reticulum (ER) is the organelle responsible for synthesis, folding and posttranslational processing of proteins destined to cell secretion or insertion into membranes, corresponding to about 50% of total protein load. The ER is also the main site of lipid metabolism, intracellular calcium control and several other functions. Growing evidences directly or indirectly implicate the ER in redox processes in physiological and pathological situations. The main ER function in this context is the redox folding nascent proteins, i.e., the insertion of disulfide bridges. This function, essential for cellular homeostasis, is performed by a family of proteins located primarily in the ER, protein disulfide isomerases (PDI), whose prototype is PDI itself (PDI or PDIA1), a dithiol redox chaperone belonging to the thioredoxin superfamily [17,24,25]. The catalysis by PDI of protein folding in the ER lumen is supported by the regeneration of its disulfide (i.e., reoxidation) via the flavin oxidase Ero1 (ER oxireductin). In turn, the reoxidation of Ero1 requires the electron transfer from its active site to molecular oxygen, generating a significant hydrogen peroxide flux. Recent evidences suggest that this peroxide can oxidize in a productive way peroxiredoxin-4 and glutathione peroxidase 7/8, which transfer their oxidant equivalents to PDI, promoting protein folding. This fact is of great importance to understand the redox equilibrium of ER. In other words, the ER has a redox-folding pathway that produces oxidants and at least two that consume this same oxidant, thus providing adequate redox balance. In each case, PDI itself and possibly other family members constitute the central regulatory hub for ER redox homeostasis [24].

Correct protein folding is an essential function to maintain cellular homeostasis. Given the importance of such mechanism, the ER has a sophisticated protein quality control system, preventing incorrect protein folding and consequent formation to aggregates, harmful to the cell. ER stress is a frequent condition, resulting from the inability of ER to adequately process recently synthesized proteins, causing accumulation of misfolded proteins. ER stress triggers a complex signaling network known as the unfolded protein response (UPR). UPR is involved in the genesis of several diseases, including cancer, neurodegenerative diseases, cardiovascular diseases such as myocardial hypertrophy and,

particularly, atherosclerosis and its risk factors such as obesity and insulin resistance [26]. UPR signaling (Fig. 10.3) involves prosurvival adaptive pathways, which induce a decrease of the protein translation load (via phosphorylation of eIF2alpha factor), expression of antioxidant genes, enhanced folding capacity (increased number of resident chaperones), degradation of un/misfold proteins by the proteasome and of protein aggregates by autophagy. Concomitantly, the UPR activates proapoptotic pathways that promote cell death when adaptive mechanisms cannot restore ER homeostasis. Several evidences suggest convergence between redox processes and UPR through a model in which the UPR triggers ROS production and at the same time the production of ROS sustains UPR signaling [24,26]. Although this convergence has been well documented, oxidative stress mechanisms during the UPR are complex and not yet clear [27]. One ROS generation pathway during UPR is the induction of the Nox4 NADPH oxidase isoform, in addition to activation of Nox2 in the initial phases of apoptosis, and connections between ER and mitochondria [17,26,27].

In parallel, the maintenance of a relatively oxidizing environment in the ER lumen, expressed as a relatively low GSH/GSSG ratio, is essential to provide an appropriate environment for the formation of protein disulfides. Thus, in vitro incubation of cells with powerful reductants, such as dithiothreitol, induces ER stress. Recently, it has been reported that in some situations, the cells produce an excess of reductive equivalents, known as "reductive stress" [28]. Reductive stress has important cellular toxicity, caused, at least in part, by a lower capacity for redox protein folding in the ER [26,27]. Mechanisms triggering reductive stress are not yet clear. In some situations, there are evidences supporting that an overcompensation of oxidative stress by the hormetic systems Nrf2 or FOXO (previously discussed) can promote reductive stress [28]. In addition to the redox environment, maintenance of high intrareticular concentrations of Ca^{+2} is essential for all the stages of the secretory process. The ER present physical interface platforms with mitochondria. These platforms are called "MAM" (mitochondrial-associated membranes) and compose signaling centers for the transport of calcium and probably lipids, in addition to having functions in protein folding [26,27]. In parallel, the chaperone function of protein quality control includes the correct incorporation of carbohydrates and traffic to the Golgi

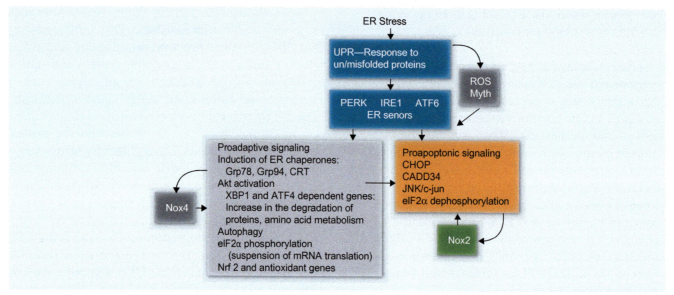

FIG. 10.3   Scheme demonstrating the main activated signaling pathways during UPR (unfolded protein response), secondary to ER stress. Endoplasmic reticulum (ER) stress, secondary to the imbalance between protein synthesis load and capacity for their processing by the ER, produces an accumulation of unmisfolded proteins in the ER lumen, triggering the UPR. The UPR is a network of signals initially promoted by un/misfolder protein sensors located at the ER (PERK kinase, IRE1 kinase/endonuclease and ATF6 transcription factor). These sensors generate a series of nuclear signals that codify genes for proadaptive responses related to the increase in protein processing capacity by the ER (ER chaperones), metabolic adaptations (amino acids), un/misfolded protein degradation by the proteasome, autophagy, antioxidant genes and survival pathways. Proadaptive signaling tends to correct stress, inhibiting the further signaling. At the same time, genes that codify proto-apoptotic genes are also transcribed. EiF2alpha phosphorylation stops proteic translation, decreasing the load of proteins as an adaptive process. However, a prolonged maintenance of such phosphorylation can cause cellular damage. At the same time, dephosphorylation of eiF2alpha (by the GADD 34 factor) can, in some instances, lead to apoptosis. Although apoptotic signaling is generally a result of inefficient adaptation, both mechanisms are coactivated during the UPR. Modified from Laurindo FR, Araujo TL, Abrahão TB. Nox NADPH oxidases and the endoplasmic reticulum. Antioxid Redox Signal 2014;20:2755–75.

and post-Golgi system. Therefore, it is not unexpected that some endogenous or exogenous ER stress induces are directed against these factors. In the case of exogenous stimulus, the most typical examples are thapsigargin, an inhibitor of Ca^{+2}-ATPase of ER, tunicamycin, an inhibitor of protein glicosylation and brefeldin A, a disruption of Golgi transport.

The main operational UPR markers are listed in Table 10.5 and are frequently used to "detect" ER stress in tissues and cells.

Another aspect related to ER involvement in several redox pathways is the effect of PDI on the regulation of Nox NADPH oxidase [24], described and characterized in detail by our group. PDI acts as a regulatory protein and maybe organizer of NADPH oxidase in vascular smooth muscle cells [29,30], as well as macrophages [31], endothelial cells and neutrophils [32]. Experiments employing PDI loss-of-function with various tools including small-interference RNA, consistently showed that PDI exerts functional modulation of the oxidase and particularly is required for angiotensin II mediated-activation. Additionally,

TABLE 10.5   Some UPR Operational Markers [26]

ER stress sensors

IRE1 Kinase/Endonuclease phosphorylation

PERK Kinase phosphorylation

Cleavage and nuclear migration of ATF6 transcription factor

UPR signaling pathways

Phosphorylation of elF2alpha factor

XBP1 transcription factor (mRNA) splicing

Nuclear migration of ATF4 factor

Nuclear migration of CHOP transcription factor

Expression of KDEL chaperones: Grp78, Grp94, calreticulin, Orp150 our data indicate that PDI colocalizes or coimmunoprecipitates at least with p22phox, Nox1, Nox2, and Nox4 subunits from NADPH oxidase, indicating close association with the enzymatic complex [24]. During oxidase activation, PDI translocates to membrane(s) fraction(s), where it seems to sustain the activity of NADPH oxidase complex via still unknown mechanisms. Acute overexpression of PDI in vascular smooth muscle cells promotes spontaneous activation of NADPH oxidase and increased Nox1, mRNA levels, in an agonist-independent fashion [30]. Moreover, PDI is essential for migration of smooth muscle cells stimulated by PDGF, a Nox1-dependant effect. Of note, PDI silencing in this model leads to decreased activities of RhoGTPases Rac1 and RhoA, which among other effects, contribute to Nox1 activation and cytoskeleton organization. In fact, the cytoskeleton organization was substantially disrupted by PDI silencing [33].

The effects of PDI on NADPH oxidase are in line with the well-known subcellular traffic of PDI to vesicles and cellular surface. PDI is involved in the subcellular transport and secretion of several proteins [24,25]. In the plasma membrane, it promotes redox modulation of surface proteins, among which integrins have been the most well-studied, in line with the reported key role of PDI in the redox control of platelet adhesion and aggregation [24,34]. PDI interacts via redox pathways with platelet integrins and is important for thrombosis generation. PDI antagonists constitute potentially innovative interventions, currently undergoing clinical tests [24]. Such PDI actions at the surface of endothelial cells and of platelets indicate new redox pathways governing vascular thrombogenicity, with significant potential therapeutic implications.

Together, these considerations suggest that the ER is a new and important component of redox signaling paradigms in physiology and pathophysiology.

REDOX PROCESSES IN VASCULAR DISEASES: VASCULAR REPAIR TO INJURY AS AN ENDOTHELIAL/ VASCULAR PATHOPHYSIOLOGY MODEL

Basically, all vascular diseases, in particular atherosclerosis, occur in association with the triad of oxidative stress, inflammation, and endothelial dysfunction, which are themselves strongly connected. Moreover, ER stress contributes to each of these

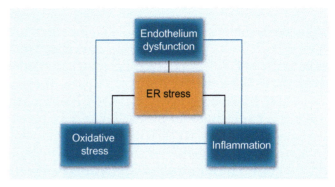

FIG. 10.4 Oxidative stress is intrinsically connected to endothelial dysfunction and inflammation. These processes permeate the natural history of genesis, evolution, and complications of essentially all the vascular diseases. Recently, several evidences indicate that endoplasmic reticulum (ER) stress contributes to these processes and it is an important component of vascular disease pathophysiology.

processes (Fig. 10.4). In this session, we discuss some evidences for the involvement of redox pathways on the pathophysiology of the vascular response to injury and atherosclerosis. Other diseases will be discussed in specific chapters and have been reviewed in the literature [14–16].

The vascular repair response to injury has a broader significance as a prototype of several common events responsible for physiological adaptations or pathological changes in a wide range of conditions, e.g., endothelial dysfunction, flow-induced remodeling, hypertension, and atherosclerosis [35,36]. Such studies have defined thrombotic and inflammatory mediators responsible for the initial inflammatory phase, which is also accompanied by massive cell loss and degradation of extracellular matrix. Several possible growth factors and cytokines responsible for the subsequent proliferative phase have been described. The late resolutive stabilization phase is governed by residual proliferation, cell migration, apoptosis, remodeling, and extracellular matrix secretion. The typical element of this injury reaction to injury, similarly to atherosclerosis, is the development of a neointimal layer. Pluripotent stem cells from several sources can give rise to the neointima, although the proportion of these cells is believed to vary according to the underlying pathophysiological process. While stem cells seem to be the main source of neointimal cells after balloon injury in the case of atherosclerosis, medial layer cells seem to be the predominant source of neointima [37].

Our group has investigated redox processes responsible for vascular response to injury for many years [35,36,38–42]. These studies and other literature reports have composed the so-called "redox hypothesis of restenosis," the basic aspects of which have been previously reviewed [35,36]. The administration of superoxide dismutase, an enzyme able to remove superoxide radicals, significantly prevents vasospasm immediately after lesion, probably involving a robust activation of NADPH oxidase(s) [40] (Fig. 10.5).

Additionally, in later phases of vascular repair, there is a sustained production of oxidants during the resolutive injury phase together with expression of vascular NADPH oxidase subunits [39,41–43]. In parallel, a decrease in SOD activity, associated with lower bioactivity of NO produced by iNOS, contributes to constrictive remodeling after injury, which is the main mechanism responsible for vascular lumen reduction. Exogenous replenishment of SOD3 (extracellular), even when started 7 days after vascular injury, restores NO production and preserves

vascular caliber via improved remodeling, without changing the neointima [41]. Thus, after an important initial production of oxidants just after injury, the neointima appears to maintain a prooxidant state in which activation of NADPH oxidase coexists with SOD underactivity. Moreover, neointimal cells are under increased secretory load, considering that more than 80% of the mature neointima is composed by extracellular matrix. In fact, oxidative stress coexists with intense ER stress, together with an extremely high expression of PDI in every vessel layer [42]. In parallel, there is also a robust increase in cell-surface PDI. Inhibition of such cell-surface PDI with neutralizing antibodies induces decrease in hydrogen peroxide production and reduces vascular caliber by constrictive remodeling [42]. This means that the surface PDI (extracellular), known to contribute to thrombosis, sustains at the same time a novel mechanism in which cytoskeleton architecture and matrix reorganization occurs in a way to sustain vascular caliber during the repair response.

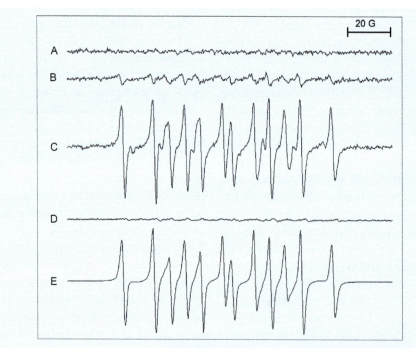

FIG. 10.5 EPR radical adduct spectra, demonstrating the production of superoxide radical after ex vivo vascular injury. Rabbit aortas were incubated in Krebs-HEPES buffer containing the spin-trap DEPMPO (5-diethoxyphosphoryl-5--methyl-1-pyrroline-N-oxide, 100 mmol/L). Little or no EPR signal was detected in the conditions shown on Panel A = Buffer with DEPMPO and on Panel B = Buffer + intact vessel. After overdistention injury, a robust EPR spectrum was detected (Panel C), which upon digital simulation (Panel E) proved to be composite of different DPMPO products. Local injury in the presence of superoxide dismutase (30 μg/mL) showed reduction in this signal (Panel D), indicating that superoxide radical is a primary species of the observed spectra. Instrument conditions: 20 mW power, 1 G amplitude modulation, 81×10^{-3} time constant; 0.466 G/s scan speed; 2×106 gain [40].

Overall, it is possible to suggest that the main redox-sensitive component of the vascular repair response is vascular remodeling, while the redox pathways that control the neointima seem to be complex and context-specific. In fact, the "Probucol and Multivitamin Trial" showed that administration of the antioxidant probucol, albeit not multivitamins, postballoon angioplasty in patients significantly reduces restenosis essentially by preventing constrictive remodeling rather than neointimal growth [44].

Redox modulation of vascular remodeling clarifies a relevant point, i.e., that postinjury oxidative stress seems to have indeed a signaling function of vascular repair and does not only represent a consequence of injury. Thus, redox-dependent signaling pathways can be identified during this process: activation of MAP kinases, protein kinase C and transcription factors, particularly NfkappaB at initial phases after injury [35]. Since NfkappaB is a factor promoting inflammatory gene responses, redox pathways, even if temporarily activated, are able to promote changes on the long-term vascular repair program. Similar phenomena are characteristic of atherosclerosis and other vascular diseases.

ATHEROSCLEROSIS, ENDOTHELIAL DYSFUNCTION, REDOX PATHWAYS, AND LIPOPROTEIN OXIDATION

The involvement of redox processes on atherosclerosis involves at least three important aspects. First, the largest source of oxidants in the arterial wall in the atherosclerotic process originates from the vessel wall itself, as second-messenger of proliferation, migration and apoptosis, among other. All the considerations at the beginning of this chapter regarding redox signaling explain this fact and provide a dimension of its magnitude. Along the same line this fact is associated with the increased complexity facing antioxidant therapies, since it involves mechanisms that go beyond the simple prooxidant imbalance. Second, inflammatory processes that characterize atherosclerosis are strongly associated with redox mechanisms (Fig. 10.3). Third, peroxidation of lipoprotein components is a relevant event in the genesis of atheroma—the so-called "oxidative theory of atherogenesis" [23,45]. In this section, we shall discuss some aspects of this theory.

In addition to all the oxidized lipoprotein effects discussed below, evidences for the oxidative theory of atherogenesis include [45]:

(a) demonstration of oxidative decomposition of LDL on the arterial wall in experimental atherosclerosis models;
(b) similarities between LDL extracted from the arterial wall in these models and oxidized LDL;
(c) demonstration of immunoreactivity for LDL modified by malondialdehyde in complicated human atheroma plaques as well as correlation between immunoreactivity for plasma oxidized LDL and atherosclerosis progression;
(d) the preventive effect of antioxidants in experimental models; and
(e) epidemiological data correlating diets and lifestyles linked to lower oxidative stress with lower incidence of coronary artery disease complications.

Several degrees of LDL oxidation can exist, varying from peroxidation of specific phospholipid targets at the particle surface (named "minimally oxidized LDL") until extensive oxidation of internal lipids and protein. Following the propagation of these processes, toxic by-products start to accumulate, e.g., malondialdehyde, 4-hydroxinonenal. These can serve as markers of the oxidation process. In certain instances, however, biological effects of minimally modified LDL can exist in the absence of changes in any of those markers [45]. Although the minimally altered LDL is still recognized by apolipoprotein E, the expression of which is negatively regulated by intracellular cholesterol levels, extensive modification of the LDL protein particle prevents its capture by these receptors, leading to altered surface charge that elicit recognition by the scavenger monocyte/macrophage receptor (CD36). This receptor is not regulated by intracellular levels of cholesterol and promotes substantial LDL uptake. In vitro, the classic trigger of LDL oxidation is exposure to transition metals, e.g., iron or copper (see below). These metals catalyze lipoperoxide (=LOOH) decomposition, according to the following reaction:

$$LOOH + Fe^{3+}[or\ Cu^{2+}] = Fe^{2+}[or\ Cu^+] + LOO.$$
$$[=peroxyl\ radical] + H^+$$

or

$$LOOH + Fe^{2+}[or\ Cu^+] = Fe^{3+}[or\ Cu^{2+}] + LO.$$
$$[=alkoxyl\ radical] + OH^-$$

Thus, from the initial peroxide (=LOOH in the reaction above), the reaction can evolve in a chain pattern due to formation of peroxyl and alkoxyl radicals. Although such metal-catalyzed pathways can also occur in vivo, there are evidences that other nonenzymatic pathways, e.g., exposure to peroxynitrite and other chlorinated species from leukocytes or enzymatic, e.g., dependent of 15-lipoxiginase activation, can be equally or more important [45]. At least, these pathways may have a role in the production (called "seeding") of the initial lipoperoxide (=LOOH in the reactions above), which could later spread via other mechanisms. The initial event of LDL oxidation is still discussed. It is unclear whether oxidation can occur in plasma, but it certainly occurs in the vascular wall, induced within endothelial cells or monocytes/macrophages. The oxidized lipoprotein can be retained in the sub endothelial space. Identification of the endothelial receptor LOX for oxidized proteins corroborates the notion that a certain degree of oxidation occurs in a physiological manner and can contribute to the elimination of LDL from the vascular wall via monocytes/macrophages [46].

Some of the main effects of the oxidized LDL are:

(a) retention in the sub endothelial space;
(b) recruitment and chemotaxis of monocytes/macrophages;
(c) activation of monocytes/macrophages and smooth muscle cells, which induce intracellular LDL uptake via scavenger receptors, giving origin to foam cells;
(d) toxicity to endothelial cells; and
(e) stimulus to excessive superoxide production of superoxide radicals by endothelial cells.

The oxidative theory of atherogenesis is intrinsically linked to the pathophysiology of endothelial dysfunction, associated with endothelial nitric oxide degradation by superoxide.

NADPH OXIDASE IN THE PATHOPHYSIOLOGY OF ATHEROSCLEROSIS

The involvement of several NADPH oxidases in atherosclerosis is complex and related to the distinctive peculiarities of each isoforms, as well as to the cell types in which they are expressed. Growth factors such as angiotensin II and PDGF activate Nox2

in endothelial cells and Nox1 in smooth muscle cells, supporting cellular proliferation, migration, and inflammatory adhesion molecule exposure. Such effects can be collectively described as cellular activation. Oxidized LDL activates Nox1 and Nox5. Contrarily, Nox4 intriguingly is able to induce hormetic transcription factors (see the beginning of this chapter) that code for antioxidant genes, such as Nrf2, and induces cell differentiation. That is, effects that can be collectively defined as protective [14,16]. Both Nox4 and Nox2 can also be indirectly associated with angiogenic pathways, inducing endothelial cell proliferation via NO or VEGF [16]. Thus, there is not an NADPH oxidase effect on atherosclerosis, but rather individual effects of each isoform. Moreover, given the complexity of effects in distinct vascular cells, integrative understanding of each Nox role is only possible upon the study of transgenic animal models or specific inhibitors (which are yet essentially nonexistent). In general, these studies support evidences that Nox1 and Nox5 contribute to atherosclerosis and its risk factors, however many such data are still indirect and focused on the risk factors and not on the plaque per se. In studies that directly analyzed murine models of atherosclerosis, the effect of Noxes was variable [14–16]. In some examples, the genetic deletion of specific Nox subunits (e.g., Nox1, Nox2, or p47phox) led to a reduced extension of atherosclerotic lesions in apoE or LDL receptor-deficient mice. Studies with Nox2 are also quite conflicting, and focus particularly on inflammation. Data about protective effects of Nox4 are suggestive, but so far there are no data specifically in atherosclerosis models. In summary, suggestive evidences for the involvement of the Nox NADPH oxidase in atherosclerosis are well-grounded in mechanistic studies and essentially based on indirect targetvariables, while studies in murine atherosclerotic models are still inconclusive. On the other hand, there is more clear evidence suggesting that Nox1 and Nox2 contribute to the growth of neointima after vascular lesion [15]. It is possible that Noxes displays cooperative effects and the inhibition of only one isoform is not sufficiently informative.

Mitochondria in the Redox Pathophysiology of Atherosclerosis

The understanding of mitochondrial function and its implications has recently advanced significantly.

In parallel, redox pathways associated with mitochondria have been studied. In the context of atherosclerosis, there are several emerging evidences for the role of such pathways, which can be summarized as follows [19,47]:

(a) There is an increased detection of mitochondrial ROS in atherosclerotic injury models.
(b) In most such cases, it is possible to detect a disruption of the mitochondrial antioxidant mechanisms.
(c) In parallel with mitochondrial dysfunction, there is evidence of defects in autophagic mechanisms that remove defective mitochondria (mitophagy).
(d) There are evidences of imbalance in the protein machinery of fusion and fission in experimental atherosclerosis, leading to changes in mitochondrial net architecture and a consequent functional loss.
(e) The uncoupling enzyme UCP2, which seemingly has protective effects regarding mitochondrial ROS production, displays altered regulation altered in the atherosclerotic murine model.

In analogy with mitochondrial dysfunction in other pathophysiological situations (e.g., metabolic syndrome), there are evidences that ROS from mitochondrial origin enhance the inflammatory response [47]. Together, the study of redox modulation by mitochondria in atherosclerosis and vascular dysfunction is an area in progress, which can reveal new information relevant to the understanding of how derangements in several metabolic pathways lead to inflammation and diseases.

ANTIOXIDANTS AND ATHEROSCLEROSIS

Few topics in pathophysiology have provoked so much controversy and discussion as the use of antioxidants to treat or prevent some diseases. Considering the aspects already discussed in this chapter, this proposal has a solid conceptual fundament. Specifically, regarding the vascular system, with respect to response to injury and atherosclerosis, the results in experimental models are extremely suggestive and encouraging. However, in the clinical sphere, results of the most well-controlled and conducted studies have been uniformly negative regarding the use of vitamins, ROS-scavenging compounds in general, and related food supplements [35,48].

For example, experimental studies investigating the effect of antioxidants after angioplasty [35] have shown a protective effect by ascorbic acid (vitamin C), alpha-tocopherol (vitamin E) and probucol in the reduction of neointima or constrictive vascular remodeling. Clinical studies [44], on the other hand, demonstrate a consistent effect of probucol, but not multivitamins, on the antagonism of constrictive remodeling but not of neointimal thickening. In the case of atherosclerosis, a number of experimental studies have documented reversion, via antioxidant therapeutic strategies, of endothelial dysfunction, plaque extension, increased superoxide radical production, macrophage infiltration, and altered signaling proteins, among other effects. In humans, administration of antioxidants consistently amplifies the beneficial effects of hypolipidemic agents in the improvement of vasomotor endothelial dysfunction and inflammation markers.

Therefore, the reasons for the negative results on clinical studies with antioxidants can be multiple [48]. These can include the following:

(a) Several oxidative processes perform transduction of physiologically protective signals.
(b) Currently available antioxidants are nonspecific and poorly efficient.
(c) Redox processes depend at the same time on oxidative and reductive events and the effect of antioxidants can be contradictory.
(d) Several antioxidants exhibit prooxidant effects depending on the concentration.
(e) Clinical studies with antioxidants may have selected events that occur in advanced plaques, the mechanism of which being so complex that the antioxidant effect becomes negligible.

In particular, it becomes clear that the study of redox mechanisms in atherosclerosis is a promising field and can definitely reveal efficient interventions, but needs further in-depth mechanistic knowledge [9]. In this context, it is important to note that the indiscriminate use of antioxidant supplements does not seem to be risk-free and at least two well-conducted experimental studies [49,50] have shown that common antioxidants can accelerate tumor growth and metastatic potential.

It is essential to emphasize that all this discussion so far is with respect to pharmacologic supplementation of antioxidants and not modulation of redox balance supported by nutrition, moderate exercise and healthy lifestyles. For those, there is substantial evidence indicating a protective effect, the mechanism of which must be multifactorial (i.e., not only redox). In relation to emerging antioxidant interventions, there are several pathways. Some of them concentrate in molecular targets such as NADPH oxidases or in the modulation of mitochondrial function by antioxidants specifically targeted to mitochondria, for example. Another study field is centered in nitric oxide releasing compounds. Also, there is great interest in the development of drugs that mimic natural products such as flavonoids and other compounds.

CONCLUSION

Vascular oxidative stress leads to inactivation of nitric oxide from the endothelium and is an important mechanistic foundation of endothelial dysfunction in atherosclerosis. Redox imbalance, inflammation, and endothelial dysfunction are closely related to the pathophysiology of atherosclerosis, neointimal proliferation, and ultimately of several vascular diseases. The genesis of oxidative stress fundamentally depends on the disequilibrium among cellular processes associated with proliferation, migration, differentiation and apoptosis. These processes involve enzymatic pathways of oxidant production, among which NADPH oxidases and mitochondria are the most studied. The oxidative change in LDL is perhaps the first consistent theory of atherogenesis to unify the cellular and lipid components of the atheroma.

Given this solid pathophysiologic foundation, it is reasonable to expect that an in-depth understanding of mechanistic aspects of these processes will lead to the future development of rational and effective therapeutic interventions. It is interesting to conclude by asking if the clinically oriented investigator can reasonably expect that pathophysiologic studies will indeed bring any contribution to patient managements. The answer can be that, similarly to analogous types of investigation, relevant advances can arise, but in an unexpected manner. Clearly, the target of such investigations is to identify specific determinants of vascular caliber loss, plaque growth and instability, and ultimately to design interventions that improve patient care. In the short term, however, the immediate goals are to understand specific

physiological and pathophysiological aspects. Discrepancies between experimental and cellular findings and clinical implications are not exclusive to the study of redox processes, nor to the cardiovascular field; rather they are part of the usual pathways of scientific discovery.

Acknowledgments

Work developed by or group has been financed by the Fundação de Amparo à Pesquisa do Estado de São Paulo (FAPESP #09/54764-6), Centro de Pesquisa, Inovação e Difusão FAPESP (CEPID "Processos Redox em Biomedicina" [Redox processes in biomedicine] #13/07937-8), Instituto Nacional de Ciência, Tecnologia e Inovação de Processos Redox em Bio-Medicina (INCT Redoxoma, CNPq), and Fundação Zerbini.

References

[1] Halliwell B, Gutteridge J. Free radicals in biology and medicine. 2nd ed. Oxford: Oxford University Press; 1999.
[2] Forman HJ, Augusto O, Brigelius-Flohe R, et al. Even free radicals should follow some rules: a guide to free radical research terminology and methodology. Free Radic Biol Med 2015;78:233–5.
[3] Augusto O, Bonini MG, Amanso AM, et al. Nitrogen dioxide and carbonate radical anion: two emerging radicals in biology. Free Radic Biol Med 2002;32:841–59.
[4] Buettner GR. The pecking order of free radicals and antioxidants: lipid peroxidation, alpha-tocopherol, and ascorbate. Arch Biochem Biophys 1993;300:535–43.
[5] Jones DP. Redefining oxidative stress. Antioxid Redox Signal 2006;8:1865–79.
[6] Winterbourn CC. Reconciling the chemistry and biology of reactive oxygen species. Nat Chem Biol 2008;4:278–86.
[7] Jones DP. Radical-free biology of oxidative stress. Am J Physiol Cell Physiol 2008;295:C849–68.
[8] Rhee SG, Woo HA, Kil IS, et al. Peroxiredoxin functions as a peroxidase and a regulator and sensor of local peroxides. J Biol Chem 2012;287:4403–10.
[9] Fernandes DC, Bonatto D, Laurindo FRM. The evolving concept of oxidative stress. In: Sauer H, Shah A, Laurindo FR, editors. Oxidative stress in clinical practice: cardiovascular diseases. Springer: New York; 2010. p. 1–41.
[10] Stralin P, Karlsson K, Johansson BO, et al. The interstitium of the human arterial wall contains very large amounts of extracellular superoxide dismutase. Arterioscler Thromb Vasc Biol 1995;15:2032–6.
[11] Howitz KT, Sinclair DA. Xenohormesis: sensing the chemical cues of other species. Cell 2008;133:387–91.
[12] Forman HJ, Traber M, Ursini F. Antioxidants: GRABbing new headlines. Free Radic Biol Med 2014;66:1–2.
[13] Ristow M, Zarse K, Oberbach A, et al. Antioxidants prevent health-promoting effects of physical exercise in humans. Proc Natl Acad Sci U S A 2009;106:8665–70.
[14] Lassègue B, San Martín A, Griendling KK. Biochemistry, physiology, and pathophysiology of NADPH oxidases in the cardiovascular system. Circ Res 2012;110:1364–90.
[15] Brandes RP, Weissmann N, Schröder K. NADPH oxidases in cardiovascular disease. Free Radic Biol Med 2010;49:687–706.
[16] Konior A, Schramm A, Czesnikiewicz-Guzik M, et al. NADPH oxidases in vascular pathology. Antioxid Redox Signal 2014;20:2794–814.
[17] Laurindo FR, Araujo TL, Abrahão TB. Nox NADPH oxidases and the endoplasmic reticulum. Antioxid Redox Signal 2014;20:2755–75.

[18] Figueira TR, Barros MH, Camargo AA, et al. Mitochondria as a source of reactive oxygen and nitrogen species: from molecular mechanisms to human health. Antioxid Redox Signal 2013;18:2029–74.

[19] Shadel GS, Horvath TL. Mitochondrial ROS signaling in organismal homeostasis. Cell 2015;163:560–9.

[20] Zhang DX, Gutterman DD. Mitochondrial reactive oxygen species-mediated signaling in endothelial cells. Am J Physiol Heart Circ Physiol 2007;292:H2023–31.

[21] Guthikonda S, Sinkey C, Barenz T, et al. Xanthine oxidase inhibition reverses endothelial dysfunction in smokers. Circulation 2003;107:416–21.

[22] Battelli MG, Polito L, Bolognesi A. Xanthine oxidoreductase in atherosclerosis pathogenesis: not only oxidative stress. Atherosclerosis 2014;237:562–7.

[23] Darley-Usmar V, Halliwell B. Blood radicals: reactive nitrogen species, reactive oxygen species, transition metal ions, and the vascular system. Pharm Res 1996;13:649–62.

[24] Laurindo FR, Pescatore LA, Fernandes D de C. Protein disulfide isomerase in redox cell signaling and homeostasis. Free Radic Biol Med 2012;52:1954–69.

[25] Hatahet F, Ruddock LW. Protein disulfide isomerase: a critical evaluation of its function in disulfide bond formation. Antioxid Redox Signal 2009;11:2807–50.

[26] Santos CX, Tanaka LY, Wosniak J, et al. Mechanisms and implications of reactive oxygen species generation during the unfolded protein response: roles of endoplasmic reticulum oxidoreductases, mitochondrial electron transport, and NADPH oxidase. Antioxid Redox Signal 2009;11:2409–27.

[27] Eletto D, Chevet E, Argon Y, et al. Redox controls UPR to control redox. J Cell Sci 2014;127:3649–58.

[28] Lloret A, Fuchsberger T, Giraldo E, et al. Reductive stress: a new concept in Alzheimer's disease. Curr Alzheimer Res 2016;13:206–11.

[29] Janiszewski M, Lopes LR, Carmo AO, et al. Regulation of NAD(P)H oxidase by associated protein disulfide isomerase in vascular smooth muscle cells. J Biol Chem 2005;280:40813–9.

[30] Fernandes DC, Manoel AH, Wosniak Jr. J, et al. Protein disulfide isomerase overexpression in vascular smooth muscle cells induces spontaneous preemptive NADPH oxidase activation and Nox1 mRNA expression: effects of nitrosothiol exposure. Arch Biochem Biophys 2009;484:197–204.

[31] Santos CX, Stolf BS, Takemoto PV, et al. Protein disulfide isomerase (PDI) associates with NADPH oxidase and is required for phagocytosis of Leishmania chagasi promastigotes by macrophages. J Leukoc Biol 2009;86:989–98.

[32] de A Paes AM, Veríssimo-Filho S, Guimarães LL, et al. Protein disulfide isomerase redox-dependent association with p47(phox): evidence for an organizer role in leukocyte NADPH oxidase activation. J Leukoc Biol 2011;90:799–810.

[33] Pescatore LA, Bonatto D, Forti FL, et al. Protein disulfide isomerase is required for platelet-derived growth factor-induced vascular smooth muscle cell migration, Nox1 NADPH oxidase expression, and RhoGTPase activation. J Biol Chem 2012;287:29290–300.

[34] Furie B, Flaumenhaft R. Thiolisomerases in thrombus formation. Circ Res 2014;114:1162–73.

[35] Azevedo LCP, Pedro MA, Souza LC, et al. Oxidative stress as a signaling mechanism of the vascular response to injury. The redox hypothesis of restenosis. Cardiovasc Res 2000;47:436–45.

[36] Laurindo FRM, Souza HP, Pedro MA, et al. Redox aspects of vascular response to injury. Methods Enzymol 2002;352:432–54.

[37] Daniel JM, Sedding DG. Circulating smooth muscle progenitor cells in arterial remodeling. J Mol Cell Cardiol 2011;50:273–9.

[38] Laurindo FRM, da Luz PL, Uint L, et al. Evidence for superoxide radical-dependent coronary vasospasm after angioplasty in intact dogs. Circulation 1991;83:1705–15.

[39] Janiszewski M, Pasqualucci CA, Souza LC, et al. Oxidized thiols markedly amplify the vascular response to balloon injury in rabbits through a redox active metal-dependent mechanism. Cardiovasc Res 1998;39:327–38.

[40] Souza HP, Souza LC, Anastacio VM, et al. Vascular oxidant stress early after balloon injury: evidence for increased NAD(P)H oxidoreductase activity. Free Radic Biol Med 2000;28:1232–42.

[41] Leite PF, Danilovic A, Moriel P, et al. Sustained decrease in superoxide dismutase activity underlies constrictive remodeling after balloon injury in rabbits. Arterioscler Thromb Vasc Biol 2003;23:2197–202.

[42] Tanaka LY, Araújo HA, Hironaka GK, et al. Peri/epicellular protein disulfide isomerase sustains vascular lumen caliber through an anti-constrictive remodeling effect. Hypertension 2016;67(3):613–22.

[43] Szöcs K, Lassègue B, Sorescu D, et al. Upregulation of nox-based NAD(P)H oxidase in restenosis after carotid injury. Arterioscler Thromb Vasc Biol 2002;22:21–7.

[44] Côté G, Tardif JC, Lespérance J, et al. Effects of probucol on vascular remodeling after coronary angioplasty. Multivitamins and Protocol Study Group. Circulation 1999;99:30–5.

[45] Navab M, Berliner JA, Watson AD, et al. The Yin and Yang of oxidation in the development of the fatty streak. A review based on the 1994 George Lyman Duff Memorial Lecture. Arterioscler Thromb Vasc Biol 1996;16:831–42.

[46] Sawamura T, Kume N, Aoyama T, et al. An endothelial receptor for oxidized low-density lipoprotein. Nature 1997;386:73–7.

[47] Wang Y, Tabas I. Emerging roles of mitochondria ROS in atherosclerotic lesions: causation or association? J Atheroscler Thromb 2014;21(5):381–90.

[48] Griendling KK, FitzGerald GA. Oxidative stress and cardiovascular injury: part II: animal and human studies. Circulation 2003;108:2034–40.

[49] Le Gal K, Ibrahim MX, Wiel C, et al. Antioxidants can increase melanoma metastasis in mice. Sci Transl Med 2015;7:308re8.

[50] Sayin VI, Ibrahim MX, Larsson E, et al. Antioxidants accelerate lung cancer progression in mice. Sci Transl Med 2014;6:221ra15.

Further Reading

[1] White CR, Darley-Usmar V, Berrington WR, et al. Circulating plasma xanthine oxidase contributes to vascular dysfunction in hypercholesterolemic rabbits. Proc Natl Acad Sci U S A 1996;93:87445–9.

CHAPTER

11

Blood Coagulation and Endothelium

Joyce M. Annichino-Bizzacchi and Erich Vinicius de Paula

INTRODUCTION

Endothelial cells form a continuous monolayer that covers the entire vascular system, maintaining blood flow inside the vessels and promoting organ perfusion. They play the role of barrier between the blood flow and tissues, and participate in several processes such as hemostasis, inflammation, and immunological mechanisms.

According to the location, such as specific organs and sites, the functions of endothelium cells are also variable. Culture of endothelial cells of several vascular, venous, or arterial sites evidence differences in protein expression, particularly in microcirculation, providing specific properties according to the location [1]. Regarding homeostasis, one example is the absence of thrombomodulin (TM) expression in cerebral microcirculation and along lower limb venous valves. The latter situation predisposes the activation of coagulation in case of venous stasis [2].

The proportion between volume of blood and endothelial area is also different in several vascular sites, contributing to the diversity of vascular functions [3].

Regarding hemostasis, physiologically the endothelium plays anticoagulant and antithrombotic roles, preventing platelet adhesion on its surface, inhibiting the complex formed by activated Factor VII (FVIIa) and tissue factor (TF), activating the anticoagulant pathways of protein C (PC), and contributing for activation of fibrinolysis by release of tissue plasminogen activator (t-PA) (Fig. 11.1).

Maintenance of blood flow is important to promote appropriate quantity of coagulation factors and platelets. The difference in blood flow influences thrombus composition, since arterial thrombi are rich in platelets, with greater amounts of fibrin and red blood cells compared to venous thrombi.

BLOOD COAGULATION

The anticoagulant and antithrombotic action of the endothelium includes receptors expression, control of fluid and macromolecules extravasation, and actions upon platelets and other blood cells.

PLATELETS

By suppressing platelet adhesion and aggregation, the endothelium prevents or limits thrombus formation. Normally, platelets do not adhere to the endothelium due to the action of glycosaminoglycans with negative charge. The ectonucleotidases present on the endothelial surface also convert ADP in adenosine, preventing platelet aggregation [4].

Activation of endothelial cells promotes cellular signaling, mediated by receptors, calcium influx, and induces phosphorylation, with generation of nitric oxide (NO), prostacyclin and prostaglandin E2, endothelin-1, and recruitment of vesicles that contain preformed proteins [5]. The release of prostacyclin and prostaglandin E2, in large vessels, is mediated by vasoactive substances, and by thrombin in small vessels [6]. Effect of prostacyclins is enhanced by NO produced at the endothelium by endothelial NO synthase (eNOS). Vasoactive agents

Endothelium and Cardiovascular Diseases
https://doi.org/10.1016/B978-0-12-812348-5.00011-8

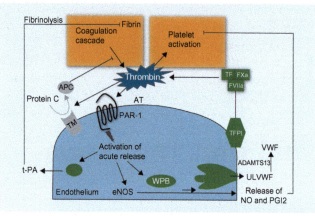

FIG. 11.1 Endothelial cells. Pathways involved in coagulation activation and inhibition and platelet activation processes. *APC*, activated protein C; *TM*, thrombomodulin; *AT*, antithrombin; *eNOS*, endothelial nitric oxide synthase; *WBP*, Weibel Palade bodies; *vWF*, von Willebrand Factor; *ULVWF*, extremely high molecular weight von Willebrand Factor; *t-PA*, tissue plasminogen activator; *FT*, tissue factor; *TFPI*, tissue factor pathway inhibitor. Adapted from van Hinsbergh VWM. Endothelium role in regulation of coagulation and inflammation. Semin Immunopathol 2012;34:93–106.

and increased vascular flow promote increased eNOS synthesis. The NO also directly inhibits platelet adhesion and aggregation, decreasing platelet cytosolic Ca^{2+} and activating membrane glycoproteins [7]. The thrombin generated, even in small quantities, activates PAR-1 that activates several intracellular signaling pathways, with increased production of NO and prostacyclins.

When the endothelium is significantly damaged, changes in balance of such processes favor thrombotic complications. Activated platelet adhere to the endothelial surface, by means of the connection between the platelet membrane glycoprotein Ibα (GpIbα) and P-selectin or von Willebrand Factor (vWF). The platelet heparanase released during activation also favors this connection, degrading proteoglycans on the endothelial surface.

After endothelial platelet adhesion, platelet membrane phospholipid receptors promote approximation of coagulation factors, favoring activation of coagulation and generating the first thrombin molecules. Thrombin has an essential role in this process, because it stimulates both coagulation and fibrin formation, and also activates control mechanisms.

VON WILLEBRAND FACTOR

vWF is synthesized by the endoplasmic reticulum of endothelial cells and megakaryocytes, and is processed in the Golgi apparatus, under the form of dimers, which polymerize in spiral multimers [8]. vWF is stored in Weibel-Palade bodies and in platelet α-granules.

The stimulus to release vWF may be represented by vasoactive substances, such as histamine, bradykinin, vasopressin, and thrombin, which increase calcium cytoplasmic concentration and activate protein kinase C. In this process, the Weibel-Palade corpusc les are fused with the cellular membrane and release vWF [9], which may be very long (ULVWF, or Ultra large vWF Multimers). The ULVWF aligns along the endothelial surface or along injured vessels, and interacts with the platelet GpIbα by means of its A1 domain. Under high flow conditions, this connection is also mediated by the vWF A3 domain. ADAMTS13 is a metalloprotease responsible for cleavage of ULVWF in smaller multimers on the endothelial surface. This lysis occurs in the specific Tyr1605-Met1606 site of the vWF A2 domain [10]. ADAMTS13 binds to the endothelial surface and promotes this proteolytic activity.

ADAMTS13 deficiency is manifested by a severe clinical picture, characterized by obstruction of microcirculation caused by the complexes formed between ULVWF and platelets, thrombocytopenia, and microangiopathic hemolytic anemia, called Thrombotic Thrombocytopenic Purpura. Therefore, the role of the endothelial surface is essential to prevent thrombotic phenomena.

Under normal flow conditions, vWF binds to GpIbα, promoting platelet adhesion, but in high shear stress situations, vWF also binds to glycoprotein IIb/IIIa, favoring platelet aggregation.

vWF also takes part in blood coagulation as carrier of factor VIII (FVIII).

COAGULATION ACTIVATION MEDIATED BY ENDOTHELIAL FACTORS

As mentioned, in normal health, the endothelium performs anticoagulant and antithrombotic activities. Depending on the stimuli, all processes are orchestrated, promoting the formation of blood clots, that contain bleeding and promote healing, without promoting unnecessary damage, limited to the local site. Therefore, coagulation activation is important to form clots according to local needs.

However, in situations in which activation level is exaggerated, as is the case of septicemia or cancer, this balance may be broken, with deleterious effects, such as massive thrombosis and disseminated intravascular coagulation, among others.

TISSUE FACTOR AND TISSUE FACTOR INHIBITOR

TF is a transmembrane protein that initiates coagulation, by binding to factor VII (FVII), which once activated acts on factor X (FX) and factor XI (FXI). Factor Xa transforms prothrombin into thrombin, the enzyme that converts fibrinogens to fibrin. Activation of FXI is very important, because factor Xa and factor VIIa are soon inactivated by the tissue factor pathway inhibitor (TFPI), expressed on the endothelial surface [11]. The activation of FXIa generates additional FXa molecules, fostering continued coagulation even if factor VIIa is inactivated.

TF is expressed on the endothelial surface and in several cells, such as macrophages, monocytes, and fibroblasts [12]. Recent studies showed the presence of TF in microparticles generated by platelets, endothelial cells, and monocytes, contributing for a more dynamic vision on the generation and expression of this factor [13]. In normal conditions, the endothelium controls the initiation of coagulation by the actions of TFPI, which limits the activity of the TF/VIIa/Xa complex [14].

On the other hand, isomerases released by the activated endothelial cells are important to promote coagulation, by the activation of TF present in microparticles [15].

CONTACT SYSTEM

The contact system includes factors XII and XI, prekallikrein (PK), high molecular weight kininogen (HK), C1-inhibitor (C1INH), and α_2-macroglobulin. This system is activated in the presence of negatively charged surfaces, leading to coagulation and generation of fibrin, by activation of factor XI [16]. Although blood coagulation may be activated through this pathway, physiologically FXI is activated directly by thrombin. The contact system also is responsible for producing bradykinin, a peptide involved in vascular permeability.

The contact system can be activated on the endothelial surface by local accumulation of its components [17]. FXII and HK bind directly to these cells, while PPK and FXI are recruited via HK. Studies suggest that changes or cellular stress evoke recruitment of contact system components.

The endothelium presents three contact system binding sites: glycosaminoglycans with negative charge [18], receptor C1q, and a cytokeratin-1 [19]. C1INH has lower capacity of inactivating FXIIa and PK in plasma when on the surface of endothelial cells, due to the presence of the glyco-saminoglycans.

FXII, as well as PPK and HK, may induce activation of plasminogen and fibrinolysis. PK may also activate urokinase, constituting an important in vivo fibrinolysis activation mechanism [20].

ANTICOAGULANT PROPERTIES OF THE ENDOTHELIUM

Antithrombin

Antithrombin (AT) is a serpin synthesized in the liver that binds to the heparan sulfate of the proteoglycans that cover the endothelial surface [21]. AT inhibits serinoproteases, the main target being thrombin, but also inactivating the activated forms of factors X, IX, VII, XI, and XII. The formation of the AT-protease complex involves specific peptides that, once formed, lead to irreversible inhibition.

The role of AT in preventing coagulation and thrombosis is evident when the severe thromboembolic state of patients with AT deficiency is considered. One of the most important anticoagulant drugs is heparin, and its mechanism of action is maximization of AT activity.

Anticoagulant Pathway of Protein C

Another pathway of extreme importance to coagulation control is the PC anticoagulant pathway. PC is a vitamin K-dependent protein synthesized in the liver, which is activated by the complex formed by thrombin/TM. Activated protein C (APC) inhibits the coagulation cascade by cleavage of factors Va and VIIIa.

At the same time that thrombin participates in fibrin formation process, by acting on fibrinogen, its connection to EGF-like domains 5 and 6 of TM turns its activity into that of an anticoagulant. There is another specific change, with effect on PC and the thrombin-activated fibrinolysis inhibitor (TAFI) [22]. The specificity of this thrombin/TM complex depends on the concentration of TM, which when elevated acts on PC and when low acts on TAFI. The expression in different tissues and vessel diameters regulate this process, which is higher in small vessels, due to the increased ration between surface and blood volume [23].

TM is synthesized almost exclusively by endothelial cells, and is expressed throughout the vascular system, except in cerebral circulation [2]. Transcription of TM is increased in situations of increased arterial flow, and decreased in the presence of inflammation [24].

The importance of TM in the complex process of coagulation regulation can be demonstrated by observing that knockout [25] mices in the beginning of the embryonic process, around the tenth day. Transgenic animals with altered TM expression only in the vasculature died in the intrauterine stage or in the first weeks after birth, showing consumption coagulopathy and formation of thrombus in several organs [26].

The endothelial receptor for PC, EPCR, has an important role in this process, by promoting the interaction of components of the PC anticoagulant pathway. PC binds to EPCR, converting into activated PC (APC) due to the action of the thrombin/TM complex. APC detaches from EPCR and acts by inhibiting factors Va and VIIIa, with protein S (PS) as a cofactor [27]. Once cleaved, these factors lose their pro-coagulant action, inhibiting the conversion of prothrombin to thrombin.

Therefore, formation of APC is limited to the generation of thrombin, exerting a negative feedback mechanism on blood coagulation. The PC cleavage process is 1000 times more efficient when thrombin forms a complex with TM, maximized by the connection between PC and EPCR [27,28].

Alterations in the PC anticoagulant pathway are clinically evidenced by thrombotic clinical pictures manifested in patients who showed PC deficiency or resistance to APC resulting from factor V gene mutation, called Factor V Leiden. Some cases of hemolytic uremic syndrome, which is characterized

by vascular obstruction, may exhibit TM gene mutations [29].

EPCR also interferes in coagulation, regardless of the PC anticoagulant pathway, due to connection with FVII or FVIIa and reduced activity of FT [30,31]. These factors are transported and stored in tissue, and have a reduced half-life [30,31].

ENDOTHELIUM AND FIBRINOLYSIS

Plasminogen is produced by the liver, and t-PA attached to fibrin is converted into plasmin, which is responsible for degrading fibrin into fibrin degradation products, avoiding blood vessel obstruction.

The endothelial cells continuously produce and release t-PA, promoting fibrinolysis effects whenever it binds to the circulating fibrin [32]. t-PA is stored in Weibel-Palade corpuscles and in specific vesicles [33], and is released by vasoactive substances or thrombin. Availability of t-PA during fibrin generation is very important to allow integration with the clot, making its action much more effective when compared to its incorporation only after formation.

Under inflammatory conditions, the urokinase plasminogen activator (u-PA) can also be synthesized and released by the endothelium [34]. However, its action is restricted to the cell surface after the connection of the urokinase receptor (UPAR) [35], and has no effects on coagulation and blood fibrinolysis.

TAFI inhibits fibrinolysis by removing the lysine resides from fibrin, which are required for the bond formation between the plasminogen and t-PA. As a consequence, plasminogen cannot be converted into plasmin, reducing the fibrinolysis potential [36]. Plasminogen activator inhibitor, PAI-1, is produced by endothelial cells and inhibits t-PA and u-PA, but its contribution in vivo is controversial.

Depending on the circulation, there is influence on the type of mechanism involved in fibrinolysis. Therefore, veins of lower limbs are prone to development of thrombosis, by stasis, but also present high fibrinolysis potential, due to the high availability of t-PA, which contributes to minimum contact between thrombus and endothelium. However, this may favor thrombus growth on the vessel lumen and embolus formation, as is the case of the most common complication in lower limb thrombosis and pulmonary embolism.

CONCLUSIONS

The endothelium has a natural anticoagulant role, but physiologically small quantities of thrombin and fibrin are generated, without pathological impacts. When this process is altered, due to activation of endothelial cells, presence of molecules or microparticles, or injury with subendothelium exposition, initiation of coagulation and exaggerated generation of thrombin may lead to serious complications, triggering thromboembolic events.

Unfortunately, in clinical practice, functional and endothelial component assessments are very limited, due to the lack of direct access to endothelial cells. Currently, we rely on indirect analysis, such as circulating TM dosage and vWF. The development of methods that allow easy and reliable analysis may contribute to a better understanding of the physiopathology of several diseases, in particular thromboembolic diseases, advancing therapeutic approaches to affected patients.

References

[1] Chi JT, Chang HY, Haraldsen G, et al. Endothelial cell diversity revealed by global expression profiling. Proc Natl Acad Sci USA 2003;100(19):10623–8.

[2] Ishii H, Salem HH, Bell CE, et al. Thrombomodulin, an endothelial anticoagulant protein, is absent from the human brain. Blood 1986;67:362–5.

[3] Aird WC. Phenotypic heterogeneity of the endothelium I. Structure, function, and mechanisms. Circ Res 2007;100:158–73.

[4] Pearson JD. Endothelial cell function and thrombosis. Best Pract Res Clin Haematol 1999;12:329–41.

[5] von Hinshergh VWM. Endothelium role in regulation of coagulation and inflammation. Semin Immunopathol 2012;34: 93–106.

[6] Gerritsen ME. Functional heterogeneity of vascular endothelial-cells—commentary. Biochem Pharmacol 1987;36:2701–11.

[7] Fulton D, Gratton JP, Mccabe TJ, et al. Regulation of endothelium-derived nitric oxide production by the protein kinase Akt. Nature 1999;399:597–601.

[8] Sadler JE. von Willebrand factor assembly and secretion. J Thromb Haemost 2009;7:24–7.

[9] Valentijn KM, Sadler JE, Valentijn JA, et al. Functional architecture of Weibel-Palade bodies. Blood 2011;117:5033–43.

[10] Turner NA, Nolasco L, Ruggeri ZM, et al. Endothelial cell ADAMTS-13 and VWF: production, release, and VWF string cleavage. Blood 2009;114:5102–11.

[11] Osterud B, Bajaj MS, Bajaj SP. Sites of tissue factor pathway inhibitor (Tfpi) and tissue factor expression under physiological and pathological conditions. Thromb Haemost 1999;73:873–5.

[12] Drake TA, Morrissey JH, Edgington TS. Selective cellular expression of tissue factor in humantissues—implications for disorders of hemostasis and thrombosis. Am J Pathol 1989;134:1087–97.

[13] Dignat-George F, Boulanger CM. The many faces of endothelial microparticles. Arterioscler Thromb Vasc Biol 2011;31:27–33.

[14] White TA, Johnson T, Zarzhevsky N, et al. Endothelial-derived tissue factor pathway inhibitor regulates arterial thrombosis but is not required for development or hemostasis. Blood 2010;116: 1787–94.

[15] Jasuja R, Furie B, Furie BC. Endothelium-derived but not platelet-derived protein disulfide isomerase is required for thrombus formation in vivo. Blood 2010;116:4665–74.

[16] Maas C, Renné T. Regulatory mechanisms of the plasma contact system. Thromb Res 2012;129(Suppl 2):S73–6.

[17] van Iwaarden F, de Groot PG, Bouma BN. The binding of high molecular weight kininogen to cultured human endothelial cells. J Biol Chem 1988;263:4698–703.

[18] Renné T, Dedio J, David G, et al. High molecular weight kininogen utilizes heparan sulfate proteoglycans for accumulation on endothelial cells. J Biol Chem 2000;275:33688–96.

[19] Joseph K, Shibayama Y, Ghebrehiwet B, et al. Factor XII-dependent contact activation on endothelial cells and binding proteins gC1qR and cytokeratin 1. Thromb Haemost 2001;85:119–24.

[20] Lynch J, Shariat-Madar Z. Physiological effects of the plasma Kallikrein-Kinin system: roles of the blood coagulation factor XII (Hageman factor). J Clinic Toxicol 2012;2:3.

[21] Bauer KA, Rosenberg RD. Role of antithrombin-III as a regulator of in vivo coagulation. Semin Hematol 1991;28:10–8.

[22] Morser J. Thrombomodulin links coagulation to inflammation and immunity. Curr Drug Targ 2012;13:421–31.

[23] Colucci M, Semeraro N. Thrombin activatable fibrinolysis inhibitor: at the nexus of fibrinolysis and inflammation. Thromb Res 2012;129:314–9.

[24] Parmar KM, Larman HB, Dai GH, et al. Integration of flow-dependent endothelial endothelial phenotypes by Kruppel-like factor 2. J Clin Investig 2006;1(16):49–58.

[25] Healy AM, Rayburn HB, Rosenberg RD, et al. Absence of the blood-clotting regulator thrombomodulin causes embryonic lethality in mice before development of a functional cardiovascular system. Proc Natl Acad Sci U S A 1995;92:850–4.

[26] Isermann B, Hendrickson SB, Zogg M, et al. Endothelium-specific loss of murine thrombomodulin disrupts the protein C anti-coagulant pathway and causes juvenile-onset thrombosis. J Clin Invest 2001;108:537–46.

[27] Rezaie AR. Regulation of the protein C anticoagulant and anti-inflammatory pathways. Curr Med Chem 2010;17:2059–69.

[28] Stearns-Kurosawa DJ, Kurosawa S, Mollica JS, et al. The endothelial cell protein C receptor augments protein C activation by the thrombin-thrombomodulin complex. Proc Natl Acad Sci U S A 1996;93(19):10212–6.

[29] Delvaeye M, Noris M, De Vriese A, et al. Thrombomodulin mutations in atypical hemolytic-uremic syndrome. New Engl J Med 2009;361:345–57.

[30] Nayak RC, Sen P, Ghosh S, et al. Endothelial cell protein C receptor cellular localization and trafficking: potential functional implications. Blood 2009;114:1974–86.

[31] Gopalakrishnan R, Hedner U, Ghosh S, et al. Biodistribution of pharmacologically administered recombinant factor VIIa (rFVIIa). J Thromb Haemost 2010;8:301–10.

[32] Medcalf RL. Fibrinolysis, inflammation, and regulation of the plasminogen activating system. J Thromb Haemost 2007;5:132–42.

[33] Emeis JJ, vanden Eijnden Schrauwen Y, vanden Hoogen CM, et al. An endothelial storage granule for tissue-type plasminogen activator. J Cell Biol 1997;139(1):245–56.

152 11. BLOOD COAGULATION AND ENDOTHELIUM

[34] vanHinsbergh VWM, Vandenberg EA, Fiers W, et al. Tumor necrosis factor induces the production of urokinase type plasminogen activator by human endothelial cells. Blood 1990;75:1991–8.

[35] Blasi F, Sidenius N. The urokinase receptor: focused cell surface proteolysis, cell adhesion and signaling. FEBS Lett 2010;584: 1923–30.

[36] Foley JH, Cook PF, Nesheim ME. Kinetics of activated thrombin-activatable fibrinolysis inhibitor (TAFIa)-catalyzed cleavage of C-terminal lysine residues of fibrin degradation products and removal of plasminogen-binding sites. J Biol Chem 2011;286:19280–6.

Further Reading

[1] Sixma JJ, vanZanten GH, Huizinga EG, et al. Platelet adhesion to collagen: an update. Thromb Haemost 1997;78:434–8.

12

Endothelium and Genetics

Riccardo Lacchini, Gustavo Henrique Oliveira de Paula,
and José Eduardo Tanus dos Santos

INTRODUCTION

When we think about diseases associated with endothelial dysfunction, it is common to identify patients that behave differently from most others, that is, with a small number of clinical risk factors develop the diseases (phenotype), while others with higher number of clinical risk factors have a certain resistance to the disease. Similarly, progression of diseases is quicker in some individuals rather than others, without an apparent cause. One of the factors that may explain this is the set of genetic characteristics (genotypes) that induce higher risk or higher resistance to a given phenotype.

A phenotype is defined as an observable trait, either directly visible, such as hair color or presence of a disease, or only measurable, such as the metabolism rate of a specific drug, response to medications, etc. Besides that, when there are genetic bases that influence the risk of a given phenotype, it is common to see cases in which the diseases repeat over generations of the same family. This is quite clear in respect to common diseases such as cancer and hypertension, among others. In fact, the estimation is that blood pressure heritability is approximately 15%–30%, and up to 60% of blood pressure variability measured in the long term is influenced by hereditary traits [1].

Currently, most of the common phenotypes are considered to have a multifactorial nature, most of them of a multigene nature [2]. This is the same as saying that there is not genetic determinism in which an individual that carries an allele (genetic characteristic) will be doomed, necessarily, to have the diseases associated with that allele [2]. In fact, these common phenotypes suffer the influence of several genetic and environmental factors that, together, determine a quantifiable risk for the development of a given phenotype (Fig. 12.1). This concept can be applied to cardiovascular diseases that involve endothelial dysfunction: in addition to clinical factors such as obesity, tobacco use, alcoholism, gender, and age [3,4], genetic variants in important components of the endothelial function may predispose an individual to or protect them against the development of the disease [3]. This obviously excludes innate metabolism errors or other rare diseases of monogenic characteristics, whose mutation lead to massive loss of capacity of a given metabolic pathway, for example, and a clear and precise association between an allele and a phenotype is noticeable [1]. This type of mutation is rare, and probably does not explain most cases of common diseases, which are more relevant from a public health perspective.

The purpose of this chapter is to provide a clinical overview of the current genetic situation involving genes expressed by the endothelium. Therefore, the objective is not to exhaust the subject, but only to comment briefly on some of the results obtained at the moment, as well as their clinical relevance. As an example, we shall focus more deeply on the gene that can be thought as the endothelium "prototype candidate gene," NOS3, which encodes endothelial nitric oxide synthesis.

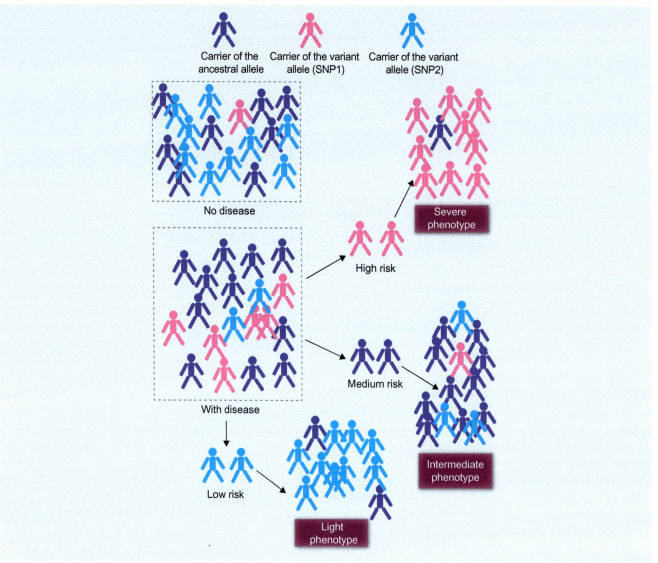

FIG. 12.1   Association of polymorphisms with the groups of phenotype carrying individuals. This figure exemplifies two polymorphisms: Single Nucleotide Polymorphism (SNP) 1 and SNP2. The examples include transporter of both ancestral alleles *(dark-blue individuals)*, transporter of variant allele SNP1 *(pink individuals)*, and transporter of variant allele SNP2 *(light blue individuals)*. To simplify the illustration, the transporters of both variant alleles are not represented. Supposing that variant allele SNP1 has harmful effects, while variant allele SNP2 has protective effects, the analysis of case-control population shows higher frequency of light blue individuals in the group without the disease in respect to the group with the disease. Similarly, the proportion of pink individuals is higher in groups with the disease when compared to the group without the disease. The same analysis can be made regarding the progression of the disease or other more specific subphenotypes (response to medications, subclassification of symptoms, among others). The idea shall be the same: if there is a genetic effect, higher proportion of protective allele carriers *(light-blue individuals)* shall be expected in populations presenting milder phenotypes, while a higher proportion of risk allele carriers *(pink individuals)* is expected in the groups of more intense phenotypes. The presence of all genotypes in the healthy group illustrates the idea that there is no genetic determinism. The fact that *pink individuals* are carriers of the risk allele does not imply that they shall develop the disease. Another interesting fact is that the analysis of each subphenotype is independent: alleles of genes not related to the disease trigger, and that therefore do not affect the risk of having the disease or not, may change other parameters such as disease progression speed or response to medications.

THIRTEEN YEARS AFTER THE CONCLUSION OF THE HUMAN GENOME PROJECT: WHAT HAS CHANGED?

One of the great recent achievements in human genetics was the analysis of the human genome, which allowed us to know ourselves in a much more comprehensive fashion [5,6]. More than a decade after the first human genome draft was disclosed, we still have no DNA tests that determine the profession, salary, and life expectancy of our children, as imagined in 1997 in the famous movie *Gattaca*. In fact, genetics of complex diseases and pharmacogenetics, which were the great promises of the Human Genome Project, did not materialize as quickly as foreseen (Fig. 12.2). The expectation that succeeded the billions of dollars spent in the Human Genome Project between 1990 and 2003 was not satisfied immediately, and much of the hype was lost. Although researchers of great reputation in the field are optimistic (Table 12.1) [2], the truth is that little regarding the clinical applicability of the huge quantity of information was in fact achieved, and a lot of time and effort are still required to realize such view [2]. Scientists had a huge "reality shock": we were able to crack the biochemical code into letters. Very well, but what words, paragraphs, chapters, and messages do these letters mean?

The truth is that the human genetic complexity is much bigger than expected. The great bottlenecks in the development of genetics of complex diseases has been the interpretation of the results obtained, experimental limitations, conclusions based on fragile associations, and the genetic nature of populations itself. The latter shows that it is not possible to directly extrapolate the data obtained from a population with well-defined ethnic origins to other populations. If a phenotype is the result of the interaction of hundreds of variants, when we analyze populations with different sets of genetic characteristics, we can understand, for example, that a decisive result in Japanese people may not be relevant for Caucasians, and so forth. Little by little, this scientific investigation field has adapted, and currently it is difficult to publish an article simply demonstrating a biological effect associated with a single polymorphism, especially if the sample is small (100 controls and 100 patients, for example). Few will give credit to such finding without a much larger number of individuals studied, and without replication of the findings in other populations. On the other hand, studies involving thousands of cases and controls with genomic approaches are more easily accepted, although until now their contribution has been relatively small in face of the high cost of execution.

Among several advances in the last decade, we have the developments of the Genome Wide Association Studies (GWAS) [2]. These wide studies conduct thousands of independent tests, examining the possible association of each common polymorphism with the phenotype studied. It is known in statistics that, if sufficient tests are conducted, at some point an association (false) is found completely due to chance. This is called error type I. To compensate for the error, we need to adjust the value of alpha (conventionally lower than 0.05 in common statistical analysis) in order to ensure that high rates of false associations are not detected. Therefore, a common value of P used in the GWAS studies is $P < 10^{-7}$ or $P < 10^{-10}$, for example, to consider an association to be significant [11]. Yet evaluating the genotypes of thousands of polymorphisms of each volunteer is not an ordinary task. To facilitate the execution of these studies, the first step was to determine the haplotype structures in different populations (HapMap Project, http://hapmap.ncbi.nlm.nih.gov). Haplotypes are sets of alleles of different polymorphisms that, by being close to each other, end up segregating together, passing from generation to generation. This is explained by the fact that gene recombination (which occurs during meiosis in gamete formation) is not a completely random event; there are DNA regions more prone to being cleaved during this process. Therefore, there is a low recombination rate within haplotype blocks, and high recombination between other blocks. After the common haplotype blocks were acknowledged, the strategy used was to evaluate genotypes of a given polymorphism that could represent which would be the set of alleles of different polymorphisms in a haplotype block (known as "tag" polymorphisms or tagSNP). This allowed mapping of a large number of frequent polymorphisms in a compelling and broad way, with only a fraction of them being actually determined experimentally.

Thus, the first GWAS were born using DNA microarray chips. These, with only a few thousand genotypes, provide information regarding the common genetic variability of the genome as a whole. In a way, this approach is also limited due to the extrapolations made, assuming homogeneity of the populations studied and representativeness of

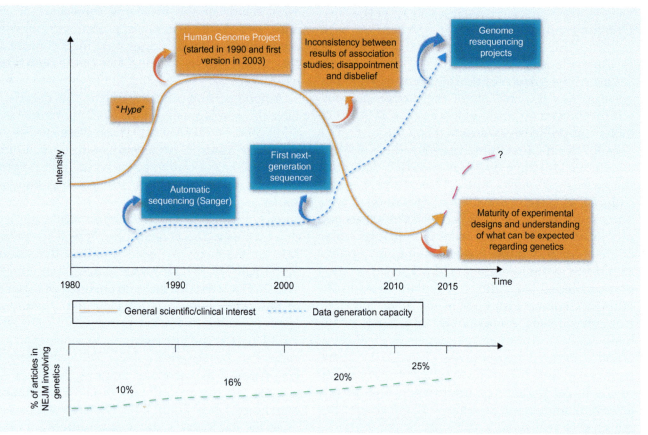

FIG. 12.2 Approximate timeline illustrating the technical capacity, clinical/scientific interest, and real impact contribution of genetics in medicine. The *blue line* represents scientific and clinical interest in genomic medicine over time, while *orange* and *black lines* represent, respectively, the technical capacity (mainly speed) to generate data and the percentage of articles in the acclaimed clinical publication New England Journal of Medicine focusing on genetic data. The 80s saw the great boom in molecular biology, creating at the time high expectations regarding the impact of molecular biology on medicine. In 1990, the Human Genome Project was created so that, in a period of 15 years, the human genetic code would be unraveled. During the 90s, huge investments were made in DNA sequencing equipment around the world, which allowed the execution of the human genome, followed by hundreds of other organisms. As the genome was obtained, the studies focused on individual markers associated with clinical phenotypes. With few exceptions, it was shown that common polymorphisms cause small effects, may times insignificant in face of environmental factors, which generated great disappointment and disbelief. The fact that thousands of articles were published with conflicting results also contributed to this disbelief, with genotype-phenotype associations obtained by some researchers that were not reproduced in other groups, or that were even rejected by others. Regardless of the disbelief by a large part of the scientific and clinical community, there was still great investment in genotyping by genome re-sequencing projects, which created a demand for the development of new sequencing technologies. With that, new techniques emerged during the 90s and in the mid-2000s, such as DNA microarrays and next-generation sequencing, respectively. These techniques demanded some time between their creation and maturity. Therefore, the real genotype generation technical capacity leap came in the mid- to late 2000s. Meanwhile in the last 5 years the growing capacity to determine genotypes increased exponentially, both by means of DNA chips (to reaching more than 1 million chip polymorphisms) and by new generation sequencers. The scientific criteria to create these types of studies also matured, which shall gradually minimize data inconsistencies, considering that technical difficulties shall be eliminated. This maturity, together with the reduced cost by sequenced DNA base, and new genomic approaches, are revitalizing clinical and scientific interest in genomic medicine. In parallel to all these developments, it is apparent that genetics is gradually consolidating in the practice/clinical fields. If in the 80s only 10% of the articles published in NEJM involved genetics, today one out of each four articles published in the same magazine involve genetic/genomic data.

TABLE 12.1 Quotes by Great Scientists Regarding Genomic in Medicine

---

FRANCIS COLLINS

- "By 2015, we will see the beginnings of a real transformation in the therapeutics of medicine, which by 2020 will have touched virtually every disorder [...]. And the drugs that we give in 2020 will for the most part be those that were based on the understanding of the genome, and the things that we use today will be relegated to the dust bin" [7]
- "[...] However, NIH will not fund any use of gene-editing technologies in human embryos. The concept of altering the human germline in embryos for clinical purposes has been debated over many years from many different perspectives, and has been viewed almost universally as a line that should not be crossed [...]." NIH statement regarding research funding using gene editing in human embryos [8]

Contributions:
- Identification of genes that cause cystic fibrosis and Huntington disease
- Leader of the Human Genome Project and Director of the National Institute of Health

---

JAMES D. WATSON

- "The ever quickening advances of science made possible by the success of the Human Genome Project will also soon let us see the essences of mental disease. Only after we understand them at the genetic level can we rationally seek out appropriate therapies for such illnesses as schizophrenia and bipolar disease."—The New York Times [9]

Contributions:
- Watson and his colleague Crick reinterpreted the results of X-ray diffraction from the 50s and proposed the double-helix model of DNA we know today

---

FRANCIS CRICK

- "Finally one should add that in spite of the great complexity of protein synthesis and in spite of the considerable technical difficulties in synthesizing polynucleotides with defined sequences it is not unreasonable to hope that all these points will be clarified in the near future, and that the genetic code will be completely established on a sound experimental basis within a few years." Francis Crick's speech when he won the Nobel Prize in Physiology and Medicine, in 1962 [10]

Contributions:
- Crick and his colleague Watson reinterpreted the results of X-ray diffraction from the 50s and proposed the double-helix model of DNA we know today

---

the assigned haplotypes. It will be obvious to the reader that this was an approach that analyzed the whole roughly, without evaluating the universe of information regarding the rarest variants of the genome, because it analyzed only a predefined set of polymorphisms. This is similar to a short-sighted doctor examining a patient without his glasses, using asbestos gloves, and asking questions in Portuguese to the patient, who only speaks German. In fact, some compelling results were found, but only a very small quantity and at exorbitantly high costs, which hindered the popularization of this experimental approach on a global scale.

More recently, the Next-Generation DNA sequencing approach was developed. This approach allows direct sequencing of an entire genome in a few weeks, determining exactly each base in each position of the DNA. Although it is still very expensive, this approach has already been used in molecular diagnose of monogenic syndromes without specific diagnoses, in some types of cancer, or research [2].

The new generation sequencing represents a great leap, which will allow investigation not only of common variants, but also of rare mutations regarding common phenotypes. By either sequencing or microarrays, the great advantage of GWAS in respect to the common candidate-gene analysis strategy is that, as this analysis is not limited by hypotheses formulated in advance, there is no limitation due to the sparse existing physiopathological knowledge.

In fact, GWAS helped to expand the network of genes and proteins involved in macroscopic phenotypes and to improve understanding the physiopathological processes [2]. Some of the "new genes" identified in GWAS have been in fact confirmed as relevant protagonists in cardiovascular diseases. On the other hand, a large part of the current set of data originating from these studies indicated associations of polymorphisms in intergenic regions or in genes with completely unknown functions. Therefore, we return to the previous question: how do we interpret these associations? If we do not know how gene "X" relates to the disease, how do we interpret the results found in GWAS? Would this be only a statistical artifact or, in fact, we are facing a new cell metabolic or signaling pathway hitherto unknown? Time will tell. Meanwhile, the reader will clearly understand that, regardless of compelling data, most relevant information regarding the genetics of complex diseases do not come from GWAS, nor from polymorphisms of candidate genes, but from the molecular clarification of monogenic diseases [1]. Now, if a mutation that generated a stop códon in gene "X" causes phenotype "Y", the relevance of subtle alterations in this same gene becomes obvious, changing more subtly the same phenotype.

A clear advance that is directly related to the genome project, but that many times is not interpreted as so, was the flood of knockout animal models for thousands of different genes that emerged in the last decade [2]. Since the installed capacity to create DNA sequences was created, why not unravel the genome of Wistar rats or BALB/C mice, and so forth? With the knowledge of DNA sequences, it is much easier to use the knockout or interference RNA technologies to modulate the gene and discover its relevance in disease models or regarding response to drugs. The great development is not in the advanced understanding of the hereditary aspects of complex diseases, but in the improved understanding of the molecular bases of the physiopathological processes and responses to drugs. This is, in fact, a great victory of genome projects that, on its own, already justifies several expenses made, overlooking the advances that will occur in decades to come.

Another extremely recent advance is the development of a technique of relatively easy execution to create lines of knockout cell cultures [12,13], and for embryonic cell alterations. This system is called CRISPR/Cas-9 (clustered regularly interspaced short palindromic repeats), and seemingly will cause the knockout model generation revolution, since it simplifies the process greatly, both to generate knockout animals as to generate immortalized knockout cell lines, or even to genetic engineering cells obtained from patients [14,15]. This technique is generating debate regarding its use in human embryos to cure monogenic diseases and possible human genetic improvement. This subject has caused negative reactions due to an ethical boundary, which cannot be crossed without a conscious decision by the scientific community and the community in general (Table 12.1) [16].

Another field that was greatly leveraged in recent years is epigenetics. It deals with factors that can also be inherited and that are related to DNA, but that are not information included in the codes of adenine, thymine, cytosine, and guanine. This field deals, for example, with DNA cytosine methylation (CpG islands), methylation and acetylation of histones (both processes regulating when a gene is available for reading or not), and production of microRNAs that act on gene expression at posttranscriptional level. This field is also in plain expansion and will, certainly, bring many encouraging results in the coming years. A beautiful revision of endothelium-related epigenetics was published in *Circulation Research* [17].

Finally, when reflecting upon why is it difficult for studies of both candidate genes and GWAS to generate reproducible results, the questioning that emerged was if we are, in fact, determining the phenotype of the patients precisely. Should a hypertensive person with excessive aldosterone and salt-sensitive hypertension be in the same group of another hypertensive person with a hyperactive sympathetic nervous system? If we understand that the physiopathology is different, why do we imagine that the genetic influence would be the same? That is, perhaps, the greatest obstacle to association studies: if we define all patients as "hypertensive" in order to achieve a number of a few thousand participants per group, how can we ensure that a study in the United Kingdom has the same proportion of salt-sensitive patients as a study in Brazil? Might the results obtained in the United Kingdom and not reproduced in Brazil be due to the population differences, or might there be a difference in phenotype composition? This is the greatest restriction of all, that in a certain way, links the advance in knowledge of genetics of complex diseases necessarily to the advance of diagnostic medicine. If in the future we separate these different subgroups of diseases, maybe the genetic association studies that in the past failed to show reliable, compelling, and reproducible evidence will begin to provide more encouraging results.

Here there is a bit of a dilemma: the need to increasingly improve the determination of patient phenotype, and thus the stratification of these into subgroups (clinical/biological reasons). On the other hand, we need to maintain a robust number of individuals in order to have enough statistical power (statistical reasons). How do we determine phenotypes and subphenotypes and how do we stratify them while keeping a huge number of individuals in each subgroup? This problem can be bypassed with two strategies, which in turn have limitations: the use multicenter studies or progressive collection of patient samples followed by a specific center along several decades. Accepting and adopting this difficulty seems to be discouraging, but it is something that can lead us in the right direction. There are catalysts that can allow the success of these endeavors: public policies that create integrated databases and integrated biobanks. This could protect and disclose the data, samples, and patient contacts for present and future studies. This strategy could help providing higher numbers of subjects, and keeping biological databases available for studies that incorporate new technologies that are becoming popular. Today it is absolutely clear that the great bottleneck in this scientific process is no longer the technical capacity to generate DNA gigabase data (already exists, although at a high cost), but, instead, in having a large number of subjects, carrying the same highly specific and well-scrutinized subphenotype, available in a short period of time.

For readers interested in learning more regarding the subject, we suggest a review article [18] and the perspective from 2011 by the National Institute of Health regarding genomic medicine [2]. Information regarding the polymorphisms and gene sequences can be found in the PubMed website (http://www.ncbi.nlm.nih.gov/ snp), on the Ensembl website (www.ensembl.org), and on the Genome Browser website (https://genome.ucsc.edu). Information regarding the allele frequencies of thousands of polymorphisms in populations of different origins is available on the HapMap website (www.hapmap.org) and Seattle SNPs website (http://pga.gs.washington.edu). Information regarding facts and figures related to human genetics can be found on the National Human Genome Research Institute website (http://www.genome.gov/10000202). Finally, information regarding pharmacogenetics can be found on the www.pharmgkb.org and www.drugbank.ca websites.

The following sections will illustrate some of the associations of genetic polymorphisms in genes expressed by the endothelium with cardiovascular diseases. In order to avoid generating an excessively long text, we shall be limited to some of the most important genes related to endothelium vascular tone control. Therefore, we shall not cover the functions of the endothelium on homeostasis or leukocyte migration during inflammatory processes, or others. In addition, we will not cover endothelium epigenetics and gene handling techniques in greater details, because regardless of being very exciting from the scientific point of view, they are still incipient in respect to clinical application.

MAIN GENES INVOLVED IN ENDOTHELIAL FUNCTIONS

One of the main genes involved in endothelial function regulation is NOS3, which encodes endothelial nitric oxide synthesis. In addition, there are genes that are not necessarily expressed in the endothelium, but that can generate products that will activate receptors there. It is interesting to expose to practitioners the critical view that, of the following associations, few examples in fact are solidified in the literature. Most data presented provide evidence of the involvement of genes that still need validation in bigger populations or populations different from the one originally studied in order to consider the use of such information in clinical practice. Furthermore, analysis regarding the cost/benefit that genetic information may add to clinical practice is warranted. For readers interested in further information regarding the genetic factors related to blood pressure, an elegant review is available in *Circulation Research* [1].

Nitric Oxide System

Endothelial Nitric Oxide Synthase (NOS3)

Endothelial nitric oxide synthase (eNOS) is the main enzyme responsible for nitric oxide (NO) synthesis in the vascular endothelium. This enzyme is encoded by gene *NOS3*, mapped in region 7q36 in human genome [19]. Since its characterization in the beginning of the 90s, a considerable number of polymorphisms has been described in this gene, including single nucleotide polymorphisms (SNPs), variable number tandem repeats (VNTRs), micro-satellites, and insertions/deletions (indel) [20]. Among these polymorphisms, the ones with functional and clinical implications stand out.

Functional Implications of Polymorphisms in Gene NOS3

In general, genetic polymorphisms of *NOS3* that are considered functional are those that affect eNOS expression or activity. Among them, two SNPs in the promoter region, one VNTR in intron 4, and one SNP in exon 7 have been widely studied [21–23]. Fig. 12.3 illustrates the functional mechanisms of these polymorphisms.

SNP g.-786 T>C (rs2070744) is characterized by the exchange of a thymine by a cytosine in position -786 in the promoter region of gene NOS3. The functional implications of this polymorphism are probably related to replication protein A1 (RPA1), which acts as transcription repressor [24]. In vitro studies have shown that this repressor protein binds with more affinity to the promoter region of gene NOS3 when allele C is present, resulting in a drastic reduction of transcriptional activity and, consequently, lower expression of eNOS in comparison to allele T [24,25]. Consistent with these findings, in vivo studies revealed lower NO bioavailability in individuals carrying allele C in respect to those carrying allele T [26], corroborating the functional role of this polymorphism.

Another important SNP located in the promoter region of gene NOS3 is g.-665C>T (rs3918226), whose functional aspect was demonstrated recently [23]. Regardless of the absence of association between this polymorphism and plasmatic levels of nitrite [27] (an important marker of endogenous NO production), in vitro studies have shown that the exchange of cytosine for thymine in position -665 of the promoter results in reduced expression of eNOS. This effect would possibly be related to changed affinity for certain transcription factors [22], which regulate the promoter, and which therefore affect genic transcription.

Different from the SNPs located in the promoter region, which modify genic transcription, the VTNR known as 4b/4a affects eNOS in the posttranscriptional level [28,29]. This VNTR is characterized by repeats of sequences of 27 base-pairs in intron 4 of gene NOS3; the most common alleles in this polymorphism are those with five (variant 4b) and four (variant 4a) copies of the 27 base pairs mentioned, although other rarer alleles have been reported [30]. The functional role of VNTR 4b/4a is related to the formation of microRNAs (microRNA). Molecular studies revealed that endothelial cells containing allele 4b have higher quantity of that specific microRNA, resulting in reduced levels of eNOS mRNA in respect to cells containing 4a allele [28,29].

Finally, the functionality of SNP Glu298Asp (rs1799983) is attributed to modified eNOS activity [31]. This polymorphism is located in exon 7 and corresponds to exchange of guanine for thymine in position 894 of gene NOS3, resulting in the replacement of the amino acid glutamine for aspartate in position 298 of the protein [19]. Cell culture studies revealed that "Asp" allele reduces the affinity between eNOS and caveolin-1, therefore decreasing the availability of the enzyme in its caveolar fraction in endothelial cells [31]. Therefore, a reduced quantity of eNOS is available for activation mediated by the calcium-calmodulin complex, resulting in lower enzymatic activity in respect to allele "Glu" [31]. These findings seem to be in fact relevant in vivo, since individuals carrying allele "Asp" showed reduced NO formation by platelets in comparison to those carrying "Glu" allele [32,33].

Clinical Implications of Polymorphisms in Gene NOS3

Arterial hypertension, pregnancy hypertensive diseases, erectile dysfunction, migraine and metabolic disturbances are examples of highly prevalent diseases that involve reduced endogenous formation of NO as a relevant physiopathological mechanism. Considering the essential role played by NO in the cardiovascular system and the effects of polymorphisms in *NOS3* gene on eNOS expression and activity, as well as on the levels of endogenous NO production markers, many studies have assessed the influence of *NOS3* polymorphisms on the diseases mentioned above. The findings of these studies shall be briefly discussed subsequently.

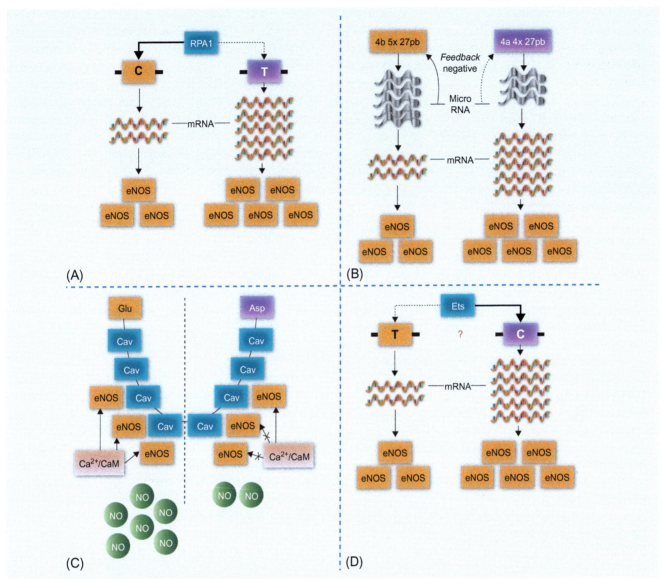

FIG. 12.3 Functional polymorphisms of gene NOS3: mechanisms by which protein expression or activity is modified: Frames A, B, C, and D represent the mechanisms of polymorphisms g.−786 T>C, Variable Number of Tandem Repeats on intron 4, g.894G>T (or Glu298Asp) e g., −665C>T, respectively. (A) The polymorphism is located in a region that, when T allele is present, configures a sequence recognized by RPA1, suppressing gene transcription, and therefore leading to reduced expression of eNOS when compared to C allele. (B) Each portion of 27 base pairs, which can be repeated five (allele 4b) or four times (allele 4a) encodes one siRNA copy that reduces the expression of the eNOS gene itself. Since allele 4b has more repeats, it generates more siNRA that, in turn, reduces mRNA of gene NOS3, and therefore lead to lower enzyme expression. (C) Allele Glu generates a protein with high affinity to caveolin 1. While caveolin 1 inactivates eNOS, this interaction serves to locate eNOS in caveolae, where it is available for activating factors. Allele Asp generates an enzyme with low affinity for caveolin 1 and, therefore, eNOS does not accumulate so efficiently on the caveolae, and therefore is not available for the eNOS activation factors. (D) It was demonstrated that polymorphism g.−665C>T is functional, so that allele T is associated with higher expression of eNOS in respect to allele C. Although functional, the mechanism itself is not known, but computer analysis suggest that changing alleles could change the affinity for transcription factors of the ETS family that could recognize this region, but this is still not definitively demonstrated. *RPA1*, replication protein A 1; *mRNA*, messenger ribonucleic acid (RNA); *eNOS*, endothelial nitric oxide synthase; *Cav*, caveolin; *Ca²⁺/CaM*, calcium-calmodulin; *NO*, nitric oxide; *MicroRNA*, naturally created interference RNA; *Ets*, E26 transformation specific (family of transcription factors).

Arterial Hypertension

As a result of many studies showing that eNOS regulation disturbances may lead to NO deficiency and cause arterial hypertension [34], several papers have evaluated if gene polymorphisms in *NOS3* are associated with susceptibility to this clinical condition. In fact, increased risk of developing arterial hypertension was observed in carriers of variant alleles for polymorphisms g.-786 T > C, g.-665C > T, 4b/4a VNTR, and Glu298Asp [23,35–38]. However, several other studies reported no association between these polymorphisms and arterial hypertension [39–41]. Such discrepancy can be related to the analysis of individual polymorphisms instead of analysis of polymorphism combinations in haplotypes, in addition to the aspects commented above.

The haplotype approach is highly recommended for studying genes that are candidate to arterial hypertension susceptibility [42]. For that matter, studies have evaluated associations between haplotypes involving polymorphisms g.-786 T > C, 4b/4a VNTR, and Glu298Asp and the risk of developing arterial hypertension [40,41,43,44]. In particular, haplotype C-4b-Glu was pointed out as a protector against the development of arterial hypertension, while haplotype C-4b-Asp increased the risk of susceptibility to the disease, in both black and white individuals [40].

Pregnancy Hypertensive Diseases

Pregnancy hypertensive diseases represent the main causes of maternal and neonatal morbidity and mortality, affecting 3%–5% of all pregnancies [45,46]. In this context, studies indicate that deficient NO formation may contribute to increase blood pressure in important pregnancy hypertensive diseases, such as gestational hypertension and preeclampsia [45]. Based on this evidence, several studies have been dedicated to investigate the effects of polymorphisms in gene NOS3 on susceptibility to pregnancy hypertensive diseases. Some of these studies noted associations between the variant alleles of polymorphisms g.-786 T > C, 4b/4a VNTR and Glu298Asp and the risk of developing preeclampsia or gestational hypertension [47–49]. However, other studies failed to demonstrate these associations [46,50]. To clarify these inconsistencies, a recent meta-analysis evaluated studies involving polymorphisms in *NOS3* gene in preeclampsia and found that C allele of SNP g.-786 T > C and 4a allele of 4b/4a VNTR increased susceptibility to this hypertensive disease [51]. In turn, polymorphism Glu298Asp did not affect the preeclampsia risk in this meta-analysis [51].

Besides analyzing the polymorphisms individually, some studies also used the haplotype approach. One of these studies, evaluating the haplotypes that included polymorphisms g.-786 T > C, 4b/4a VNTR, and Glu298Asp, noted an association between haplotype C-4b-Asp and preeclampsia [48]. On the other hand, another study noted higher frequency of haplotype C-4b-Glu in healthy pregnant women when compared to preeclampsia [46]. Curiously, the same haplotype was associated with increased bioavailability of NO in healthy pregnant women, which suggests that haplotype C-4b-Glu may protect against preeclampsia by increasing the formation of NO [46].

Migraine

Migraine is an important neurovascular disease of mutifactorial etiology and complex physiopathology. Being an important mediator to control cerebral blood flow and contribute to the activation of nociceptors, NO plays an important role in migraine genesis [52]. In this context, polymorphisms in *NOS3* gene seem to contribute to susceptibility to this disease.

Variant alleles for polymorphisms g.-786 T > C and Glu298Asp were associated with increased risk of developing migraine [53,54]. Besides that, genotype "AspAsp" for polymorphism Glu298Asp was associated with three times greater risk of occurrence of aura [54] (transient neurological symptoms that precede migraine attacks). The same polymorphism may also influence pain intensity and the age in which migraine begins [53]. On the other hand, other studies reported absence of association between polymorphisms in *NOS3* gene and migraine [55,56], which reinforces the idea that polymorphisms may not be as effective to evidence genetic effects as the haplotype approach. In fact, a study involving haplotypes, which included polymorphisms g.-786 T > C, g.-665C > T, 4b/4a VNTR, Glu298Asp, and tagSNP rs743506, noted that individuals carrying haplotypes C-C-4a-Glu-G and C-C-4b-Glu-G had increased risk of occurrence of aura [56], which suggests that the haplotypes in *NOS3* gene may affect susceptibility to occurrence of such symptoms in patients with migraine.

Metabolic Diseases

Metabolic diseases, such as *diabetes mellitus* and obesity, are highly prevalent and have affected a large number of individuals around the world [57]. Considering that the physiopathology of both diseases involves reduced bioavailability of NO, studies have evaluated the contribution of polymorphisms in *NOS3* gene for the susceptibility to such diseases.

For that matter, strong evidence suggests that allele 4a for VNTR 4b/4a is associated with susceptibility to *diabetes mellitus*, both type 1 and type 2 [58–60], and this effect would probably be related to the endothelial dysfunction associated with this allele in diabetic patients [61]. Allele "Asp" for Glu298Asp polymorphism also seems to be associated with increased risk for developing *diabetes mellitus* [60,62], and this association is particularly important in obese individuals [63]. A haplotype analysis involving polymorphisms g.-786 T > C, 4b/4a VNTR and Glu298Asp evidenced haplotype C-4b-Glu as protector against the development of *diabetes mellitus* [64], indicating that the combination of polymorphisms in *NOS3* gene may also predict susceptibility to this disease.

Polymorphisms in *NOS3* gene also have been associated with obesity and related factors. In fact, individuals carrying allele "Asp" in Glu298Asp polymorphism showed higher body mass indexes and greater abdominal circumference [65], which suggests that this polymorphism may influence genetic susceptibility to obesity. In addition, genotype 4a4a in 4b/4a VNTR was associated with obesity in children and teenagers [66]. In this context, considering that obesity frequently develops during childhood, it is essential to evaluate the genetic susceptibility to this disease in children and teenagers.

Erectile Dysfunction

Reduced bioavailability of NO is an important factor that contributes to erectile dysfunction, a disorder characterized by the inability to obtain and maintain an erection that allows sexual intercourse [67]. Therefore, several studies have been dedicated to evaluate the effects of polymorphisms in *NOS3* gene on susceptibility to this disorder. The "Asp" allele related to Glu298Asp SNP has been constantly associated with increased risk of erectile dysfunction in several studies [68–70]. In addition, CC genotype related to g.-786 T > C SNP also has been associated with erectile dysfunction, being considered an independent risk factor in the absence of other factors [71]. Although until now there are no studies evaluating the influence of haplotypes in gene NOS3 on susceptibility to erectile dysfunction, a recent study showed that haplotypes in this gene may affect the response to sildenafil [72], a medication frequently prescribed to treat this disorder. These findings suggest important pharmacogenetic implications of polymorphisms in *NOS3* gene, as shall be discussed ahead.

Pharmacogenetic Implications of Polymorphisms in NOS3 Gene

Polymorphisms in *NOS3* gene are the main candidates to affect responses to medications that act on the signaling pathway mediated by NO, such as statins. Besides inhibiting cholesterol synthesis, statins have pleiotropic effects such as eNOS activation and increased expression, resulting in high NO bioavailability [21]. In this context, endothelial cells carrying CC genotype for g.-786 T > C polymorphism treated with statins showed higher levels of eNOS mRNA than cells carrying TT genotype [73]. This effect probably results from the increased transcriptional rate, improved eNOS mRNA stability, and decreased expression of repressor protein RPA1 promoted by statin. This would happen more pronouncedly in cells carrying CC genotype [73]. Furthermore, healthy volunteers transporting the rare allele and treated with atorvastatin showed increased NO bioavailability [26], corroborating with in vitro findings. This data suggests that statins may restore impaired endogenous NO production, possibly associated with CC genotype. ACE-I also interferes with NO production by eNOS, which supports the notion that responses to these agents are influenced by *NOS3* gene. In fact, hypertensive patients carrying TC or CC genotypes, and C allele for g.-786 T > C polymorphism showed greater anti-hypertensive response to ACE-I enalapril [74]. Sildenafil is a drug with effects modulated by NO bioavailability, with therapeutic responses modified by polymorphisms in *NOS3* gene [75]. Therefore, individuals with erectile dysfunction, carrying C allele for g.-786 T > C SNP, and 4a allele for 4b/4a VNTR showed better response to treatment with sildenafil when compared to patients carrying T and 4b alleles [72,76]. These findings are consistent with a recent study that observed that lower NO bioavailability in patients with erectile dysfunction is associated with better therapeutic response to sildenafil [77]. Finally, haplotypes including polymorphisms g.-786 T > C, 4b/4a VNTR, and Glu298Asp also influenced responses to sildenafil [72], suggesting that haplotype analysis may help defining the genetic contributions for drug responses.

Arginase I (ARG1) and Arginase II (ARG2)

Arginase I and Arginase II are homologous enzymes that transform L-arginine into urea and L-ornithine [78,79]. Although these enzymes are expressed in the whole body, they are more expressive in the liver, endothelium, smooth muscle cells (Arginase I), and mitochondria in kidney cells, prostate cells, gastrointestinal tract cells, and vessels (Arginase II) [79]. The main role of Arginase I in the liver is to eliminate the nitrogen generated by the metabolism of amino acids and nucleotides by the urea cycle [79,80]. The role of Arginase II is not well known yet, regardless of its involvement in regulation of L-arginine homeostasis [79].

What is interesting regarding arginase is that these enzymes compete with eNOS for the same substrate, L-arginine, and in some tissues they are co-localized with eNOS. In fact, it is believed that unbalanced expression and activity of arginase and eNOS may be one of the causes of endothelial dysfunction [79]. Several factors affect Arginase expression regulation, including inflammatory factors (lipopolysaccharides, tumor necrosis factor (TNF-α), interferon-γ [79,81], interleukins 4, 10, 13 [82], among others), by-products of oxidative stress (H_2O_2 [83], peroxinitrite [84,85]), hypoxia [86,87], and angiotensin II [88], among others [79]. The interaction between arginase and eNOS seems important, mainly in cardiovascular diseases [79,89,90] and diabetes [79,91,92].

Apparently, there is reduced NO production due to deficient L-arginine substrate. In fact, it has been shown that increased arginase activity is associated with endothelial dysfunction in several contexts, such as hypertension, atherosclerosis [93], diabetes [94,95], and aging. Polymorphisms in arginase I and II were widely studied regarding allergic responses and asthma [96–98], with sparse results showing an association of polymorphisms in these genes with blood pressure [99], myocardial infarction risk [100], and carotid intima media thickness

[100]. This is a gene that requires more attention in future studies, due to the great potential of affecting endothelial function.

Vascular Endothelial Growth Factor (VEGF)

VEGF is one of the main citokynes involved in angiogenesis stimulation [101–103]. This citokyne is also related to other effects, such as endothelial eNOS upregulation and increased production of nitric oxide by the endothelium [104–109]. At least three polymorphisms were widely studied due to their functional aspect, that is, ability to change VEGF expression VEGF [110,111]: g.-2578C > A (rs699947), g.-1154G > A (rs1570360), and g.-634G > C (rs2010963). These polymorphisms were associated with several cardiovascular diseases, for example: preeclampsia [112], migraine [113], heart hypertrophy [114], child obesity [115], atherosclerosis [116], and coronary disease [117]. Although interesting, most of the studies mentioned had limitations, for example, small number of individuals included or mixed population. This does not disqualify the findings, but they must be validated in larger populations to allow extrapolating the data to clinical practice.

Asymmetrical Dimethylarginine (ADMA) and Its Metabolism Enzyme, Dimethylarginine Dimethylamino-Hydrolase (DDAH1 and 2)

ADMA is a natural form of L-arginine that derives from proteolysis of methylated proteins [118]. ADMA acts as endogenous inhibitor of three subtypes of NO synthase (eNOS, nNOS, and iNOS). Therefore, when higher concentrations of ADMA are present, there is a reduction in enzymatic NO production. In fact, ADMA was implied in the pathogenesis of cardiovascular diseases due to reduction of NO synthesis [119]. Besides that, there is clear increase in plasmatic ADMA levels in patients with heart failure [120], congenital heart disease [121], arrhythmia [122], arterial hypertension [123], among others. Several studies show that the plasmatic levels of ADMA may predict deleterious events and cardiovascular mortality [118]. There are some enzymes that metabolize ADMA, the most important being dimethylarginine dimethylamino-hydrolase types 1 and 2. These enzymes are encoded by genes *DDAH1* and *DDAH2*, and are responsible for the metabolism of 80% of the ADMA generated intracellularly [124]. Recent clinical studies have shown that polymorphisms in *DDAH1* gene are able to modulate the plasmatic levels of ADMA [125–127], and these levels, when increased, also increase the risk of cardiovascular outcomes [128–130]. The association between polymorphisms in *DDAH1* and cardiovascular outcomes has already been demonstrated in Chinese people regarding infarction and coronary disease [131], but requires expansion to and validation in other populations.

Endothelin System

The endothelin system involves at least six main genes: *EDN1* and *EDN2*, which encode respectively endothelins 1 and 2; genes *EDNRA* and *EDNRB* that encode endothelin receptors A and B, and the genes of Endothelin Converting Enzymes *ECE1* and *ECE2*. Endothelins 1 and 2 are synthesized as larger and inactive forms, which are subsequently cleaved into active forms. When these peptides act on ET_a and ET_b receptors in vessel smooth muscle cells, they lead to intense vasoconstriction by increasing the cytoplasmic calcium in these cells [132]. On the other hand, the ET_b receptor expressed in endothelial cells leads to endothelium nitric oxide production and also works as a circulating endothelin removal and internalization system. The importance of this system is illustrated by the effectiveness of Bosentan and Darusentan, antagonists mainly of ET_a receptor in reducing blood pressure [133,134]. Changes in the endothelin system seem to be particularly important in renal function and salt-sensitive hypertension genesis [132]. Some studies show associations between polymorphisms in endothelin system genes and cardiovascular phenotypes. Polymorphisms in the receptor gene, *EDNRA*, were associated with idiopathic pulmonary hypertension [135] and worst correlated phenotypes. Polymorphisms in genes of both receptors (*EDNRA* and *EDNRB*) were associated with changes in pulse wave speed and carotid intima media thickness in hypertensive subjects [136]. Polymorphisms in gene *EDN1* were associated with endothelial dysfunction in children [137]. Polymorphisms in gene of Endothelin Converting Enzyme 1, *ECE1*, were associated with higher blood pressure values in Japanese hypertensive women [138]. Finally, a polymorphism next to *EDN3* and *GNAS* genes (the latter encodes the alpha subunit of G protein, which associates with several membrane receptors, including adrenergic receptors), was associated with differences in blood pressure drop in response

to hydrochlorothiazide in a large GWAS [139]. In fact, the last study does not clarify if there is a relation with the endothelin system, requiring more studies to validate this association.

Inflammatory Mediators

E Selectin (SELE)

The expression of this protein is increased in the endothelial membrane after pro-inflammatory stimuli [140]. Once expressed, the function of E selectin is to promote leukocyte adhesion, allowing these cells to execute their inflammatory and/or immune responses. The relation with hypertension is because these immunological responses may cause endothelial damage that, in turn, may lead to hypertension and other cardiovascular diseases. A large study with Asian subjects followed 1768 patients for 9 years (2003–12) in order to check blood pressure increase. Among the several candidate genes tested, four *TAGSNPs* in *SELE* gene were associated with blood pressure progression and risk of developing hypertension in this sample [141]. These results are consistent with another work with Caucasians, also with a very robust number of patients followed by 11 years, in which polymorphisms in *SELE* were also associated with blood pressure changes [142]. Some factors make these studies very robust: first, the large sample size; second, the phenotype was not followed only in a single moment, but in reality was evaluated repeatedly along time, showing a consistent increase pattern. This reduces the chances of random fluctuations influencing the results reported. The same study with Asiatic people found associations (less decisive) in respect to genes *DDAH1* and *EDNRA* (mentioned in other sections herein).

WHAT DOES GENOMIC MEDICINE HOLD FOR THE FUTURE OF ENDOTHELIAL DISEASE TREATMENT?

Currently, in terms of applicability of genomic medicine to treat cardiovascular diseases, we still have relatively incipient data and evidence. On the other hand, there is still a lot to study and discover, which gives us the idea that it is possible to extrapolate here all the excitement seen in other fields of medicine (especially oncology), with the advances in genomics. The understanding of how genetic factors affect diseases may allow us to select patients that benefit from personalized pharmacological treatments. For example, carriers of a panel of high-risk alleles in important genes may benefit from early treatment or from more rigid blood pressure control goals. Drawing upon genetic data, the physician can decide for the use of an optimized dose of the best medication for a given patient, minimizing the need for the trial-and-error approach. Genomics will bring improvements in understanding of cardiovascular biology and pathology, which may result in new therapeutic targets and new drugs. The use of nucleic acids in the form of gene therapy, interference RNAs, or even CRISPR/Cas9 system and other approaches may be employed in the pharmacological treatment of these diseases. Today, we are in a situation that is analogous to the beginning of the 70s, when it was thought that it would be fantastic to use an antibodies to treat a disease (today is a reality, regardless of the costs). Obviously, there is a series of barriers that must be overcome besides the acquisition of knowledge that we still do not possess. The main barrier is still the cost per base analyzed (genotyped). The cost is lowering over the years (http://genome.gov/sequencingcosts), starting at 10 dollars per base (in 1990), found today to be a little below 10,000 dollars per genome. It is clear that upscaling of current technologies per se will not be able to reach genome sequencing at sufficiently low costs. We still have a long road to develop a technique with relatively low cost that allows genome sequencing, for example, in neonates, for predictive trial purposes (eventually replacing the neonatal heel prick test). There is no denying of the several advantages and benefits that may come from advancements in genomic medicine. We cannot know, however, if we will be alive to see the full use of these potential, and if it will in fact revolutionize cardiovascular disease treatment.

References

[1] Padmanabhan S, Caulfield M, Dominiczak AF. Genetic and molecular aspects of hypertension. Circ Res 2015;116(6):937–59.

[2] Green ED, Guyer MS. Charting a course for genomic medicine from base pairs to bedside. Nature 2011;470(7333):204–13.

[3] Rosendorff C, Lackland DT, Allison M, et al. Treatment of hypertension in patients with coronary artery disease: a scientific statement from the American Heart Association, American College of Cardiology, and American Society of Hypertension. Circulation 2015;131(19):e435–70.

[4] [VI Brazilian guidelines on hypertension]. Arq Bras Cardiol 2010;95 (1 Suppl):1–51.

[5] Lander ES, Linton LM, Birren B, et al. Initial sequencing and analysis of the human genome. Nature 2001;409(6822):860–921.

[6] Venter JC, Adams MD, Myers EW, et al. The sequence of the human genome. Science 2001;291(5507):1304–51.

[7] Lesko LJ, Zineh I. DNA, drugs and chariots: on a decade of pharmacogenomics at the US FDA. Pharmacogenomics 2010;11 (4):507–12.

[8] Collins FS. Statement on NIH funding of research using geneediting technologies in human embryos. [Internet]. Available at: http://www.nih.gov/about/director/04292015_statement_gene_editing_technologies.htm [accessed 20.06.16].

[9] Watson J. Statement by James D Watson. The New York Times. [Internet]. Available at: http://www.nytimes.com/2007/10/25/science/26wattext.html?_r=0 [accessed 20.06.16].

[10] Crick F. On the genetic code—nobel lecture. [Internet]. Available at: http://www.nobelprize.org/nobel_prizes/medicine/laureates/1962/crick-lecture.html [accessed 20.06.16].

[11] Link E, Parish S, Armitage J, et al. SLCO1B1 variants and statin-induced myopathy—a genomewide study. N Engl J Med 2008; 359(8):789–99.

[12] Hendel A, Bak RO, Clark JT, et al. Chemically modified guide RNAs enhance CRISPR-Cas genome editing in human primary cells. Nat Biotechnol 2015;33(9):985–9.

[13] Hale CR, Majumdar S, Elmore J, et al. Essential features and rational design of CRISPR RNAs that function with the Cas RAMP module complex to cleave RNAs. Mol Cell 2012;45(3):292–302.

[14] Sander JD, Joung JK. CRISPR-Cas systems for editing, regulating and targeting genomes. Nat Biotechnol 2014;32(4):347–55.

[15] Shalem O, Sanjana NE, Zhang F. High-throughput functional genomics using CRISPR-Cas9. Nat Rev Genet 2015;16(5):299–311.

[16] Baltimore D, Berg P, Botchan M, et al. Biotechnology. A prudent path forward for genomic engineering and germline gene mo-dification. Science 2015;348(6230):36–8.

[17] Matouk CC, Marsden PA. Epigenetic regulation of vascular endothelial gene expression. Circ Res 2008;102(8):873–87.

[18] Lander ES. Initial impact of the sequencing of the human genome. Nature 2011;470(7333):187–97.

[19] Marsden PA, Heng HH, Scherer SW, et al. Structure and chromosomal localization of the human constitutive endothelial nitric oxide synthase gene. J Biol Chem 1993;268(23):17478–88.

[20] Cooke GE, Doshi A, Binkley PF. Endothelial nitric oxide synthase gene: prospects for treatment of heart disease. Pharmacogenomics 2007;8(12):1723–34.

[21] Lacchini R, Silva PS, Tanus-Santos JE. A pharmacogenetics-based approach to reduce cardiovascular mortality with the prophylactic use of statins. Basic Clin Pharmacol Toxicol 2010;106(5):357–61.

[22] Salvi E, Kutalik Z, Glorioso N, et al. Genomewide association study using a high-density single nucleotide polymorphism array and case-control design identifies a novel essential hypertension susceptibility locus in the promoter region of endothelial NO synthase. Hypertension 2012;59(2):248–55.

[23] Salvi E, Kuznetsova T, Thijs L, et al. Target sequencing, cell experiments, and a population study establish endothelial nitric oxide synthase (eNOS) gene as hypertension susceptibility gene. Hypertension 2013;62(5):844–52.

[24] Miyamoto Y, Saito Y, Nakayama M, et al. Replication protein A1 reduces transcription of the endothelial nitric oxide synthase gene containing a −786 T−>C mutation associated with coronary spastic angina. Human Mol Gen 2000;9(18):2629–37.

[25] Nakayama M, Yasue H, Yoshimura M, et al. T-786—>C mutation in the 5′-flanking region of the endothelial nitric oxide synthase gene is associated with coronary spasm. Circulation 1999; 99(22):2864–70.

[26] Nagassaki S, Sertório JT, Metzger IF, et al. eNOS gene T-786C polymorphism modulates atorvastatin-induced increase in blood nitrite. Free Radic Biol Med 2006;41(7):1044–9.

[27] Luizon MR, Metzger IF, Lacchini R, et al. Endothelial nitric oxide synthase polymorphism rs3918226 associated with hyperten-sion

does not affect plasma nitrite levels in healthy subjects. Hypertension 2012;59(6):e52. author reply e53.

[28] Zhang MX, Zhang C, Shen YH, et al. Effect of 27 nt small RNA on endothelial nitric-oxide synthase expression. Mol Biol Cell 2008; 19(9):3997–4005.

[29] Zhang MX, Zhang C, Shen YH, et al. Biogenesis of short intronic repeat 27-nucleotide small RNA from endothelial nitric-oxide synthase gene. J Biol Chem 2008;283(21):14685–93.

[30] Tanus-Santos JE, Desai M, Flockhart DA. Effects of ethnicity on the distribution of clinically relevant endothelial nitric oxide variants. Pharmacogenetics 2001;11(8):719–25.

[31] Joshi MS, Mineo C, Shaul PW, et al. Biochemical consequences of the NOS3 Glu298Asp variation in human endothelium: altered caveolar localization and impaired response to shear. FASEB J 2007;21(11):2655–63.

[32] Tanus-Santos JE, Desai M, Deak LR, et al. Effects of endothelial nitric oxide synthase gene polymorphisms on platelet function, nitric oxide release, and interactions with estradiol. Pharmacogenetics 2002;12(5):407–13.

[33] Godfrey V, Chan SL, Cassidy A, et al. The functional consequence of the Glu298Asp polymorphism of the endothelial nitric oxide synthase gene in young healthy volunteers. Cardiovasc Drug Rev 2007;25(3):280–8.

[34] Thomas GD, Zhang W, Victor RG. Nitric oxide deficiency as a cause of clinical hypertension: promising new drug targets for refractory hypertension. JAMA 2001;285(16):2055–7.

[35] Uwabo J, Soma M, Nakayama T, et al. Association of a variable number of tandem repeats in the endothelial constitutive nitric oxide synthase gene with essential hypertension in Japanese. Am J Hypertens 1998;11(1 Pt 1):125–8.

[36] Hyndman ME, Parsons HG, Verma S, et al. The T-786->C mutation in endothelial nitric oxide synthase is associated with hypertension. Hypertension 2002;39(4):919–22.

[37] Pereira TV, Rudnicki M, Cheung BM, et al. Three endothelial nitric oxide (NOS3) gene polymorphisms in hypertensive and normotensive individuals: meta-analysis of 53 studies reveals evidence of publication bias. J Hypertens 2007;25(9):1763–74.

[38] De Miranda JA, Lacchini R, Belo VA, et al. The effects of endothelial nitric oxide synthase tagSNPs on nitrite levels and risk of hypertension and obesity in children and adolescents. J Hum Hypertens 2015;29(2):109–14.

[39] Kato N, Sugiyama T, Morita H, et al. Lack of evidence for association between the endothelial nitric oxide synthase gene and hypertension. Hypertension 1999;33(4):933–6.

[40] Sandrim VC, Coelho EB, Nobre F, et al. Susceptible and protective eNOS haplotypes in hypertensive black and white subjects. Atherosclerosis 2006;186(2):428–32.

[41] Sandrim VC, Yugar-Toledo JC, Desta Z, et al. Endothelial nitric oxide synthase haplotypes are related to blood pressure elevation, but not to resistance to antihypertensive drug therapy. J Hypertens 2006;24(12):2393–7.

[42] Yagil Y, Yagil C. Candidate genes, association studies and haplotype analysis in the search for the genetic basis of hypertension. J Hypertens 2004;22(7):1255–8.

[43] Sandrim VC, de Syllos RW, Lisboa HR, et al. Endothelial nitric oxide synthase haplotypes affect the susceptibility to hypertension in patients with type 2 diabetes mellitus. Atherosclerosis 2006;189(1):241–6.

[44] Vasconcellos V, Lacchini R, Jacob-Ferreira AL, et al. Endothelial nitric oxide synthase haplotypes associated with hypertension do not predispose to cardiac hypertrophy. DNA Cell Biol 2010;29(4):171–6.

[45] Sandrim VC Palei AC, Metzger IF, et al. Nitric oxide formation is inversely related to serum levels of antiangiogenic factors soluble fms-like tyrosine kinase-1 and soluble endogline in preeclampsia. Hypertension 2008;52(2):402–7.

[46] Sandrim VC, Palei AC, Sertorio JT, et al. Effects of eNOS polymorphisms on nitric oxide formation in healthy pregnancy and in pre-eclampsia. Mol Hum Reprod 2010;16(7):506–10.

[47] Tempfer CB, Dorman K, Deter RL, et al. An endothelial nitric oxide synthase gene polymorphism is associated with pre-eclampsia. Hypertens Pregnancy 2001;20(1):107–18.

[48] Serrano NC, Casas JP, Díaz LA, et al. Endothelial NO synthase genotype and risk of preeclampsia: a multicenter case-control study. Hypertension 2004;44(5):702–7.

[49] Seremak-Mrozikiewicz A, Drews K, Barlik M, et al. The significance of -786 T->C polymorphism of endothelial NO synthase (eNOS) gene in severe preeclampsia. J Matern Fetal Neonatal Med 2011;24(3):432–6.

[50] Landau R, Xie HG, Dishy V, et al. No association of the Asp298 variant of the endothelial nitric oxide synthase gene with preeclampsia. Am J Hypertens 2004;17(5 Pt 1):391–4.

[51] Dai B, Liu T, Zhang B, et al. The polymorphism for endothelial nitric oxide synthase gene, the level of nitric oxide and the risk for pre-eclampsia: a meta-analysis. Gene 2013;519(1):187–93.

[52] Olesen J. Nitric oxide-related drug targets in headache. Neurotherapeutics 2010;7(2):183–90.

[53] Eröz R, Bahadir A, Dikici S, et al. Association of endothelial nitric oxide synthase gene polymorphisms (894G/T, -786 T/C, G10T) and clinical findings in patients with migraine. Neuromolecular Med 2014;16(3):587–93.

[54] Borroni B, Rao R, Liberini P, et al. Endothelial nitric oxide synthase (Glu298Asp) polymorphism is an independent risk factor for migraine with aura. Headache 2006;46(10):1575–9.

[55] Toriello M, Oterino A, Pascual J, et al. Lack of association of endothelial nitric oxide synthase polymorphisms and migraine. Headache 2008;48(7):1115–9.

[56] Goncalves FM, Martins-Oliveira A, Speciali JG, et al. Endothelial nitric oxide synthase haplotypes associated with aura in patients with migraine. DNA Cell Biol 2011;30(6):363–9.

[57] Golden SH, Robinson KA, Saldanha I, et al. Clinical review: prevalence and incidence of endocrine and metabolic disorders in the United States: a comprehensive review. J Clin Endocrinol Metab 2009;94(6):1853–78.

[58] Galanakis E, Kofteridis D, Stratigi K, et al. Intron 4 a/b polymorphism of the endothelial nitric oxide synthase gene is associated with both type 1 and type 2 diabetes in a genetically homogeneous population. Hum Immunol 2008;69(4-5):279–83.

[59] Mehrab-Mohseni M, Tabatabaei-Malazy O, Hasani-Ranjbar S, et al. Endothelial nitric oxide synthase VNTR (intron 4 a/b) polymorphism association with type 2 diabetes and its chronic complications. Diabetes Res Clin Pract 2011;91(3):348–52.

[60] Jia Z, Zhang X, Kang S, et al. Association of endothelial nitric oxide synthase gene polymorphisms with type 2 diabetes mellitus: a meta-analysis. Endocr J 2013;60(7):893–901.

[61] Komatsu M, Kawagishi T, Emoto M, et al. ecNOS gene polymorphism is associated with endothelium-dependent vasodilation in type 2 diabetes. Am J Physiol Heart Circ Physiol 2002;283(2):H557–61.

[62] Monti LD, Barlassina C, Citterio L, et al. Endothelial nitric oxide synthase polymorphisms are associated with type 2 diabetes and the insulin resistance syndrome. Diabetes 2003;52(5):1270–5.

[63] Bressler J, Pankow JS, Coresh J, et al. Interaction between the NOS3 gene and obesity as a determinant of risk of type 2 diabetes: the Atherosclerosis Risk in Communities study. PloS ONE 2013;8(11):e79466.

[64] De Syllos RW, Sandrim VC, Lisboa HR, et al. Endothelial nitric oxide synthase genotype and haplotype are not associated with diabetic retinopathy in diabetes type 2 patients. Nitric Oxide 2006;15(4):417–22.

[65] Podolsky RH, Barbeau P, Kang HS, et al. Candidate genes and growth curves for adiposity in African- and European-American youth. Int J Obes 2007;31(10):1491–9.

I. STRUCTURE AND FUNCTIONS

[66] Souza-Costa DC, Belo VA, Silva OS, et al. eNOS haplotype associated with hypertension in obese children and adolescents. Int J Obes 2011;35(3):387–92.

[67] Hatzimouratidis K, Amar E, Eardley I, et al. Guidelines on male sexual dysfunction: erectile dysfunction and premature ejaculation. Eur Urol 2010;57(5):804–14.

[68] Hermans MP, Ahn SA, Rousseau MF. eNOS [Glu298Asp] polymorphism, erectile function and ocular pressure in type 2 diabetes. Eur J Clin Invest 2012;42(7):729–37.

[69] Lee YC, Huang SP, Liu CC, et al. The association of eNOS G894T polymorphism with metabolic syndrome and erectile dysfunction. J Sex Med 2012;9(3):837–43.

[70] Wang JL, Wang HG, Gao HQ, et al. Endothelial nitric oxide synthase polymorphisms and erectile dysfunction: a meta-analysis. J Sex Med 2010;7(12):3889–98.

[71] Sinici I, Güven EO, Serefoglu E, et al. T-786C polymorphism in promoter of eNOS gene as genetic risk factor in patients with erectile dysfunction in Turkish population. Urology 2010;75(4):955–60.

[72] Muniz JJ, Lacchini R, Rinaldi TO, et al. Endothelial nitric oxide synthase genotypes and haplotypes modify the responses to sildenafil in patients with erectile dysfunction. Pharmacogenomics J 2013;13(2):189–96.

[73] Abe K, Nakayama M, Yoshimura M, et al. Increase in the transcriptional activity of the endothelial nitric oxide synthase gene with fluvastatin: a relation with the –786 T->C polymorphism. Pharmacogenet Genomics 2005;15(5):329–36.

[74] Silva PS, Fontana V, Luizon MR, et al. eNOS and BDKRB2 genotypes affect the antihypertensive responses to enalapril. Eur J Clin Pharmacol 2013;69(2):167–77.

[75] Lacchini R, Tanus-Santos JE. Pharmacogenetics of erectile dysfunction: navigating into uncharted waters. Pharmacogenomics 2014; 15(11):1519–38.

[76] Peskircioglu L, Atac FB, Erdem SR, et al. The association between intron 4 VNTR, E298A and IVF 23+10 G/T polymorphisms of ecNOS gene and sildenafll responsiveness in patients with erectile dysfunction. Int J Impot Res 2007;19(2):149–53.

[77] Muniz JJ, Lacchini R, Sertório JT, et al. Low nitric oxide bioavailability is associated with better responses to sildenafil in patients with erectile dysfunction. Naunyn Schmiedebergs Arch Pharmacol 2013;386(9):805–11.

[78] Durante W, Johnson FK, Johnson RA. Arginase: a critical regulator of nitric oxide synthesis and vascular function. Clin Exp Pharmacol Physiol 2007;34(9):906–11.

[79] Pernow J, Jung C. Arginase as a potential target in the treatment of cardiovascular disease: reversal of arginine steal? Cardiovasc Res 2013;98(3):334–43.

[80] Masuda H. Significance of nitric oxide and its modulation mechanisms by endogenous nitric oxide synthase inhibitors and arginase in the micturition disorders and erectile dysfunction. Int J Urol 2008;15(2):128–34.

[81] Nelin LD, Wang X, Zhao Q, et al. MKP-1 switches arginine metabolism from nitric oxide synthase to arginase following endotoxin challenge. Am J Physiol Cell Physiol 2007;293(2):C632–40.

[82] Munder M. Arginase: an emerging key player in the mammalian immune system. Br J Pharmacol 2009;158(3):638–51.

[83] Thengchaisri N, Hein TW, Wang W, et al. Upregulation of arginase by H$_2$O$_2$ impairs endothelium-dependent nitric oxide-media-ted dilation of coronary arterioles. Arterioscler Thromb Vasc Biol 2006;26(9):2035–42.

[84] Sankaralingam S, Xu H, Davidge ST. Arginase contributes to endothelial cell oxidative stress in response to plasma from women with preeclampsia. Cardiovasc Res 2010;85(1):194–203.

[85] Chandra S, Romero MJ, Shatanawi A, et al. Oxidative species increase arginase activity in endothelial cells through the RhoA/ Rho kinase pathway. Br J Pharmacol 2012;165(2):506–19.

[86] Prieto CP, Krause BJ, Quezada C, et al. Hypoxia-reduced nitric oxide synthase activity is partially explained by higher

[87] Chen B, Calvert AE, Cui H, et al. Hypoxia promotes human pulmonary artery smooth muscle cell proliferation through induction of arginase. Am J Physiol Lung Cell Mol Physiol 2009;297(6):L1151–9.

[88] Toque HA, Romero MJ, Tostes RC, et al. p38 Mitogen-activated protein kinase (MAPK) increases arginase activity and contributes to endothelial dysfunction in corpora cavernosa from angiotensin-II-treated mice. J Sex Med 2010;7(12):3857–67.

[89] Quitter F, Figulla HR, Ferrari M, et al. Increased arginase levels in heart failure represent a therapeutic target to rescue microvascular perfusion. Clin Hemorheol Microcirc 2013;54(1):75–85.

[90] Porembska Z, Kedra M. Early diagnosis of myocardial infarction by arginase activity determination. Clin Chim Acta 1975;60(3):355–61.

[91] Ogino K, Takahashi N, Takigawa T, et al. Association of serum arginase I with oxidative stress in a healthy population. Free Radic Res 2011;45(2):147–55.

[92] Kashyap SR, Lara A, Zhang R, et al. Insulin reduces plasma arginase activity in type 2 diabetic patients. Diabetes Care 2008; 31(1):134–9.

[93] Ryoo S, Lemmon CA, Soucy KG, et al. Oxidized low-density lipoprotein-dependent endothelial arginase II activation contributes to impaired nitric oxide signaling. Circ Res 2006;99(9):951–60.

[94] Romero MJ, Platt DH, Tawfik HE, et al. Diabetes-induced coronary vascular dysfunction involves increased arginase activity. Circ Res 2008;102(1):95–102.

[95] Bivalacqua TJ, Hellstrom WJ, Kadowitz PJ, et al. Increased expression of arginase II in human diabetic corpus cavernosum: in diabetic-associated erectile dysfunction. Biochem Biophys Res Commun 2001;283(4):923–7.

[96] Li H, Romieu I, Sienra-Monge JJ, et al. Genetic polymorphisms in arginase I and II and childhood asthma and atopy. J Allergy Clin Immunol 2006;117(1):119–26.

[97] Vonk JM, Postma DS, Maarsingh H, et al. Arginase 1 and arginase 2 variations associate with asthma, asthma severity and beta2 agonist and steroid response. Pharmacogenet Genomics 2010; 20(3):179–86.

[98] Salam MT, Bastain TM, Rappaport EB, et al. Genetic variations in nitric oxide synthase and arginase influence exhaled nitric oxide levels in children. Allergy 2011;66(3):412–9.

[99] Meroufel D, Dumont J, Médiène-Benchekor S, et al. Characterization of arginase 1 gene polymorphisms in the Algerian popu-lation and association with blood pressure. Clin Biochem 2009;42(10-11):1178–82.

[100] Dumont J, Zureik M, Cottel D, et al. Association of arginase 1 gene polymorphisms with the risk of myocardial infarction and common carotid intima media thickness. J Med Genet 2007; 44(8):526–31.

[101] Ferrara N. Vascular endothelial growth factor. Arterioscler Thromb Vasc Biol 2009;29(6):789–91.

[102] Holmes DI, Zachary I. The vascular endothelial growth factor (VEGF) family: angiogenic factors in health and disease. Genome Biol 2005;6(2):209.

[103] Giordano FJ, Gerber HP, Williams SP, et al. A cardiac myocyte vascular endothelial growth factor paracrine pathway is required to maintain cardiac function. Proc Natl Acad Sci U S A 2001; 98(10):5780–5.

[104] Hood JD, Meininger CJ, Ziche M, et al. VEGF upregulates ecNOS message, protein, and NO production in human endothelial cells. Am J Physiol 1998;274(3 Pt 2):H1054–8.

[105] Van Der Zee R, Murohara T, Luo Z, et al. Vascular endothelial growth factor/vascular permeability factor augments nitric oxide release from quiescent rabbit and human vascular endothelium. Circulation 1997;95(4):1030–7.

[106] Lin CS, Ho HC, Chen KC, et al. Intracavernosal injection of vascular endothelial growth factor induces nitric oxide synthase isoforms. BJU Int 2002;89(9):955–60.

arginase-2 activity and cellular redistribution in human umbilical vein endothelium. Placenta 2011;32(12):932–40.

[107] Shen BQ, Lee DY, Zioncheck TF. Vascular endothelial growth factor governs endothelial nitric-oxide synthase expression via a KDR/Flk-1 receptor and a protein kinase C signaling pathway. J Biol Chem 1999;274(46):33057–63.

[108] Musicki B, Kramer MF, Becker RE, et al. Inactivation of phosphorylated endothelial nitric oxide synthase (Ser-1177) by O-GlcNAc in diabetes-associated erectile dysfunction. Proc Natl Acad Sci USA 2005;102(33):11870–5.

[109] Musicki B, Kramer MF, Becker RE, et al. Age-related changes in phosphorylation of endothelial nitric oxide synthase in the rat penis. J Sex Med 2005;2(3):347–55. discussion 355-7.

[110] Lambrechts D, Storkebaum E, Morimoto M, et al. VEGF is a modifier of amyotrophic lateral sclerosis in mice and humans and protects motoneurons against ischemic death. Nat Genet 2003;34(4):383–94.

[111] Shahbazi M, Fryer AA, Pravica V, et al. Vascular endothelial growth factor gene polymorphisms are associated with acute renal allograft rejection. J Am Soc Nephrol 2002;13(1):260–4.

[112] Sandrim VC, Palei AC, Cavalli RC, et al. Vascular endothelial growth factor genotypes and haplotypes are associated with pre-eclampsia but not with gestational hypertension. Mol Hum Reprod 2009;15(2):115–20.

[113] Goncalves FM, Martins-Oliveira A, Speciali JG, et al. Vascular endothelial growth factor genetic polymorphisms and haplotypes in women with migraine. DNA Cell Biol 2010;29(7):357–62.

[114] Lacchini R, Luizon MR, Gasparini S, et al. Effect of genetic polymorphisms of vascular endothelial growth factor on left ventricular hypertrophy in patients with systemic hypertension. Am J Cardiol 2014;113(3):491–6.

[115] Belo VA, Souza-Costa DC, Luizon MR, et al. Vascular endothelial growth factor haplotypes associated with childhood obesity. DNA Cell Biol 2011;30(9):709–14.

[116] Howell WM, Ali S, Rose-Zerilli MJ, et al. VEGF polymorphisms and severity of atherosclerosis. J Med Genet 2005;42(6):485–90.

[117] Biselli PM, Guerzoni AR, de Godoy MF, et al. Vascular endothelial growth factor genetic variability and coronary artery disease in Brazilian population. Heart Vessels 2008;23(6):371–5.

[118] Bouras G, Defereos S, Tousoulis D, et al. Asymmetric Dimethylarginine (ADMA): a promising biomarker for cardiovascular disease? Curr Top Med Chem 2013;13(2):180–200.

[119] Pope AJ, Karuppiah K, Cardounel AJ. Role of the PRMT-DDAH-ADMA axis in the regulation of endothelial nitric oxide production. Pharmacol Res 2009;60(6):461–5.

[120] Kielstein JT, Bode-Böger SM, Klein G, et al. Endogenous nitric oxide synthase inhibitors and renal perfusion in patients with heart failure. Eur J Clin Invest 2003;33(5):370–5.

[121] Tutarel O, Denecke A, Bode-Böger SM, et al. Asymmetrical dimethylarginine—more sensitive than NT-proBNP to diagnose heart failure in adults with congenital heart disease. PLoS ONE 2012;7(3):e33795.

[122] Yang L, Xiufen Q, Shugin S, et al. Asymmetric dimethylarginine concentration and recurrence of atrial tachyarrythmias after catheter ablation in patients with persistent atrial fibrillation. J Interv Card Electrophysiol 2011;32(2):147–54.

[123] Tousoulis D, Bouras G, Antoniades C, et al. Methionine-induced homocysteinemia impairs endothelial function in hypertensives: the role of asymmetrical dimethylarginine and antioxidant vitamins. Am J Hypertens 2011;24(8):936–42.

[124] Teerlink T. ADMA metabolism and clearance. Vasc Med 2005;(10 Suppl 1):S73–81.

[125] Anderssohn M, McLachlan S, Lüneburg N, et al. Genetic and environmental determinants of dimethylarginines and association with cardiovascular disease in patients with type 2 diabetes. Diabetes Care 2014;37(3):846–54.

[126] Seppala I, Kleber ME, Lyytikäinen LP, et al. Genome-wide association study on dimethylarginines reveals novel AGXT2 variants associated with heart rate variability but not with overall mortality. Eur Heart J 2014;35(8):524–31.

[127] Abhary S, Burdon KP, Kuot A, et al. Sequence variation in DDAH1 and DDAH2 genes is strongly and additively associated with serum ADMA concentrations in individuals with type 2 diabetes. PLoS ONE 2010;5(3):e9462.

[128] Krzyzanowska K, Mittermayer F, Wolzt M, et al. Asymmetric dimethylarginine predicts cardiovascular events in patients with type 2 diabetes. Diabetes Care 2007;30(7):1834–9.

[129] Lu TM, Chung MY, Lin MW, et al. Plasma asymmetric dimethylarginine predicts death and major adverse cardiovascular events in individuals referred for coronary angiography. Int J Cardiol 2011;153(2):135–40.

[130] Boger RH, Sullivan LM, Schwedhelm E, et al. Plasma asymmetric dimethylarginine and incidence of cardiovascular disease and death in the community. Circulation 2009;119(12):1592–600.

[131] Ding H, Wu B, Wang H, et al. A novel loss-of-function DDAH1 promoter polymorphism is associated with increased susceptibility to thrombosis stroke and coronary heart disease. Circ Res 2010;106(6):1145–52.

[132] Boesen EI. Endothelin receptors, renal effects and blood pressure. Curr Opin Pharmacol 2005;21:25–34.

[133] Krum H, Viskoper RJ, Lacourciere Y, et al. The effect of an endothelin-receptor antagonist, bosentan, on blood pressure in patients with essential hypertension. Bosentan Hypertension Investigators. N Engl J Med 1998;338(12):784–90.

[134] Nakov R, Pfarr E, Eberle S. Darusentan: an effective endothelinA receptor antagonist for treatment of hypertension. Am J Hypertens 2002;15(7 Pt 1):583–9.

[135] Calabro P, Limongelli G, Maddaloni V, et al. Analysis of endothelin-1 and endothelin-1 receptor A gene polymorphisms in patients with pulmonary arterial hypertension. Intern Emerg Med 2012;7(5):425–30.

[136] Yasuda H, Kamide K, Takiuchi S, et al. Association of single nucleotide polymorphisms in endothelin family genes with the progression of atherosclerosis in patients with essential hypertension. J Hum Hypertens 2007;21(11):883–92.

[137] Chatsuriyawong S, Gozal D, Kheirandish-Gozal L, et al. Genetic variance in nitric oxide synthase and endothelin genes among children with and without endothelial dysfunction. J Transl Med 2013;11:227.

[138] Banno M, Hanada H, Kamide K, et al. Association of genetic polymorphisms of endothelin-converting enzyme-1 gene with hypertension in a Japanese population and rare missense mutation in proproendothelin-1 in Japanese hypertensives. Hypertens Res 2007;30(6):513–20.

[139] Turner ST, Boerwinkle E, O'Connell JR, et al. Genomic association analysis of common variants influencing antihypertensive response to hydrochlorothiazide. Hypertension 2013;62(2):391–7.

[140] Kansas GS. Selectins and their ligands: current concepts and controversies. Blood 1996;88(9):3259–87.

[141] Liu F, He J, Gu D, et al. Associations of endothelial system genes with blood pressure changes and hypertension incidence: the GenSalt study. Am J Hypertens 2015;28(6):780–8.

[142] Sass C, Pallaud C, Zannad F, et al. Relationship between E-selectin L/F554 polymorphism and blood pressure in the Stanislas cohort. Hum Genet 2000;107(1):58–61.

13

Epigenetic Regulation of Endothelial Function: With Focus on MicroRNAs

Fernanda Roberta Roque, Clara Nobrega, Tiago Fernandes, and Edilamar Menezes de Oliveira

INTRODUCTION

How do endothelial cells know they are different from other cells in the system? How do these cells know they must divide themselves in cells of the same type? Although classic genetics answers the first question by means of protein transcription factors that truly guide some genes to be activated while others remain silenced, the second question is still not answered, leading us to study the epigenetic mechanisms that regulate endothelial function and its cell identity.

The term "epigenetics" was used for the first time by Conrad Waddington, in 1946, when he defined it as "a field of biology that studies the causal interaction between genes and their products" [1]. Although very broad, this definition created interest on epigenetics as modulator of gene transcription that, with the growing relevance of the studies, resulted in the currently most accepted definition, which says that epigenetics studies modifications in gene expression, transmissible by meiosis and/or mitosis, that is, hereditary changes in chromatin, which do not depend or result in changes in the DNA sequence [2]. Different characteristics in monozygotic twins, progressive changes in chromatic function during the aging process, inactivation of the X chromosome in women, genomic imprinting, and tissue-specific gene expression are some examples of phenotypic differences inherited nonmutated DNA sequences [3,4].

Due to the dynamic character of the epigenetic code, it is understood that susceptibility to diseases, mainly those related to metabolic dysfunctions, cancer, and cardiovascular diseases, from random environmental influences, exposure to chemical reactants, and diversified behavioral patterns suffered by previous generations is prominent, and may influence the health of its descendants from differences in gene expression attributed to epigenetic variations in genome encoding or not encoding regions. Therefore, knowing the mechanisms that lead to these events and their consequences can be the key to develop methods able to prevent the accumulation of epigenetic changes that are detrimental to the health of future generations, regardless of the influences and environmental exposure suffered by the current generation [3–6]. Thus, we need to know the three main epigenetic mechanism regulation groups: DNA methylation, histone modification, and regulation of genes associated with RNA (Fig. 13.1).

DNA Methylation

DNA methylation consists of adding one methyl group to carbon 5 (5-mC) on the cytosine-phosphate-guanine dinucleotide (CpGs) sites, with this reaction being well conserved in prokaryots and eukaryots. The CpGs islands are genome regions associated with ~50% of the genic promoters that are generally not methylated, being methylated only for genic regulation. The cytokines may be methylated in the promoter regions, CpGs, or even in intergenic regions, which are

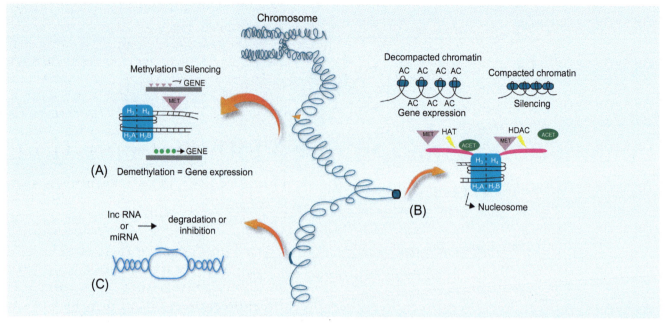

FIG. 13.1 Epigenetics regulatory mechanisms. (A) DNA methylation in CpGs islands promotes gene silencing. (B) Modification of histones from HATs and HDACs promotes chromatic remodeling. By adding an acetyl group, the HATs promote chromatic relaxation, decompacting it and allowing connection to transcriptional regulators. On the other hand, the HDACs remove the acetyl group, compacting the chromatin, which prevents connection to these regulators and promotes gene silencing. (C) Epigenetic regulation associated with RNA occurs by means of the connection between long noncoding RNAs (lncRNA) or microRNAs (miRNAs) with their target genes, promoting gene silencing. *Source: Fernanda Roberta Roque, Tiago Fernandes, Clara Nobrega, and Edilamar Menezes de Oliveira.*

nonencoding regions. When they occur in CpGs, they result in gene silencing due to genic transcription suppression, inhibiting the connection between the transcription factors and their promoter sites. Although DNA methylation plays an important role during the embryonic stage, establishing properties that lead to cell identity, induction of hypermethylation by means of environmental factors has been related to silencing of repressors of tumor growth, with deregulation of gene expression in diseases such as lupus, multiple sclerosis, and heart failure. The mechanisms regulating the balance methylation/demethylation are of paramount importance to maintain physiological homeostasis, such as the balance between enzymes that catalyze the DNA methylation reactions, methyltransferases (DNMT) 1, 3A, and 3B, and the enzymes that promote DNA demethylation, the ten-eleven methylcytosine dioxygenase (TET) [7,8].

Modification of Histones

In eukaryotes, DNA is compacted in chromatin, whose base unit is the nucleosome, which in turn has two copies of histones H2A, H2B, H3, and H4 in its core, and is surrounded by ~146 base pairs. Modification of histones regulates gene transcription, as they change the chromatin condensation stage, an essential phenomenon to allow or prevent transcription. The most studied histone modifications are acetylation, methylation, phosphorylation, and ubiquitination, with main focus on the first one. The histone acetyltransferase enzymes (HAT) are able to add acetyl groups to the lysines located on histone tails, especially H3 and H4, in a process that promotes chromatin decondensation, which is then known as euchromatin and is associated with DNA demethylation. On the other hand, histone deacetylases class I and II (HDAC I and II) remove acetyl groups, promoting chromatin condensation, called heterochromatin, which is associated with histone and DNA methylation. Therefore, while euchromatin is related to gene activation, since it allows more access to transcription factors, heterochromatin is related to gene repression, as it does not allow transcription factors to bind to promoter regions [9,10].

Gene Regulation Associated With RNA: lncRNA and MicroRNA

RNAs derived from noncoding sequences may play an important role in regulating gene expression by means of small interference RNAs, long noncoding RNAs (lncRNA) or even microRNAs (miRNAs). Most lncRNA have more than 200 nucleotides, which can interact in several cell levels such as structure, conformation, and cell activation and repression mechanisms [11]. In the vascular system, the lncRNAs expressed on the endothelium and smooth muscle cells (SMC) were recently demonstrated as regulators of both endothelial growth and function regulations, regarding the contractile phenotype of the SMC, respectively [12]. Yet, studies show that lncRNAs can be regulated by the shear forces imposed by the blood flow on the endothelium, or even reduced effects of metalloproteinase AMZ2 by means of the connection between the lncRNA to the chromatic repression site [13]. On the other hand, miRNAs are small endogenous RNA molecules that regulate posttranscriptional gene expression. When the miRNAs bind to their target genes, they promote total degradation of inhibition of these protein coding genes [14]. In studies of endothelial functions, miRNAs play a fundamental role in regulating several pathways and mechanisms involved with this system, but also play an important role in epigenetic regulation itself, since these molecules are able to regulate DNA methylation and chromatic condensation, bleaching DNMT 1, 3A and 3B, and HDAC [15]. Given the great regulatory potential of miRNAs, we shall analyze them along this chapter, from their biogenesis up to physiological and pathological role in vascular function.

MicroRNA

The miRNAs are single-strand, noncoding ribonucleic acid (RNA) molecules with ~22 nucleotides in size, which act mainly by repressing protein translation in plants and animals [16,17]. Bioinformatics studies showed that one-third of all protein coding genes and, essentially, all biological pathways are under the control of miRNAs [18]. Therefore, it is not surprising that this class of small RNAs acts on vascular disease progression. miRNAs regulate a series of biological processes that began to be investigated as a result of the discovery of lin-4 participation in the larval development of

Caenorhabditis elegans (nematodes) [19,20] Later, it was demonstrated that lin-4 acted on the posttranscriptional level and showed partial complementarity with region 3'-UTR (nontranslated region) of mRNA in protein lin-13 [14,21]. The second miRNA, called let-7, was only discovered 7 years after lin-4. It then became clear that let-7 also acted on posttranscriptional level and showed partial complementarity with the 3'-UTR region of mRNA in protein lin-41 [22,23]. This discovery originated the hypothesis that small RNAs may be present and perform similar functions in other species beside nematodes. It was proven that lin-7 and lin-41 are evolutionarily conserved in metazoans, with counterparts that were promptly detected in flies, mice, and humans. This evolutionary conservation indicated a more general role of small RNAs in regulating development, which opened perspectives to discover new miRNAs and other biological processes that included its regulation mechanism [24].

Since then, new miRNAs have been discovered and today there are more than 35,000 sequence of mature miRNAs, of more than 220 types of different organisms, listed in the bioinformatics database (miRBase) [25,26]. Until the beginning of 2015, according to the miRBase 21.0 site, more than 2500 miRNAs were identified in the human species, compared to the more than 1900 found in mice and 700 in rats, indicating that the number for miRNAs seems to correlate with the greatest complexity of the organism [27]. It is possible that this numbers are even larger and that this class is one of the largest endogenous gene regulators found now [28].

In humans, approximately one-third of miRNAs are organized in clusters. It is probable that a given cluster is a single transcriptional unit, suggesting coordinated regulation of miRNAs in clusters. In silico analysis revealed that more than half of the clusters have two or more similar miRNA sequences. However, it is very rare to have mature miRNAs showing identical sequences being repeated in a cluster. This genomic organization grants simultaneous expression of similar miRNAs, possibly leading to the combinational diversity and synergy between biological effects. However, all miRNAs derived from the same cluster are not expressed in equal levels, suggesting that the miRNAs are also regulated at posttranscriptional level. In addition to that, a significant part of the miRNAs is located in the intronic region of protein coding genes, and may act in synergy with it [29].

The miRNAs perform their regulatory roles by binding to the 3'-UTR region of target mRNAs. This mechanism of action allows reducing the protein levels of the target genes, rarely affecting he transcriptional expression level [16]. Since the miRNAs have small sequences and act without complete pairing, a single miRNA can regulate several target mRNAs, besides cooperating to control a single mRNA [17,30]. Some studies indicate that one miRNA can regulate 200 mRNAs showing totally different functions. Therefore, miRNAs constitute a huge and complex cell signaling regulatory network. In plants, miRNA regulation occurs mainly due to its perfect interaction with mRNA, leading to its degradation (siRNA mechanism). However, there are also examples of occurrence of this type of gene silencing in mammals [17,31,32].

Recent transcriptome and bioinformatics studies suggest the existence of thousands of other types of nonprotein coding RNAs, but the number of those encoded in the human genome is unknown. Differently from miRNA, many noncoding RNAs identified recently were not validated for its function; it is possible that many are not functional [33]. lncRNAs are part of noncoding RNAs that interact with the main cell growth, proliferation, differentiation, and survival pathways. Recently discovered, the lncRNAs have been described for regulating gene expression, and may act as mirNA sponges to reduce their levels [34,35]. Therefore, the knowledge regarding mechanisms regulated by different noncoding RNAs revealed a promising investigation field that may contribute to future understanding of molecular processes induced by physiological and pathological conditions.

Biogenesis and MicroRNA Function

Most of the miRNAs are transcribed by RNA polimerase II, generating a primary transcript called pri-miRNA [36]. These miRNAs were transcribed as if they were protein coding genes, and therefore have the cap 5' structure and poly(A) tail [37]. Mature miRNA generation is preceded by cleavage events executed by enzyme complexes [38].

The miRNA formation process starts from the pri-miRNA, which has hairpin type double helix (~300 nucleotides), being processed in the nucleus by a type III ribonuclease called Drosha and its co-factor DGCR8 (DiGeorge syndrome critical region gene 8). The resulting molecule of the Drosha processing is called miRNA precursor (pre-miRNA), which is exported from the nucleus to the cytoplasm by means of exporting 5, a nuclear exportation protein depending on Ran-GTp as co-factor. In cytoplasm, pre-miRNA suffers the action of another type III ribonuclease called Dicer, and originates a small and imperfect RNA duplex (double strand) that contains both the mature miRNA strand and its antisense strand [16,26,28,39].

The Dicer product is incorporated by a multimeric complex called RISC (RNA-induced silence complex), which contains proteins of the Argonaut family as main catalysis actuation components [36]. Only one of the double strands of the miRNA remains in the RISC complex to control posttranscriptional expression of target genes. In general, the duplex part with lower stability if selected to be the RISC complex element [40–42]. The miRNA expressed quantity can be controlled in pri-miRNA transcription, during the biogenesis and also mature miRNA turnover steps [43].

The gene endogenous expression can be regulated in two ways by the mature miRNAs. The first is cleavage or total degradation of the miRNA, with the mRNA induced by perfect pairing of miRNA nucleotides and target mRNA [44]. The second regulation form, more common in animals, occurs via translational repression, by means of partial complementarity [45]. In this type of regulation, only the miRNA region that covers from the second to eighth nucleotide is totally complementary of the target mRNA. This sequence is called seed region, and is responsible for miRNA specificity [42,46]. Animals have seed complementarity regions in the 3'-UTR regions of the target mRNAs, and this mRNA region is the preferred region for miRNA coupling and execution of its negative regulation mechanism [21,42]. The mature single-strand miRNA, incorporated to the RISC protein complex, guide the miRNA to the complementarity region in the 3'-UTR region of the target mRNA, in order to perform gene expression negative regulation by means of translational repression (Fig. 13.2) [43].

In mammals, the biological relevance of the miRNAs was first demonstrated by an animal model knockout for Dicer [47] and DGCR8 [48], both involved in the miRNA biogenesis process, which resulted in embryonic lethality and therefore showed that complete absence of miRNAs was incompatible with life. However, over the last two decades, several studies conducted with cell

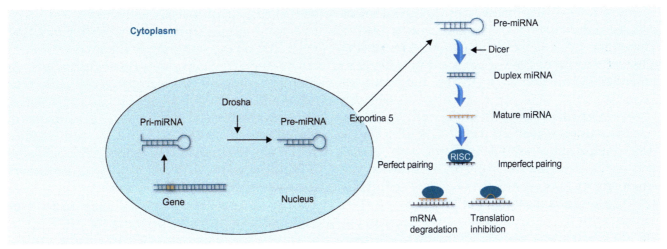

FIG. 13.2   miRNA biogenesis and mechanisms of action. *Adapted from Song XW, Shan DK, Chen J, et al. miRNAs and lncRNAs in vascular injury and remodeling. Sci China Life Sci 2014;57:826–35.*

cultures or in animal models, demonstrated that there were <10% of the miRNAs that were individually required for development and feasibility processes, but most of them acted together to regulate a variety of targets, therefore modulating the homeostasis and disease processes [49,50].

In the vascular system, growing evidence suggests that the actions of several miRNAs are important for signaling and vascular function, mediated by their involvements in regulating several cellular processes such as proliferation, differentiation, migration, and apoptosis, and therefore may contribute to endothelial dysfunction, angiogenesis, vascular rarefaction, and vascular remodeling [18,51].

Mechanisms of MicroRNAs Regulation of Vascular Function: Physiological and Pathological Aspects

The vascular system is responsible for distributing blood flow to several tissues, regulating vascular resistance and blood pressure, in order to allow important processes such as gas exchange, offer of nutrients, and removal of metabolic residues, among others. This system is classified primarily according to its function, that is, conductance arteries, resistance arteries, exchange vessels, and capacitance vessels. The basic vascular structure includes three separate and interconnected layers: the more internal layer is called intima, the central layer is called media, and the more outer layer is called adventitia, besides the extracellular matrix, which shows varied composition according to each type of vascular segment and respective functions.

Different physiological or pathological characteristic stimuli may induce adaptive processes on the vascular wall structure, called vascular remodeling. Examples of physiological vascular remodeling include the embryonic angiogenesis and arteriogenesis processes, and also angiogenesis induced by physical activity. Examples of pathological vascular remodeling include restenosis, which occurs due to accumulated vascular smooth muscle cells (VSMCs) forming on the neointima, hypertrophy of the vessel medial layer in arterial hypertension, or even atherosclerosis, which occurs as a result of inflammatory response, among others. Therefore, endothelial vascular cells, SMC, and fibroblasts, in addition to inflammatory cells and extracellular matrix proteins themselves, play an important role in the vascular remodeling process. Recent research is trying to identify miRNAs that may regulate the vascular remodeling process and, therefore, be used as target to treat vascular diseases [18,51–53].

MicroRNAs and Endothelial Cells

Integrity of the endothelial layer is essential for vascular system homeostasis; once damaged, the endothelium may contribute for the pathogenesis of several vascular diseases, such as arterial hypertension and atherosclerosis. The endothelial cells play a very important role in the development, maintenance, and remodeling of the vascular networks required by angiogenesis, a process that involved endothelial cells proliferation, migration, and differentiation [25]. A remarkable characteristic of vascular diseases is the loss of functional micro-vessels and impaired angiogenic process [54–56]. Therefore, improvement of endothelial function and correction of microvascular rarefaction are potential therapeutic strategies for these diseases [54–59].

Several studies suggest that the miRNA biogenesis regulators, as well as specific miRNAs, may be important in endothelial dysfunction pathogenesis and reduced angiogenic capacity in vascular diseases. In fact, specific miRNAs regulate endothelial cell functions, and these miRNAs are involved mainly with the regulation of the vascular apoptosis, angiogenesis, and inflammation processes [25,60,61].

The importance of miRNAs in endothelial cells was demonstrated by studies with Dicer knockout models, which observed that the presence of this enzyme is required to embryonic angiogenesis and mice normal development [62]. Similarly, the knockout model for the Dicer enzyme resulted in reduced proliferation of endothelial cells and vascular formation [63]. Besides that, the knockout model for both Dicer and Drosha resulted in impaired capillary development and formation of endothelial cell tubes [64]. Therefore, the results showed the decisive role of miRNAs on endothelial cell regulation and postnatal angiogenesis due to its processing by the Drosha and Dicer enzymes.

The miRNA expression profiles in endothelial cells was documented, and several highly expressed miRNAs were related to pro- and antiangiogenic factors (proangiogenic miRNAs: -17-92 *cluster*, -23/-27, -126, -130a, -210, -296, -378, and let-7f; antiangiogenic miRNAs: -15b, -16, -20, -21, 92a, -221, -222, and -328) [25,60,61,65], according to the target miRNAs predicted by algorithms. However, specific validated targets and endothelial cell functions related to angiogenesis, inflammation, apoptosis, and vascular function have been characterized by only a few of these miRNAs. Among the miRNAs involved in these processes, miRNAs-16, -21, -126, -155, -22, and -222 have an important role [25,60,61,66–70].

MiRNA-126 is apparently, the only specifically expressed in endothelial cells and hematopoietic progenitor cells. With abundant expression, miRNA-126 was described for regulating migration of inflammatory cells, formation of capillary networks, and cell survival, and therefore is involved in vascular dysfunction, inflammation, and

rarefaction in different pathologies [51,71–73]. Important studies that used animals knockdown for miRNA-126 have shown impaired endothelial cell migration during vessel growth, vascular lumen rupture, compromised formation of endothelial tubes, and bleeding processes [51,66,71,72]. Molecular analysis revealed that miRNA-126 inhibits the Sprout-related protein 1 (SPRED1) and regulator subunit 2 of phosphatidylinositol 3,4,5 triphosphate (PIK3R2, also known as p85-beta), which negatively regulate signaling of the vascular endothelial growth factor (VEGF) by inhibiting MAPK and PI3K/Akt pathways, respectively [66,71,73].

Previous studies conducted by our group have showed reduced expression of microvascular miRNA-126 in spontaneously hypertensive rats (SHR) and increased expression of the PI3KR2 target when compared to normotensive animals (WKY). The increased PI3KR2 leads to inhibition of PI3K and the signaling pathway regulated by it, such as Akt/eNOS. In fact, reduced expression of endothelial nitric oxide synthase (eNOS) was observed in these animals, indicating the possible role of this pathway in capillary rarefaction in hypertensive animals [74].

Interestingly, VEGF and antiapoptotic protein Bcl-2 were identified in bioinformatics approaches and validated as targets for miRNA-16 in endothelial cells, showing that equivalents of this miRNA promoted reduced expression of VEGF and Bcl-2, while the specific antagonists increased expression [25,60,61,69,70]. Furthermore, studies showed that miRNA-16 superexpression promoted smaller endothelial cell proliferation, migration and tube formation in vitro, and superexpression of miRNA-16 by *lentiviruses* reduced the capacity of endothelial cells to form blood vessels in vivo [75,76]. Supporting this argument, miRNA-21 is also an apoptotic miRNA that bleaches Bcl-2, suggesting its important role in regulating the intrinsic angiogenic activity of the cell [25,60,61,67]. Curiously, we observed that arterial hypertension was also associated with increased expression of miRNAs-16 and peripheral-21 in parallel with decreased expression of target genes VEGF and Bcl-2 [74].

Studies also show that miRNA-21 can influence endothelial progenitor cell function and migration in coronary arterial diseases [77]. Similarly, angiogenesis induced by ischemia in adult tissues can be promoted or inhibited by the use of oligonucleotides antisense for miRNA-92a [78] and miRNA-126

[79], respectively. MiRNA-16 may also influence susceptibility to atherosclerosis by changing the function of endothelial cells [80].

Several studies have shown, by means of bioinformatics prediction, that miRNAs-221 e -222 bleach eNOS and c-kit, a stem cell marker also present in endothelial progenitor cells [25,56,60,61]. Interestingly, Sun et al. [70]. have shown that miRNA-155 is also an essential regulator of eNOS expression and endothelium-dependent vasodilation. Therefore, the authors suggest that miRNA-155 inhibition may be a new therapeutic approach to improve endothelial dysfunction during the development of cardiovascular diseases. NO released by the endothelium plays an important role in maintaining vessel baseline tone, blood pressure regulation, and blood flow distribution, contributing significantly for vasodilatory properties and maintenance of vascular structure. Endothelial dysfunction refers to reduced endothelium-derived vasoactive properties, mainly detected by diminished NO bioavailability, regularly observed in arterial hypertension [81,82].

Although the subject is exciting and with great potential regarding understanding of several cardiovascular diseases, there are few studies involving pathological and physiological aspects. Based on descriptions of the studies mentioned, evidence suggests that the action of specific miRNAs is very important to maintain microvascular homeostasis, since they are involved in regulation of several cellular processes such as proliferation, differentiation, migration, inflammation, and apoptosis, and may therefore contribute to the endothelial dysfunction, reduced angiogenic capacity, and vascular remodeling widely observed in vascular diseases [39,51,56,83].

MicroRNAs and VSMC

VSMCs are the main component of the vessel medial layer, whose main function in adult blood vessels is adjusting vascular tone due tis contraction and relaxation capacity, different from what is observed in vascular development states, in which their main functions are high proliferation and migration capacity, and ability to synthesize extracellular matrix components, including collagen, elastin, integrins, cadherins, and proteoglycans. Plasticity is one of the characteristics of VSMCs, and their differentiation process may show a variety

of different phenotypes that oscillate between contractile or synthetic, occurring mainly during the development stages, but also in adult organisms in response to several conditions, for example, stimuli related to physiological or pathological situations.

Differences in morphology, proliferation and migration potential, and even VSMC marker gene expression, may be observed in the contractile and synthetic phenotypes. Higher expression of contractile phenotype proteins, among them, smooth muscle myosin heavy chain (SM-MHC), alpha-smooth muscle actin (α-SMA), SM-22α, Calponin-h1, and lower proliferation and migration ratio are characteristics of contractile phenotype VSMCs, that is, differentiated VSMCs. Just like reduced expression of contractile phenotype marker proteins and increased expression of some proteins such as embryonic smooth muscle myosin heavy chain (SMemb), cellular retinol binding protein (CRBP-1), associated with increased proliferation and migration ratio, and production of extracellular matrix proteins, are characteristics of synthetic phenotype VSMCs. Phenotypic modulation, that is, capacity to transition between phenotypes, is considered a key process in vascular injury recovery; however, the possibility of altering VSMC growth/apoptosis, contraction/relaxation, migration, and differentiation processes, impaired production/degradation of the extracellular matrix and stimulation of inflammatory responses my result in structural remodeling, and lead to vascular disease progression [84,85].

The process of transition between the VSMCs contractile and synthetic phenotypes depends on the integration of several factors that act locally, and many studies have demonstrated the role of specific miRNAs acting on the target genes involved in phenotypic modulation. Several miRNAs have been considered important for the VSMC differentiation process, including miRNAs-1, -10a, -21, -24, -26a, -31, -100, -133, -143, -145, -146a, -204, -208, -221, -222, and let-7 [51,86]. Some of the most studied miRNAs involved in promotion of the contractile phenotype (differentiated) or synthetic phenotype (proliferative) shall be discussed below.

The critical role of miRNAs-143/145 in VSMCs was demonstrated by the fact that mice double knockout for these miRNAs showed a phenotype very similar to the model knockout for Dicer, although not lethal and less severe, which showed reduced blood pressure, significant reduced number of contractile VSMCs, and significant increased number of proliferative cells, displaying therefore a synthetic morphology associated significant reduction in contractile markers [87]. These miRNAs have higher expression of VSMCs, and their function is to repress multiple factors that normally produce a more synthetic phenotype, and therefore are important to maintain the VSMC contractile phenotype. Studies that showed, for example, that the target of miRNA-145 is a network of transcription factors such as factors 4 and 5 of the Kruppel type (KLF-4, KLF-5). KLF4 represses myocardin-induced activation of SMC genes and myocardin expression itself [88]. When repressed, myocardin, which is a positive regulator of the contractile phenotype, contributes for the proliferation process. The activation of miRNA-145 inhibits its targets, KLF-4 and KLF-5, allowing a positive effect on myocardin activity, which associates to the serum response factor (SRF) collaborating for a contractile phenotype. In fact, superexpression of miRNA-145 increased the expression of contractile genes such as SM-MHC, α-SMA, and calponin.

Although microRNA-143 seems to have a smaller regulatory effect than miRNA-145, it also collaborates to maintain the contractile phenotype. Studies have shown that mice with deficient miRNAs-143/145 need both miRNAs in order to have transition between the phenotypes. MiRNA-143 acts by repressing Elk-1, which competes with myocardin to bind to SRF, or even, by repression of tropomyosin 4 (TPM-4), a structural protein with high expression of VSMCs with synthetic phenotype [89,90]. As mentioned, the VSMCs of animal models with miRNA143/145$^{-/-}$ showed reduced expression of contractile genes, reduced blood pressure, and more synthetic characteristics. In contrast, high expression of miRNA-145 was observed in lung tissue and VSMCs of patients with pulmonary arterial hypertension [91]. Therefore, it seems that reduced miRNAs-143/145 may collaborate with a more synthetic phenotype as observed in vascular injury situations, as well as an increase as observed in arterial hypertension, which seems to be harmful as it collaborates more with a hypertensive picture. This data indicate that the miRNA-143/145 family has its expression increased or reduced according to different pathological stimuli, and handling the expression of these miRNAs may be fundamentally important to change VSMC phenotypes, for artery contractility, and blood pressure control.

The miRNAs-1/133 family is considered specifically expressed in cardiac and skeletal muscle and cell differentiation and proliferation mediators [92]. However, a more recent study has shown that VSMC proliferation was also inhibited by the expression of miRNA-1 induced by myocardin, with this response being mediated by the action of the miRNA on its target, the proto-oncogene serine-threonine kinase protein (Pim-1). Additionally, it was noted that the myocardin-miRNA-1-Pim-1 signaling cascade was also involved in regulating in vivo vascular proliferation, and a model with noeointimal injury induced by ligation of the carotid artery showed reduced expression of myocardin and miRNA-1 associated with increased Pim-1 [93]. In the same year, another study demonstrated that miRNA-1 expression is required for the embryonic stem cell differentiation process in SMCs in vitro, and the use of a miRNA-1 (antagomir) inhibitor reduces the expression of specific contractile SMC markers and the SMC population. The authors observed that the repression of KLF-4 by miRNA-1 took part in the differentiation process [94]. Regardless of the previous results, there is still controversy, and Torella et al. [95] have shown that not miRNA-1, but miRNA-133 instead, regulated VSMC growth inhibiting in vitro proliferation and migration, besides showing reduced expression of VSMC, in proliferation state, due to the vascular injury caused by in vivo angioplasty. In these studies, the authors concluded that the miRNA-133 pathway also included the KLF4-myocardin axis be means of repression of the Sp-1 transcription factor, already known for regulating this axis [95].

Still regarding KLF-4, it is important to mention the role of miRNA-146a in regulating the phenotype of VSMCs. The expression of miRNA-146a is increased in rat carotid arteries in a model of vascular injury induced by a balloon. This model allows observing VSMC proliferation and neointimal hyperplasia [96,97]. Therefore, expression of miRNA-146a is increased in proliferative cells, interestingly reducing KLF-4 expression and increasing expression of proliferation marker PCNA (proliferating cell nuclear antigen). Additionally, reduced expression of KLF4 induced by miRNA-146a also inhibits the transcription of the miRNA itself, therefore forming a negative feedback cycle in which one adjusts the expression of the other. However, KLF4 competes with KLF5 to regulate miRNA-146a transcription, showing an opposite effect of both, that is, while superexpression of KLF4 decreases the level of expression of miRNA-146a, superexpression of KLF5 increases the expression of the same miRNA [97].

miRNA-21 has been extensively studied in cancer, and therefore is called an oncomiR, but also in many other pathological conditions, such as cardiovascular and pulmonary diseases [98]. Expression of miRNA-21 is abundantly increased in rat carotid arteries after vascular injury caused by angioplasty [96]. In addition, this study also showed that reduction in the aberrant expression of miRNA-21 in this model decreased the neointimal formation, and its inhibition decreases in vitro VSMC proliferation and increases apoptosis, signaling the involvement of a tumor suppressor, phosphatase and tensin homolog (PTEN) and antiapoptotic protein Bcl-2 [96]. A more recent study observed that SRF regulates the expression of PTEN in VSMCs by means of the miRNA-143-FRA-1-miRNA-21 axis, showing communication between signaling pathways and miRNAs [99]. On the other hand, there is still controversy regarding the role of microRNA-21, with some studies indicating miRNA-21 as inducer of contractile protein synthesis. This is the case of studies which observed that miRNA-21 reduced the expression of the VSMC contractile gene negative regulator, PDCD4 (programmed cell death 4) [100]. More recently, the same group published a new miRNA-21 target, members of the DOCK family (dedicator of cytokinesis), which due to reduced expression of miRNA-21 inhibited cell migration and increased the expression of contractile genes, such as α-SMA and calponin [101].

MiRNAs-221/222 are expressed in VSMCs, endothelial cells, and hematopoietic cells, by means of a common set of genes in chromosome X. The function of these miRNAs on the cardiovascular system is completely dependent of the cell type. In VSMCs, we observed that these miRNAs regulate the cell differentiation process, since their expression inhibits differentiation by repressing the transcription of specific smooth muscle contractile genes. Therefore, reduced expression of these miRNAs increases the expression of differentiation markers in VSMCs, blocks the proliferation and migration effects, and increased expression of miRNAs 221/222 promotes VSMC proliferative effect. These proliferative effects are mediated by reduced expression of the protein tyrosine kinase c-kit (receptor of the stem cell factor (SCF)), which in turn reduces the level of expression

of myocardin and therefore inhibits VSMC contractile gene expression [102], and also due to reduced p27(kip1) and p57(kip2) (members of the cyclin inhibitor family, involved in the cell cycle process) [103]. In fact, this effect can be clearly observed in vivo, in a model of vascular injury followed by angioplasty, and also in vitro, in VSMC in proliferative state stimulated by the platelet-derived growth factor (PDGF), when there is elevated expression of miRNAs-221/222 [103].

In short, as observed in this brief discussion on miRNAs, it is possible to confirm their important role in regulating the VSMC differentiation, proliferation, and migration processes, unquestionably contributing to homeostasis and vascular function, and for many pathologies that involve the vascular system. Table 13.1 shows microRNAs of endothelial cells and SMC: target genes and vascular function.

MicroRNAs as Vascular Injury Biomarkers

miRNAs show high stability in bodily fluids such as plasma, platelets, and red blood cells, urine, and saliva, and may be used as molecular biomarkers for several diseases. Therefore, several miRNAs circulating in plasma were identified as biomarkers for several diseases, including cardiovascular diseases [104–106] and, have been performing increasingly important roles in clinical applications, such as diagnosis of diseases and monitoring therapeutic effects. Specifically, circulating miRNAs have many characteristics of good biomarkers:

(1) Are stable in circulation and resistant to digestion with RNAse and other adverse conditions, such as extreme pH, boiling, extended storage, and several freezing/thawing cycles;
(2) Many sequences are conserved between species;
(3) Changes in circulating levels have been associated with several diseases; and
(4) Can be easily determined by several methods [107,108].

Curiously, circulating miRNAs are protected from the activity of endogenous RNAse [109], and evidence shows that this protection is reached by means of packaging the circulating miRNAs into microparticles such as exosomes, microvesicles, or apoptotic bodies [110,111]; by connection of proteins bound to the RNA, such as Argonaut 2 e

TABLE 13.1   MicroRNAs of Endothelial Cells and Smooth Muscle Cells: Target Genes and Vascular Function

MirNa	Location	Target gene	Function	Reference
miRNA-126	Endothelial cell	SPRED1, PIK3R2	Angiogenesis, vascular integrity	[66,71,72]
miRNA-16	Endothelial cell	VEGF, Bcl-2	Angiogenesis, vascular integrity	[69,75,76]
miRNA-21	Endothelial cell	SOD2, Sprouty2, Bcl-2	Angiogenesis, apoptosis	[25,77]
miRNA-92a	Endothelial cell	Integrin a5	Angiogenesis	[78]
miRNA-221/222	Endothelial cell	eNOS, c-kit	Endothelial function, angiogenesis	[25,60]
miRNA-155	Endothelial cell	eNOS	Vascular function	[70]
miRNA-145	Smooth muscle cell	KLF4, KLF5	Differentiation, proliferation	[88–90]
miRNA-143	Smooth muscle cell	Elk-1, TPM-4	Differentiation, proliferation	[89,90]
miRNA-1	Smooth muscle cell	Pim-1	Differentiation, proliferation	[93]
miRNA-133	Smooth muscle cell	Sp-1	Proliferation	[95]
miRNA-146a	Smooth muscle cell	KLF4	Proliferation	[97]
miRNA-21	Smooth muscle cell	PTEN, Bcl-2 PDCD4, DOCK	Proliferation, apoptosis Differentiation	[96,99–101]
miRNA-221/222	Smooth muscle cell	c-Kit, p27(kip1), p57(kip2)	Differentiation, proliferation	[102,103]

Bcl-2, antiapoptotic protein Bcl-2; *c-kit*, protein tyrosine kinase c-kit receptor; *DOCK*, family of dedicator of cytokinesis proteins; *Elk-1*, member of the family of transcription specific factors with ETD domain E-26; *eNOS*, endothelial nitric oxide synthase; *KLF4*, KLF5, Kruppel factors 4 and 5; *p27(kip1) and p57(kip2)*, members of the family of cycling inhibitors.; *PDCD4*, programmed cell death 4; *PIK3R2*, phosphatidylinositol 3,4,5 triphosphate regulator subunit 2; *Pim-1*, proto-oncogene serine-threonine kinase protein; *PTEN*, phosphatase and tensin homolog tumor suppressor; *SOD2*, superoxide dismutase 2; *SP-1*, transcription factor SP-1; *SPRED1*, Sprout-related protein 1; *TPM-4*, tropomyosin 4; *VEGF*, vascular endothelial growth factor.

nucleophosmin 1 [112]; or by connection with high-density lipoproteins (HDL) [113] (Fig. 13.3).

The exosomes are small vesicles (50–100 nm) originated from the endosome and released by cells when the multivesicular bodies fuse with plasma membrane. Microvesicles are membranous vesicles that are bigger (0.1–1 μM) than exosomes and are released by cells by formation of bubbles in the plasma membrane. Apoptotic bodies are bigger than microparticles (0.5–2 μM) and are eliminated from cells during apoptosis [105]. Kosaka et al. [114] have shown that miRNAs are released to circulation by means of a ceramide-dependent secreting apparatus. Besides that, the authors showed that channels and receptors associated with the cell membrane allow the passage of the Argonaut2-miRNAs protein complex.

The presence of miRNAs in microparticles leads to the intriguing idea that circulating miRNAs may have a function in cell to cell communication. This suggests that miRNAs are selectively guided

for cell secreting and absorbed by a distant target cell, possibly to regulate gene expression. This is a field of intensive investigation, and the fist studies recently revealed that miRNAs can effectively work as cell to cell communication mediators [105,111,115]. In fact, it has been reported that apoptotic bodies are released in circulation, inhibiting atherosclerosis progression. A study conducted by Zernecke et al. [111] proposed that miRNA-126 is released by apoptotic bodies responsible for this protective effect by inducing chemokine CXCL12. Besides that, the authors showed that apoptotic bodies of control mice protected against atherosclerosis of mice with deleted apolipoprotein E submitted to atherogenic diet, as evidenced by the reduced infiltration of macrophages.

Distinct patterns of circulating miRNAs were found for myocardial infarction, heart failure, atherosclerotic disease's *diabetes melittus* type 2 (DM2), and arterial hypertension [105,116–118]. Although these studies indicated that plasmatic

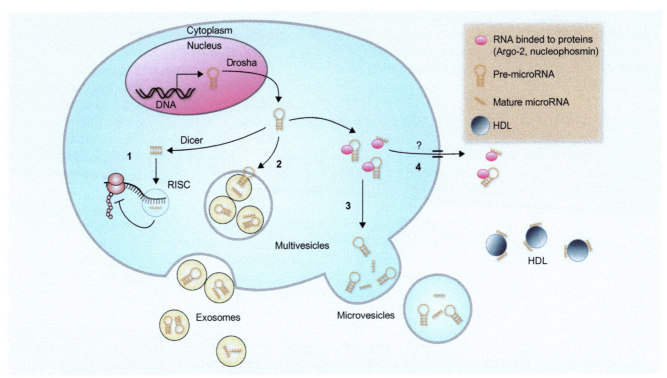

FIG. 13.3 miRNA release mechanism and extracellular transportation system. One of the miRNA duplex strands can bind to the RISC complex, guiding it to specific mRNA targets to avoid gene translation into protein (1). The other miRNA duplex strand can be degraded or released by the cell by means of the exportation mechanisms described below. In the cytoplasm, pre-miRNAs can also be incorporated into small vesicles called exosomes and released by the cells when the multivesicular bodies fuse to plasma membrane (2). Cytoplasmic miRNAs (pre-miRNA or mature miRNA) can also be released by microvesicles that are liberated by cells through plasma membrane vesicles (3). The miRNAs are also found in circulation in the free form of microparticles. These miRNAs can be associated with high-density lipoproteins (HDL) or connect to RNA binding proteins such as Argo-2. These miRNAs can be released passively, as by-products of dead cells, or actively, by means of the interaction with the membrane channels or specific proteins (4). *Adapted from Creemers EE, Tijsen AJ, Pinto YM. Circulating microRNAs: novel biomarkers and extracellular communicators in cardiovascular disease? Circ Res 2012;110:483–95.*

levels of specific miRNAs correlate to several forms of cardiovascular diseases, they currently do not show if the plasmatic levels of miRNAs correspond to worsening or improvement of these pathologies.

Several groups have studied the hypothesis that specific cardiac miRNAs are released in circulation during acute myocardial infarction (AMI) and may be used to detect and monitor myocardium injury. Four cardiac miRNAs, such as miRNAs-208a, -499, -1, and -133 are found consistently high in the plasma of patients with AMI within hours after the beginning of the infarction [105,108,116].

Tijsen et al. [117] was one of the first groups to investigate differential expression of plasmatic miRNAs in heart failure, including patients with dyspnea and dyspnea without heart failure. Six miRNAs were confirmed in these patients, such as miRNAs-18b, -423-5p, -622, -675, and -1254.

miRNA-423-5p showed an important role by differentiating patients with heart failure from healthy individuals and patients with dyspnea without heart failure, also correlating positively with the levels of brain natriuretic peptide (BNP) and negatively with the left ventricular ejection fraction. A recent study has confirmed increased expression of miRNA-423-5p and of a set of circulating miRNAs, such as miRNAs-16, -20b, -93, -106b, and -223, in hypertension-induced heart failure in salt-sensitive rats (Dahl). Curiously, these alterations were attenuated with treatment with antimiR-208a and inhibitor of the angiotensin converting enzyme [119].

Fukushima et al. [120] confirmed a negative correlation between miRNA-126 expression and age, BNP, and functional class in patients with heart failure of ischemic ethiology in comparison to healthy individuals. Curiously, altered expression of miRNA-126 in the endothelial progenitor cells of 106 patients with heart failure was confirmed and associated with mortality [121]. Alterations in the circulating levels of miRNA-126 also were found in patients with atherosclerosis [118] and DM 2 [122], and may reflect the status of endothelial cell in heart failure.

A signature of plasmatic miRNAs has been identified in essential arterial hypertension [123]. After the initial analysis of the miRNA microarray in plasma of hypertensive and control patients, they confirmed different levels of three miRNAs (hcmv-miRNA-UL112, let-7e, and miRNA-296-5p). Interestingly, one of the miRNAs validated seems to be a promoter of the human cytomegalovirus (HCMV), which suggests a new bond between HCMV infection and essential hypertension.

In brief, identification of stable circulating miRNAs challenges a series of concepts and introduces a new generation of potential biomarkers. However, fundamental questions regarding its transport, distant actions, and feedback on these actions must be answered before the idea that circulating miRNAs are part of a mobile message system based on miRNAs can be shared even more.

POTENTIAL ROLE OF MicroRNAs IN VASCULAR THERAPY

Role of miRNAs as Pharmacological Therapy in Vascular Diseases

As a single miRNA is able to interact with several targets, related to a given phenotype, its modulation can have more potent effects when compared to the current treatments that act on a single target, and therefore has been indicated as an attractive treatment for cardiovascular diseases, among others. Of the tools available to modulate the actions of miRNAs on their targets, chemically modified oligonucleotides that complement the sequence of mature miRNAs (antimiR) can reduce the levels of miRNAs with increased expression. Conversely, sequences that are analogous to the miRNA (miR-mimic) may elevate the levels of miRNAs with reduced expression. As the miRNAs act as genic expression inhibitors, the effect of adding specific miR-mimics in a system promotes reduced expression of the target miRNAs. Inversely, the effect of the antimiR attenuates the inhibition of

target genes. Therefore, the main effect of miRNA inhibitors is activation of gene expression and the main effect of miR-mimics is gene suppression [124]. The studies that use silencing or mimic miRNA functions have already been investigating several signaling pathways and also several pathologies; however, most of these studies have been conducted in vitro, and then their relevance is subsequently investigated in vivo in mice. However, to test the therapeutic potential and also the safety of using these tools, some studies have proposed investigation using animals that are more complex than rodents. Advancing even more, recently an important clinical study was published demonstrating the safety and efficiency of the therapeutic use of miRNA-122 in patients with chronic hepatitis C [125].

Several vascular alterations are the focus of many studies that investigate the therapeutic role of miRNAs in arterial hypertension. This is characterized by persistent elevation of systemic blood pressure, and is considered an important risk factor for the development of coronary arterial diseases, heart failure, and cerebrovascular accident. In the recent review by Shi et al. [126] we can clearly see the role of several miRNAs, as well as their target genes, whose function is to control VSMC proliferation, differentiation, and vascular tone, regulate the endothelial function and act on vascular development, vascular tone regulation, and VSMC phenotype itself, among others, and that have altered expression in case of arterial hypertension, therefore representing an interesting therapeutic target. To focus on the vascular system, among the several miRNAs that are altered with arterial hypertension showed in the recent literature review conducted by Shi et al. [126], we can mention those that are related to VSMC alterations, such as miRNAs-9, -17, -21, -130/301, -145, -193, -204, -206, and -328, and those with alterations in endothelial cells, such as miRNAs-17/92, -21, -103/301, -126, -155, -204, -424/503, and -708. It is important to emphasize that, among the different types of hypertension, for example: pulmonary, genetics, gestational, induced by salt, induced by angiotensin II, among others, the miRNA expression profile is differentiated, and therefore the therapeutic targets shall also be specific according to each situation, even if within a hypertensive condition [126,127].

In atherosclerosis, a chronic inflammatory disease of the arterial wall, which induces important reduction of vascular lumen and may cause death by AMI or cerebrovascular accident due to ruptured thrombosis plaques, a recent revision work also discusses how the miRNAs can contribute to endothelial cell loss of quiescence and barrier function, among them miRNAs-221/222, -155, -126, -21, -10a, or for phenotypic alterations in VSMCs, such as miRNAs-221/222, 143/145, -146a, -133, -21, for the extracellular matrix synthesis and degradation processes, such as miRNAs-29 and -204, and also for alterations caused by monocytes and macrophages, such as miRNAs-33a/33b, -125a-5p, -146a, and -155, which are commonly observed in this disease [128].

Other recent and interesting publications have been demonstrating the importance of miRNAs in several cardiovascular diseases [129], among them, AMI [130], aortic aneurysm [131], and heart failure [132]. The last two decades have shown quick advances in understanding the molecular processes regulated by miRNAs, and how they can act in pathological processes, which represents an important step for therapeutic medicine to seek tools to treat and prevent diseases.

Modulation or MicroRNAs by Physical Activity: A Nonpharmacological Therapy to Treat Cardiovascular Diseases

Aerobic physical activity is considered one of the most important nonpharmacological measures to prevent and treat cardiovascular diseases. Recent studies have collaborated to prove that physical activity is able to modulate miRNAs; however, little is still known regarding its effects on regulation. We recently revisited several miRNAs that may be promoting physiological cardiac remodeling induced by aerobic physical activity [133]. Heart physiological hypertrophy induced by physical activity involves the regulation of miRNA-27a and -27b, as well as miRNA-143 [134]. Our work showed that these miRNAs can promote positive balance in the angiotensin (1–7) vasodilation formation pathway in detriment of the angiotensin II vasoconstriction pathway. In addition, we observed increased expression or miRNA-29 correlated to reduction of collagen type I and type III, collaborating to improved ventricular compliance in female rats subjected to different aerobic protocols [135] and miRNA-126 in regulating cardiac angiogenesis induced by physical activity by inhibiting target

genes SPRED-1 and PI3KR2 [73]. However, not only aerobic physical activity, but also resistance physical activity, can modulate a miRNA and regulate cardiac function, as demonstrated by the reduction of miRNA-214 and parallel increased expression of SERCA2a, a calcium pump of the sarcoplasmic reticulum [136].

Since aerobic physical activity of low to moderate intensity is the most well-established type of training recommended to treat cardiovascular diseases, some studies are already investigating the effects on modulation of miRNAs associated with diseases. Specifically in vascular diseases such as arterial hypertension, we can observe several miRNAs that can be modulated by physical activity and revert the alterations observed with this pathology, therefore showing therapeutic potential [137]. As already mentioned, our group demonstrated that arterial hypertension was associated with reduced expression of microvascular miRNA-126 in parallel with increased expression of its target, PI3KR2, besides increased peripheral expression of miRNAs-16 and -21, in parallel with decreased expression of target genes VEGF and Bcl-2. Interestingly, aerobic physical activity reestablished the expression of these parameters in arterial hypertension, recovering the expression of miRNAs-126, -16, and -21, and in parallel of its target genes. This response was followed by correction of capillary rarefaction in trained hypertensive animals, indicating that the formation of new capillaries may depend on the balance between the proangiogenic/survival factors and the antiangiogenic/apoptotic factors of the miRNA action pathway [74]. In atherosclerosis, a recent study showed that physical activity, as well as treatment with statins, may induce the expression of miRNA-146a and reduce expression of miRNA-155, demonstrating the probable protective effects of physical activity in vascular diseases [138]

CONCLUSIONS

With the new discoveries of genetics in the last decades, miRNAs have gained considerable prominence, as they have been identified as one of the largest classes of gene regulators. This new class of small RNAs regulates several cellular processes, taking part in a wide spectrum of human diseases. Therefore, the use or miRNA microarrays complemented with bio-informatics prediction software to identify miRNA groups, related signaling networks, and target genes contributed significantly to understand the biological importance of miRNAs.

As evidenced along the review, miRNAs work in orchestrated fashion to control a common pathways or biological function; this unique characteristic of miRNAs makes them efficient tools to determine specific pathways involved in diseases or biological processes. Therefore, we observed that there are specific miRNA profiles responsible for endothelial cell and VSMC differentiation, proliferation, and migration processes, contributing greatly for homeostasis and vascular function, and also for many pathologies that involve the vascular system.

Regardless of the therapeutic options currently available to treat cardiovascular diseases, alterations in vascular function and structure persist in a large number of patients, leading to injury in target organs, such as the heart and skeletal muscle, representing very significant complications. Therefore, the search for new therapies, such as the use or miRNAs, shall be stimulated. Thus, identification of specific miRNA profiles for a given disease and specific tissue opens perspectives for future studies, using miRNAs as clinical diagnosis biomarkers and therapeutic agents by means of subexpression or inhibition.

References

[1] Armstrong L. Epigenetics. New York, NY: Garland Science; 2014.

[2] Zimmer P, Bloch W. Physical exercise and epigenetic adaptations of the cardiovascular system. Herz 2015;40:353–60.

[3] Webster AL, Yan MS, Marsden PA. Epigenetics and cardiovascular disease. Can J Cardiol 2013;29:46–57.

[4] Jiang YZ, Manduchi E, Jiménez JM, et al. Endothelial epigenetics in biomechanical stress: disturbed flow-mediated epigenomic plasticity in vivo and in vitro. Arterioscler Thromb Vasc Biol 2015;35:1317–26.

[5] Bird A. Perceptions of epigenetics. Nature 2007;447:396–8.

[6] Turgeon PJ, Sukumar AN, Marsden PA. Epigenetics of cardiovascular disease: a new "beat" in coronary artery disease. Med Epigenet 2014;2:37–52.

[7] Udali S, Guarini P, Moruzzi S, et al. Cardiovascular epigenetics: from DNA methylation to microRNAs. Mol Asp Med 2013;34:883–901.

[8] Nührenberg T, Gilsbach R, Preissl S, et al. Epigenetics in cardiac development, function, and disease. Cell Tissue Res 2014;3: 585–600.

[9] Han P, Hang CT, Yang J, et al. Chromatin remodeling in cardiovascular development and physiology. Circ Res 2010;108:378–96.

[10] Dunn J, Simmons R, Thabet S, et al. The role of epigenetics in the endotelial cell shear stress response and atherosclerosis. Int J Biochem Cell Biol 2015;67:167–76.

[11] Lee JT. Epigenetic regulation by long noncoding RNAs. Science 2012;338:1435–9.

[12] Uchida S, Dimmeler S. Long noncoding RNAs in cardiovascular diseases. Circ Res 2015;116:737–50.

[13] Hartung CT, Michalik KM, You X, et al. Regulation and function of the laminar flow-induced non-coding RNAs in endothelial cells. Circulation 2014;130:A18783.

[14] Ambros V. The functions of animal microRNAs. Nature 2004;431: 350–5.

[15] Chuang JC, Jones PA. Epigenetics and microRNAs. Pediatr Res 2007;61:24R–9R.

[16] Kim VN. MicroRNA biogenesis: coordinated cropping and dicing. Nat Rev Mol Cell Biol 2005;6:376–85.

[17] Guo H, Ingolia NT, Weissman JS, et al. Mammalian microRNAs predominantly act to decrease target mRNA levels. Nature 2010;466:835–40.

[18] Hartmann D, Thum T. MicroRNAs and vascular (dys)function. Vasc Pharmacol 2011;55:92–105.

[19] Lee RC, Feinbaun RL, Ambros V. The C. elegans heterochronic gene lin-4 encodes small RNAs with antisense complementarity to lin-14. Cell 1993;75:843–54.

[20] Wightman B, Ha I, Ruvkun G. Posttranscriptional regulation of the heterochronic gene lin-14 by lin-4 mediates temporal pattern formation in C. elegans. Cell 1993;75:855–62.

[21] Lau NC, Lim LP, Weinstein EG, et al. An abundant class of tiny RNAs with probable regulatory roles in Caenorhabditis elegans. Science 2001;294:858–62.

[22] Reinhart BJ, Slack FJ, Basson M, et al. The 21-nucleotide let-7 TNA regulates developmental timing in Caenorhabditis elegans. Nature 2000;403:901–6.

[23] Lee RC, Ambros V. An extensive class of small RNAs in Caenorhabditis elegans. Science 2001;294:862–4.

[24] Pasquinelli AE, Reinhart BJ, Slack F, et al. Conservation of the sequence and temporal expression of let-7 heterochronic regulatory RNA. Nature 2000;408:86–9.

[25] Urbich C, Kuehbacher A, Dimmeler S. Role of microRNAs in vascular diseases, inflammation, and angiogenesis. Cardiovasc Res 2008;79:581–8.

[26] Kingwell K. Cardiovascular disease: microRNA protects the heart. Nat Rev Drug Discov 2011;10:98.

[27] Liu N, Williams AH, Maxeiner JM, et al. MicroRNA-206 promotes skeletal muscle regeneration and delays progression of Du-chenne muscular dystrophy in mice. J Clin Invest 2012;122:2054–65.

[28] Olena AF, Patton JG. Genomic organization of microRNAs. J Cell Physiol 2010;222:540–5.

[29] Lee YS, Dutta A. MicroRNAs in cancer. Annu Rev Pathol 2009;4: 199–227.

[30] Brennecke J, Stark A, Russell RB, et al. Principles of microRNA-target recognition. PLoS Biol 2005;3:e85.

[31] Hammond SM. Dicing and slicing: the core machinery of de RNA interference pathway. FEBS Lett 2005;579:5822–9.

[32] Valencia-Sanches MA, Liu J, Hannon GJ, et al. Control of translation and mRNA degradation by miRNAs and siRNAs. Genes Dev 2006;20:515–24.

[33] Hangauer MJ, Vaughn IW, McManus MT. Pervasive transcription of the human genome produces thousands of previously unidentified long intergenic noncoding RNAs. PLoS Genet 2013;9(6)e1003569.

[34] Hansen TB, Jensen TI, Clausen BH, et al. Natural RNA circles function as efficient microRNA sponges. Nature 2013;495:384–8.

[35] Wang K, Liu F, Zhou LY, et al. The long noncoding RNA CHRF regulates cardiac hypertrophy by targeting miR-489. Circ Res 2014;114:1377–88.

[36] Czech B, Hannon GJ. Small RNA sorting: matchmaking for Argonautes. Nat Rev Genet 2011;12:19–31.

[37] Williams AH, Liu N, van Rooij E, et al. MicroRNA control of muscle development and disease. Curr Opin Cell Biol 2009;21:461–9.

[38] He L, Hannon GJ. MicroRNAs: small RNAs with a big role in gene regulation. Nat Rev Genet 2004;5:522–31.

[39] Thum T. MicroRNA therapeutics in cardiovascular medicine. EMBO Mol Med 2012;4:3–14.

[40] Khvorova A, Reynolds A, Jayasena SD. Functional siRNAs and miRNAs exhibit strand bias. Cell 2003;115:209–16.

[41] Schwarz DS, Hutvágner G, Du T, et al. Asymmetry in the assembly of the RNAi enzyme complex. Cell 2003;115:199–208.

[42] Vasques LR, Schoof CRG, Botelho EL. MicroRNAs: a new paradigm for gene regulation. In: Gaur RK, Gafni Y, Sharma P, Gupta VK, editors. RNAi technology. 1st ed., vol. 1. New Hampshire: Science Publishers, Enfield; 2011. p. 135–53.

[43] Schwarz S, Grande AV, Bujdoso N, et al. The microRNA regulated SBP-box genes SPL9 and SPL15 control shoot maturation in Arabidopsis. Plant Mol Biol 2008;67:183–95.

[44] Rhoades MW, Reinhart BJ, Lim LP, et al. Prediction of plant microRNA targets. Cell 2002;1(10):513–20.

[45] Doench JG, Sharp PA. Specificity of microRNA target selection in translational repression. Genes Dev 2004;18:504–11.

[46] Lewis BP, Burge CB, Bartel DP. Conserved seed pairing, often flanked by adenosines, indicates that thousands of human genes are microRNA targets. Cell 2005;120:15–20.

[47] Bernstein E, Kim SY, Carmell MA, et al. Dicer is essential for mouse development. Nat Genet 2003;35:215–7.

[48] Wang Y, Medvid R, Melton C, et al. DGCR8 is essential for microRNA biogenesis and silencing of embryonic stem cell self-renewal. Nat Genet 2007;39:380–5.

[49] Vidigal JA, Ventura A. The biological function of miRNAs: lessons from in vivo studies. Trends Cell Biol 2015;25:137–47.

[50] Hammond SM. An overview of microRNAs. Adv Drug Deliv Rev 2015;87:3–14.

[51] Nazari-Jahantigh M, Wei Y, Schober A. The role of microRNAs in arterial remodelling. Thromb Haemost 2012;107:611–8.

[52] Neth P, Nazari-Jahantigh M, Schober A, et al. MicroRNAs in flow-dependent vascular remodeling. Cardiovasc Res 2013;99: 294–303.

[53] Wey Y, Schober A, Weber C. Pathogenic arterial remodeling: the good and bad of microRNAs. Am J Physiol Heart Circ Physiol 2013;304:H1050–9.

[54] Levy BI, Ambrosio G, Pries AR, et al. Microcirculation in hypertension: a new target for treatment? Circulation 2001;104:735–40.

[55] Feihl F, Liaudet L, Waeber B, et al. Hypertension: a disease of the microcirculation? Hypertension 2006;48:1012–7.

[56] Bátkai S, Thum T. MicroRNAs in hypertension: mechanisms and therapeutic targets. Curr Hypertens Rep 2012;14:79–87.

[57] Laurent S, Boutouyrie P. The structural factor of hypertension: large and small artery alterations. Circ Res 2015;116:1007–21.

[58] Coffman TM. Under pressure: the search for the essential mechanisms of hypertension. Nat Med 2011;17:1402–9.

[59] Roque FR, Briones AM, García Redondo AB, et al. Aerobic exercise reduces oxidative stress and improves vascular changes of small mesenteric and coronary arteries in hypertension. Br J Pharmacol 2013;168:686–703.

[60] Suárez Y, Sessa WC. MicroRNAs as novel regulators of angiogenesis. Circ Res 2009;104:442–54.

[61] Quintavalle C, Garofalo M, Croce CM, et al. "ApoptomiRs" in vascular cells: their role in physiological and pathological angiogenesis. Vasc Pharmacol 2011;55:87–91.

[62] Yang WJ, Yang DD, Na S, et al. Dicer is required for embryonic angiogenesis during mouse development. J Biol Chem 2005;280:9330–5.

[63] Suárez Y, Fernández-Hernando C, Pober JS, et al. Dicer dependent microRNAs regulate gene expression and functions in human endothelial cells. Circ Res 2007;100:1164–73.

[64] Kuehbacher A, Urbich C, Zeiher AM, et al. Role of Dicer and Drosha for endothelial microRNA expression and angiogenesis. Circ Res 2007;101:59–68.

[65] Poliseno L, Tuccoli A, Mariani L, et al. MicroRNAs modulate the angiogenic properties of HUVECs. Blood 2006;108:3068–71.

[66] Wang S, Aurora AB, Johnson BA, et al. The endothelial-specific microRNA miR-126 governs vascular integrity and angiogenesis. Dev Cell 2008;15:261–71.

[67] Sen CK, Gordillo GM, Khanna S, et al. Micromanaging vascular biology: tiny microRNAs play big band. J Vasc Res 2009;46:527–40.

[68] Weber M, Baker MB, Moore JP, et al. MiR-21 is induced in endothelial cells by shear stress and modulates apoptosis and eNOS activity. Biochem Biophys Res Commun 2010;393:643–8.

[69] Chamorro-Jorganes A, Araldi E, Penalva LO, et al. MicroRNA-16 and microRNA-424 regulate cell-autonomous angiogenic functions in endothelial cells via targeting vascular endothelial growth factor receptor-2 and fibroblast growth factor receptor-1. Arterioscler Thromb Vasc Biol 2011;31:2595–606.

[70] Sun HX, Zeng DY, Li RT, et al. Essential role of microRNA-155 in regulating endothelium-dependent vasorelaxation by targeting endothelial nitric oxide synthase. Hypertension 2012;60:1407–14.

[71] Fish JE, Santoro MM, Morton SU, et al. miR-126 regulates angiogenic signaling and vascular integrity. Dev Cell 2008;15:272–84.

[72] Staszel T, Zapala B, Polus A, et al. Role of microRNAs in endothelial cell pathophysiology. Pol Arch Med Wewn 2011;121:361–6.

[73] Da Silva Jr ND, Fernandes T, Soci UP, et al. Swimming training in rats increases cardiac MicroRNA-126 expression and angiogenesis. Med Sci Sports Exerc 2012;44:1453–62.

[74] Fernandes T, Magalhães FC, Roque FR, et al. Exercise training prevents the microvascular rarefaction in hypertension balancing angiogenic and apoptotic factors. Role of microRNAs-16, -21, and -126. Hypertension 2012;59:513–20.

[75] Cimmino A, Calin GA, Fabbri M, et al. miR-15 and miR-16 induce apoptosis by targeting BCL2. Proc Natl Acad Sci U S A 2005;102:13944–9.

[76] Dejean E, Renalier MH, Foisseau M, et al. Hypoxia-microRNA-16 downregulation induces VEGF expression in anaplastic lymphoma kinase (ALK)-positive anaplastic large-cell lymphomas. Leukemia 2011;25:1882–90.

[77] Fleissner F, Jazbutyte V, Fiedler J, et al. Asymmetric dimethylarginine impairs angiogenic progenitor cell function in patients with coronary artery disease through a microRNA-21-dependent mechanism. Circ Res 2010;107:138–43.

[78] Bonauer A, Carmona G, Iwasaki M, et al. MicroRNA-92a controls angiogenesis and functional recovery of ischemic tissues in mice. Science 2009;324:1710–3.

[79] van Solingen C, Seghers L, Bijkerk R, et al. Antagomir-mediated silencing of endothelial cell specific microRNA-126 impairs ischemia-induced angiogenesis. J Cell Mol Med 2009;13:1577–85.

[80] Harris TA, Yamakuchi M, Ferlito M, et al. MicroRNA-126 regulates endothelial expression of vascular cell adhesion molecule 1. Proc Natl Acad Sci U S A 2008;105:1516–21.

[81] Landmesser U, Drexler H. Endothelial function and hypertension. Curr Opin Cardiol 2007;22:316–20.

[82] Gkaliagkousi E, Douma Ø, Zamboulis C, et al. Nitric oxide dysfunction in vascular endothelium and platelets: role in essential hypertension. J Hypertens 2009;27:2310–20.

[83] Dangwal S, Bang C, Thum T. Novel techniques and targets in cardiovascular microRNA research. Cardiovasc Res 2012;93:545–54.

[84] Owens GK, Kumar MS, Wamhoff BR. Molecular regulation of vascular smooth muscle cell differentiation in development and disease. Physiol Rev 2003;84:767–801.

[85] Rensen SS, Doevendans PA, van Eys GJ. Regulation and characteristics of vascular smooth muscle cell phenotypic diversity. Neth Hear J 2007;15:100–8.

[86] Kang H, Hata A. MicroRNA regulation of smooth muscle gene expression and phenotype. Curr Opin Hematol 2012;19:224–31.

[87] Elia L, Quintavalle M, Zhang J, et al. The knockout of miR-143 and miR-145 alters smooth muscle cell maintenance and vascular homeostasis in mice: correlates with human disease. Cell Death Differ 2009;16:1590–8.

[88] Liu Y, Sinha S, McDonald OG, et al. Kruppel-like factor 4 abrogates myocardin-induced activation of smooth muscle gene expression. J Biol Chem 2005;280:9719–27.

[89] Cordes KR, Sheehy NT, White M, et al. miR-145 and miR-143 regulate smooth muscle cell fate decisions. Nature 2009;460:705–10.

[90] Rangrez AY, Massy ZA, Metzinger-Le Meuth VM, et al. miR-143 and miR-145. Molecular keys to switch the phenotype of vascular smooth muscle cells. Circ Cardiovasc Genet 2011;4:197–205.

[91] Caruso P, Dempsie Y, Stevens HC, et al. A role for miR-145 in pulmonary arterial hypertension: evidence from mouse models and patients samples. Circ Res 2012;111:290–300.

[92] Townley-Tilson WH, Callis TE, Wang D. MicroRNAs 1, 133, and 206: critical factors of skeletal and cardiac muscle development, function, and disease. Int J Biochem Cell Biol 2010;42:1252–5.

[93] Chen J, Yin H, Jiang Y, et al. Induction of microRNA-1 by miocardin in smooth muscle cells inhibits cell proliferation. Arterioscler Thromb Vasc Biol 2011;31:368–75.

[94] Xie C, Huang H, Sun X, et al. MicroRNA-1 regulates smooth muscle cell differentiation by repressing Kruppel-like factor 4. Stem Cells Dev 2011;20:205–10.

[95] Torella D, Iaconetti C, Catalucci D, et al. MicroRNA-133 controls vascular smooth muscle cell phenotypic switch in vitro and vascular remodeling in vivo. Circ Res 2011;109:880–93.

[96] Ji R, Cheng Y, Yue J, et al. MicroRNA expression signature and antisense-mediated depletion reveal an essential role of microRNA in vascular neointimal lesion formation. Circ Res 2007;100:1579–88.

[97] Sun SG, Zheng B, Han M, et al. miR-146a and Kruppel-like factor 4 form a feedback loop to participate in vascular smooth muscle cell proliferation. EMBO Rep 2011;12:56–62.

[98] Kumarswamy R, Volkmann I, Thum T. Regulation and function of miRNA-21 in health and disease. RNA Biol 2011;8:706–13.

[99] Horita HN, Simpson PA, Ostriker A, et al. Serum response factor regulates expression of PTEN through a micro-RNA network in vascular smooth muscle cells. Arterioscler Thromb Vasc Biol 2011;31:2909–19.

[100] Davis BN, Hilyard AC, Lagna G, et al. SMAS proteins control DROSHA-mediated microRNA maturation. Nature 2008;454:56–61.

[101] Kang H, Davis-Dusenbery BN, Nguyen PH, et al. Bone morphogenetic protein 4 promotes vascular smooth muscle contractility by activating microRNA-21 (miR-21), which down-regulates expression of family of dedicator of cytokinesis (DOCK) proteins. J Biol Chem 2012;287:3976–86.

[102] Davis BN, Hilyard AC, Nguyen PH, et al. Induction of microRNA-221 by platelet-derived growth factor signaling is critical for modulation of vascular smooth muscle phanotype. J Biol Chem 2009;284:3728–38.

[103] Liu X, Cheng Y, Zhang S, et al. A necessary role of miR-221 and miR-222 in vascular smooth muscle cell proliferation and neointimal hyperplasia. Circ Res 2009;104:476–87.

[104] Gupta SK, Bang C, Thum T. Circulating microRNAs as biomarkers and potential paracrine mediators of cardiovascular disease. Circ Cardiovasc Genet 2010;3:484–8.

[105] Creemers EE, Tijsen AJ, Pinto YM. Circulating microRNAs: novel biomarkers and extracellular communicators in cardiovascular disease? Circ Res 2012;110:483–95.

[106] Khalyfa A, Gozal D. Exosomal miRNAs as potential biomarkers of cardiovascular risk in children. J Transl Med 2014;12:162.

[107] Park NJ, Zhou H, Elashoff D, et al. Salivary microRNA: discovery, characterization, and clinical utility for oral cancer detection. Clin Cancer Res 2009;15:5473–7.

[108] Corsten MF, Dennert R, Jochems S, et al. Circulating microRNA-208b and microRNA-499 reflect myocardial damage in cardiovascular disease. Circ Cardiovasc Genet 2010;3:499–506.

[109] Mitchell PS, Parkin RK, Kroh EM, et al. Circulating microRNAs as stable blood-based markers for cancer detection. Proc Natl Acad Sci U S A 2008;105:10513–8.

[110] Valadi H, Ekstrom K, Bossios A, et al. Exosome-mediated transfer of mRNAs and microRNAs is a novel mechanism of genetic exchange between cells. Nat Cell Biol 2007;9:654–9.

[111] Zernecke A, Bidzhekov K, Noels H, et al. Delivery of microRNA-126 by apoptotic bodies induces CXCL12-dependent vascular protection. Sci Signal 2009;2: ra81.

[112] Arroyo JD, Chevillet JR, Kroh EM, et al. Argonaute2 complexes carry a population of circulating microRNAs independent of vesicles in human plasma. Proc Natl Acad Sci U S A 2011;108:5003–8.

[113] Vickers KC, Palmisano BT, Shoucri BM, et al. MicroRNAs are transported in plasma and delivered to recipient cells by high-density lipoproteins. Nat Cell Biol 2011;13:423–33.

[114] Kosaka N, Iguchi H, Yoshioka Y, et al. Secretory mechanisms and intercellular transfer of microRNAs in living cells. J Biol Chem 2010;285:17442–52.

[115] Zhang Y, Liu D, Chen X, et al. Secreted monocytic miR-150 enhances targeted endothelial cell migration. Mol Cell 2010;39:133–44.

[116] Wang GK, Zhu JQ, Zhang JT, et al. Circulating microRNA: a novel potential biomarker for early diagnosis of acute myocardial infarction in humans. Eur Heart J 2010;31:659–66.

[117] Tijsen AJ, Creemers EE, Moerland PD, et al. MiR423-5p as a circulating biomarker for heart failure. Circ Res 2010;106:1035–9.

[118] Fichtlscherer S, De Rosa S, Fox H, et al. Circulating microRNAs in patients with coronary artery disease. Circ Res 2010;107:677–84.

[119] Dickinson BA, Semus HM, Montgomery RL, et al. Plasma microRNAs serve as biomarkers of therapeutic efficacy and disease progression in hypertension-induced heart failure. Eur J Heart Fail 2013;15:650–9.

[120] Fukushima Y, Nakanishi M, Nonogi H, et al. Assessment of plasma miRNAs in congestive heart failure. Circ J 2011;75:336–40.

[121] Qiang L, Hong L, Ningfu W, et al. Expression of miR-126 and miR-508-5p in endothelial progenitor cells is associated with the prognosis of chronic heart failure patients. Int J Cardiol 2013;168:2082–8.

[122] Zampetaki A, Kiechl S, Drozdov I, et al. Plasma microRNA profiling reveals loss of endothelial miR-126 and other microRNAs in type 2 diabetes. Circ Res 2010;107:810–7.

[123] Li S, Zhu J, Zhang W, et al. Signature microRNA expression profile of essential hypertension and its novel link to human cytomegalovirus infection. Circulation 2011;124:175–84.

[124] Van Rooij E, Marshall WS, Olson EN. Toward microRNA-based therapeutics for heart disease: the sense in antisense. Circ Res 2008;103:919–28.

[125] Van der Ree MH, van der Meer AJ, de Bruijne J, et al. Long-term safety and efficacy of microRNA-targeted therapy in chronic hepatitis C patients. Antivir Res 2014;111:53–9.

[126] Shi L, Liao J, Liu B, et al. Mechanisms and therapeutic potential of microRNAs in hypertension. Drug Discov Today 2015;20 (10):1188–204.

[127] Gupta S, Li L. Modulation of microRNAs in pulmonary hypertension. Int J Hypertens 2015;2015:169069.

[128] Araldi E, Chamorro-Jorganes A, van Sollingen C, et al. Therapeutic potential of modulating microRNAs in atherosclerotic vascular disease. Curr Vasc Pharmacol 2015;13:291–304.

[129] Nishiguchi T, Imanishi T, Akasaka T. MicroRNAs and cardiovascular diseases. Biomed Res Int 2015;2015:682857.

[130] Boon RA, Dimmeler S. MicroRNAs in myocardial infarction. Nat Rev Cardiol 2015;12:135–42.

[131] Fu XM, Zhou YZ, Cheng Z, et al. MicroRNAs: novel players in aortic aneurysm. Biomed Res Int 2015;2015:831641.

[132] Duygu B, de Windt LJ, da Costa Martins PA. Targeting microRNAs in heart failure. Trends Cardiovasc Med 2016;26(2):99–100.

[133] Fernandes T, Barauna VG, Negrão CE, et al. Aerobic exercise training promotes physiological cardiac remodeling involving a set of microRNAs. Am J Physiol Heart Circ Physiol 2015;309(4): H543–52.

[134] Fernandes T, Hashimoto N, Magalhães FC, et al. Aerobic exercise training-induced left ventricular hypertrophy involves regulatory microRNAs, decreased angiotensin-converting enzyme-angiotensin II, and synergistic regulation of angiotensin-converting enzyme 2-angiotensin (1–7). Hypertension 2011;58:182–9.

[135] Soci UP, Fernandes T, Hashimoto NY, et al. MicroRNAs 29 are involved in the improvement of ventricular compliance promoted by aerobic exercise training in rats. Physiol Genomics 2011;43: 665–73.

[136] Melo SF, Barauna VG, Junior MA, et al. Resistance training regulates cardiac function through modulation of microRNA-214. Int J Mol Sci 2015;16:6855–67.

[137] Neves VJ, Fernandes T, Roque FR, et al. Exercise training in hypertension: role of microRNAs. World J Cardiol 2014;6:713–27.

[138] Wu XD, Zeng K, Liu WL, et al. Effect of aerobic exercise on miRNA-TLR4 signaling in atherosclerosis. Int J Sports Med 2014;35:344–50.

Further Reading

[1] Song XW, Shan DK, Chen J, et al. miRNAs and lncRNAs in vascular injury and remodeling. Sci China Life Sci 2014;57:826–35.

14

Adhesion Molecules and Endothelium

Juliana Carvalho Tavares and Marcelo Nicolás Muscará

INTRODUCTION

In the inflammatory process, migration of circulating leukocytes to the focus of lesion is a central event of the process that follows local vasodilation. It not only allows a greater influx of leukocytes to the lesion site, but also facilitates plasma extravasation in tissues (edema).

In these events, endothelial cells participate in various forms, either by releasing vasodilator mediators (such as nitric oxide—NO and prostacyclin—PGI_2, as well as other inflammatory mediators (cytokines and other lipids derived from arachidonic acid) with chemotactic characteristics for leukocytes. Thus, adhesion of these to the microcirculation endothelial layer precedes their migration into tissue (diapedesis). Endothelial layer continuity rupture in this process promotes increased vascular permeability, resulting in plasma and proteins output in inflammatory focus (edema).

Favoring circulating leukocytes adhesion to endothelial surface is mediated by a large increase in expression of molecules that mediate this interaction in both leukocytes and endothelial cells, in response to a wide variety of inflammatory mediators. For example, histamine and thrombin (mediators released in acute phase of inflammatory process) promote, in endothelial cells, short-term (minutes) displacement of adhesion molecules stored in granules to the cell surface, while stimuli such as endotoxin, IL-1β and TNF-α stimulate medium-/long-term (hours/days) de novo synthesis of these molecules.

ENDOTHELIAL BARRIER

Barrier properties that characterize the endothelium in vascular wall are due to expression of proteins with adhesive properties and which determine the various junctions described between endothelial cells: adherens junctions, gap junctions and tight junctions. Water movement and various solutes between blood and interstitial fluid are regulated by these junctions and thus alterations in protein expression or disruption of physical integrity of the junctions cause increase in vascular permeability that occurs in the inflammatory process.

Contact between adjacent endothelial cells is mediated by adherent junctions through binding of cadherin protein molecules present in both cells. This binding results in endothelial cytoskeleton reorganization through interaction with actin filaments. Other proteins (such as α-catenin, β-catenin, p120 catenin, vinculin, and α-actinin) participate in this structure.

Connexin family proteins form gap junctions, which allow communication between adjacent endothelial cells through channels, thus connecting cytoplasm and allowing, in this way, the passage of ions and small molecules (less than 1 kDa). These junctions participate in other cellular processes, such as apoptosis and differentiation, and allow communication between endothelial cells and the smooth muscle of the vascular wall.

Tight junctions are characteristic of cerebral vasculature (which results in the blood-brain barrier), and cause adjacent endothelial cells to be virtually

189

merged, therefore reinforcing the restriction of water and plasma solute passage to the interstitial space. These junctions are formed by a wide variety of proteins, such as occludin, members of claudin and *zonula occludens* families, and other junctional adhesion molecules. For a deeper approach on endothelial junctions and proteins involved in vascular permeability increase in inflammatory processes, see the review by Rodrigues and Granger [1].

PROTEINS INVOLVED IN LEUCOCYTE-ENDOTHELIUM INTERACTIONS

Selectins are adhesion receptors expressed by both endothelial and leukocyte cells (or hematopoietic in general). There are three types of selectins: P-, E-, and L-selectins, with selectin ligands comprising a variety of sialylated and fucosylated carbohydrates. Selectins participate in a wide range of physiologically important processes, including hematopoietic stem cell interactions with bone marrow microenvironment, recruitment of lymphocytes to endothelial venules, migration of leukocytes to inflammation areas, and cancer cell metastasis. In general, selectins mediate rapid interactions, whereas another group of adhesion receptors, integrins, is involved in stable high affinity bonds with immunoglobulin-like cell adhesion molecules (IgCAM). Thus, selectins participate in initial interaction between hematopoietic cells in motion and other cells, whether at rest, such as endothelial cells, or in motion, such as other circulating blood cells. Shear stress can be critical for activation of some events mediated by selectins, which can then lead to adhesion of one cell to another short-term cell (*tethering*). In interactions involving endothelial cells, the rolling phenomenon can be observed, which consists of repeated short term interactions of hematopoietic cells in motion with endothelium surface. After these transient interactions, there are other firmer interactions mediated by other types of adhesion receptors and ligands. Selectins are thus responsible for the rolling of circulatory cells on endothelial surfaces, whereas integrins and IgCAMs mediate a more intense and longer-lasting interaction with such cells.

Selectins

In general, selectins have at the extreme N-terminal region a lectin domain responsible for carbohydrate binding and whose activity is dependent on calcium ion. In the membrane direction, an epidermal growth factor (EGF) domain is linked to a variable number of single consensus repeats for each selectin: two in L-, six in E-, and nine in P-selectin. The remaining molecules consist of a transmembrane domain and a short C-terminal cytoplasmic domain [2].

E-selectin, also called type 1 endothelium-leukocyte adhesion molecule, CD62E, or type 2 leukocyte-endothelium adhesion molecule, is poorly expressed in resting endothelial cells, but its expression increases considerably in response to shear stress and proinflammatory (de novo synthesis via activation of *SELE* gene). There is also E-selectin expression in skin and bone marrow. Mononuclear and polymorphonuclear leukocytes express ligands for E-selectin that mediate interactions of these cells with endothelium during the inflammatory process.

L-selectin (or CD62L) is constitutively expressed on mononuclear and polymorphonuclear leukocytes as well as in naïve and central memory type T lymphocytes, thus demonstrating the importance of these molecules for immune cell trafficking. The interaction of L-selectin with its ligands results in the activation of several signaling pathways (src tyrosine kinase $p56^{lck}$, Ras, MAP kinase, Rac2) that lead to increased production of superoxide anion (O_2^-). In neutrophils and lymphocytes, L-selectin expression is regulated by endometalloproteinases (such as ADAM-17), which cleave the extracellular end of L-selectin when cells are activated. In response to the loss of L-selectin, increased expression of other adhesion molecules occurs in leukocytes and the consequent increase of the circulating concentrations of soluble L-selectin (SL-selectin) exerts a negative feedback on leukocyte adhesion to endothelium.

In the case of P-selectin, its expression on the surface of platelets and endothelial cells needs to be activated. In platelets, a rapid translocation of P-selectin stored in dense granules to the cell surface is induced by thrombin; in endothelial cells, histamine, phorbol esters, components of the complement system, calcium ionophore, thrombin, hydrogen peroxide, hypoxia and compounds containing heme group stimulate the rapid translocation of P-selectin stored in Weibel-Palade bodies to the cell surface. This expression is of short duration, falling considerably in a matter of minutes [3].

Integrins

Integrins, with some membrane proteoglycans, are generally the major cell receptors for binding to most extracellular matrix proteins (ECMs, including various types of laminins, collagen, and fibronectin), which allow the communication between ECM and cell cytoskeleton.

They constitute a large family of transmembrane receptors and, in the case of blood cells, also serve as adhesion molecules between cells. As with selectins and other cell adhesion molecules, integrins have much lower affinities for the ligands than those usually found in pharmacological receptors; however, on the cell surface, they are approximately 10–100 times more abundant than the latters, and also activate intracellular signaling pathways.

Structurally, they are glycoproteins heterodimers (termed alpha and beta subunits), noncovalently associated. Binding of integrins to ligands depends on extracellular Ca^{2+} or Mg^{2+} cations (depending on the integrin), which may also influence its binding affinity and specificity. In humans, integrin heterodimers result from combination of 24 types of alpha subunits and 9 types of beta subunits, and the number of this large variety of possible heterodimers can be further increased by alternative splicing of some mRNAs.

The same integrin molecule may have different binding specificities, depending on the cell; thus, specific factors of each cell type are likely to interact with integrins and modulate their activity. On the other hand, different integrins can bind to the same protein; for example, fibronectin is a ligand of at least eight different integrins, and laminin of at least five.

As already discussed, integrins not only mediate cell interaction with ECM, but also with other cells. For example, β_2 subunits form dimers with at least four alpha subunit types and are exclusively expressed on the surface of circulating inflammatory cells. These integrins mediate interaction with specific endothelial cell ligands (members of the adhesion immunoglobulin superfamily, as discussed below) to firmly attach to infection sites and fight the invading agent. Thus, humans with the genetic disease "leukocyte adhesion deficiency" suffer from repeated bacterial infections because of their inability to synthesize β_2 subunits.

β_3 integrins mediate, among other actions, binding of circulating platelets to fibrinogen during blood clotting; thus, patients with Glanzmann's disease are characterized by excessive bleeding secondary to $\beta 3$ subunit genetic deficiency.

The relationship between intracellular signaling and integrin expression and/or activity is bidirectional. On one hand, in response to certain stimuli the cell may increase or decrease integrin expression or, more often, their affinity for the ligand. For example, before integrins can mediate platelet or white blood cell adhesion to endothelium, they must be activated either by phosphorylation of the cytoplasmic portion or by association of that portion with cytoplasmic proteins. In addition, this activation occurs, for example, in platelets, after contact with a vascular lesion or in response to soluble signaling molecules. Stimulus triggers intracellular signaling pathways, which result in the rapid activation (conformational alteration of extracellular domain) of a β_3 integrin on platelet membrane, which allows a high affinity binding with fibrinogen, thus favoring fibrinolysis and the consequent formation of a platelet plug.

On the other hand, integrins bind ECM molecules to the cytoskeleton (specifically, actin filaments), regulating cell shape, orientation and movement. Intracellular signaling pathways may also be activated by mechanisms similar to those associated with various types of pharmacological receptors, such as changes in gene expression or activation of tyrosine kinase proteins (such as FAK—focal adhesion kinase).

Furthermore, integrins and pharmacological receptors can act together. For example, growth and proliferation of certain cells in culture will not occur in response to extracellular growth factors if cells are not bound to ECM molecules via integrins. Endothelial cells enter apoptosis when they lose contact with ECM via integrins. This mandatory binding to ECM, as a condition of survival and proliferation, is perhaps an indication that cells survive and proliferate only when they are in the proper position [4]

Immunoglobulin-Like Cell Adhesion Molecules—IgCAM

Migration of leukocytes through endothelium is a complex process, which occurs sequentially and involves active functions by both the adherent (e.g., leukocyte) and endothelium, the latter playing an important role in promoting efficient diapedesis. Signal transduction mechanisms within endothelial cells are mediated by immunoglobulin type cell adhesion molecules (IgCAM), which affect not only leukocyte-endothelium interactions, but also those between endothelial cells.

As described, transendothelial migration firstly involves a short-term adhesion (tethering) to endothelium, which is followed by rolling along the endothelial lining until the firm adhesion. The leukocyte then spreads and effectively migrates through the endothelium (diapedesis) to the lesion focus. Firm adhesion of leukocytes to the endothelium is mediated by integrins and IgCAMs, such as intercellular adhesion molecule-1 (ICAM-1).

Under resting conditions, the endothelial expression of ICAM-1 is low; however, it increases significantly in response to proinflammatory stimuli and thus facilitates adhesion and transmigration of leukocytes mediated by integrins. The structure of ICAM-1 consists of five immunoglobulin (Ig)-like domain replicates in the extracellular portion, a transmembrane portion and a short intracellular domain of 28 amino acids involved in the activation of signal transduction pathways and transendothelial migration. Integrin-mediated endothelial ICAM-1 reuptake is the onset of intracellular signaling involving the activation of small RhoA GTPases (such as RhoG), cytoskeletal alterations and Src-kinase activation, among others. These pathways act synergistically, transiently reducing endothelial integrity and, consequently, facilitating an efficient transendothelial leukocyte migration. A similar mechanism occurs in the case of the vascular cell adhesion molecule-1 (VCAM-1), a β_1 integrin ligand, which also involves the production of reactive oxygen species that contribute to an efficient transendothelial migration by decreasing endothelial junctions via regulation of extracellular metalloproteinase activity.

In addition to the known leukocyte migration that occurs through junctions between adjacent (paracellular) endothelial cells, there is also trans-cellular migration (i.e., through endothelial cell bodies) whose mechanisms of regulation are unknown. PECAM-1 (platelet-endothelial cell adhesion molecule-1) is one of IgCAMs with greater possibilities of involvement in this process. Leukocytes and endothelial cells express PECAM, and these molecules, expressed in both cell types, facilitate diapedesis by establishing a homotypical adhesion between them. In many cases PECAM acts as a negative regulator affecting function of other membrane proteins, as for example, toll-like receptor 4 (TLR4), α_{2b}-β_3 integrin and ICAM-1 [5].

EXAMPLES OF ADHESION MOLECULES PARTICIPATION IN CARDIOVASCULAR DISEASES

Atherosclerosis

Endothelial cells exhibit innate heterogeneity, expressed by differences in structure and function, according to the vessel and/or tissue in which they reside. This endothelial heterogeneity explains blood vessel-specific pathologies, such as the development of arterial atherosclerotic plaques in areas prone to injury, venous thrombosis and vascular extravasation in venules. Modulation of endothelial cell constitutive functions, such as selective permeability and biosynthesis capacity, occurs under conditions of hyperglycemia and/or hyperlipidemia. Endothelial cell lining from regions resistant or susceptible to atherosclerosis found in human carotid and aorta have unique structural, molecular and functional differences that help to explain, at least in part, atherogenic vs. atheroprotective phenotypes [6].

Atherosclerotic vascular disease is the underlying cause of myocardial infarction, cerebrovascular accident (CVA), unstable angina, and sudden cardiac death [7]. Collectively these diseases are the leading cause of death worldwide, and their incidence continues to increase as a result of the global epidemics of obesity and type 2 diabetes, which are major risk factors for atherosclerosis, in addition to hyperlipidemia caused by genetic and/or environmental factors. The disease is initiated by the deposition of apolipoprotein B (apoB)-containing lipoproteins (LPs) in the subendothelial region by interaction of apoB with ECM proteoglycans in focal artery areas, particularly in regions where laminar flow is disturbed by curves or branching points in arteries [8]. In fact, there is a positive correlation between low shear stress areas, LDL accumulation sites and injury onset [9]. Several molecular modifications of retained LP, such as oxidation, lipolysis, proteolysis, or aggregation [10], probably mimic pathogen-associated molecular patterns (PAMPs) and/or tissue damage molecular patterns (DAMPS), and trigger a low-intensity inflammatory response.

Over the last decade, chronic disease of the arterial wall has been associated with an inflammatory response mediated by the immune system, since the interaction of cells of this system with the vascular endothelium is an essential event. A few days after atherosclerotic plaque formation, an increase in leukocyte recruitment is triggered; these leukocytes adhere and migrate through the endothelial cell monolayer and subsequently differentiate into macrophages (Fig. 14.1). Macrophages are cells responsible for phagocytizing lipoproteins, causing the formation of foam cells [11]. The lipid content from dead foam cells contribute to the formation of the necrotic lesion core, with accumulation of smooth muscle cells secreting fibrotic elements which contribute to the expansion of the fibrotic plaque [12]. The lesion continues to grow due to migration of circulating mononuclear cells. In addition, there is an increase in cell proliferation and ECM formation [13].

Thus, it is evident that monocyte recruitment is an important step in atherosclerotic plaque development. In the leukocyte recruitment mechanism associated with the inflammatory response, the first step is leukocyte rolling on the surface of endothelial cells, and as already discussed, this mechanism is mediated by selectins.

In an experimental model of atherosclerosis, P- and E-selectin-deficient mice showed a significant decrease in atherosclerosis development, thus evidencing the role of these adhesion proteins in the pathogenesis

of the disease [14]. After the rolling step, leukocytes must adhere firmly to endothelium in order to allow the transmigration step to take place through the endothelial cell layer toward the intima [14]. The firm adhesion of monocytes and T cells on endothelium is mediated by the interaction of IgCAMs, such as ICAM-1, VCAM-1 and PECAM-1 present on the endothelial surface, with integrins VLA4, CD11/CD18, or $\alpha_4\beta_7$ present on the leukocytes' surface [15].

In addition to the adhesion molecules, different chemokines and cytokines regulate leukocyte recruitment. Chemokines induce chemotaxis through activation of G protein-coupled receptors, and recruit circulating monocytes or lymphocytes to the plaque site. Studies with animals deficient in chemokine CCL2/MCP1 (CCL2$^{-/-}$) and/or the CCR2 receptor (CCR2$^{-/-}$) demonstrated that these molecules play a key role in monocyte and T cell migration toward the intima layer [16].

Proinflammatory cytokines, such as interleukins IL-1β, IL-6, IL-8 and TNF-α, produced by cells present in the intima, in response to lipoproteins, may also activate circulating monocytes [17]. Endothelial growth factors and inflammatory mediators, such as NO derived from inducible NO synthase (iNOS) and

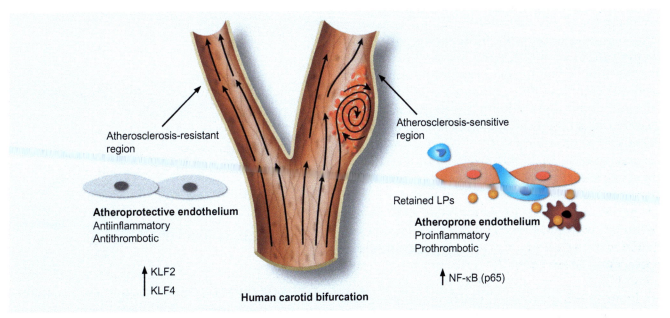

FIG. 14.1 Endothelial cell and early development of atherosclerotic lesions. Endothelial cells display an atherogenic phenotype, which promotes a proinflammatory environment by activating NF-κB signaling pathway, which is then perpetuated in response to subendothelial apoB-containing LPs. Activation of NF-κB promotes entry of blood monocytes (blue cells), through endothelial cell junctions (orange cells), into the intima, where they differentiate into macrophages (red cells). In contrast, arterial geometries that are exposed to uniform laminar flow present an atheroprotective endothelial cell phenotype (acronyms, see text).

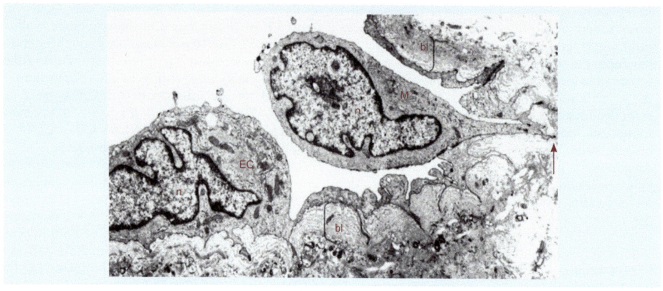

FIG. 14.2 Monocyte (M) diapedesis through an interendothelial junction (*arrow*) in lesion-prone aorta area, which shows hyperplastic basal lamina (bl), and accumulation of modified lipoproteins. N, nucleus; *ECs*, endothelial cells; *l*, lumen. *Reproduced from Simionescu M, Popov D, Sima A, et al. Pathobiochemistry of combined diabetes and atherosclerosis studied on a novel model. The hyperlipemic-hyperglycemic hamster. Am J Pathol 1996;148:997–1014.*

angiotensin I and II, also contribute to regulate the adhesive properties of vascular endothelium [18,19].

Endothelium-derived NO is a critical molecule for normal vessel function, as well as for the progression/reversal of atherosclerosis. Decreased NO production/bioavailability from endothelial NO synthase (eNOS) is characteristic of endothelial dysfunction associated with increased production of superoxide anion (O_2^-). In fact, within the atherosclerotic plaque, inflammatory cells and smooth muscle cells are a source of superoxide anion [20,21]. Inactivation of NO by superoxide anion leads to formation of highly oxidizing products (peroxynitrite anion—$ONOO^-$, hydroxyl radical—OH), which will result in lipid peroxidation and protein nitration, directly affecting the endothelial cell membrane, and causing depletion of the antioxidant defense reserves [22,23].

Accumulation of oxidized LDL can directly stimulate endothelial cells to increase adhesion molecule expression, and the subsequent migration of circulating monocytes toward the intima layer (Fig. 14.2).

During atherosclerosis, platelet activation also occurs, which promotes interaction between monocytes and endothelial cells mediated by the proinflammatory chemokine CCL5 and increased expression of P-selectin [24,25]. On the other hand, the injection of platelets expressing P-selectin in animals that do not express apolipoprotein E (ApoE$^{-/-}$)

accelerates the formation of atherosclerotic disease lesions, whereas the injection of P-selectin-deficient platelets causes only minor lesions [26].

Monocyte diapedesis occurs through endothelial cell junctions, specifically in focal areas with deposition of modified lipoproteins, in hyperplastic basal lamina (Fig. 14.3).

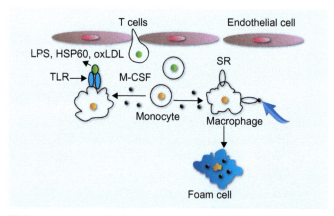

FIG. 14.3 Newly formed macrophages are characterized by presenting increased expression of patterns recognition receptors (PRRs), such as scavenger receptors (SR, a protein family that includes CD36, CD68, CXCL16, SR-A, and SR-B1) and *toll*-like receptors (TLR). SR can bind to oxLDL in intima, which leads to the formation of foam cells. T cells migrate into the intima layer and secrete cytokines that stimulate monocyte differentiation.

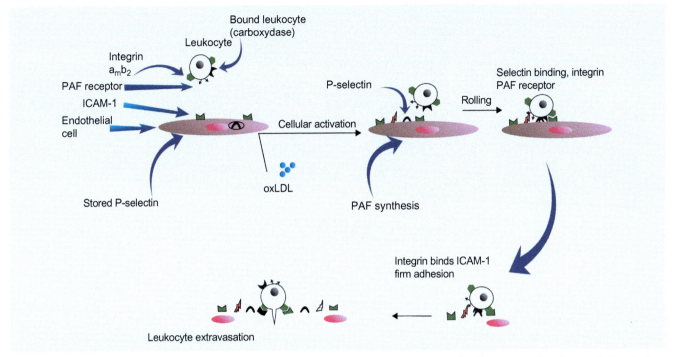

FIG. 14.4 Oxidized LDL (oxLDL) triggers an increased expression of endothelial adhesion molecules, and circulating leukocytes and lymphocytes adhere to endothelial cells through adhesion molecules. P-selectin surface expression leads to a weak interaction with circulating leukocytes, which reduce their speed, but do not stop. For firm adhesion to take place, integrins β_2 should be activated. Binding of PAF to its receptor on immune cells activates β_2 integrins via Rho. Once activated, integrins bind ICAMs expressed on endothelial cell surface, allowing a strong adhesion.

Monocytes retained in the intima layer differentiate into macrophages (Fig. 14.4), and macrophage colony stimulating factor (M-CSF) has a key role in triggering this differentiation [27]. In fact, M-CSF-deficient animals develop late lesions and have a lower accumulation of macrophages [28,29]. T lymphocytes, which also migrate into intima, secrete cytokines that promote differentiation of monocytes into macrophages.

In addition to differentiating into macrophages, monocytes can also differentiate into dendritic cells (DC), which act as antigen presenting cells within the intima layer. In the subendothelial layer, antigens presented by dendritic cells or macrophages can interact with T cells and, consequently, activate them. Activation of T cells culminates with the production of the proinflammatory cytokines interferon gamma (IFNγ) and TNF-α, which stimulate macrophage activation and the release of other inflammatory mediators [30]. It is believed that the preactivation of T lymphocytes takes place at the lymph node region, once antigen-presenting cells can move from the plaque to the lymph nodes [31].

The newly formed macrophages are characterized by increased expression of pattern recognition receptors (PRRs), such as scavenger receptors (SR, a protein family that includes CD36, CD68, CXCL16, SR-A, and SR-B1) and toll-like receptors (TLR). TLRs can be activated by lipopolysaccharide (LPS) or the HSP60 heat shock protein. This signaling may trigger the production of proinflammatory molecules, such as cytokines, chemokines, matrix metalloproteinases (MMP), and iNOS-derived NO [32] (Fig. 14.5).

Formation of Foam Cells and Fibrotic Plaques

In the arterial wall, macrophages contribute to the formation of the fibrous plaque, via metabolization of subendothelial components [33]. Lipoproteins infiltrated into the intima are retained before undergoing chemical changes (enzymatic or nonenzymatic oxidation), which are responsible for producing oxLDLs. On the other hand, oxLDL present in the intima can bind to SR receptors present on macrophages, leading to formation of these foam cells [31] (Fig. 14.6).

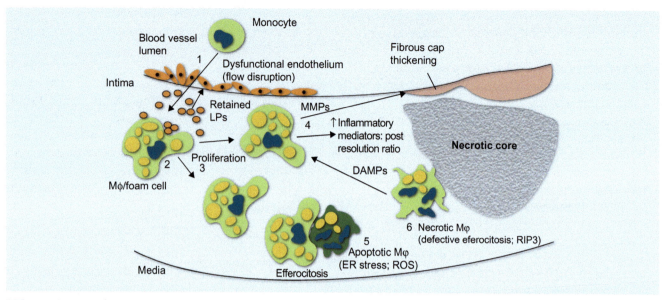

FIG. 14.5 Proatherogenic roles of macrophages present in lesions. (1) Endothelial dysfunction and accumulation of apolipoprotein B triggers the entry of proinflammatory monocytes into the subendothelial intima (*red arrows* represent endothelial dysfunction triggered by LPs retained in the intima). (2) Macrophages phagocyte retained LPs and transform into lipid loaded foam cells. (3) The advanced lesions present in macrophages can proliferate. (4) The macrophages promote plaque progression, amplifying the increased inflammatory response. Moreover, matrix metalloproteinases (MMP) secreted by proinflammatory macrophages can lead to a thinning of the fibrous cap and plaque rupture. (5) Environmental factors in advanced lesions promote macrophage apoptosis, for example, as a result of prolonged endoplasmic reticulum (ER) and/or oxidative stress. The apoptotic cell death may not be problematic if efficiently removed by macrophages of the lesion (eferocytosis). (6) However, in advanced atherosclerosis, the process leads to postapoptotic necrosis. Necrotic cells, which can also develop through RIP3 activation (primary necrosis), release damage-associated molecular pattern (DAMP) receptors which amplify the inflammatory process. These cells may also merge into areas, called necrotic cores, which promote breaking of the plaque and the subsequent thrombosis.

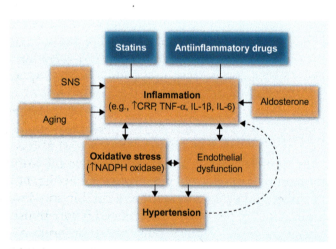

FIG. 14.6 Illustrative diagram of the interaction between inflammation and hypertension, and factors that may be associated. Antiinflammatory drugs and statins may be effective antihypertensive agents due to their antiinflammatory actions.

On the other hand, reactive oxygen species produced by endothelial cells and enzymes such as myeloperoxidase, phospholipase A2, and sphingomyelinase, are all involved in the process of LDL modification within atherosclerotic lesions in humans [34,35]

Apolipoprotein E produced by macrophages may inhibit the transformation of macrophages into foam cells. Chimeras obtained from bone marrow transplantation of apo-E-deficient animals into control animals promoted the development of large atherosclerotic lesions in transplanted animals [36].

The pronounced aggregation of foam cells may lead to the formation of a necrotic core (Fig. 14.6). At the center of the core, the foam cells die and start to accumulate extracellular debris and lipids. The more the plaque grows, the more the necrotic core grows, and consequently protease release occurs through the foam cells, which facilitates plaque destruction.

Flibrotic plaque is comprised by the necrotic core, containing lipids and dead cell debris, with a fibrotic layer that coats the core. The fibrotic cover consists of smooth muscle cells (SMCs), a growing mass of extracellular lipids and extracellular matrix (ECM) derived from SMC cells [13]. Immune cells such as macrophages, T cells and mast cells are found in fibrotic plate [31]. Inflammatory mediators such as cytokines and growth factors secreted by immune cells are essential elements of the plaques inflammatory response and vascular function [37,38].

The fibrous cap is coated with endothelial cells which sequester peripheral blood from the subendothelial lesion before the plaque rupture takes place. During the plaque rupture process, cytokines produced by T cells, mast cells, some B cells and even natural killer cells (NK), leadng to endothelial cell death [39,40]. Moreover, collagenases and other proteases produced by macrophages degrade the extracellular matrix of SMC cells and trigger the plaque rupture [41]. On the other hand, plaque growth and rupture can be inhibited by antiinflammatory cytokines, such as IL-10 and transforming growth factor beta (TGF-β), which stimulate collagen production and help to increase fibrous cap stability [42].

Advanced Lesion and Thrombosis

Matrix degradation within the fibrous cap mediated by different proteases (collagenases, gelatinases, and stromelysin), makes plaque vulnerable and more susceptible to breakage [41]. The presence of tissue factor in the lesion's lipid core is an indispensable element when initiating thrombosis. This factor is produced by activated endothelial cells and macrophages. When plaque rupture takes place, fibrin present in the necrotic core is exposed to circulating platelets, causing platelet adhesion and aggregation; this process induces the thrombosis mechanism [43]. Other molecules, such as the plasminogen activating factor, are also important thrombosis mediators, since they contribute to the degradation of the fibrin mesh [13].

Hypertension

Hypertension is the most common cardiovascular risk factor [44], and contributes to increase global morbidity and mortality rates [45]. An interaction between hypertension and inflammatory response

has been demonstrated; however, it is not yet clear whether inflammation is a cause or a consequence of hypertension.

Among the current hypertension therapies are angiotensin II (Ang II) AT1 receptor (AT1R) antagonists, angiotensin converting enzyme (ACE) inhibitors, diuretics, calcium channel blockers, and beta-adrenergic blockers. Moreover, even when blood pressure targets are achieved, many hypertensive patients still remain at risk of a cardiovascular event, which may be due to the underlying inflammation.

Inflammatory Response Associated With Hypertension

Inflammation is a defensive response aimed to stop the invasion of pathogens or to repair an injury. It is a complex process characterized by vascular alterations, and a cascade of molecular and cellular events, being of great relevance the leukocyte recruitment into the tissue, which involves leukocyte-endothelial interactions mediated by cell adhesion molecules, migration across vascular endothelium, and the release of various inflammatory mediators. Immune cells infiltrated into the target tissue are to eliminate the offending agent and cooperate with the repair process of the lesion site. However, an exacerbated inflammatory response may have harmful effects and contribute to the progression of chronic diseases such as atherosclerosis [7], rheumatoid arthritis [46], and systemic lupus erythematosus [47].

The acute phase protein, C-reactive protein (CRP), is involved in the innate immune responses and has functions that include phagocytosis potencialization [48], stimulation of the synthesis and secretion of proinflammatory cytokines, such as IL-6 IL-1β and TNF-α [49] by monocytes, and expression of the adhesion molecules ICAM-1 and VCAM-1 on endothelial cells [50]. Curiously, hypertensive patients usually have elevated circulating concentrations of CRP [51], and thus CRP is considered an inflammatory marker associated with hypertension (Fig. 14.7).

Immune cells have been implicated in the development of hypertension. Patients with hypertensive nephrosclerosis have higher renal infiltration of CD4+ and CD8+ T cells than normotensive controls [52]. Knock-out mice for RAG-1 gene (deficient in T and B cells) did not develop hypertension in response to either Ang II, treatment with deoxycorticosterone+saline overload (DOCA-salt) [53] or

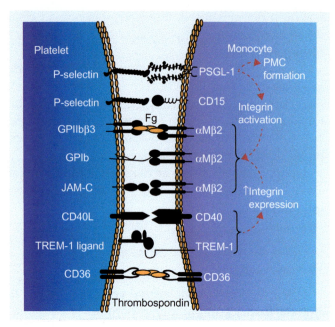

FIG. 14.7 Molecular platelet-monocyte interaction. P-selectin expressed in platelets mediates the initial contact with monocytes via PSGL-1 and CD15. Other interactions occur via CD40L-CD40, TREM-1 ligand-TREM-1 and CD36-CD36 binding via thrombospondin (acronyms, see text).

chronic stress [54]. Furthermore, it is important to note that T cells express AT_1 receptor, ACE, angiotensinogen, renin, and renin receptor, being all these important components of the renin-angiotensin-aldosterone system (RAAS) [55].

Endothelial Dysfunction Induced by Inflammation in Hypertension

Inflammation can modulate the availability of vasoconstrictor and vasodilator mediators, including NO, an important signaling molecule in the regulation of vascular tone. Inflammation has been shown to negatively regulate eNOS activity. In contrast, gene therapy to increase eNOS activity attenuates the expression of leukocyte adhesion molecules and hypercholesterolemia-induced monocyte infiltration [56]. Chronic inhibition of NOS increases endothelium-dependent contraction of rat aorta, induces COX-2 expression and increases endothelium-derived constricting factor (EDCFs) production [57]. The complex mechanism of increased blood pressure in NO-deficient hypertensive rats involves increased sympathetic and renin-angiotensin

system activation, and oxidative stress [58]. Therefore, endothelial dysfunction may exacerbate vascular inflammation, which can, in turn, contribute to the development of hypertension.

Oxidative stress is a major cause of endothelial dysfunction, especially by reducing NO bioavailability. Chronic inflammation can also cause oxidative stress, which has been associated with hypertension [59]. Cells of innate immune system, such as neutrophils and macrophages, produce reactive oxygen species (ROS), such as superoxide and hydrogen peroxide, as microbicide mechanism. The nicotinamide adenine dinucleotide phosphate (NADPH) oxidase enzyme complex is a major source of ROS in both immune cells and in the vasculature. It is interesting to point out that passive transfer of T cells deficient in NADPH oxidase not only resulted in lower superoxide production but also attenuated Ang II-induced hypertension.

Proinflammatory Effects of Aldosterone

Aldosterone is a steroid hormone, synthesized by the adrenal cortex and secreted into blood. In the kidney, aldosterone enters main distal tubule cells and binds to its cytoplasmic mineralocorticoid receptor (MR), modulating renal tubular sodium reabsorption. An increase in sodium reabsorption promotes increased osmolarity, stimulating thirst. Increased water intake increases blood volume and, consequently, blood pressure.

Until recently, it was believed that aldosterone actions were restricted to kidney. However, other tissues associated with control of arterial pressure, including brain [60], vessels [61] and heart [62], were characterized as aldosterone targets. Some proinflammatory effects of aldosterone have been reported. For example, exogenous administration of aldosterone in animals has led to increased levels of ICAM-1, MCP-1 and TNF-α in coronary arteries [63], as well as vascular infiltration of macrophages and lymphocytes [64]. In the heart, vascular aldosterone increased the expression of ICAM-1, MCP-1, osteopontin and COX-2, effects that can be blocked by the MR antagonist eplerenone [62]. In addition, there is an association between inflammation and aldosterone in essential hypertensive patients, where high plasma aldosterone levels have been correlated with elevated circulating levels of CRP and leucocytes [65].

Platelet-Monocyte Complexes in Cardiovascular Diseases

In addition to the known functions of platelets in the control of coagulation and healing, these cells may contribute to endothelial integrity and participate of the inflammatory processes. Under resting conditions, the nonactivated functional endothelium prevents platelet adhesion to the vessel wall due to its antithrombotic properties, involving the release of substances that inhibit platelet activation, such as NO and cyclo-oxigenase-2-derived prostacyclin [66]. However, under inflammatory conditions, the endothelial phenotype can be changed to a prothrombotic profile, by stimulating the release of platelet activating factors, such as ADP and von Willebrand factor (vWF), and modulate the expression of tissue factors and adhesion molecules [67]. The adhesion of platelets is further stimulated when the ECM proteins, collagen and vWF -potent platelet ligands, are exposed onto an injured vessel wall. Upon platelet activation, a rapid platelet adhesion to the ECM proteins is the primary event in thrombus formation.

Platelet activation also leads to increased circulating platelet-leukocyte aggregates. Platelet-monocyte complexes have been described in different clinical conditions such as peripheral vascular disease, hypertension [68], acute and stable coronary syndromes [69], stroke [70], and diabetes; [71] increases in the levels of monocyte-platelet complex have been used as an early marker of acute myocardial infarction [72]. On the other hand, high intake of omega-3 unsaturated fatty acids leads to decreased number of activated platelets and platelet-monocyte complexes [73]. However, the presence of these complexes is not only a sensitive marker of in vivo platelet activation and cardiovascular disease, but is also considered as a cardiovascular risk factor [74]. The importance of activated platelets and platelet-monocyte complexes in vascular diseases is reinforced by several studies that show that inhibition of platelet adhesion to monocytes, mediated by blockade of platelet P-selectin binding to the P-selectin glycoprotein ligand-1 (PSGL-1) constitutively expressed in monocytes, reduces inflammation. The complex PSGL1-P-selectin induces activation of integrin in monocytes, thus increasing molecular interactions, such as CD40L-CD40, TREM1 ligand-TREM1, and CD36-CD36 binding via thrombospondin (Fig. 14.7).

References

[1] Rodrigues SF, Granger DN. Blood cells and endothelial barrier function. Tissue Barriers 2015;3(1–2)e978720.

[2] Telen MJ. Cellular adhesion and the endothelium: E-selectin, L-selectin, and pan-selectin inhibitors. Hematol Oncol Clin North Am 2014;28:341–54.

[3] Kutlar A, Embury SH. Cellular adhesion and the endothelium: P-selectin. Hematol Oncol Clin North Am 2014;28:323–39.

[4] Alberts B, Johnson A, Lewis J, et al. Integrins. In: Molecular biology of the cell. 4th ed. New York: Garland Science; 2002.

[5] van Buul JD, Hordijk PL. Endothelial signalling by Ig-like cell adhesion molecules. Transfus Clin Biol 2008;15(1–2):3–6.

[6] Gimbrone Jr. MA, Garcia-Cardena G. Vascular endothelium, hemodynamics, and the pathobiology of atherosclerosis. Cardiovasc Pathol 2013;22:9–15.

[7] Lusis AJ. Atherosclerosis. Nature 2000;407:233–41.

[8] Williams KJ, Tabas I. The response-to-retention hypothesis of early atherogenesis. Arterioscler Thromb Vasc Biol 1995;15:551–61.

[9] Zand T, Hoffman AH, Savilonis BJ, et al. Lipid deposition in rat aortas with intraluminal hemispherical plug stenosis. A morphological and biophysical study. Am J Pathol 1999;155:85–92.

[10] Goldstein JL, Brown MS. Binding sites on macrophages that mediate uptake and degradation of aceylated low density lipoprotein, producing massive cholesterol deposition. Proc Natl Acad Sci U S A 1979;76:3337.

[11] Ross R. The pathogenesis of atherosclerosis: a perspective for the 1990s. Nature 1993;362:8019.

[12] Tamminen M, Mottino G, Qiao H, et al. Ultrastructure of early lipid accumulation in apoE-deficient mice. Arterioscler Thromb Vasc Biol 1999;19:84753.

[13] Lusis AJ. Atherosclerosis. Nature 2000;407:23341.

[14] Cybulsky M, Gimbrone MA. Endothelial expression of a mononuclear leukocyte adhesion molecule during atherosclerosis. Science 1991;251:78891.

[15] Colling RG. P-selectin or intercellular adhesion molecule (ICAM1) deficiency substantially protects against atherosclerosis in apolipoprotein E-deficient mice. J Exp Med 2000;191:18994.

[16] Boring L, Gosling J, Cleary M, et al. Decreased lesion formation in CCR2/mice reveals a role for chemokines in the initiation of atherosclerosis. Nature 1998;394:8947.

[17] Bazan JF, Bacon KB, Hardiman G, et al. A new class of membrane bound chemokine with a CX3C motif. Nature 1997;385:6404.

[18] Daugherty A, Webb NR, Rateri DL, et al. Thematic review series: the immune system and atherogenesis. Cytokine regulation of macrophage functions in atherosclerosis. J Lipid Res 2005;46:181222.

[19] Zernecke A, Weber C. Inflammatory mediators in atherosclerotic vascular disease. Basic Res Cardiol 2005;100:93101.

[20] Heistad DD. Oxidative stress and vascular disease. Arterioscler Thromb Vasc Biol 2006;26;689–95.

[21] Manea A, Raicu M, Simionescu M. Expression of functionally phagocyte-type NAD(P)H oxidase in pericytes: effect of angioten-sin II and high glucose. Biol Cell 2005;97:723–34.

[22] Henning B, Chow CK. Lipid peroxidation and endothelial cell injury: implications in atherosclerosis. Free Radic Biol Med 1988;4:9–106.

[23] Rubanyi GM, Vanhoutte PM. Oxygen-derived free radicals, endothelium, and responsiveness of vascular smooth muscle cells. Am J Physiol 1986;250:H815–21.

[24] Huo Y, Ley KF. Role of platelets in the development of atherosclerosis. Trends Cardiovasc Med 2004;14:18–22.

[25] von Hundelshausen P, Weber KS, Huo Y, et al. RANTES deposition by platelets triggers monocyte arrest on inflamed and atherosclerotic endothelium. Circulation 2001;103:1772–7.

[26] Simionescu M, Popov D, Sima A, et al. Pathobiochemistry of combined diabetes and atherosclerosis studied on a novel model.

The hyperlipemic-hyperglycemic hamster. Am J Pathol 1996;148:997–1014.

[27] Simionescu M. Implications of early structural-functional changes in the endothelium for vascular disease. Arterioscler Thromb Vasc Biol 2007;27(2):266–74.

[28] Rosenfeld ME, Yla-Herttuala S, Lipton BA, et al. Macrophage colony-stimulating factor mRNA and protein in atherosclerotic lesions of rabbits and humans. Am J Pathol 1992;140:291300.

[29] Yuri VB. Monocyte recruitment and foam cell formation in atherosclerosis. Micron 2006;37:20822.

[30] Chi Z, Melendez AJ. Role of cell adhesion molecules and immune-cell migration in the initiation, onset and development of atherosclerosis. Cell Adh Migr 2007;1(4):171–5.

[31] Quehenberger O. Molecular mechanisms regulating monocyte recruitment in atherosclerosis. J Lipid Res 2005;46:158290.

[32] Goran KH, Peter L. The immune response in atherosclerosis: a doubleedged sword. Nat Rev Immunol 2006;6:50819.

[33] Miller YI, Viriyakosol S, Binder CJ, et al. Minimally modified LDL binds to CD 14, induces macrophage spreading via TLR4/MD2 and inhibits phagocytosis of apoptotic cells. J Biol Chem 2003;278:15618.

[34] Itabe H. Oxidized low density lipoproteins: what is understood and what remains to be clarified. Biol Pharm Bull 2003;26:19.

[35] Ivandic B, Castellani LW, Wang XP, et al. Role of group II secretory phospholipase A2 in atherosclerosis: 1. Increased atherogenesis and altered lipoproteins in transgenic mice expressing group IIa phospholipase A2. Arterioscler Thromb Vasc Biol 1999;19:128490.

[36] Marathe S, Kuriakose G, Williams KJ, et al. Sphingomyelinase, an enzyme implicated in atherogenesis, is present in atherosclerotic lesions and binds to specific components of the subendothelial extracellular matrix. Arterioscler Thromb Vase Biol 1999;19:264858.

[37] Accad M, Smith SJ, Newland DL, et al. Massive xanthomatosis and altered composition of atherosclerotic lesions in hyperlipidemic mice lacking acyl CoA: cholesterol acyltransferase 1. J Clin Invest 2000;105:7119.

[38] Moreno PR, Falk E, Palacios IF, et al. Macrophage infiltration in acute coronary syndromes: implications for plaque rupture. Circulation 1994;90:7758.

[39] Weber C. Platelets and chemokines in atherosclerosis, partners in crime. Circ Res 2005;96:6126.

[40] Bobryshev YV, Lord RS. A S100 positive cells in human arterial intima and in atherosclerotic lesions. Cardiovas Res 1995;29:68996.

[41] Friesel R, Komoriya A, Maciag T. Inhibition of endothelial cell proliferation by INFg. J Cell Biol 1987;104:68996.

[42] Libby P. Changing concepts of atherosclerosis. J Intern Med 1999;247:34958.

[43] AitOufella H, Salomon BL, Potteaux S, et al. Natural regulatory T cells control the development of atherosclerosis in mice. Nat Med 2006;12:17880.

[44] Mach F, Schönbeck U, Bonnefoy JY, et al. Activation of monocyte/macrophage functions related to acute atheroma complication by ligation of CD40: induction of collagenases, stromelysin, and tissue factor. Circulation 1997;96:3969.

[45] Ventura HO, Taler SJ, Strobeck JE. Hypertension as a hemodynamic disease: the role of impedance cardiography in diagnostic, prognostic, and therapeutic decision making. Am J Hypertens 2005;18(2):26S–43S.

[46] Kearney PM, Whelton M, Reynolds K, et al. Global burden of hypertension: analysis of worldwide data. Lancet 2005;365 (9455):217–23.

[47] Sweeney SE, Firestein GS. Rheumatoid arthritis: regulation of synovial inflammation. Int J Biochem Cell Biol 2004;36(3):372–8.

[48] Asanuma YE. Accelerated atherosclerosis and inflammation in systemic lupus erythematosus. Jpn J Clin Immunol 2012;35 (6):470–80.

[49] Mortensen RF, Zhong W. Regulation of phagocytic leukocyte activities by C-reactive protein. J Leukoc Biol 2000;67(4):495–500.

[50] Ballou SP, Lozanski G. Induction of inflammatory cytokine release from cultured humanmonocytes by C-reactive protein. Cytokine 1992;4(5):361–8.

[51] Pasceri V, Willerson JT, Yeh ET. Direct proinflammatory effect of C-reactive protein on human endothelial cells. Circulation 2000;102(18):2165–8.

[52] Dinh QN, Drummond GR, Sobey CG, et al. Roles of inflammation, oxidative stress, and vascular dysfunction in hypertension. Biomed Res Int 2014;2014:406960.

[53] Youn J, Yu HT, Lim BJ, et al. Immunosenescent CD8+ T cells and C-X-C cemokine receptor type 3 chemokines are increased in human hypertension. Hypertension 2013;62(1):126–33.

[54] Guzik TJ, Hoch NE, Brown KA, et al. Role of the T cell in the genesis of angiotensin II-induced hypertension and vascular dysfunction. J Exp Med 2007;204(10):2449–60.

[55] Marvar PJ, Vinh A, Thabet S, et al. T lymphocytes and vascular inflammation contribute to stress-dependent hypertension. Biol Psychiatry 2012;71(9):774–82.

[56] Jurewicz M, McDermott DH, Sechler JM, et al. Human T and natural killer cells possess a functional renin-angiotensin system: further mechanisms of angiotensin II-induced inflammation. J Am Soc Nephrol 2007;18(4):1093–102.

[57] Qian H, Neplioueva V, Shetty GA, et al. Nitric oxide synthase gene therapy rapidly reduces adhesion molecule expression and inflammatory cell infiltration in carotid arteries of cholesterol-fed rabbits. Circulation 1999;99(23):2979–82.

[58] Qu C, Leung SWS, Vanhoutte PM, et al. Chronic inhibition of nitric-oxide synthase potentiates endothelium-dependent contractions in the rat aorta by augmenting the expression of cyclooxygenase-2. J Pharmacol Exp Ther 2010;334(2):373–80.

[59] Tomita H, Egashira K, Kubo-Inoue M, et al. Inhibition of NO synthesis induces inflammatory changes and monocyte chemoattractant protein-1 expression in rat hearts and vessels. Arterioscler Thromb Vasc Biol 1998;18(9):1456–64.

[60] Crowley SD. The cooperative roles of inflammation and oxidative stress in the pathogenesis of hypertension. Antioxid Redox Signal 2014;20(1):102–20.

[61] Zhang Z, Yu Y, Kang Y, et al. Aldosterone acts centrally to increase brain renin-angiotensin system activity and oxidative stress in normal rats. Am J Physiol Heart Circ Physiol 2008;294(2): H1067–74.

[62] Leibovitz E, Ebrahimian T, Paradis P, et al. Aldosterone induces arterial stiffness in absence of oxidative stress and endothelial dysfunction. J Hypertens 2009;27(11):2192–200.

[63] Rocha R, Rudolph AE, Frierdich GE, et al. Aldosterone induces a vascular inflammatory phenotype in the rat heart. Am J Physiol Heart Circ Physiol 2002;283(5):H1802–10.

[64] Sun Y, Zhang J, Lu L, et al. Aldosterone-induced inflammation in the rat heart: role of oxidative stress. Am J Pathol 2002;161 (5):1773–81.

[65] Kasal DA, Barhoumi T, Li MW, et al. T regulatory lymphocytes prevent aldosterone-induced vascular injury. Hypertension 2012;59 (2):324–30.

[66] Tzamou V, Vyssoulis G, Karpanou E, et al. Aldosterone levels and inflammatory stimulation in essential hypertensive patients. J Human Hypertens 2013;27(9):535–8.

[67] Jin RC, Voetsch B, Loscalzo J. Endogenous mechanisms of inhibition of platelet function. Microcirculation 2005;12:247–58.

[68] Van Hinsbergh VW. The endothelium: vascular control of hemostasis. Eur J Obstet Gynecol Reprod Biol 2001;95:198–201.

[69] Nomura S, Kanazawa S, Fukuhara S. Effects of efonidipine on platelet and monocyte activation markers in hypertensive patients with and without type 2 diabetes mellitus. J Hum Hypertens 2002;16:539–47.

[70] Neumann FJ, Marx N, Gawaz M, et al. Induction of cytokine expression in leukocytes by binding of thrombin-stimulated platelets. Circulation 1997;95:2387–94.

[71] McCabe DJ, Harrison P, Mackie IJ, et al. Platelet degranulation and monocyte-platelet complex formation are increased in the acute and convalescent phases after ischemic stroke or transient ischemic attack. Br J Haematol 2004;125:777–87.

[72] Elalamy I, Chakroun T, Gerotziafas GT, et al. Circulating platelet-leukocyte aggregates: a marker of micro vascular injury in diabetic patients. Thromb Res 2008;121:843–8.

[73] Li G, Sanders JM, Bevard MH, et al. CD40 ligand promotes Mac-1 expression, leukocyte recruitment, and neointima formation after vascular injury. Am J Pathol 2008;172:1141–52.

[74] Atarashi K, Hirata T, Matsumoto M, et al. Rolling of Th1 cells via P-selectin glycoprotein ligand-1 stimulates LFA-1-mediated cell binding to ICAM-1. J Immunol 2005;174:1424–32.

Further Reading

[1] Tabas I, García-Cardeña G, Owens GK. Recent insights into the cellular biology of atherosclerosis. J Cell Biol 2015;209(1):13–22.

[2] Huo Y, Schober A, Forlow SB, et al. Circulating activated platelets exacerbate atherosclerosis in mice deficient in apolipoprotein E. Nat Med 2003;9:61–7.

[3] Smith JD, Trogan E, Ginsberg M, et al. Decreased atherosclerosis in mice deficient in both macrophage colonystimulating factor (op) and apolipoprotein E. Proc Natl Acad Sci U S A 1995;92:82648.

[4] Stumpf C, Jukic J, Yilmaz A, et al. Elevated VEGF-plasma levels in young patients with mild essential hypertension. Eur J Clin Invest 2009;39(1):31–6.

[5] Lippi G, Montagnana M, Salvagno GL, et al. Risk stratification of patients with acute myocardial infarction by quantification of circulating monocyte-platelet aggregates. Int J Cardiol 2007;115:101–2.

[6] van Gils JM, Zwaginga JJ, Hordijk PL. Molecular and functional interactions among monocytes, platelets, and endothelial cells and their relevance for cardiovascular diseases. J Leukoc Biol 2009;85 (2):195–204.

15

Endothelium and the Renin-Angiotensin System

Walkyria Oliveira Sampaio and Robson Augusto Souza dos Santos

INTRODUCTION

Over the years, most publications related to hemodynamic actions of the renin-angiotensin system (RAS) emphasize the vascular actions of angiotensins. The last two decades saw the important expansion of the concept that defined RAS only as a linear formation cascade of angiotensin (Ang II), constituted by angiotensinogen-renin-angiotensin I—angiotensin converting enzyme (ACE)—angiotensin II—AT$_1$ receptor. Identification of biological activity of other angiotensins, the discovery of new receptors and enzymatic pathways, in particular in the angiotensin-(1-7)/ACE2/receptor axis and, more recently, of alamandine and receptor MrgD, led to the recognition that a complex system with intrinsic counter-regulatory characteristics, in which the angiotensins formed, their receptors, and signalling pathways exercise modulatory influence among themselves. Therefore, as shall be discussed herein, there is an important counterpoint between the actions of these axes in modulating vascular tone and endothelial function, since the vascular responses to angiotensin-(1-7) (Ang-(1-7)) are significantly different from the ones induced by angiotensin II (Ang II). Ang II is classically known for its vasoconstrictor, pro-oxidative, proliferative, and thrombotic effects. On the other hand, the axis formed by Ang-(1-7) and alamandine induces vasodilation, inhibits vascular smooth muscle cell proliferation, and reduces the formation of thrombus.

EXPRESSION OF RAS COMPONENTS ON VASCULATURE

The presence of the RAS (Fig. 15.1) in vasculature is well established. Both the circulating and local systems perform important functions in vessels, related mainly to maintaining endothelial functionality, and regulating vascular growth and proliferation [1]. The mapping of the distribution of the main components of the RAS in normal and altered vessels brought the morphological basis to understand the RAS vascular functions on physiology and physiopathology. Vascular ACE is present predominantly in the endothelium and adventitious layer, and in a smaller quantity in smooth muscle cells [2–4]. ACE2, an enzyme homologous to ACE, especially involved in the formation of heptapeptide Ang-(1-7) from Ang II, is expressed in the endothelium and in smooth muscle [5]. AT receptors are expressed in large quantity in smooth muscle cells of large vessel and at a lower proportion in the adventitious layer. High levels of this receptor also occur in smooth muscle cells associated with advanced vascular injuries [6]. Receptor AT$_2$ is the main and practically the only Ang II receptor expressed in fetal aorta, although its expression is reduced significantly after birth and during the maturation process [7]. In contrast to AT$_1$ receptor, AT$_2$ receptor is expressed mainly in the adventitious layer of the human renal artery [8] and seems to mediate important Ang II actions on renal vasculature, since animals with deleted AT$_2$ receptors showed worse

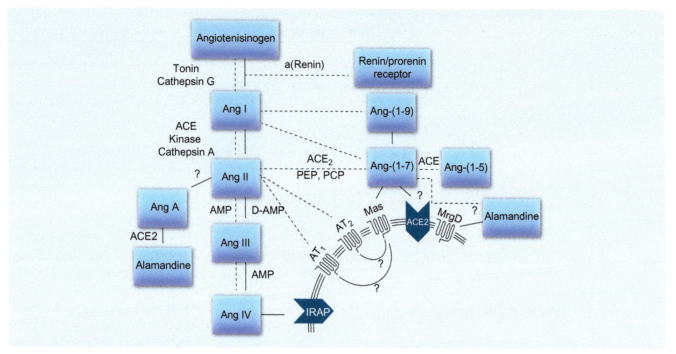

FIG. 15.1   Updated and simplified view of the renin-angiotensin system. *ACE*, angiotensin converting enzyme; *AMP*, aminopeptidase; *Ang*, angiotensin; *AT1*, angiotensin receptor type 1; *AT2*, angiotensin receptor type 2; *D-Amp*, dipeptidyl aminopeptidase I-III; *IRAP*, insulin-regulated aminopeptidase; *Mas*, Mas receptor; *MrgD*, Mas-related G-protein-coupled receptor D; *NEP*, neutral endopeptidase; *PCP*, prolyl carboxypeptidase; *PEP*, prolyl endopeptidase. *From Santos RAS. Angiotensin-(1–7). Hypertension 2014;63(6):1138–47.*

glomerular injury, altered renal function, and albuminuria [8,9]. Such changes are probably secondary in respect to nitric oxide synthesis reduction [10]. Endothelial cells have AT_1 and AT_2 receptors, both in direct contact with circulating Ang II [11]. The MAS receptor is also expressed both in the endothelium and vascular smooth muscle cells, mediating the counter-regulatory actions of Ang-(1-7) [12–14]. Recently, a new peptide that composes the system was discovered, alamandine. The structure of this is very similar to Ang-(1-7), differing only in terms of the presence of alanine residue replacing the aspartate residue in the amino-terminal portion of the peptide. In addition, the same authors described that alamandine is a binder of the MrgD (Mas-related receptor D) receptor. This receptor is expressed in the endothelium and in vascular smooth muscle cells, and mediates vasodilation and antiproliferative effects [15,16].

Vascular Actions of RAS and Cellular Mechanism Involved

Beyond the classic vasoconstrictor effect, which contributes to increase peripheral vascular resistance and maintain normal levels of blood pressure and angiotensin II if inappropriately increased, it develops/maintains high blood pressure levels. In addition, a great deal of evidence demonstrates the role of Ang II in endothelial injury. Ang II, by means of AT_1 receptor, is a potent activator of oxidative and inflammatory cascades, main mediators of endothelial dysfunction and lipid peroxidation. Simultaneously, the actions of Ang II on vascular smooth muscle cells lead to proliferation, migration, and phenotypic changes, arising from production of growth factors and extracellular matrix [17,18]. Excessive production of reactive oxygen and nitrogen species (RONS) is the primordial precursor stage of several deleterious vascular effects of Ang II, considering that these species, besides injuring the cell membrane, act as intracellular and intercellular signaling molecules, activating proliferation and inflammation related cascades. RONS have several origins: mitochondria, xanthine oxidase, NADPH oxidase, lipoxygenase, and uncoupled nitric oxide synthase (NOS), with NADPH oxidase being the main source of ROS in the vascular wall [18,19]. Ang II is considered the main stimulus that activates NADPH oxidase in vessels and, therefore, increases RONS production. The signaling pathways by means of which Ang II stimulates NADPH oxidase involve PLD, PKC, PLA2, PI3K, and thiol-oxidoreductases. In vascular smooth muscle cells, tyrosine kinase c-Src regulates the activity of Ang II-induced NADPH oxidase, stimulating phosphorylation and translocation of subunit p47phox. Additionally, Src is essential for the effects of Ang II in the synthesis of NADPH subunits, stimulating increased expression of gp91phox, p22phox, p47phox, and p67phox. Furthermore, stimulating NADPH oxidase, Ang II also activates other RONS generation processes. Among these, we can emphasize NOS uncoupling, characterized by arginine and tetrahydrobiopterin deficiency, which often leads NOS to generate O_2. NOS uncoupling occurs in several pathological conditions with increased SRA activity (hypertension, atherosclerosis, and diabetes) [18–22]. The enzymes of the mitochondrial chain, stimulated by Ang II, may also contribute to increased O_2 production in the vasculature. Intracellular targets of RONS encompass transcription factors (NFkB, Ap-1, HIF-1), tyrosine kinases (PDGFR, EGRF, Src, JAK2, Pyk2, and Akt), MAPKs (p38MAPK), ion channels, and phospholipases. In addition, RONS can modify the activity of enzymes as phosphatases, inactivating them through oxidation. Thus, increased RONS production feedbacks deleterious mechanisms stimulated by Ang II [21,22].

Actions of Ang II via AT_2 receptors are, in part, different from those mediated by AT_1 receptor. The effects of stimulation of AT_2 receptors on blood pressure are still controversial; however, a recent study has described that stimulation of the AT_2 receptor resulted in vasodilation response in several vascular beds, such as mesenteric, renal, coronary, cerebral, cutaneous, and uterine artery of rodents [23]. The stimulated AT_2 receptor activates at least three different classic signaling pathways: cGMP/nitric oxide, phosphatases, and phospholipase A2. This receptor also can mediate antiinflammatory actions, by inhibiting NF-κB, antifibrotic effects, reducing the expression of metalloproteinase, and antiproliferative effects, by means of phosphatase SHP-1 and the protein associated with the receptor (ATIP) [24]. However, the role of AT_2 receptors in apoptosis is controversial, since it mediates proapoptotic or antiapoptotic actions. It has been demonstrated that this receptor can mediate an inactivation effect of the MAP kinase regulated by extracellular

signal-regulated kinases (ERK) 1/2, which results in dephosphorylation of Bcl-2 and proapoptotic effects induced by bax. On the other hand, stimulation of the AT_2 receptor also stimulates transcription of GATA-6 by activation of ERK1/2 and c-Jun N-terminal kinase (JNK). GATA-6 activates the promoter FasL, and consequently induces apoptosis via caspase [24].

In turn, the vascular responses of Ang-(1-7) are significantly different from those induced by angiotensin II (Ang II). In blood vessels. The Ang-(1-7) axis has vasodilatory, antiproliferative, and antithrombotic actions [1]. Blood vessels of several species, including humans, are able to generate Ang-(1-7) and express the MAS receptor, both in endothelial cells and vascular smooth muscle cells [25–27]. Vasodilation is the best characterized effect of Ang-(1-7). As examples of wide vasodilation action, Ang-(1-7) produces relaxation of the aortic annulus of Sprague-Dawley rats and mren-2 rats [28], coronary arteries of pigs [29,30], middle cerebral artery of dogs [31], pial arterioles [32], systemic vasculature of cats [33], afferent arterioles of rabbits [34], and mesenteric microvessels of rats [35,36]. Besides that, normotensive animals and hypertensive animals, Ang-(1-7) maximizes the vasodilation effect of bradykinin in several vascular beds, including coronaries [37] of dogs and rats [29,36], and renal and mesenteric arteries of rats [35,38].

The selectivity of the vasodilatory effect of Ang-(1-7)/MAS contrasts with the more generalized vasoconstriction described for Ang II/AT_1. In normotensive rats, Ang-(1-7) produces remarkable changes on the regional blood flow, increasing mesenteric vascular conductance, as well as cerebral, cutaneous, and renal territory conductance. In addition, Ang-(1-7) increased cardiac output (CO) by 30%, and simultaneously, reduces total peripheral resistance (TPR) by 26%. These opposite changes in total peripheral resistance and cardiac output justify the absence of substantial changes in blood pressure during administration of Ang-(1-7) [39]. Likewise, transgenic rats that have increased circulating levels of Ang-(1-7)-(TGR L-3292) showed accentuated alteration in regional blood flow, resulting in increased vascular conductance in the kidneys, lungs, adrenal glands, spleen, brain, testicles, and brown adipose tissue [40]. In contrast, mice with deleted MAS receptor showed high vascular resistance in several territories, such as the kidney, lungs, adrenal glands, mesenteric, spleen, and brown tissue. In this model, parallel increase of total peripheral resistance and reduction of cardiac index were also observed [41]. In addition to its actions on vascular tone, Ang-(1-7) inhibits vascular growth and has protective effects against thrombosis [26,42]. It is important to emphasize that this antiproliferative potential was equally observed in cardiac fibroblasts [43] and tumor cells [44,45].

The main underlying effect of the vascular effects of Ang-(1-7) consists of nitric oxide (NO) release. Human endothelial cells express the receptor coupled to protein G, MAS, by means of which Ang-(1-7) mediates the activation of endothelial nitric oxide synthase (eNOS) and production of NO. NO release stimulated by Ang-(1-7) depends on the phosphatidylinositol-3-phosphate kinase (PI3K)/AKT pathway. With that, Ang-(1-7) phosphorylates Akt into S473 residue that, in turn, phosphorylates eNOS synthase. The effect of Ang-(1-7) on NO, involves regulated coordination of eNOS phosphorilation of the amino acid Ser 1177 and dephosphorilation of Thr 495 [14]. In resting conditions, eNOS is phosphorylated into residue Thr495 and only weakly phosphorylated into Ser1177. Therefore, the main mechanism of action of Ang-(1-7) on eNOS is phosphorylation of Sr1177, via Akt and simultaneous dephosphorylation of Thr495. These posttranslational effects stimulate the catalytic activity of the enzyme and NO production. The evidence of participation of the MAS receptor was obtained by treating cells with MAS antagonist, A-779, which blocked the effects of Ang-(1-7) on the PI3K/Akt/eNOS pathway and NO production [14]. Fig. 15.2 summarize the main signaling cascades that mediates the modulatory actions of the SRA axes. Alamandine also has vasodilatory action on the isolated aorta, including in vessels removed from mice with deleted MAS receptor but blocked by the Mas/MrgD receptor antagonist D-Pro^7-Ang-(1-7), showing that this actions results from the interaction between the peptide and its specific receptor, MrgD. The vasodilatory action of alamandine is mediated by the release of nitric oxide [15].

However, regardless of the stimulating effects of Ang-(1-7) on PI3K/Akt—known as an important intracellular pathway that produces antiapoptotic and proliferative effects—it has recently been demonstrated that Ang-(1-7) is also able to activate other more peripheral intracellular components, such as transcription factor FOXO1, a potent tumor suppressor [46]. Such paradoxical findings indicate that the modulatory effect of Ang-(1-7) is complex and probably involves multiple effectors. These, by means of refined intracellular mechanisms, are able to coordinate negative feedbacks, resulting in specific effects demonstrated in different cells. The examples of specificity include the different actions of Ang-(1-7) on the endothelium (antiapoptotic) and vascular smooth muscle cells (antiproliferative).

SRA ON ENDOTHELIAL DYSFUNCTION, ATHEROSCLEROSIS, AND THROMBOSIS

In addition to being one of the most important peptides in arterial hypertension genesis, Ang II also has a direct participation in endothelial dysfunction that triggers atherosclerotic disease. Atherosclerosis is a chronic inflammatory process, orchestrated by systemic (liver, kidney, heart, lungs, adipose tissue, adrenal glands, pancreas, and sexual glands) and local (leukocyte, endothelium, and vascular smooth muscle cells) production of inflammatory mediators [47]. Ang II also participates in atherosclerosis pathogenesis as a potent proinflammatory agent. In vessels, Ang II regulates the expression of adhesion molecules (VCAM-1, ICAM-1, and p-selectin), stimulates the classic inflammatory mediators, and regulates the activation of complement system proteins. Effectively, the ACE/Ang II/AT$_1$ axis shares intracellular signaling cascades with other cytokines (TNF-α and IL-1b), such as the activation of RONS and pathways PTK, PKC, MAPK, and AP-1 [47,48]. As a consequence, some authors classify Ang II as a real cytokine. Furthermore, there is evidence of participation of Ang II in increased vascular permeability, leukocyte infiltration, and tissue remodeling, events that represent the initial stages of the inflammatory process. The effects of Ang II include increased expression of cytokines (IL-6, IL-1, IL-18Ra), chemokines (MCP-1), leukocyte adhesion molecules (selectins P, E, and L, integrins α1 and β2, VCAM, and ICAM) and LDLox receptors in endothelial cells, increased oxidation of LDL-c in macrophages, and lipid peroxidation. Furthermore, Ang II favors intra-plaque recruitment of monocytes and lymphocytes, and directly increases the expression TNF-α and COX-2 in atherosclerotic arteries [47,48].

The main mediator of the proinflammatory effects of Ang II is activation of the NF-κB protein complex, which acts as transcription factor. The inactive

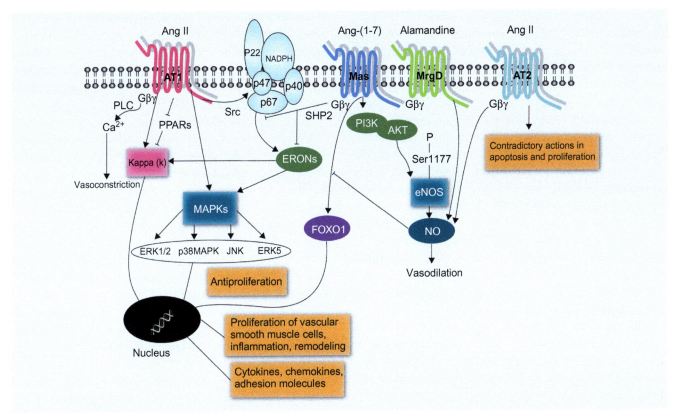

FIG. 15.2 Main signaling pathways involved in the mediation of vascular effects of Ang II and Ang-(1-7). The signaling pathways that mediate the actions of the Ang II/AT1 axis include, among others: activation of NADPH oxidase, via Src, increasing the production of RONS; activation of MAPKs (ERK1/2, p28MAPK, JNK, ERK5), leading to remodeling and fibrosis processes; activation of NF-κB, main mediator of inflammatory processes; inhibition of the PPAR antiinflammatory pathway; and increased calcium concentration. On the other hand, the pathways activated by the Ang-(1-7)/Mas axis lead to increased production of nitric oxide (NO), via PI3K/Akt; activation for FOXO1, inhibition of MAPKs (ERK1/2, p38 MAPK); and inhibition of NADPH oxidase, by means of Src dephosphorylation, probably by phosphatase SHP2. Alamandine stimulates NO production by means of receptor MrgD. Regardless of stimulating NO production, the actions of Ang II mediated by the AT2 receptor in apoptosis and proliferation are controversial. *PLC*, phospholipase C; *PPAR*, peroxisome proliferator-activated receptor; *Src*, tyrosine kinase; *RONS*, reactive oxigen and nitrogen species; *eNOS*, endothelial nitric oxide synthase; *SHP2*, tyrosine phosphatase homologous to Src; *PI3K*, phosphatidylinositol-3-kinase; *Akt*, serine-threonine kinase; *FOXO1*, forkhead box protein O1; *NF-κB*, nuclear factor kappa B; *MAPK*, mitogen-activated protein kinase; *ERK1*, extracellular signal-regulated kinase; *JNK*, janus kinase. *From Walkyria Sampaio and Robson Augusto Souza dos Santos.*

cytoplasmic form of NF-κB results from the connection with inhibiting kB proteins (IkBs). Ang II stimulates, by means of the AT_1 and AT_2 receptors, the phosphorylation of subunit p65 of the NF-κB complex and its nuclear translocation, leading to synthesis of inflammatory molecules and angiotensinogen. Activation of NF-κB also inhibits PPARs, which represent an important modulatory pathway due to attenuated expression of proinflammatory genes [49]. Ang II also stimulates inflammatory mediators involved in the coagulation cascade. In particular, the ACE/Ang II/AT_1 axis inhibits fibrinolysis and

increases thrombosis, resulting in high production of PAI-1 in endothelial cells and smooth muscle cells [50].

Ang II also acts on platelets, stimulating the release of thromboxane A2 and the growth factor derived from such cells [50]. There is evidence that the inflammatory effects of Ang II on the vasculature may participates directly in the endothelial dysfunction that involve the mediation of the AT_2 receptor [49]. Notwithstanding, several studies showed the effects of Ang II in vascular remodeling. In vascular smooth muscle cells, Ang II, via AT_1, activates tyrosine-kinases (Src, FAK, PI3K, and JAK2), which, in turn, regulate cellular cascades such as MEKs/MAPKs (ERK1/2, p38MAPK, JNK, and ERK5) and growth factors (bFGF, PDGF, VEGF, TGF-beta) that result in proliferation, apoptosis, inflammation, and cell migration [19–21].

Additionally, Ang II stimulates collagen synthesis and changes the activity of metalloproteinases, leading to deposition of ECM and vascular stiffening. The oxidative and inflammatory events triggered by Ang II are directly linked to cellular remodeling, since they activate molecules involved and act as process feedback pathways [20–47]. Among the actions of Ang II that reinforce vascular dysfunction, it is worth mentioning stimulation of expression of other vasoactive substances, such as endothelin and aldosterone, and increased sympathetic activity [18,19].

On the other hand, actions of Ang-(1-7) via MAS counter-regulate the vascular growth process, by reducing ERK1/2 activity and stimulating production of prostacyclin in vascular smooth muscle cells [51]. It has also been shown that the neointimal formation was reduced after stents were implanted in rats treated with Ang-(1-7) [52]. Similarly, it has been recently shown that alamandine receptor MrgD is expressed in atherosclerotic plaques and seems to have vascular protective effects, by means of vasodilation and antiproliferative actions [16].

In human endothelial cells, Ang-(1-7) also counter-regulates Ang II signaling, by reducing phosphorylation of c-Src and ERK1/2, as well as activation of NAD (P) H oxidase by Ang II. This modulatory effect is mediated by activation of phosphatase SHP-2. A-779, antagonist of MAS receptor, inhibits these actions, therefore indicating that they are mediated by this receptor [53]. According to this concept, compromised endothelial function found in two different genetic lines with deleted MAS (C57BL/6 and FVB/N), indicates a crucial role of this receptor on the endothelium [54,55].

In the FVB/N line, endothelial dysfunction is associated with increased blood pressure [54], while C57BL/6 mice did not show such pressure alteration [55]. Besides that, animals with deleted MAS receptor showed aggravated hypertension resulting from the Goldblatt model (2 kidneys-1 clip) [56]. On the other hand, short-term infusion of Ang-(1-7) promoted improvement of the endothelial function in normotensive rats, increasing significantly the hypotensive effect of intra-arterial administration

of acetylcholine [57]. Oral administration of alamandine, included in HP-beta cyclodextrin, reduced blood pressure in spontaneously hypertensive rats, indicating that the vasodilatory action of the peptide is broad and probably decreases the total peripheral vascular resistance [15].

The Ang-(1-7)/MAS axis also has been acknowledged by its antithrombotic and antiproliferative effects. For that matter, mice with deleted receptor showed increased thrombus size and reduced bleeding time [58]. Opposite changes were observed with the active formulation of Ang-(1-7) included in HP-cyclodextrin, administered orally [59]. The antithrombotic activity also was found in mice with deleted bradykinin receptors B2 (BDKRB2 − / −), in which the antagonist of receptor MAS A-779, reduced arterial thrombus formation time as well as tail bleeding [59]. The mechanisms involved in these antithrombogenic actions include NO release and increased prostacyclin in platelets [60].

The antiproliferative action of Ang-(1-7) on the vasculature is observed in vascular smooth muscle cells (VSMCs), in which this peptide reduced the activity of MAP kinase ERK1/2 induced by Ang II [51]. The findings also described increased production of prostacyclin (PGI2) in smooth muscle cells treated with Ang-(1-7). Furthermore, in rats with vascular calcification, Ang-(1-7) reverted, in VSMCs, the reduction in markers such as (SM) alpha-actin, SM22alpha, calpolin, and smoothelin, and delayed the osteogenic transition of these cells [61].

The actions of Ang-(1-7) on the human vasculature still require additional investigation to confirm the potent vasodilation effect described in rodents. Although the first investigations focusing on human vessels showed some controversy, the contrasting results obtained in humans may be justified by methodological discrepancies, racial differences, or differences in sensitivity to Ang-(1-7) due to different vascular territories. Infusion of Ang-(1-7) in patients chronically treated with ACE inhibitors did not result in altered forearm blood flow, while infusion of bradykinin produced vasodilation [62]. This observation was used as evidence to justify the lack of relevance of Ang-(1-7) on ACE inhibitors hemodynamic effects. However, such conclusion did not consider that ACE inhibition significantly increases the circulating levels of Ang-(1-7) [63]. Therefore, it is evident that it would have been more appropriate to use the antagonist of the MAS receptor instead of Ang-(1-7) to evaluate the effect of the peptide on the experimental

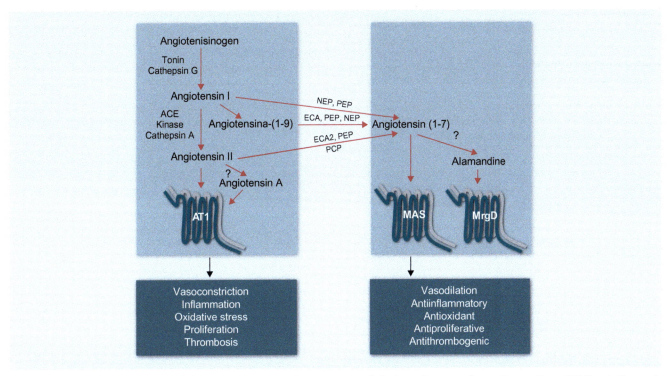

FIG. 15.3 Modulatory axes of the renin-angiotensin system and their main vascular actions. *From Walkyria Sampaio and Robson Augusto Souza dos Santos.*

model mentioned. However, authors such as Sasaki et al. [64] noted dose-dependent vasodilation in the forearm circulation of normotensive patients, as well as essential hypertension. Similarly, Ueda et al. [65] reported that Ang-(1-7) maximizes, in dose-dependent fashion, the vasodilation induced by bradykinin in forearm resistance vessels of normotensive individuals. This finding represents an important example of the classic bradykinin maximizing activity of Ang-(1-7) described in several models [29,35,36].

Ang-(1-7) also attenuated the vasoconstrictor effect of Ang II in the forearm of normotensive patients, as well as thoracic artery in vitro [64,66]. Another study described that Ang-(1-7) antagonizes Ang II in renal vessels in vitro [67]. Van Twist et al. [68] observed significant dose-dependent increase of blood flow to the kidney during intra-renal infusion of Ang-(1-7) in hypertensive patients. Curiously, similar studies showed similar results in renal vessels of rats [69]. This effect was attenuated in patients using sodium restrictive diet, probably due to the fact that this restriction increases circulating Ang-(1-7) and, consequently, its endogenous action.

In short, the RAS, either circulating or local, importantly influences vascular function, contributing to maintain normal blood pressure levels. However, via Ang II, when inappropriately high, SRA contributes significantly with the endothelial dysfunction that follows arterial hypertension and other cardiovascular diseases. On the other hand, Ang-(1-7), and possibly alamandine, contribute to counter-regulate the deleterious effects of Ang II both on the endothelium and on vascular smooth muscle cells. Fig. 15.3 lists the main vascular effects resulting from the activation of the SRA axes. Ongoing studies suggest that the therapeutic use of ACE2/Ang-(1-7)/MAS agonists may contribute to treat endothelial dysfunction and its consequences.

References

[1] Santos RAS. Angiotensin-(1–7). Hypertension 2014;63(6):1138–47.
[2] Cadwell PR, Segal BC, Hsu KC, et al. Angiotensin converting enzyme: vascular endothelial localization. Science 1976;191 (4231):1050–1.
[3] Rogerson FM, Chai SY, Schlawe I, et al. Presence of angiotensin converting enzyme in the adventitia of large vessels. J Hypertens 1992;7:615–20.
[4] Arnal JF, Battle T, Rasetti C, et al. ACE in three tunicae of rat aorta: expression in smooth muscle and effect of renovascular hypertension. Am J Physiol 1994;267(5 Pt 2):H1777–84.
[5] Shenoy V, Qi Y, Katovich MJ, et al. ACE2, a promising therapeutic target for pulmonary hypertension. Curr Opin Pharmacol 2011;11 (2):150–5.

[6] Allen AM, Zhuo J, Mendelsohn FAO. Localization and function of angiotensin AT1 receptors. Am J Hypertens 2000;13:31S–38S.

[7] Carey RM. Update on the role of AT2 receptor. Curr Opin Nephrol Hypertens 2005;14(1):67–71.

[8] Ozono R, Wang ZQ, Moore AF, et al. Expression of the subtype 2 angiotensin (AT2) receptor protein in rat kidney. Hypertension 1997;30(5):1238–46.

[9] Carey RM, Wang ZQ, Siragy HM. Role of the angiotensin type-2 receptor in the control of blood pressure and renal function. Hypertension 2000;35:155–63.

[10] Benndorf RA, Krebs C, Hirsch-Hoffmann B, et al. Angiotensin II type 2 receptor deficiency aggravates renal injury and reduces survival in chronic kidney disease in mice. Kidney Int 2009;75 (10):1039–49.

[11] Ardaillou R. Angiotensin II receptors. J Am Soc Nephrol 1999;10: S30–9.

[12] Tallant EA, Lu X, Weiss RB, et al. Bovine aortic endothelial cells contain an angiotensin-(1–7) receptor. Hypertension 1997;29(1 Pt 2):388–93.

[13] Santos RA, Simões e Silva AC, Maric C, et al. Angiotensin-(1–7) is an endogenous ligand for the G protein-coupled receptor Mas. Proc Natl Acad Sci U S A 2003;100(14):8258–63.

[14] Sampaio WO, Santos RAS, Faria-Silva R, et al. Angiotensin-(1–7) through receptor Mas mediates endothelial nitric oxide synthase activation via Akt-dependent pathways. Hypertension 2007;49 (1):185–92.

[15] Lautner RQ, Villela DC, Fraga-Silva RA, et al. Discovery and characterization of alamandine: a novel component of the renin-angiotensin system. Circ Res 2013;112(8):1104–11.

[16] Habiyakare B, Alsaadon H, Mathai ML, et al. Reduction of angiotensina and alamandine vasoactivity in the rabbit model of atherogenesis: differential effects of alamandine and Ang(1-7). Int J Exp Pathol 2014;95(4):290–5.

[17] Billet S, Aguilar F, Baudry C, et al. Role of angiotensin II AT1 receptor activation in cardiovascular diseases. Kidney Int 2008;74:1379–84.

[18] Mehta PK, Griendling KK. Angiotensin II cell signaling: physiological and pathological effects in the cardiovascular system. Am J Physiol Cell Physiol 2007;292:C82–97.

[19] Touyz RM, Schiffrin EL. Signal transduction mechanisms mediating the physiological and pathophysiological actions of angiotensin II in vascular smooth muscle cells. Pharmacol Rev 2000;52 (4):639–72.

[20] Nguyen Dinh Cat A, Touyz RM. Cell signaling of angiotensin II on vascular tone: novel mechanisms. Curr Hypertens Rep 2011;13 (2):122–8.

[21] Briones AM, Touyz RM. Oxidative stress and hypertension: current concepts. Curr Hypertens Rep 2010;12(2):135–42.

[22] Touyz RM. Reactive oxygen species and angiotensin II signaling in vascular cell—implications in cardiovascular disease. Braz J Med Biol Res 2004;37(8):1263–73.

[23] Danyel LA, Schmerler P, Paulis L, et al. Impact of AT2-receptor stimulation on vascular biology, kidney function, and blood pressure. Integr Blood Press Control 2013;6:153–61.

[24] Namsolleck P, Recarti C, Foulquier S, et al. AT(2) receptor and tissue injury: therapeutic implications. Curr Hypertens Rep 2014;16 (2):416.

[25] Santos RA, Brosnihan KB, Jacobsen DW, et al. Production of angiotensin-(1–7) by human vascular endothelium. Hypertension 1992;19:II56–61.

[26] Sampaio WO, Santos RAS, Faria-Silva R, et al. Angiotensin-(1–7) through receptor Mas mediates endothelial nitric oxide synthase activation via Akt-dependent pathways. Hypertension 2007;49:185–92.

[27] Freeman EJ, Chisolm GM, Ferrario CM, et al. Angiotensin-(1–7) inhibits vascular smooth muscle cell growth. Hypertension 1996;28:104–8.

[28] Lemos VS, Cortes SF, Silva DM, et al. Angiotensin-(1–7) is involved in the endothelium-dependent modulation of phenylefrine-induced contraction in the aorta of m-Ren transgenic rats. Br J Pharmacol 2002;135(7):1743–8.

[29] Brosnihan KB, Li P, Ferrario CM. Angiotensin-(1–7) dilates canine coronary arteries through kinins and nitric oxide. Hypertension 1996;27(3 Pt 2):523–8.

[30] Pörsti I, Bara AT, Busse R, et al. Release of oxide nitric by angiotensin-(1–7) from porcine coronary endothelium: implications for a novel angiotensin receptor. Br J Pharmacol 1994;111 (3):652–4.

[31] Feterik K, Smith L, Katusic ZS. Angiotensin-(1–7) causes endothelium-dependent relaxatation in canine middle cerebral artery. Brain Res 2000;873:75–82.

[32] Meng W, Busija DW. Comparative effects of angiotensin-(1–7) and angiotensin II on piglet pial arterioles. Stroke 1993;24:2041–5.

[33] Osei SY, Ahima RS, Minkes RK, et al. Differential responses to Ang-(1–7) in the feline mesenteric and hindquarter vascular beds. Eur J Pharmacol 1993;234:35–42.

[34] Ren Y, Garvin JL, Carretero OA. Vasodilator action of angiotensin-(1–7) on isolated rabbit afferent arterioles. Hypertension 2002;39:799–802.

[35] Fernandes L, Fortes ZB, Nigro D, et al. Potentiation of bradykinin by angiotensin-(1–7) on arterioles of spontaneously hypertensive rats studies in vivo. Hypertension 2001;37(2 Pt 2):703–9.

[36] Oliveira MA, Fortes ZB, Santos RAS, et al. Synergistic effect of angiotensin-(1–7) on bradykinin arteriolar dilation in vivo. Peptides 1999;20:1195–201.

[37] Almeida AP, Fabregas BC, Madureira MM, et al. Angiotensin-(1–7) potentiates the coronary vasodilatory effect of bradykinin in the isolated rat heart. Braz J Mol Biol Med 1999;33:709–13.

[38] Santos RAS, Passaglio KT, Pesquero JB, et al. Interactions between kinins and angiotensin-(1–7) in kidney and blood vessels. Hypertension 2001;38(3 Pt 2):660–4.

[39] Sampaio WO, Nascimento AA, Santos RA. Systemic and regional hemodynamic effects of angiotensin-(1–7) in rats. Am J Physiol Heart Circ Physiol 2003;284(6):H1985–94.

[40] Botelho-Santos GA, Sampaio WO, Reudelhuber TL, et al. Expression of an angiotensin-(1–7)-producing fusion protein in rats induced marked changes in regional vascular resistance. Am J Physiol Heart Circ Physiol 2007;292(5):H2485–90.

[41] Botelho-Santos GA, Bader M, Alenina N, et al. Altered regional blood flow distribution in Mas-deficient mice. Ther Adv Cardiovasc Dis 2012;6(5):201–11.

[42] Fraga-Silva RA, Da Silva DG, Montecucco F, et al. The angiotensin-converting enzyme 2/angiotensin-(1–7)/Mas receptor axis: a potential target for treating thrombotic diseases. Thromb Haemost 2012;108(6):1089–96.

[43] McCollum LT, Gallagher PE, Tallant EA. Angiotensin-(1–7) abrogatesmitogen-stimulated proliferation of cardiac fibroblasts. Peptides 2012;34:380–8.

[44] Gallagher PE, Tallant EA. Inhibition of human lung cancer cell growth by angiotensin-(1–7). Carcinogenesis 2004;25:2045–52.

[45] Ni L, Feng Y, Wan H, et al. Angiotensin-(1–7) inhibits the migration and invasion of A549 human lung adenocarcinoma cells through inactivation of the PI3K/AktandMAPK signaling pathways. Oncol Rep 2012;27(3):783–90.

[46] Verano-Braga T, Schwämmle V, Sylvester M, et al. Time-resolved quantitative phosphoproteomics: new insights into angiotensin-(1-7) signaling networks in human endothelial cells. J Proteome Res 2012;11(6):3370–81.

[47] Marchesi C, Paradis P, Schiffrin EL. Role of the renin-angiotensin system in vascular inflammation. Trends Pharmacol Sci 2008;29 (7):367–74.

[48] Montecucco F, Pende A, Mach F. The renin-angiotensin system modulates inflammatory processes in atherosclerosis: evidence from basic research and clinical studies. Mediators Inflamm 2009;2009:752406.

[49] Ruiz-Ortega M, Lorenzo O, Rupérez M, et al. Angiotensin II activates nuclear transcription factor kappaB through AT(1) and AT(2) in vascular smooth muscle cells: molecular mechanisms. Circ Res 2000;86(12):1266–72.

[50] Celi A, Cianchetti S, Dell'Omo G, et al. Angiotensin II, tissue factor and the thrombotic paradox of hypertension. Expert Rev Cardiovasc Ther 2010;8(12):1723–9.

[51] Tallant EA, Clark MA. Molecular mechanisms of inhibition of vascular growth by angiotensin-(1–7). Hypertension 2003;42 (4):574–9.

[52] Langeveld B, Van Gilst WH, Tio RA, et al. Angiotensin-(1–7) attenuates neointimal formation after stent implantation in the rat. Hypertension 2005;45(1):138–41.

[53] Sampaio WO, Castro CH, Santos RA, et al. Angiotensin-(1–7) counterregulates angiotensin II signaling inhuman endothelial cells. Hypertension 2007;50:1093–8.

[54] Xu P, Costa-Goncalves AC, Todiras M, et al. Endothelial dysfunction and elevated blood pressure in MAS gene-deleted mice. Hypertension 2008;51:574–80.

[55] Rabelo LA, Xu P, Todiras M, et al. Ablation of angiotensin-(1–7) receptor Mas in C57Bl/6 mice causes endothelial dysfunction. J Am Soc Hypertens 2008;2:418–24.

[56] Rakušan D, Bürgelová M, Vanéčková I, et al. Knockout of angiotensin-(1-7) receptor Mas worsens the course of two-kidney, one-clip Goldblatt hypertension: roles of nitric oxide deficiency and enhanced vascular responsiveness to angiotensin II. Kidney Blood Press Res 2010;33:476–88.

[57] Faria-Silva R, Duarte FV, Santos RA. Short-term angiotensin(1–7) receptor MAS stimulation improves endothelial function in normotensive rats. Hypertension 2005;46:948–52.

[58] Fraga-Silva RA, Pinheiro SVB, Gonçalves ACC, et al. The antithrombotic effect of angiotensin-(1–7) involves Mas-mediated NO release from platelets. Mol Med 2008;14(1–2):28–35.

[59] Fraga-Silva RA, Costa-Fraga FP, De Sousa FB, et al. An orally active formulation of angiotensin-(1–7) produces an antithrombotic effect. Clinics 2011;66(5):837–41.

[60] Fang C, Stavrou E, Schmaier AA, et al. Angiotensin-(1–7) and Mas decrease thrombosis in Bdkrb2−/− mice by increasing NO and prostacyclin to reduce platelet spreading and glycoprotein VI activation. Blood 2013;121(15):3023–32.

[61] Sui YB, Chang JR, Chen WJ, et al. Angiotensin-(1–7) inhibits vascular calcification in rats. Peptides 2013;42:25–34.

[62] Wilsdorf T, Gainer JV, Murphey LJ, et al. Angiotensin-(1–7) does not affect vasodilator or TPA responses to bradykininin human forearm. Hypertension 2001;37:1136–40.

[63] Davie AP, McMurray JJ. Effect of angiotensin-(1–7) and bradykinin inpatients with heart failure treated with an ACE inhibitor. Hypertension 1999;34:457–60.

[64] Sasaki S, Higashi Y, Nakagawa K, et al. Effects of angiotensin-(1–7) on forearm circulation in normotensive subjects and patients with essential hypertension. Hypertension 2001;38(1):90–4.

[65] Ueda S, Masumori-Maemoto S, Wada A, et al. Angiotensin(1–7) potentiates bradykinin-induced vasodilatation in man. J Hypertens 2001;19:2001–9.

[66] Roks AJ, Van Geel PP, Pinto YM, et al. Angiotensin-(1-7) is a modulator of the human renin-angiotensin system. Hypertension 1999;34(2):296–301.

[67] Roks AJ, Nijholt J, Van Buiten A, et al. Low sodium diet inhibits the local counter-regulator effect of angiotensin-(1–7) on angiotensin II. J Hypertens 2004;22:2355–61.

[68] Van Twist DJ, Houben AJ, de Haan MW, et al. Angiotensin-(1–7)-induced renal vasodilation in hypertensive humans is attenuated by low sodium intake and angiotensin II co-infusion. Hypertension 2013;62:789–93.

[69] Van der Wouden EA, Ochodnický P, Van Dokkum RP, et al. The role of angiotensin(1–7) in renal vasculature of the rat. J Hypertens 2006;24:1971–8.

I. STRUCTURE AND FUNCTIONS

ENDOTHELIAL DYSFUNCTION AND CLINICAL SYNDROMES

METHODS OF INVESTIGATION

16

Methods to Investigate Endothelial Function in Humans

Valéria Costa-Hong, Keyla Yukari Katayama, and Fernanda Marciano Consolim-Colombo

The endothelium, a layer of specialized cells that covers the luminal surface of blood vessels, works as an organ that is spatially distributed along the entire body and is highly active from the metabolic point of view [1].

It is well established that the endothelium has several physiological functions, with the overall purpose of maintaining vascular health [1]. Besides that, it is involved in several pathological processes, including development of atherosclerosis and cancer [2].

Although of the same embryological origin, endothelial cells do not represent a homogeneous set of cells, because they are different regarding morphology and function along the several segments of the circulatory system. Genetic and biochemical aspects, and participation of biomechanical forces, contribute for this significant characteristic, which has only recently been addressed [3]. The phenotypic heterogeneity of endothelial cells provides at least two advantages: first, it allows the endothelium to act according to the several needs of the underlying tissues along the entire body; second, it favors cell adaptation capacity to different microenvironments, for example, the hyperosmolar and hypoxic environment of the internal medulla of the kidney, and the well-oxygenated environment of pulmonary alveoli [3].

There are properties and functions that are common to all endothelial cells. Endothelium-dependent vasodilation was defined with the discovery of the endothelium-derived relaxation factor (EDRF) in 1980 by R. Furchgott [4]. This significant function is responsible for autoregulation of tissue blood flows by microcirculation and changes in diameter as a response to shear stress in conductance vessels. With the advance of knowledge in endothelial physiology, its other functions began to be described: coagulation and anticoagulation regulation, platelet aggregation prevention, and modulation of the vascular inflammatory response. Furthermore, it became evident that endothelial dysfunction has deep and complex implications in several diseases, such as arterial hypertension, *diabetes mellitus*, dyslipidemia, atherosclerosis, acute coronary syndromes, sepsis, and systemic inflammatory syndrome, with the latter exposing the importance of endothelium dysfunction in microcirculation alterations [5–9].

The endothelium can be studied in different ways, from cellular and molecular aspects to physiological preparations in experimental animals and isolated vessels/organs, and may also be inferred, even if indirectly, in humans. Each investigation method is appropriate to elucidate a given type of scientific question, from molecular intimacy of the intra- and intercellular signaling mechanisms, encompassing circulation and coagulation homeostasis mechanisms, inflammation physiopathology, and vascular repair mechanisms, up to the clinical implications in diagnosis, prognosis, and treatment of a given pathology.

EVALUATION OF THE ENDOTHELIAL FUNCTION IN HUMANS

Regardless of the giant leaps in comprehending the biology of endothelial cells, little perception of this organ arrive at clinical practice. During patient

217

evaluation, endothelium health is not a routine question, nor a situation in which specific therapeutic correction is sought. Many factors corroborate for this situation. The endothelium is not available for inspection, palpation, percussion, and/or auscultation. Besides, the tests that measure endothelial function available at the moment, either biochemical or functional, have limitations regarding standardization, ease of execution, reproducibility, and variability [1,10]. Therefore, the invisible and diffuse nature of the cell layer, the complexity of the system, and its adaptability are characteristics that hinder the incorporation of endothelial function evaluation in clinical practice. Therefore, the huge potential of the endothelium as a diagnosis, preventive, and therapeutic target remains widely unexplored [1].

Currently, there are some tests that provide information, even if indirectly, regarding the endothelial function in humans, and therefore they are able to detect endothelial injury [10,11]. These tests consist of functional evaluations that interfere with the vessel wall structure, and dosage of blood markers.

Functional tests allow estimating vasomotor reactivity, that is, the blood vessel dilation/contraction capacity or blood flow increase/reduction of a segment, in face of several stimuli. Vascular dilation that follows physiological and/or pharmacological stimulation, by administering drugs that act specifically on endothelial cell receptors, for example, acetylcholine, bradykinin, and substance P, is understood as dilation that depends on endothelial function integrity [11]. Methods that evaluate vessel reactivity are considered invasive in cases of infusion of vasoactive drugs directly inside arteries or veins. Methods are considered noninvasive when the vessel is stimulated indirectly, by means of maneuvers, such as reactive hyperemia, hand grip, or mental stress. Vessel diameter variation or blood flow volume, in response to several stimuli, can be quantified by means of different techniques, such as plethysmography or vascular ultrasound [11–16].

INVASIVE METHODS

Coronary Arteries—Cineangiography

In patients subjected to diagnostic cineangiography, biplane quantitative angiography of a coronary epicardial segment and use of intracoronary Doppler ultrasound probe (Flow Wire) allow the evaluation of vessel diameter and coronary flow before and after intracoronary injection of acetylcholine and bradykinin, which are endothelium-dependent vasodilators. The response observed in normal individuals is vessel dilation and increased blood flow. Patients with coronary atherosclerotic disease (CAD) show, both on the affected vessel and areas without evident obstructive injuries, significant reduction of dilation or even vasoconstrictor response when infused with vasodilators [17]. The expression "endothelial dysfunction" was coined in 1986 based on this observation. Several studies demonstrated that patients without CAD, but carrying risk factors such as *diabetes mellitus*, dyslipidemia, hypertension or tobacco use, may show vasodilation and coronary flow alterations. On the other hand, control of diabetes, hypercholesterolemia, arterial hypertension, and quitting tobacco use may revert this endothelial dysfunction. In addition to demonstrating the level of endothelial dysfunction in the arterial territory, the reactivity of coronaries evaluated in invasive fashion shows correlation with the risk of future adverse coronary events [18].

Brachial Artery—Plethysmography

Plethysmography with venous occlusion is a technique used for approximately 50 years to evaluate the peripheral blood flow [19]. As it was developed by Whitney, the variations in volume of a segment, for example, the forearm, are automatically transformed in percentage variations of the blood flow into the region (mL/100 mL of tissue/min).

The equipment that composes the assembly required to perform plethysmography includes stretching sensors (mercury in sylastic strain-gauge), wrist and arm pressure cuffs, air compressor and blower, and signal transformed, for example, the plethysmograph itself. The installation of the equipment is briefly described as follows: a strain-gauge with appropriate diameter for the forearm is placed 5 cm below the cubital fold and the arm must be maintained elevated 10 cm from the level of the heart by means of a support (Fig. 16.1). Hand circulation is excluded by insufflating the cuff placed on the wrist with supra-systolic pressure (200 mmHg), 1 minute before determining the flow volume. Venous return occlusion is intermittent, applying pressure of 35–40 mmHg on the cuff located on the middle third of the arm. Occlusion

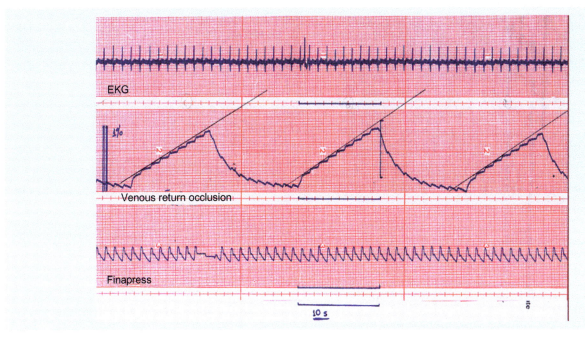

FIG. 16.1 Chart records. The tracing shows three channels: *ECG*, a derivation of eletroccardiogram; *Flow*, plethysmograph curves; *Finapress*, noninvasive, beat-by-beat blood pressure curves. *Source*: Hypertension Unit Lab, InCor; Laboratório de Hipertensão Arterial, InCor.

may last 7–10 seconds, and release may last another 7–10 seconds. This provides records of three flow curves per minute (Fig. 16.1).

Blood flow may be calculated using manual or semimanual analysis, with the aid of a computer program. Quantification of forearm blood flow by means of plethysmography during infusion of vasoactive drugs in the brachial artery is considered the gold standard of endothelial function evaluation. To execute this pharmacological test, the artery must be punctured and infused with drug doses adjusted to the volume of the segment to be analyzed, so that there are no systemic effects that may interfere with the interpretation of the results. Usually, increasing doses of the medication selected are infused to obtain the dose-response curve; (e.g., vasodilation response curve) [19]. The procedure, although simple to execute in clinical research centers, has minimal complication risks due to arterial puncture, such as artery spasm, thrombosis, bleeding, or hematoma.

In brief, initially the endothelial function study protocol by means of brachial artery drug infusion quantifies the forearm volume with the purpose of adapting drug infusion; then, the individuals are positioned in a stretcher in the supine position; in aseptic conditions and with local superficial

anesthesia, the brachial artery of the nondominant arm is punctured (2 cm from the cubital fold) using a specific catheter; the catheter is fixed at the site and connected to a pathway that maintains continuous infusion of saline solution with heparin to maintain system permeability and another pathway to infuse vasoactive drugs (Fig. 16.2). After puncture, the accessories of the plethysmography assembly (Hokannson, Inc., llevue, WA, USA) are positioned to obtain the forearm blood flow curves. The minimum waiting time before starting drug infusion is 30 minutes. With the purpose of studying endothelium-dependent dilation, at least three incremental doses of acetylcholine (0.75: 5 and 15 mg/100 mL of tissue/min) are infused, with each infusion lasting 2 minutes. After a 15 minutes break, the same sequence is repeated, now infusing sodium nitroprusside (doses of 1, 2, and 4 mg/100 mL of tissue/min, also 2 minutes each dose), which is a medication that causes vessel dilation due to direct action on smooth muscles (evaluation of endothelium-dependent vasodilation). Total infusion volume shall not exceed 2 mL/min, in order to avoid the effect that a higher volume might have on the endothelium (increased shear stress). At the end of the examination, the artery puncture location is compressed for 10–15 minutes [20].

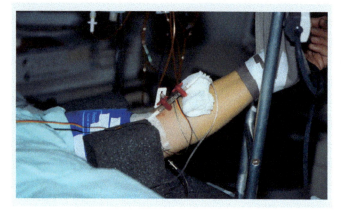

FIG. 16.2 Plethysmography. *Source*: Laboratory of Arterial Hypertension, InCor.

Dorsal Hand Vein—Compliance Technique (Dorsal Hand Vein Technique—DHV)

The method to evaluate endothelial function in the venous territory considers the same principles of intravascular infusion of growing doses of vasodilator drugs with local action, but on an easily accessible territory with smaller possibility of complications: the dorsal hand vein.

In brief, the method can be described as follows: the patient is placed in the supine position with the left forearm resting on a support with an angle of 30 degree in respect to the horizontal position. A 23G needle is inserted in the dorsal vein of the left hand, and then a physiological solution of 0.9% is infused by means of a Harvard infusion pump (Harvard Apparatus Inc., South Natick, MA) for 30 minutes at 0.3 mL/min, in order to allow recovery from the vasoconstriction caused by inserting the needle. A tripod with central transducer (LVDT, Shaevitz Engineering, Pennsauken, NJ) is fixed on the back of the hand, 10 mm above the needle insertion point (Fig. 16.3). The transducer detects small linear displacements on the dorsal surface of the left hand. The variation in vein diameter promotes proportional vertical movement of the transducer, which is recorded and transformed into a signal to create the curve [12]. Then, incremental doses of acetylcholine, sodium nitroprusside, or other drugs to be investigated are infused, always respecting the pharmacokinetics of the drugs and the use of doses and concentrations with pure local effects [21].

There are studies that demonstrate a good correlation between the endothelial function of the arterial territory and endothelial function of the venous territory [12] evaluated with different

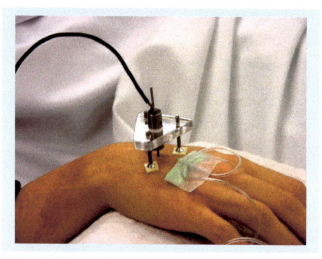

FIG. 16.3 Transducer positioning. *Source*: Laboratory of Arterial Hypertension, InCor.

techniques, even if it is possible to quantity therapeutic responses in similar fashion in these vascular territories [22].

Therefore, selection of the method to be used in each protocol depends on the questions to be answered, the facilities available in the services, and the experience of the researchers with the different techniques.

NONINVASIVE METHODS

Artery Diameters—High Resolution Vascular Ultrasound

Peripheral arterial flow and diameter may be measured using vascular ultrasound equipment with high resolution transducer [23–25]. This technique can be used in resting situations and during maneuvers that cause increased blood flow and vessel dilation (reactive hyperemia) or systemic administrative of several vasodilatory drugs, including sublingual nitrate. Therefore, this method permits to evaluate vessel diameter changes in response to endothelium-dependent and independent stimuli and provides information regarding the endothelial function of the segment studied.

The evaluation of arterial dilation that depends on increased flow after reactive hyperemia is called flow-mediated dilation (FMD) and is one of the most used techniques not only for diagnosis of endothelial function, but also as cardiovascular prognosis marker [25].

It is known that when an arterial segment is occluded, ischemic tissue released factors or molecules, such as adenosine, H+ ions, and EDHF, among others, which cause microcirculation dilation, in an attempt to restore blood perfusion. When the occluded segment is opened, there is immediate great blood supply to the ischemic region, causing tissue hyperemia, the so called reactive hyperemia. Increased flow depends initially on blood pressure and quantity of substance produced by the endothelium in microcirculation and represents the functional state of these cells. Increased speed and blood flow causes increased shear stress on the arteries that nourish the previously ischemic tissue (conductance arteries), and this is a potent stimulus for endothelial cells to produce vasodilation substances. Shear stress stimulates membrane potassium channels, leading to endothelial cell hyperpolarization, with consequent increased intracellular calcium, activation of the endothelial nitric oxide synthase (eNOS) enzyme, and release of nitric oxide (NO). As a result, the underlying vascular smooth muscle relaxes and the arterial diameter increases. This process occurs in a short period of time, and therefore FMD can be quantified in great arteries minutes after the occlusion/ischemia is cleared from the irrigated segment [26,27].

Smaller FMD capacity characterizes endothelial dysfunction in conductance arteries. Since the FMD mechanism depends on NO [26,27], endothelial function can be considered a noninvasive marker of NO bioavailability, a significant factor that opposes the onset and/or progression of atherosclerosis [28].

The use of endothelium-dependent vasodilating drugs, such as sublingual nitroglycerin, allows checking if the vessel smooth muscle layer is working normally or not in response to direct stimulation. Smaller FMD associated with preserved vasodilation independent from the endothelium proves the presence of dysfunction in the endothelial layer alone. The importance of smooth muscle cell dysfunction in the stratification of cardiovascular disease (CVD) risk is still not properly established.

The brachial artery ultrasound FMD technique is the most common method used to evaluate endothelial dysfunction in humans [29,30]. This method is employed to investigate patients with CVDs, carriers of cardiovascular risk factors, asymptomatic individuals, and even children. FMD correlates to the endothelial function of coronary arteries

evaluated in noninvasive fashion [31], and is a predictor of cardiovascular events [32–35].

Due to the good accuracy and reproducibility, the technique must be executed by expert professionals, using standardized equipment, and according to an established protocol. In 2002, Corretti et al. [36] published the first guideline to execute the FMD technique in the brachial artery territory. Other guidelines appeared over the years, with the purpose of incorporating discoveries regarding how physiological variables and variables of the technique itself can influence vasodilation response [23,37,38].

We shall cover some points that are significant in the FMD technique, and also for other techniques to evaluate endothelial function. To increase accuracy and reproducibility of the test, and reduce conditions that influence vascular reactivity, such as diet, tobacco use, physical activity, medication, and menstrual cycle, among others, certain principles must be observed and standardized (Table 16.1).

TABLE 16.1 Recommendations to Evaluate FMD in Conductance Arteries [23,37,38]

- Vitamin supplements must be withdrawn at least 72 h before evaluation
- Caffeine must be avoided at least 12 h before evaluation
- No smoking or exposure to cigarette smoke for at least 12 h before evaluation
- Women in the fertile phase must always be evaluated in the same stage of the menstrual cycle
- In studies focusing the differences between genders, the evaluation must be conducted between the 1st and 7th days of the menstrual cycle
- Whenever possible, evaluations must be conducted with the patient fasting for at least 6 h
- Tests with repeated measurements must always be redone on the same day
- Abstinence of physical activity at least 12 h before the test
- Whenever possible, vasoactive medication must be avoided at least 4 half-lives before the evaluation; when this is not possible, the role of these medications must be considered when interpreting the results
- Nonsteroid antiinflammatories and aspirin must be withdrawn 1 and 3 days, respectively, before
- The examination must be conducted in a calm environment, with controlled temperature between 20°C and 25°C

The patient must remain in the supine position but selection of the arm to be evaluated is indifferent (neutral in respect of some specified physical property: endothelial function). The pressure cuff used to obstruct blood circulation may be placed on the limb, in the forearm. The image of the longitudinal section of the brachial artery must be captured above the elbow fold and keeping an anatomical landmark as reference, in order to use the same artery image in successive records. The depth and gain are optimized to identify the vessel lumen and wall in each study, and the Doppler probe is positioned at 60 in respect to the center of the vessel. The diameter of the brachial artery is defined by the distance between the anterior wall intima layer and the posterior wall, quantified manually or semiautomatically, with the aid of a specific *software* program, in order to avoid observer interference. For reading purposes, the images captured must contain at least six cardiac cycles, coinciding with the R wave in the ECG, and when an automatic analysis software is used, the diameter is calculated during a complete heart cycle.

After evaluating the artery diameter in the resting position, the reactive hyperemia maneuver is started. Therefore, a pneumatic tourniquet is placed around the forearm and inflated up to 50 mmHg above the systolic pressure value, for 5 minutes. After deflating the cuff, the first five flows (Doppler) are recorded and the artery image is captured to calculate the diameter, for a period of 10 seconds before releasing the cuff up to 180 seconds after releasing the cuff (Fig. 16.4). There is a waiting period of 20 minutes for the artery to return to normal conditions, and then a second resting moment is recorded. Then, nitroglycerin spray or sublingual nitrate (5 mg) is administered. After 5 minutes of use, the diameter and flow are analyzed and the brachial artery dilation percentage is calculated. The nitrate shall not be administered in hypotensive patients or patients with clinically significant bradycardia. In this method patients with brachial artery diameter smaller than 2.5 mm must be excluded due to difficulty in making the measurements, as well as patients with diameter bigger than 5 mm due to smaller vasodilation [39].

FMD is expressed by the percentage of brachial artery diameter change after stimulus in respect to the baseline diameter, according to the following formula:

$$FMV = \frac{HR \text{ diameter} - Baseline \text{ diameter}}{Baseline \text{ diameter}} \times 100$$

Next, we have the formula to calculate the percentage of diameter change after the use of nitrate:

$$FFMV = \frac{Prenitrate \text{ diameter} - Postnitrate \text{ diameter}}{Prenitrate \text{ diameter}} \times 100$$

Maximum vasodilation does not always occur 60 seconds after releasing the pressure cuff, as shown by Black et al. [40]. These authors evaluated the time in which maximum vasodilation occurred in three groups: youngsters, sedentary elders, and physically active elders; the group of youngsters reached the maximum vasodilation peak before the groups of elderly patients. Regarding FMD, if analyzed in traditional fashion, for example, 60 seconds after releasing the cuff, there would be no differences between the three groups, but continuous analysis allowed observing that FMD was similar

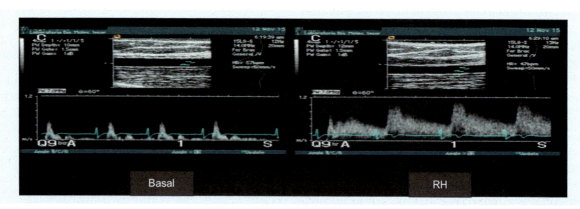

FIG. 16.4 Brachial artery diameter and flow in the baseline and reactive hyperemia stages. *Source:* Laboratory of Arterial Hypertension, InCor.

among youngsters and physically active elders; sedentary elders, on the other hand, showed worse FMD when compared to the group of young patients.

There are software programs that capture flow velocity and artery diameter simultaneously and continuously, making it possible to observe blood flow values before releasing the cuff and continue the evaluation for up to 120 seconds, in cases of brachial arteries. This methodology allows evaluating the blood flow peak velocity, maximum dilation value, and then moment in which vasodilation occurred. In the absence of this technology, the diameter should be measured 1 minute after the cuff is released.

Vessel baseline diameter is inversely proportional to the level of vasodilation obtained after reactive hyperemia stimulus (major arteries dilate less that smaller arteries) and there is variability in arterial diameter from individual to individual. Therefore, it is recommended that studies provide baseline and absolute diameter variability together with the value of FMD. In comparative studies, it is significant that the baseline diameter value of the groups are the same, and in the case of significant difference, statistical tests must be used for correction purposes.

If these guidelines are observed, reproducibility of the FMD technique increases, and variation observed in studies that evaluated technique reproducibility can be explained by lack of standardization, use of transducer with inappropriate frequency, location, cuff time and pressure, patient preparation regarding diet, physical activity, menstrual cycle, circadian rhythm, and analysis method, and all of the factors listed interfere with examination reproducibility and must be rigorously controlled. Welsch et al. [41] did not observe significant differences in the baseline diameter and maximum diameter after hyperemia in three repeated measurements; however, they noted a difference when they compared the measurements made between two performers. The recommendations also state that the tests must be conducted by a single well-trained performer. A recent study conducted in men with coronary disease evaluated the variability of the FMD technique. The variability observed in repeated measurements, with a time interval of 30 minutes, showed a coefficient of 10%, and between two sequential days of 11%, proving that FMD is a safe tool to evaluate the action of interventions in patients with CVD [30].

Evaluation of FMD in Other Conductance Arteries

NO synthesis on the arterial tree is heterogeneous, and the role of NO on DMF in different beds is variable. Therefore, the technique used to study FMD in the brachial artery cannot be extrapolated to other territories. Among the territorial studies, it is known that FMD of the superficial femoral artery reflects predominantly the endothelium-dependent vasodilation action mediated by NO [27], when the cuff is placed distal to the occluder site and inflation is maintained for 5 minutes. Another significant point is the time to reach the diameter peak after HR and postnitrate. Smaller arteries reach the peak faster than bigger arteries. As shown in the study conducted by Thijssen [42], diameter evaluation must be prolonged for a longer period of time depending on the vessel analyzed, for example, after 3 minutes, 100% of the patients have reached the maximum dilation peak when studying the brachial artery, but for the same period in the superficial femoral artery, only 58% of the individuals would have dilated. When the femoral artery is evaluated, the analysis time must be extended to 6 minutes.

Peripheral Arterial Tonometry

Peripheral arterial tonometry (PAT) is a noninvasive technique, independent from the observer, which captures, by means of beat by beat plethysmography, the digital pulse wave amplitude (PWA), which corresponds to digital volume variation [17]. The PWA evaluation follows the same recommendations for FMD described in Table 16.1.

Methodology

The EndoPAT device (Endo-PAT2000; Itamar Medical, Caesarea, Israel) consists of two digital probes with an inflatable internal membrane system. When inflated, these membranes apply counter-pressure of 70 mmHg on the distal phalanges of the index fingers. The pressure signal variations are filtered, amplified, and stored by the device for future analysis [43].

PWA is evaluated before and during the reactive hyperemia maneuver. The pressure cuff is placed in the proximal part of the arm (study arm) and the contralateral arm works as control.

The PAW baseline measurement is determined by plethysmography by the probes positioned on the

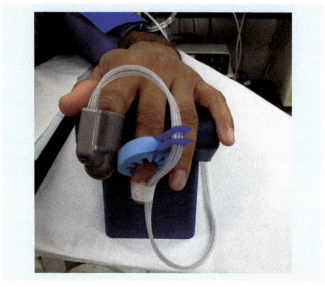

FIG. 16.5 Probe positioning. *Source*: Laboratory of Arterial Hypertension, InCor.

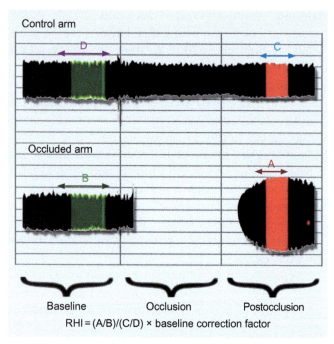

$$RHI = (A/B)/(C/D) \times \text{baseline correction factor}$$

FIG. 16.6 RHI calculation by endoPAT [45]. *Source*: Laboratório de Hipertensão Arterial, InCor.

index fingers of both hands for a period of 5 minutes (Fig. 16.5). Hyperemia begins with the occlusion of the brachial artery blood flow for 5 minutes, with pressure of 50 mmHg above the systolic pressure [44], and arterial flow occlusion is confirmed by reducing the outline to zero (Fig. 16.5). The PWA signal is recorded for 5 minutes after the pressure cuff is released. The data are analyzed automatically by the device software.

The reactive hyperemia index (RHI) is the result of the rate between the averages of PWA of the post- and preocclusion values. To calculate the index, we use the PAW average during 1 minute, 60 seconds after releasing the pressure cuff, divided by the average PWA during 210 seconds preocclusion (Fig.16.6). The RHI is normalized by the measurements of the opposing arm, which serves as control of the systemic effect of reactive hyperemia [43].

This index is considered an endothelial function marker; [29] however, the pulse amplitude after reactive hyperemia is complex, reflecting flow changes and digital microcirculation dilation, and is partially dependent on NO [18]. Studies have shown that the endothelial function evaluated by the EndoPAT method correlates with microvascular function of the coronaries in patients in early atherosclerosis stages [46] and is a predictor of cardiovascular events [47].

The RHI measurement is based on the same principle as the FMD technique, but the Framingham study showed that there is no significant correlation between RHI and FMD, contributing for the idea that the techniques reflect different and complementary aspects of the vascular function [17,48].

Quantification of Substances Produced by the Endothelium in Blood

In addition to vessel dilation capacity evaluation, it is possible to measure substances produced by the endothelium.

In general, the presence of endothelial dysfunction is considered when there is smaller production of vasodilation substances and/or higher production of vasoconstrictor substances, such as inflammatory pathway activation markers, thrombogenic factors, oxidative stress markers, and cell activation markers. More recently, the quantification of circulating endothelial progenitor cells (EPC) has been used to estimate the endothelium recovery capacity. Thus, in hypertensive and prehypertensive patients [49], as well as other cardiovascular risk situations, there is a smaller number and activity of EPC, associated with smaller endothelium-dependent vasodilation and worsened cardiovascular prognosis.

The EPCs include an extremely rare cellular group of nonhematopoietic cells that can be recruited from bone marrow by several stimuli, such as cytokines (VEGF, SDF-1), drugs such as statins, estrogens, erythropoietin, and physical activity, among others, and play a fundamental role in maintaining endothelial integrity. They can be identified and characterized by means of sensitive techniques such as flow cytometry [47].

The quantification of these cells in peripheral blood may be considered a predictor of atherosclerosis risk and extension, because individuals with reduced number of EPCs in blood flow have higher risk of developing endothelial dysfunctions, since vasculogenesis is compromised [50].

The advanced in endothelial function evaluation, by means of peripheral blood markers, shall be based on the development of platforms with assay combinations for doses of soluble mediators and/or microparticles (MP) derived from endothelial cells [51]. There are high levels of these MP in several pathological situations [52], and the circulating level is associated with endothelial dysfunction. The term microparticles is used to described small vesicles released by different cell types after activation or apoptosis, containing cellular material such as proteins, mRNAs, lipoproteins, and debris [53]. Endothelial cells are rich in phosphatidylserine, which has potent procoagulant activity, indicating that endothelial microparticles (EMP) may determine increased thrombogenicity of the atherosclerotic plaque [54].

The EPCs associated with the MP can be considered as useful CVD biomarkers, as they are directly related to endothelial homeostasis (Fig. 16.7) [53].

Recruitment of EPCs suggests a compensatory vascular repair mechanism that contributes to restore endothelial integrity. Increased MPs is directly related to endothelial dysfunction and, consequently, to atherosclerosis progression. Therefore, the MP/EPC ratio can indicate the level of unbalance between endothelial damage and repair capacity [53]. Chapter 17 contains more information on this topic.

CONCLUSIONS

In summary, among the tests available for endothelial evaluation, few were systematically assessed regarding the possibility of adding value to patient prognosis. Regardless of the clinical data, due the dynamic nature of the endothelium and its responses to environmental factors, a single methodology to evaluate endothelial function may not reflect the real function. For example, environmental factors affect endothelial function and may only be a transient state that does not represent a pathological entity.

Stratification of the risk of CVDs and prescription of therapeutic measures, such as lifestyle changes or pharmacotherapy, may be more refined with the incorporation of methods that evaluate endothelial function with the purpose of reducing the incidence of CVDs.

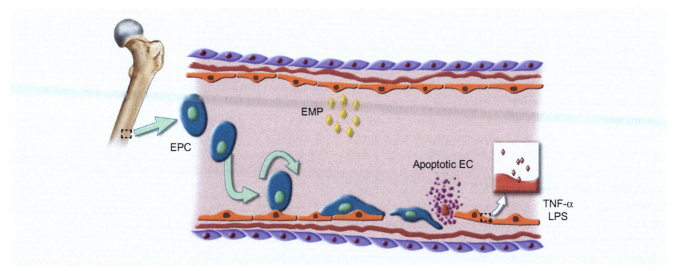

FIG. 16.7 Replacement of endothelial cells (CEs) that suffer apoptosis on the vascular wall by recruiting endothelial progenitor cells (EPC) from bone marrow. The ECs in apoptosis release endothelial microparticles (EMP), which are also formed after activation by different stimuli (TNF-α, LPS, among others—in detail). *TNF*, tumoral necrosis factor; *LPS*, lipopolysaccharides.

It is unquestionable that detection of endothelial dysfunction is an early marker of vascular dysfunction and atherosclerosis development. In addition, several studies correlated endothelial dysfunction, either in conductance vessels or microvasculature, with cardiovascular risk, classical risk factors, changes in lifestyle and response to pharmacological interventions [25,29].

However, the fundamental role of the methods for evaluation of endothelial function in research has not yet been translated into clinical practice. At this moment, taking into account limitations of each technique, there are no strong recommendations for adoption of endothelial evaluation in asymptomatic patients [55].

On the other hand, in patients with CVD, no recovery of endothelial function under optimized pharmacological treatment permits identification of a subpopulation at higher or recurrent events [55].

Forearm pletismography with infusion of vasoactive drugs in the brachial artery may be considered the gold standard for endothelial function evaluation, albeit its invasive nature limits its use [25].

FMD with ultrasound is frequently used in clinical studies given its noninvasive nature, its ability to predict cardiovascular events and its correlation with coronary endothelial function. A disadvantage is that it requires high resolution equipment, and is therefore highly operator dependent. Yet reclassification of patients at risk was demonstrated in studies with large number of participants [29,31,35]. In the MESA study [56], it was shown that the combination of FMD with Framingham score was superior to either alone in risk evaluation.

Flow variation recorded by EndoPAT is indicative of microvascular functions and resistant arterias; hence EndoPAT evaluates endothelial function in different ways than FMD. EndoPAT is easy to perform and operator independent. However, digital probes are expensive, reproducibility is not uniform, and predictive value for cardiovascular is not as well established as for FMD [16,17,47].

Recently a study compared three techniques—pletismography, FMD, and Endo PAT—in 50-year-old individuals, and correlated the findings with the Framingham risk score [57]. No differences were observed among tests of endothelium dependent dilation, suggesting that each technique contributes with unique information regarding different vascular territories. There was significant correlation of cardiovascular risk with FMD and vasoactive drugs

infused in pletismography, indicating that reactivity in certain vascular beds (conductance arteries) correlates with cardiovascular risk in the age range studied.

Based on other studies, FMD seems to be more useful in individuals with atherosclerosis while evaluations by EndoPAT is better for younger individuals in whom microvascular function predicts early cardiovascular risk [29].

Better standardization of the techniques, specific softwares for data analysis, and training are necessary to reduce biases in these methods [23]. It is hoped that future clinical studies will fill up the gaps in information relative to the use of endothelial function as an important tool in the assessment of patients for primary and secondary prevention.

References

[1] Aird WC. Endothelium as an organ system. Crit Care Med 2004;32 (5 Suppl.):S271–9.

[2] Feletou M, Vanhoutte PM. Endothelial dysfunction: a multifaceted disorder (The Wiggers Award Lecture). Am J Physiol Heart Circ Physiol 2006;291(3):H985–H1002.

[3] Aird WC. Phenotypic heterogeneity of the endothelium: I. Structure, function, and mechanisms. Circ Res 2007;100(2):158–73.

[4] Furchgott RF, Zawadzki JV. The obligatory role of endothelial cells in the relaxation of arterial smooth muscle by acetylcholine. Nature 1980;288(5789):373–6.

[5] Alexander RW. Oxidized LDL, autoantibodies, endothelial dysfunction, and transplant-associated arteriosclerosis. Arterioscler Thromb Vasc Biol 2002;22(12):1950–1.

[6] Levi M, ten CH, van der Poll T. Endothelium: interface between coagulation and inflammation. Crit Care Med 2002;30(5 Suppl.): S220–4.

[7] Reinhart K, Bayer O, Brunkhorst F, et al. Markers of endothelial damage in organ dysfunction and sepsis. Crit Care Med 2002;30 (5 Suppl.):S302–12.

[8] Bertoluci MC, Ce GV, da Silva AM, et al. Endothelial dysfunction as a predictor of cardiovascular disease in type 1 diabetes. World J Diabetes 2015;6(5):679–92.

[9] Libby R. Current concepts of the pathogenesis of the acute coronary syndromes. Circulation 2001;104(3):365 72.

[10] Deanfield JE, Halcox JP, Rabelink TJ. Endothelial function and dysfunction: testing and clinical relevance. Circulation 2007;115 (10):1285–95.

[11] Pedro MA, Coimbra SR, Colombo FMC. Métodos de investigação do endotélio. In: Luz PL, Laurindo FRM, Chagas ACP, editors. Endotélio e doenças cardiovasculares. 1st ed. Rio de Janeiro: Atheneu; 2003. p. 53–68.

[12] Rubira MC, Consolim-Colombo FM, Rabelo ER, et al. Venous or arterial endothelium evaluation for early cardiovascular dysfunction in hypertensive patients? J Clin Hypertens (Greenwich) 2007;9(11):859–65.

[13] Inoue T, Matsuoka H, Higashi Y, et al. Flow-mediated vasodilation as a diagnostic modality for vascular failure. Hypertens Res 2008;31 (12):2105–13.

[14] Wilkinson IB, Webb DJ. Venous occlusion plethysmography in cardiovascular research: methodology and clinical applications. Br J Clin Pharmacol 2001;52(6):631–46.

[15] Kura N, Fujikawa T, Tochikubo O. New finger-occlusion plethysmograph for estimating peripheral blood flow and vascular resistance. Circ J 2008;72(8):1329–35.

[16] Matsuzawa Y, Guddeti RR, Kwon TG, et al. Secondary prevention strategy of cardiovascular disease using endothelial function testing. Circ J 2015;79(4):685–94.

[17] Poredos P, Jezovnik MK. Testing endothelial function and its clinical relevance. J Atheroscler Thromb 2013;20(1):1–8.

[18] Fukuda D, Yoshiyama M, Shimada K, et al. Relation between aortic stiffness and coronary flow reserve in patients with coronary artery disease. Heart 2006;92(6):759–62.

[19] Benjamin N, Calver A, Collier J, et al. Measuring forearm blood flow and interpreting the responses to drugs and mediators. Hypertension 1995;25(5):918–23.

[20] Lima SM, Aldrighi JM, Consolim-Colombo FM, et al. Acute administration of 17beta-estradiol improves endothelium-dependent vasodilation in postmenopausal women. Maturitas 2005;50(4):266–74.

[21] Rabelo ER, Rohde LE, Schaan BD, et al. Bradykinin or acetylcholine as vasodilators to test endothelial venous function in healthy subjects. Clinics (Sao Paulo) 2008;63(5):677–82.

[22] de Sousa MG, Yugar-Toledo JC, Rubira M, et al. Ascorbic acid improves impaired venous and arterial endothelium-dependent dilation in smokers. Acta Pharmacol Sin 2005;26(4):447–52.

[23] Thijssen DH, Black MA, Pyke KE, et al. Assessment of flow-mediated dilation in humans: a methodological and physiological guideline. Am J Physiol Heart Circ Physiol 2011;300(1):H2–H12.

[24] Frolow M, Drozdz A, Kowalewska A, et al. Comprehensive assessment of vascular health in patients; towards endothelium-guided therapy. Pharmacol Rep 2015;67(4):786–92.

[25] Anderson TJ, Phillips SA. Assessment and prognosis of peripheral artery measures of vascular function. Prog Cardiovasc Dis 2015;57(5):497–509.

[26] Joannides R, Haefeli WE, Linder L, et al. Nitric oxide is responsible for flow-dependent dilatation of human peripheral conduit arteries in vivo. Circulation 1995;91(5):1314–9.

[27] Kooijman M, Thijssen DH, de Groot PC, et al. Flow-mediated dilatation in the superficial femoral artery is nitric oxide mediated in humans. J Physiol 2008;586(4):1137–45.

[28] Celermajer DS, Sorensen KE, Gooch VM, et al. Non-invasive detection of endothelial dysfunction in children and adults at risk of atherosclerosis. Lancet 1992;340(8828):1111–5.

[29] Flammer AJ, Anderson T, Celermajer DS, et al. The assessment of endothelial function: from research into clinical practice. Circulation 2012;126(6):753–67.

[30] Onkelinx S, Cornelissen V, Goetschalckx K, et al. Reproducibility of different methods to measure the endothelial function. Vasc Med 2012;17(2):79–84.

[31] Anderson TJ, Uehata A, Gerhard MD, et al. Close relation of endothelial function in the human coronary and peripheral circulations. J Am Coll Cardiol 1995;26(5):1235–41.

[32] Kitta Y, Obata JE, Nakamura T, et al. Persistent impairment of endothelial vasomotor function has a negative impact on outcome in patients with coronary artery disease. J Am Coll Cardiol 2009;53(4):323–30.

[33] Karatzis EN, Ikonomidis I, Vamvakou GD, et al. Long-term prognostic role of flow-mediated dilatation of the brachial artery after acute coronary syndromes without ST elevation. Am J Cardiol 2006;98(11):1424–8.

[34] Gokce N, Keaney Jr. JF, Hunter LM, et al. Risk stratification for postoperative cardiovascular events via noninvasive assessment of endothelial function: a prospective study. Circulation 2002;105(13):1567–72.

[35] Inaba Y, Chen JA, Bergmann SR. Prediction of future cardiovascular outcomes by flow-mediated vasodilatation of brachial artery: a meta-analysis. Int J Cardiovasc Imaging 2010;26(6):631–40.

[36] Corretti MC, Anderson TJ, Benjamin EJ, et al. Guidelines for the ultrasound assessment of endothelial-dependent flow-mediated vasodilation of the brachial artery: a report of the International Brachial Artery Reactivity Task Force. J Am Coll Cardiol 2002;39(2):257–65.

[37] Deanfield J, Donald A, Ferri C, et al. Endothelial function and dysfunction. Part I: methodological issues for assessment in the different vascular beds: a statement by the Working Group on Endothelin and Endothelial Factors of the European Society of Hypertension. J Hypertens 2005;23(1):7–17.

[38] Harris RA, Nishiyama SK, Wray DW, et al. Ultrasound assessment of flow-mediated dilation. Hypertension 2010;55(5):1075–85.

[39] Stout M. Flow-mediated dilatation: a review of techniques and applications. Echocardiography 2009;26(7):832–41.

[40] Black MA, Cable NT, Thijssen DH, et al. Importance of measuring the time course of flow-mediated dilatation in humans. Hypertension 2008;51(2):203–10.

[41] Welsch MA, Allen JD, Geaghan JP. Stability and reproducibility of brachial artery flow-mediated dilation. Med Sci Sports Exerc 2002;34(6):960–5.

[42] Thijssen DH, Dawson EA, Black MA, et al. Heterogeneity in conduit artery function in humans: impact of arterial size. Am J Physiol Heart Circ Physiol 2008;295(5):H1927–34.

[43] Kuvin JT, Patel AR, Sliney KA, et al. Assessment of peripheral vascular endothelial function with finger arterial pulse wave amplitude. Am Heart J 2003;146(1):168–74.

[44] Higashi Y. Assessment of endothelial function. History, methodological aspects, and clinical perspectives. Int Heart J 2015;56(2):125–34.

[45] McCrea CE, Skulas-Ray AC, Chow M, et al. Test–retest reliability of pulse amplitude tonometry measures of vascular endothelial function: implications for clinical trial design. Vasc Med 2012;17(1):29–36.

[46] Bonetti PO, Pumper GM, Higano ST, et al. Noninvasive identification of patients with early coronary atherosclerosis by assessment of digital reactive hyperemia. J Am Coll Cardiol 2004;44(11):2137–41.

[47] Rubinshtein R, Kuvin JT, Soffler M, et al. Assessment of endothelial function by non-invasive peripheral arterial tonometry predicts late cardiovascular adverse events. Eur Heart J 2010;31(9):1142–8.

[48] Hamburg NM, Palmisano J, Larson MG, et al. Relation of brachial and digital measures of vascular function in the community: the Framingham heart study. Hypertension 2011;57(3):390–6.

[49] Giannotti G, Doerries C, Mocharla PS, et al. Impaired endothelial repair capacity of early endothelial progenitor cells in prehypertension: relation to endothelial dysfunction. Hypertension 2010;55(6):1389–97.

[50] Werner N, Kosiol S, Schiegl T, et al. Circulating endothelial progenitor cells and cardiovascular outcomes. N Engl J Med 2005;353(10):999–1007.

[51] Dignat-George F, Boulanger CM. The many faces of endothelial microparticles. Arterioscler Thromb Vasc Biol 2011;31(1):27–33.

[52] Martinez MC, Tesse A, Zobairi F, et al. Shed membrane microparticles from circulating and vascular cells in regulating vascular function. Am J Physiol Heart Circ Physiol 2005;288(3):H1004–9.

[53] França C, Izar MC, Amaral J, et al. Micropartículas e célulasprogenitoras: novosmarcadores da disfunçãoendotelial. Rev Soc Cardiol Estado de São Paulo 2013;23(4):33–9.

[54] Mallat Z, Benamer H, Hugel B, et al. Elevated levels of shed membrane microparticles with procoagulant potential in the peripheral circulating blood of patients with acute coronary syndromes. Circulation 2000;101(8):841–3.

[55] Greenland P, Alpert JS, Beller GA, et al. 2010 ACCF/AHA guideline for assessment of cardiovascular risk in asymptomatic adults: executive summary: a report of the American College of Cardiology Foundation/American Heart Association Task Force on Practice Guidelines. Circulation 2010;122(25):2748–64.

[56] Ganz P, Hsue Y. Individualized approach to the management of coronary heart disease: identifying the nonresponders before it is too late. J Am Coll Cardiol 2009;53:331–3.

[57] Yeboah J, Folsom AR, Burke GL, et al. Predictive value of brachial flow-mediated dilation for incident cardiovascular events in a population-based study: the multi-ethnic study of atherosclerosis. Circulation 2009;120(6):502–9.

Further Reading

[1] Lind L. Relationship between three different tests to evaluate endothelium-dependent vasodilation and cardiovascular risk in a middle-aged sample. J Hypertens 2013;31:1570–4.

17

Endothelial Biomarkers

Francisco Antonio Helfenstein Fonseca and Maria Cristina O. Izar

INTRODUCTION

Blood vessels provide oxygen and nutrients to tissues, while lymphatic vessels filter and absorb the interstitial fluids from them. During the last three decades, modifications in vascular endothelium properties have been associated with several metabolic changes and cardiovascular diseases. The quick change in functional and metabolic properties that the vascular endothelium assumes are notable, ranging from a quiescent and physiological state to dysfunctional state [1]. The endothelium has the same general characteristics in all territories, and therefore the exposure to different concentrations of oxygen and nutrients (as in arterial circulation and pulmonary microcirculation), as well as high cerebral metabolic needs, also are associated with important differences in endothelial metabolism, such as quantity of mitochondria, glycolysis rate, and oxidative metabolism [1]. Regardless of these heterogeneities, the search for new functional endothelium state biomarkers has been promising and seems essential to better understand cardiovascular diseases and event prevention.

ENDOTHELIAL PROGENITOR CELLS AND VASCULAR REPAIR

In 1997, Asahara et al. [2] described progenitor endothelial cells in blood circulation and their potential in differentiating from the vascular endothelium, form new vessels and to repair senescent or apoptotic cells. In the following years, there was great interest in characterization and demonstration of endothelial progenitor cell subpopulations in several diseases and health states, but due to the very small number of such progenitor cells in circulation, their role in vascular repair and angiogenesis began to be questioned, and new evidence regarding contribution of resident endothelial cells was more recently appreciated [3]. It is possible that resident endothelial cells maintain their differentiation capacity in new endotheliocyte, just as endothelial progenitor cells, in more immature form in circulation, also contribute to this process. Many initial studies involving endothelial progenitor cell quantification with flow cytometry and some more characteristic DCs of these cells showed promising results, and a reduced number of circulating endothelial progenitor cells has been associated with cardiovascular outcomes [4,5]. A smaller rate of circulating endothelial progenitor cells was also found in HIV positive patients associated with endothelial dysfunction, a condition that seems to contribute to the severity and precocity of cardiovascular diseases in these patients [6]. Therefore, a reduced number of circulating endothelial progenitor cells, as well as smaller differentiation capacity of resident endothelial cells, seem to contribute to lower tissue reparation ability and exposure of the vascular intimate structure to blood thrombotic factors.

Microparticles

Microparticles are enucleated phospholipid cell fragments, of diameter between 100 and 1000 nm [7] and also usually quantified by flow cytometry and use of antibodies markers of this cell type. Endothelial microparticles are considered new endothelial injury and vascular dysfunction biomarkers, and their serum concentration is related to cardiovascular, inflammatory, and metabolic diseases [8].

The number of endothelial microparticles seems to reflect the balance between cell proliferation and death [9] and many inflammatory cytokines may influence their serum levels [10], as well as the intensity of oxidative stress [11].

Not only endothelial microparticles have been postulated as vascular dysfunction markers but also have been associated with prediction of cardiovascular outcomes [6,12,13]. In fact, these microparticles modulate inflammation, coagulation, adhesion, and leukocyte recruitment, potentially contributing to atherosclerotic plaque formation and cardiovascular outcomes. However, more recently, the physiological role of these microparticles has been further elucidated, and in certain concentrations they seem to act as cell mechanisms of antiinflammatory and cytoprotective actions [14]. It is interesting that the use of some drugs with known cardiovascular protection, such as statins and aspirin, have the ability to reduce the circulating levels of endothelial microparticles [15]. On the other hand, in the presence of comorbidities, such as end-stage renal failure, increased endothelial microparticles in circulation has been described as an independent mortality predictor [16]. Regardless of some controversies, more recent studies and reviews do in fact suggest that increased plasma endothelial microparticles concentration is associated with higher risk of cardiovascular events [15–18].

During vascular injury, microparticles derived from endothelium may contain tissue factor and activate coagulation. Specifically, in tumor necrosis factor alpha (TNF-α), lipopolysaccharide or oxidized LDL, an increased endothelial microparticle release containing tissue factor has been described [19,20]. On the other hand, microparticles derived from endothelium containing tissue factor pathway inhibitor (TFPI) can counterbalance or completely eliminate this effect [21]. Platelet microparticles are among the most numerous circulating microparticles, and have been considered as new platelet activation biomarkers. Increased platelet microparticles have been related to atherosclerosis, arterial hypertension, acute coronary disease, and cerebrovascular accidents (CVA) [19–21]. More recently, studies described that the number of platelet microparticles is associated with the volume of cerebrovascular accident [22]. Interestingly, in coronary disease patients, regardless of maintenance of stable antiplatelet doses, withdrawal of statin increased the number of circulating platelet microparticles significantly [23].

Oxidized Low-Density Lipoprotein and Immune Response

In parallel to the discovery of nitric oxide (NO) and its relevance in vascular function, studies regarding LDL oxidation also revealed that its increased plasma concentration has antagonistic action, consisting of an important vascular dysfunction biomarker [24]. LDL oxidation promotes endothelial dysfunction and favors the development of atherosclerosis. The mechanism involved in endothelial dysfunction seems to involve smaller production of NO and smaller expression of cyclic guanosine monophosphate (GMPc) [25]. Furthermore, recent studies have shown that increased production of antioxidized LDL antibodies has antiatherogenic action, representing one of the main defense mechanisms of our organism against the insult of oxidized LDL [26–28]. In fact, subtype B lymphocytes (B1) produce antibodies against oxidized LDL (IgM) that neutralize and avoid exposure to dendritic cells, preventing presentation of lymphocyte T antigens, reducing, therefore, the infiltration of these cells in the vascular intimate structure [29,30]. Oxidized LDL also promotes increased expression of its receptor (LOX-1) in endothelial cells, as well as in smooth muscle cells and macrophages. Internalization of oxidized LDL determines profound modification in endothelial cells, causing increased activation, smaller vasodilation, higher apoptosis rates, cell proliferation, inflammatory infiltrate, and platelet activation [31].

The central role of lymphocytes in the evolution of atherosclerosis has been studied extensively in recent years, turning endothelial function immunomodulation and recovery strategies very promising avenues in cardiovascular disease prevention or therapy. This was demonstrated experimentally in genetic models deficient in B and T lymphocytes. In these experiments, atherosclerosis was significantly attenuated in animals deficient in lymphocyte T, but again expressed when lymphocytes T CD4+ were transferred to these animals, promoting atherosclerosis and artery inflammation [32]. Currently, cunning studies have sought atherosclerosis reduction by increasing production of protective interleukins (such as IL-10) or regulatory T cells, by means of stimulation with peptides derived from Apolipoprotein B, or apoB100, combined with subunit B of the cholera toxin, through a vaccine [33–35]. Initial results were promising, but further investigations are required to define the real importance of such an approach.

MicroRNAs AND ENDOTHELIAL DYSFUNCTION

MicroRNAs are small RNA sequences (approximately 22 nucleotides) that regulate messenger RNA expression at the posttranscriptional level by degradation or translational repression. More than 2500 microRNAs have been described in human beings, and each one of them may interact with several hundreds of messenger RNA targets. MicroRNAs influence all stages of atherosclerosis, including endothelial dysfunction [36].

MiR-126, one of the most studied microRNAs in cardiovascular disease, is reduced in patients with *diabetes mellitus* type 2 (DM2), and not only relates to the development of insulin resistance by inhibition of substrate 1 of the insulin receptor, but also regulates the vascular endothelium growth factor (VEGF) pathway, reducing angiogenesis and vascular repair [36–38]. In addition, miR-126 is expressed in endothelial microparticles, assuming an important information role in vascular repair, which is reduced in patients with diabetes [39–41]. Several other microRNAs affect endothelial function, either because of increased transcription of vasoconstrictor factors, such as endothelin-1, or due to higher inflammatory activity [36].

ICAM, VCAM, AND SELECTINS

The intercellular adhesion molecule type 1 (ICAM 1) is a protein greatly expressed in endothelial cells in the presence of cytokine stimulation, such as with interleukin 1-β (IL-1 b) or tumor necrosis factor alpha (TNF-α). The relevance of this protein relates to its ability to act as binder of integrin LFA-1, a leukocyte receptor. Therefore, ICAM-1 participates in leukocyte migration to the vascular intima, an event of great relevance in atherosclerosis. In addition, ICAM-1 has also been implicated in viral internalization, and therefore participates actively in the cell inflammatory and infectious micro environment

[42]. The vascular cell adhesion molecule type 1 (VCAM-1) can also be expressed in the vascular endothelium under the same inflammatory cytokine stimuli, and is equally involved in leukocyte transmigration to the vascular intimate structure. Selectins (selectins L, E, and P) have major structural homology, but are different because they are expressed in different cell types. L-selectins are expressed in leukocytes, P-selectins in platelets, and E-selectins in the vascular endothelium. E-selectins are expressed under inflammatory stimuli by interleukins, and are related to the leukocyte rolling on the endothelial surface, preceding cell transmigration. It is interesting that classic risk factors, such as diabetes, hypertension, or hypercholesterolemia, are associated with increased expression of E-selectins, suggesting an important link in atherosclerosis [43].

CONCLUSIONS

The age of a person may be better estimated by the age of the person's vessels. The field of biomarkers seeks to monitor aging, including parameters that reflect progressive endothelial dysfunction, associated with accelerated development of atherosclerosis and complications such as myocardial infarction and cerebrovascular accident. A European Commission is presently examining a large number of biomarkers that may more precisely predict the cardiovascular risk associated with aging, as well as improve the quality of life of a population with progressively increased life expectancy [44]. Unfortunately, profound endothelial cell phenotype changes occur with aging, but the comprehensiveness of this dysfunction seems more associated with traditional cardiovascular risk factors [45,46], emphasizing the need for controlling such classic factors. However, currently significant advances occurred related to molecular mechanisms linked to the development of atherosclerosis and vascular dysfunction, new biomarkers and treatment targets, which may attenuate are continuously being described. In this context, another fascinating aspect of endothelium is its phenotypic change affects the balance of factors that modulate thrombosis. Indeed, an increase in early recurrence of acute thromboembolic events may be related to greater and transitory gene expression of prothrombogenic factors [47], in addition to acute infiltration of

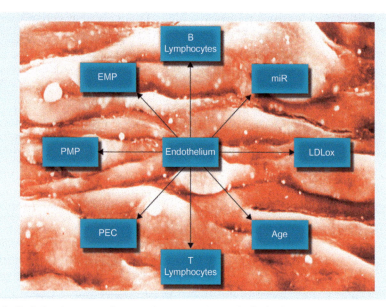

FIG. 17.1   New endothelial dysfunction biomarkers. *EMP*, endothelial microparticles; *PMP*, platelet microparticles; *PEC*, progenitor endothelial cells; *miR*, microRNAs; *oxLDL*, oxidized LDL. Increased microparticles, reduced endothelial progenitor cells, reduction of some miRNAs and increased oxidized LDL are associated endothelial dysfunction. Aging is associated with progressive endothelial dysfunction, mainly in the presence of uncontrolled risk factors.

inflammatory cells in vascular intimate structure [48]. This inflammatory infiltration includes some receptors that seem to modulate the magnitude of inflammatory response; among these CD40 or receptor 5 of superfamily TNF-α, which is activated by its bind (CD40L), has received particular attention since it is expressed not only in lymphocytes, but in macrophages as well [49]. Fig. 17.1 summarizes the main new endothelial dysfunction biomarkers.

References

[1] Eelen G, de Zeeuw P, Simons M, et al. Endothelial cell metabolism in normal and diseased vasculature. Circ Res 2015;116(7):1231–44.

[2] Asahara T, Murohara T, Sullivan A, et al. Isolation of putative progenitor endothelial cells for angiogenesis. Science 1997;275 (5302):964–7.

[3] Zhang M, Malik AB, Rehman J. Endothelial progenitor cells and vascular repair. Curr Opin Hematol 2014;21(3):224–8.

[4] Werner N, Kosiol S, Schiegl T, et al. Circulating endothelial progenitor cells and cardiovascular outcomes. N Engl J Med 2005;353 (10):999–1007.

[5] Schmidt-Lucke C, Rössig L, Fichtlscherer S, et al. Reduced number of circulating endothelial progenitor cells predicts future cardiovascular events: proof of concept for the clinical importance of endogenous vascular repair. Circulation 2005;111(22):2981–7.

[6] da Silva EF, Fonseca FA, França CN, et al. Imbalance between endothelial progenitors cells and microparticles in HIV-infected patients naive for antiretroviral therapy. AIDS 2011;25(13): 1595–601.

[7] Berezin A, Zulli A, Kerrigan S, et al. Predictive role of circulating endothelial-derived microparticles in cardiovascular diseases. Clin Biochem 2015;48(9):562–8.

[8] Mause SF, Weber C. Microparticles: protagonists of a novel communication network for intercellular information exchange. Circ Res 2010;107(9):1047–57.

[9] Spencer DM, Mobarrez F, Wallén H, et al. The expression of HMGB1 on microparticles from Jurkat and HL-60 cells undergoing apoptosis in vitro. Scand J Immunol 2014;80(2):101–10.

[10] Winner M, Koong AC, Rendon BE, et al. Amplification of tumor hypoxic responses by macrophage migration inhibitory factor-dependent hypoxia-inducible factor stabilization. Cancer Res 2007;67(1):186–93.

[11] Mezentsev A, Merks RM, O'Riordan E, et al. Endothelial microparticles affect angiogenesis in vitro: role of oxidative stress. Am J Physiol Heart Circ Physiol 2005;289(3):H1106–14.

[12] Meziani F, Tesse A, Andriantsitohaina R. Microparticles are vectors of paradoxical information in vascular cells including the endothelium: role in health and diseases. Pharmacol Rep 2008;60(1):75–84.

[13] Chironi GN, Boulanger CM, Simon A, et al. Endothelial microparticles in diseases. Cell Tissue Res 2009;335(1):143–51.

[14] Morel O, Morel N, Jesel L, et al. Microparticles: a critical component in the nexus between inflammation, immunity, and thrombosis. Semin Immunopathol 2011;33(5):469–86.

[15] Schiro A, Wilkinson FL, Weston R, et al. Endothelial microparticles as conveyors of information in atherosclerotic disease. Atherosclerosis 2014;234(2):295–302.

[16] Amabile N, Guérin AP, Tedgui A, et al. Predictive value of circulating endothelial microparticles for cardiovascular mortality in end-stage renal failure: a pilot study. Nephrol Dial Transplant 2012;27(5):1873–80.

[17] Werner N, Wassmann S, Ahlers P, et al. Circulating CD31+/ annexin V+ apoptotic microparticles correlate with coronary endothelial function in patients with coronary artery disease. Arterioscler Thromb Vasc Biol 2006;26(1):112–6.

[18] Sinning JM, Losch J, Walenta K, et al. Circulating CD31+/Annex-inV+ microparticles correlate with cardiovascular outcomes. Eur Heart J 2011;32(16):2034–41.

[19] Nadaud S, Poirier O, Girerd B, et al. Small platelet microparticle levels are increased in pulmonary arterial hypertension. Eur J Clin Invest 2013;43(1):64–71.

[20] Lukasik M, Rozalski M, Luzak B, et al. Enhanced platelet-derived microparticle formation is associated with carotid atherosclerosis in convalescent stroke patients. Platelets 2013;24(1):63–70.

[21] Michelsen AE, Brodin E, Brosstad F, et al. Increased level of platelet microparticles in survivors of myocardial infarction. Scand J Clin Lab Invest 2008;68(5):386–92.

[22] Chen Y, Xiao Y, Lin Z, et al. The role of circulating platelets microparticles and platelet parameters in acute ischemic stroke patients. J Stroke Cerebrovasc Dis 2015;24(10):2313–20.

[23] Pinheiro LF, França CN, Izar MC, et al. Pharmacokinetic interactions between clopidogrel and rosuvastatin: effects on vascular protection in subjects with coronary heart disease. Int J Cardiol 2012;158(1):125–9.

[24] Gradinaru D, Borsa C, Ionescu C, et al. Oxidized LDL and NO synthesis – biomarkers of endothelial dysfunction and ageing. Mech Ageing Dev 2015;151:101–13.

[25] Vicinanza R, Coppotelli G, Malacrino C, et al. Oxidized low-density lipoproteins impair endothelial function by inhibiting non-genomic action of thyroid hormone-mediated nitric oxide production in human endothelial cells. Thyroid 2013;23(2):231–8.

[26] Brandão SA, Izar MC, Fischer SM, et al. Early increase in autoantibodies against human oxidized low-density lipoprotein in hypertensive patients after blood pressure control. Am J Hypertens 2010;23(2):208–14.

[27] da Fonseca HA, Fonseca FA, Monteiro AM, et al. Inflammatory environment and immune responses to oxidized LDL are linked to systolic and diastolic blood pressure levels in hypertensive subjects. Int J Cardiol 2012;157(1):131–3.

[28] Izar MC, Fonseca HA, Pinheiro LF, et al. Adaptive immunity is related to coronary artery disease severity after acute coronary syndrome in subjects with metabolic syndrome. Diab Vasc Dis Res 2013;10(1):32–9.

[29] Ait-Oufella H, Sage AP, Mallat Z, et al. Adaptive (T and B cells) immunity and control by dendritic cells in atherosclerosis. Circ Res 2014;114(10):1640–60.

[30] Sage AP, Murphy D, Maffia P, et al. MHC class II-restricted antigen presentation by plasmacytoid dendritic cells drives proatherogenic T cell immunity. Circulation 2014;130(16):1363–73.

[31] Pirillo A, Norata GD, Catapano AL. LOX-1, OxLDL, and atherosclerosis. Mediators Inflamm 2013;2013:152786.

[32] Zhou Y, Robertson AK, Hjerpe C, et al. Adoptive transfer of CD4+ T cells reactive to modified low-density lipoprotein aggravates atherosclerosis. Arterioscler Thromb Vasc Biol 2006;26(4):864–70.

[33] Hermansson A, Johansson DK, Ketelhuth DF, et al. Immunotherapy with tolerogenicapolipoprotein B-100-loaded dendritic cells

attenuates atherosclerosis in hypercholesterolemic mice. Circulation 2011;123(10):1083–91.

[34] Klingenberg R, Lebens M, Hermansson A, et al. Intranasal immunization with an apolipoprotein B-100 fusion protein induces antigen-specific regulatory T cells and reduces atherosclerosis. Arterioscler Thromb Vasc Biol 2010;30(5):946–52.

[35] Geng YJ, Jonasson L. Linking immunity to atherosclerosis: implications for vascular pharmacology – a tribute to Göran K. Hansson. Vascul Pharmacol 2012;56(1–2):29–33.

[36] Nishiguchi T, Imanishi T, Akasaka T. MicroRNAs and cardiovascular diseases. Biomed Res Int 2015;2015:682857.

[37] Fish JE, Santoro MM, Morton SU, et al. miR-126 regulates angiogenic signaling and vascular integrity. Dev Cell 2008;15(2):272–84.

[38] Wang S, Aurora AB, Johnson BA, et al. The endothelial-specific microRNA miR-126 governs vascular integrity and angiogenesis. Dev Cell 2008;15(2):261–71.

[39] Jansen F, Yang X, Hoelscher M, et al. Endothelial microparticle-mediated transfer of microRNA-126 promotes vascular endothelial cell repair via SPRED1 and is abrogated in glucose-damaged endothelial microparticles. Circulation 2013;128(18):2026–38.

[40] Wei Y, Nazari-Jahantigh M, Neth P, et al. MicroRNA-126, -145, and -155: a therapeutic triad in atherosclerosis? Arterioscler Thromb Vasc Biol 2013;33(3):449–54.

[41] Schober A, Nazari-Jahantigh M, Wei Y, et al. MicroRNA-126-5p promotes endothelial proliferation and limits atherosclerosis by suppressing Dlk1. Nat Med 2014;20(4):368–76.

[42] Adamson P, Etienne S, Couraud PO, et al. Lymphocyte migration through brain endothelial cell monolayers involves signaling through endothelial ICAM-1 via a rho-dependent pathway. J Immunol 1999;162(5):2964–73.

[43] Roldán V, Marín F, Lip GY, et al. Soluble E-selectin in cardiovascular disease and its risk factors. A review of the literature. Thromb Haemost 2003;90(6):1007–20.

[44] Bürkle A, Moreno-Villanueva M, Bernhard J, et al. MARK-AGE biomarkers of ageing. Mech Ageing Dev 2015;151:2–12.

[45] Camici GG, Sudano I, Noll G, et al. Molecular pathways of aging and hypertension. Curr Opin Nephrol Hypertens 2009;18(2):134–7.

[46] Stampfli SF, Akhemedov A, Gebhard C, et al. Aging induces endothelial dysfunction while sparing arterial thrombosis. Arterioscler Thromb Vasc Biol 2010;30(10):1960–7.

[47] Sayed S, Cockerill GW, Torsney E, et al. Elevated tissue expression of thrombomodulatory factors correlates with acute symptomatic carotid plaque phenotype. Eur J Vasc Endovasc Surg 2009;38(1):20–5.

[48] Dutta P, Courties G, Wei Y, et al. Myocardial infarction accelerates atherosclerosis. Nature 2012;487(7407):325–9.

[49] Jansen MF, Hollander MR, van Royen N, et al. CD40 in coronary artery disease: a matter of macrophages? Basic Res Cardiol 2016;111:38.

AGING AND COGNITIVE FUNCTIONS

18

Endothelial Alterations in Aging

Mauricio Wajngarten, Amit Nussbacher, Paulo Magno Martins Dourado, and Antonio Carlos Palandri Chagas

INTRODUCTION

The prevalence of atherosclerosis increases with aging. The incidence of acute myocardial infarction (AMI), stroke and peripheral vascular (CVA) insufficiency is high in elderly individuals, even after control for other cardiovascular risk factors such as hypercholesterolemia, smoking, and hypertension [1,2].

The blood vessel of the elderly is less prepared to defend itself against aggression, as it occurs in patients with atherosclerotic disease. Similarly to this, many of the changes observed in aging and its consequent vulnerability to cardiovascular diseases result from endothelial dysfunction that develops over the years.

Endothelium consists of a thin layer of cells that covers the inner surface of the vessels, acts as a barrier between circulating medium and arterial wall, secreting vasodilatory substances (EDRF) and vasoconstricting substances (EDCF), and this way regulates arterial vasomotricity, protects against blood clotting, and is antiproliferative. It exerts active endocrine function in response to humoral, neural and mechanical stimuli, synthesizing and releasing vasoactive substances that modulate blood flow, tone, and vascular caliper.

This chapter will discuss the main evidences regarding changes that lead to endothelial dysfunction related to aging. In addition, data from studies that seek to analyze relationships between dietary restriction, vascular function, and longevity will be presented.

IMPAIRMENT OF ENDOTHELIUM DEPENDENT VASODILATION ASSOCIATED WITH AGING

The most widely used method to evaluate endothelial function is response to flow endothelium-mediated dilation (FMD). Age is a predictor of impairment of endothelium-dependent vasodilation of coronary epicardial arteries, resistance coronary vessels, and peripheral arteries [3–8]. It has been shown that age-associated decrease of FMD response is progressive and linear [9], while endothelium-independent vasodilation does not change with age, as can be seen by the response to sodium nitroprusside infusion, a direct stimulator smooth muscle cGMP (Fig. 18.1).

Therefore, it is not surprising that age is a strong predictor of endothelium-dependent vasodilatation exerting greater influence than classic risk factors for cardiovascular disease, such as serum levels of total cholesterol, LDL-cholesterol, or even blood pressure levels.

STRUCTURAL CHANGES RELATED TO CELL REPLICATION/REPLACEMENT— TELOMERE FUNCTION

Telomeres, terminal parts of chromosomes, suffer shortening of their length with each replicative cycle of somatic cells in culture [10,11].

This process also occurs in vivo, since there is an inverse relationship between telomere length in

237

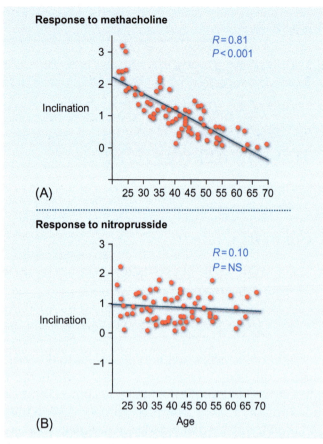

FIG. 18.1 (A) Only the endothelium-dependent vasodilation decays with aging. (B) Vasodilation induced by the direct relaxant of vascular smooth muscle, sodium nitroprusside does not change with age. Adapted from Gerhard M, Roddy MA, Creager SJ, et al. Aging progressively impairs endothelium-dependent vasodilation in forearm resistance vessels of humans. Hypertension 1996;27:849–53.

replicating somatic cells and age of people who donated cells [10–13]. Therefore, replicative history of somatic cells is an important determinant of telomere length.

Recent experimental data suggest that telomeres can serve as biological clocks, marking not only age at cellular level but also aging at systemic level. This data demonstrate that:

- telomere shortening prevention by forced expression in somatic cells in culture of telomerase catalytic portion, reverse transcriptase that adds telomere repeats to chromosome terminal portion and postpones the replicative senescence [14,15]; and
- telomerase knockout in mouse amplifies some characteristics associated with systemic aging in

subsequent generations of mice presenting substantially shortened telomere length [16].

On the other hand, observations in humans imply that telomere shortening in vivo may contribute to pathogenesis of vascular changes associated with aging. In fact, it has been observed progressive telomere shortening in human arteries in regions susceptible to atherosclerosis [17], and that telomere length is inversely correlated with pulse pressure and atherosclerosis degree [18,19].

It has been demonstrated, in an experimental model [20] that loss of telomere function induces endothelial dysfunction observed in aged arteries, while inhibition of telomere shortening suppresses these changes. To investigate whether endothelin cell senescence causes dysfunction, senescence was induced in human aortic endothelial cells, inhibiting telomere function. Consequently, there were phenotypic characteristics of cellular senescence, such as enlargement of cell shape accompanied by increased expression of intracellular adhesion molecule-1 (ICAM-1) and reduction of endothelial nitric oxide (NOS) synthesis activity, changes involved in atherogenesis. On the other hand, the reverse phenomenon was observed by introducing catalyst component telomerase activity. This intervention extended lifespan and inhibited functional changes associated with senescence of endothelial cells from human aorta, with a reduction in ICAM-1 levels and an increase in NOS levels, indicating that promotion of telomerase activity conferred protection against endothelial dysfunction associated with replicative senescence (Fig. 18.2).

Decreased Availability of NO—The Main Mechanism Responsible for Compromising Aging-Related Endothelium-Dependent Vasodilation

The main mechanism responsible for aging-related endothelial dysfunction appears to be reduction of NO availability. Several studies, both in experimental and human models, suggest that nitric oxide release or activity is reduced in aging [9,21,22].

NO is synthesized in endothelial cells from its precursor, L-arginine, by the enzyme NOS and induces vascular relaxation through activation of guanylate cyclase (cGMP) in vascular smooth muscle, which leads to vascular smooth muscle cell hyperpolarization and consequent relaxation (vasodilatation).

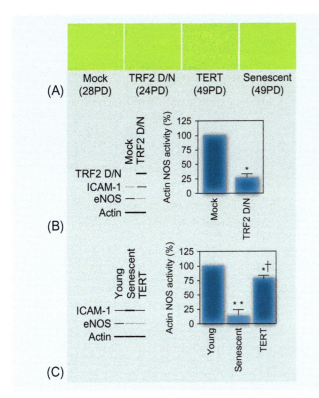

FIG. 18.2 (A) Cell morphology of HAECs infected with Mock (LPCX, 28PD), TRF2D/N (24PD) and TERT (49PD), and senescent HAECs (49PD). Original magnification was ×40. (B) Western blot analysis of ICAM-1, eNOS, and TRF2D/N and NOS activity in HAECs. The NOS activity in HAECs infected with LPCX is set at 100% and compared with that in HAECs infected with TRF2D/N (graph, $n = 4$; *$P<0.001$, paired-t test). (C) Western blot analysis of ICAM-1 and eNOS, and NOS activity in parental young HAECs (20PD), senescent HAECs (49PD), and HAECs infected with TERT (49PD). The NOS activity in parental HAECs is set at 100% and compared with that in senescent HAECs and HAECs infected with TERT. Graph: $n = 4$. *$P<0.05$ versus parental, **$P<0.001$ versus parental, †$P<0.01$ versus senescent by ANOVA. Similar results were observed in three independent Western blot analyses. Adapted from Minamino T, Miyauchi H, Yoshida T, et al. Endothelial cell senescence in human atherosclerosis: role of telomere in endothelial dysfunction. Circulation 2002;105:1541–4.

In fact, its reduction harms homeostasis and vascular protection against aggression, insofar as it leads to compromised actions such as: regulation of vascular tone [23,24], inhibition of platelet aggregation, inhibition of molecular adhesion to endothelial surface and inhibition of proliferation of vascular smooth muscle proliferation [23–27]. Therefore, reducing its availability may contribute to atherogenesis in the elderly.

In addition to nitric oxide, endothelial cell produces other factors that cause vasodilation, among which prostaglandin I_2, a derivative of arachidonic acid, cyclooxygenase pathway, and hyperpolarizing factor agent derived from the endothelium (FHDE). In subjects with hypercholesterolemia, exercise-induced vasodilation, predominantly mediated by FHDE, is impaired [28]. The effect of aging on these factors is less known.

Intracellular availability and/or mobilization of L-arginine is an early determinant of endothelial function in aging and hypertension, as described for dyslipidemia and atherosclerosis. Infusion of L-arginine, precursor of NO, potentiates vasodilator response to acetylcholine. In hypertensive individuals, potentiation of vasodilator response to acetylcholine by L-arginine is already noticeable even in individuals younger than 30 years (Fig. 18.3) [29]. A trial using venous occlusion plethysmography to study forearm blood flow in humans [30] assessed whether reduced availability of NO related to aging depends on reduction of its production, by compromising L-arginine-NO pathway, or increased degradation by oxidative stress. Vasodilator response to isolated acetylcholine was analyzed and in the presence of L-NMMA (Ng monomethyl L-arginine, a nonspecific inhibitor of NO synthase), of an antioxidant agent (vitamin C), or both.

In normotensive individuals, the inhibitory effect of L-NMMA on the vasodilator response to acetylcholine decreased in parallel with age and administration of agent that reduces oxidative stress (vitamin C) potentiated vasodilator response to acetylcholine only in elderly. In hypertensive individuals, vitamin C potentiated vasodilation to acetylcholine in younger subjects and counteracted the inhibitory effect of L-NMMA [30].

Therefore, the progressive impairment of age-related endothelium-dependent vasodilatation is caused by a change in the L-arginine-NO pathway. Only in the elderly (>60 years) does the influence of oxidative stress appear, leading to complete impairment of NO availability. On the other hand, in hypertensive patients, involvement of the L-arginine pathway is already observed even in young individuals. Production of oxidative stress appears decades before it does in normotensive.

This change in L-arginine-NO pathway, associated to age, could be related to decreased availability of substrate [31] or the presence of endogenous inhibitor of NOS, such as ADMA (asymmetric

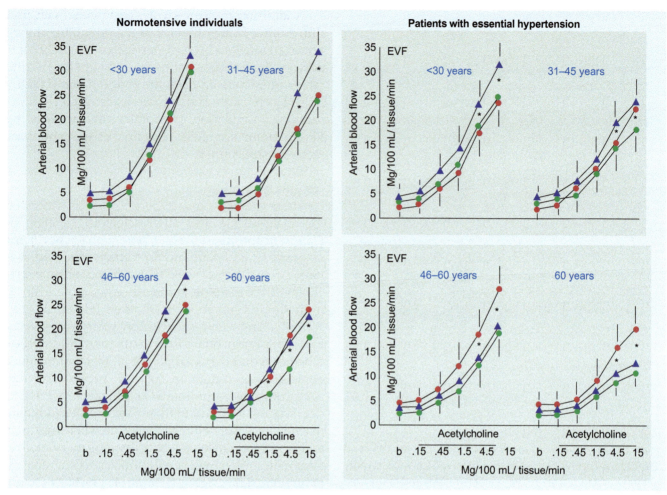

FIG. 18.3   Response of endothelial vasodilator function to infusion of L-arginine and indomethacin (-indomethacin, -L-arginine, -saline infusion [control]). Response of endothelial vasodilator function to infusion of L-arginine and indomethacin (-saline infusion, -L-arginine, and -indomethacin). Data is mean ± SD expressed as absolute values. *Significant difference between saline infusion and L-arginine or indomethacin ($P < 0.05$). *EVF*, endothelial vasodilator function. Adapted from Taddei S, Virdis A, Mattei P, et al. Hypertension causes premature aging of endothelial function in humans. Hypertension 1997;29(3):736–43.

dimetilarginine), known as an endothelial dysfunction and atherosclerosis mediator. ADMA circulating levels are related to cardiovascular risk factors such as hypercholesterolemia, hypertension, *diabetes mellitus*, hyperhomocysteinemia, age, and smoking. High serum concentrations of ADMA have been associated with carotid atherosclerotic lesions evaluated by measurement of the intima-media thickness and presence of plaques and may represent a marker of asymptomatic carotid atherosclerosis in the elderly [32].

Regarding possible sources of oxidative stress both in aging and hypertension, experimental evidence indicates that some systems could be responsible for increased production of reactive oxygen species, including studies with rat aorta rings that evaluated both NADH system/NADPH oxidase [33], as well as tetrahydrobiopterin [34]; on the other hand, a study on brachial artery of humans showed that the cyclooxygenase pathway contributes to the inhibition of vasodilator response to acetylcholine via production of prostanoids vasoconstrictors.

The vascular response to acetylcholine depends on its binding to muscarinic receptors on the surface of endothelial cells, which induces increased intracellular calcium and elevation of NO. With intact endothelium vasodilatation is observed; however, in the presence of endothelial lesions, paradoxical vasoconstriction takes place [35].

INCREASE IN PRODUCTION OF VASOCONSTRITOR MEDIATORS

In addition to vasodilator mediators, endothelium also produces vasoconstrictors via the cyclooxygenase pathway, such as thromboxane A2 and prostaglandin H2 (vasoconstrictor prostanoids). Its exacerbated production could contribute to aging-related endothelial dysfunction.

To investigate the contribution of vasoconstrictor production to aging-related endothelial dysfunction, a study using venous occlusion plethysmography to mediate forearm blood flow in humans evaluated whether inhibition of cyclooxygenase pathway by indomethacin, with consequent decrease in production of vasoconstrictor prostanoids, could improve vasodilatory response to acetylcholine. In fact, this occurred only in individuals over 60 years of age. On the other hand, in hypertensive patients, indomethacin potentiated the response to acetylcholine at an earlier age than in normotensive ones (Fig. 18.3) [29].

Interrelation Between Arterial Stiffening and Endothelial Dysfunction

The main vascular alteration related to aging is arterial stiffness [36]. This is responsible for elevation of systolic and pulse pressure related to aging and is primarily responsible for cardiac function impairment [37], predisposition to stroke and coronary artery disease in the elderly [38]. Although this process is due to structural changes in the middle layer of large and medium-sized arteries, including increased collagen content, reduction, and fracture of elastin fibers and calcification, an important two-way interaction between these abnormalities and those that occur in the endothelium. Arterial rigidity is determined not only by structural changes in the middle layer, but also by vascular endothelial regulation of vascular smooth muscle. This interrelationship between middle-layer and endothelial changes is bidirectional because disturbances in mechanical properties of the vascular wall are a strong predisposed to atherosclerosis development, which affects endothelial function. On the other hand, endothelial dysfunction contributes to vascular enhancement through modulation of vascular smooth muscle tonus. As endothelium-mediated vasodilation is compromised with aging, an important blood pressure regulating mechanism is lost, and blood pressure rises with aging due to arterial stiffness, allowing the elevation of pulse pressure and exposing the elderly to an increased risk of cardiovascular events. Thus, changes in the middle layer, resulting in arterial stiffening, and endothelial function, with vasodilatation impairment, culminate in a vascular aging process that facilitates the development of cardiovascular diseases and promotes dysfunction in vascular aging [39].

Dietary Restriction, Vascular Function, and Longevity

In recent years, there has been a growing search for knowledge about nutrition and health. In this sense, experimental studies evaluated relationship between food restriction and longevity. Caloric restriction (CR) is the only nongenetic interference that has proven to increase longevity in several experimental models from yeasts to primates. Abundant, experimental evidence indicates that the effect of dietary restriction on stimulating vasculature preservation associates various metabolic pathways and stress resistance. Identification of molecules that can prevent occurrence of age-related diseases is a crucial issue in development of aging-related medical research [40–43].

Downstream effects of various pathways with potential influence on aging include reduction in cellular damage induced by oxidative stress, increased efficiency of mitochondrial functions, thus attenuating age-related declines.

Among these mechanisms are highlighted those involving sirtuins, *AMP* kinase, the insulin/insulin-like growth factor 1, and protein-kinase TOR.

Sirtuins

Sirtuin name originates from the gene responsible for cellular regulation in *yeast* (*silent information regulation 2*).

Sirtuins are class II proteins of deacetylases NAD+ dependent histone (Nicotinamide adenine dinucleotide) family, responsible for regulating a variety of cellular functions, such as response to stress, regulation of metabolic energy and caloric restriction. They act on ADP-ribosyl transferase and exert deacetylation catalytic activity, both dependent on NAD+ and can regulate the activity of several transcriptional factors (Fig. 18.4) [41].

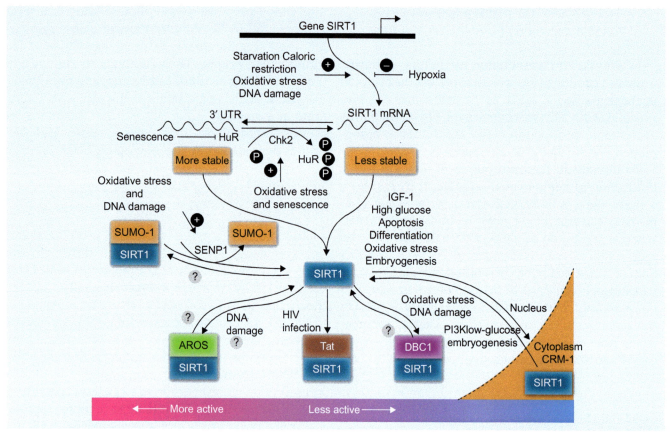

FIG. 18.4   Sirtuin regulatory pathways. *SIRT1*, Sirtuin 1; *Gene SIRT1*, Sirtuin 1 gene; *DNA damage*, DNA damage; *MRNA*, messenger ribonucleic acid; *Hur*, Hu Antigen R; *Tat*, transactivation active region; JUICE 1, small ubiquitin-like modifier; *SENP1*, Sentrin-specific protease 1; CRM-1, chromosomal maintenance 1; *DBC1*, deleted in bladder cancer protein 1; *AROS*, active regulator of SIRT1. Adapted from Kwon HS, Ott M. The ups and downs of SIRT1. Trends Biochem Sci 2008;33(11):517–25.

The increase of these positive and negative regulators can clarify the action mechanisms of these enzymes in various pathologies, thus opening paths to establish therapeutic potential of these proteins [40–43].

Sirtuin Sir2 was able to extend the lifespan by inhibiting the formation of toxic circles of extracromosomic rhabdomomal DNA, but may have other yet unknown functions that contribute to longevity [42]. Sir2 extends life through maintenance of gene silencing at the telomeres during aging [42]. This pathway could potentially affect metazoans, which have rDNA circles. In other species, overexpression of SIR-2.1 sirtuin gene extends life by activating protein complex DAF-16/forkhead box (FOXO) [40]. SIR-2.1 is able to activate DAF-16 directly, by desacetilation and mammals SIRT1 is known to deacetylate FOXO proteins in response to oxidative stress [40]. The discovery that sirtuin can deacetylate FOXO proteins directly as well as

the fact that the mutant insulin/IGF pathway-1 (insulin growth factor-1) do not need Sir-2.1 for increased longevity suggests that they may influence sirtuins through DAF-16/FOXO complex and the independent pathway of insulin and IGF-1 signaling in some species [40]. Sirtuin overexpression has not been able to extend the lifespan of mammals [43].

Resveratrol was the only one among the various investigated substances that effectively increased longevity of several species in the laboratory [43] except mammals. Thus, this substance has been exhaustively studied with regard to its therapeutic capacity in the prevention or reduction of diseases related to aging and, in relation to its mechanism of action related to a powerful activation of sirtuins, especially SIRT1, exerting a similar action to the RC [43]. This conclusion does not necessarily mean that sirtuins may not extend longevity in mammals, because the relationship between resveratrol and sirtuins is not fully understood. Resveratrol cannot actually activate sirtuins in a simple way. It stimulates SIRT1 in mammals by deacetylating a fluorescent substrate, but not the native SIRT1 substrates that have been tested [41,43]. As NAD+ and NADH (nicotinamide adenine dinucleotide dehydrogenase) are important regulators of metabolism, sirtuins are good candidates to represent proteins that respond to dietary restriction. However, whatever the effects of food restriction on the levels of NAD+ and NADH that occur, it is clear that dietary restriction does not increase sirtuin activity in yeast [41]. In flies, only one way of dietary restriction mode was tested, and here sirtuins were necessary for life extension [40]. Similarly, in rats, chronic dietary restriction cannot increase lifespan in the absence of SIRT1 [44]. Consistent with sirtuin sensors being nutrients, they regulate a wide variety of metabolic pathways and stress in response to dietary restriction in mice [44].

Some studies show that plants such as *Lithospermum erythrorhizon*, *Panax ginseng*, *Ginkgo biloba*, and *Rhodiola rosea* can exert beneficial effects on cell senescence and longevity, which are associated with aging process by a mechanism that involves SIRT1 [45–48] and one study evaluated the role of Japanese Humulus Emulsion (EHJ) in this process [49]. The function of sirtuins in modulating life in yeast has been recognized for over a decade; however, the ability of sirtuins to extend life expectancy in other organisms remains controversial. There are seven homologues of sirtuin (SIRT1-7) in mammals, of which SIRT1 is the most extensively studied. In mammals, it has been shown that antiaging mechanism from CR (caloric restriction) involves activation of SIRT1 in numerous tissues [50]. Thus, increased expression of SIRT1 in mice results in phenotypes that resemble life-prolonging RC effects [51]. Since it was discovered that SIRT1 plays a key role in life modulation, the protein has attracted increasingly more attention as a potential drug target capable of slowing onset of aging and prolonging life. For example, the polyphenol resveratrol related to SIRT1 exerts a beneficial effect on lifespan [52,53]. In addition, low levels of SIRT1 were observed in the heart tissue of elderly rats [54]. Therefore, it is possible that life-prolongation effect by EHJ is mediated by SIRT1 regulation. However, the effect exerted by EHJ on lifespan of higher order organisms is yet to be fully elucidated.

AMP Kinase

AMP kinase is a nutrient and energy sensor that activates catabolic pathways and inhibits anabolic pathways when AMP/ATP cellular ratio increases. Overexpression of AMP kinase prolongs life in *Caenorhabditis elegans* [55], and metformin, an antidiabetic drug, which activates AMP kinase, can extend the lifespan of mice [56]. AMP kinase is also required for changes in insulin/IGF-1 capable of extending the lifespan of worms [55], but exactly how it fits in this pathway is not known. AMP kinase can also extend lifespan in response to food restriction (Fig. 18.5) [55].

In *C. elegans*, extension of life span when food is limited is initiated in middle age and requires the participation of AMP kinase, which appears to act by phosphorylating and directly activating DAF-16/FOXO [57]. This AMP kinase pathway is not required for extending the longevity due to a reduced availability of food [57] and, on the other hand, genes involved in extension of life related to food reduction are not required when feed restriction is initiated in middle age [58].

One trial [49] showed that EHJ, whose active components are quercetin and luteolin, displays antioxidant and antiaging action via modulation of the AMPK pathway (AMP activated protein kinase)—SIRT1. EHJ activates AMPK in human fibroblast cells. It has been demonstrated that AMPK plays a key role in the process of aging and the determination of life time [59]. Overexpression of AMPK has been associated with prolonged lifespan in *C. elegans* and *Drosophila* fruit fly [60]. Thus, several studies have shown that activation of AMPK is involved in lifespan [59,60,57,61]. Moreover, other authors have demonstrated that presence of AMPK is essential for lifespan extension by RC in *C. elegans* by phosphorylation of FOXO transcription factor [57]. Remarkably, AMPK has been reported as capable of phosphorylating FOXO3 in mammalian cells, suggesting that the modulation of AMPK by FOXO can be preserved among species [61]. Results of EHJ use were consistent with previous studies, which demonstrated that small molecules such as chicoric acid and metformin, are able to prolong life of worms through modulation of AMPK expression [62,63]. In addition, previous studies have reported that luteolin and quercetin, activate AMPK, indicating that these substances may contribute to EHJ effect on AMPK expression levels

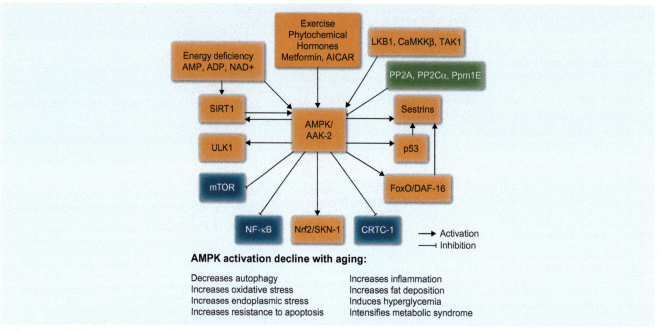

FIG. 18.5 Schematic view of the target signaling pathways of AMPK activation. AMP kinase activity can be stimulated by energy deficiency and various physiological and chemical agents, for example metformin and many phytochemicals. *AMPK*, AMP-activated protein kinase; *NAD+*, nicotine adenine dinucleotide; *SIRT1*, Sirtuin 1; *FOXO*, forkhead box; *ADP*, adenosine diphosphate; *p53*, protein 53; *MTOR*, mammalian target of rapamycin; *NF-κB*, nuclear transcription factor; *CRTC-1*, CREB regulated transcription coactivator 1; *AAK-2*, AMP activated kinase AAK-2. Adapted from Lee H, Cho JS, Lambacher N, et al. The *Caenorhabditis elegans* AMP-activated protein kinase AAK-2 is phosphorylated by LKB1 and is required for resistance to oxidative stress and for normal motility and foraging behavior. J Biol Chem 2008;283(22):14988–93.

[64–66]. The results of this study [66] showed that EHJ was able to prolong the lifespan of yeast cells. Other experiments have shown that EHJ unregulates the proteins associated with sirtuin 1 longevity and AMP activated protein kinase and that effectively inhibited production of reactive oxygen species (ROS). In addition, antioxidant potential of the active components of EHJ, including luteolin, luteolin-7 glycoside, quercetin and quercitrin, was evaluated and results demonstrated that these flavonoids were able to remove cell ROSs and in intracellular systems. In summary, results revealed that EHJ has potential for antioxidant activity; however, more research in vivo is needed in order to develop antiaging drug agents safely and effectively.

Insulin/Insulin-Like Growth Factor

Data obtained using the commonly accepted models to study longevity demonstrates that reduction of insulin/IGF-1 signal receptor pathway leads to increased lifespan. It would act as a the longevity module regulator in the prolongation of life in humans [67]. Inhibition of insulin/IGF-1 signaling alter longevity due to changes in gene expression by the DAF-16, a FOXO transcription factor, heat shock transcription factor (HSF-1), and SKN-1 [68], a response factor to xenobiotics similar to Nrf (nuclear factor erythroid-2 related factor). These transcription factors, in turn, regulate positively or negatively several genes that act to produce effects on longevity [69]. Disturbance of insulin/IGF-1 pathway activity appears to increase longevity in humans [40] (Fig. 18.6).

Mutations known to impair IGF-1 receptor function are over-represented in a cohort of Ashkenazi [70] centenarian Jews and DNA variants at insulin receptor gene linked to longevity in a Japanese cohort [71]. AKT variants and FOXO3A [72] have been associated to longevity in different populations throughout the world. Among Germans [72,73], FOXO3A variants are larger in centenarians than in the 90-year-old group, reinforcing the hypothesis that these variants increase lifetime. The FOXO1 gene variants are also linked to increased longevity in Americans and Chinese [73,74]. It is impressive

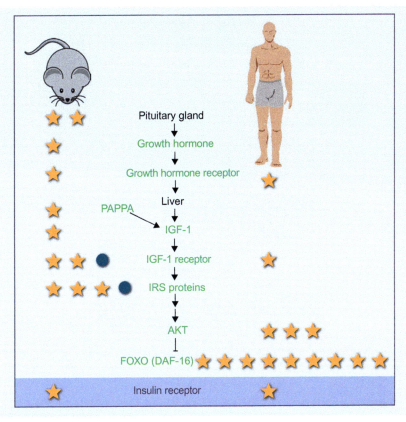

FIG. 18.6 Insulin/IGF-1 and FOXO signaling pathways affect life in mice and humans, since plasma IGF-1 is produced by the liver in response to growth hormone secreted from pituitary gland. Furthermore, PAPPA (pregnancy associated plasma protein), a metalloproteinase which inactivates IGF-1 binding proteins available increases IGF-1 levels. (Inactivation of PAPPA reduces IGF-1 signaling). In response to IGF-1, IGF-1 activates a downstream receptor signaling pathway that contains many proteins, which showed to affect life, such as IRS proteins (insulin receptor substrate protein 1 and 2) and AKT (a kinase that phosphorylates and inactivates FOXO transcription factors). Each *star* represents a long-term mutant strain of mouse (left) or human cohort that DNA variants are associated with increased longevity (right). Insulin receptor studies are shown below the line. The *star mouse* IGF-1 is a study in which the low IGF-1 levels were correlated to longevity among 31 strains of mice, providing strong evidence that this hormone affects longevity. *Circles* represent studies that similar mutants were examined, but the life extension was not observed. Adapted from Kenyon CJ. The genetics of ageing. Nature 2010;464(7288):504–12 [Review]. Erratum in: Nature 2010;467(7315):622.

how FOXO variants are consistently associated with longevity. This is possibly why FOXO proteins act in many pathways that affect longevity [75] (Fig. 18.7).

TOR Protein Kinase

TOR protein kinase is an amino acid and nutrients sensor that stimulates growth and blocks pathways salvage, such as autophagy when food is abundant. Inhibiting TOR pathway increases lifespan in many species, from yeast to mice [76–78].

TOR inhibition increases resistance to environmental stress [79] and, in certain species, such inhibition appears to activate a pathway that is distinct from insulin/IGF-1, since it increases life spam regardless of DAF- 16/FOXO [79,80]. TOR inhibition also has effects on the translation that involves breathing in response to longevity and dietary restriction [81]. When nutrient levels and TOR activity fall, transcription levels also fall, affecting lifespan [77,78,82–84].

Elderly knockout mice for S6 kinase increased body oxygen consumption rates [85], suggesting that this life extension mechanism can be maintained. TOR inhibition also stimulates autophagy, which, as in the mutant insulin/IGF-1 pathway, is necessary to extend lifespan [86,87].

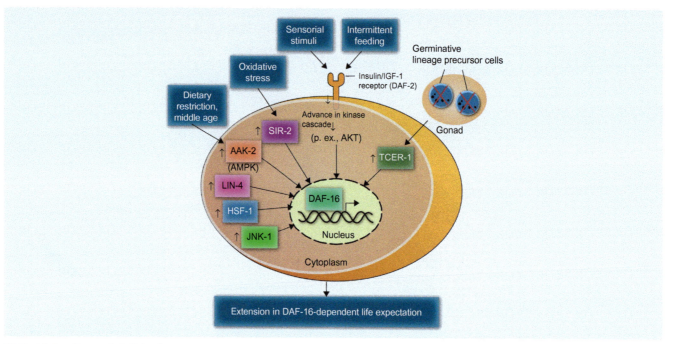

FIG. 18.7 *Caenorhabditis elegans* transcription factor DAF-16/FOXO promotes longevity in response to many stimuli. *Arrows* to the left of the protein depict increased or decreased gene expression. Overexpression of SIR-2 sirtuin, the heat shock transcription factor HSF-1, the development timing micro-RNA LIN-4 AAK-2 (kinase subunit AMP) kinase, Jun 1 (JNK-1) or the transcription of the REER-1 elongation factor increases life. Inhibition of DAF-2 insulin/IGF-1 receptor, or components of a kinase cascade also extends downstream life. In each case, extension of lifespan is dependent on DAF-16. Signals that activate these pathways include dietary restriction, oxidative stress, sensory cues and ablation of germ cell precursors in the developing gonad (indicated by *red crosses*). Some of the proteins listed separately in the figure can act together on the same path. Adapted from Lunetta KL, D'Agostino Sr RB, Karasik D, et al. Genetic correlates of longevity and selected age-related phenotypes: a genome-wide association study in the Framingham Study. BMC Med Genet 2007;8(Suppl. 1):S13.

Of all the nutrients detection pathways, TOR pathway has been most consistently linked to food restriction. TOR inhibition mimics the physiological effects of food restriction, and in animal studies, the lifespan extension produced by TOR inhibition was not increased by dietary restriction [76,77,79] in worms as, in flies, chronic dietary restriction extends lifespan, as at least in part, to stimulate breathing [81]. The subjacent mechanism depends not only on translational control, since it requires the SKN-1 transcription factor.

Mutant S6 kinase mice display gene expression patterns and durability similar to those triggered by dietary restriction [88], suggesting that the TOR/S6 kinase pathway also influences the response to dietary restriction in mammals.

TOR in animals of the mammalian class operates through a complex called mTORC1 (rapamycin target 1 complex). When there is food in abundance, which causes increased production of insulin and proteins, mTORC1 acts by stimulating cellular

components synthesis, leading to growth and cell division and the same time reversing autophagy; however, when food is scarce, mTORC1 acts causing cells to self-preserve avoiding replication, promoting autophagy to supply necessary substances for cell repair and energy generation. After maturation, continuous mTORC1 activity leads to excessive protein synthesis and formation of destructive protein aggregates destructive and can also lead to undue production of smooth muscle cells, which may favor the development of atherosclerosis and may also cause the decline of cell function and senescence, promoting the development of toxicity by suppressing autophagy, allowing the material to remain in cells damaged condition, so the inhibition of mTORC1 becomes a therapeutic target for interfering with these mechanisms, and thereby delaying aging [86,87]. TOR complex 1 (TORC1) is sensitive to rapamycin and is the central element of an integrated TOR signaling network (Fig. 18.8). It controls and integrates a diverse array of intracellular and

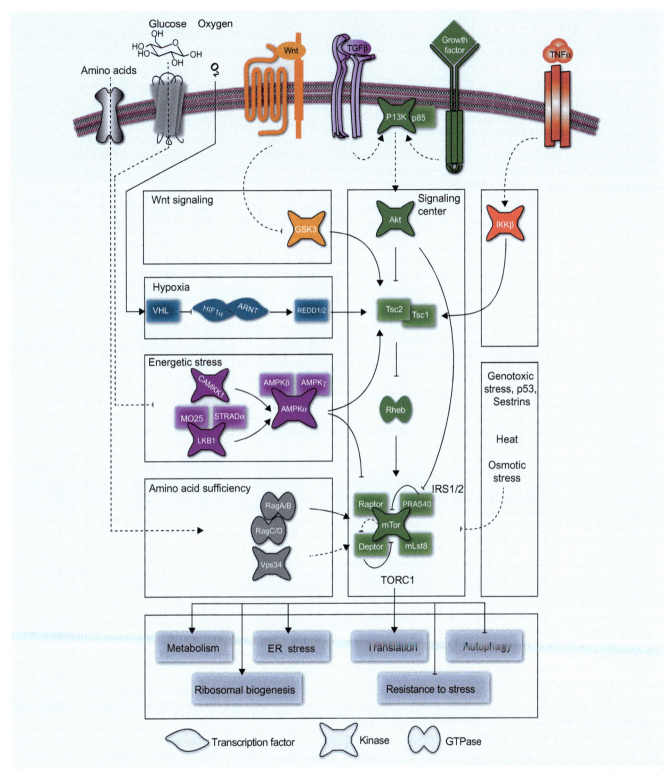

FIG. 18.8 TOR signaling pathways. TORC1 (transducer of regulated CREB activity 1) integrates intra and extracellular environmental stimuli through signaling modules that detect and transmit multiple entries for a central "signaling core" *(in green)*. This figure summarizes biochemical evidence from several studies that have identified TORC1 inputs and outputs. Multiple outputs *(in orange)* that keep cell growth in balance with the environment are regulated by TORC1. Adapted from Selman C, Tullet JM, Wieser D, et al. Ribosomal protein S6 kinase 1 signaling regulates mammalian life span. Science 2009;326:140–4.

extracellular parameters and controls cell size of cell proliferation and lifespan by a variety of downstream pathways. It consists of the serine/threonine kinase TOR and their associated proteins Raptor (regulatory associated protein TOR), mLst8, PRAS40, and (in mammals) Deptor. Moreover, the TOR 2 complex (TORC2) is insensitive to rapamycin; it controls serum kinase activity, is induced by glucocorticoids (SGK), and contributes to the full activation of Akt. It contains TOR, Rictor (partner rainsensitive pamicina mTOR), SIN1, Proctor/PRR5L, mLst8, and (again, only in mammals) Deptor [89].

Process Changes—Perspectives

Vascular changes specific of aging, including arterial stiffness and endothelial dysfunction with consequent elevation of systolic arterial pressure and pulse pressure, precede clinically manifested cardiovascular disease and vascular aging make vascular aging the greatest risk factor for the development of atherosclerotic disease, hypertension, and stroke.

Lifestyle has a major impact on the speed and magnitude with which this process of vascular aging takes place. Interventions on lifestyle, with the purpose of slowing down the vascular aging process, should be adopted before installation of clinically manifested disease. A study compared endothelial function in individuals of a rural population in China to an urban Australian population. The prevalence of coronary artery disease is about 20% lower in Chinese rural population than in industrialized countries. This study showed that Chinese presented less endothelial dysfunction than urban center Australians [90].

A low-salt diet is associated with lower aging-related arterial stiffness [91]. There is a huge prevalence of physical inactivity in the elderly [92]. Aging increases the risk of cardiovascular diseases, particularly due to the increased stiffness of large arteries and development of endothelial dysfunction. In contrast, regular exercise acts against development of arterial stiffness and endothelial dysfunction associated with increasing age [93,94]. Fitness improves function of baroreceptors [95] and endothelial function in elderly subjects [96]. Therefore, the vascular aging process can be slowed down with lifestyle changes such as regular physical activity and healthy diet, with consequent improvement in endothelial function and lower arterial stiffness.

There are few controlled studies on the effects of calorie restriction in humans. CALERIE study group [97] assessed feasibility, safety and effects of CR on longevity predictors, risk factors and quality of life in nonobese humans. A total of 218 subjects aged 21–51 years were randomized to a 2-year intervention designed to achieve 25% of CR or an uncontrolled diet. Variables assessed were change on resting metabolic rate (adjusted for weight change and body temperature); tri-iodothyronine (T3), and plasma concentrations of tumor necrosis factor-alpha as well as physiological and psychological exploratory measures. The protocol was completed by 82% of the CR group and 95% of the group with free diet. Weight reduction was observed only in CR group: this was 10.4%. The adjusted resting metabolic rate was higher in the CR group at 12 months ($P = 0.04$), but not at 24 months. Body temperature did not differ between groups. T3 level decreased more in the CR group at 12 and 24 months ($P < 0.001$), while tumor necrosis factor-alpha significantly decreased only at 24 months ($P = 0.02$). CR group showed greater reduction in cardiometabolic risk factors and daily energy expenditure adjusted for the change in weight without adverse effects on quality of life. The authors concluded that sustained CR is feasible in nonobese humans. The effects of CR observed in this study suggest potential benefits related to aging, and these will be further elucidated with ongoing studies.

Dietary restriction mitigates many negative effects of aging, consequently promotes health, and increases longevity. Although over the last few years, extensive research has been dedicated to understanding the biology of aging, the precise mechanistic aspects of dietary restriction are yet to be resolved. However, literature also accumulates conflicting evidence as to how dietary restriction improves mitochondrial performance and if this is enough to slow cell deterioration and the body dependent on age [98].

CONCLUSIONS

Age is an important factor modifying structure and endothelial function in humans. Mechanisms involved in reduction of endothelium-dependent vasodilation related to age are focused, above all, on a primary change in the route of L-arginine-NO. Oxidative stress in arterial wall plays an important role in elderly because it compromises NO availability. Impairment of endothelium-dependent vasodilation in hypertension appears to represent an acceleration of the changes seen in aging. Endothelial dysfunction contributes in fundamental ways to make vascular aging a strong risk factor for developing cardiovascular diseases.

References

[1] Yeap BB, McCaul KA, Flicker L, et al. Diabetes, myocardial infarction and stroke are distinct and duration-dependent predictors of subsequent cardiovascular events and all-cause mortality in older men. J Clin Endocrinol Metab 2015;100(3):1038–47.

[2] Wilsgaard T, Loehr LR, Mathiesen EB, et al. Cardiovascular health and the modifiable burden of incident myocardial infarction: the Tromsø Study. BMC Public Health 2015;15:221.

[3] Ong P, Athanasiadis A, Hill S, et al. Coronary microvascular dysfunction assessed by intracoronary acetylcholine provocation testing is a frequent cause of ischemia and angina in patients with exercise-induced electrocardiographic changes and unobstructed coronary arteries. Clin Cardiol 2014;37(8):462–7.

[4] Tarhouni K, Freidja ML, Guihot AL, et al. Role of estrogens and age in flow-mediated outward remodeling of rat mesenteric resistance arteries. Am J Physiol Heart Circ Physiol 2014;307(4):H504–14.

[5] Feher A, Broskova Z, Bagi Z. Age-related impairment of conducted dilation in human coronary arterioles. Am J Physiol Heart Circ Physiol 2014;306(12):H1595–601.

[6] Donato AJ, Morgan RG, Walker AE, et al. Cellular and molecular biology of aging endothelial cells. J Mol Cell Cardiol 2015;89(Pt B):122–35.

[7] Conti V, Corbi G, Simeon V, et al. Aging-related changes in oxidative stress response of human endothelial cells. Aging Clin Exp Res 2015;27(4):547–53.

[8] Sena CM, Pereira AM, Seiça R. Endothelial dysfunction—a major mediator of diabetic vascular disease. Biochim Biophys Acta 2013;1832(12):2216–31.

[9] Gerhard M, Roddy MA, Creager SJ, et al. Aging progressively impairs endothelium-dependent vasodilation in forearm resistance vessels of humans. Hypertension 1996;27:849–53.

[10] Greider CW. Telomeres and senescence: the history, the experiment, the future. Curr Biol 1998;8:R178–81.

[11] Borghini A, Giardini G, Tonacci A, et al. Chronic and acute effects of endurance training on telomere length. Mutagenesis 2015;30 (5):711–0.

[12] Srettabunjong S, Satitsri S, Thongnoppakhun W, et al. The study on telomere length for age estimation in a Thai population. Am J Forensic Med Pathol 2014;35(2):148–53.

[13] Tower J. Programmed cell death in aging. Ageing Res Rev 2015;23 (Pt A):90–100.

[14] Bodnar AG, Quellete M, Frolkis M, et al. Extension of life-span by introduction of telomerase into normal human cells. Science 1998;279:349–52.

[15] Chiodi I, Belgiovine C, Zongaro S, et al. Super-telomeres in transformed human fibroblasts. Biochim Biophys Acta 2013;1833 (8):1885–93.

[16] Rudolph KL, Chang S, Lee HW, et al. Longevity, stress response, and cancer in aging telomerase-deficient mice. Cell 1999;96:701–12.

[17] Babizhayev MA, Vishnyakova KS, Yegorov YE. Oxidative damage impact on aging and age-related diseases: drug targeting of telomere attrition and dynamic telomerase activity flirting with imidazole-containing dipeptides. Recent Pat Drug Deliv Formul 2014;8(3):163–92.

[18] Jeanclos E, Schork NJ, Kyvik KO, et al. Telomere length inversely correlates with pulse pressure and is highly familial. Hypertension 2000;36:195–200.

[19] Hunt SC, Kimura M, Hopkins PN, et al. Leukocyte telomere length and coronary artery calcium. Am J Cardiol 2015;116(2):214–8.

[20] Minamino T, Miyauchi H, Yoshida T, et al. Endothelial cell senescence in human atherosclerosis: role of telomere in endothelial dysfunction. Circulation 2002;105:1541–4.

[21] Ramezani TF, Behboudi-Gandevani S, Ghasemi A, et al. Association between serum concentrations of nitric oxide and transition to menopause. Acta Obstet Gynecol Scand 2015;94 (7):708–14.

[22] Spier SA, Delp MD, Meininger CJ, et al. Effects of ageing and exercise training on endothelium-dependent vasodilatation and structure of rat skeletal muscle arterioles. J Physiol 2004;556(Pt 3):947–58.

[23] Arora DP, Hossain S, Xu Y, et al. Nitric oxide regulation of bacterial biofilms. Biochemistry 2015;54(24):3717–28.

[24] Alfieri A, Ong AC, Kammerer RA, et al. Angiopoietin-1 regulates microvascular reactivity and protects the microcirculation during acute endothelial dysfunction: role of eNOS and VE-cadherin. Pharmacol Res 2014;80:43–51.

[25] Radomski MW, Palmer RMJ, Moncada S. An L-arginine-nitric oxide pathway present in human platelets regulates aggregation. Proc Natl Acad Sci U S A 1990;87:5193–7.

[26] Cockrell A, Laroux FS, Jourd'heuil D, et al. Role of inducible nitric oxide synthase in leukocyte extravasation in vivo. Biochem Biophys Res Commun 1999;257(3):684–6.

[27] Shoker AS, Yang H, Murabit MA, et al. Analysis of the in vitro effect of exogenous nitric oxide on human lymphocytes. Mol Cell Biochem 1997;171(1–2):75–83.

[28] Ozkor MA, Hayek SS, Rahman AM, et al. Contribution of endothelium-derived hyperpolarizing factor to exercise-induced vasodilation in health and hypercholesterolemia. Vasc Med 2015;1:14–22.

[29] Taddei S, Virdis A, Mattei P, et al. Hypertension causes premature aging of endothelial function in humans. Hypertension 1997;29 (3):736–43.

[30] Taddei S, Virdis A, Ghiadoni L, et al. Age-related reduction of NO availability and oxidative stress in humans. Hypertension 2001;38 (2):274–9.

[31] Risbano MG, Gladwin MT. Therapeutics targeting of dysregulated redox equilibrium and endothelial dysfunction. Handb Exp Pharmacol 2013;218:315–49.

[32] Riccioni G, Scotti L, D'Orazio N, et al. ADMA/SDMA in elderly subjects with asymptomatic carotid atherosclerosis: values and site-specific association. Int J Mol Sci 2014;15(4):6391–8.

[33] Rajagopalan S, Kurz S, Munzel T, et al. Angiotensin II-mediated hypertension in the rat increases vascular superoxide production via membrane NADH/NADPH oxidase activation: contribution to alterations of vasomotor tone. J Clin Invest 1996;97:1916–23.

[34] Wang Q, Yang M, Xu H, et al. Tetrahydrobiopterin improves endothelial function in cardiovascular disease: a systematic review. Evid Based Complement Alternat Med 2014;2014:850312.

[35] Alley H, Owens CD, Gasper WJ, et al. Ultrasound assessment of endothelial-dependent flow-mediated vasodilation of the brachial artery in clinical research. J Vis Exp 2014;92:e52070.

[36] Sun Z. Aging, arterial stiffness, and hypertension. Hypertension 2015;65(2):252–6.

[37] Cunha PG, Cotter J, Oliveira P, et al. Pulse wave velocity distribution in a cohort study: from arterial stiffness to early vascular aging. J Hypertens 2015;33(7):1438–45.

[38] Villella E, Cho JS. Effect of aging on the vascular system plus monitoring and support. Surg Clin North Am 2015;95(1):37–51.

[39] Taddei S, Bruno RM. Endothelial dysfunction in hypertension: implications for treatment. J Hypertens 2015;33(6):1137–8.

[40] Kenyon CJ. The genetics of ageing. Nature 2010;464(7288):504–12. [Review]. Erratum in: Nature 2010;467(7315):622.

[41] Kwon HS, Ott M. The ups and downs of SIRT1. Trends Biochem Sci 2008;33(11):517–25.

[42] Winogradoff D, Echeverria I, Potoyan DA, et al. The acetylation landscape of the H4 histone tail: disentangling the interplay between the specific and cumulative effects. J Am Chem Soc 2015;137(19):6245–53.

[43] Beher D, Wu J, Cumine S, et al. Resveratrol is not a direct activator of SIRT1 enzyme activity. Chem Biol Drug Des 2009;74:619–24.

[44] Li L, Sun Q, Li Y, et al. Overexpression of SIRT1 Induced by resveratrol and inhibitor of miR-204 suppresses activation and proliferation of microglia. J Mol Neurosci 2015;56(4):858–67.

[45] Yoo HG, Lee BH, Kim W, et al. Lithospermum erythrorhizon extract protects keratinocytes and fibroblasts against oxidative stress. J Med Food 2014;17:1189–96.

[46] Hwang E, Lee TH, Park SY, et al. Enzyme-modified Panax ginseng inhibits UVB-induced skin aging through the regulation of procollagen type I and MMP-1 expression. Food Funct 2014;5:265–74.

[47] Kampkötter A, Pielarski T, Rohrig R, et al. The Ginkgo biloba extract EGb761 reduces stress sensitivity, ROS accumulation and expression of catalase and glutathione S-transferase 4 in Caenorhabditis elegans. Pharmacol Res 2007;55:139–47.

[48] Gospodaryov DV, Yurkevych IS, Jafari M, et al. Lifespan extension and delay of age-related functional decline caused by Rhodiola rosea depends on dietary macronutrient balance. Longev Healthspan 2013;2:5.

[49] Sung B, Chung JW, Bae HR, et al. Humulus japonicus extract exhibits antioxidative and anti-aging effects via modulation of the AMPK-SIRT1 pathway. Exp Ther Med 2015;9(5):1819–26.

[50] Cohen HY, Miller C, Bitterman KJ, et al. Calorie restriction promotes mammalian cell survival by inducing the SIRT1 deacetylase. Science 2004;305:390–2.

[51] Bordone L, Cohen D, Robinson A, et al. SIRT1 transgenic mice show phenotypes resembling calorie restriction. Aging Cell 2007;6:759–67.

[52] Valenzano DR, Terzibasi E, Genade T, et al. Resveratrol prolongs lifespan and retards the onset of age-related markers in a short-lived vertebrate. Curr Biol 2006;16:296–300.

[53] Baur JA, Pearson KJ, Price NL, et al. Resveratrol improves health and survival of mice on a high-calorie diet. Nature 2006;444:337–42.

[54] Tong C, Morrison A, Mattison S, et al. Impaired SIRT1 nucleocytoplasmic shuttling in the senescent heart during ischemic stress. FASEB J 2013;27:4332–42.

[55] Lee H, Cho JS, Lambacher N, et al. The Caenorhabditis elegans AMP-activated protein kinase AAK-2 is phosphorylated by LKB1 and is required for resistance to oxidative stress and for normal motility and foraging behavior. J Biol Chem 2008;283(22):14988–93.

[56] Anisimov VN, Popovich IG, Zabezhinski MA, et al. Sex differences in aging, life span and spontaneous tumorigenesis in 129/Sv mice neonatally exposed to metformin. Cell Cycle 2015;14(1):46–55.

[57] Greer EL, Dowlatshahi D, Banko MR, et al. An AMPK-FOXO pathway mediates longevity induced by a novel method of dietary restriction in C. elegans. Curr Biol 2007;17:1646–56.

[58] Castelein N, Cai H, Rasulova M, et al. Lifespan regulation under axenic dietary restriction: a close look at the usual suspects. Exp Gerontol 2014;58:96–103.

[59] Burkewitz K, Zhang Y, Mair WB. AMPK at the nexus of energetics and aging. Cell Metab 2014;20:10–25.

[60] Apfeld J, OConnor G, McDonagh T, et al. The AMP-activated protein kinase AAK-2 links energy levels and insulin-like signals to lifespan in C. elegans. Genes Dev 2004;18:3004–9.

[61] Greer EL, Oskoui PR, Banko MR, et al. The energy sensor AMP-activated protein kinase directly regulates the mammalian FOXO3 transcription factor. J Biol Chem 2007;282:30107–19.

[62] Schlernitzauer A, Oiry C, Hamad R, et al. Chicoric acid is an antioxidant molecule that stimulates AMP kinase pathway in L6 myotubes and extends lifespan in Caenorhabditis elegans. PLoS One 2013;8:e78788.

[63] Onken B, Driscoll M. Metformin induces a dietary restriction-like state and the oxidative stress response to extend C. elegans healthspan via AMPK, LKB1, and SKN-1. PLoS One 2010;5:e8758.

[64] Yin Y, Li W, Son YO, et al. Quercitrin protects skin from UVB-induced oxidative damage. Toxicol Appl Pharmacol 2013;269:899.

[65] Eid HM, Martineau LC, Saleem A, et al. Stimulation of AMP-activated protein kinase and enhancement of basal glucose uptake in muscle cells by quercetin and quercetin glycosides, active principles of the antidiabetic medicinal plant Vaccinium vitis-idaea. Mol Nutr Food Res 2010;54:991–1003.

[66] Liu JF, Ma Y, Wang Y, et al. Reduction of lipid accumulation in HepG2 cells by luteolin is associated with activation of AMPK and mitigation of oxidative stress. Phytother Res 2011;25:588–96.

[67] Villa F, Carrizzo A, Spinelli CC, et al. Genetic analysis reveals a longevity-associated protein modulating endothelial function and angiogenesis. Circ Res 2015;117(4):333–45.

[68] Tullet JM, Hertweck M, An JH, et al. Direct inhibition of the longevity-promoting factor SKN-1 by insulin-like signaling in C. elegans. Cell 2008;132(6):1025–38.

[69] Goudeau J, Bellemin S, Toselli-Mollereau E, et al. Fatty acid desaturation links germ cell loss to longevity through NHR-80/HNF4 in C. elegans. PLoS Biol 2011;9(3)e1000599.

[70] Suh Y, Atzmon G, Cho MO, et al. Functionally significant insulin-like growth factor I receptor mutations in centenarians. Proc Natl Acad Sci U S A 2008;105:3438–42.

[71] Kojima T, Kamei H, Aizu T, et al. Association analysis between longevity in the Japanese population and polymorphic variants of genes involved in insulin and insulin-like growth factor 1 signaling pathways. Exp Gerontol 2004;39:1595–8.

[72] Flachsbart F, Caliebe A, Kleindorp R, et al. Association of FOXO3A variation with human longevity confirmed in German centenarians. Proc Natl Acad Sci U S A 2009;106:2700–5.

[73] Li Y, Wang WJ, Cao H, et al. Genetic association of FOXO1A and FOXO3A with longevity trait in Han Chinese populations. Hum Mol Genet 2009;18:4897–904.

[74] Lunetta KL, D'Agostino Sr. RB, Karasik D, et al. Genetic correlates of longevity and selected age-related phenotypes: a genome-wide association study in the Framingham Study. BMC Med Genet 2007;8(Suppl. 1):S13.

[75] Kenyon C. The plasticity of aging: insights from long-lived mutants. Cell 2005;120:449–60.

[76] Stracka D, Jozefczuk S, Rudroff F, et al. Nitrogen source activates TOR (target of rapamycin) complex 1 via glutamine and independently of Gtr/Rag proteins. J Biol Chem 2014;289(36):25010–20.

[77] Zhang X, Camprecíós G, Rimmelé P, et al. FOXO3-mTOR metabolic cooperation in the regulation of erythroid cell maturation and homeostasis. Am J Hematol 2014;89(10):954–63.

[78] Harrison DE, Strong R, Sharp ZD, et al. Rapamycin fed late in life extends lifespan in genetically heterogeneous mice. Nature 2009;460:392–5.

[79] Hansen M, Taubert S, Crawford D, et al. Lifespan extension by conditions that inhibit translation in Caenorhabditis elegans. Aging Cell 2007;6:95–110.

[80] Yano T, Ferlito M, Aponte A, et al. Pivotal role of mTORC2 and involvement of ribosomal protein S6 in cardioprotective signaling. Circ Res 2014;114(8):1268–80.

[81] Zid BM, Rogers AN, Katewa SD, et al. 4E-BP extends lifespan upon dietary restriction by enhancing mitochondrial activity in Drosophila. Cell 2009;139:149–60.

[82] Amiel E, Everts B, Fritz D, et al. Mechanistic target of rapamycin inhibition extends cellular lifespan in dendritic cells by preserving mitochondrial function. J Immunol 2014;193(6):2821–30.

[83] Steffen KK, MacKay VL, Kerr EO, et al. Yeast life span extension by depletion of 60s ribosomal subunits is mediated by Gcn4. Cell 2008;133:292–302.

[84] Syntichaki P, Troulinaki K, Tavernarakis N. eIF4E function in somatic cells modulates ageing in *Caenorhabditis elegans*. Nature 2007;445:922–6.

[85] Um SH, Frigerio F, Watanabe M, et al. Absence of S6K1 protects against age- and diet-induced obesity while enhancing insulin sensitivity. Nature 2004;43:200–5.

[86] Tóth ML, Sigmond T, Borsos E, et al. Longevity pathways converge on autophagy genes to regulate life span in *Caenorhabditis elegans*. Autophagy 2008;4:330–8.

[87] Bjedov I, Toivonen JM, Kerr F, et al. Mechanisms of life span extension by rapamycin in the fruit fly *Drosophila melanogaster*. Cell Metab 2010;11:35–46.

[88] Selman C, Tullet JM, Wieser D, et al. Ribosomal protein S6 kinase 1 signaling regulates mammalian life span. Science 2009;326:140–4.

[89] Kapahi P, Chen D, Rogers AN, et al. With TOR, less is more: a key role for the conserved nutrient-sensing TOR pathway in aging. Cell Metab 2010;11(6):453–65.

[90] Woo KS, McCrohon JA, Chook P, et al. Chinese adults are less susceptible than whites to age-related endothelial dysfunction. J Am Coll Cardiol 1997;30:113–8.

[91] Hu J, Jiang X, Li N, et al. Effects of salt substitute on pulse wave analysis among individuals at high cardiovascular risk in rural China: a randomized controlled trial. Hypertens Res 2009;32(4):282–8.

[92] DeVan AE, Seals DR. Vascular health in the ageing athlete. Exp Physiol 2012;97(3):305–10.

[93] Santos-Parker JR, LaRocca TJ, Seals DR. Aerobic exercise and other healthy lifestyle factors that influence vascular aging. Adv Physiol Educ 2014;38(4):296–307.

[94] Gutierrez J, Marshall RS, Lazar RM. Indirect measures of arterial stiffness and cognitive performance in individuals without traditional vascular risk factors or disease. JAMA Neurol 2015;72(3):309–15.

[95] Tomiyama H, Matsumoto C, Kimura K, et al. Pathophysiological contribution of vascular function to baroreflex regulation in hypertension. Circ J 2014;78(6):1414–9.

[96] Suboc TB, Knabel D, Strath SJ, et al. Associations of reducing sedentary time with vascular function and insulin sensitivity in older sedentary adults. Am J Hypertens 2016;29(1):46–53.

[97] Ravussin E, Redman LM, Rochon J, et al. A 2-year randomized controlled trial of human caloric restriction: feasibility and effects on predictors of health span and longevity. J Gerontol A Biol Sci Med Sci 2015;70(9):1097–104.

[98] Ruetenik A, Barrientos A. Dietary restriction, mitochondrial function and aging: from yeast to humans. Biochim Biophys Acta 2015;1847(11):1434–47.

II. ENDOTHELIAL DYSFUNCTION AND CLINICAL SYNDROMES

19

Vascular Function and Cognitive Decline

Ivan Aprahamian, Fabiano Vanderlinde, and Marina Maria Biella

INTRODUCTION

The prevalence of dementia and other cognitive impairments has increased in recent years. The population of older people advanced very quickly mainly in developed countries. In 2000, in these countries, approximately 600 million people were 60 years old or more. The estimates indicate that in 2025 the number of older people shall reach 1.2 billion, and in 2050, 2 billion [1]. The calculations indicate that Vascular Dementia (VD) occurs in 1.2%–4.2% of adults above 65 years old [2], showing a longitudinal increase of incidence with aging [3], and may manifest in up to 30% of the population above 80 years old. Among the types of dementia, VD occupies the second place regarding prevalence. In the Americas and in Europe, VD answers for 10%–20% of all dementia cases [4].

Interestingly, these patients undergo a reasonable period of mild cognitive impairment without showing social or labor dysfunctions. This pre-dementia state, of high lability in its clinical evolution, is called Mild Vascular Cognitive Impairment (MVCI), which identifies clear vascular disorder as the main cause of the problem.

RISK FACTORS

The main risk factors for VD and MVCI are related to the physiopathology of the atherosclerotic vascular disease, especially cerebrovascular diseases (Table 19.1). However, there is certain scientific controversy regarding the negative influence of several factors, such as hypertension, diabetes, and dyslipidemia, among others. These differences are partially explained by different control group selections. In a large epidemiological study [5], metabolic syndrome, which incorporates a series of classic factors that precipitate vascular diseases, was not significantly associated with VD, while another similar study found a strong association [6]. Demographic factors, such as male gender and advanced age, besides the factors associated with neuroplasticity, such as low educational level and little physical and mental activity, also are closely related to higher chance of developing MVCI and VD.

PHYSIOPATHOLOGY OF VASCULAR DEMENTIA

The pathology and underlying mechanisms of VD are still not completely known (Fig. 19.1). Macroscopically, three events closely related to cognitive impairment and dementia associated with vascular dysfunction are observed [7]: infarctions related to the major cerebral arteries with cortical and/or subcortical impairment, smaller subcortical or lacunar stroke following the distribution of penetrating arteries (thalamus, basal ganglia, internal capsule, cerebellum, and brainstem), and finally, periventricular chronic cerebral ischemia, causing chronic endothelial dysfunction. Chronic hypoperfusion reduces cerebrovascular blood flow, causing hypoxia, oxidative stress, and inflammation, with the hippocampus, periventricular white matter, and basal ganglia being most affected [8]. Normally, small artery cerebrovascular diseases involve both chronic microangiopathy and lacunar strokes, and are originated by the lipohyalinosis and microatheromatosis processes [9]. This process has a

253

TABLE 19.1 MCVI and VD Risk Factors

Demographic and environmental factors	• Age • Male gender • Low educational level • Little physical and mental activity
Genetic factors	• Related to cardiovascular and metabolic risk factors • Related to other diseases (e.g., CADASIL, sickle cell anemia)
Cardiovascular factors	• Cerebrovascular disease • Coronary disease • Previous stroke • Atrial fibrillation • Chronic kidney disease • Low cardiac output • Peripheral arterial disease • Arterial hypertension • Tobacco use • Inflammation: increased CRP, IL-1, IL-6, TNF-α • Dyslipidemia
Metabolic factors	• Diabetes mellitus type 2 • Resistance to insulin • Obesity • Metabolic syndrome

Stroke, *cerebrovascular accident*; CRP, *C-Reactive Protein*; IL-1, *interleukin-1*; IL-6, *interleukin-6*; TNF-α, *Alpha Tumoral Necrosis Factor*.

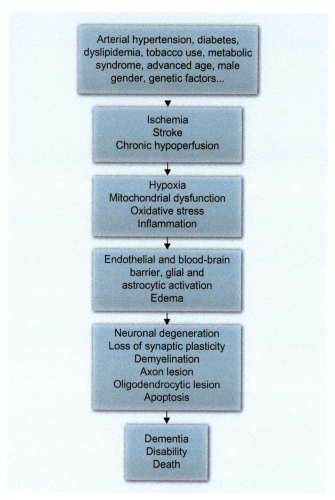

FIG. 19.1 VD physiopathology.

common vascular origin and is common in hypertensive, diabetic, and older people. The 3T magnetic resonance technology added cerebral microhemorrhages to this physiopathology. The bleeding was associated with both small artery cerebrovascular disease and arterial hypertension [10].

Hypoxia produces micro-vascular and neurovascular lesions, resulting in blood-brain barrier (BBB) dysfunction, swelling, and astrocytic and glial activation. The oxidative stress induced by hypoxia causes mitochondrial dysfunction, promoting neuronal death and apoptosis by released free radicals, reactive oxygen species, nitric oxide synthase, and malondialdehyde synthase. Increased BBB permeability allows infiltration of inflammatory factors, such as interleukins-1 and -6, alpha tumoral necrosis factor, C-reactive protein, and matrix metalloproteinase. Once inside the brain tissue, these inflammatory factors increase neuroglial inflammation, causing white matter lesions with demyelination, axon loss, oligodendrocyte neurodegeneration, loss of synaptic and dendritic plasticity and apoptosis [8].

Chronically, in all three events related to cognitive decline, there is endothelial dysfunction. The atherosclerosis process triggers a continued inflammatory and hypoperfusion condition, which reduces nitric oxide production and the anti-inflammatory protection of the endothelial wall. With the reduction in vasodilatation, we can infer the existence of progressive neuronal lesion and the development of apoptosis mechanisms, resulting in cognitive decline and subsequently dementia. The entire process continues with progressive endothelial dysfunction, breakage of the concomitant blood-brain barrier, and penetration of free radicals and other neurotoxic products (abnormal concentrations of cytokines, chemokines, and neurotransmitters) in the brain. This sequence of chronic hypoperfusion, hypoxia, inflammation, oxidative stress, and neuronal lesion gradually leads to dementia, causing disability and finally death [11].

Some exceptions to this scenario must be indicated, in which, regardless of vascular damage, the term dementia is not used due to the long clinical stability of the condition, despite substantial damage. Examples can be observed in subarachnoid hemorrhage, cognitive decline after cardiac bypass surgery and encephalopathy after prolonged cardiac arrest. Another important point is the frequent observation of cerebrovascular disease in small vessels in patients with Alzheimer's dementia, whose exact physiopathological role is still poorly established. Finally, location of vascular damage is by far more important than its volume, since strategic infarctions (hippocampus, angular gyrus, and cingulum gyrus, among others) may be disastrous and evolve into dementia right after the event.

CLINICAL PRESENTATION

The basic characteristic of VD is the development of cognitive impairment in multiple domains such as memory, aphasia, apraxia, agnosia, or executive dysfunction, which is sufficient to cause serious decline in daily activities in respect to a previous levels of performance [12,13]. Due to the large number of lesions in the frontal lobe, changes in humor and behavior are frequent, and may lead to depression, anxiety, and disorientation. The etiology of vascular lesion can vary a significantly, making its presentation very heterogeneous, which may be consequence of both cerebrovascular accident in major vessels and micro-vascular lesions. Clinical symptoms depend on the location and extension of the lesions, which may be focal, multifocal, or diffuse, especially involving the subcortex. The onset is generally sudden, according to the vascular lesion, and the course of decline is gradual, with periods of worsening followed by periods of stability and even certain improvement, in a pattern known as "step-by-step" deterioration. This fluctuating evolution is due to the occurrence of multiple cerebral infarctions over time. However, when lesions occur in the microvasculature, especially in white matter, basal ganglia, or thalamus, the onset of may be particularly more subtle and with slower progress. The acute and gradual character is the main difference between VD and Alzheimer's disease (AD), since the latter shows more insidious onset, with slow and progressive evolution, affecting mainly memory in the beginning. However, this distinction is not so clear in clinical practice [14]. Another point of interest to distinguish VD from other dementias

is its specific evolution. The first frequently shows long stability periods. For example, we can mention hypothetically two identical patients: one with AD and the other with VD. After 3 years of evolution, the AD patient already shows cognitive impairment, while the VD patient may be relatively stable.

Furthermore, many risk factors are common to both diseases and part of their physiopathological processes is interconnected. The blood-brain barrier dysfunction caused by vascular lesion, for example, may lead to increased deposition of beta-amyloid protein in the brain. Therefore, a neurodegenerative process associated with a cerebrovascular pathology in not uncommon. These cases show what is called Mixed Dementia (MD), the most common association between VD and AD. The symptomatology of MD is variable, with clinical manifestations related to both conditions.

CLINICAL PRESENTATION OF VASCULAR DEMENTIA

Post-stroke VD

The cortico-subcortical lesion after ischemic cerebrovascular accident is the easiest characterized situation in clinical practice and results from occlusion of a major vessel. The dimension of the affected area varies widely and is related to the arterial obstruction location. VD after hemorrhagic cerebrovascular accident exhibit manifestations that frequently resemble the ischemic form, with varied cognitive decay. Several forms and extensions of intraparenchymatous lesion may occur. When the bleeding occurs in the subarachnoid space, ischemic lesion may also occur due to secondary arterial vasospasm [13]. It is estimated that, after a cerebrovascular accident, 20%–30% of patients are diagnosed with dementia [12].

VD due to Bordering Territory Infarction

Hypoperfusion and ischemia in bordering territories cause several brain lesions, which may affect the cerebral cortex and basal ganglia.

VD due to Strategic Infarction

Strategic infarctions are characterized by small vessel lesions that affect regions of broad functional relevance. They may cause VD states with acute onset. The cognitive impairment varies according to the affected area. There may be cases of amnesia, Wernicke aphasia, Broca aphasia, attention deficit, spatial disorientation, executive dysfunction, and behavioral changes, among others.

Multiple VD Infarction

The symptoms are diversified and any of them may cause cognitive impairment, functional impairment, and behavioral changes. Usually, it is characterized by attention, executive, and amnesic deficit, followed by apathy and other behavioral changes. The presentation is similar to VD due to subcortical infarctions.

VD due to Subcortical Ischemia

Subcortical Ischemic Vascular Dementia (SIVD) is one of the most common clinical presentations [15], and may be responsible for up to 50% of the cases of VD [13]. It is essentially a small vessels disease, with microangiopathy associated with atherosclerosis, arterial hypertension, *diabetes mellitus*, and other vascular risk factors. Another cause of SIVD is Cerebral Amyloid Angiopathy (CAA), which is characterized by progressive deposition of beta-amyloid protein in small arteries [15,16]. As it affects small vessels, ischemia results in subcortical lesions that cause lacunar strokes (smaller than 15 mm), as well as white matter lesions, where it promotes microangiopathic demyelination, also known as leukoaraiosis [13]. SIVD is also present in Binswanger Subcortical Arteriosclerotic Encephalopathy (Fig. 19.2), a form of vascular dementia associated with systemic arterial hypertension, characterized by a diffuse process of white matter demyelination and axon loss, initially periventricular, which then spreads to the entire subcortex [17].

VD AND GENETIC ALTERATIONS

Although rare, two genetic alteration associated with VD must be emphasized due to their importance in our population: Cerebral Autosomal Dominant Arteriopathy with Subcortical Infarcts and Leucoencephalopaty (CADASIL) and Sickle Cell Anemia [13].

CADASIL (Fig. 19.3) is a small vessel cerebral disease, of hereditary character, which is the main cause

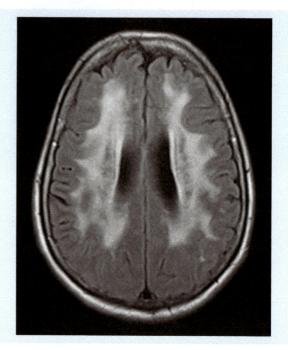

FIG. 19.2 Axial-FLAIR transversal magnetic resonance image of the encephalon showing extensive hypersignal in almost all subcortical white matter, characterizing severe microangiopathy. Multiple lacunar strokes are observed inside the affected area.

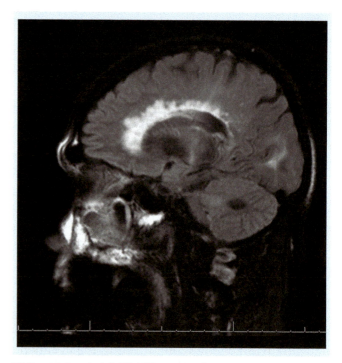

FIG. 19.3 Axial-FLAIR sagittal magnetic resonance image of the encephalon showing extensive hypersignal in almost all subcortical white matter with nodular areas.

TABLE 19.2 Major or Mild Vascular Neurocognitive Disorder (DSM-5)

DIAGNOSIS CRITERIA

A. The major or mild neurocognitive disorder criteria are met

B. Clinical aspects are consistent with vascular etiology, as suggested by one of the following:

  1. The onset of cognitive deficit is temporarily related to one or more cerebrovascular events
  2. Impairment evidence is emphasized in complex attention (including processing speed) and frontal executive function

C. There is evidence of cerebrovascular disease from the historic, physical examination, and/or neuroimaging considered sufficient to explain the cognitive impairment

D. The symptoms are not better explained by other mental or systemic disorders

Probable neurocognitive vascular disorder is diagnosed if either of the following is present; otherwise, possible neurocognitive vascular disorder should be considered:

  1. The clinical criteria are supported by neuroimaging of significant parenchymal lesion, attributed to cerebrovascular disease (with neuroimaging support)
  2. The neurocognitive syndrome is temporarily related to one or more documented cerebrovascular events
  3. Presence of clinical and genetic evidence (i.e., cerebral autosomal dominant arteriopathy, with subcortical infarctions and leukoencephalopathy) of cerebrovascular disease

Possible neurocognitive disorder is diagnosed when the clinical criteria are met, but there is no neuroimaging available, and the temporal relation between the neurocognitive syndrome and one or more cerebrovascular events is not established

Adapted from the Diagnostic and Statistical Manual for Mental Disorders 5th edition (DSM-5).

of cerebral infarctions of genetic origin in adults. The main characteristic is the occurrence of successive subcortical infarctions, with neuroimaging showing white matter alterations, which become more diffuse over time [18]. Cognitive impairment tends to appear late on, and is generally accompanied by migraine and progressive mood swings and disability [19].

Sickle cell anemia is a recessive autosomal disorder in which abnormal hemoglobin leads to deformed red blood cell structure. There are frequent vascular obstruction episodes caused by sickle cells, which may occur spontaneously or due to infection, dehydration, or hypoxia. Cerebral vessel lesions may cause ischemic or bleeding events, causing progressive cognitive impairment.

DIAGNOSTIC CRITERIA

Due to its heterogeneity, VD has a broad variety of diagnosis criteria with different sensitivities and specificities. Currently the most used criteria are the *Diagnostic and Statistical Manual for Mental Disorders,*

5th edition (DSM-5), published by the American Psychiatric Association [12] (Table 19.2), and the International Statistical Classification of Diseases (ICD-10) [20] (Table 19.3).

The criteria in DSM-5 are divided in Major Neurocognitive Disorders (Major NCDs) and Mild Neurocognitive Disorder (mNCD). The Major NCD criteria are: evidence of important cognitive impairment from the previous level of performance in one or more cognitive domains (complex attention, executive function, learning and memory, language, perceptual-motor, or social cognition), based on patient concern or information from a knowledgeable person or physician; substantial impairment in cognitive performance, preferably documented by standardized neuropsychological testing or, in its absence, another quantified clinical assessment; the cognitive impairment interferes with independence in everyday activities (i.e., at a minimum, requiring assistance with complex instrumental activities of daily living such as paying bills or managing medications); the cognitive deficits do not occur exclusively in the context of a delirium, and

TABLE 19.3 Vascular Dementia (ICD-10)

- Cognitive and functional impairment is asymmetrical, and there may be memory loss, intellectual impairment, and focal neurological signs. Critical thinking and judgment may be relatively preserved. An abrupt onset and progressive deterioration, but with intervals of stability, as well as the presence of focal neurological symptoms and signs, increase the probability of the diagnosis; in some cases, it may be only confirmed with the use of neuroimaging or, finally, by means of neuropathological examination

- Symptoms must be present for at least 6 months in order to confirm the diagnosis and the *delirium* clinical picture must be excluded

Adapted from ICD-10, classification of mental and behavioral disorders, clinical descriptions, and diagnosis instructions; WHO, 1993.

are not explained by another mental disorder (e.g., major depressive disorder, schizophrenia).

The mNCD criteria are: evidence of modest cognitive decline from a previous level of performance in one or more cognitive domains (complex attention, executive function, learning and memory, language, perceptual-motor, or social cognition) based on concern of the individual, a knowledgeable informant, or the physician recognition; modest impairment in cognitive performance, preferably documented by standardized neuropsychological testing or, in its absence, another quantified clinical assessment; the cognitive deficits do not interfere with capacity for independence in everyday activities (i.e., complex instrumental activities of daily living such as paying bills or managing medications are preserved, but greater effort, compensatory strategies, or accommodation may be required); the cognitive deficits do not occur exclusively in the context of a delirium, and are not better explained by another mental disorder (e.g., major depressive disorder, schizophrenia).

In 2011, the American Heart Association/American Stroke Association (AHA/ASA) also published their own criteria (Table 19.4), with a broader concept defined as Vascular Cognitive Impairment (VCI). These also include criteria for vascular dementia, as well as Mild Vascular Cognitive Impairment (MCVI), which in turn is divided in probable, possible, and unstable [1].

There are also the criteria defined by the National Institute of Neurological Disorders and Stroke—*Association Internationale pour la Recherche et l'Enseignement en Neurosciences* (NINDS-AIREN) [21], which are long-established and broadly used in research.

It is common to use scores to estimate the probability of ischemic lesions in patients with cognitive disorders, with the Hachinski Ischemic Score

(HIS) [22] being the most employed (Table 19.5). The score is based entirely on clinical data, and is useful to differentiate between VD and AD. Scores higher than 7 suggest VD, while scores lower than 4 suggest AD. Between 4 and 7, there is higher probability of mixed dementia.

COGNITIVE TRACKING TESTS

The cognitive tracking tests are greatly relevant in clinical practice, because they allow to detect new cases, help with diagnosis, and monitor dementia. Neuropsychometric tests are simpler, brief and widely used, such as mini-mental examination [23,24]. The neuropsychological test denomination is reserved for neuropsychological assessment, administered by a qualified neuro-psychologist. Early identification of individuals with probable dementia is one of the main measures that may benefit the patient. Early diagnosis is vital for immediate identification of possible reversible causes of dementia, to control the risk factors involved, and to begin drug treatment in order to slow the course of the process.

Most of the cognitive tracking tests are applied quickly and can be easily executed by the physician himself at his office or in hospitals. The tests may be more or less comprehensive, evaluating one or more cognitive domains. It is advantageous that the physician know the most common tests, in order to promptly refer to specialized treatment. Bellow are some of the most commonly used tests in clinical practice.

Mini-Mental State Examination (MMSE)

The MMSE is the cognitive tracking test most commonly used in clinical practice. Besides helping with the diagnosis, it may be used as longitudinal monitoring method. The test assesses several

TABLE 19.4 Vascular Cognitive Impairment (AHA/ASA)

1. VCI includes all forms of cognitive deficit from vascular dementia to mild cognitive impairment

2. These criteria cannot be used in individuals with active abusive alcohol use or drug addiction. Individuals must be free from such substances for at least 3 months

3. These criteria shall not be used for individuals with *delirium*

VASCULAR DEMENTIA (VD)

1. The diagnosis of dementia must be based on functional cognitive impairment in relation to previous state and on permanent deficit in two or more cognitive domains that are sufficient to affect everyday activities

2. The diagnosis of dementia must be based on cognitive tests, and at least four cognitive domains must be assessed: executive functions, memory, language, and visual-spatial functions

3. The deficits in everyday activities must be independent from the sensitive and motor sequel of vascular events

PROBABLE VD

1. Existence of cognitive impairment and evidence of cerebrovascular disease in neuroimaging tests and

   (a) existence of a clear relation between a vascular event (for example, stroke) and the onset of cognitive deficits, or

   (b) existence of a clear relation between the severity and pattern of the cognitive impairment and the presence of diffuse subcortical cerebrovascular pathology

   (c) Possible VD

THERE IS COGNITIVE IMPAIRMENT AND EVIDENCE OF CEREBROVASCULAR DISEASE BUT

1. There is temporal relation, relation of severity, or cognitive pattern between the vascular disease (e.g., silent infarction, small vessel cerebral disease) and cognitive impairment

2. There is not sufficient information to diagnose VD (e.g., clinical symptoms suggest the presence of vascular disease, but there are no CT or MRN tests available)

3. The severity of aphasia does not allow appropriate cognitive assessment. However, individuals with documented evidence of normal cognitive function before the clinical event that caused aphasia may be classified as having probable VD

4. There is evidence of other neurodegenerative diseases or conditions, besides the cerebrovascular disease, that may affect cognition, such as:

   (a) history of other neurodegenerative diseases (e.g., Parkinson's disease, Lewy body dementia)

   (b) presence of Alzheimer's disease confirmed by biomarkers (e.g., PET, CSF, amyloid ligands)

   (c) history of active cancer or metabolic or mental diseases that may affect cognitive functions

MILD VASCULAR COGNITIVE IMPAIRMENT (MVCI)

1. MVCI includes the four sub-types proposed for mild cognitive impairment: amnesic, amnesic plus other domains, non-amnesic with a single domain, non-amnesic with multiple domains

2. The MVCI rating must be based on cognitive tests and at least four cognitive domains must be assessed: executive functions, memory, language, and visual-spatial functions. The classification must be based on subjective assessment of cognitive decline in respect to a previous state and impairment of at least one cognitive domain

3. Instrumental functions of everyday life may be normal or mildly altered, regardless of the presence or motor or sensitive symptoms

Continued

TABLE 19.4   Vascular Cognitive Impairment (AHA/ASA)—cont'd

PROBABLE MVCI

---

1. There is cognitive impairment and imaging evidence of cerebrovascular disease and

    (a) existence of clear temporal relation between the vascular event (e.g., stroke) and the beginning of cognitive deficits, or

    (b) existence of a clear relation between severity and pattern of the cognitive impairment and the presence of diffuse subcortical cerebrovascular pathology

2. History of progressive and gradual cognitive deficit before or after the stroke, which suggests the presence of non-vascular neurodegenerative disorder

---

POSSIBLE MVCI

There is cognitive impairment and evidence of cerebrovascular disease, but:

1. there is no clear temporal relation, relation of severity, or cognitive pattern between the vascular disease (e.g., silent infarction, small vessel subcortical disease) and the beginning of cognitive deficits

2. there is no sufficient information to diagnose MVCI (e.g., clinical symptoms suggest the presence of vascular disease, but there are no CT or MRN tests available)

3. the severity of aphasia does not allow appropriate cognitive assessment. However, individuals with documented evidence of normal cognitive function before the clinical event that caused aphasia may be classified as having probable MVCI

4. there is evidence of other neurodegenerative diseases or conditions, besides the cerebrovascular disease, that may affect cognition, such as:

    (a) history of other neurodegenerative diseases (e.g., Parkinson's disease, Lewy body dementia)

    (b) presence of Alzheimer's disease confirmed by biomarkers (e.g., PET, CSF, amyloid ligands)

    (c) history of active cancer or metabolic or mental diseases that may affect cognitive functions

---

UNSTABLE MVCI

Individuals with diagnosed probable or possible MVCI, whose symptoms return to normal, must be classified as having unstable MVCI

Stroke, *cerebrovascular accident;* PET, *positron emission tomography;* CT, *computerized tomography;* MRN, *nuclear magnetic resonance;* CSF, *cerebrospinal fluid.*
From Gorelick FB, Scuteri A, Black SE, et al. Vascular contributions to cognitive impairment and dementia: a statement for healthcare professionals from the American Heart Association/American Stroke Association. Stroke 2011;42(9):2672–713.

cognitive functions, such as spatial and temporal orientation, immediate memory, attention and calculation, evocation memory, language, and constructional apraxia.

Montreal Cognitive Assessment (MOCA)

The MOCA [25] is a cognitive tracking test used mainly to detect mild cognitive deficits, especially in patients with high level of education and whose MMSE performance is normal. The MOCA is also a test of multiple cognitive domains, such as visual-spatial and executive abilities, designation, memory, attention, language, abstraction, evocation memory, and orientation.

Semantic Verbal Fluency (sVF)

The sVF tests [26] are easily applicable and widely used in clinical practice to assess executive function and semantic memory. The test consists of requesting the patient to say the highest number of words of a given category in 1 min. In general, the patient is requested to name as many animals as possible. There is also another application mode in which the patient is requested to say the highest number of words they can that start with a given letter.

Clock Drawing Test (CDT)

The easily applied CDT [27] is also widely used in clinical practice. The patient is requested to draw a

TABLE 19.5 Hachinski Ischemic Scale

Finding	Score
Abrupt onset	2
"Step-by-step" deterioration	1
Fluctuating course	2
Nocturnal confusion	1
Relative preservation of reality	1
Depression	1
Somatic complaints	1
Emotional lability	1
History of hypertension	1
History of strokes	2
Evidence of associated atherosclerosis	1
Focal neurological symptoms	2
Focal neurological sings	2

TABLE 19.6 Pfeffer Scale to Assess Everyday Activities

1. Does he/she handle their own money?
2. Is he/she able to buy clothes, food, and household products by himself?
3. Is he/she able to heat water to make coffee and put out the fire?
4. Is he/she able to cook food?
5. Is he/she able to keep up to date with community or neighborhood events?
6. Is he/she able to pay attention, understand, and discuss radio or TV shows, newspapers, or magazines?
7. Is he/she able to remember commitments, family events, holidays?
8. Is he/she able to handle his own medication?
9. Is he/she able to take a walk in the neighborhood and find the way back home?
10. Is he/she able to leave the house alone safely?

Note: 0—normal or never done it but could; 1—does it with difficulty or never done it and now would have difficulty; 2—need help; 3—does not do it.

clock on a white sheet of paper, with pointer positions indicating the time requested (e.g., 11:10 a.m.). The test also evaluates several cognitive domains, such as understanding, concentration, abstraction, planning, visual memory, execution, and motor programming.

EVERYDAY ACTIVITIES

The loss of ability to execute everyday activities (EA) is part of dementia syndrome diagnosis criteria, which makes its appreciation mandatory when assessing these patients. There are several scales that can be used to determine the patient's degree of dependence. Basic activities (such as going to the restroom, feeding, etc.) and instrumental activities (such as ability to handle money, go shopping, etc.). One of the most widely used scales is the Pfeffer [28] scale or Functional Activities Questionnaire (Table 19.6). The scores go from 0 (normal) to 3 (unable) for each question, totaling a maximum of 30 points. Elders with 5 points or more are already considered dependent. The higher the dependency of the patient to perform everyday activities, more advanced the stage of the disease.

NEUROIMAGING

The preferred method to diagnose VD is nuclear magnetic resonance (NMR). White Matter Hypersignals (WMH) are frequent findings in elders, and are closely related to VD. The prevalence of this type of signal may vary around 11%–21% in elders with approximately 65 years old, and 94% in elders with approximately 80 years old [29]. However, it is important to emphasize that these findings may also be present in other types of dementia, such as AD, or simply have no clinical relation whatsoever. Generally, these images represent demyelination, astrogliosis, microglia activation, and axon lesion processes, and are consequences, in most cases, of a chronic ischemic process in small vessels (microangiopathy) [30].

The hypersignals are seen in T2-weighted NMR imaging, in T2-weighted-Fluid-Attenuated Inversion Recovery (FLAIR), and proton density-weighted imaging, without prominent hyposignals in T1 weighted images. They may also be seen in Diffusion Tensor Imaging (DTI) even before they appear in conventional NMR; however, this exam is not commonly used in clinical practice. In Computerized Tomography (CT), the lesions are generally shown as hypo-attenuation areas.

In functional images, Positron Emission Tomography (PET) may help differentiating AD and VD. In the first case, the exam evidences lower cerebral metabolism in the temporoparietal region, while the second case shows higher cerebral damage in the frontal region. Single-Photon Emission Computerized Tomography (SPECT) can be useful to demonstrate areas of cerebral hypoperfusion secondary to ischemic lesion. Usually, the collective neuroimaging data set is used to establish the VD diagnosis, in contrast to AD, in which biological markers (for example, hippocampal atrophy or temporoparietal hypometabolism) may individually be of great complementary diagnosis value.

Normally, WMH begins with small points around the frontal and occipital lobes of the lateral ventricles and, with disease progression, begin to form a small "margin" around the lateral walls of these ventricles. Afterwards, the lesions tend to get bigger, becoming confluent, and may reach the subcortical and deep white matter areas.

Several scales were proposed to classify the level of white matter lesion in images. One of the most widely used is the Modified Fazekas Scale [31–33] (Table 19.7), which uses T2-weighted NMR and FLAIR imaging. The score goes from 0 to 3 according to the extension of the lesions (Fig. 19.4).

Due to lack of access to NMR in many places, recently a new scale has been proposed that may also be used in images obtained by means of CT: the Age-Related White Matter Changes Rating Scale for MRI and CT—ARWMC Scale [34] (Table 19.8).

TABLE 19.7 Modified Fazekas Scale

Hypersignals in white matter	Score
Absent	0
Punctual or focal images	1
Onset of confluence	2
Confluent or irregular periventricular hyperintensities	3

TREATMENT

The treatment must focus on preventing cerebrovascular events, with rigorous control of risk factors, because, once dementia is present, the tendency is that the patient evolves with gradual decline of cognitive functions. Most of the MVCI and VD risk factors are not different from the other risk factors of atherosclerotic vascular diseases, which have well-established treatment, control, and prevention means, and that can be performed in primary healthcare levels.

Drug treatment of dementia is done using anticholinesterase. The two most used drugs are donepezil (class IIa, evidence level A), and galantamine, which seem to show more benefits in cases of mixed DV/AD dementia (class IIa, evidence level A). The benefits of using rivastigmine and memantine are not well established yet. Drugs with a vasodilatation effect, such as nimodipine and vimpocetine, as well as other compounds such as diphosphate-choline (citicoline), piracetam, and huperzine-A, have not shown convincing data up to now [1].

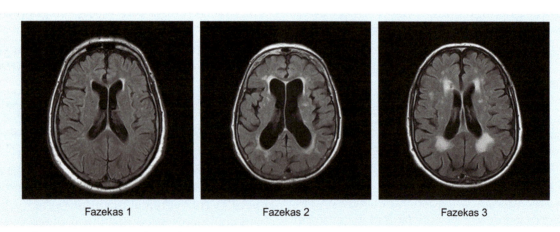

Fazekas 1   Fazekas 2   Fazekas 3

FIG. 19.4 Illustrative images of Axial FLAIR magnetic resonance regarding the Fazekas rating.

TABLE 19.8 ARWMC Scale

WHITE MATTER LESIONS

0 No lesions (including symmetrical, well-defined caps or bands)

1 Focal lesions

2 Beginning confluence of lesions

3 Diffuse involvement of the entire region, with or without involvement of U fibers

BASAL GANGLIA LESIONS

0 No lesions

1 Focal lesion (≥5 mm)

2 >1 focal lesion

3 Confluent lesions

CONCLUSIONS

Similarly to other alterations resulting from atherosclerotic vascular lesions, MVCI and VD are diseases of great relevance to public health, because, due to the aging population, their social and economic burden has increased. Therefore, it is very important that health professionals recognize these diseases as early as possible in order to provide prompt treatment and monitoring; this will also lead to better quality of life for caregivers and families.

References

[1] Gorelick FB, Scuteri A, Black SE, et al. Vascular contributions to cognitive impairment and dementia: a statement for healthcare professionals From the American Heart Association/American Stroke Association. Stroke 2011;42(9):2672–713.

[2] Hébert R, Brayne C. Epidemiology of vascular dementia. Neuroepidemiology 1995;14(5):240.

[3] Jorm AF, Jolley D. The incidence of dementia: a meta-analysis. Neurology 1998;51(3):728.

[4] Lobo A, Launer LJ, Fratiglioni L, et al. Prevalence of dementia and major subtypes in Europe: a collaborative study of population-based cohorts. Neurologic Diseases in the Elderly Research Group. Neurology 2000;54(11 Suppl 5):S4–9.

[5] Kalmijn S, Foley D, White L, et al. Metabolic cardiovascular syndrome and risk of dementia in Japanese-American elderly men. The Honolulu-Asia aging study. Arterioscler Thromb Vasc Biol 2000;20(10):2255.

[6] Solfrizzi V, Scafato E, Capurso C, et al. Italian Longitudinal Study on Ageing Working Group. Metabolic syndrome and the risk of vascular dementia: the Italian Longitudinal Study on Ageing. J Neurol Neurosurg Psychiatry 2010;81(4):433.

[7] Kalaria RN. Cerebrovascular disease and mechanisms of cognitive impairment: evidence from clinicopathological studies in humans. Stroke 2012;43(9):2526–34.

[8] Venkat P, Chopp M, Chen J. Models and mechanisms of vascular dementia. Exp Neurol 2015;272:97–108.

[9] Rost NS, Rahman RM, Biffi A, et al. White matter hyperintensity volume is increased in small vessel stroke subtypes. Neurology 2010;75(19):1670–7.

[10] Qiu C, Cotch MF, Sigurdsson S, et al. Cerebral microbleeds, retinopathy, and dementia: the AGES-Reykjavik Study. Neurology 2010;75(24):2221.

[11] Luz PL, Fialdini RC, Nishiyama M. Red wine, resveratrol and vascular aging: implications for dementia and cognitive decline. In: Martin CR, Preedy VR, editors. Diet and nutrition in dementia and cognitive decline. Cambridge: Academic Press; 2014. p. 944–5.

[12] Manual Diagnóstico e Estatístico de Doenças Mentais 5a edição (DSM-5), da Associação Americana de Psiquiatria, tradução portuguesa.

[13] Engelhardt E, Tocquer C, André C, et al. Demência vascular. Critérios diagnósticos e exames complementares. Dement Neuropsychol 2011;5(1):49–77.

[14] Wiesmann M, Kiliaan AJ, Claassen JAHR. Vascular aspects of cognitive impairment and dementia. J Cereb Blood Flow Metab 2013;33(11):1696–706.

[15] Thal DR, Grinberg LT, Attems J. Vascular dementia: different of vessel disorders contribute to the development of dementia in the elderly brain. Exp Gerontol 2012;47(11):816–24.

[16] Pantoni L. Cerebral small vessel disease: from pathogenesis and clinical characteristics to therapeutic challenge. Lancet Neurol 2010;9:689–701.

[17] Oliveira ASB, Massaro AR, Campos CJR, et al. Encefalopatia subcortical arteriosclerótica de Binswanger. Forma especial de demência associada à hipertensão arterial sistêmica. Arq Neuropsiquiatr 1986;44(3):255–62.

[18] Joutel A, Corpechot C, Ducros N, et al. Notch3 mutations in CADASIL, a hereditary adult-onset condition causing stroke and dementia. Nature 1996;383(6602):707–10.

[19] Pantoni L, Pescini F, Nannucci S, et al. Comparison of clinical, familial, and MRI features of CADASIL and NOTCH3-negative patients. Neurology 2010;74:57–63.

[20] World Health Organization. The ICD-10 classification of mental and behavioural disorders. Geneva: WHO; 1993.

[21] Román GC, Tatemichi TK, Erkinjuntti T, et al. Vascular dementia: diagnostic criteria for research studies: Report of the NINDS-AIREN International Workshop. Neurology 1993;43:250–60.

[22] Hachinski VC, Iliff LD, Zilhka E, et al. Cerebral blood flow in dementia. Arch Neurol 1975;32:632–7.

[23] Folstein MF, Folstein SE, McHugh PR. Mini-mental state: a practical method for grading the cognitive state of patients for the clinician. J Psychiatr Res 1975;12:189–98.

[24] Brucki SMD, Nitrini R, Caramelli P, et al. Sugestões para o uso do Mini-Exame do Estado Mental no Brasil. Arq Neuropsiquiatr 2003;61:777–81.

[25] Bertolucci PHF, Sarmento ALR, Wajman JR. Montreal cognitive assessement. Versão experimental brasileira. [Internet]. Disponível em: www.mocatest.org; [Acesso em 25.06.16].

[26] Brucki SMD, Malheiros SMF, Okamoto IH, et al. Dados normativos para o teste de fluência verbal categoria animais em nosso meio. Arq Neuropsiquiatr 1997;55:56–61.

[27] Aprahamian I, Martinelli JE, Neri AL, et al. The Clock Drawing Test: a review of its accuracy in screening for dementia. Dement Neuropsychol 2009;3:74–80.

[28] Pfeffer RI, Kurosaki TT, Harrah Jr. CH, et al. Measurement of functional activities in older adults in the community. J Gerontol 1982;37(3):323–9.

[29] Debette S, Markus HS. The clinical importance of white matter hyperintensities on brain magnetic resonance imaging: systematic review and meta-analysis. BMJ 2010;341:c3666.

[30] Prins ND, Scheltens P. White matter hyperintensities, cognitive impairment and dementia: an update. Nat Rev Neurol 2015;11:157–65.

[31] Fazekas F, Chawluk JB, Alavi A, et al. MR signal abnormalities at 1.5T in Alzheimer's dementia and normal aging. AJR 1987;149:351–6.

[32] Schmidt R, Fazekas F, Kleinert G, et al. Magnetic resonance imaging signal hyperintensities in the deep and subcortical white matter: a comparative study between stroke patients and normal volunteers. Arch Neurol 1992;49:825–7.

[33] Mäntylä R, Erkinjuntti T, Salonen O, et al. Variable agreement between visual rating scales for White matter hyperintensities on MRI. Stroke 1997;28:1614–23.

[34] Wahlund LO, Barkhof F, Fazekas F. A new rating scale for age-related white matter changes applicable to MRI and CT. Stroke 2001;32:1318–22.

20

Emotional Stress and Influences on Endothelium

Mayra Luciana Gagliani, Elaine Marques Hojaij, and Protásio Lemos da Luz

INTRODUCTION

The intimate mind/body relationship has been known since Hippocrates' time. However, Hanz Selye [1] made the classic description of stress syndrome, which includes several organic disorders such as gastrointestinal bleeding and adrenal lesions. He defined the term "General Adaptation Syndrome." Recently, Dimsdale [2] found about 40,000 citations on psychological stress. Here, we shall use the concept "stress" as the set of body reactions to a challenge, a new circumstance that provokes organic functional emotions or responses of several levels, and may include hemodynamic, biochemical or metabolic processes.

Stress can exist in an isolation form, in individuals with mental health preserved or associated with mental disorders. This is the case of depression and anxiety; they may appear in sufficient intensity to interfere with quality of life, including impairing physical health. In addition to being an isolated risk factor for cardiovascular disease, mental stress is closely linked to other factors associated with lifestyle, such as smoking, sedentary lifestyle, diabetes, excessive alcohol use, hypertension, and poor diet [3].

This chapter will address general aspects of mental illness as risk factors for cardiovascular diseases, physiopathological mechanisms, and the specific endothelium involvement when the organism is exposed to stress situations, as well as possible therapeutic interventions.

MENTAL DISORDERS AND CARDIOVASCULAR DISEASES

As Table 20.1 shows, there are a number of studies involving large numbers of participants associating generalized anxiety disorder, depression, posttraumatic stress, panic attacks, anxiety, social phobia, sleep deprivation, unemployment, stress in the workplace, hostility and anger to cardiovascular and all-cause mortality, cardiovascular events, stroke, and prevalence of coronary disease. The studies include meta-analysis, reviews, and group prospecting [4].

The *Interheart* Study [5] evaluated the relationship between coronary risk factors and acute myocardial infarction. In this study, 11,119 individuals with a history of acute infarction were compared to another 13,648 from a control group. The results evidenced that patients with first infarction showed greater prevalence of some kind of mental stress than controls, i.e., chronic stress at work or at home, financial worries, depression, or stressful events in daily life. The effect of psychosocial stress on coronary disease was decisive in patients from all regions, of different ages and ethnic groups, and both genders. Mental stress accounted for approximately one-third of myocardial infarction risk.

The *Framingham* study [6] found that psychosocial stress was a stronger risk factor that diabetes, smoking, obesity, poor eating habits, and sedentary lifestyle (Fig. 20.1).

265

TABLE 20.1   Psychological Factors and Cardiovascular Outcomes [4]

Author	Factor	Type of study	Participants	Association
Sofi et al. (2014)	Insomnia	Meta-analysis	12,250	CV events
Capuccio et al. (2011)	Duration of sleep	Meta-analysis	474,684	CHD, Stroke
Nicholson et al. (2006)	Depression	Meta-analysis	146,538	CHD
Roest et al. (2010)	Anxiety	Meta-analysis	249,846	CHD, Cardiac mortality
Roest et al. (2010)	Anxiety after myocardial infarction	Meta-analysis	5750	Mortality due to different causes, cardiac mortality, CV events
Edmondson et al. (2013)	Post traumatic stress disorder	Systematic review and meta-analysis	402,274	CHD
Chida et al. (2009)	Anger and hostility	Meta-analytic review of prospective evidence	2770	CHD
Kivimäki et al. (2012)	Psychosocial stress (work stress)	Meta-analysis	197,473	CHD
Roelfs et al. (2011)	Unemployment	Meta-analysis and meta-regression	More than 20 million	Mortality due to different causes
Holt-lunstad et al. (2010)	Social relationships	Meta-analysis	308,849	Mortality due to different causes
Martens et al. (2010)	Generalized anxiety disorder	Prospective Cohort	1015	CV events
Roest et al. (2012)	Generalized anxiety disorder	Naturalistic Cohort	438	Mortality due to different causes, CV events
Smoller et al. (2007)	Panic attacks	Prospective Cohort	3369	CHD, stroke
Brummett et al. (2006)	Scores on the optimism-pessimism scale	Observational Cohort	6958	Mortality due to different causes

CHD, *coronary heart disease; CV, cardiovascular.*
Adapted from Rozanski, A. Behavioral cardiology: current advances and future directions. J Am Coll Cardiol 2014;64:100–10.

The observational study *Women's Health Initiative* (WHI) [7] showed that depressive disorders were present in 15.8% of women, which significantly increased the likelihood of stroke, angina, and death.

About 374 young people, aged between 18 and 30, underwent a psychological test battery in the study *Coronary Artery Risk Development in Young Adults* (Cardia) [8]. Ten years later, those with hostility score above average were twice as likely to have coronary atherosclerotic disease as documented by the calcium score in coronary computed tomography.

The *Japan Collaborative Cohort Study* (JACC) [9] included 73,424 individuals in Japan. Out of these, 30,180 were men and 43,244 women aged between 40 and 79 years old. None had a history of stroke, heart disease, or cancer. After participants completed a questionnaire regarding their own perception of their mental health, the results showed that women with a higher stress level were twice as likely to have a heart attack or stroke when compared to others with lower levels.

In the *Whitehall II Study* [10], held in London and which included 10,308 civilian workers, it was

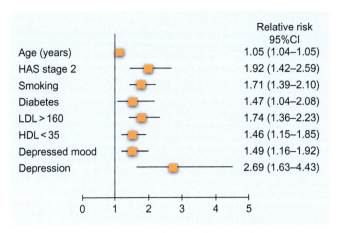

FIG. 20.1 Risk factors for cardiovascular events in Framingham study [6]. *Adapted from Rozanski A, Blumenthal JA, Davidson KW, et al. The epidemiology, pathophysiology, and management of psychosocial risk factors in cardiac practice: the emerging field of behavioral cardiology. J Am Coll Cardiol 2005;45:637–51.*

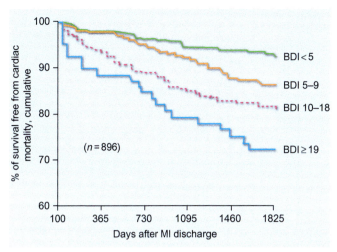

FIG. 20.2 Postinfarct mortality, according to Beck Depression Inventory (BDI) [14]. *Adapted from Lespérance F, Frasure-Smith N, Talajic M, et al. Five-year risk of cardiac mortality in relation to initial severity and one-year changes in depression symptoms after myocardial infarction. Circulation 2002;105:1049–53.*

observed that the worst conditions of life are associated with increased heart rate even at rest, in addition to a low variation in heart rate and increased risk of coronary heart disease over 5.3 years. The data obtained also suggest that the increase in risk is proportional to the low purchasing power, in part, is related to the losses that stress causes in the autonomic nervous system. Tense working conditions are also associated with a generalized increase in inflammation.

In the *Multiple Risk Factor Intervention Trial* (MRFIT) [11], men with hostile behavior were followed for 16 years and had a higher risk of death from cardiovascular disease when compared to low-hostility men. It was also observed that early traumatic experiences, such as physical or sexual abuse, neglect, and dysfunctional homes, may predispose individuals to coronary diseases after a few decades.

Rugulies [12] conducted a meta-analysis of 11 studies which investigated the impact of depression on the development of coronary heart disease in healthy individuals. In general, the relative risk of developing disease was 1.64% when the individual was depressed. *Nippon Data 80* [13], a prospective study conducted in Japan, showed that the relative risk of death from coronary artery disease between individuals with cholesterol levels between 240 and 259 mL was 1.8% compared to others whose levels were 160–179 mL. These findings indicate that the association between mental stress and

cardiovascular disease risk is similar to that observed in high cholesterol rates.

In addition, depression has a negative impact on the prognosis of patients with heart disease. There is evidence that in postinfarction, depression is associated with higher mortality (Fig. 20.2). Depression is more prevalent in patients with coronary artery disease (CAD) than in the general population; in addition, 45% of patients with acute myocardial infarction were depressed [14,15]. It is complex to demonstrate the relationship between depression and cardiopathy; however, one may predispose patients to develop the other, while the disease alone may depress the patient's mental state. Added to this, mental stress results in deterioration and progression of heart disease, creating a vicious circle. Figs. 20.3 and 20.4 illustrate the impacts of depression, physical inactivity, and lack of purpose in life.

Therefore, numerous studies in different populations and regions of the world, of both genders, have shown that several forms of psychological stress are associated with the development of cardiopathies, especially coronary disease.

PREVALENCE OF MENTAL DISORDERS

Mental disorders represent at least 12% of all cardiovascular disease occurrences, and are expected to

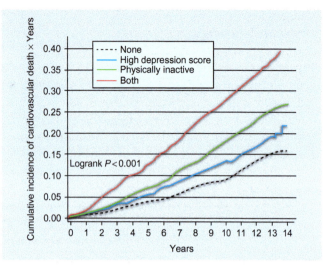

FIG. 20.3 We observed the cumulative effect of the association "depression and physical inactivity" on cardiovascular mortality over 14 years in elderly adults in The Cardiovascular Health Study [4]. *Adapted from Rozanski, A. Behavioral cardiology: current advances and future directions. J Am Coll Cardiol 2014;64:100–10.*

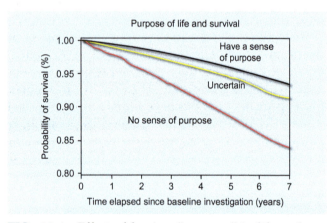

FIG. 20.4 Effect of having "purpose" in life and not having it under the probability of survival over 7 years, according to the Japanese study *Ohsaki Study [4]. Adapted from Rozanski, A. Behavioral cardiology: current advances and future directions. J Am Coll Cardiol 2014;64:100–10.*

reach 15% by 2020. It is estimated that one in four people in the world will be affected by these disorders at some time in their lives [16–18]. In the United States, 61.5 million Americans (about 25% of the adult population) are already suffering some mental disease. The forecast for 2030 is that, worldwide depression should be the second cause of nontransmissible illness. Bipolar affective disorders also have a high prevalence in the world population: 60 million people [16–18].

In the Brazilian adult population, the prevalence of mental disorders ranges from 20% to 56% depending on the population studied [19]. The 25 studies reviewed by Santos and Siqueira [19] point out that women are more likely to anxiety and mood disorders; men tend to have disorders related to substance use. In the city of São Paulo, in a sample of just over 5000 participants, 29.6% were diagnosed of at least one occurrence per year, with an emphasis on anxiety at 19.9% and mood alterations 11% [20]. A study carried out on 829 children in Salvador showed a prevalence of 23.2% of psychiatric cases [21].

COSTS OF MENTAL DISORDERS

These costs can be educational, social, familial, and financial. In Brazil, the total expenditure of Ministry of Health with public health actions and services between 2001 and 2009 increased by 55% [22]. The common mental disorders remain the third main cause of social security benefits granting of sick pay for work absence. The average annual impact is 186 million for the social security system, with an average increase of 7.1% per year in the amount of new sick pay for mental disorders [23]. In developed countries, costs reach up to 4% of GDP. In 2006, about 36.2 million Americans had mental health expenses of an amount equal to that spent on cancer therapies, behind only treatments for cardiovascular disease and trauma [23].

Considering only depression, the costs are around US$10 billion a year, and much of it is related to job loss or low productivity [24]. It is therefore not just a clinical problem; the high costs of mental illness represent serious problem for all countries.

ENDOTHELIUM AND CARDIOVASCULAR DISEASES

In addition to depression and other psychopathologies, there are different forms of stress that cause changes in the endothelium. Endothelial dysfunction precedes atherosclerosis, in addition to being a fundamental element in its development and evolution, characterizing a cardiovascular risk marker [25]. Emotional disorders directly influence its course, especially in coronary disease.

STRESS AND ENDOTHELIAL DYSFUNCTION IN GENERAL POPULATION

Mental stress induced by a test with color differences causes increased platelet activity, as measured by beta-thromboglobulin levels and endothelial activity [26]. This was observed in healthy and hypertensive patients. The stress caused by any type of arithmetic activity induces reduction of flow-mediated dilation (FMD) in both healthy men and women.

In our laboratory, we subjected healthy elderly individuals to the Stroop Color Test, and measured FMD-mediated dilation in the brachial artery, which presented a significant reduction. They were then treated with ginkgo biloba (80 mg/day) for 30 days, and the test was repeated. Ginkgo biloba ingestion normalized endothelial function, which did not occur in nontreated control-individuals [27] (Fig. 20.5).

Stress caused by anger and hostile personality traits attenuates the brachial artery FMD in normal or high cholesterol individuals. Signs of hostility are directly linked to increased adverse effects caused by mental stress on endothelial function. In postmenopausal women, with normal angina and coronary arteries, stress generated by anger was able to cause myocardial ischemia associated with endothelial dysfunction in only 5 min [4,15,26].

Another type of acute stress is triggered by natural disasters, such as earthquakes, wars, suicide attacks, etc [15,28]. Both sudden death and acute myocardial infarction were more common in survivors of Taiwan's earthquakes during the first week after the event, and in Israelis immediately after missile attacks on the first day after the Gulf War. Residents of New York who had cardiac fibrillation experienced two to three times more tachycardia and ventricular fibrillations for 1 month after the September 11th attacks.

Intense emotions can also cause cardiac function deterioration. Wittstein et al. [29] observed that stress can reduce ventricular function immediately after a very strong emotional impact, such as the death of a loved one. These individuals had normal coronary arteries, but had signs of ischemia on the electrocardiogram. Plasma catecholamine levels in acute stress situations are about 30 times greater than normal, and approximately four to five times higher than in patients with acute myocardial infarction.

Moreover, Schmidt et al. [30] conducted a double-blind study in 75 volunteers, with an average age of 26, who were separated at random: a control-group and the other exposed to noise (simulating the sound of an aircraft) in two frequencies. After assessing endothelial function, sleep quality, and plasma adrenaline concentration, Schmidt et al. observed that noise impaired sleep quality, reduced dose-dependent endothelial function, and increased adrenaline production. Endothelial dysfunction was reversed with vitamin C, suggesting mediation by oxidative stress.

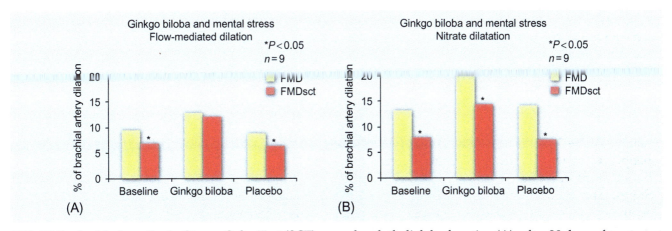

FIG. 20.5 In elderly patients, Stroop Color Test (SCT) caused endothelial dysfunction (A); after 30 days of treatment with ginkgo biloba, the same test did not cause endothelial dysfunction, indicating that endothelium was protected, whereas control individuals remained vulnerable. Endothelium-independent dilation (B) was not affected by ginkgo biloba. *FMD*, flow-mediated dilation; *yellow* represents FMD without SCT; *FMDsct*, in *red*, flow-mediated dilation under SCT effect. Atherosclerosis Unit. Heart Institute (InCor)-USP.

On the other hand, young people were asked to watch a short, funny movie, and also a thrilling movie [31]. The cheerful movie caused increased FMD, while the thrilling one caused depression (Fig. 20.6). The same authors have shown that listening to animated music also improves FMD; such effects were attributed to endorphin release [32].

Adolescents exposed to chronic and negative stressors, which worsen over time, increase their chances of having some cardiovascular manifestation. Healthy male students who had sleep deprivation for four weeks and were under great pressure due to academic activities showed a significant decrease in blood flow-mediated vasodilation [26].

In medical students who had a high level of stress, smoked, and had a sedentary life and poor dietary habits, the main factor that caused an imbalance of endothelial function was emotional stress and cigarette smoking [26]. Low social status is linked to FMD reduction of the brachial artery in healthy adults. It was observed that the low social status assessed by MacArthur Scale of Subjective Social Status may be related to cardiovascular disease due to weakening in vasodilatation. These findings explain, at least in part, the findings of the *Whitehall Study* [10], in which low purchasing power and impossibility of self-determination were associated with higher mortality among British workers.

According to Mausbach et al. [33], the chronic stress of elderly caregivers is associated with endothelial dysfunction and may be a potential mechanism related to increased risk of cardiovascular disease in this group. In healthy adults, an increase in Profile of Mood States Scale, i.e., depression/discouragement, tension/anxiety, anger/hostility, fatigue/inertia, and confusion/disorientation, is related to endothelial dysfunction [34]. Mood disorders may contribute to cardiovascular disease by decreasing vasodilation. There is a significant interaction between carotid artery thickness and exhaustion degree, characterized by internal fatigue and irritability only for males. No interaction was found in the vasodilatation flow and exhaustion, in both genders. Hostile behaviors are also related to increased adverse effects of mental stress on endothelial function [26,28,34].

Aging effects on endothelial function have been reported in several circumstances. Depression, exhaustion, and social isolation are often associated with cardiovascular events in elderly people [35]. Major and minor depression give a risk of 3% and 1.6% compared to individuals without the disease, according to Penninx et al.'s [36] study of 2847 men and women aged 55–85 years old. Other factors, such as negative affectivity, anger, comorbidities, and poor social support, also increased the

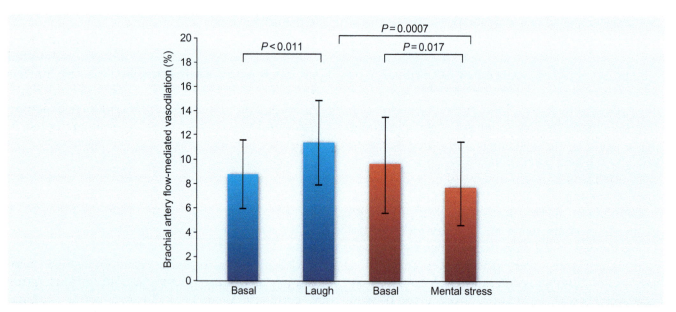

FIG. 20.6 Effect of 15–30-min movie in young, one causing laugh, and the other mental stress on brachial artery FMD [31]. *Adapted from Miller M, Mangano C, Park Y, et al. Impact of cinematic viewing on endothelial function. Heart 2006;92:261–2.*

risk of postinfarct cardiac mortality compared to individuals with adequate social support [37–40].

The mechanisms responsible for increased morbidity/mortality include prothrombotic, proinflammatory factors, leucocyte and platelet counts, factors VII and VIII, and fibrinogen. Changes in sympathetic and parasympathetic systems, as well as production of steroids, also contribute to the condition evolution.

Treatments with antidepressants, rehabilitation, or cognitive behavioral therapy improve depressive symptoms, although the risk of cardiovascular events, such as death and nonfatal heart attacks, has not been reduced [35,41].

Aging is associated with morphological and functional changes in the vasculature in general. Egashira et al. [42] studied 18 patients aged between 23 and 70 years old, with atypical chest pain and who did not present risk factors for coronary artery disease or any cardiac disease. Increased coronary flow, in response to acetylcholine, decreased significantly in older patients. On the other hand, blood flow in response to papaverine (an endothelium-independent smooth muscle dilator) had subtle changes over the years.

We also analyzed [43] the mechanisms contributing to vascular aging associated with cognitive decline that is observed during the process: oxidative stress, arterial stiffness, mitochondrial dysfunction, apoptosis, inflammation, reduction in cellular replication and endothelial progenitor cells generation, and endothelial dysfunction. Celermajer et al. [44] compared the influence of aging on endothelial function of 103 healthy men and 135 healthy women, nonsmokers, aged between 15 and 72 years old. Aging was associated with the progressive loss of FMD, more so in men than in women. Celermajer et al. concluded that the greatest decline in endothelial function in men related to advancing age is justified by the alteration of sex hormones production, since estrogens protect women from vascular function weakening.

Chauhan et al. [45] studied 34 patients of both genders, aged between 27 and 73 years old, who presented atypical chest pain, normal exercise tests, and angiography, who were nonsmokers and had no risk factors. Chauhan et al. found that intracoronary administration of L-arginine, a precursor of nitric oxide, could reverse endothelial dysfunction caused by aging. According to a study by Woo et al. [46], curiously, the elderly Chinese are less susceptible to endothelial dysfunction than the elderly white. The authors hypothesized that such differences are due to the high consumption of vegetables, fish, and green tea by the Chinese, because of the antioxidant and flavonoid effects.

Chronic inflammation is a hallmark of Alzheimer's disease. Kelleher and Souza [47] found 15 studies on endothelial dysfunction in patients with the disease; 10 of them had evidence of this disorder. It is concluded that irrespective of any vascular risk factor, endothelial dysfunction is present in this disease, suggesting a vascular component, although there is no direct method of functional evaluation of cerebral circulation.

Females may show special vascular function characteristics depending on age, menstrual cycle, or associated pathologies. Wagner et al. [48] studied 39 women in postmenopause, without cardiovascular disease; one group with a history of psychiatric illness and another with no symptoms of the disease. None of them had an active disease and had not used antidepressants for at least a year. Those with prior depression had lower brachial FMD compared to those who had never had depression. There was also a direct relationship between the number of episodes of depression and FMD. Thus, even after years of absence of depression, it appeared to negatively influence endothelial function.

Can physical activity restore endothelial function in both animals and humans? [49,50] The practice of sports and emotional stress have some common characteristics, such as increased activity of sympathetic system, heart rate, cardiac output and blood pressure. Thus, Hambrecht et al. [49] submitted 10 patients with CAD to controlled exercise for four weeks and compared the coronary flow under acetylcholine action, to nine controls. This conditioning significantly increased coronary flow compared to controls.

ENDOTHELIAL DYSFUNCTION AND EMOTIONAL STRESS IN CARDIAC DISEASES

Alterations induced by emotional stress have been systematically documented in various forms of coronary disease, including stable angina, postmyocardial infarction, and silent ischemia [51–55]. Emotional stress can induce myocardial ischemia, sudden death, and severe left ventricular dysfunction, as well as causing myocardial spasm. Moreover, this phenomenon also has prognostic value, as noted by Bairey et al. [55], identifying patients at increased risk of future cardiovascular events. However, endothelial function was not analyzed in most of the cited studies; only possible vasoconstrictive effect of emotional stress was mentioned. Yeung et al. [56] were perhaps the first to identify endothelial dysfunction under emotional stress in CAD patients (Fig. 20.7). They observed that nervousness in solving arithmetic tests caused vasoconstriction in stenosed coronary segments, but did not affect normal ones. Studies in patients with coronary atherosclerosis have also shown that stenosed coronary segments, as well as neighboring segments, respond with vasoconstriction to acetylcholine infusion instead of vasodilation, as would be expected in coronary arteries [57].

Lavoie et al. [58] studied 323 patients with major or minor depression degrees, with coronary artery disease or risk of developing the disease. FMD of the forearm was assessed when they were subjected to myocardial scintigraphy. Individuals with depression, both major and minor, had worse endothelial function compared to nondepressive ones.

On the other hand, in hypertensive people, endothelial dysfunction has been systematically observed [59–61], and attributed to oxidative stress dependent on NADPH oxidase vascular and mitochondrial sources. For example, by over expressing thioredoxin, which is an important peroxidase in the conversion of hydrogen peroxide to water, transgenic mice become resistant to A-II hypertension, oxidative stress, and endothelial dysfunction.

With respect to psychological stress and hypertension, Gasperin et al. [61] conducted a meta-analysis that included 34,556 patients with medium segment in 7.5 years. They concluded that individuals with more intense responses to psychological stress tests were 21% more likely to develop hypertension than those with more attenuated responses. Spruill [60] also reviewed the influence of chronic stress, i.e., occupational, stressful aspects of everyday life, low social status, and racial discrimination on the occurrence of hypertension. In general, the relationship between chronic psychological stress and hypertension is well established. However, obesity is associated with the production of inflammatory and noninflammatory adipocytokines by adipose tissue and endothelial dysfunction; perivascular adipose tissue regulates in part vascular homeostasis. Obesity and hypertension are two important components of the metabolic syndrome, which, in turn, has a strong connection with the development of coronary disease. Therefore, local and systemic vascular inflammation is also related to hypertension.

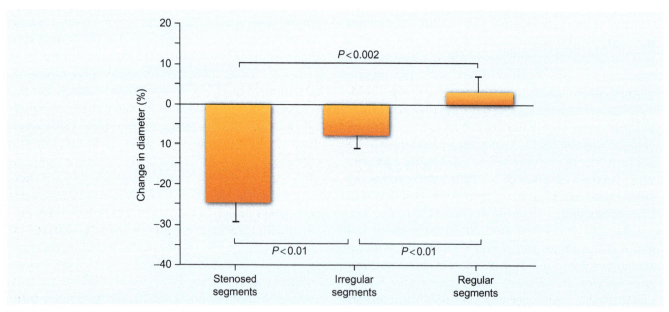

FIG. 20.7 Divergent effects of mental stress on coronary arteries. Psychological tests produced vasodilation in normal coronary segments and vasoconstriction in stenosed segments [56]. *Adapted from Yeung AC, Vekshtein VI, Krantz DS, et al. The effect of atherosclerosison the vasomotor response of coronary arteries to mentalstress. N Engl J Med 1991;325:1551–6.*

Occupational stress can be evaluated by the model of Karasek et al. [62], which is based on two variables: work demand and decision freedom, i.e., control that the employee has on occupational methods and strategies. Basically, it is the relationship between responsibility and authority; when someone is held accountable, but does not have autonomy to manage work, is subject to a considerable degree of stress. For example, in the CARDIA study [8], increasing demands at work were predictive of future hypertension in 3200 young and healthy over 8 years. The same was observed in 8395 workers in Canada for 7.5 years; in this case, the employees with lower degrees of support suffered more. Likewise, insecurity at work, unemployment, and low productivity aggravated the symptoms [59–62].

Social isolation and marital status are also associated with hypertension. In general, married individuals have lower rates of cardiovascular complications and lower mortality than singles. Naturally, this also depends on marriage quality; in happy relationships, problems are smaller than in those unhappy, according to various evaluations, both in acute or chronic conditions. On the other hand, social isolation rated by family support, living alone, number of friends and participation in community or religious activities influence the occurrence of hypertension [60].

Racial discrimination has recently emerged as a possible factor of hypertension. Thus, African-Americans are more hypertensive than whites, and part of that is attributed to racism [60,61]. Several studies recorded that cardiovascular reactivity tests applied to African-Americans were more positive compared to whites, which may also reflect a genetic component or eating habits.

Finally, belonging to a lower class and be rated by education level, occupation type, remuneration, social group and residence place, as documented in the *MESA* study [63,64], has been associated with higher prevalence of hypertension. In general, regions with fewer economic resources and few (or no) education opportunities make the individual more vulnerable. This is mainly explained by common lifestyles in communities.

The mechanisms associated with emotional stress, and hypertension are mainly dependent on the activation of the sympathetic nervous system (SNS) with catecholamines release, which increase heart rate, cardiac output, vascular resistance, and arterial pressure. Repeated negative thoughts, such as those that occur in paranoia, also appear to contribute. Regarding the therapeutic measures, the transcendental meditation was the only intervention that reduced blood pressure; this will be discussed later.

HYPERCHOLESTEROLEMIA

Experimental and clinical studies indicate that high levels of cholesterol in blood affect endothelial function, and psychological factors, when influencing life style, especially by consuming a fat-rich diet or sedentary lifestyle, raise cholesterol, and consequently reduce endothelial function [3]. Therefore, these systems are interconnected.

The oxidized LDL particles induce production of adhesion molecules on endothelium and monocytes attraction, have cytotoxic effects and increase proinflammatory gene activation and growth factors on endothelial cells and vascular wall, all leading to endothelial dysfunction; this, in turn, promotes platelet aggregation, expression of metalloproteinases, thrombogenesis, and forming foam cells. Such processes increase oxidative stress and generation of reactive oxygen species (ROS_2), creating a vicious spiral. Several studies have shown that LDL oxidized particles inhibit endothelium dependent relaxation by nitric oxide inactivation, inactivation of induced nitric oxide synthase (iNOS), or reduced arginine availability, the synthesis substrate for oxide generation. Thus, an increase in LDL is accompanied by a drop in flow-mediated dilation [65–68].

LDL also affects endothelial nitric oxide synthase (eNOS) by raising caveolin-1 expression and stabilization of heterocomplex of its connection with the synthase. It is known that caveolin-1, complex protein of caveolae layer, turns synthase inactive. And this effect is proportional to the increase of intracellular cholesterol, which modulates gene transcription of caveolin-1 through regulatory protein of sterol binding (SREBP). Thus, nitric oxide bioavailability may decrease independently of nitric oxide synthase concentration [65–68].

Nitric oxide bioavailability is also affected by ADMA (asymmetrical dimethy L-arginine) ratio with cholesterol. ADMA release through endothelial cells is increased by the presence of native or oxidized LDL, possibly mediated by super-regulation of S-adenoyilmethionine-dependent methyltransferases, with subsequent dimethyl L-arginine proteolysis. ADMA levels are higher in CAD patients and are markers of increased risk of cardiovascular events [69–72].

Unlike LDL, HDL activates eNOS via SR-BI (scavenger receptor BI) by a process requiring apo A-I binding. This action, in addition to its antioxidant and antiinflammatory activities, makes it protective against atherosclerosis. However, under certain conditions, such as diabetes and coronary heart disease, HDL may lose these beneficial activities by inactivating paraoxanase-1 and other oxidative modifications, becoming blocker of nitric oxide synthase via LOX1 activation (lectin-like receptor of oxidized LDL) and PKC beta II (C beta II protein kinase) [73–76].

There are also direct cholesterol actions in ionic potassium channels, among them the rectifier channels of cell intruder chain. Kir2, Kir4, and Kir6 are highlighted. The Kir2 channel is suppressed by cholesterol elevation in the cell membrane and potentiated by its depletion. It has been demonstrated by Fang et al. [77] that hypercholesterolemia leads to depolarization of endothelial cell membrane with impairment of endothelium-mediated vasodilation.

The highest levels of cholesterol, even if they are still in the normal range, lead to decreased flow-mediated vasodilation and increased sympathetic tone. Finally, it was observed that cholesterol reduction by statins restores endothelial function [67,78]. However, there are no controlled studies of the effect of psychological interventions on plasma cholesterol levels and endothelial function.

STUDIES WITH ANIMALS SUBJECTED TO MENTAL STRESS

Some studies in animals deserve attention. For example, in hypertensive rats isolated coronary artery, exposed to chronic stress by jets of air during 2 h per day for 2–10 days, it was noted that the dilator response to acetylcholine was lower in the stressed ones than in the calm ones, but only in aged rats with borderline hypertension [79]. However, the relaxation induced by sodium nitroprusside was also mitigated, but the response to isoproterenol had no effect. In another experiment performed with ovariectomized rats, subjected to high levels of stress due to box exchange, there was also

endothelial dysfunction, but the responses were attenuated by estrogen replacement [80]. Chronic stress modifies the rat's sexual behavior, probably from changes in sexual hormones in endocrine factors and nitric oxide. Thus, mice of both sexes, exposed to a moderate level of stress for 6 weeks, experienced different oxidative stress and compensatory responses. This is probably due to different oxidants/antioxidants underlying mechanisms. The responses to chronic stress in females were accompanied by lower levels of soluble type 1 intercellular molecules, which suggest a less impaired endothelium in females [80].

In monkeys subjected to high cholesterol diet for 36 months, the group undergoing a permanent stressful environment showed FMD by acetylcholine, significantly reduced in the iliac artery in relation to animals on the same diet, but without current stress. Others who had experienced stress, but were not under stress at the time of experience, had a normal response [81]. With respect to nitroglycerin, both groups were similar. This experiment suggests that monkeys exposed to acute, nonchronic stress experience reduced endothelium-dependent vasodilation.

MENTAL STRESS AND ENDOTHELIAL DYSFUNCTION—MECHANISMS

The various forms of mental stress mentioned, such as depression, social chronic stress, or acute stress caused by public speaking (in people), or tail shock and immobilization (in animals), cause endothelial dysfunction through various common mechanisms, which are activated differently depending on the circumstances.

From a biochemical point of view, the mental stress acts primarily through the activation of two systems: the pathway of hypothalamic/pituitary/adrenal (HPA) axis and sympathetic nervous system (SNS). Fig. 20.8 illustrates schematically the various mechanisms involved in the process of mental stress, endothelial dysfunction, and atherosclerosis. Psychological stress influences behavior, as will be discussed below [4,15,26,28].

The mental stress is felt by the cerebral cortex, which sends signals to hypothalamic paraventricular nucleus, which induces CRF (corticotropin-releasing factor) release in pituitary portal circulation; CRF binds to specific receptors on the pituitary;

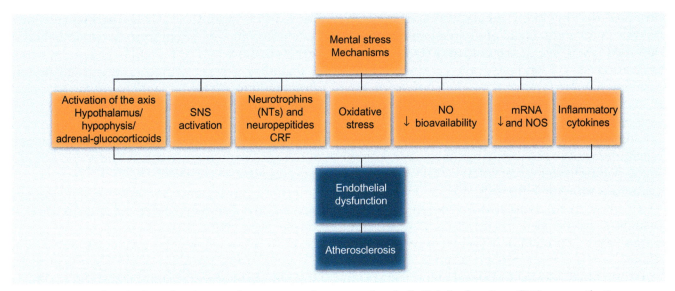

FIG. 20.8 Biochemical mechanisms acting on mental stress and endothelial dysfunction. *SNS*, sympathetic nervous system; *CRF*, corticotropin-releasing fator; *NO*, nitric oxide; *NOS*, nitric oxide synthase.

this, in turn, releases ACTH (adrenocorticotrofic hormone). ACTH releases glucocorticoid from adrenals (cortisol in humans and corticosterone in rodents) [82], a substance which mitigates flow-mediated arterial dilation in men. Thus, glucocorticoid treatment reduced FMD in 20 patients treated with prednisone in the long run for autoimmune chronic diseases [83]. Moreover, in HUVEC (human umbilical vein endothelial cells), the production of hydrogen peroxide was significantly enhanced by dexamethasone at the same time it increased the amount of peroxynitrite and decreased the amount of nitric oxide; this study also demonstrated that the generation of ROS$_2$ was due to mitochondrial electron transport chain, NADPH oxidase and xanthine oxidase. As is well known, nitric oxide is responsible for approximately 80% of arterial dilation. It originates from L-arginine under activation of eNOS gene NADPH enzyme action, and various cofactors such as thiols, flavins and tetrahydrobiopterin; the entire process results in production of L-citrulline and nitric oxide [77,78]. Another known phenomenon is ROS nitric oxide inactivation, which reduces its bioavailability. ROS includes superoxide, hydrogen peroxide and peroxynitrite radical. Therefore, Iuri et al.'s [76] data indicate that excess glucocorticoid leads to ROS production with consequent nitric oxide inactivation in vascular endothelium, thus representing a mechanism of vascular injury in patients with excess glucocorticoids [83].

Furthermore, the lipid and glucose metabolism are exacerbated by the production of glucocorticoids

by HPA-axis activation. For example, Pan et al. [84] reviewed 29 clinical studies and documented a significant association between depression and metabolic syndrome with a relative risk of 1.34%.

On the other hand, mental stress causes sympathetic nervous system (SNS) activation, releasing catecholamines in circulation; catecholamines and glucocorticoid are known as stress hormones. For example, Lambert et al. [85] analyzed SNS activation pattern in patients with metabolic syndrome and hypertension. The authors noted a high incidence of multiple spikes of sympathetic nerves associated with depressive symptoms.

The SNS activation increases vascular tone, myocardial oxygen consumption, platelet production, and activation of rennin angiotensin system (RAS). One RAS activation product is angiotensin-II (A-II), which, in turn, is potent stimulator of vascular oxidase NADPH, which is the main source of ROS in cardiovascular system [86]. Therefore, this pathway also causes oxidative stress and consequent inactivation or decrease of nitric oxide bioavailability and, consequently, endothelial dysfunction. In addition, there are additional humoral factors that respond to emotional stress. These include neurotrophins (NTs), such as urocortin 1, 2, and 3. NTs form a polypeptides family that includes NGF (nerve growth factor), BDNF (brain derived neurotrophic factor) NT3, NT4, and NT5 in humans. NTs' secretion by the hypothalamus, pituitary, and peripheral nerves is markedly altered in psychological stress conditions [87]. In line with this concept, it has been proposed that BDNF dysregulation plays an essential role in the mechanism of depression. It was also observed that BDNF causes oxidative stress, since its expression was found associated with increased expression of NADPH oxidase and superoxide production in rats aortic with spontaneous hypertension (SHR) [88].

Moreover, neuropeptides of the corticotropin-releasing factor (CFR) family members are involved in the pathophysiology of behavior and psychological stress through HPA axis regulation. CRF is a 41-amino acids neuropeptide; it represents a family including urocortins 1–3, CRF1, CRF2 which bind intracellular G protein and specific receptors. This system seems critical to the beginning of response to stress. Thus, Neufeld-Cohen et al. [89] showed that urocortin is essential in the recovery process from stress in mice knockout for the three urocortin genes. It was also observed that urocortin is expressed in endothelial cells and has a powerful antioxidant effect [90].

Seo et al. [91] reported that NADPH oxidase in the brain plays a key role in depressive behavior through p47phox and p67phox subunits. Several observations suggest that depressive behavior occurs via up-regulation of cerebral NADPH oxidase [28].

There is evidence that endothelin-1 (ET-1) secretion is altered in situations of psychological stress. ET-1 is a peptide with 21 amino acids, which has high vasoconstrictor power. Thus, Mangiafico et al. [92] demonstrated that in patients with intermittent claudication, ET-1 levels were higher at baseline situation than in individuals without claudication. Especially when subjected to arithmetic stresses, patients with intermittent claudication increased ET-1 levels significantly more than controls. Therefore, the increased ET-1 release may be a trigger for cardiovascular events or acceleration of atherosclerosis.

There are also numerous experimental and clinical evidences indicating that psychological stresses cause inflammatory cytokines production, such as interleukins (IL-1, IL-6), TNF-α, and interferon-gamma, in both the blood and the brain [93]. Cytokines are bioactive mediators produced by several cell types, particularly macrophages and lymphocytes, which act synergistically or antagonistically, in a complex mode, and are associated with inflammation, immune response, differentiation or cell death. Psychological stresses are clearly related to such stimulation, causing a true inflammatory condition. A consequence of increased cytokines production is the induction of oxidative stress, with consequent deleterious effects on endothelium-dependent dilation, which has been observed, for example, in patients with depression and bipolar disorder. In experimental animals undergoing foot shocks or immobilization, an increase in IL-6 plasma concentration was also noticed [94].

Another consequence of acute stress in animals, induced by immobilization, is the release of nuclear factor NF-kB. In men undergoing "forced speech," it was observed to be the same with increased catecholamines and glucocorticoids [95]. NF-kB plays a key role in the stimulation of many genes critical for vascular function, including ERO$_2$ generation.

In conclusion, from the biochemical point of view, psychological stresses cause HPA axis activation with glucocorticoid production, SNS with catecholamine release, inflammatory cytokines and NF-kB release, an increase in ET-1 production and oxidative stress. These factors are responsible for the decreased bioavailability of nitric oxide, decreased FMD and increased vascular tone, with consequent activation of endothelium, whose vasodilator function is impaired to cause clinical syndromes such as angina and left ventricular dysfunction.

Other aspects germane to endothelial dysfunction are related to behaviors and lifestyle (Fig. 20.9). Thus, emotional disorders tend to facilitate unhealthy habits, such as smoking, eating fatty foods, and sedentary lifestyle, all associated with endothelial dysfunction [45]. Fig. 20.10 exemplifies practices promoting health as opposed to others that cause disease. The mechanisms associated with this type of endothelial dysfunction contribute to atherosclerosis development.

TREATMENT OPTIONS FOR MENTAL DISEASES AND ENDOTHELIUM FUNCTION

The classic way of treating emotional problems such as depression and other forms of stress include the use of medication, psychological therapies and lifestyle changes. An extensive review on the use of medications escapes from the purpose of this chapter. Therefore, we shall mention just a few most

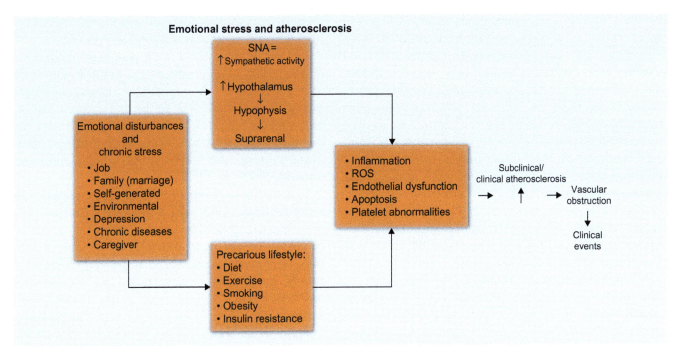

FIG. 20.9 Dependent mechanisms of psychological stress that trigger biochemical processes that change lifestyle; both lead to atherosclerosis development. *ANS*, autonomic nervous system [3]. *Adapted from Da Luz PL, Nishiyama M, Chagas AC. Drugs and lifestyle for the treatment and prevention of coronary artery disease: comparative analysis of the scientific basis. Braz J Med Biol Res 2011;44:973–91.*

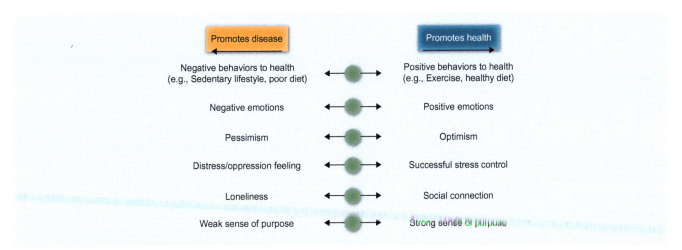

FIG. 20.10 Healthy and harmful behaviors to health [4]. *Adapted from Rozanski, A. Behavioral cardiology: current advances and future directions. J Am Coll Cardiol 2014;64:100–10.*

common medications. These include antidepressants as follows:

- Monoamine oxidase inhibitors (e.g., tranylcypromine);
- tricyclic (e.g., amitriptyline, imipramine, and clomipramine);
- selective serotonin reuptake inhibitors (e.g., fluoxetine, paroxetine, sertraline, citalopram, and fluvoxamine);

- serotonin and noradrenaline reuptake inhibitors (e.g., duloxetine and venlafaxine);
- serotonin receptor antagonists (e.g., trazodone and mirtazapine); and
- dopamine and noradrenaline reuptake inhibitors (bupropion).

It can be said that lifestyle changes are important but difficult to implement. For example, facing family problems that may require divorce, fight drug use by children, or changing jobs to avoid stress at work may be necessary, but they require huge behavioral changes. However, it has been shown that regular exercise practice may be as effective as antidepressant medication and prevents more recurrences than medications [96]. However, in practice, convince sedentary and depressed people to take regular exercise is notoriously difficult. There are even unsolvable situation in the short and medium term, as a chronic disease of some family member who is under the person's care, or circumstances of poverty, lack of social or family support which clearly interfere with the individual's spiritual peace. The number of situations that can cause emotional stress is almost infinite.

There are also circumstances beyond the person's autonomy, such as political instability in a country, global economic crises, environmental aggression, violence, war, terrorism, and deficits in the pension and health system, all of which may deeply affect individuals. The complicating factor is that the individual suffers the consequences but has no way to directly affect the course of events.

Excluding these possibilities, we are left with those in which people have direct responsibility and, therefore, can influence outcomes. Since these require in-depth psychological analyses and corresponding actions, so-called behavioral medicine has been gaining more and more relevance.

Meditation, Prayer, Yoga

Wang et al. [97] studied 10 experienced meditators, measuring cerebral blood flow by functional magnetic resonance imaging and associated responses during and after meditation. It turned out that frontal regions, anterior cingulate, limbic system, and parental lobes were affected during practice. A strong correlation was also observed between the depth of meditation and neural activity in the frontal areas of the left lower brain, including the insula, inferior frontal cortex, and temporal pole. There were persistent changes in the left anterior insula and precentral turning, even after meditation had been stopped. The study revealed that changes occur in the brain during meditation and these changes are associated with practitioners' subjective experiences.

Cahn and Polich's review [98] shows that neuroimaging studies indicate an increase in cerebral blood flow measurements during meditation, which seems to reflect changes in the anterior cingulate cortex and prefrontal and dorsolateral areas.

In Lou et al.'s study, PET was used to measure brain activity in meditative yoga practitioners at different stages. In all meditation stages, a bilateral increase was observed in hippocampus, parietal and sensory occipital; however, in orbitofrontal, prefrontal dorsolateral, anterior cingulate cortex, thalamus, bridge, and cerebellum, a reduction was noticed.

In Lazar et al.'s study [99], using a specific yoga practice (Kundalini), which consists of the repetition of a mantra associated with breathing exercises, cerebral activity was evaluated by means of functional magnetic resonance imaging (fMRI) measurements. It was observed that this practice increases the activity in the putamen, midbrain, pregenual anterior cingulate cortex, and para-hippocampal formation, as well as areas within the frontal and parietal cortex. Therefore, it is assumed that with increasing time of meditation, individuals produce altered brain states, which can change their state of consciousness as they continue the meditative activity. The main increased activity areas aid attention (frontal and parietal cortex, particularly the dorsolateral prefrontal cortex) and autonomic control (limbic regions, midbrain and pregenual anterior cingulate cortex). Thus, there is variability in the findings as a result of the different styles of meditative yoga.

Zen is a kind of contemplative meditation. Some practitioners were subjected to magnetic resonance imaging (MRI) exams and showed activation in the areas of hippocampus, left frontal, right temporal, and anterior cingulate cortex, with deactivations in the visual cortex and left frontal lobe [98]. The increased activity in dorsolateral prefrontal cortex may contribute to cerebral function self-regulation, because it has been seen as a self-regulatory of emotional reactions. At the same time, the decrease in emotional reactivity is reported in meditators individuals [100,101].

Christian practitioners of prayer were analyzed in some studies. Azari et al. [102] compared religious individuals with nonreligious ones while listening to the stories of a children's poetry and a phone book. The religious group reported achieving spiritual elevation while reading, and found significant activation of dorsolateral prefrontal lobe, right medial parietal, and dorsomedial prefrontal cortex when compared to those who were nonreligious.

The increased dorsolateral and dorsomedial prefrontal cortex was significantly higher for all comparisons. In contrast, the nonreligious reported a state of happiness while listening to poetry, which was associated with left amygdale activation, directly related to affective states [98].

Moreover, while Franciscan nuns prayed, they were subjected to single-photon emission computed tomography (SPECT) mapping, in Newberg et al.'s study [103]. In comparison to the baseline values, the scans during the prays showed increased blood flow in prefrontal cortex (7.1%), which was lower in parietal lobes (6.8%) and higher in frontal lobes (9%), and also a strong correlation between changes in blood flow in prefrontal and ipsilateral cortex, and in the superior parietal lobe. The results suggest that the meditative-spiritual experiences are in part mediated by a change in the superior parietal lobe, which helps generate the normal spatial awareness sense [103].

Mindfulness is a derivation of meditation that has been widely used for psychological symptoms treatment and control. A systematic review, including nine articles [104] in 2014, evaluated the effects of this method in patients with vascular disease, including CAD, angina, IAM, stroke, and peripheral vascular disease, as well as diabetes, hypertension, and hypercholesterolemia, but who had not yet developed vascular disease. Psychological and physical outcomes of the interventions were evaluated which, for the most part, lasted 8 weeks. There was a significant improvement of psychological aspects, such as symptoms of anxiety, depression, and stress, compared to parameters verified at the beginning of the studies. There were no positive effects of mindfulness on the physical part, but the authors believe that the intervention time is insufficient to achieve benefits.

Pargaonkar et al. [105], from the University of Stanford, observed similar results in the study of women with angina, but without obstructive coronary artery disease, undergoing 8weeks of mindfulness associated with the use of medications. When compared to women who received only medication, they showed improvements in psychological conditions.

In conclusion, it is important to understand that there is a considerable discrepancy when the results of different studies are compared. This is probably due to the absence of a standardized model for evaluating the effects of meditation, as well as the lack of appropriate techniques to apply in new studies. More importantly, neurological exam results are beginning to show some consistency in the location of meditative practices. The frontal and prefrontal areas appear as relatively high demand activation sites. Finally, a significant number of reports—in both psychological and medical clinic—suggest significant effects, offering significant correlations between meditation and brain activity.

Cognitive Behavioral Therapy (CBT)

Behavioral and emotional reactions are strongly influenced by our thoughts, which determine how we perceive things. Thus, our emotions are consequences not of the situation itself, but of our perceptions, expectations and interpretations. This is the basic notion of Aaron T. Beck's theory [106]. Based on this theory, Beck developed cognitive behavioral therapy (CBT), whose goal is to aid the identification and evaluation of higher order thoughts and beliefs in order to encourage more realistic thoughts, generating more functional behaviors which will achieve psychological well-being [106].

CBT is effective in reducing symptoms and recurrence rates, with or without medication, in a wide variety of psychiatric disorders, including depression, suicidal tendencies, anxiety disorders and phobias, among others [107–109]. Studies using neuroimaging have proven their efficacy. For example, Goldapple et al. [110] evaluated two groups of patients with depression: one treated with paroxetine and the other treated with CBT. They concluded that CBT promotes clinical recovery by modulating the functioning of specific sites in limbic and cortical system regions.

In addition to the cognitive restructuring, CBT uses several complementary techniques such as relaxation, breathing techniques, meditation/mindfulness. In short, as mentioned by Rozansky [4], CBT today is an important tool in clinical cardiology when one wants to implement changes in the patients' lifestyle. Our clinical experience fully supports this notion.

References

[1] Selye H, Fortier C. Adaptive reactions to stress. Res Publ Assoc Res Nerv Ment Dis 1949;29:3–18.
[2] Dimsdale JE. Psychological stress and cardiovascular disease. J Am Coll Cardiol 2008;51:1237–46.

[3] Da Luz PL, Nishiyama M, Chagas AC. Drugs and lifestyle for the treatment and prevention of coronary artery disease: comparative analysis of the scientific basis. Braz J Med Biol Res 2011;44:973–91.

[4] Rozanski A. Behavioral cardiology: current advances and future directions. J Am Coll Cardiol 2014;64:100–10.

[5] Rosengren A, Hawken S, Ounpuu S, et al. Association of psychosocial risk factors with risk of acute myocardial infarction in 11 119 cases and 13 648 controls from 52 countries (the INTER-HEART study): case-control study. Lancet 2004;364:953–62.

[6] Rozanski A, Blumenthal JA, Davidson KW, et al. The epidemiology, pathophysiology, and management of psychosocial risk factors in cardiac practice: the emerging field of behavioral cardiology. J Am Coll Cardiol 2005;45:637–51.

[7] Kim CK, McGorray SP, Bartholomew BA, et al. Depressive symptoms and heart rate variability in postmenopausal women. Arch Intern Med 2005;165:1239–44.

[8] Iribarren C, Sidney S, Bild DE, et al. Association of hostility with coronary artery calcification in young adults: the CARDIA study. Coronary Artery Risk Development in Young Adults. JAMA 2000;283:2546–51.

[9] Isso H, Date C, Yamamoto A, et al. Perceived mental stress and mortality from cardiovascular disease among Japanese men and women: the Japan Collaborative Cohort Study for Evaluation of Cancer Risk Sponsored by Monbusho (JACC Study). Circulation 2002;106:1229–36.

[10] Bosma H, Peter R, Siegrist J, et al. Two alternative job stress models and the risk of coronary heart disease. Am J Public Health 1998;88:68–74.

[11] Matthews KA, Gump BB, Harris KF, et al. Hostile behaviors predict cardiovascular mortality among men enrolled in the Multiple Risk Factor Intervention Trial. Circulation 2004;109:66–70.

[12] Rugulies R. Depression as a predictor for coronary heart disease. E review and meta-analysis. Am J Prev Med 2002;23:51–61.

[13] Okamura T, Tanaka H, Miyamatsu N, et al. The relationship between serum total cholesterol and all-cause or cause-specific mortality in a 17.3-year study of a Japanese cohort. Atherosclerosis 2007;190:216–23.

[14] Lespérance F, Frasure-Smith N, Talajic M, et al. Five-year risk of cardiac mortality in relation to initial severity and one-year changes in depression symptoms after myocardial infarction. Circulation 2002;105:1049–53.

[15] Das S, O'Keefe JH. Behavioral cardiology: recognizing and addressing the profound impact of psychosocial stress on cardiovascular health. Curr Atheroscler Rep 2006;8:111–8.

[16] Gonçalves RW, Vieira FS, Delgado PGG. Política de Saúde Mental no Brasil: evolução do gasto federal entre 2001 e 2009. Rev Saúde Pública 2012;46:51–8.

[17] Academy of Medical Sciences. Challenges and priorities for global mental health research in low- and middle-income countries. Londres: Symposium report; 2008.

[18] The National Alliance on Mental Illness. Mental Illness: facts and numbers. [Internet]. Available at: http://www2.nami.org/factsheets/mentalillness_factsheet.pdf; 2013 [accessed 24.06.16].

[19] Santos EG, Siqueira MM. Prevalência dos transtornosmentaisnapopulaçãoadultabrasileira: umarevisãosistemática de 1997 a 2009. J Bras Psiquiatr 2010;593:238–46.

[20] Andrade LH, Wang YP, Andreoni S, et al. Mental disorders in megacities: findings from the São Paulo megacity mental health survey, Brazil. PLoS ONE 2012;7:e31879.

[21] Feitosa HN, Ricou M, Rego S, et al. A saúde mental das crianças e dos adolescentes: consideraçõesepidemiológicas, assistenciais e bioéticas. Rev Bioet 2011;19:259–75.

[22] Silva-Junior JS, Fischer FM. Adoecimento mental incapacitante: benefíciosprevidenciários no Brasil entre 2008-2011. Rev Saúde Pública 2014;48:186–90.

[23] Razzouk D. Economia da saúdeaplicada à saúde mental. In: Mateus MD, editor. Políticas de saúde mental: baseado no cursoPolíticaspúblicas de saúde mental, do CAPS Luiz R. Cerqueira. São Paulo: Instituto de Saúde; 2013. p. 230–51.

[24] Wang PS, Simon G, Kessler RC. The economic burden of depression and the cost-effectiveness of treatment. Int J Methods Psychiatr Res 2003;12:22–33.

[25] Da Luz PL, Favarato D. A disfunçãoEndotelialcomoÍndicePrognóstico e AlvoTerapêuticoemEndotélio&Doençcas Cardio-vasculares de PL Da Luz, FRM Laurindo, ACP Chagas. São Paulo: Atheneu; 2005. p. 203–20.

[26] Toda N, Nakanishi-Toda M. How mental stress affects endothelial function. Pflugers Arch 2011;462:779–94.

[27] Coimbra S, Da Luz PL. Efeito de Ginkobilobasobredilataçãoendotéliodependenteemidosossadios. Dados nãopublicados.

[28] Inoue N. Stress and atherosclerotic cardiovascular disease. J Atheroscler Thromb 2014;21:391–401.

[29] Wittstein IS, Thieman DR, Lima JA, et al. Neurohumoral features of myocardial stunning due to sudden emotional stress. N Engl J Med 2005;352:539–48.

[30] Schmidt FP, Basner M, Kröger G, et al. Effect of nighttime aircraft noise exposure on endothelial function and stress hormone release in healthy adults. Eur Heart J 2013;34:3508–14.

[31] Miller M, Mangano C, Park Y, et al. Impact of cinematic viewing on endothelial function. Heart 2006;92:261–2.

[32] Miller M, Fry WF. The effect mirthful laughter on the human cardiovascular system. Med Hypotheses 2009;73:636–9.

[33] Mausbach BT, Roepke SK, Ziegler MG, et al. Association between chronic caregiving stress and impaired endothelial function in the elderly. J Am Coll Cardiol 2010;55:2599–606.

[34] Harris CW, Edwards JL, Baruch A, et al. Effects of mental stress on brachial artery flow-mediated vasodilation in healthy normal individuals. Am Heart J 2000;139:405–11.

[35] Zieman SJ, Malasky BR. Cardiovascular risk factors in the elderly—evaluation and intervention. In: Cardiovascular disease in the elderly of Gerstenblith G. New York: Humana Press; 2010. p. 79–102.

[36] Penninx BW, Beekman AT, Honing A, et al. Depression and cardiac mortality: results from a community-based longitudinal study. Arch Gen Psychiatry 2001;58:221–7.

[37] Berkman LF, Leo-Summers L, Horwitz RI. Emotional support and survival after myocardial infarction. A prospective population-based study of the elderly. Ann Intern Med 1992;117:1003–9.

[38] Frasure-Smith N, Lesperance F. Depression and other psychological risks following myocardial infarction. Arch Gen Psychiatry 2003;60:627–36.

[39] Watkins LL, Scheiderman N, Blumenthal JA, et al. Cognitive and somatic symptoms of depressions are associated with medical comorbity in patients after acute myocardial infarction. Am Heart J 2003;146:48–54.

[40] Shiotani I, Sato H, Kinjo K, et al. Depressive symptoms predict 12-month prognosis in elderly patients with acute myocardial infarction. J Cardiovasc Risk 2002;9:153–60.

[41] Berkman LF, Blumenthal J, Burg M, et al. Effects of treating depression and low perceived social support on clinical events after myocardial infarction: the Enhancing Recovery in Coronary Heart Disease Patients (ENRICHD) Randomized Trial. JAMA 2003;289:3106–16.

[42] Egashira K, Inou T, Hirooka Y, et al. Effects of age on endothelium-dependent vasodilation of resistance coronary artery by acetylcholine in humans. Circulation 1993;88:77–81.

[43] Da Luz PL, Fialdini RC, Nishiyama M. Red wine, resveratrol and vascular aging: implications for dementia and cognitive decline in diet and nutrition in dementia and cognitive decline. In: Martin CR, Preedy VR, editors. Diet and nutrition in dementia cognitive decline. Rio de Janeiro: Elsevier; 2015. p. 943.

[44] Celemajer DS, Sorensen KE, Spiegelhalter DJ, et al. Aging is associated with endothelial dysfunction in healthy men years before the age-related decline in women. J Am Coll Cardiol 1994;24:471–6.

[45] Chauhan A, More RS, Mullins PA. Aging-associated endothelial dysfunction is humans is reversed by L-arginine. J Am Coll Cardiol 1996;28:1796–804.

[46] Woo KS, Jane A, Chook P, et al. Chinese adults are less susceptible than whites to age-related endothelial dysfunction. J Am Coll Cardiol 1997;30:113–8.

[47] Kelleher RJ, Soiza RL. Evidence of endothelial dysfunction in the development of Alzheimers disease: is Alzheimers a vascular disorder? Am J Cardiovasc Dis 2013;3:197–226.

[48] Wagner JA, Tennen H, Mansoor GA, et al. History of major depressive disorder and endothelial function in postmenopausal women. Psychosom Med 2006;68(1):80–6.

[49] Hambrecht R, Wolf A, Gielen S, et al. Effect of exercise on coronary endothelial function in patients with coronary artery disease. N Engl J Med 2000;342:454–60.

[50] Rush JWE, Denniss SG, Graham DA. Vascular nitric oxide and oxidative stress: determinants of endothelial adaptations to cardiovascular disease and to physical activity. Can J Appl Physiol 2005;30:442–74.

[51] Jain D, Shaker SM, Burg M, et al. Effects of mental stress on left ventricular and peripheral vascular performance in patients with coronary artery disease. J Am Coll Cardiol 1998;31:1314–22.

[52] Rozanski A, Krantz DS, Bairey CN. Ventricular responses to mental stress testing in patients with coronary artery disease. Pathophysiological implications. Circulation 1991;83:II137.

[53] Goldberg AD, Becker LC, Bonsall R, et al. Ischemic, hemodynamic, and neurohormonal responses to mental and exercise stress. Experience from the psychophysiological investigations of myocardial ischemia study (PIMI). Circulation 1996;94:2402–9.

[54] Okano Y, Utsunomiya T, Yano K. Effect of mental stress on hemodynamics and left ventricular diastolic function in patients with ischemic heart disease. Jpn Circ J 1998;62:173–7.

[55] Bairey CN, Krantz DS, Rozanski A. Mental stress as an acute trigger of ischemic left ventricular dysfunction and blood pressure elevation in coronary artery disease. Am J Cardiol 1990;66:28G–31G.

[56] Yeung AC, Vekshtein VI, Krantz DS, et al. The effect of atherosclerosis on the vasomotor response of coronary arteries to mental stress. N Engl J Med 1991;325:1551–6.

[57] Ludmer PL, Selwyn AP, Shook TL, et al. Paradoxical vasoconstriction induced by acetylcholine in atherosclerosis coronary arteries. N Engl J Med 1986;315:1046–51.

[58] Laroia KT, Pelletier R, Arsenault A, et al. Association between clinical depression and endothelial function measured by forearm hyperemic reactivity. Psychosom Med 2010;72:20–6.

[59] Cardillo C, Kilcoyne CM, Cannon RO, et al. Impairment of the nitric oxide-mediated vasodilator response to mental stress in hypertensive but not in hypercholesterolemic patients. J Am Coll Cardiol 1998;32:1207–13.

[60] Spruill TM. Chronic psychosocial stress and hypertension. Curr Hypertens Rep 2010;12:10–6.

[61] Gasperin D, Netuveli G, Dias-da-Costa JS, et al. Effect of psychological stress on blood pressure increase: a meta-analysis of cohort studies. Cad Saude Publica 2009;25:715–26.

[62] Karasek RA, Baker D, Marxer F, et al. Job decision latitude, job demands, and cardiovascular disease: a prospective study of Swedish men. Am J Public Health 1981;71:694–705.

[63] Ranjit N, Diez-Roux AV, Shea S, et al. Psychosocial factors and inflammation in the multi-ethnic study of atherosclerosis. Arch Intern Med 2007;167:174–81.

[64] Ranjit N, Diez-Roux AV, Shea S, et al. Socioeconomic position, race/ethnicity, and inflammation in the multi-ethnic study of atherosclerosis. Circulation 2007;116:2383–90.

[65] Michel JB, Feron O, Sase K, et al. Caveolin versus calmodulin. Counterbalancing allosteric modulators of endothelial nitric oxide synthase. J Biol Chem 1997;272:25907–12.

[66] Feron O, Dessy C, Moniotte S, et al. Hypercholesterolemia decreases nitric oxide production by promoting the interaction of caveolin and endothelial nitric oxide synthase. J Clin Invest 1999;103:897–905.

[67] Feron O, Dessy C, Desager JP, et al. Hydroxymethylglutaryl-coenzyme A reductase inhibition promotes endothelial nitric oxide synthase activation through a decrease in caveolin abundance. Circulation 2001;103:113–8.

[68] Pelat M, Dessy C, Massion P, et al. Rosuvastatin decreases caveolin-1 and improves nitric oxide-dependent heart rate and blood pressure variability in apolipoprotein E-/- mice in vivo. Circulation 2003;107:2480–6.

[69] Boger RH, Sydow K, Borlak J, et al. LDL cholesterol upregulates synthesis of asymmetrical dimethylarginine in human endothelial cells: involvement of S-adenosylmethionine-dependent methyltransferases. Circ Res 2000;87:99–105.

[70] Perticone F, Sciacqua A, Maio R, et al. Asymmetric dimethylarginine, L-arginine, and endothelial dysfunction in essential hypertension. J Am Coll Cardiol 2005;46:518–23.

[71] Melikian N, Wheatcroft SB, Ogah OS, et al. Asymmetric dimethylarginine and reduced nitric oxide bioavailability in young Black African men. Hypertension 2007;49:873–7.

[72] Juonala M, Viikari JSA, Alfthan G, et al. Brachial artery flow mediated dilation and asymmetrical dimethylarginine in the cardiovascular risk in young Finns study. Circulation 2007;116:1367–73.

[73] Yuhanna IS, Zhu Y, Cox BE, et al. High-density lipoprotein binding to scavenger receptor-BI activates endothelial nitric oxide synthase. Nat Med 2001;7:853–7.

[74] Besler C, Heinrich K, Rohrer L, et al. Mechanisms underlying adverse effects of HDL on eNOS-activating pathways in patients with coronary artery disease. J Clin Invest 2011;121:2693–708.

[75] Xu S, Ogura S, Chen J, et al. LOX-1 in atherosclerosis: biological functions and pharmacological modifiers. Cell Mol Life Sci 2013;70:2859–72.

[76] Undurti A, Huang Y, Lupica JA, et al. Modification of high density lipoprotein by myeloperoxidase generates a pro-inflammatory particle. J Biol Chem 2009;284:30825–35.

[77] Fang Y, Mohler ER, Hsieh III E, et al. Hypercholesterolemia suppresses inwardly rectifying k+ channels in aortic endothelium in vitro and in vivo. Circ Res 2006;98:1064–71.

[78] Steinberg HO, Bayazeed B, Hook G, et al. Endothelial dysfunction is associated with cholesterol levels in the high normal range in humans. Circulation 1997;96:3287–93.

[79] Fuchs LC, Landas SK, Johnson AK. Behavioral stress alters coronary vascular reactivity in borderline hypertensive rats. J Hypertens 1997;15:301–7.

[80] Morimoto K, Kurahashi Y, Shintani-Ishida K, et al. Estrogen replacement suppresses stress-induced cardiovascular responses in ovariectomized rats. Am J Physiol Heart Circ Physiol 2004;287:H1950–6.

[81] Strawn WB, Bondjers G, Kaplan JR, et al. Endothelial dysfunction in response to psychosocial stress in monkeys. Circ Res 1991;68:1270–9.

[82] Alkadhi K. Brain physiology and pathophysiology in mental stress. ISRN Physiol 2013;2013. https://doi.org/10.1155/2013/806104.

[83] Iuchi T, Akaike M, Mitsui T, et al. Glucocorticoid excess induces superoxide production in vascular endothelial cells and elicits vascular endothelial dysfunction. Circ Res 2003;92:81–7.

[84] Pan A, Keum N, Okereke OI, et al. Bidirectional association between depression and metabolic syndrome: a systematic review and meta-analysis of epidemiological studies. Diabetes Care 2012;35:1171–80.

II. ENDOTHELIAL DYSFUNCTION AND CLINICAL SYNDROMES

[85] Lambert E, Dawood T, Straznicky N, et al. Association between the sympathetic firing pattern and anxiety level in patients with the metabolic syndrome and elevated blood pressure. J Hypertens 2010;28:543–50.

[86] Fernandes DC, Bonatto D, Laurindo FRM. The evolving concept of oxidative stress. In: Sauer H, Shah A, Laurindo FR, editors. Oxidative stress in clinical practice: cardiovascular diseases. New York: Springer; 2010.

[87] Naert G, Ixart G, Maurice T, et al. Brain-derived neurotrophic factor and hypothalamic-pituitary-adrenal axis adaptation processes in a depressive-like state induced by chronic restraint stress. Mol Cell Neurosci 2011;46:55–66.

[88] Amoureux S, Lorgis L, Sicard P, et al. Vascular BDNF expression and oxidative stress during aging and the development of chronic hypertension. Fundam Clin Pharmacol 2012;26:227–34.

[89] Neufeld-Cohen A, Tsoory MM, Evans AK, et al. A triple urocortin knockout mouse model reveals an essential role for urocortins in stress recovery. Proc Natl Acad Sci U S A 2010;107:19020–5.

[90] Lassègue B, San Martín A, Griendling KK. Biochemistry, physiology, and pathophysiology of NADPH oxidases in the cardiovascular system. Circ Res 2012;110:1364–90.

[91] Seo JS, Park JY, Choi J, et al. NADPH oxidase mediates depressive behavior induced by chronic stress in mice. J Neurosci 2012;32:9690–9.

[92] Mangiaflco RA, Malatino LS, Attinà T, et al. Exaggerated endothelin release in response to acute mental stress in patients with intermittent claudication. Angiology 2002;53:383–90.

[93] García-Bueno B, Caso JR, Leza JC. Stress as a neuroinflammatory condition in brain: damaging and protective mechanisms. Neurosci Biobehav Rev 2008;32:1136–51.

[94] LeMay LG, Vander AJ, Kluger MJ. The effects of psychological stress on plasma interleukin-6 activity in rats. Physiol Behav 1990;47:957–61.

[95] Bierhaus A, Wolf J, Andrassy M, et al. A mechanism converting psychosocial stress into mononuclear cell activation. Proc Natl Acad Sci U S A 2003;100:1920–5.

[96] Dunn AL, Trivedi MH, Kampert JB, et al. Exercise treatment for depression efficacy and dose response. Am J Prev Med 2005;28:1–8.

[97] Wang DJJ, Rao H, Korczykowski M, et al. Cerebral blood flow changes associated with different meditation practices and perceived depth of meditation. Psychiatry Res 2011;191:60–7.

[98] Cahn BR, Polich J. Meditation states and traits: EEG, ERP, and neuroimaging studies. Psychol Bull 2006;132:180–211.

[99] Lazar SW, Rosman IS, Vangel M, et al. Functional brain imaging of mindfulness and mantra-based meditation. In: Paper presented at the meeting of the Society for Neuroscience, New Orleans; 2003.

[100] Goleman DJ. The meditative mind: varieties of meditative experience. New York: Penguin Putnam; 1977/1988.

[101] Wallace RK. Physiological effects of transcendental meditation. Science 1970;167:1751–4.

[102] Azari NP, Nickel J, Wunderlich G. Neural correlates of religious experience. Eur J Neurosci 2003;13:1649–52.

[103] Newberg A, Pourdehnad M, Alavi A, et al. Cerebral blood flow during meditative prayer: preliminary findings and methodological issues. Percept Mot Skills 2003;97:625–30.

[104] Abbott RA, Whear R, Rodgers LR, et al. Effectiveness of mindfulness-based stress reduction and mindfulness based cognitive therapy in vascular disease: a systematic review and meta-analysis of randomised controlled trials. J Psychosom Res 2014;76:341–51.

[105] Pargaonkar V, Goldner J, Edwards KS, et al. The effect of mindfulness-based stress reduction on angina and vascular function in women with non-obstructive coronary artery disease. JACC 2015;65(10S):A1658.

[106] Hofmann SG. Introdução à terapia cognitive-comportamentalcontemporânea. Porto Alegre: Ed. Artmed; 2014. p. 236.

[107] Hofmann SG, Asnaani A, Vonk IJJ, et al. The efficacy of cognitive behavioral therapy: a review of meta analyses. Cognit Ther Res 2012;36:427–40.

[108] Lam D, Watkins E, Hayward P, et al. A randomized controlled trial of cognitive therapy of relapse prevention for bipolar disorder: outcome of the first year. Arch Gen Psychiatry 2003;60:145–52.

[109] Scoot J, Paykel E, Morris R, et al. Cognitive-behavioural therapy for severe and recurrent bipolar disorders. Randomised controlled trial. Br J Psychiatry 2006;188:313–20.

[110] Goldapple K, Segal Z, Garson C, et al. Modulation of cortical-limbic pathways in major depression—treatment-specific effects of cognitive behavior therapy. Arch Gen Psychiatry 2004;61:34–41.

Further Reading

[1] Lou HC, Kjaer TW, Friberg L. A 15O-H2O PET study of meditation and the resting state of normal consciousness. Hum Brain Mapp 1999;7:98–105.

LIPOPROTEINS

21

Lipids and Lipoprotein Mediators of Endothelial Function and Dysfunction

Helena Coutinho Franco de Oliveira

INTRODUCTION

This chapter focuses on the role of plasmatic lipoproteins, as well as associated molecules, upon endothelial function, presenting recent experimental, observational, and clinical scientific findings. Some types of lipoproteins, especially their lipid components, may contribute to preserve endothelial function or, in contrast, trigger endothelial dysfunction, contributing significantly to the atherogenesis process. Knowledge regarding the interaction of the systemic lipid mediators with the endothelium may indicate potential relevant preventive or therapeutic intervention targets in case of risk of cardiovascular diseases.

In its physiological condition, the endothelium is continuously exposed to several systemic circulatory molecules that may have significant effects on its function. Plasmatic lipoproteins are very heterogeneous molecular macroaggregates in terms of size, lipid and protein composition, density, and electrical charge, acknowledged as the main transport vehicle for the endogenous and exogenous lipids with energetic, structural, and regulatory functions. Lipoproteins interact directly with vascular endothelial cells, by several types of proteins, receptors, and transporters located in the cell plasma membrane, either inside or outside special regions called lipid rafts and caveolae. Lipid rafts are firmly-packed microdomains, rich in glycosphingolipids, cholesterol, and proteins, that compartmentalize cellular processes related to signal transduction and transport through the membrane. Caveolae are a special type of lipid rafts that contain a protein called caveolin-1 and several receptors and molecules involved in signal transduction pathways (Fig. 21.1) [1]. Caveolae are particularly abundant in endothelial cells, where they regulate functions such as vascular tone, permeability, transendocytosis, angiogenesis, and response to several types of stress. Caveolae contain several endothelial function modulating molecules, such as endothelial nitric oxide synthase (eNOS), which regulates nitric oxide (NO) production, and scavenger receptors class B type I (SR-BI) and CD36, which interact with plasmatic lipoproteins and directly or indirectly regulate eNOS activity [2]. Plasmatic lipoproteins may also induce endothelial signaling due to their interaction with proteins in other regions of the endothelial membrane, such as transporters ABCA1/G1 (ATP binding cassette transporters) and LOX-1 receptors (lectin-type oxidized LDL receptor 1), as will be discussed later.

LOW DENSITY LIPOPROTEINS

The interactions between plasmatic lipoproteins and the endothelium may result in beneficial or deleterious effects, depending on the chemical nature of the lipoprotein. The most abundant lipoproteins are low density lipoproteins (LDL), the main carriers of cholesterol in human blood stream. LDL is rich in polyunsaturated fatty acids (PUFA), which are constituents of its complex lipids. PUFA are the main substrate for lipid peroxidation in oxidative stress conditions associated with metabolic disorders, such as hyperlipidemia, hyperglycemia, insulin

Endothelium and Cardiovascular Diseases
https://doi.org/10.1016/B978-0-12-812348-5.00021-0

285

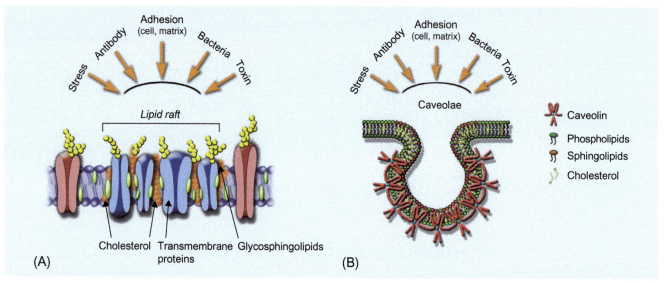

FIG. 21.1 Schematic representation of plasma membrane microdomains called *lipid rafts* (A) and caveolae (B), abundant in endothelial cells. Lipid rafts are firmly-packed microdomains, rich in glycosphingolipids, cholesterol, and proteins that compartmentalize cellular processes normally related to signal transduction and transport through the membrane. Panel A: from Wikibooks (https://en.wikibooks.org/wiki/File:Lipid_Raft*.png) and Panel B: adapted from Razani B, Woodman SE, Lisanti MP. Caveolae: from cell biology to animal physiology. Pharmacol Rev 2002;54 (3):431–467.

resistance, diabetes, and atherosclerosis. Therefore, excessive LDL, susceptibility of LDL to oxidation, the presence of oxidized LDL (LDLox), or anti-LDLox antibodies are indications of vascular oxidative stress [3–8]. On the other hand, other lipoproteins, notably high density lipoproteins (HDL), are relatively protected and protective against lipid peroxidation because they transport several enzymatic and nonenzymatic antioxidants [9,10]. LDLox is recognized as the main mediator of several events, including endothelial dysfunction, which leads to atherosclerosis initiation and progression [11,12]. Currently, there is solid evidence demonstrating that increased plasmatic concentrations of LDL-cholesterol induces oxidative stress and endothelial dysfunction [12], while reduced LDL-cholesterol improves endothelial function [13]. Such evidences are part of the basis of the atherogenesis oxidative hypothesis [11].

In the last three decades, both LDL oxidation and NO synthesis have been intensively and widely studied, and seem to have antagonistic actions in the vascular endothelial microenvironment, significantly influencing atherogenesis [14]. The main antiatherogenic effects of NO and the proatherogenic actions of LDLox involved in endothelial dysfunction are represented in Fig. 21.2. LDLox antagonize the atheroprotective effects of NO by inducing: (1) vasoconstriction (inhibition of NO

synthesis and induction of endothelin-1); (2) apoptosis; (3) inflammation; and (4) adhesiveness (expression of adhesion molecules); and by perpetuating the reactive oxygen species (ROS) generation vicious cycle. Regarding the latter, it is important to emphasize that the ROS produced by vessel wall cells oxidize LDL, which, when oxidized, promotes most of its deleterious effects by inducing more intracellular ROS generation, as will be discussed later on.

LDLox acts on the endothelium by binding to several scavenger receptors, including SR-A, SR-BI, CD36, and LOX-1 [15]. The LOX-1 receptor was initially identified as the main LDLox receptor in endothelial cells, but it is also expressed in macrophages and smooth muscle cells [16]. LDLox, LOX-1, and ROS are considered inseparable factors in the endothelial damage process [17]. When LDLox interacts with LOX-1, there is positive feedback of several pathways, resulting in increased expression of LOX-1 itself, increased ROS, reduced NO generation, and increased endothelial apoptosis [18]. When LDLox binds to LOX-1, NADPH oxidase (an enzymatic complex that constitutes the main source of ROS in endothelial cells) is activated due to induced translocation of specific subunits of this enzyme to endothelial cell membranes, leading to a fast increase in superoxide and hydrogen peroxide production [19,20]. The superoxide quickly reacts with NO, forming peroxynitrite radical and decreasing NO availability, increasing

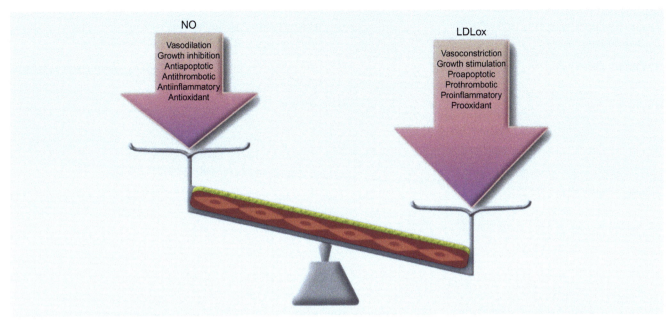

FIG. 21.2 Main antagonistic actions of nitric oxide (NO) (atheroprotective) and oxidized LDL (atherogenic) involved in endothelial function and dysfunction. Modified from Gradinaru D, Borsa C, Ionescu C, et al. Oxidized LDL and NO synthesis—biomarkers of endothelial dysfunction and ageing. Mech Ageing Dev 2015;151:101–13.

LOX-1 expression, and, therefore, additionally increasing ROS production [16]. On the other hand, increased ROS activates inflammatory and cell death pathways [21,22]. In addition, LDLox induces the so-called eNOS uncoupling, by means of which eNOS itself increases superoxide production and reduces NO generation [23]. Another redox enzyme that is involved with ROS production, specifically of mitochondrial origin, is p66^{Shc}. Exposure of endothelial cells to LDLox leads to p66^{Shc} phosphorylation, which can be dependent or independent from LOX-1 [24], but dependent on the eNOS uncoupling [25].

LDLox reduces NO availability by several mechanisms besides inducing the formation of superoxide (via NADPH oxidase), including eNOS displacement from caveolae and activation of arginase II, which compete with eNOS for the arginine substrate [26–29]. LDLox also contributes to alterations

of vascular tone by stimulating the generation of the vasoconstrictor endothelin-1 and the expression of the angiotensin converting enzyme [30,31].

High LDLox concentrations induce cell death both by necrosis and by apoptosis. Apoptosis is triggered by several mechanisms, including increased ROS levels, activation of NF-κB pathway, alteration of intracellular calcium homeostasis, alteration of anti- and proapoptotic proteins expression, activation of caspases, and increased expression of Fas death receptor [22,32–34]. Endothelial cell apoptosis results in increased vascular permeability to molecules, macromolecules, and cells, also increasing coagulation and proliferation of smooth muscle cells, contributing for the development of atherosclerotic lesions and complications.

Most LDLox actions described before seem to be mediated by LOX-1, since the use of anti-LOX-1

TABLE 21.1 Role of LOX-1 in Endothelial Dysfunction

LOX-1 inducers	LOX-1 ligands	Results of LOX-1 activation
• Proinflammatory cytokines (TNF-α, IFNg, LPS, RPC) • Modified lipoproteins (LDLox, HDLox) • Hypertension (AgII, ET-1, shear stress) • Hyperglycemia (AGEs)	Modified LP: • LDLox (copper) • Delipidated LDLox • Glycoxidized LDL • LDLox (15-lipoxygenase) • HDLox (15-lipoxygenase) • HDL modified by HOCl	↑ Endothelial activation: • ↑ ROS • NF-κB activation • ↑ expression of adhesion molecules (E-selectin, P-selectin, VCAM, ICAM) • ↑ secretion of MCP-1 (via MAP kinase) and other cytokines

Continued

TABLE 21.1 Role of LOX-1 in Endothelial Dysfunction—cont'd

LOX-1 inducers	LOX-1 ligands	Results of LOX-1 activation
• Others (homocysteine, free radicals)	Others: • Apoptotic cells • Activated platelets, AGEs	↓ Endothelium-dependent vasorelaxation: • ↓ NO • ↑ ET-1 • ↑ ACE ↑ Apoptosis: • ↑ FAS expression • ↓ antiapoptotic proteins (Bcl-2, c-IAP-1) • ↑ Caspases-3/9

LP, lipoproteins; TNF-α, tumor necrosis factor alpha; IFNg, interferon gamma; LPS, lipopolysaccharides; RPC, reactive protein C; AGEs, advanced glycation end products; NF-κB, nuclear factor kappa B; VCAM, vascular cell adhesion molecule; ICAM, intercellular adhesion molecule; MCP-1, monocyte chemoattractant protein-1; ET-1, endothelin-1; ACE, angiotensin converting enzyme; FAS, receptor of the cell death receptor family, or CD95; Bcl-2, B cell lymphoma 2; c-IAP-1, cellular inhibitor of apoptosis protein 1.

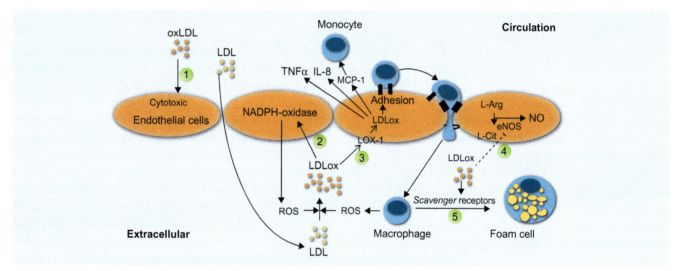

FIG. 21.3   Schematic representation of the main deleterious actions of excessive LDL and LDLox upon the endothelium: 1. cytotoxic for endothelial cells, 2. activation of endothelial NADPH oxidase, 3. induction of adhesion molecules expression and proinflammatory cytokines synthesis, 4. inhibition of NO synthesis, 5. stimulation of foam cell formation (see text for abbreviations).

antibodies inhibits or attenuates the effects of LDLox [16]. The various conditions that induce expression of LOX-1, its ligands, and the main consequences of its activation that result in endothelial dysfunction are summarized in Table 21.1.

The main deleterious actions of excessive LDL and LDLox upon the endothelium are represented in Fig. 21.3.

LIPOPROTEINS OF THE POSTPRANDIAL PERIOD

The lipoproteins present in plasma during the postprandial period have also been associated with endothelial dysfunction. Chylomicrons and their remnants are able to increase endothelium permeability, are cytotoxic, and inhibit endothelium-dependent relaxation in isolated aorta [35]. The contribution of lipoproteins of the postprandial state to atherosclerosis has been suggested long ago [36], but only more recent human studies have demonstrated strong and consistent associations between postprandial lipemia and endothelial dysfunction, and between postprandial lipemia and oxidative stress [37]. Several works have demonstrated an occurrence of an acute oxidative stress associated with the lipoproteins of this fed state, and it seems to be the most plausible mechanism to explain endothelial dysfunction in such a state [8], although there is still controversy [38,39]. The current thinking is that triglyceride rich lipoproteins are implicated in atherogenesis because, besides being the vehicle for diet cholesterol, they may induce the formation of free radicals, inflammation, leukocyte activation, and endothelium activation [40]. These effects seem to be related to their products of intravascular lipolysis and are proportional to the magnitude of postprandial lipemia, evidenced mainly in secondary hyperlipidemias, such as in diabetes and metabolic syndrome [41,42], although there are other important factors in these cases besides lipemia per se.

The products of postprandial lipoprotein lipolysis generated on the endothelial surface, where lipoprotein lipase (LPL) is anchored, may be toxic for the endothelium. Lipolysis of the triglyceride rich lipoproteins releases several types of fatty acids, lysophospholipids, and oxidized lipids. At least in vitro, many of them show atherogenic properties such as induction of adhesion molecule expression, reduction of NO generation, and increased ROS production. On the other hand, lipolysis provides long-chain polyunsaturated fatty acids, which are PPAR (peroxisome proliferator activated receptors) ligands, transcription factors that trigger antiinflammatory responses [43]. High local concentration of free fatty acids derived from lipolysis of triglyceride rich lipoproteins may also break the endothelial barrier, significantly increasing its permeability and, therefore, favoring the entrance of not only the fatty acids themselves, but also LDL and remnant lipoproteins [43].

Fatty acid uptake by endothelial cells is not completely elucidated and may involve both receptor-mediated transport and nonspecific transport. The most well-known transporters are CD36/FAT (fatty acid translocase) and the FATP (fatty acid transport proteins) family. The site in which CD36 binds with fatty acids overlaps with the LDLox binding site [44]. Endothelial cells express mainly FATP1 and FATP4 [45], and other members of the FATP family have less expression. Intracellularly, fatty acids transport is mediated by FABP (fatty acid binding proteins), proteins that also have tissue-specific distribution [46].

Recent evidence reveals that VEGF-B (vascular endothelial growth factor-B) also targets endothelium fatty acid transport and promotes its storage in underlying tissues, such as skeletal muscle, cardiac muscle, and adipose tissue [47]. The binding between VEGF-B and its receptor (VEGFR1 and NRP1) in endothelial cells results in increased expression of FATP3 and FATP4. These studies were confirmed by genetic manipulation models of

VEGF-B and their receptors specifically in endothelial cells, and showed that inhibiting this signaling pathway was relevant to the attenuation of undesirable fat accumulation in skeletal muscle and cardiac muscle in the context of fat enriched diets and diabetes [48,49]. However, before considering VEGF-B inhibition as possible therapeutic target, it is important to consider its angiogenic and neurogenic effects, especially relevant in vascular ischemia and diabetic neuropathy conditions.

Previous studies with endothelial cell cultures using PPARg pharmacological activators (glitazones) showed that the activation of these nuclear receptors, which are also activated by unsaturated fatty acids, such as oleic acid, linoleic acid, EPA (eicosapentaenoic acid), and DHA (docosahexaenoic acid), result in reduced inflammatory response, including inhibition of NF-κB and reduced expression of chemokines and adhesion molecules [50–52]. On the other hand, specific deletion of PPARg in endothelial cells reduces the expression of CD36 and intracellular chaperone FABP4 (fatty acid binding protein, also known as AP2) in endothelial cells, decreasing the fatty acid influx through the endothelium. This genetic manipulation reduces fat accumulation and improves skeletal muscle sensitivity to insulin, but impairs endothelium-mediated vasodilation, suggesting the participation of independent pathways for the metabolic and vascular effects [53,54]. Therefore, both VEGF-B and PPARg seem to promote fatty acid uptake by the endothelium. However, the effects of VEGF-B are more restricted to tissues that express such factor (skeletal muscle, heart, and brown adipose tissue), and its inhibition redirects the fatty acid flux to white adipose tissue. On the other hand, endothelial PPARg increased fatty acid uptake more extensively, but its persistent activation in vivo (e.g., glitazones) could have deleterious effects on the endothelium itself. [47]

Saturated fatty acids may be ligands or activators of another family of membrane receptors, the so-called toll like receptors (TLR), especially TLR2 and TLR4 [55–57], which are responsible for activating inflammatory pathways. In vitro, saturated fatty acids such as palmitate (C16:0) and stearate (C18:0) induce the generation and secretion of several proinflammatory molecules of endothelial cells, for example, cytokines of the CCL and CXCL family and IL-6, which may be involved in the recruitment of monocytes and other leukocytes [58,59]. This response is probably mediated through the NF-κB and endoplasmic reticulum pathways [59]. In several systems, potentially toxic saturated fatty acids may be neutralized by adding oleic acid (monounsaturated), perhaps by inducing higher palmitate esterification into triacylglycerol [60]. Saturated fatty acids seem to induce endothelial cell apoptosis independently from TLR4 [58]. These effects are neutralized in the presence of unsaturated fatty acids or by suppressing the stearoyl-CoA desaturase (SCD), enzyme that converts C16:0 and C18:0 into C16:1 and C18:1, respectively [61].

Other possible saturated fatty acid inflammatory mechanism is the excessive generation of ceramides. Serine-palmitoil-transferase (SPT) catalyzes the first step in the synthesis of ceramide, sphingomyelin, and sphingosine-1-phosphate (S1P). While some sphingolipids regulate cell survival (e.g., S1P), others induce cell death (e.g., ceramides) [62]. Ceramides are mediators implicated in cell death and eNOS inhibition [63]. SPT inhibition protects the endothelium from the defective endothelium-dependent vasorelaxation induced by palmitate. In addition, reduced SPT activity decreases sphingo-myelin synthesis, disrupting lipid rafts formation, as well as the function of proteins associated with these structures, such as TLR4 [64].

Regarding unsaturated fatty acids, there are many studies demonstrating their effects, generally beneficial, especially omega-3 fatty acids, such as EPA and DHA [65–68]. However, there is still controversy regarding the antiatherogenic role of these molecules in clinical studies [69]. It is important to consider that the endothelium is never exposed to a single type of fatty acid in vivo, and that the combinations of saturated and unsaturated acids that mimic those observed in human plasma do not induce the inflammatory and apoptotic effects observed for saturated fatty acids in endothelial cells [43,58]. The possible actions of fatty acids derived from postprandial lipoproteins in endothelial cells discussed herein are illustrated in Fig. 21.4.

High Density Lipoproteins

HDL is an antiatherogenic lipoprotein [10]. This property has been attributed mainly because of its ability to remove cholesterol from tissue membranes, including from macrophages of the vascular athero-sclerotic lesions, and transport them to the liver, from where cholesterol can be excreted from the body, a process known as reverse cholesterol transport.

Besides this well-established function demonstrated in experimental and clinical studies, HDL has other protective roles related to the endothelium, such as antioxidant, antiapoptotic, antiinflammatory, and antithrombotic functions. The mechanisms by which HDL performs these functions are related to its size and lipid and protein composition [70].

Vasodilation via NO production is an HDL well-characterized action upon endothelial cells. HDL induces NO production by several mechanisms [71], as described below:

1. HDL regulates membrane eNOS stability. eNOS is located in the caveolae, and HDL regulates caveolae cholesterol content. Lipoproteins that are able to work as cholesterol acceptors or donors can disturb caveolae structure and eNOS functions. For example, LDLox causes caveolae cholesterol depletion via CD36, leading to intracellular eNOS redistribution and reduced enzyme activity [26]. However, HDL maintains caveolae cholesterol content by supplying cholesteryl ester to the cell via SR-BI, and therefore preserves eNOS localization in the caveolae [72]. It is important to emphasize that different types of HDL promote distinct cell cholesterol fluxes by interacting with SR-BI and with ABCA1/G1 membrane transporters. When interacting with SR-BI, cholesteryl ester enriched HDL promotes cell cholesterol influx and cholesteryl ester poor HDL promotes cell cholesterol efflux from cell membranes (Fig. 21.5A). On the other hand, the interaction between HDL and ABC transporters always results in cholesterol efflux from the membrane to the lipoprotein. Small HDL and lipid poor HDL (prebeta HDL) interact with ABCA1 and spherical HDL containing esterified cholesterol (HDL2 and 3) interact with ABCG1 transporters (Fig. 21.5B).
2. Based on studies with ApoAI mimetics, HDL has been postulated to prevent the so-called eNOS uncoupling induced by LDLox, which results in higher production of superoxide, decreasing the NO availability, particularly in oxidative stress and hypercholesterolemia conditions [73].
3. HDL triggers signaling initiated in the membrane that results in increased eNOS activity [74]. The connection between SR-BI and HDL via ApoAI causes a rapid activation of the tyrosine-kinase Src, leading to activation of downstream kinases,

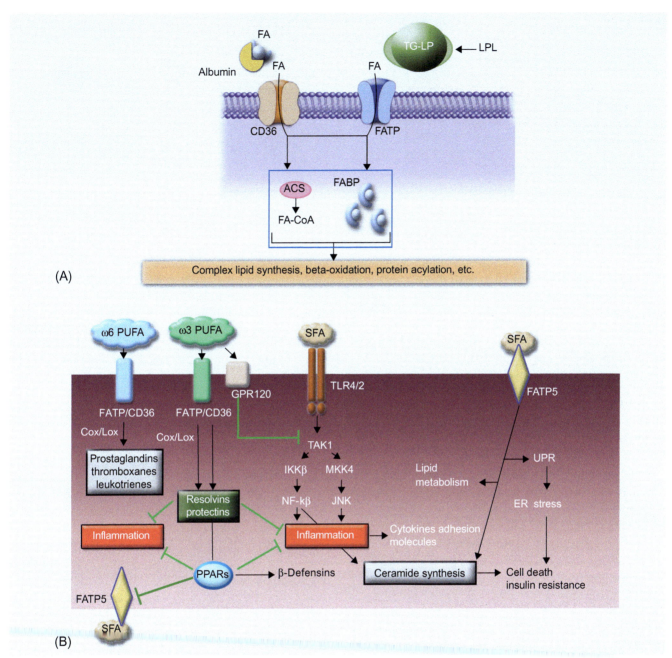

FIG. 21.4 Effects of fatty acids derived from lipolysis of triglyceride rich lipoproteins (TG-LP) in endothelial cells. (A) Fatty acids (FA) are released by TG-LP due to the action of the lipoprotein lipase (LPL) anchored on the endothelium luminal surface. FA are taken up by endothelial cells by the FATP and CD36 transporter proteins. Once inside the cells, FA bind to chaperones called FABP or are converted into acyl-CoA by acyl-CoA synthase (ACS) and used in complex lipid synthesis or beta-oxidation and protein acylation, among other processes. (B) Saturated fatty acids (SFA) activate *toll-like receptors* (TLR4/2), which results in proinflammatory responses. SFAs also inhibit the antiinflammatory responses triggered by the activation of PPAR (peroxisome proliferator activated receptors) transcription factors, and may induce the so-called "unfolded protein response" (UPR) in the endoplasmic reticulum (ER), which may lead to ER stress and apoptosis. Omega-3 polyunsaturated fatty acids (w3 PUFA) and its derivatives may induce PPAR activation, inhibiting proinflammatory genes expression and activating antiinflammatory genes. w3 PUFA have direct antiinflammatory action, as they generate eicosanoids called resolvins and protectins, while w6 PUFA generate proinflammatory eicosanoids (prostaglandins, thromboxanes, and leukotrienes). *TAK1*, TGF beta activated kinase 1; *IKKb*, inhibitor of kappa B kinase; *NF-κB*, nuclear factor kappa beta; *MKK4*, mitogen-activated protein kinase kinase 4; *JNK-c*, Jun N-terminal kinase.

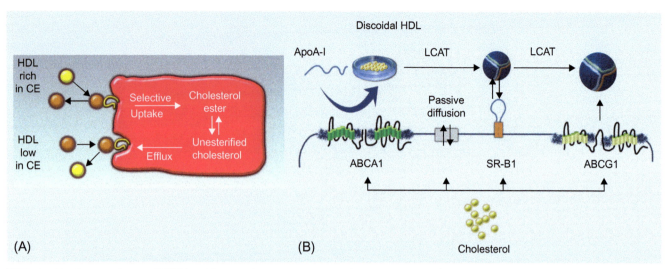

FIG. 21.5   Interactions of HDL with membrane receptors that may result in supply (influx) or removal (efflux) of cellular cholesterol. (A) Interactions between HDL and SR-BI receptors: cholesteryl ester (CE) enriched HDL promotes cell cholesterol influx by a process named selective uptake. CE poor HDL promotes membrane cholesterol efflux. (B) Interactions of HDL with the ABC transportes results in membrane cholesterol efflux. Lipid poor discoidal HDL interact with ABCA1 while spherical CE containing HDL interact with ABCG1 transporters. Discoidal and small HDL particles are converted into CE enriched HDL by the action of the lecithin cholesterol acyl transferase (LCAT). *CE*, cholesterylester; *SR-BI*, scavenger receptor class B type I; *ABCA1/G1*, ATP binding cassette transporter A1 and G1. Panel A: partial reproduction from: Krieger M. Charting the fate of the "good cholesterol": identification and characterization of the high-density lipoprotein receptor SR-BI. Ann Rev Biochem 1999;68:1–1068. Panel B: partial reproduction from: Hafiane A, Genest J. HDL, atherosclerosis, and emerging therapies. Cholesterol 2013;2013, Article ID 891403.

such as PI3K, Akt, and MAP kinases, which increase eNOS activity [75–77]. Although ApoAI and HDL phospholipids are required to activate HDL-mediated signaling, there is also involvement of lysophospholipids, sphingosylphosphorylcholine (SPC), sphingosine-1-phosphate (S1P), and lysosulfatides (LSF), all transported in HDL, which cause eNOS-dependent relaxation in precontracted rat aortic rings. These species act through the lysophospholipid receptor S1P3 present in the membrane of endothelial cells [78].

4. HDL regulates eNOS mass. The binding between HDL and SR-BI or S1P G coupled protein receptors results not only in increased NO production, but also increased eNOS abundance. The mechanism responsible for increasing eNOS mass seems to result from its decreased degradation [79,80].

The four mechanisms by which HDL preserves the endothelium-dependent vasodilation (eNOS) are represented in Fig. 21.6.

It has been postulated that intracellular signaling triggered by HDL depends on endothelial membrane cholesterol efflux, since cholesterol acceptor methyl-beta-cyclodextrin (MBCD) mimics the actions of HDL [74,81]. However, particles composed by ApoAI and phospholipids, but without cholesterol, activate eNOS, while particles loaded with cholesterol do not possess such effect. On the other hand, the SR-BI receptor is required for the direct actions of ApoAI and HDL upon the endothelium, sometimes in cooperation with ABCA1 and G1 [75,82,83]. The activation of eNOS both by HDL and by MBCD depend on SR-BI [84]. Considering that the effects of MBCD do not require any type of protein on the cell surface, these findings suggest that eNOS activation by HDL requires: (1) cholesterol efflux; (2) sufficient ApoAI and HDL phospholipids to start intracellular signaling; and (3) SR-BI working as a membrane cholesterol movement sensor [74].

It is important to emphasize that HDL oxidative modifications impair its protective role in vascular endothelium. After HDL incubation with 15-lipoxygenase, HDL3 no longer activates eNOS and NO generation, both by reducing its affinity

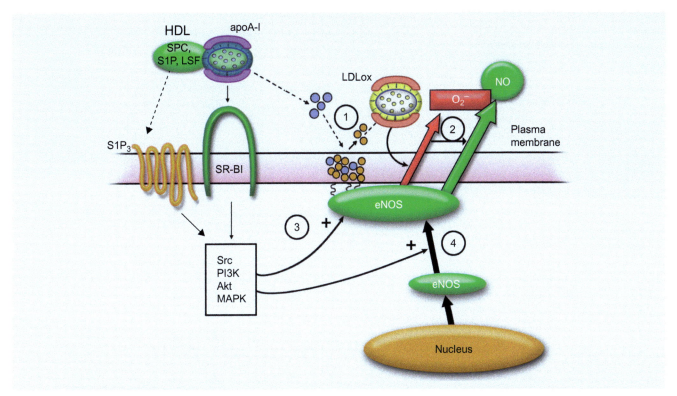

FIG. 21.6 Mechanisms by which HDL performs its endothelium-dependent vasodilation action, modulating endothelial nitric oxide synthase (eNOS): 1. regulates membrane eNOS stability; 2. prevents the so-called "eNOS uncoupling" induced by LDLox; 3. triggers signaling via SR-BI and S1P receptor, which results in increased eNOS activity; 4. regulates eNOS abundance in cells. Adapted from Mineo C, Deguchi H, Griffin JH, et al. Endothelial and antithrombotic actions of HDL. Circ Res 2006;98(11):1352–64.

to SR-BI and by reducing the expression of this receptor after endothelial cells are exposed to the modified HDL [85].

S1P is a lipid mediator produced by sphingolipid metabolism, and it is present at high concentrations in plasma, particularly bound to HDL (HDL-S1P) and in low quantities in tissues. This vascular gradient is essential to modulate HDL-S1P-mediated vascular permeability [86,87]. S1P found in HDL is bound to the apolipoprotein M. S1P-ApoM may activate S1P receptors in endothelial cells, while ApoM deficiency abolishes the presence of S1P in HDL. ApoM deficient rats show endothelial dysfunction manifested especially in the lungs [88].

HDL also acts by reducing endothelial activation. This has antithrombotic consequences and occurs mainly because of the reduced expression of tissue factor, P-selectin, and E-selectin, and by means of the already mentioned increased NO production [71].

HDL also maintains endothelial monolayer integrity by reducing apoptosis and increasing endothelial proliferation and migration. HDL carries out antiapoptotic actions that include preventing persistent intracellular calcium increase induced by proapoptotic agents, such as LDLox, and inhibiting the activation of caspases 3 and 9. Endothelial cell proliferation and migration stimulated by HDL depends on calcium and is mediated by multiple kinase cascades involving PI3K, p38 and p42/44 MAPK, Rho kinase, and small GTPases Rac [71].

Since phospholipids and derivatives represent the main bioactive components of HDL [89], the phosphosphingolipidome of the different subclasses of HDL was characterized and related to the HDL functionalities [90]. The main biological activities, such as cholesterol efflux and antioxidant, antithrombotic, antiinflammatory, and antiapoptotic activities, were predominantly associated with small, dense, protein-rich HDL (HDL3). The biological activities were also correlated to multiple phosphosphingolipidome components. In particular, the phosphatidylserine content was correlated to all HDL functionality measurements [90].

TABLE 21.2 Main Lipid Mediators of Endothelial Function and Dysfunction

Lipoproteins	Components	Responses
LDL	Oxidized LDL: peroxidized lipids (phospho- and lysophospholipids), but also oxidation of the ApoB protein component.	• Oxidative stress • Vasoconstriction • Inflammation • Permeability • Cell death
Triglyceride rich LP	Products of lipolysis: saturated fatty acids, peroxidized fatty acids and lysophospholipids. Substrates for synthesis of ceramide.	• Possible acute oxidative stress • Vasoconstriction • Inflammation • Permeability • Cytotoxicity
Triglyceride rich LP	Products of lipolysis: mono- and polyunsaturated fatty acids (especially omega-3, such as EPA and DHA). Substrates for synthesis of Sphingosine-1-phosphate (S1P).	• Activation of antiinflammatory and cell survival pathways
HDL	HDL (especially small and dense, HDL3): ApoA1, phospholipids (phosphatidylserine), lysophospholipids, sphingosylphosphorylcholine (SPC), sphingosine-1-phosphate (S1P), and lysosulfatides (LSF).	• Cholesterol efflux • Vasodilator • Antioxidant • Antiinflammatory • Antithrombotic • Antiapoptotic

EPA, *eicosapentaenoic acid*; DHA, *docosahexaenoic acid*.

In summary, this chapter calls attention to the molecular mechanisms by means of which plasmatic lipid and lipoprotein mediators contribute to maintain endothelial homeostasis or trigger poorly adaptive cell responses that break such homeostasis. The main components emphasized and discussed are summarized in Table 21.2.

References

[1] Head BP, Patel HH, Insel PA. Interaction of membrane/lipid rafts with the cytoskeleton: impact on signaling and function: membrane/lipid rafts, mediators of cytoskeletal arrangement and cell signaling. Biochim Biophys Acta 2014;1838(2):532–45.
[2] Frank PG, Woodman SE, Park DS, et al. Caveolin, caveolae, and endothelial cell function. Arterioscler Thromb Vasc Biol 2003;23(7):1161–8.
[3] Fang JC, Kinlay S, Behrendt D, et al. Circulating autoantibodies to oxidized LDL correlate with impaired coronary endothelial function after cardiac transplantation. Arterioscler Thromb Vasc Biol 2002;22(12):2044–8.
[4] Mizuno T, Matsui H, Imamura A, et al. Insulin resistance increases circulating malondialdehyde-modified LDL and impairs endothelial function in healthy young men. Int J Cardiol 2004;97(3):455–61.
[5] Yin WH, Chen JW, Tsai C, et al. L-Arginine improves endothelial function and reduces LDL oxidation in patients with stable coronary artery disease. Clin Nutr 2005;24(6):988–97.
[6] Woodman RJ, Watts GF, Playford DA, et al. Oxidized LDL and small LDL particle size are independently predictive of a selective defect in microcirculatory endothelial function in type 2 diabetes. Diabetes Obes Metab 2005;7(5):612–7.
[7] Delporte C, Van Antwerpen P, Vanhamme L, et al. Low-density lipoprotein modified by myeloperoxidase in inflammatory pathways and clinical studies. Mediators Inflamm 2013;2013:971579.
[8] Le NA. Lipoprotein-associated oxidative stress: a new twist to the postprandial hypothesis. Int J Mol Sci 2014;16(1):401–19.
[9] Mineo C, Shaul PW. Novel biological functions of high-density lipoprotein cholesterol. Circ Res 2012;111(8):1079–90.
[10] Camont L, Chapman MJ, Kontush A. Biological activities of HDL subpopulations and their relevance to cardiovascular disease. Trends Mol Med 2011;17(10):594–603.
[11] Steinberg D, Witztum JL. Oxidized low-density lipoprotein and atherosclerosis. Arterioscler Thromb Vasc Biol 2010;30(12):2311–6.
[12] Hermida N, Balligand JL. Low-density lipoprotein-cholesterol-induced endothelial dysfunction and oxidative stress: the role of statins. Antioxid Redox Signal 2014;20(8):1216–37.
[13] Delles C, Dymott JA, Neisius U, et al. Reduced LDL-cholesterol levels in patients with coronary artery disease are paralleled by improved endothelial function: An observational study in patients from 2003 and 2007. Atherosclerosis 2010;211(1):271–7.
[14] Gradinaru D, Borsa C, Ionescu C, et al. Oxidized LDL and NO synthesis—biomarkers of endothelial dysfunction and ageing. Mech Ageing Dev 2015;151:101–13.
[15] Levitan I, Volkov S, Subbaiah PV. Oxidized LDL: diversity, patterns of recognition, and pathophysiology. Antioxid Redox Signal 2010;13(1):39–75.
[16] Pirillo A, Norata GD, Catapano AL. LOX-1, OxLDL, and atherosclerosis. Mediators Inflamm 2013;2013:152786.
[17] Lubrano V, Balzan S. LOX-1 and ROS, inseparable factors in the process of endothelial damage. Free Radic Res 2014;48(8):841–8.
[18] Mollace V, Gliozzi M, Musolino V, et al. Oxidized LDL attenuates protective autophagy and induces apoptotic cell death of

endothelialcells: role of oxidative stress and LOX-1 receptor expression. Int J Cardiol 2015;184:152–8.

[19] Rueckschloss U, Galle J, Holtz J, et al. Induction of NAD(P)H oxidase by oxidized low-density lipoprotein in human endothelial cells: antioxidative potential of hydroxymethylglutaryl coenzyme A reductase inhibitor therapy. Circulation 2001;104(15):1767–72.

[20] Rueckschloss U, Duerrschmidt N, Morawietz H. NADPH oxidase in endothelial cells: impact on atherosclerosis. Antioxid Redox Signal 2003;5(2):171–80.

[21] Cominacini L, Pasini AF, Garbin U, et al. Oxidized low density lipoprotein (ox-LDL) binding to ox-LDL receptor-1 in endothelial cells induces the activation of NF-kappaB through an increased production of intracellular reactive oxygen species. J Biol Chem 2000;275 (17):12633–8.

[22] Chen XP, Xun KL, Wu Q, et al. Oxidized low density lipoprotein receptor-1 mediates oxidized low density lipoprotein-induced apoptosis in human umbilical vein endothelial cells: role of reactive oxygen species. Vascul Pharmacol 2007;47(1):1–9.

[23] Fleming I, Mohamed A, Galle J, et al. Oxidized low-density lipoprotein increases superoxide production by endothelial nitric oxide synthase by inhibiting PKCalpha. Cardiovasc Res 2005;65(4):897–906.

[24] Shi Y, Cosentino F, Camici GG, et al. Oxidized low-density lipoprotein activates p66Shc via lectin-like oxidized low-density lipoprotein receptor-1, protein kinase C-beta, and c-Jun N-terminal kinase kinase in human endothelial cells. Arterioscler Thromb Vasc Biol 2011;31(9):2090–7.

[25] Shi Y, Lüscher TF, Camici GG. Dual role of endothelial nitric oxide synthase in oxidized LDL-induced, p66Shc-mediated oxidative stress in cultured human endothelial cells. PLoS One 2014;9(9).e107787.

[26] Blair A, Shaul PW, Yuhanna IS, et al. Oxidized low density lipoprotein displaces endothelial nitric-oxide synthase (eNOS) from plasmalemmal caveolae and impairs eNOS activation. J Biol Chem 1999;274(45):32512–9.

[27] Cominacini L, Rigoni A, Pasini AF, et al. The binding of oxidized low density lipoprotein (ox-LDL) to ox-LDL receptor-1 reduces the intracellular concentration of nitric oxide in endothelial cells through an increased production of superoxide. J Biol Chem 2001;276(17):13750–5.

[28] Ryoo S, Lemmon CA, Soucy KG, et al. Oxidized low-density lipoprotein-dependent endothelial arginase II activation contributes to impaired nitric oxide signaling. Circ Res 2006;99(9):951–60.

[29] Förstermann U, Sessa WC. Nitric oxide synthases: regulation and function. Eur Heart J 2012;33(7):829–37. 837a–d.

[30] Sakurai K, Cominacini L, Garbin U, et al. Induction of endothelin-1 production in endothelial cells via co-operative action between CD40 and lectin-like oxidized LDL receptor (LOX-1). J Cardiovasc Pharmacol 2004;44(Suppl. 1):S173–80.

[31] Montecucco F, Pende A, Mach F. The renin-angiotensin system modulates inflammatory processes in atherosclerosis: evidence from basic research and clinical studies. Mediators Inflamm 2009;2009:752406.

[32] Salvayre R, Auge N, Benoist H, et al. Oxidized low-density lipoprotein-induced apoptosis. Biochim Biophys Acta 2002;1585 (2–3):213–21.

[33] Imanishi T, Hano T, Sawamura T, et al. Oxidized low density lipoprotein potentiation of Fas-induced apoptosis through lectin-like oxidized-low density lipoprotein receptor-1 in human umbilical vascular endothelial cells. Circ J 2002;66(11):1060–4.

[34] Chen J, Mehta JL, Haider N, et al. Role of caspases in Ox-LDL-induced apoptotic cascade in human coronary artery endothelial cells. Circ Res 2004;94(3):370–6.

[35] Jagla A, Schrezenmeir J. Postprandial triglycerides and endothelial function. Exp Clin Endocrinol Diabetes 2001;109(4):S533–47.

[36] Zilversmit DB. Atherogenic nature of triglycerides, postprandial lipidemia, and triglyceride-rich remnant lipoproteins. Clin Chem 1995;41(1):153–8.

[37] Wallace JP, Johnson B, Padilla J, et al. Postprandial lipaemia, oxidative stress and endothelial function: a review. Int J Clin Pract 2010;64(3):389–403.

[38] Sodré FL, Paim BA, Urban A, et al. Reduction in generation of reactive oxygen species and endothelial dysfunction during postprandial state. Nutr Metab Cardiovasc Dis 2011;21(10):800–7.

[39] Muntwyler J, Sütsch G, Kim JH, et al. Post-prandial lipaemia and endothelial function among healthy men. Swiss Med Wkly 2001;131(15–16):214–8.

[40] Botham KM, Wheeler-Jones CP. Postprandial lipoproteins and the molecular regulation of vascular homeostasis. Prog Lipid Res 2013;52(4):446–64.

[41] Norata GD, Grigore L, Raselli S, et al. Post-prandial endothelial dysfunction in hypertriglyceridemic subjects: molecular mechanisms and gene expression studies. Atherosclerosis 2007;193(2): 321–7.

[42] Anderson RA, Evans ML, Ellis GR, et al. The relationships between post-prandial lipaemia, endothelial function and oxidative stress in healthy individuals and patients with type 2 diabetes. Atherosclerosis 2001;154(2):475–83.

[43] Goldberg IJ, Bornfeldt KE. Lipids and the endothelium: bidirectional interactions. Curr Atheroscler Rep 2013;15(11):365.

[44] Kuda O, Pietka TA, Demianova Z, et al. Sulfo-N-succinimidyl oleate (SSO) inhibits fatty acid uptake and signaling for intracellular calcium via binding CD36 lysine 164: SSO also inhibits oxidized low density lipoprotein uptake by macrophages. J Biol Chem 2013;288(22):15547–55.

[45] Sandoval A, Fraisl P, Arias-Barrau E, et al. Fatty acid transport and activation and the expression patterns of genes involved in fatty acid trafficking. Arch Biochem Biophys 2008;477(2):363–71.

[46] Furuhashi M, Hotamisligil GS. Fatty acid-binding proteins: role in metabolic diseases and potential as drug targets. Nat Rev Drug Discov 2008;7(6):489–503.

[47] Mehrotra D, Wu J, Papangeli I, et al. Endothelium as a gatekeeper of fatty acid transport. Trends Endocrinol Metab 2014;25 (2):99–106.

[48] Hagberg CE, Falkevall A, Wang X, et al. Vascular endothelial growth factor B controls endothelial fatty acid uptake. Nature 2010;464(7290):917–21.

[49] Hagberg CE, Mehlem A, Falkevall A, et al. Targeting VEGF-B as a novel treatment for insulin resistance and type 2 diabetes. Nature 2012;490(7420):426–30.

[50] Marx N, Mach F, Sauty A, et al. Peroxisome proliferator-activated receptor-gamma activators inhibit IFN-gamma-induced expression of the T cell-active CXC chemokines IP-10, Mig, and I-TAC in human endothelial cells. J Immunol 2000;164(12):6503–8.

[51] Pasceri V, Wu HD, Willerson JT, et al. Modulation of vascular inflammation in vitro and in vivo by peroxisome proliferator-activated receptor-gamma activators. Circulation 2000;101(3):235–8.

[52] Jackson SM, Parhami F, Xi XP, et al. Peroxisome proliferator-activated receptor activators target human endothelial cells to inhibit leukocyte-endothelial cell interaction. Arterioscler Thromb Vasc Biol 1999;19(9):2094–104.

[53] Kanda T, Brown JD, Orasanu G, et al. PPAR gamma in the endothelium regulates metabolic responses to high-fat diet in mice. J Clin Invest 2009;119(1):110–24.

[54] Goto K, Iso T, Hanaoka H, et al. Peroxisome proliferator-activated receptor-g in capillary endothelia promotes fatty acid uptake by heart during long-term fasting. J Am Heart Assoc 2013;2(1). e004861.

[55] Erridge C. Endogenous ligands of TLR2 and TLR4: agonists or assistants? J Leukoc Biol 2010;87(6):989–99.

[56] Wong SW, Kwon MJ, Choi AM, et al. Fatty acids modulate Toll-like receptor 4 activation through regulation of receptor dimerization and recruitment into lipid rafts in a reactive oxygen species-dependent manner. J Biol Chem 2009;284(40):27384–92.

[57] Cheng AM, Handa P, Tateya S, et al. Apolipoprotein A-I attenuates palmitate-mediated NF-κB activation by reducing Toll-like receptor-4 recruitment into lipid rafts. PLoS One 2012;7(3). e33917.

[58] Li X, Gonzalez O, Shen X, et al. Endothelial acyl-CoA synthetase 1 is not required for inflammatory and apoptotic effects of a saturated fatty acid-rich environment. Arterioscler Thromb Vasc Biol 2013;33(2):232–40.

[59] Krogmann A, Staiger K, Haas C, et al. Inflammatory response of human coronary artery endothelial cells to saturated long-chain fatty acids. Microvasc Res 2011;81(1):52–9.

[60] Listenberger LL, Han X, Lewis SE, et al. Triglyceride accumulation protects against fatty acid-induced lipotoxity. Proc Natl Acad Sci U S A 2003;100(6):3077–82.

[61] Peter A, Weigert C, Staiger H, et al. Induction of stearoyl-CoA desaturase protects human arterial endothelial cells against lipotoxicity. Am J Physiol Endocrinol Metab 2008;295(2):E339–49.

[62] Young MM, Kester M, Wang HG. Sphingolipids: regulators of crosstalk between apoptosis and autophagy. J Lipid Res 2013;54(1):5–19.

[63] Symons JD, Abel ED. Lipotoxicity contributes to endothelial dysfunction: a focus on the contribution from ceramide. Rev Endocr Metab Disord 2013;14(1):59–68.

[64] Chakraborty M, Lou C, Huan C, et al. Myeloid cell-specific serine palmitoyltransferase subunit 2 haploinsufficiency reduces murine atherosclerosis. J Clin Invest 2013;123(4):1784–97.

[65] De Caterina R, Massaro M. Omega-3 fatty acids and the regulation of expression of endothelial pro-atherogenic and pro-inflammatory genes. J Membr Biol 2005;206(2):103–16.

[66] Dessì M, Noce A, Bertucci P, et al. Atherosclerosis, dyslipidemia, and inflammation: the significant role of polyunsaturated Fatty acids. ISRN Inflamm 2013;2013:191823.

[67] Miyoshi T, Noda Y, Ohno Y, et al. Omega-3 fatty acids improve postprandial lipemia and associated endothelial dysfunction in healthy individuals—a randomized crossover trial. Biomed Pharmacother 2014;68(8):1071–7.

[68] Ishida T, Naoe S, Nakakuki M, et al. Eicosapentaenoic acid prevents saturated fatty acid-induced vascular endothelial dysfunction: involvement of long-chain acyl-CoA synthetase. J Atheroscler Thromb 2015;22(11):1172–85.

[69] Bosch J, Gerstein HC, Dagenais GR, et al. n-3 fatty acids and cardiovascular outcomes in patients with dysglycemia. N Engl J Med 2012;367(4):309–18.

[70] Tran-Dinh A, Diallo D, Delbosc S, et al. HDL and endothelial protection. Br J Pharmacol 2013;169(3):493–511.

[71] Mineo C, Deguchi H, Griffin JH, et al. Endothelial and antithrombotic actions of HDL. Circ Res 2006;98(11):1352–64.

[72] Uittenbogaard A, Shaul PW, Yuhanna IS, et al. High density lipoprotein prevents oxidized low density lipoprotein-induced inhibition of endothelial nitric-oxide synthase localization and activation in caveolae. J Biol Chem 2000;275(15):11278–83.

[73] Ou Z, Ou J, Ackerman AW, et al. L-4F, an apolipoprotein A-1 mimetic, restores nitric oxide and superoxide anion balance in low-density lipoprotein-treated endothelial cells. Circulation 2003;107(11):1520–4.

[74] Mineo C, Shaul PW. Regulation of signal transduction by HDL. J Lipid Res 2013;54(9):2315–24.

[75] Yuhanna IS, Zhu Y, Cox BE, et al. High-density lipoprotein binding to scavenger receptor-BI activates endothelial nitric oxide synthase. Nat Med 2001;7(7):853–7.

[76] Mineo C, Yuhanna IS, Quon MJ, et al. High density lipoprotein-induced endothelial nitric-oxide synthase activation is mediated by Akt and MAP kinases. J Biol Chem 2003;278(11):9142–9.

[77] Drew BG, Fidge NH, Gallon-Beaumier G, et al. High-density lipoprotein and apolipoprotein AI increase endothelial NO synthase activity by protein association and multisite phosphorylation. Proc Natl Acad Sci U S A 2004;101(18):6999–7004.

[78] Nofer JR, van der Giet M, Tölle M, et al. HDL induces NO-dependent vasorelaxation via the lysophospholipid receptor S1P3. J Clin Invest 2004;113(4):569–81.

[79] Ramet ME, Ramet M, Lu Q, et al. High-density lipoprotein increases the abundance of eNOS protein in human vascular endothelial cells by increasing its half-life. J Am Coll Cardiol 2003;41:2288–97.

[80] Campbell S, Genest J. HDL-C: clinical equipoise and vascular endothelial function. Expert Rev Cardiovasc Ther 2013;11(3):343–53.

[81] Prosser HC, Ng MK, Bursill CA. The role of cholesterol efflux in mechanisms of endothelial protection by HDL. Curr Opin Lipidol 2012;23(3):182–9.

[82] Seetharam D, Mineo C, Gormley AK, et al. High-density lipoprotein promotes endothelial cell migration and reendothelialization via scavenger receptor-B type I. Circ Res 2006;98(1):63–72.

[83] Kimura T, Tomura H, Mogi C, et al. Role of scavenger receptor class B type I and sphingosine 1-phosphate receptors in high density lipoprotein-induced inhibition of adhesion molecule expression in endothelial cells. J Biol Chem 2006;281(49):37457–67.

[84] Assanasen C, Mineo C, Seetharam D, et al. Cholesterol binding, efflux, and a PDZ-interacting domain of scavenger receptor-BI mediate HDL-initiated signaling. J Clin Invest 2005;115(4):969–77.

[85] Cutuli L, Pirillo A, Uboldi P, et al. 15-Lipoxygenase-mediated modification of HDL3 impairs eNOS activation in human endothelial cells. Lipids 2014;49(4):317–26.

[86] Xiong Y, Hla T. S1P control of endothelial integrity. Curr Top Microbiol Immunol 2014;378:85–105.

[87] Wilkerson BA, Argraves KM. The role of sphingosine-1-phosphate in endothelial barrier function. Biochim Biophys Acta 2014;1841 (10):1403–12.

[88] Christoffersen C, Nielsen LB. Apolipoprotein M: bridging HDL and endothelial function. Curr Opin Lipidol 2013;24(4):295–300.

[89] Spijkers LJ, Alewijnse AE, Peters SL. Sphingolipids and the orchestration of endothelium-derived vasoactive factors: when endothelial function demands greasing. Mol Cells 2010;29 (2):105–11.

[90] Camont L, Lhomme M, Rached F, et al. Small, dense high-density lipoprotein-3 particles are enriched in negatively charged phospholipids: relevance to cellular cholesterol efflux, antioxidative, antithrombotic, anti-inflammatory, and antiapoptotic functionalities. Arterioscler Thromb Vasc Biol 2013;33(12):2715–23.

22

HDL and Endothelium

Raul Cavalcante Maranhão, Antonio Casela Filho, Gilbert Alexandre Sigal,
Antonio Carlos Palandri Chagas,
and Protásio Lemos da Luz

INTRODUCTION

The concept of the role of plasmatic lipids in atherogenesis consists of two main pillars with a firm foundation on epidemiologic studies. The first is that the plasmatic fraction of low density lipoproteins (LDL), mediated by the cholesterol contained in this fraction (LDL-cholesterol), correlates positively with the incidence of atherosclerotic cardiovascular disease, in particular coronary arterial disease. The second is that high density lipoproteins (HDL), measured by HDL-cholesterol, correlate negatively with the incidence of such diseases [1]. The meta-analysis of four studies, totaling more than 15,000 individuals, estimated that for each 1 mg/dL less in HDL-cholesterol levels, there were 2% more cases of coronary arterial disease in men and 3% more in women [2].

Very low density lipoprotein (VLDL) and its catabolism products, such as intermediate density lipoprotein (IDL) and LDL, as well as chylomicrons and their remnants, contain apolipoprotein (apo) B. In cytoplasm organelles, apo B is the element that gathers lipids, such as triglycerides, and is therefore the intracellular synthesis center for both chylomicrons, in the intestine, and VLDL, in the liver. Chylomicrons and VLDL leave enterocytes and hepatocytes into circulation with a maximum triglyceride load, which is hydrolyzed in the capillary endothelium by lipoprotein lipase, going from the lipoproteins to the tissue where they are stored as fat tissue and muscle. In this catabolic cascade, the lipoproteins become progressively smaller, depleted from their triglyceride load, and are finally captured by the tissues, mainly the liver.

HDL constitutes a specific lipid transport circuit in circulation, and does not contain apo B. HDL formation is centered in apo A-I, a protein with very distinctive properties that, by interacting with other cellular and plasmatic systems, allows removing cholesterol from tissues and promotes cholesterol homeostasis in the plasmatic compartment and organism.

Dietary measures, such as reduction in saturated fat percentage, and current drugs available, such as statins, are very effective in reducing the plasmatic concentration of LDL-cholesterol. These means provided epidemiological evidence that LDL-cholesterol reduction is beneficial both in primary and secondary prevention of cardiovascular diseases [3].

The means to raise HDL-cholesterol levels are more indirect, less specific, and less efficient. One of them depends on the presence of hypertriglyceridemia: such treatment results in increased HDL-cholesterol, the so-called "teeter-totter effect;" this will be discussed later on.

Tobacco use reduces HDL-cholesterol, and quitting the habit may increase this fraction [4]. Likewise, physical activity, preferably more intense, and moderate consumption of alcoholic beverages, raise HDL-cholesterol levels [5,6]. However, these responses of HDL-cholesterol may be attributed, at least in part, to triglyceridemia: physical training reduces and tobacco use increases triglycerides, which may indirectly affect HDL due to the "teeter-totter effect" [7].

Drugs whose therapeutic target is HDL-cholesterol increase, such as CETP (cholesteryl ester transfer protein) inhibitors, were only developed recently. In clinical tests, although these medications significantly increase HDL-cholesterol concentration, so far they have been unable to reduce the incidence of cardiovascular disease complications. It is interesting to note that CETP inhibitors may also reduce LDL-cholesterol [8,9]. In contrast, several lipid-lowering drugs, such as statins and fibrates, also increase HDL-cholesterol. Could this increase in HDL-cholesterol also be related to reduction in cardiovascular events, achieved by treatment with these drugs?

Thus, although the negative correlation between HDL-cholesterol and coronary arterial diseases is a fact, there is lack of clinical evidence that HDL-cholesterol increase per se is beneficial to prevent these diseases, in terms of primary prevention and secondary prevention as well. On the other hand, a growing number of endothelium-protective functions, antiatherogenic functions, and other functions, related to HDL, have been demonstrated [10]. These findings suggest that HDL-cholesterol measurement, which translates only the quantity of HDL, is only one part of the issue and the evaluation of multifaceted qualitative aspects is required to understand the broader significance of this lipoprotein. Among the questions and answers lies the immense complexity of HDL.

The goal of this chapter is to show some aspects of HDL metabolism that are more relevant to understand how HDL contributes to protect and repair the endothelium and the current drug interventions that modify HDL-cholesterol. An additional purpose is to raise possibilities to find new biomarkers associated with cardiovascular physiopathology and new therapeutic instruments to maintain endothelium integrity and therefore achieve longer and healthier longevity.

HDL STRUCTURE AND METABOLISM, ESTERIFICATION, AND REVERSE CHOLESTEROL TRANSPORT

HDL is the lipoprotein fraction with the smallest particles (approximately 10 nm). HDL is constituted by spherical or discoid particles varying from 1.063 to 1.21 g/mL, due to the high protein content (>30%) in comparison to other classes of lipoproteins [11]. Discoidal HDLs are the smallest HDL particles, consisting of apo A-I with monolayer of phospholipids and free cholesterol. Spherical HDLs are larger and with a hydrophobic core formed by cholesterol esters and small quantities of triglycerides surrounded by a monolayer of phospholipids and nonesterified cholesterol [11,12]. Several apolipoproteins and many other proteins bond to the surface of HDL fraction particles.

More than 100 different proteins and enzymes in constant movement, evidenced by proteomics techniques, prove the HDL fraction functional characteristics [11]. The functions of many of these proteins and associated enzymes are still unknown, and only some of them are directly related to lipid metabolism. The others are related to the protease inhibition, complement regulation, and acute phase response processes, among others.

In contrast to lipoproteins that contain apo B, HDL formation begins with apo A-I lipidation that occurs mainly in the plasmatic compartment and not in cytoplasm, as happens with lipoproteins that contain apo B [13]. Apo A-I is produced mainly by the liver, but also by the intestine. It receives phospholipids in the endoplasmic reticulum and, after extrusion to the plasmatic compartment, continues to receive phospholipids and nonesterified cholesterol, proceeding from cells in which the lipids are removed and incorporated to HDL by means of ABCA1 (ATP-binding cassette transporter A1) and ABCG1 (ATP-binding cassette transporter G1) transporters [14]. It also receives cholesterol and phospholipids from other plasmatic lipoproteins assuming discoid format, the nascent HDLs. As seen, chylomicron and VLDL lipolysis reduces the size of those lipoproteins; the excessive phospholipids and cholesterol on the surface after lipolysis decouples from the VLDL and chylomicrons; any may be incorporated to apo A-I and therefore contribute to HDL formation [15].

The nonesterified form of cholesterol is continuously esterified by the catalytic action of LCAT (lecithin-cholesterol-acyl-transferase), of which apo A-I is a co-factor. Cholesterol esterification in nascent HDL leads to progressive disc rounding, in a process called lipoprotein maturation. Cholesterol esters are more hydrophobic than nonesterified cholesterol, and therefore, after esterification, cholesterol is displaced from the particle surface to the nucleus, where it remains isolated from the water medium surrounding the lipoprotein [16].

Cholesterol esterification and sequestering inside lipoproteins is essential to stabilize the cholesterol plasmatic pool. The concentration of nonesterified free cholesterol must be maintained according to a strict limit: excessive free cholesterol changes cell membrane permeability and affects its functions. Regarding the endothelium, excessive cholesterol may have important consequences in triggering atherogenesis processes. Therefore, circulation has an immense cholesterol esterification machine based mainly on the HDL fraction, although it may occur to a lesser extent in other lipoprotein fractions. Each HDL particle that receives free cholesterol, consisting of LCAT and its co-factor apo A-I, composes a minuscule unit of this machine. Cholesterol esterification also creates the gradient between the cell membrane and HDL that allows reverse cholesterol transport [17].

HDL Subfractions

The density interval that corresponds to the HDL fraction (1.63–1.210 g/mL) has continuous particles of growing size and density [18]. Usually, two large subclasses that can be separated by different methods are distinguished in the HDL fraction: HDL2, which is bigger, lighter, and richer in lipids (d 1.63–1.125) and HDL3, which is smaller, denser, and richer in protein (d 1.125–1.210 g/mL) [18].

HDL can be separated in a much higher number of subfractions, using methods such as one-dimensional or two-dimensional electrophoresis, ionic mobility, or more discriminative ultra centrifugation methods. With these approaches, HDL2 and HDL3 may be separated in 5 subpopulations by means of gel gradient electrophoresis, according to size: HDL3c (7.2–7.8 nm), HDL3b (7.8–8.2 nm), HDL3a (8.2–8.8 nm), HDL2a (8.8–9.7 nm), and HDL2b (9.7–12.9 nm) [19]. Using two-dimensional electrophoresis, HDL can be separated by load into more than 10 subclasses. Based on protein composition, HDL can also be separated in subpopulations using the immunoaffinity method [19]. Agarose gel electrophoresis allows the separating an HDL subfraction with slower prebeta mobility (pre-β HDL), due to the weak negative charge load on its surface related to the lack of neutral lipid core; this mobility is different from most HDL particles, which have alpha mobility (α-HDL). Pre-β HDL has a discoid, flattened format, and would be the subfraction that initially receives free cholesterol proceeding from the cells. Approximately 60% of the cell cholesterol efflux depends on this subfraction [20].

The HDL subclasses profile reflect the dynamics of the complex metabolism of this lipoprotein in the plasmatic compartment. Several factors continuously interfere with this metabolism. Lipid transfers among lipoprotein classes, activity of several enzymes that interfere in lipid metabolism, in general, and HDL metabolism in particular, such as LCAT, kinetics and concentration of several lipoproteins, among many other factors, may change this profile. This process is frequently called HDL remodeling [21].

In the case of LDL subfractions, it is interesting to note that it is established that small and dense LDL subfractions have higher atherogenicity when compared to the subfraction of larger and less dense particles [22]. It is also possible to determine a profile of LDL subfractions associated with gender and age, in which larger and less dense fractions are more prominent in women, whose atherosclerotic manifestations are at a later stage. However, LDL is the final product of catabolism, in which larger biochemical events does not occur, besides lipoprotein removal from the plasmatic compartment by LDL receptors [22]. For HDL, metabolism in the plasmatic compartment is much more complex and dynamic that for LDL, and maybe due to this reason the association of HDL subclass profiles with cardiovascular events is controversial, with no safe evidence that a given subclass is less protective than others. A recent meta-analysis postulated that there is no different regarding cardio-protection between the two major classes, HDL2 and HDL3 [23]. A recent report indicates that in 2414 cases of acute myocardial infarction (AMI), there is increased risk of cardiovascular events associated with low HDL3 and not HDL2 [24]. However, the authors did not find a correlation between the events and HDL-cholesterol, which contradicts the literature [3].

HDL and Gender

All parameters related to HDL plasmatic concentration are higher in women in comparison with men [25]. This includes HDL-cholesterol and the HDL2 and HDL3 subfractions. Apo A-I concentration is also higher in women than in men [26]. The onset of menopause does not change the levels of HDL-cholesterol or apo A-I [27], but the use of contraceptives or estrogen replacement therapy increases HDL-cholesterol and apo A-I levels in plasma. Adjustment by age, body mass index, tobacco use, or alcoholism does not change these gender differences [26].

Apo A-I and Apo A-II

The two main apos in HDL are apo A-I and A-II, with 243 and 154 amino acids, respectively. The amphipathic molecular structure of apo A-I includes an alpha-helix N-terminal segment (aa 1–184) and a C-terminal (aa 185–243) domain without a defined structure. The latter is responsible for initiating the bond and aggregation of lipids to apo A-I, and for the efflux of cholesterol from cells to HDL. The first 43 residues stabilized the lipid-free apo A-I structure [28]. From these findings, it is clear that HDL functions depend largely on the properties of apo A-I.

Apo A-II plays an important role in HDL structure conformation and stability, granted by its higher hydrophobicity. It also collaborates to HDL stability by inhibiting the action of lipases upon the lipoprotein, as suggested in studies with transgenic animals [29].

Injected in the delipidated form in individuals with low HDL-cholesterol, apo A-I showed half-life in plasma of 15–54 h; mature discoid HDL has longer half-life [30].

LCAT

In its mature format, LCAT is a 416-amino acid protein that in circulation tends to associate with the HDL fraction. In plasma, it is the only enzyme able to esterify cholesterol, which occurs by means of a transesterification reaction. In this reaction, LCAT cleaves a fatty acid in the sn-2 position of lecithin, due to its activity as phospholipase A2. Then, in the second step, LCAT acts as acyltransferase, and the cleaved fatty acid is transesterified to the cholesterol hydroxyl group, in carbon 3 of the steroid ring [31]. Although apo A-I is the best LCAT activator, apo E may also activate this enzyme in plasma, which may occur in apolipoproteins that contain apo B. Apo A-II, apo A-IV, apo C-I, and apo C-III are very weak enzyme activators [31].

Reverse Cholesterol Transport

Reverse cholesterol transport consists of cellular cholesterol transported from peripheral tissues to the liver, from where it is eliminated in feces as bile acid, cholesterol, and other catabolism products. The cholesterol excreted can also be recycled after intestinal resorption.

In the beginning of the process, which involves several stages, discoid apo A-I particles with low levels of phospholipids and cholesterol (HDL pre-beta1 subfraction) interact with the ABCA1 transporter, with efflux of cholesterol accumulated on the cell membrane to HDL [32]. A second mechanism involves cholesterol efflux to mature HDL particles, which interact with the cell membrane by means of ABCG1 transporters [33]. Cholesterol may also be transferred from the membrane to HDL particles by means of passive diffusion. When the free cholesterol esterified in HDL becomes very hydrophobic, it is pushed to the core of the lipoprotein, away from contact with the water medium. From here on, it may take two flow paths. In the first one, it remains in the HDL particle until it is finally collected by the liver by means of SR-BI receptors. In the second path, it is transferred to other lipoprotein classes, such as VLDL or LDL, and is finally collected by the liver as one of their components, by means of LDL receptors [33].

A more direct specific aspect of participation of HDL-mediated reverse transport in antiatherogenic defense consists of removal of cholesterol deposited in macrophages in the arterial intimal layer, by means of ABCA1 and ABCG1 transporters. This process may contribute to stabilize or even revert atherosclerotic lesions [34].

Fig. 22.1 illustrates cholesterol reverse transport.

Lipid Transfers

Lipid transfers are mediated by transfer proteins CETP and PLTP (phospholipid transfer protein). CETP favors the transference of cholesterol esters, triglycerides, and phospholipids, while PLTP promotes the transference of phospholipids and influences the transference of nonesterified cholesterol. Lipid transfers between lipoprotein classes are bidirectional, but may enrich the specific class of a given lipid and deplete another lipid [35]. For that matter, exchanges between VLDL and HDL tend to have a higher transference of cholesterol esters from HDL to VLDL and more triglycerides from VLDL to HDL. In hypertriglyceridemia, due to accumulated VLDL, by the law of mass action, HDL becomes triglyceride-rich and

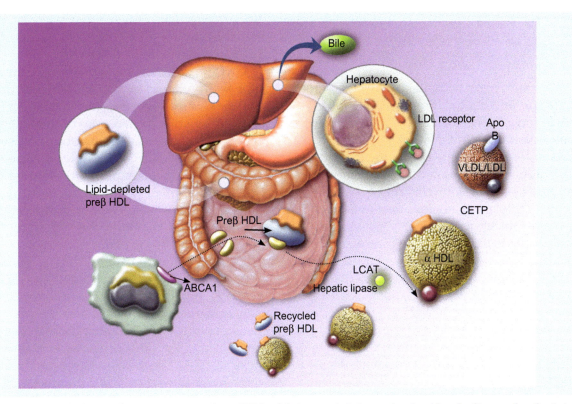

FIG. 22.1 Reverse cholesterol transport—pre-beta HDL, rich in apo A-I, is synthesized by the liver or by the intestinal mucosa and released in circulation, where by promoting the transference of the excessive free cholesterol in macrophages it increases in size and transforms into HDL3 and HDL2. These are transported to the liver, where they are processed. Adapted from Ashen MD, Blumenthal RS. Clinical practice. Low HDL cholesterol levels. N Engl J Med 2005;353:1252–60.

cholesterol-ester-depleted. Thus, when HDL suffers hepatic lipase attack, it destabilizes and is removed more quickly from circulation, resulting in reduced HDL-cholesterol concentration. In this "teeter-totter," triglycerides go up as HDL-cholesterol goes down; triglycerides go down as HDL-cholesterol goes up [36].

ABCA1 and ABCG1 Transporters

ABC proteins (ATP-binding cassette) are universally present in prokaryotes and eukaryotes. In the organism, they perform the first step in reverse transport: moving cell cholesterol to HDL [37]. Most organism cells are not able to catabolize cholesterol, and therefore cell cholesterol efflux is essential for homeostasis.

Macrophages, which are cells with essential importance in atherogenesis, have four pathways through which nonesterified cholesterol flows to the extracellular medium. Passive processes consist of simple diffusion by means of water medium and diffusion facilitated by SR-BI [37]. Active transport is mediated by the ABCA1 and ABCG1 transporters, which are membrane lipid translocases. Efflux of phospholipids and nonesterified cholesterol to apo A-I promoted by ABCA1 is essential for cholesterol homeostasis.

Caveolin-1 (CAV1), a protein that organizes and concentrates certain signaling molecules and receptors in cell plasma membrane caveolae, regulates cholesterol efflux mediated by ABCA1 and ABCG1 [37].

An important aspect of ABC transporters is their action upon proliferation control of stem cell of the hematopoietic system and multipotent progenitor cells in bone marrow, and mobilization of such cells. Activation of the cholesterol efflux paths by HDL infusion results in suppressed mobility of hematopoietic system stem cells, leading in inhibited production of monocytes and neutrophils, and atherosclerosis in apo E knockout mice [38]. ABCG4, a transporter similar to ABCG1 that mediates cholesterol efflux in megakaryocyte precursor cells, controls platelet production and thrombosis [38].

The ABCA1 transporter is located in the cell plasma membrane. Apo A-I and other apolipoproteins, such as apo E, bind directly to the transporter. Apo A-I binds to cholesterol with little efficiency, so that phospholipid efflux into the protein is very important for cholesterol efflux. After binding to ABCA1, apo A-I is lipidated and assumes discoid shape, with each discoid containing two apo A-I molecules, which promotes decoupling of the ABCA1 nascent lipoprotein [32]. ABCA1 is apparently also able to move from the plasma membrane to the interior and act on cholesterol traffic, promoting efflux from intracellular compartments. ABCA1 was discovered originally in patients with Tangier disease. This disease, in which the transporter has genetic defects, shows the importance of the transporter in HDL formation: in the Tangier homozygous forms, HDL-cholesterol is almost absent due to ABCA1 deficiency [39,40].

Differently from ABCA1, the ABCG1 transporter does not perform cholesterol efflux by means of apo A-I still without lipids, but by interaction with more mature forms of HDL [41]. It also promotes efflux of oxysteroids that may be toxic to cells, such as 7-ketocholesterol [42].

ABCG1 knockout mice show deficient cholesterol efflux from macrophages, with formation of foam cells in several tissues, mainly in the lung [43]. However, there is little change in the lipid profile of ABCG1 knockout mice, which may be explained by low hepatic expression of the transporter. In addition to ABCA1, ABCG1 has additional effects on macrophages and plays a role in cholesterol intracellular traffic [38].

CETP

CETP, a glycosylated protein with 476 aa and 75 kDa, is mainly secreted by the liver, fat tissue, enterocytes, and spleen [44] and circulates in plasma mainly associated with HDL [45].

CETP only promotes lipid transfers between lipoprotein subclasses that have different cholesterol ester/triglyceride mass coefficients. Therefore, CETP mediates HDL cholesterol ester transfer to lipoproteins that contain apo B in exchange for triglycerides, and transfer of triglycerides from lipoproteins that contain apo B in exchange for cholesterol ester [46–48]. CETP also favors cholesterol ester transfers between HDL subfractions [46].

CETP activity depends on its concentration and ability to interact with lipoproteins. Interaction may be stimulated by free fatty acids generated during hydrolysis of triglycerides proceeding from food or inhibited by specific apolipoproteins, such as apo C-I [49]. Most plasma cholesterol esters originate from the esterification reaction catalyzed by LCAT, which occurs mainly in the HDL fraction [50]. Triglycerides in circulation are present mostly in chylomicrons and VLDL before the hydrolysis performed by lipoprotein lipase [45].

As one of the cholesterol flow regulators of the reverse cholesterol transport system, CETP can be seen as potentially having both proatherogenic and antiatherogenic properties. In its proatherogenic action, CETP-mediated cholesterol ester transfer may reduce flow of cholesterol from HDL to hepatic SR-BI, with consequent increased cholesterol mass transported by the atherogenic VLDL, IDL, and LDL of arterial wall, increasing LDL levels [51,52]. After that, cholesterol deposits tend to increase in peripheral tissues and the arterial wall [52]. In addition to that, CETP also interacts with triglyceride lipases, generating small and dense LDL that is the most atherogenic LDL subclass [53,54]. A reduction in HDL particle size mediated by CETP is followed by dissociation of lipid-poor apo A-I from the particle [55,56].

On the other hand, CETP can also have antiatherogenic effects by promoting cholesterol ester flow to the liver by means of the indirect reverse cholesterol transport pathway, with predominant cholesterol ester hepatic collection by means of the antiatherogenic pathways of the LDL receptor [46].

The CETP activity level is decisive to divide the plasmatic cholesterol ester pool between LDL and HDL. Excessive CETP activity increases bidirectional transfer between HDL and LDL without resulting in important changes in cholesterol ester distribution between these two fractions [47]. When VLDL levels are normal, cholesterol CETP-mediated ester transfer from HDL is directed preferably to LDL, but when VLDL is accumulated in plasma, HDL cholesterol ester is preferably transferred by CETP to VLDL particles [57].

Some hypotheses have been suggested regarding the mechanisms by means of which CETP acts on lipid transfers. CETP transports cholesterol ester between donor and acceptor lipoproteins by means of the water medium [58]. Alternatively, in a tunnel-like mechanism, CETP connects these two lipoproteins. HDL binds to the N-terminal and LDL or VLDL to the C-terminal end. A transient ternary complex is formed, with neutral lipids flowing from the donor lipoprotein to the acceptor lipoprotein by means of the CETP molecule [59,60]. In a modified tunnel hypothesis, the lipids would be tunneled in a CETP dimer [45].

The reduced HDL levels and antiatherogenic functions and increased proatherogenic factors that occur in metabolic syndrome and *diabetes mellitus* type 2 (DM2) coexist with elevated CETP concentrations in plasma [61,62]. Treatment with pioglitazone reduces the hepatic triglyceride content and increases HDL-cholesterol levels, while simultaneously reducing CETP [63]. In individuals with normal weight, there is a reverse correlation between CETP levels and visceral fat gain and between CETP levels and body mass index (BMI) [64].

HDL ATHEROPROTECTIVE FUNCTIONS

In addition to cholesterol esterification and reverse transport, several other antiatherogenic functions have been attributed to HDL. These functions are performed by apolipoproteins, in particular apo A-I, and by several other proteins found in the HDL fraction, presumably on the surface of lipoprotein particles. It can be supposed that, depending on the roster of proteins on the particle surface, the protective actions of each HDL subclass vary in type and intensity.

Recent years have seen the development of the concept that HDL-cholesterol level, although important marker and predictor of coronary arterial disease, is not the entirely protector potential of this lipoprotein. May be the most remarkable observation regarding this fact was the discovery of apo A-I$_{Milano}$, a polymorphism prevailing in the population in the north of Italy. This apo A-I isoform determines lower levels of HDL-cholesterol; nevertheless, the incidence of cardiovascular disease in that population is lower [65]. On the other hand, very high levels of HDL-cholesterol may not grant antiatherogenic protection. It was also observed that higher cell cholesterol removal by HDL, the first step in reverse transport, is correlated only partially

(40%) with the HDL-cholesterol plasmatic concentration [65].

The several functions that have been attributed to HDL, from cholesterol esterification by LCAT in HDL and involvement in reverse cholesterol transport up to antiapoptotic, vasodilation, antioxidant, antiinflammatory, and antithrombotic functions, among others, compete to repair the effects of endothelial injury, the first event in the atherogenesis process, and to attenuate the onset and development of the process and its clinical manifestations.

Endothelium Reparation Function, Vasodilation, and Endothelium Antiapoptosis

The primary function of HDL is to protect endothelium integrity and function. Lower HDL-cholesterol concentrations are associated with endothelial dysfunction. HDL tends to increase the number of progenitor cells in the serum compartment and in regions with endothelial injuries, suggesting direct endothelium reparation action [66].

An aspect of great importance is HDL's ability to promote vasodilation. HDL binds to apo A-I of SR-BI receptors and this activates eNOS (endothelial nitric oxide synthase), since the presence of apo A-I allows eNOS to couple to the SR-BI receptor. Besides that, HDL also contributes to vasodilation by means of SR-BI, by inducing expression of cyclooxygenase 2 and production of prostacyclin (PGI2) by endothelial cells. Beside its vasodilation function, PGI2 also inhibits platelet aggregation.

HDL may inhibit endothelial cell apoptosis induced by oxidized LDL and TNF-α. Possibly, the HDL antiapoptotic effect is due to sphingolipids and apo A-I, inhibiting intracellular generation of reactive oxygen species (ROS), mitochondrial apoptotic pathways, and caspase-independent apoptotic pathways [67,68].

Fig. 22.2 illustrates the bond between HDL and scavenger receptor-BI (SR-BI) activating the eNOS enzyme.

Antioxidant Function

One of HDL's antioxidant effects is inhibiting the production of oxidized phospholipid in LDL. Apo A-I is able to remove oxidized phospholipids from oxidized LDL and cells binding these molecules and forming biologically active phospholipids in

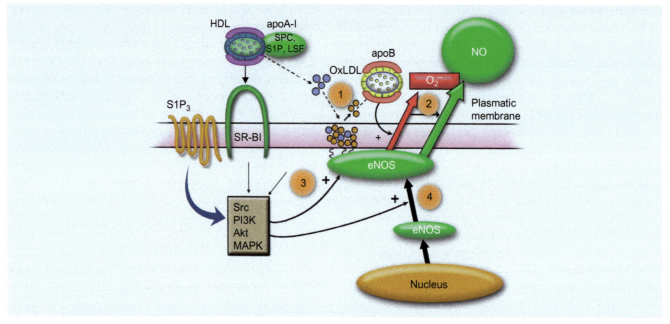

FIG. 22.2   Bond between HDL and scavenger receptor BI (SR-BI) activating the eNOS enzyme (see text for abbreviations). Adapted from Mineo C, Shaul PW. Role of high-density lipoprotein and scavenger receptor B type I in the promotion of endothelial repair. Trends Cardivasc Med 2007;17:156–61.

LDL. Besides that, it inactivates lipid hydroperoxides [69,70]. Other apolipoproteins present in the HDL fraction, such as A-II, C, E, A-IV, J, D, and M, may also protect LDL from oxidation induced by free radicals.

Antioxidant enzymes present in the HDL fraction, such as paraoxonase 1 (PON1) and PAF-AH (platelet activating factor acetylhydrolase) that catalyze oxidized, proinflammatory phospholipid hydrolysis, converge into the HDL antioxidant activity.

PON1 is an esterase and lactonase that catalyzes peroxides and lactones hydrolysis. It is synthesized and secreted by the liver, and full activation depends on apo A-I. PON1 associates exclusively with HDL by means of hydrophobic interactions. It is located in HDL subfractions that contain apo J (clusterin) and apo A-I [71,72].

The best PON1 transporters are small HDL particles contained in the HDL3 subfraction. HDL supposedly transfers PON1 to tissues [73]. The possible actions of PON1 include inhibiting lipid oxidation, delaying LDL aggregation and formation of oxidized LDL, and preventing lipid peroxide accumulation in this lipoprotein [74]. This enzyme also promotes breakdown of specific oxidized lipids in oxidized LDL, reduces cholesterol synthesis, decreases macrophage oxidized LDL collection, stimulates HDL-mediated cholesterol efflux from macrophages, and suppresses monocyte differentiation into macrophages [74].

PON1 has been demonstrated to promote lipid peroxide hydrolysis in coronary arteries and carotid atherosclerotic lesions in men [71,72,75]. PON1 serum activity is inversely related to the risk of cardiovascular disease [76], and low PON1 activity has been observed in patients with atherosclerosis, hypercholesterolemia, and *diabetes mellitus* [77,78].

HDL's antioxidant activity is higher in smaller and denser subfractions. This difference in antioxidant activity between HDL subfractions is due, probably, to uneven distribution of apolipoproteins and enzymes over the subfraction spectrum. The potent antioxidative protective activity observed in subfractions with smaller and denser particles is probably due to synergy of several enzymatic mechanisms (PON, LCAT, and PAF-AH) and nonenzymatic upon oxidized lipid inactivation [71].

Antiinflammatory Functions

HDL has antiinflammatory activities related to protection against cardiovascular disease. When exposed to HDL of healthy individuals, endothelial cells and leukocytes attenuate the expression of proinflammatory molecules. This effect is attributed to sphingosine-1-phosphate (S1P), which is part of HDL lipids, or equally to proteins associated with HDL, such as apoA-I, PON1, or clusterin [79].

Apo A-I, mediated by SR-BI, is able to reduce the expression of adhesion molecules, such as VCAM-1, ICAM-1, and selectin-E, and the production of IL-8 and IL-1 [80,81]. HDL and apo A-I reduce neutrophil migration and diffusion and neutrophil-platelet interaction [81]. HDL attenuates the activation of monocytes/macrophages and neutrophils, and inhibits linkage between T cell microparticles and monocytes, therefore reducing the production of proinflammatory cytokines [82].

The presence of HDL reduces the expression of MCP-1 (monocyte chemoattractant protein-1) and this weakens the bond between inflammatory cells and the dysfunctional endothelium, and reduces displacement of inflammatory cells to the subendothelial space. HDL also increases production of antiinflammatory cytokine IL-10, which is an important protection against the development of advanced atherosclerotic lesions [82]. HDL also inhibits ICAM-1 and VCAM-1 superexpression induced by oxidized LDL on the surface of endothelial cells. This contributes to reduce leukocyte bond and infiltration, and free radical production [82].

Antithrombotic Functions

HDL has antithrombotic and fibrinolysis properties. Possibly, HDL's most important antithrombotic effects are due to its vascular effects. For that matter, HDL reduces the expression of adhesion molecules on the endothelial surface and modulates blood flow by affecting NO production. HDL also reduces thrombin-mediated tissue factor production in endothelium cells [83] and inhibits factor X.

In addition to these actions on the vasculature, which are indirect antithrombotic effects, HDL also inhibits platelet aggregation. This occurs due to the inhibition of thrombin, adenosine diphosphate, and adrenalin-dependent mechanisms [84]. HDL also inhibits platelet activation. Apolipoprotein apo E present in HDL may induce NO production by platelets, and this inhibits platelet activation [85].

Regarding HDL fibrinolysis activity, higher levels of HDL-cholesterol are associated with improved permeability and fibrin clot lysis [86]. HDL may be associated with increased fibrinolysis resulting from increased plasmin generation [87]. Besides that, HDL-cholesterol is inversely correlated to the plasmatic levels of plasminogen inhibitor-I (PAI-I), which is the inhibitor of the tissue plasminogen activator and the urokinase-type plasminogen. Both are secreted by endothelial cells, and positive correlation has been evidenced between HDL-cholesterol and D-dimer levels, one of the products of fibrin degradation [88,89].

Antiinfective and Cytoprotective Activities

HDL also has antiinfective activity. HDL-cholesterol is reduced during infections, and endotoxemia modifies HDL composition by depleting phospholipids and apo A1 and increasing serum amyloid A (SAA) and PAF-AH [90]. The number of HDL particles remains unchanged, but the number of small and medium-sized particles decreases. Due to its ability to bind to lipopolysaccharides (LPS) in gram-negative bacteria and to lipoteichoic acid (LTA) in gram-positive bacteria, HDL neutralizes and increases plasma bleaching of these substances. Due to direct interaction with apo A-I [91], the lipoprotein binds and removes endotoxins from circulation by bile excretion, reducing the production of endotoxins induced by TNF-α and CD14 expression in monocytes. It has also been suggested that HDL protects against viral and parasitic infections [92–94]. The implications of the infectious processes, such as endothelium aggression factors, have been widely studied [95].

HDL and Sphingosine-1-Phosphate

Sphingolipids are cell membrane components and also important signaling molecules. S1P is a sphingolipid synthesized and released by phosphorylation of sphingosine by sphingosine kinase [96]. Blood cells are the biggest sources of S1P, in particular erythrocyte, but vascular and lymphatic endothelial cells also synthesize and release S1P. Endothelial cells express S1P receptors that are stimulated by thrombin and in hypoxia conditions [97].

S1P levels in tissues are low, but are elevated in plasma. The S1P in plasma binds to lipoproteins, 70%–90% to HDL, preferably HDL3. Transfer of S1P from cell membrane to HDL requires the HDL to contact the cell membrane. Possibly ABCA1 is the transporter in S1P efflux from cell to plasma, and apo A-I would mediate S1P migration to apoM of HDL. S1P binds to HDL with great affinity, and the S1P-HDL bond is biologically active [98].

The effects of S1P are linked to endothelial cell NO production, due to eNOS activation, NO-dependent vasodilation, and angiogenesis. Akt, a specific serine/threonine-protein kinase, and extracellular regulation pathways are also influenced by S1P, and play a role in some aspects of HDL's antioxidant, antiapoptotic, and antiinflammatory actions [99].

It is interesting to note that S1P content in the HDL of patients with coronary arterial diseases is lower than in healthy individuals. The opposite occurs with S1P out of the HDL fraction, which is increased in those patients when compared to healthy individuals [100].

HDL and MicroRNAs

MicroRNAs (miRNAs) are small RNA chains, with 18–25 nucleotides, which regulate gene expression [101]. The miRNAs are no encoding RNAs, but they regulate posttranscriptional gene expression, by inhibiting translation of promoting mRNA degradation. By this time, more than 2500 miRNA sequences have been described. They circulate in bodily fluids, including blood, in a highly stable fashion, associated with microparticles, apoptotic bodies, exosomes, protein complexes, and lipoproteins, our specific interest. This association makes them resistant to degradation outside cells by serum nucleases. The miRNAs are one of the most important biological discoveries in recent years [102].

HDL transports miRNAs and delivers them to different cell types. The expression of several genes associated with HDL metabolism, such as ABCA1, ABCG1, and SR-BI receptor genes, is influenced by miRNAs [103]. Two different miRNAs, miR-486 and miR-92a, indicated as stable or unstable coronary disease markers, respectively, are transported in HDL [104]. The most abundant miRNA bonded to HDL is miR-223. HDL particles have the ability of releasing miR-223 to recipient cells and mediating reduction of expression of certain genes involved in the SR-BI pathway [105,106]. Transferred to endothelial cells, HDL miR-223 reduced the expression of ICAM-1 (intracellular adhesion molecule 1) and, therefore, reduced monocyte adhesion and inflammation. This could be a mechanisms in the set of HDL antiinflammatory effects [107]. Therefore, miRNAs carried in HDL deserve research attention focus. The discovery that ABCA1 expression is widely regulated by miRNAs [103,108] was

essential, and opens the perspective of new therapeutic targets for atherosclerosis.

HDL and Islets of Langerhans β-Cells

HDL and apo A-I have antidiabetogenic properties, increasing pancreatic beta cell functions and sensitivity to insulin. HDL and apo A-I increase insulin synthesis and secretion by beta cells. This mechanism involves the activation of the Gαs subunit of protein G on the surface of islet beta cells. This leads to the activation of a transmembrane adenylyl cyclase and increases adenosine monophosphate and intracellular calcium level, with the activation of protein kinase A. This activates insulin synthesis, excluding FoxO1 from the nucleus of beta cells, which increases insulin synthesis and secretion by pancreatic cells and releases insulin gene transcription [109].

Although this HDL property is not directly related to the endothelium, it may protect it since the installation of insulin resistance and DM2 trigger several proatherogenic factors, including dyslipidemia.

Fig. 22.3 illustrates HDL proteomics.

DYSFUNCTIONAL HDL AND PROATHEROGENIC HDL

Conditions such as tobacco use, infections, surgery, or diseases such as chronic systemic inflammation, atherosclerosis, *diabetes mellitus*, metabolic syndrome, rheumatic disease, coronary arterial disease, acute coronary syndrome, or chronic renal dysfunction, may change HDL composition and structure. Finally, this may reduce HDL function, even though HDL-cholesterol serum levels remain normal or even high. This dysfunctional HDL may even have proinflammatory and prooxidative activity and become a proatherogenic lipoprotein.

Acute phase reactions such as those that occur in infections and after surgeries may turn HDL into a proinflammatory lipoprotein. Serum amyloid A (SAA) and ceruloplasmin, which are elevated in the acute phase response, bind to HDL, and replace apo A-I in the lipoprotein structure. In addition to that, apo A-I synthesis is reduced in inflammatory states.

Dysfunctional HDL loses its vascular protection effects and, consequently, its antiinflammatory

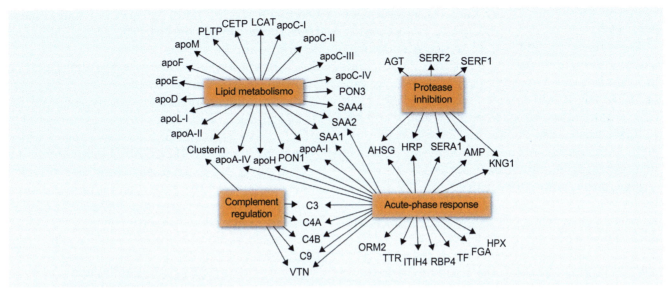

FIG. 22.3 HDL proteomics: identified proteins associated with HDL are involved in different functions. *Apo*, apolipoprotein; *AGT*, angiotensinogen; *AHSG*, α-2-HS-glycoprotein; *AMP*, bikunin; *CETP*, cholesterol ester transfer protein; *FGA*, fibrinogen; *HPX*, hemopexin; *HRP*, haptoglobin-related protein; *ITIH4*, inter-α-trypsin inhibitor heavy chain H4; *KNG1*, kininogen-1; *LCAT*, lecithin-cholesterol acyltransferase; *ORM2*, α-1-acid glycoprotein 2; *PLTP*, phospholipid transfer protein; *PON*, paraoxonase; *RBP4*, retinol binding protein; *SAA*, serum amyloid A; *SERA1*, α-1-antitrypsin; *SERF1*, serpin peptidase inhibitor (clade F, member 1); *SERF2*, α-2-antiplasmin; *TF*, transferrin; *TTR*, transthyretin; *VTN*, vitronectin [110]. Adapted from Vaisar T, Pennathur S, Green PS, et al. Shotgun proteomics implicates protease inhibition and complement activation in the anti-inflammatory properties of HDL. J Clin Invest 2007;117(3):746–56.

effects, and therefore the HDL endothelial reparation ability may be affected. Dysfunctional HDL does not stimulate endothelial cell NO release and even inhibits it by inhibiting eNOS activation. Besides that, inhibition of endothelial VCAM-1 expression and inhibition of the adhesion of white blood cells to activated endothelial cells occurs. MCP-1 production activity by human arterial wall cells is increased, and consequently monocyte migration into the arterial wall is increased as well. Finally, dysfunctional HDL loses its antiapoptotic effects upon endothelial cells.

The activity of the antioxidant enzymes associated with HDL is decreased when HDL becomes dysfunctional. Possibly, reduced PON1 activity is due to changes in lipoprotein composition, in particular apo A-I [111], and also due to reduced expression of the PON1 gene in the liver that occurs during acute phase response. Other enzymes with antioxidant functions associated with HDL, such as Lp-PLA2 and LCAT, may also be affected in case of inflammation. Another possibility raised is that dysfunctional HDL may have prothrombotic effects [112,113].

Metalloproteinases (MPO) have an interesting aspect. They are a source of ROS in inflammatory processes, and may oxidize apo A-I in HDL, inhibiting the lipoprotein's atheroprotective functions. Recently, the formation of a complex with MPO and PON1 in HDL has been described, in which PON1 partially inhibits MPO activity, which, in turn, inactivates PON1. This complex seems to perform reciprocal modulation between these two enzymes during inflammation, in which HDL serves as support for both [114].

Methods to Evaluate Dysfunctional HDL

Reproducible clinical tests, such as cholesterol efflux, monocyte chemotaxis levels, endothelial inflammation, oxidation, NO production, and thrombosis, have been proposed to access HDL functionality [115].

An important part of reverse cholesterol transport, which is cholesterol efflux from cells, has been estimated by incubating donor cells in medium containing acceptors such as human serum. Cholesterol efflux can be estimated by measuring the quantity of cholesterol released by donor cells into the medium [116].

Another methodology allows measuring cholesterol efflux from macrophages to HDL, which is more specifically related to reverse cholesterol transport [117].

The effects of HDL in preventing endothelial lesion may be accessed by measuring HDL ability in inhibiting cytokine-mediated protein expression of VCAM-1, ICAM-1, and E-selectin [118]. These effects may also be evaluated in vivo and the reduced expression of VCAM-1 and ICAM-1 after HDL infusion in surgical carotid lesion has been demonstrated in animal models [115].

HDL proinflammatory actions may be accessed performing tests that estimate monocyte chemotaxis activity by measuring cell migration into the subendothelial space in the presence and absence of HDL. Increased migration observed in endothelial aortic cells co-culture has been the basis to demonstrate HDL's proinflammatory activity [115].

HDL's antioxidant activity may be accessed incubating HDL with phospholipids commonly found in oxidized LDL and measuring HDL's effects on reducing oxidize phospholipid formation. LDL oxidation inhibited by HDL or oxidation of PAPC (1-1-palmitoyl-2-arachidonoyl-sn-glycero-3-phosphoryl-choline) or oxidized PAPC inactivation have been commonly employed. The use of spectroscopy and a fluorescent phospholipid oxidation marker allows estimating HDL oxidant activity quantitatively, and therefore demonstrate if HDL has anti- or prooxidant activity [115,119,120].

The effects of HDL in improving endothelial functions can be estimated by measuring NO production by endothelial cells stimulated by HDL. This test is performed by incubating endothelial cells with L-arginine and one of the eNOS antagonists and measuring the quantity of L-citrulline produced from the L-arginine [121] or nitrite and nitrate production. On the other hand, incubating endothelial cells with eNOS inhibitor and measuring oxygen free radical concentration allows testing agents used to restore endothelial capacity. Finding of reduced oxygen free radical concentration may suggest that the agent tested has antiatherogenic effects [115].

To access HDL's antiplatelet and antithrombotic activity effects, serotonin release inhibition and platelet preparation aggregation have been studied [115,122].

Proatherogenic HDL

In some conditions such as inflammatory states, *diabetes mellitus*, or metabolic syndrome, apo A-I may be modified by reactive oxygen species [123].

Oxidative modifications of apo A-I may prevent HDL from acting on cell cholesterol efflux by impairing the interaction between apo A-I and the ABCA1 complex, which pumps the cholesterol out of the cells [123]. The antioxidative enzymes associated with HDL may also be inactivated, and oxidized proteins and lipids may accumulated in HDL, therefore impairing the lipoprotein antiinflammatory function [124,125]. The replacement of SAA by apo A-I in HDL structure may revert its functions, making the lipoprotein proinflammatory and proatherogenic. This may occur in conditions associated with high SAA serum levels, as is the case of metabolic syndrome, *diabetes mellitus*, and chronic renal disease [125–127].

Lipid Transfers to HDL and Antiatherosclerosis Protection

Instituto do Coração (InCor-FMUSP) conducted an in vitro test to evaluate lipid transfers to HDL. This method measures simultaneous transfer of radioactive lipids—phospholipids, triglycerides, esterified cholesterol, and nonesterified cholesterol—from the donor lipid emulsion to HDL. It demonstrated that esterified cholesterol transfer was reduced in the presence of coronary arterial disease, both in nondiabetic patients and those with DM 2 [128]. Besides that, in clinical conditions that favor the development of coronary arterial disease, as is the case of heart transplant, familial hypercholesterolemia, and sedentarism, transfer of nonesterified cholesterol to HDL was reduced [21,129]. These results demonstrated the importance of phenomena related to HDL metabolism and reverse transport, such as proatherogenic factors.

HDL in Metabolic Syndrome and *Diabetes Mellitus* Type 2

Combined with overweight or obesity, glucose intolerance, systemic arterial hypertension, and hypertriglyceridemia, low HDL-cholesterol concentration is one of the characteristics that close the concept of metabolic syndrome [130]. It is important to note that the concept of syndrome is an attempt to typify individuals with higher risk of developing DM2 and cardiovascular diseases.

In metabolic syndrome, insulin resistance is usually higher as a result from overweight and obesity. This is manifested as hypertriglyceridemia and low HDL-cholesterol, which constitute the typical dyslipidemia pattern of insulin resistance and DM 2, and therefore these two lipid alterations are included in the diagnosis criterion. Coronary arterial disease is the most frequent fatal complication in DM 2, with risk increased 2–4 times, and is responsible for 60% of diabetic patient deaths. In patients with metabolic syndrome, moderate physical activity for a period of 3 months was not able to increase HDL-cholesterol, but improved HDL and PON1 antioxidant activity [131].

Among several and important steps in lipid metabolism, insulin regulates triglyceride-rich lipoprotein degradation by lipoprotein lipase and hormone-sensitive lipase that catalyze the release of fatty acids from peripheral tissues such as muscle and fatty tissue, and VLDL synthesis. With the tendency of increased triglycerides in metabolic syndrome and DM 2, HDL-cholesterol concentration would tend to decrease. However, resistance to insulin also has direct effects on HDL synthesis and metabolism [132].

Therapy with niacin in patients with DM type 2 improved several HDL functional properties, including those related to endothelium functions. HDL isolated from diabetic patients after therapy with niacin demonstrated higher capacity in stimulating endothelial NO production and improving endothelium-dependent vasodilation [133].

Effects of Lipid-Lowering Drugs Upon HDL and Endothelial Function

The usual lipid-lowering agent with the highest increased in HDL-cholesterol is niacin (nicotinic acid). HDL-cholesterol increased approximately 20%–30%, at the same time in which LDL-cholesterol is reduced by a similar proportion, 20%–30%, and triglycerides by 35%–45% [134].

It is interesting that, in patients with low HDL-cholesterol and without other CAD risk factors, treatment with niacin for 3 months was able to improve endothelial function without increasing HDL-cholesterol (Fig. 22.4).

Probucol, an LDL-cholesterol reducer and powerful antioxidant drug, was taken from the market because it reduced HDL-cholesterol, although this was not proved to be harmful.

In hypercholesterolemic patients with normal HDL-cholesterol, treatment with statins increased HDL-cholesterol approximately 5%–10%. This effect was independent on the type of statin and dose employed. In patients with low HDL-cholesterol

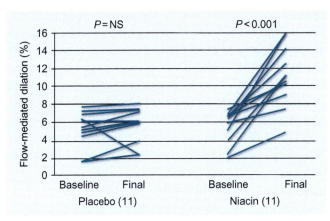

FIG. 22.4 Variation of flow-mediated dilation in the placebo and niacin groups. Adapted from Benjo AM, Maranhão RC, Coimbra S, et al. Accumulation of chylomicron remnants and impaired vascular reactivity occur in subjects with isolated low HDL cholesterol:effects of niacin treatment. Atherosclerosis 2006;187:116–22.

levels, below 35 mg/dL, there were different responses depending on the statin employed. The effects of the maximum dose of simvastatin, 80 mg/dL, upon apo A-I and HDL-cholesterol were higher than the effects of an equivalent dose of atorvastatin [135]. Cases of low HDL-cholesterol associated with hypertriglyceridemia showed 20%–30% increase in HDL-cholesterol with the use of rosuvastatin [136].

Fibrates increase HDL-cholesterol by stimulating PPAR-alpha upon expression of apo A-I and apo A-II, which increases HDL-cholesterol [137]. The effect upon HDL-cholesterol is more important in patients with hypertriglyceridemia, when the "teeter-totter" effect occurs, because it reduced the passage of triglycerides to HDL and therefore accelerated lipoprotein catabolism. Besides that, there is increased HDL formation from products of triglyceride-rich lipoproteins lipolysis: since the treatment intensifies lipolysis, there is more lipidation of apo A-I and formation of initial HDL. Fenofibrates has more potent action than gemfibrozil to increase HDL-cholesterol [138].

Ezetimibe, and have little effect both upon triglycerides and HDL-cholesterol [139,140]. Combined with statins, medication scheme that is frequently recommended, it produced significant HDL-cholesterol increase when compared to isolated ezetimibe of statin [141].

Two new medications in the process of being introduced in clinical practice lead to accentuated LDL-cholesterol reduction. Mipomersen is a complementary antisense nucleotide specific for the mRNA of apo B100, which inhibits apo B100 synthesis by the liver. Mipomersen does not alter HDL-cholesterol levels significantly [142]. PCSK9 (proprotein convertase subtilisin/kexin type 9) inhibitors are monoclonal antibodies that block PCSK9, which degrades the LDL receptor; with that, they increase LDL removal from circulation [143]. PCSK9 inhibitors *evolocumab* and *alirocumab* only increase HDL-cholesterol moderately [144].

TREATMENT WITH HDL, HDL MIMETICS, AND CETP INHIBITORS

In the mid-1990s, studies observed that apo A-I has antiinflammatory properties that inhibit expression of cell adhesion molecules induced by cytokines such as VCAM-1, ICAM-1, and E-selectin, which play key role in diapedese of immunocompetent cells from circulation to the arterial wall [145]. Subsequent research demonstrated that the interaction between apo A-I and the ABCA1 complex, by means of which cholesterol is transferred from cells to HDL, promotes the activation of the signal transducer and activator of transcription (STAT-3) by janus kinase 2 (JAK2). Finally, this abolishes the lipopolysaccharides (LPS) induced in the proinflammatory response, involving the production of IL-1β, IL-6 e TNF-α by monocytes [146]. This occurred without affecting cholesterol efflux from cells to HDL, the first step in reverse cholesterol transport.

Recently, another antiinflammatory mechanism related to HDL has been described, involving the activation of transcription factor 3 (ATF3), a transcriptional modulator that inhibits immune signaling or TLR (toll-like receptor) [147,148]. Previous treatment with native HDL or reconstituted HDL particles inhibited the production of cytokines in vivo and also in human mononuclear cells of peripheral blood.

Experimental treatment with intravenous infusions of HDL containing apo A-1$_{Milano}$ [149], the mutation of the apolipoprotein with incremented antiatherosclerotic properties, or with apo A-I mimetic synthetic peptides, has been tested with promising expectations. Some apo A-I mimetic peptides showed antiinflammatory effects in experimental systems. Therefore, the interference of atherogenesis proinflammatory mechanisms is also implied in this strategy to treat the disease [150–153].

Recently, it has been demonstrated that the bond between HDL and the SR-BI receptor activates the eNOS enzyme by means of intracellular mobilization of Ca++ and phosphorylation of eNOS into Ser1177, induced and measured by means of the phosphatidylinositol-3 (PI3K) and Akt kinase enzymes, promoting NO release by endothelial cells [154]. The vasoactive effects seem to relate to lysophospholipids carried by HDL and represent an interesting aspect of the atherogenic function of these biomolecules [154].

Clinically proving this data, it has been demonstrated that acute infusion of reconstituted HDL in hypercholesterolemic individuals was effective to improve the endothelial dysfunction that generally follows these cases. This effect was obtained by increasing NO bioavailability, clearly indicating that HDL has endothelium-protective effects by means of quick mechanisms [155].

Fig. 22.5A shows that endothelium-dependent arterial dilation is reduced in hypercholesterolemic individuals when compared to control individuals, but is reverted after recombinant HDL infusion. Fig. 22.5B shows that intraarterial infusion of L-NMMA (NG-monomethyl-L-arginine), an NO synthesis inhibitor, blocks the effects of recombinant HDL upon flow-mediated dilation (FMD), identifying higher availability of NO as the mechanism responsible for improving endothelial dysfunction. Fig. 22.5C shows FMD before and after infusion of recombinant HDL [155].

CETP Inhibitors and HDL

Increased serum concentration of CETP tends to reduce CETP deficiency to increase HDL-cholesterol, which makes this protein the therapeutic target to expand the HDL fraction. Several CETP chemical inhibitors were developed, including torcetrapib (Pfizer, USA), dalcetrapib (Roche, Switzerland), anacetrapibe (Merck, USA) and evacetrapibe (Lilly, USA). Torcetrapib, a 3,5-bis(trifluoromethyl) phenyl derivative, is a potent CETP activity inhibitor able to increase HDL-cholesterol levels by 72% and reduce LDL-cholesterol by 25%. However, despite of the lipid changes, it was unable to reduce carotid intima-media thickness in patients with familial hypercholesterolemia ad mixed dyslipidemia [156,157].

In 2006, the ILLUMINATE study was prematurely terminated due to adverse effects of the drugs, excessive deaths, and cardiovascular disease [158].

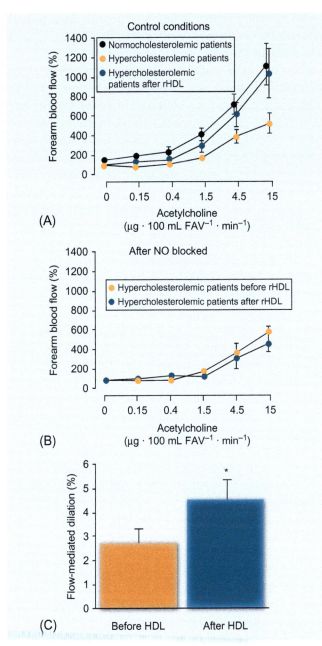

FIG. 22.5 (A) Endothelium-dependent arterial dilation is reduced in hypercholesterolemic patients when compared to control individuals, but is reverted after recombinant HDL infusion. (B) Intraarterial infusion of L-NMMA (NG-monomethyl-L-arginine), an NO synthesis inhibitor, blocks the effects of recombinant HDL upon flow-mediated dilation (FMD), identifying higher availability of NO as the mechanism responsible for improving endothelial dysfunction. (C) FMD before and after recombinant HDL infusion. *Adapted from Spieker LE, Sudano I, Hürlimann D, et al. High-density lipoprotein restores endothelial function in hypercholesterolemic men. Circulation 2002;105:1399–402.*

Further analysis of the study data revealed other undesirable effects of torcetrapib, including increased systemic blood pressure and sodium, bicarbonate, and aldosterone plasmatic levels, and reduced potassium levels [159,160]. Hypertension was a consequence of increased production of adrenal steroids such as aldosterone and cortisol.

Dalcetrapib, a benzenethiol derivative, was the first small molecule to show CETP inhibiting action, besides providing antiatherogenic effect in vivo [161]. It binds irreversibly to CETP, but differently from torcetrapib, dalcetrapib does not seem to induce the formation of CETP-HDL complex at therapeutic plasmatic concentrations [162]. Dalcetrapib is less potent that torcetrapib regarding HDL-cholesterol increase. In healthy individuals, daily treatment with 600 mg increased HDL-cholesterol by 23% after 4 weeks [163], and 28% in patients with familial hypercholesterolemia using pravastatin. reducing LDL-cholesterol levels by approximately 7% [164]. After 24 weeks, in daily doses of 900 mg, dalcetrapib was able to increase HDL-cholesterol levels by 33% in patients using atorvastatin, but LDL-cholesterol levels did not change [165]. Despite of the effects on HDL-cholesterol, dalcetrapib did not reduce cardiovascular events, and the clinical tests were interrupted in 2012.

Anacetrapib is another 3,5-bis(trifluoromethyl) phenyl derivative used as CETP inhibitor, forming a firm but reversible bond with the protein [166]. Anacetrapib inhibits cholesterol ester transference from HDL to LDL and HDL3 do HDL2 [166]. In normolipidemic individuals and dyslipidemic patients treated with atorvastatin, daily dose of 300 mg of anacetrapib were able to increased HDL-cholesterol levels by 130%, apo A-I by 47%, and reduced LDL-cholesterol levels by 40%. Treatment with anacetrapib did not increased systemic blood pressure nor aldosterone synthesis [167,168].

Anacetrapib can also improve HDL functions. HDL in patients treated with anacetrapib showed increased cholesterol efflux in foam cell cultures, regardless of HDL-cholesterol levels, and also maintained HDL antiinflammatory activity [169].

Phase III of the ongoing REVEAL study was designed to test if anacetrapib reduces the incidence of coronary events in 30,000 patients with established cardiovascular disease using statins, and its conclusion is expected for 2017 [170,171].

Evacetrapib, the most recent CETP inhibitor, is undergoing clinical tests. This is a new benzazepine compound with selective and potent CETP inhibiting activity [172]. In monotherapy, the drug increases HDL-cholesterol levels dose-dependent from 54% to 130%, and reduces LDL-cholesterol levels from 14% to 36% [173]. Combined with statin therapy, evacetrapib also increased HDL-cholesterol and reduced LDL-cholesterol levels by 83% and 13%, respectively. These tests did not find changes in blood pressure nor aldosterone and mineralocorticoid concentrations [173]. Evacetrapib clinical tests are currently being conducted.

CONCLUSIONS

As we have seen, while the main role of chylomicrons and VLDL is transporting fat absorbed in the intestine or produced by the liver in plasma for further lipolysis on the endothelium surface, and then storing it in tissues as adipose tissue, the role of HDL is mainly related to reverse cholesterol transport and homeostasis in plasma and the organism. It can be assumed that lipoproteins that contain apo B tend to stress the endothelium, while HDL has endothelial pavement reparation effects.

In principle, considering ideal plasmatic lipid regulation, the better- adapted organism would be the one able to remove lipoproteins containing apo B from circulation more quickly, maintaining them at low concentrations. Simultaneously, this organism would be able to maintain HDL-cholesterol at higher concentration, removing HDL more slowly or increasing lipoprotein formation.

As described, HDL protects the endothelium against almost all atherogenic processes identified so far, such as cholesterol deposit in macrophages, lipoprotein oxidation, inflammation, thrombogenesis, endothelial cell apoptosis, and others. Apo A-I, apo A-II, and several other proteins associated with the HDL fraction grant it several endothelium and organism protective properties.

The set of systems and processes that involve HDL and that we only recently began to understand leaves several questions open. For example, to what extent HDL-cholesterol increase favors each HDL action? For a given function of the lipoprotein, more HDL particles in circulation means the function is more amplified? In the first step of reverse transport, the cell cholesterol efflux, what is most important,

ABCA1 transporter efficiency or apo and nascent HDL properties in interacting with the transporter? And about the functional interfaces of HDL as CETP or PLTP, how therapeutic interventions in these transfer proteins, could be able to increase reverse cholesterol transport and other related functions? Or, on the contrary, could they impair these functions and arterial intimal layer protection? What about the possibility of increasing expression of several proteins whose functions are related to HDL protective properties, such as PON1? Or using miRNAs as a tool, since they are transported in HDL, and besides that, are able to modulate functions attributed to the lipoprotein?

Another range of questions refers to the state of HDL in hypertriglyceridemia and other dysfunctions originated in the circuit of lipoproteins that contain apo B that are reflected in HDL, leading to functional deficit. Another crucial question that also involves the entire lipid transport system,. including HDL, is related to insulin resistance states. Also intriguing is what happens to HDL when apo A-I is interchanged with SAA in infectious diseases and also when the lipoproteins becomes openly proatherogenic.

Considering that some drugs that reduce cholesterol and triglycerides, such as statins and fibrates, also increase HDL-cholesterol, the notorious cardiovascular benefit promoted by them may be, at least in part, due to the effects on HDL and related metabolic events, and that are not evaluated by measuring HDL-cholesterol. On the other hand, in patients with low HDL-cholesterol, we noted reduced removal of chylomicron remnants, which are atherogenic lipoproteins, suggesting chylomicron involvement in increased cardiovascular events associated with low HDL levels [134].

HDL can be seen as a kind of endogenous drug whose beneficial effects are more appreciated at every new function attributed to it. From another perspective, HDL is part of a much wider and multifaceted entity, a gigantic plasmatic pool where several proteins and miRNAs converge, with multiple functions able to protect the endothelium and the organism and that interlace with the chylomicron and VLDL circuit, from which they suffer crucial metabolic influences.

With its multiple actions, HDL becomes one of the attractive biochemical pathways that are open to research for longer and healthier survival. Certainly, strategies can only be tested if safe therapeutic means are obtained to achieve the several effects possible upon HDL. However, we cannot forget that it is part of a larger system that also encompasses lipoproteins that contain apo B. Meanwhile, quitting tobacco use, regular physical activity, reverting overweight and obesity, moderate alcohol consumption, and dietary and drug treatment of hyperlipidemia increase HDL-cholesterol and, most importantly, knowingly promote endothelium integrity and cardiovascular health.

References

[1] Subedi BH, Joshi PH, Jones SR, et al. Current guidelines for high-density lipoprotein cholesterol in therapy and future directions. Vasc Health Risk Manag 2014;10:205–16.

[2] Gordon DJ, Probstfield JL, Garrison RJ, et al. High-density lipoprotein cholesterol and cardiovascular disease. Four prospective American studies. Circulation 1989;79:8–15.

[3] Morris PB, Ballantyne CM, Birtcher KK, et al. Review of clinical practice guidelines for the management of LDL-related risk. J Am Coll Cardiol 2014;64:196–206.

[4] Chelland Campbell S, Moffatt RJ, Stamford BA. Smoking and smoking cessation—the relationship between cardiovascular disease and lipoprotein metabolism: a review. Atherosclerosis 2008;201:225–35.

[5] Gordon B, Chen S, Durstine JL. The effects of exercise training on the traditional lipid profile and beyond. Curr Sports Med Rep 2014;13:253–9.

[6] Matsumoto C, Miedema MD, Ofman P, et al. An expanding knowledge of the mechanisms and effects of alcohol consumption on cardiovascular disease. J Cardiopulm Rehabil Prev 2014;34:159–71.

[7] Tenenbaum A, Klempfner R, Fisman EZ. Hypertriglyceridemia: a too long unfairly neglected major cardiovascular risk factor. Cardiovasc Diabetol 2014;13:159.

[8] Barter PJ, Rye KA. Targeting high-density lipoproteins to reduce cardiovascular risk: what is the evidence? Clin Ther 2015;37:2716–31.

[9] Mabuchi H, Nohara A, Inazu A. Cholesteryl ester transfer protein (CETP) deficiency and CETP inhibitors. Mol Cells 2014;37:777–84.

[10] Arora S, Patra SK, Saini R. HDL-A molecule with a multi-faceted role in coronary artery disease. Clin Chim Acta 2016;452:66–81.

[11] Martin SS, Jones SR, Toth PP. High-density lipoprotein subfractions: current views and clinical practice applications. Trends Endocrinol Metab 2014;25:329–36.

[12] Segrest P, Harvey SC, Zannis V. Detailed molecular model of apolipoproteins A-I on the surface of high-density lipoproteins and its functional implications. Trends Cardiovasc Med 2000;10:246–52.

[13] Maric J, Kiss RS, Franklin V, et al. Intracellular lipidation of newly synthesized apolipoprotein A-I in primary murine hepatocytes. J Biol Chem 2005;280:39942–9.

[14] Castro GR, Fielding CJ. Early incorporation of cell-derived cholesterol into pre beta-migrating high-density lipoprotein. Biochemistry 1988;27:25–9.

[15] Tall AR, Sammett D, Vita GM, et al. Lipoprotein lipase enhances the cholesteryl ester transfer protein-mediated transfer of cholesteryl ester from high density lipoproteins to very low density lipoproteins. J Biol Chem 1984;259:9587–94.

[16] Czarnecka H, Yokoyama S. Regulation of cellular cholesterol efflux by lecithin: Cholesterol acyl transferase reaction through nonspecific lipid exchange. J Biol Chem 1996;271:2023–8.

[17] Rousset X, Vaisman B, Amar M, et al. Lecithin: cholesterol acyltransferase—from biochemistry to role in cardiovascular disease. Curr Opin Endocrinol Diabetes Obes 2009;16:163–71.

[18] Rosenson RS, Brewer Jr. HB, Ansell B, et al. Translation of high-density lipoprotein function into clinical practice: current prospects and future challenges. Circulation 2013;128:1256–67.

[19] Rosenson RS, Brewer HB, Chapman MJ, et al. HDL measures, particles heterogeneity, proposed nomenclature and relation to atherosclerotic cardiovascular events. Clin Chem 2011;57:392–410.

[20] Wróblewska M. The origin and metabolism of a nascent pre-ß high density lipoprotein involved in cellular cholesterol efflux. Acta Biochim Pol 2011;58:275–85.

[21] Rye KA, Barter PJ. Regulation of high-density lipoprotein metabolism. Circ Res 2014;1(14):143–56.

[22] Diffenderfer MR, Schaefer EJ. The composition and metabolism of large and small LDL. Curr Opin Lipidol 2014;25:221–6.

[23] Superko HR, Pendyala L, Williams PT, et al. High-density lipoprotein subclasses and their relationship to cardiovascular disease. J Clin Lipidol 2012;6:496–523.

[24] Martin SS, Khokhar AA, May HT, et al. HDL cholesterol subclasses, myocardial infarction, and mortality in secondary prevention: the Lipoprotein Investigators Collaborative. Eur Heart J 2015;36:22–30.

[25] Davis CE, Williams DH, Oganov RG, et al. Sex difference in high density lipoprotein cholesterol in six countries. Am J Epidemiol 1996;143:1100–6.

[26] Gardner CD, Tribble DL, Young DR, et al. Population frequency distributions of HDL, HDL(2), and HDL(3) cholesterol and apolipoproteins A-I and B in healthy men and women and associations with age, gender, hormonal status, and sex hormone use: the Stanford Five City Project. Prev Med 2000;31:335–45.

[27] Mascarenhas-Melo F, Sereno J, Teixeira-Lemos E, et al. Markers of increased cardiovascular risk in postmenopausal women: focus on oxidized-LDL and HDL subpopulations. Dis Markers 2013;35:85–96.

[28] Mei X, Atkinson D. Lipid-free apolipoprotein A-I structure: insights into HDL formation and atherosclerosis development. Arch Med Res 2015;46:351–60.

[29] Tailleux A, Duriez P, Fruchart JC, et al. Apolipoprotein A-II, HDL metabolism and atherosclerosis. Atherosclerosis 2002;164:1–13.

[30] Nanjee MN, Crouse JR, King JM, et al. Effects of intravenous infusion of lipid-free apo A-I in humans. Arterioscler Thromb Vasc Biol 1996;16:1203–14.

[31] Kunnen S, Van Eck M. Lecithin: cholesterol acyltransferase: old friend or foe in atherosclerosis? J Lipid Res 2012;53:1783–99.

[32] Wang S, Smith JD. ABCA1 and nascent HDL biogenesis. Biofactors 2014;40:547–54.

[33] Phillips MC. Molecular mechanisms of cellular cholesterol efflux. J Biol Chem 2014;289:24020–9.

[34] Fisher EA, Feig JE, Hewing B, et al. High-density lipoprotein function, dysfunction, and reverse cholesterol transport. Arterioscler Thromb Vasc Biol 2012;32:2813–20.

[35] Barter PJ, Rye KA. Cholesteryl ester transfer protein inhibition as a strategy to reduce cardiovascular risk. J Lipid Res 2012;53:1755–66.

[36] Maranhão RC, Freitas FR. HDL metabolism and atheroprotection: predictive value of lipid transfers. Adv Clin Chem 2014;65:1–41.

[37] Ashen MD, Blumenthal RS. Clinical practice. Low HDL cholesterol levels. N Engl J Med 2005;353:1252–60.

[38] Westerterp M, Bochem AE, Yvan-Charvet L, et al. ATP-binding cassette transporters, atherosclerosis, and inflammation. Circ Res 2014;114:157–70.

[39] Fredrickson DS. The inheritance of high density lipoprotein deficiency (Tangier disease). J Clin Invest 1964;43:228.

[40] Rust S, Rosier M, Funke H, et al. Tangier disease is caused by mutations in the gene encoding ATP-binding cassette transporter 1. Nat Genet 1999;22:352–5.

[41] Wang N, Lan D, Chen W, et al. ATP-binding cassette transporters G1 and G4 mediate cellular cholesterol efflux to high-density lipoproteins. Proc Natl Acad Sci U S A 2004;101:9774–9.

[42] Terasaka N, Wang N, Yvan-Charvet L, et al. High-density lipoprotein protects macrophages from oxidized low-density lipoprotein-induced apoptosis by promoting efflux of 7-ketocholesterol via ABCG1. Proc Natl Acad Sci U S A 2007;104:15093–8.

[43] Kennedy MA, Barrera GC, Nakamura K, et al. ABCG1 has a critical role in mediating cholesterol eâiux to HDL and preventing cellular lipid accumulation. Cell Metab 2005;1:121–31.

[44] Drayna D, Jarnagin AS, McLean J, et al. Cloning and sequencing of human cholesteryl ester transfer protein cDNA. Nature 1987;327:632–4.

[45] Tall AR. Plasma lipid transfer proteins. Annu Rev Biochem 1995;64:235–57.

[46] Barter PJ. CETP and atherosclerosis. Arterioscler Thromb Vasc Biol 2000;20:2029–31.

[47] Barter PJ, Hopkins CJ, Calver GD. Transfers and exchanges of esterified cholesterol between plasma lipoproteins. Biochem J 1982; 208:1–7.

[48] Marcel YL, McPherson M, Hogue H, et al. Distribution and concentration of cholesteryl ester transfer protein in plasma of normolipemic subjects. J Clin Invest 1990;85:10–7.

[49] Gautier T, Masson D, de Barros JP, et al. Human apolipoprotein CI accounts for the ability of plasma high density lipoproteins to inhibit the cholesteryl ester transfer protein activity. J Biol Chem 2000;275:37504–9.

[50] Savel J, Lafitte M, Pucheu Y, et al. Molecular cloning low levels of HDL-cholesterol and atherosclerosis, a variable relationship—a review of LCAT deficiency. Vasc Health Risk Manag 2012;8:357–61.

[51] Chapman MJ, Le Goff W, Guerin M, et al. Cholesteryl ester transfer protein: at the heart of the action of lipid-modulating therapy with statins, fibrates, niacin, and cholesteryl ester transfer protein inhibitors. Eur Heart J 2010;31:149–64.

[52] Williams KJ, Tabas I. The response-to-retention hypothesis of early atherogenesis. Arterioscler Thromb Vasc Biol 1995;15:551–61.

[53] Chung BH, Segrest JP, Franklin F. In vitro production of beta-very low density lipoproteins and small, dense low density lipoproteins in mildly hypertriglyceridemic plasma: role of activities of lecithin: cholesterol acyl transferase, cholesteryl ester transfer protein and lipoprotein lipase. Atherosclerosis 1998;141:209–25.

[54] Newnham HH, Barter PJ. Synergistic effects of lipid transfers and hepatic lipase in the formation of very small high density lipoproteins during incubation of human plasma. Biochim Biophys Acta 1990;1044:57–64.

[55] Liang HQ, Rye KA, Barter PJ. Dissociation of lipid-free apolipoprotein A-I from high density lipoproteins. J Lipid Res 1994;35:1187–99.

[56] Rye KA, Hime NJ, Barter PJ. Evidence that CETP-mediated reductions in reconstituted high density lipoprotein size involve particle fusion. J Biol Chem 1997;272:5953–60.

[57] Mann CJ, Yen FT, Grant AM, et al. Mechanism of cholesteryl ester transfer in hypertriglyceridemia. J Clin Invest 1991;88:2059–66.

[58] Barter PJ, Jones ME. Kinetic studies of the transfer of esterified cholesterol between human-plasma low and high-density lipoproteins. J Lipid Res 1980;21:238–49.

[59] Ihm J, Quinn DM, Busch SJ, et al. Kinetics of plasma protein-catalyzed exchange of phosphatidylcholine and cholesteryl ester between plasma lipoproteins. J Lipid Res 1982;23:1328–41.

[60] Zhang L, Yan F, Zhang S, et al. Structural basis of transfer between lipoproteins by cholesteryl ester transfer protein. Nat Chem Biol 2012;8:342–9.

[61] Gomez Rosso L, Benitez MB, Fornari MC, et al. Alterations in cell adhesion molecules and other biomarkers of cardiovascular disease in patients with metabolic syndrome. Atherosclerosis 2008;199:415–23.

[62] Coniglio RI, Merono T, Montiel H, et al. HOMA-IR and non-HDL-C as predictors of high cholesterol ester transfer protein activity in patients at risk for type 2 diabetes. Clin Biochem 2012;45:566–70.

[63] Jonker JT, Wang Y, de Haan W, et al. Pioglitazone decreases plasma cholesteryl ester transfer protein mass, associated with a decrease in hepatic triglyceride content, in patients with type 2 diabetes. Diabetes Care 2010;33:1625–8.

[64] Oliveira HC, de Faria EC. Cholesteryl ester transfer protein: the controversial relation to atherosclerosis. IUBMB Life 2011;63:248–57.

[65] Khera AV, Cuchel M, de la Llera-Moya M, et al. Cholesterol efflux capacity, high-density lipoprotein function, and atherosclerosis. N Engl J Med 2011;364:127–35.

[66] Mineo C, Shaul PW. Role of high-density lipoprotein and scavenger receptor B type I in the promotion of endothelial repair. Trends Cardiovasc Med 2007;17:156–61.

[67] Tran-Dinh A, Diallo D, Delbosc S, et al. HDL and endothelial protection. Br J Pharmacol 2013;169:493–511.

[68] de Souza JA, Vindis C, Nègre-Salvayre A, et al. Small, dense HDL 3 particles attenuate apoptosis in endothelial cells: pivotal role of apolipoprotein A-I. J Cell Mol Med 2010;14:608–20.

[69] Navab M, Berliner JA, Subbanagounder G, et al. HDL and the inflammatory response induced by LDL-derived oxidized phospholipids. Arterioscler Thromb Vasc Biol 2001;21:481–8.

[70] Annema W, von Eckardstein A. High-density lipoproteins. Multifunctional but vulnerable protections from atherosclerosis. Circ J 2013;77:2432–48.

[71] Kontush A, Chantepie S, Chapman MJ. Small, dense HDL particles exert potent protection of atherogenic LDL against oxidative stress. Arterioscler Thromb Vasc Biol 2003;23:1881–8.

[72] Carreón-Torres E, Rendón-Sauer K, Monter Garrido M, et al. Rosiglitazone modifies HDL structure and increases HDL-apo AI synthesis and catabolism. Clin Chim Acta 2009;401:37–41.

[73] Deakin SP, Bioletto S, Bochaton-Piallat ML, et al. HDL-associated paraoxonase-1 can redistribute to cell membranes and influence sensitivity to oxidative stress. Free Radic Biol Med 2011;50:102–9.

[74] Précourt LP, Amre D, Denis MC, et al. The three-gene paraoxonase family: physiologic roles, actions and regulation. Atherosclerosis 2011;214:20–36.

[75] Aviram M, Hardak E, Vaya J, et al. Human serum paraoxonases (PON1) Q and R selectively decrease lipid peroxides in human coronary and carotid atherosclerotic lesions: PON1 esterase and peroxidase-like activities. Circulation 2000;101:2510–7.

[76] Kunutsor SK, Bakker SJ, James RW, et al. Serum paraoxonase-1 activity and risk of incident cardiovascular disease: the PREVEND study and meta-analysis of prospective population studies. Atherosclerosis 2015;245:143–54.

[77] Sozer V, Himmetoglu S, Korkmaz GG, et al. Paraoxonase, oxidized low density lipoprotein, monocyte chemoattractant protein-1 and adhesion molecules are associated with macrovascular complications in patients with type 2 diabetes mellitus. Minerva Med 2014;105:237–44.

[78] Zhu Y, Huang X, Zhang Y, et al. Anthocyanin supplementation improves HDL-associated paraoxonase 1 activity and enhances cholesterol efflux capacity in subjects with hypercholesterolemia. J Clin Endocrinol Metab 2014;99:561–9.

[79] Riwanto M, Rohrer L, Roschitzki B, et al. Altered activation of endothelial anti- and proapoptotic pathways by high-density lipoprotein from patients with coronary artery disease: role of high-density lipoprotein-proteome remodeling. Circulation 2013;127:891–904.

[80] Vuilleumiera N, Dayer JM, Eckardstein A, et al. Pro- or anti-inflammatory role of apolipoprotein A-1 in high-density lipoproteins? Swiss Med Wkly 2013;143:1–12.

[81] Brewer Jr HB. The evolving role of HDL in the treatment of high-risk patients with cardiovascular disease. J Clin Endocrinol Metab 2011;96:1246–57.

[82] Mineo C, Shaul PW. Novel biological functions of high-density lipoprotein cholesterol. Circ Res 2012;111:1079–90.

[83] Ossoli A, Remaley AT, Vaisman B, et al. Plasma-derived and synthetic high-density lipoprotein inhibit tissue factor in endothelial cells and monocytes. Biochem J 2016;473:211–9.

[84] Nofer JR, Brodde MF, Kehrel BE. High-density lipoproteins, platelets and the pathogenesis of atherosclerosis. Clin Exp Pharmacol Physiol 2010;37:726–35.

[85] Brodde MF, Korporaal SJ, Herminghaus G, et al. Native high-density lipoproteins inhibit platelet activation via scavenger receptor BI: role of negatively charged phospholipids. Atherosclerosis 2011;215:374–82.

[86] Zqbczyk M, Hondo L, Krzek M, et al. High-density cholesterol and apolipoprotein AI as modifiers of plasma fibrin clot properties in apparently healthy individuals. Blood Coagul Fibrinolysis 2013;24:50–4.

[87] Kaba NK, Francis CW, Moss AJ, et al. Effects of lipids and lipid-lowering therapy on hemostatic factors in patients with myocardial infarction. J Thromb Haemost 2004;2:718–25.

[88] Asselbergs FW, Williams SM, Hebert PR, et al. Gender-specific correlations of plasminogen activator inhibitor-1 and tissue plasminogen activator levels with cardiovascular disease-related traits. J Thromb Haemost 2007;5:313–20.

[89] van der Stoep M, Korporaal SJ, Van Eck M. High-density lipoprotein as a modulator of platelet and coagulation responses. Cardiovasc Res 2014;103:362–71.

[90] Van Lenten BJ, Hama SY, de Beer FC, et al. Anti-inflammatory HDL becomes pro-inflammatory during the acute phase response. Loss of protective effect of HDL against LDL oxidation in aortic wall cell cocultures. J Clin Invest 1995;96:2758–67.

[91] Ma J, Liao XL, Lou B, et al. Role of apolipoprotein A-I in protecting against endotoxin toxicity. Acta Biochim Biophys Sin 2004;36:419–24.

[92] Pajkrt JE, Doran F, Koster Lerch PG, et al. Anti-inflammatory effects of reconstituted high density lipoprotein during human endotoxemia. J Exp Med 1996;184:1601–8.

[93] Stoll LL, Denning GM, Weintraub NL. Potential role of endotoxin as a proinflammatory mediator of atherosclerosis. Arterioscler Thromb Vasc Biol 2004;24:2227–36.

[94] Rizzo M, Otvos J, Nikolic D, et al. Subfractions and subpopulations of HDL: an update. Curr Med Chem 2014;21:2881–91.

[95] Campbell LA, Rosenfeld ME. Infection and atherosclerosis development. Arch Med Res 2015;46:339–50.

[96] Karliner JS. Sphingosine kinase and sphingosine 1-phosphate in the heart: a decade of progress. Biochim Biophys Acta 1831;2013:203–12.

[97] Sattler K, Levkau B. Sphingosine-1-phosphate as a mediator of high-density lipoprotein effects in cardiovascular protection. Cardiovasc Res 2009;82:201–11.

[98] Levkau B. HDL-S1P: cardiovascular functions, disease-associated alterations, and therapeutic applications. Front Pharmacol 2015;6:243.

[99] Zhang QH, Zu XY, Cao RX, et al. An involvement of SR-B1 mediated PI3K-Akt-eNOS signaling in HDL-induced cyclooxygenase 2 expression and prostacyclin production in endothelial cells. Biochem Biophys Res Commun 2012;420:17–23.

[100] Sorci-Thomas MG, Thomas MJ. High density lipoprotein biogenesis, cholesterol efflux, and immune cell function. Arterioscler Thromb Vasc Biol 2012;32:2561–5.

[101] Bartel DP. MicroRNAs: genomics, biogenesis, mechanism, and function. Cell 2004;1(16):281–97.

[102] Friedman RC, Farh KK, Burge CB, et al. Most mammalian mRNAs are conserved targets of microRNAs. Genome Res 2009;19:92–105.

[103] Canfrán-Duque A, Ramírez CM, Goedeke L, et al. microRNAs and HDL life cycle. Cardiovasc Res 2014;103:414–22.

[104] Niculescu LS, Simionescu N, Sanda GM, et al. miR-486 and miR-92a identified in circulating HDL discriminate between stable and vulnerable coronary artery disease patients. PLoS One 2015;10:e0140958.

[105] Creemers EE, Tijsen AJ, Pinto YM. Circulating microRNAs: novel biomarkers and extracellular communicators in cardiovascular disease? Circ Res 2012;110:483–95.

[106] Vickers KC, Palmisano BT, Shoucri BM, et al. MicroRNAs are transported in plasma and delivered to recipient cells by high-density lipoproteins. Nat Cell Biol 2011;13:423–33.

[107] Tabet F, Vickers KC, Cuesta Torres LF, et al. HDL-transferred microRNA-223 regulates ICAM-1 expression in endothelial cells. Nat Commun 2014;5:3292.

[108] Rayner KJ, Suárez Y, Dávalos A, et al. miR-33 contributes to the regulation of cholesterol homeostasis. Science 2010;328:1570–3.

[109] Cochran BJ, Bisoendial RJ, Hou L, et al. Apolipoprotein A-I increases insulin secretion and production from pancreatic b-cells via a G-protein-cAMP-PKA-FoxO1-dependent mechanism. Arterioscler Thromb Vasc Biol 2014;34:2261–7.

[110] Vaisar T, Pennathur S, Green PS, et al. Shotgun proteomics implicates protease inhibition and complement activation in the anti-inflammatory properties of HDL. J Clin Invest 2007;117(3):746–56.

[111] Eren E, Yilmaz N, Aydin O. Functionally defective high-density lipoprotein and paraoxonase: a couple for endothelial dysfunction in atherosclerosis. Cholesterol 2013;2013:792090.

[112] Annema W, von Eckardstein A, Kovanen PT. HDL and athero-thrombotic vascular disease. Handb Exp Pharmacol 2015;224:369–403.

[113] Otocka-Kmiecik A, Mikhailidis DP, Nicholls SJ, et al. Dysfunctional HDL: a novel important diagnostic and therapeutic target in cardiovascular disease? Prog Lipid Res 2012;51:314–24.

[114] Huang Y, Wu Z, Riwanto M, et al. Myeloperoxidase, paraoxonase-1, and HDL form a functional ternary complex. J Clin Invest 2013;123:3815–28.

[115] deGoma EM, deGoma RL, Rader DJ. Beyond high-density lipoprotein cholesterol levels evaluating high-density lipoprotein function as influenced by novel therapeutic approaches. J Am Coll Cardiol 2008;51:2199–211.

[116] de la Llera Moya M, Atger V, Paul JL, et al. A cell culture system for screening human serum for ability to promote cellular cholesterol efflux. Relations between serum components and efflux, esterification, and transfer. Arterioscler Thromb 1994;14:1056–65.

[117] Zhang Y, Zanotti I, Reilly MP, et al. Overexpression of apolipoprotein A-I promotes reverse transport of cholesterol from macrophages to feces in vivo. Circulation 2003;108:661–3.

[118] Clay MA, Pyle DH, Rye KA, et al. Time sequence of the inhibition of endothelial adhesion molecule expression by reconstituted high density lipoproteins. Atherosclerosis 2001;157:23–9.

[119] Ragbir S, Farmer JA. Dysfunctional high-density lipoprotein and atherosclerosis. Curr Atheroscler Rep 2010;12:343–8.

[120] Riwanto M, Landmesser U. High-density lipoprotein structure, function, and metabolism high density lipoproteins and endothelial functions: mechanistic insights and alterations in cardiovascular disease. J Lipid Res 2013;54:3227–43.

[121] Yuhanna IS, Zhu Y, Cox BE, et al. High-density lipoprotein binding to scavenger receptor-BI activates endothelial nitric oxide synthase. Nat Med 2001;7:853–7.

[122] Knetsch ML, Aldenhoff YB, Koole LH. The effect of high-density-lipoprotein on thrombus formation on and endothelial cell attachment to biomaterial surfaces. Biomaterials 2006;27:2813–9.

[123] Yu R, Yekta B, Vakili L, et al. Proatherogenic high-density lipoprotein, vascular inflammation, and mimetic peptides. Curr Atheroscler Rep 2008;10:171–6.

[124] Shah PK, Chyu KY. Apolipoprotein A-I mimetic peptides: potential role in atherosclerosis management. Trends Cardiovasc Med 2005;15:291–6.

[125] Navab M, Anantharamaiah GM, Reddy ST, et al. Mechanisms of disease: proatherogenic HDL—an evolving field. Nat Clin Pract Endocrinol Metab 2006;2:504–11.

[126] Tolle M, Huang T, Schuchardt M, et al. High-density lipoprotein loses its anti-inflammatory capacity by accumulation of pro-inflammatory-serum amyloid A. Cardiovasc Res 2012;94:154–62.

[127] Dullaart RP, de Boer JF, Annema W, et al. The inverse relation of HDL anti-oxidative functionality with serum amyloid A is lost in metabolic syndrome. Obesity 2013;21:361–6.

[128] Sprandel MC, Hueb WA, Segre A, et al. Alterations in lipid transfers to HDL associated with the presence of coronary artery disease in patients with type 2 diabetes mellitus. Cardiovasc Diabetol 2015;14:1–9.

[129] Maranhão RC, Freitas FR, Strunz CM, et al. Lipid transfers to HDL are predictors of precocious clinical coronary heart disease. Clin Chim Acta 2012;413:502–5.

[130] World Health Organization. Definition, diagnosis and classification of diabetes mellitus and its complications. Report of a WHO consultation. Geneva: World Health Organization; 1999. p. 1.

[131] Casella-Filho A, Chagas AC, Maranhão RC, et al. Effect of exercise training on plasma levels and functional properties of high-density lipoprotein cholesterol in the metabolic syndrome. Am J Cardiol 2011;107:1168–72.

[132] Borggreve SE, De Vries R, Dullaart RP. Alterations in high-density lipoprotein metabolism and reverse cholesterol transport in insulin resistance and type 2 diabetes mellitus: role of lipolytic enzymes, lecithin: cholesterol acyltransferase and lipid transfer proteins. Eur J Clin Invest 2003;33:1051–69.

[133] Sorrentino SA, Besler C, Rohrer L, et al. Endothelial-vasoprotective effects of high-density lipoprotein are impaired in patients with type 2 diabetes mellitus but are improved after extended-release niacin therapy. Circulation 2010;121:110–22.

[134] Benjo AM, Maranhão RC, Coimbra S, et al. Accumulation of chylomicron remnants and impaired vascular reactivity occur in subjects with isolated low HDL cholesterol: effects of niacin treatment. Atherosclerosis 2006;187:116–22.

[135] Crouse JR, Kastelein J, Isaacsohn J, et al. A large, 36 week study of the HDL-C raising effects and safety of simvastatin versus atorvastatin. Atherosclerosis 2000;151:8–9.

[136] Hunninghake DB, Stein EA, Bays HE, et al. Rosuvastatin improves the atherogenic and atheroprotective lipid profiles in patients with hypertriglyceridemia. Coron Artery Dis 2004;15:115–23.

[137] Kersten S, Desvergne B, Wahli W. Roles of PPARs in health and disease. Nature 2000;405:421–4.

[138] Rotllan N, Llaverias G, Julve J, et al. Differential effects of gemfibrozil and fenofibrate on reverse cholesterol transport from macrophages to feces in vivo. Biochim Biophys Acta 2011;1811:104–10.

[139] Dujovne CA, Ettinger MP, McNeer JF, et al. Efficacy and safety of a potent new selective cholesterol absorption inhibitor, ezetimibe, in patients with primary hypercholesterolemia. Am J Cardiol 2002;90:1092–7.

[140] Araújo RG, Casella Filho A, Chagas AC. Ezetimibe—pharmacokinetics and therapeutics. Arq Bras Cardiol 2005;85:20–4.

[141] Goldberg RB, Guyton JR, Mazzone T, et al. Ezetimibe/simvastatin vs atorvastatin in patients with type 2 diabetes mellitus and hypercholesterolemia: the VYTAL study. Mayo Clin Proc 2006;81:1579–88.

[142] Bell DA, Hooper AJ, Watts GF, et al. Mipomersen and other therapies for the treatment of severe familial hypercholesterolemia. Vasc Health Risk Manag 2012;8:651–9.

[143] Seidah NG, Awan Z, Chrétien M, et al. PCSK9: a key modulator of cardiovascular health. Circ Res 2014;114:1022–36.

[144] Reiner Ž. PCSK9 inhibitors—past, present and future. Expert Opin Drug Metab Toxicol 2015;11:1517–21.

[145] Cockerill GW, Rye KA, Gamble JR, et al. High-density lipoproteins inhibit cytokine-induced expression of endothelial cell adhesion molecules. Arterioscler Thromb Vasc Biol 1995;15:1987–94.

[146] Tang C, Liu Y, Kessler PS, et al. The macrophage cholesterol exporter ABCA1 functions as an anti-inflammatory receptor. J Biol Chem 2009;284:32336–43.

[147] De Nardo D, Labzin LI, Kono H, et al. High-density lipoprotein mediates anti-inflammatory reprogramming of macrophages via the transcriptional regulator ATF3. Nat Immunol 2014;15:152–60.

[148] Moore KJ, Fisher EA. High-density lipoproteins put out the fire. Cell Metab 2014;19:175–6.

[149] Chyu KY, Shah PK. HDL/ApoA-1 infusion and ApoA-1 gene therapy in atherosclerosis. Front Pharmacol 2015;6:187.

[150] Stoekenbroek RM, Stroes ES, Hovingh GK. ApoA-I mimetics. Handb Exp Pharmacol 2014;224:631–48.

[151] Chenevard R, Hürlimann D, Spieker L, et al. Reconstituted HDL in acute coronary syndromes. Cardiovasc Ther 2012;30:e51–7.

[152] Uehara Y, Chiesa G, Saku K. High-density lipoprotein-targeted therapy and apolipoprotein A-I mimetic peptides. Circ J 2015;79:2523–8.

[153] Ibanez B, Giannarelli C, Cimmino G, et al. Recombinant HDL(Milano) exerts greater anti-inflammatory and plaque stabilizing properties than HDL(wild-type). Atherosclerosis 2012;220:72–7.

[154] Nofer JR, van der Giet M, Tölle M, et al. HDL induces NO-dependent vasorelaxation via the lysophospholipid receptor S1P3. J Clin Invest 2004;1(13):569–81.

[155] Spieker LE, Sudano I, Hürlimann D, et al. High-density lipoprotein restores endothelial function in hypercholesterolemic men. Circulation 2002;105:1399–402.

[156] Bots ML, Visseren FL, Evans GW, et al. Torcetrapib and carotid intima-media thickness in mixed dyslipidaemia (RADIANCE 2 study): a randomised, double-blind trial. Lancet 2007;370:153–60.

[157] Kastelein JJ, van Leuven SI, Burgess L, et al. Effect of torcetrapib on carotid atherosclerosis in familial hypercholesterolemia. N Engl J Med 2007;356:1620–30.

[158] Barter PJ, Caulfield M, Eriksson M, et al. Effects of torcetrapib in patients at high risk for coronary events. N Engl J Med 2007;357:2109–22.

[159] Clerc RG, Stauffer A, Weibel F, et al. Mechanisms underlying off-target effects of the cholesteryl ester transfer protein inhibitor torcetrapib involve L-type calcium channels. J Hypertens 2010;28:1676–86.

[160] Hu X, Dietz JD, Xia C, et al. Torcetrapib induces aldosterone and cortisol production by an intracellular calcium-mediated mechanism independently of cholesteryl ester transfer protein inhibition. Endocrinology 2009;150:2211–9.

[161] Okamoto H, Yonemori F, Wakitani K, et al. A cholesteryl ester transfer protein inhibitor attenuates atherosclerosis in rabbits. Nature 2000;406:203–7.

[162] Niesor EJ, von der Marck E, Brousse M, et al. Inhibition of cholesteryl ester transfer protein (CETP): different in vitro characteristics of RO4607381/JTT-705 and torcetrapib (TOR). Atherosclerosis 2008;199:231.

[163] de Grooth GJ, Kuivenhoven JA, Stalenhoef AF, et al. Efficacy and safety of a novel cholesteryl ester transfer protein inhibitor, JTT-705, in humans: a randomized phase II dose-response study. Circulation 2002;105:2159–65.

[164] Kuivenhoven JA, de Grooth GJ, Kawamura H, et al. Effectiveness of inhibition of cholesteryl ester transfer protein by JTT-705 in combination with pravastatin in type II dyslipidemia. Am J Cardiol 2005;95:1085–8.

[165] Stein EA, Roth EM, Rhyne JM, et al. Safety and tolerability of dal-cetrapib (RO4607381/JTT-705): results from a 48-week trial. Eur Heart J 2010;31:480–8.

[166] Ranalletta M, Bierilo KK, Chen Y, et al. Biochemical characterization of cholesteryl ester transfer protein inhibitors. J Lipid Res 2010;51:2739–52.

[167] Krishna R, Anderson MS, Bergman AJ, et al. Effect of the cholesteryl ester transfer protein inhibitor, anacetrapib, on lipoproteins in patients with dyslipidaemia and on 24-h ambulatory blood pressure in healthy individuals: two double-blind, randomized placebo-controlled phase I studies. Lancet 2007;370:1907–14.

[168] Bloomfield D, Carlson GL, Sapre A, et al. Efficacy and safety of the cholesteryl ester transfer protein inhibitor anacetrapib as monotherapy and coadministered with atorvastatin in dyslipidemic patients. Am Heart J 2009;157:352–60.

[169] Yvan-Charvet L, Kling J, Pagler T, et al. Cholesterol efflux potential and antiinflammatory properties of high-density lipoprotein after treatment with niacin or anacetrapib. Arterioscler Thromb Vasc Biol 2010;30:1430–8.

[170] Gutstein DE, Krishna R, Johns D, et al. Anacetrapib, a novel CETP inhibitor: pursuing a new approach to cardiovascular risk reduction. Clin Pharmacol Ther 2012;91:109–22.

[171] Shinkai H. Cholesteryl ester transfer-protein modulator and inhibitors and their potential for the treatment of cardiovascular diseases. Vasc Health Risk Manag 2012;8:323–31.

[172] Cao G, Beyer TP, Zhang Y, et al. Evacetrapib is a novel, potent, and selective inhibitor of cholesteryl ester transfer protein that elevates HDL cholesterol without inducing aldosterone or increasing blood pressure. J Lipid Res 2011;52:2169–76.

[173] Nicholls SJ, Brewer HB, Kastelein JJ, et al. Effects of the CETP inhibitor evacetrapib administered as monotherapy or in combination with statin on HDL and LDL cholesterol. A randomized controlled trial. JAMA 2011;306:2099–109.

Further Reading

[1] Wang F, Gu HM, Zhang DW. Caveolin-1 and ATP binding cassette transporter A1 and G1-mediated cholesterol efflux. Cardiovasc Hematol Disord Drug Targets 2014;14:142–8.

[2] Nair DR, Nair A, Jain A. HDL genetic defects. Curr Pharm Des 2014;20:6230–7.

23

Artificial Lipoproteins in Endothelial Dysfunction and Atherosclerosis

Raul Cavalcante Maranhão

INTRODUCTION

One of the ways of studying nature is simulating it. Lipoproteins are, in fact, natural emulsions consisting of lipids and proteins. The strategy used as a starting point to explore lipid metabolism in plasma was to produce artificial emulsions that mimic the composition and structure of lipoproteins.

These artificial emulsions, produced in a laboratory without the protein part of lipoproteins, have an invaluable advantage: a single preparation, marked with radioisotopes of other methods, allows conducting safe plasmatic kinetics studies in a large number of individuals. This is important because, if natural lipoproteins isolated from the blood of one person were injected in another individual, there is a risk of transmitting viruses such as HIV or hepatitis and causing immune reactions. Therefore, this procedure is ethically forbidden. On the other hand, injecting the fractions obtained from the serum of the person itself is a very laborious and nonfunctional process.

The study of chylomicron metabolism exemplifies the application of this methodological approach. Chylomicrons are lipoproteins that transport diet fats in the lymph and blood. Obtaining samples from experimental animals requires lymph cannulation and long harvest period, a laborious and complex process. In humans, it is difficult to isolate chylomicrons in plasma after fatty meals, since chylomicrons have physical-chemical characteristics that overlap those of very low-density lipoproteins (VLDL). Since artificial chylomicrons are injected

intravenously, in a pool, the intestinal absorption component is suppressed, which ensures great advantage over other methods based on ingesting a fat load followed by postprandial blood sample collection, which requires a longer observation period and suffers interference from intestinal absorption speed variations from person to person.

ARTIFICIAL CHYLOMICRONS

Chylomicrons constitute one of the four great classes of lipoproteins, together with VLDL, low-density lipoproteins (LDL), and high-density lipoproteins (HDL). Chylomicron and VLDL particles have very similar composition and are known as the triglyceride-rich proteins (TRLP). Chylomicrons are synthesized in the intestine, from lipids such as diet fats, phospholipids, and cholesterol absorbed by enterocytes. VLDLs are synthesized in the liver [1] (Fig. 23.1).

In systemic circulation, TRLPs in contact with capillary endothelial surface are degraded by lipoprotein lipase, as this enzyme is more commonly known. The bond between chylomicrons and VLDL with lipases is made by apolipoprotein (apo) CII, one of the proteins present on the surface of lipoprotein particles. Besides binding with the enzyme (apo), CII also stimulates TRLP triglyceride hydrolysis. After suffering lipase action, degraded TRLPs decouple from the enzyme, binding again, downstream, with another lipase molecule, in a continuous lipolysis process. The resulting particles,

319

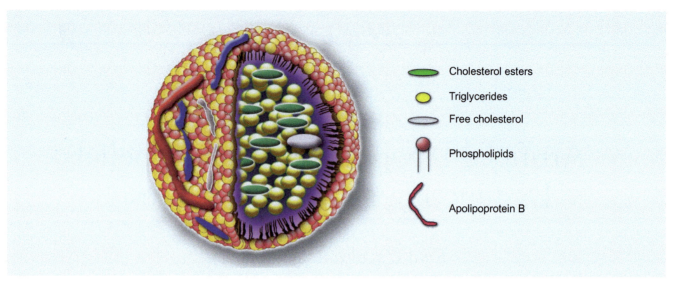

FIG. 23.1   Structure of triglyceride-rich lipoproteins (TRLP—chylomicrons and VLDL). One monolayer of phospholipids, with free cholesterol, involves the triglycerides, which constitute most of the particle. Cholesterol ester molecules are also in the core part of TRLPs.

therefore depleted from triglycerides, are called chylomicron remnants and, in the case of VLDL, intermediate density lipoproteins (IDL) and LDL. LDL is the final product of VLDL degradation, in which the proportion of triglycerides in particles is only residual. The chylomicron remnants are finally captured by cells, mainly in the liver, by means of cell membrane receptors that recognize apo E, and by other mechanisms [2–5] (Fig. 23.2).

TRLP lipolysis by lipoprotein lipase on the endothelium surface is essential for the organism. By means of this mechanism, diet fats absorbed in the intestine and transported in the lymph and then in blood, packed in chylomicron particles, are broken

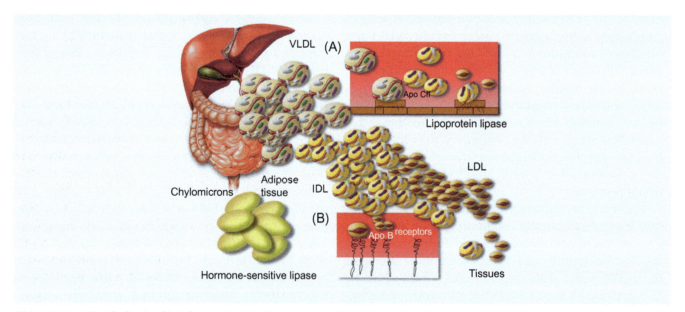

FIG. 23.2   Metabolism of triglyceride-rich lipoproteins (TRLP—chylomicrons and VLDL). After entering systemic circulation, the lipoproteins synthesized in the intestine (chylomicrons) or in the liver (VLDL) go through two main processes: (A) lipolysis, or particle triglycerides breakdown, carried out by lipoprotein lipase, an enzyme bounded to the endothelium by proteoglycans and that is stimulated by the apo CII present in TRLPs. (B) Chylomicron or VLDL (LDL) remnants, which are generated, decouple from the enzyme after substantial depletion of the triglyceride content, and are taken up by the liver, mainly by receptors that recognize apo E or apo B, and other mechanisms.

on the endothelium surface. Triglyceride hydrolysis is complete, until it generates free fatty acids (FFAs) and glycerol. FFAs and glycerol are absorbed and stored in adipose tissue and muscle. In cytoplasm, triglycerides are resynthesized from recently absorbed FFAs and glycerol. These constitute large organism energy stores, which allow survival for extended fasting periods. During fasting, the organism begins to break adipose tissue and muscle fats to generate energy. This occurs due to activation of hormone-sensitive lipase, an intracellular enzyme. Cell FFAs are transported to the liver bound to albumin.

Chylomicron remnants, generated by lipoprotein lipase activity, may take part in atherogenesis, a hypothesis that was explained by Zilversmit in the postprandial atherosclerosis theory [6].

In the laboratory, we synthesized lipid emulsions consisting of nanospheres with a chemical composition similar to chylomicrons, but without the apolipoproteins. The method used to prepare the emulsions was based on ultrasonic radiation and gradient centrifugation of lipid mixtures containing phospholipids, both cholesterol forms (esterified and nonesterified), and triglycerides in defined proportions. Emulsion particles consist predominantly of triglycerides, involved by a monolayer of phospholipids [1,2].

When the emulsions were injected in rats, we observed that they acquired all apolipoproteins in contact with natural lipoproteins, except for apo B. This occurs because apo B is a very large and hydrophobic molecule that cannot move from the natural to the artificial particle. One of the apo acquired was apo CII, which after adhering to the surface of artificial particles is now able to bind to the lipoprotein lipase on the endothelium surface and stimulate artificial chylomicron triglyceride breakdown by the enzyme. Another apo acquired, apo E, allows artificial chylomicron remnants depleted from triglycerides to bind to hepatic receptors, such as LDL receptor and others. After binding to the receptors, the remnants are captured by the cells, mainly by hepatocytes [1,2,7–10].

It is interesting to note that the natural and artificial chylomicron remnant particles are depleted from triglycerides, since most of the compounds is removed from the particles by the action of lipoprotein lipase. However, the cholesterol content, mainly in the form of cholesterol esters, is maintained inside the particles. Therefore, when we produced the emulsion with cholesterol esters and radioactively marked triglycerides and injected in experimental animal circulation, collecting samples at predetermined time intervals, up to 1 hour, we observed that the triglyceride curves were quicker than those of cholesterol esters marked in artificial chylomicrons [1].

This observation allowed the conceiving of a new method to evaluate chylomicron metabolism based on artificial emulsions. When patients were injected with a given diseases and their controls were injected with double emulsion markers, we standardized for a new, practical, and specific approach to study chylomicron metabolism in humans, avoiding the technical difficulties of other previous study methods. The radioactive dose injected in patients results in even lower exposure that a simple chest X-ray, ensuring procedural safety.

Using the new kinetic method, we explored several aspects of chylomicron metabolism and physiopathology and, by inference, of VLDLs, since they are TRLP with intravascular metabolism analogous to chylomicrons.

We demonstrated that, in patients with coronary arterial disease (CAD), both the plasmatic clearance of triglycerides, which represents the lipolysis process carried out by lipoprotein lipase, and of cholesterol esters, which represents remnant removal by the liver, are reduced [11]. We have demonstrated that the plasmatic clearance of chylomicrons was reduced in patients with CAD, and directly correlated to the intensity of the coronary bed lesion [12] (Fig. 23.3).

In prospective studies, we demonstrated that slower plasmatic removal of triglycerides from circulation was associated with the development of CAD [13]. Therefore, in a broader scenario, extrapolating the results described, it is possible to imagine two large groups of individuals in the general population. In the first group, after a fatty meal, the circulating remnant lipolysis and liver removal processes are quicker and more efficient. In the second group, these postprandial processes are slower and less efficient, chylomicron lipolysis is slower, and the remnants circulate in the bloodstream for a longer period. These results converged with other works that, using different methods, substantiated the role played by chylomicron metabolism in atherogenesis [14–16].

After establishing the main relationship between the metabolization speed of artificial chylomicrons and the presence of CAD, we started studying three

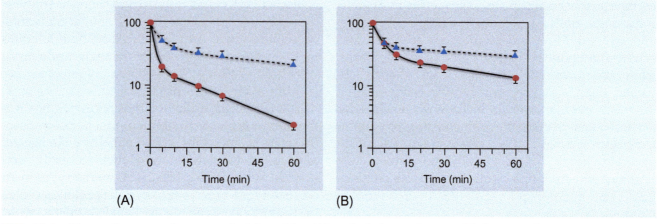

FIG. 23.3 Plasmatic removal curves of triglycerides (A) and cholesterol esters (B) from artificial chylomicrons in patients with coronary arterial disease (CAD, *dotted line*) and individuals without the disease *(continuous line)*. The chylomicrons are produced with marked radioactive lipids and are injected intravenously. Plasma is sampled for up to 1 hour in predetermined time intervals to count radioactivity. Both the removal of triglycerides, which represents chylomicron lipolysis that occurs in capillary endothelium due to lipoprotein lipase activity, and removal of cholesterol esters, which represents capture of remnants by the liver, are slower in patients with CAD [11].

aspects of the chylomicron-atherosclerosis relationship: the state of chylomicrons in diseases with higher incidence of cardiovascular complications, the effects of lipid-lowering medication (i.e., drugs that reduce cholesterol and plasmatic triglycerides), and diet influences.

Chronic inflammatory diseases are frequently followed by the risk of developing cardiovascular diseases, in particular CAD. In case of systemic lupus erythematosus (SLE), the risk of CAD is increased 50 times, recalling that the disease occurs mainly in young women, where CAD is unlikely [17]. In patients with SLE, we documented that both lipolysis and remnant removal were significantly reduced [18]. Besides the increased plasmatic triglycerides and reduced HDL-cholesterol observed in these patients, the chylomicron metabolism disorder shall contribute to predisposition of developing atherosclerotic cardiovascular diseases.

Remnant removal is reduced in obese individuals [19,20]. Chylomicrons are also altered in both metabolic syndrome and *diabetes mellitus* (DM) type 2 [21]. This alteration was expected to be more intense in diabetes than in metabolic syndrome. However, this did not happen and the disorder was equally intense in both conditions, with the same observed regarding the lipid profile. In polycystic ovary syndrome, which also shows insulin resistance, lipolysis was not altered, but remnant removal decreased [22]. In contrast, in patients with malignant arterial hypertension, lipolysis was reduced

but remnants were removed normally [23]. In heart transplantation, the accelerated form of atherosclerosis developed is the main cause of procedural failure after the first year from the surgery [24]. In patients with heart transplantation, both lipolysis and remnant removal were reduced [25], which can contribute to the development of graft vasculopathy. All of these diseases show chronic inflammation that causes or interacts with the lipid alterations to develop the atherogenic process. Hematologic neoplasia also alters chylomicron metabolism, as observed in leukemias [26], multiple myeloma [27], and Hodgkin and non-Hodgkin lymphoma [28]. This finding is possibly related to neoplastic cell secretion of factors, such as TNF-α, which act on lipid metabolism.

Another aspect we explored was the action of lipid-lowering drugs upon chylomicron metabolism. Statins, such as simvastatin and atorvastatin, accelerated both triglyceride and chylomicron remnant removal [29,30]. Fibrates, such as gemfibrozil and etofibrate, also accelerated chylomicron metabolism [31,32]. Hypercholesterolemic patients treated with a more recent class of drugs, which act on cholesterol intestinal absorption, called ezetimibe, also accelerated chylomicron metabolism [33]. However, the association between ezetimibe and simvastatin, commercially used as Vitorin, had no additional effects [33].

Patients with low HDL-cholesterol also showed deficient removal of chylomicrons from circulation.

Treatment with niacin, a drug that increases the circulating levels of HDL-cholesterol, with slow-release, low-flush presentation and a dose of 1.5 g/day, did not change chylomicron metabolism [34]. However, it is worth mentioning that the treatment improved vascular reactivity [34].

Another range of influences on chylomicron metabolism that we investigated consisted of dietary interventions. For that matter, we visited a classic nutritional debate again: is ingestion of eggs deleterious for lipid plasmatic metabolism? Eggs are a quite interesting class of food, due to the protein content and high nutritional value. However, egg yolk has one of the highest cholesterol contents per gram among several foods consumed by humans. Therefore, egg is considered to have a proatherogenic character, although saturated fats are much more potent than cholesterol as dietary factor able to increased LDL-cholesterol levels. Unfortunately, in the prospective study that assessed the effects of ingesting three eggs per day, we found out that remnant removal was reduced by egg consumption, although triglyceride lipolysis was not altered [35]. These results corroborate with the directives of the American Heart Association (AHA) regarding egg consumption. Another great nutrition and dietary landmark was also explored in our works: the most extreme vegetarian pattern, veganism, which excludes ingestion of any food products of animal origin. Vegans were compared with individuals with lacto ovo and omnivorous diets. We observed that the individuals in the vegan group were more efficient in metabolizing artificial chylomicrons, followed by individuals with lacto ovo diet, with individuals with omnivorous diet having the lowest remnant plasmatic removal rates [36,37]. However, lipolysis is not different among those who followed vegan, lacto ovo, and omnivorous diets [20,21].

A toxicological subject important for public health was also covered: the use of anabolic-androgenic steroids by bodybuilders, notoriously detrimental to health, resulted in decreased plasmatic removal of chylomicron remnants [38].

Another aspect of TRLP physiology and physiopathology that was also covered in our research was the relationship between common chylomicron and LDL removal mechanisms. The positive correlation found between the plasmatic clearance of chylomicron remnants and LDL-cholesterol concentration and the fact that chylomicron and remnant

removal is reduced in patients with heterozygous familial hypercholesterolemia, with diagnosed mutations of the LDL receptors, clearly demonstrated the role of LDL receptors in removing remnants from circulation [39,40].

Therefore, our research accumulated considerable experience in chylomicron metabolism clinical studies, using these artificial lipoproteins to demonstrate pathophysiological aspects, metabolic changes due to drug and dietary interventions, and the influence of TRLP metabolism on CAD evolution prognosis.

Although there is no practical method to assess chylomicron metabolism in clinical laboratories, it is important to have a clear notion that, dietary guidance or prescription of statins or fibrates not only reduces LDL-cholesterol or triglyceride concentrations. In fact, the entire process of removing fat from blood circulation and cholesterol absorbed from food is being accelerated. The speed and efficiency of chylomicron plasmatic catabolism translates into protection against the atherogenesis process.

ARTIFICIAL LDL OR LDE

In 1987, a new model based on artificial emulsions was developed in our laboratory, with the purpose of acquiring a tool to explore LDL plasmatic metabolism. Therefore, an artificial LDL was created as an emulsion consisting of nanoparticles with structure similar to LDL. These emulsions are formed by the same compounds that constitute artificial chylomicrons, but in very different proportions. Just like LDL, the proportion of triglycerides is very small and the proportion of cholesterol esters is predominant in the core of the particles. A minimum proportion of free cholesterol is also present, and the particle is covered, as all lipoproteins are, by a monolayer of phospholipids (Fig. 23.4).

"Artificial LDL" is produced without protein, by means of technologies based on extended ultrasonic irradiation followed by differential centrifugation or high-pressure microfluidization. We observed that, in contact with plasma, they acquire several apolipoproteins present in circulating lipoproteins. One of these apos, apo E, is recognized by the LDL receptors and allows the nanoemulsion to be taken up by cells by means of the same process as LDLs, LDL receptor-mediated endocytosis. Therefore, after acquiring apolipoproteins in contact with plasma, the artificial LDL starts to mimic the

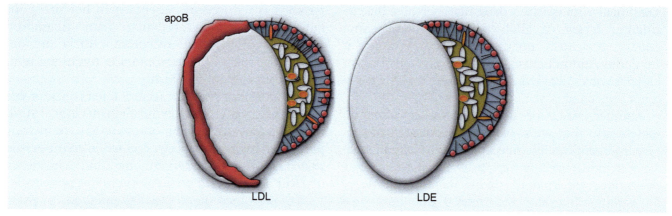

FIG. 23.4 Chemical structure of low-density lipoproteins (LDL) and LDE: the cholesterol ester nucleus is covered by a monolayer of phospholipids. The monolayer also has a small percentage of free cholesterol and the nucleus has a small percentage of triglycerides. On the surface of large protein molecules (>300 kDa), LDL has apo B100, which binds the lipoprotein to its cell membrane receptors. LDE is prepared without protein.

metabolism of native LDL [41,42]. We frequently refer to nanoemulsions as LDE to remind us of the structural similarity with LDL and apo E, which is the ligand that binds the nanoparticles and lipoprotein receptors.

At the time, there was still strong controversy regarding the value of LDL cholesterol concentrations as a CAD risk factor. We expected that, when LD was injected in the circulation of two groups of patients with similar LDL-cholesterol levels, one of them without CAD and the other with CAD, LDE removal in the group with the disease would be slower. If the difference in removal kinetics was very sharp with reduced standard-deviation, we would undoubtedly have a new and potent disease biomarker.

However, the results of the experiment were disappointing: the LDE cholesterol removal curves were similar [43]. Further on, we demonstrated that patients with hypercholesterolemia showed reduced LDE removal, which, however, was increased with the use of statins [44].

Using LDE as LDL metabolism investigation instrument, we demonstrated, for example, that physical activity, a factor that reduces the risk of CAD, accelerates LDE plasmatic removal and, by extension, native LDL [45]. These results are very important, since they imply that physical activity increases LDL turnover. Therefore, the renewed LDL is less exposed to the modification and oxidation processes that lead to uptake by macrophages and atherogenesis. In amateur cyclists who practice daily aerobic exercises, the plasmatic removal rates

were five times higher than in sedentary individuals [45] (Fig. 23.5), despite their LDL cholesterol not being different.

It is interesting to note that hypercholesterolemic patients also benefit from the effects of physical activity: their LDL clearance is also increased [46]. Similar to aerobic exercises, resistance exercises performed by individuals with normal cholesterol

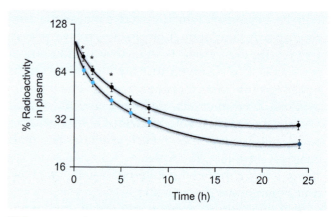

FIG. 23.5 Plasmatic removal of LDE with EV-injected radioactive markers in cyclists *(blue circles)* and sedentary individuals *(black circles)*. Both groups showed equal LDL cholesterol concentrations. As LDE removal was much quicker in athletes, it is assumed that the liver produces more LDL-cholesterol, therefore with higher LDL turnover in athletes. Therefore, the LDL in athletes is "newer" than the one in sedentary individuals and less susceptible to oxidation and other modification processes involved with atherogenesis. This is one of the mechanisms by means of which physical activity has antiatherosclerotic action [45].

levels also increase LDE clearance and, by extrapolation, LDL clearance [47]. Another important issue, due to the fact that advanced age is critical for the appearance of atherosclerosis manifestations, we demonstrated that LDL removal, studied by means of LDE plasmatic kinetics, is reduced in elder individuals. This had already been demonstrated before; however, in our work, when we injected the standard preparation, in this case LDE and not LDL of each individual reinjected in the patients themselves, it became clear that the removal mechanisms determine LDL removal efficiency loss associated with age instead of lipoprotein composition modifications [48].

We also studied LDE kinetics in patients with heart transplants, noting that these transplants are followed by lower LDL removal [49].

Studying the kinetics of cholesterol forms (esterified and nonesterified) in LDE, we noted that nonesterified cholesterol tends to dissociate from LDE in patients with CAD [43]. In other experiments, we have injected in CAD patients radioactively marked LDE in both cholesterol forms. The analysis of the vessel fragments discarded by the surgeon showed that the percentage of LDE nonesterified cholesterol accumulated was higher than LDE esterified cholesterol [50]. With this data, we established the hypothesis of a new atherogenesis mechanism in which nonesterified cholesterol decouples from the lipoprotein and accumulates directly in the arterial endothelium. The supposed accumulation of the nonesterified form on the artery could lead to endothelial dysfunction, considering that the excessive cholesterol in the plasma membrane could lead to membrane structural and fluidity alterations that strongly influence the enzymatic reactions and other processes carried out in cells.

Since apo E, present in LDE but not in LDL, has a much higher affinity than apo B100, the only apo in LDL, LDE is removed much more quickly than natural LDL [42]. This favors the execution of kinetic tests, since the plasmatic removal follow-up time, that is, time required to collect a series of blood samples, is much shorter. However, extrapolation of LDE results to natural LDL metabolism must be careful. In some circumstances, LDE is expected not to mimic the natural LDL metabolism, and this must be taken into account when interpreting the results.

As a lesson taken from these works for practitioners regarding the use of LDE to explore LDL metabolism, there are phenomena translated in LDL plasmatic kinetics that are not necessarily expressed by LDL-cholesterol concentration. The examples include the accelerated LDL clearance in athlete circulation, decreased clearance in sedentary individuals, and CAD cholesterol free form kinetics, as described before. Although there are not clinical routine exams to measure these events, such knowledge is important to understand the atherosclerosis predisposing phenomena and the strategies required to deal with them.

TRANSFER OF LIPIDS TO HDL

In works published on 2008 onward, after the incursions related to chylomicron and LDL transportation cycles, we focused on HDL metabolism. Mimicking its general structure to conduct in vivo kinetic studies, as conducted with chylomicrons and LDL, did not seem a good strategy due to the difficulty in producing such small particles ranging from 8 to 12 nm in diameter. This would require the presence of apo A-I in the reparation process, which is the main HDL apo and an important lipoprotein structuring element. Isolating apo A-I from the serum or producing it by means of genetic engineering was always the complicating factor of the experiments. At the time, we decided to link our HDL approach to lipid transfer, a subject that has fascinated us for a very long time. This was the subject of our habilitation thesis, presented in 1987 [51]. The interest in lipid transfers was only just beginning at the time.

Lipid transfers are an intriguing aspect of lipid metabolism in plasma. In circulation, molecules of several lipid species, such as cholesterol, triglycerides, and phospholipids, pass from one lipoprotein to another, and vice versa. This is like a frantic molecular ping-pong between lipoprotein particles, which is favored by specialized proteins, the so-called transfer proteins. Namely, the proteins are cholesteryl ester transfer protein (CETP) [52] and phospholipids transfer protein (PLTP) [53]. The first facilitates mainly the transfer of cholesterol esters and triglycerides, while PLTP favors the transfer of phospholipids, with some action upon free form (nonesterified) cholesterol transfer (Fig. 23.6).

There is no consensus regarding higher or lower action of CETP or PLTP results in higher or lower propensity to develop atherosclerosis. Lipid transfer

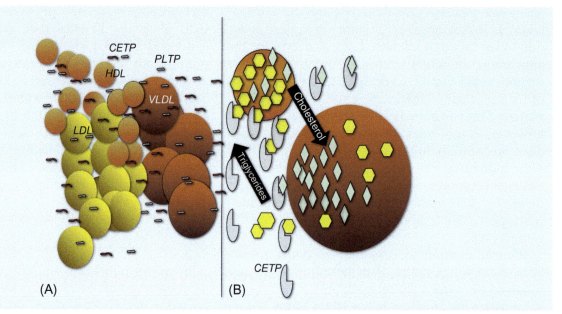

FIG. 23.6   (A) Lipid transfers between lipoproteins: the lipids that compose the several classes of lipoproteins, such as cholesterol, in both forms (free and esterified), triglycerides, and phospholipids, leap from one lipoprotein particle to another. Transfer is promoted by transfer proteins CETP (cholesteryl ester transfer protein) and PLTP (phospholipid transfer protein). (B) Lipid transfer movement is bidirectional, but CETP tends to favor triglyceride removal from VLDL to HDL, and removal of cholesterol esters from HDL to VLDL. Therefore, when VLDL accumulates in plasma, by the law of mass action, VLDL receives more cholesterol esters and HDL receives more triglycerides. The consequences of this exchange are unfavorable for HDL. The excessive triglycerides in the structure makes this lipoprotein class suffers the action of hepatic lipase, resulting in accelerated HDL catabolism and low concentration of HDL-cholesterol. This causes the "teeter-totter" effect observed in clinical practice routine: triglycerides (VLDL) go up, and HDL-cholesterol goes down.

between lipoproteins also depends on other factors, such as concentration and composition of several lipoprotein classes, among others. It primarily affects HDL, a lipoprotein formed mainly in the plasmatic compartment, instead of others, formed in enterocytes and hepatocytes. HDL formation depends on lipid transfers, mainly cholesterol and phospholipids from other lipoproteins and cells to apo A-I. The apo A-I lipidation process forms phospholipid and cholesterol discs around the protein. The entrance of free cholesterol in these discs and esterification due to the action of lecithin-cholesterol acyltransferase (LCAT) makes emerging HDL particles increasingly round [54]. Lipid transfer is directly connected to HDL functions such as cholesterol reverse transportation and stabilization in the plasmatic compartment (Fig. 23.7).

Considering that lipid transfer between lipoproteins mainly affects HDL and is an essential process for the metabolism and functions of these lipoproteins, we have imagined and developed a method focusing on assessing lipid transfer to HDL [55].

In this new approach, consisting of in vitro tests in which LDE marked with radioactive lipids—free and esterified cholesterol, phospholipids, and triglycerides—is incubated with total plasma, LDE serves as donor of lipids for the lipoproteins. The lipids transferred to HDL are quantified after radioactive counting of the HDL fraction isolated by means of chemical precipitation [56].

This new methodological approach allowed us to obtain a new CAD marker: in patients with early CAD, we observed reduced transfer of free cholesterol and cholesterol ester to HDL. In that specific case, the plasmatic lipid profile of patients with CAD was not different from the control individuals, without the disease, and the transfer of cholesterol to HDL was the only biochemical marker found in the study. Furthermore, in patients with DM 2 who developed CAD, the transfer of both forms of cholesterol was reduced [57]. Another aspect is that reduced transfer associated with CAD occurs both in men and women, either diabetic or not (unpublished data). The gender factor is very important,

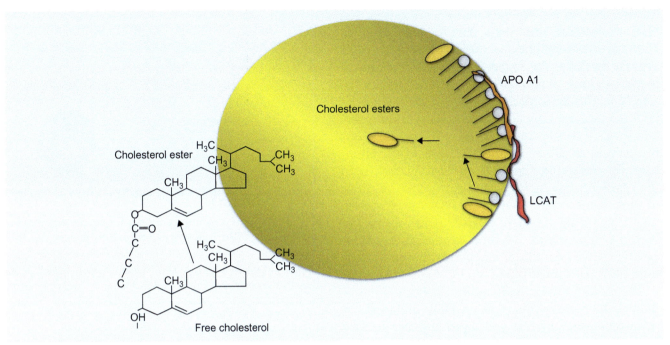

FIG. 23.7 HDL cholesterol esterification. Free (nonesterified) cholesterol molecules are located in the phospholipid monolayer that covers HDL. In the esterification process, cholesterol receives an acyl radical from the phospholipids by means of a reaction catalyzed by acyl-cholesterol acyl-transferase, which is activated by apolipoprotein A-I (apo AI), the main apo in HDL. The recently esterified cholesterol, which is more hydrophobic, moves from the HDL surface layer to the nucleus, were it remains sequestered and is only removed by other lipoproteins by means of CETP action. The esterification process stabilizes cholesterol in the plasmatic compartment and promotes reverse cholesterol transportation. *LCAT*, lecitin cholesterol acyl transferase.

considering that HDL-cholesterol concentration is higher in women, who have a lower propensity to develop CAD. Incidentally, advent of menopause did not change the flow of lipid transfer to HDL [58]. The presence of DM2 is also an important factor, considering that these patients have a higher propensity to develop CAD.

Besides the main issue of how lipid transfer to HDL occurs in CAD, we also investigated the influence of other diseases, clinical situations, dietary factors, and physical activity upon these parameters. We observed that DM 2, and not DM type 1, modifies the transfers [59]. In hypercholesterolemia, the transfer of nonesterified cholesterol to HDL is reduced, which may contribute to aggravate the proatherogenesis state resulting from LDL-cholesterol accumulated in blood [60].

We also documented reduced transfer of cholesterol to HDL in patients with heart transplantation [61], who, as already mentioned, suffer rapidly developing atherosclerosis, known as graft vasculopathy. In patients with subclinical hypothyroidism, who have a higher possibility of developing atherosclerosis, reduced transfer of phospholipids to HDL was the only altered lipid marker [62].

Physical activity [63–65], both when applied to individuals with metabolic syndrome [63] and when applied to elder individuals [64], may alter lipid transfers. Marathon runners show transfer of all four lipids much higher than sedentary individuals; however, during the marathons, transfer is inhibited [65]. Dietary factors may also change the transfer of lipids to HDL [36,37,66,67].

Differently from plasmatic kinetic tests based on artificial chylomicron models and LDE, the in vitro test of transfer of lipids to HDL may be adaptable and operational in clinical laboratory routines. The test is comprehensive, since it simultaneously assesses the transfer of the four main lipids present in lipoproteins and brings most of the factors that operate in the process to the test tube, such as transfer proteins and lipoproteins, in the same concentrations in which they are found in plasma.

From the teachings and up to clinical practice that may have brought the data obtained in the lipid transfer test so far, the fact that stands out is that HDL-cholesterol only tells part of the story of HDL as a potent endogenous drug that protects against the long atherogenesis process and clinical manifestations of the disease. Several functional aspects of this lipoprotein may be independent from HDL-cholesterol concentration, at least from a certain level of concentration, as demonstrated by our results and results of previous studies in the literature. By observing the HDL-cholesterol results of their patients, physicians can be aware that in front of them lies a very important risk factor. However, much regarding the state of HDL is not translated by these figures that, if estimated in a clinical laboratory could constitute independent risk factors with their own meanings.

The lipid transfer test has two aspects. One of them is to demonstrate that an HDL metabolic or functional step, in this case lipid transfer to said lipoprotein, is altered in atherosclerosis. The other is that lipid transfers, as a complex phenomenon that occurs in several classes of lipoproteins but is measured herein in only a specific vector—donor lipoprotein lipids going to the receptor HDL fraction—are also altered in this disease.

The approach of both aspects herein is pioneering [68]. To the best of our knowledge, this is the first biochemical, practical, in vitro, comprehensive test of the transfer phenomenon.

The fact that free cholesterol (nonesterified) transfer is reduced in conditions associated with atherosclerosis completes the description of our hypothesis that there is a new additional mechanism of lipid accumulation on arterial walls [68]. There is surely still a long way to go to prove this hypothesis (Fig. 23.8).

USE OF LDE IN DRUG TARGETING

The greatest progress in our line of investigation of artificial lipoproteins was, without a doubt, the discovery that LDE can be used as a vehicle able to concentrate drugs in their action sites (drug targeting). This has been for a long time one of the great objectives of therapeutics, formulated initially by Paul Ehrlich (1854–1915), Nobel Medicine Prize winner in 1908, in his "magic bullets" concept. When treating cancer, due to the low therapeutic levels and high toxicity of chemotherapeutic agents, this strategy may be particularly useful to try to increase

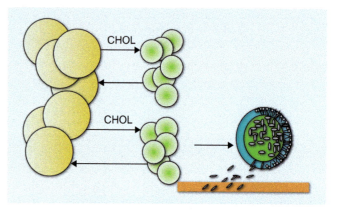

FIG. 23.8 Direct accumulation of free (nonesterified) cholesterol in the endothelium: a hypothesis. In this hypothesis, transfer of free cholesterol from other lipoproteins to HDL, where it is esterified, is deficient. Therefore, the excessive free cholesterol would decouple from the lipoprotein and, due to its low solubility in plasma, accumulate on vessel surfaces, causing endothelial function disorders [43,48,68].

pharmacological action and reduce toxicity of antineoplastic agents. The first drug delivery studies were conducted with liposomes, spherical vesicles formed basically by phospholipid bilayers with an aqueous medium inside. Besides the liposome systems, micelles, dendrimers, solid lipid particles or lipid nanoemulsions, metallic nanoparticles, semiconductor nanoparticles, or polymeric nanoparticles have also been studied as drug vehicles, as a new discipline was arranged: nanotechnology applied to biomedical sciences, with the potential of revolutionizing the therapeutic and diagnosis fields.

Emulsion systems have been used for a long time in pharmacies as drug vehicles for parenteral use without, however, having drug targeting effects. In articles published since 1992 [69,70], we have demonstrated that LDE is able to concentrate on neoplastic cells and therefore carry the chemotherapeutic agents directed to those cells. This was a pioneering discovery in the field of nanotechnology applied to medicine, since it was the first time that solid particles were directed to the action site artificially, and not by liposomes. The discovery was the object of patents granted by the U.S. Department of Commerce.

The expression of LDL receptors is increased in neoplastic cells. This phenomenon, described by Ho and two winners of the Nobel Medicine Prize in 1985, Michael Brown and Joseph Goldstein [71], results from accelerated mitosis of neoplastic cells, which demand lipids for the synthesis of new membranes required due to cell duplication. This results in uptake increased of LDL by these cells, and also reduces the serum concentration of LDL-cholesterol [72]. LDE that binds to the same receptors, but with even more affinity than native LDL itself since, due to the fact that it binds via apo E, it is also avidly captured by neoplastic cells (Fig. 23.9).

In patients with acute myelocytic leukemia, LDE removal from plasma was much quicker than in individuals without the disease [69,70]. When the patients were treated, intake was reduced up to control levels, and at the same time LDL cholesterol level increased [70]. By means of calculations based on plasmatic kinetics, we estimated that the concentration of LDE in acute myelocytic leukemia cells was approximately 50 times higher. In another study, LDE was marked with Technetium 99m to generate nuclear medicine imaging of patients with breast cancer. We demonstrated that LDE is not concentrated only in the primary tumor, but also in metastasis [73]. LDE intake was increased by five and ten times in breast and ovarian cancer, respectively [74,75] (Fig. 23.10).

Up to now, we have developed five formulations of chemotherapeutic agents associated with LDE: carmustine (BCNU) and derivatives of paclitaxel, etoposide, methotrexate, and daunorubicin. The purpose of deriving the four latter chemotherapeutic agents to associate them to LDE was to increase their lipophilicity and, therefore, the efficiency of the association and to obtain stable preparations. In cell culture and pharmacokinetics studies in animals, and later in patients, several chemotherapeutic agent-LDE preparations proved in fact to be stable, and the drug was transported by the nanoparticles up to the cells, which is fundamental to obtain the drug targeting effect [75–82].

Animal testing showed that, based on the classic pharmacology parameters, such as lethal dose calculation, this drug vehiculation process drastically reduces chemotherapeutic agent toxicity [79,80]. Animal testing with oncological models showed that association to LDE also increases the therapeutic action of the drugs [79,80] and extends survival time (Figs. 23.11 and 23.12).

Clinical testing in patients with advanced malignant neoplasia resistant to many drugs showed that the preparations were safe and confirmed LDE's

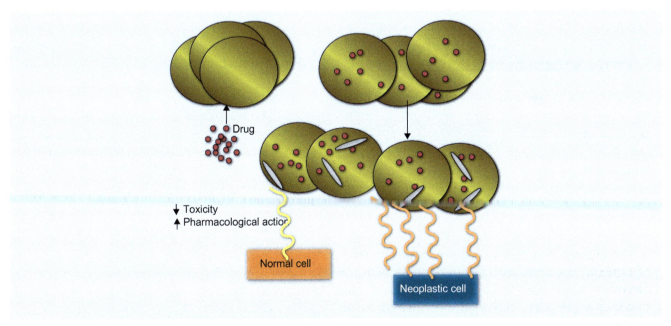

FIG. 23.9 Using the LDE system to treat cancer: anticancer drugs are associated with LDE nanoparticles and injected in circulation. In contact with plasma, the nanoparticles containing the drug acquire apolipoprotein E from plasmatic lipoproteins, such as VLDL, chylomicrons, or HDL. LDE binds to the lipoprotein receptors, mainly LDL receptors that recognize apo E on the surface of nanoparticles. Then, LDE is captured by cells by means of receptor-mediated endocytosis, pouring its drug load into the cytoplasm. Since the expression of LDL receptors is very high in neoplastic tissues, LDE is captured by these tissues much more than by normal tissues, whose cells have less LDL receptors. This is the mechanism that grants LDE with its drug targeting properties [69,70].

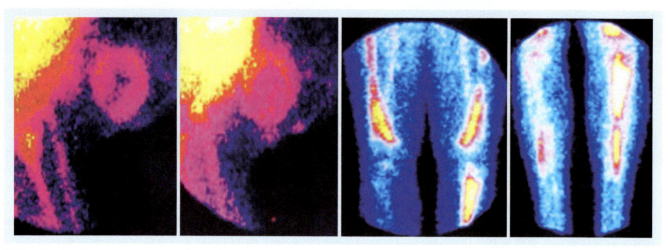

FIG. 23.10 Proof-of-concept of the ability of LDE to concentrate on malignant tumor tissues: LDE was marked with Technetium 99m, a radioisotope able to generate nuclear medicine images. After intravenous injection and imaging of patients with breast cancer, it is demonstrated that LDE is captured more intensely by both the primary tumor and metastases, in this case, bone cancer.

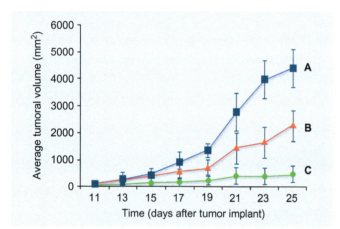

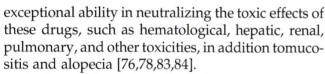

FIG. 23.11 Treatment of rats with B16 melanoma implant using a single dose of: (A) saline solution (control); (B) etoposide in its commercial presentation; (C) etoposide associated with LDE. The curves represent tumor size evolution after the treatments [79].

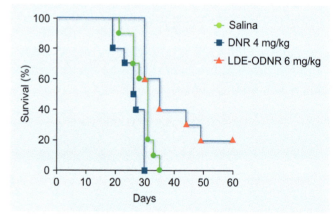

FIG. 23.12 Kaplan Meier survival plots of rats with melanoma B16 implant treated with a single dose of saline solution (controls), with the commercial preparation of daunorubicin (DNR) or with DNR associated with LDE.

exceptional ability in neutralizing the toxic effects of these drugs, such as hematological, hepatic, renal, pulmonary, and other toxicities, in addition tomucositis and alopecia [76,78,83,84].

During a phase 1 study with LDE-carmustine, responses to the treatment that would not be expected by conventional chemotherapy were present in several patients. This finding also allowed conducting extended treatment with these high-toxicity chemotherapeutic agents for periods longer than 2 years without significant toxicity, in at least two patients, which would be unthinkable in conventional therapy (unpublished data).

It is easy to understand the remarkable reduced toxicity that is achieved by associating chemotherapeutic agents to LDE. Certainly, it cannot be exclusively attributed to the concentration of the drugs in the action sites. This must be the result of the new biodistribution of the chemotherapeutic agents created by the association to LED, half-life, protection granted in circulation by covering the drugs with nanoparticles, among other factors.

Such significant reduced toxicity allows (Table 23.1):

(a) chemotherapy of disabled or very old patients, for which there were no alternatives to chemotherapeutic agents;

TABLE 23.1 Toxicity Study of the Carmustine (BCNU) Preparation Associated With LDE: The Groups of Patients, With a Total Number of 46, Were Treated With Increasing Doses of Chemotherapeutic Agents (Dose Escalation Protocol)

| | Carmustine dose (mg/m² of bodily surface) | | | | | | | | | |
| | 150 | | 190 | | 240 | | 300 | | 350 | |
	Grade 1	Grade 2	Grade 1	Grade 2	Grade 1	Grade 2	Grade 1	Grade 2	Grade 1	Grade 2
Nausea	25	0	0	0	18	0	14	0	14	0
Vomiting	13	0	20	0	0	0	0	0	24	0
Local pain	31	0	60	0	14	0	40	0	59	0
Arterial hypertension	0	0	0	0	0	0	3	0	7	0
Fever	0	0	0	0	0	0	0	0	0	0
Dyspnea	0	0	0	0	0	0	0	0	0	0
Alopecia	0	0	0	0	0	0	3	0	0	0
Anemia	0	0	0	0	0	0	3	0	17	0
Leukopenia	0	0	0	0	0	0	3	0	24	0
Thrombocytopenia	0	0	0	0	7	0	6	0	0	0
Hepatic										
AST	0	0	0	0	14	0	6	3	17	0
ALP	0	0	0	0	0	0	6	0	34	0
Bilirubin	0	0	0	0	0	0	6	0	5	0
Renal										
Urea	0	0	0	0	0	0	6	0	0	0
Creatine	0	0	0	0	0	0	0	0	0	0
Alkaline phosphatase	0	0	10	0	10	0	20	0	34	7

Toxicity is classified from level zero (0), absence of toxicity, to level four (4), very high toxicity. Only toxicity levels 1 and 2 were observed, and even so only at a small percentage. Note that the commercial preparation of carmustine is a chemotherapeutic agent known for its high toxicity, which makes its current use very restricted [78].

(b) chemotherapy for extended periods, since cumulative toxicity that forces treatment interruption no longer exists; and

(c) increased chemotherapeutic agent doses, which increases the anticancer efficiency of the treatment, as the drugs have dose-response behavior. These points have been demonstrated in our experiments so far.

Using groups of seven patients with multiple myeloma, we clearly demonstrated the efficiency of the LDE-carmustine preparation to improve clinical state and serum concentration of the disease marker, gamma globulin M [83]. In several patients with solid tumors, we observed disease improvement and extended tumor progression-free time. However, the issue of therapeutic efficiency in patients with malignant neoplasia and precise indications of cancer treatment with chemotherapeutic agents vehiculated buy LDE is complex, and still requires extensive work.

Nonneoplastic proliferative processes also make them develop superexpression of LDL receptors. In patients with thalassemia minor, in which accelerated destruction of red blood cells leads to accelerated proliferation of hematopoietic cells, LDE plasmatic removal rates also increased [85].

In 2008, we reported that LDE also concentrates in atherosclerotic lesions induced in rabbits by means of cholesterol-rich diet [86]. According to the knowledge consolidated from the work conducted by Ross [87] that led to the current pathophysiological concept, atherosclerosis is basically a chronic inflammatory-proliferative process [88,89].

Since the drugs used in anticancer treatment are the most potent antiproliferative agents in the therapeutic arsenal, we raised the hypothesis that they could be used, vehiculated by LDE, to treat atherosclerosis. This opening is possible because the association with LDE neutralizes the toxicity of the chemotherapeutic agents. Thus, the highly toxic treatment that would be inconceivable from the cardiology perspective becomes perfectly tolerable.

In fact, treatment of rabbits with atherosclerosis using LDE-paclitaxel resulted in atherosclerotic lesion reduction of 60%–70% [86], also achieve with LDE-etoposide [90] and LDE-methotrexate [91,92]. Macrophagic invasion of the intima, as well as proliferation of smooth muscle cells proceeding from the medial layer and identified by immunohistochemistry, were significantly reduced [86,91,92]. We demonstrated that these effects were achieved due to the inhibiting action of the drugs upon these proinflammatory factors, as well as the stimulation effect upon antiinflammatory cytokines [91,92]. Treatment of rabbits with atherosclerosis by means of association of antiproliferative drugs vehiculated by LDE, such as LDE-methotrexate and LDE-paclitaxel, reduced lesions even more [93] (Figs. 23.13–23.15).

The large reduction in toxicity and remarkable antiproliferative and antiinflammatory activity observed in experimental atherosclerosis led us to open a new and fascinating avenue of LDE applications: the use of the anticancer arsenal—drugs such as paclitaxel, etoposide, and methotrexate to treat cardiovascular diseases and other chronic degenerative diseases.

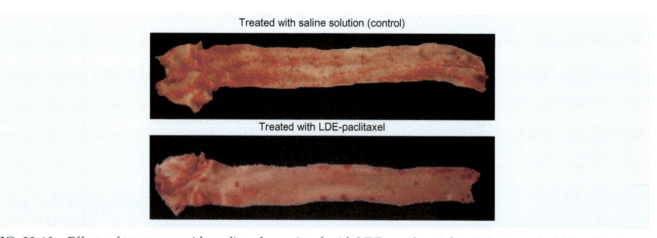

FIG. 23.13   Effects of treatment with paclitaxel associated with LDE on atherosclerotic lesions of rabbit with atherosclerosis induced by cholesterol-rich diet: *on the top*, aortas of control rabbits treated with saline solution injection; *on the bottom*, aortas of rabbits treated with paclitaxel associated with LDE [86].

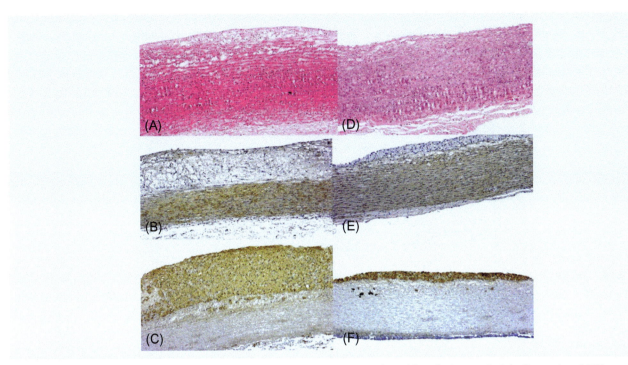

FIG. 23.14 Effects of treatment of rabbits with atherosclerosis induced by cholesterol-rich diet using LDE-paclitaxel. Microscopy of the aorta wall with hematoxylin-eosin coloration *(top images)* Immunohistochemistry to mark smooth muscle cells *(middle images)* and macrophages *(bottom images)*. To the left (A, B, C), rabbits treated with saline solution *(controls)*; to the right (D, E, F) rabbits treated with LDE-paclitaxel [86].

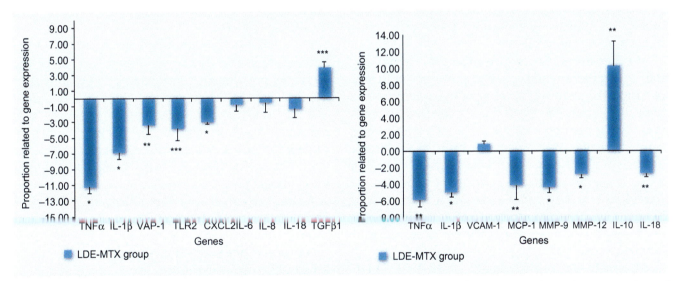

FIG. 23.15 The chemotherapeutic agents associated with LDE, such as methotrexate, reduce the expression of proinflammatory factors, such as TNF-α, IL1b, and others, in rabbit aorta with atherosclerosis induced by cholesterol-rich diet [92].

In order to explore these possibilities, in experiments conducted in rabbit models with induced rheumatoid arthritis, intraarticular injection of LDE-methotrexate preparation drastically reduced the synovial membrane inflammatory process, reducing the number of inflammatory cells in the synovial liquid [94]. The commercial preparation had no effect on both synovial inflammation or the number of inflammatory cells in the intraarticular fluid. This is probably due to the fact that, in commercial preparations, methotrexate intake is very low, increasing approximately 90 times when

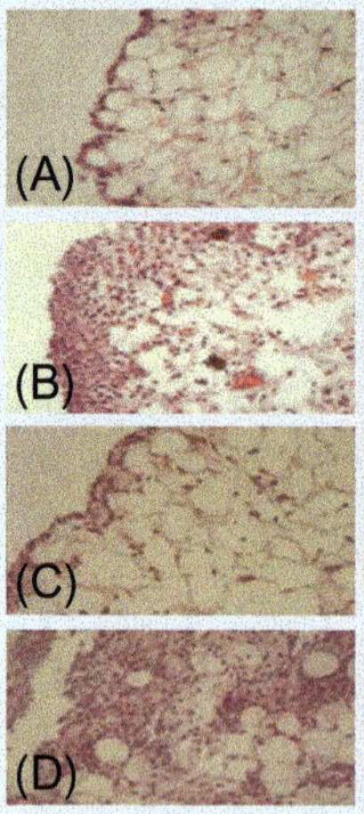

Joint	Cells/100 mm^2
Saline	0.41 ± 0.01
AIA	23.46 ± 2.6
AIA + LDE-MTX	2.96 ± 0.44*
AIA + MTX commercial	19.1 ± 1.2

FIG. 23.16   Effect of methotrexate associated with LDE in models of arthritis induced in rabbits by means of methylated albumin: microscopy of joints after: (A) control 1, normal synovium; (B) control 2, rabbit synovium with induced arthritis, no treatment. (C) Treated with preparation of methotrexate associated with LDE. (D) Treated with commercial methotrexate. *To the right*, number of inflammatory cells in the synovial liquid collected from the joints [94]. The images make it clear that treatment with LDE-MTX drastically reduced the inflammatory process. This was also confirmed by the cell count in the synovial liquid, in which treatment with LDE-MTX drastically reduced the number of inflammatory cells. This effect was not achieved by commercial MTX.

incorporated to LDE. Treatment of rabbits with rheumatoid arthritis using intravenous injection of LDE-methotrexate was also clearly superior to commercial methotrexate regarding therapeutic effectiveness [94] (Fig. 23.16).

In other investigation, we raised the hypothesis that chemotherapeutic agent preparations with anti-cancer actions could play a role in the graft immunological rejection and vasculopathy process, priority issue when treating patients after heart transplantation [95]. Paclitaxel vehiculated by LDE was tested in heterotopic heart transplant models in rabbits [96]. These tests are described in Chapter 49. Recently, a review article was published covering the new therapeutic strategies based on drug delivery to treat cardiovascular diseases originated from atherosclerosis [97].

A recent pilot-experiment tested a LDE-methotrexate preparation administered intraperitoneally in rats with acute myocardial infarction (AMI) induced by ligature of the anterior descending coronary artery. In the evaluation conducted 6 weeks after the AMI event, by comparing with untreated controls, the treatment showed very significant improvement of echocardiographic parameters. The histopathological analysis of infarcted hearts showed significant reduction of the infarction area, as well as inflammatory infiltrate, and replacement of the scar tissue for muscle tissue (unpublished data, presented in the Congress of Cardiology of the Brazilian Society of Cardiology, 2016).

In a pilot-study conducted in Instituto Dante Pazzanese, in São Paulo, we tested the use of the paclitaxel preparation vehiculated by LDE in elderly patients with extensive aortic atheroma. The dose and chemotherapeutic scheme of paclitaxel was the same used in oncological treatment, that is, 175 mg/m^2 of bodily surface every 3 weeks, administered for six cycles. The toxicity observed during the treatment period was practically irrelevant. Of eight patients in whom lesions were observed by means of angiotomography before and after the treatment, four showed reduced plaque volumes, three showed no changes, and one showed increased volume (unpublished data, presented in the International Society of Atherosclerosis Congress, Amsterdam, 2015).

CONCLUSIONS

As shown, with the use of original methodological approaches based on two types of artificial lipoproteins, artificial chylomicrons and LDE, we have investigated the main lipid transport cycles in circulation and the phenomenon of lipid transfer among lipoproteins.

In the assessment of this exploratory effort, we emphasize the confirmation that reduced chylomicron catabolism is associated with the presence and development of CAD. The original demonstration that clinical conditions such as SLE, metabolic syndrome, and heart transplant are followed by chylomicron catabolism disorders that contribute to explain the cause of increased atherosclerosis and its complications in these patients. In addition to that, our observations regarding the action of lipid-lowering drugs such as statins, fibrates, and dietary factors have important pathophysiological and therapeutic implications in promoting higher chylomicron catabolism efficiency.

Among the studies using LDE, we emphasize the results that suggest accumulation of nonesterified cholesterol in arteries as a possible proatherogenic factor. Considering that this cholesterol form is decisive in membrane fluidity and in the activation state of the enzymes present on the cellular surface, excessive levels may cause endothelial function disorders. Still in the same line of work, the demonstration that physical activity accelerates plasmatic removal of LDE and, by analogy, probably increases the expression of LDL receptors in peripheral tissues, provides an entirely new insight in the field of physical activity, lipids, and atherosclerosis.

LDE, used as a lipid-donor to HDL, allowed standardizing what seems to be the first in vitro test to assess lipid transfers between lipoproteins, in addition to constituting a test to evaluate HDL metabolism. The alterations in cholesterol transfer to HDL found in patients with CAD may be, in fact, the first finding associating lipid transfers to atherosclerosis, since up to now it has not been demonstrated that CETP or PLTP activities are related to the presence of CAD.

As seen, after the pioneering demonstration that LDE is able to concentrate anticancer drugs in tumors and drastically reduce their toxicity, we opened a new frontier by demonstrating the possibility of applying this vast and potent chemotherapeutic arsenal in cardiac therapeutics. Since atherosclerosis is a chronic process of slow progression, which lasts from youth up to senescence, the potential of this new noninvasive treatment would be directed to cases in which disease aggravation requires more energetic and quick pharmacological action.

However, regardless of the promising results exposed above, full confirmation that the association of chemotherapeutic agents to LDE can be used in human atherosclerosis is still required and demands extensive clinical research efforts.

Acknowledgments

The author thanks Thauany Martins Tavoni, Pharmacist-Biochemist for the text revision. The researches described in this chapter were carried out with financial support from the Foundation for Research Support of the State of São Paulo (FAPESP), National Council for Scientific and Technological Development (CNPq), and Financier of Studies and Projects (FINEP).

References

[1] Redgrave TG, Maranhao RC. Metabolism of protein-free lipid emulsion models of chylomicrons in rats. Biochim Biophys Acta 1985;835:104–12.

[2] Maranhao RC, Tercyak AM, Redgrave TG. Effects of cholesterol content on the metabolism of protein-free emulsion models of lipoproteins. Biochim Biophys Acta 1986;875:247–55.

[3] Redgrave TG. Chylomicron metabolism. Biochem Soc Trans 2004;321:79–82.

[4] Sacks FM. The crucial roles of apolipoproteins E and C-III in apoB lipoprotein metabolism in normolipidemia and hypertriglyceridemia. Curr Opin Lipidol 2015;26:56–63.

[5] Huang Y, Mahley RW. Apolipoprotein E: structure and function in lipid metabolism, neurobiology, and Alzheimer's diseases. Neurobiol Dis 2014;72:3–12.

[6] Zilversmit DB. Atherogenesis: a postprandial phenomenon. Circulation 1979;60:473–85.

[7] Karpe F, Bickerton AS, Hodson L, et al. Removal of triacylglycerols from chylomicrons and VLDL by capillary beds: the basis of lipoprotein remnant formation. Biochem Soc Trans 2007;35:472–6.

[8] Hirata MH, Oliveira HC, Quintão EC, et al. The effects of Triton WR-1339, protamine sulfate and heparin on the plasma removal of emulsion models of chylomicrons and remnants in rats. Biochim Biophys Acta 1987;917:344–6.

[9] Oliveira HC, Hirata MH, Redgrave TG, et al. Competition between chylomicrons and their remnants for plasma removal: a study with artificial emulsion models of chylomicrons. Biochim Biophys Acta 1988;958:211–7.

[10] Redgrave TG, Maranhao RC, Tercyak AM, et al. Uptake of artificial model remnant lipoprotein emulsions by the perfused rat liver. Lipids 1988;23:101–5.

[11] Maranhão RC, Feres MC, Martins MT, et al. Plasma kinetics of a chylomicron-like emulsion in patients with coronary artery disease. Atherosclerosis 1996;126:15–25.

[12] Sposito AC, Ventura LI, Vinagre CG, et al. Delayed intravascular catabolism of chylomicron-like emulsions is an independent predictor of coronary artery disease. Atherosclerosis 2004;176 (2):397–403.

[13] Sposito AC, Lemos PA, Santos RD, et al. Impaired intravascular triglyceride lipolysis constitutes a marker of clinical outcome in patients with stable angina undergoing secondary prevention treatment: a long-term follow-up study. J Am Coll Cardiol 2004;43:2225–32.

[14] Chacra AP, Santos RD, Amâncio RF, et al. Clearance of a 3H-labeled chylomicron-like emulsion following the acute phase of myocardial infarction. Int J Cardiol 2004;93:181–7.

[15] Groot PH, van Stiphout WA, Krauss XH, et al. Postprandial lipoprotein metabolism in normolipidemic men with and without coronary artery disease. Arterioscler Thromb 1991;11:653–62.

[16] Patsch JR, Miesenböck G, Hopferwieser T, et al. Relation of triglyceride metabolism and coronary artery disease. Studies in the postprandial state. Arterioscler Thromb 1992;12:1336–45.

[17] Nurmohamed MT, Heslinga M, Kitas GD. Cardiovascular comorbidity in rheumatic diseases. Nat Rev Rheumatol 2015;11 (12):693–704.

[18] Borba EF, Bonfá E, Vinagre CG, et al. Chylomicron metabolism is markedly altered in systemic lupus erythematosus. Arthritis Rheum 2000;43:1033–40.

[19] Oliveira MR, Maranhão RC. Plasma kinetics of a chylomicron-like emulsion in normolipidemic obese women after a short-period weight loss by energy-restricted diet. Metabolism 2002;51:1097–103.

[20] Oliveira MR, Maranhão RC. Relationships in women between body mass index and the intravascular metabolism of chylomicron-like emulsions. Int J Obes Relat Metab Disord 2004;28:1471–8.

[21] Silva VM, Vinagre CG, Dallan LA, et al. Plasma lipids, lipoprotein metabolism and HDL lipid transfers are equally altered in metabolic syndrome and in type 2 diabetes. Lipids 2014;49:677–84.

[22] Rocha MP, Maranhão RC, Seydell TM, et al. Metabolism of triglyceride-rich lipoproteins and lipid transfer to high-density lipo-protein in young obese and normal-weight patients with polycystic ovary syndrome. Fertil Steril 2010;93:1948–56.

[23] Bernardes-Silva H, Toffoletto O, Bortolotto LA, et al. Malignant hypertension is accompanied by marked alterations in chylomicron metabolism. Hypertension 1995;26:1207–10.

[24] Schmauss D, Weis M. Cardiac allograft vasculopathy. Recent developments. Circulation 2008;117:2131–41.

[25] Vinagre CG, Stolf NA, Bocchi E, et al. Chylomicron metabolism in patients submitted to cardiac transplantation. Transplantation 2000;69:532–7.

[26] Sakashita AM, Bydlowski SP, Chamone DA, et al. Plasma kinetics of an artificial emulsion resembling chylomicrons in patients with chronic lymphocytic leukemia. Ann Hematol 2000;79:687–90.

[27] Hungria VT, Brandizzi LI, Chiattone CS, et al. Metabolism of an artificial emulsion resembling chylomicrons in patients with multiple myeloma. Leuk Res 1999;23:637–41.

[28] Gonçalves RP, Hungria VT, Chiattone CS, et al. Metabolism of chylomicron-like emulsions in patients with Hodgkin's and with non-Hodgkin's lymphoma. Leuk Res 2003;27:147–53.

[29] Santos RD, Sposito AC, Ventura LI, et al. Effect of pravastatin on plasma removal of a chylomicron-like emulsion in men with coronary artery disease. Am J Cardiol 2000;85:1163–6.

[30] Sposito AC, Santos RD, Amâncio RF, et al. Atorvastatin enhances the plasma clearance of chylomicron-like emulsions in subjects with atherogenic dyslipidemia: relevance to the in vivo metabolism of triglyceride-rich lipoproteins. Atherosclerosis 2003;166:311–21.

[31] Santos RD, Ventura LI, Spósito AC, et al. The effects of gemfibrozil-upon the metabolism of chylomicron-like emulsions in patients with endogenous hypertriglyceridemia. Cardiovasc Res 2001;49:456–65.

[32] Spósito AC, Maranhão RC, Vinagre CG, et al. Effects of etofibrate upon the metabolism of chylomicron-like emulsions in patients with coronary artery disease. Atherosclerosis 2001;154:455–61.

[33] Mangili OC, Moron Gagliardi AC, Mangili LC, et al. Favorable effects of ezetimibe alone or in association with simvastatin on the removal from plasma of chylomicrons in coronary heart disease subjects. Atherosclerosis 2014;233:319–25.

[34] Benjó AM, Maranhão RC, Coimbra SR, et al. Accumulation of chylomicron remnants and impaired vascular reactivity occur in subjects with isolated low HDL cholesterol: effects of niacin treatment. Atherosclerosis 2006;187:116–22.

[35] César TB, Oliveira MR, Mesquita CH, et al. High cholesterol intake modifies chylomicron metabolism in normolipidemic young men. J Nutr 2006;136:971–6.

[36] Vinagre JC, Vinagre CC, Pozzi FS, et al. Plasma kinetics of chylomicron-like emulsion and lipid transfers to high-density lipoprotein (HDL) in lacto-ovo vegetarian and in omnivorous subjects. Eur J Nutr 2014;53:981–7.

[37] Vinagre JC, Vinagre CG, Pozzi FS, et al. Metabolism of triglyceride-rich lipoproteins and transfer of lipids to high-density lipoproteins (HDL) in vegan and omnivore subjects. Nutr Metab Cardiovasc Dis 2013;23:61–7.

[38] Morikawa AT, Maranhão RC, Alves MJ, et al. Effects of anabolic androgenic steroids on chylomicron metabolism. Steroids 2012;77:1321–6.

[39] Sposito AC, Santos RD, Hueb W, et al. LDL concentration is correlated with the removal from the plasma of a chylomicron-like emulsion in subjects with coronary artery disease. Atherosclerosis 2002;161:447–53.

[40] Carneiro MM, Miname MH, Gagliardi AC, et al. The removal from plasma of chylomicrons and remnants is reduced in heterozygous familial hypercholesterolemia subjects with identified LDL receptor mutations: study with artificial emulsions. Atherosclerosis 2012;221:268–74.

[41] Maranhão RC, Cesar TB, Pedroso-Mariani SR, et al. Metabolic behavior in rats of a nonproteinmicroemulsion resembling low-density lipoprotein. Lipids 1993;28:691–6.

[42] Hirata RD, Hirata MH, Mesquita CH, et al. Effects of apolipoprotein B-100 on the metabolism of a lipid microemulsion model in rats. Biochim Biophys Acta 1999;1437:53–62.

[43] Santos RD, Hueb W, Oliveira AA, et al. Plasma kinetics of a cholesterol-rich emulsion in subjects with or without coronary artery disease. J Lipid Res 2003;44(3):464–9.

[44] Santos RD, Chacra AP, Morikawa A, et al. Plasma kinetics of free and esterified cholesterol in familial hypercholesterolemia: effects of simvastatin. Lipids 2005;40:737–43.

[45] Vinagre CG, Ficker ES, Finazzo C, et al. Enhanced removal from the plasma of LDL-like nanoemulsioncholesteryl ester in trained men compared with sedentary healthy men. J Appl Physiol 2007;103:1166–71.

[46] Ficker ES, Maranhão RC, Chacra AP, et al. Exercise training accelerates the removal from plasma of LDL-like nanoemulsion in moderately hypercholesterolemic subjects. Atherosclerosis 2010;212:230–6.

[47] da Silva JL, Vinagre CG, Morikawa AT, et al. Resistance training changes LDL metabolism in normolipidemic subjects: a study with a nanoemulsion mimetic of LDL. Atherosclerosis 2011;219:532–7.

[48] Pinto LB, Wajngarten M, Silva EL, et al. Plasma kinetics of a cholesterol-rich emulsion in young, middle-aged, and elderly subjects. Lipids 2001;36:1307–11.

[49] Puk CG, Vinagre CG, Bocchi E, et al. Plasma kinetics of a cholesterol-rich microemulsion in patients submitted to heart transplantation. Transplantation 2004;78:1177–81.

[50] Couto RD, Dallan LA, Lisboa LA, et al. Deposition of free cholesterol in the blood vessels of patients with coronary artery disease: a possible novel mechanism for atherogenesis. Lipids 2007;42:411–8.

[51] Maranhão RC, Roland IA, Hirata MH. Effects of Triton WR 1339 and heparin on the transfer of surface lipids from triglyceride-rich emulsions to high density lipoproteins in rats. Lipids 1990;25 (11):701–5.

[52] Charles MA, Kane JP. New molecular insights into CETP structure and function: a review. J Lipid Res 2012;53(8):1451–8.

[53] Albers JJ, Vuletic S, Cheung MC. Role of plasma phospholipid transfer protein in lipid and lipoprotein metabolism. Biochim Biophys Acta 1821;2012:345–57.

[54] Rye KA, Barter PJ. Cardioprotective functions of HDLs. J Lipid Res 2014;55:168–79.

[55] Lo Prete AC, Dina CH, Azevedo CH, et al. In vitro simultaneous transfer of lipids to HDL in coronary artery disease and in statin treatment. Lipids 2009;44:917–24.

[56] Maranhão RC, Freitas FR, Strunz CM, et al. Lipid transfers to HDL are predictors of precocious clinical coronary heart disease. Clin Chim Acta 2012;413:502–5.

[57] Sprandel MC, Hueb WA, Segre A, et al. Alterations in lipid transfers to HDL associated with the presence of coronary artery disease in patients with type 2 diabetes mellitus. Cardiovasc Diabetol 2015;14:107–16.

[58] Giribela AH, Melo NR, Latrilha MC, et al. HDL concentration, lipid transfer to HDL, and HDL size in normolipidemicnonobese menopausal women. Int J Gynaecol Obstet 2009;104:117–20.

[59] Feitosa AC, Feitosa-Filho GS, Freitas FR, et al. Lipoprotein metabolism in patients with type 1 diabetes under intensive insulin treatment. Lipids Health Dis 2013;12:15.

[60] Martinez LR, Santos RD, Miname MH, et al. Transfer of lipids to high-density lipoprotein (HDL) is altered in patients with familial hypercholesterolemia. Metabolism 2013;62:1061–4.

[61] Puk CG, Bocchi EA, Lo Prete AC, et al. Transfer of cholesterol and other lipids from a lipid nanoemulsion to high-density lipo-protein in heart transplant patients. J Heart Lung Transplant 2009;28:1075–80.

[62] Sigal GA, Medeiros-Neto G, Vinagre JC, et al. Lipid metabolism in subclinical hypothyroidism: plasma kinetics of triglyceride-rich lipoproteins and lipid transfers to high-density lipoprotein before and after levothyroxine treatment. Thyroid 2011;21:347–53.

[63] Casella-Filho A, Chagas AC, Maranhão RC, et al. Effect of exercise training on plasma levels and functional properties of high-density lipoprotein cholesterol in the metabolic syndrome. Am J Cardiol 2011;107:1168–72.

[64] Bachi AL, Rocha GA, Sprandel MC, et al. Exercise training improves plasma lipid and inflammatory profiles and increases cholesterol transfer to high-density lipoprotein in elderly women. J Am Geriatr Soc 2015;63:1247–9.

[65] Vaisberg M, Bachi AL, Latrilha C, et al. Lipid transfer to HDL is higher in marathon runners than in sedentary subjects, but is acutely inhibited during the run. Lipids 2012;47:679–86.

[66] Gagliardi AC, Maranhão RC, de Sousa HP, et al. Effects of margarines and butter consumption on lipid profiles, inflammation markers and lipid transfer to HDL particles in free-living subjects with the metabolic syndrome. Eur J Clin Nutr 2010;64:1141–9.

[67] Cesar TB, Aptekmann NP, Araujo MP, et al. Orange juice decreases low-density lipoprotein cholesterol in hypercholesterolemic subjects and improves lipid transfer to high-density lipoprotein in normal and hypercholesterolemic subjects. Nutr Res 2010;30:689–94.

[68] Maranhão RC, Freitas FR. HDL metabolism and atheroprotection: predictive value of lipid transfers. Adv Clin Chem 2014;65:1–41.

[69] Maranhão RC, Garicochea B, Silva EL, et al. Increased plasma removal of microemulsions resembling the lipid phase of low-density lipoproteins (LDL) in patients with acute myeloid leukemia: a possible new strategy for the treatment of the disease. Braz J Med Biol Res 1992;25:1003–7.

[70] Maranhão RC, Garicochea B, Silva EL, et al. Plasma kinetics and biodistribution of a lipid emulsion resembling low-density lipoprotein in patients with acute leukemia. Cancer Res 1994;54:4660–6.

[71] Ho YK, Smith RG, Brown MS, et al. Low-density lipoprotein (LDL) receptor activity in human acute myelogenous leukemia cells. Blood 1978;52:1099–114.

[72] Vitols S, Angelin B, Ericsson S, et al. Uptake of low density lipoproteins by human leukemic cells in vivo: relation to plasma lipoprotein levels and possible relevance for selective chemotherapy. Proc Natl Acad Sci U S A 1990;87:2598–602.

[73] Graziani SR, Igreja FA, Hegg R, et al. Uptake of a cholesterol-rich emulsion by breast cancer. Gynecol Oncol 2002;85(3):493–7.

[74] Ades A, Carvalho JP, Graziani SR, et al. Uptake of a cholesterol-rich emulsion by neoplastic ovarian tissues. Gynecol Oncol 2001;82:84–7.

[75] Dias ML, Carvalho JP, Rodrigues DG, et al. Pharmacokinetics and tumor uptake of a derivatized form of paclitaxel associated to a cholesterol-rich nanoemulsion (LDE) in patients with gynecologic cancers. Cancer Chemother Pharmacol 2007;59:105–11.

[76] Pires LA, Hegg R, Valduga CJ, et al. Use of cholesterol-rich nanoparticles that bind to lipoprotein receptors as a vehicle to paclitaxel in the treatment of breast cancer: pharmacokinetics, tumor uptake and a pilot clinical study. Cancer Chemother Pharmacol 2009;63:281–7.

[77] Rodrigues DG, Covolan CC, Coradi ST, et al. Use of a cholesterol-rich emulsion that binds to low-density lipoprotein receptors as a vehicle for paclitaxel. J Pharm Pharmacol 2002;54(6):765–72.

[78] Maranhão RC, Graziani SR, Yamaguchi N, et al. Association of carmustine with a lipid emulsion: in vitro, in vivo and preliminary studies in cancer patients. Cancer Chemother Pharmacol 2002;49:487–98.

[79] Rodrigues DG, Maria DA, Fernandes DC, et al. Improvement of paclitaxel therapeutic index by derivatization and association to a cholesterol-rich microemulsion: in vitro and in vivo studies. Cancer Chemother Pharmacol 2005;55(6):565–76.

[80] Valduga CJ, Fernandes DC, Lo Prete AC, et al. Use of a cholesterol-rich microemulsion that binds to low-density lipoprotein receptors as vehicle for etoposide. J Pharm Pharmacol 2003;55:1615–22.

[81] Almeida CP, Vital CG, Contente TC, et al. Modification of composition of a nanoemulsion with different cholesteryl ester molecular species: effects on stability, peroxidation, and cell uptake. Int J Nanomedicine 2010;5:679–86.

[82] Kretzer IF, Maria DA, Maranhão RC. Drug-targeting in combined cancer chemotherapy: tumor growth inhibition in mice by association of paclitaxel and etoposide with a cholesterol-rich nanoemulsion. Cell Oncol 2012;35(6):451–60.

[83] Hungria VT, Latrilha MC, Rodrigues DG, et al. Metabolism of a cholesterol-rich microemulsion (LDE) in patients with multiple myeloma and a preliminary clinical study of LDE as a drug vehicle for the treatment of the disease. Cancer Chemother Pharmacol 2004;53(1):51–60.

[84] Pinheiro KV, Hungria VT, Ficker ES, et al. Plasma kinetics of a cholesterol-rich microemulsion (LDE) in patients with Hodgkin's and non-Hodgkin's lymphoma and a preliminary study on the toxicity of etoposide associated with LDE. Cancer Chemother Pharmacol 2006;57:624–30.

[85] Naoum FA, Gualandro SF, Latrilha MdaC, et al. Plasma kinetics of a cholesterol-rich microemulsion in subjects with heterozygous beta-thalassemia. Am J Hematol 2004;77(4):340–5.

[86] Maranhão RC, Tavares ER, Padoveze AF, et al. Paclitaxel associated with cholesterol-rich nanoemulsions promotes atherosclerosis regression in the rabbit. Atherosclerosis 2008;197:959–66.

[87] Ross R, Glomset JA. The pathogenesis of atherosclerosis (first of two parts). N Engl J Med 1976;295:369–77.

[88] Libby R. Inflammation in atherosclerosis. Arterioscler Thromb Vasc Biol 2012;32:2045–51.

[89] Hansson GK, Libby P, Tabas I. Inflammation and plaque vulnerability. J Intern Med 2015;278(5):483–93.

[90] Tavares ER, Freitas FR, Diament J, et al. Reduction of atherosclerotic lesions in rabbits treated with etoposide associated with cholesterol-rich nanoemulsions. Int J Nanomedicine 2011;6:2297–304.

[91] Bulgarelli A, Martins Dias AA, Caramelli B, et al. Treatment with methotrexate inhibits atherogenesis in cholesterol-fed rabbits. J Cardiovasc Pharmacol 2012;59(4):308–14.

[92] Bulgarelli A, Leite Jr AC, Dias AA, et al. Anti-atherogenic effects of methotrexate carried by a lipid nanoemulsion that binds to LDL receptors in cholesterol-fed rabbits. Cardiovasc Drugs Ther 2013;27:531–9.

[93] Leite Jr. AC, Solano TV, Tavares ER, et al. Use of combined chemotherapy with etoposide and methotrexate, both associated to lipid nanoemulsions for atherosclerosis treatment in cholesterol-fed rabbits. Cardiovasc Drugs Ther 2015;29:15–22.

II. ENDOTHELIAL DYSFUNCTION AND CLINICAL SYNDROMES

[94] Mello SB, Tavares ER, Bulgarelli A, et al. Intra-articular methotrexate associated to lipid nanoemulsions: anti-inflammatory effect upon antigen-induced arthritis. Int J Nanomedicine 2013;8:443–9.

[95] Lourenço-Filho DD, Maranhão RC, Méndez-Contreras CA, et al. An artificial nanoemulsion carrying paclitaxel decreases the transplant heart vascular disease: a study in a rabbit graft model. J Thorac Cardiovasc Surg 2011;141:1522–8.

[96] Arora S, Gullestad L. The challenge of allograft vasculopathy in cardiac transplantation. Curr Opin Organ Transplant 2014;19:508–14.

[97] Maranhão RC, Tavares ER. Advances in non-invasive drug delivery for atherosclerotic heart disease. Expert Opin Drug Deliv 2015;12:1135–47.

Further Reading

[1] Ross R, Glomset JA. The pathogenesis of atherosclerosis (second of two parts). N Engl J Med 1976;295:420–5.

DIET AND ENDOTHELIUM

Influence of Diet on Endothelial Dysfunction

Ana Maria Lottenberg, Maria Silvia Ferrari Lavrador, Milessa Silva Afonso, and Roberta Marcondes Machado

INTRODUCTION

Several causes, among them genetics and lifestyle, in which the quality of the diet plays a crucial role, are known to be involved in the genesis of cardiovascular disease (CVD). The relationship between diet and coronary artery disease was one of the main common points in health research for almost half a century [1]. The pioneer study was published by Keys et al., who demonstrated in the Seven Countries Study the strong association between cardiovascular risk with fat consumption greater than 30% of total calories [2]. In subsequent years, the strong association between the effect of nutrients on cardiovascular risk was demonstrated in a broad variety of epidemiological, clinical and experimental studies [3–7]. All these findings, along with the advances in nutrigenomics and molecular biology, motivated the development of research protocols with the objective of evaluating the effect of micro and macronutrients on several steps enrolled in endothelial dysfunction, which, along with artery thickening, is characterized as an early predictor of CVD [8,9]. The results of these studies have strengthened the nutrition guidelines which advocate a diet free of trans fatty acids and the consumption of up to 7% of saturated fatty acids [10–12].

Although nutritional recommendation is based primarily on the percentage of macronutrients (carbohydrates, proteins, and fats), recently the same international guidelines began to prioritize the recommendation of healthy eating pattern [10], with the appreciation of the type and variety of foods consumed in the diet. This new approach is based on the outcome of important epidemiological and interventional studies, such as DASH (Dietary Approach to Stop Hypertension) [13], Interheart [14], and Predmed (*Prevención con Dieta Mediterránea*) [15]. The DASH study is considered the best clinical investigation involving the relevance of diet in controlling systemic blood pressure (SBP). In this study, it was shown that the consumption of fruits and vegetables enhances the effect of salt reduction on hypertension [3]. It is noteworthy that this effect was even more pronounced with the adequate consumption of fats and with the inclusion of low fat dairy products. In Interheart Study (case-control) [14], dietary consumption of the population from 52 countries was evaluated and 3 dietary standards used throughout the world were identified. The first, characterized by a normal amount of fat and small amount of fruits, was defined as Oriental. The second, called Western, presented a high content of fat in the diet and low amount of fruits and vegetables. The other pattern, defined as Prudent, was similar to Oriental pattern in regard to fat content, but has a higher amount of fruits and vegetables. The latter was associated with lower risk for acute myocardial infarction (AMI) and granted cardiovascular protection. In addition, an important systematic review conducted by a team from the University of Toronto [1] assessed the results of randomized and controlled studies, and demonstrated that the Mediterranean nutrition standard was clearly associated with lower cardiovascular risk.

Hence, it becomes evident the strong influence of the diet on all steps associated with the development of atherosclerosis, i.e., from the retention of LDL-cholesterol particle, recruitment and differentiation of immune cells in the subendothelial space until its subsequent interaction between endothelial cells, platelets, and smooth muscle cells [4,16,17], conditions that alter the endothelial function.

Endothelium is capable of recognizing changes in hemodynamic forces and respond with synthesis and secretion of vasoactive substances, like nitric oxide (NO), which is able to promote relaxation of the smooth muscle, inducing vasodilation and bronchodilation [18]. Therefore, in response to different stimuli, such as shear stress or hypoxia, the endothelial cell secrets vasoactive molecules which coordinate the vasomotor function promoting vasoconstriction or vasodilation. The main vasoconstrictor factors are prostaglandin H2 (PGH2) and thromboxane A2 (TXA2) (derived from *arachidonic* acid), endothelin-1 (ET-1), and reactive oxygen species (ROS), while the main vasodilators

are NO and endothelium-derived hyperpolarizing factor (EDHF) [19,20].

Different stimuli can trigger endothelial activation, including the presence of oxidized LDL (OxLDL), lipopolysaccharide (LPS), cytokines and free radicals, dyslipidemia, hyperhomocysteinemia, hyperglycemia, hyperinsulinemia and smoking. Once activated, the endothelium expresses adhesion molecules and releases cytokines and chemokines [21,22] (Fig. 24.1).

Reducing plasma LDL concentrations is one of the main goals for cardiovascular risk reduction. This is due to the fact that high LDL concentrations, especially small and dense particles, favor their retention in the subendothelial space, where they are modified, whether by oxidative or glycation processes [22,23].

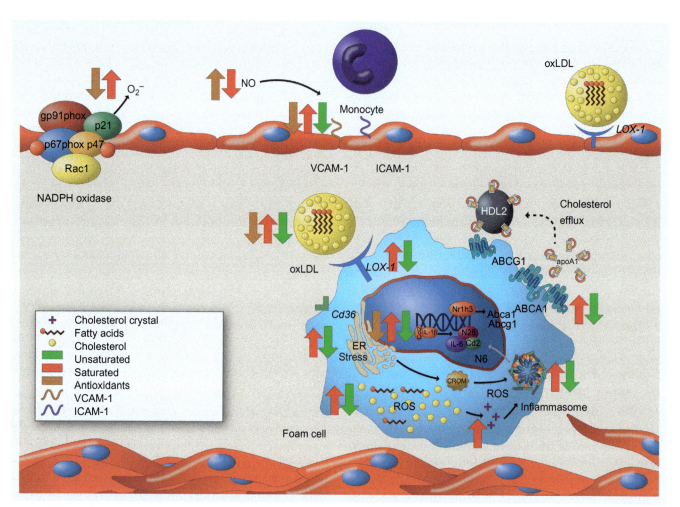

FIG. 24.1 *Endothelial dysfunction and atherosclerosis development. ABCA1,* ATP-A1 binding cassette transporter; *ABCG1,* ATP-G1 binding cassette transporter; *Ccl2/MCP-1,* monocyte chemoattractive protein; *CD36,* differentiation cluster 36; *ER,* endoplasmic reticulum; *HDL,* high-density lipoprotein; *ICAM-1,* intercellular adhesion molecule-1; *IL-1β,* interleukin-1β; *IL-6,* interleukin-6; *LOX-1,* oxidized LDL receptor-1; *NF-κB,* nuclear factor-κB; *NO,* nitric oxide; *Nr1h3/LXR-,* liver X receptor; *O2⁻,* anion superoxide; *oxLDL,* oxidized LDL; *ROS,* reactive oxygen species; *VCAM-1,* vascular cell adhesion molecule [23–27].

Cholesterol homeostasis in arterial macrophages is not only governed by the uptake of LDL particles, but also by the intracellular removal of cholesterol in a mechanism known as reverse cholesterol transport (RCT). In this process, HDL particle removes cholesterol from peripheral tissues and macrophages present in the intima of the vessel and send it to the liver to be excreted in the bile. Thus, apolipoprotein A-I (Apo A-I), the main HDL protein, removes cholesterol through the interaction with ATP-binding cassette transporter A1 (ABCA1), while HDL interacts with ABCG1 to remove cellular cholesterol [28,29]. This process is fundamental to cholesterol homeostasis in peripheral cells, since they are not able to degrade it.

Macrophages contained within the atherosclerotic plaque are capable of secreting proinflammatory cytokines [23,30], as well as growth factors and metalloproteinases, that induce cell proliferation and degradation of the matrix, which can change the stability of the atherosclerotic plaque [31]. Endothelial cell injury leads to the activation of different cell types (endothelial cells, platelets, and leukocytes), releasing inflammatory mediators, such as cytokines and growth factors, which will induce phenotypic alterations in smooth muscle cells (SMC). SMC exist in different phenotypes, the most common is the quiescent or contractile subtype which has the main function of regulating the vessel tonus. In response to deleterious stimuli, SMC is by the synthetic phenotype. Unlike the contractile phenotype, the synthetic has little control on the regulation of the contractility of the vessel; on the other hand, it has a greater capacity to generate extracellular matrix proteins and to promote migration to the intima of the vessel and cell proliferation [32]. Furthermore, once activated, SMCs are able to express receptors for lipoproteins and promote the uptake of lipids, such as modified LDL, leading to the formation of foam cells in a similar manner to macrophages, contributing to the lesion progression [33,34].

Choi et al. [35] demonstrated that the SMC in the intima of the vessel presents lower ABCA1 expression and thus has less binding capacity to Apo A-I. This leads to lower HDL formation and reduction of cellular cholesterol efflux, further favoring intracellular lipid accumulation.

Due to the fundamental importance of the diet in almost all these steps described above, this chapter aims to discuss the most important findings in the literature related to the role of micro and macronutrients in lipid metabolism, as well as on innate and adaptive immunity, conditions closely involved in the genesis of endothelial dysfunction.

DIETARY FATS

It has already been established that both the amount and the type of fatty acids exert direct influence on the concentration of lipids and plasma lipoproteins, insulin resistance, and consequently on cardiovascular risk [36–38]. This is due to the type of fatty acid present in the fat, which can modulate several regulatory mechanisms at the cellular level [39].

In general, the investigations suggest a protective role of polyunsaturated fat on the development of CVD [7]; accordingly its consumption was stimulated by the American Heart Nutrition Committee [40]. In regard to monounsaturated fatty acids, when compared to the consumption of saturated fat, they are capable of reducing plasma cholesterol concentration [41]. In addition, it is well documented that Mediterranean populations, which consume high quantities of oleic acid, have lower prevalence of obesity, metabolic syndrome, type 2 diabetes mellitus (T2DM) and cardiovascular events [42]. On the other hand, high amount of saturated fatty acids [43] and the intake of trans fatty acids are consensually appointed as atherogenic, increasing the cardiovascular risk [4,44].

POLYUNSATURATED FATTY ACIDS

Polyunsaturated fatty acids have two or more double bonds in the carbon chain and the location of the first double bond, from the methyl terminal, is identified by the letter ω and determine the fatty acid series. Thus, these fatty acids are classified into omega 3 and omega 6.

Alpha linolenic acid (ALA; C18:3, omega 3), as well as linoleic acid (C18:2, omega 6), are found mostly in seeds and their oils. Linoleic acid is found in all vegetable oils (soybean, canola, sunflower and corn), while ALA is found in significant quantities only in canola and soybean oils. As they are not synthesized by the human organism, they are considered essential and must be provided by the diet. Docosahexanoic acid (DHA; C22:6, omega 3), eicosapentaenoic acid (EPA; C20:5, omega 3), and arachidonic acid (AA; C20:4, omega 6), can be synthesized from essential fatty acids through successive steps that involve the action of delta-6 desaturase and elongase enzymes [18]. However, most of the tissue and circulating content of EPA and DHA comes from diet, since this conversion is quite limited in humans [45].

Greater attention has been given to the effects of different classes of polyunsaturated fatty acids in CVD prevention, especially those from omega 3 series. Although some studies support a cardio-protective action from ALA [46], the beneficial action of the polyunsaturated fatty acids seems to be more associated with the consumption of EPA and DHA present in fish of very cold and deep water [45,47,48].

The effects of omega 3 fatty acids series upon cardiovascular outcome include: reduction of triglyceridemia [49], improvement of the endothelial function [50] and antiinflammatory action, with reduction of inflammatory biomarkers such as interleukin-1β (IL-1β), tumor necrosis factor-α (TNF-α), and interleukin 6 (IL-6) [50], and antithrombotic and antiarrhythmic effects. Furthermore, in vitro and in vivo studies suggest a direct action of omega-3 fatty acids in the electrophysiology of atrial and ventricular myocytes, which would lead to a reduction of arrhythmia episodes [51].

The mechanisms by which polyunsaturated fatty acids reduce cholesterolemia involve: (1), alteration in the spatial configuration of the lipids into which they are incorporated due to the angles present in the carbon chain of the polyunsaturated fatty acids; as a result they occupy more space within LDL particles reducing the volume available to transport cholesterol [52]; (2) reduction on hepatic synthesis of apolipoprotein B (apo B)-rich particles [53,54] associated with increased LDL catabolism, which may be involved in the increase in LDL receptor (LDLr) activity, as well as increased particle fluidity granted by cholesterol and triglycerides (TG) contents, which may affect its surface properties, influencing the catabolism of the particle [55]; (3) decreased transfer of cholesterol esters from HDL to VLDL/LDL particles mediated by cholesterol ester transfer protein (CETP) [56].

The reduction of triglyceridemia mediated by polyunsaturated fatty acids is related to their action on sterol regulatory element-binding protein (SREBP), a transcription factor that regulates the transcription of genes related to fatty acid, TG, and cholesterol biosynthesis. SREBP transcription is dependent on the liver X receptor (LXR), which, once activated by cholesterol oxides, forms heterodimers with retinoic X receptor (RXR), binds to the sterol responsive element (SRE), in the nucleus, activating the transcription of their target genes [57]. Polyunsaturated fatty acids act as LXR antagonists, preventing its activation and, consequently, SREBP transcription [58].

In addition, the incorporation of polyunsaturated fatty acids by the phospholipids (PL) of the endoplasmatic reticulum membrane interferes with membrane compartmentalization, and also inhibits the degradation of insulin-induced gene protein (INSIG), which maintains the complex SREBP cleavage-activating protein (SCAP)-SREBP anchored in the endoplasmic reticulum membrane, preventing the maturation and translocation of SREBP to the nucleus, thus reducing its transcriptional activity [59–61].

Other effects of unsaturated fatty acids (UFA) on lipids metabolism involve their capability of reducing ABCA1 and ABCG1 content on the plasma membrane, reducing the ability to remove cellular cholesterol. The mechanisms described involve suppression of transporters transcription, as well as post translational modulations that induce ABCA1 degradation [62–64]. Nonetheless, it is important to emphasize that unsaturated fatty acids reduce plasma cholesterol concentrations as well as atherogenic lipoproteins [53], and also diminish the expression of receptors involved in OxLDL by macrophages, reducing intracellular cholesterol accumulation [65,66]. In this context, the effect of UFA on reducing the expression of proteins involved in cholesterol removal does not grant atherogenic potential.

It has already been demonstrated that omega 3 fatty acids reduce the expression of chemoattractant and adhesion molecules, and promote plaque stability [18]. This effect is more pronounced the greater the number of double bonds present in the fatty acid carbon chain [67–69]. Treatment of endothelial cells with DHA prevented their activation even when stimulated with cytokines, inhibited the expression of adhesion molecule (vascular cellular adhesion molecule 1—VCAM-1, E-selectin, and intercellular adhesion molecule 1—ICAM-1) in a dose-dependent manner and reduced the secretion of proinflammatory cytokines (IL-6 and interleukin-8, or IL-8) as well as monocyte adhesion. These findings correlate positively with the incorporation of DHA in membrane phospholipids (PL) and inversely with the incorporation of omega 6 [70].

In another study, it was observed that EPA increased NO production by endothelial cells even in the presence of high-glucose concentrations [71]. It is known that in high-concentrations glucose is, in part, converted to sorbitol in a reaction catalyzed by aldose reductase, that uses NADH as a cofactor, which is also necessary for the activity of endothelial nitric oxide synthase (eNOS). This competition for the cofactor may lead to a decreased NO production by endothelial cells. In addition, intracellular accumulation of sorbitol can directly contribute to endothelial cell dysfunction [72]. The addition of EPA to the culture medium enriched with glucose prevented the decrease in NO production induced by glucose, and increased intracellular calcium concentration. The study also suggests that EPA action in the increase of NO production is dependent on calcium-calmodulin system, since the effect in increasing NO production was abolished by the use of calmodulin inhibitor [71]. The eNOS enzyme, constitutively expressed in endothelial cells is calcium-calmodulin dependent. In addition, treatment with EPA has increased the concentration of this fatty acid in endothelial cell plasma membrane, increasing the molar ratio EPA/AA. This alteration in the composition of membrane PL of the endothelial cells may reflect in the type of prostanoids and prostaglandins to be formed from the fatty acid released from the PL [71].

A recent study in humans showed that omega 3 supplementation (2 g; 46% EPA and 38% DHA) for 12 weeks reduced IL-6 plasma concentrations, increased plasminogen activator inhibitor (PAI-1) and improved the lipid profile of individuals with metabolic syndrome. The improvement of metabolic profile favored endothelial function, leading to a linear increase in flow-mediated dilatation [73].

Fatty acid content of LDL reflects the type of fat in the diet [74]. Thus, a diet rich in polyunsaturated fats can increase LDL susceptibility to oxidation, when compared to those rich in saturated and monounsaturated [75,76], since the unsaturations in the carbon chain of polyunsaturated fatty acids make them less stable and more susceptible to peroxidation [77]. Once oxidized, LDL particles are taken up by macrophage scavenger receptors, which may aggravate the atherosclerotic process [77,78]. However, contrary to these results, Calzada et al. [79]. showed in normolipidemic individuals that DHA supplementation with doses between 200 and 800 mg/day, conferred a protective and antioxidant effect to LDL. The discrepancy between the results may be attributed to different reasons, such as gender, age, and lipid profile of participants, as well as different doses of omega 3 utilized in the studies.

Other molecules derived from omega 3 fatty acids EPA e DHA are involved in the inflammation resolution. These compounds, known as resolvins and protectins promote a series of actions that seek to restore the homeostasis of the affected tissue. Antiinflammatory bioactivity of these mediators includes: reduction of inflammatory cell migration to the inflammation site, reduction of the proinflammatory cytokine synthesis, and clearance of apoptotic cells [80,81].

In addition to the classic antiinflammatory effects of omega 3 polyunsaturated fatty acids, related to their ability to synthesize eicosanoids from the odd series, more recently, another pathway by which omega 3, more specifically DHA and EPA, exert antiinflammatory action was described. These fatty acids are ligands of G-protein couple receptor 120 (GPR120), which are broadly expressed in adipocytes and inflammatory macrophages [82,83]. Once activated by DHA, this receptor recruits beta-arrestin 2, blocking the activity of proteins upstream of nuclear factor kappa B (NF-κB) signaling pathway and, therefore, preventing the secretion of TNF-α and IL-6 by macrophages stimulated with LPS [82]. Recent studies have shown that polymorphisms in GPR120 with loss of function were associated with obesity and insulin resistance, both in animal models and humans [84,85]. Additionally, omega 3 fatty acids activate the AMPK/SIRT1 (protein kinase activated by AMP/sirtuin-1) pathway and repress NF-κB transcriptional activity in macrophage stimulated with LPS, confirming the antiinflammatory role of omega 3 fatty acids [86].

These antiinflammatory mechanisms mediated by polyunsaturated fatty acids contribute to the reduction of inflammasome assembly, since NF-κB activation presensitizes macrophages for its formation. The inflammasome is a multiprotein complex, an innate immune system component, which triggers inflammatory cascade by the activation of caspase-1, leading to the secretion of interleukin-1β (IL-1β) and interleukin 18 (IL-18) [87]. It was demonstrated that the nutrient-responsive component, known as NLR family pyrin domain-containing 3 (NLRP3), can be activated by saturated fatty acids leading to IL-1β secretion. On the other hand, the unsaturated fatty acids prevent inflammasome assembly induced by saturated fatty acids in human monocytes/macrophages, counteracting IL-1β secretion [88].

The progression of atherosclerotic lesion is directly associated with macrophage infiltration [6,89]. However, classification of the macrophage subtype present in the lesion depends on the identification of their polarization, which can be either M1 (classic activation), associated with proinflammatory cytokines (TNF-α and IL-6) production, or M2 (alternative activation), related to the tissue repair process and the production of antiinflammatory cytokines (IL-4, transforming growth factor beta—TGF-β). Initial lesions show a predominance of M2 subtype, however with the lesion progression an increase of M1 inflammatory macrophages also takes place [89].

Polyunsaturated fatty acids can modulate macrophage polarization. LDLr-KO mice fed an omega 3 enriched diet had lower number of Ly6C^{hi} monocytes, associated with M1 subtype, when compared to animals consuming palm oil [90]. The macrophages incubated with DHA presented higher efferocytosis capability, as well as production of antiinflammatory cytokines and markers associated with M2 polarization. These DHA effects were dependent on the activation of peroxisome proliferator-activated receptor-γ (PPAR-γ) [91], which plays a key role in programming macrophage alternative activation [92].

The excess of free cholesterol is toxic to the cells, thus the enzyme acyl-coenzyme A (CoA):cholesterol acyltransferase (ACAT) esterifies the cholesterol so it may be stored in structures known as lipid droplets [93]. The release of cholesterol from lipid droplets can occur by the action of neutral cholesterol ester hydrolase (nCEH) [93], or by autophagy, in a process known as macrolipophagy, in which occurs the fusion of the lipid droplets with lysosomes, which contains acid lipases [94]. It has been shown that during atherosclerosis progression, autophagy becomes progressively deffective, favoring the accumulation of lipid droplets and keeping the inflammasome complex activated [95], since autophagy is required for cytosolic protein degradation. Omega 3 fatty acids also reduce inflammasome complex activation by promoting its degradation in an autophagy-mediated process [96].

MONOUNSATURATED FATTY ACIDS

Monounsaturated fatty acids have a single double bond, being oleic acid (18:1, omega 9) the main

representative in the diet and the major sources are olive and canola oil. The concept that these fatty acids offer protection against CVD comes especially from studies carried out in populations of Mediterranean regions, which present low CVD prevalence and whose main fat component in the diet is olive oil, rich in monounsaturated fatty acids [97,98].

Oleic acid has a neutral effect on cholesterolemia, especially because it is the main substrate of the cholesterol-esterifying enzyme, ACAT, reducing the intracellular pool of free cholesterol, which contributes to the increase in the activity of hepatic LDL receptors [99,100]. When concentrations of free cholesterol rise, ACAT esterifies cholesterol so it can be stored in lipid droplets [93]. While this effect may be considered beneficial because it decreases free cholesterol toxicity, the accumulation of esterified cholesterol in lipid droplets can hinder cellular cholesterol efflux mediated by ABCA1 and ABCG1 receptors, thus favoring cellular accumulation of lipids [101].

It has already been demonstrated that the type of fatty acid is able to influence LDL susceptibility to oxidation. Thus, when comparing LDL from individuals who consume a diet rich in oleic acid, with LDL rich in omega 6 polyunsaturated fatty acids, it was observed that omega 6 rich particles were more susceptible to oxidation and capable of stimulating inflammatory pathways associated with macrophage trafficking and expression of genes that encode adhesion molecules [102]. It is noteworthy that, unlike omega 3, the omega 6 polyunsaturated fatty acids, such as AA, are known for their proinflammatory capability [103], especially because they are substrates for the synthesis of prostaglandin E2 (PGE2), TXA2 and leukotriene B4 (LTB4), which are associated with vasoconstriction, platelet aggregation and inflammation [17]. It is known that the cell membrane lipids composition reflects the fatty acids in the diet and that this composition promotes differences in membrane fluidity, which can interfere with protein activity and also cell signaling. Thus, in addition to its greater susceptibility to oxidation, the incorporation of omega 6 fatty acids in the plasma membrane of inflammatory cells can modulate its response to different stimuli [17].

Besides being more stable and less susceptible to oxidation, oleic acid is also capable of reducing endothelial cell activation. Massaro et al. [104] demonstrated that treatment with oleic acid reduced endothelial cells activation induced by LPS and

IL-1, prevented depletion of intracellular glutathione pool, as well as NF-κB activation, and inhibited the expression of VCAM-1 and secretion of macrophage colony stimulating factor (MCSF) in the culture medium. It is worth noting that the effects of oleic acid observed in this study were directly linked to the increase of their incorporation in cellular lipids.

SATURATED FATTY ACIDS

Saturated fatty acids are naturally present in the diets, mainly as palmitic acid (16:0) and, in lesser quantities, as myristic acid (14:0), stearic acid (18:0) and lauric acid (12:0), which is barely found in the diet. In addition to animal fats (palmitic), vegetable fats such as coconut (lauric), palm (palmitic), and cocoa (stearic) are food sources of saturated fatty acids.

There is a consensus in the literature that a diet rich in saturated fat increases plasma cholesterol, when compared to polyunsaturated fat, and several mechanisms of action have been described:

(1) Saturated fatty acids present straight carbon chain and therefore are linearly pack in the lipoprotein core, allowing these particles to transport larger quantities of cholesterol ester [52].

(2) Lower LDLr expression and/or activity reduces the removal of LDL particles from plasma [55,105,106]; in this sense, saturated fat is capable of reducing LDLr expression [107,108], and activity when consumed in association with cholesterol, possibly due to the alteration in membrane fluidity due to the incorporation of saturated fatty acids into PL [109].

(3) Saturated fatty acids induce hepatic expression of peroxisome-proliferator-activated receptor-gamma coactivator 1β (PGC-1β), which induces the expression of transcription factors related to hepatic lipid synthesis, increasing synthesis, production and secretion of VLDL rich in cholesterol and TG [110].

An experimental study performed in LDL receptor knockout mice (LDLr-KO) fed a cholesterol-rich diet and supplemented with palmitic, omega 3, or omega 6 fatty acids, has shown that saturated fatty acid induced the development of a greater area of atherosclerotic lesion and increased recruitment of inflammatory cells to the arterial intima, a fact that was prevented by the consumption of omega 3 and omega 6 fatty acids [90].

Wang et al. demonstrated that LDLr-KO mice fed a diet rich in saturated fats supplemented with omega 6: omega 3 had smaller lesion area, less accumulation of cholesterol in macrophages and lower expression of inflammatory markers (IL-6, TNF-α and monocyte chemotactic protein 1, or MCP-1) when they received the smaller ratio of omega 6: omega 3 when compared with those who received only omega 6 supplementation [103]. In another study, LDLr-KO mice fed with different hyperlipidemic diets showed that a diet rich in omega 6 induced more pronounced inflammatory response, compared to a diet rich in saturated fat, with higher concentrations of IL-6 and TNF-α in plasma and in the culture medium of LPS-stimulated peritoneal macrophages; however, animals that consumed a diet rich in omega 6 showed lower plasma cholesterol concentrations and smaller atherosclerosis lesion areas [6]. These studies demonstrate that even with unfavorable inflammatory profile, plasma lipid concentrations are preponderant for the atherosclerotic lesion progression, confirming the lipid hypothesis [24].

It is known that saturated fatty acids, such as palmitic and stearic, precursors of ceramides, are capable of activating macrophage inflammatory response increasing TNF-α concentrations in the culture medium [111–113]. Part of the inflammatory events caused by saturated fatty acids is mediated by Toll-like receptors family (TLR2 and TLR4), which trigger a signaling cascade mediated by serine kinases c-Jun N-terminal (JNK) and IkB kinase (IKK), proteins upstream of NF-κB and activator protein-1 (AP-1) [25].

The activation of NF-κB mediated by saturated fatty acids favors synthesis and secretion of proinflammatory cytokines, such as TNF-α, IL-6 and MCP-1, contributing to greater recruitment of inflammatory cells and accumulation of cholesterol in the lesion area [103]. It was also demonstrated that NF-κB activation by saturated fatty acids induces the expression of cyclooxygenase-2 (COX-2), interleukin 1α (IL-1α), and inducible nitric oxide synthase (iNOS) [114].

In another study, it was demonstrated that saturated fatty acid promotes maturation of dendritic cells, increasing the expression of class II histocompatibility molecules (MHC-II) and costimulator molecules inducing proinflammatory cytokine secretion

(IL-12p70 and IL-6), leading to naive T cell activation. This regulation is, in part, modulated by TLR 4 and inhibited by long chain polyunsaturated fatty acids [115].

Other inflammatory effects of saturated fatty acids include the presensitization of macrophages via transcriptional activation of NF-κB, and activation of inflammasome complex assembly, inducing caspase-1 cleavage and raising the secretion of IL-1β and IL-18 [116,117]. Furthermore, by increasing LDL particle uptake, saturated fatty acids induce intracellular accumulation of free cholesterol leading to the formation of cholesterol crystals, which, in turn, stimulate the inflammasome activation [118].

Recruitment of inflammatory cells, especially monocytes to the subendothelial space, is a *sine qua non* condition for atherosclerotic lesion formation. In this environment, MCSF converts monocytes into macrophages, which express receptors responsible for the uptake of modified LDL, such as oxidized LDL receptor-1 (LOX-1), cluster of differentiation 36 (CD36), and scavenger receptors class A and B (SRA and SR-BI, respectively) [23,26]. Saturated fatty acids induce MCSF-mediated differentiation and increase the expression of receptors involved in modified LDL uptake when compared to unsaturated fatty acids [65,66,104,119].

One of the possible mechanisms by which saturated fatty acids induce scavenger receptor expression involves the activation of endoplasmic reticulum (ER) stress [120]. ER stress is characterized by accumulation of unfolded or misfolded proteins in the reticulum, exceeding the capacity of the organelle and inducing an adaptive response known as UPR (unfolded protein response), which aims to restore reticulum homeostasis. In this process, the cell first reduces protein synthesis, in order to diminish the accumulation of proteins in the reticulum. The reticulum ability to process unfolded proteins is then increased, and finally, if reticulum homeostasis is not restored, apoptosis ensues [27]. Recent studies have shown that chronic conditions such as inflammation, dyslipidemia, and hyperglycemia, associated with obesity and diabetes, can induce ER stress [121].

It has been demonstrated that palmitic acid promotes the activation of ER stress in macrophages, leading to increased LOX-1 expression, an effect that was mitigated by unsaturated fatty acids (oleic and linoleic), since they have suppressed ER stress induced by palmitic acid [66]. A study from the same group also showed that the increased LOX-1 expression in macrophages, induced by palmitic acid, led to an increased OxLDL uptake [65], which favors the accumulation of cholesterol in macrophages, leading to foam cell formation and contributing to atherosclerosis development.

Giving induction of ER stress, as well as inflammatory stress and oxidative stress, saturated fatty acids contribute to macrophage apoptosis, which can be cleared by other phagocytic cells, in a process known as efferocytosis [122]. It has been reported that in advanced plaques this mechanism becomes inefficient, and larger concentration of saturated fatty acids in the plasma membrane can undermine the process further. On the other hand, omega 3 polyunsaturated fatty acids can reestablished efferocytosis in obese LDLr-KO mice by a phosphatidyl inositol 3-kinase (PI3K) dependent mechanism [123].

It is known that high-glucose concentrations can activate both T cells and endothelial cells, turning them responsive to insulin. Incubation with palmitic acid activate both cells lineages, inducing insulin receptor expression, glucose transporter 4 (GLUT 4), and increased production of reactive oxygen species and proinflammatory cytokins secretion (TNF-α, IL-2, IL-6, and IL-1β). These results indicate that chronic activation of both cells either by hyperglycemia or by elevated palmitic acid concentrations can contribute to CVD progression [124].

Reaffirming all the mechanisms described above, the replacement of 5% of the caloric value provided in the form of saturated by unsaturated fatty acids reduces cardiovascular risk in 42%, and replacement of 2% of calories in the form of trans by unsaturated fatty acids can reduce ~53% of the cardiovascular risk [98].

TRANS FATTY ACIDS

Trans fatty acids are geometric isomers of cis fatty acids. They can be produced through the biohydrogenation of fat by microbial action in ruminant animals [125] or by industrial hydrogenation of vegetable oils [126]. In the diet, trans fatty acid is mainly found in the form of elaidic acid (trans 18:1, n-9) and basically comes from the consumption of industrialized products prepared with hydrogenated fat.

Trans fatty acids have deleterious effects and their consumption increase TG, total cholesterol and

LDL-c and VLDL-c plasma concentrations. In addition, when compared to saturated fatty acids, trans fatty acids reduce HDL-c concentrations [127,128]. Trans consumption increases apo AI catabolism, leading to lower prebeta HDL and HDL formation, and decreases apolipoprotein B100 (Apo B100) catabolism, increasing atherogenic particles plasma concentration [129]. Furthermore, trans intake increases small and dense LDL particles concentration, which are more atherogenic as they infiltrate more easily into artery walls [130].

It has been shown that a trans-fat rich-diet increased the hepatic expression of genes involved in lipids synthesis, fatty acid synthase, SREBP, microsomal transfer protein for triglycerides (MTP), and apo B100. In addition, trans diet intake reduced vitamin E plasma concentrations by 30%, altering the defense mechanisms against oxidative stress, which may explain the increment in mRNA expression of glucose-regulated protein-78 (Grp78), an endoplasmic reticulum chaperone whose expression is elevated during oxidative stress [131].

Another deleterious action of elaidic acid involves the reduction of cellular cholesterol efflux mediated by ABCA-1, an effect attributed to alterations in cell membrane fluidity [132]. Therefore, this fatty acid contributes to cholesterol accumulation in macrophages and, consequently, to atherosclerotic lesion progression.

The Nurses' Health showed that dietary replacement of saturated and trans fatty acids by mono and polyunsaturated is more effective to prevent CVD than total fat reduction in the diet [7]. Using data from the same population, the relationship between consumption of different types of fat and risk for T2DM was evaluated; it was observed that the consumption of saturated and monounsaturated fatty acids was not associated with risk of T2DM; however, trans fatty acids consumption increased and of polyunsaturated fatty acids reduced the risk for T2DM [133].

In a prospective analysis on a subgroup of the Women's Health Study (WHS), the association between the consumption of different types of fat and the risk of developing hypertension was evaluated. After a multivariate analysis with adjustment for factors related to obesity, only trans fatty acids consumption showed a positive association with hypertension risk [134].

In addition to raising the incidence of cardiovascular disease by direct action on risk factors, trans fatty acids also do so through the induction of endothelial injury. In vitro study in human umbilical vein endothelial cells (HUVEC) showed that elaidic acid can cause cell death through the activation of caspase pathway [135]. In another study, in human aortic endothelial cells (HAEC), the trans fatty acid induced NF-κB activation, mediated by an increase in ROS production, which led to an increase in the expression of VCAM-1 and ICAM-1 and, consequently, an increase in the leukocyte adhesion [136].

Human studies corroborate the results of experimental studies. Baer et al. [137] reported increased plasma concentrations of E-selectin and C reactive protein (CRP) in men consuming a diet enriched with trans fatty acids for 5 weeks.

With respect to the inflammatory response, trans fatty acids intake can increase TNF-α and IL-6 production in humans [138]. In subjects with established CVD, trans fatty acids were positively associated with systemic inflammation, represented by the increase of proinflammatory cytokines IL-1β, IL-6, TNF-α, and MCP-1 [139]. These studies demonstrate that, in addition to adverse effects on lipid metabolism, trans fatty acid is capable of inducing proinflammatory profile further aggravating its harmful effects to health.

Because transfatty acids have the greatest atherogenic potential in comparison to all fatty acids in the diet, their consumption is not recommended by any international guidelines [10–12].

SUGARS

The American Heart Association (AHA) and the World Health Organization (WHO) recommend, that the consumption of sugar should be limited to 10% of daily total carbohydrate intake, and that an intake below 5% of the total caloric value could grant greater health benefits (WHO) [140,141].

In the 1960s, Yudkin et al. demonstrated that high-sugar consumption was associated with increased CVD [142,143]. Later, a follow-up study of 10 years conducted in 75,521 American women, showed that the consumption of a high glycemic index diet rich in sugar, increased the risk for CVD development, regardless of other risk factors [144].

Results of several epidemiological and experimental studies have demonstrated an association between high carbohydrate [145–147] and sugars [148–150] intake, particularly fructose [151–154] with increased risk for CVD.

SUCROSE

Sucrose is a disaccharide composed of 50% glucose and 50% fructose, and its main sources are honey, sugar cane and its products, such as sugar cane juice, brown sugar, white sugar, and, some fruits. It has already been demonstrated that the high consumption of sucrose is strongly related to a high risk for CVD [155].

A longitudinal study conducted in 88,520 women for 24 years demonstrated a significant increase in the incidence of CVD among women who consumed two servings of sweetened beverages per day [149]. An increase in cardiovascular risk factors was also observed in a study performed with teenagers, whose average sugar consumption was higher than 20% of the daily total calories [150]. Regardless of other variables, including body weight, high-sugar consumption was inversely associated with plasma HDL-c concentrations and directly associated with plasma LDL-c concentration.

Although there were several evidences that sucrose can raise blood pressure [156], clinical study results are still very controversial. A study by Van der Schaaf et al. [157] showed that the consumption of a sucrose-rich diet (40% of total calories) for 6 weeks, did not induce blood pressure alterations in hypertensive individuals and adults with polycystic kidney disease. The same results were observed in young and healthy subjects consuming isocaloric diets, for 6 weeks, with 10% or 25% of total calories as sucrose [158]. However, there was an elevation in blood pressure with the consumption of beverages sweetened with sucrose (28% of total calories) for 10 weeks as compared to the consumption of beverages containing artificial sweeteners [159]. Differently from previous studies, the deleterious effect observed in this investigation, may have been due to the significant difference in daily total caloric consumption between sucrose and artificial sweetener groups, which also induced differences in weight gain and adipose tissue content.

Regarding epidemiological studies, the association between sugar consumption and increased blood pressure was demonstrated. Data from National Health and Nutrition Examination Survey

(NHANES 1999–2004) showed that high consumption of sweetened beverages was associated with elevated blood pressure in 4,867 American adolescents (aged 12–18 years) [160]. The results were adjusted for CVD risk factors, both from clinical and nutritional variables. The Framingham Heart Study [161] showed that the consumption of at least one serving of soft drinks (350 mL) is associated with an increase in CVD markers, such as prevalence and incidence of hypertension, hypertriglyceridemia as well as reduction in HDL-c plasma concentrations. Sucrose and glucose can raise blood pressure effect by action, with higher sodium retention [162]. Sugar consumption can also increase sympathetic nervous system activity, as demonstrated in healthy individuals after an oral glucose administration which led to increased levels of circulating norepinephrine [163].

FRUCTOSE

The main sources of fructose in the diet are fruits, and also sucrose and corn syrup, which contains equal amounts of fructose and glucose. The consumption of fructose in the United States increased significantly after 1967, due to the enrichment of juices and soft drinks with corn syrup [153,154]. Several authors have demonstrated that fructose could be the critical component in the relationship between increased sugar consumption and increased incidence of CVD. A longitudinal study conducted in 88,520 women for 24 years showed an increase in CVD incidence among women who consumed two servings of beverages sweetened with sugar per day [149]. Welsh et al. [150] demonstrated that the consumption of sweetened beverages by American adolescents was positively associated with an increase in CVD risk.

Fructose consumption can increase cardiovascular risk, in part, due to its effect on the elevation of plasma TG concentration, which can be explained by two different mechanisms. First, in the liver, fructose uptake is insulin independent, and its conversion is rapidly processed by fructokinase. Fructose metabolism lacks feedback mechanisms, ignoring the energy status of the cell, therefore the intracellular concentrations of adenosine triphosphate (ATP) or citrate do not promote negative feedback, as seen in glucose metabolism. Thus, fructose can enter into the glycolytic and gluconeogenic pathways in a continuous and uncontrolled manner, producing large amounts of lactate and pyruvate (glycerol precursors) and glycogen [164]. Additionally, high-fructose consumption induces the expression of lipogenic enzymes in the liver [165].

The second mechanism associated with the atherogenic potential of fructose is involved with its effect on blood pressure rise, probably due to endothelial dysfunction [166]. In order to evaluate the effects of sugar on acute cardiovascular response, healthy individuals received 60 g of fructose, sucrose, or glucose, and an increase of hemodynamic responses with fructose was observed. These alterations resulted in significant rise in blood pressure, ascribed to cardiac sympathetic activation, increased cardiac output and lower contractility, however, without peripheral resistance alteration [154]. Regarding glucose consumption, an increase in cardiac output with concomitant peripheral vasodilation was observed, without any alterations in blood pressure [154].

An experimental study conducted in mice, compared the intake of a fructose-rich diet (corn syrup) to a high-fat diet, and revealed that both diets led to hyperinsulinemia, hyperglycemia, and hyperlipidemia when compared to a normocaloric diet. However, only the fructose-rich diet induced a rise in blood pressure, i.e., fructose consumption causes metabolic effects similar to the intake of a high-fat diet [167].

Studies conducted in both animals and humans, showed that fructose increased plasma uric acid concentration, induced hyperinsulinemia and impaired insulin sensitivity [166,168]. Uric acid is an independent risk factor for CVD and plays an important role in hypertension [169,170].

The intermediate step of fructose metabolism generates uric acid and induces ROS accumulation as a result of ATP degradation in adenosine diphosphate (ADP) [171–173]. Therefore, hyperuricemia leads to an increase in ROS content which in turn reduces NO bioavailability, leading to endothelial dysfunction and blood pressure elevation [166,174]. Another uric acid mechanism of action in hypertension is related to the activation of the renin-angiotensin-aldosterone system and inhibition of neural NO synthase [175] which also leads to endothelial dysfunction, as shown in experimental studies [174]. This same investigation has also shown that uric acid reduces the synthesis of NO induced by vascular endothelial growth factor in endothelial cells from bovine aorta [174].

In vitro studies demonstrated that the treatment of HAEC cells, with physiological concentrations of fructose increased ICAM-1 expression in a time and dose-dependent manner, possibly due to the reduction in intracellular ATP concentrations, leading to the reduction of NO synthesis and inducing endothelial dysfunction. These changes were independent from the activity of NF-κB [173]. In animals, fructose consumption was associated with greater ICAM-1 expression and plasma concentration in renal endothelial cells [173].

Data from Framingham Heart Study's third generation which evaluate, in 5,887 adults, the association between endothelial function and the adherence to the Dietary Guidelines for Americans (2010). It was demonstrated that greater adherence to the guide was associated with greater blood flow speed and arterial wave, i.e., better dietary quality is associated with vascular health [176].

On the other hand, the consumption of fructose, in the context of a diet rich in vegetables and fruits, is inversely associated with endothelial dysfunction biomarkers, such as adhesion molecules, CRP, and other inflammatory biomarkers [177,178]. However, higher fructose consumption in a Western diet pattern is positively related to endothelial dysfunction, possibly due to the high-fructose content present in corn syrup-sweetened beverages [179], as well as higher concentrations of saturated and trans fatty acids, which contribute to endothelium damage as well [4,44].

Although there is not a consensus regarding the direct role of fructose/sucrose on blood pressure and, consequently, in endothelial function, it is known that high consumption of these sugars is associated with poor dietary habits and increased prevalence of obesity, especially when consumed as juices, teas or soft drinks because they induce lower satiety [180,181]. Thus, an adequacy of fructose and sucrose consumption is recommended to prevent hypertension. Supporting this recommendation, dietary hypertension treatment is based on the DASH diet, which recommends low sucrose/fructose consumption [149].

SODIUM

According to the latest family budget survey (POF—*Pesquisa de Orçamentos Familiares*) [182], the average consumption of sodium in the Brazilian population is 3,200 mg/day, exceeding WHO [183] and Dietary Guidelines for Americans (2010) [184] recommendations, which advocate the consumption of 2,000 mg/day and 2,300 mg/day, respectively. A reduction in sodium intake is linked to a drop in blood pressure [185].

The WHO had established a goal for a 30% reduction in the mean population intake of sodium. This is one of the nine global targets implemented by WHO, projected for the year of 2025, in order to reduce the prevalence of nontransmissible chronic diseases. Hendriksen et al. [186] showed that a 30% reduction in salt consumption led to a decline in the prevalence of acute myocardial infarction between 6.4% and 13.5%, and in the ischemic heart disease, between 4% and 9%, in nine European countries.

The majority of sodium consumed in Brazilian families comes from table salt and sodium-based spices (74%); however, data from the last POF survey (2008 to 2009) showed significant increase in sodium intake due to higher consumption of ultra-processed foods, when compared to POF 2002–2003 [187]. The same was observed in the United States, where ultra-processed foods, along with restaurant dining, contribute with more than 75% of the total dietary sodium [188].

Excessive sodium consumption is one of the main risk factors for hypertension [189], and other cardiovascular diseases, since sodium can also induces independent effects on endothelial function [190]. In this context, excessive sodium consumption among salt-sensitive subjects may negatively influence the sympathetic nervous system, increase angiotensin II, catecolamines, and other factors such as inflammatory cytokines and prooxidant species, inducing oxidative stress [191].

Sodium balance in the organism depends on hormones and systems like the renin-angiotensin-aldosterone system, sympathetic nervous system, atrial natriuretic peptides, and others [192]. Seals et al. [193] demonstrated, in postmenopause women, that moderate dietary sodium restriction, for 3 months, diminished blood pressure and pulse pressure, changes that can be, in part, explained by decrease in the stiffness of the large elastic arteries [193]. One of the mechanisms associated with reduced elasticity in the arterial wall is lower plasma concentration of C-natriuretic peptide (CNP) as demonstrated in 35 hypertensive individuals [192]. CNP is a peptide stored by endothelial cells, that promote vasodilation of the vascular smooth muscle

[193] and can be found in low concentrations in healthy individuals [194].

Clinical and experimental studies have demonstrated the effect of sodium on endothelial function, regardless of blood pressure [190,192]. Vascular tonus is determined by the balance between vasoconstrictor and vasodilator agents in the endothelium and sodium presents a significant role in endothelium-dependent vasodilation, by reducing NO bioavailability [195,196]. NO is the first compound responsible for vasodilation in the arteries [197]; in addition, it inhibits platelet aggregation and SMC proliferation, modulates the leukocyte-endothelium interaction by altering the expression of cell adhesion molecules and reducing monocyte adhesion [198].

NO reduction has strong association with increased ROS levels, generated by NADPH oxidase, xanthine oxidase or by reduced eNOS activity within the vascular wall, leading to NO scavenging, and also an interruption in some signaling pathways that are involved in NO production [195]. A study conducted in experimental models of hypertension showed that excessive sodium consumption reduces NO production and endothelium-dependent vasodilation with impairment in the insulin /PI3K-AKT/eNOS signaling pathway [196].

The consumption of a sodium enriched diet by salt-sensitive subjects increased S-selectin, E-selectin and endothelin-1 plasma concentrations, as well as urinary albumin excretion. In contrast, there was no difference in adhesion molecule concentrations (ICAM-1 and VCAM-1) when compared to salt-resistant subjects [199].

The higher intake of sodium by 14 salt-resistant subjects reduced angiotensin II and aldosterone plasma concentrations and renin activity as compared to individuals on a low-sodium diet [190]. However, excessive salt consumption brought on deleterious effects on endothelium-dependent dilatation. An experimental study showed that a high-salt diet increased systolic blood pressure and reduced angiotensin II plasma concentration and plasma renin activity both in normotensive and spontaneously hypertensive rats [200].

Renin-angiotensin system components are locally synthesized in cardiac tissue [201] and distributed in cardiomyocytes, endothelial and vascular SMC which are independetly regulated from circulatory levels [202]. Angiotensin II increases ROS production in the endothelium, reducing NO [203]

bioavailability and inducing oxidative stress [196]. As evidenced in experimental studies with animal models for hypertension, excessive salt intake increases mRNA and protein levels of angiotensin II receptor (AT-1) and ROS production in aortas [196]; it also increases the synthesis of prorenin receptor, AT-1 receptors and angiotensinogen in myocardium inducing interstitial fibrosis and cardiomyocyte hypertrophy [204]. In another study, high-salt intake increased the expression and content of renin and prorenin receptors, and increased the activity of angiotensinogen and AT-1 receptors in cardiac tissue. This led to activation of kinase regulated by extracellular signal (ERK 1/2) and protein-kinase activated by mitogen (MAPK38p) signaling pathways, inducing TGF-1β secretion, which led to interstitial cardiac fibrosis and perivascular fibrosis, as well as cardiomyocyte hypertrophy [200].

PHYTOSTEROLS

Phytosterols (PS) are natural constituents of plants, found in seeds and vegetable oils; being beta-sitosterol the most abundant, followed by campesterol, and stigmasterol. PS and cholesterol have similar chemical structure, differing only on the carbon side chain, by the presence of a methyl or a ethyl group [205]. PS are obtained exclusively from diet, and as they are minimally absorbed [206], their plasma concentrations (0.3–1.7 mg/dL) are much lower than cholesterol concentration (150–260 mg/dL).

A regular diet provides an average of 150–400 mg/day of PS [207] and a relevant review of approximately 40 studies noted that a daily dose of 2 g of PS resulted in a 10% reduction on LDL-c plasma concentration, however higher doses did not increase this action [207]. Most clinical studies showed that PS reduce plasma cholesterol concentrations in ~10%–15% [208]. Based on this information, the National Cholesterol Education Program [11] included the daily consumption of 2 g of PS in the nutritional recommendations for the treatment of hypercholesterolemia. However, it is important to note that PS supplementation is indicated for moderated hypercholesterolemia treatment, and its efficiency should be individually tested, since heterogeneous responses were observed [209].

The hypocholesterolemic effect of PS can be explained by two mechanisms: competition with cholesterol during absorption and stimulation of transintestinal cholesterol excretion (TICE). During the absorption process competition between PS and cholesterol occurs on the micelles, which are structures containing bile acids, fatty acids, diglycerides, monoglycerides, phospholipids, lysophospholipids, cholesterol, PS, and lipid-soluble vitamins [210]. Since PS and phytostanols are very hydrophobic, they have higher physicochemical affinity by the micelles in intestinal lumen, thus interfering with cholesterol solubility in this structure, favoring the permanence of PS in micelle interior on the detriment of cholesterol, resulting in less cholesterol absorption [211]. Hence, cholesterol is removed from micelles, increasing its excretion in feces (Fig. 24.2).

Although sterol absorption has not been completely elucidated, several molecular mechanisms involved in this step have already been proposed [208]. Niemann-Pick-C1-like-1 protein (NPC1L1) is the main protein in the brush border membrane responsible for sterols absorption [213]. Inside the enterocyte, cholesterol is mostly esterified by ACAT2, and subsequently incorporated into

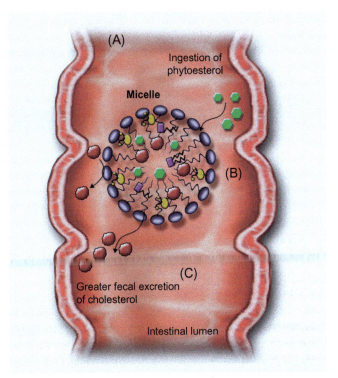

FIG. 24.2 Structure of micelles and competition mechanism between cholesterol and PS within this structure: (A) PS nutricional supplementation; (B) competitive mechanism within the micelle: PS has greater physicochemical affinity with micelle, hampering cholesterol solubilization and incorporation; (C) greater fecal excretion of cholesterol [211,212].

chylomicrons in the basolateral membrane, culminating in the secretion of these particles to the lymphatic system. A small amount of nonesterified cholesterol returns to the intestinal lumen through specific transporters, known as ATP-binding cassette transporter G5 (ABCG5) and the ATP-binding cassette transporter G8 (ABCG8) (Fig. 24.3) [214].

Compared to cholesterol, the proportion of PS esterification is much lower [215]. Furthermore, ABCG5/G8 excrete sterols to the intestinal lumen [216]. For these reasons, NPC1L1 receptor is much less efficient in PS absorption than in cholesterol [217]. A very small amount of PS can be esterified by ACAT2 and further transported by lipoproteins and excreted by the bile (Fig. 24.3) [218].

Another important explanation for PS hypocholesterolemic effect is the stimulation of cholesterol excretion by intestinal mucosa, known as transintestinal cholesterol excretion (TICE), which occurs independently from the bile pathway [219]. TICE is an active system present in the proximal portion of small intestine and involves cholesterol transport through enterocytes apical and basolateral membranes back into the intestinal lumen. It was concluded that ABCG5/ABCG8 heterodimer is involved in facilitating this flow of cholesterol induced by PS [220].

PS supplementation is safe for human consumption [207], reduces plasma cholesterol concentration, does not accumulate in the artery wall, and in addition, prevent atherosclerotic lesion development, as shown in LDLr-KO mice [5]. Although there are no studies on the effect of PS on cardiovascular outcomes, a recent review of clinical studies has shown that PS does not impair endothelial function [212].

Antioxidants

Antioxidants are substances that retard or inhibit the oxidation of substrates when present in low concentrations in regard to oxidizable products [221]. They differ in molecular structure, and therefore

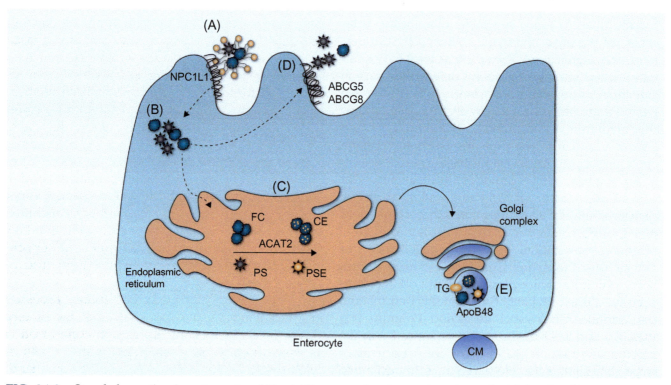

FIG. 24.3　Sterol absorption in enterocytes: (A) micelle approximation on enterocyte brush border; (B) nonesterified cholesterol and PS (FC and PS), enter the cell through the sterol transporter protein NPC1L1; (C) FC is esterified by ACAT2 in the endoplasmic reticulum (ER); a small part of the PS can also be esterified by means of that same enzyme; (D) Most PS returns to the intestinal lumen through transporter proteins ABCG5 and ABCG8; EC, along with PSE, TG and apoB 48, form the Chylomicrons (CM), which are transported by the lymphatic system; (E) Cholesteryl ester (CE), phytoesterol ester (PSE), triglycerides (TG) and apolipoprotein B-48 (Ape B48) are processed in the Golgi apparatus to form chylomicrons [212].

exhibit distinct mechanism of action, which aim different targets, modulating endothelial function [222]. They can be classified as endogenous antioxidants (glutathione peroxidase (GPx), glutathione reductase (GR), superoxide dismutase (SOD) and catalase (CAT)) and exogenous, which can be obtained from the diet. Vitamins C and E, selenium, and nicotinic acid are the most studied regarding their possible action on cardiovascular risk reduction. Antioxidants improve endothelial function, attenuate inflammatory processes and make LDL more resistant to oxidation [222].

Oxidative stress and inflammation are conditions that jointly induce endothelial dysfunction. Particularly the first is characterized by an imbalance between oxidants and antioxidants, with prevalence of oxidants, such as anion superoxide (O_2^-), peroxynitrite $(ONOO^-)$, hydroxyl (OH^-), and hydrogen peroxide (H_2O^2) [222,223]. The generation of these compounds is a continuous and physiological process [224,225], however in situations in which the generation of these radicals exceeds the degradation capacity, oxidative stress is established [226].

On the other hand, it has already been demonstrated that antioxidants naturally present in the diet may act on signaling pathways involved in oxidative and inflammatory processes, improving endothelial function. One of the main pathways of antioxidant involves the phosphorylation and nuclear translocation of a nuclear factor derived from erythrocyte 2 (Nrf2), a transcription factor that binds to the antioxidant responsive element (ARE) inducing the transcription of genes encoding SOD, GPx, CAT, glutathione-S-transferase (GST), and reduced glutathione (GSH) [227,228]. Moreover, antioxidant compounds are capable to reduce NF-κB inhibitor kinase (IkBa) phosphorilation, inhibiting NF-κB translocation and, consequently, the expression of genes regulated by this transcription factor, such as adhesion molecules VCAM-1 and ICAM-1, P-selectin, monocyte chemotatic protein (MCP-1), and proinflammatory cytokines [229,230]. All these factors, combined with an increase in eNOS activity, contribute to the prevention of endothelial dysfunction [231,232].

Vitamin C acts on redox signaling, reducing oxidative stress by acting as a ROS sequester, thus decreasing secretion of proinflammatory cytokines and, consequently, improving endothelial function [229].

Polyphenols (flavonoids, theaflavina, and epicatechin) found in diffuse form in vegetables and olive oil can also act on redox signaling, reducing ROS and increasing heme oxygenase-1 (HO-1), as well as decreasing LTB4, plasma P-selectin, VCAM, ICAM, and MCP-1, and the expression of NF-κB [233,234]. In the endothelium, flavonoids increase eNOS production and prevent endothelial dysfunction and vascular remodeling [231,232].

Present in vegetable oils and seeds such as nuts and almonds, vitamin E decreases the expression of adhesion molecules and CD-36 [235], and increases Nrf2 [236] phosphorylation, and nitric oxide plasma concentration, contributing to improve endothelial function [237].

The elucidation of these mechanisms prompted several researchers to investigate potential benefits of antioxidant supplementation on CVD prevention and treatment in experimental, epidemiological studies and clinical studies. A clinical study performed in patients with coronary artery disease (CAD) showed that acute consumption of 2 g of ascorbic acid improved endothelium-dependent vasodilation [229]. Studies evaluating the effect of diet on cardiovascular risk did not find any association between dietary intake of vitamins C or E with CAD incidence [238]. However, this meta-analysis also shows that among individuals with supplementation exceeding 700 mg/day of vitamin C, lower CAD risk was found (RR=0.75; IC 95%: 0.60–0.93). The association between vitamin C supplementation and CAD risk persisted after adjustments for risk factors not related to diet (smoking, diabetes) and for dietary factors (fiber and saturated fat intake). The authors concluded that, due to the effects of high-antioxidants intake not being fully understood, the results did not provide sufficient evidence to recommend high doses of vitamin C.

The analysis of 16 prospective studies discussed in a recent meta-analysis showed that the dietary intake of vitamin C and its plasma concentration as well were inversely associated with the risk of cerebrovascular accident (CVA). On the other hand, vitamin supplementation did not influence the risk of this vascular event [239]. Another important meta-analysis with clinical ($n=78,296,707$ participants) evaluated the effect of antioxidant supplementation (vitamin A, C, E, or selenium) on cardiovascular mortality [240]. The average participant age was 63 years and all had controlled diseases from various etiologies, including cardiovascular diseases [240]. Interestingly both beta-carotene and vitamin E significantly increased mortality, while vitamin A, vitamin C, and selenium did not influence mortality [240].

The effect of vitamin E supplementation associated with vitamin C, on alternate days for a period of 8 years, was evaluated in 14,641 American doctors aged over 50. Vitamin E or vitamin C supplementation did not reduce the incidence infarction, total stroke, and cardiovascular mortality. However, vitamin E supplementation was associated with increased hemorrhagic stroke incidence [241].

Hyperglycemia is one of the main inducers of endothelial dysfunction observed in diabetes, for promoting the pathological activation of polyol and hexosamine pathways, protein kinase C and the shunt of the pentose pathway [242], in addition to increasing superoxide anion with decreased nitric oxide production [243]. Glycemic variation maximizes oxidative stress in endothelial cells [244] and, in this context, diet plays a central role in preventing damages induced by oxidative stress.

Regarding the benefits of vitamin supplementation to improve endothelial function in diabetes, study results are conflicting. A recent review concluded that supplementation with ascorbic acid, taurine and nicotinamida could improve endothelial function in children with type 1 diabetes [245]. The explanation for this results was the fact that endothelial function remains less sensitive to glycemia variations. Despite this conclusion favoring vitamin supplementation, the majority of the studies discussed in this review were performed in cells and animals, and few clinical studies were included. Due to the lack of clinical studies showing effectiveness of vitamin supplementation on endothelial function in diabetes, the American Diabetes Association persists with the recommendation that there is no need for vitamin supplementation in diabetic individuals [12].

Although synthetic micronutrients may have chemical structures similar to those found in foods, fruits and vegetables have other nutrients and phytochemicals that enhance the bioavailability of vitamins and other antioxidant compounds [222]. Thus, although the plasma concentration of antioxidants from diet is associated with endothelial function improvement, megadoses of antioxidants proved ineffective to prevent or reduce cardiovascular mortality [246].

CONCLUSIONS

Nutritional recommendations for cardiovascular prevention, with emphasis on improving endothelial function, are based on the adequacy of dietary calorie consumption, the removal of trans fatty acids, and reduction of saturated fatty acids, sugar, and salt, as well as an increase of monounsaturated fatty acids intake [247]. The Dietary Guide for the Brazilian Population (2014) [248] as well as the Dietary Guidelines for Americans (2015) [184], emphasizes the importance of consuming minimally processed foods and reaffirms the importance of including fruits, vegetables, grains, nuts, dairy products, and lean meats. These nutritional recommendations contemplate some of the nine global targets projected for the year 2025 by the World Health Organization [249], with the purpose of preventing chronic diseases, including cardiovascular disease.

References

[1] Mente A, de Koning L, Shannon HS, et al. A systematic review of the evidence supporting a causal link between dietary factors and coronary heart disease. Arch Intern Med 2009;169(7):659–69.

[2] Keys A. Diet and the epidemiology of coronary heart disease. J Am Med Assoc 1957;164(17):1912–9.

[3] Appel LJ, Moore TJ, Obarzanek E, et al. A clinical trial of the effects of dietary patterns on blood pressure. DASH Collaborative Research Group. N Engl J Med 1997;336(16):1117–24.

[4] Mensink RP, Zock PL, Kester AD, et al. Effects of dietary fatty acids and carbohydrates on the ratio of serum total to HDL cholesterol and on serum lipids and apolipoproteins: a meta-analysis of 60 controlled trials. Am J Clin Nutr 2003;77(5):1146–55.

[5] Bombo RP, Afonso MS, Machado RM, et al. Dietary phytosterol does not accumulate in the arterial wall and prevents atherosclerosis of LDLr-KO mice. Atherosclerosis 2013;231(2):442–7.

[6] Machado RM, Nakandakare ER, Quintao EC, et al. Omega-6 polyunsaturated fatty acids prevent atherosclerosis development in LDLr-KO mice, in spite of displaying a pro-inflammatory profile similar to trans fatty acids. Atherosclerosis 2012;224:66–74.

[7] Hu FB, Stampfer MJ, Manson JE, et al. Dietary fat intake and the risk of coronary heart disease in women. N Engl J Med 1997;337:1491–9.

[8] Vlachopoulos C, Aznaouridis K, Stefanadis C. Prediction of cardiovascular events and all-cause mortality with arterial stiffness: a systematic review and meta-analysis. J Am Coll Cardiol 2010;55 (13):1318–27.

[9] Yeboah J, Folsom AR, Burke GL, et al. Predictive value of brachial flow-mediated dilation for incident cardiovascular events in a population-based study: the multi-ethnic study of atherosclerosis. Circulation 2009;120(6):502–9.

[10] Eckel RH, Jakicic JM, Ard JD, et al. 2013 AHA/ACC guideline on lifestyle management to reduce cardiovascular risk: a report of the American College of Cardiology/American Heart Association Task Force on Practice Guidelines. Circulation 2014;129(25 Suppl. 2):S76–99.

[11] National Cholesterol Education Program (NCEP) Expert Panel on Detection, Evaluation, and Treatment of High Blood Cholesterol

in Adults (Adult Treatment Panel III). Third Report of the National Cholesterol Education Program (NCEP) Expert Panel on Detection, Evaluation, and Treatment of High Blood Cholesterol in Adults (Adult Treatment Panel III) final report. Circulation 2002;106(25):3143–421.

[12] American Diabetes Association. Standards of medical care in diabetes. Diabetes Care 2014;37(1):S14–80.

[13] Sacks FM, Appel LJ, Moore TJ, et al. A dietary approach to prevent hypertension: a review of the Dietary Approaches to Stop Hypertension (DASH) Study. Clin Cardiol 1999;22(7 Suppl.):III6–III10.

[14] Yusuf S, Hawken S, Ounpuu S, et al. Effect of potentially modifiable risk factors associated with myocardial infarction in 52 countries (the INTERHEART study): case-control study. Lancet 2004;364(9438):937–52.

[15] Estruch R, Martínez-González MA, Corella D, et al. Effects of a Mediterranean-style diet on cardiovascular risk factors: a randomized trial. Ann Intern Med 2006;145(1):1–11.

[16] Marin C, Ramirez R, Delgado-Lista J, et al. Mediterranean diet reduces endothelial damage and improves the regenerative capacity of endothelium. Am J Clin Nutr 2011;93(2):267–74.

[17] Calder PC. Fatty acids and inflammation: the cutting edge between food and pharma. Eur J Pharmacol 2011;668(Suppl. 1):S50–8.

[18] Massaro M, Scoditti E, Carluccio MA, et al. Omega-3 fatty acids, inflammation and angiogenesis: nutrigenomic effects as an explanation for anti-atherogenic and anti-inflammatory effects of fish and fish oils. J Nutrigenet Nutrigenomics 2008;1(1–2):4–23.

[19] Rubanyi GM. The discovery of endothelin: the power of bioassay and the role of serendipity in the discovery of endothelium-derived vasocative substances. Pharmacol Res 2011;63:448–54.

[20] Vanhoutte PM. Regeneration of the endothelium in vascular injury. Cardiovasc Drugs Ther 2010;24:299–303.

[21] Lu H, Daugherty A. Atherosclerosis. Arterioscler Thromb Vasc Biol 2015;35:485–91.

[22] Witztum JL, Steinberg D. Role of oxidized low density lipoprotein in atherogenesis. J Clin Invest 1991;88:1785–92.

[23] Ross R. Atherosclerosis—an inflammatory disease. N Engl J Med 1999;340:115–26.

[24] Steinberg D. Thematic review series: the pathogenesis of atherosclerosis. An interpretive history of the cholesterol controversy, part V: the discovery of the statins and the end of the controversy. J Lipid Res 2006;47:1339–51.

[25] Lee JY, Zhao L, Youn HS, et al. Saturated fatty acid activates but polyunsaturated fatty acid inhibits Toll-like receptor 2 dimerized with Toll-like receptor 6 or 1. J Biol Chem 2004;279:16971–9.

[26] Rocha VZ, Libby R. Obesity, inflammation, and atherosclerosis. Nat Rev Cardiol 2009;6:399–409.

[27] Ron D, Walter R. Signal integration in the endoplasmic reticulum unfolded protein response. Nat Rev Mol Cell Biol 2007;8:519–29.

[28] Lewis GF, Rader DJ. New insights into the regulation of HDL metabolism and reverse cholesterol transport. Circ Res 2005; 96(12):1221–32.

[29] Tall AR, Yvan-Charvet L, Terasaka N, et al. HDL, ABC transporters, and cholesterol efflux: implications for the treatment of atherosclerosis. Cell Metab 2008;7(5):365–75.

[30] Lusis AJ. Atherosclerosis. Nature 2000;407(6801):233–41.

[31] Barter P. The inflammation: lipoprotein cycle. Atheroscler Suppl 2005;6(2):15–20.

[32] Rudijanto A. The role of vascular smooth muscle cells on the pathogenesis of atherosclerosis. Acta Med Indones 2007;39(2):86–93.

[33] Rong JX, Shapiro M, Trogan E, et al. Transdifferentiation of mouse aortic smooth muscle cells to a macrophage-like state after cholesterol loading. Proc Natl Acad Sci U S A 2003; 100(23): 13531–6.

[34] Doran AC, Meller N, McNamara CA. Role of smooth muscle cells in the initiation and early progression of atherosclerosis. Arteriscler Thromb Vasc Biol 2008;28:812–9.

[35] Choi HY, Rahmani M, Wong BW, et al. ATP-binding cassette transporter A1 expression and apolipoprotein A-I binding are impaired in intima-type arterial smooth muscle cells. Circulation 2009;119:3223–31.

[36] Nicolosi RJ, Wilson TA, Rogers EJ, et al. Effects of specific fatty acids (8:0, 14:0, cis-18:1, trans-18:1) on plasma lipoproteins, early atherogenic potential, and LDL oxidative properties in the hamster. J Lipid Res 1998;39:1972–80.

[37] Bray GA, Lovejoy JC, Smith SR, et al. The influence of different fats and fatty acids on obesity, insulin resistance and inflammation. J Nutr 2002;132:2488–91.

[38] Haag M, Dippenaar NG. Dietary fats, fatty acids and insulin resistance: short review of a multifaceted connection. Med Sci Monit 2005;11:359–67.

[39] Manco M, Calvani M, Mingronr G. Effects of dietary fatty acids on insulin sensitivity and secretion. Diabetes Obes Metab 2004;6:402–13.

[40] Harris WS, Mozaffarian D, Rimm E, et al. Omega-6 fatty acids and risk for cardiovascular disease: a science advisory from the American Heart Association Nutrition Subcommittee of the Council on Nutrition, Physical Activity, and Metabolism; Council on Cardiovascular Nursing; and Council on Epidemiology and Prevention. Circulation 2009;119:902–7.

[41] Reaven PD, Grasse BJ, Tribble DL. Effect of linoleate-enriched and oleate-enriched diets in combination with a-tocopherol on the susceptibility of LDL and LDL subfractions to oxidative modification in humans. Arterioscler Thromb 1994;14:557–66.

[42] De Lorgeril M, Salen P. The Mediterranean-style diet for the prevention of cardiovascular diseases. Public Health Nutr 2006;9:118–23.

[43] Nicolosi RJ, Stucchi AF, Kowala MC, et al. Effect of dietary fat saturation and cholesterol on LDL composition and metabolism. In vivo studies of receptor and nonreceptor-mediated catabolism of LDL in cebus monkeys. Arteriosclerosis 1990;10:119–28.

[44] Warensjö E, Sundström J, Vessby B, et al. Markers of dietary fat quality and fatty acid desaturation as predictors of total and cardiovascular mortality: a population-based prospective study. Am J Clin Nutr 2008;88:203–9.

[45] Mozaffarian D, Wu JH. Omega-3 fatty acids and cardiovascular disease: effects on risk factors, molecular pathways, and clinical events. J Am Coll Cardiol 2011;58:2047–67.

[46] Campos H, Baylin A, Willett WC. Alpha-linolenic acid and risk of nonfatal acute myocardial infarction. Circulation 2008; 1(18):339–45.

[47] Lavie CJ, Milani RV, Mehra MR, et al. Omega-3 polyunsaturated fatty acids and cardiovascular diseases. J Am Coll Cardiol 2009;54:585–94.

[48] Skeaff CM, Miller J. Dietary fat and coronary heart disease: summary of evidence from prospective cohort and randomised controlled trials. Ann Nutr Metab 2009;55:173–201.

[49] Harris WS, Bulchandani D. Why do omega-3 fatty acids lower serum triglycerides? Curr Opin Lipidol 2006;17:387–93.

[50] Dangardt F, Osika W, Chen Y, et al. Omega-3 fatty acid supplementation improves vascular function and reduces inflammation in obese adolescents. Atherosclerosis 2010;212:580–5.

[51] McLennan PL. Myocardial membrane fatty acids and the antiarrhythmic actions of dietary fish oil in animal models. Lipids 2001;36(Suppl.):S111–4.

[52] Spritz N, Mishkel MA. Effects of dietary fats on plasma lipids and lipoproteins: an hypothesis for the lipid-lowering effect of unsaturated fatty acids. J Clin Invest 1969;48:78–86.

[53] Chan DC, Watts GF, Mori TA, et al. Randomized controlled trial of the effect of n-3 fatty acid supplementation on the metabolism of apolipoprotein B-100 and chylomicron remnants in men with visceral obesity. Am J Clin Nutr 2003;77:300–7.

II. ENDOTHELIAL DYSFUNCTION AND CLINICAL SYNDROMES

[54] Ouguerram K, Maugeais C, Gardette J, et al. Effect of n-3 fatty acids on metabolism of apoB100-containing lipoprotein in type 2 diabetic subjects. Br J Nutr 2006;96:100–6.

[55] Woollett LA, Spady AK, Dietschy JM. Saturated and unsaturated fatty acids independently regulate low density lipoprotein receptor activity and production rate. J Lipid Res 1992;33:77–88.

[56] Lottenberg SA, Lottenberg AM, Nunes VS, et al. Plasma cholesteryl ester transfer protein concentration, high-density lipoprotein cholesterol esterification and transfer rates to lighter density lipoproteins in the fasting state and after a test meal are similar in Type II diabetics and normal controls. Atherosclerosis 1996;127:81–90.

[57] Brown MS, Goldstein JL. The SREBP pathway: regulation of cholesterol metabolism by proteolysis of a membrane-bound transcription factor. Cell 1997;89:331–40.

[58] Ou J, Tu H, Shan B, et al. Unsaturated fatty acids inhibit transcription of the sterol regulatory element-binding protein-1c (SREBP-1c) gene by antagonizing ligand-dependent activation of the LXR. Proc Natl Acad Sci U S A 2001;22:6027–32.

[59] Gale SE, Westover EJ, Dudley N, et al. Side chain oxygenated cholesterol regulates cellular cholesterol homeostasis through direct sterol-membrane interactions. J Biol Chem 2009;284:1755–64.

[60] Jelinek D, Castillo JJ, Richardson LM, et al. The Niemann-Pick C1 gene is downregulated in livers of C57BL/6 J mice by dietary fattyacids, but not dietary cholesterol, through feedback inhibition of the SREBP pathway. J Nutr 2012;142:1935–42.

[61] Kim SK, Seo G, Oh E, et al. Palmitate induces RIP1-dependent necrosis in RAW 264.7 cells. Atherosclerosis 2012;225:315–21.

[62] Uehara Y, Miura S, von Eckardstein A, et al. Unsaturated fatty acids suppress the expression of the ATP-binding cassette transporter G1 (ABCG1) and ABCA1 genes via an LXR/RXR responsive element. Atherosclerosis 2007;191:11–21.

[63] Wang Y, Oram JF. Unsaturated fatty acids inhibit cholesterol eâlux from macrophages by increasing degradation of ATP-binding cassette transporter A1. J Biol Chem 2002;277:5692–7.

[64] Wang Y, Oram JF. Unsaturated fatty acids phosphorylate and destabilize ABCA1 through a protein kinase C delta pathway. J Lipid Res 2007;48:1062–8.

[65] Ishiyama J, Taguchi R, Yamamoto A, et al. Palmitic acid enhances lectin-like oxidized LDL receptor (LOX-1) expression and promotes uptake of oxidized LDL in macrophage cells. Atherosclerosis 2010;209:118–24.

[66] Ishiyama J, Taguchi R, Akasaka Y, et al. Unsaturated FAs prevent palmitate-induced LOX-1 induction via inhibition of ER stress in macrophages. J Lipid Res 2011;52:299–307.

[67] De Caterina R, Bernini W, Carluccio MA, et al. Structural requirements for inhibition of cytokine-induced endothelial activation by unsaturated fatty acids. J Lipid Res 1998;39:1062–70.

[68] Jinno Y, Nakakuki M, Kawano H, et al. Eicosapentaenoic acid administration attenuates the pro-inflammatory properties of VLDL by decreasing its susceptibility to lipoprotein lipase in macrophages. Atherosclerosis 2011;219:566–72.

[69] Cawood AL, Ding R, Napper FL, et al. Eicosapentaenoic acid (EPA) from highly concentrated n-3 fatty acid ethyl esters is incorporated into advanced atherosclerotic plaques and higher plaque EPA is associated with decreased plaque inflammation and increased stability. Atherosclerosis 2010;212:252–9.

[70] De Caterina R, Cybulsky MI, Clinton SK, et al. The omega-3 fatty acid docosahexaenoate reduces cytokine-induced expression of proatherogenic and proinflammatory proteins in human endothelial cells. Arterioscler Thromb 1994;14:1829–36.

[71] Okuda Y, Kawashima K, Sawada T, et al. Eicosapentaenoic acid enhances nitric oxide production by cultured human endothelial cells. Biochem Biophys Res Commun 1997;232:487–91.

[72] Stevens MJ, Dananberg J, Feldman EL, et al. The linked roles of nitric oxide, aldose reductase, and (Na+, K+)-ATPase in the slowing of nerve conduction in the streptozotocin diabetic rat. J Clin Invest 1994;94(2):853–9.

[73] Tousoulis D, Plastiras A, Siasos G, et al. Omega-3 PUFAs improved endothelial function and arterial stiffness with a parallel antiinflammatory effect in adults with metabolic syndrome. Atherosclerosis 2014;232:10–6.

[74] Callow J, Summers LK, Bradshaw H, et al. Changes in LDL particle composition after the consumption of meals containing different amounts and types of fat. Am J Clin Nutr 2002;76:345–50.

[75] Mata P, Varela O, Alonso R, et al. Monounsaturated and polyunsaturated n-6 fatty acid-enriched diets modify LDL oxidation and decrease human coronary smooth muscle cell DNA synthesis. Arterioscler Thromb Vasc Biol 1997;17:2088–95.

[76] Vecera R, Skottová N, Vána P, et al. Antioxidant status, lipoprotein profile and liver lipids in rats fed on high-cholesterol diet containing currant oil rich in n-3 and n-6 polyunsaturated fatty acids. Physiol Res 2003;52:177–87.

[77] Mazière C, Dantin F, Conte MA, et al. Polyunsaturated fatty acid enrichment enhances endothelial cell-induced low-density lipoprotein peroxidation. Biochem J 1998;336:57–62.

[78] Parthasarathy S, Steinberg D, Witztum JL. The role of oxidized low-density lipoproteins in the pathogenesis of atherosclerosis. Annu Rev Med 1992;43:219–25.

[79] Calzada C, Colas R, Guillot N, et al. Subgram daily supplementation with docosahexaenoic acid protects low-density lipoproteins from oxidation in healthy men. Atherosclerosis 2010;208:467–72.

[80] Serhan CN, Savill J. Resolution of inflammation: the beginning programs the end. Nat Immunol 2005;6:1191–7.

[81] Levy BD. Resolvins and protectins: natural pharmacophores for resolution biology. Prostaglandins Leukot Essent Fat Acids 2010;82:327–32.

[82] Oh DY, Talukdar S, Bae EJ, et al. GPR120 is an omega-3 fatty acid receptor mediating potent anti-inflammatory and insulinsensitizing effects. Cell 2010;142:687–98.

[83] Sykaras AG, Demenis C, Case RM, et al. Duodenal enteroendocrine I-cells contain mRNA transcripts encoding key endocannabinoid and fatty acid receptors. PLoS One 2012;7:e42373.

[84] Oh DY, Olefsky JM. Omega 3 fatty acids and GPR120. Cell Metab 2012;15:564–5.

[85] Ichimura A, Hirasawa A, Poulain-Godefroy O, et al. Dysfunction of lipid sensor GPR120 leads to obesity in both mouse and human. Nature 2012;483:350–4.

[86] Xue B, Yang Z, Wang X, et al. Omega-3 polyunsaturated fatty acids antagonize macrophage inflammation via activation of AMPK/SIRT1 pathway. PLoS One 2012;7:e45990.

[87] Stienstra R, van Diepen JA, Tack CJ, et al. Iníiaminasome is a central player in the induction of obesity and insulin resistance. Proc Natl Acad Sci U S A 2011;108:15324–9.

[88] L'homme L, Esser N, Riva L, et al. Unsaturated fatty acids prevent activation of NLRP3 inflammasome in human monocytes/macrophages. J Lipid Res 2013;54:2998–3008.

[89] Khallou Laschet J, Varthaman A, Fornasa G, et al. Macrophage plasticity in experimental atherosclerosis. PLoS One 2010;5:e8852.

[90] Brown AL, Zhu X, Rong S, et al. Omega-3 fatty acids ameliorate atherosclerosis by favorably altering monocyte subsets and limiting monocyte recruitment to aortic lesions. Arterioscler Thromb Vasc Biol 2012;32:2122–30.

[91] Chang HY, Lee HN, Kim W, et al. Docosahexaenoic acid induces M2 macrophage polarization through peroxisome proliferator activated receptor g activation. Life Sci 2015;120:39–47.

[92] Chawla A. Control of macrophage activation and function by PPARs. Circ Res 2010;106:1559–69.

[93] Ikonen E. Cellular cholesterol trafficking and compartmentalization. Nat Rev Mol Cell Biol 2008;9:125–38.

[94] Singh R, Kaushik S, Wang Y, et al. Autophagy regulates lipid metabolism. Nature 2009;458:1131–5.

[95] Razani B, Feng C, Coleman T, et al. Autophagy links inflammasomes to atherosclerotic progression. Cell Metab 2012;15:534–44.

[96] Williams-Bey Y, Boularan C, Vural A, et al. Omega-3 free fatty acids suppress macrophage inflammasome activation by inhibiting NF-κB activation and enhancing autophagy. PLoS One 2014;9: e97957.

[97] Menotti A, Keys A, Aravanis C, et al. First 20-year mortality data in 12 cohorts of six countries. Ann Med 1989;21:175–9.

[98] Menotti A, Kromhout D, Blackburn H, et al. Food intake patterns and 25-year mortality from coronary heart disease: cross-cultural correlations in the Seven Countries Study. Eur J Epidemiol 1999;15:507–15.

[99] Rumsey SC, Galeano NF, Lipschitz B, et al. Oleate and other long chain fatty acids stimulate low density lipoprotein receptor activity by enhancing acyl coenzyme A:cholesterol acyltransferase activity and altering intracellular regulatory cholesterol pools in cultured cells. J Biol Chem 1995;270:10008–16.

[100] Yu-Poth S, Yin D, Kris-Etherton PM, et al. Long-chain polyunsaturated fatty acids upregulate LDL receptor protein expression in fibroblasts and HepG2 cells. J Nutr 2005;135:2541–5.

[101] Ouimet M, Marcel YL. Regulation of lipid droplet cholesterol efflux from macrophage foam cells. Arterioscler Thromb Vasc Biol 2012;32:575–81.

[102] Tsimikas S, Philis-Tsimikas A, Alexopoulos S, et al. LDL isolated from Greek subjects on a typical diet or from American subjects on an oleate-supplemented diet induces less monocyte chemotaxis and adhesion when exposed to oxidative stress. Arterioscler Thromb Vasc Biol 1999;19:122–30.

[103] Wang S, Wu D, Matthan NR, et al. Reduction in dietary omega-6 polyunsaturated fatty acids: eicosapentaenoic acid plus docosahexaenoic acid ratio minimizes atherosclerotic lesion formation and inflammatory response in the LDL receptor null mouse. Atherosclerosis 2009;204:147–55.

[104] Massaro M, Carluccio MA, Paolicchi A, et al. Mechanisms for reduction of endothelial activation by oleate: inhibition of nuclear factor-kappaB through antioxidant effects. Prostaglandins Leukot Essent Fat Acids 2002;67:175–81.

[105] Spady DK, Dietschy JM. Dietary saturated triacylglycerols suppress hepatic low density lipoprotein receptor activity in the hamster. Proc Natl Acad Sci U S A 1985;82:4526–30.

[106] Mustad VA, Ellsworth JL, Cooper AD, et al. Dietary linoleic acid increases and palmitic acid decreases hepatic LDL receptor protein and mRNA abundance in young pigs. J Lipid Res 1996;37:2310–23.

[107] Fox JC, McGill Jr HC, Carey KD, et al. In vivo regulation of hepatic LDL receptor mRNA in the baboon. Differential effects of saturated and unsaturated fat. J Biol Chem 1987;262:7014–20.

[108] Bennett AJ, Billett MA, Salter AM, et al. Modulation of hepatic apolipoprotein B, 3-hydroxy-3-methylglutaryl-CoA reductase and low-density lipoprotein receptor mRNA and plasma lipoprotein concentrations by defined dietary fats. Comparison of trimyristin, tripalmitin, tristearin and triolein. Biochem J 1995;311:167–73.

[109] Srivastava RA, Ito H, Hess M, et al. Regulation of low density lipoprotein receptor gene expression in HepG2 and Caco2 cells by palmitate, oleate, and 25-hydroxycholesterol. J Lipid Res 1995;36:1434–46.

[110] Lin J, Yang R, Tarr PT, et al. Hyperlipidemic effects of dietary saturated fats mediated through PGC-1 coactivation of SREBP. Cell 2005;120:261–73.

[111] de Lima-Salgado TM, Alba-Loureiro TC, do Nascimento CS, et al. Molecular mechanisms by which saturated fatty acids modulate TNF-α expression in mouse macrophage lineage. Cell Biochem Biophys 2011;59:89–97.

[112] Holland WL, Bikman BT, Wang LP, et al. Lipid-induced insulin resistance mediated by the proinflammatory receptor TLR4 requires saturated fatty acid-induced ceramide biosynthesis in mice. J Clin Invest 2011;121:1858–70.

[113] Schilling JD, Machkovech HM, He L, et al. Palmitate and lipopolysaccharide trigger synergistic ceramide production in primary macrophages. J Biol Chem 2013;288:2923–32.

[114] Lee HU, Lee HJ, Park HY, et al. Effects of hemeoxygenase system on the cyclooxygenase in the primary cultured hypothalamic cells. Arch Pharm Res 2001;24:607–12.

[115] Weatherill AR, Lee JY, Zhao L, et al. Saturated and polyunsaturated fatty acids reciprocally modulate dendritic cell functions mediated through TLR4. J Immunol 2005;174:5390–7.

[116] Wen H, Gris D, Lei Y, et al. Fatty acid-induced NLRP3-ASC inflammasome activation interferes with insulin signaling. Nat Immunol 2011;12:408–15.

[117] Csak T, Ganz M, Pespisa J, et al. Fatty acid and endotoxin activate inflammasomes in mouse hepatocytes that release danger signals to stimulate immune cells. Hepatology 2011;54:133–44.

[118] Duewell P, Kono H, Rayner KJ, et al. NLRP3 inflammasomes are required for atherogenesis and activated by cholesterol crystals. Nature 2010;464:1357–61.

[119] Gao D, Pararasa C, Dunston CR, et al. Palmitate promotes monocyte atherogenicity via de novo ceramide synthesis. Free Radic Biol Med 2012;53:796–806.

[120] Cao J, Dai DL, Yao L, et al. Saturated fatty acid induction of endoplasmic reticulum stress and apoptosis in human liver cells via the PERK/ATF4/CHOP signaling pathway. Mol Cell Biochem 2012;364(1–2):115–29.

[121] Ozcan U, Cao Q, Yilmaz E, et al. Endoplasmic reticulum stress links obesity, insulin action, and type 2 diabetes. Science 2004;306:457–61.

[122] Liao X, Sluimer JC, Wang Y, et al. Macrophage autophagy plays a protective role in advanced atherosclerosis. Cell Metab 2012;4:545–53.

[123] Li S, Sun Y, Liang CP, et al. Defective phagocytosis of apoptotic cells by macrophages in atherosclerotic lesions of ob/ob mice and reversal by a fish oil diet. Circ Res 2009;105:1072–82.

[124] Stentz FB, Kitabchi AE. Palmitic acid-induced activation of human T-lymphocytes and aortic endothelial cells with production of insulin receptors, reactive oxygen species, cytokines, and lipid peroxidation. Biochem Biophys Res Commun 2006;346:721–6.

[125] Wolff RL, Precht D, Nasser B, et al. Trans- and cis-octadecenoic acid isomers in the hump and milk lipids from Camelus dromedarius. Lipids 2001;36:1175–8.

[126] Block JM, Barrera-Arellano D. Hydrogenated products in Brazil: trans isomers, physico-chemical characteristics and fatty acid composition. Arch Latinoam Nutr 1994;44:281–5.

[127] Lichtenstein AH, Ausman LM, Jalbert SM, et al. Effects of different forms of dietary hydrogenated fats on serum lipoprotein cholesterol levels. N Engl J Med 1999;340:1933–40.

[128] Grundy SM, Denke MA. Dietary influences on serum lipids and lipoproteins. J Lipid Res 1990;31:1149–72.

[129] Matthan NR, Welty FK, Barrett PH, et al. Dietary hydrogenated fat increases high-density lipoprotein apoA-I catabolism and decreases low-density lipoprotein apoB-100 catabolism in hypercholesterolemic women. Arterioscler Thromb Vasc Biol 2004;24:1092–7.

[130] Mauger JF, Lichtenstein AH, Ausman LM, et al. Effect of different forms of dietary hydrogenated fats on LDL particle size. Am J Clin Nutr 2003;78:370–5.

[131] Cassagno N, Palos-Pinto A, Costet P, et al. Low amounts of trans 18:1 fatty acids elevate plasma triacylglycerols but not cholesterol and alter the cellular defence to oxidative stress in mice. Br J Nutr 2005;94:346–52.

[132] Fournier N, Attia N, Rousseau-Ralliard D, et al. Deleterious impact of elaidic fatty acid on ABCA-1-mediated cholesterol efflux from mouse and human macrophages. Biochim Biophys Acta 1821;2012:303–12.

II. ENDOTHELIAL DYSFUNCTION AND CLINICAL SYNDROMES

[133] Salmerón J, Hu FB, Manson JE, et al. Dietary fat intake and risk of type 2 diabetes in women. Am J Clin Nutr 2001;73:1019–26.

[134] Wang L, Manson JE, Forman JP, et al. Dietary fatty acids and the risk of hypertension in middle-aged and older women. Hypertension 2010;56:598–604.

[135] Zapolska-Downar D, Kosmider A, Naruszewicz M. Trans fatty acids induce apoptosis in human endothelial cells. J Physiol Pharmacol 2005;56:611–25.

[136] Bryk D, Zapolska-Downar D, Malecki M, et al. Trans fatty acids induce a proinflammatory response in endothelial cells through ROS-dependent nuclear factor-KB activation. J Physiol Pharmacol 2011;62:229–38.

[137] Baer DJ, Judd JT, Clevidence BA, et al. Dietary fatty acids affect plasma markers of inflammation in healthy men fed controlled diets: a randomized crossover study. Am J Clin Nutr 2004;79:969–73.

[138] Han SN, Leka LS, Lichtenstein AH, et al. Effect of hydrogenated and saturated, relative to polyunsaturated, fat on immune and inflammatory responses of adults with moderate hypercholesterolemia. J Lipid Res 2002;43:445–52.

[139] Mozaffarian D, Rimm EB, King IB, et al. Trans fatty acids and systemic inflammation in heart failure. Am J Clin Nutr 2004;80:1521–5.

[140] Johnson RK, Appel LJ, Brands M, et al. American Heart Association Nutrition Committee of the Council on Nutrition, Physical Activity, and Metabolism and the Council on Epidemiology and Prevention. Dietary sugars intake and cardiovascular health: a scientific statement from the American Heart Association. Circulation 2009;120(11):1011–20.

[141] World Health Organization. Guideline: sugars intake for adults and children. Geneva: World Health Organization; 2015.

[142] Yudkin J. Dietary factors in arteriosclerosis: sucrose. Lipids 1978;13:370–2.

[143] Yudkin J. Sugar and ischaemic heart disease. Practitioner 1967;198:680–3.

[144] Liu S, Willett WC, Stampfer MJ, et al. A prospective study of dietary glycemic load, carbohydrate intake, and risk of coronary heart disease in US women. Am J Clin Nutr 2000;71(6):1455–61.

[145] Parks EJ, Hellerstein MK. Carbohydrate-induced hypertriacylglycerolemia: historical perspective and review of biological mechanisms. Am J Clin Nutr 2000;71:412–33.

[146] Merchant AT, Anand SS, Kelemen LE, et al. Carbohydrate intake and HDL in a multiethnic population. Am J Clin Nutr 2007;85:225–30.

[147] Vartanian LR, Schwartz MB, Brownell KD. Effects of soft drink consumption on nutrition and health: a systematic review and meta-analysis. Am J Public Health 2007;97:667–75.

[148] Frayn KN, Kingman SM. Dietary sugars and lipid metabolism in humans. Am J Clin Nutr 1995;62(1 Suppl.):250S–61S.

[149] Fung TT, Malik V, Rexrode KM, et al. Sweetened beverage consumption and risk of coronary heart disease in women. Am J Clin Nutr 2009;89:1037–42.

[150] Welsh JA, Sharma AS, Cunningham SA, et al. Consumption of added sugars and indicators of cardiovascular disease risk among US adolescents. Circulation 2011;123:249–57.

[151] Bantle JP, Raatz SK, Thomas W, et al. Effects of dietary fructose on plasma lipids in healthy subjects. Am J Clin Nutr 2000;72:1128–34.

[152] Havel PJ. Dietary fructose: implications for dysregulation of energy homeostasis and lipid/carbohydrate metabolism. Nutr Rev 2005;63:133–57.

[153] Johnson RJ, Segal MS, Sautin Y, et al. Potential role of sugar (fructose) in the epidemic of hypertension, obesity and the metabolic syndrome, diabetes, kidney disease, and cardiovascular disease. Am J Clin Nutr 2007;86:899–906.

[154] Brown CM, Dulloo AG, Yepuri G, et al. Fructose ingestion acutely elevates blood pressure in healthy young humans. Am J Physiol Regul Integr Comp Physiol 2008;294:R730–7.

[155] Elliott SS, Keim NL, Stern JS, et al. Fructose, weight gain, and the insulin resistance syndrome. Am J Clin Nutr 2002;76(5):911–22.

[156] Xi B, Huang Y, Reilly KH, et al. Sugar-sweetened beverages and risk of hypertension and CVD: a dose-response meta-analysis. Br J Nutr 2015;113(5):709–17.

[157] Van der Schaaf MR, Koomans HA, Joles JA. Dietary sucrose does not increase twenty-four-hour ambulatory blood pressure in patients with either essential hypertension or polycystic kidney disease. J Hypertens 1999;17:453–4.

[158] Black RN, Spence M, McMahon RO, et al. Effect of eucaloric high- and low-sucrose diets with identical macronutrient profile on insulin resistance and vascular risk: a randomized controlled trial. Diabetes 2006;55(12):3566–72.

[159] Raben A, Vasilaras TH, Moller AC, et al. Sucrose compared with artificial sweeteners: different effects on ad libitum food intake and body weight after 10 weeks of supplementation in overweight subjects. Am J Clin Nutr 2002;76:721–9.

[160] Nguyen S, Choi HK, Lustig RH, et al. Sugar-sweetened beverages, serum uric acid, and blood pressure in adolescents. J Pediatr 2009;154:807–13.

[161] Dhingra R, Sullivan L, Jacques PF, et al. Soft drink consumption and risk of developing cardiometabolic risk factors and the metabolic syndrome in middle-aged adults in the community. Circulation 2007;116:480–8 [Published correction appears in Circulation 2007;116:e557].

[162] Rebello T, Hodges RE, Smith JL. Short-term effects of various sugars on antinatriuresis and blood pressure changes in normotensive young men. Am J Clin Nutr 1983;38(1):84–94.

[163] Rowe JW, Young JB, Minaker KL, et al. Effect of insulin and glucose infusions on sympathetic nervous system activity in normal man. Diabetes 1981;30(3):219–25.

[164] Spruss A, Bergheim I. Dietary fructose and intestinal barrier: potential risk factor in the pathogenesis of nonalcoholic fatty liver disease. J Nutr Biochem 2009;20(9):657–62.

[165] Hirahatake KM, Meissen JK, Fiehn O, et al. Comparative effects of fructose and glucose on lipogenic gene expression and intermediary metabolism in HepG2 liver cells. PLoS One 2011;6(11):e26583.

[166] Nakagawa T, Hu H, Zharikov S, et al. A causal role for uric acid in fructose-induced metabolic syndrome. Am J Physiol Ren Physiol 2006;290(3):F625–31.

[167] Singh VP, Aggarwal R, Singh S, et al. Metabolic syndrome is associated with increased oxo-nitrative stress and asthma-like changes in lungs. PLoS One 2015;10(6)e0129850.

[168] Stanhope KL, Schwarz JM, Keim NL, et al. Consuming fructose-sweetened, not glucose-sweetened, beverages increases visceral adiposity and lipids and decreases insulin sensitivity in overweight/obese humans. J Clin Invest 2009;119(5):1322–34.

[169] Niskanen LK, Laaksonen DE, Nyyssonen K, et al. Uric acid level as a risk factor for cardiovascular and all-cause mortality in middle-aged men: a prospective cohort study. Arch Intern Med 2004;164:1546–51.

[170] Gagliardi AC, Miname MH, Santos RD. Uric acid: a marker of increased cardiovascular risk. Atherosclerosis 2009;202(1):11–7.

[171] Hallfrisch J. Metabolic effects of dietary fructose. FASEB J 1990;4(9):2652–60.

[172] Strazzullo P, Puig JG. Uric acid and oxidative stress: relative impact on cardiovascular risk? Nutr Metab Cardiovasc Dis 2007;17(6):409–14.

[173] Glushakova O, Kosugi T, Roncal C, et al. Fructose induces the inflammatory molecule ICAM-1 in endothelial cells. J Am Soc Nephrol 2008;19(9):1712–20.

[174] Khosla UM, Zharikov S, Finch JL, et al. Hyperuricemia induces endothelial dysfunction. Kidney Int 2005;67(5):1739–42.

[175] Mazzali M, Hughes J, Kim YG, et al. Elevated uric acid increases blood pressure in the rat by a novel crystal-independent mechanism. Hypertension 2001;38(5):1101–6.

[176] Sauder KA, Proctor DN, Chow M, et al. Endothelial function, arterial stiffness and adherence to the 2010 Dietary Guidelines for Americans: a cross-sectional analysis. Br J Nutr 2015;113(11): 1773–81.

[177] Oude Griep LM, Wang H, Chan Q. Empirically-derived dietary patterns, diet quality scores, and markers of inflammation and endothelial dysfunction. Curr Nutr Rep 2013;2(2):97–104.

[178] Lowndes J, Sinnett S, Yu Z, et al. The effects of fructose-containing sugars on weight, body composition and cardiometabolic risk factors when consumed at up to the 90th percentile population consumption level for fructose. Nutrients 2014;6(8):3153–68.

[179] Bray GA. Fructose: should we worry? Int J Obes 2008;32(Suppl. 7): S127–31.

[180] Gibson SA. Dietary sugars intake and micronutrient adequacy: a systematic review of the evidence. Nutr Res Rev 2007;20(2):121–31.

[181] Frary CD, Johnson RK, Wang MQ. Children and adolescents' choices of foods and beverages high in added sugars are associated with intakes of key nutrients and food groups. J Adolesc Health 2004;34(1):56–63.

[182] Instituto Brasileiro de Geografla e Estatística. Pesquisa de Orçamentos Familiares 2008/2009: antropometria e estado nutricional de crianças, adolescentes e adultos no Brasil. Rio de Janeiro: Instituto Brasileiro de Geografia e Estatística; 2010.

[183] World Health Organization. Global strategy on diet, physical activity and health: a framework to monitor and evaluate implementation, Geneva: World Health Organization; 2006. http:// www.who.int/dietphysicalactivity/Indicators%20English.pdf [accessed June 21, 2016].

[184] U.S. Department of Agriculture, U.S. Department of Health and Human Services. Dietary Guidelines for Americans, 2015. 8th ed. Washington, DC: U.S. Government Printing Office; 2015.

[185] Sacks FM, Svetkey LP, Vollmer WM, et al. Effects on blood pressure of reduced dietary sodium and the Dietary Approaches to Stop Hypertension (DASH) diet. DASH-Sodium Collaborative Research Group. N Engl J Med 2001;344(1):3–10.

[186] Hendriksen MA, Van Raaij JM, Geleijnse JM, et al. Health gain by salt reduction in Europe: a modelling study. PLoS One 2015;10(3): e0118873.

[187] Sarno F, Claro RM, Levy RB, et al. Estimated sodium intake for the Brazilian population, 2008–2009. Rev Saude Publica 2013;47(3): 571–8.

[188] Centers for Disease Control and Prevention. Sodium's role in processed food, Atlanta, GA: Centers for Disease Control and Prevention; 2012, http://www.cdc.gov/salt/pdfs/sodium_role_ processed.pdf [accessed June 22, 2016].

[189] Zhao D, Qi Y, Zheng Z, et al. Dietary factors associated with hypertension. Nat Rev Cardiol 2011;8(8):456–65.

[190] DuPont JJ, Greaney JL, Wenner MM, et al. High dietary sodium intake impairs endothelium-dependent dilation in healthy salt-resistant humans. J Hypertens 2013;31(3):530–6.

[191] Majid DS, Prieto MC, Navar LG. Salt-sensitive hypertension: perspectives on intrarenal mechanisms. Curr Hypertens Rev 2015; 11(1):38–48.

[192] Todd AS, Macginley RJ, Schollum JB, et al. Dietary salt loading impairs arterial vascular reactivity. Am J Clin Nutr 2010;91 (3):557–64.

[193] Seals DR, Tanaka H, Clevenger CM, et al. Blood pressure reductions with exercise and sodium restriction in postmenopausal women with elevated systolic pressure: role of arterial stiffness. J Am Coll Cardiol 2001;38(2):506–13.

[194] Stingo AJ, Clavell AL, Heublein DM, et al. Presence of C-type natriuretic peptide in cultured human endothelial cells and plasma. Am J Physiol 1992;263(4 Pt 2):H1318–21.

[195] Boegehold MA. The effect of high salt intake on endothelial function: reduced vascular nitric oxide in the absence of hypertension. J Vasc Res 2013;50(6):458–67.

[196] Zhou MS, Schulman IH, Raij L. Role of angiotensin II and oxidative stress in vascular insulin resistance linked to hypertension. Am J Physiol Heart Circ Physiol 2009;296:H833–9.

[197] Moncada S, Higgs A. The L-arginine-nitric oxide pathway. N Engl J Med 1993;329:2002–12.

[198] Tsao PS, Buitrago R, Chan JR, et al. Fluid flow inhibits endothelial adhesiveness: nitric oxide and transcriptional regulation of VCAM-1. Circulation 1996;94:1682–9.

[199] Ferri C, Bellini C, Desideri G, et al. Clustering of endothelial markers of vascular damage in human salt-sensitive hypertension: influence of dietary sodium load and depletion. Hypertension 1998;32(5):862–8.

[200] Hayakawa Y, Aoyama T, Yokoyama C, et al. High salt intake damages the heart through activation of cardiac (pro) renin receptors even at an early stage of hypertension. PLoS One 2015;10(3): e0120453.

[201] Baker KM, Booz GW, Dostal DE. Cardiac actions of angiotensin II. Role of an intracardiac renin-angiotensin system. Annu Rev Physiol 1992;54:227–41.

[202] Ferreira DN, Katayama IA, Oliveira IB, et al. Salt-induced cardiac hypertrophy and interstitial fibrosis are due to a blood pressure-independent mechanism in Wistar rats. J Nutr 2010;140:1742–51.

[203] Laursen JB, Rajagopalan S, Galis Z, et al. Role of superoxide in angiotensin II-induced but not catecholamine-induced hypertension. Circulation 1997;95(3):588–93.

[204] Katayama IA, Pereira RC, Dopona EP, et al. High-salt intake induces cardiomyocyte hypertrophy in rats in response to local angiotensin II type 1 receptor activation. J Nutr 2014;144(10): 1571–8.

[205] Ostlund Jr. RE. Phytosterols in human nutrition. Annu Rev Nutr 2002;22:533–49.

[206] de Jong A, Plat J, Mensink RP. Metabolic effects of plant sterols and stanols. J Nutr Biochem 2003;14(7):362–9.

[207] Katan MB, Grundy SM, Jones P, et al. Efficacy and safety of plant stanols and sterol in the management of blood cholesterol levels. Mayo Clin Proc 2003;78(8):965–78.

[208] Patel MD, Thompson PD. Phytosterols and vascular disease. Atherosclerosis 2006;186(1):12–9.

[209] Lottenberg AM, Nunes VS, Nakandakare ER, et al. Food phytosterol ester efficiency on the plasma lipid reduction in moderate hypercholesterolemic subjects. Arq Bras Cardiol 2002;79(2): 139–42.

[210] Law M. Plant sterol and stanol margarines and health. BMJ 2000;320(7238):861–4.

[211] Ikeda I, Tanaka K, Sugano M, et al. Inhibition of cholesterol absorption in rats by plant sterols. J Lipid Res 1988;29(12):1573–82.

[212] Lottenberg AM, Bombo RP, Ilha A, et al. Do clinical and experimental investigations support an antiatherogenic role for dietary phytosterols/stanols? IUBMB Life 2012;64(4):296–306.

[213] Garcia-Calvo M, Lisnock J, Bull HG, et al. The target of ezetimibe is Niemann-Pick C1-Like 1 (NPC1L1). Proc Natl Acad Sci U S A 2005;102(23):8132–7.

[214] Yu L, Hammer RE, Li-Hawkins J, et al. Disruption of Abcg5 and Abcg8 in mice reveals their crucial role in biliary cholesterol secretion. Proc Natl Acad Sci U S A 2002;99(25):16237–42.

[215] Lin DS, Steiner RD, Merkens LS, et al. The effects of sterol structure upon sterol esterification. Atherosclerosis 2010;208(1):155–60.

[216] Sanclemente T, Marques-Lopes I, Fajó-Pascual M, et al. A moderate intake of phytosterols from habitual diet affects cholesterol metabolism. J Physiol Biochem 2009;65(4):397–404.

[217] Ge L, Wang J, Qi W, et al. The cholesterol absorption inhibitor ezetimibe acts by blocking the sterol-induced internalization of NPC1L1. Cell Metab 2008;7(6):508–19.

[218] von Bergmann K, Sudhop T, Lütjohann D. Cholesterol and plant sterol absorption: recent insights. Am J Cardiol 2005;96(1A): 10D–14D.

[219] Van Der Velde AE, Brufau G, Groen AK. Transintestinal choles-terol efflux. Curr Opin Lipidol 2010;21:167–71.

[220] Brufau G, Kuipers F, Lin Y, et al. A reappraisal of the mechanism by which plant sterols promote neutral sterol loss in mice. PLoS One 2011;6(6):e21576.

[221] Halliwell B, Gutteridge JC. The definition and measurement of antioxidants in biological systems. Free Radic Biol Med 1995;18:125.

[222] Siti HN, Kamisah Y, Kamsiah J. The role of oxidative stress, anti-oxidants and vascular inflammation in cardiovascular disease. Vasc Pharmacol 2015;71:40–56.

[223] Sies H. Oxidative stress: oxidants and antioxidants. Exp Physiol 1997;82:291–5.

[224] Murphy MP, Holmgren A, Larsson NG, et al. Unraveling the bio-logical roles of reactive oxygen species. Cell Metab 2011; 13(4):361–6.

[225] Schieber M, Chandel NS. ROS function in redox signaling and oxi-dative stress. Curr Biol 2014;24(10):R453–62.

[226] Sies H, Stahl W, Sevanian A. Nutritional, dietary and postprandial oxidative stress. J Nutr 2005;135(5):969–72.

[227] Ramyaa P, Krishnaswamy R, Padma W. Quercetin modulates OTA-induced oxidative stress and redox signalling in HepG2 cells—up regulation of Nrf2 expression and down regulation of NF-κB and COX-2. Biochim Biophys Acta 2014;1840(1): 681–92.

[228] Cho BO, Ryu HW, Jin CH, et al. Blackberry extract attenuates oxi-dative stress through up-regulation of Nrf2-dependent antioxi-dant enzymes in carbon tetrachloride-treated rats. J Agric Food Chem 2011;59(21):11442–8.

[229] Uzun A, Yener U, Cicek OF, et al. Does vitamin C or its combina-tion with vitamin E improve radial arteryendothelium-dependent vasodilatation in patients awaiting coronary artery bypass sur-gery? Cardiovasc J Afr 2013;24:255–9.

[230] Sahu BD, Kumar JM, Sistla R. Bioflavonoid, prevents cisplatin-induced acute kidney injury by up-regulating antioxidant defenses and down-regulating the MAPKs and NF-κB pathways. PLoS One 2015;10(7):e0134139.

[231] Da Costa CA, de Oliveira PR, de Bem GF, et al. Euterpeoleracea Mart-derived polyphenols prevent endothelial dysfunction and vascular structural changes in renovascular hypertensive rats: role of oxidative stress. Naunyn Schmiedeberg's Arch Pharmacol 2012;385(12):1199–209.

[232] Widmer RJ, Freund MA, Flammer AJ, et al. Beneficial effects of polyphenol-rich olive oil in patients with early atherosclerosis. Eur J Nutr 2013;52(3):1223–31.

[233] Loke WM, Proudfoot JM, Hodgson JM, et al. Specific dietary poly-phenols attenuate atherosclerosis in apolipoprotein e-knockout mice by alleviating inflammation and endothelial dysfunction. Arterioscler Thromb Vasc Biol 2010;30(4):749–57.

[234] Scoditti E, Calabriso N, Massaro M, et al. Mediterranean diet poly-phenols reduce inflammatory angiogenesis through MMP-9 and COX-2 inhibition in human vascular endothelial cells: a poten-tially protective mechanism in atherosclerotic vascular disease and cancer. Arch Biochem Biophys 2012;527(2):81–9.

[235] Meydani M, Kwan P, Band M, et al. Long-term vitamin E supplementation reduces atherosclerosis and mortality in Ldlr−/− mice, but not when fed Western style diet. Atherosclerosis 2014;233(1):196–205.

[236] Bozaykut P, Karademir B, Yazgan B, et al. Effects of vitamin E on peroxisome proliferator-activated receptor g and nuclear factorer-ythroid 2-related factor 2 in hypercholesterolemia-induced ath-erosclerosis. Free Radic Biol Med 2014;70:174–81.

[237] Leong XF, Najib MN, Das S, et al. Intake of repeatedly heated palm oil causes elevation in blood pressure with impaired vasor-elaxation in rats. Tohoku J Exp Med 2009;219:71–8.

[238] Knekt P, Ritz J, Pereira MA, et al. Antioxidant vitamins and cor-onary heart disease risk: a pooled analysis of 9 cohorts. Am J Clin Nutr 2004;80(6):1508–20.

[239] Chen GC, Lu DB, Pang Z, et al. Vitamin C intake, circulating vita-min C and risk of stroke: a meta-analysis of prospective studies. J Am Heart Assoc 2013;2(6):e000329.

[240] Bjelakovic G, Nikolova D, Gluud LL, et al. Antioxidant supple-ments for prevention of mortality in healthy participants and patients with various diseases. Cochrane Database Syst Rev 2012;3:CD007176.

[241] Sesso HD, Buring JE, Christen WG, et al. Vitamins E and C in the prevention of cardiovascular disease in men: the Physicians' Health Study II randomized controlled trial. JAMA 2008; 300(18):2123–33.

[242] Giannini C, Mohn A, Chiarelli F, et al. Macrovascular angiopathy in children and adolescents with type 1 diabetes. Diabetes Metab Res Rev 2011;27:436–60.

[243] Loomans CJM, De Koning EJP, Staal FJT, et al. Endothelial pro-genitor cell dysfunction in type 1 diabetes: another consequence of oxidative stress? Antioxid Redox Signal 2005;7:1468–75.

[244] Hirsch IB, Brownlee M. Should minimal blood glucose variability become the gold standard of glycemic control? J Diabetes Compli-cat 2005;19:178–81.

[245] Hoffman RP. Vascular endothelial dysfunction and nutritional compounds in early type 1 diabetes. Curr Diabetes Rev 2014; 10(3):201–7.

[246] Halliwell B. Free radicals and antioxidants: updating a personal view. Nutr Rev 2012;70(5):257–65.

[247] Santos RD, Gagliardi AC, Xavier HT, et al. First guidelines on fat consumption and cardiovascular health. Arq Bras Cardiol 2013;100(1 Suppl. 3):1–40.

[248] Brasil. Ministério da Saúde. Secretaria de Atenção à Saúde. Depar-tamento de Atenção Básica. Guia alimentar para a população bra-sileira/Ministério da Saúde, Secretaria de Atenção à Saúde, Departamento de Atenção Básica. 2nd ed. Brasília: Ministério da Saúde; 2014. p. 156.

[249] WHO. NCD global monitoring framework; 2011. Disponível em: http://www.who.int/nmh/global_monitoring_framework/en [Internet] [Acesso em July 22, 2016].

Effects of Mediterranean Diet on Endothelial Function

Jordi Merino, Richard Kones, and Emilio Ros

INTRODUCTION

Concept of Dietary Patterns as the Best Measures of Exposure in Relation to Disease Outcomes in Nutritional Science

Foods are consumed in a number of combinations, providing a range of nutrients and other dietary components, which interact in complex ways. The effects of diet are the results of a web-like network of myriads of chemicals within foods reacting with an even larger network of human biochemical pathways, a complex exposure variable that calls for multiple approaches to examine the relationship between diet and disease risk. In the 1980s, Schwerin and coworkers [1] first attempted to examine the prevalence, magnitude and distribution of malnutrition and related health problems within the United States by using a theoretical concept proposed at the White House Conference on Food, Nutrition and Health in 1969. At the time, the nascent hypothesis emphasized the importance of examining "the relationship of food consumption and patterns of eating to the health of [the] American population." The findings of this study illustrated the value of using an eating pattern model to explore the complex association between food consumption and health. Today, this methodology has emerged as an alternative and complementary approach to examining the relationship between diet and chronic disease risk. Conceptually, dietary patterns represent a broader picture of food and nutrient consumption, and may thus be more predictive of complex disease risk than individual foods or nutrients [2]. An examination of dietary patterns also closely parallels events in the real world, in which nutrients and foods are consumed in combination; hence, their joint effects may best be investigated by considering the entire eating pattern.

Dietary patterns are not directly measurable, a reason why various techniques have been used to characterize them from collected dietary information. Two methods are widely used: information is obtained from the dietary data collected to define patterns by a posteriori statistical techniques (factor or cluster analyses) [2,3] or dietary indices are defined a priori based on previous knowledge of what constitutes a healthy diet, such as the Dietary Approaches to Stop Hypertension (DASH) diet [4], the Mediterranean diet (MedDiet) [5], or other healthy dietary indices [6,7]. Studying dietary patterns has important public health implications because the overall patterns of food consumption in their multiple combinations are easier for the public to understand and implement. In fact, current dietary guidelines (e.g., Dietary Guidelines for Americans 2010 and 2013 American Heart Association/American College of Cardiology Guideline on Lifestyle Management to Reduce Cardiovascular Risk) recommend a dietary pattern approach for reducing chronic disease risk [8,9]. One pattern, originally described as the traditional diet and lifestyle of olive-growing inhabitants of Crete, Greece, southern Italy, and Spain in the 1950s–1960s, has roots dating back to Crete, 2700–1450 BC. This pattern, the MedDiet, has attracted increasing interest by both nutrition investigators and the general public [6].

Definition of the Mediterranean Diet

The ancient Greek word *diaita*, from which the word diet derives, means *balanced lifestyle*, and this is exactly what the MedDiet is—much more than a nutritional pattern. The MedDiet is a lifestyle, not simply a collection of foods and nutrients. Rather, the foods are grown locally and prepared using recipes and cooking methods unique to each locale; meals are shared, along with celebrations and traditions, and this is coupled with moderate physical activity, favored by a welcoming climate and adequate time outdoors in the sunshine [10], which completes a lifestyle that modern science now recommends to optimize long-term health [11]. The MedDiet lifestyle model is characterized by abundant foods of plant origin such as vegetables, legumes, fruits, whole-grain cereals, and nuts; the use of olive oil as the main source of culinary fat, moderate consumption of fish, seafood, poultry, dairy products (yogurt, cheese), and eggs, as well as small amounts of red meat, and a daily moderate intake of wine generally at meals. The pattern involves a low intake of saturated fatty acids (SFA), high content of monounsaturated fatty acids (MUFA), complex carbohydrates, and fiber, as well as abundant antioxidants [10]. Health benefits are attributed to dietary *balance, variety, and moderation*, but go beyond foods and nutrients to apply these same principles to a way of life. In addition to variety, simplicity and sustainability are key features. A unique characteristic of the MedDiet is that has been, and remains, an evolutionary, dynamic, and vital cultural heritage.

SCIENTIFIC EVIDENCE LINKING THE MEDITERRANEAN DIET TO CARDIOVASCULAR PREVENTION

Overview of Epidemiological Studies

The first scientific evidence on the health properties of the MedDiet stemmed from the Seven Countries study, conducted in the 1950s [12]. This ecological study, comparing dietary habits in seven countries worldwide, reported that both the overall diet and the type of fat consumed in the Mediterranean regions was associated with low cardiovascular disease (CVD) mortality rates. Subsequent evidence came from the MONICA project, whereby the lower rates of coronary heart disease (CHD) and

its risk factors observed in Southern European Countries were also attributed to the cardioprotective effect of the MedDiet [13].

Prospective cohort studies have consistently found that a MedDiet-like pattern is associated with a decreased risk of CHD and mortality [14,15]. In the landmark Nurse's Health Study (NHS), a greater adherence to the MedDiet was associated with a 29% lower risk of incident CHD and 27% lower risk of stroke [16]. In the European Prospective Investigation into Cancer and Nutrition (EPIC) Spanish cohort study, the MedDiet was associated with 27% lower risk of CHD [17]. EPIC has also provided evidence that a 2-point increase in the MedDiet score was associated with a 25% reduced risk of all-cause mortality in a Greek population [18] and a 14% similarly lowered risk in elderly subjects [19]. Moreover, a meta-analysis including seven prospective cohort studies and 200,000 subjects concluded that a 2-point increase in a 9-point score of adherence to the MedDiet was associated with a significant 8% reduction of overall mortality and 10% reduction of CVD incidence and mortality [15].

Clinical Trial Evidence on Hard Outcomes: The PREDIMED Study

Intervention studies have helped establish causality regarding the protective role of the MedDiet on cardiovascular health. The Lyon Diet Heart Study was a randomized controlled trial (RCT) of secondary CHD prevention evaluating the effect of a modified MedDiet on the risk of recurrent myocardial infarction after intervention for 46 months [20]. In this RCT, dietary recommendations emphasized the consumption of more whole grains, root and green vegetables and fish, less beef, lamb and pork (replaced with poultry), no day without fruit, and butter and cream replaced with margarine high in α-linolenic acid (ALA), the vegetable n-3 fatty acid. However, use of olive oil was not advised, hence a critical component of the MedDiet was missing. Compared to the control group, those who were randomized to the MedDiet group had a 47% reduced risk of myocardial infarction and CVD mortality [20]. In the GISSI-Prevenzione secondary prevention trial with 11,000 patients who had suffered a myocardial infarction, participants were advised to increased consumption of foods from the traditional MedDiet, such as extra-virgin olive oil (EVOO), fish, fruit, raw and cooked vegetables, and nuts. After an

6-year follow-up, participants who reported better adherence to the MedDiet benefited from a 49% reduced risk of all-cause death compared to those who were not following the diet [21]. Until recently, there was no evidence of the effect of the MedDiet in the primary prevention of CVD. The PREDIMED (PREvención con DIeta MEDiterránea) study is a multicenter RCT of primary CVD prevention conducted in Spain [22]. This study included 7447 participants (aged 55–80 years, of whom 57% were women) at high-CVD risk, but no CVD at enrollment. Participants were randomized to one of three intervention groups: (i) a MedDiet supplemented with EVOO, (ii) a MedDiet supplemented with mixed nuts, or (iii) a control diet (advice to follow a low-fat diet). The participants received quarterly nutrition education in individual and group sessions and, depending on the assigned group, free provision of EVOO, mixed nuts (walnuts, hazelnuts and almonds), or nonfood gifts. The primary objective of PREDIMED was to evaluate the effect of the MedDiet on the risk of major CVD events with a primary endpoint composed of myocardial infarction, stroke or CVD death. After a mean follow-up of 4.8 years, 288 participants suffered a CVD event and the results showed that, compared to participants in the control diet group, those randomized to the MedDiet supplemented with EVOO and the MedDiet supplemented with nuts had both lower CVD risk by nearly 30% [22].

Mechanisms of Protection

The cardioprotective effect of the MedDiet can be explained by a beneficial effect on classical and emerging cardiovascular risk factors, including inflammation insulin resistance and oxidative stress [23]. In proof, other PREDIMED reports have demonstrated that both supplemented MedDiets are associated with lower blood pressure, improved lipid profiles, reduced concentrations of circulating inflammatory makers, and decreased insulin resistance and subsequent risk of type-2 diabetes (T2D) and metabolic syndrome (MetS) [24–27]. Although the underlying mechanisms of protection against CVD by the MedDiet are not fully understood, its richness in beneficial bioactive compounds is likely to be highly relevant.

The antiinflammatory effects of the MedDiet have been tested in various PREDIMED substudies. In a 3-month pilot study conducted in the first 772 participants recruited into the trial, plasma concentrations of interleukin-6 (IL-6), vascular cell adhesion molecule (VCAM)-1 and endothelial intercellular adhesion molecule (ICAM)-1 significantly decreased in the MedDiet intervention groups, while plasma concentrations of high-sensitive C-reactive protein (CRP) only decreased in the MedDiet with EVOO [24]. In contrast, plasma concentrations of VCAM-1 and ICAM-1 increased in the control diet group. An additional study examined changes in peripheral blood mononuclear cell expression of cell surface inflammatory mediators at 3 months [28]. Again, the PREDIMED MedDiets exerted beneficial effects on adhesion molecules and CD40 expression on T lymphocytes and monocytes. Moreover, there were significant reductions in CD49d in peripheral T-lymphocytes and CD11b, CD49d (a crucial adhesion molecule for leukocyte homing) and CD40 (a proinflammatory ligand) in monocytes, suggesting a mechanistic pathway by which the MedDiet might influence the inflammatory status [28]. An additional PREDIMED substudy with data collected at 3 months demonstrated that the MedDiets had a beneficial effect on the expression of genes involved in vascular inflammation, foam cell formation and plaque stability [29]. A further 12-month PREDIMED substudy involving 164 participants reported that participants in the MedDiet supplemented with nuts had a significant 34% reduction in CD40 expression on monocyte surfaces compared to those in the control diet [30]. In addition, inflammatory biomarkers related to plaque instability such as CRP and IL-6 were reduced by 45% and 35% and 95% and 90% in the MedDiet with EVOO and nuts, respectively, compared with the control diet. Likewise, compared with the control diet group, soluble ICAM-1 and P-selectin were also reduced by 50% and 27%, respectively, in the MedDiet with EVOO group, and P-selectin by 19% in the MedDiet with nuts group. The conclusion was that a MedDiet supplemented with EVOO or nuts modifies the process of adhesion of circulating monocytes and T-lymphocytes to endothelial cells during inflammation, a crucial early step in the initiation and progression of atherosclerosis [30]. Therefore, in patients at high risk of CVD, MedDiets downregulate cellular and circulating inflammatory biomarkers believed to be involved in the pathogenesis of CVD. A recent PREDIMED report shows that the inflammatory potential of the diet, more pronounced when conforming less to the MedDiet, relates to the occurrence of CVD events [31].

Inflammation is an important factor linking diet with visceral adiposity and insulin resistance, a central feature of the MetS. Several studies have focused on the association between the MedDiet and insulin sensitivity. As the ratio of dietary MUFA/SFA rises, both insulin sensitivity and pancreatic β-cell function improve in the postprandial state [32]. In a Greek adult population, an inverse association between MedDiet adherence and indices of glucose homeostasis and insulin resistance was found in nondiabetic subjects [33]. In a seminal RCT testing various diets during 24 months for weight loss, insulin resistance evaluated by the Homeostasis Model Assessment Method (HOMA-IR) was improved by a MedDiet compared with a low-fat diet [34]. In 215 newly diagnosed T2D patients randomized to a MedDiet or a low-fat diet, a similar inverse association was observed between the MedDiet and the HOMA-IR index after 1 year of dietary intervention [35]. In contrast, there was no difference in HOMA-IR indices at 1 year in 116 patients with T2D assigned to either a MedDiet or the dietary pattern suggested by the American Diabetes Association (ADA) [36]. Substituting fewer and lower glycemic index carbohydrates in a traditional MedDiet appeared to maximize clinical control in a PREDIMED substudy of 191 diabetic participants followed for 1 year, as increased values of both adiponectin/leptin ratio and adiponectin/HOMA-IR ratio, and significantly lower waist circumference were observed in all three diet arms; in both MedDiet groups, but not in the low-fat diet group, these results were associated with significant reductions in body weight [37].

Oxidative stress is another critical link to cardiometabolic disorders and CVD that can be modulated by the MedDiet. The LIPGENE study recently showed that a diet high in MUFA enhances postprandial glutathione levels and the reduced/oxidized glutathione ratio compared with a low-fat diet and a diet high in SFA [38]. These results concur with those of a 3-month PREDIMED substudy showing the two intervention MedDiets to be associated with lower plasma oxidized LDL levels [39] and with data collected in a MedDiet intervention study in Canadian women [40].

The high-EVOO content of the MedDiet may explain in part the antioxidant and antiatherogenic properties of this dietary pattern. At least in vitro, polyphenols from EVOO suppressed the reactive oxygen species (ROS)-mediated activation of NF-κB, as well as the expression of proatherogenic molecules matrix metalloproteinase (MMP)-9 and cyclooxygenase-2 [41]. Other potent polyphenols found in olives and EVOO are hydroxytyrosol and oleuropein; the biology of these compounds suggests a possible role in the prevention of CVD, although the evidence is only preclinical [42]. One of the secondary outcomes of the PREDIMED study was incident atrial fibrillation, a condition linked to an enhanced inflammatory status. After a follow-up of 4.7 years, the Mediterranean diet with EVOO significantly reduced the risk of atrial fibrillation, with a hazard ratio of 0.62 [95% confidence interval (CI), 0.45–0.85] compared with the control diet group, a result that is congruent with the antiinflammatory properties of the MedDiet in general and EVOO polyphenols in particular [43]. A recent prospective study from Italy conducted in patients with atrial fibrillation showed that those with best adherence to the MedDiet had reduced CVD event rates compared with less adherent participants by an antioxidant mechanism, as shown by concomitant downregulation of Nox2 and decreased excretion of F2-isoprostanes [44].

MEDITERRANEAN DIET AND ENDOTHELIAL FUNCTION

The Whole Dietary Pattern and Endothelial Function

Assessment Using Circulating Molecules

Endothelial dysfunction comprises a specific state of endothelial activation that is characterized by enhanced expression and release into the circulation of inflammatory cytokines and adhesion molecules. Different dietary patterns were related to improvement in markers of inflammation and endothelial dysfunction, including CRP, IL-6, ICAM-1, and VCAM-1 in the NHS I cohort. In a group of 1900 women with MetS, those with close adherence to a MedDiet pattern had reduced serum concentrations of CRP, IL-6, IL-7 and IL-18, decreased insulin resistance, and improved endothelial function compared with those with poorer adherence [45]. In addition, a prudent dietary pattern similar to the MedDiet was inversely associated with plasma CRP and E-selectin concentrations, whereas a Western dietary pattern, with higher intake of red meat, sweets and refined grains, was positively associated with levels of CRP, IL-6, E-selectin, ICAM-1 and VCAM-1 [46]. Participants in the Greek Attica Study who were highly adherent to the traditional MedDiet had lower plasma concentrations of CRP, IL-6, homocysteine and fibrinogen, as well as a lower white blood cell count and a borderline decrease of TNF-α and amyloid A levels compared with those who were least adherent [47]. In agreement with these studies, a recent systematic review and meta-analysis of 17 RCTs examining the effects of the MedDiet versus control diets on endothelial function in 2300 study subjects showed that the MedDiet significantly improved markers of inflammation such as CRP and IL-6 and of endothelial function such as ICAM-1 [48].

Measurements in the Arterial Bed

The effects of the MedDiet on endothelial function measured in the arterial bed have been evaluated in several RCTs using different validated methodologies (Table 25.1). Studies have examined the effects

of the MedDiet by using flow mediated dilation (FMD) ultrasonographically determined in the brachial artery [49–51], peripheral arterial tonometry (PAT) in small arteries [52–55], or venous occlusion plethysmography (VOP) in the forearm [56]. In one of the first RCTs assessing a whole dietary pattern for outcomes of endothelial function, Fuentes et al. used FMD in 22 hypercholesterolemic men consuming a MedDiet or a low-fat diet for 4 weeks each with a crossover design [49]. The results showed a significant improvement of FMD only with the MedDiet, which was enriched with EVOO. The same authors used PAT to assess fasting and postprandial endothelial function after two high-fat diets, enriched in EVOO (MedDiet) or SFA, and a low-fat diet enriched with ALA [52]. Compared with the other two diets, the MedDiet improved both fasting and postprandial endothelial function while reducing circulating ICAM-1 and increasing nitrous oxide (NO). An 8-week RCT conducted in Greece in

TABLE 25.1 Randomized Clinical Trials Assessing the Effect of the Mediterranean Diet on Endothelial Function

Author, year (reference)	Study population	Intervention	EF assessment	Main result
Fuentes et al. (2001) [49]	20 Men with HC	MedDiet vs low-fat diet (4-wk, crossover)	FMD	Improved EF with MedDiet
Rallidis et al. (2009) [50]	90 Abdominally obese individuals	Enhanced MedDiet vs control MedDiet (8-wk, parallel group)	FMD	EF improved with enhanced MedDiet
Buscemi et al. (2009) [51]	20 Overweight or obese women	MedDiet vs very-low CHO diet, both hypocaloric (8-wk, parallel group)	FMD	Low CHO impaired EF compared with MedDiet
Fuentes et al. (2008) [52]	20 Healthy men	MedDiet enriched with EVOO vs SFA diet vs low-fat, ALA-enriched diet (4-wk, crossover)	Fasting and postprandial PAT	Improvement of both fasting and postprandial EF with MedDiet
Pérez-Martínez et al. (2010) [53]	75 Subjects with metabolic syndrome	4 Diets differing in SFA, MUFA and n-3 PUFA content (12-wk, parallel group)	Postprandial PAT	EF improved with high-MUFA diet
Marin et al. (2011) [54]	20 Healthy subjects	MedDiet vs SFA diet vs low-fat diet (4-wk, crossover)	PAT	SFA diet had lower EF than MedDiet and low-fat diet
Fernández et al. (2012) [55]	45 Subjects with metabolic syndrome	MedDiet with physical exercise vs MeDiet alone (12-wk, parallel group)	PAT	Improved EF with MedDiet plus exercise
Ambring et al. (2004) [56]	22 Healthy subjects	MedDiet enriched with n-3 PUFA and phytosterols vs control diet (4-wk, crossover)	VOP	No differences in EF

ALA, *alpha-linolenic acid*; CHO, *carbohydrate*; EF, *endothelial function*; wk, *week*; FMD, *flow mediated dilation*; HC, *hypercholesterolemia*; MUFA, *monounsaturated fatty acids*; PAT, *peripheral arterial tonometry*; PUFA, *polyunsaturated fatty acids*; SFA, *saturated fatty acids*; VOP, *venous occlusion plethysmography*.

90 subjects with abdominal obesity showed that, compared with counsel on the MedDiet (control), intervention with an enhanced Mediterranean-style diet improved FMD by a mean 2.05% in concert with a significant fall in diastolic blood pressure [50]. In an Italian parallel-group weight loss study, low-calorie diets (MedDiet and very-low carbohydrate) given for 8 weeks were compared for effects on FMD at 7 days and 2 months in 20 overweight or obese women [51]. At 7 days, FMD was reduced in the low-carbohydrate diet group and increased in the MedDiet group, but at 2 months it returned to baseline in both groups.

In addition to the study of Fuentes et al. [52], three additional RCTs have tested the MedDiet for effects on endothelial function measured by PAT. Within the frame of the LIPGENE study, Pérez-Martínez et al. evaluated postprandial endothelial function by PAT in subjects with MetS allocated to four diets differing in fat quantity and quality during 12 weeks [53]. The postprandial reactive hyperemia index (RHI) after test meals with a composition similar to the chronic diet was higher after a MedDiet high in MUFA from olive oil than after diets high in SFA or two low-fat diets (one supplemented with n-3 PUFA). Lower postprandial ICAM-1 levels and higher NO were also observed after the high-MUFA diet. In a RCT conducted in Spain, Marin et al. randomized 20 healthy elderly subjects to three diets, MedDiet, high-SFA diet and low-fat, ALA-enriched diet for 4 weeks each in a crossover design [54]. The RHI was higher with the MedDiet and low-fat diet compared with the high-SFA diet, but only the MedDiet reduced the number of circulating endothelial cell microparticles and increased the number of circulating endothelial progenitor cells, two relevant markers of vascular function. Another study from the same Spanish group using PAT randomized 45 subjects with MetS to a control MedDiet or a MedDiet with intensive physical exercise training for 12 weeks each in a parallel-group design [55]. The RHI improved only after the MedDiet with exercise, which also increased the number of circulating endothelial progenitor cells. On the other hand, Ambring et al. found no beneficial effect of a Mediterranean-style diet given for 1 month to healthy subjects on endothelial function measured invasively by VOP [56].

A recent cross-sectional study from Italy evaluated 95 subjects referred for CVD risk assessment and found a significant correlation between adherence to the MedDiet and the RHI measured by PAT [57]. The cited meta-analysis of RCTs [48] showed that, in comparison with control interventions, the MedDiet augmented FMD values (weighted mean differences: 1.86%; 95% CI, 0.23–3.48) along with a rise in adiponectin. As shown in another recent meta-analysis [58], each 1% increase in FMD is associated with a 13% decrease in the risk of future CVD events. Thus, improved FMD following the MedDiet translates into a significant CVD risk reduction.

Effects of Key Mediterranean Diet Components

The key components of the MedDiet most studied for effects on endothelial function are three foods particularly rich in bioactive polyphenols, namely virgin olive oil, nuts and red wine. Seafood is another critical food of the MedDiet and its main nutrients, long chain (LC) n-3 polyunsaturated fatty acids (PUFA), have long been evaluated for effects on endothelial function.

Olive Oil

MUFA-rich olive oil is the main source of fat in the MedDiet. Virgin olive oil, produced by mechanically pressing ripe olives, contains multiple bioactive compound beyond MUFA, such as polyphenols, phytosterols and vitamin E [59]. Epidemiologic evidence suggests that olive oil consumption is inversely associated with CVD risk and all-cause and CVD mortality [60]. Recently, the PREDIMED trial revealed that a MedDiet enriched with EVOO decreased CVD risk by 30% [22]. Similarly, a beneficial effect has been demonstrated for intermediate phenotypes such as blood lipids, insulin sensitivity, glycemic control, and blood pressure [24,25]; use of an EVOO-enriched MedDiet was also inversely associated with new-onset T2D [26]. In a PREDIMED substudy, baseline total olive oil consumption, especially the EVOO variety, was associated with a significantly lower risk of CVD events (39%) and CVD mortality (48%), and for each increase of 10 g/day (two teaspoons) in EVOO intake, CVD and mortality risk decreased by 10% and 7%, respectively [61]. A recent meta-analysis concluded that epidemiologic studies consistently found an inverse association between olive oil consumption and stroke, but there were inconsistencies among studies regarding olive oil intake and CHD as the end-point [60].

Several studies have examined the antiinflammatory and vasculoprotective effects of olive oil compounds. In vitro studies have shown that oleic acid prevents endothelial activation by inhibiting the expression of leukocyte adhesion molecules [62], scavenging intracellular reactive oxygen species [63], or interfering with the activation of NF-κB, a key modulator of the inflammatory response [64]. In in vitro studies, Carluccio et al. showed that incubation of endothelial cells with oleic acid increased the proportion of oleate in total cell lipids while diminishing the relative proportions of SFA in association with endothelial antiinflammatory actions [62]. Oleic acid was able to reduce the inflammatory effects of SFA on human aortic endothelial cells by suppressing the incorporation of stearic acid into phospholipids [65]. Human LDL enriched in oleic acid lowered monocyte chemotaxis by 52% and reduced monocyte adhesion by 77%, compared with linoleic acid-enriched LDL, which increased oxidative stress [65]. Isolated LDL from healthy subjects who had consumed an oleic acid-rich diet for 8 weeks promoted a decrease in the expression of ICAM-1 [66]. Inflammatory markers, such as CRP, IL-6 and ICAM-1, were lower after both short-term (3 months) and long-term (2 years) consumption of olive oil-rich diets [67,68]. After consumption of EVOO (containing 1125 mg polyphenols/kg and 350 mg tocopherols/kg), as compared with refined olive oil (containing no polyphenols or tocopherols), there was a postprandial reduction in inflammatory mediators derived from arachidonic acid, such as thromboxane B2 and 6-keto-prostaglandin F1α [69], with a decrease in serum levels of ICAM-1 and VCAM-1 [70]. In patients with stable CHD, the daily consumption of 50 mL of refined olive oil with different doses of phenolic compounds for 3 weeks improved other inflammatory markers, such as CRP or IL-6 [71]. Several PREDIMED reports confirm the antiinflammatory effect of the MedDiet supplemented with EVOO in comparison with the control diet [24,28–30].

The role of dietary fat in insulin resistance and T2D has been of clinical interest for decades. In general, intake of MUFA or enrichment of membrane lipids with these fatty acids has been found to be neutral regarding either diabetes risk in epidemiological studies [72] or glycemic control in diabetic patients in RCTs comparing MUFA to carbohydrate-rich diets [73]. However, recent data from subjects of a Mediterranean country with high-dietary MUFA intake in the form of olive oil show a significant inverse association between the serum phospholipid proportions of oleic acid, the main MUFA, and insulin resistance assessed by the HOMA method [74]. The KANWU study was a parallel-arm feeding trial in 162 healthy subjects who were given diets with 37% energy from fat, either a high-SFA diet (17% SFA, 14% MUFA) or a high-MUFA diet (8% SFA, 23% MUFA) [75]. The main finding was that substitution of MUFA for SFA improved insulin sensitivity, which was impaired on the SFA diet (−10%) but did not change on the MUFA diet. Another important finding was that subjects with total fat intake >37% of energy obtained no benefit from MUFA. While this is consistent with results from a controlled lifestyle intervention trial where changes in estimated desaturase activities (derived from plasma fatty acid composition) were related to changes in insulin sensitivity only in subjects with a total fat intake below 35.5% of energy intake [76], it does not concur with findings from the PREDIMED study, whereby the MedDiet with EVOO reduced T2D risk in spite of having a total fat intake of 42% of daily energy [26]. In line with the KANWU study, a controlled short-term trial in 59 healthy subjects reported impaired insulin sensitivity in those who consumed a SFA-enriched diet compared with those who consumed a MUFA-rich diet, both diets having a total fat content of 38% of energy [77].

Results from the EUROLIVE study have confirmed the in vivo antioxidant properties of olive oil polyphenols in humans [78]. The EUROLIVE was a large, crossover, multicenter clinical trial performed in 200 individuals from five European countries. Participants were randomly assigned to receive 25 mL/day of three similar olive oils, but with different phenolic content, in intervention periods of 3 weeks. The results showed that all olive oils increased HDL-cholesterol and the ratio between reduced and oxidized forms of glutathione and decreased triglycerides, total cholesterol: HDL-cholesterol ratios, and DNA oxidative damage [78,79]. Consumption of medium- and high-phenolic content olive oil also reduced circulating oxidized LDL and other biomarkers of oxidation. In another RCT conducted in 25 healthy men, consumption of olive oil with a modestly elevated (but real life) polyphenol content versus a low-polyphenol one induced a significant decrease in apolipoprotein B levels, total LDL particle number, and small dense LDL particles [80]. Thus, beyond oleic acid, the polyphenols in virgin varieties of olive oil are presumably responsible for many of its beneficial cardiometabolic effects, including preservation of endothelial function.

Some RCTs have tested the effects of consumption of different types of olive oil on endothelial function in either chronic or acute studies. Concerning chronic studies, a small nonrandomized study in 11 diabetic patients showed that an olive oil-enriched diet attenuated the FMD-determined endothelial dysfunction present during consumption of a baseline diet high in PUFA, while at the same time reducing insulin resistance; no details on the type of olive oil used were provided [81]. In another chronic study, a crossover RCT of dietary supplementation with 30 mL of polyphenol-enriched olive oil versus a polyphenol-depleted olive oil during 4 months in 24 women with mild hypertension, the polyphenol-rich oil improved endothelial function as measured by PAT, together with a decrease in blood pressure and reduction in levels of CRP and oxidized LDL [82].

Several acute studies have examined the effects of olive oil meals on endothelial function measured in the postprandial state. High-fat meals impair postprandial endothelial function, which is a relevant outcome in dietary studies. The first postprandial FMD study using olive oil was carried out by Vogel et al. in 10 healthy volunteers who were given five different meals containing 50 g fat each [83]. FMD was measured at baseline and 3 h postprandially. A meal with 50 mL EVOO and bread impaired postprandial FMD, while similar meals with canola oil and bread, salmon with cereals, EVOO with bread and vitamins, and EVOO with bread, vinegar and vegetables did not, implying that vascular reactivity was preserved due to antioxidants in vegetables. That antioxidants (polyphenols) are important to preserve postprandial endothelial function was demonstrated by Ruano et al. in postprandial studies performed in 21 hypercholesterolemic subjects who were given test meals containing bread and 40 mL virgin olive oil either enriched or depleted in polyphenols [84]. Improvement of the RHI (measured with PAT) via reduced oxidative stress and increased NO metabolites was reported after the intake of the phenol-rich olive oil in comparison with the low-phenol one. Another acute study showing improved postprandial FMD in healthy subjects after acute consumption of both 250 mL

red wine and 50 mL "green" (virgin) olive oil, two main components of the MedDiet, supports the importance of polyphenols for improved vascular reactivity [85]. However, not all studies are consistent in showing a beneficial effect of EVOO. For instance, in another acute study, 37 healthy volunteers were randomized to receive 50 mL of maize oil, cod liver oil, soya oil, EVOO or water as an isolated meal and endothelial function was measured by VOP at baseline and 3 h postprandially [86]. No changes of postischemic forearm blood flow, serum VCAM-1 or total lipid peroxidation were observed after EVOO. Nevertheless, the consumption of a sizeable amount of oil without any other food is far from a customary meal, which may explain the discrepancy with other studies. Also, not all olive oils labeled as "extra-virgin" contain comparable amounts of polyphenols.

Nuts

The term nuts encompasses tree nuts, such as almonds, brazil nuts, hazelnuts, macadamias, pine nuts, pistachios, and walnuts, but also peanuts, which botanically are legumes but have a nutrient profile similar to tree nuts. Nuts have traditionally been avoided because of their high-fat content. However, their fatty acid profile is favorable, being rich in MUFAs (most nuts) and PUFAs (walnuts), and also contain substantial amounts of beneficial nutrients, such as dietary fiber, minerals (e.g., potassium, magnesium, and calcium), vitamins (e.g., folate, vitamin E), and other bioactive compounds, such as phytosterols and polyphenols [87–89]. Such an optimal nutrient composition predicts the health benefits of frequent nut consumption [90]. Indeed, there is a large body of scientific evidence supporting the cardioprotective properties of nuts. Several large prospective studies have reported on incident CHD in relation to frequency of nut consumption, including peanuts and peanut butter. Attesting to the interest of the topic for the nutrition community, three meta-analyses of prospective studies were recently published in a single issue of a leading nutrition journal [91–93]. Studies consistently reported a protective effect of nut consumption on fatal and nonfatal CHD, resulting in an inverse association with fatal CHD [relative risk (RR) 0.76; CI 0.69–0.84] and nonfatal CHD (RR 0.78; CI 0.67–0.92) per four servings of nuts/week (one serving equals 28.4 g). When the results were expressed for each serving/day, the pooled RR for CHD (fatal and nonfatal) was 0.72 (CI 0.64–0.81). A dose-response relationship between nut consumption and reduced CHD outcomes was described in all studies. The consistency of findings in all prospective studies strongly suggests a causal association between nut consumption and CHD protection. In the PREDIMED trial, participants randomized to the MedDiet supplemented with nuts displayed a 30% decreased risk of CVD [22].

The consumption of nuts also related inversely to T2D, with an RR per 4 weekly servings of 0.87 (CI 0.81–0.94) [91]. On the other hand, these meta-analyses do not suggest an association between nut consumption and stroke risk [91,92]. The results of the PREDIMED trial, however, indicate that the MedDiet supplemented with nuts reduced the risk of stroke [22]. The meta-analysis of prospective studies in which exposure to nuts was related to incident hypertension also showed a protective effect [93]. It also deserves to be mentioned that pooled RRs for each serving/day were 0.83 (CI 0.76–0.91) for all-cause mortality, ascertained in five studies [92,93]. Two additional meta-analyses of epidemiological studies focused on mortality support the inverse association between nut consumption and overall and cause-specific mortality, particularly CVD and cancer [94,95].

Many feeding trials have clearly shown that regular consumption of all kinds of nuts has a cholesterol lowering effect, even in the context of healthy diets [96–98]. The pooled analysis of Sabaté et al. of 25 RCTs using various nuts indicated a consistent cholesterol-lowering effect, with a mean 7.4% LDL-cholesterol reduction for an average consumption of 67 g (2.4 oz) of nuts, which was independent of the type of nut tested [98]. Nut diets also reduced triglycerides when they were elevated at baseline. As indicated in a recent meta-analysis of 21 RCTs, nut consumption also has modest blood pressure-lowering properties in individuals without T2DM, pistachios having the strongest effect [99].

Acute studies using test meals with a high-glycemic index with or without nuts have shown reduced postprandial glucose responses with nut meals, suggesting that nuts could be useful in diabetic control [100]. A recent meta-analysis of 12 nut feeding trials in patients with T2D suggests that, compared to control diets, nut diets modestly lower HbA1c and fasting glucose but have no effect on fasting insulin or HOMA-IR [101]. Another meta-analysis of 49 RCTs with nuts reporting on at least one component of the MetS concludes that nut diets also reduce fasting glucose in nondiabetic individuals [102]. The other MetS criterion that was significantly reduced in this meta-analysis was serum triglycerides. In the PREDIMED trial, the MedDiet with nuts was associated with increased reversion of MetS, and this effect was largely due to favorable effects on blood glucose and waist circumference [27].

Cohort studies have also reported an association between nut consumption and reduced circulating levels of inflammatory biomarkers [90]. In a cross-sectional analysis of data from 6000 participants in the prospective Multi-Ethnic Study of Atherosclerosis (MESA) [103], concentrations of soluble inflammatory markers (CRP, IL-6 and fibrinogen) decreased across increasing categories of nut and seed consumption. A cross-sectional substudy of the PREDIMED cohort involving 772 participants reported decreasing serum concentrations of inflammatory markers ICAM-1 and VCAM-1, but not those of CRP or IL-6, across increasing tertiles of nut consumption. When examined as secondary outcomes in nut feeding trials, inflammatory molecules variably decreased in comparison with control diets, as shown in various reports of the PREDIMED study [24,28–30].

Given that nuts are a rich source of antioxidants, it is not surprising that their consumption has been associated with improved oxidative status. This has been reported for MUFA-rich nuts, such as almonds, pistachios, and hazelnuts [104], but not for PUFA-rich walnuts, albeit no deleterious effects on oxidation have been described [105].

The optimal nutrient composition of nuts and their consistent beneficial effects on CVD, T2DM and cardiometabolic risk factors argue in favor of a salutary effect of nuts on vascular reactivity. Several RCTs have examined the effect of chronic nut consumption on endothelial function (Table 25.2), as assessed by brachial artery FMD [106–111] or small artery PAT [112]. Three studies evaluated postprandial endothelial function in the vascular bed after nut or control meals [113–115]. Some of these studies also assessed circulating molecules related to endothelial activation. The first RCT of this kind was performed by Ros et al. in 21 hypercholesterolemic subjects following a MedDiet who received walnuts (40–65 g/day, ≈18% of total energy) or an isoenergetic control diet for 4 weeks

TABLE 25.2 Randomized Clinical Trials Assessing the Effect of Nut Consumption on Endothelial Function

Author, year (reference)	Study population	Interventions	EF assessment	Main results
Ros et al. (2004) [106]	21 HC subjects	MedDiet with or without walnuts (4-wk, cross-over)	FMD	Walnut diet improved EF
Ma et al. (2010) [107]	24 Patients with type-2 diabetes	Walnut diet vs control diet (8-wk, crossover)	FMD	Walnut diet improved EF
West et al. (2010) [108]	12 HC subjects	Western diet vs walnut+walnut oil diet vs walnut+walnut oil+flaxseed oil diet (6-wk, crossover)	FMD	Walnut+walnut oil+flaxseed oil diet improved EF
Katz et al. (2012) [109]	46 Overweight subjects	Walnut diet vs control diet (8-wk, crossover)	FMD	Walnut diet improved EF
Orem et al. (2013) [110]	21 HC subjects	Control-hazelnut-control diets (4-wk, sequential)	FMD	Hazelnut diet improved EF
Kasliwal et al. (2015) [111]	60 Dyslipidemic subjects	Pistachio diet vs control diet (12-wk, parallel group)	FMD	Pistachio diet improved EF
López-Uriarte et al. (2010) [112]	50 Subjects with metabolic syndrome	Healthy diet with or without mixed nuts (12-wk, parallel group)	PAT	No significant differences in EF
Cortés et al. (2006) [113]	12 HC and 12 healthy subjects	High-fat meals with walnuts or olive oil (control) (crossover)	Postprandial FMD	Walnut meal improved postprandial EF
Berryman et al. (2013) [114]	15 HC subjects	Various test meals with walnuts, nut skins, defatted nut meat or walnut oil (cross-over)	Postprandial PAT	Walnut oil preserved postprandial EF
Kendall et al. (2014) [115]	20 Subjects with metabolic syndrome	Various tests meals with bread, and/or butter and cheese, with or without pistachios (cross-over)	Postprandial PAT	Pistachio meals preserved postprandial EF

EF, *endothelial function*; FMD, *flow mediated dilation*; HC, *hypercholesterolemia*; PAT, *peripheral arterial tonometry*; wk, *week*.

each in a crossover design [106]. Compared with the control diet, the walnut diet was associated with a mean increase of 2.3% in FMD and reduced serum VCAM-1. Three subsequent chronic RCTs used walnut diets versus control diets in FMD studies conducted in patients with T2D [107], hypercholesterolemic subjects [108], and overweight adults with visceral obesity [109], and the results of all three studies confirmed a beneficial effect of walnuts on vascular reactivity. Improved FMD has also been shown for hazelnuts in hypercholesterolemic subjects [110] and pistachios in a mildly dyslipidemic population [111]. Serum concentrations of oxidized LDL, PCR and VCAM-1 were also lower with the hazelnut diet compared with the control diet [110].

The single chronic study that assessed endothelial function by PAT failed to show any effect of a mixed nut diet (30 g/day) on endothelial function [112]. Concerning postprandial endothelial dysfunction after a fatty meal, the study of Cortés et al. showed that, compared with soaking the bread in a saturated fat-rich sandwich with common (not virgin) olive oil, adding walnuts to a similar sandwich preserved 4-h postprandial FMD in healthy subjects and increased it by 1% in those with hypercholesterolemia [113]. This occurred in the absence of changes in oxidative stress but in association with decreased circulating levels of E-selectin, an adhesion molecule involved in the early steps of monocyte recruitment to the endothelium. Another acute study assessed postprandial endothelial function by PAT in hypercholesterolemic subjects after test meals consisting of whole walnuts (85 g), nut skins (5.6 g), defatted nut meat (34 g), or walnut oil (51 g) [114]. Only walnut oil had the favorable effect of preserving postprandial vascular reactivity. This study reported increased cholesterol efflux from cholesterol-laden macrophages exposed to serum collected after the walnut or walnut oil meals. Cholesterol efflux from macrophages is an important step in the antiatherogenic reverse cholesterol transport pathway. This is a novel and important finding that needs to be confirmed in further studies; it suggests that walnut consumption improves the functionality of HDL, the main lipoprotein particles involved in reverse cholesterol transport. A third acute study also used postprandial PAT to assess changes after various meals with or without added pistachios and reported preserved postprandial endothelial function with the pistachio meals [115].

A recent RCT focusing on circulating endothelial activation molecules reported that a walnut diet reduced E-selectin levels in comparison with a diet rich in fatty fish [116]. This study and 6 of 10 RCTs examining endothelial function in the arterial bed used walnuts and all reported beneficial effects. Among nuts, walnuts have the best nutrient composition for improving endothelial function because, besides having high-polyphenol content, they are rich in ALA, the vegetable n-3 fatty acid, and both are strong bioactive molecules [90].

Red Wine

Excessive alcohol consumption is a major global risk factor for morbidity and mortality, but drinking in moderation, mainly in the form of wine or beer, is cardioprotective [117]. Moderate consumption of alcoholic beverages is considered an integral part of a healthful lifestyle, and drinking wine in moderation with meals is a key component of the Mediterranean way of life [10]. The association of exposure to alcoholic beverages and CVD outcomes has been examined in many observational studies. A meta-analysis of 84 prospective studies concluded that, in comparison to abstention from alcohol, light-to-moderate consumption of alcoholic beverages reduced fatal and nonfatal CHD by close to 30% and overall mortality by 13%, but had no effect on stroke [118]. Moderate alcohol consumption protects from incident T2D as well, as shown in a meta-analysis of 20 cohort studies [119].

A meta-analysis of RCTs assessing the effects of moderate alcohol drinking on intermediate CVD risk markers shows that increased HDL-cholesterol is a universal effect observed with any type of alcoholic beverage, that is, it may be ascribed to ethanol itself [120]; as HDL-cholesterol relates inversely to CVD risk, this has traditionally been considered the main mechanism for the CVD protection afforded by alcoholic beverages. In addition, alcohol consumption has been associated with a reduced risk of venous thrombosis and lower fibrinogen levels [120]. A meta-analysis of studies using different alcoholic beverages suggests that moderate daily intake of wine and beer may confer higher protection against CVD than moderate intake of spirits, which can be attributed to the higher polyphenolic content of fermented alcoholic beverages [121].

A recent RCT showed that moderate doses of dealcoholized red wine decreased systolic and diastolic blood pressure while increasing plasma NO concentrations [122]. Red wine tended to have effects similar to those of dealcoholized red wine, but the changes did not achieve statistical significance, and gin had no effect. These results suggest that the salutary effects of red wine on vascular function are attributable to polyphenols rather than alcohol. The cardiovascular benefit of moderate wine intake has been related to beneficial effects on oxidative status and arterial wall inflammation. Although alcohol itself induces oxidative stress, wine and beer polyphenols are strong antioxidants that might counteract the prooxidant properties of ethanol. RCTs have shown that red wine increases plasma antioxidant capacity, suppresses reactive oxygen species, and decreases LDL oxidation and oxidative DNA damage [123]. Postprandial reduction of oxidative stress has also been observed after red wine consumption, an interesting effect that supports the Mediterranean way of drinking wine with meals [124].

One of the mechanisms of CVD protection by red wine and its polyphenols lies in reduction of low-grade inflammation and blunting of endothelial activation. An RCT comparing the effects of red wine and white wine on inflammatory pathways showed that either beverage decreased CRP, ICAM-1, and IL-6, while only red wine reduced VCAM-1 and E-selectin [125]. Besides, both wines blunted cell adhesion molecule expression by mononuclear cells and reduced monocyte adhesion to stimulated endothelial cells, with a greater effect of red wine. Another

RCT from the same group testing red wine and dealcoholized red wine showed reduced plasma and monocyte inflammatory biomarkers related to early stages of atherosclerosis after both drinks, suggesting that wine polyphenols (not alcohol) exert an antiinflammatory effect on the vascular endothelium [126]. Other studies have supported the beneficial properties of red wine on inflammation and endothelial function, both measured as circulating endothelial activation molecules and in the arterial bed after chronic studies or following acute ingestion for assessment of postprandial changes, as reviewed [117,127,128]. These topics will not be discussed further here because an entire chapter of this book is dedicated to the effects of red wine and polyphenols on endothelial function (Chapter 26).

Seafood

Fish (used hereafter to refer to finfish and shellfish) is another key food of the MedDiet that is believed to contribute to its health benefits [10]. There is a large body of evidence on the cardioprotective properties of LC n-3 PUFA, mainly eicosapentaenoic (C20:5n3, EPA) and docosahexaenoic (C22:6n3, DHA) acids [129]. The flesh of fatty fish (such as mackerel, herring, salmon, tuna or sardines) is the major food source of these fatty acids, while in lean fish (such as cod, snapper, flounder, or rockfish) they are confined to the liver, which is an important natural source of fish oil. Moreover, fish contains vitamin D, taurine, selenium, additional minerals, and other bioactive components that could contribute to the health benefits attributed to fish consumption. The salutary effects of fish, however, may be counteracted by pollutant contamination, especially polychlorinated biphenyls and heavy metals such as mercury [130].

Evidence from multiple lines of research, including in vitro studies, animal experiments, observational studies, and RCTs, supports the cardiovascular benefits of LC n-3 PUFA. Consistent evidence from epidemiological studies suggests that consumption of fish or fish oil supplements protects from CHD, particularly sudden cardiac death (SCD) due to a well-established antiarrhythmic effect that can be observed at intakes as low as 250 mg/day of EPA+DHA [130,131]. This is easily achievable by meeting the American Heart Association recommendations to consume at least two servings of fish per week, preferably fatty fish [132]. Recent meta-analyses have confirmed the beneficial effect of fish consumption against CVD [131,133–136]. A meta-analysis of 17 prospective studies in primary prevention cohorts reported that, compared to subjects with the lowest consumption, those who consumed fish once per week had a 16% lower risk of fatal CHD [133]. Also, in a dose-response analysis of data from eight prospective studies, each additional 100-g weekly serving of fish was associated with a 5% reduced risk of acute coronary syndrome [134]. Regarding fish consumption and stroke risk, a meta-analysis of data from 19 cohorts provided evidence of a modest beneficial association, in particular against ischemic stroke [135]. An increment of two servings/week of any fish type was associated with a mean 4% reduced risk of cerebrovascular disease in another meta-analysis [136]. There is also consistent evidence from epidemiological studies that fish consumption protects from heart failure [137]. However, according to a recent meta-analysis of observational studies, fish/seafood or EPA+DHA intake is not associated with T2D risk [138].

Regarding RCTs with outcomes on CVD events, an important issue to consider is that fish has rarely been the main food in a nutrition intervention study, a notable exception being the DART trial, which showed a beneficial effect on cardiac death and overall mortality of a diet containing two servings/week of fatty fish for 2 years in patients with stable angina [139]. Most RCTs have used fish oil supplements, usually at doses higher than those that can be obtained from regular fish consumption. Remarkably, in the last 5 years the results of several large RCTs have been published showing little effect of LC n-3 PUFA supplements on CHD mortality [140–144], thus leading to a pessimistic view on the cardioprotective role of LC n-3 PUFA. However, methodological limitations in these trials, such as the background diet, low statistical power, length of intervention, and particularly the type of participants, mostly patients with prior CHD or multiple risk factors receiving up-to-date cardiac treatment, preclude drawing firm conclusions, as recently discussed [145]. Another important consideration on the issue of fish versus fish oil supplements is that the whole food has other beneficial nutrients that are not present in supplements, although the latter are presumably devoid of heavy metals and other pollutants.

As extensively reviewed by Mozaffarian and Wu [131], LC n-3 PUFA are bioactive molecules that have a beneficial effect on several intermediate CVD risk factors, including hypertriglyceridemia, blood pressure, resting heart rate, thrombosis, low-grade inflammation, and lipid peroxidation. Some of these effects are observed with low doses of LC n-3 PUFA, as found with customary fish consumption, for example, lowering of blood pressure, while others like triglyceride reduction, even if dose-related, are only relevant at "pharmacologic" doses >3–4 g/day. Concordant with the null effect of LC n-3 PUFA on T2D risk, meta-analyses of RCTs assessing effects of fish oil supplements (0.9–18 g/day) on glycemic control in patients with T2D found no effects on fasting glucose or hemoglobin A1c [146,147].

The mechanisms of action of LC n-3 PUFA involve diverse effects at the molecular level, including modifications in membrane structure and function, tissue metabolism, and genetic regulation, well known for the peroxisome proliferator-activated receptor gamma (PPARγ), an antiinflammatory, insulin-sensitizing, hypolipidemic nuclear receptor that is activated when bound to LC n-3 PUFA. Experimental data have confirmed the ability of EPA and DHA to modulate cell and organelle membrane structure and function, ion channels and cellular electrophysiology, regulation of nuclear receptors and transcription factors; and competition of their metabolites with proinflammatory arachidonic acid-derived eicosanoids [131]. The latter is a biological pathway of LC n-3 PUFA that has been extensively studied at the experimental level: EPA and DHA are precursor for a family of antiinflammatory lipid mediators known as resolvins and protectins, which are inflammation resolving [148,149]. There is, however, scanty evidence of these effects in humans. Of note, in a RCT conducted in patients awaiting carotid endarterectomy, carotid plaques retrieved from subjects supplemented with EPA for several weeks before surgery had a lower arachidonic acid-to-EPA ratio and increased stability (reduced foam cells and reduced activity of inflammatory macrophages) compared to plaques from patients receiving placebo, suggesting a direct antiatherosclerotic effect of LC n-3 FA related to their antiinflammatory action [150]. The influence of dietary fish consumption or usual fish oil supplement doses on levels of these inflammation-resolving mediators and their clinical relevance are promising areas of research.

The antiinflammatory properties of LC n-3 PUFA support a beneficial effect on endothelial function.

There is also experimental and clinical evidence that these fatty acids promote the synthesis of NO, a molecule strongly linked to endothelial health [151]. The bulk of clinical evidence on the effects of LC n-3 PUFA on inflammation and endothelial function stems from RCTs using supplemental doses of fish oil rather than natural fish products. A recent systematic review of 26 RCTs using various doses of EPA and DHA in healthy individuals, patients with CVD or T2DM, or acutely ill patients concludes that LC n-3 PUFA intake is associated with circulating biomarker levels, reflecting lower levels of inflammation and endothelial activation [152]. A meta-analysis of 16 fish oil RCTs with outcomes on endothelial function assessed in the arterial bed indicated that, compared to placebo, LC n-3 PUFA supplements significantly increased fasting FMD by 2.30% (CI, 0.89–3.72) at doses ranging from 0.45 to 4.5 g/day over a median of 56 days [153].

Regarding scientifically sound studies testing endothelial function in relation to consumption of fish or fish products, the results of nine cross-sectional studies [154–162] and eight RCTs [83,116,163–168] that examined effects on variables other than CRP alone have been published (Table 25.3). The exposure variable in cross-sectional studies was fish consumption with the usual diet. Five of these studies used circulating levels of inflammatory/endothelial activation molecules as outcome variables, and in all of them fish consumption had a beneficial effect [154–158]. On the other hand, endothelial function measured in the arterial bed improved with increasing fish consumption only in two of four cross-sectional studies using FMD or PAT [159–162]. The findings of six RCTs assessing the chronic effects of fatty fish-enriched diets versus control diets on circulating biomarkers were mixed, as four studies showed a beneficial effect and two studies showed no effect [116,163–167]. Surprisingly, few RCTs have sought to evaluate whether a fish-based diet intervention, instead of LC n-3 PUFA supplements, can improve endothelial function as measured directly in the arterial bed. A classical study testing various meals for effects on postprandial endothelial function found that a salmon meal prevented the endothelial dysfunction that follows a fatty meal [83]. A single chronic study conducted in Japanese women with T2D examined the effects of a diet rich in fatty fish versus a control diet and reported improved endothelial function measured by VOP with the fish diet [168] (Table 25.3).

TABLE 25.3 Human Studies Assessing the Effects of Fish on Endothelial Function

Author, year (reference)	Study design	Population	Intervention	EF assessment	Main findings
Pischon et al. (2003) [154]	Cross-sectional	405 Healthy men and 454 healthy women	No intervention (habitual diet)	Inflammation/endothelial activation molecules	Inverse association of LC n-3 PUFA intake from fish with sTNF-R1 and sTNF-R2, somewhat less for CRP
Lopez-Garcia et al. (2004) [155]	Cross-sectional	727 Healthy women	No intervention (habitual diet)	Endothelial activation molecules	Inverse association of LC n-3 PUFA intake from fish with VCAM-1 and ICAM-1
Zampelas et al. (2005) [156]	Cross-sectional	1514 Healthy men and 1528 healthy women	No intervention (habitual diet)	Inflammation/endothelial activation molecules	Greater intake of fish related to lower levels of CRP, IL-6 and TNFα
He et al. (2009) [157]	Cross-sectional	5677 Healthy men and women	No intervention (habitual diet)	Inflammation/endothelial activation molecules	Greater intake of LC n-3 PUFA from fish inversely associated with levels of CRP, IL-6 and MMP-3
van Bussel et al. (2011) [158]	Cross-sectional	301 Healthy men and women	No intervention (habitual diet)	Inflammation/endothelial activation molecules	Fish consumption and LC n-3 PUFA inversely associated with vWF, TM, VCAM-1, E-selectin and IL-8
Anderson et al. (2010) [159]	Cross-sectional	3045 Healthy men and women	No intervention (habitual diet)	FMD	No association of fish consumption with EF
Petersen et al. (2010) [160]	Cross-sectional	40 Healthy subjects	No intervention (habitual diet)	FMD	No association of fish consumption with EF
Buscemi et al. (2014) [161]	Cross-sectional	54 Healthy subjects	No intervention (habitual diet)	FMD	EF improved in high-fish consumers
Merino et al. (2014) [162]	Cross-sectional within dietary intervention RCT	108 Patients at high risk for CVD	Intensive dietary advice for 1-year	PAT and inflammation/endothelial activation molecules	High intake of fish-derived LC n-3 PUFA associated with improved EF by PAT and lower CRP, TNF-α, ICAM-1 and VCM-1

Study	Design	Subjects	Diets	Outcome	Results
Seierstad et al. (2005) [163]	RCT (6-wk, parallel groups)	60 Patients with CHD	3 Diets: salmon fed fish oil; salmon fed rapeseed oil; salmon fed 50% fish; and rapeseed oils	Endothelial activation molecules	Salmon fed fish oil reduced VCAM-1 and IL-6
De Mello et al. (2009) [164]	RCT (8-wk, parallel groups)	27 Patients with CHD	3 Diets: fatty fish; lean fish; and control	PMBC expression of inflammatory and endothelial function genes	No effect of fatty fish or lean fish
Lindqvist et al. (2009) [165]	RCT (6-wk, crossover)	35 Overweight men	2 Diets: herring vs control	Inflammation/endothelial activation molecules	No effect of fish diet on CRP, IL-6, IL-18, or ICAM-1
van den Elsen et al. (2011) [166]	RCT ($\simeq$6-month, parallel group)	123 Pregnant women	2 Diets: fatty fish (salmon) vs control	Endothelial activation molecule expression in cultivated umbilical cord endothelial cells	Fish diet reduced expression of ICAM-1
de Mello et al. (2011) [167]	RCT (12-wk, parallel groups)	104 Subjects with MetSyn	3 Diets: high-fish, berries, and wholegrain; wholegrain; control	Inflammation/endothelial activation molecules	High-fish diet reduced CRP and E-selectin vs control diet
Chiang et al. (2012) [116]	RCT (4-wk, crossover)	25 Healthy or hyperlipidemic subjects	3 Diets: salmon; walnuts; control diet	Endothelial activation molecules	High-fish diet reduced sICAM-1 vs control diet
Vogel et al. (2000) [83]	RCT (crossover)	10 Healthy or hyperlipidemic subjects	5 Meals with 50 g fat: canned red salmon; olive oil; olive oil with vegetables; olive oil with vitamins; canola oil	Postprandial FMD	Salmon meal preserved postprandial EF
Kondo et al. (2014) [168]	RCT (4-wk, crossover)	23 Women with T2D	2 Diets: high-fatty fish vs control	VOP	High-fish diet improved EF

CRP, C-reactive protein; EF, endothelial function; FMD, flow mediated dilation; ICAM, intercellular adhesion molecule; IL, interleukin; MMP, metalloproteinase; PAT, peripheral arterial tonometry; RCT, randomized controlled trial; sTNF-R, soluble tumor necrosis factor receptor; TM, thrombomodulin; TNF, tumor necrosis factor; VCAM, vascular cell adhesion molecule; VOP, venous occlusion plethysmography; vWF, von Willebrand factor; wk, week.

II. ENDOTHELIAL DYSFUNCTION AND CLINICAL SYNDROMES

Additional cross-sectional studies have shown that consumption of fish and LC n-3 PUFA with the habitual diet is associated with lower CRP levels [169,170] and that increased proportions of EPA and DHA determined in plasma lipid fractions as reliable markers of intake also relate to lower levels of inflammatory/endothelial activation molecules [171–173]. Thus, although not all studies have provided evidence of benefit, the general picture is that, by way of their antiinflammatory properties, fish and fish products do indeed improve endothelial function.

ADDITIONAL CONSIDERATIONS ON THE MEDITERRANEAN DIET AND HEALTH

Our earlier brief overview on the salutary properties of the MedDiet was focused on CVD risk and intermediate biomarkers in order to introduce its effects on endothelial function. However, there are additional dimensions to the healthful effects of the MedDiet that are worth summarizing to grasp a more encompassing picture of its manifold beneficial properties. They include protection from SCD, lack of weight gain despite its high-fat content, and beneficial effects beyond cardiometabolic health on chronic devastating conditions, such as dementia and cancer. Logically given the above beneficial effects, adherence to the MedDiet is associated with longevity and a healthy lifespan. The optimal nutrient adequacy of the MedDiet will also be discussed. Finally, two topics of increasing interest and ongoing research, nutrigenomic effects and gut microbiota in relation to the MedDiet, will be briefly reviewed.

Sudden Cardiac Death

Among the cardiovascular benefits of the MedDiet, one that deserves to be mentioned in the light of the above discussed antiarrhythmic properties of fish and its main constituents LC n-3 PUFA, key components of the MedDiet, is protection against SCD. As SCD occurs within minutes, with no time to intervene, primary prevention is critical. Chiuve et al. examined the association between an alternate MedDiet score and SCD risk in women of the prospective NHS and reported that higher scores related to a reduced risk (RR 0.60; CI, 0.43–0.84) [174]. A recent report from the large cohort of the Women's Health Initiative study also suggests an inverse relationship between a MedDiet score and SCD, with a hazard ratio of 0.64 (CI, 0.43–0.94), but not when using a high-DASH diet score [175]. In that study consumption of total fish was individually associated with lower risk of SCD, with a hazard ratio of 0.63 (CI, 0.42–0.95) for quintile five compared with quintile 1 [175]. Thus, increasing fish consumption within the MedDiet probably contributes to protection against SCD. A critical MedDiet component, moderate alcohol (wine) intake, may help reduce SCD as well [118]. The Lyon Diet Heart Study also reported a beneficial effect of the MedDiet on SCD, but there were only eight cases [20]. There were also too few SCD events in the PREDIMED trial to draw conclusions on the effects of the intervention MedDiets.

Body Weight

Despite the high-fat content of the MedDiet, a consistent finding in both epidemiological studies [176–178] and RCTs [178,179] is the lack of weight gain over time or the low risk of developing overweight or obesity reported in individuals with better adherence to this dietary pattern. The MedDiet has also been examined for effects on weight loss, but the conclusion of a recent meta-analysis of short-term RCTs is that body weight is reduced only when the diet is calorie restricted [179]. A cross-sectional analysis of the PREDIMED cohort also showed an inverse association between MedDiet adherence and adiposity indexes [180]. The lack of weight gain with the MedDiet is not unexpected, since it is enriched in foods that have not been associated with long-term deleterious effects on adiposity, such as nuts, vegetables, fruits, and whole grains, and it is poor in foods and beverages that relate to long-term weight gain, such as fast-foods, sweets and deserts, butter, red meat and processed meat, and sugar-sweetened beverages [181].

The lack of a fattening effect of the high-fat MedDiet has practical implications, as the fear of weight gain in an era of obesity pandemic need no longer be an obstacle to adherence to a dietary pattern known to provide much clinical and metabolic benefit. This observation is are also relevant for public health, as it supports unrestricted fat intake as appropriate for body weight maintenance and overall cardiometabolic health, as recently acknowledged by the Dietary Guidelines for Americans 2015 Advisory Committee [182].

Cognitive Decline and Dementia

The progressive lifespan increase in recent decades has resulted in a greatly increased frequency of age-related diseases, including neurodegenerative disorders such as Alzheimer's disease, the most common type of dementia [183]. The social and economic burden of caring for persons with dementia is rising exponentially and to date pharmacologic agents have proven ineffective to prevent or treat the disease [184]. Most likely, to be successful an intervention would need to be started in the preclinical stage, a common situation in the elderly population worldwide [185].

There is emerging evidence suggesting an association between dietary habits and cognitive performance [186]. As oxidative stress is believed to play a major role in cognitive decline and dementia, it is plausible that antioxidant-rich foods or dietary patterns like the MedDiet might afford protection. Indeed, several large prospective studies have related long-term dietary exposures to cognitive impairment or dementia and, as recently reviewed and meta-analyzed, the results suggest that the MedDiet also protects from age-related cognitive deterioration, Alzheimer disease and other types of dementia [187–189]. Recent novel findings from the PREDIMED trial in a subcohort of participants undergoing sequential neurocognitive tests indicate that, as opposed to the control diet, the intervention MedDiets supplemented with EVOO and nuts were able to counteract age-related cognitive decline [190]. An earlier cross-sectional evaluation at baseline of the same PREDIMED subcohort showed that increasing consumption of polyphenol-rich foods (olive oil, walnuts, wine, and coffee) was associated with better memory scores, reinforcing the notion that phenolic compounds protect brain function [191]. Regarding the two key foods of the MedDiet that were supplemented in the PREDIMED trial, there is scanty epidemiological evidence of an association between consumption of olive oil [192] and nuts [193,194] and better cognition at older age.

The lack of effective therapies for cognitive decline and associated neurodegenerative disorders underlines the need of preventive strategies to delay the presentation and/or reduce the burden of these devastating conditions. The results of the PREDIMED trial with the MedDiet, the first demonstrating a beneficial effect of a dietary pattern on cognition, are encouraging but further research is clearly indicated.

Cancer

Due to the long inception period of many cancers and their multifactorial etiology, diet effects are difficult to detect in epidemiological studies and even more in RCTs. The meta-analysis of observational studies of Sofi et al. reporting on the beneficial effects of adherence to the MedDiet on several chronic conditions concluded that a modest protective effect against total cancer incidence and mortality existed [15]. A more recent meta-analysis of epidemiological studies including 21 prospective cohort studies and 12 case-control studies focused exclusively on the effects of the MedDiet on overall cancer and type-specific cancer [195]. This systematic review provides convincing evidence that high adherence to the MedDiet is associated with a reduced risk of overall cancer mortality/incidence (10%) as well as colorectal cancer (14%), probable evidence of an association with a reduced risk of prostate cancer (4%), and probable evidence of no effects on breast and gastric cancer risk. Regarding RCT evidence, the results of the Lyon Diet Heart Study with 605 participants showed a 61% reduction of total cancer risk in the group assigned the cardioprotective Mediterranean-type diet used in this study compared with the control group after follow-up for 4 years, but these results were based only on a total of 14 cancers [196]. Cancer incidence was a prespecified secondary outcome of the PREDIMED trial. Recently, the first PREDIMED report on this condition assessed 4282 women and showed a 68% reduction in breast cancer risk in those allocated to the MedDiet with EVOO group compared with the control group after follow-up for 4.8 years [197]. Again, these results are based on few confirmed cases of breast cancer ($n=33$).

Concerning exposure to olive oil and cancer risk, a recent systematic review and meta-analysis of 19 observational studies suggested a 34% lower odds of having any type of cancer when comparing the highest category of olive oil consumption with the lowest figure [198].

Nuts were also associated with decreased risk of cancer mortality when comparing highest with lowest categories of consumption in a meta-analysis of three large prospective studies (RR 0.86; CI, 0.75–0.98) [94]. A more recent meta-analysis including one additional prospective study reports similar protection from nut consumption against cancer mortality (RR 0.85; CI, 0.77–0.93) [95]. Finally, a recent report from the large prospective The

Moli-sani study from Italy also suggests reduced cancer mortality with increasing nut consumption [199]. However, no data on nut consumption and cancer incidence are available.

Longevity

The consistent evidence of a beneficial effect of the MedDiet on incidence and mortality from chronic prevalent conditions such as CVD, dementia, and some types of cancer, together with a protective effect on obesity and T2D, indicate that adherence to this dietary pattern is associated with increased longevity and a healthy lifespan [200]. The meta-analysis of Sofi et al. [15] already concluded that increased MedDiet adherence decreased overall mortality, a finding that has been confirmed in subsequent prospective studies in the United States [201] and Northern Europe [202]. A recent report from the prospective NHS also suggests that greater adherence to the MedDiet at midlife is strongly linked to greater health and wellbeing (no major chronic diseases or impairments in cognition, physical function or mental health) in women surviving to age 70 or older [203].

A significant contributor to accelerated aging, particularly in insulin-resistant individuals, is endothelial dysfunction, senescence, and impaired repair [204]. Hence, the discussed salutary effects of the MedDiet on endothelial function are likely to underlie in part the increased longevity observed in populations adhering to this dietary pattern.

A critical biomarker of aging is telomere length, as shorter telomeres are associated with a decreased life expectancy and increased rates of developing age related chronic diseases. Telomeres are repetitive DNA sequences located at the ends of chromosomes that undergo attrition each time a somatic cell divides, a process that is accelerated by oxidative stress and inflammation, thus being modifiable [205]. Given the protective effects of the MedDiet on oxidative stress and chronic inflammation, it comes as no surprise that greater adherence to this dietary pattern was associated with longer telomeres in two large cross-sectional studies of older persons conducted in the United States [206,207] and a smaller study in Italy [208]. These results support the benefit of adherence to the MedDiet for promoting longevity while pointing to a relevant biologic mechanism for its antiaging effects.

Nutritional Adequacy

Unfortunately the customary use of the word "diet" has evolved to mean "weight loss diet" rather than the correct meaning, the totality of what is ingested on a regular basis. As a result, the MedDiet has been frequently discussed in relation to effects on adiposity in comparison to weight-loss diets, distorting its proper place as a health-supporting dietary pattern. In a meaningful way, the MedDiet conforms to the new public health paradigm of addressing health promotion and chronic disease prevention rather than disease-driven acute care. However, as a therapeutic diet as well, the MedDiet is now arguably the most scientifically credited, particularly regarding CVD and its risk factors. Among the merits of the MedDiet, there is a unique advantage in the wide spectrum of nutrients it contains. Nutrient adequacy in the customary diet is important because deficient dietary micronutrient content has been associated with CVD, lower resistance to infection, surgical complications, birth defects, cancer, obesity, and osteoporosis and it is acknowledged that micronutrient deficiencies persist in developed regions of the world such as America [209] and the European Union [210]. Changes in dietary patterns and macronutrient ratios influence dietary micronutrients. For instance, rises in the quantity together with diminished quality of carbohydrates are associated with poorer micronutrient intake [211], as are high-dietary glycemic loads [212].

That the MedDiet supports the concept of dietary variety while fulfilling nutrient requirements for optimal health has been examined in several studies. A report from the SUN cohort in Spain divided participants into those best adhering to the MedDiet or to a Western dietary pattern [213]. Probabilistic assessment of nutrient adequacy was performed for ≥19 vitamins and minerals, a greater number than the usual "nutrients of concern," and higher adherence to the MedDiet was clearly associated with a better micronutrient profile. In a French study using diet models, nutrient adequacy was analyzed while translating nutrient recommendations into combinations of real food choice [214]. The investigators varied macronutrients, seeking to minimize alterations in other foods to achieve nutrient adequacy, using computer-generated personalized diets. By using eight models, it was observed that foods typical of the MedDiet were needed for balance and were the most direct way to reach overall nutrient adequacy. A recent review of available evidence on the nutritional properties of the MedDiet indicated that this dietary pattern was associated with adequate micronutrient intake in both adults and children and concluded that it could be used in public health nutrition policies to prevent micronutrient deficiencies in vulnerable population groups [215].

As one consequence of long-term dietary micronutrient (and protein) inadequacy in the elderly is frailty (decreased strength, endurance capacity and overall physiological functioning, with loss of lean body mass and increased vulnerability to disease and death), that increased adherence to the MedDiet was associated with reduced frailty in a prospective Spanish cohort study of 1815 community-dwelling individuals aged ≥60 years further supports its nutritional adequacy [216]. Other large prospective studies in older persons from Mediterranean countries revealed that adherence to the MedDiet was associated with a slower decline in lower extremity mobility [217] or with a delayed onset of disability in activities of daily living [218], although in this study the beneficial effect was limited to women.

Ongoing Research: Nutrigenomics and Gut Microbiota

The discipline of nutritional genomics has developed since the 1990s with the aim of gaining knowledge on the interaction between dietary factors and the genetic background and the role they have in the modulation of both phenotypic traits and disease risk [219]. Thus nutrigenomics attempts to grasp the genetic basis for the known interindividual responses to diet (a classical example is blood cholesterol changes after egg consumption) and the reasons for the often dissimilar clinical phenotypes observed in carriers of the same genetic variant. Regarding the identification of gene-diet interactions determining CVD risk or related metabolic traits, some PREDIMED studies have investigated whether the effects of the MedDiet differ depending on the individual's genetic profile [220–224]. A genetic score, including the melanocortin four receptor (MC4R) rs17782313 and fat mass and obesity (FTO) rs9939609 variants, both strongly associated with BMI, had no statistically significant interaction with the MedDiet, but greater adherence to this dietary pattern significantly reduced BMI in

genetically susceptible individuals [220]. An additional assessment of these obesity-related variants focusing on the diabetes phenotype showed significant interactions between them and the MedDiet in determining T2D risk, in such a way that greater adherence reduced risk in susceptible individuals [221]. The transcription factor 7-like 2 (TCF7L2) gene is strongly associated with T2D but controversially with plasma lipids and CVD. Adherence to the MedDiet reduced the adverse effect of the TCF7L2 rs7903146 (C>T) polymorphism on CVD risk factors (fasting glucose and lipids) and, importantly, on stroke incidence [222]. Another PREDIMED study focused on the rs3812316 variant previously associated with lower triglycerides of the Max-like protein X interacting protein-like (MLXIPL) gene, which encodes the carbohydrate response element binding protein. The results showed that the MedDiet enhances the triglyceride-lowering effect of this variant while strengthening its protective effect on incident myocardial infarction [223]. Genetic and epigenetic effects combined focusing on microRNA target site polymorphisms were also analyzed. The MedDiet interacted with the gain-of-function microRNA-410 rs13702 T>C polymorphism in the LPL gene, known to be associated with lower triglyceride levels, by enhancing its triglyceride-lowering effect and reducing stroke incidence as well [224]. Thus far, the important lesson from these nutrigenomic studies is that adherence to the MedDiet counteracts the genetic susceptibility to develop detrimental cardiometabolic phenotypes and associated CVD events.

The human gut is home to a vast number of microorganisms, mostly bacteria and fungi, known as the intestinal microbiota, which profoundly influence physiology and metabolism. The microbiota plays a critical role in the metabolism of xenobiotics and various nutrients, particularly the indigestible components of complex carbohydrates (fiber) leading to production of short-chain fatty acids through fermentation [225], but it also has a dynamic nature, especially in response to changes in diet, which may result in beneficial or detrimental metabolic changes for the host, particularly affecting the susceptibility to CVD and obesity. This is a hot topic of contemporary research [226,227]. That the taxonomy and properties of the gut microbiome is shaped by diet was elegantly proven by a comparative study of children living in a rural African village and those living in Europe [228]. The microbiota of African children had an overall greater microbial richness and produced higher levels of short-chain fatty acids than that of European children. Compared with the Western European diet (high in animal protein, sugar, starch, and fat and low in fiber), the rural African diet (rich in complex carbohydrate and fiber from fruits and vegetables and poor in animal protein) must have had a prominent role in the observed differences in gut microbiota, rendering it less proinflammatory. Other clinical studies have associated diets higher in fruits, vegetables and fiber with increased microbiome richness, which in turn related to better health, while lower bacterial richness was associated with obesity, insulin resistance, dyslipidemia, and inflammatory disorders [229,230]. There is also important emerging evidence that dietary effects on the microbiota contribute to development of disease, particularly CVD. This is epitomized by the metabolism of choline, an amino acid produced by the hydrolysis of phosphatidylcholine (abundant in eggs) by the intestinal microbiota, resulting in the formation of trimethylamine, which is transformed by the liver into trimethylamine oxide (TMAO) [231], a small molecule shown in a seminal prospective study to be strongly associated with an increased CHD risk [232]. A similar microbial pathway has been identified for the conversion of dietary carnitine, abundant in red meat, into TMAO [233]. Last but not least, the bioavailability and biologic effects of dietary polyphenols depend largely on their transformation by the gut microbiota in a reciprocal relationship whereby polyphenols and their metabolites modulate microbial balance in favor of beneficial organisms, exerting a prebiotic-like action [234]. Although thus far there are no published studies relating adherence to the MedDiet to changes in the intestinal microbiome or derived metabolites, it is not difficult to envision that such a plant-based diet, rich in complex carbohydrate, fiber, and polyphenols and low in eggs, meat and processed meat products that contain the TMAO precursors choline and carnitine, should promote a healthy microbiome. The results of ongoing studies on this critical topic are eagerly awaited.

CONCLUSIONS

Although possessing an inherent country-specific variability, the MedDiet can be defined overall by a high consumption of plant foods (vegetables, fruits, whole grains, beans, nuts, and spices, all good sources of fiber and polyphenols), moderate consumption of wine with meals, low intake of meat, low-moderate intake of dairy products and eggs, fish and seafood at least twice per week, and a high-total fat, unsaturated fat intake (~35–45% of energy) chiefly due to the nonrestricted use of olive oil in the kitchen and at the table. Indeed the MedDiet has a high ratio of MUFA to other fats and greater amounts of plant and marine LC n-3 PUFA than most other healthy dietary patterns. Also, the MedDiet is reputed for its variety, balance, palatability, satiating power, and the nurturing social milieu in which foods are consumed [10]. Importantly, the MedDiet does not involve processed foods in bags, boxes, or cans, "fast foods," confectionery, or sweetened beverages, thus being the opposite of the reputedly unhealthy Western dietary pattern. Importantly, the MedDiet was tested for its ability to lower chronic disease burden and increase longevity as an experiment of nature, long before its "discovery" by Keys et al. [12]. It was not an artificially constructed diet designed by an expert committee based upon anticipated characteristics of a mixture of nutrients. Therefore, there was no deliberate plan to add a minimum number of servings of produce, or molding of contents to assure a low sodium intake, low glycemic load, or concern about processing to delay nutrient degradation. Hence, intrinsically, there is a greater volume of vegetables and fruits than in other diet patterns, especially within Mediterranean countries and the EU [10]. The base of the MedDiet pattern was always predominantly fresh foods of plant origin, containing a high ratio of potassium-to-sodium, and inherently richer in fiber, magnesium and other beneficial micronutrients, compared to Western or common contemporary dietary patterns. Another often overlooked consequence of the predominance of real food grown locally in the MedDiet is a crowding-out of processed foods and a high degree of sustainability [235].

Decades of nutritional epidemiological research and food and dietary pattern based RCTs have provided high-quality evidence on the power of the MedDiet and its principal components, such as EVOO, nuts, red wine and seafood, to improve endothelial function, a critical component of cardiovascular health and biological pathways linked to longevity. As discussed, the MedDiet appears to be optimal for the prevention of CVD and T2D and, while the evidence is not as strong, it also has a salutary effect on other chronic conditions, such as hypertension, cognitive dysfunction, dementia, and cancer. Adherence to the MedDiet also relates to healthy aging in part because of the provision of a high level of nutrient adequacy and the evidence from nutrigenomic studies that it might counteract the genetic susceptibility to disease. Knowledge of the overall benefit of the MedDiet may have an especially important role in promoting its adherence: maintaining physical, cognitive, and mental health with aging should be a more powerful driving force for dietary change than simply prolonging life or avoiding a single chronic disease. The dietary approach to health is necessarily cost-effective compared with standard pharmacotherapy, which could be avoided by sizeable segments of the population if diet-focused strategies are implemented.

While there is no "perfect" dietary pattern suited for everyone and individuals are usually advised to choose a healthy diet that is adapted to their cultural background and which they can follow life-long, there are certain evidence-based principles for diets to be considered healthy [236]. By any standard, the MedDiet ranks among the top few reputedly healthy dietary patterns while satisfying many of the criteria listed by major international societies. For instance, constituents of the MedDiet fulfill those in the healthy diet defined by the American Heart Association (AHA) Strategic Committee among behaviors that define ideal cardiovascular health [237], and those in the 2013 AHA/American College of Cardiology Lifestyle Guideline [9]. Remarkably, based on the findings of the PREDIMED trial [22], the recent AHA/American Stroke Association guidelines for the primary prevention of stroke specifically state that a MedDiet supplemented with nuts may be considered in lowering the risk of stroke [238]. More importantly, the MedDiet, a dietary pattern translatable the world over with few modifications [239], has a prominent place in the Scientific Report of the 2015 Dietary Guidelines Advisory Committee that sets forth updated dietary recommendations for Americans [209]. Having stood the test of time for 5 millennia and now the tests of modern science, the reward to risk ratio of following the MedDiet is one of the best in medical sciences. This knowledge needs to be translated to the public, as dietary quality is still far from optimal in large segments of the population, due in part to socio-economic inequalities, as a healthy diet is associated with financial and educational status. Thus, besides extolling the MedDiet, public health policy must focus on the most vulnerable segments of society in order to improve the quality of their diet and, consequently, their health status.

Acknowledgments

All authors have read and approved the final manuscript. CIBERDEM and CIBEROBN are initiatives of ISCIII, Spain.

References

[1] Schwerin HS, Stanton JL, Riley AM, et al. Food eating patterns and health: a reexamination of the Ten-State and HANES I surveys. Am J Clin Nutr 1981;34:568–80.
[2] Hu FB. Dietary pattern analysis: a new direction in nutritional epidemiology. Curr Opin Lipidol 2002;13:3–9.
[3] Millen BE, Quatromoni PA, Gagnon DR, et al. Dietary patterns of men and women suggest targets for health promotion: the Framingham Nutrition Studies. Am J Health Promot 2015;11:42–52 [discussion 52-3].
[4] Appel LJ, Moore TJ, Obarzanek E, et al. A clinical trial of the effects of dietary patterns on blood pressure. N Engl J Med 1997;336:1117–24.
[5] Shen J, Wilmot KA, Ghasemzadeh N, et al. Mediterranean dietary patterns and cardiovascular health. Annu Rev Nutr 2015;35:425–49.
[6] Chiuve SE, Fung TT, Rimm EB, et al. Alternative dietary indices both strongly predict risk of chronic disease. J Nutr 2012;142:1009–18.
[7] Harmon BE, Boushey CJ, Shvetsov YB, et al. Associations of key diet-quality indexes with mortality in the Multiethnic Cohort: the Dietary Patterns Methods Project. Am J Clin Nutr 2015;101:587–97.
[8] US Department of Agriculture, US Department of Health and Human Services. Dietary guidelines for Americans 2010. 7th ed. Washington, DC: US Government Printing Office; 2010.
[9] Eckel RH, Jakicic JM, Ard JD, et al. 2013 AHA/ACC guideline on lifestyle management to reduce cardiovascular risk: a report of the American College of Cardiology/American Heart Association Task Force on practice guidelines. J Am Coll Cardiol 2014;63:2960–84.
[10] Bach-Faig A, Berry EM, Lairon D, et al. Mediterranean diet pyramid today. Science and cultural updates. Public Health Nutr 2011;14:2274–84.
[11] Hoffman R, Gerber M. Evaluating and adapting the Mediterranean diet for non-Mediterranean populations: a critical appraisal. Nutr Rev 2013;71:573–84.
[12] Keys A, Menotti A, Karvonen MJ, et al. The diet and 15-year death rate in the seven countries study. Am J Epidemiol 1986;124:903–15.
[13] Tunstall-Pedoe H, Kuulasmaa K, Mähönen M, et al. Contribution of trends in survival and coronary-event rates to changes in coronary heart disease mortality: 10-year results from 37 WHO MONICA project populations. Monitoring trends and determinants in cardiovascular disease. Lancet 1999;353:1547–57.
[14] Dilis V, Katsoulis M, Lagiou P, et al. Mediterranean diet and CHD: the Greek European Prospective Investigation into Cancer and Nutrition cohort. Br J Nutr 2012;108:699–709.
[15] Sofi F, Abbate R, Gensini GF, et al. Accruing evidence on benefits of adherence to the Mediterranean diet on health: an updated systematic review and meta-analysis. Am J Clin Nutr 2010;92:1189–96.

[16] Fung TT, Rexrode KM, Mantzoros CS, et al. Mediterranean diet and incidence of and mortality from coronary heart disease and stroke in women. Circulation 2009;119:1093–100.

[17] Guallar-Castillón P, Rodríguez-Artalejo F, Tormo MJ, et al. Major dietary patterns and risk of coronary heart disease in middle-aged persons from a Mediterranean country: the EPIC-Spain cohort study. Nutr Metab Cardiovasc Dis 2012;22:192–9.

[18] Trichopoulou A, Costacou T, Bamia C, et al. Adherence to a Mediterranean diet and survival in a Greek population. N Engl J Med 2003;348:2599–608.

[19] Bamia C, Trichopoulos D, Ferrari P, et al. Dietary patterns and survival of older Europeans: the EPIC-Elderly Study (European Prospective Investigation into Cancer and Nutrition). Public Health Nutr 2007;10:590–8.

[20] De Lorgeril M, Salen P, Martin JL, et al. Mediterranean diet, traditional risk factors, and the rate of cardiovascular complications after myocardial infarction: final report of the Lyon Diet Heart Study. Circulation 1999;99:779–85.

[21] Barzi F, Woodward M, Marfisi RM, et al. Mediterranean diet and all-causes mortality after myocardial infarction: results from the GISSI-Prevenzione trial. Eur J Clin Nutr 2003;57:604–11.

[22] Estruch R, Ros E, Salas-Salvadó J, et al. Primary prevention of cardiovascular disease with a Mediterranean diet. N Engl J Med 2013;368:1279–90.

[23] Bulló M, Lamuela-Raventós R, Salas-Salvadó J. Mediterranean diet and oxidation: nuts and olive oil as important sources of fat and antioxidants. Curr Top Med Chem 2011;11:1797–810.

[24] Estruch R, Martínez-González MA, Corella D, et al. Effects of a Mediterranean-style diet on cardiovascular risk factors: a randomized trial. Ann Intern Med 2006;145:1–11.

[25] Doménech M, Roman P, Lapetra J, et al. Mediterranean diet reduces 24-hour ambulatory blood pressure, blood glucose, and lipids: one-year randomized, clinical trial. Hypertension 2014;64:69–76.

[26] Salas-Salvadó J, Bulló M, Estruch R, et al. Prevention of diabetes with Mediterranean diets: a subgroup analysis of a randomized trial. Ann Intern Med 2014;160:1–10.

[27] Babio N, Toledo E, Estruch R, et al. Mediterranean diets and metabolic syndrome status in the PREDIMED randomized trial. CMAJ 2014;186:E649–57.

[28] Mena MP, Sacanella E, Vázquez-Agell M, et al. Inhibition of circulating immune cell activation: a molecular anti-inflammatory effect of the Mediterranean diet. Am J Clin Nutr 2009;89:248–56.

[29] Llorente-Cortés V, Estruch R, Mena MP, et al. Effect of Mediterranean diet on the expression of pro-atherogenic genes in a population at high cardiovascular risk. Atherosclerosis 2010;208:442–50.

[30] Casas R, Sacanella E, Urpí-Sardà M, et al. The effects of the Mediterranean diet on biomarkers of vascular wall inflammation and plaque vulnerability in subjects with high risk for cardiovascular disease. A randomized trial. PLoS One 2014;9:e100084.

[31] Garcia-Arellano A, Ramallal R, Ruiz-Canela M, et al. Dietary inflammatory index and incidence of cardiovascular disease in the PREDIMED Study. Nutrients 2015;7:4124–38.

[32] López S, Bermúdez B, Pacheco YM, et al. Distinctive postprandial modulation of beta cell function and insulin sensitivity by dietary fats: monounsaturated compared with saturated fatty acids. Am J Clin Nutr 2008;88:638–44.

[33] Panagiotakos DB, Tzima N, Pitsavos C, et al. The association between adherence to the Mediterranean diet and fasting indices of glucose homoeostasis: the ATTICA Study. J Am Coll Nutr 2007;26:32–8.

[34] Shai I, Schwarzfuchs D, Henkin Y, et al. Weight loss with a low-carbohydrate, Mediterranean, or low-fat diet. N Engl J Med 2008;359:229–41.

[35] Esposito K, Maiorino MI, Ciotola M, et al. Effects of a Mediterranean-style diet on the need for antihyperglycemic drug therapy in patients with newly diagnosed type 2 diabetes: a randomized trial. Ann Intern Med 2009;151:306–14.

[36] Elhayany A, Lustman A, Abel R, et al. A low carbohydrate Mediterranean diet improves cardiovascular risk factors and diabetes control among overweight patients with type 2 diabetes mellitus: a 1-year prospective randomized intervention study. Diabetes Obes Metab 2010;12:204–9.

[37] Lasa A, Miranda J, Bulló M, et al. Comparative effect of two Mediterranean diets versus a low-fat diet on glycaemic control in individuals with type 2 diabetes. Eur J Clin Nutr 2014;68:767–72.

[38] Peña-Orihuela P, Camargo A, Rangel-Zuñiga OA, et al. Antioxidant system response is modified by dietary fat in adipose tissue of metabolic syndrome patients. J Nutr Biochem 2013;24:1717–23.

[39] Fitó M, Guxens M, Corella D, et al. Effect of a traditional Mediterranean diet on lipoprotein oxidation: a randomized controlled trial. Arch Intern Med 2007;167:1195–203.

[40] Lapointe A, Goulet J, Couillard C, et al. A nutritional intervention promoting the Mediterranean food pattern is associated with a decrease in circulating oxidized LDL particles in healthy women from the Québec City metropolitan area. J Nutr 2005;135:410–5.

[41] Shukla S, Gupta S. Suppression of constitutive and tumor necrosis factor alpha-induced nuclear factor (NF)-kappaB activation and induction of apoptosis by apigenin in human prostate carcinoma PC-3 cells: correlation with down-regulation of NF-kappaB-responsive genes. Clin Cancer Res 2004;10:3169–78.

[42] Efentakis P, Iliodromitis EK, Mikros E, et al. Effects of the olive tree leaf constituents on myocardial oxidative damage and atherosclerosis. Planta Med 2015;81:648–54.

[43] Martínez-González MA, Toledo E, Arós F, et al. Extra-virgin olive oil consumption reduces risk of atrial fibrillation: the PREDIMED trial. Circulation 2014;130:18–26.

[44] Pastori D, Carnevale R, Bartimoccia S, et al. Does Mediterranean diet reduce cardiovascular events and oxidative stress in atrial fibrillation? Antioxid Redox Signal 2015;23:682–7.

[45] Fargnoli JL, Fung TT, Olenczuk DM, et al. Adherence to healthy eating patterns is associated with higher circulating total and high-molecular-weight adiponectin and lower resistin concentrations in women from the Nurses' Health Study. Am J Clin Nutr 2008;88:1213–24.

[46] Lopez-Garcia E, Schulze MB, Fung TT, et al. Major dietary patterns are related to plasma concentrations of markers of inflammation and endothelial dysfunction. Am J Clin Nutr 2004;80:1029–35.

[47] Chrysohoou C, Panagiotakos DB, Pitsavos C, et al. Adherence to the Mediterranean diet attenuates inflammation and coagulation process in healthy adults: the ATTICA Study. J Am Coll Cardiol 2004;44:152–8.

[48] Schwingshackl L, Hoffmann G. Mediterranean dietary pattern, inflammation and endothelial function: a systematic review and meta-analysis of intervention trials. Nutr Metab Cardiovasc Dis 2014;24:929–39.

[49] Fuentes F, López-Miranda J, Sánchez E, et al. Mediterranean and low-fat diets improve endothelial function in hypercholesterolemic men. Ann Intern Med 2001;134:1115–9.

[50] Rallidis LS, Lekakis J, Kolomvotsou A, et al. Close adherence to a Mediterranean diet improves endothelial function in subjects with abdominal obesity. Am J Clin Nutr 2009;90:263–8.

[51] Buscemi S, Verga S, Tranchina MR, et al. Effects of hypocaloric very-low-carbohydrate diet vs. Mediterranean diet on endothelial function in obese women. Eur J Clin Investig 2009;39:339–47.

[52] Fuentes F, López-Miranda J, Pérez-Martínez P, et al. Chronic effects of a high-fat diet enriched with virgin olive oil and a low-fat diet enriched with α-linolenic acid on post-prandial endothelial function in healthy men. Br J Nutr 2008;100:159–65.

[53] Perez-Martinez P, Moreno-Conde M, Cruz-Teno C, et al. Dietary fat differentially influences regulatory endothelial function during the postprandial state in patients with metabolic syndrome: from the LIPGENE study. Atherosclerosis 2010;209:533–8.

[54] Marin C, Ramirez R, Delgado-Lista J, et al. Mediterranean diet reduces endothelial damage and improves the regenerative capacity of endothelium. Am J Clin Nutr 2011;93:267–74.

[55] Fernández JM, Rosado-Álvarez D, Da Silva Grigoletto ME, et al. Moderate-to-high-intensity training and a hypocaloric Mediterranean diet enhance endothelial progenitor cells and fitness in subjects with the metabolic syndrome. Clin Sci (Lond) 2012;123:361–73.

[56] Ambring A, Friberg P, Axelsen M, et al. Effects of a Mediterranean-inspired diet on blood lipids, vascular function and oxidative stress in healthy subjects. Clin Sci (Lond) 2004;106:519–25.

[57] Cioni G, Boddi M, Fatini C, et al. Peripheral-arterial tonometry for assessing endothelial function in relation to dietary habits. J Investig Med 2013;61:867–71.

[58] Inaba Y, Chen JA, Bergmann SR. Prediction of future cardiovascular outcomes by flow-mediated vasodilatation of brachial artery: a meta-analysis. Int J Cardiovasc Imaging 2010;26:631–40.

[59] López-Miranda J, Pérez-Jiménez F, Ros E, et al. Olive oil and health: summary of the II international conference on olive oil and health consensus report, Jaén and Córdoba (Spain) 2008. Nutr Metab Cardiovasc Dis 2010;20:284–94.

[60] Martínez-González MA, Dominguez LJ, Delgado-Rodríguez M. Olive oil consumption and risk of CHD and/or stroke: a meta-analysis of case-control, cohort and intervention studies. Br J Nutr 2014;112:248–59.

[61] Guasch-Ferré M, Hu FB, Martínez-González MA, et al. Olive oil intake and risk of cardiovascular disease and mortality in the PREDIMED Study. BMC Med 2014;12:78.

[62] Carluccio MA, Massaro M, Bonfrate C, et al. Oleic acid inhibits endothelial activation: a direct vascular antiatherogenic mechanism of a nutritional component in the Mediterranean diet. Arterioscler Thromb Vasc Biol 1999;19:220–8.

[63] Massaro M, Basta G, Lazzerini G, et al. Quenching of intracellular ROS generation as a mechanism for oleate-induced reduction of endothelial activation and early atherogenesis. Thromb Haemost 2002;88:335–44.

[64] Perez-Martinez P, Lopez-Miranda J, Blanco-Colio L, et al. The chronic intake of a Mediterranean diet enriched in virgin olive oil, decreases nuclear transcription factor kappaB activation in peripheral blood mononuclear cells from healthy men. Atherosclerosis 2007;194:e141–6.

[65] Harvey KA, Walker CL, Xu Z, et al. Oleic acid inhibits stearic acid-induced inhibition of cell growth and pro-inflammatory responses in human aortic endothelial cells. J Lipid Res 2010;51:3470–80.

[66] Tsimikas S, Philis-Tsimikas A, Alexopoulos S, et al. LDL isolated from Greek subjects on a typical diet or from American subjects on an oleate-supplemented diet induces less monocyte chemotaxis and adhesion when exposed to oxidative stress. Arterioscler Thromb Vasc Biol 1999;19:122–30.

[67] Yaqoob P, Knapper JA, Webb DH, et al. Effect of olive oil on immune function in middle-aged men. Am J Clin Nutr 1998;67:129–35.

[68] Esposito K, Marfella R, Ciotola M, et al. Effect of a Mediterranean-style diet on endothelial dysfunction and markers of vascular inflammation in the metabolic syndrome: a randomized trial. JAMA 2004;292:1440–6.

[69] Bogani P, Galli C, Villa M, et al. Postprandial anti-inflammatory and antioxidant effects of extra virgin olive oil. Atherosclerosis 2007;190:181–6.

[70] Pacheco YM, Bemúdez B, López S, et al. Minor compounds of olive oil have postprandial anti-inflammatory effects. Br J Nutr 2007;98:260–3.

[71] Fitó M, Cladellas M, de la Torre R, et al. Anti-inflammatory effect of virgin olive oil in stable coronary disease patients: a randomized, crossover, controlled trial. Eur J Clin Nutr 2008;62:570–4.

[72] Risérus U, Willett WC, Hu FB. Dietary fats and prevention of type 2 diabetes. Prog Lipid Res 2009;48:44–51.

[73] Ros E. Dietary cis-monounsaturated fatty acids and metabolic control in type 2 diabetes. Am J Clin Nutr 2003;78:617S–625S.

[74] Sala-Vila A, Cofán M, Mateo-Gallego R, et al. Inverse association between serum phospholipid oleic acid and insulin resistance in subjects with primary dyslipidaemia from a Mediterranean population. Clin Nutr 2011;30:590–2.

[75] Vessby B, Uusitupa M, Hermansen K, et al. Substituting dietary saturated for monounsaturated fat impairs insulin sensitivity in healthy men and women: the KANWU Study. Diabetologia 2001;44:312–9.

[76] Corpeleijn E, Feskens EJM, Jansen EHJM, et al. Improvements in glucose tolerance and insulin sensitivity after lifestyle intervention are related to changes in serum fatty acid profile and desaturase activities: the SLIM study. Diabetologia 2006;49:2392–401.

[77] Pérez-Jiménez F, López-Miranda J, Pinillos MD, et al. A Mediterranean and a high-carbohydrate diet improve glucose metabolism in healthy young persons. Diabetologia 2001;44:2038–43.

[78] Covas M-I, Nyyssönen K, Poulsen HE, et al. The effect of polyphenols in olive oil on heart disease risk factors: a randomized trial. Ann Intern Med 2006;145:333–41.

[79] Machowetz A, Poulsen HE, Gruendel S, et al. Effect of olive oils on biomarkers of oxidative DNA stress in Northern and Southern Europeans. FASEB J 2007;21:45–52.

[80] Hernáez Á, Remaley AT, Farràs M, et al. Olive oil polyphenols decrease LDL concentrations and LDL atherogenicity in men in a randomized controlled trial. J Nutr 2015;145:1692–7.

[81] Ryan M, McInerney D, Owens D, et al. Diabetes and the Mediterranean diet: a beneficial effect of oleic acid on insulin sensitivity, adipocyte glucose transport and endothelium dependent vasoreactivity. QJM 2000;93:85–91.

[82] Moreno-Luna R, Muñoz-Hernandez R, Miranda ML, et al. Olive oil polyphenols decrease blood pressure and improve endothelial function in young women with mild hypertension. Am J Hypertens 2012;25:1299–304.

[83] Vogel RA, Corretti MC, Plotnick GD. The postprandial effect of components of the Mediterranean diet on endothelial function. J Am Coll Cardiol 2000;36:1455–60.

[84] Ruano J, Lopez-Miranda J, Fuentes F, et al. Phenolic content of virgin olive oil improves ischemic reactive hyperemia in hypercholesterolemic patients. J Am Coll Cardiol 2005;46:1864–8.

[85] Karatzi K, Papamichael C, Karatzis E, et al. Postprandial improvement of endothelial function by red wine and olive oil antioxidants: a synergistic effect of components of the Mediterranean diet. J Am Coll Nutr 2008;27:448–53.

[86] Tousoulis D, Papageorgiou N, Antoniades C, et al. Acute effects of different types of oil consumption on endothelial function, oxidative stress status and vascular inflammation in healthy volunteers. Br J Nutr 2010;103:43–9.

[87] Ros E, Mataix J. Fatty acid composition of nuts. Implications for cardiovascular health. Br J Nutr 2006;96:S29–35. Erratum in: Br J Nutr 2008;99:447–8.

[88] Segura R, Javierre C, Lizarraga MA, et al. Other relevant components of nuts: phytosterols, folates and minerals. Br J Nutr 2006;96: S36–44. Erratum in: Br J Nutr 2008;99:447–8.

[89] Salas-Salvadó J, Bulló M, Pérez-Heras A, et al. Dietary fibre, nuts and cardiovascular disease. Br J Nutr 2006;96:S45–51. Erratum in: Br J Nutr 2008;99:447–8.

[90] Ros E. Health benefits of nut consumption. Nutrients 2010;2:652–82.

[91] Afshin A, Micha R, Khatibzadeh S, et al. Consumption of nuts and legumes and risk of incident ischemic heart disease, stroke, and diabetes: a systematic review and meta-analysis. Am J Clin Nutr 2014;100:278–88.

[92] Luo C, Zhang Y, Ding Y, et al. Nut consumption and risk of type 2 diabetes, cardiovascular disease, and all-cause mortality: a systematic review and meta-analysis. Am J Clin Nutr 2014;100:256–69.

[93] Zhou D, Yu H, He F, et al. Nut consumption in relation to cardiovascular disease risk and type 2 diabetes: a systematic review and meta-analysis of prospective studies. Am J Clin Nutr 2014;100:270–7.

[94] Grosso G, Yang J, Marventano S, et al. Nut consumption on all-cause, cardiovascular, and cancer mortality risk: a systematic review and meta-analysis of epidemiologic studies. Am J Clin Nutr 2015;101:783–93.

[95] van den Brandt PA, Schouten LJ. Relationship of tree nut, peanut and peanut butter intake with total and cause-specific mortality: a cohort study and meta-analysis. Int J Epidemiol 2015;44:1038–49.

[96] Mukuddem-Petersen J, Oosthuizen W, Jerling JC. A systematic review of the effects of nuts on blood lipid profiles in humans. J Nutr 2005;135:2082–9.

[97] Griel AE, Kris-Etherton PM. Tree nuts and the lipid profile: a review of clinical studies. Br J Nutr 2006;96:S68–78.

[98] Sabaté J, Oda K, Ros E. Nut consumption and blood lipids: a pooled analysis of 25 intervention trials. Arch Intern Med 2010;170:821–7.

[99] Mohammadifard N, Salehi-Abargouei A, Salas-Salvadó J, et al. The effect of tree nut, peanut, and soy nut consumption on blood pressure: a systematic review and meta-analysis of randomized controlled clinical trials. Am J Clin Nutr 2015;101:966–82.

[100] Kendall CW, Josse AR, Esfahani A, et al. Nuts, metabolic syndrome and diabetes. Br J Nutr 2010;104:465–73.

[101] Viguiliouk E, Kendall CW, Blanco Mejia S, et al. Effect of tree nuts on glycemic control in diabetes: a systematic review and meta-analysis of randomized controlled dietary trials. PLoS One 2014;9:e103376.

[102] Blanco Mejia S, Kendall CW, Viguiliouk E, et al. Effect of tree nuts on metabolic syndrome criteria: a systematic review and meta-analysis of randomised controlled trials. BMJ Open 2014;4: e004660.

[103] Jiang R, Jacobs Jr DR, Mayer-Davis E, et al. Nut and seed consumption and inflammatory markers in the multi-ethnic study of atherosclerosis. Am J Epidemiol 2006;163:222–31.

[104] López-Uriarte P, Bulló M, Casas-Agustench P, et al. Nuts and oxidation: a systematic review. Nutr Rev 2009;67:497–508.

[105] Banel DK, Hu FB. Effects of walnut consumption on blood lipids and other cardiovascular risk factors: a meta-analysis and systematic review. Am J Clin Nutr 2009;90:56–63.

[106] Ros E, Núñez I, Pérez-Heras A, et al. A walnut diet improves endothelial function in hypercholesterolemic subjects: a randomized crossover trial. Circulation 2004;109:1609–14.

[107] Ma Y, Njike VY, Millet J, et al. Effects of walnut consumption on endothelial function in type 2 diabetic subjects: a randomized controlled crossover trial. Diabetes Care 2010;33:227–32.

[108] West SG, Krick AL, Klein LC, et al. Effects of diets high in walnuts and flax oil on hemodynamic responses to stress and vascular endothelial function. J Am Coll Nutr 2010;29:595–603.

[109] Katz DL, Davidhi A, Ma Y, et al. Effects of walnuts on endothelial function in overweight adults with visceral obesity: a randomized, controlled, crossover trial. J Am Coll Nutr 2012;31:415–23.

[110] Orem A, Yucesan FB, Orem C, et al. Hazelnut-enriched diet improves cardiovascular risk biomarkers beyond a lipid-lowering effect in hypercholesterolemic subjects. J Clin Lipidol 2013;7:123–31.

[111] Kasliwal RR, Bansal M, Mehrotra R, et al. Effect of pistachio nut consumption on endothelial function and arterial stiffness. Nutrition 2015;31:678–85.

[112] López-Uriarte P, Nogués R, Saez G, et al. Effect of nut consumption on oxidative stress and the endothelial function in metabolic syndrome. Clin Nutr 2010;29:373–80.

[113] Cortés B, Núñez I, Cofán M, et al. Acute effects of high-fat meals enriched with walnuts or olive oil on postprandial endothelial function. J Am Coll Cardiol 2006;48:1666–71.

[114] Berryman CE, Grieger JA, West SG, et al. Acute consumption of walnuts and walnut components differentially affect postprandial lipemia, endothelial function, oxidative stress, and cholesterol efflux in humans with mild hypercholesterolemia. J Nutr 2013;143:788–94.

[115] Kendall CW, West SG, Augustin LS, et al. Acute effects of pistachio consumption on glucose and insulin, satiety hormones and endothelial function in the metabolic syndrome. Eur J Clin Nutr 2014;68:370–5.

[116] Chiang YL, Haddad E, Rajaram S, et al. The effect of dietary walnuts compared to fatty fish on eicosanoids, cytokines, soluble endothelial adhesion molecules and lymphocyte subsets: a randomized, controlled crossover trial. Prostaglandins Leukot Essent Fat Acids 2012;87:111–7.

[117] Chiva-Blanch G, Arranz S, Lamuela-Raventos RM, et al. Effects of wine, alcohol and polyphenols on cardiovascular disease risk factors: evidences from human studies. Alcohol Alcohol 2013;48:270–7.

[118] Ronksley PE, Brien SE, Turner BJ, et al. Association of alcohol consumption with selected cardiovascular disease outcomes: a systematic review and meta-analysis. BMJ 2011;342:d671.

[119] Baliunas DO, Taylor BJ, Irving H, et al. Alcohol as a risk factor for type 2 diabetes: a systematic review and meta-analysis. Diabetes Care 2009;32:2123–32.

[120] Brien SE, Ronksley PE, Turner BJ, et al. Effect of alcohol consumption on biological markers associated with risk of coronary heart disease: systematic review and meta-analysis of interventional studies. BMJ 2011;342:d636.

[121] Costanzo S, Di Castelnuovo A, Donati MB, et al. Wine, beer or spirit drinking in relation to fatal and non-fatal cardiovascular events: a meta-analysis. Eur J Epidemiol 2011;26:833–50.

[122] Chiva-Blanch G, Urpi-Sardá M, Ros E, et al. Dealcoholized red wine decreases systolic and diastolic blood pressure and increases plasma nitric oxide: short communication. Circ Res 2012;111:1065–8.

[123] Estruch R, Sacanella E, Mota F, et al. Consumption of red wine, but not gin, decreases erythrocyte superoxide dismutase activity: a randomised cross-over trial. Nutr Metab Cardiovasc Dis 2011;21:46–53.

[124] Covas MI, Gambert P, Fitó M, et al. Wine and oxidative stress: up-to-date evidence of the effects of moderate wine consumption on oxidative damage in humans. Atherosclerosis 2010;208:297–304.

[125] Sacanella E, Vázquez-Agell M, Mena MP, et al. Anti-inflammatory effects of moderate wine consumption in women. Am J Clin Nutr 2007;86:1463–9.

[126] Chiva-Blanch G, Urpi-Sarda M, Llorach R, et al. Differential effects of polyphenols and alcohol of red wine on the expression of adhesion molecules and inflammatory cytokines related to atherosclerosis: a randomized clinical trial. Am J Clin Nutr 2012;95:326–34.

[127] Karatzi K, Karatzis E, Papamichael C, et al. Effects of red wine on endothelial function: postprandial studies vs clinical trials. Nutr Metab Cardiovasc Dis 2009;19:744–50.

[128] Mangoni AA, Stockley CS, Woodman RJ. Effects of red wine on established markers of arterial structure and function in human studies: current knowledge and future research directions. Expert Rev Clin Pharmacol 2013;6:613–25.

[129] De Caterina R. n-3 fatty acids in cardiovascular disease. N Engl J Med 2011;364:2439–50.

[130] Mozaffarian D, Rimm EB. Fish intake, contaminants, and human health: evaluating the risks and the benefits. JAMA 2006;296:1885–99.

[131] Mozaffarian D, Wu JHY. Omega-3 fatty acids and cardiovascular disease: effects on risk factors, molecular pathways, and clinical events. J Am Coll Cardiol 2011;58:2047–67.

[132] Kris-Etherton PM, Harris WS, Appel LJ, et al. Omega-3 fatty acids and cardiovascular disease: new recommendations from the

American Heart Association. Arterioscler Thromb Vasc Biol 2003;23:151–2.

[133] Zheng J, Huang T, Yu Y, et al. Fish consumption and CHD mortality: an updated meta-analysis of seventeen cohort studies. Public Health Nutr 2012;15:725–37.

[134] Leung Yinko SS, Stark KD, Thanassoulis G, et al. Fish consumption and acute coronary syndrome: a meta-analysis. Am J Med 2014;127:848–57.

[135] Xun P, Qin B, Song Y, et al. Fish consumption and risk of stroke and its subtypes: accumulative evidence from a meta-analysis of prospective cohort studies. Eur J Clin Nutr 2012;66:1199–207.

[136] Chowdhury R, Stevens S, Gorman D, et al. Association between fish consumption, long chain omega 3 fatty acids, and risk of cerebrovascular disease: systematic review and meta-analysis. BMJ 2012;345:e6698.

[137] Djoussé L, Akinkuolie AO, Wu JH, et al. Fish consumption, omega-3 fatty acids and risk of heart failure: a meta-analysis. Clin Nutr 2012;31:846–53.

[138] Wu JH, Micha R, Imamura F, et al. Omega-3 fatty acids and incident type 2 diabetes: a systematic review and meta-analysis. Br J Nutr 2012;107:S214–27.

[139] Burr ML, Fehily AM, Gilbert JF, et al. Effects of changes in fat, fish, and fibre intakes on death and myocardial reinfarction: diet and reinfarction trial (DART). Lancet 1989;2:757–61.

[140] Kromhout D, Giltay EJ, Geleijnse JM. n-3 fatty acids and cardiovascular events after myocardial infarction. N Engl J Med 2010;363:2015–26.

[141] Rauch B, Schiele R, Schneider S, et al. OMEGA, a randomized, placebo-controlled trial to test the effect of highly purified omega-3 fatty acids on top of modern guideline-adjusted therapy after myocardial infarction. Circulation 2010;122:2152–9.

[142] Galan P, Kesse-Guyot E, Czernichow S, et al. Effects of B vitamins and omega 3 fatty acids on cardiovascular diseases: a randomised placebo controlled trial. BMJ 2010;341:c6273.

[143] ORIGIN Trial Investigators, Gerstein HC, Bosch J, et al. n-3 fatty acids and cardiovascular outcomes in patients with dysglycemia. N Engl J Med 2012;367:309–18.

[144] Risk and Prevention Study Collaborative Group, Roncaglioni MC, Tombesi M, et al. n-3 fatty acids in patients with multiple cardiovascular risk factors. N Engl J Med 2013;368:1800–8.

[145] Harris WS. Are n-3 fatty acids still cardioprotective? Curr Opin Clin Nutr Metab Care 2013;16:141–9.

[146] Montori VM, Farmer A, Wollan PC, et al. Fish oil supplementation in type 2 diabetes: a quantitative systematic review. Diabetes Care 2000;23:1407–15.

[147] Hartweg J, Perera R, Montori V, et al. Omega-3 polyunsaturated fatty acids (PUFA) for type 2 diabetes mellitus. Cochrane Database Syst Rev 2008;CD003205.

[148] Serhan CN, Chiang N, Van Dyke TE. Resolving inflammation: dual anti-inflammatory and pro-resolution lipid mediators. Nat Rev Immunol 2008;8:349–61.

[149] Calder PC. The role of marine omega-3 (n-3) fatty acids in inflammatory processes, atherosclerosis and plaque stability. Mol Nutr Food Res 2012;56:1073–80.

[150] Cawood AL, Ding R, Napper FL, et al. Eicosapentaenoic acid (EPA) from highly concentrated n-3 fatty acid ethyl esters is incorporated into advanced atherosclerotic plaques and higher plaque EPA is associated with decreased plaque inflammation and increased stability. Atherosclerosis 2010;212:252–9.

[151] Balakumar P, Taneja G. Fish oil and vascular endothelial protection: bench to bedside. Free Radic Biol Med 2012;53:271–9.

[152] Rangel-Huerta OD, Aguilera CM, Mesa MD, et al. Omega-3 long-chain polyunsaturated fatty acids supplementation on inflammatory biomakers: a systematic review of randomised clinical trials. Br J Nutr 2012;107:S159–70.

[153] Wang Q, Liang X, Wang L, et al. Effect of omega-3 fatty acids supplementation on endothelial function: a meta-analysis of randomized controlled trials. Atherosclerosis 2012;221:536–43.

[154] Pischon T, Hankinson SE, Hotamisligil GS, et al. Habitual dietary intake of n-3 and n-6 fatty acids in relation to inflammatory markers among US men and women. Circulation 2003;108:155–60.

[155] Lopez-Garcia E, Schulze MB, Manson JE, et al. Consumption of (n-3) fatty acids is related to plasma biomarkers of inflammation and endothelial activation in women. J Nutr 2004;134:1806–11.

[156] Zampelas A, Panagiotakos DB, Pitsavos C, et al. Fish consumption among healthy adults is associated with decreased levels of inflammatory markers related to cardiovascular disease: the ATTICA study. J Am Coll Cardiol 2005;46:120–4.

[157] He K, Liu K, Daviglus ML, et al. Associations of dietary long-chain n-3 polyunsaturated fatty acids and fish with biomarkers of inflammation and endothelial activation (from the Multi-Ethnic Study of Atherosclerosis [MESA]). Am J Cardiol 2009;103:1238–43.

[158] van Bussel BC, Henry RM, Schalkwijk CG, et al. Fish consumption in healthy adults is associated with decreased circulating biomarkers of endothelial dysfunction and inflammation during a 6-year follow-up. J Nutr 2011;141:1719–25.

[159] Anderson JS, Nettleton JA, Herrington DM, et al. Relation of omega-3 fatty acid and dietary fish intake with brachial artery flow-mediated vasodilation in the Multi-Ethnic Study of Atherosclerosis. Am J Clin Nutr 2010;92:1204–13.

[160] Petersen MM, Eschen RB, Aardestrup I, et al. Flow-mediated vasodilation and dietary intake of n-3 polyunsaturated acids in healthy subjects. Cell Mol Biol (Noisy-le-Grand) 2010;56:38–44.

[161] Buscemi S, Vasto S, Di Gaudio F, et al. Endothelial function and serum concentration of toxic metals in frequent consumers of fish. PLoS One 2014;9:e112478.

[162] Merino J, Sala-Vila A, Kones R, et al. Increasing long-chain n-3 PUFA consumption improves small peripheral artery function in patients at intermediate-high cardiovascular risk. J Nutr Biochem 2014;25:642–6.

[163] Seierstad SL, Seljeflot I, Johansen O, et al. Dietary intake of differently fed salmon; the influence on markers of human atherosclerosis. Eur J Clin Investig 2005;35:52–9.

[164] de Mello VD, Erkkilä AT, Schwab US, et al. The effect of fatty or lean fish intake on inflammatory gene expression in peripheral blood mononuclear cells of patients with coronary heart disease. Eur J Nutr 2009;48:447–55.

[165] Lindqvist HM, Langkilde AM, Undeland I, et al. Herring (Clupea harengus) intake influences lipoproteins but not inflammatory and oxidation markers in overweight men. Br J Nutr 2009;101:383–90.

[166] van den Elsen LW, Noakes PS, van der Maarel MA, et al. Salmon consumption by pregnant women reduces ex vivo umbilical cord endothelial cell activation. Am J Clin Nutr 2011;94:1418–25.

[167] de Mello VD, Schwab U, Kolehmainen M, et al. A diet high in fatty fish, bilberries and wholegrain products improves markers of endothelial function and inflammation in individuals with impaired glucose metabolism in a randomised controlled trial: the Sysdimet study. Diabetologia 2011;54:2755–67.

[168] Kondo K, Morino K, Nishio Y, et al. A fish-based diet intervention improves endothelial function in postmenopausal women with type 2 diabetes mellitus: a randomized crossover trial. Metabolism 2014;63:930–40.

[169] Madsen T, Skou HA, Hansen VE, et al. C-reactive protein, dietary n-3 fatty acids, and the extent of coronary artery disease. Am J Cardiol 2001;88:1139–42.

[170] Niu K, Hozawa A, Kuriyama S, et al. Dietary long-chain n-3 fatty acids of marine origin and serum C-reactive protein

concentrations are associated in a population with a diet rich in marine products. Am J Clin Nutr 2006;84:223–9.

[171] Fernandez-Real JM, Broch M, Vendrell J, et al. Insulin resistance, inflammation, and serum fatty acid composition. Diabetes Care 2003;26:1362–8.

[172] Klein-Platat C, Drai J, Oujaa M, et al. Plasma fatty acid composition is associated with the metabolic syndrome and low-grade inflammation in overweight adolescents. Am J Clin Nutr 2005;82:1178–84.

[173] Ferrucci L, Cherubini A, Bandinelli S, et al. Relationship of plasma polyunsaturated fatty acids to circulating inflammatory markers. J Clin Endocrinol Metab 2006;91:439–46.

[174] Chiuve SE, Fung TT, Rexrode KM, et al. Adherence to a low-risk, healthy lifestyle and risk of sudden cardiac death among women. JAMA 2011;306:62–9.

[175] Bertoia ML, Triche EW, Michaud DS, et al. Mediterranean and dietary approaches to stop hypertension dietary patterns and risk of sudden cardiac death in postmenopausal women. Am J Clin Nutr 2014;99:344–51.

[176] Beunza JJ, Toledo E, Hu FB, et al. Adherence to Mediterranean diet, long-term weight change, and incident overweight or obesity: the Seguimiento Universidad de Navarra (SUN) cohort. Am J Clin Nutr 2010;92:1484–93.

[177] Romaguera D, Norat T, Vergnaud AC, et al. Mediterranean dietary patterns and prospective weight change in participants of the EPIC-PANACEA project. Am J Clin Nutr 2010;92:912–21.

[178] Buckland G, Bach A, Serra-Majem L. Obesity and the Mediterranean diet: a systematic review of observational and intervention studies. Obes Rev 2008;9:582–93.

[179] Esposito K, Kastorini CM, Panagiotakos DB, et al. Mediterranean diet and weight loss: meta-analysis of randomized controlled trials. Metab Syndr Relat Disord 2011;9:1–12.

[180] Martinez-Gonzalez MA, García-Arellano A, Toledo E, et al. A 14-item Mediterranean assessment tool and obesity index among high-risk subjects: the PREDIMED trial. PLoS One 2012;7:e43134.

[181] Mozzafarian D, Hao T, Rimm EB, et al. Changes in diet and lifestyle and long-term weight gain in women and men. N Engl J Med 2011;364:2392–404.

[182] Mozaffarian D, Ludwig DS. The 2015 US Dietary Guidelines. Lifting the ban on total dietary fat. JAMA 2015;313:2421–2.

[183] Ritchie K, Lovestone S. The dementias. Lancet 2002;360:1759–66.

[184] Iqbal K, Liu F, Gong CX. Alzheimer disease therapeutics: focus on the disease and not just plaques and tangles. Biochem Pharmacol 2014;88:631–9.

[185] Vos SJ, Xiong C, Visser PJ, et al. Preclinical Alzheimer's disease and its outcome: a longitudinal cohort study. Lancet Neurol 2013;12:957–65.

[186] Otaegui-Arrazola A, Amiano P, Elbusto A, et al. Diet, cognition, and Alzheimer's disease: food for thought. J Nutr 2014;53:1–23.

[187] Lourida I, Soni M, Thompson-Coon J, et al. Mediterranean diet, cognitive function and dementia: a systematic review. Epidemiology 2013;24:479–89.

[188] Psaltopoulou T, Sergentanis TN, Panagiotakos DB, et al. Mediterranean diet and stroke, cognitive impairment, depression: a meta-analysis. Ann Neurol 2013;74:580–91.

[189] Singh B, Parsaik AK, Mielke MM, et al. Association of Mediterranean diet with mild cognitive impairment and Alzheimer's disease: a systematic review and meta-analysis. J Alzheimers Dis 2014;39:271–82.

[190] Valls-Pedret C, Sala-Vila A, Serra-Mir M, et al. Mediterranean diet and age-related cognitive decline: a randomized clinical trial. JAMA Intern Med 2015;175:1094–103.

[191] Valls-Pedret C, Lamuela-Raventós RM, Medina-Remón A, et al. Polyphenol-rich foods in the Mediterranean diet are associated with better cognitive function in elderly subjects at high cardiovascular risk. J Alzheimers Dis 2012;29:773–82.

[192] Berr C, Portet F, Carriere I, et al. Olive oil and cognition: results from the three-city study. Dement Geriatr Cogn Disord 2009;28:357–64.

[193] Nooyens AC, Bueno-de-Mesquita HB, van Boxtel MP, et al. Fruit and vegetable intake and cognitive decline in middle-agedmen and women: the Doetinchem Cohort Study. Br J Nutr 2011;106:752–61.

[194] O'Brien J, Okereke O, Devore E, et al. Long-term intake of nuts in relation to cognitive function in older women. J Nutr Health Aging 2014;18:496–502.

[195] Schwingshackl L, Hoffmann G. Adherence to Mediterranean diet and risk of cancer: a systematic review and meta-analysis of observational studies. Int J Cancer 2014;135:1884–97.

[196] de Lorgeril M, Salen P, Martin JL, et al. Mediterranean dietary pattern in a randomized trial: prolonged survival and possible reduced cancer rate. Arch Intern Med 1998;158:1181–7.

[197] Toledo E, Salas-Salvadó J, Donat-Vargas C, et al. Mediterranean diet and invasive breast cancer risk among women at high cardiovascular risk in the PREDIMED trial. A randomized clinical trial. JAMA Intern Med 2015;175 [Epub ahead of print] PMID: 26365989.

[198] Psaltopoulou T, Kosti RI, Haidopoulos D, et al. Olive oil intake is inversely related to cancer prevalence: a systematic review and a meta-analysis of 13,800 patients and 23,340 controls in 19 observational studies. Lipids Health Dis 2011;10:127.

[199] Bonaccio M, Di Castelnuovo A, De Curtis A, et al. Nut consumption is inversely associated with both cancer and total mortality in a Mediterranean population: prospective results from the Molisani study. Br J Nutr 2015;114:804–11.

[200] Trichopoulou A. Traditional Mediterranean diet and longevity in the elderly: a review. Public Health Nutr 2004;7:943–7.

[201] Lopez-Garcia E, Rodriguez-Artalejo F, Li TY, et al. The Mediterranean-style dietary pattern and mortality among men and women with cardiovascular disease. Am J Clin Nutr 2014;99:172–80.

[202] Tognon G, Nilsson LM, Lissner L, et al. The Mediterranean diet score and mortality are inversely associated in adults living in the subartic region. J Nutr 2012;142:1547–53.

[203] Samieri C, Sun Q, Townsend MK, et al. The association between dietary patterns at midlife and health in aging: an observational study. Ann Intern Med 2013;159:584–91.

[204] Avogaro A, de Kreutzenberg SV, Federici M, et al. The endothelium abridges insulin resistance to premature aging. J Am Heart Assoc 2013;2:e000262.

[205] Blasco MA. Telomeres and human disease: ageing, cancer and beyond. Nat Rev Genet 2005;6:611–22.

[206] Crous-Bou M, Fung TT, Prescott J, et al. Mediterranean diet and telomere length in Nurses' Health Study: population based cohort study. BMJ 2014;349:g6674.

[207] Gu Y, Honig LS, Schupf N, et al. Mediterranean diet and leukocyte telomere length in a multi-ethnic elderly population. Age (Dordr) 2015;37:24.

[208] Boccardi V, Esposito A, Rizzo MR, et al. Mediterranean diet, telomere maintenance and health status among elderly. PLoS One 2013;8:e62781.

[209] Scientific Report of the 2015 Dietary Guidelines Advisory Committee. Washington, DC: Office of Disease Prevention and Health Promotion. Available at: http://health.gov/dietaryguidelines/2015-scientific-report.

[210] Roman Viñas B, Ribas Barba L, Ngo J, et al. Projected prevalence of inadequate nutrient intakes in Europe. Ann Nutr Metab 2011;59:84–95.

[211] Zazpe I, Sánchez-Taínta A, Santiago S, et al. Association between dietary carbohydrate intake quality and micronutrient intake adequacy in a Mediterranean cohort: the SUN (Seguimiento Universidad de Navarra) Project. Br J Nutr 2014;1–10 [Epub ahead of print].

[212] Louie JCY, Buyken AE, Brand-Miller JC, et al. The link between dietary glycemic index and nutrient adequacy. Am J Clin Nutr 2012;95:694–702.

[213] Serra-Majem L, Bes-Rastrollo M, Román-Viñas B, et al. Dietary patterns and nutritional adequacy in a Mediterranean country. Br J Nutr 2009;101:S21–8.

[214] Maillot M, Issa C, Vieux F, et al. The shortest way to reach nutritional goals is to adopt Mediterranean food choices: evidence from computer-generated personalized diets. Am J Clin Nutr 2011;94:1127–37.

[215] Castro-Quezada I, Román-Viñas B, Serra-Majem L. The Mediterranean diet and nutritional adequacy: a review. Nutrients 2014;6:231–48.

[216] León-Muñoz LM, Guallar-Castillón P, López-García E, et al. Mediterranean diet and risk of frailty in community-dwelling older adults. J Am Med Dir Assoc 2014;15:899–903.

[217] Milaneschi Y, Bandinelli S, Corsi AM, et al. Mediterranean diet and mobility decline in older persons. Exp Gerontol 2011;46:303–8.

[218] Feart C, Peres K, Samieri C, et al. Adherence to a Mediterranean diet and onset of disability in older persons. Eur J Epidemiol 2011;26:747–56.

[219] Corella D, Ordovas JM. Nutrigenomics in cardiovascular medicine. Circ Cardiovasc Genet 2009;2:637–51.

[220] Corella D, Ortega-Azorín C, Sorlí JV, et al. Statistical and biological gene-lifestyle interactions of MC4R and FTO with diet and physical activity on obesity: new effects on alcohol consumption. PLoS One 2012;7:e52344.

[221] Ortega-Azorín C, Sorlí JV, Asensio EM, et al. Associations of the FTO rs9939609 and the MC4R rs17782313 polymorphisms with type 2 diabetes are modulated by diet, being higher when adherence to the Mediterranean diet pattern is low. Cardiovasc Diabetol 2012;11:137.

[222] Corella D, Carrasco P, Sorlí JV, et al. Mediterranean diet reduces the adverse effect of the TCF7L2-rs7903146 polymorphism on cardiovascular risk factors and stroke incidence: a randomized controlled trial in a high-cardiovascular-risk population. Diabetes Care 2013;36:3803–11.

[223] Ortega-Azorín C, Sorlí JV, Estruch R, et al. Amino acid change in the carbohydrate response element binding protein is associated with lower triglycerides and myocardial infarction incidence depending on level of adherence to the Mediterranean diet in the PREDIMED trial. Circ Cardiovasc Genet 2014;7:49–58.

[224] Corella D, Sorlí JV, Estruch R, et al. MicroRNA-410 regulated lipoprotein lipase variant rs13702 is associated with stroke incidence and modulated by diet in the randomized controlled PREDIMED trial. Am J Clin Nutr 2014;100:719–31.

[225] Sekirov I, Russell SL, Antunes LCM, et al. Gut microbiota in health and disease. Physiol Rev 2010;90:859–904.

[226] Power SE, O'Toole PW, Stanton C, et al. Intestinal microbiota, diet and health. Br J Nutr 2014;111:387–402.

[227] Albenberg LG, Wu GD. Diet and the intestinal microbiome: associations, functions, and implications for health and disease. Gastroenterology 2014;146:1564–72.

[228] De Filippo C, Cavalieri D, Di Paola M, et al. Impact of diet in shaping gut microbiota revealed by a comparative study in children from Europe and rural Africa. Proc Natl Acad Sci U S A 2010;107:14691–6.

[229] Claesson MJ, Jeffery IB, Conde S, et al. Gut microbiota composition correlates with diet and health in the elderly. Nature 2012;488:178–84.

[230] Cotillard A, Kennedy SP, Kong LC, et al. Dietary intervention impact on gut microbial gene richness. Nature 2013;500:585–8.

[231] Wang Z, Klipfell E, Bennett BJ, et al. Gut flora metabolism of phosphatidylcholine promotes cardiovascular disease. Nature 2011;472:57–63.

[232] Tang WH, Wang Z, Levison BS, et al. Intestinal microbial metabolism of phosphatidylcholine and cardiovascular risk. N Engl J Med 2013;368:1575–84.

[233] Koeth RA, Wang Z, Levison BS, et al. Intestinal microbiota metabolism of L-carnitine, a nutrient in red meat, promotes atherosclerosis. Nat Med 2013;19:576–85.

[234] Cardona F, Andrés-Lacueva C, Tulipani S, et al. Benefit of polyphenols on gut microbiota and implications in human health. J Nutr Biochem 2013;24:1415–22.

[235] Medina FX. Food consumption and civil society: Mediterranean dietas a sustainable resource for the Mediterranean area. Public Health Nutr 2011;14:2346–9.

[236] Willett WC, Stampfer MJ. Current evidence on healthy eating. Annu Rev Public Health 2013;34:77–95.

[237] Huffman MD, Capewell S, Ning H, et al. Cardiovascular health behavior and health factor changes (1988–2008) and projections to 2020: results from the National Health and Nutrition Examination Surveys. Circulation 2012;125:2595–602.

[238] Meschia JF, Bushnell C, Boden-Albala B, et al. Guidelines for the primary prevention of stroke: a statement for healthcare professionals from the American Heart Association/American Stroke Association. Stroke 2014;45:3754–832.

[239] Trichopoulou A, Martínez-González MA, Tong TY, et al. Definitions and potential health benefits of the Mediterranean diet: views from experts around the world. BMC Med 2014;12:112.

26

Action of Red Wine and Polyphenols Upon Endothelial Function and Clinical Events

Protásio Lemos da Luz, Desidério Favarato, and Otavio Berwanger

INTRODUCTION

Alcohol, and in particular polyphenols found in red wine, have potential cardioprotective actions, including vasodilation, antioxidation, and platelet antiaggregation, in addition to HDL increase. Mechanistic studies indicate that flavonoids can affect the formation and evolution of atherosclerotic plaques, but the mechanisms involved in such protective effect have only recently began to be explained. Red wine flavonoids are believed to reduce endothelin-1 (ET-1) production, block NF-κB expression and increase nitric oxide secretion by endothelial cells [1–3].

The cardioprotective effect suggested by mechanistic studies is also corroborated in epidemiological studies involving different populations. In general, large-scale observational studies consistently demonstrate the protective effect of moderate alcohol consumption against the development of atherosclerotic disease and cardiovascular morbimortality, with an estimated 20%–40% risk reduction. That is consistent with historical observations; thus, in 1819, Samuel Black observed a high incidence of CAD in autopsies performed in Ireland and a low incidence in procedures in France and Mediterranean countries [4].

A recent meta-analysis of 84 prospective cohort studies in adults without manifested cardiovascular disease (CVD) evaluated the association between alcohol consumption, major cardiovascular events and total mortality [5]. Among the studies, the median follow-up time was 11 years, and 85% of them monitored the participants for more than 5 years. The main results are in Table 26.1, where the protective effects of alcohol consumption in relation to major fatal and nonfatal cardiovascular events were observed.

In addition to the protective effect demonstrated upon major cardiovascular events, the relative risk adjusted for total mortality was (0.87 CI 95%, 0.83–0.92) comparing consumers and nonconsumers. That finding is important, as we compute deaths secondary to traffic accidents and hepatic cirrhosis. However, as suggested by many individual studies and confirmed by the meta-analysis of studies included in Table 26.2, the protective effect is stronger in lower doses. The risk adjusted for cerebral vascular accident (CVA) was significantly increased with daily consumption above 15 g.

The findings of this systematic review are consistent with observational evidence from large-scale individual studies. A recent survey involved 245,207 adults participating in the US National Health Interview Survey (NHIS) [6], an annual survey of nationally representative samples of adults in the USA, between 1987 and 2000. This survey includes detailed questions about alcohol consumption. Participants were divided into abstemious, light consumers (3 drinks or less per week), moderate consumers (4–7 drinks per week for women and 4–14 drinks per week for men) and heavy consumers (more than 7 or 14 drinks per week for women and men, respectively). Mortality was verified by combining the NHIS database with the National Death Index in 2002. Results were adjusted for several

TABLE 26.1   Relative Risk (RR) of Cardiovascular Events Among Alcohol Consumers Compared to Nonconsumers— Meta-analysis of 84 Cohort Studies [5]

Outcome	Number of studies	RR 95% CI
Cardiovascular mortality	21	0.75 (0.70–0.80)
CAD	29	0.71 (0.66–0.77)
Mortality by CAD	31	0.75 (0.68–0.81)
CVA	17	0.98 (0.91–1.06)
Mortality by CVA	10	1.06 (0.91–1.23)

CAD, coronary artery disease; *CVD*, cerebral vascular accident; *RR*, relative risk; *CI*, confidence interval.
Adapted from Ronksley PE, Brien SE, Turner BJ, et al. Association of alcohol consumption with selected cardiovascular disease outcomes: a systematic review and meta-analysis. BMJ 2011;342:d671.

covariables and information that were also recorded in the NHIS questionnaire. During the total of 1,987,439 person-years of follow-up, there were 10,670 cardiovascular deaths, including 6135 related to coronary disease and 1758 related to CVA. The results showed that, in general, moderate consumption was associated with lower cardiovascular mortality and light consumption was also associated with a better result than abstinence, while excessive consumption was not clearly related to higher or lower risk (Table 26.3).

There was little difference in risk between life-long abstemious participants, rare/sporadic life-long consumers and former consumers. In addition, the results did not show any frequency standard for alcohol consumption or excessive alcohol consumption. Additionally, the results provided some of the strongest evidence so far that the

TABLE 26.2   Relative Risk (RR) Adjusted for Outcomes by Daily Consumption vs Nonconsumption—Meta-analysis of 84 Cohort Studies [5]

Outcome	<2.5 g/d, RR 95% CI	2.5–14.9 g/d[a], RR 95% CI	15–29.9 g/d, RR 95% CI
Cardiovascular mortality	0.71 (0.57–0.89)	0.77 (0.71–0.83)	0.75 (0.70–0.80)
CAD	0.71 (0.57–0.89)	0.96 (0.86–1.06)	0.66 (0.59–0.75)
Mortality by CAD	0.92 (0.80–1.06)	0.79 (0.73–0.86)	0.79 (0.71–0.88)
CVA	0.81 (0.74–0.89)	0.80 (0.74–0.87)	0.92 (0.82–1.04)
Mortality by CVA	1.00 (0.75–1.34)	0.86 (0.73–0.99)	1.15 (0.86–1.54)

[a] *Around one drink per day.*
CAD, coronary artery disease; *CVD*, cerebral vascular accident; *RR*, relative risk; *CI*, confidence interval.
Adapted from Ronksley PE, Brien SE, Turner BJ, et al. Association of alcohol consumption with selected cardiovascular disease outcomes: a systematic review and meta-analysis. BMJ 2011;342:d671.

TABLE 26.3   Adjusted Risk Rates* 95% CI for Cardiovascular Diseases, Coronary Disease, Cerebral Vascular Accident and Mortality According to Alcohol Consumption [6]

Alcohol consumption	Nonconsumers	Light	Moderate	High
Cardiovascular deaths	1.00	0.76 (0.68–0.85)	0.67 (0.59–0.77)	0.89 (0.73–1.10)
Cardiovascular deaths adjusted for diet/physical activity	1.00	0.77 (0.69–0.85)	0.69 (0.61–0.80)	0.90 (0.73–1.10)
Mortality by coronary disease	1.00	0.75 (0.66–0.84)	0.67 (0.57–0.79)	0.80 (0.61–1.05)
Mortality by CVA	1.00	0.80 (0.61–1.05)	0.76 (0.58–0.99)	1.25 (0.92–1.70)

* *Adjusted for age, sex, race, smoking, marital status, education, region, urbanization, body mass index (BMI), and general health index.*
CAD, coronary artery disease; *CVD*, cerebral vascular accident; *RR*, relative risk; *CI*, confidence interval.
Adapted from Schoenborn CA, Adams PF, Barnes PM, et al. Health behaviors of adults: United States, National Health Interview Surveys (NHIS). Vital Health Stat 2004;219:1–79.

associations observed regarding the inverted relationship between light/moderate alcohol consumption and cardiovascular mortality can be generalized for the US population and are not restricted to thoroughly monitored volunteer groups [7].

Atherosclerosis Risk in Communities (ARIC) [8], a prospective epidemiological study in men and women aged between 45 and 64 years old, in four communities of the United States, identified 7697 participants who were abstemious at the baseline. Among them, 6% began to drink moderate amounts of alcohol, such as two drinks a day or less for men and one drink a day or less for women, during the initial 6-year evaluation. The entire group was monitored for another 4 years, and those that began to drink were compared to those that continued without drinking alcohol. The results showed that the new moderate consumers had a 38% lower chance of developing CVD defined as cardiovascular death, infarction, diagnosed coronary cardiac disease, procedure related to coronary cardiac disease, or definitive or probable CVA than their opposite persistent abstemious participants, a difference that remained after the adjustment for demographic and cardiovascular risk factors. However, there was no difference in all-cause mortality between new consumers and nonconsumers. Table 26.4 shows the risk of cardiovascular risk and mortality in new consumers compared to abstemious participants.

New consumers also showed slight improvement in HDL-cholesterol levels and no adverse effects on arterial pressure. Researchers say that this data supports the idea that taking up alcohol in one's midlife can have a positive global impact on cardiovascular health and that, for carefully selected individuals, a "heart healthy diet" can include limited alcohol consumption even among individuals who did not previously imbibe alcohol [9].

Cohort studies based on the follow-up of patients included in randomized clinical trials have been evaluating the effects of alcohol consumption upon major cardiovascular events, such as the Physicians' Health Study [10], originally a randomized study with aspirin and beta-carotene. The study included more than 88,000 male physicians, who provided self-referred information on their alcohol consumption. In this cohort's database, researchers identified 14,125 men at the baseline with a history of current or past treatment of hypertension, and who were free from AMI, cancer, CVA or hepatic disease. They were followed-up in relation to the occurrence of all mortality causes or deaths that were considered to be due to CVD. During 75,710 person-years of follow-up, there were 1018 deaths, 579 by CVD. Compared to those who classified themselves as rare consumers or abstemious participants, the individuals who reported monthly, weekly, or daily alcohol consumption had a decreasing risk of death, both for total mortality ($P < 0.001$ for a linear trend) and for mortality by CVD ($P < 0.001$ for a linear trend). Table 26.5 shows an adjusted relative risk of mortality by CVD versus alcohol consumption according to the Physicians' Health Study.

Results observed in the North American population are also consistent in other locations. In the EPIC study [11], for example, researchers evaluated the consumption of alcohol among 15,630 men and 25,808 women in Spain for a 10-year median based on their answers to a diet history validated questionnaire. The participants reported the amount of beer, cider, wine, liqueurs, and other distilled drinks they consumed per day or per week during the 12 months prior to being recruited by the EPIC. Total alcohol ingestion was calculated using the average alcohol

TABLE 26.4 Risk of Cardiovascular Disease and Mortality in New Consumers vs Abstemious Participants [8]

Outcome	OR	95% CI
Cardiovascular disease	0.62	0.40–0.95
Mortality	0.71	0.31–1.64

OR, odds ratio.
Adapted from Fuchs FD, Chambless LE, Folsom AR, et al. Association between alcoholic beverage consumption and incidence of coronary heart disease in whites and blacks: the Atherosclerosis Risk in Communities Study. Am J Epidemiol 2004;160:466–74.

TABLE 26.5 Physicians Health Study: Adjusted Relative Risk of Mortality by CVD vs Alcohol Consumption [10]

Alcohol consumption	Relative risk	95% CI
RARELY OR NEVER		
Monthly	0.83	0.62–1.13
Weekly	0.61	0.49–0.77
Daily	0.56	0.44–0.71

CAD, coronary artery disease; CVD, cerebral vascular accident; RR, relative risk; CI, confidence interval.
Adapted from Berger K, Ajani UA, Kase CS, et al. Light-to-moderate alcohol consumption and risk of stroke among U.S. male physicians. N Engl J Med 1999;341:1557–64.

content in a standard glass of any kind of drink consumed. In Spain, a "standard drink" or "glass of drink" is estimated to contain around 10 g of alcohol. The participants also provided information about lifestyle, including exercise, smoking, and other risk factors for CVDs, such as obesity and high cholesterol levels. All were free from coronary disease at the baseline. During the 10-year follow-up, there were 609 coronary events (481 in men and 128 in women) for an incidence rate of 30,056 in 100,000 person-years for men and 4793 in 100,000 person-years for women. In a multivariate analysis, the researchers found an inverted association between alcohol consumption and coronary disease in men. Moderate, high and very high alcohol consumption was associated with a reduced coronary disease risk. The HR adjusted risk rate was 0.90 (95% CI, 0.56–1.44) for former consumers—0.90 (95% CI, 0.56–1.4); light consumers—0.65 (95% CI, 0.41–1.04); moderate consumers—0.49 (95% CI, 0.32–0.75); high dose consumers—0.46 (95% CI, 0.30–0.71); and very high dose consumers—0.50 (95% CI, 0.29–0.85). Women also benefited from alcohol intake, but the effects were not statistically significant, perhaps because of the small number of coronary events they experienced [11].

Many authors argue if the potential cardioprotective effect of moderate alcohol consumption cannot be explained by the confusion bias, in particular the proper adherence to lifestyle changes in patients who moderately consume alcohol. This hypothesis was tested by a recent analysis of the Health Professionals Follow-up Study (HPFS) [12], which suggests that, even after correction for confounding factors, the effect remains unaltered. HPFS was a prospective study of 42,847 male health professionals, aged between 40 and 75 years old, who in 1986 answered the diet and clinical history questionnaire and, afterward, regularly reported follow-up information about diet, health, and lifestyle. The study cohort consisted of 8867 participants who were initially free from CVD, cancer and diabetes, and who reported the presence of four healthy lifestyle characteristics or behaviors: body mass index <25, "moderate to vigorous" physical activity of at least 30 minutes a day, abstinence from smoking, and a high diet score. Such higher scores, according to the group, reflect a large intake of vegetables, fruit, "cereal fiber," fish, chicken proteins, vegetables polyunsaturated fats and low intake of trans fats and red meat. In this analysis, a lower cardiovascular risk was observed among men who reported daily alcohol consumption of 5–30 g, corresponding to around 1–2 drinks (Table 26.6) [12].

New HPFS data suggest that men with hypertension and who drink moderately may not need to give up alcohol consumption due to cardiovascular heart risk. In this study, Beulens et al. [13] analyzed alcohol consumption, assessing every four years the incidence of nonfatal AMI, fatal coronary disease, and CVA events in 11,711 men with hypertension between 1986 and 2002. Altogether, 653 events occurred during the study period. Reasons for AMI risk were reduced with an increase in alcohol consumption, compared with people who abstained from alcohol, varying from 1.09 in individuals who drank 0.1–4.9 g of alcohol per day to 0.41 in patients who consumed 50 g or more alcohol per day

TABLE 26.6 A Relative AMI Risk in 16 Years According to the Alcohol Consumption Level in 8867 Men With Healthy Levels With Healthy Levels of Four Lifestyle Factors: Weight, Activity, Smoking, and Diet [12]

Parameter	0	0.1–4.9 g/d	5.0–14.9 g/d	15.0–29.9 g/d	>30.0 g/d
No	1.889	2.252	2.730	1.282	714
AMI prevalence%	1.5	1.5	1.0	0.6	1.3
RR 95% CI[a]	1.00	0.98 0.55–1.74	0.59 0.33–1.07	0.38 0.16–0.89	0.86 0.36–2.05
RR 95% CI[b]	1.00	0.92 0.51–1.65	0.52 0.28–0.96	0.32 0.13–0.75	0.70 0.29–1.70

[a] Adjustment for age, family AMI history, regular use of aspirin, hypertension, and hypercholesterolemia: $P < 0.04$ for the trend.
[b] Adjusted for individual body mass levels, diet physical activity, and smoking: $P < 0.04$ for the trend. AMI, acute myocardial infarction. RR, relative risk.
Adapted from Pai JK, Mukamal KJ, Rimm EB. Long-term alcohol consumption in relation to all-cause and cardiovascular mortality among survivors of myocardial infarction: the Health Professionals Follow-up Study. Eur Heart J 2012;33:1598–605.

($P < 0.001$ for the trend). The risk of stroke, however, at least in patients who consumed 10–29.9 g of alcohol per day, was not different from that of abstemious participants, with a risk rate of 1.40 and 1.55, respectively. There were no significant statistical differences between the type of beverage and AMI risk [13].

Can the improvement in surrogate biomarkres explain the potential beneficial effects of alcohol consumption? This issue was addressed by a systematic review with a meta-analysis of 44 observational studies assessing the effect of alcohol consumption upon the biomarkers [14]. No significant effect of alcohol use was observed on total cholesterol levels and LDL, triglycerides, or Lpa levels. However, "statistically significant changes in the levels of HDL, fibrinogen and adiponectin after alcohol consumption had relevant pharmacological magnitude." The effect on HDL increased with the level of alcohol intake, an increase of 0.072 mmol/L in one or more drinks per day up to 0.14 mmol/L in more than four drinks per day, $P = 0.013$ for the trend. The effects were independent from the fact of alcohol having been consumed via beer, wine, or distilled drinks [5,13]. Table 26.7 shows biomarker levels during alcohol use periods versus abstinence periods.

In women, the evidence regarding the benefits and risks associated with moderate consumption remains controversial. Despite observational studies suggesting the reduction of cardiovascular events, there is the suggestion of increase in neoplasia incidence. The risk of sudden cardiac death (SCD) was significantly reduced: 36% among women initially free from cardiac disease who reported consuming 5–15 g of alcohol per day, in comparison to those who considered themselves to be abstemious, throughout 25 years of follow-up in the Nurses' Health Study (Table 26.8) [15]. The SCD risk increased with higher levels of alcohol consumption. The prospective analysis included 85,067 women, initially without apparent chronic disease whose medical history, cardiovascular risk factors, and lifestyle, including alcohol consumption, were monitored through questionnaires every 4 years between 1980 and 2006. The relationship observed between alcohol intake and SCD risk, adjusted for several cardiovascular risk factors, was U-shaped ($P < 0.02$), while the relationship with the risk of fatal and nonfatal coronary cardiac disease was more linear ($P < 0.001$).

According to the authors, one of the study's limitations is that there was not a great number of women who drank higher doses of alcohol, thus, only 4% of the sample drank two or more drinks per day, and less than 1% drank higher amounts. Therefore, we cannot draw definitive conclusions about the association between high alcohol consumption and SCD risk.

On the other hand, according to the Million Women Study [16], conducted in the United

TABLE 26.7 Biomarker Levels During Alcohol Use Periods vs Nonconsumption Periods [14]

Biomarker	Number of studies	Average of difference 95% CI
HDL (mmol/L)	33	0.094 0.064–0.123
Apolipoprotein A1 (g/L)	16	0.101 0.073–0.129
Fibrinogen (g/L)	7	−0.20 −0.29 to −0.11
Adiponectin (mg/L)	4	0.56 0.39–0.72

Modified from Brien ES, Ronksley PE, Turner BJ, et al. Effect of alcohol consumption on biological markers associated with risk of coronary heart disease: systematic review and meta-analysis of interventional studies. BMJ 2011;342:d636.

TABLE 26.8 Relative Sudden Cardiac Death Risk for Total Alcohol Intake in Comparison With the Report of Nonconsumption of Alcohol [15]

Consumption range	Model 1[a]	Model 2[b]
Former consumers	0.78 (0.54–1.12)	0.79 (0.55–1.14)
0.1–4.9 g/day	0.67 (0.49–0.91)	0.77 (0.57–1.06)
5.0–14.9 g/day	0.54 (0.36–0.80)	0.64 (0.43–0.95)
15.0–29.9 g/day	0.58 (0.32–1.05)	0.68 (0.38–1.23)
≥30.0 g/day	1.01 (0.62–1.64)	1.15 (0.70–1.87)

[a] *Adjusted for age, caloric consumption, smoking, body mass index, family AMI history, menopause status, use of postmenopause hormonal therapy, aspirin use, vitamin supplement consumption, physical activity, ratio of polyunsaturated/saturated fat consumption, omega 3, alpha-linoleic acid, and trans-fat consumption (P = 0.002 for trend).*
[b] *Adjusted also for coronary disease diagnosis, cerebral vascular accident, diabetes, high blood pressure, and high cholesterol (P = 0.02 for trend).*
Modified from Chiuve S, Rimm E, Mukamal K, et al. Light to moderate alcohol consumption and risk of sudden cardiac death in women. Heart Rhythm 2010;7:1374–80.

Kingdom, even for low to moderate alcohol consumption, there is a significant increase in the risk of cancer, both globally and site-specific. The women in this study were 55 years old on average, and 75% said they drank alcohol, consuming an average of one drink per day (10 g of alcohol). Few drank more than three drinks per day, and there was no difference between wine and other drinks, such as distilled drinks, although most women drank wine. Consuming one drink per day increased the incidence of neoplasia in 13% in relation to that expected in the general population. Most of the events were related to breast cancer, but there was also an increase in the incidence of hepatic, intestinal, and head and neck neoplasia in women who also smoked. It is noteworthy that, much like similar studies in the field, alcohol consumption was self-reported. The researchers did not use abstemious participants as a reference group because they suspected there was a "relatively high risk in that group." The data collected did not differentiate those who had never drunk alcohol and those not drinking currently, but did so in the past. The researchers suspected that many abstemious participants were actually former consumers that had given up drinking because of disease. They found an increased cancer risk in abstemious women compared to the group that drank two drinks per week or less. Additionally, trend analyses were performed concerning the association between the amount of alcohol consumed and the risk of cancer. All estimates were adjusted for potential confusion factors, such as age, smoking, use of oral contraceptives, and hormone replacement therapy. Each additional beverage consumed per day was associated with 11 additional breast neoplasia for every 1000 women up to 75 years old; with an additional oral cavity and pharynx cancer and an additional rectum cancer; and 0.7 for other types of neoplasia, for example, esophagus, larynx, and hepatic [16].

In the CASCADE study [17], 224 participants with type 2 diabetes were randomized to receive 150 mL of mineral water ($n=83$), white wine ($n=68$) or red wine ($n=73$) during dinner, for 2 years, and all of them followed the Mediterranean diet without caloric restriction. At the end of observation, red wine drinkers had an increase of HDL (2.0 mg/dL) and polyprotein A (0.03 g/L), and reduction of the total cholesterol/HDL ratio (0.27), all statistically significant. These modifications did not occur in the other groups. Within the genetic point of view, only the slow ethanol metabolites with ADH1B*1 alcoholic dehydrogenase alleles experienced effects on glycemic control compared to the fast ethanol metabolites. In addition, the quality of sleep was improved in both wine drinking groups. There was no difference in relation to arterial pressure, adiposity, hepatic function, symptoms, drug therapeutics or quality of life, except sleep improvement.

Altogether, red wine drinkers experienced a reduction in the number of metabolic syndrome components. Therefore, this original, randomized study showed that wine, especially red wine, in well-controlled diabetic patients, is safe as part of a healthy diet and reduces the cardiometabolic risk. Genetic data also suggests that ethanol affects glucose metabolism. To date, this is the only randomized study in patients that shows the beneficial effects of red wine on metabolic syndrome components.

Di Castelnuovo et al. [18] published a meta-analysis of 26 studies that specifically evaluated wine and beer in relation to cardiovascular risk. In 13 studies with wine, in 209,418 people, the relative risk was 0.68 in relation to nondrinkers. A J-curve and a statistically significant inverted relationship were observed between the amount consumed and the risk up to 150 mL/day. In 15 studies with beer, the relative risk was 0.78, but there was no relationship with the amount ingested. The authors concluded that wine protects against cardiovascular risk. As for beer, such protection was considered uncertain, since there was no relationship with the amount ingested. Also in the Inter-Heart study [19], in which 27,000 patients from 52 countries were analyzed in relation to factors associated with myocardium infarction, it was observed that regular consumption of alcoholic beverages, in moderate doses, reduced the incidence of infarction in both sexes and in all age groups.

Trichopoulou et al. [20], in the Greek study EPIC, analyzed the contribution of different Mediterranean diet components that were associated with mortality reduction in 23,349 male and female individuals, followed for 8.5 years. They calculated that the regular intake of a moderate quantity of wine corresponded to 23.5% of the total protective effect and represented the greatest isolated contribution toward the positive effect; the combination of moderate ethanol use, low consumption of meat and byproducts, and high intake of vegetables, fruit, nuts, and olive oil was associated with protection against global mortality.

The PREDIMED study [21], carried out in Spain and monitoring 5505 individuals, men and women, for up to 7 years, documented the reduction in the incidence of depression in moderate red wine consumers, while heavy drinkers were at higher risk.

In our group [22], we performed an observational study in 101 healthy men, regular red wine drinkers, for 18 years in average, comparing them to 104 abstemious participants of the same age group. Against our expectations, the number of lesions on the anterior descending, right coronary and circumflex arteries analyzed by computed angiotomography was not different in relation to abstemious participants. However, red wine drinkers had significantly higher HDL (46.9 ± 10.9 vs 39.5 ± 9 mg/dL) and lower glycemia (97.6 ± 18.2 vs 105.9 ± 32.0 mg/dL) than abstemious participants. Associating the coronary calcium score of another population (data not shown), we observed that chronic red wine drinkers had a significantly higher coronary Ca^{++} score than abstemious participants, as shown in Fig. 26.1.

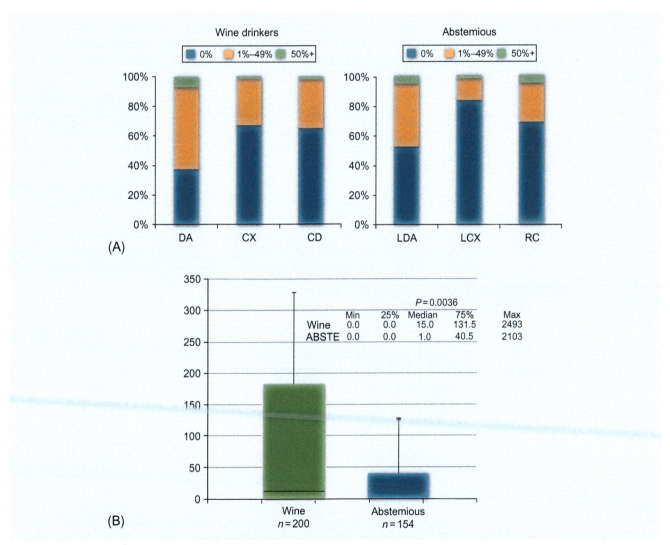

FIG. 26.1 (A) Frequency of coronary lesions by computed tomography in 101 regular red wine drinkers and 104 abstemious participants. There is no significant difference between coronary lesions in the anterior descending, circumflex, or right coronary arteries between red wine drinkers and abstemious participants. (B) This panel illustrates the coronary calcium score data in 354 individuals. There is a significant increase in the calcium score among chronic wine drinkers in comparison to abstemious participants [22]. Adapted from Da Luz PL, Coimbra S, Favarato D. Coronary artery plaque burden and calcium scores in healthy men adhering to longterm wine drinking or alcohol abstinence. Braz J Med Biol Res 2014;47:697–705.

This apparently paradoxical finding corresponds to the observations of Puri et al., published in 2015 [23]. These authors studied many individuals in eight studies treated with high doses ($n=1545$), and low doses of statins ($n=1726$), or with no statins ($n=224$). They noticed that patients treated with high statin doses showed reduction in the volume of atherosclerotic plaques, but an increase of the coronary Ca^{++} score was also noticed. The highest calcium increases correspond to the higher statin doses and plaque regressions. They interpreted such findings as an indication that, in these cases, the calcification results from the "scarring" or lesion stabilization phenomenon.

We interpreted our findings regarding red wine with a similar rationale; that is, red wine most likely leads to coronary lesion stabilization. This partially explains the several clinical documentations that moderate long-term red wine consumption reduces cardiovascular events.

EXPERIMENTAL STUDIES

Caloric restriction is known to be the only intervention capable of increasing the life span of several species, including worms, fish and fungi [24–26]. Red wine resveratrol was shown to follow the same

metabolic pathway of RC by stimulating sirtuins (SIRT2, SIRT1), which are members of the NAD-dependent deacetylase family, which participate in several metabolic processes. This process leads to the stability and repair of DNA, transcriptional silencing, and P53 regulation, resulting in increased lifespan [24].

Thus, Baur et al. [27] tested the effects of resveratrol on the afterlife of mice subjected to a hypocaloric diet. These animals had a clearly reduced survival, while those that received resveratrol along with the diet had a survival like that of controls. Therefore, resveratrol antagonized the damaging effects of the hypercaloric diet to the point of increasing life span.

We also studied the action of red wine and nonalcoholic wine, nonalcoholic wine products (NAWP) upon the formation of atherosclerotic plaques in the aorta of rabbits fed with a hypercholesterolemic diet, using Sudan IV staining [28]. Animals that received red wine in their water along with a fatty diet developed less plaques than the controls; those that received NAWP also developed less plaques than controls, but this effect was somewhat smaller than with red wine. Curiously, plaque formation reduction was not followed by plasmatic lipid reduction, suggesting that red wine and NAWP protective actions do not depend on lipids (Fig. 26.2).

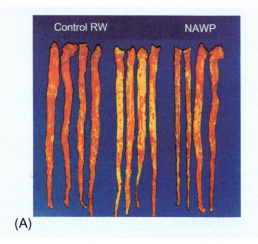
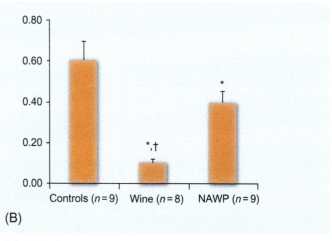

(A)　　　　(B)

FIG. 26.2　Red wine (RW) and nonalcoholic wine products' (NAWP) protective action in rabbits subjected to a hypercholesterolemic diet for three months. (A) Representative specimens of rabbit aortas stained with Sudan IV. The control group shows a virtually complete infiltration of the entire aorta surface; in contrast, the animals that received the diet plus RW in their drinking water, or diet and NAWP, showed a clear reduction of atherosclerosis plaques. (B) Analysis of the aorta intima/media thickness in the control groups, RW and NAWP. A statistically significant reduction is noticed on both groups treated in comparison to control, without significant difference between both treatment groups. $*P < 0.001$, controls vs wine; $^{\dagger}P < 0.005$, wine vs NAWP [28]. Adapted from Da Luz PL, Serrano CV, Chacra AP, et al. The effect of red wine on experimental atherosclerosis: lipid-independent protection. Exp Mol Path 1999;65:150–9.

We also analyzed the effects of red wine, and low and high doses of resveratrol in normal rats subjected to exercise [29]. Red wine and resveratrol in any dosage did not affect afterlife, but red wine and low doses of resveratrol improved the animals' physical capacity, increased the endothelium-dependent dilation of the aorta rings, and reduced the expression of P53 and P16; concurrently, they induced an increase of the telomere length in aorta homogenized components, which was associated with an increase in telomerase activity (Figs. 26.3–26.5). Therefore, red wine and resveratrol in low doses improved vascular function and reduced aging rates.

In a similar manner, Pearson et al. [30] observed in mice on normal diets that resveratrol induces the expression of genes that are similar to those induced by caloric restriction; resveratrol also reduced the

expression of aging rates, such as albuminuria, inflammation, and apoptosis in the vascular endothelium, it increased aorta elasticity, improved motor coordination, and reduced the formation of cataracts. However, survival did not increase. Thus, the authors concluded that mice fed with a normal diet, and which received resveratrol since they were 12 months old, obtained several functional benefits, but not an increase in life span.

Barger et al. [31] also showed, in mice, that resveratrol mimics caloric restriction in terms of gene transcription in the heart, skeletal muscle, and brain. Specifically, dietetic resveratrol mimicked the effects of caloric restriction in glucose capture mediated by insulin in the muscle. They concluded that resveratrol can retard aging by alterations in the chromatin structure and in transcription.

Other many beneficial effects of red wine and polyphenols were documented in bench studies or animal experimentation, in different species [32–35].

In summary, many experimental studies in several models and species, and experiments in humans as well, suggest cardiovascular benefits of red wine and especially polyphenols, among which resveratrol stands out. The mechanisms involved in these phenomena will be discussed later on.

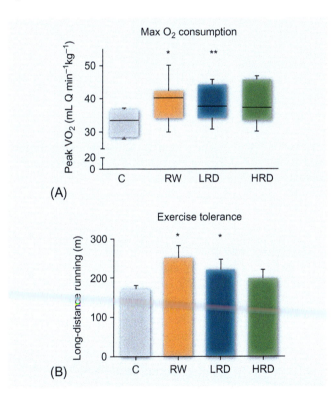

(A)

(B)

FIG. 26.3 Aerobic capacity and response to exercise in rats that consumed red wine (RW) and resveratrol in low (LRD) and high (HRD) doses (A) Max VO_2 was significantly increased in animals that received RW and both doses of resveratrol in comparison to controls. (B) Tolerance to exercise significantly increased only with RW and low doses of resveratrol [29]. Adapted from Da Luz PL, Tanaka L, Brum PC, et al. Red wine and equivalent oral pharmacological doses of resveratrol delay vascular aging but do not extend life span in rats. Atherosclerosis 2012;224:136–42.

RESVERATROL BIOAVAILABILITY

This is a major and quite controversial issue [36]. Several in vitro and in vivo observations point to the fact that bioavailability is dose-dependent. Experimental studies show that in vitro effects are obtained with much larger doses than what would have been obtained with in vivo oral doses [35]. On the other hand, there is a discrepancy between bioavailability, which is low, and the uniformly beneficial experimental effects [37]. While attempting to explain this seeming paradox and increase the bioavailability, many alternatives were tested, including the increase of intestinal absorption, with combinations such as multiple polyphenols (food and drink and liquid formulations); these combinations clearly increased bioavailability when compared to ingestion alone.

Different formulations, such as the use of nanoparticles and intravenous infusion, were also tested. The latter clearly increases bioavailability, but it is not practical. In parallel, there is sufficient research showing that small doses of resveratrol are beneficial.

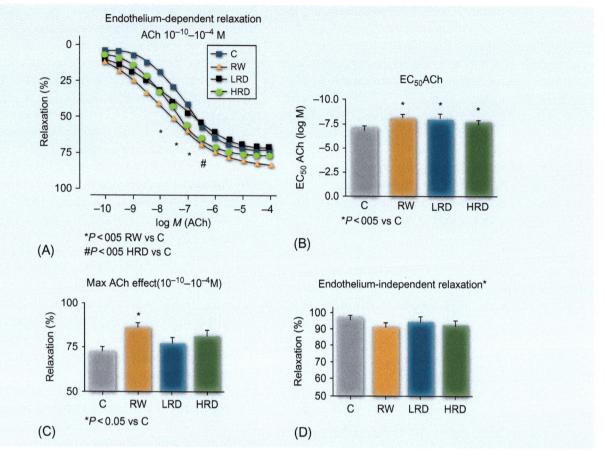

FIG. 26.4 Vascular function in aorta rings of normal rats. (A) Flow mediated dilation significantly increased after RW and low doses of resveratrol. (B) E50 increased with RW and with both doses of resveratrol. (C) Max relaxation increased only with RW. (D) Endothelium-independent dilation was not affected by any treatment [29]. Adapted from Da Luz PL, Tanaka L, Brum PC, et al. Red wine and equivalent oral pharmacological doses of resveratrol delay vascular aging but do not extend life span in rats. Atherosclerosis 2012;224:136–42.

For example, resveratrol increases eNOS activity from 0.1 to 1 µM in human endothelial cells with only 2 min of incubation. Resveratrol also increases the activity of AMPK via SIRT-1 to a 3-µM concentration [35]. Another study in men documented an increase of brain flow in resveratrol doses of 5.65–14.4 ng/dL, and 0.025 and 0.061 µM [38].

The low bioavailability and beneficial effects paradox was also addressed by Andreadi et al. [39]. Resveratrol ingested is extensively metabolized in sulfate and glucuronide conjugates, which limits its plasmatic bioavailability. The authors showed that resveratrol sulfate actually comprises an intracellular resveratrol reservoir. They studied cancerous human colorectal cells incubated with resveratrol doses compatible with clinical doses. They noticed that larger sulfate concentrations occurred in the HT-29 line; in these they also observed the greater antiproliferative effect, which

occurred by autophagy. Sulfate also increased acid beta-galactosidase associated with the senescence in these cells, indicating that they were undergoing an aging process.

These authors concluded that resveratrol itself, and not its metabolites, is responsible for the autophagy and senescence of cancer cells. Thus, the findings could also explain the beneficial effects of resveratrol, even with low plasmatic bioavailability.

POLYPHENOLS VS ALCOHOL

Several studies show the benefits of alcoholic beverages, especially red wine, upon cardiovascular risk. Red wine supremacy is especially attributed to its polyphenolic compounds, especially resveratrol. Polyphenols also explain why red wines are

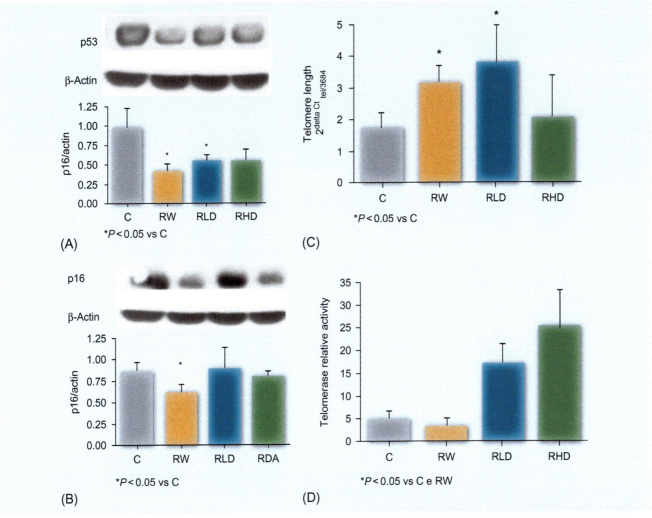

FIG. 26.5 Cell aging biomarkers in aorta homogenized components of normal rats. (A) P53 was reduced by RW and resveratrol in low dosage. (B) P16 was also reduced by RW, but not by resveratrol. (C) Telomere length increased with RW and low dosage of resveratrol, but not with high doses. (D) Telomerase was increased by both doses of resveratrol, but not by RW [28]. C, controls; RW, red wine; RLD, resveratrol, low dose; RHD, resveratrol, high dose [29]. Adapted from Da Luz PL, Tanaka L, Brum PC, et al. Red wine and equivalent oral pharmacological doses of resveratrol delay vascular aging but do not extend life span in rats. Atherosclerosis 2012;224:136–42.

better than white (in these they are found in smaller amounts). The studies of Gronbaek et al. [40], Klasky et al., Renaud et al. [41], and Dr. Castelnuevo's meta-analysis [18] showed the supremacy of wine over other alcoholic beverages. However, an American study [42] does not record differences.

Alcohol itself has some beneficial effects, among which the most important seems to be HDL increase [22]; alcohol could also increase the intestinal absorption of red wine polyphenols. Thus, clinical studies are altogether inconclusive [32].

Nonetheless, when mechanistic studies are analyzed, the data favors polyphenols. For example,

resveratrol mimics caloric restriction; polyphenols are inhibitor of platelet aggregation and vasodilators by stimulating the production of eNOS and nitric oxide; they stimulate the production of sirtuins and modulate glucose metabolism [24,26,34]. Data analysis becomes complicated because red wine drinkers can have a healthier lifestyle, with a healthy diet, more exercise, and less smoking.

In summary, the set of mechanistic and clinical evidence seems to favor polyphenols as the main agents responsible for the positive effects; but a supporting effect of alcohol cannot be neglected.

WINE, RESVERATROL, AND COGNITIVE FUNCTION

Reduction of the cognitive function is a natural consequence of advanced age, and it typically starts at around 60 years [43]. With the population's progressive aging, this phenomenon takes on considerable proportions, with major personal and socioeconomic consequences. The current predominant concept is that, although this decrease is a natural phenomenon, there is room for such reduction to be decreased; physical and mental exercises seem to be the interventions that best protect. Resveratrol, due to its vasodilator actions, endothelial function improvement, mitochondrial biogenesis induction, and effects on glycemic metabolism have chemical characteristics capable of protecting the brain against cognitive function deterioration.

There is experimental evidence that, in mice with induced brain lesion, resveratrol increases the brain flow to the affected areas and improves functional recovery [39]. in vitro studies documented that resveratrol and epigallocatechin modify the reticular structure of an amyloid particulate, transforming them into complexes that are not absorbed by neurons, and would therefore protect against Alzheimer's [42]. In addition, wine has been clinically observed to protect against Alzheimer's and dementia [44–46].

A study in elderly women [47] documented that light consumption of alcoholic beverages partially antagonized the decrease in cognitive function. That is corroborated by the findings of Mukamal et al. [48], who observed in moderate drinking elders a reduction in cardiovascular events, plasmatic fibrinogen reduction, and other thrombotic factors in comparison to nondrinkers. They also noticed preservation of the cerebral vasculature and reduction of subclinical CVAs. Mukamal et al. [45] noticed a "sharp reduction of vascular dementia and Alzheimer's among people who consumed 1 to 6 drinks/week."

In turn, a study by Kennedy et al. [38] in normal individuals, who received resveratrol, showed an increase of cerebral flow to the frontal region, but without modifications of the cognitive function examined in standard psychological tests.

We studied cognitive function in individuals with dementia and regular red wine drinkers, comparing them to abstemious participants, both subjected to standardized physical exercise during 3 months.

Individuals from both groups were evaluated by functional cerebral magnetic resonance (FCMR) and by standardized psychological tests. We observed in FCMR that there is a different BOLD signal pattern (magnetic resonance signal dependent on cerebral blood oxygen concentration) among the groups, a finding that can represent a different cerebral activation pattern. In the preliminary evaluation of results in relation to operational memory, a greater BOLD signal was observed in individuals who had lower systemic VO_2 during the cardiopulmonary test (unpublished data).

Altogether, current evidence does not allow concluding that red wine or nutrition polyphenols effectively protect against cognitive function deterioration in men. However, this field certainly deserves further investigation.

J CURVE

The relationship between amount of alcoholic beverage ingestion and general mortality or cardiovascular events has been constantly seen as a J curve. That is, abstemious individuals and those who ingest more than 30 g of alcohol/day experience greater mortality than those that consume small/moderate amounts of alcoholic beverages. Consuming more than 30 g of alcohol/day also causes increases of arterial pressure and hepatic enzymes; and a heavier intake increases mortality in direct proportion to the dose [32].

In contrast, drinking 1–2 drinks/day reduces the risk in CAD, especially mortality, hospitalization and angina, in both sexes [18]. Small daily doses seem to have a higher protective effect than a single dose equal to the day. Individuals who have had an infarction are also protected by moderate intake for having reduction in new coronary events. The J curve is independent from the type of beverage consumed.

In a careful review, O'Keefe et al. [49] observed a clear J relationship for all-cause mortality, in both men and women, and the same was observed for CVA. The authors also noticed a protective effect of consuming 1.50–29.9 g alcohol/day in 8867 middle-aged men that had a healthy lifestyle; but lower or higher intakes did not induce protection.

In summary, the many clinical studies demonstrated that those who do not consume any alcoholic beverages, and at the other extreme, those that do so in excess, do not obtain advantages. Specifically, excess alcohol is associated with the reduction of the left ventricle's ejection fraction, left ventricular progressive hypertrophy, greater risk of hemorrhagic cerebral accident, and dementia.

ADVERSE EFFECTS OF ALCOHOL

Unlike the beneficial effects of wine, alcohol can cause several adverse effects, depending on the dosage and individual susceptibility. The causes of increased mortality in heavy alcohol consumers include liver cancer, cirrhosis, cancers of the digestive and respiratory tracts, mouth, esophagus, larynx and pharynx, suicidal tendencies, accidents, homicides, and cardiovascular events.

Harmful effects of alcohol should not be underestimated. Mortality is not the only problem to be considered. Alcohol consumption can cause gastrointestinal neoplasia, both fatal and nonfatal, cardiac arrhythmias, such as atrial fibrillation and ventricular extrasystoles in susceptible individuals. Car accidents, industrial accidents, chronic alcoholism, psychosis, unemployment and sexual harassment have also been documented [50,51]. A special case is that of teenagers that can easily exaggerate in drinking, particularly due to the "zebra" type effect. Thus, individuals that excessively consume alcohol can be part of a specific socioeconomic context, with smoking, drug use, and inadequate diet; in this situation, it is not easy to single out the effects of alcohol. Some people should refrain: those who have a family history of alcoholism, hepatic diseases, uncontrolled hypertension, pregnant women and individuals that use medication that interferes in alcohol metabolism (sedatives, antidepressants). Contraindications also include cardiac arrhythmias, myocardiopathies, ventricular dysfunction, cardiac insufficiency, decompensated diabetes, alcoholism, and hypertriglyceridemia. As seen previously, Allen et al. [16] observed an increase in the incidence of several cancers, even with low intake of alcoholic beverages; they calculated that approximately 13% of breast, digestive tract, liver, and rectum cancers could be attributed to alcohol. Thus, in an editorial along with the report, Lauer and Sorlie [52] recommend caution in the interpretation, especially because the study has important limitations. For example, the study participants were all seen in a clinic for cancer diagnosis; therefore, they cannot represent the general population; also, the data is all based on questionnaires, and not from objective determinations; finally, there is no information about total mortality or cardiovascular events. Therefore, the study raises a potentially important problem, but conclusions cannot be taken as definitive ones.

RESVERATROL AND LONGEVITY

Resveratrol increases the life span of several species, including worms, insects, fungi, and fish [24,25]. In mammals such as mice and rats, this polyphenol positively affects factors that protect against vascular aging, inducing P53 reduction, telomere length, and telomerase increase. The issue of human aging is more complex [52]. For example, a study in the Greek island of Ikaria [53] documented that 13% of 1420 people examined were more than 80 years old, and that 6–10 individuals aged more than 90 years old were physically active. This data is above the European population average. Therefore, longevity stands out not only by the number of years, but also in the excellent quality of life of long-lived people. Several factors are believed to contribute to such longevity, including a diet rich in fruit and vegetables, the regular exercise of the people that inhabit this mountainous region, air quality, close family ties and the behavior itself of people that seem to deal with low levels of emotional stress; such habits include after-lunch rest, the traditional "siesta." Among dietary factors, wine consumption is included. In Brazil, the studies of Moriguchi et al. [54] in the town of Veranopolis also pointed out the large number of 80-year-olds with a great quality of life. These authors studied 213 individuals over the age of 80. They observed that several traditional risk factors are not associated with cardiovascular mortality, while arterial pressure identified people with a higher risk of cardiovascular death. Alcoholic beverage intake did not correlate with total or cardiovascular mortality.

In summary, clinical studies available so far do not warrant the assertion that consuming any alcoholic beverage is associated with greater longevity. The presence of several factors that interfere in longevity does not allow us to single out any of them as the responsible one. It seems that the factors in combination are responsible for the increase in survival.

BASIC ACTION MECHANISMS OF WINE AND POLYPHENOLS

Actions in the Endothelium

Endothelial dysfunction can be assessed via a noninvasive method by the measurement of flow-mediated vasodilation in upper limb arteries by ultrasound, through the technique of transient occlusion, via artery upstream pneumatic compression. A dysfunctional endothelium is already detected in the mere presence of risk factors, such as arterial hypertension, obesity, hypercholesterolemia, low HDL, diabetes and smoking, even before clinical manifestations of atherosclerosis. Endothelial dysfunction is not only a coronary disease marker, but also a good indicator of future cardiovascular events [55–59].

The efficacy of red wine in the improvement of flow-mediated dilation in hypertensives [60,61], smokers [62,63], diabetics [64], metabolic syndrome [65], and hypercholesterolemics [66] has been well demonstrated.

Chen et al. [67] demonstrated that polyphenols, resveratrol, and quercetin, found in red wine and red grape juice, induced vasodilation in the arterial aortas, which was reverted by L-nitroarginine, a nitric oxide synthase inhibitor, only when polyphenols were added in low doses, without however blocking it in high concentrations, assuming indirect and direct vasodilator effects of these substances.

Andriambelon et al. [68] and Cishek et al. [69] also showed such effect in rat aorta rings. The same was demonstrated by Flesh et al. in human coronaries in vitro [70].

Stein et al. [71] analyzed the effects of consuming grape juice for two weeks by CAD patients and observed an improvement of FMD and an increase in the resistance to LDL oxidation. Acute ingestion of wine or its dealcoholized product also increased FMD in men without coronary disease. Likewise, chronic ingestion of fermented alcoholic beverages increased FMD in that same type of patient, as shown by several authors, including Teragawa [72], Whelan [73], and Lekakis [74], with their respective collaborators.

Among us, Coimbra et al. [66] demonstrated that consuming grape juice or wine for 2 weeks improves flow-mediated vasodilation, endothelium-dependent vasodilation, in hypercholesterolemic patients.

Suzuki et al. revealed that individuals who had from one dose per month to two daily doses of an alcoholic beverage had higher flow-mediated dilation than abstemious participants and heavier drinkers [75]. Several others proved such efficacy.

Next, we shall analyze the mechanisms by which wine and polyphenols lead to the improvement of FMD and other dysfunctional endothelium actions.

Fig. 26.6 shows the polyphenols' actions upon the cellular signaling systems.

Mechanism of Action Upon the Endothelial System

ET-1 was initially described as a powerful vasoconstrictor, its super production is related to vascular disease and atherosclerosis [76]. In experimental models, its inhibition prevents the start of atherogenesis, by avoiding endothelial dysfunction and the formation of a fatty stria; even in established atherosclerosis its blockage reduces the incidence of infarction. In ACD patients, the local super production of ET-1 is known to reduce coronary flow sharply [77,78].

ET-1 acts in two receptors: ETA (vasoconstrictor and growth inducer) and ETB (vasodilator and cell growth inhibitor) and, also, in the clearance of ETA receptors from the membrane surface, both of them activating families of signaling proteins G. The ET-1 gene is expressed in endothelial cells, cardiomyocytes, hepatocytes, renal collector duct cells, neurons, and keratinocytes. Receptors are expressed in smooth muscle cells, cardiomyocytes, hepatocytes, neurons, osteoblasts, keratinocytes, and adipocytes, while the ETB receptor is expressed in endothelial cells, smooth muscle cells, cardiomyocytes, hepatocytes, renal collector duct cells, neurons, osteoblasts, keratinocytes, and adipocytes [79,80].

ET-1 contributes to the pathogenesis of salt-sensitive hypertension in animals and humans [81] and secondary hypertension with low renin [82]. These effects can be due to the direct vasoconstrictor effect and due to the increase in superoxide production via ETA receptor and NADPH oxidase [83,84]. ET-1 is increased not only in arterial hypertension, but also in hypercholesterolemia, hyperglycemia, and metabolic syndrome [85–87].

Ethanol alone can reduce the levels of ET-1 [88]. However, polyphenols in wine, green tea, and

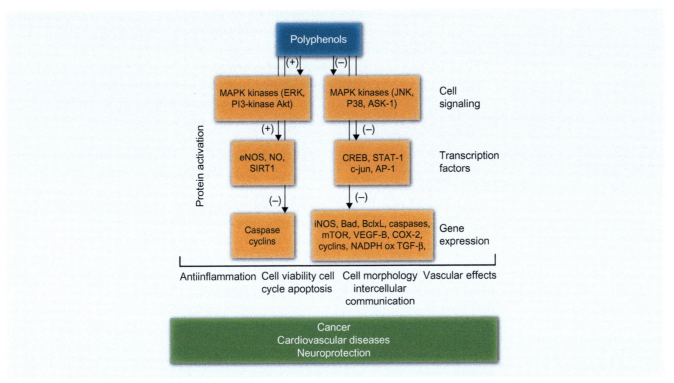

FIG. 26.6 Metabolic pathways of polyphenol actions that can affect clinical situations. *AP-1*, activator protein 1; *ASK 1*, apoptosis signal-regulating kinase 1; *Bad*, death promoter associated with Bcl-XL/Bcl-2, *Bcl*, antiapoptotic protein family; *COX-2*, cyclooxygenase 2; *CREB*, cyclic AMP responsive element binding protein; *eNOS*, endothelial nitric oxide-synthase; *ERK*, extracellular signaling regulated-kinase; *iNOS*, inducible nitric oxide-synthase; *JNK*, c-Jun amino-terminal kinase; *MAPK*, mitosis activator protein kinases; *mTOR*, rapamycin target in mammals; *NADPH ox*, NADPH oxidase; *NO*, nitric oxide; *STAT*, signal transducer and transcription activator; *TGF-b*, growth and transformation factor b; *SIRT-1*, sirtuin 1.

extra-virgin olive oil also reduce ET-1 synthesis in endothelial cells by ET-1 gene transcription suppression and increase in nitric oxide synthase activity [89–91].

Nicholson et al. [92] demonstrated that diet polyphenols reduce endothelin gene expression; other authors demonstrated that wine polyphenols (purified quercetin and epicatechin) reduce the ET-1 concentrations by action upon the cell's redox balance [89–91].

Corder [93] and Khan [94], with collaborators, showed such effects in bovine aorta endothelial cells. According to Khan, these actions were caused by signaling modifications of tyrosine kinase. Resveratrol inhibits the secretion of endothelin-1, the levels of ET-1 mRNA and endothelin promotor gene activity for interfering with the ERK1/2 pathway and diminishing the formation of reactive oxygen species [95–97]. In addition, quercetin and epicatechin inhibit the release of ET-1 [91,96–98]. Isoflavones, such as genistein, also improve FMD in menopausal women and reduce the levels of endothelin-1, improving the nitric oxide/ET-1 ratio [99].

This hypertension model induced by DOCA (deoxycorticosterone acetate) and salt, resveratrol an apocynin reduced the plasma levels of ET-1 and the hyperexpression of the p22phox gene in the aorta with FMD improvement [100].

Actions on NO Synthase and Nitric Oxide

Endothelial nitric oxide synthase (eNOS), in addition to being expressed in endothelial cells, is also expressed in cardiomyocytes, platelets, certain brain neurons, in the syncytiotrophoblasts of human placenta, and renal tubular epithelial cells [101,102]. Calmodulin activated by calcium is an important eNOS activity regulator and increases its activity. Several other proteins also interact with eNOS and regulate its activity, such as heat shock protein 90 (HSP 90), which binds allosterically and activates it through recoupling [103].

The eNOS fraction located on the caveolae interacts with the protein that covers it, caveolin-1, becoming inactive. Calmodulin and HSP90 bind to caveolin-1 and dislocate eNOS, activating-it [104].

Nonetheless, there are ways to activate eNOS that are nondependent of calcium concentration, such as shear stress that activates the enzyme via phosphorylation [105]. That mechanism is also exerted by estrogen and by the vascular endothelium growth factor (VEGF).

Nitric oxide is one of the main vasoprotector molecules, for in addition to being a vasodilator, it has antiatherosclerotic activities, such as inhibition of platelet aggregation, of leukocyte adhesion, of smooth muscle cell proliferation and atherogenic gene expression, such as the chemoattractive protein-1 (MCP-1), vascular cellular adhesion molecule-1 (VCAM-1), and intercellular adhesion molecule-1 (ICAM-1) [106–109]. Such adhesive molecules are related to leukocyte adhesion and migration through the vascular wall. In addition, endothelial permeability reduction decreases the LDL flow into the arterial wall, it reduces its oxidation and, therefore, adds more antiatherogenic effects [110–119].

Hypercholesterolemia, diabetes mellitus, arterial hypertension and smoking are associated both to nitric oxide synthesis reduction or degradation increase. This nitric oxide bioavailability reduction is followed by endothelial dysfunction with an alteration in vasomotricity and the proatherogenic state.

Red wine polyphenols sharply increase the expression and activity of nitric oxide synthase and, hence, nitric oxide release [120]. Resveratrol increases the eNOS promoter activity (transcriptional effect) and stabilizes eNOS' mRNA (posttranscriptional effect) [121,122].

Resveratrol, besides increasing nitric oxide production, by incrementing eNOS expression, reduces NADPH oxidase activity on arterial walls. The mechanism suggested is the activation of structures related to the cell membrane, such as estrogen receptors that trigger a cascade of signaling pathways whose target is the proteins' AMPK kinase activated by AMP and eNOS activation by serine phosphorylation 1177 [123]. This pathway could also activate SIRT1, causing acetylation decrease and also activating eNOS. Insulin increases nitric oxide bioavailability via this pathway [124–126].

Wine polyphenols relax aorta rings by strengthening nitric oxide synthase and increasing eNOS

expression, and not by the increase of nitric oxide's biological effectiveness or by protecting it from superoxide action [126,127]. Huang et al. compared the effects of water, red wine, beer, and vodka on endothelial function, determined by flow mediated dilation. Only red wine improved endothelial function and increased nitric oxide levels in the plasma [128–130].

The extract of red wine polyphenols causes the vasodilation of aorta rings previously retracted by norepinephrine due to a sharp nitric oxide increase, an effect shared by anthocyanin delphinidin, but not by malvidin, cyanidin, quercetin, catechin, and apicatechin [131]. Red wine extract also elevates intracellular concentration of ion calcium, which is the main signaling pathway of nitric oxide production by wine polyphenols [132,133].

Several authors replicated these polyphenols' effects from several sources, such as different wines, cocoa, tea, hawthorn, and maritime pine bark, in isolated animal or human vessels. All of them produced endothelium-dependent vasodilation with an increase of cyclic GMP and blocked by eNOS inhibitors [70,134–142]. In addition, Burns et al. found that this effect was strongly related to polyphenols' concentration in wine [143].

The administration of wine phenolic compounds reduced arterial pressure in rats. This effect was due to an increase in the expression of nitric oxide synthase and cyclooxygenase genes in the arterial wall [144].

The observation that white wine, poor in resveratrol, also has cardiovascular protective effects, lead some researchers to investigate the role of caffeic acid and tyrosol, abundant in white wine, in flow-mediated vasodilation [145–147]. Migliorini et al. demonstrated in human endothelial cells that caffeic acid, but not tyrosol, increased the production of nitric oxide under acetylcholine stimulation [148]. Resveratrol increases nitric oxide production by increasing eNOS expression and reducing NADPH oxidase on arterial walls. The mechanism suggested is the activation of structures related to the cell membrane, such as estrogen receptors that trigger a cascade of signaling pathways whose target is the AMPK protein kinase activated by AMP and eNOS activation by serine phosphorylation 1177 [123]. This pathway could also activate SIRT1, causing acetylation decrease and, also, activating eNOS. This is the same pathway through which insulin increases nitric oxide availability [124–126].

Epigallocatechin increases the concentration of calcium in cytosol, which activates several calcium-dependent enzymes, including calmodulin-dependent protein-kinase II and CaMKKb calcium/calmodulin-dependent protein kinase kinase [149,150], the latter one of the initiators of the AMP/AMPK dependent cascade, an enzyme that has a key role in energetic metabolism [151–153]. This increase of cytosolic calcium leads calmodulin to release caveolae nitric oxide synthase and increase nitric oxide synthesis [154–157].

ENDOTHELIAL PROGENITOR CELLS

Endothelial progenitor cells are mononuclear cells originated from the bone marrow and have several reparative functions for the dysfunctional endothelium, in the neovasculogenesis of ischemic tissues and in the tumoral microenvironment [158–160]. Endothelial progenitor cells are also susceptible to oxidative stress [161–166].

Hill et al. [167–169] showed that progenitor endothelial cells have several endothelial characteristics, such as the expression of CD31, TIE2 receptor of tyrosine-kinase, and receptor 2 of vascular growth endothelial growth factor [167–169].

Hill et al. [167] revealed that the number of endothelial progenitor cells was reduced in hypercholesterolemia, hypertension, and diabetes.

Others also found this inverted correlation between risk factors and number of circulating endothelial progenitor cells and associated this with the prognosis in this situation [170–173]. Low resveratrol concentrations increase the proliferation, migration, and adhesion of progenitor cells in culture and increased the expression of eNOS mRNA [174].

Balestrieri et al. [175] revealed an increase of endothelial progenitor cells with an addition of red wine to the diet of mice subjected to physical exercise, which was also observed in young adults with red wine [176,177] and in smokers, with the use of green tea [62]. The most accepted mechanism is via the increase of nitric oxide bio-availability [177].

Resveratrol reduces the senescence of the endothelial progenitor cell through the increase of telomerase activity [178].

All polyphenols, such as puerarin [179,180], wine resveratrol [174,175,181,182], ginkgo biloba [183], berberine [184], salvianolic acids [185], and ginsenosides [186], increased the endothelial progenitor cells bioactivity.

COAGULATION AND PLATELETS

While moderate alcoholic beverage consumption is followed by the decrease of several factor VII coagulation factors, tissue factor, fibrinogen and von Willebrand factor, viscosity and increase of the fibrinolytic capacity, its heavy consumption has the opposite effects. Consumption of beer and distilled drinks, but not wine, increases the PAI-1/tPA ratio. Some authors showed that alcohol and some polyphenols, such as catechin and quercetin, increase t-PA transcription. These actions can be associated with the observations of venous thrombosis reduction with moderate use of wine [187].

Due to its eNOS activator and COX-2 inhibitor activities, wine consumption has antiplatelet activity [187].

Resveratrol inhibits platelet aggregation induced by thrombin, collagen, PAF (platelet activation factor) and ADP [108]. Resveratrol hinders platelet aggregation by thromboxane B2 due to its inhibition of protein kinase C pathway [188]. Quercetin also hinders aggregation induced by thrombin and ADP, and platelet activation by peptide agonists-6 (thrombin receptor activator), arachidonic acid, ADP, epinephrine, collagen, and ristocetin, immobilization by calcium, granule secretion and fibrinogen binding [189]. In addition, quercetin inhibits platelet adhesion to vascular endothelial cells and platelet aggregation stimulated by collagen by inhibiting the glycoprotein VI signaling pathway [190–204].

ANTIINFLAMMATORY EFFECTS

In experimental studies, alcohol suppresses proinflammatory cytokines synthesis, such as TNF-α, IL-1b, IL-6, IL-8, and MCP-1 in alveolar macrophages and monocytes [205–208]. Such effects have also been observed by red wine use in individuals with high risk of developing CVD, which increased interleukin-10, which is antiinflammatory, and reduced interleukin-6, which is inflammatory [209].

The intake of wine, beer or distilled beverages has been associated with reductions in reactive C protein, fibrinogen, viscosity, and leukocyte count in elderly individuals [210].

The expression of adhesion molecules VCAM-1 and ICAM-1 is also reduced with moderate red wine ingestion [211].

Inflammatory stimuli, such as TNF-α and lipopolysaccharides and adhesion molecules, promote IKB phosphorylation (nuclear factor kappa B inhibitor), molecules that maintain NF-κB inactive in cytoplasm, enabling is ubiquitination and degradation, releasing the NF-κB subunits, p50 and p65, the latter being the most active, to migrate to the nucleus and begin proinflammatory gene transcription, related to apoptosis, cellular cycle regulation, cellular invasion, and metastatic growth [212]. In regard to atherosclerosis, NF-κB promotes the transcription of adhesion molecules, growth factors, and intercellular matrix metalloproteinases.

NF-κB is regulated by the redox state, the inflammatory stimuli augment the generation of reactive oxygen species, which activate the phosphorylation pathways that lead to IKB ubiquitination and NF-κB release to the nucleus.

N-acetylcysteine inhibits phosphorylation in serine 536 of unit P56, without acting upon the IKK/NF-κB complex, while red wine does not act in this site, blocking IKBα serine 32 phosphorylation, with the same final effects of blocking NF-kB activation by TNF-α [213,214].

LIPIDIC EFFECTS

Among the several phenolic compounds and resveratrol effects are: cholesterol efflux increase, HDL augmentation, oxidized LDL reduction, and decrease in foam cell formation. Meta-analyses revealed that moderate alcoholic beverage intake increases HDL between 8% and 9%, reduces LDL in 11%, and does not alter total cholesterol nor triglycerides. There is increase of apolipoprotein A-I between 1% and 7% [215,216].

Some authors found difference in the impact of the type of beverage on the lipid profile. Ethanol alone reduces ApoB concentration, while red wine, but not gin, increase ApoA-I and ApoA-II in healthy volunteers. However, other authors that evaluated a

high-risk population for CVD found that ethanol increases not only HDL, but also ApoA-I and ApoA-II [217–222].

Among the mechanisms proposed for HDL increase are the increase of the transportation rate and augmentation of the lipoproteic lipase activity [223–225]. Another of wine and polyphenols' action mechanism is the inhibition of intestinal cholesterol absortion by the Nieman-Pick C1-like cholesterol transporter, the same action site of ezetimibe [226].

In addition, there is also an increase in the expression of LDL receptors in the hepatocytes by proteolytic activation of ligand proteins to the SREBPs sterols regulator element [227].

GLYCEMIC METABOLISM

Two meta-analyses demonstrated that moderate alcoholic beverage consumption has a protective effect against diabetes mellitus type 2 [228,229].

Fig. 26.7 shows the action pathways of polyphenols that influence glycidic metabolism.

This beneficial effect is related to insulin sensitivity improvement with a corresponding insulinemia reduction [230,231].

Diabetes and hyperglycemia are related to premature atherosclerotic disease and accelerated by the

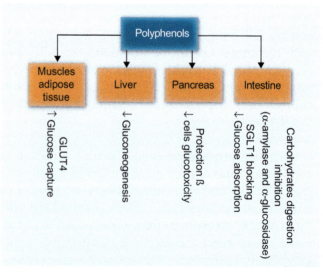

FIG. 26.7 Schematic representation of the action pathways of polyphenols that influence glycidic metabolism. *GLUT4*, glucose transporter type 4; *SGLT1*, glucose transporter protein coupled to Na⁺.

lipid alterations that follow them—VLDL concentration augmentation and HDL decrease and qualitative alteration of lipoproteins by their oxidation. Lipooxidation is exacerbated in diabetes by several prooxidant systems, including the augmented formation of advanced glycation end products (AGEs), protein kinase C delta (PKC δ) activation and increase of the NADPH oxidase activation in macrophages [232]. Along with these oxidative alterations, cellular depletion of antioxidants occurs, such as the hydrolyzer enzyme of oxidized lipids, paraoxanase 1 PON1, reduced glutathione GSH, and vitamins C and E [233–235].

In diabetes, there is a key alteration in the insulin cellular signaling sequence. Instead of the normal phosphorylation pathway in IRS-1 tyrosine that activates the PI3 kinase pathway, which increases glucose capture, the synthesis of glycogen, protein and lipids, and eNOS phosphorylation, there is phosphorylation in serine that leads to the activation of the MAP kinase pathway, which promotes growth signaling, proliferation, differentiation, inflammation, and gene expression [236–239].

AGEs are prooxidants formed by the elevation of glucose concentration and by the nonenzymatic reaction of sugars reduction and amines of proteins, aminolipids, and nucleic acids [240]. This process decisively alters proteins of the connective tissue, plasmatic lipoproteins, membrane phospholipids, and DNA. The activation of AGEs' receptors (RAGEs) leads to the oxidative process amplification [241–244].

The AGEs, both in normal individuals and in diabetics, increase oxidative stress, increase TNF-α, VCAM-1, and PCR inflammation, increase the resistance to insulin, aggravate vascular dysfunction, reduce adiponectin, and increase deacotylation of subunit p65 of NF-κB, that is, its activation [245–258].

Wine polyphenols offset this abnormal pathway, inhibiting it and activating the physiological pathway [259]. In animal models and some studies in humans, polyphenols and drinks rich in polyphenols reduced fasting hyperglycemia and the postprandial glycemic peak, and improved acute secretion and sensitivity to insulin. The mechanisms considered are the inhibition of carbohydrates' digestion and glucose absorption in the intestine, stimulation to insulin secretion, hepatic neoglucogenesis modulation, receptor activation, and

glucose capture in the insulin-sensitive tissues and the modulation of signaling pathways and cellular gene expression.

In metabolic syndrome and type 2 diabetes mellitus, several nutrition sources rich in polyphenols have been studied, among them, isoflavone-rich soy [260]; epigallocatechin-rich tea [261,262]; phenolic-rich coffee [263]; grapes(particularly due to resveratrol) [264]; apples (due to flavonoids) [265]; and several other herbs [266].

Several polyphenols inhibit alpha-amylase and alpha-glycosidase, including flavonoids—anthocyanins, catechins, flavanones, flavonoids, flavones, and isoflavones; the phenolic acids and tannins—proanthocyanins and ellagitannins. This effect was verified by the fact that polyphenols consistently reduce glycemia after maltose and amide ingestion, but not all of them inhibit hyperglycemia after glucose ingestion [267–282].

Several polyphenols also inhibit glucose inhibition. SGLT-1 is inhibited by the chlorogenic, ferulic, caffeic, and tannic acids [283]; quercetin [284], catechins [285–287], and naringerin [288].

GLUT-2 transport, in turn, is inhibited by quercetin, myricetin, apigenin, and catechins [287,289]. Epigallocatechin increases glucose capture in skeletal muscles by increasing GLUT-4 translocation to the cell membrane; since there is increase in AMPK phosphorylation, this signaling pathway is believed to be responsible for GLUT-4 translocation [290].

Resveratrol also increases glucose absorption by skeletal muscle cells via AMPK activation in the presence of active insulin, on the PI3k-Akt signaling pathway; [291] however, in its absence, AMPK activation occurs by sirtuin (SIRT1) [292–295].

Kaempferol and quercetin improve glucose capture in adipose tissue only in the presence of insulin, suggesting they act as insulin peripheral sensitizers by acting in the PPAR-g, as the glytazones [296].

In a study with db/db mice, Wolfram et al. [297] demonstrated that epigallocatechin reduced glycemia, and an increase of glucokinase hepatic expression and a reduction of the gluconeogenic enzyme phosphoenolpyruvate carboxykinase [298], that is, there was a displacement of the gluconeogenesis prevalence state to a glycogen generator one.

In summary, red wine, especially via polyphenol effects, acts on several metabolic pathways that can reduce mortality, as shown in Fig. 26.8.

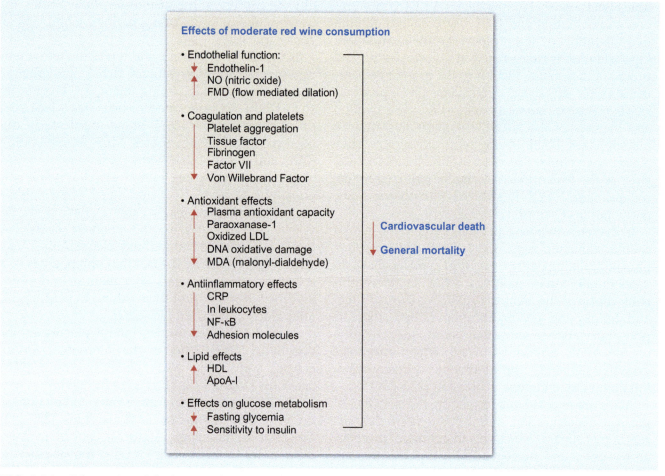

FIG. 26.8   Schematics of the basic red wine action mechanisms.

CONCLUSIONS

Several clinical studies that suggest protective actions of wine against global and cardiovascular mortality are observational. Therefore, although very compelling, they do not offer solid evidence of this potential beneficial effect, because confounding factors interfere to an extent that is not quantifiable with certainty.

Thus, healthy lifestyles, including appropriate diets, especially of the Mediterranean type, regular physical activity, nonsmoking, proper body weight, lower stress levels, close family bonds, after-lunch rest, and a proper amount of sleep (including siesta) are associated with light/moderate wine consumption to provide cardiovascular protection.

In addition, mechanistic studies on the actions of red wine and polyphenols provide pathophysiological support to the clinical findings. For example, arterial vasodilation, endothelial function protection via eNOS and nitric oxide induction, antiplatelet action, endothelin production inhibition, and NF-κB are beneficial and protective effects against atherosclerosis.

Therefore, concerning practical recommendations, it may be said that light/moderate red wine consumption can be beneficial in atherosclerosis protection, if there are no contraindications.

References

[1] Da Luz PL, Coimbra SR. Wine, alcohol and atherosclerosis: clinical evidences and mechanisms. Braz J Med Biol Res 2004;37:1275–95.
[2] Da Luz PL, Nishiyama M, Chagas ACP. Drugs and lifestyle for the treatment and prevention of coronary artery disease a comparison. Braz J Med Biol Res 2011;44:973–99.
[3] Renaud S, de Lorgeril M. Wine, alcohol, platelets, and the French paradox for coronary heart disease. Lancet 1992;339:1523–6.
[4] Evans A. Dr Black's favourite disease. Br Heart J 1955;74:696–7.
[5] Ronksley PE, Brien SE, Turner BJ, et al. Association of alcohol consumption with selected cardiovascular disease outcomes: a systematic review and meta-analysis. BMJ 2011;342:d671.

[6] Schoenborn CA, Adams PF, Barnes PM, et al. Health behaviors of adults: United States, National Health Interview Surveys (NHIS). Vital Health Stat 2004;219:1–79.

[7] Mukamal KJ, Chen CM, Rao SR, et al. Alcohol consumption and cardiovascular mortality among US adults, 1987 to 2002. J Am Coll Cardiol 2010;55:1328–35.

[8] Fuchs FD, Chambless LE, Folsom AR, et al. Association between alcoholic beverage consumption and incidence of coronary heart disease in whites and blacks: the Atherosclerosis Risk in Communities Study. Am J Epidemiol 2004;160:466–74.

[9] King DE, Mainous AG, Geesey ME. Adopting moderate alcohol consumption in middle age: subsequent cardiovascular events. Am J Med 2008;121:201–36.

[10] Berger K, Ajani UA, Kase CS, et al. Light-to-moderate alcohol consumption and risk of stroke among U.S. male physicians. N Engl J Med 1999;341:1557–64.

[11] Arriola L, Martinez-Camblor P, Larrañaga N, et al. Alcohol intake and the risk of coronary heart disease in the Spanish EPIC cohort study. Heart 2009;96(2):124–30.

[12] Pai JK, Mukamal KJ, Rimm EB. Long-term alcohol consumption in relation to all-cause and cardiovascular mortality among survivors of myocardial infarction: the Health Professionals Follow-up Study. Eur Heart J 2012;33:1598–605.

[13] Beulens JWJ, Rimm EB, Ascherio A, et al. Alcohol consumption and risk for coronary heart disease among men with hypertension. Ann Intern Med 2007;146:10–9.

[14] Brien ES, Ronksley PE, Turner BJ, et al. Effect of alcohol consumption on biological markers associated with risk of coronary heart disease: systematic review and meta-analysis of interventional studies. BMJ 2011;342:d636.

[15] Chiuve S, Rimm E, Mukamal K, et al. Light to moderate alcohol consumption and risk of sudden cardiac death in women. Heart Rhythm 2010;7:1374–80.

[16] Allen NE, Beveral V, Casabonne D, et al. Moderate alcohol intake and cancer incidence in women. J Natl Cancer Inst 2009;101:296–305.

[17] Gepner Y, Golan R, Harman-Boehm I, et al. Effects of initiating moderate alcohol intake on cardiometabolic risk in adults with type 2 diabetes. Ann Intern Med 2015;163:569–79.

[18] Di Castelnuovo A, Rotondo S, Iacoviello L, et al. Meta-analysis of wine and beer consumption in relation to vascular risk. Circulation 2002;105:2836–44.

[19] Yusuf S, Hawken S, Ounpuu S, et al. Effect of potentially modifiable risk factors associated with myocardial infarction in 52 countries (the INTERHEART study): case-control study. Lancet 2004;364:937–52.

[20] Trichopoulou A, Bamia C, Trichopoulos D. Anatomy of healthy effects of Mediterranean diet: Greek EPIC prospective cohort study. BMJ 2009;338:b2337.

[21] Gea A, Beunza JJ, Estruch R, et al. Alcohol intake, wine consumption and the development of depression: the PREDIMED study. BMC Med 2013;11:192.

[22] Da Luz PL, Coimbra S, Favarato D. Coronary artery plaque burden and calcium scores in healthy men adhering to longterm wine drinking or alcohol abstinence. Braz J Med Biol Res 2014;47:697–705.

[23] Puri R, Nicholls SJ, Shao M, et al. Impact of statins on serial coronary calcification during atheroma progression and regression. J Am Coll Cardiol 2015;65:1273–82.

[24] Howitz KT, Bitterman KJ, Cohen HY, et al. Small molecule activators of situins extend *Saccharomyces cerevisiae* lifespan. Nature 2003;425:191–6.

[25] Lin S-j, Desfossez PA, Guarente L. Requirement of NAD and SIR2 for life-span extendion by colorie restriction in *Saccharomyces cerevisiae*. Science 2000;289:2126–8.

[26] Wood JG, Rogina B, Lavu S, et al. Sirtuin activators mimic caloric restriction and delay ageing in metazoans. Nature 2004;430:686–9.

[27] Baur JA, Pearson KJ, Price NL, et al. Resveratrol improves health and survival of mice on a high-calorie diet. Nature 2006;444:337–42.

[28] Da Luz PL, Serrano CV, Chacra AP, et al. The effect of red wine on experimental atherosclerosis: lipid-independent protection. Exp Mol Pathol 1999;65:150–9.

[29] Da Luz PL, Tanaka L, Brum PC, et al. Red wine and equivalent oral pharmacological doses of resveratrol delay vascular aging but do not extend life span in rats. Atherosclerosis 2012;224:136–42.

[30] Pearson KJ, Baur JA, Lewis KN, et al. Resveratrol delays age-related deterioration and mimics trscriptional aspects of dietary restriction without extending lifespan. Cell Metab 2008;8:157–68.

[31] Barger JL, Kayo T, Vann JM, et al. A low dose of dietary resveratrol partially mimics caloric restriction and retards aging parameters in mice. PLoS One 2008;3:e2264.

[32] Opie LH, Lecour S. The red wine hypothesis: from concepts to protective signaling molecules. Eur Heart J 2007;28:1683–93.

[33] Stoclet J-C, Chataigneau T, Ndiaye M, et al. Vascular protection by dietary polyphenols. Eur J Pharmacol 2004;500:299–313.

[34] Wallerath T, Deckert G, Ternes T, et al. Resveratrol, a polyphenolic phytoalexin present in red wine, enchances expression and activity of endothelial nitric oxide synthase. Circulation 2002;106:1652–8.

[35] Sinclair DA, Guarent L. Small-molecule allosteric activators of sirtuins. Annu Rev Pharmacol Toxicol 2014;54:363–80.

[36] Smoliga JM, Blanchard O. Enchancing the delivery of resveratrol in humans: if low bioavailability is the problem, what is the solution? Molecules 2014;19:17154–72.

[37] Goldberg DM, Yan J, Soleas GJ. Absorption of three wine-related polyphenols in three diferent matrices by healthy subjects. Clin Biochem 2003;36:79–87.

[38] Kennedy DO, Wightman EL, Reay JL, et al. Effects of resveratrol on cerebral blood flow variables and cognitive performance in humans: a double-blind, placebo-controlled, crossover investigation. Am J Clin Nutr 2010;91:1590–7.

[39] Andreadi C, Britton RG, Patel KR. Resveratrol-sulfates provide an intracellular reservoir for generation of parent resveratrol, which induces autophagy in cancer cells. Autophagy 2014;10:524–5.

[40] Gronbaek M, Becker U, Johansen D, et al. Type of alcohol consumed and mortality from all causes, coronary heart disease, and cancer. Ann Intern Med 2000;133:411–9.

[41] Reanud SC, Gueguen R, Siest G, et al. Wine, beer, and mortality in middle-aged men from eastern France. Arch Intern Med 1999;159:1865–70.

[42] Mukamal KJ, Conigrave KM, Mittleman MA, et al. Roles of drinking pattern and type of alcohol consumed in coronary heart disease in men. N Engl J Med 2003;348:109–18.

[43] Da Luz PL, Fialdini RC, Nishiyama M, et al. Red wine resveratrol and vascular aging: implications for dementia and cognitive decline. In: Diet and nutrition in dementia and cognitive decline. Massachusetts: Academic Press; 2015. p. 943.

[44] Truelsen T, Thudium D, Gronbaek M. Amount and type of alcohol and risk of dementia: the Copenhagen City Heart Study. Neurology 2002;59:1313–9.

[45] Mukamal KJ, Kuller LH, Fitzpatrick AL, et al. Prospective study of alcohol consumption and risk of dementia in older adults. JAMA 2003;289:1405–13.

[46] Orgogozo JM, Dartigues JF, Lafont S, et al. Wine consumption and dementia in the elderly: a prospective community study in the Bordeaux area. Rev Neurol (Paris) 1997;153:185–92.

[47] Stampfer MJ, Kang JH, Chen J, et al. Effects of moderate alcohol consumption on cognitive function in women. N Engl J Med 2005;352:245–53.

[48] Mukamal KJ, Longstreth Jr WT, Mittleman MA, et al. Alcohol consumption and subclinical findings on magnetic resonance imaging of the brain in older adults: the cardiovascular health study. Stroke 2001;32:1939–46.

[49] OKeefe JH, Bybee KA, Lavie CJ. Alcohol and cardiovascular health: the razor-sharp double-edged sword. J Am Coll Cardiol 2007;50:1009–14.

[50] Castelli WP. How many drinks a day? JAMA 1979;242:2000.

[51] Fernandez-Sola J, Estruch R, Grau JM, et al. The relation of alcoholic myopathy to cardiomyopathy. Ann Intern Med 1994;120:529–36.

[52] Lauer MS, Sorlie R. Alcohol, cardiovascular disease, and cancer: treat with caution. J Natl Cancer Inst 2009;101:282–3.

[53] Panagiotakos DB, Chryssohoou C, Siasos G, et al. Sociodemographic and lifestyle statistics of oldest old people (>80 years) living in Ikara Island: the Ikara study. Cardiol Res Pract 2010;2011:679187.

[54] Werle MH, Moriguchi E, Fuchs SC, et al. Risk factors cardiovascular disease in the very elderly: results of a cohort study in a city in southern Brazil. Eur J Cardiovasc Prev Rehabil 2011;18:369–77.

[55] Caramori PR, Zago AJ. Endothelial dysfunction and coronary artery disease. Arq Bras Cardiol 2000;75:163–82.

[56] Ceravolo R, Maio R, Pujia A, et al. Pulse pressure and endothelial dysfunction in never treated hypertensive patients. J Am Coll Cardiol 2003;41:1753–8.

[57] Shechter M, Issachar A, Marai I, et al. Long-term association of brachial artery flow mediated vasodilation and cardiovascular events in middle aged subjects with no apparent heart disease. Int J Cardiol 2009;134:52–8.

[58] Yeboah J, Crouse JR, Hsu FC, et al. Brachial flow-mediated dilation predicts incident cardiovascular events in older adults: the Cardiovascular Health Study. Circulation 2007;1(15):2390–7.

[59] Rossi R, Nuzzo A, Origliani G, et al. Prognostic role of flow-mediated dilation and cardiac risk factors in postmenopausal women. J Am Coll Cardiol 2008;51:997–1002.

[60] Nogueira LP, Knibel MP, Torres MR, et al. Consumption of high-polyphenol dark chocolate improves endothelial function in individuals with stage 1 hypertension and excess body weight. Int J Hypertens 2012;2012:147321.

[61] Grassi D, Necozione S, Lippi C, et al. Cocoa reduces blood pressure and insulin resistance and improves endothelium-dependent vasodilation in hypertensives. Hypertension 2005;46:398–405.

[62] Kim W, Jeong MH, Cho SH, et al. Effect of green tea consumption on endothelial function and circulating endothelial progenitor cells in chronic smokers. Circ J 2006;70(8):1052–7.

[63] Heiss C, Kleinbongard P, Dejam A, et al. Acute consumption of flavanol-rich cocoa and the reversal of endothelial dysfunction in smokers. J Am Coll Cardiol 2005;46:1276–83.

[64] Machha A, Achike FI, Mustafa AM, et al. Quercetin, a flavonoid antioxidant, modulates endothelium-derived nitric oxide bioavailability in diabetic rat aortas. Nitric Oxide 2007;16:442–7.

[65] Rizza S, Muniyappa R, Iantorno M, et al. Citrus polyphenol hesperidin stimulates production of nitric oxide in endothelial cells while improving endothelial function and reducing inflammatory markers in patients with metabolic syndrome. J Clin Endocrinol Metab 2011;96:E782–92.

[66] Coimbra SR, Lage SH, Brandizzi L, et al. The action of red wine and purple grape juice on vascular reactivity is independent of plasma lipids in hypercholesterolemic patients. Braz J Med Biol Res 2005; 38(9):1339–47.

[67] Chen CK, Pace-Asciak CR. Vasorelaxing activity of resveratrol and quercetin in isolated rat aorta. Gen Pharmacol 1996;27 (2):363–6.

[68] Andriambeloson E, Stoclet JC, Andriantsitohaina R. Mechanism of endothelial nitric oxide-dependent vasorelaxation induced by wine polyphenols in rat thoracic aorta. J Cardiovasc Pharmacol 1999;33:248–54.

[69] Cishek MB, Galloway MT, Karim M, et al. Effect of red wine on endothelium-dependent relaxation in rabbits. Clin Sci 1997;93:507–11.

[70] Flesch M, Schwarz A, Bohm M. Effects of red and white wine on endothelium-dependent vasorelaxation of rat aorta and human coronary arteries. Am J Phys 1998;275:H1183–90.

[71] Stein JH, Keevil JG, Wiebe DA, et al. Purple grape juice improves endothelial function and reduces the susceptibility of LDL cholesterol to oxidation in patients with coronary artery disease. Circulation 1999;100:1050–5.

[72] Teragawa H, Fukuda Y, Matsuda K, et al. Effect of alcohol consumption on endothelial function in men with coronary artery disease. Atherosclerosis 2002;165(1):145–52.

[73] Whelan AP, Sutherland WH, McCormick MP, et al. Effects of white and red wine on endothelial function in subjects with coronary artery disease. Intern Med J 2004;34:224–8.

[74] Lekakis J, Rallidis LS, Andreadou I, et al. Polyphenolic compounds from red grapes acutely improve endothelial function in patients with coronary heart disease. Eur J Cardiovasc Prev Rehabil 2005;12:596–600.

[75] Suzuki K, Elkind MS, Boden-Albala B, et al. Moderate alcohol consumption is associated with better endothelial function: a cross sectional study. BMC Cardiovasc Disord 2009;9:8.

[76] Yanagisawa M, Kunihara H, Kimura S, et al. A novel potent vasoconstrictor peptide produced by vascular endothelial cells. Nature 1988;332:411–5.

[77] Corder R. Handbook of experimental pharmacology. In: Warner TD, editor. Endothelin and its inhibitors. Berlin: Springer; 2001. p. 35–67.

[78] Caligiuri G, Levy B, Pernow J, et al. Myocardial infarction mediated by endothelin receptor signaling in hypercholesterolemic mice. Proc Natl Acad Sci U S A 1999;96:6920–4.

[79] Rubanyi GM, Polokoff MA. Endothelins: molecular biology, biochemistry, pharmacology, physiology, and pathophysiology. Pharmacol Rev 1994;46:325–415.

[80] Xu D, Emoto N, Giaid A, et al. ECE-1: a membrane-bound metalloprotease that catalyzes the proteolytic activation of big endothelin-1. Cell 1994;78:473–85.

[81] Schiüin EL. Vascular endothelium in hypertension. Vasc Pharmacol 2005;43:19–29.

[82] Elijovich F, Laffer CL, Amador E, et al. Regulation of plasma endothelin by salt in salt-sensitive hypertension. Circulation 2001;103:263–8.

[83] Bohm F, Pernow J. The importance of endothelin-1 for vascular dysfunction in cardiovascular disease. Cardiovasc Res 2007;76:8–18.

[84] Bouallegue A, Daou GB, Srivastava AK. Endothelin-1-induced signaling pathways in vascular smooth muscle cels. Curr Vasc Phamacol 2007;5:45–52.

[85] Haak T, Marz W, Jugmann E, et al. Elevated endothelin levels in patients with hyperlipoproteinemia. J Clin Investig 1999;72:580–4.

[86] Manea AS, Todirita A, Manea A. High glucose-induced expression of endothelin-1 in human endothelial cells is mediated by activated CCAAT/enhacer-binding proteins. PLoS One 2013;8 (12)e84170.

[87] Yu AP, Tam BT, Yau WY, et al. Association of endothelin-1 and matrix metalloproteinase-9 with metabolic syndrome in middle-aged and older adults. Diabetol Metab Syndr 2015;7:111.

[88] Bau PF, Bau CH, Rosito GA, et al. Alcohol consumption, cardiovascular health, and endothelial function markers. Alcohol 2007;41:479–88.

[89] Li M, Ma G, Han L, et al. Regulating effect of tea polyphenols on endothelin, intracellular calcium concentration, mitochondrial membrane potentials in vascular cells injured by angiotensin II. Ann Vasc Surg 2014;28(4):1016–22.

[90] Storniolo CE, Rosello-Catafau J, Pinto X, et al. Polyphenol fraction of extra virgin olive oil protects against endothelial dysfunction induced by high glucose and free fatty acids through modulation of nitric oxide and endothelin-1. Redox Biol 2014;2:971–7.

[91] Loke WM, Hodgson JM, Proudfoot JM, et al. Pure dietary flavonoids quercetin and (−)-epicatechin augment nitric oxide products and reduce endothelin-1 acutely in healthy men. Am J Clin Nutr 2008;88:1018–25.

[92] Nicholson SK, Tucker GA, Brameld JM. Physiological concentrations of dietary polyphenolsnregulate vascular endothelial cell expression of genes important in cardiovascular health. Br J Nutr 2010;103(10):1398–403.

[93] Corder R, Douthwaite JA, Lee DM, et al. Endothelin-1 synthesis reduced by wine. Nature 2001;414:863–4.

[94] Khan NQ, Lees DM, Douthwaite JA, et al. Comparison of red wine extract and polyphenol constituents on endothelin-1 synthesis by cultured endothelial cells. Clin Sci 2002;103(Suppl. 48): 72S–75S.

[95] Reiter CEN, Kim J, Quon MJ. Green tea polyphenol epigallocatechin gallate reduces endothelin-1 expression and secretion in vascular endothelial cells: roles for AMP-activated protein kinase, Akt, and FOXO1. Endocrinology 2010;151:103–14.

[96] Zhao X, Gu Z, Attele AS, et al. Effects of quercetin on the release of endothelin, prostacyclin and tissue plasminogen activator from human endothelial cells in culture. J Ethnopharmacol 1999;67:279–85.

[97] El Mowafy AM, White RE. Resveratrol inhibits MAPK activity and nuclear translocation in coronary artery smooth muscle: reversal of endothelin-1 stimulatory effects. FEBS Lett 1999;451:63–7.

[98] Liu JC, Chen JJ, Chan P, et al. Inhibition of cyclic strain-induced endothelin-1 gene expression by resveratrol. Hypertension 2003;42:1198–205.

[99] Squadrito F, Altavilla D, Morabito N, et al. The effect of the phytoestrogen gensitein on plasma nitric oxide concentration, endothelin-1 levels and endothelium dependent vasodilatation in postmenopausal women. Atherosclerosis 2002;162:339–47.

[100] Jimenez R, Lopes-Sepulveda R, Kadmiri M, et al. Polyphenols restore endothelial function in DOCA-salt-hypertension: role of endothelin-1 and NADPH oxidase. Free Radic Biol Med 2007; 43(3):462–73.

[101] Forstermann U, Closs EI, Pollock JS, et al. Nitric oxide synthase isozymes. Characterization, puriflcation, molecular cloning, and functions. Hypertension 1994;23:1121–31.

[102] Forstermann U. Regulation of nitric oxide synthase expression and activity. In: Mayer B, editor. Handbook of experimental pharmacology-nitric oxide. Berlin: Springer; 2000. p. 71–91.

[103] Pritchard Jr KA, Ackerman AW, Gross ER, et al. Heat shock protein 90 mediates the balance of nitric oxide and superoxide anion from endothelial nitric-oxide synthase. J Biol Chem 2001;276:17621–4.

[104] Song Y, Cardounel AJ, Zweier JL, et al. Inhibition of superoxide generation from neuronal nitric oxide synthase by heat shock protein 90: implications in NOS regulation. Biochemistry 2002;41:10616–22.

[105] Sowa G, Pypaert M, Sessa WC. Distinction between signaling mechanisms in lipid rafts vs. caveolae. Proc Natl Acad Sci U S A 2001;98:14072–7.

[106] Zeiher AM, Fisslthaler B, Schray Utz B, et al. Nitric oxide modulates the expression of monocyte chemoattractant protein 1 in cultured human endothelial cells. Circ Res 1995;76:980–6.

[107] Tsao PS, Wang B, Buitrago R, et al. Nitric oxide regulates monocyte chemotactic protein-1. Circulation 1997;96:934–40.

[108] Arndt H, Smith CW, Granger DN. Leukocyte ± endothelial cell adhesion in spontaneously hypertensive and normotensive rats. Hypertension 1993;21:667–73.

[109] Garg UC, Hassid A. Nitric oxide-generating vasodilators and 8-bromo-cyclic guanosine monophosphate inhibit mitogenesis and proliferation of cultured rat vascular smooth muscle cells. J Clin Invest 1989;83:1774–7.

[110] Nakaki T, Nakayama M, Kato R. Inhibition by nitric oxide and nitric oxide-producing vasodilators of DNA synthesis in vascular smooth muscle cells. Eur J Pharmacol 1990;189:347–53.

[111] Nunokawa Y, Tanaka S. Interferon-gamma inhibits proliferation of rat vascular smooth muscle cells by nitric oxide generation. Biochem Biophys Res Commun 1992;188:409–15.

[112] Hogan M, Cerami A, Bucala R. Advanced glycosylation end products block the antiproliferative effect of nitric oxide. Role in the vascular and renal complications of diabetes mellitus. J Clin Invest 1992;90:1110–5.

[113] Kubes P, Suzuki M, Granger DN. Nitric oxide: an endogenous modulator of leukocyte adhesion. Proc Natl Acad Sci U S A 1991;88:4651–5.

[114] Davenpeck KL, Gauthier TW, Lefer AM. Inhibition of endothelial-derived nitric oxide promotes P-selectin expression and actions in the rat microcirculation. Gastroenterology 1994;107:1050–8.

[115] Gauthier TW, Scalia R, Murohara T, et al. Nitric oxide protects against leukocyte ± endothelium interactions in the early stages of hypercholesterolemia. Arterioscler Thromb Vasc Biol 1995;15:1652–9.

[116] De Caterina R, Libby P, Peng HB, et al. Nitric oxide decreases cytokine-induced endothelial activation. Nitric oxide selectively reduces endothelial expression of adhesion molecules and proinflammatory cytokines. J Clin Invest 1995;96:60–8.

[117] Tsao PS, Buitrago R, Chan JR, et al. Fluid flow inhibits endothelial adhesiveness. Nitric oxide and transcriptional regulation of VCAM-1. Circulation 1996;94:1682–9.

[118] Cardona-Sanclemente LE, Born GV. Effect of inhibition of nitric oxide synthesis on the uptake of LDL and fibliurinogen by arterial walls and other organs of the rat. Br J Pharmacol 1995;114:1490–4.

[119] Draijer R, Atsma DE, van der Laarse A, et al. cGMP and nitric oxide modulate thrombin-induced endothelial permeability. Regulation via different pathways in human aortic and umbilical vein endothelial cells. Circ Res 1995;76:199–208.

[120] Wallerath T, Poleo D, Li H, et al. Red wine increases the expression of human endothelial nitric oxide synthase: a mechanism that may contribute to its beneflcial cardiovascular effects. J Am Coll Cardiol 2003;4:471–8.

[121] Wallerath T, Deckert G, Ternes T, et al. Resveratrol, a polyphenolic phytoalexin present in red wine, enhances expression and activity of endothelial nitric oxide synthase. Circulation 2002;106:1652–8.

[122] Wallerath T, Li H, Godtel-Ambrust U, et al. A blend of polyphenolic compounds explains the stimulatory effect of red wine on human endothelial NO synthase. Nitric Oxide 2005;12:97–104.

[123] Spanier G, Xu H, Xia N, et al. Resveratrol reduces endothelial oxidative stress by modulating the gene expression of superoxide dismutase 1 (SOD1), glutathione peroxidase 1 (GPx1) and NADPH oxidase subunit (Nox4). J Physiol Pharmacol 2009;60 (Suppl. 4):111–6.

[124] Yang J, Wang N, Li J, et al. Effects of resveratrol on NO secretion stimulated by insulin and its dependence on SIRT1 in high glucose cultured endothelial cells. Endocrine 2010;37:365–72.

[125] Arunachalam G, Yao H, Sundar IK, et al. SIRT1 regulates oxidant- and cigarette smoke-induced eNOS acetylation in endothelial cells: role of resveratrol. Biochem Biophys Res Commun 2010;393:66–72.

[126] Gresele P, Pignatelli P, Guglielmini G, et al. Resveratrol, at concentrations attainable with moderate wine consumption, stimulates human platelet nitric oxide production. J Nutr 2008;138:1602–8.

[127] Leikert JF, Rathel TR, Wohlfart P, et al. Red wine polyphenols enhance endothelial nitric oxide synthase expression and subsequent nitric oxide release from endothelial cells. Circulation 2002;106:1614–7.

[128] Huang PH, Chen YH, Tsai HY, et al. Intake of red wine increases the number and functional capacity of circulating endothelial progenitor cells by enhancing nitric oxide bioavailability. Arterioscler Thromb Vasc Biol 2010;30:869–77.

[129] Fitzpatrick DF, Bing B, Rohdewald R. Endothelium-dependent vascular effects of Pycnogenol. J Cardiovasc Pharmacol 1998;32(4):509–15.

[130] Yamakoshi J, Kataoka S, Koga T, et al. Proanthocyanidin-rich extract from grape seeds attenuates the development of aortic atherosclerosis in cholesterol-fed rabbits. Atherosclerosis 1999;142(1):139–49.

[131] Stoclet JC, Kleschyov A, Andriambeloson E, et al. Endothelial NO release caused by red wine polyphenols. J Physiol Pharmacol 1999;50(4):535–54.

[132] Martin S, Andriambeloson E, Takeda K, et al. Red wine polyphenols increase calcium in bovine aortic endothelial cells: a basis to elucidate signaling pathways leading to nitric oxide production. Br J Pharmacol 2002;136(6):1579–87.

[133] Fitzpatrick DF, Fleming RC, Bing B, et al. Isolation and characterization of endothelium-dependent vasorelaxing compounds from grape seeds. J Agric Food Chem 2000;48:6384–90.

[134] Andriambeloson E, Magnier C, Haan-Archipoff G, et al. Natural dietary polyphenolic compounds cause endothelium-dependent vasorelaxation in rat thoracic aorta. J Nutr 1998;128:2324–33.

[135] Chen ZY, Zhang ZS, Kwan KY, et al. Endothelium-dependent relaxation induced by hawthorn extract in rat mesenteric artery. Life Sci 1998;63:1983–91.

[136] Duarte J, Jimenez R, Villar IC, et al. Vasorelaxant effects of the bioflavonoid chrysin in isolated rat aorta. Planta Med 2001;67:567–9.

[137] Fitzpatrick DF, Hirschfeld SL, Ricci T, et al. Endothelium-dependent vasorelaxation caused by various plant extracts. J Cardiovasc Pharmacol 1995;26:90–5.

[138] Karim M, McCormick K, Kappagoda CT. Effects of cocoa extracts on endothelium-dependent relaxation. J Nutr 2000;130:2105S–2108S.

[139] Kim SH, Kang KW, Kim KW, et al. Procyanidins in crataegus extract evoke endothelium-dependent vasorelaxation in rat aorta. Life Sci 2000;67:121–31.

[140] Lemos VS, Freitas MR, Muller B, et al. Dioclein, a new nitric oxide- and endothelium-dependent vasodilator flavonoid. Eur J Pharmacol 1999;386:41–6.

[141] Lorenz M, Wessler S, Follmann E, et al. A constituent of green tea, epigallocatechin-3-gallate, activates endothelial nitric oxide synthase by a phosphatidylinositol-3-OH-kinase-, cAMP-dependent protein kinase-, and Akt-dependent pathway and leads to endothelial-dependent vasorelaxation. J Biol Chem 2004;279:6190–5.

[142] Taubert D, Berkels R, Klaus W, et al. Nitric oxide formation and corresponding relaxation of porcine coronary arteries induced by plant phenols: essential structural features. J Cardiovasc Pharmacol 2002;40:701–13.

[143] Burns J, Gardner PT, O'Neil J, et al. Relationship among antioxidant activity, vasodilation capacity, and phenolic content of red wines. J Agric Food Chem 2000;48:220–30.

[144] Diebolt M, Bucher B, Andriantsitohaina R. Wine polyphenols decrease blood pressure, improve NO vasodilatation, and induce gene expression. Hypertension 2001;38:159–65.

[145] Cui J, Tosaki A, Cordis GA, et al. Cardioprotective abilities of white wine. Ann N Y Acad Sci 2002;957:308–16.

[146] Samuel SM, Thirunavukkarasu M, Penumathsa SV, et al. Akt/FOXO3a/SIRT1-mediated cardioprotection by n-tyrosol against ischemic stress in rat in vivo model of myocardial infarction: switching gears toward survival and longevity. J Agric Food Chem 2008;56:9692–8.

[147] Thirunavukkarasu M, Penumathsa SV, Samuel SM. White wine induced cardioprotection against ischemia-reperfusion injury is mediated by life extending Akt/FOXO3a/NFkappaB survival pathway. J Agric Food Chem 2008;56:6733–9.

[148] MIgliorini M, Cantaluppi V, Mannari C, et al. Caffeic acid, a phenol found in white wine modulates endothelial nitric oxide production and protects from oxidative stress-associated endothelial cell injury. PLoS One 2015;10(4)e0117530.

[149] Kargacin ME, Emmett TL, Kargacin GJ. Epigallocatechin-3-gallate has dual, independent effects on the cardiac sarcoplasmic reticulum/endoplasmic reticulum Ca^{2+} ATPase. J Muscle Res Cell Motil 2011;32:89–98.

[150] Soler F, Asensio MC, Fernández-Belda F. Inhibition of the intracellular Ca^{2+} transporter SERCA (Sarco-Endoplasmic Reticulum Ca^{2+}-ATPase) by the natural polyphenol epigallocatechin-3-gallate. J Bioenerg Biomembr 2012;44:597–605.

[151] O'Neill LA, Hardie DG. Metabolism of inflammation limited by AMPK and pseudo-starvation. Nature 2013;493:346–55.

[152] Ruderman NB, Carling D, Prentki M, et al. AMPK, insulin resistance, and the metabolic syndrome. J Clin Invest 2013;123:2764–72.

[153] Towler MC, Hardie DG. AMP-activated protein kinase in metabolic control and insulin signaling. Circ Res 2007;100:328–41.

[154] Hellermann R, Solomonson LP. Calmodulin promotes dimerization of the oxygenase domain of human endothelial nitric-oxide synthase. J Biol Chem 1997;272:12030–4.

[155] Hong Byun E, Fujimura EY, Yamada K, et al. TLR4 signaling inhibitory pathway induced by green tea polyphenol epigallocat-echin-3-gallate through 67-kDa laminin receptor. J Immunol 2010;185:33–45.

[156] Hotta Y, Huang L, Muto T, et al. Positive inotropic effect of purified green tea catechin derivative in guinea pig hearts: the mea-surements of cellular Ca2þ and nitric oxide release. Eur J Pharmacol 2006;552:123–30.

[157] Michel JB, Feron O, Sacks D, et al. Reciprocal regulation of endothelial nitric-oxide synthase by Ca^{2+}-calmodulin and caveolin. J Biol Chem 1997;272:15583–6.

[158] Urbich C, Dimmeler S. Endothelial progenitor cells: characterization and role in vascular biology. Circ Res 2004;95:343–53.

[159] Rafii S, Lyden D. Therapeutic stem and progenitor cell transplantation for organ vascularization and regeneration. Nat Med 2003;9:702–12.

[160] Khakoo AY, Finkel T. Endothelial progenitor cells. Annu Rev Med 2005;56:79–101.

[161] Dernbach E, Urbich C, Brandes RP, et al. Antioxidative stress associated genes in circulating progenitor cells: evidence for enhanced resistance against oxidative stress. Blood 2004;104:3591–7.

[162] Wang X, Chen J, Tao Q, et al. Effects of ox-LDL on number and activity of circulating endothelial progenitor cells. Drug Chem Toxicol 2004;27:243–55.

[163] Ma FX, Zhou B, Chen Z, et al. Oxidized low density lipoprotein impairs endothelial progenitor cells by regulation of endothelial nitric oxide synthase. J Lipid Res 2006;47:1227–37.

[164] Zhou B, Ma FX, Liu PX, et al. Impaired therapeutic vasculogenesis by transplantation of OxLDL-treated endothelial progenitor cells. J Lipid Res 2007;48:518–27.

[165] Di Santo S, Diehm N, Ortmann J, et al. Oxidized low density lipoprotein impairs endothelial progenitor cell function by down-regulation of E-selectin and integrin alpha(v) beta5. Biochem Biophys Res Commun 2008;373:528–32.

[166] Wu Y, Wang Q, Cheng L, et al. Effect of oxidized low-density lipoprotein on survival and function of endothelial progenitor cell mediated by p38 signal pathway. J Cardiovasc Pharmacol 2009;53:151–6.

[167] Hill JM, Zalos G, Halcox JPJ, et al. Circulating endothelial progenitor cells, vascular function, and cardiovascular risk. N Eng J Med 2003;348:593–600.

[168] Asahara T, Murohara T, Sullivan A, et al. Isolation of putative endothelial progenitor cells for angiogenesis. Science 1997;275:964–7.

[169] Ito H, Rovira II, Bloom ML, et al. Endothelial progenitor cells as putative targets for angiostatin. Cancer Res 1999;59:5875–7.

[170] Schmidt-Lucke C, Rossig L, Fichtlscherer S, et al. Reduced number of circulating endothelial progenitor cells predicts future cardiovascular events: proof of concept for the clinical importance of endogenous vascular repair. Circulation 2005;111(22):2981–7.

[171] Werner N, Kosiol S, Schiegl T, et al. Circulating endothelial progenitor cells and cardiovascular outcomes. N Engl J Med 2005;353(10):999–1007.

[172] Cuadrado-Godia E, Regueiro A, Nunez J, et al. Endothelial progenitor cells predict cardiovascular events after atherothombotic stroke and acute myocardial infarction. A PROCELL substudy. PLoS One 2015;10(9)e013245.

[173] Vasa M, Fichtlscherer S, Aicher A, et al. Number and migratory activity of circulating endothelial progenitor cells inversely correlate with risk factors for coronary artery disease. Circ Res 2001;89(1):E1–7.

[174] J G, Cq W, Hh F, et al. Effects of resveratrol on endothelial progenitor cells and their contributions to reendothelialization in intima-injured rats. J Cardiovasc Pharmacol 2006;47:711–21.

[175] Balestrieri ML, Fiorito C, Crimi E, et al. Effect of red wine antioxidants and minor polyphenolic constituents on endothelial progenitor cells after physical training in mice. Int J Cardiol 2008;126:295–7.

[176] Hamed S, Alshiek J, Aharon A, et al. Red wine consumption improves in vitro migration of endothelial progenitor cells in young healthy individuals. Am J Clin Nutr 2010;92:161–9.

[177] Huang PH, Chen YC, Tsai HY, et al. Intake of red wine increases the number and functional capacity of circulating endothelial progenitor cells by enhancing nitric oxide bioavaiability. Arterioscler Thromb Vasc Biol 2010;30:869–77.

[178] Xia L, Wang XX, Hu XS, et al. Resveratrol reduces endothelial progenitor cells senescence through augmentation of telomerase activity by Akt-dependent mechanisms. Br J Pharmacol 2008;155:387–94.

[179] Zhu JH, Wang XX, Chen JZ, et al. Effects of puerarin on number and activity of endothelial progenitor cells from peripheral blood. Acta Pharmacol Sin 2004;25(8):1045–51.

[180] Zhu J, Wang X, Shang Y, et al. Puerarin reduces endothelial progenitor cells senescence through augmentation of telomerase activity. Vascul Pharmacol 2008;49(2-3):106–10.

[181] Lefèvre J, Michaud SE, Haddad P, et al. Moderate consumption of red wine (cabernet sauvignon) improves ischemiainduced neovascularization in ApoE-deficient mice: effect on endothelial progenitor cells and nitric oxide. FASEB J 2007;21(14):3845 52.

[182] Wang XB, Huang J, Zou JG, et al. Effects of resveratrol on number and activity of endothelial progenitor cells from human peripheral blood. Clin Exp Pharmacol Physiol 2007;34(11):1109–15.

[183] Dong XX, Hui ZJ, Xiang WX, et al. Ginkgo biloba extract reduces endothelial progenitorcell senescence through augmentation of telomerase activity. J Cardiovasc Pharmacol 2007;49(2):111–5.

[184] Xu MG, Wang JM, Chen L, et al. Berberine-induced upregulation of circulating endothelial progenitor cells is related to nitric oxide production in healthy subjects. Cardiology 2009;112(4):279–86.

[185] Li YJ, Duan CL, Liu JX, et al. Pro-angiogenic actions of Salvianolic acids on in vitro cultured endothelial progenitor cells and chick embryo chorioallantoic membrane model. J Ethnopharmacol 2010;131(3):562–6.

[186] He W, Wu WK, Wu YL, et al. Ginsenoside-Rg1 mediates microenvironment-dependent endothelial differentiation of human mesenchymal stem cells in vitro. J Asian Nat Prod Res 2011;13(1):1–11.

[187] Volpato S, Pahor M, Ferrucci L, et al. Relationship of alcohol intake with inflammatory markers and plasminogen activator inhibitor-1 in well-functioning older adults: the Health, Aging, and Body Composition study. Circulation 2004;109(5):607–12.

[188] Grenett HE, Aikens ML, Torres JA, et al. Ethanol transcriptionally upregulates t-PA and u-PA gene expression in cultured human endothelial cells. Alcohol Clin Exp Res 1998;22:849–53.

[189] Abou-Agag LH, Aikens ML, Tabengwa EM, et al. Polyphyenolics increase t-PA and u-PA gene transcription in cultured human endothelial cells. Alcohol Clin Exp Res 2001;25:155–62.

[190] Mukamal KJ, Jadhav PP, D'Agostino RB, et al. Alcohol consumption and hemostatic factors: analysis of the Framingham Offspring cohort. Circulation 2001;104(12):1367–73.

[191] Mukamal KJ, Cushman M, Mittleman MA, et al. Alcohol consumption and inflammatory markers in older adults: the Cardio-vascular Health Study. Atherosclerosis 2004;173(1):79–87.

[192] Toth A, Sandor B, Papp J, et al. Moderate red wine consumption improves hemorheological parameters in healthy volunteers. Clin Hemorheol Microcirc 2014;56(1):13–23.

[193] Tousoulis D, Ntarladimas I, Antoniades C, et al. Acute effects of different alcoholic beverages on vascular endothelium, inflammatory markers and thrombosis fibrinolysis system. Clin Nutr 2008;27:594–600.

[194] Djousse L, Pankow JS, Arnett DK, et al. Alcohol consumption and plasminogen activator inhibitor type 1: the National Heart, Lung, and Blood Institute Family Heart Study. Am Heart J 2000;139(4):704–9.

[195] Rimm EB, Williams P, Fosher K, et al. Moderate alcohol intake and lower risk of coronary heart disease: meta-analysis of effects on lipids and hemostatic factors. BMJ 1999;319(7224):1523–8.

[196] Pace-Asciak CR, Hahn S, Diamandis EP, et al. The red wine phenolics trans-resveratrol and quercetin block human platelet aggregation and eicosanoid synthesis: implications for protection against coronary heart disease. Clin Chim Acta 1995;235:207–19.

[197] Olas B, Wachowicz B, Saluk-Juszczak J, et al. Effect of resveratrol, a natural polyphenolic compound, on platelet activation induced by endotoxin or thrombin. Thromb Res 2002;107:141–5.

[198] Wang Z, Huang Y, Zou J, et al. Effects of red wine and wine polyphenol resveratrol on platelet aggregation in vivo and in vitro. Int J Mol Med 2002;9:77–9.

[199] Fragopoulou E, Nomikos T, Antonopoulou S, et al. Separation of biologically active lipids from red wine. J Agric Food Chem 2000;48:1234–8.

[200] Yang Y, Wang X, Zhang L, et al. Inhibitory effects of resveratrol on platelet activation induced by thromboxane a(2) receptor agonist in human platelets. Am J Chin Med 2011;39:145–59.

[201] Oh WJ, Endale M, Park SC, et al. Dual roles of quercetin in platelets. phosphoinositide-3-kinase and MAP kinases inhibition and cAMP-dependent vasodilator-stimulated phosphoprotein stimulation. Evid Based Complement Alternat Med 2012;2012:485262.

[202] Janssen K, Mensink RP, Cox FJ, et al. Effects of the flavonoids quercetin and apigenin on hemostasis in healthy volunteers: results from an in vitro and a dietary supplement study. Am J Clin Nutr 1998;67:255–62.

[203] Fan PS, Gu ZL, Liang ZQ. Effect of quercetin on adhension of platelets to microvascular endothelial cells in vitro. Acta Pharmacol Sin 2001;22:857–60.

[204] Hubbard GP, Stevens JM, Cicmil M, et al. Quercetin inhibits collagen-stimulated platelet activation through inhibition of multiple components of the glycoprotein VI signaling pathway. J Thromb Haemost 2003;1(5):1079–88.

[205] Nelson S, Bagby GJ, Bainton BG, et al. The effects of acute and chronic alcoholism on tumor necrosis factor and the inflammatory response. J Infect Dis 1989;160:422–9.

[206] Kolls JK, Xie J, Lei D, et al. Differential effects of in vivo ethanol on LPS-induced TNF and nitric oxide production in the lung. Am J Physiol 1995;268:L991–8.

[207] Szabo G, Mandrekar P, Catalano D. Inhibition of superantigenin-duced T cell proliferation and monocyte IL-1 beta, TNF-alpha, and IL-6 production by acute ethanol treatment. J Leukoc Biol 1995;58:342–50.

[208] Verma BK, Fogarasi M, Szabo G. Down-regulation of tumor necrosis factor alpha activity by acute ethanol treatment in human peripheral blood monocytes. J Clin Immunol 1993;13:8–22.

[209] Chiva-Blanch G, Urpi-Sarda M, Llorach R, et al. Differential effects of polyphenols and alcohol of red wine on the expres-sion of adheion molecules and inflammatory cytokines related to athero-sclerosis: a randomized clinical trial. Am J Clin Nutr 2012;95:326–34.

[210] Wannamethee SG, Lowe GD, Shaper G, et al. The effects of different alcoholic drinks on lipids, insulin and haemostatic and inflamma-tory markers in older men. Thromb Haemost 2003;90:1080–7.

[211] Estruch R, Sacanella E, Badia E, et al. Different effects of red wine and gin consumption on inflammatory biomarkers of atheroscle-rosis: a prospective randomized crossover trial; effects of wine on inflammatory markers. Atherosclerosis 2004;175:117–23.

[212] Grefen FR, Karin M. The IKK/NFKB activation pathway—a tar-get for prevention and treatment of cancer. Cancer Lett 2004;206:193–9.

[213] Schubert SY, Neeman I, Resnick N. A novel mechanism for the inhibition of NF-KB activation in vascular endothelial cells by nat-ural antioxidants. FASEB J 2002;16(14):1931–3.

[214] Martinez N, Casos K, Simonetti P, et al. De-alchoolized white and red wines decreases inflammatory makers and NF-kB in atheroma plaques in apoE-deficient mice. Eur J Nutr 2013;52(2):737–47.

[215] Rimm EB, Williams P, Fosher K, et al. Moderate alcohol intake and lower risk of coronary heart disease: meta-analysis of effects on lipids and haemostatic factors. BMJ 1999;319(7224):1523–8.

[216] Brien SE, Ronksley PE, Turner BJ, et al. Effect of alcohol consump-tion on biological markers associated with risk of coronary heart disease: systematic review and meta-analysis of interventional studies. Br Med J 2011;342:d636.

[217] Schafer C, Parlesak A, Ekoldt J, et al. Beyond HDL-cholesterol increase: phospholipid enrichment and shift from HDL3 to HDL2 in alcohol consumers. J Lipid Res 2007;48:1550–8.

[218] Avellone G, Di Garbo V, Campisi D, et al. Effects of moderate Sicilian red wine consumption on inflammatory biomarkers of atherosclerosis. Eur J Clin Nutr 2006;60:41–7.

[219] Droste DW, Iliescu C, Vaillant M, et al. A daily glass of red wine associated with lifestyle changes independently improves blood lipids in patients with carotid atherosclerosis: results from a ran-domized controlled trial. Nutr J 2013;12:147.

[220] Rifler JP, Lorcerie F, Durand P, et al. A moderate red wine intake improves blood lipid parameters and erythrocytes membrane flu-idity in post myocardial infarct patients. Mol Nutr Food Res 2011;56:345–51.

[221] Estruch R, Sacanella E, Mota F, et al. Moderate consumption of red wine, but not gin, decreases erythrocyte superoxide dismutase activity: a randomised cross-over trial. Nutr Metab Cardiovasc Dis 2011;21:46–53.

[222] Chiva-Blanch G, Urpi-Sarda M, Ros E, et al. Effects of red wine polyphenols and alcohol on glucose metabolism and the lipid pro-file: a randomized clinical trial. Clin Nutr 2013;32(2):200–6.

[223] De Oliveira E, Silva ER, Foster D, et al. Alcohol consumption raises HDL cholesterol levels by increasing the transport rate of apolipoproteins A-I and A-II. Circulation 2000;102:2347–52.

[224] Nishiwaki M, Ishikawa T, Ito T, et al. Effects of alcohol on lipopro-tein lipase, hepatic lipase, cholesteryl ester transfer protein, and lecithin: cholesterol acyltransferase in high-density lipoprotein cholesterol elevation. Atherosclerosis 1994;111:99–109.

[225] Taskinen MR, Nikkila EA, Valimaki M, et al. Alcohol-induced changes in serum lipoproteins and in their metabolism. Am Health J 1987;113:458–64.

[226] Leifert WR, Abeywardena MY. Grape seed and wine polyphenol extracts inhibit cellular cholesterol uptake, cell proliferation, and 5-lipoxygenase acitivity. Nutr Res 2008;28:842–50.

[227] Yashiro T, Nanmoku M, Shimizu M, et al. Resveratrol increases the expression and activity of the low density lipoprotein receptor in hepatocytes by the proteolytic activation of the sterol regulatory element-binding proteins. Atherosclerosis 2012;220(2):369–74.

[228] Koppes LL, Dekker JM, Hendriks HF, et al. Moderate alcohol con-sumption lowers the risk of type 2 diabetes: a meta-analysis of prospective observational studies. Diabetes Care 2005;28:719–25.

[229] Baliunas DO, Taylor BJ, Irving H, et al. Alcohol as a risk factor for type 2 diabetes: a systematic review and meta-analysis. Diabetes Care 2009;32:2123–32.

[230] Kim SH, Abbasi F, Lamendola C, et al. Effect of moderate alco-holic beverage consumption on insulin sensitivity in insulin-resistant, nondiabetic individuals. Metabolism 2009;58:3872.

[231] Napoli R, Cozzolino D, Guardasole V, et al. Red wine consump-tion improves insulin resistance but not endothelial function in type 2 diabetic patients. Metabolism 2005;54:306–13.

[232] Cai W, Torreggiani M, Zhu L, et al. AGER1 regulates endothelial cell NADPH oxidase-dependent oxidant stress via PKC-delta: implications for vascular disease. Am J Physiol Cell Physiol 2009;298:C624–34.

[233] Kota SK, Meher LK, Kota SK, et al. Implications of serum paraox-anase activity in obesity, diabetes mellitus, and dyslipidemia. Indian J Endocrinol Metab 2013;17(3):402–12.

[234] Bachetti T, Masciangelo S, Armeni T, et al. Glycation of human high density lipoprotein by methoxilglyoxal: effects on HDK-paraoxanase activity. Metabolism 2014;63(3):307–11.

[235] Shen Y, Ding FH, Sun JT, et al. Association of elevated apoA-I gly-cation and reduced HDL-associated paraoxanase1, 3 activity, and their interaction with angiographic severity of coronary disease in patients with type 2 diabetes mellitus. Cardiovasc Diabetol 2015;14:52.

[236] Taniguchi CM, Emanuelli B, Kahn CR. Critical nodes in signalling pathways: insights into insulin action. Nat Rev Mol Cell Biol 2006;7:85–96.

[237] Cusi K, Maezono K, Osman A, et al. Insulin resistance differen-tially affects the PI 3-kinase- and MAP kinase-mediated signaling in human muscle. J Clin Invest 2000;105:311–20.

[238] Das Evcimen N, King GL. The role of protein kinase C activation and the vascular complications of diabetes. Pharmacol Res 2007;55:498–510.

[239] Rask-Madsen C, King GL. Proatherosclerotic mechanisms involv-ing protein kinase C in diabetes and insulin resistance. Arterios-cler Thromb Vasc Biol 2005;25:487–96.

[240] Brownlee M. Biochemistry and molecular cell biology of diabetic complications. Nature 2001;414:813–20.

[241] Fu MX, Requena JR, Jenkins AJ, et al. The advanced glycation end product, Nepsilon-(carboxymethyl))lysine, is a product of both lipid peroxidation and glycoxidation reactions. J Biol Chem 1996;271:9982–6.

[242] Schalkwijk CG, Stehouwer CD, van Hinsbergh VW. Fructose-mediated non-enzymatic glycation: sweet coupling or bad modi-fication. Diabetes Metab Res Rev 2004;20:369–82.

[243] Zhang Q, Ames JM, Smith RD, et al. A perspective on the Maillard reaction and the analysis of protein glycation by mass spectrom-etry: probing the pathogenesis of chronic disease. J Proteome Res 2009;8:754–69.

[244] Monnier VM, Bautista O, Kenny D, et al. Skin collagen glycation, glycoxidation, and crosslinking are lower in subjects with long-term intensive versus conventional therapy of type 1 diabetes: rel-evance of glycated collagen products versus HbA1c as markers of

diabetic complications. DCCT Skin Collagen Ancillary Study Group Diabetes Control and Complications Trial. Diabetes 1999;48:870–80.

[245] Semba RD, Nicklett EJ, Ferrucci L. Does accumulation of advanced glycation end products contribute to the aging phenotype? J Gerontol A Biol Sci Med Sci 2010;65:963–75.

[246] Kilhovd BK, Juutilainen A, Ronemaa T, et al. Increased serum levels of advanced glycation end products predict total, cardiovascular and coronary mortality in women with type 2 diabetes: a population-based 18 year follow-up study. Diabetologia 2007;50:1409–17.

[247] Samuel VT, Petersen KF, Shulman GI. Lipid-induced insulin resistance: unravelling the mechanism. Lancet 2010;375:2267–77.

[248] Bucala R, Makita Z, Vega G, et al. Modification of low density lipoprotein by advanced glycation end products contributes to the dyslipidemia of diabetes and renal insufficiency. Proc Natl Acad Sci U S A 1994;91:9441–5.

[249] Verma N, Manna SK. Advanced glycation end products (AGE) potently induce autophagy through activation of RAF kinase and NF-KB. J Biol Chem 2016;291(3):1481–91.

[250] Adachi T, Inoue M, Hara H, et al. Relationship of plasma extracellular-superoxide dismutase level with insulin resistance in type 2 diabetic patients. J Endocrinol 2004;181:413–7.

[251] Liang F, Kume S, Koya D. SIRT1 and insulin resistance. Nat Rev Endocrinol 2009;5:367–73.

[252] Yoshizaki T, Milne JC, Imamura T, et al. SIRT1 exerts anti-inflammatory effects and improves insulin sensitivity in adipocytes. Mol Cell Biol 2009;29:1363–74.

[253] de Kreutzenberg SV, Ceolotto G, Papparella I, et al. Downregulation of the longevity-associated protein sirtuin 1 in insulin resistance and metabolic syndrome: potential biochemical mechanisms. Diabetes 2010;59:1006–15.

[254] Migliaccio E, Giorgio M, Mele S, et al. The p66shc adaptor protein controls oxidative stress response and life span in mammals. Nature 1999;402:309–13.

[255] Bierhaus A, Shiekofer S, Schwaninger M, et al. Diabetes-associated sustained activation of the transcription factor nuclear factor-KB. Diabetes 2001;50:2792–808.

[256] Basta G, Schmidt AM, De Caterina R. Advanced glycation end products and vascular inflammation: implications for accelerated atherosclerosis in diabetes. Cardiovasc Res 2004;63:582–92.

[257] Borra MT, Smith BC, Denu JM. Mechanism of human SIRT1 activation by resveratrol. J Biol Chem 2005;280:17187–95.

[258] Langley E, Pearson M, Faretta M, et al. Human SIR2 deacetylates p53 and antagonizes PML/p53-induced cellular senescence. EMBO J 2002;21:2383–96.

[259] Sadowska-Bartoz I, Galiniak S, Barttoz G. Polyphenols protect against protein glycation. Free Radic Biol Med 2014;75(Suppl. 1):S47.

[260] Orgaard A, Jensen L. The effects of soy isoflavones on obesity. Exp Biol Med 2008;233:1066–80.

[261] Wolfram S. Effects of green tea and EGCG on cardiovascular and metabolic health. J Am Coll Nutr 2007;26:373S–388S.

[262] Jang HJ, Ridgeway SD, Kim J. Effects of the green tea polyphenol epigallocatechin-3-gallate on high-fat diet-induced insulin resistance and endothelial dysfunction. Am J Physiol Endocrinol Metab 2013;305(12):E1444–51.

[263] van Dam RM, Hu FB. Coffee consumption and risk of type 2 diabetes: a systematic review. JAMA 2005;294:97–104.

[264] Zunino S. Type 2 diabetes and glycemic response to grapes or grape products. J Nutr 2009;139:1794S–1800S.

[265] Boyer J, Liu RH. Apple phytochemicals and their health benefits. Nutr J 2004;3:5.

[266] Hui H, Tang G, Go VL. Hypoglycemic herbs and their action mechanisms. Chin Med 2009;4:11.

[267] Iwai K, Kim MY, Onodera A, et al. Alpha-glucosidase inhibitory and antihyperglycemic effects of polyphenols in the fruit of Viburnum dilatatum Thunb. J Agric Food Chem 2006;54:4588–92.

[268] Tadera K, Minami Y, Takamatsu K, et al. Inhibition of alpha-glucosidase and alpha amylase by flavonoids. J Nutr Sci Vitaminol 2006;52:149–53.

[269] Lo Piparo E, Scheib H, Frei N, et al. Flavonoids for controlling starch digestion: structural requirements for inhibiting human alpha-amylase. J Med Chem 2008;51:3555–61.

[270] Kim JS, Kwon CS, Son KH. Inhibition of alpha-glucosidase and amylase by luteolin, a flavonoid. Biosci Biotechnol Biochem 2000;64:2458–61.

[271] Funke I, Melzi MF. Effect of different phenolic compounds on alpha-amylase activity: screening by microplate-reader based kinetic assay. Pharmazie 2005;60:796–7.

[272] Narita Y, Inouye K. Kinetic analysis and mechanism on the inhibition of chlorogenic acid and its components against porcine pancreas alpha-amylase isozymes I and II. J Agric Food Chem 2009;57:9218–25.

[273] McDougall GJ, Shpiro F, Dobson P, et al. Different polyphenolic components of soft fruits inhibit alpha-amylase and alpha-glucosidase. J Agric Food Chem 2005;53:2760–6.

[274] Lee YA, Cho EJ, Tanaka T, et al. Inhibitory activities of proanthocyanidins from persimmon against oxidative stress and digestive enzymes related to diabetes. J Nutr Sci Vitaminol 2007;53:287–92.

[275] Adisakwattana S, Charoenlertkul P, Yibchok-Anun S. Alpha-glucosidase inhibitory activity of cyanidin-3-galactoside and syner-gistic effect with acarbose. J Enzyme Inhib Med Chem 2009;24:65–9.

[276] Adisakwattana S, Ngamrojanavanich N, Kalampakorn K, et al. Inhibitory activity of cyanidin-3-rutinoside on alpha-glucosidase. J Enzyme Inhib Med Chem 2004;19:313–6.

[277] Matsui T, Ueda T, Oki T, et al. Alpha-glucosidase inhibitory action of natural acylated anthocyanins. 2. Alpha-glucosidase inhibition by isolated acylated anthocyanins. J Agric Food Chem 2001;49:1952–6.

[278] Matsui T, Ueda T, Oki T, et al. Alpha-glucosidase inhibitory action of natural acylated anthocyanins. Survey of natural pigments with potent inhibitory activity. J Agric Food Chem 2001;49:1948–51.

[279] Lee DS, Lee SH. Genistein, a soy isoflavone, is a potent alpha-glucosidase inhibitor. FEBS Lett 2001;501:84–6.

[280] Adisakwattana S, Chantarasinlapin P, Thammarat H, et al. A series of cinnamic acid derivatives and their inhibitory activity on intestinal alpha-glucosidase. J Enzyme Inhib Med Chem 2009;24:1194–200.

[281] Chauhan A, Gupta S, Mahmood A. Effect of tannic acid on brush border disaccharidases in mammalian intestine. Indian J Exp Biol 2007;45:353–8.

[282] Schafer A, Hogger P. Oligomeric procyanidins of French maritime pine bark extract (Pycnogenol) effectively inhibit alpha-glucosidase. Diabetes Res Clin Pract 2007;77:41–6.

[283] Welsch CA, Lachance PA, Wasserman BP. Dietary phenolic compounds: Inhibition of Na+-dependent D-glucose uptake in rat intestinal brush border membrane vesicles. J Nutr 1989;119:1698–704.

[284] Cermak R, Landgraf S, Wolffram S. Quercetin glucosides inhibit glucose uptake into brushborder-membrane vesicles of porcine jejunum. Br J Nutr 2004;91:849–55.

[285] Kobayashi Y, Suzuki M, Satsu H, et al. Green tea polyphenols inhibit the sodium-dependent glucose transporter of intestinal epithelial cells by a competitive mechanism. J Agric Food Chem 2000;48:5618–23.

[286] Shimizu M, Kobayashi Y, Suzuki M, et al. Regulation of intestinal glucose transport by tea catechins. Biofactors 2000;13:61–5.

II. ENDOTHELIAL DYSFUNCTION AND CLINICAL SYNDROMES

[287] Johnston K, Sharp P, Clifford M, et al. Dietary polyphenols decrease glucose uptake by human intestinal Caco-2 cells. FEBS Lett 2005;579:1653–7.

[288] Li JM, Che CT, Lau CB, et al. Inhibition of intestinal and renal Na$^+$-glucose cotransporter by naringenin. Int J Biochem Cell Biol 2006;38:985–95.

[289] Song J, Kwon O, Chen S, et al. Flavonoid inhibition of sodium-dependent vitamin C transporter 1 (SVCT1) and glucose transporter isoform 2 (GLUT2), intestinal transporters for vitamin C and glucose. J Biol Chem 2002;277:15252–60.

[290] Zhang ZF, Li Q, Liang J, et al. Epigallocatechin-3-O-gallate (EGCG) protects the insulin sensitivity in rat L6 muscle cells exposed to dexamethasone condition. Phytomedicine 2010;17:14–8.

[291] Park CE, Kim MJ, Lee JH, et al. Resveratrol stimulates glucose transport in C2C12 myotubes by activating AMP-activated protein kinase. Exp Mol Med 2007;39:222–9.

[292] Guarente L. Sirtuins, aging, and medicine. N Eng J Med 2013;364:2235–44.

[293] Deng JY, Hsieh PS, Huang JP, et al. Activation of estrogen receptor is crucial for resveratrol-stimulating muscular glucose uptake via both insulin-dependent and -independent pathways. Diabetes 2008;57:1814–23.

[294] Breen DM, Sanli T, Giacca A, et al. Stimulation of muscle cell glucose uptake by resveratrol through sirtuins and AMPK. Biochem Biophys Res Commun 2008;374:117–22.

[295] Lagouge M, Argmann C, Gerhart-Hines Z, et al. Resveratrol improves mitochondrial function and protects against metabolic disease by activating SIRT1 and PGC-1alpha. Cell 2006;127:1109–22.

[296] Fang XK, Gao J, Zhu DN. Kaempferol and quercetin isolated from *Euonymus alatus* improve glucose uptake of 3T3-L1 cells without adipogenesis activity. Life Sci 2008;82:615–22.

[297] Wolfram S, Raederstorff D, Preller M, et al. Epigallocatechin gallate supplementation alleviates diabetes in rodents. J Nutr 2006;136:2512–8.

[298] Collins QF, Liu HY, Pi J, et al. Epigallocatechin-3-gallate (EGCG), a green tea polyphenol, suppresses hepatic gluconeogenesis through 50-AMP-activated protein kinase. J Biol Chem 2007;282:30143–9.

Further Reading

[1] Klatsky AL, Friedman GD, Armstrong MA, et al. Wine, liquor, beer and mortality. Am J Epidemiol 2003;158:585–95.

[2] Romero M, Jimenez R, Sanchez M, et al. Quercetin inhibits vascular superoxide production induced by endothelin-1: role of NADPH oxidase, uncoupled eNOS and PKC. Atherosclerosis 2009;202:58–67.

[3] Lopez-Sepulveda R, Gomez-Guzman M, Zarzuelo MJ, et al. Red wine polyphenols prevent endothelial dysfunction induced by endothelin-1 in rat aorta: role of NADPH oxidase. Clin Sci 2011;120:321–33.

[4] Hemmens B, Mayer B. Enzymology of nitric oxide synthases. Methods Mol Biol 1998;100:1–32.

KIDNEY AND HYPERTENSION

27

Kidney and Endothelium

Jose Jayme Galvão De Lima

INTRODUCTION

The endothelium is one of the most extensive organs of the body; it plays important roles in blood pressure regulation, vascular tonus, hemodynamics, neurotransmission and coagulation, among others. For these reasons, endothelial dysfunction has consequences that impact almost the entire organism (Table 27.1). The purpose of this chapter is to discuss endothelial dysfunction from the perspective of its relations with the kidney.

RENAL ENDOTHELIUM

Morphological and functional renal endothelium characteristics vary markedly in different regions of the kidney, reflecting the multiplicity of the organ's functions (Table 27.2). These differences are more evident when we compare the cortical and medular regions. Thus, endothelial cells coating glomerular capillaries, mainly associated with glomerular filtration are fenestrated, that is, discontinuous, which facilitates glomerular filtration. These cells secrete nitric oxide, endothelin-1, angiotensin, and TNF-a, they also express histocompatibility antigens class II (HLA-DR) and receptors for the endothelial growth factor (VEGF). The latter is produced mainly by podocytes, it increases permeability, promotes the formation of fenestrae and is essential in the endothelium repair processes. In contrast, the endothelium that covers the vasa recta presents water and urea transport channels, and only the one associated with the ascending portion of the loop of Henle is fenestrated.

In basal conditions, the endothelium maintains vessels in a state of relative dilation and releases substances that oppose platelet aggregation and fibrosis. Endothelial aggression by conditions such as hypertension, diabetes, and increased free radicals turn the endothelium vasoconstrictor and prothrombotic. Endothelial dysfunction may evolve to estructural endothelium damage and reduction in vascular network, reducing the supply of oxygen and nutrients to tissues. Compromise of endothelial integrity plays a central role in establishing renal lesions in both chronic and acute diseases. In fact, endothelial dysfunction is currently considered one of the key processes in the evolution of all-cause nephropathies.

Due to its strategic location between the blood and parenchyma, renal vascular endothelium is the primary target of several morbid processes, such as glomerulonefrites, vasculitis, lupus nephritis, preeclampsia, hemolytic uremic syndrome (HUS), acute renal failure (ARF), chronic graft disease (CGD), and in the progression of kidney disease. In all these conditions, endothelial damage plays an important pathogenetic role. Diversity of renal vascular endothelium partially explains renal clinical and pathological manifestations in the face of several aggression factors. These aspects will be discussed in detail later on.

NITRIC OXIDE AND THE KIDNEY

Nitric oxide is one of the main endothelial factors that control vascular homeostasis. The endothelium, by means of nitric oxide release, promotes vasodilation and inhibits inflammation, thrombosis, and cell

421

TABLE 27.1 Vascular Alterations Associated With Endothelial Dysfunction

---

THROMBOEMBOLISM

Platelet aggregation
Coagulation factors
Endothelial adhesion molecules

VASOCONSTRICTION

Vascular tonus
Vascular reactivity

TROPHIC EFFECTS

Hypertrophy
Hyperplasia

VASCULAR LESION

Fibrosis
Atherogenesis
Oxidative stress

INFLAMMATION

TABLE 27.2 Morphological and Functional Characteristics of Renal Endothelium

Vase	Endothelium	Main function
Afferent arterioles	Nonfenestrated	Regulation of the glomerular blood flow
Glomerular capillary	Fenestrated, without diaphragm	Selective filtration of macromolecules
Efferent arteriole	Nonfenestrated	Filtration fraction regulation
Peritubular plexus	Fenestrated	Isosmotic absorption of water and electrolytes
Descending vasa recta	Nonfenestrated	Exchanger with counter-current
Ascending vasa recta	Fenestrated with diaphragm	Exchanger and multiplier with counter-current

proliferation. Nitric oxide regulates renal hemodynamics. In strains of animals in which the gene of the constitutive enzyme nitric oxide-synthase (NOS) is abolished or blocked by the competitive inhibitor L-arginine *N*-nitro methyl ester (L-NAME) a consistent increase in systemic blood pressure

and in glomerular capilar pressure, as well as, a reduced glomerular filtration rate, filtration coefficient, and renal plasma flow are observed [1]. These anomalies are associated with an increase in plasma renin, proteinuria, and, in later stages, by glomerular sclerosis [2], and are totally or partially reversed by the administration of L-arginine [3,4]. These findings suggest that factors capable of interfering with the synthesis and release of nitric oxide can cause functional changes and even irreversible lesions in the kidneys.

ENDOTHELIUM AND NEPHROPATHY PROGRESSION

The integrity of endothelial cells is essential for survival of other cells as result of the endothelium role as an intermediary in the provision of oxygen and nutrients to tissues. One of the most complex problems in nephrology concerns the tendency of several renal diseases, of the most various causes, to evolve to terminal uremia even after the apparent control or extinction of trigger agents. The classic explanation for this phenomenon indicates that hemodynamic adjustments in response to the reduction of nephron population cause increased intraglomerular pressure, which, in turn, lead to progressive destruction of surviving nephrons. Recently, new evidence suggests that in addition to the hemodynamic factor there is an ischemic component due to the continuous endothelium destruction and reduction in the number of vessels, leading to vascular rarefaction. In this sense, VEGF deficiency caused by interference of proinflammatory factors, such as IL-1β, IL-6, and TNF-a, seems to play an important role. VEGF deficiency prevents endothelial cell restoration and contributes to vascular rarefaction. This latter phenomenon adds to the hemodynamic aggression an ischemia component that accelerates the destruction of remaining nephrons.

Nitric oxide reduction that follows renal disease results from decrease in synthesis or increase in destruction, or both. In several animal models of renal failure there is glomerular hypertension, proteinuria, and decreased renal plasmatic flow, similar to the abnormalities observed in the examples of nitric oxide deficient animals already discussed. Correction of this deficiency by prolonged L-arginine administration results in improved renal plasma flow and proteinuria, and reduction of intraglomerular pressure on some models [3,5–7]. In diabetic rats, L-arginine, by favoring nitric oxide synthesis, also reduces glomerular pressure and proteinuria [8]. What is perhaps most important is the verification that nitric oxide metabolism improvement in some models comes with the attenuation of inflammatory reaction and interstitial fibrosis. As we know, interstitial fibrosis, in particular, is one of the most sensitive markers of unfavorable renal disease evolution and, in general, it is preceded by local inflammation. There is currently great interest in determining the factors that interfere in the production and/or destruction of nitric oxide in nephropathies and in seeking means to reverse these abnormalities. One of the candidates is oxidative stress mediated by free radicals.

OXIDATIVE STRESS AND CHRONIC RENAL DISEASE

Oxidative stress results from the imbalance between the formation of free radicals and defense antioxidants mechanisms. Markers of oxidative stress were described in chronic renal failure patients [9]. Free radicals inactivate nitric oxide, causing functional renal alterations as those described in the previous item, in addition to favoring atherogenesis, apoptosis and cell proliferation [10]. Free radicals are generated by various cells and tissues. In the kidney, its main sources are endothelial, glomerular, and interstitial cells. The generation of free radicals is high when endothelium becomes dysfunctional. Inflammatory cells, especially macrophages [11], are also important generation sites of reactive oxygen species. These cells are present in interstitial infiltrates in most nephropathies.

Oxidative stress can be due not only to increased free radical generation, but also to reduce antioxidant factors. Nrf2 (nuclear factor-erythroid-2-related factor 2) controls the expression of several genes involved in antioxidant activity. Deficiency of this factor has been widely documented in chronic renal disease [11] and probably contributes to the increase of oxidative stress in this condition.

In chronic renal disease, in addition to oxidative stress, other factors contribute to the reduction of nitric oxide availability. Thus, nitric oxide generation is compromised in renal disease due to competitive

inhibition of the NOS enzyme by asymmetric dimethylarginine (ADMA). This compound is usually produced by several cells, including endothelial cells from protein metabolism, catabolism being inhibited in renal failure. This is due to impairment in production of the enzyme dimethylarginine dimethylaminohydrolase (DDAH) which hydrolyses ADMA. At the same time, in renal disease, ADMA excretion is reduced leading to increased systemic concentration. Accumulation of ADMA in the blood stream is correlated with several alterations characteristic of nitric oxide deficiency, such as hypertension and atherosclerosis, and it is an independent risk factor for death and cardiovascular events in patients with chronic renal disease. In humans, ADMA infusion increases systemic vascular resistance and blood pressure, in addition to reducing the renal plasma flow and renal sodium excretion. Finally, ADMA favors free radicals generation, possibly via local angiotensin production, and reduces the number of endothelium progenitor cells in the circulation.

Another important aspect is the relationship between endothelial dysfunction and cardiovascular disease in chronic renal patients. Chronic uremia is associated with high morbidity and cardiovascular mortality, while in the general population, the oxidative stress correlates to cardiovascular disease [12]. The relationship between oxidative stress and cardiovascular complications in these individuals remains a matter of debate in literature [13]. The use of antioxidants seems to reduce cardiovascular complications in patients with advanced renal failure [14]. In patients with less severe renal failure, there is no information in this regard. Interventions intended to reduce oxidative stress can interfere in the progression of chronic nephropathies. In this sense, inhibitors of the renin-angiotensin system have shown promise as drugs that can reduce oxidative stress [10].

OTHER FACTORS INVOLVED IN NEPHROPATHY PROGRESSION

Endothelial cells of animals and patients with chronic nephropathy are subjected to mechanical stress resulting from straining and increased hydrostatic pressure. These alterations make the endothelium dysfunctional and collaborate to stimulate the production, by the endothelium and by other local or infiltrative cells, of a wide variety of profibrotic factors, such as TGF-β, endothelin (ET), angiotensinogen, fibronectin, and laminin [15]. Angiotensin-II generated in renal circulation, in addition to promoting intraglomerular pressure increase, stimulates cell proliferation, attracts inflammatory cells and promotes fibrosis possibly mediated by TGF-β. The fact that angiotensin-II is capable of generating free radicals is of interest to pathogenesis of endothelial dysfunction in nephropathies [16]. The activity of endothelial cyclooxygenase is increased in renal failure models, favoring the production of prostacyclin, thromboxane, and prostaglandin F2-a [15]. These factors attract inflammatory cells. Finally, nitric oxide derived from the inducible NOS form (iNOS), released by inflammatory cells, exerts toxic effects and takes part in the pathogenesis of experimental nephropathies [17].

ET is a 21-amino acid peptide produced in three isoforms, ET-1, ET-2, and ET-3, by endothelium cells in the kidneys, brain, and lungs [18]. The physiological effects of ET are mediated by receptors ETA and ETB. ETA stimulation causes vasoconstriction and cell proliferation, while the bind to ETB increases nitric oxide production by the endothelium followed by vasodilation. ET is also implicated as an inflammatory response mediator. In humans, intravenous ET-1 infusion, in doses that do not alter blood pressure, results in a significant renal vasoconstriction with increase of vascular resistance [19]. These results indicate that the kidney is particularly sensitive to the action of ET. It is of interest that in studies in animal models with renal disease blocking of ET receptors resulted in symptom attenuation.

Positive results with the use of ET blockers were observed in animals with Heymann nephritis, autoimmune spontaneous nephritis, cyclosporin nephropathy, and experimental diabetes, in the model of renal ablation, and in several types of ARF [18]. In these models, blocking ET reduced proteinuria, the degree of interstitial fibrosis and blood pressure, and improved glomerular filtration rate at variable intensities. There are few clinical studies with the use ET receptors blockers. In healthy volunteers, blocking ET has caused a drop in blood pressure and systemic vascular resistance [20]. Other investigations are underway in patients with essential hypertension. Thrombospontin-1 (TSP-1) is an antiangiogenic factor inhibitor of endothelial cell proliferation and promoter of apoptosis present in macrophages, in other inflammatory cells and in renal parenchyma cells in animals with experimental nephropathy. In the normal kidney, the production of TSP-1 is small, but it is much increased in several experimental nephropathy models, especially by mesangial cells [6]. TSP-1 promotes renal vascular rarefaction and ischemia; its presence correlates with the progression of renal disease.

Another angiogenesis inhibitor associated with the development of chronic renal disease is the protein SPARC (secreted protein, acidic, cysteine-rich [osteonectin]) that inhibits endothelial proliferation. It is reduced in models of renal hypertrophy, as in diabetes, and increased in models that accompany fibrosis and vascular rarefaction [6].

The deficiency of renoprotective factors is also part of renal disease progression. Erythropoietin (EPO) is a factor of renal origin that regulates hematopoiesis. Other important EPO actions include inhibition of apoptosis and protection against ischemia by means of JAK2 and STAT5 and PI3K/AKT pathway activation [7]. These effects are observed with the use of reduced EPO doses that do not interfere in erythropoiesis. Endothelial cells express receptors for EPO, which, in turn, stimulate nitric oxide production via nitric oxide-synthase (eNOS) and increases the number of endothelial progenitor cells. Since EPO is reduced in renal failure, it is speculated that this deficiency can contribute to the progression of nephropathy. However, investigations in humans have not yet conclusively demonstrated the renoprotective effect of EPO.

As mentioned, receptors for VEGF are present in the endothelium. This is an angiogenic factor constitutively expressed by visceral epithelial cells of the glomerular basal membrane (podocytes) and in renal tubular cells. VEGF increases eNOS activity in endothelium and protects the endothelium against several types of vascular attacks; it is reduced in renal disease and this deficiency contributes to impair the regeneration of damaged endothelial cells. Inflammatory cytokines, such as IL-1β, TNF-a, and IL-6, reduce VEGF expression by renal cells.

BIOMARKERS OF ENDOTHELIAL FUNCTION

Due to the key role of endothelial dysfunction in the pathogenesis of nephropathies, methods that allow early detection of endothelium compromise, especially renal, have potential clinical importance.

Although not yet included in current practice, these markers could be used in the future to assess the severity and progression of nephropathies, in addition to verifying the effectiveness of therapeutic interventions. Among the most promising are the determination of ADMA levels, endothelial microparticles, circulating endothelium progenitor cells, and inflammatory cytokines. These markers, however, are not specific for renal diseases, and may be altered in systemic endothelial dysfunction.

ENDOTHELIUM IN SOME TYPES OF RENAL DISEASE

As we have seen, endothelium participates in the progression of several types of nephropathies. There are some diseases in which endothelial dysfunction plays a central role in pathogenesis. Among these are diabetic nephropathy, HUS and preeclampsia. Table 27.3 summarizes the main endothelial alterations found in human pathology and in experimental models, and the therapeutic approaches used. Diabetes can be considered a prototype of the disease in which endothelium compromise is almost constant, given the large number of factors capable of injuring this tissue. Dyslipidemia, hypertension, hyperglycemia, increased free radicals, and mechanical stress [21] contribute to damage and alter renal endothelial function. In the glomeruli, these alterations lead to increased permeability, with leakage of proteins and other macromolecules to the urinary space; mesangial proliferation with consequent glomerular sclerosis; stimulation of factors favoring fibrosis such as ET and angiotensin-II. Hyperglycemia and tissue glycation products are particularly harmful, since they inactivate nitric oxide and reduce eNOS expression in the endothelium [21].

TABLE 27.3 Alterations of the Endothelium in Experimental and Clinical Conditions

Condition	Alteration	Treatment	Reference
L-NAME administration (rat)	Reduction of nitric oxide, hypertension, proteinuria, glomerulosclerosis, hyperreninemia	L-Arginine	[1,2]
Renal ablation (rat)	Reduction of nitric oxide, interstitial fibrosis, proteinuria, uremia	L-Arginine	[7]
Obstructive uropathy (rat)	Reduction of nitric oxide, interstitial fibrosis	L-Arginine	[3]
Spontaneous diabetes (rat)	Reduction of nitric oxide, increased renal angiotensin generation	L-Arginine, ACE inhibitors	[4,8]
Human CRF	Increased oxidative stress	ACE inhibitors	[9,10,13]
Endothelin EV (man)	Renal vasoconstriction		[19]
Experimental and spontaneous nephritis, diabetes and renal ablation (rat)	Hypertension, proteinuria, interstitial fibrosis, uremia	Blocking of ET_A/ET_B	[18]
Diabetes	Increased angiotensin renal generation, free radicals, reduction of nitric oxide	ACE inhibitors	[4,21]
Hemolytic uremic syndrome	Endothelium destruction, nitric oxide reduction, hypertension, uremia, hypercoagulation/fibrinolysis	Fresh plasma (protease?)	[22]
Malignant nephrosclerosis (man)	Endothelium lesion, nitric oxide reduction, hypertension, uremia, hypercoagulation/fibrinolysis, hyperreninemia	ACE inhibitors	[22,23]
Preeclampsia	Endothelium lesions, hypertension, proteinuria	Aspirin	[23]

See text for abbreviations.
CRF, *chronic renal failure.*

The pathogenesis of HUS and of its clinical variant thrombotic thrombocytopenic purpura have not yet been fully clarified, but there is no doubt of renal endothelium participation in the syndrome's main expressions, which includes ARF, arterial hypertension and microangiopathic hemolytic anemia. Recent evidence suggests that a protease deficiency or inactivation, probably of endothelial origin, capable of depolymerizing molecules of high molecular weights of the von Willebrand factor, is critical to trigger the syndrome [22]. Von Willebrand's factor of high molecular weight, when found in high concentrations in plasma, causes platelet aggregation, thromboembolism, and disseminated endothelial injury, specially in the kidneys. These alterations almost always come with other signs of endothelial dysfunction, such as increased oxygen-free radicals and of vasoconstrictor and prothrombotic factors. In the hemolytic-uremic syndrome, caused by *Escherichia coli* strains, which express verotoxin, higher susceptibility of renal cortical vessels can be explained by the abundance of receptors for verotoxin in the cortical endothelium in comparison to other kidney regions.

Malignant nephrosclerosis has several points in common with hemolytic-uremic syndrome: hypertension, elevation of renal vascular resistance, microangiopathic hemolytic anemia, hypercoagulability, and fibrinolysis. As occurs in the hypertension model induced by L-NAME [1], plasma renin is markedly increased. These alterations are in large part due to the destruction of the renal endothelium.

Preeclampsia is another condition in which the participation of the renal endothelium seems to be relevant. It is a multisystemic disease peculiar to pregnancy. It is characterized by hypertension, proteinuria, renal dysfunction, and microangiopathy. As in the case of hemolytic uremic syndrome, pathogenesis is controversial. However, a likely role of endothelial dysfunction is indicated by the reduction of VEGF production generated by an inhibitor originating from the placenta, causing endotheliosis, proteinuria, and decreased glomerular filtration rate [23].

Endothelial lesions are also observed in acute renal lesions (ARL) caused by ischemia or sepsis, and seem to be associated with activation of inflammatory factors and increased expression in endothelial cells of molecules that facilitate the adhesion of inflammatory cells to the endothelium.

Changes in renal microvascular endothelium are relevant characteristics of chronic allograft nephropathy (CAN) an important cause of renal transplant function loss. In that case, the presence of histocompatibility antigens class II in the cortical renal endothelium plays an important role. An important characteristic of CAN is the presence of the complement system activation marker C4d in the endothelium and this finding is an independent predictor of interstitial fibrosis and graft loss. Data from our laboratory suggest that secondary hyperparathyroidism has a negative effect on endothelial function in hemodialysis patients [24] and that parathyroidectomy reduces cardiovascular mortality in that setting [25]. In one study, vitamin D deficiency has been associated with endothelial dysfunction in chronic renal failure patients [26].

THERAPEUTIC IMPLICATIONS

General measures that benefit the cardiovascular system also improve endothelial function. Among them are regular exercises, abstention of tobacco, control of weight, dyslipidemia, diabetes, and hypertension. The value of more specific interventions for endothelium protection, however, still lacks solid experimental evidence. The best option is the use of renin-angiotensin inhibitors that increase nitric oxide bioavailability by reducing oxidative stress. Other drugs that may reduce oxidative stress, such as vitamin E and *N*-acetylcysteine (NAC), have therapeutic potential and are under evaluation. Folic acid participates as a cofactor in ADMA catabolism and is commonly prescribed for chronic renal patients, but their effectiveness is still debated. Another approach is the use of drugs that stimulate renal endothelium preservation and repair. Thiazolidinediones are drugs used in diabetes treatment and they exert their effects by increased expression of PPAR isoform γ (PPAR-γ) in the membrane of several tissues, including the endothelium. In animals and in vitro studies, receptor PPAR-γ activation stimulates endothelial cell regeneration by factors such as VEGF, in addition to increasing nitric oxide production and inhibiting endothelin-1 synthesis [27]. Drugs that stimulate angiogenesis have great therapeutic potential in theory, but there are problems with their use due to the possibility of an increased incidence of tumors and worsening of retinopathy in diabetic patients. The value of EPO is still under investigation. In experimental models and in humans, NAC, an oxidative stress blocker, reduces the incidence of acute renal injury. Recently, our group demonstrated that NAC in maximum doses reduces by about 50% the incidence of acute renal lesion in patients with chronic renal disease undergone coronary intervention [28]. Other authors have also demonstrated the importance of endothelium in postoperative acute renal injury [29]. In patients with type 2 diabetes, bardoxolone methyl, an inflammatory response modulator and antioxidant that acts as a Keap1-Nrf2 [11] pathway activator was recently tested in humans with encouraging results such as preservation of glomerular filtration rate [30]. Atrasentan, an ET blocker, reduced proteinuria in patients with diabetes in one study [31]. In rats, blocking trombospondin-1 (TSP-1) activity reduces proteinuria in diabetic animals [32], but this treatment has not yet been evaluated. A large number of clinical studies are still necessary in this quite promising area.

CONCLUSIONS

Renal endothelium plays crucial roles in blood pressure regulation, resistance, and vascular tonus. It is not surprising, therefore, that the kidney is particularly sensitive to endothelial dysfunction, which is found in virtually all patients with renal disease, in whom it plays a significant role in evolution and prognosis. Furthermore, in certain renal pathologies, the endothelial dysfunction seems to be in the center of disorders. From the clinical point of view, there are still few methods for intervention upon the endothelium capable of influencing the course of nephropathies. We still lack methods that allow direct evaluation of renal endothelium. This is an important point, since systemic endothelial function may not necessarily correlate with local endothelial function in different organs. It is also evident that more information on the biology, phenotypic, and genotypic characteristics of endothelial cells are still required in order to generate approaches capable of affecting the development of nephropathies. These procedures should focus on endothelial lesion prevention and on means to accelerate its recovery.

References

[1] Baylis C, Mitruka B, Deng A. Chronic blockade of nitric oxide synthesis in the rat produces systemic hypertension and glomerular damage. J Clin Invest 1992;90:278–81.
[2] Ribeiro MO, Antunes E, de Nucci G, et al. Chronic inhibition of nitric oxide synthesis: a new model arterial hypertension. Hypertension 1992;20:298–303.
[3] Cherla G, Jaimes EA. Role of L-arginine in the pathogenesis and treatment of renal disease. J Nutr 2004;134:2801S–2806S.
[4] Klahr S. The role of nitric oxide in hypertension and renal disease progression. Nephrol Dial Transplant 2001;16(Suppl. 1):60–2.
[5] Malyzco J. Mechanism of endothelial dysfunction in chronic kidney disease. Clin Chim Acta 2010;411:1412–20.
[6] Kang DH, Kanellis J, Hugo C, et al. Role of microvascular endothelium in progressive renal disease. J Am Soc Nephrol 2002;13:806–16.
[7] Filser D. Perspectives in renal disease progression: the endothelium as a treatment target in chronic kidney disease. J Nephrol 2010;23:369–76.
[8] Reys A, Karl IE, Kissane J, et al. L-Arginine administration prevents glomerular hyperfiltration and decreases proteinuria in diabetic rats. J Am Soc Nephrol 1993;4:1039–45.
[9] Small DM, Coombes JS, Bennett N, et al. Oxidative stress, antioxidant therapies and chronic kidney disease. Nephrology (Carlton) 2012;17:311–21.
[10] Cachofeiro V, Goicochea M, de Vinuesa SG, et al. Oxidative stress and inflammation, a link between chronic kidney disease and cardiovascular disease. Kidney Int 2008;74:S4–9.
[11] Ruiz S, Pergola PE, Zager RA, et al. Targeting the transcription factor Nrf2 to ameliorate oxidative stress and inflammation in chronic kidney disease. Kidney Int 2013;83:1029–41.
[12] Csányi G, Miller Jr FJ. Oxidative stress in cardiovascular disease. Int J Mol Sci 2014;15:6002–8.
[13] Jun M, Venkataraman V, Razavian M, et al. Antioxidants for chronic kidney disease. Cochrane Database Syst Rev 2012;10:CD008176.
[14] Boaz M, Smetana S, Weintein T, et al. Secondary prevention with antioxidants of cardiovascular disease in end stage renal disease (SPACE): randomized placebo-controlled trial. Lancet 2000;356:1213–8.
[15] Zatz R, Fujihara CK. Mechanisms of progressive renal disease: role of angiotensin II, cyclooxigenase products and nitric oxide. J Hypertens 2002;20(Suppl. 3):S37–44.
[16] Zhong JC, Guo D, Chen CB, et al. Prevention of angiotensin II-mediated renal oxidative stress, inflammation, and fibrosis by angiotensin converting enzyme 2. Hypertension 2011;57:314–22.
[17] Wang Y, Harris DCH. Macrophages in renal disease. J Am Soc Nephrol 2011;22:21–7.
[18] Maguire JJ, Davenport AP. Endothelin@25—new agonists, antagonists, inhibitors and emerging research frontiers: IUPHAR review. Br J Pharmacol 2014;171:5555–72.
[19] Sorensen SS, Madsen JK, Pedersen EB. Systemic and renal effect of intravenous infusion of endothelin-1 in healthy human volunteers. Am J Physiol 1994;266:F411–8.
[20] Haynes WG, Ferro CJ, O'Kane KPJ, et al. Systemic endothelin receptor blockade decreases peripheral vascular resistance and blood pressure in humans. Circulation 1996;93:1860–70.
[21] Tabit CE, Chung WB, Hamburg NM, et al. Endothelial dysfunction in diabetes mellitus: molecular mechanisms and clinical implications. Rev Endocr Metab Disord 2010;11:61–74.
[22] Boyer O, Niaudet P. Hemolytic uremic syndrome: new developments in pathogenesis and treatment. Int J Nephrol 2011;2011:908407.
[23] Noori M, Donald AE, Angelakopoulou A, et al. Prospective study of placental angiogenic factors and maternal vascular function

before and after preeclampsia and gestational hypertension. Circulation 2010;122:478–87.

[24] Bortolotto LA, Costa-Hong V, Jorgetti V, et al. Vascular changes in chronic renal disease patients with secondary hyperparathyroidism. J Nephrol 2007;20:1–7.

[25] Costa-Hong V, Jorgetti V, Gowdak LHW, et al. Parathyroidectomy reduces cardiovascular events and mortality in renal hyperparathyroidism. Surgery 2007;142:699–703.

[26] Chitalia N, Recio-Mayoral A, Kaski JC, et al. Vitamin D deficiency and endothelial dysfunction in non-dialysis chronic kidney disease patients. Atherosclerosis 2012;220:265–8.

[27] Westerweel PE, Verhaar M. Protective actions of PPAR-g activation in renal endothelium. PPAR Res 2008;10:1–9.

[28] Santana-Santos E, Gowdak LH, Gaiotto FA, et al. High dose of N-acetylcystein prevents acute kidney injury in chronic kidney disease patients undergoing myocardial revascularization. Ann Thorac Surg 2014;97:1617–23.

[29] Xu C, Chang A, Hack BK, et al. TNF-mediated damage to glomerular endothelium is an important determinant of acute kidney injury in sepsis. Kidney Int 2014;85:72–81.

[30] Pergola PE, Raskin PR, Toto RT, et al. Bardoxolone methyl and kidney function in CKD with type 2 diabetes. N Engl J Med 2011;365: 327–36.

[31] Kohan DE, Pritchett Y, Molitch M, et al. Addition of atrasentan to renin-angiotensin system blockade reduces albuminuria in diabetic nephropathy. J Am Soc Nephrol 2011;22:763–72.

[32] Lu A, Miao M, Schoeb TR, et al. Blockade of TSP1-dependent TGF- activity reduces renal injury and proteinuria in a murine model of diabetic nephropathy. Am J Pathol 2011;178:2573–86.

28

Endothelium and Arterial Hypertension

Fernanda Marciano Consolim-Colombo and Luiz Aparecido Bortolotto

INTRODUCTION

Systemic arterial hypertension (SAH) is considered one of the most important risk factors for development of cardiovascular diseases such as coronary syndromes, stroke and congestive heart failure [1]. Considering the current values that define arterial hypertension, sustained blood pressure levels equal or above 140/90 mmHg, it is estimated that 25%–30% of the population older than 18 years is hypertensive. However, there is a progressive and linear increase in cardiovascular mortality as blood pressure reaches values above 115 mmHg for systolic and above 75 mmHg for diastolic [2]. The development of hypertension takes place when there is malfunction in different mechanisms that normally act to maintain pressure within a normal range. Several systems are important, such as the sympathetic nervous system, renin-angiotensin-aldosterone system, and sodium and water retention mechanisms controlled by the kidneys [3]. Since recognition of endothelium as an active endocrine system, of great importance in regulation of vascular tone, the role of endothelial dysfunction in development of hypertension and development of target organ injury, has been widely studied [4].

ENDOTHELIAL DYSFUNCTION

Thus, inability of a vascular segment (e.g., coronary artery) to dilate in response to direct stimuli to endothelial cells such as acetylcholine infusion was defined as endothelial dysfunction [5]. In parallel to the increasing knowledge on the role of endothelial cells in vessel dynamics and in blood components, this concept was extended. Currently, "endothelial dysfunction" is understood as a functional alteration of the endothelium, which starts to have pro-thrombotic, pro-inflammatory, and pro-constrictive phenotype [6].

Presence of endothelial dysfunction is frequently detected in animal models of hypertension and in hypertensive patients, in very early stages of the disease [7,8]. In fact, this dysfunction has been systematically demonstrated in primary (Fig. 28.1), secondary, and even in offspring of hypertensive individuals [9–11]. However, despite a lot of research, it has not yet been clarified whether endothelial dysfunction is a cause or consequence of hypertension [12]. Recent findings on the role of endothelial dysfunction in the pathophysiologic mechanisms associated with SAH indicate that oxidative stress and inflammatory processes are crucial both to start and to maintain the functional and structural changes of vessels that characterize this pathology [13].

Use of various classes of antihypertensive drugs was effective in reducing blood pressure and thus reduce the high morbidity and mortality associated with SAH. However, there is a provocative hypothesis that some classes are able to improve endothelial function, regardless of their effect on blood pressure values, and may add beneficial effects on the cardiovascular prognosis of hypertensive patients [14]. Thus, it is evident the importance of understanding, in the light of current knowledge, the relationship among endothelial function, SAH, cardiovascular morbidity and mortality.

Endothelium has many physiological functions, which act together in order to maintain vascular health [15]. As endothelial cells are in a strategic position in the vessel wall, they receive

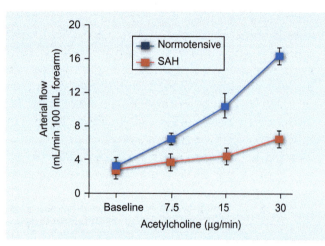

FIG. 28.1 Endothelial function assessed by vasodilator response induced by acetylcholine in patients with primary hypertension (SAH) and altered in comparison to normotensive individuals. From Arterial Hypertension Team, InCor.

hemodynamic and humoral signals; in response to this signaling, the endothelium synthesize substances that affect not only the vessel cells themselves but also circulating cells in the blood. They function, therefore, as effectors of local adaptive responses. Generally, factors produced by endothelium regulate vascular tone, monocyte proliferation, local lipid metabolism, cell growth and migration, and integration with extracellular matrix. In addition, the endothelium produces adhesion proteins and can act as a critical trigger for inflammatory response, mediating the passage of inflammatory cells through the vascular wall [16].

VASODILATORS: NITRIC OXIDE

Substances that regulate vascular tonus act in vessels smooth muscle, causing dilation (e.g., nitric oxide, prostacyclins (PGI-2), endothelial hyperpolarizing factor, and bradykinin) and contraction (endothelin, thromboxane A2, Prostaglandin H2, angiotensin-II, and superoxide anions). Under physiological conditions, there is balance in the release of vasodilators and vasoconstrictors by the endothelium. When there is an imbalance in the release of these factors, due to increase in the release of vasoconstrictors or reduction of vasodilators availability, it is called endothelial dysfunction. Increased constriction of microcirculation vessels increase peripheral vascular resistance, raising blood pressure values. On the other hand, endothelial dysfunction

in large arteries (coronary, renal, brachial, femoral) limits vasodilatation of these arterial segments, and is directly related to remodeling process of the arterial wall and to development of atheromatous disease [17]. More recently, we demonstrated venous endothelial dysfunction in hypertensive patients [18]. Functional impact of the poor vasodilation of this territory seems to be due a reduction of capacity to accommodate venous volume, that is, two-thirds of the total blood volume of the body.

Cellular and molecular mechanisms involved in the development of hypertension may include lower bioavailability of nitric oxide and increased oxidative stress. Nitric oxide is an important vasodilator produced by endothelium, but it has several other functions [14]. It is a signaling molecule that regulates several intracellular pathways of the endothelium itself and interferes with the metabolism of adjacent cells of other tissues. Because of their actions, nitric oxide is considered the main vasodilator trigger, and also have anti-inflammatory, antiproliferative and antithrombotic properties. The vasodilatation and all effects linked to this oxide depend on the quantity available in tissues (bioavailability), which depends on the relation between synthesis and degradation of nitric oxide. Thus, the bioavailability of nitric oxide can be reduced, either if the synthesis decrease or if the degradation increased. Different animal models of arterial hypertension have less nitric oxide bioavailability, but they present different synthesis and degradation capacities. In addition, chronic inhibition of nitric oxide synthase (NOS), through drugs or genetic manipulation, causes hypertension in experimental animals, while increase in nitric oxide levels lowers blood pressure in these models [19].

Studies in humans have shown decreased nitric oxide bioavailability, reduction of nitric oxide *per se* and lower vasodilating effect at different stages of hypertension development [20]. The NOS can also serve as a source of active oxygen species when the flow of electrons is uncoupled from nitric oxide synthesis and not linked to superoxide production [21,22]. Loss of the NOS cofactor tetrahydrobiopterin (BH4) has been identified as the main cause of the disconnection, and oxidative stress can also deplete BH4 levels [23]. Increased BH4 levels lower blood pressure in animal models and improves vasodilation in femoral arteries in spontaneously hypertensive animals [24] and in the forearm of subjects with hypertension [25]. Supplementation with BH4, aiming to reverse the uncoupling of NOS, is currently approved by the Food and Drug Administration (FDA) for phenylketonuria and has been used to treat hypertension in clinical trials [26,27]. In fact, oral supplementation of BH4 also decreases BP in patients with poorly controlled hypertension [28].

Evidence accumulated in the last decade strongly supports the notion that reactive oxygen species (ROS) are generated in the vasculature primarily by NADPH oxidase, by a mechanism that is dependent on angiotensin-II [29,30]. Activation of this enzyme leads to production of superoxide and uncoupling of endothelial nitric oxide synthase (eNOS), which maintains oxidative stress, increasing levels of peroxynitrite that is harmful to the tissues. The latter process may result in vascular dysfunction. Formation of NADPH-dependent ROS, in particular H_2O_2, may also contribute to vascular injury by maintaining activation of NADPH oxidase, which promotes expression of inflammatory genes, reorganization of extracellular matrix and growth (hypertrophy/hyperplasia) of vascular smooth muscle cells. Effect of ROS appears to be mediated by redox targets, such as tyrosine kinases and phosphatases, mitogen-activated protein kinases, transcription factors, matrix metalloproteinases, activated peroxisome proliferator, alpha-receptor, poly (ADP-ribose) polymerase-1, Ca (2+) signaling mechanisms and secreted factors, such as cyclophilin A and heat shock protein 90-alpha [31]. Redox targets seem to play a central role in normal vascular function, but can also lead to remodeling of the vascular wall, and improve vascular reactivity and hypertension. Polymorphisms of the p22phox promoter gene could determine susceptibility to oxidative stress mediated by NADPH oxidase in humans and animals with hypertension. Although ROS are strongly implicated in the etiology of hypertension, clinical trials with antioxidants are inconclusive in relation to their efficacy in treating the disease. New drugs with antihypertensive action and antioxidant properties (celiprolol, carvedilol) offer promising results in hypertension management.

Another signaling pathway that, when stimulated, may interrupt normal functioning of the cell, triggering oxidative stress or acting on cell survival pathways, is endoplasmic reticulum stress (ERS) [32]. Accumulation of unfolded proteins in the reticulum triggers an intracellular activation sequence designated UPR (unfolded protein response), to restore cellular homeostasis [33]. Prolonged activation of this cellular response results in ERS, and recent studies have linked the ERS in different sites and the development of hypertension [34]. Interestingly, ERS in subfornical organ precedes development of oxidative stress induced by angiotensin-II in mice, and that ERS inhibition blocks the increase of blood pressure induced by angiotensin-II [35]. These data corroborate other observations that inhibition of ERS decreases oxidative stress markers in animals vessels [36]. These studies raise the possibility that ERS may be the most promising therapeutic target in blocking oxidative stress increase, if it can be possible to determine that ERS is a modulator which precedes this process (upstream modulator).

Nevertheless, if we consider the possibility that endothelial dysfunction associated with hypertensive state may be due to increase in synthesis and release of vasoconstricting substances by the endothelium, we must highlight the role of endothelin-1 (ET-1). The ET-1 is one of the most potent vasoconstrictors produced by the human body, and has a natriuretic action in the kidneys [37]. Molecular and biochemical characteristics of ET-1 are well defined, but its importance in cardiovascular and renal regulation is still being evaluated. The ET-1 can cause an increase in blood pressure, activating type A specific receptors (ETA), or produce antihypertensive effects by activating the pathway that starts with stimulation of type B receptors in the kidneys. Thus, ET-1 can influence BP regulation in different ways depending on the region where it is being produced and what type of receptor is activated. Thus, ET-1, by activating ETA receptors, produces systemic and renal vasoconstriction, alters the pressure-natriuresis curve, reduces glomerular filtration rate, and induces cell proliferation in several diseases, including hypertension [37]. Despite participation of ET-1 in animal hypertension models, its role in human hypertension is still obscure. Clinical use of a drug that inhibits ETA receptors has not been shown to be effective in the treatment of hypertension. New drugs with greater selectivity for ET-1 receptors are being studied [38].

IMMUNOLOGICAL MECHANISMS AND ARTERIAL HYPERTENSION

Despite extensive research, etiology of primary hypertension remains undefined. Over the last few years, more and more studies have addressed the role of immunity in cardiovascular disease [39]. "Low-grade" or subclinical inflammation is currently a recognized characteristic of hypertension, and there is a growing number of evidences regarding the role of humoral factors (cytokines) and inflammatory cells in different animal models of hypertension and also in clinical trials. There are several excellent reviews describing the immune system in hypertension [40–43]. Much of the data presented is derived from experimental and clinical observational studies since trials evaluating interventional approaches on this factor in clinical practice are still ongoing.

A new paradigm is thus revealed, which includes active participation of different components of the immune system, including innate (e.g., macrophages) and adaptive (effector T lymphocytes and regulatory T lymphocytes) immunity in mechanisms associated with hypertension pathophysiology. In addition, some studies have demonstrated the beneficial effect of immune therapy, which was able to improve or prevent experimental hypertension in animal models [40,41,43]. In this scenario we shall briefly present the most current data on immune mechanisms and hypertension.

Subclinical inflammation is increasingly recognized as an integral part of the developmental pathophysiology and progression of vascular disease [39]. Studies started four decades ago have suggested a relationship between the immune system and hypertension, in models of renal infarction, hypertension induced by mineralocorticoid, and spontaneously hypertensive rats. After a period of limited information, literature in the past decade has added information on immunological basis of hypertension, mediated by inflammatory cytokines and immune system cells [29].

Guzik et al. have recognized a role for the adaptive immune response in promoting blood pressure elevation [44]. RAG-1 −/− mice (animals genetically deficient in T and B lymphocytes) were resistant to development of hypertension and protected from vascular injury induced by angiotensin II infusion. Adoptive transfer of T cells, but not B, restored hypertensive response. These researchers also showed a characteristic pattern of T cell infiltration into the adventitia of blood vessels, whose expression of NADPH oxidated and cytokines, such as interleukin-17 (IL-17), of T cells, were required for maximum elevation of blood pressure [44]. Crowley et al. conducted an experiment in mice with combined immunodeficiency, which do not develop T or B lymphocytes (in a manner similar to RAG-1 −/− mice), and also observed resistance to

blood pressure elevation and lower degrees of ventricular hypertrophy, cardiac fibrosis and albuminuria in these animals after angiotensin-II infusion [45].

Harrison et al. advanced the hypothesis that a hypertensive stimulus leads to renal injury, formation of antigens (neoantigens), and activation of T lymphocytes in the kidneys. Cytokines released by T lymphocytes [41] promotes the entry of other inflammatory cells such as macrophages, in the kidney, and in perivascular fat leading to renal vasoconstriction and increased sodium reabsorption, increasing the severity of hypertension [41].

Specifically, Th17 lymphocytes contribute to increases in BP and other vascular lesions, whereas regulatory T lymphocytes (Tregs) are immunosuppressive and limit increases in blood pressure [46]. Animal studies showed that administration of immunonosuppressors reduces blood pressure of spontaneously hypertensive animals, associated with an increase in Treg cells and reduction of Th17 cells [47]. In addition, Treg lymphocytes reduce oxidative stress and vascular endothelial dysfunction induced by angiotensin II, thereby reducing the vascular injury [48]. These cells also diminish heart damage (hypertrophy, fibrosis, inflammation) in response to angiotensin-II infusion through anti-inflammatory action [49].

Among the questions raised by these studies, one should investigate whether T lymphocyte response reflects immunity to specific antigens, or is simply a nonspecific response to tissue injury. Recent studies by Harrison et al. suggest that antigen-specific T cell response leads to hypertension [45]. Another point of interest is the participation of central nervous system areas such as immune activation circuit, causing angiotensin-II-dependent hypertension [50]. These findings are provocative, but it is not known how much of this information can be translated to the pathogenesis of essential hypertension in humans.

THE ROLE OF CYTOKINES IN HYPERTENSION

Cytokines are the major components that regulate T lymphocytes, and are produced by cells of immune system and other tissues. There are proinflammatory (e.g., IL-6, IFN-γ) and anti-inflammatory (e.g., IL-10) cytokines. The importance of certain cytokines has been explored in studies with different models of hypertension: gestational hypertension (IL-10), ischemic renal injury (IL-10, IL-17); Angiotensin-II infusion (IL-10, IL-17, TNF-α); fructose overload (TNF-α) [42]. It is important to note that the final result does not depend solely on a specific cytokine, but on a set of several cytokines present in the site that is being evaluated. Thus, an environment rich in proinflammatory cytokines will promote activation of immune cells, which produce more cytokines and exacerbate inflammatory cascade, and which, under certain circumstances, may contribute to mechanisms that lead to hypertension and target organs lesions.

Several clinical studies have demonstrated increased inflammatory markers and vascular lesion markers in patients with hypertension [41]. The us-CRP is the most studied marker of vascular inflammation, and the more robust and reproducible in hypertension. It has been demonstrated in hypertensive patients, that us-CRP continuously and gradually increased, and also elevated levels of this protein in normotensive and prehypertensive patients are predictors of hypertension development.

Data regarding the association of other inflammatory markers (IL-6 and IL-1b cytokines, TNF-α) exist, but is not as consistent. The list of vascular lesion markers involved in inflammation and hypertension has been continuously increased (fibrinogen, VWF, VCAM-1, ICAM-1, P-selectin, E-selectin, D-dimer, PAI-1, endothelin, MP-2, CD40 linker) [29].

In clinical context, inflammation causes endothelial dysfunction and, consequently, changes in vasodilators and vasoconstrictors synthesis. In this context, endothelial dysfunction may represent the basis upon which chronic inflammation leads to hypertension, but mechanisms are not fully elucidated [51]. All demonstrated changes in vessels, including remodeling, stiffening, calcification, and inflammation, share the ability to increase total peripheral resistance (TPR) [40], and consequently lead to arterial hypertension.

ENDOTHELIUM AND ARTERIAL STIFFNESS IN HYPERTENSION

Increased arterial stiffness results from various phenomena involving vessel structure, such as wall fibrosis, elastin fiber rupture, collagen accumulation, inflammation, calcification, macromolecule diffusion into arterial wall and endothelial dysfunction, and can be influenced by physiological conditions, genetic profile, and cardiovascular risk factors [52–54].

In subjects with arterial hypertension, the main arterial wall modification is middle layer hypertrophy [55]. In young hypertensive, alterations in mechanical properties result mainly from hemodynamic effect per se (pressure increase), because decrease in carotid arterial distensibility disappears in isobaric conditions [54,55].

However, in other territories, such as femoral artery or thoracic aorta, intrinsic changes in stiffness (i.e., increased stiffness under isobaric conditions) can also be observed [55]. In hypertensive patients, active mechanisms within the arterial wall must be involved, because in predominantly muscular peripheral artery such as the radial artery, the diameter is unchanged despite the high pressure whereas in central arteries, predominatly elastic, the diameter increases according to pressure [55].

Among older hypertensive patients, wall hypertrophy is associated with increased development of extracellular matrix of the middle layer and adventitia. This pattern is associated with reduced arterial compliance and distensibility, regardless of BP level [55]. These changes are observed in central arteries, but not in peripheral ones [56].

In addition to the isolated hemodynamic effect on functional and structural properties of large vessels, arterial hypertension can modify these properties through modifications of different systems involved in blood pressure control. Thus, an increase in renin-angiotensin-aldosterone system activity may play an important role in the development of arterial stiffness in hypertensive patients, since angiotensin II stimulates vascular smooth muscle cell hypertrophy and collagen accumulation, whereas aldosterone promotes growth of extracellular matrix by fibroblasts; both changes promote reporcussions on large arteries function [53]. On the other hand, low degree chronic inflammation is associated with infiltration of smooth muscle cells, macrophages and mononuclear cells; increased cytokine and matrix metalloproteinases content; calcifications; changes in composition of proteoglycans and cellular infiltration around *vasa vasorum*, leading to vascular ischemia [57]. Furthermore, senescence and endothelial cell dysfunction can lead to structural changes that lead to increased intrinsic stiffness of the arterial wall [58].

Modifications of the functional and structural properties of large arteries in hypertensive individuals also

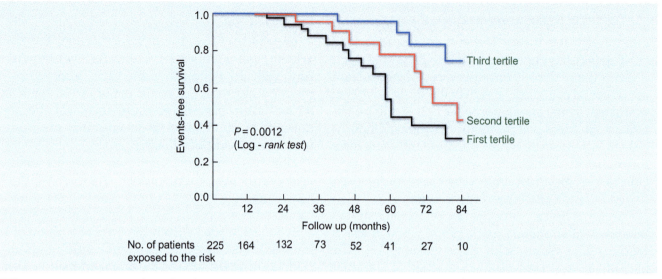

FIG. 28.2 Prognostic value of endothelial dysfunction in patients with hypertension. Patients of the first tertile (lower vasodilatation capacity) had shorter events-free survival than those of the third tertile (greater vasodilation capacity). Adapted from Perticone F, Ceravolo R, Pujia A, et al. Prognostic significance of endothelial dysfunction in hypertensive patients. Circulation 2001;104(2):191–6.

appear to be genetically mediated [59]. Studies evaluating the genetic determinants of arterial stiffness in hypertensive patients demonstrated that polymorphism of elastin gene [60] determined changes in stiffness of central arteries. Increase in arterial stiffness in hypertensive patients may also be influenced by other frequently associated comorbidities, such as atherosclerosis, diabetes, chronic kidney failure and obstructive sleep apnea [61].

PROGNOSTIC VALUE OF ENDOTHELIAL DYSFUNCTION IN ARTERIAL HYPERTENSION

According to some authors, endothelial dysfunction that promotes poorer coronary dilatation can predict progression of atherosclerotic disease and occurrence of cardiovascular events in the long term follow-up [62]. Thus, coronary endothelial vasoreactivity can provide prognostic and diagnostic information in patients at risk for coronary artery disease.

Patients with arterial hypertension have also demonstrated a relationship between endothelial dysfunction and cardiovascular risk. In one of these evidence, patients with hypertension and endothelial dysfunction assessed by arterial vasodilation capacity induced by acetylcholine showed a higher risk of cardiovascular events [63] (Fig. 28.2).

Endothelial dysfunction assessed by fluxomedial dilatation was also associated with increased cardiovascular risk in postmenopausal women, and the authors suggest that this evaluation in postmenopausal women with risk factors, especially hypertension, may be an effective strategy for measuring the risk in this population [64]. In a recent meta-analysis, Matsuzawa observed, in 35 studies involving nearly 18,000 participants, in whom prognostic value of flow-mediated vasodilation was evaluated, that endothelial dysfunction was an excellent predictor of cardiovascular events [65].

TREATMENT OF ARTERIAL HYPERTENSION AND ENDOTHELIUM

In relation to antihypertensive treatment, several evidences with angiotensin converting enzyme (ACE) inhibitors, angiotensin-II receptor blockers (ARBs), calcium channel blockers (CCB) and certain beta-blockers, particularly in the group containing nebivolol molecule, showed beneficial effects on endothelial function. Thus, ACE inhibitors enhance endothelium function by reducing the effects of angiotensin II. In addition, ACE inhibitors promote bradykinin stabilization, which induces nitric oxide and prostacyclin release, and reduces production of free radicals through vascular NADPH oxidase,

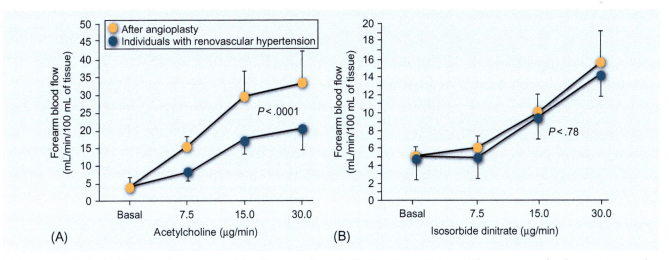

FIG. 28.3 Endothelial function assessed by flow-mediated dilation in patients with renovascular hypertension by fibro muscular dysplasia before and after renal artery angioplasty. Graph (A) endothelium dependent. Graph (B) modified endothelium-independent vasodilation. Adapted from Higashi Y, Sasaki S, Nakagawa K, et al. Endothelial function and oxidative stress in renovascular hypertension. N Engl J Med 2002;346:1954–62.

which is stimulated by angiotensin-II [66]. In the Trend [67] study, 6 months of treatment with ACE inhibitors (quinapril) has promoted improvement in endothelial dysfunction in normotensive patients with coronary artery disease. These benefits probably occur due to attenuation of vasoconstricting effects and generation of superoxide by angiotensin-II and by augmented release of cellular endothelial nitric oxide secondary to the reduction of bradykinin breakdown. A recent meta-analysis has shown that ACE inhibitors improve endothelial function in patients with endothelial dysfunction caused by various conditions and are superior to BCC and beta-blockers. There was no significant difference between the effects of ACE inhibitors and ARBs on peripheral endothelial function [68].

On the other hand, studies with ARB have demonstrated a positive effect on endothelial function, which endorses the important role of angiotensin-II in the development of atherosclerosis [69]. On the Island study, ARBs have improved endothelial function and reduced inflammatory markers, which suggests an important role of these factors in the atherosclerosis pathogenesis [69].

Regarding beta-blockers, there was a historical evolution in relation to endothelial function. Unlike the first and second generations of beta-blockers, third-generation drugs, such as carvedilol [70] and nebivolol [71], have favorable effects on endothelial function. Both drugs stimulate β3 receptors, which

activate eNOS, have antioxidant effects, and increase nitric oxide release [70,72]. In a recent randomized study, it was observed that, in comparison with metoprolol, carvedilol significantly improved endothelial function in patients with hypertension and type 2 *diabetes mellitus* when administered for five months, in addition to usual medications [73].

Concerning the BCCs, they reduce calcium entry into voltage-dependent L-type channels in vascular muscle cells, promoting peripheral and coronary arterial vasodilation. In addition, some BCCs either activate eNOS or have antioxidant properties, therefore increasing bioavailability of this oxide [74]. ENCORE-1 [75] and ENCORE-2 [76] studies have shown that long-acting nifedipine improves endothelial coronary function in patients with stable CAD, an effect that persisted even after drug discontinuation.

In other forms of hypertension, such as hypertension secondary to renal artery stenosis, treatment of the cause may also provide improvement of endothelial function. In an elegant work published by Higashi et al. [77], patients with renovascular hypertension submitted to renal artery angioplasty presented improved flow-mediated vasodilation and oxidative stress markers, suggesting that excessive oxidative stress is involved, at least in part, in endothelial dysfunction in patients with renal artery stenosis, probably due to an increase in renin-angiotensin-aldosterone system activity (Fig. 28.3).

In addition to antihypertensive drugs, other medications often used by patients with arterial hypertension due to the association of comorbidities have proven effective in improving endothelial function [78]. Thus, hypertensive patients with hypercholesterolemia who received statins improved endothelial function, assessed by ultrasound, with a reduction in cholesterol levels and no significant change in blood pressure [78]. Dyslipidemic hypertensive patients also showed improved flow-mediated vasodilation after aspirin added to statins. Obese hypertensive patients had improvement in endothelial function with weight reduction achieved with orlistat, effect independent from reduction of blood pressure achieved [78].

Experimental drugs which act on advanced glycation end-products of administered in hypertensive patients improved endothelial function regardless of effect on pressure [78].

References

[1] Perkovic V, Huxley R, Wu Y, et al. The burden of blood pressure-related disease: a neglected priority for global health. Hypertension 2007;50(6):991–7.
[2] Lawes CM, Vander Hoorn S, Rodgers A. Global burden of blood-pressure-related disease, 2001. Lancet 2008;371(9623):1513–8.
[3] Kaplan NM. Primary hypertension: pathogenesis. In: Kapklan NM, Victor R, editors. Clinical hypertension. Baltimore, MD: Williams & Wilkins; 2009. p. 41–99.
[4] Hall JE. Textbook of medical physiology. Philadelphia, PA: Elsevier; 2011.
[5] Gokce N, Keaney Jr. JF, Vita JA. Endotheliopathies: clinical manifestations of endothelial dysfunction. In: Loscalzo J, Shafer AI, editors. Thrombosis and hemorrhage. Baltimore, MD: Williams & Wilkins; 1998. p. 901–24.
[6] Widlansky ME, Gokce N, Keaney Jr. JF, et al. The clinical implications of endothelial dysfunction. J Am Coll Cardiol 2003;42:1149–60.
[7] Treasure CB, Manoukian SV, Klein JL, et al. Epicardial coronary artery responses to acetylcholine are impaired in hypertensive patients. Circ Res 1992;71:776–81.
[8] Panza JA, Quyyumi AA, Callahan TS, et al. Effect of antihypertensive treatment on endothelium-dependent vascular relaxation in patients with essential hypertension. J Am Coll Cardiol 1993;21(5):1145–51.
[9] Panza JA, Casino PR, Kilcoyne CM, et al. Role of endothelium-derived nitric oxide in the abnormal endothelium-dependent vascular relaxation of patients with essential hypertension. Circulation 1993;87:1468–74.
[10] Panza JA, Garcia CE, Kilcoyne CM, et al. Impaired endothelium-dependent vasodilation in patients with essential hypertension: evidence that nitric oxide abnormality is not localized to a single signal transduction pathway. Circulation 1995;91:1732–8.
[11] Shimbo D, Muntner P, Mann D, et al. Endothelial dysfunction and the risk of hypertension: the Multi- Ethnic Study of Atherosclerosis. Hypertension 2010;55:1210–6.
[12] Quyyumi AA, Patel RS. Endothelial dysfunction and hypertension: cause or effect? Hypertension 2010;55:1092–4.
[13] Escobales N, Crespo MJ. Oxidative-nitrosative stress in hypertension. Curr Vasc Pharmacol 2005;3(3):231–46.
[14] McVeigh GE, Plumb R, Hughes S. Vascular abnormalities in hypertension: cause, effect, or therapeutic target? Curr Hypertens Rep 2004;6(3):171–6.
[15] Moncada S, Palmer RM, Higgs EA. Nitric oxide: physiology, pathophysiology, and pharmacology. Pharmacol Rev 1991;43(2):109–42.
[16] Aroor AR, Demarco VG, Jia G, et al. The role of tissue Renin-Angiotensin-aldosterone system in the development of endothelial dysfunction and arterial stiffness. Front Endocrinol 2013;4:161.
[17] Feletou M, Vanhoutte PM. Endothelial dysfunction: a multifaceted disorder (The Wiggers Award Lecture). Am J Physiol Heart Circ Physiol 2006;291(3):H985–H1002.
[18] Rubira MC, Consolim-Colombo FM, Rabelo ER, et al. Venous or arterial endothelium evaluation for early cardiovascular dysfunction in hypertensive patients? J Clin Hypertens (Greenwich) 2007;9(11):859–65.
[19] Baylis C. Nitric oxide synthase derangements and hypertension in kidney disease. Curr Opin Nephrol Hypertens 2012;21(1):1–6.
[20] Panza JA, Quyyumi AA, Brush Jr JE, et al. Abnormal endothelium-dependent vascular relaxation in patients with essential hypertension. N Engl J Med 1990;323(1):22–7.
[21] Kietadisorn R, Juni RP, Moens AL. Tackling endothelial dysfunction by modulating NOS uncoupling: new insights into its pathogenesis and therapeutic possibilities. Am J Physiol Endocrinol Metab 2012;302(5):E481–95.
[22] Michel T. NO way to relax: the complexities of coupling nitric oxide synthase pathways in the heart. Circulation 2010;121(4):484–6.
[23] Roe ND, Ren J. Nitric oxide synthase uncoupling: a therapeutic target in cardiovascular diseases. Vascul Pharmacol 2012;57(5–6):168–72. Epub 2012/03/01.
[24] Noguchi K, Hamadate N, Matsuzaki T, et al. Improvement of impaired endothelial function by tetrahydrobiopterin in stroke-prone spontaneously hypertensive rats. Eur J Pharmacol 2010;631(1–3):28–35.
[25] Higashi Y, Sasaki S, Nakagawa K, et al. Tetrahydrobiopterin enhances forearm vascular response to acetylcholine in both normotensive and hypertensive individuals. Am J Hypertens 2002;15(4 Pt 1):326–32. Epub 2002/05/07.
[26] Moens AL, Kietadisorn R, Lin JY, et al. Targeting endothelial and myocardial dysfunction with tetrahydrobiopterin. J Mol Cell Cardiol 2011;51(4):559–63.
[27] Porkert M, Sher S, Reddy U, et al. Tetrahydrobiopterin: a novel antihypertensive therapy. J Hum Hypertens 2008;22(6):401–7.
[28] Fortepiani LA, Reckelhoff JF. Treatment with tetrahydrobiopterin reduces blood pressure in male SHR by reducing testosterone synthesis. Am J Physiol Regul Integr Comp Physiol 2005;288(3):R733–6.
[29] Boos CJ, Lip GY. Is hypertension an inflammatory process? Curr Pharm Des 2006;12(13):1623–35.
[30] Montezano AC, Touyz RM. Molecular mechanisms of hypertension-reactive oxygen species and antioxidants: a basic science update for the clinician. Can J Cardiol 2012;28(3):288–95.
[31] Montezano AC, Touyz RM. Oxidative stress, Noxs, and hypertension: experimental evidence and clinical controversies. Ann Med 2012;44(Suppl 1):S2–S16.
[32] Laurindo FRM, Pescatore LA, Fernandes DC. Protein disulfide isomerase in redox cell signaling and homeostasis. Free Radic Biol Med 2012;52(9):1954–69.
[33] Santos CXC, Tanaka LY, Wosniak Jr J, et al. Mechanisms and implications of reactive oxygen species generation during the unfolded protein response: roles of endoplasmic reticulum oxidoreductases mitochondrial electron transport, and NADPH oxidase. Antioxid Redox Signal 2009;1(1):2409–27.
[34] Hasty AH, Harrison DG. Endoplasmic reticulum stress and hypertension—a new paradigm? J Clin Invest 2012;122(11):3859–61.
[35] Young CN, Cao X, Guruju MR, et al. ER stress in the brain subfornical organ mediates angiotensin-dependent hypertension. J Clin Invest 2012;122(11):3960–4.

[36] Kassan M, Galan M, Partyka M, et al. Endoplasmic reticulum stress is involved in cardiac damage and vascular endothelial dysfunction in hypertensive mice. Arterioscler Thromb Vasc Biol 2012;32(7):1652–61.

[37] Schiffrin EL. Vascular endothelin in hypertension. Vascul Pharmacol 2005;43(1):19–29.

[38] Krum H, Viskoper RJ, Lacourciere Y, et al. The effect of an endothelin-receptor antagonist, bosentan, on blood pressure in patients with essential hypertension. Bosentan Hypertension Investigators. N Engl J Med 1998;338(12):784–90.

[39] Packard RR, Lichtman AH. Libby R Innate and adaptive immunity in atherosclerosis. Semin Immunopathol 2009;31(1):5–22.

[40] Coffman TM. Under pressure: the search for the essential mechanisms of hypertension. Nat Med 2011;17(11):1402–9.

[41] Harrison DG, Guzik TJ, Lob HE, et al. Inflammation, immunity, and hypertension. Hypertension 2011;57(2):132–40.

[42] Leibowitz A, Schiffrin EL. Immune mechanisms in hypertension. Curr Hypertens Rep 2011;13(6):465–72.

[43] Schiffrin EL. The immune system: role in hypertension. Can J Cardiol 2013;29(5):543–8.

[44] Guzik TJ, Hoch NE, Brown KA, et al. Role of the T cell in the genesis of angiotensin II induced hypertension and vascular dysfunction. J Exp Med 2007;204(10):2449–60.

[45] Vinh A, Chen W, Blinder Y, et al. Inhibition and genetic ablation of the B7/CD28 T-cell costimulation axis prevents experimental hypertension. Circulation 2010;122(24):2529–37.

[46] Madhur MS, Lob HE, McCann LA, et al. Interleukin 17 promotes angiotensin II-induced hypertension and vascular dysfunction. Hypertension 2010;55(2):500–7.

[47] Tipton AJ, Baban B, Sullivan JC. Female spontaneously hypertensive rats have greater renal anti-inflammatory T lymphocyte infiltration than males. Am J Physiol Regul Integr Comp Physiol 2012;303(4):R359–67.

[48] Barhoumi T, Kasal DA, Li MW, et al. T regulatory lymphocytes prevent angiotensin II-induced hypertension and vascular injury. Hypertension 2011;57(3):469–76.

[49] Kvakan H, Kleinewietfeld M, Qadri F, et al. Regulatory T cells ameliorate angiotensin II-induced cardiac damage. Circulation 2009;119(22):2904–12.

[50] Marvar PJ, Thabet SR, Guzik TJ, et al. Central and peripheral mechanisms of T-lymphocyte activation and vascular inflammation produced by angiotensin II-induced hypertension. Circ Res 2010;107(2):263–70.

[51] Savoia C, Sada L, Zezza L, et al. Vascular inflammation and endothelial dysfunction in experimental hypertension. Int J Hypertens 2011;2011:281240.

[52] Laurent S, Briet M, Boutouyrie R. Large and small artery cross-talk and recent morbidity mortality trials in hypertension. Hypertension 2009;54(2):388–92.

[53] Lacolley P, Safar M, Regnault V, et al. Angiotensin II, mechanotransduction, and pulsatile arterial hemodynamics in hypertension. Am J Physiol Heart Circ Physiol 2009;297(5):H1567–75.

[54] Safar M, Struijker-Boudier H. Cross-talk between macro- and microcirculation. Acta Physiol (Oxf) 2010;198(4):417–30.

[55] Safar M. Hypertension, systolic blood pressure, and large arteries. Med Clin North Am 2009;93(3):605–19.

[56] Bortolotto LA, Hanon O, Franconi G, et al. The aging process modifies the distensibility of elastic but not muscular arteries. Hypertension 1999;34(4 Pt 2):889–92.

[57] Laurent S, Boutouyrie P. The structural factor of hypertension: large and small artery alterations. Circ Res 2015;116(6):1007–21.

[58] Donato AJ, Morgan RG, Walker AE, et al. Cellular and molecular biology of aging endothelial cells. J Mol Cell Cardiol 2015;89(Pt B):122–35.

[59] Laurent S, Boutouyrie P, Lacolley P. Structural and genetic bases of arterial stiffness. Hypertension 2005;45(6):1050–5.

[60] Hanon O, Luong V, Mourad J, et al. Aging, carotid artery distensibility, and the Ser422Gly elastin gene polymorphism in humans. Hypertension 2001;38(5):1185–9.

[61] Zieman SJ, Melenovsky V, Kass D. Mechanisms, pathophysiology and therapy of arterial stiffness. Arterioscler Thromb Vasc Biol 2005;25:932–43.

[62] Suwaidi JA, Hamasaki S, Higano ST, et al. Long-term follow-up of patients with mild coronary artery disease and endothelial dysfunction. Circulation 2000;101:948–54.

[63] Perticone F, Ceravolo R, Pujia A, et al. Prognostic significance of endothelial dysfunction in hypertensive patients. Circulation 2001;104(2):191–6.

[64] Rossi R, Nuzzo A, Origliani G, et al. Prognostic role of flow-mediated dilation and cardiac risk factors in post-menopausal women. J Am Coll Cardiol 2008;51:997–1002.

[65] Matsuzawa Y, Kwon TG, Lennon RJ, et al. Prognostic value of flow-mediated vasodilation in brachial artery and fingertip artery for cardiovascular events: a systematic review and meta-analysis. J Am Heart Assoc 2015;4(11):pii. e002270.

[66] Rajagopalan S, Harrison DG. Reversing endothelial dysfunction with ACE inhibitors. A new trend. Circulation 1996;94(3):240–3.

[67] Mancini GB, Henry GC, Macaya C, et al. Angiotensin-converting enzyme inhibition with quinapril improves endothelial vasomotor dysfunction in patients with coronary artery disease. The TREND (Trial on Reversing ENdothelial Dysfunction) Study. Circulation 1996;94(3):258–65.

[68] Shahin Y, Khan JA, Samuel N, et al. Angiotensin converting enzyme inhibitors effect on endothelial dysfunction: a meta-analysis of randomised controlled trials. Atherosclerosis 2011;216(1):7–16.

[69] Sola S, Mir MQ, Cheema FA, et al. Irbesartan and lipoic acid improve endothelial function and reduce markers of inflammation in the metabolic syndrome: results of the Irbesartan and Lipoic Acid in Endothelial Dysfunction (ISLAND) study. Circulation 2005;111(3):343–8.

[70] Kalinowski L, Dobrucki LW, Szczepanska-Konkel M, et al. Third-generation beta-blockers stimulate nitric oxide release from endothelial cells through ATP eâlux: a novel mechanism for antihypertensive action. Circulation 2003;107(21):2747–52.

[71] Toblli JE, DiGennaro F, Giani JF, et al. Nebivolol: impact on cardiac and endothelial function and clinical utility. Vasc Health Risk Manag 2012;8:151–60.

[72] Khan MU, Zhao W, Zhao T, et al. Nebivolol: a multifaceted antioxidant and cardioprotectant in hypertensive heart disease. J Cardiovasc Pharmacol 2013;62:445–51.

[73] Bank AJ, Kelly AS, Thelen AM, et al. Effects of carvedilol versus metoprolol on endothelial function and oxidative stress in patients with type 2 diabetes mellitus. Am J Hypertens 2007;20:777–83.

[74] Tang EH, Vanhoutte PM. Endothelial dysfunction: a strategic target in the treatment of hypertension? Pflugers Arch 2010;459:995–1004.

[75] ENCORE Investigators. Evaluation of nifedipine and cerivastatin on recovery of coronary endothelial function. Circulation 2003;107(3):422–8.

[76] Lüscher TF, Pieper M, Tendera M, et al. A randomized placebo-controlled study on the effect of nifedipine on coronary endothelial function and plaque formation in patients with coronary artery disease: the ENCORE II study. Eur Heart J 2009;30:1590–7.

[77] Higashi Y, Sasaki S, Nakagawa K, et al. Endothelial function and oxidative stress in renovascular hypertension. N Engl J Med 2002;346:1954–62.

[78] Miyamoto M, Kotani K, Taniguchi N. Effect of non-antihypertensive drugs on endothelial function in hypertensive subjects evaluated by flow-mediated vasodilation. Curr Vasc Pharmacol 2015;13(1):121–7.

29

Endothelial Alterations in Pulmonary Hypertension

Mariana Meira Clavé and Antonio Augusto Lopes

INTRODUCTION

Pulmonary vascular disease comprises an extensive set of abnormalities of diverse etiology. Pulmonary hypertension, a late hemodynamic consequence of these abnormalities, is now classified into five categories according to the pathophysiological similarities among the diseases [1]. The first category corresponds to pulmonary arterial hypertension (PAH), the most widely studied from the genetic and biochemical points of view, through experimental animal, clinical, therapeutic, and prognostic models. In this category, idiopathic PAH, hereditary PAH, as well as forms associated with connective tissue diseases, congenital heart diseases, schistosomiasis, HIV infection, and ingestion of anorexigenics and toxic substances are classified. In the second category are the diseases of the left side of the heart, with post capillary, or mixed, pre- and post-capillary pulmonary hypertension. The third category includes together respiratory disorders. Fourth is thromboembolic disease. Finally, the fifth category combines etiologies with pathophysiological mechanisms that have not yet been completely clarified. In all of these situations, pulmonary vascular remodeling is present in variable extension. Vascular changes are observed in all segments (from arterial to venous) and in all layers of the vessel wall [2].

The pulmonary vascular endothelium has a central role in vessels dynamics under physiological conditions and is markedly involved in pathological situations. Vascular tonus regulation and antithrombotic protection are critical functions of the pulmonary endothelium. Particularly in small vessels, physiological and pathological events are expressed through intense and extensive cross-talk between endothelial cells and circulating blood elements on the luminal face, as well as cellular and matrix components on the abluminal side. For this reason, there has been a tendency to replace the term "endothelial dysfunction" with "microvascular dysfunction," considering the complexity of the interactive phenomena.

ENDOTHELIUM IN PULMONARY VASCULAR DYSFUNCTION: DECREASE IN CAPACITY TO MEDIATE VASODILATION

Pulmonary hypertension is established by the progressive loss of pulmonary vascular bed (notably the sum of the cross-sectional areas of the resistance vessels) due to vasoconstriction and vaso-occlusion (intimate proliferative lesions). The hemodynamic result is elevation of pulmonary arterial pressure at levels above 25 mmHg at rest, generally associated with an increase in pulmonary vascular resistance above 3 Wood units per square meter. In typical PAH, wedge pulmonary pressure is less than 15 mmHg, and transpulmonary gradient (mean pressure—wedge) and diastolic gradient (diastolic pressure—wedge) higher than 12 and 7 mmHg, respectively.

There is experimental and clinical evidence on the imbalance between natural vasodilating and vasoconstricting factors in PAH. Prostacyclin synthesis (prostaglandin I2), as well as urinary excretion of metabolite, 2,3-dinor-6-keto-PGF$_{1\alpha}$ are reduced [3]. The reduced ability to promote vasodilation is aggravated by decrease in nitric oxide levels (partly due to reduced expression in endothelial nitric oxide synthase [eNOS]), as well as the second messenger in the same signaling pathway, cyclic guanosine monophosphate (cyclic GMP) [4–6]. Increase in asymmetric dimethyl-arginine levels (an endogenous inhibitor of nitric oxide synthase) may be an additional factor [7,8].

Taken together, these abnormalities in the generation of nitric oxide and prostacyclin production result, besides the reduction in the ability to promote vasodilation, also defective endothelium behavior in its properties of inhibiting platelet aggregation and cell proliferation. Drugs have been developed to compensate for these defects. Examples are prostanoids (epoprostonol, iloprost, and treprostinil), phosphodiesterase-5 inhibitors that can preserve intracellular levels of cyclic GMP (sildenafil, tadalafil, vandernafil) and, more recently, the possibility of direct activation of guanylate cyclase (nitric oxide receptor) by means of riociguat. In addition, nitric oxide is administered by inhalation in several acute conditions that occur with increased arterial resistance in pulmonary territory.

In contrast, there is an excess of vasoconstrictors stimuli especially represented by endothelin-1 [9,10], thromboxane [11] and 05-hydroxytryptamine [12]. Endothelin-1 is the most potent endogenous vasoconstrictor known, producing effects from its binding to ET$_A$ and ET$_B$ receptors both of type seven transmembrane domains. These receptors are coupled to Protein G, with effects derived from production of mediators such as diaglicerol (C protein kinase pathway inducer—PKC) and inositol triphosphate (intracellular calcium mobilizer). Furthermore, intracellular signaling from endothelin coupling to its receptor results in activation of MAP kinase pathway, as well as expression of fast response genes, such as c-fos and c-jun. [13] The result of these processes, in addition to vasoconstriction, is induction of cell proliferation, with endothelin-1 being considered as mitogenic.

Under physiological conditions, lungs remove endothelin from circulation, so that concentration in pulmonary veins is inferior to arterial one. In PAH, the situation is reversed, with expression and release of endothelin in the circulation. Therapeutic approaches used to reduce their effects are represented by endothelin receptor antagonists (bosentan, ambrisentan, macitentan), widely used in PAH. An interesting observation, with therapeutic implication, is reduction of exhaled nitric oxide under conditions of low expiratory flow in patients with PAH, and the demonstration that this abnormality can be reversed by bosentan administration. Observation suggests that nitric oxide suppression may be related in part to increased endothelin expression in these patients [14].

PARTICIPATION IN VASCULAR REMODELING PROCESS

First, one must consider that vascular tonus modulators, nitric oxide, prostacyclin, and endothelin-1 are also cell growth modulators. Nitric oxide is an inhibitor of smooth muscle cell growth, leukocyte adhesion and platelet aggregation. Prostacyclin, the most potent endogenous antiplatelet agent, is able to inhibit DNA synthesis and smooth muscle cells proliferation. As mentioned, endothelin-1 is considered a mitogenic peptide, given the activation of several signaling pathways that result in cell replication.

Loss of endothelial function as a "barrier" between circulating blood and vessel wall elements is an important event in the development of pulmonary vascular disease, and a whole line of research has been developed in this regard. Physical forces of shear stress and stretch probably induce appearance of failures in integrity and continuity between endothelial cells, allowing substances in serum (apolipoprotein-A1 is a candidate) to contact the subendothelial region. A succession of molecular events results in the activation of endovascular elastase (EVE). It is a serine protease produced and released by smooth muscle cells (precursors or mature cells), with properties to promote not only elastin fragmentation, but also pericellular proteolysis, which results in the release and availability of growth factors (fibroblast growth factor (FGF), transforming growth factor beta (TGF-β)) from their deposits in the proteoglycan matrix. The interaction of muscle cells with the altered matrix (especially modified collagen) promotes expression and re-binding to other extracellular proteins, such as tenascin-C, all these integrin-mediated mechanisms. The result is the organization of the cytoskeleton (smooth muscle cells), the formation of focal adhesion complexes with various signaling molecules and activation receptors for growth factors, which form clusters (e.g., EGF receptor *epidermal growth factor*), able to start signaling processes that culminate in cell replication (smooth muscle and myofibroblasts). All these mechanisms have been extensively studied in animal models, cell culture, and human disease [15].

Another important aspect, widely explored of the past decades is cellular dysfunction due to changes in BMPR2 receptors (for bone phogenetic protein). These are TGF-β superfamily proteins, whose normal functioning tend to maintain cells in the quiescent state, ie outside the replication cycle. This type of receptor constitutes one of the dimeric receptor chains. Numerous mutations have been described in PAH as human disease, both in transcribed and non-transcribed regions of their gene, representing the genetic / hereditary aspect of the disease. Mutations in receptors of the same family, such as endoglin and ALK-1, are also widely reported in PAH. Thus, TGF-β system dysfunction, not only of genetic origin but also in terms of signaling pathways, both in PAH in humans and in experimental models, constitutes an important topic among pathophysiological aspects. In pulmonary arterial endothelial cells, the loss of BMPR2 function is related to apelin protein repression [16]. As a result, there is a reduction in a microRNA that normally suppress FGF-2 growth factor [17]. BMPR2 dysfunction in vascular cells also is associated with increased expression of interleukin-6 (IL-6) and increase GM-CSF (granulocyte-macrophage colony-stimulating factor) secretion in response to TNF-α (tumor necrosis factor alpha). Finally, changes in signaling pathways from BMPR2 result in Foxp3⁺ Treg regulatory system (T cells) dysfunction, which in normal conditions is critical for suppressing autoimmune response associated with development of human and experimental pulmonary vascular disease [18].

There is still considerable controversy regarding proliferation of endothelial cells and the role of VEGF (vascular endothelial growth factor) and its receptors in the pathophysiology of pulmonary vascular alterations. At first, angiogenesis, from endothelial cells, would be a restorative phenomenon; it is also part of the normal developmental process of pulmonary circulation. It is well known that in infants and young children with severe PAH associated with cardiac septal defects, there is a rarefaction of small intra-acinary vessels with increase in the

ratio between the number of alveoli and arteries [19]. Accordingly, VEGF and FGF-2 (Basic FGF or FGFb) and their receptors, among other molecules, play a central role in the physiological sense. It is noteworthy that in experimental conditions, inhibition of VEGF receptors with animals exposed to hypoxia constitutes a classic model of severe pulmonary vascular disease [20], while VEGF hyperexpression attenuates the development of pulmonary hypertension [21]. Furthermore; it is known that endothelial proliferation is part of the pulmonary vascular alterations. The current concept is that apoptosis of endothelial cells would be an early event, which would favor selection of apoptosis-resistant and hyperproliferative cells. Ultimately, these cells would contribute to the development of so-called plexiform lesions, which are characteristic of advanced pulmonary vascular disease. An interesting and critical observation from the pathophysiological point of view: cells that make up plexiform lesions have growth patterns with a monoclonal characteristic [22]. In plexiform lesions of patients with idiopathic PAH, somatic mutations in genes encoding regulatory proteins such as proliferation and apoptosis TGF-β RII and Bax were observed conferring growth instability to endothelial cells present in plexuses characterizing neoplastic type behavior, analogous to what is observed in colon cancer. In the pathogenesis of pulmonary vascular disease, a second hit could then be represented by hypoxia or viral infections, use of anorexigens, or additional gene mutations [23].

The so-called PPARs (*peroxisome proliferator-activated receptors*) are a group of transcription factors involved in physiological events such as lipogenesis and inflammation. PPAR-γ is expressed in many cell types, including endothelium and vascular smooth muscle. It can be activated by natural agonists and synthetic substances such as thiazolidinediones (troglitazone, rosiglitazone) used in the treatment of type 2 *diabetes mellitus*. The role of PPAR-γ in the pathogenesis of PAH has been widely demonstrated. Mice spontaneously develop pulmonary hypertension by conditional knock out of PPAR-γ in endothelial or smooth muscle cells [24]. Activation of PPAR-γ results in reversal of pulmonary hypertension in an experimental model of insulin resistance [25]. Studies using PPAR-γ antagonism show that PPAR-γ itself is necessary to inhibit proliferation of smooth muscle cells mediated by BMP-2 [26]. Activation of PPAR-γ reduces levels of two critical factors in the pathogenesis of PAH: endothelin-1 and asymmetric dimethylarginine (endogenous antagonist of nitric oxide synthase [NOS]) [27,28]. Thus, this signaling pathway becomes a potential target for therapeutic measures to be developed and tested in the future.

ENDOTHELIUM, THROMBOSIS AND INFLAMMATION IN PULMONARY VASCULAR DISEASE

The first point to be considered is recognition of thrombosis as a relevant pathophysiological element in pulmonary vascular disease. On the one hand, it is emphasized that chronic pulmonary thromboembolism constitutes Group 4 in the current diagnostic classification of pulmonary hypertension [29]. On the other hand, it is important to remember that pulmonary thrombosis (including central, involving the main pulmonary arteries, right and left) is a critical complication in other forms of the disease. Patients with PAH associated with congenital heart diseases, especially adults, with advanced disease (Eisenmenger syndrome, with cyanosis, erythrocytosis, and hyperviscosity) may present this type of complication, which implies a poor prognosis [30–33]. Moreover, thrombotic complication is part of the pathophysiological scenario of both idiopathic and hereditary PAH, which are together with congenital heart defects in Group 1 of the diagnostic classification.

Even before thromboembolic or thrombotic events can be observed as filling defects in large pulmonary arteries recognized through angiographic or angiotomographic examination, occlusion of small subsegmental and intra-acinar vessels is likely to evolve silently and be responsible for worsening clinical evolution without an apparent cause. In pulmonary vascular disease, as in other related processes, thrombosis and inflammation are unsociable pathophysiological components. For example, high circulating levels of D-dimer constitute a diagnostic marker in thromboembolic disease. However, D-dimer levels may be elevated in the absence of clinical thromboembolism, and have been associated with inflammatory processes [34–36]. In patients with idiopathic PAH, D-dimer circulating levels have implications with severity of the disease and is associated with functional class, hemodynamic parameters and survival [37].

Increased circulating D-dimer levels presume generation of thrombin in association with pulmonary vascular abnormalities [38,39]. It is a serine protease with significant actions on the behavior of platelets, endothelial cells and smooth muscle, with a central role in the interchange between inflammation and thrombosis. Thrombin activates platelets, and other cellular elements, by cleavage of the PAR-type receptors (protease activated receptors belongs to the superfamily of type seven transmembrane domains coupled to G protein receptors), promoting aggregation, in addition to surface expression and secretion of various molecules. In PAH patients at the Heart Institute of the University of São Paulo School of Medicine, we had the opportunity to observe an increase in the density of PAR-1 receptors associated with an increase in the expression of P-selectin after platelet stimulation in vitro by thrombin [40]. In addition, PAH progresses with increased CD40 expression and its binding protein, CD40L (or CD154), which enables interaction between platelets and endothelial cells [41]. These latter also have the PAR-1 type receptors that respond to stimulation by thrombin with mobilization of Weibel-Palade bodies and fusion with the plasma membrane, formation of secretion pores and release of various molecules, including P-selectin and von Willebrand factor. Platelets, on the other hand, have several receptors for von Willebrand factor (mainly glycoproteins Ib and IIb/IIIa), whereas granulocytes and monocytes can be recruited by binding between P-selectin and PSGL-1. Thus, P-selectin is an important player in the interaction among endothelial cells, platelets and neutrophils, especially in hypoxia [42]. Therefore, thrombin, signaling from PAR-type receptors, P-selectin expression and interactive events among endothelial cells, leukocytes and platelets, involving CD40-CD40L complex, become central in the unification of thrombosis and inflammation processes, widely studied in pulmonary vascular disease, in a similar way to atherosclerosis.

Abnormalities related to inflammation and immunity have been extensively studied in pulmonary vascular disease. Events involving inflammatory elements and mediators are described in all vessel wall layers, from endothelium to adventitia, in different forms of the disease, in human condition and in all kinds of laboratory preparations [43]. Endothelial cells produce and/or are stimulated by a number of cytokines and growth factors, including interleukin 1 (IL-1β), 4, 6 and 8, TNF-α and interferon-gamma (IFN-γ). Among other effects, cytokines are capable of inducing the genetic expression of tissue factor in endothelial cells and monocytes/macrophage via NF-κB transcription factor, creating a local pro-coagulant condition. Increased circulating levels of IL-1 and IL-6 are found in patients with pulmonary hypertension [44]. Endothelial cells also express chemokines like IL-8 (or NAP-1, neutrophil attractant protein 1), MCP-1 (monocyte chemoattractant protein 1) and the CC subfamily receptors (CCRs to RANTES—Regulated on activation, Normal T cell Expressed and secreted and MCP-1, for example) and CXC (CXCRs where GROs—growth regulated oncogene, NAP-1 and others bind). Chemokines act through its receptors type seven transmembrane domains, to induce expression of integrins that interact with adhesion molecules of ICAM-1 (intercellular adhesion molecule 1) and VCAM-1 (vascular cellular adhesion molecule 1) type, promoting cellular recruitment. Studies have demonstrated increased expression of fractalkine (FKN) and its receptor CX3CR1 in patients with PAH, emphasizing the importance of such complexes (FKN / CX3CR1) in leukocyte adhesion [45,46]. Finally, interactive events are highlighted, for example, from endothelial cells and platelets, resulting in tendency to vasoconstriction and cellular proliferation. Platelets express TGF-β and RANTES, among other factors [47]. It has been demonstrated that by means of these two cytokines, platelets are able to induce endothelin-1 expression in endothelial cells [48,49].

Abnormalities involving coagulation factors may make patients with PAH predisposed to bleeding episodes. This is particularly important in adolescents and adults with chronic hypoxemia associated with Eisenmenger syndrome. The frailty of small pulmonary vessels that are a series of dilated lesions favors ruptures, which are aggravated by alterations in coagulation or the use of anticoagulants, which are manifested by hemoptysis, which is sometimes copious. On the other hand, alterations in the endothelium in the natural and fibrinolytic anticoagulant systems, which predispose to chronic intravascular coagulation and aggravation of the global vaso-occlusive condition, are of concern. Thrombomodulin is an endothelial-surface proteoglycan, rich in chondroitin sulfate capable of binding to newly formed thrombin, with subsequent activation of protein C and inactivation of coagulation factors V and VIII. Levels of thrombomodulin can be easily

measured in plasma, and reflect proteolytic fragments present in circulation. Although these levels are markedly elevated in acute situations that occur with widespread proteolysis [50], in chronic situations they tend to be decreased, reflecting reduced production. It is classically known that hypoxia suppresses endothelial expression of thrombomodulin at the transcriptional level [51], favoring, by this mechanism, fibrin deposition in animal models [52]. In PAH, circulating levels of thrombomodulin are reduced [53,54], a finding that is compatible with defective endothelial expression that may predispose especially to patients with chronic hypoxemia (Eisenmenger syndrome), intravascular coagulation, and thrombosis. It should be emphasized that endothelial dysfunction in the various forms of PAH is not restricted to pulmonary circulation.

PLATELETS IN PULMONARY VASCULAR DISEASE

It is practically impossible to discuss pulmonary vascular endothelium without mentioning the participation of platelets, especially in the microvascular dysfunction/lesion process. Platelets and endothelial cells have great structural and functional proximity. Both have storage granules (Weibel-Palade bodies in endothelial cells) that store the same molecules (P-selectin, von Willebrand factor) and secrete them by similar stimuli (thrombin, via PAR-like receptors). Both have integrin type receptors ($\alpha_{IIb}/\beta3, \alpha_v/\beta3$ and others), selectins and their binding proteins (PSGL), cellular adhesion molecules of the immunoglobulin family, as well as their binding proteins, and CD40-CD40L complex elements, which makes them capable of interacting both with each other and with leukocytes. Both have active cyclooxygenase pathway, with preferential production of thromboxane (platelets) or prostacyclin (endothelium).

It has long been known that part of the platelets are generated from megakaryocytes outside the bone marrow, constituting extramedullary thrombopoiesis. There seems to be no more doubt in this regard, except the percentage of the total number of platelets they represent. It is believed that about 20% of platelets are so generated in the pulmonary circulation (capillary bed) under physiological conditions, and this percentage may be higher in the disease. In addition, drugs that mimic thrombopoietin are able to increase the traffic of megakaryocytes from the medullary compartment towards pulmonary microvasculature [55–57]. It is understood, therefore, that platelets play an important role in pulmonary vascular biology, both in physiological and pathological situations. Pulmonary hypertension is a known phenomenon in family disease called platelet storage pool disease, in which platelets are unable to store their granular content (especially dense granules) resulting in increased circulating serotonin [58]. It is known that serotonin and their receptors play a critical pathophysiological role in the development of pulmonary vascular disease, not only towards vasoconstriction but also on proliferation of smooth muscle cells.

Except for familial disease with a storage defect, there is no evidence that platelets may be primarily involved in the genesis of pulmonary vascular disease. However, due to their interactions with endothelial cells and leukocytes, in addition to other elements, evidence indicates their pathophysiological involvement in progression and complications of the disease. Platelets store and release TGF-β and RANTES, important cytokines in induction of endothelin-1 endothelial expression [47–49]. In addition to these, several other molecules are stored in platelets with significant effect on the surrounding microenvironment, especially PDGF (platelet derived growth factor with important mitogenic role), FGF-2 (basic) and VEGF A, in addition to other cytokines. Platelets also have TLR type receptors (Toll-like receptors), in particular TLR-2 and TLR-4. These receptors are activated by histones or histone-DNA complexes released by activated or apoptotic leukocytes (mainly neutrophils), resulting in the development of proinflammatory and prothrombotic responses, with release of P-selectin, cytokines, and chemokines [59]. It should be noted that interactions between platelets and leukocytes are ascribed to several other intermediate actions. In activated platelets, surface-mobilized P-selectin binds to PSGL-1 (neutrophils, monocytes). This binding signals for expression, in leukocytes, of Mac-1 (α-chain of integrin $\alpha_M\beta2$, which together with $\alpha L\beta2$, or LFA-1 corresponds to ICAM-1 receptor), capable of binding to platelet glycoprotein GPIbα. This second interaction amplifies activation of integrin $\alpha_M\beta2$, which indirectly binds to platelet integrin $\alpha_{IIb}/\beta3$ using fibrinogen molecule as a bridge. This "secondary recapture" is often associated with interactions between leukocytes and endothelial cells, followed by release of inflammatory and prothrombotic

mediators [60]. Some of these molecular events are shown in Fig. 29.1.

Despite the recognized participation of platelets in several pathophysiological events that contribute to worsening of pulmonary vascular disease and occurrence of clinical events, there are still no randomized trials, involving a large number of patients, with the objective of testing the possible benefits of antiplatelet drugs in the treatment of this cardiopathy. In clinical practice, especially in the management of advanced PAH associated with congenital heart disease with chronic hypoxemia and blood hyperviscosity due to erythrocytosis, reduction in the number of circulating platelets may be of concern (below 50,000 platelets/mm^3). This phenomenon is associated with platelet hypoaggregability, suggesting that when examining ex vivo platelet, these are exhausted, as a result of chronic endogenous activation. Initial studies in our institution show that platelets of patients with PAH have hyperphosphorylated proteins on tyrosine, reduced half-life (increased turnover) and form aggregates in circulation, with other cellular elements [61–63]. Therefore, platelets and all systems operating jointly contributing to the tendency to thrombosis are recognized as potential targets for therapeutic intervention in PAH and other forms of the disease, while clinical trials are expected in this regard.

MARKERS OF MICROVASCULAR DYSFUNCTION IN PULMONARY VASCULAR DISEASE

Initially, these molecules were characterized as markers of endothelial dysfunction/lesion. However, after knowing interactions among endothelium, leukocytes, platelets, and circulating soluble elements that make up the pathophysiological scenario biomarkers, they are better designated as markers of microvascular events, with the caveat that in advanced disease, the pathological process is not restricted to microcirculation. Identification of biomarkers in pulmonary hypertension becomes central to understand the disease and its evolution. For example, as studies progressed, two curious facts were observed. The application of specific therapies (prostanoids, endothelin receptors antagonists, phosphodiesterase inhibitors, and other drugs), in randomized studies, provides improvement in functional capacity, hemodynamic parameters, and quality of life. Impact on survival has also been

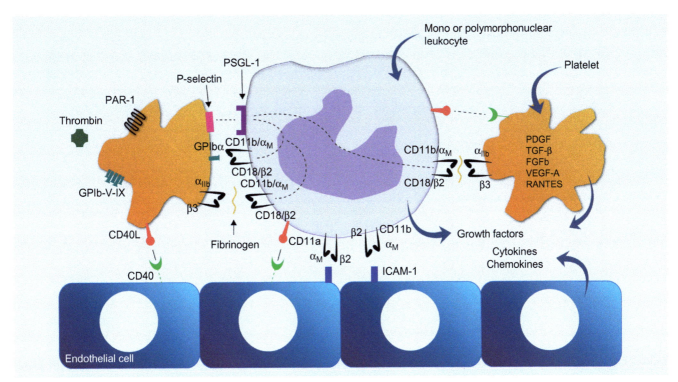

FIG. 29.1 Illustration of some molecular mechanisms that operate in interaction among endothelial cells, platelets and leukocytes, characterizing inflammation and thrombosis as a single process. Having thrombin generation in circulation, this serine protease acts via its receptor PAR-1 (protease activated receptor-1), which results in P-selectin expression on the platelet membrane. The first interaction with leukocytes occurs by binding of P-selectin to PSGL-1 (P-selectin glycoprotein ligand-1). Signaling in leukocyte follows the integrin expression of type β2 (CD11b / CD18 or α_M β2), which interact with platelet glycoprotein Ibα and subsequently with the glycoprotein α_{IIb}/β3, the fibrinogen molecule acting as a "bridge." The same class of β2 integrins, in the leukocyte (α_Lβ2 and α_Mβ2) serves as a receptor for ICAM-1 (intercellular adhesion molecule-1) molecules on endothelial cells. Interactions between platelets, leukocytes and endothelium also occur via the CD40 molecule and its binding protein, CD154, often termed CD40L. Finally, results from this multiple interaction, the release of various cytokines, chemokines and growth factors, acting on all cell types, particularly the endothelium. *PDGF*, platelet-derived growth factor; *TGF-β*, transforming growth factor-beta; *FGFb*, basic fibroblast growth factor; *VEGF-A*, Vascular endothelial growth factor A; *RANTES*, regulated on activation, normal, T cell expressed and secreted.

demonstrated, and more recent studies address morbidity and mortality as outcomes. However, there is a relatively small number of studies seeking to connect clinical improvement with parameters that define biological behavior. Occasionally, there is improvement in biochemical markers in the course of specific therapies [64]. However, the persistence of biochemical abnormalities, or no change in these parameters despite treatments, suggests that effectiveness of these new drugs is often characterized by improvement of symptoms and hemodynamic variables, although the disease continues progressing. Indeed, it is known that patients with PAH are hardly maintained stable for more than 2 years under monotherapies, especially oral drugs. In addition, unpublished data from clinical observations suggest that even patients maintained stable for years under

prostacyclin (e.g., intravenously), when evolving to death; their lung tissue examined by autopsy showed serious and extensive vascular lesions in the pulmonary circulation (plexiform lesions, for example, considered to be advanced in disease progression). Therefore, it should be emphasized with respect to prognostic markers in pulmonary hypertension, that although has evolved in terms of functional parameters (indices of physical performance, cardiac output, right atrial pressure, echocardiographic measurements and magnetic resonance imaging of the right ventricular function, natriuretic peptides), there is still a long way in relation to the role of biological markers in disease evolution.

Hence, progressive incorporation is needed in clinical practice. Fig. 29.2 shows some microvascular dysfunction/lesion markers that have been

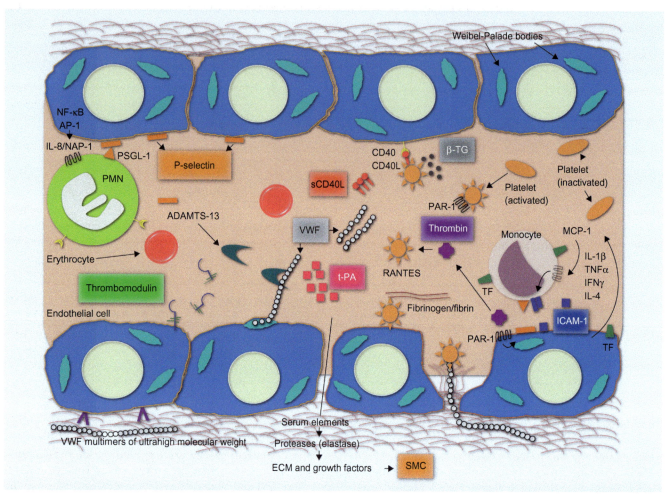

FIG. 29.2 Illustration of some biological events that occur in endothelial cell activation process, introducing elements that are considered and investigated as microvascular dysfunction markers in pulmonary vascular disease. Various stimuli such as inflammatory mediators (cytokines, chemokines), thrombin, hypoxia and physical forces like shear stress and stretch acting on the endothelium, promoting mobilization of Ca ++ and fusion of Weibel-Palade bodies with the plasma membrane, followed by secretion of its content (P-selectin, von Willebrand factor (VWF), tissue—type plasminogen activator (t-PA), among other molecules). P-selectin participates in the recruitment of leukocytes (polymorphonuclear above all in hypoxic conditions), an event facilitated by chemotaxis from NAP-1 chemokine (neutrophil attractant protein-1, also known as interleukin 8—IL-8). The interaction via P-selectin is through its binding protein, PSGL-1 (P-selectin glycoprotein ligand-1). The VWF once secreted, may circulate or remain on surface, binding to P-selectin, participating in recruitment of platelets. Once in circulation, the VWF is physiologically cleaved by ADAMTS-13 metalloprotease (a desintegrin and metalloproteinase with thrombospondin type 1 motif, member 13), operating as a carrier for clotting factor VIII. Loss of endothelium integrity exposes VWF molecules of very high molecular weight stored in the subendothelial region, that recruit platelets through interaction with the platelet glycoprotein complex GPIb-V-IX. The same loss of integrity allows serum factors coming into contact with the subendothelium, participating in the activation of endovascular elastase, followed by a cascade of events that result in release of growth factors and proliferation of smooth muscle cells and myofibroblasts. In endothelial cells and monocytes, stimulation via inflammatory mediators (IL-1β, interleukin-1β, TNF-α, tumor necrosis factor-alpha, IFN-γ, interferon gamma, IL-4, interleukin-4) induces expression of tissue factor (TF), with activation of the extrinsic pathway of the coagulation system resulting in thrombin generation. This acts through its PAR-1 receptor (protease-activated receptor-1) promoting, in endothelial cells, the secretion of Weibel-Palade bodies content. Platelets activated by thrombin release beta-thromboglobulin (β-TG), CD40L, adhesion molecules, growth factors and chemokines (RANTES, regulated on activation, Normal T cell Expressed and secreted). Activated endothelial cells express adhesion molecules, including ICAM-1 from the immunoglobulin family (intercellular adhesion molecule-1), which interact with leukocyte by means of class β2 integrins induced by MCP-1 chemokine type (monocyte chemoattractant protein-1). In acute situations, the protease activity results in breakdown and release into circulation of thrombomodulin fragments, whose concentration increases. In situations of chronic endothelial dysfunction, the plasma concentration is reduced as a result of defective endothelial expression. In pulmonary vascular disease, VWF, P-selectin, thrombomodulin, t-PA, β-TG, and the soluble CD40L have been measured, along with cytokines, chemokines and other adhesion molecules, as markers of endothelial and microvascular dysfunction.

investigated in pulmonary vascular disease. It is known that plasma thrombomodulin is an important marker of microvascular processes. What is measured in circulation are proteolytic fragments of this surface proteoglycan.

Older studies, but absolutely applicable to present days, show that in acute situations like disseminated intravascular coagulation, fulminant hepatitis, venous thrombosis, thrombophlebitis, and certain types of leukemia, circulating levels of thrombomodulin are elevated [50]. In general, these situations occur with activation of multiple proteases. In chronic patients with PAH, however, plasma thrombomodulin is typically reduced [53,54], reflecting, probably, reduced endothelial expression. It is possible that plasma thrombomodulin constitutes a biochemical marker of cellular response to therapeutic agents [64]. Unpublished data from our experience suggests that oral administration of therapies for PAH can also act favorably on thrombomodulin levels.

The von Willebrand factor is an important marker of microvascular dysfunction in acute and chronic diseases. It is a multimeric protein of high molecular mass, expressed only by megakaryocytes (platelets), and endothelial cells. It lies in the subendothelium as ultrahigh molecular weight forms, anchoring endothelial cells by binding to vitronectin receptors (integrin $\alpha_V\beta3$) and serving to platelet adhesion in injury situations, and in circulation, in which, among other functions, it acts as a carrier of VIII clotting factor. It is assumed that plasma pool is representative of protein derived essentially from the endothelium. Platelets secreted factor would be readily used in surface, adhesion and aggregation processes, contributing little to the plasma pool (Fig. 29.2). Data from our institution (and other authors) shows that in PAH, the von Willebrand factor is elevated in circulation, with several structural and functional changes and with respect to its sialic acid content, that gives greater adhesiveness to protein [65–67]. In some early studies of clinical correlation, we observed that von Willebrand factor was seen as prognostic indicator in the short term (1 year of evolution) in patients with PAH [68–70]. More recently, a group from Columbia University identified von Willebrand factor as an independent predictor of survival in the long-term in patients with PAH, either idiopathic, familial or associated with the ingestion of anorexigenics [71]. Later, in a 4-year survival study, we characterized such protein as an independent predictor of survival in patients with PAH associated with congenital heart defects [72]. Finally, data not yet published from our group confirms the association between high levels of von Willebrand factor and reduction of expectancy survival in PAH in congenital heart defects, this time in a 9 years follow-up (Fig. 29.3).

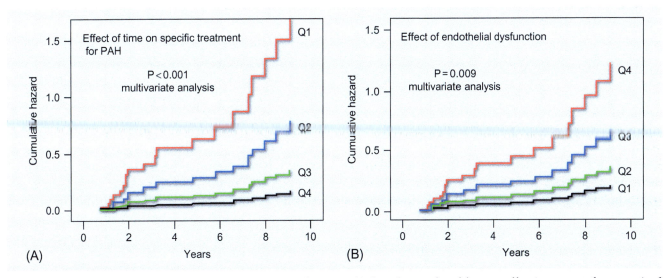

FIG. 29.3   Cox Proportional Hazards regression analysis applied to the study of factors affecting event-free survival (clinical worsening accompanied by low cardiac output) in patients with advanced pulmonary hypertension associated with congenital heart defects. (A) Exposure time to specific drugs for pulmonary arterial hypertension (years) are analyzed as quartiles; Q1 and Q4, respectively, lower and higher treatment periods. (B) plasma von Willebrand factor at baseline and its concentrations (U/dL) analyzed as quartiles; Q1 and Q4, respectively, lower and higher concentrations (abbreviations, see text).

P-selectin is another important marker of endothelial dysfunction, although it is also expressed in other cell types, (megakaryocytes and platelets), stored together with von Willebrand factor in Weibel-Palade bodies in endothelial cells, is secreted by various stimuli (hypoxia, PAR receptor activation by thrombin etc.). It is considered an important mediator of inflammation, since it participates, as adhesion molecules, of mono and polymorphonuclear leukocyte recruitment by binding to its receptor PSGL-1 (Figs. 29.1 and 29.2). Increased circulating levels of P-selectin are considered as endothelial and platelet activation indicators, and thus are described in various diseases including thrombotic thrombocytopenic purpura, acute lung injury, hypercholesterolemia and pulmonary hypertension (in which respond to certain specific therapies) [64,73–75]. In a placebo-controlled study conducted in our institution, we observed that medium-term treatment (6 months) with rosuvastatin was able to reduce significantly plasma level of P-selectin in patients with PAH [76].

Tissue-type plasminogen activator (t-PA), expressed in endothelial cells, is a serine protease involved in fibrinolysis but acting not only in coagulation process. It is known that t-PA operates in different pericellular proteolysis events, with important role in cell growth, including neoplastic. Pericellular proteolysis allows release of growth factors (TGF-β, FGF-2) from their deposits in the matrix. In physiological conditions, in terms of fibrinolysis, response to systemic venous compression is carried out by local release of t-PA and suppression of its inhibitor PAI-1 (plasminogen activator inhibitor-1). In pathological conditions, systemic venous circulation responds in a defective form in the release of t-PA without PAI-1 suppression [77,78]. In advanced pulmonary vascular disease, in resting conditions, without venous tourniquet, an increase in plasma concentration of t-PA is observed, here emerging as a marker of endothelial dysfunction, presumably pulmonary (Fig. 29.2). In this sense, plasma t-PA changes in the same direction as von Willebrand factor and P-selectin (increase) and in contrast to what is observed with thrombomodulin (which is reduced in circulation) [54,79]. Recent data, not yet published by our group, suggests that the plasma level of t-PA, initially increased, undergoes reduction in the short to medium term by specific treatments for PAH with vasodilators.

Many other elements have been investigated and their concentrations in circulating blood measured as markers of microvascular dysfunction and lesion in pulmonary hypertension (Figs. 29.1 and 29.2). Among them, cytokines, chemokines, cell adhesion molecules and other mediators of inflammation are highlighted [80–82]. The CD40-CD40L complex has been investigated for its role in the interactions among endothelium, leukocytes and platelets [41]. Endothelial cells may be present in circulation, in pathological conditions including pulmonary vascular disease; they are detected by flow cytometry, and its number is related to severity [83,84]. In tissue preparations, endothelial cells apoptosis has been investigated as sign of dysfunction, and correlated with disease severity [85]. Endothelial dysfunction has also been studied in terms of genetic aspects. Polymorphisms have been investigated, for example, in the gene encoding eNOS; replacement of Glu298Asp, in the transcription region, is related to defective expression associated with elevated pulmonary pressures in children with congenital heart disease undergoing surgical treatment [86]. Finally, many mutations have been described for the genes encoding TGF-β family receptors, many of which have profound implications with the performance of endothelial cells, especially BMPR2 as mentioned above.

THE FUTURE OF BIOMARKERS IN PULMONARY HYPERTENSION

In pulmonary vascular disease, similar to other vascular disorders, vascular injury markers are inserted within a larger context of biomarkers. The characterization of these substances, whether recoverable in peripheral blood samples, tissue or represented by genetic material implies specific questions and several steps. Firstly, one needs to know which phenomenon that particular substance or molecule "marks." In this sense, correlations are established with major events and outcomes, such as survival, clinical, deterioration and other markers already characterized (functional, hemodynamic, metabolic, etc.). The next step is to identify whether a specific marker is part of a chain ("cascade" signaling pathway, etc.) already known, or if it represents a finding, occasionally secondary. Is it an evolution marker or a marker of complication? Next, it is sought to verify its response to treatments. Biomarkers that remain altered in the course of successful treatment may mean insidious progression of the disease despite transient clinical improvement. One also seeks to check whether it is an independent marker, and for that, comparative analysis with several possible confounders becomes imperative, often with use of exhaustive statistical tools. Finally, particularly in the pediatric population, it is necessary to investigate the behavior of certain markers during normal growth and development, and only later to characterize the pathological deviations.

Thus, it is clear that biomarkers identification is a difficult task, which should not be limited to simple correlation analysis with facts known in the disease. In pulmonary hypertension, the ground is wide, with numerous research possibilities in the direction of questions to be answered. For the future, it is expected that the list of biomarkers will become progressively more consistent, and to help us identify new, more specific targets for therapies as well as subtle changes in the disease course to enable us to modify treatment strategies accordingly.

References

[1] Simonneau G, Gatzoulis MA, Adatia I, et al. Updated clinical classification of pulmonary hypertension. J Am Coll Cardiol 2013;62 (25 Suppl):D34–41.

[2] Stenmark KR, Frid MG. Pulmonary vascular remodeling: cellular and molecular mechanisms. In: Yuan JX-J, Garcia JGN, Hales CA, et al., editors. Textbook of pulmonary vascular disease. New York: Springer Science+Business Media; 2011. p. 759–77.

[3] Christman BW, McPherson CD, Newman JH, et al. An imbalance between the excretion of thromboxane and prostacyclin metabolites in pulmonary hypertension. N Engl J Med 1992;327(2):70–5.

[4] Giaid A, Saleh D. Reduced expression of endothelial nitric oxide synthase in the lungs of patients with pulmonary hypertension. N Engl J Med 1995;333(4):214–21.

[5] Ghofrani HA, Osterloh IH, Grimminger F. Sildenafil: from angina to erectile dysfunction to pulmonary hypertension and beyond. Nat Rev Drug Discov 2006;5(8):689–702.

[6] Wharton J, Strange JW, Møller GM, et al. Antiproliferative effects of phosphodiesterase type 5 inhibition in human pulmonary artery cells. Am J Respir Crit Care Med 2005;172(1):105–13.

[7] Leiper J, Nandi M, Torondel B, et al. Disruption of methylarginine metabolism impairs vascular homeostasis. Nat Med 2007;13(2):198–203.

[8] Pullamsetti S, Kiss L, Ghofrani HA, et al. Increased levels and reduced catabolism of asymmetric and symmetric dimethylarginines in pulmonary hypertension. FASEB J 2005;19(9):1175–7.

[9] Giaid A, Yanagisawa M, Langleben D, et al. Expression of endothelin-1 in the lungs of patients with pulmonary hypertension. N Engl J Med 1993;328(24):1732–9.

[10] Rubens C, Ewert R, Halank M, et al. Big endothelin-1 and endothelin-1 plasma levels are correlated with the severity of primary pulmonary hypertension. Chest 2001;120(5):1562–9.

[11] Farber HW, Loscalzo J. Pulmonary arterial hypertension. N Engl J Med 2004;351(16):1655–65.

[12] Launay JM, Hervé P, Peoc'h K, et al. Function of the serotonin 5-hydroxytryptamine 2B receptor in pulmonary hypertension. Nat Med 2002;8(10):1129–35.

[13] Jeffery TK, Morrell NW. Molecular and cellular basis of pulmonary vascular remodeling in pulmonary hypertension. Prog Cardiovasc Dis 2002;45(3):173–202.

[14] Girgis RE, Champion HC, Diette GB, et al. Decreased exhaled nitric oxide in pulmonary arterial hypertension: response to bosentan therapy. Am J Respir Crit Care Med 2005;172(3):352–7.

[15] Rabinovitch M. Pulmonary hypertension and the extracellular matrix. In: Yuan JX-J, Garcia JGN, Hales CA, et al., editors. Textbook of pulmonary vascular disease. 1st ed. New York: Springer Science+Business Media; 2011. p. 801–9.

[16] Alastalo TP, Li M, Perez Vde J, et al. Disruption of PPARg/b-catenin-mediated regulation of apelin impairs BMP-induced mouse and human pulmonary arterial EC survival. J Clin Invest 2011;121(9):3735–46.

[17] Kim J, Kang Y, Kojima Y, et al. An endothelial apelin-FGF link mediated by miR-424 and miR-503 is disrupted in pulmonary arterial hypertension. Nat Med 2013;19(1):74–82.

[18] Taraseviciene-Stewart L, Nicolls MR, Kraskauskas D, et al. Absence of T cells confers increased pulmonary arterial hypertension and vascular remodeling. Am J Respir Crit Care Med 2007;175(12):1280–9.

[19] Rabinovitch M, Haworth SG, Castaneda AR, et al. Lung biopsy in congenital heart disease: a morphometric approach to pulmonary vascular disease. Circulation 1978;58(6):1107–22.

[20] Taraseviciene-Stewart L, Kasahara Y, Alger L, et al. Inhibition of the VEGF receptor 2 combined with chronic hypoxia causes cell death-dependent pulmonary endothelial cell proliferation and severe pulmonary hypertension. FASEB J 2001;15(2):427–38.

[21] Farkas L, Farkas D, Ask K, et al. VEGF ameliorates pulmonary hypertension through inhibition of endothelial apoptosis in experimental lung fibrosis in rats. J Clin Invest 2009;119(5):1298–311.

[22] Tuder RM, Chacon M, Alger L, et al. Expression of angiogenesis-related molecules in plexiform lesions in severe pulmonary hypertension: evidence for a process of disordered angiogenesis. J Pathol 2001;195(3):367–74.

[23] Yeager ME, Halley GR, Golpon HA, et al. Microsatellite instability of endothelial cell growth and apoptosis genes within plexiform lesions in primary pulmonary hypertension. Circ Res 2001;88(1):E2–E11.

[24] Guignabert C, Alvira CM, Alastalo TP, et al. Tie2-mediated loss of peroxisome proliferator-activated receptor-gamma in mice causes PDGF receptor-beta-dependent pulmonary arterial muscularization. Am J Physiol Lung Cell Mol Physiol 2009;297(6):L1082–90.

[25] Hansmann G, Wagner RA, Schellong S, et al. Pulmonary arterial hypertension is linked to insulin resistance and reversed by peroxisome proliferator-activated receptor-gamma activation. Circulation 2007;115(10):1275–84.

[26] Hansmann G, de Jesus Perez VA, Alastalo TP, et al. An antiproliferative BMP-2/PPARgamma/apoE axis in human and murine SMCs and its role in pulmonary hypertension. J Clin Invest 2008;118(5):1846–57.

[27] Martin-Nizard F, Furman C, Delerive P, et al. Peroxisome proliferator-activated receptor activators inhibit oxidized low-density lipoprotein-induced endothelin-1 secretion in endothelial cells. J Cardiovasc Pharmacol 2002;40(6):822–31.

[28] Wakino S, Hayashi K, Tatematsu S, et al. Pioglitazone lowers systemic asymmetric dimethylarginine by inducing dimethylarginine dimethylaminohydrolase in rats. Hypertens Res 2005;28(3):255–62.

[29] Kim NH, Delcroix M, Jenkins DP, et al. Chronic thromboembolic pulmonary hypertension. J Am Coll Cardiol 2013;62(25 Suppl):D92–9.

[30] Perloff JK, Hart EM, Greaves SM, et al. Proximal pulmonary arterial and intrapulmonary radiologic features of Eisenmenger syndrome

and primary pulmonary hypertension. Am J Cardiol 2003;92(2): 182–7.

[31] Silversides CK, Granton JT, Konen E, et al. Pulmonary thrombosis in adults with Eisenmenger syndrome. J Am Coll Cardiol 2003;42 (11):1982–7.

[32] Broberg C, Ujita M, Babu-Narayan S, et al. Massive pulmonary artery thrombosis with haemoptysis in adults with Eisenmenger's syndrome: a clinical dilemma. Heart 2004;90(11):e63.

[33] Caramuru L, Maeda N, Bydlowski S, et al. Age-dependent likelihood of In situ thrombosis in secondary pulmonary hypertension. Clin Appl Thromb Hemost 2004;10(3):217–23.

[34] Eggebrecht H, Naber CK, Bruch C, et al. Value of plasma flbrin D-dimers for detection of acute aortic dissection. J Am Coll Cardiol 2004;44(4):804–9.

[35] Monaco C, Rossi E, Milazzo D, et al. Persistent systemic inflammation in unstable angina is largely unrelated to the atherothrombotic burden. J Am Coll Cardiol 2005;45(2):238–43.

[36] Rajappa M, Goswami B, Balasubramanian A, et al. Interplay between inflammation and hemostasis in patients with coronary artery disease. Indian J Clin Biochem 2015;30(3):281–5.

[37] Shitrit D, Bendayan D, Bar-Gil-Shitrit A, et al. Significance of a plasma D-dimer test in patients with primary pulmonary hypertension. Chest 2002;122(5):1674–8.

[38] Tournier A, Wahl D, Chaouat A, et al. Calibrated automated thrombography demonstrates hypercoagulability in patients with idiopathic pulmonary arterial hypertension. Thromb Res 2010;126(6): e418–22.

[39] Ataga KI, Moore CG, Hillery CA, et al. Coagulation activation and inflammation in sickle cell disease-associated pulmonary hypertension. Haematologica 2008;93(1):20–6.

[40] Maeda NY, Carvalho JH, Otake AH, et al. Platelet protease-activated receptor 1 and membrane expression of P-selectin in pulmonary arterial hypertension. Thromb Res 2010;125(1):38–43.

[41] Damås JK, Otterdal K, Yndestad A, et al. Soluble CD40 ligand in pulmonary arterial hypertension: possible pathogenic role of the interaction between platelets and endothelial cells. Circulation 2004;110(8):999–1005.

[42] Polanowska-Grabowska R, Wallace K, Field JJ, et al. P-selectin-mediated platelet-neutrophil aggregate formation activates neutrophils in mouse and human sickle cell disease. Arterioscler Thromb Vasc Biol 2010;30(12):2392–9.

[43] Rabinovitch M, Guignabert C, Humbert M, et al. Inflammation and immunity in the pathogenesis of pulmonary arterial hypertension. Circ Res 2014;115(1):165–75.

[44] Budhiraja R, Tuder RM, Hassoun PM. Endothelial dysfunction in pulmonary hypertension. Circulation 2004;109(2):159–65.

[45] Balabanian K, Foussat A, Dorfmüller P, et al. CX(3)C chemokine fractalkine in pulmonary arterial hypertension. Am J Respir Crit Care Med 2002;165(10):1419–25.

[46] Dorfmüller P, Perros F, Balabanian K, et al. Inflammation in pulmonary arterial hypertension. Eur Respir J 2003;22(2):358–63.

[47] von Hundelshausen P, Weber KS, Huo Y, et al. RANTES deposition by platelets triggers monocyte arrest on inflamed and atherosclerotic endothelium. Circulation 2001;103(13):1772–7.

[48] Humbert M, Morrell NW, Archer SL, et al. Cellular and molecular pathobiology of pulmonary arterial hypertension. J Am Coll Cardiol 2004;43(12 Suppl S):13S–24S.

[49] Molet S, Furukawa K, Maghazechi A, et al. Chemokine- and cytokine-induced expression of endothelin 1 and endothelin-converting enzyme 1 in endothelial cells. J Allergy Clin Immunol 2000;105(2 Pt 1):333–8.

[50] Takahashi H, Ito S, Hanano M, et al. Circulating thrombomodulin as a novel endothelial cell marker: comparison of its behavior with von Willebrand factor and tissue-type plasminogen activator. Am J Hematol 1992;41(1):32–9.

[51] Ogawa S, Gerlach H, Esposito C, et al. Hypoxia modulates the barrier and coagulant function of cultured bovine endothelium. Increased monolayer permeability and induction of procoagulant properties. J Clin Invest 1990;85(4):1090–8.

[52] Healy AM, Hancock WW, Christie PD, et al. Intravascular coagulation activation in a murine model of thrombomodulin deficiency: effects of lesion size, age, and hypoxia on fibrin deposition. Blood 1998;92(11):4188–97.

[53] Cacoub P, Karmochkine M, Dorent R, et al. Plasma levels of thrombomodulin in pulmonary hypertension. Am J Med 1996;101(2): 160–4.

[54] Lopes AA. Pathophysiological basis for anticoagulant and antithrombotic therapy in pulmonary hypertension. Cardiovasc Hematol Agents Med Chem 2006;4(1):53–9.

[55] Aliberti G, Proietta M, Pulignano I, et al. The lungs and platelet production. Clin Lab Haematol 2002;24(3):161–4.

[56] Weyrich AS, Zimmerman GA. Platelets in lung biology. Annu Rev Physiol 2013;75:569–91.

[57] Léon C, Evert K, Dombrowski F, et al. Romiplostim administration shows reduced megakaryocyte response-capacity and increased myelofibrosis in a mouse model of MYH9-RD. Blood 2012;119 (14):3333–41.

[58] Herve P, Drouet L, Dosquet C, et al. Primary pulmonary hypertension in a patient with a familial platelet storage pool disease: role of serotonin. Am J Med 1990;89(1):117–20.

[59] Beaulieu LM, Freedman JE. Inflammation & the platelet histone trap. Blood 2011;118(7):1714–5.

[60] Kroll MH, Afshar-Kharghan V. Platelets in pulmonary vascular physiology and pathology. Pulm Circ 2012;2(3):291–308.

[61] Lopes AA, Maeda NY, Ebaid M, et al. Effect of intentional hemodilution on platelet survival in secondary pulmonary hypertension. Chest 1989;95(6):1207–10.

[62] Lopes AA, Maeda NY, Almeida A, et al. Circulating platelet aggregates indicative of in vivo platelet activation in pulmonary hypertension. Angiology 1993;44(9):701–6.

[63] Maeda NY, Bydlowski SP, Lopes AA. Increased tyrosine phosphorylation of platelet proteins including pp 125(FAK) suggests endogenous activation and aggregation in pulmonary hypertension. Clin Appl Thromb Hemost 2005;11(4):411–5.

[64] Sakamaki F, Kyotani S, Nagaya N, et al. Increased plasma P-selectin and decreased thrombomodulin in pulmonary arterial hypertension were improved by continuous prostacyclin therapy. Circulation 2000;102(22):2720–5.

[65] Lopes AA, Maeda NY, Aiello VD, et al. Abnormal multimeric and oligomeric composition is associated with enhanced endothelial expression of von Willebrand factor in pulmonary hypertension. Chest 1993;104(5):1455–60.

[66] Lopes AA, Maeda NY. Abnormal degradation of von Willebrand factor main subunit in pulmonary hypertension. Eur Respir J 1995;8(4):530–6.

[67] Lopes AA, Ferraz de Souza B, Maeda NY. Decreased sialic acid content of plasma von Willebrand factor in precapillary pulmonary hypertension. Thromb Haemost 2000;83(5):683–7.

[68] Lopes AA, Maeda NY. Circulating von Willebrand factor antigen as a predictor of short-term prognosis in pulmonary hypertension. Chest 1998;114(5):1276–82.

[69] Lopes AA, Maeda NY, Bydlowski SP. Abnormalities in circulating von Willebrand factor and survival in pulmonary hypertension. Am J Med 1998;105(1):21–6.

[70] Lopes AA, Maeda NY, Gonçalves RC, et al. Endothelial cell dysfunction correlates differentially with survival in primary and secondary pulmonary hypertension. Am Heart J 2000;139(4):618–23.

[71] Kawut SM, Horn EM, Berekashvili KK, et al. von Willebrand factor independently predicts long-term survival in patients with pulmonary arterial hypertension. Chest 2005;128(4):2355–62.

[72] Lopes AA, Barreto AC, Maeda NY, et al. Plasma von Willebrand factor as a predictor of survival in pulmonary arterial hypertension associated with congenital heart disease. Braz J Med Biol Res 2011;44(12):1269–75.

[73] Katayama M, Handa M, Araki Y, et al. Soluble P-selectin is present in normal circulation and its plasma level is elevated in patients with thrombotic thrombocytopenic purpura and haemolytic uraemic syndrome. Br J Haematol 1993;84(4):702–10.

[74] Sakamaki F, Ishizaka A, Handa M, et al. Soluble form of P-selectin in plasma is elevated in acute lung injury. Am J Respir Crit Care Med 1995;151(6):1821–6.

[75] Davì G, Romano M, Mezzetti A, et al. Increased levels of soluble P-selectin in hypercholesterolemic patients. Circulation 1998;97 (10):953–7.

[76] Barreto AC, Maeda NY, Soares RPS, et al. Rosuvastatin and vascular dysfunction markers in pulmonary arterial hypertension: a placebo-controlled study. Braz J Med Biol Res 2008;41(8):657–63.

[77] Binotto M, Maeda N, Lopes A. Evidence of endothelial dysfunction in patients with functionally univentricular physiology before completion of the Fontan operation. Cardiol Young 2005;15(1): 26–30.

[78] Binotto MA, Maeda NY, Lopes AA. Altered endothelial function following the Fontan procedure. Cardiol Young 2008;18 (1):70–4.

[79] Caramuru L, Lopes A, Maeda N, et al. Long-term behavior of endothelial and coagulation markers in Eisenmenger syndrome. Clin Appl Thromb Hemost 2006;12(2):175–83.

[80] Brun H, Holmstrøm H, Thaulow E, et al. Patients with pulmonary hypertension related to congenital systemic-to-pulmonary shunts are characterized by inflammation involving endothelial cell activation and platelet-mediated inflammation. Congenit Heart Dis 2009;4(3):153–9.

[81] Sungprem K, Khongphatthanayothin A, Kiettisanpipop P, et al. Serum level of soluble intercellular adhesion molecule-1 correlates with pulmonary arterial pressure in children with congenital heart disease. Pediatr Cardiol 2009;30(4):472–6.

[82] Oguz MM, Oguz AD, Sanli C, et al. Serum levels of soluble ICAM-1 in children with pulmonary artery hypertension. Tex Heart Inst J 2014;41(2):159–64.

[83] Smadja DM, Gaussem P, Mauge L, et al. Circulating endothelial cells: a new candidate biomarker of irreversible pulmonary hypertension secondary to congenital heart disease. Circulation 2009;119(3):374–81.

[84] Levy M, Bonnet D, Mauge L, et al. Circulating endothelial cells in refractory pulmonary hypertension in children: markers of treatment efficacy and clinical worsening. PLoS ONE 2013;8(6):e65114.

[85] Lévy M, Maurey C, Celermajer DS, et al. Impaired apoptosis of pulmonary endothelial cells is associated with intimal proliferation and irreversibility of pulmonary hypertension in congenital heart disease. J Am Coll Cardiol 2007;49(7):803–10.

[86] Loukanov T, Hoss K, Tonchev P, et al. Endothelial nitric oxide synthase gene polymorphism (Glu298Asp) and acute pulmonary hypertension post cardiopulmonary bypass in children with congenital cardiac diseases. Cardiol Young 2011;21(2):161–9.

CORONARY DISEASE AND ATHEROSCLEROSIS

Endothelial Alterations in Chronic Coronary Disease

Aline Alexandra Iannoni de Moraes, Antonio Carlos Palandri Chagas, José Rocha Faria Neto, and Protásio Lemos da Luz

INTRODUCTION

Chronic coronary atherosclerotic disease (CAD) is characterized by reversible episodes of myocardial ischemia related to imbalance between supply and demand of oxygen [1]. Coronary obstruction by stable atherosclerotic plaque is the main pathophysiological mechanism of CAD. However, functional alterations of the epicardial vessels and/or coronary microcirculation may also act as inducers of myocardial ischemia, both individually and in association with atherosclerosis [1]. Ischemia episodes may be related to chest discomfort (*angina pectoris*), but there are cases of asymptomatic disease and patients who develop acute coronary syndrome without ever before having presented symptoms of angina [1].

The broad symptomatic and pathophysiologic spectrum of CAD makes it difficult to estimate its prevalence, but the most common symptom, *angina pectoris*, affects between 5% and 14% of individuals, being more frequent in men and elderly [1]. Equally broad is its prognosis, with annual mortality rates ranging between 0.63% and 3.8%, depending on factors such as degree and site of coronary obstruction, and associated morbidities [2].

Although CAD and *angina pectoris* have been described more than two centuries ago by William Heberden [3], demonstration of endothelium's active participation in atheromatous process was only possible 40 years ago. In the early 1970s, Ross and Glomset demonstrated that the mechanical removal of endothelial cells accelerating progression of atherosclerosis

[4], and, at the end of the same decade, Moncada and Vane identified prostacyclin [5]. Twenty years later, in the late 1990s, characterization of nitric oxide as a coronary vasodilator agent gave Furchgott, Ignarro, and Murad the Nobel Prize in medicine [6]. Even more recent are the findings of endothelin, reactive oxygen species and NADPH oxidase, peroxynitrite and the role of mitochondrial respiration function as a regulator of endothelial cells [3,7].

With the understanding of the endothelium's importance in the atheromatous process, several methods to analyze endothelial function have been elaborated. These exams allowed a better understanding of the endothelium's role in clinical practice, prognostic evaluation of endothelial dysfunction in CAD, as well as analysis of the impact of different treatments on the endothelium.

In this chapter, we shall discuss the participation of endothelium in CAD, methods for measuring endothelial dysfunction in CAD, and the available evidence on the impact of different treatments on the endothelium. We shall start with a brief description of the pathophysiological basis for CAD.

PHYSIOPATHOLOGICAL BASES FOR CAD

Coronary Flow in the Normal Heart

Accuracy of the coronary flow adjustment is essential for the myocardial muscle. Unlike skeletal musculature, capable to reduce its O_2 extraction rate to 30%–40% at rest, the heart keeps pace at

60–70 beats/min even at rest. The high demand of O_2 is compensated by the extraction rate of 70%–80% [8]. With the high rate of extraction at rest, increase in coronary flow is the main mechanism to provide the necessary energy supply in situations such as physical activity [9]. Coronary flow is able to increase 4–6 times in response to vigorous physical activity from baseline at rest of 0.5–1.5 mL/min/g of myocardium [9].

The coronary arterial system can be understood as three compartments with different functions, although without a clear anatomical distinction [10]. The major epicardial coronary arteries comprise the proximal compartment, their diameter varies between 500 and 2–5 mm, and they have conductivity as the main function, motivating the denomination of "conductance arteries" [10]. Follows compartment of the prearterioles, still in the extramyocardial position and with a diameter of 100–500 µm [10]. Prearterioles function is to maintain constant shear stress by vasodilation or vasoconstriction in response to changes in pressure or flow independent from endothelium, what is also performed, less importantly, by the epicardial coronary arteries [10].

The distal compartment is represented by intramural arterioles with a diameter <100 µm and the important function is to regulate the coronary flow according to myocardial consumption of 10. Intramural arteries present high tonus at rest and dilate in response to release of metabolites by myocardium as a result of increased O_2 consumption and mediated by the endothelium [10]. Arteriolar vasodilation ends up reducing resistance across the network, and the consequent increase in shear force induces proximal prearterioles and epicardial vessels to vasodilation [10]. Thus, an increase in oxygen consumption by the myocardium promotes vasodilation of the entire coronary system to ensure adequate energy intake in situations of increased demand (e.g., during exercise) [10]. Fig. 30.1 summarizes the functional anatomy of the coronary system.

Role of the Endothelium in Regulating Coronary Flow in the Normal Heart

The endothelium is essential in the regulation of coronary flow. In response to increased or decreased myocardial metabolic demand, endothelial cell produces several vasodilating and vasoconstrictor agents, ensuring an adequate O_2 supply, and preventing myocardial ischemia [9]. Fig. 30.2 summarizes the action of each agent described in sequence.

The main vasodilator agent produced by the endothelium is nitric oxide. Its vasodilatory action occurs by increasing cGMP in smooth muscle cells, which results in activation of K_{Ca} and K_{ATP} channels, and cell hyperpolarization, with consequent muscle relaxation [9]. Another nitric oxide action pathway through cGMP is the activation of GMPc (PKG)-dependent protein-kinase. PKG reduces intracellular calcium concentration by activating myosin light chain phosphorylation (MLCP), which in turn reduces the sensitivity of myofilaments to calcium, culminating in vasodilation [3]. Production of nitric oxide in the endothelium takes place by action of oxide nitric-synthase on L-arginine and the process may be modulated by stimulation of receptors (such as alpha 2 adrenergic receptors and ET_B) and by mechanical deformation resulting from shear forces imposed by blood flow [9]. Thus, its action is mild during rest and increases in situations of greater O_2 demand (such as physical activity) due to stimuli such as increased shear force and the activation of alpha 2 adrenergic receptors in the endothelium [9].

Other vasodilator agents produced by endothelium include prostanoids and endothelial hyperpolarizing factors (EDHF) [9]. Prostacyclin and other prostanoids are produced from arachidonic acid through cyclooxygenase 1 and promote muscle relaxation by opening of K_{ATP} channels via cAMP. EDHF include arachidonic acid metabolism products by cytochrome P-450 [11] and hydrogen peroxide [12] and its action on smooth muscle cell also takes place through the opening of K_{Ca} channels, with consequent hyperpolarization and muscle relaxation.

Another vasodilator is worth mentioning, although there is evidence that it is produced more by myocytes than by the endothelium: adenosine. At rest, its production takes place in the extracellular environment and the molecule is used by the cardiomyocyte to form AMP [9]. When there is an increase in myocardial metabolism, increased ATP hydrolysis results in an increase of free ADP in an intracellular medium [9]. Adenylate cyclase converts ADP to AMP and this, in turn, is converted to adenosine by AMP 5′-nucleotidase. Adenosine produced by cardiomyocyte is then released into

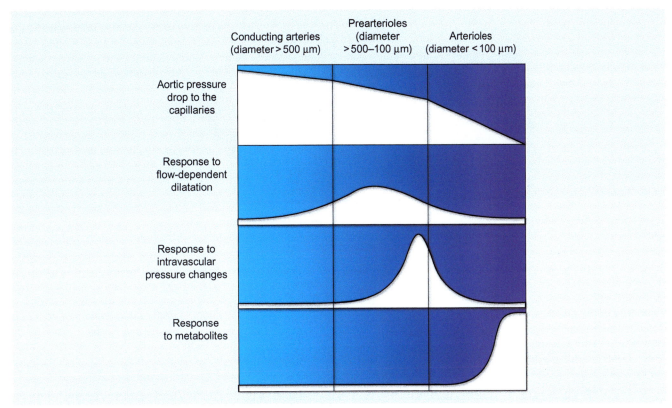

FIG. 30.1 Functional anatomy of the coronary artery system. Epicardial arteries, that is, conductance arteries, prearterioles, and arterioles form the functional subdivisions of the coronary artery system. The pressure fall is mild in the conductance arteries, more pronounced in the prearterioles and greater in the arterioles. The conductance arteries and, to a greater degree, the proximal arterioles are responsive to coronary flow. Distal prearterioles are more responsive than other segments to changes in intravascular pressure and are the main responsible for self-regulation of coronary artery flow. Arterioles are responsive to changes in the intramyocardial concentration of metabolites and are the main vessels responsible for the adequacy of the blood flow to the metabolic needs of the myocardium. Prearterioles, by definition, are not exposed to myocardial metabolites, since they are located in the epicardium (acronyms, see text). Adapted from Camici PG, Crea F. Coronary microvascular dysfunction. N Engl J Med 2007;356(8):830–40.

the extracellular medium, in which it promotes arteriolar vasodilation [9]. The mechanism through which adenosine causes muscle relaxation appears to involve the opening of K_{ATP} channels after stimulation of A_1 and A_2 receptors [9].

Endothelin-1 (ET-1), produced by the endothelium after cleavage of its precursors, exerts different functions on the endothelium and on smooth muscle cells [9]. In the endothelium, its binding to ET receptor promotes nitric oxide and prostacyclin production, inducing vasodilation [9]. In contrast, binding of ET-1 to ET_A and ET_B receptors in smooth muscle cells results in vasoconstriction [9]. Studies in animals suggest a mild vasoconstrictor action of ET-1 in rest situation, and that this action is compensated by increased release of nitric oxide by endothelium during situations of increased demand

for O_2 (such as physical activity), culminating in predominance of vasodilation rather than vasoconstriction [9].

Finally, the bloodstream, as well as erythrocytes and platelets, also produces or brings with it vasodilator factors, such as acid pH, ATP, and nitric oxide, and vasoconstrictors, such as circulating catecholamines, angiotensin, histamine, and thromboxane A2 [9]. Serotonin has the interesting property of causing microvascular vasodilation and, at the same time, induces vasoconstriction of epicardial vessels [13]. In the normal heart, peripheral vasodilation effect predominates over vasoconstriction [14]. The heart's sympathetic and parasympathetic systems are also able to induce vasodilation and coronary vasoconstriction [9], exemplifying the complexity of the coronary self-regulation system to ensure

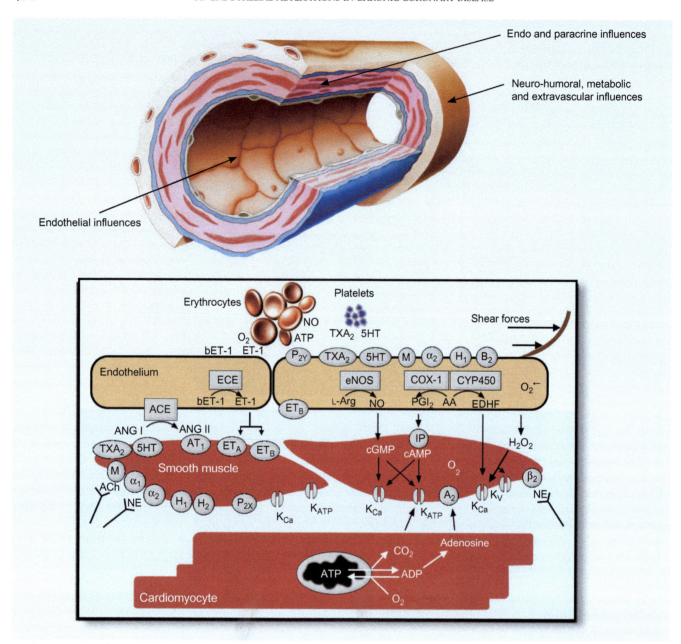

FIG. 30.2　Coronary arteriole and the various influences that determine vasomotor tone. PO_2, oxygen tension; TxA_2, thromboxane A_2 and its receptor; *5HT*, serotonin and its receptor; *P* and P_{2X2Y}, purinergic subtypes 2X and 2Y receptors that mediate, respectively, vasoconstriction and vasodilation induced by ATP; *Ach*, acetylcholine; *M*, muscarinic receptor; M_1 and M_2, histamine type 1 and 2; B_2, bradykinin receptor subtype 2; *ANG-I and ANG-II*, angiotensin-I and II; AT_1, angiotensin-II receptor subtype 1; *ET-1*, endothelin-1; ET_A and ET_B, the endothelin receptor subtypes and B; A_2, adenosine receptor subtype 2; *B*, beta 2 adrenergic receptor; *A and α*, alpha adrenergic receptors subtypes 1 and 2; *NO*, nitric oxide; *eNOS*, endothelial nitric oxide synthase; PGI_2, prostacyclin; *IP*, prostacyclin receptor; *COX-1*, cyclooxygenase-1; *EDHF*, endothelial hyperpolarizing factor; *CYP450*, cytochrome 450 2C9; K_{Ca}, K sensitive to Ca^{++}; K_{ATP}, K channel sensitive to ATP; K_v, K voltage-sensitive channel; *AA*, arachidonic acid; *L-Arg*, L-arginine; O_2^-, superoxide. The receptors are represented by *circular or oval symbols* and enzymes by *rectangular symbols*. Adapted from Duncker DJ, Bache RJ. Regulation of coronary blood flow during exercise. Physiol Rev 2008;88(3):1009–86.

supply of O_2 appropriate to myocardial metabolic demands in physiological situations.

Coronary Flow in CAD

In CAD, myocardial ischemia is mainly induced by coronary stenosis, but modulated by imbalance between O_2 supply and demand [1]. Coronary blood flow has certain characteristics. First, it occurs place mainly in diastole. Second, it is pressure-dependent, and perfusion pressure corresponds to the difference between the diastolic pressure in the aortic root and right atrial pressure. Third, there is the autoregulation phenomenon, whereby sudden changes in blood pressure are soon automatically antagonized, so that coronary flow remains constant between approximately 140 and 60 mmHg; this property allows coronary flow to be maintained even in face of large pressure variations, such as occurs in normal life or in arterial hypertension. However, from approximately 60 mmHg down, myocardial perfusion is entirely pressure-dependent (Fig. 30.3).

The major cause of chronic CAD is coronary obstruction due to atherosclerotic plaque. Plaque develops and progresses slowly. Initially, resistance in conductance arteries is minimal and flow does not change. However, when the obstruction reaches 70% or more of arterial lumen, arterial resistance (R_1) at the lesion level increases exponentially in proportion to increase in stenosis. Simultaneously, resistance (R_2) in microcirculation decreases gradually due to the accumulation of metabolites secondary to ischemic process. Coronary flow is initially preserved, but decreases progressively with progression of stenosis and R_1 because perfusion pressure downstream of the stenosis gradually falls, causing ischemia. On the other hand, coronary flow's capacity to increase by reactive hyperemia also reduces until reaching a minimum point, and the process of myocardial ischemia becomes persistent [15].

Fig. 30.4A shows influence of stenosis degree on resistance and coronary flow, and Fig. 30.4B shows influence of coronary stenosis under baseline conditions (interrupted line), and under vasodilator stimulus. Another important determinant of the ischemia degree is collateral circulation; this may completely compensate for the effects of coronary stenosis when fully developed.

Involvement of the endothelium in chronic CAD has been analyzed in several respects. Ludmer et al. [16] have made fundamental observation that in

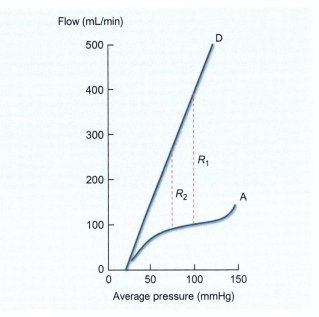

FIG. 30.3 Diagram demonstrating the self-regulation phenomenon in the normal left ventricle (line A) and with maximum vasodilation (line D). R_1 and R_2 coronary flow reserve under perfusion pressure of 75 and 100 mmHg, with constant aortic pressure and heart rate. Adapted from Hoffman JL. Maximal coronary flow and the concept of coronary vascular reserve. Circulation 1984;70:153–9.

men with CAD intracoronary injection of acetylcholine causes vasoconstriction at the site of atherosclerotic plaque as well as on adjacent segments to the lesion, rather than the expected vasodilation in segments without atherosclerosis. This clearly showed that endodontic vasodilator function is grossly impaired in CAD. Other researchers have also confirmed the presence of endothelial dysfunction in CAD, as well as its correlation with future clinical events [17]. Even individuals with only a family history of CAD, but without clinically documented CAD, present endothelial dysfunction [18].

Another essential characteristic is that endothelial dysfunction precedes clinical manifestations of CAD, and is, therefore, a risk marker. Confirming endothelium participation, treatment of dyslipidemia, and other risk factors improve endothelial function in a short period of time [19], too narrow to allow plaque regression, therefore indicates functional recovery of endothelium.

On the other hand, at least part of myocardial ischemia occurs as a consequence of coronary microvascular dysfunction (CMD) induced by cardiovascular risk factors such as hypertension, diabetes, and

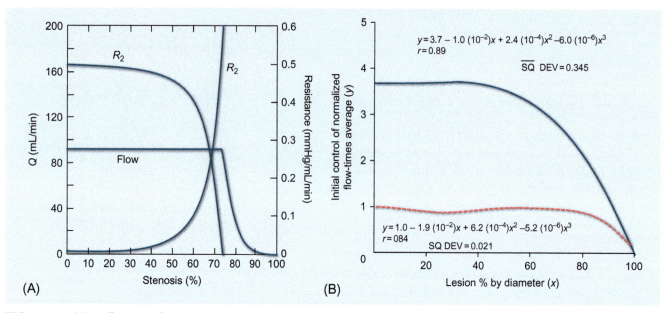

FIG. 30.4 (A) Influence of stenosis degree on coronary resistance and coronary flow increased epicardial coronary resistance (R_1) and decreased arteriolar resistance as the stenosis degree increases. When R_2 reserve is exhausted, subsequent increases in R_1 cause reduced flow. (B) Influence of coronary stenosis at baseline *(interrupted line)* and under vasodilator stimulus. The vasodilatory capacity is progressively reduced as stenosis evolves; it is noted that the vasodilator-induced flow is modified with smaller lesions than those that alter basal flow. When the lesion is larger, both flows are reduced. Adapted from Epstein SE, Cannon RO 3rd, Talbot TL. Hemodynamic principles in the control of coronary blood flow. Am J Cardiol 1985;56:4E–10E.

smoking [3]. In the presence of CMD, vasodilation induced by nitric oxide may not be sufficient due to imbalance in the formation of oxide release stimulating factors (such as bradykinin [20]) and oxide action inhibitors (such as asymmetric dimethylarginine (ADMA) [3]).

Parallel to the insufficient action of nitric oxide, there is an increase in the release of ET-1 due to stimulation of angiotensin-II and LDL oxidized by NADP oxidase [3,15]. ET-1 promotes development of atheromatosis plaque by stimulating proliferation and migration of smooth muscle cell [3,21], acting as a chemotactic factor for macrophages [22] and inducing formation of fibronectin [23] and proteins in the extracellular matrix [24]. Its proinflammatory effect mediated by increased expression of adhesion molecules [25], production of proinflammatory cytokines (such as interleukin-1 beta and TNF-α) [21] and growth factors (such as vascular endothelial [26]), and leads to the formation of plaques with greater lipid core and intraplate angiogenesis [3,21]. Release of peroxynitrite and increased platelet aggregation induced by ET-1 [3,21] become even greater propensity to form unstable atheroma plaques. Small ruptures of the fibrous plaque and blood

vessels within the plaque, followed by scarring, provide rapid atheroma growth and increased propensity to instability and acute coronary syndrome. Finally, endothelin has, by itself, vasoconstrictor effect [27] and inhibitor of nitric oxide release [21], completing its important role in all stages of formation and destabilization of coronary atheromatous plaque. Fig. 30.5 summarizes the main factors of coronary atherosclerosis.

In an attempt to prevent myocardial ischemia, compensatory mechanisms to promote chronic vasodilation develop in patients with CAD. There is an increase in the release of nitric oxide by erythrocytes [28,29] and by the endothelium [30] stimulated by low O_2 tension and activation of endothelial adrenergic alpha 2 receptors [31]. Production of adenosine by the endothelium and myocytes [9,32] is stimulated by higher concentration of ADP [9]. The effect of vasodilatory prostaglandins also appears to be enhanced in patients with chronic ischemia [33,34] in comparison to physiological situations [9]. The effect of serotonin in patients with CAD differs from that exerted on normal hearts. Since other compensatory mechanisms are already causing vasodilation, the vasoconstricting effect of

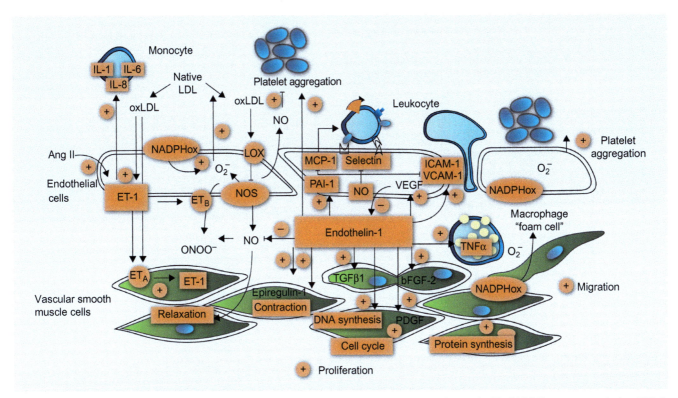

FIG. 30.5 Role of nitric oxide and endothelin-1 in atheromatosis. *Ang-II*, angiotensin-II; *ONOO⁻*, peroxynitrite; *ET-1*, endothelin-1; *ET$_A$*, endothelin receptor subtype A; *ET$_B$*, endothelin receptor subtype B; *NADHPox*, NADPH oxidase; *NO*, nitric oxide; *NOS*, nitric oxide synthase; *MCP-1*, monocyte-1 chemotactic protein; *ICAM-1*, intracellular adhesion molecule-1; *VCAM-1*, vascular adhesion molecule-1; *O$_2$*, superoxide anion; *LDL*, low density lipoprotein; *OxLDL*, oxidized low density lipoprotein; *IL-1*, interleukin-1; *IL-6*, interleukin-6; *IL-8*, interleukin-8; *TNF-α*, tumor necrosis factor alpha; *TGFb-1*, growth transforming factor beta-1; *PDGF*, platelet-derived growth factor; *BFGF-2*, basic fibroblast growth factor; *VEGF*, vascular endothelial growth factor; (+) indicates stimulation and (−) indicates inhibition (acronyms, see text). Adapted from Luscher TF, Barton M. Endothelins and endothelin receptor antagonists: therapeutic considerations for a novel class of cardiovascular drugs. Circulation 2000;102(19):2434–40.

epicardial vessels prevails, contributing to an increase in myocardial ischemia. The remaining [9] coronary vasodilation mechanisms previously described are also present and may even allow further adjustment to an increased metabolic demand (e.g., physical activity) [9].

Unfortunately, the action of adaptive mechanisms to ischemia may be compromised in patients with CAD. The vasodilator response may be limited by coronary microvascular degree (CMD) [9,35,36] and smooth muscle cells can enter a state of hyperresponsiveness which make them prone to vasoconstriction [1]. The causes of coronary muscle hyperreactivity are not well understood, but appear to be associated with changes of K$_{ATP}$ channels and Na$^+$H$^+$ pump [37]. Due to the dynamic nature of the smooth muscle cells response, the compensation for ischemia varies from person to person and even in

the same person [1]. Fig. 30.6 summarizes the action of vasodilator and vasoconstrictor agents in CAD.

Endothelium is also fundamental in another mechanism of coronary flow compensation: development of collateral circulation [38]. In the formation of collateral circulation via preformed vessels in the embryonic period, increase in blood flow takes place by the substantial increase of the arterioles and capillaries lumen diameter, and by differentiation of smooth muscle cells [39,40]. Another possible mechanism of collateral circulation formation is angiogenesis, although its degree of importance in the DAC is not well determined [38]. In this process, the capillaries or postcapillary venules endothelial cells are activated by a hypoxia inducible factor 1-α (HIF1-α) [41,42]. Endothelial activation induces formation of cytoplasmatic protrusions [38]. Then comes degradation of the

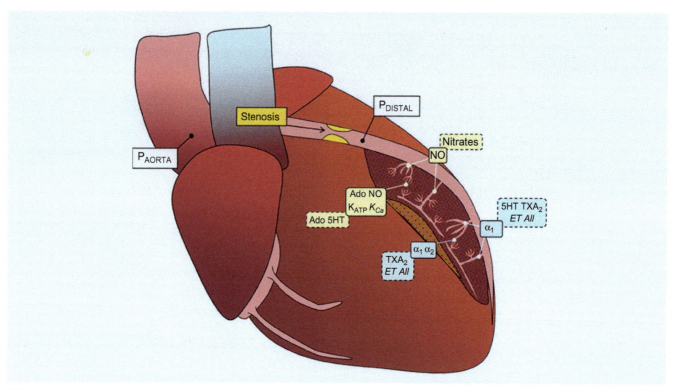

FIG. 30.6 Scheme of vasodilator (*yellow*) and vasoconstrictor (*blue*) factors that affect various segments of the microvasculature (small intramural arteries A and B and arterioles) in the presence of coronary stenosis during exercise. *Closed boxes* are endogenous substances that contribute to vasomotor control; in *boxes with broken lines*, externally administered vasoactive substances. Regular sources refer to mechanisms demonstrated in waken animals, in exercise; in *italics*, mechanisms demonstrated in anesthetized animals with open thorax. *TAX$_2$*, thromboxane A$_2$; *5HT*, serotonin; *AII*, angiotensin-II; *ET*, endothelin; *α_1 and α_2*, adrenergic receptors; *Ado*, adenosine; *NO*, nitric oxide; *K_{ATP}*, K+ channel sensible to ATP; *K_{Ca}*, K$^+$ channel sensitive to CA^{++}. Reproduced with permission from Duncker DJ, Bache RJ. Regulation of coronary blood flow during exercise. Physiol Rev 2008;88(3):1009–86.

basement membrane, direct migration of endothelial cells towards an angiogenic stimulus and cell multiplication [38].

Endothelial cells elongate and lineup, establishing tubular formations [38]. Anastomosis of tubular formations and basement membrane formation establish blood flow [42].

Increase of the shear stress on collateral circulation vessels promotes genetic alterations in endothelial cells [38]. Cytokines, growth factors, and adhesion molecules stimulate monocytes adhesion, which migrate to the vessel wall [43]. Here, macrophages secrete cytokines and growth factors that stimulate the proliferation and differentiation of smooth muscle cells, so that vessels before with only one or two layers of muscle cells increase up to 20 times their diameter and up to 50 times this mass [44]. Blood vessels resulting from collateral circulation are often tortuous; their anastomosis with the distal part of the occluded coronary is usually

angular [38] and the supply may be insufficient to ensure myocardial demand.

In short, various mechanisms involving coronary vasodilatation and the development of collateral circulation attempt to increase O$_2$ supply to territories where there is stenosis of the epicardial coronary artery. When these mechanisms are insufficient to overcome myocardial metabolic demand, myocardial ischemia ensues.

Consequences of Myocardial Ischemia

Myocardial ischemia appears in the presence of stenosis in epicardial artery of such magnitude that overcome the coronary adaptive mechanisms and limits O$_2$ supply. The most typical symptom of myocardial ischemia is *angina pectoris*. This is pain or precordial discomfort referred to as oppressive or burning, occasionally with irradiation to the left upper limb or mandible, and which may be associated with other symptoms, such as dyspnea and fatigue [1]. Angina associated with chronic CAD, in which the coronary plaque is intact, typically occurs with physical efforts or emotions due to the incapacity of adaptive mechanisms to increase O_2 supply in situations of increased myocardial demand [1,9]. Endothelial dysfunction and the state of vascular hyperactivity present in CAD can cause fluctuations in the degree of physical effort required to generate symptoms or even result in episodes of angina at rest [1]. Differentiation between episodes of stable angina at rest and episodes of unstable angina or acute myocardial infarction (associated with the rupture of coronary plaque) can be difficult, but the rapid relief of pain with the use of sublingual nitrates increases the chance of being a stable condition [1]. Silent ischemia, in which patients have documented ischemia in the absence of symptoms, is another form of clinical presentation [1].

Ischemia can also lead to the reduction of O_2 consumption by the myocardium. In myocardial segments irrigated by a stenosed coronary, cardiomyocytes reduce their contractility in an attempt to adjust their demand to the O_2 supply available. This condition is called hibernating myocardium, and may be evidenced by echocardiography through hypo or akinesia of myocardial segments [1]. Administration of dobutamine in echo stress can return contractility of the akinetic region, which differentiates viable myocardium (hibernating) from akinesia area consistent with previous infarctions and fibrosis [1,45].

When the area of myocardial ischemia is very large, which may occur in, for example, cases of proximate stenosis of the anterior descending artery, the area of hibernating myocardium may be so large that it may result in global ventricular dysfunction [1]. Clinically, these patients present symptoms of heart failure, such as dyspnea on effort, fatigue and systemic congestion, with or without angina [1].

Intercellular communication between cardiomyocytes ensures propagation of depolarization and maintains appropriate cardiac rhythm. Largely responsible for ensuring proper intercellular communication are gap junctions formed by connexins [46]. Conductance of gap junctions is modulated by H^+, Ca^{2+}, Mg^{2+}, arachidonic acid, and ATP metabolites that accumulate or are consumed during myocardial ischemia [46]. The change in gap junctions conductance can generate inappropriate coupling between cardiomyocytes and may result in arrhythmias and

sudden death [46]. Other mechanisms for the development of ventricular arrhythmias and sudden death include reentry myocardial ischemic or fibrotic areas, as well as aneurysm in the left ventricular tip after apical infarction [47].

In short, consequences of myocardial ischemia vary with the myocardium area at risk, which, in turn, depends on the location and degree of coronary stenosis, as well as capacity of the compensatory mechanisms, especially collateral circulation, to limit ischemia. Patients may present with effort angina, occasional episodes at rest, have clinical cardiac failure, arrhythmias, or may even be asymptomatic.

EVALUATION OF CORONARY MICORVASCULAR DYSFUNCTION

It is necessary to differentiate the methods used for CAD diagnosis from those employed in diagnosis of CMD. A detailed description of the complementary examinations for the diagnosis of CAD is beyond the scope of this chapter. However, a brief description is necessary for better understanding of the diagnostic methods for CMD.

The gold standard method for recognition and degree of obstruction caused by coronary artery plaque is cineangiocoronariography. However, due to adaptation mechanisms of the coronary circulation and on cardiac myocytes, the severity of coronary artery stenosis does not necessarily reflect the severity of ischemia caused by the lesion. For this reason, several methods that assess the presence and degree of myocardial ischemia have been developed and include exercise testing, imaging methods (echocardiography, myocardial perfusion scintigraphy and nuclear magnetic resonance) associated with physical or pharmacological stress, and fractional flow reserve (FFR) [1].

Assessing endothelial function is still a challenge. There are no methods that allow direct visualization of coronary microcirculation in humans. Hence, indirect measures that quantify blood flow through coronary circulation are used [10]. Invasive methods are the most widely evaluated experimentally, but have limited if any clinical application the need for coronary angiography. Noninvasive tests have been developed and it is expected that the evaluation of CMD gains popularity for diagnostic purposes and evaluation of new therapeutic interventions [48].

Invasive Methods

The method most commonly used in clinical studies to evaluate CMD is the determination of coronary flow reserve (CFR) using intracoronary Doppler ultrasound [48]. CFR measures the change of intravascular blood flow obtained after maximum vasodilation achieved with the application of adenosine and dipyridamole [48]. The rationale is that, in the presence of coronary artery stenosis or CMD related to other pathologies, there is chronic release of vasodilator agents, including adenosine itself. Thus, application of vasodilator agents will not be able to increase blood flow to the myocardium in the required ratio, that is, CFR is low. CFR is calculated using a ratio between coronary flow under the action of vasodilators agents and coronary flow at rest [48].

While CFR is an accepted method in identifying CMD related to cardiomyopathies, iatrogenic and in CMD not related to cardiomyopathies or CAD, the method loses in accuracy in measuring CMD related to CAD [42]. This is because epicardial coronary lesions may compromise measurement of CFR, so that quantify CMD in these patients is an added challenge [10,48]. Nevertheless, the extent of CFR is one of the most studied methods for quantifying CMD in DAC [48].

Low CFR values have been demonstrated in angiographically normal arteries of patients with CAD [49,50] and correlated with worse prognosis [50]. In a study involving 100 patients with acute myocardial infarction, a flow reserve <2.1 in the normal coronary angiography was associated with increased cardiac mortality in 4.09 times over 10 years of follow-up [50].

CRF also proved useful in predicting postangioplasty events in chronic CAD. In 225 patients undergoing angioplasty for unstable angina, those with CRF > 2.5 after the procedure had lower recurrence of symptoms (23% vs. 47%, $P = 0.005$), restenosis (16% vs. 41%, $P = 0.002$) and need for reoperation (16% vs. 34%, $P = 0.24$) [51].

One of the challenges in clinical application of CFR is the lack of definition of its cutoff. As studies have shown that CFR values lower than 2.1–2.3 relate to increased risk of cardiac death up to four times over 5–10 years [50,52], the value <2 is the most accepted. However, more studies are needed to establish cutoff values, which still need to be corrected for age and sex [10].

Before measurement of coronary flow in vivo in humans, the intracoronary pressure was calculated, and this measure was used as a way to infer CFR [53,54]. The method has similarities with CFR: at the coronary stenosis site adenosine is applied for vasodilation and consequent maximum myocardial hyperinflation [54]. Then, using a catheter with a pressure sensor, pressure distal to the lesion and pressure in the aorta can be measured [54]. Ratio between pressure distal to the lesion and the pressure in the aorta during vasodilation with maximal hyperemia is called FFR [54]. It is important to consider that for FFR to be able to reflect CFR, the resistance exerted by other coronary segments should be minimal, that is, maximum vasodilatation must be ensured [54]. Adenosine is used in an attempt to achieve this condition, called maximal hyperemia. However, the complexity of the self-regulation mechanism in coronary could compromise the effect of adenosine and more than that: its effect may vary from person to person, depending on factors such as medication use, smoking, comorbidities or caffeine consumption [54]. The variability of the state of maximal hyperemia limits FFR's ability to reflect CFR and is one of the reasons why FFR and CFR measures can be discordant [54–56].

FFR is a measure validated and applied in clinical practical to define the presence of ischemia determined by coronary stenosis [1]. The cutoff value of FFR ≤ 0.8 may be used in clinical practice to aid in angioplasty indication [1,57].

Another strategy to evaluate the coronary flow is using TIMI (thrombolysis in myocardial infarction) flow. This technique analyzes the degree of opacity in myocardial tissue (blush) reached after contrast injection in epicardial coronary [10]. The more intense the myocardial blush, and faster for their disappearance, the better myocardial perfusion. The scale goes from zero to three. Although indirect, it is a method often used in daily practice because of its wide availability and technical facility [10].

Noninvasive Methods

The best noninvasive method for measurement of CFR is PET-scan [52]. This was studied in over 14,000 patients in the last 25 years [58]. In CAD, the average CFR detected by PET-scan is 2.02, value similar to the cutoff proposed for intravascular Doppler [58]. PET-scan has additional benefit of being able to calculate blood flow per unit mass of the myocardium (MBF—expressed as volume of blood per minute per gram of tissue) [58]. The cutoff values for CFR and MBF in PET-scan are not yet well defined. The larger study involving 1674 patients demonstrated that a maximum in stress of 0.91 cc/min/g in CFR of 1.74 were the cut values that better identified the group of patients with significant ischemia [59]. Several studies have shown a correlation between low CFR and cardiovascular events [58]. The largest one evaluated 2783 patients with suspected CAD and subject to PET-scan. Individuals with CFR <1.5 had a mortality rate 5.6 times higher cardiac death compared to those with CFR >2.0 (CI 2.5–12.4, $P < 0.0001$) [60].

Doppler echocardiography permits to assess CFR in anterior descending coronary artery (LAD) after application of intravenous adenosine and dipyridamole [61]. In experienced hands, has good correlation with intracoronary Doppler [62], and the application of microbubbles and Doppler harmonic second can improve image clarity [62]. It was observed that the method is an independent predictor of death in a study involving 1620 patients with stable angina. CFR in ADA ≤1.8 was an independent predictor of death, myocardial infarction and revascularization at 19 months of follow-up [63]. The method is promising because it is not invasive; however, there is considerable intra- and interobserver variability, so that more studies are needed to determine the role of the method in clinical practice [48].

Transthoracic echocardiography also permits to evaluate MBF when using contrast with microbubbles. The studies are small, but MBF calculated from echocardiogram is promising. It showed good correlation with CFR calculated by intracoronary Doppler [64] and higher than TIMI blush in predicting ventricular remodeling 6 months after IAMI [65].

Analysis of myocardial perfusion, CFR, and MBF by MRI are rapidly developing. One of the largest limiting factors is the need to correct for image artifacts. However, studies in noncoronary patients such as MESA [66] and WISE [67], already demonstrated the method's ability to predict cardiovascular events.

CMD AND CAD TREATMENT

In the presence of a coronary stenosis, a treatment able to restore normal blood flow may be the most intuitive therapy. Development of coronary angiography and bypass in the 1950s made surgical myocardial revascularization possible [68]. In this mode of revascularization, coronary flow is restored by the implantation of saphenous vein grafts or arterial grafts distally to the stenotic lesion. The first successful surgical revascularization of myocardium was performed in 1960 by Goetz [69], but it was Rene Favaloro who made the surgery a success by employing saphenous vein anastomoses [70]. Since then, technical improvements made the surgery the choice revascularization procedure in situations such as the treatment of patients with complex CAD [1,71].

Another modality of myocardial revascularization, coronary angioplasty was performed for the first time in 1977 [72]. In angioplasty, arterial blood flow is restored to the native arteries by inflating an intracoronary balloon and it may or may not be followed by implantation of a stent—a device that became available for human use in 1986 [73]. Angioplasty is now the revascularization modality of choice in situations such as the treatment of patients who remain symptomatic despite drug therapy and who have coronary anatomy favorable to the procedure [1,71].

Although return of blood flow to the ischemic area is the most intuitive therapy, one must keep in mind that the objectives of the treatment of patients with CAD are to reduce the risk of death and morbidity (acute myocardial infarction, arrhythmias, heart failure) and improve symptoms. Drug treatment and lifestyle changes are critical, and in some patients may be more important than the surgical myocardial revascularization and coronary angioplasty in obtainment of these objectives [1,74,75]. Lifestyle changes and drugs control angina's symptoms; reduce the progression of atheromatous plaque; stabilize the plaque due to control of inflammation; and if the plaque rupture occurs, drugs decrease the chance of a quick coronary obstruction due to thrombus formation, that is, reduces the risk of acute coronary syndrome [1]. The main drugs involved in the treatment of CAD include beta-blockers, calcium channel blockers, nitrates, drugs inhibiting renin-angiotensin system (RAS), statins and acetyl-salicylic acid [1].

It is noteworthy that drugs associated with reduction of cardiovascular events have favorable action on the stabilization of endothelial dysfunction and CMD. In the late 1990s, the concept of "endothelial therapy" was developed as an approach to preserving or restoring health of the endothelium and consequently control atherosclerosis, reducing cardiovascular events [76]. Within this concept, it is possible to see that RAS inhibitor drugs beta-blockers, statins, acetylsalicilic acid and physical activity, measures that are proven to reduce cardiovascular events in CAD, have their effect partially due to improvement in CMD they provide.

RAS Inhibitor Drugs

SRA central effector agent is angiotensin-II [77]. Angiotensin-II is formed from the conversion of angiotensin-I by angiotensin converting enzyme (ACE) [78] which appears in large amounts on the surface of endothelial cells [20] and its performance in cell level has proinflammatory, proliferative, and vasoconstrictor effects that contribute to the onset and progression of atherosclerosis. Thus, it is expected that drugs that reduce the formation of angiotensin in (as ACE—iECA inhibitors) or prevent its performance (as angiotensin receptor blockers—BRA) have favorable effect on remodeling vascular and endothelial dysfunction [77]. iECA drugs also have an additional effect compared to the BRA: increase of bradykinin tissue levels [20], and peptide produced locally, which promotes coronary vasodilation to promote the release of nitric oxide and prostacyclin. ECA promotes bradykinin degradation [20]. Thus, ACE inhibition increases bradykinin activity, and results in improved endothelial function [20].

In fact, ACE inhibitors are able to improve endothelial function [77]. Studies have shown reduction in ADMA levels and consequent increase of nitric oxide, and decreased levels of von Willebrand's factor in patients with syndrome vs. treated with enalapril [79]. Quinapril administered for 16 weeks reduced CFR and angina [60] in women with chest pain and CFR ≤ 2.5, but without CAD [80]. Ramipril increases platelet responsiveness to nitric oxide, which can justify its antiplatelet action [81]. Thus, it is possible that the benefit of reducing cardiovascular events shown in clinical studies involving iACE [82] and BRA [83] is, at least partially, a result of improved endothelial function that the drug promotes [84].

Aliskiren, a renin inhibitor drug, is able to block the RAA system in the initial phase of its activation [77]. Its effect not only reduces levels of angiotensin-II, but also angiotensin-I and renin peptides also have deleterious effects on microcirculation [85].

There are no clinical studies examining the effects of this drug in patients with CAD; however, the Altitude study [86] included patients with cardiovascular disease and associated aliskiren with ACE inhibitors or ARBs in diabetic patients with chronic kidney and/or cardiovascular disease [86]. The expected result was that the double blockade of RAA system culminated in reduction of cardiovascular events. What was observed, however, was increased risk of hyperkalemia and hypotension and association between aliskiren and ACEI or ARB was banned in this scenario [87]. Results of the Altitude study are a warning that although an action may seem logical, not necessarily it is of benefit to the patient and demonstrate the importance of evaluating the results of clinical trials before admitting a therapeutic strategy [77].

Aldosterone is the final mediator of the RA system. Although its main impact is on the kidneys, the hormone has many effects on blood vessels. Animal studies suggest that aldosterone may affect coronary endothelial function, favoring inflammation and progression of fibrinoid necrosis [88]. However, there are no large clinical studies evaluating effects of aldosterone antagonists in patients with CAD, so it is not yet clear whether the benefits of CMD are reversed in reducing cardiovascular events in humans [77].

Beta-Blockers

Beta-blockers are known to be beneficial in the treatment of CAD due to various facets of their action [89]. The main effect is reduction in O_2 consumption by the myocardium to reduce myocardial contractility, blood pressure, and heart rate [89]. By inhibiting beta-1 adrenergic receptors, beta-blockers also reduce the release of renin and consequently reduce the levels of angiotensin II [89]. Inhibition of angiotensin-II brings the additional effect of decreased oxidative stress, as previously establishe [89].

Beta-blockers are a heterogeneous class of drugs with effects that differ according to the receptor blocked. In usual situations, β1 receptors in the myocardium have chrono and positive inotropic action, and β2 receptors in bronchial musculature have bronchodilator effect [89]. In blood vessels the activation of receptor β1, β2, and β3 adrenergic stimulate the production of nitric oxide by the endothelium. Thus, the first generation blockers (nonselective) and second generation (selective β1) have no vasodilating properties, while nebivolon and carvediol (both third-generation beta-blockers) present peripheral vasodilator effect [89]. This profile of drugs difference is reflected clinically in the fact that third generation beta-blockers exert more favorable effect on the lipid and glycemic profile compared to traditional BB [90].

Third-generation beta-blockers possibly also play role on improvement of CMD [89]. In the case of nebivolol, vasodilatation is induced due to increased production and availability of nitric oxide. Increased production of oxide is induced by nebivolol through activation of vascular β3 receptors, that activate NOS and greater bioavailability due to inhibition of ADMA [89]. Carvedilol is also able to reduce ADMA plasma levels and this effect seems to be associated with improvement in ejection fraction in patients with ventricular dysfunction [91]. Studies in rats have demonstrated the nebivolol's ability to preserve ventricular function after AMI [92] and studies in rabbits demonstrated a reduction in the development of atherosclerotic plaque [93].

Other Antihypertensive Drugs

The use of calcium channel blockers (CCB) in CAD initially was limited due to the evidence of increased mortality with the use of rapid action nifedipine [94]. Subsequently, it was demonstrated that long-acting agents did not cause this harm. Amlodipine proved effective in reducing hospitalization [95] and nifedipine GITS reduced the need for new coronary procedures in patients with CAD [96]. The benefits of dihydropyridine CCB result from its antihypertensive and coronary vasodilator effects, being able therefore to reduce O_2 consumption and increase supply [97]. Another possible explanation for the beneficial effects of long-acting CCB in CAD is its antioxidant effect [97]. Study on endothelial cells of rabbits showed increased production of nitric oxide and superoxide reduction after treatment with nifedipine [98] and use of nifedipine for up to 2 years in CAD patients showed a reduction in coronary vasoreactivity [99].

Nitrates are very effective antianginal agents in reducing myocardial ischemia [100]. Their efficacy stems from its ability to reduce the preload of the left ventricle (reduced O_2 consumption) associated with the vasodilatory effect on coronary arteries and collateral circulation (increased of O_2 supply). Chronic

use of nitrates, however, is limited by the development of tolerance and this phenomenon is related to the CMD by increased oxidative stress and activation of the RA system [100]. The mechanism of the increased oxidative stress associated the nitrate is not yet clear, but appears to be related to the release of O_2 during the process of biotransformation of nitroglycerin or activation of NADPH oxidase by RA system [100]. The process culminates in increased oxidative stress, vasoconstriction, and intravascular volume expansion. There is evidence that treatment with ACEI, ARB, BB, and statins reduce the risk of tolerance and CMD related to nitrate [100,101], and the guidelines recommend their use in acute angina and those chronic patients who keep symptoms despite treatment with aspirin, statins, BB, and BCC [1]. There is, to date, no prospective clinical trials testing the benefits of using nitrates on cardiovascular events in long run, but retrospective studies and subanalyzes of large studies [102,103] reported increased risk of cardiovascular events in patients with CAD in chronic use of nitrates.

Statins

Statins limit the progression of atherosclerosis and may even regress plates and are first choice in the treatment of dyslipidemia in the context of CAD [1]. The main statin mechanism of action to reduce the occurrence of cardiovascular events is lower LDL cholesterol lowering. However, they are more effective in preventing cardiovascular events than other hypolipidemic probably by their pleiotropic effects [104].

The pleiotropic effects of statins are closely related to inhibition of isopropenoids formation, which are formed during the synthesis of cholesterol by HMG-CoA reductase [104]. Isoprenoids are important anchors of cytoplasmic membrane, particularly GTPases of the Rho family, such as Rac 1 or RhoA. Reduction of these "anchors" results in many effects. Decreased interactions between the cytoskeleton provides RhoA and increased nitric oxides synthase expression from its RNA messenger, resulting in an increase of nitric oxide released by the endothelium and all have beneficial effects previously described [104].

The antiinflammatory effect of statins is related to the limitation of the connection between the membrane and Rac1, which prevents the activation of

NADPH oxidase and diminishes lipid peroxidation [104]. In addition to antiinflammatory effect, inhibition of NADPH oxidase provides reduced activity and platelet recruitment, justifying the anticoagulant effect of statins [104].

Another justification for the antiinflammatory effects of statins appear to be increased transcription of Krüppler-like 2 (KLF-2) factor. In the endothelium, this factor reduces cell proliferation [105,106], and in cells of acquired and innate immune system, increased KLF-2 activity has an antiinflammatory effect [107]. Statins also improve angiogenic response [108] probably due to increase of circulating monocytes with angiogenic potential by means of signaling changes in bone marrow level [108].

The lipid-lowering effect of statins already starts at low doses and takes weeks to occur. In contrast, the pleiotropic effects require higher doses and begin a few hours after drug intake [89,109]. Withdrawal of statins can lead to a rebound effect of NADPH oxidase activation. With the inability of GTPases to bind to Rac1, there is a lack of negative feedback and consequent increase in GTPases activity. With withdrawal of statins and increase in the isoprenoid, action of GTPases becomes exaggerated, resulting in excessive activation of NDPH oxidase, suppression of NOS and consequent increase in oxidative stress and proinflammatory state [110].

Acetylsalicylic Acid

Although atherosclerosis is the major cause of cardiovascular disease, thrombosis is a major factor in the development of cardiovascular events [111]. Acetylsalicylic acid (ASA) is one of the mainstays of CAD treatment, promoting reduction by 22% of new cardiovascular events after myocardial infarction [112]. The main mechanism by which ASA confers this degree of protection is its antithrombotic effect. ASA inhibits production of thromboxane A2 through inhibition of the cyclooxygenase pathway, which results in an irreversible antiplatelet effect [111]. More recently, other ASA action mechanisms related to an increase in nitric oxide production have been studied. The activation of NOS activation of cyclic GMP pathway related to nitric oxide in endothelial cells and platelets, as well as the reduction of ADMA activity seem to appear to be related to ASA. In a randomized study of 37 patients with CAD using ASA for 12 weeks, there was reduction of plasma ADMA concentrations and increase in levels of hemeoxigenase (a metabolite of nitric oxide), indicating increased production of nitric oxide promoted by ASA [113]. The effects of ASA on oxide release may be partially responsible for the benefits of the drug.

Lifestyle Changes

Lifestyle changes are the mainstay of treatment of atherosclerosis and cardiovascular risk factors. Regular physical activity improves exercise capacity and reduces perfusion defects, indicating a possible role for the improvement of myocardial perfusion [114–116].

The benefit in patients with CAD may be related to improvement of physical activity-induced microcirculation. A small study showed that 10 minutes of activity, six times a day, under 80% of maximum heart rate for 4 weeks was able to increase CRF in 29% of 10 patients with CHD. The control group of 9 patients kept under physical inactivity, did not get improvement in the measurements [117].

The weight loss associated with physical activity also may play an important role in improving CMD in patients with CAD [3]. In addition to weight reduction provide improved control of hypertension and diabetes [3], obesity itself results in CMD by increased sympathetic activity and ET-1, with consequent vasoconstriction and the inflammatory activation [3]. Small studies have shown that weight loss associated with moderate physical activity (3 hours/week) and vegetable-based, low-fat diet (10% of calories) improved endothelial dysfunction [117,118].

CONCLUSIONS

Endothelial dysfunction plays a central role in the pathophysiology of coronary artery disease, but its measurement is still a challenge. Analyses of endothelial function via MBF and CFR, particularly employing PET-scan or intracoronary Doppler, bring prognostic information in patients with CAD. However, it is not yet clear the gain in risk stratification and modification of treatment that these methods can provide.

Changes in lifestyle and the drugs that have proven benefit in treatment of CAD owe their effectiveness, at least in part, to their effect on stabilization of endothelial function. However, not all actions that demonstrate beneficial profile on endothelial dysfunction were able to demonstrate reduction of cardiovascular events in clinical studies. The pathophysiological complexity of atheromatosis is a challenge in developing methods and complementary treatments directed to endothelial dysfunction. A greater understanding of the endothelium's role in DAC is a key element for the progress of researches and has control over one of the largest causes of death worldwide: cardiovascular disease.

References

[1] Montalescot G, Sechtem U, Achenbach S, et al. 2013 ESC guidelines on the management of stable coronary artery disease: the Task Force on the management of stable coronary artery disease of the European Society of Cardiology. Eur Heart J 2013;34 (38):2949–3003.

[2] Steg PG, Bhatt DL, Wilson PW, et al. One-year cardiovascular event rates in outpatients with atherothrombosis. J Am Med Assoc 2007;297(11):1197–206.

[3] Barton M. Prevention and endothelial therapy of coronary artery disease. Curr Opin Pharmacol 2013;13(2):226–41.

[4] Ross R, Glomset JA. Atherosclerosis and the arterial smooth muscle cell: proliferation of smooth muscle is a key event in the genesis of the lesions of atherosclerosis. Science (New York, NY) 1973;180 (4093):1332–9.

[5] Moncada S, Gryglewski R, Bunting S, et al. An enzyme isolated from arteries transforms prostaglandin endoperoxides to an unstable substance that inhibits platelet aggregation. Nature 1976;263 (5579):663–5.

[6] SoRelle R. Nobel prize awarded to scientists for nitric oxide discoveries. Circulation 1998;98(22):2365–6.

[7] Kluge MA, Fetterman JL, Vita JA. Mitochondria and endothelial function. Circ Res 2013;112(8):1171–88.

[8] Feigl EO. Coronary physiology. Physiol Rev 1983;63(1):1–205.

[9] Duncker DJ, Bache RJ. Regulation of coronary blood flow during exercise. Physiol Rev 2008;88(3):1009–86.

[10] Camici PG, Crea F. Coronary microvascular dysfunction. N Engl J Med 2007;356(8):830–40.

[11] Quilley J, Fulton D, McGiff JC. Hyperpolarizing factors. Biochem Pharmacol 1997;54(10):1059–70.

[12] Jones CJ, Kuo L, Davis MJ, et al. Regulation of coronary blood flow: coordination of heterogeneous control mechanisms in vascular microdomains. Cardiovasc Res 1995;29(5):585–96.

[13] Lamping KG, Kanatsuka H, Eastham CL, et al. Nonuniform vasomotor responses of the coronary microcirculation to serotonin and vasopressin. Circ Res 1989;65(2):343–51.

[14] Bache RJ, Stark RP, Duncker DJ. Serotonin selectively aggravates subendocardial ischemia distal to a coronary artery stenosis during exercise. Circulation 1992;86(5):1559–65.

[15] Epstein SE, Cannon 3rd RO, Talbot TL. Hemodynamic principles in the control of coronary blood flow. Am J Cardiol 1985;56:4E–10E.

[16] Ludmer PL, Selwyn AP, Shook TL, et al. Paradoxical vasoconstriction induced by acetylcholine in atherosclerotic coronary arteries. N Engl J Med 1986;315:1046–51.

[17] Suwaidi JA, Hamasaki S, Higano ST, et al. Long-term follow-up of patients with mild coronary artery disease and endothelial dysfunction. Circulation 2000;101:948–54.

[18] Clarkson P, Celermajer DS, Powe AJ, et al. Endothelium-dependent dilatation is impaired in young healthy subjects with a family history of premature coronary disease. Circulation 1997;96:3378–83.

[19] O'Driscoll G, Green D, Taylor RR. Simvastatin, an HMG-coenzyme A reductase inhibitor, improves endothelial function within 1 month. Circulation 1997;95:1126–31.

[20] Prasad A, Husain S, Quyyumi AA. Abnormal flow-mediated epicardial vasomotion in human coronary arteries is improved by angiotensin-converting enzyme inhibition: a potential role of bradykinin. J Am Coll Cardiol 1999;33(3):796–804.

[21] Luscher TF, Barton M. Endothelins and endothelin receptor antagonists: therapeutic considerations for a novel class of cardiovascular drugs. Circulation 2000;102(19):2434–40.

[22] Haller H, Schaberg T, Lindschau C, et al. Endothelin increases $[Ca^{2+}]_i$, protein phosphorylation, and $O_2^{-\bullet}$ production in human alveolar macrophages. Am J Physiol 1991;261(6 Pt 1):L478–84.

[23] Marini M, Carpi S, Bellini A, et al. Endothelin-1 induces increased fibronectin expression in human bronchial epithelial cells. Biochem Biophys Res Commun 1996;220(3):896–9.

[24] Guidry C, Hook M. Endothelins produced by endothelial cells promote collagen gel contraction by fibroblasts. J Cell Biol 1991; 115(3):873–80.

[25] Lopez Farre A, Riesco A, Espinosa G, et al. Effect of endothelin-1 on neutrophil adhesion to endothelial cells and perfused heart. Circulation 1993;88(3):1166–71.

[26] Matsuura A, Yamochi W, Hirata K, et al. Stimulatory interaction between vascular endothelial growth factor and endothelin-1 on each gene expression. Hypertension 1998;32(1):89–95.

[27] Yanagisawa M, Kurihara H, Kimura S, et al. A novel potent vasoconstrictor peptide produced by vascular endothelial cells. Nature 1988;332(6163):411–5.

[28] Singel DJ, Stamler JS. Chemical physiology of blood flow regulation by red blood cells: the role of nitric oxide and S-nitrosohemoglobin. Annu Rev Physiol 2005;67:99–145.

[29] Stamler JS, Jia L, Eu JP, et al. Blood flow regulation by S-nitrosohemoglobin in the physiological oxygen gradient. Science (New York, NY) 1997;276(5321):2034–7.

[30] Pohl U, Busse R. Hypoxia stimulates release of endothelium-derived relaxant factor. Am J Physiol 1989;256(6 Pt 2):H1595–600.

[31] Jones CJ, DeFily DV, Patterson JL, et al. Endothelium-dependent relaxation competes with alpha 1- and alpha 2-adrenergic constriction in the canine epicardial coronary microcirculation. Circulation 1993;87(4):1264–74.

[32] Sparks Jr. HV, Bardenheuer H. Regulation of adenosine formation by the heart. Circ Res 1986;58(2):193–201.

[33] Duffy SJ, Castle SF, Harper RW, et al. Contribution of vasodilator prostanoids and nitric oxide to resting flow, metabolic vasodilation, and flow-mediated dilation in human coronary circulation. Circulation 1999;100(19):1951–7.

[34] Friedman PL, Brown Jr. EJ, Gunther S, et al. Coronary vasoconstrictor effect of indomethacin in patients with coronary-artery disease. N Engl J Med 1981;305(20):1171–5.

[35] Gambuccti G, Marzilli M, Marracchi P, et al. Coronary vasoconstriction during myocardial ischemia induced by rises in metabolic demand in patients with coronary artery disease. Circulation 1997;95(12):2652–9.

[36] Zeiher AM, Krause T, Schachinger V, et al. Impaired endothelium-dependent vasodilation of coronary resistance vessels is associated with exercise-induced myocardial ischemia. Circulation 1995; 91(9):2345–52.

[37] Lanza GA, Careri G, Crea F. Mechanisms of coronary artery spasm. Circulation 2011;124(16):1774–82.

[38] Zimarino M, D'Andreamatteo M, Waksman R, et al. The dynamics of the coronary collateral circulation. Nat Rev Cardiol 2014; 11(4):191–7.

[39] Cai W, Schaper W. Mechanisms of arteriogenesis. Acta Biochim Biophys Sin (Shanghai) 2008;40(8):681–92.

[40] Carmeliet P. Mechanisms of angiogenesis and arteriogenesis. Nat Med 2000;6(4):389–95.

[41] Semenza GL. Angiogenesis in ischemic and neoplastic disorders. Annu Rev Med 2003;54:17–28.

[42] Fraisl P, Mazzone M, Schmidt T, et al. Regulation of angiogenesis by oxygen and metabolism. Dev Cell 2009;16(2):167–79.

[43] van Royen N, Piek JJ, Buschmann I, et al. Stimulation of arteriogenesis; a new concept for the treatment of arterial occlusive disease. Cardiovasc Res 2001;49(3):543–53.

[44] Heil M, Schaper W. Influence of mechanical, cellular, and molecular factors on collateral artery growth (arteriogenesis). Circ Res 2004;95(5):449–58.

[45] Bonow RO, Maurer G, Lee KL, et al. Myocardial viability and survival in ischemic left ventricular dysfunction. N Engl J Med 2011;364(17):1617–25.

[46] Cascio WE, Yang H, Muller-Borer BJ, et al. Ischemia-induced arrhythmia: the role of connexins, gap junctions, and attendant changes in impulse propagation. J Electrocardiol 2005;38(4 Suppl.):55–9.

[47] Priori SG, Blomstrom-Lundqvist C, Mazzanti A, et al. 2015 ESC guidelines for the management of patients with ventricular arrhythmias and the prevention of sudden cardiac death: The Task Force for the Management of Patients with Ventricular Arrhythmias and the Prevention of Sudden Cardiac Death of the European Society of Cardiology (ESC) Endorsed by: Association for European Paediatric and Congenital Cardiology (AEPC). Eur Heart J 2015;36(41):2793–867.

[48] Camici PG, d'Amati G, Rimoldi O. Coronary microvascular dysfunction: mechanisms and functional assessment. Nat Rev Cardiol 2015;12(1):48–62.

[49] Uren NG, Marraccini P, Gistri R, et al. Altered coronary vasodilator reserve and metabolism in myocardium subtended by normal arteries in patients with coronary artery disease. J Am Coll Cardiol 1993;22(3):650–8.

[50] van de Hoef TP, Bax M, Meuwissen M, et al. Impact of coronary microvascular function on long-term cardiac mortality in patients with acute ST-segment-elevation myocardial infarction. Circ Cardiovasc Interv 2013;6(3):207–15.

[51] Serruys PW, di Mario C, Piek J, et al. Prognostic value of intracoronary flow velocity and diameter stenosis in assessing the short- and long-term outcomes of coronary balloon angioplasty: the DEBATE Study (Doppler Endpoints Balloon Angioplasty Trial Europe). Circulation 1997;96(10):3369–77.

[52] Pepine CJ, Anderson RD, Sharaf BL, et al. Coronary microvascular reactivity to adenosine predicts adverse outcome in women evaluated for suspected ischemia results from the National Heart, Lung and Blood Institute WISE (Women's Ischemia Syndrome Evaluation) study. J Am Coll Cardiol 2010;55(25):2825–32.

[53] De Bruyne B, Baudhuin T, Melin JA, et al. Coronary flow reserve calculated from pressure measurements in humans. Validation with positron emission tomography. Circulation 1994;89(3):1013–22.

[54] van de Hoef TP, Meuwissen M, Piek JJ. Fractional flow reserve-guided percutaneous coronary intervention: where to after FAME 2? Vasc Health Risk Manag 2015;11:613–22.

[55] van de Hoef TP, van Lavieren MA, Damman P, et al. Physiological basis and long-term clinical outcome of discordance between fractional flow reserve and coronary flow velocity reserve in coronary stenoses of intermediate severity. Circ Cardiovasc Interv 2014;7(3):301–11.

[56] Echavarria-Pinto M, Escaned J, Macias E, et al. Disturbed coronary hemodynamics in vessels with intermediate stenoses evaluated with fractional flow reserve: a combined analysis of epicardial and microcirculatory involvement in ischemic heart disease. Circulation 2013;128(24):2557–66.

[57] De Bruyne B, Pijls NH, Kalesan B, et al. Fractional flow reserve-guided PCI versus medical therapy in stable coronary disease. N Engl J Med 2012;367(11):991–1001.

[58] Gould KL, Johnson NP, Bateman TM, et al. Anatomic versus physiologic assessment of coronary artery disease. Role of coronary flow reserve, fractional flow reserve, and positron emission tomography imaging in revascularization decision-making. J Am Coll Cardiol 2013;62(18):1639–53.

[59] Johnson NPGK. Physiologic basis for angina and ST change: PET-verified thresholds of quantitative stress myocardial perfusion and coronary flow reserve. J Am Coll Cardiol Img 2011;4:990–8.

[60] Murthy VL, Naya M, Foster CR, et al. Improved cardiac risk assessment with noninvasive measures of coronary flow reserve. Circulation 2011;124(20):2215–24.

[61] Hozumi T, Yoshida K, Akasaka T, et al. Noninvasive assessment of coronary flow velocity and coronary flow velocity reserve in the left anterior descending coronary artery by Doppler echocardiography: comparison with invasive technique. J Am Coll Cardiol 1998;32(5):1251–9.

[62] Caiati C, Montaldo C, Zedda N, et al. Validation of a new noninvasive method (contrast-enhanced transthoracic second harmonic echo Doppler) for the evaluation of coronary flow reserve: comparison with intracoronary Doppler flow wire. J Am Coll Cardiol 1999;34(4):1193–200.

[63] Cortigiani L, Rigo F, Gherardi S, et al. Implication of the continuous prognostic spectrum of Doppler echocardiographic derived coronary flow reserve on left anterior descending artery. Am J Cardiol 2010;105(2):158–62.

[64] Lepper W, Hoffmann R, Kamp O, et al. Assessment of myocardial reperfusion by intravenous myocardial contrast echocardiography and coronary flow reserve after primary percutaneous transluminal coronary angioplasty [correction of angiography] in patients with acute myocardial infarction. Circulation 2000;101(20):2368–74.

[65] Galiuto L, Garramone B, Scara A, et al. The extent of microvascular damage during myocardial contrast echocardiography is superior to other known indexes of post-infarct reperfusion in predicting left ventricular remodeling: results of the multicenter AMICI study. J Am Coll Cardiol 2008;51(5):552–9.

[66] Wang L, Jerosch-Herold M, Jacobs Jr. DR, et al. Coronary risk factors and myocardial perfusion in asymptomatic adults: the Multi-Ethnic Study of Atherosclerosis (MESA). J Am Coll Cardiol 2006;47(3):565–72.

[67] Doyle M, Weinberg N, Pohost GM, et al. Prognostic value of global MR myocardial perfusion imaging in women with suspected myocardial ischemia and no obstructive coronary disease: results from the NHLBI-sponsored WISE (Women's Ischemia Syndrome Evaluation) study. JACC Cardiovasc Imaging 2010;3(10):1030–6.

[68] Dallan LA, Jatene FB. Myocardial revascularization in the XXI century. Rev Bras Cir Cardiovasc 2013;28(1):137–44.

[69] Goetz RH, Rohman M, Haller JD, et al. Internal mammary-coronary artery anastomosis. A nonsuture method employing tantalum rings. J Thorac Cardiovasc Surg 1961;41:378–86.

[70] Forrester JS. When the pampas came to Cleveland. The Heart Healers. St Martin's Press, Macmillan; 2015. p. 143.

[71] Kappetein AP, Feldman TE, Mack MJ, et al. Comparison of coronary bypass surgery with drug-eluting stenting for the treatment of left main and/or three-vessel disease: 3-year follow-up of the SYNTAX trial. Eur Heart J 2011;32(17):2125–34.

[72] Singh IM, Holmes Jr. DR. Myocardial revascularization by percutaneous coronary intervention: past, present, and the future. Curr Probl Cardiol 2011;36(10):375–401.

[73] Sigwart U, Puel J, Mirkovitch V, et al. Intravascular stents to prevent occlusion and restenosis after transluminal angioplasty. N Engl J Med 1987;316(12):701–6.

[74] Boden WE, O'Rourke RA, Teo KK, et al. Optimal medical therapy with or without PCI for stable coronary disease. N Engl J Med 2007;356(15):1503–16.

[75] Group BDS, Frye RL, August P, et al. A randomized trial of therapies for type 2 diabetes and coronary artery disease. N Engl J Med 2009;360(24):2503–15.

[76] Barton M, Haudenschild CC. Endothelium and atherogenesis: endothelial therapy revisited. J Cardiovasc Pharmacol 2001;38 (Suppl. 2):S23–5.

[77] Sheppard RJ, Schiffrin EL. Inhibition of the renin-angiotensin system for lowering coronary artery disease risk. Curr Opin Pharmacol 2013;13(2):274–9.

[78] Hoffman JL. Maximal coronary flow and the concept of coronary vascular reserve. Circulation 1984;70:153–9.

[79] Chen JW, Hsu NW, Wu TC, et al. Long-term angiotensin-converting enzyme inhibition reduces plasma asymmetric dimethylarginine and improves endothelial nitric oxide bioavailability and coronary microvascular function in patients with syndrome X. Am J Cardiol 2002;90(9):974–82.

[80] Pauly DF, Johnson BD, Anderson RD, et al. In women with symptoms of cardiac ischemia, nonobstructive coronary arteries, and microvascular dysfunction, angiotensin-converting enzyme inhibition is associated with improved microvascular function: a double-blind randomized study from the National Heart, Lung and Blood Institute Women's Ischemia Syndrome Evaluation (WISE). Am Heart J 2011;162(4):678–84.

[81] Willoughby SR, Rajendran S, Chan WP, et al. Ramipril sensitizes platelets to nitric oxide: implications for therapy in high-risk patients. J Am Coll Cardiol 2012;60(10):887–94.

[82] Fox KM. Efficacy of perindopril in reduction of cardiovascular events among patients with stable coronary artery disease: randomised, double-blind, placebo-controlled, multicentre trial (the EUROPA study). Lancet 2003;362(9386):782–8.

[83] Investigators O, Yusuf S, Teo KK, et al. Telmisartan, ramipril, or both in patients at high risk for vascular events. N Engl J Med 2008;358(15):1547–59.

[84] Mangiacapra F, Peace AJ, Di Seraflno L, et al. Intracoronary enalaprilat to reduce MICR ovascular damage during percutaneous coronary intervention (ProMicro) study. J Am Coll Cardiol 2013;61 (6):615–21.

[85] Riccioni G, Vitulano N, Zanasi A, et al. Aliskiren: beyond blood pressure reduction. Expert Opin Investig Drugs 2010;19 (10):1265–74.

[86] Parving HH, Brenner BM, McMurray JJ, et al. Cardiorenal end points in a trial of aliskiren for type 2 diabetes. N Engl J Med 2012;367(23):2204–13.

[87] Mancia G, Fagard R, Narkiewicz K, et al. 2013 ESH/ESC guidelines for the management of arterial hypertension: the Task Force for the Management of Arterial Hypertension of the European Society of Hypertension (ESH) and of the European Society of Cardiology (ESC). Eur Heart J 2013;34(28):2159–219.

[88] Shah NC, Pringle S, Struthers A. Aldosterone blockade over and above ACE inhibitors in patients with coronary artery disease but without heart failure. J Renin Angiotensin Aldosterone Syst 2006;7(1):20–30.

[89] Vanhoutte PM, Gao Y. Beta blockers, nitric oxide, and cardiovascular disease. Curr Opin Pharmacol 2013;13(2):265–73.

[90] Fonseca VA. Effects of beta-blockers on glucose and lipid metabolism. Curr Med Res Opin 2010;26(3):615–29.

[91] Alfieri AB, Briceno L, Fragasso G, et al. Differential long-term effects of carvedilol on proinflammatory and antiinflammatory cytokines, asymmetric dimethylarginine, and left ventricular function in patients with heart failure. J Cardiovasc Pharmacol 2008;52 (1):49–54.

[92] Sorrentino SA, Doerries C, Manes C, et al. Nebivolol exerts beneficial effects on endothelial function, early endothelial progenitor cells, myocardial neovascularization, and left ventricular dysfunction early after myocardial infarction beyond conventional beta1-blockade. J Am Coll Cardiol 2011;57(5):601–11.

[93] de Nigris F, Mancini FP, Balestrieri ML, et al. Therapeutic dose of nebivolol, a nitric oxide-releasing beta-blocker, reduces atherosclerosis in cholesterol-fed rabbits. Nitric Oxide 2008;19 (1):57–63.

[94] Furberg CD, Psaty BM, Meyer JV. Nifedipine. Dose-related increase in mortality in patients with coronary heart disease. Circulation 1995;92(5):1326–31.

[95] Pitt B, Byington RP, Furberg CD, et al. Effect of amlodipine on the progression of atherosclerosis and the occurrence of clinical events. PREVENT Investigators. Circulation 2000;102(13):1503–10.

[96] Poole-Wilson PA, Lubsen J, Kirwan BA, et al. Effect of long-acting nifedipine on mortality and cardiovascular morbidity in patients with stable angina requiring treatment (ACTION trial): randomised controlled trial. Lancet 2004;364(9437):849–57.

[97] Cooper-DeHoff RM, Chang SW, Pepine CJ. Calcium antagonists in the treatment of coronary artery disease. Curr Opin Pharmacol 2013;13(2):301–8.

[98] Brovkovych VV, Kalinowski L, Muller-Peddinghaus R, et al. Synergistic antihypertensive effects of nifedipine on endothelium: concurrent release of NO and scavenging of superoxide. Hypertension 2001;37(1):34–9.

[99] Luscher TF, Pieper M, Tendera M, et al. A randomized placebo-controlled study on the effect of nifedipine on coronary endothelial function and plaque formation in patients with coronary artery disease: the ENCORE II study. Eur Heart J 2009;30(13): 1590–7.

[100] Munzel T, Gori T. Nitrate therapy and nitrate tolerance in patients with coronary artery disease. Curr Opin Pharmacol 2013;13 (2):251–9.

[101] Knorr M, Hausding M, Kroller-Schuhmacher S, et al. Nitroglycerin-induced endothelial dysfunction and tolerance involve adverse phosphorylation and S-glutathionylation of endothelial nitric oxide synthase: beneficial effects of therapy with the AT1 receptor blocker telmisartan. Arterioscler Thromb Vasc Biol 2011;31(10):2223–31.

[102] Ishikawa K, Kanamasa K, Ogawa I, et al. Long-term nitrate treatment increases cardiac events in patients with healed myocardial infarction. Secondary Prevention Group. Jpn Circ J 1996;60 (10):779–88.

[103] Nakamura Y, Moss AJ, Brown MW, et al. Long-term nitrate use may be deleterious in ischemic heart disease: a study using the databases from two large-scale postinfarction studies. Multicenter Myocardial Ischemia Research Group. Am Heart J 1999;138(3 Pt 1):577–85.

[104] Babelova A, Sedding DG, Brandes RP. Anti-atherosclerotic mechanisms of statin therapy. Curr Opin Pharmacol 2013;13(2):260–4.

[105] Parmar KM, Nambudiri V, Dai G, et al. Statins exert endothelial atheroprotective effects via the KLF2 transcription factor. J Biol Chem 2005;280(29):26714–9.

[106] Sen-Banerjee S, Mir S, Lin Z, et al. Kruppel-like factor 2 as a novel mediator of statin effects in endothelial cells. Circulation 2005;112 (5):720–6.

[107] Bu DX, Tarrio M, Grabie N, et al. Statin-induced Kruppel-like-factor 2 expression in human and mouse T cells reduces inflammatory and pathogenic responses. J Clin Invest 2010;120 (6):1961–70.

[108] Zhou Q, Liao JK. Pleiotropic effects of statins—basic research and clinical perspectives. Circ J 2010;74(5):818–26.

[109] Liu PY, Liu YW, Lin LJ, et al. Evidence for statin pleiotropy in humans: differential effects of statins and ezetimibe on rho-associated coiled-coil containing protein kinase activity, endothelial function, and inflammation. Circulation 2009;119(1): 131–8.

[110] Jasinska-Stroschein M, Owczarek J, Wejman I, et al. Novel mechanistic and clinical implications concerning the safety of statin discontinuation. Pharmacol Rep 2011;63(4):867–79.

II. ENDOTHELIAL DYSFUNCTION AND CLINICAL SYNDROMES

[111] Hennekens CH, Dalen JE. Aspirin in the treatment and prevention of cardiovascular disease: past and current perspectives and future directions. Am J Med 2013;126(5):373–8.

[112] Antithrombotic Trialists C. Collaborative meta-analysis of randomised trials of antiplatelet therapy for prevention of death, myocardial infarction, and stroke in high risk patients. BMJ 2002;324 (7329):71–86.

[113] Hetzel S, DeMets D, Schneider R, et al. Aspirin increases nitric oxide formation in chronic stable coronary disease. J Cardiovasc Pharmacol Ther 2013;18(3):217–21.

[114] Ehsani AA, Heath GW, Hagberg JM, et al. Effects of 12 months of intense exercise training on ischemic ST-segment depression in patients with coronary artery disease. Circulation 1981;64 (6):1116–24.

[115] Schuler G, Hambrecht R, Schlierf G, et al. Myocardial perfusion and regression of coronary artery disease in patients on a regimen of intensive physical exercise and low fat diet. J Am Coll Cardiol 1992;19(1):34–42.

[116] Da Luz PL, Nishyiama M, Chagas ACP. Drugs and lifestyle for the treatment and prevention of coronary artery disease—comparative analysis of the scientific basis. Braz J Med Biol Res 2011;44:973–91.

[117] Hambrecht R, Wolf A, Gielen S, et al. Effect of exercise on coronary endothelial function in patients with coronary artery disease. N Engl J Med 2000;342(7):454–60.

[118] Dod HS, Bhardwaj R, Sajja V, et al. Effect of intensive lifestyle changes on endothelial function and on inflammatory markers of atherosclerosis. Am J Cardiol 2010;105(3):362–7.

31

Molecular Mechanisms of the Arterial Wall in Acute Coronary Syndromes

Breno Bernardes de Souza, Haniel Alves Araújo, Viviane Zorzanelli Rocha Giraldez, Peter Libby, and Roberto Rocha C.V. Giraldez

INTRODUCTION

Over the last two decades, the vascular endothelium that covers the blood vessel luminal surface has emerged as the main regulator of vascular homeostasis. Initially considered a mere inert barrier, semipermeable to blood macromolecule diffusion into interstitial space, the endothelium plays an essential role in macro and microcirculation blood vessel reactivity control, thrombosis, and intravascular coagulation, as well as inflammatory processes that involve the vessel wall. The discovery of the endothelium's decisive role in the maintenance of vascular homeostasis has shown that risk factors for the development of atherosclerosis, such as dyslipidemia, hypertension, and diabetes, or in the presence of smoking, can adversely alter impair endothelial function [1].

These risk factors activate the endothelium, and threaten its ability to maintain vascular homeostasis. Endothelial dysfunction may then directly interfere with vascular health, reducing the endothelium's basal ability to inhibit inflammatory processes and smooth muscle cell proliferation in the arterial intima, predisposing the patient to the onset and progression of atherosclerotic disease.

In addition to interfering with chronic processes, endothelial dysfunction may induce acute changes in vascular equilibrium, facilitating vasospasm or even intravascular thrombosis. The endothelium's regulatory role seems especially critical in coronary circulation, a frequent site of atherosclerosis. Acute forms of coronary artery disease and its complications are a leading cause of death in developed countries and contribute increasingly to the global burden of disease. Thus, it is essential to understand endothelial involvement in acute complications of atherosclerosis; that is, in what way the endothelium may influence the evolution of the atherothrombotic process associated with disruptions of atherosclerotic plaque. In addition, it is important to evaluate how we can intervene in this process, reestablishing the physiological function of the endothelium and allowing a more favorable outcome of the acute condition. It is also necessary to understand the endothelium's involvement in acute coronary syndromes (ACS) that are not related to atherosclerotic disease; the pathophysiology of this type of ACS, however, which is less frequent, remains poorly understood [2].

ENDOTHELIAL DYSFUNCTION IN ACUTE CORONARY SYNDROMES

Acute coronary syndromes have different pathophysiological mechanisms (Fig. 31.1) [3], and endothelial dysfunction appears to participate, whether in an acute or chronic form, in at least one step of all major different pathogenic pathways. Endothelial dysfunction contributes to the onset as well as to the progression and complications of atherosclerotic disease. The entry of inflammatory cells into arterial intima depends on adhesion molecules expressed in the endothelium, and the activated

473

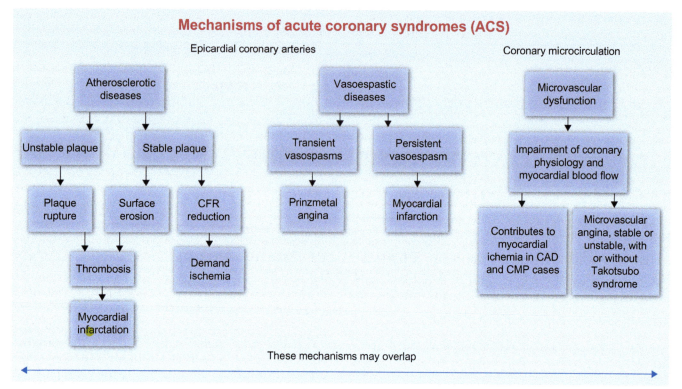

FIG. 31.1 Mechanisms that lead to acute coronary syndromes (ACS). *CMP*, cardiomyopathy; *CAD*, coronary artery disease; *CFR*, coronary flow reserve. *Note*: rare ACS mechanisms that are apparently unrelated to endothelial function (such as spontaneous coronary dissection, arteritis, and myocardial bridges) were not included in this chapter. Fig. 31.1 Adapted from Crea F, Camici PG, Bairey Merz CN. Coronary microvascular dysfunction: an update. Eur Heart J 2014;35(17):1101–11.

endothelium plays a decisive role in atherosclerotic plaque progression, stability, and persistence. This chronic role of the endothelium in atherosclerosis is discussed in Chapters 33 and 39.

The two main mechanisms of atherosclerotic plaque acute complications are plaque rupture and surface erosion. Since the endothelium does not participate directly in the acute phase of the first mechanism, except in cases of plaque rupture due to endothelium-associated paradoxical vasoconstriction, plaque rupture will not be discussed here. Endothelial dysfunction, however, plays a fundamental role in the pathogenesis of superficial erosion, which is a form of plaque disruption with increasing incidence, and will be detailed below.

The endothelium also participates in ACS that result from vasospastic conditions of epicardial coronary arteries and from ACS related to microvascular coronary dysfunction. The understanding of types of ACS unrelated to obstructive atherosclerotic disease is limited, but endothelial involvement seems to be important, mainly due to alterations in nitric oxide production or action in the endothelial layer [2].

The endothelium also participates in events that occur after ACS. Endothelial cells contribute to both reperfusion of injury after myocardial ischemia and in the healing of vessels after the ischemic event. Endothelial function may rapidly change and a healthy endothelium likely contributes to improved outcomes in cases of acute coronary events. The benefit of therapeutic measures currently used after ACS also likely improves endothelial function. For example, statin administration in the acute phase of coronary disease may not merely reduce LDL plasma levels, but also ameliorate endothelial function beyond effects on cholesterolemia. These non-LDL-dependent effects of statins may enhance the return of vascular homeostasis and consequently reduce morbidity and mortality.

Surface Erosion

Physical disruptions of atherosclerotic plaques cause most ACS. Rupture of the fibrous cap has received much attention as the main mechanism of plaque disruption leading to ACS [4,5]. Pathophysiological mechanisms that cause rupture of thin fibrous cap atheroma generally do not involve abnormalities of endothelial function [4]. Surface erosion, however, a second form of disruption which can cause ACS, indubitably involves endothelial dysfunction [5].

Previously considered responsible for 20%–25% of acute coronary syndromes, superficial erosion as a cause of ACS may be on the rise [6]. We have argued that the increasing use of statins, smoking cessation, and other preventive measures, especially those that address high levels of LDL, have caused plaque rupture to wane [7]. Surface erosion will likely become more prominent as a mechanism of coronary thrombosis given the high penetrance of statin therapy and with the introduction of newer LDL-lowering agents such as cholesterol absorption inhibitors and modulators of PCSK9, which achieve lower LDL levels than those achieved by statin mono-therapy. The demographic profile attributed to surface erosion in postmortem studies reflects the populations that have an increasing share of ACS in the era of statins: younger patients, females, diabetics, or those with histories of insulin resistance and dyslipidemia characterized by high triglyceride levels [8,9].

The causes and pathophysiology of surface erosion have received relatively little attention compared to the research surrounding the pathophysiology of fibrous cap plaque rupture [4]. The mechanisms underlying surface erosion probably differ significantly from those that cause rupture of the fibrous cap (Fig. 31.2). Lesions that cause death due to superficial erosion generally contain few chronic inflammatory cells. Yet these plaques contain abundant smooth muscle cells, proteoglycans, and glycosaminoglycans [10].

We hypothesized, at the beginning of the millennium, the role of endothelial abnormalities in surface erosion pathogenesis [11]. We proposed that endothelial desquamation occurs due to jeopardized adhesion to the underlying basement membrane and increased cellular death. In fact, we have proposed that a weakened tether of the basal (abluminal) surface of endothelial cells to the underlying basement promotes cell death (anoikis, or anchorage-dependent cell death). In early initial experiments we verified that minimally modified LDL, which can accumulate below the endothelial

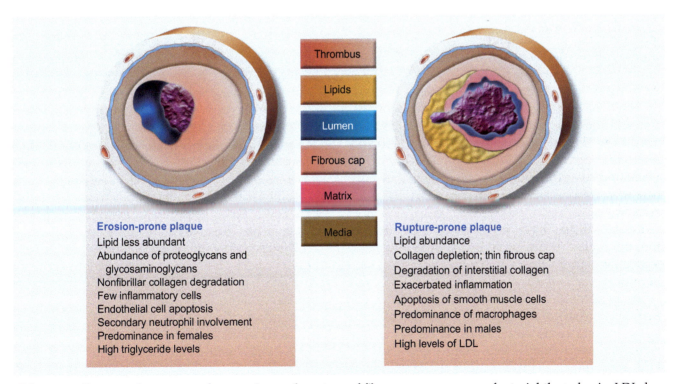

FIG. 31.2 Contrasts between surface erosion and rupture of fibrous cap as causes of arterial thrombosis. *LDL,* low density lipoprotein.Fig. 31.2 Adapted from Libby P, Pasterkamp G. Requiem for the "vulnerable plaque." Eur Heart J 2015;36(43):2984–7.

monolayer during atherogenesis, can activate matrix metalloproteinase 14 (MMP-14, or MT1-MMP) in endothelial cells. This enzyme localizes in the cell membrane rather than the extracellular matrix, like most enzymes in MMP family [12]. MMP-14 activates the zymogen precursor of another matrix metalloproteinase (MMP-2), also known as type IV collagenase [13]. MMP-2-degrades nonfibrillar collagens, such as collagen type IV, that abound in the basement membrane, to which endothelial cells adhere. These experiments demonstrate a relationship between oxidized lipoproteins and the weakening of endothelial cell anchorage in their underlying matrix. A recent genome-wide association study (GWAS) implicated variants in the COL4A1/CO-L4A2 locus (locus related to type IV collagen expression) in the risk of coronary disease [14].

We also hypothesized the existence of a role for hypochlorous acid (HOCl) in surface erosion. A subpopulation of macrophages in atherosclerotic plaques contains myeloperoxidase (MPO), an enzyme that generates HOCl [15–17]. in vitro experiments showed concentration-dependent apoptosis in endothelial cells induced by HOCl, and the HOCl concentrations used conformed to those expected in inflammatory sites [18]. Furthermore, the endothelial cells undergoing death in response to HOCl increased their production of tissue factor, a potent procoagulant. Subsequent studies by Professor Filippo Crea's group have shown that patients with ACS due to superficial erosion, as assessed by optical coherence tomography, have higher serum concentrations of MPO [19]. Altogether, these laboratory and clinical observations support the hypothesis that oxidative stress mediated by HOCl generated by MPO increases endothelial dysfunction in patients prone to ACS secondary to surface erosion.

Recent studies have investigated other triggers for endothelial dysfunction in surface erosion. Sites with altered blood flow distal to stenoses show signs of endothelial cell apoptosis more frequently than those in arterial regions with laminar flow [20]. Regions with disturbed flow in atherosclerotic arteries of mice express higher levels of the pattern recognition receptor *Toll-like receptor* 2 (TLR2) [21]. Disturbed flow in vitro enhances TLR2 expression in cultured human endothelial cells [22]. These observations suggest that TLR2 participates in cell signaling pathways that pertain to surface erosion.

Hyaluronic acid, a glycosaminoglycan that accumulates in plaques that undergo erosion, may be one endogenous activator of TLR2 [23]. In vitro studies showed that culture of human endothelial cells on a surface coated with hyaluronic acid leads to the activation of several proinflammatory signaling pathways, including expression of IL-8, a potent granulocyte chemoattractant [24]. The exposure of human endothelial cells to hyaluronic acid activates caspase-3, a protease important in cell death by apoptosis. Exposure to hyaluronan also breaks intercellular junctions of VE-cadherin between endothelial cells, which contribute to the integrity of the endothelial monolayer. In vitro, TLR2 agonists promote greater susceptibility to monolayer desquamation. The presence of polymorphonuclear leukocytes (PMN) greatly enhances many of the effects of TLR2 activation in endothelial cells. These experiments suggest that TLR2 activation participates in both desquamation and endothelial cell death. We have also demonstrated that hyaluronic acid can act as an endogenous ligand of TLR2 in plaques with morphology associated with superficial erosion [24].

In addition, synergy in the pathophysiology of endothelial dysfunction, between granulocyte recruitment and activation of TLR2, suggests a "two-step" model for surface erosion (Fig. 31.3). Initially, a few endothelial cells would slough and/or undergo apoptosis (left side of Fig. 31.3). Triggering stimuli, such as the activation of TLR2, could impair the recovery of small "wounds" in the endothelial monolayer, which usually have regenerative capacity. In the context of endothelial dysfunction, granulocyte chemoattractants, such as IL-8, could recruit such acute inflammatory cells to the lesion site. Triggering events could initiate a second stage (right side of Fig. 31.3) in which events such as the generation of hypochlorous acid by myeloperoxidase, an abundant neutrophilic enzyme, and the recruitment of platelets due to basement membrane type IV collagen exposure and tissue factor-induced thrombosis, could—together with other potential pathways—amplify thrombotic and inflammatory responses on eroding plaque.

Granulocytes generally undergo apoptosis after their recruitment to inflammatory sites. Histones, which normally keep the DNA tightly packed into chromatin, may lose their positive charges conferred by arginine due to the action of peptidyl arginine deiminase type 4 (PAD4). This enzyme converts positively charged arginine residues into uncharged

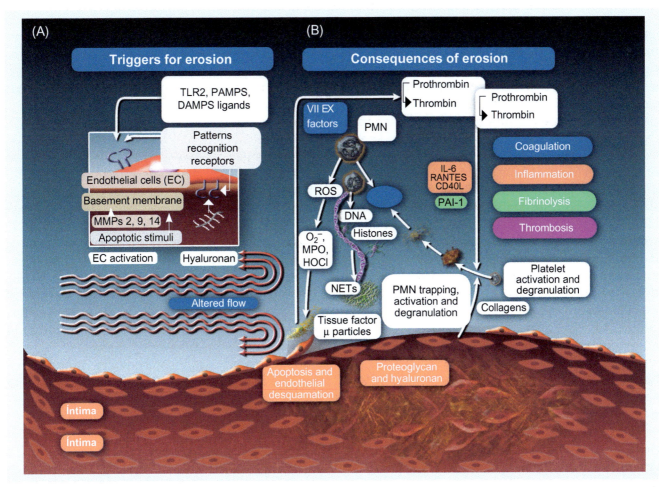

FIG. 31.3 Potential pathophysiological pathway by which surface erosion causes thrombotic complications of atherosclerotic plaques: "two-step" model. The lower part of the diagram shows a longitudinal section of a proteoglycan-rich atherosclerotic plaque-affected artery. The darker brownish hue indicates accumulation of proteoglycans, such as hyaluronic acid. (A) Some of the possible triggers for endothelial damage, which are considered underlying causes of surface erosion. Such triggers include pathogen-associated molecular patterns (PAMPs), molecular damage-associated patterns (DAMPs), and other binding inhibitors of innate immunity, including TLR2. These mediators interact with pattern recognition receptors present on the surface of the endothelial cell. Hyaluronic acid, a common constituent of plaques affected by superficial erosion, may act as a ligand for TLR2. In addition, various apoptotic stimuli derived from inflammatory cells from plaques, as well as from modified lipoproteins, may promote apoptosis of endothelial cells. Matrix-degrading enzymes, such as matrix metalloproteinases (MMPs), may compromise basement membrane constituents, which provide a substrate for endothelial cell adhesion by integrins and other adhesion molecules present on the basal surface of the endothelial cell. Nonfibrillar collagenases MMP-2 and MMP-9, and the MMP-2 activator MMP-14 are enzymes that are frequently and abnormally expressed on plaques and can damage the anchorage of endothelial cells on the surface of the intima. (B) shows some of the consequences of erosion. When undergoing desquamation (as illustrated by the damaged endothelial cell with a pycnotic nucleus), the agonal endothelial cell can release microparticles that have tissue factor, which accentuates the activity of factors VII and X, increasing the formation of thrombin and ultimately, the conversion of fibrinogen into fibrin produces a thrombus. Subendothelial matrix exposure may provide substrate for granulocyte adhesion, activation and degranulation. These polymorphonuclear leukocytes (PMN) are the source of reactive oxygen species (ROS) such as hypochlorous acid (HOCl), a product of myeloperoxidase (MPO), and also the superoxide anion (O_2). According to this schematic diagram, the granulocytes reach the lesion site after the initial desquamation of the endothelial monolayer. Granulocytes may also secrete MRP-8/14, a member of the calgranulin family that is involved with inflammation and other aspects of atherothrombosis. Dying granulocytes release DNA and histones into the extracellular medium, forming "extracellular neutrophil traps" (NETs). The DNA fibers of these structures form a network that can contribute to increase the extent of thrombosis and the trapping of leukocytes from the bloodstream, amplifying the local inflammatory response. The exposure of the macromolecules of the subendothelial extracellular matrix can activate platelets, resulting in their degranulation with release of proinflammatory mediators such as interleukin-6 and RANTES. Activated platelets may also secrete plasminogen activator inhibitor type 1 (PAI-1), an important inhibitor of endogenous fibrinolytic enzymes. PAI-1 can therefore reduce fibrinolysis and increase clot stability. Fig. 31.3 Adapted from Quillard T, Araujo HA, Franck G, et al. TLR2 and neutrophils potentiate endothelial stress, apoptosis and detachment: implications for superficial erosion. Eur Heart J 2015;36(22):1394–404.

citrulline, leading to the dissociation of ionic bonds that keep DNA (with negative charges) tightly twisted around histone oligomers to form nucleosomes. The granulocytes undergoing cell death can release long fibers of DNA into extracellular medium, forming neutrophil extracellular traps (NETs) [25]. The NET generation process, known as NETosis, consists of a recently recognized type of cell death (distinct from apoptosis) in which granulocytes engage in a suicide mission and form these structures, and thus contribute to innate immunity [26]. These DNA traps become decorated with constituents of neutrophil granules including its bactericidal enzyme myeloperoxidase, and also bind the procoagulant tissue factor [27]. Considerable evidence suggests that NETs provide a surface that together with fibrin networks, platelets and leucocytes, can exacerbate both thrombosis and acute inflammatory local responses. Analyses of human atherosclerotic plaques have revealed that lesions with morphological characteristics associated with superficial erosion show apoptotic endothelial cells with NETs [24].

Taken together, these observations corroborate the fundamental role of endothelial dysfunction in superficial erosion pathogenesis as a triggering factor for thrombosis and acute coronary syndromes. As superficial erosion may account for a growing proportion of acute coronary syndromes in the present era, we should expand our concept of endothelial dysfunction beyond the defective vasomotor responses to encompass proinflammatory and prothrombotic aspects. The study of surface erosion on a mechanistic level is just beginning. This aspect of endothelial dysfunction will require considerably greater attention from a pathophysiological perspective. Such studies may suggest new therapeutic strategies to combat acute coronary syndromes, such as treatments with deoxyribonuclease (DNAse) that promotes NETs degradation.

VASOSPASTIC AND CORONARY MICROVASCULAR DYSFUNCTION

Although atherosclerosis most commonly underlies the occurrence of ACS, some affected patients do not have evidence of obstructive epicardial coronary artery disease as determined by invasive and noninvasive diagnostic methods [3,28]. In such cases, pathogenetic mechanisms may include epicardial coronary artery spasms or dysfunction of coronary microcirculation (Fig. 31.1) [3]. In either case, endothelial cells likely contribute to the pathogenesis.

Coronary artery spasms were suggested in 1959 by Prinzmetal et al. [29] Locally increased vascular tone accounted for the symptoms of patients with recurrent attacks of chest pain at rest associated with transient elevation of the ST segment. Sites of such spasm can manifest minimal plaque, as seen by arteriography. The causes involve hyperreactivity to vasoconstrictor stimuli and/or exposure to these stimuli (common in patients with abuse of stimulant substances, such as cocaine). These causal factors converge to modify the normal physiology of smooth muscle cells, rendering them hyperreactive.

Yet, endothelial dysfunction also likely contributes to this clinical condition [30]. Kaneda et al. [31] induced spasm with intracoronary acetylcholine infusion in healthy patients and in patients with angina without evidence of stenosis in coronary angiography. Genetic evaluation in both groups showed a correlation between occurrence of spasm after infusions and a higher prevalence of mutations in the endothelial nitric oxide synthase (eNOS) gene that impair nitric oxide production. Evidence of alteration in endothelial nitric oxide-mediated vasodilation also can occur in peripheral arteries [32] and in the nonspastic regions of coronary arteries [33] of patients with a history of coronary spasm. Thus, nitric oxide deficiency may affect the entire vascular system of these patients. Yet, some studies [34,35] did not identify endothelial dysfunction in this population, and only a third of cases have polymorphisms in the eNOS gene [36]. The uncertainly regarding endothelial involvement in this condition may reflect a less consistent role of the endothelium than smooth muscle cells in some cases of coronary arterial spasm [30]. Studies that demonstrate symptom reduction with statins [37] or vitamin E [38], substances known to improve endothelial vasodilator function, support endothelial involvement in the disease process.

In addition to the two classic mechanisms that lead to ACS (atherosclerotic and vasospastic diseases), dysfunction of the coronary microcirculation has emerged as a third potential mechanism for ACS [3]. Alterations in coronary microvascular homeostasis can compromise blood supply to the myocardium in the absence of epicardial stenosis or spasm [39]. In patients with coronary artery disease or cardiomyopathies, microvascular dysfunction may arise secondarily [40]. These very ischemic events may further aggravate microvasculature homeostasis by microembolization after ACS (no-reflow phenomenon, discussed in the next section). In individuals without coronary artery disease and without cardiomyopathies, however, coronary microvascular dysfunction may independently cause unstable or stable angina, or the apical ballooning characteristic of Takotsubo disease (many currently consider Takotsubo syndrome a form of microvascular coronary disease) [3,41].

The concept of microvascular dysfunction as a primary cause of ACS is very broad can encompass cases of unstable or stable angina [40,42]. Diagnosis requires exclusion, and applies to patients with angina-like pain who have no evidence of changes in epicardial coronary arteries [3]. Many practice environments lack access to validated methods that access coronary microvascular function [39]. The causes and pathophysiology of coronary microvascular dysfunction remain obscure and a complete review is beyond the scope of this chapter (for more details see Crea et al., 2014) [2,3,43]. Yet, endothelial vasodilator and smooth muscle cell dysfunction probably contribute to the pathogenesis, together with arteriolar remodeling, and autonomic dysfunction [3,39]. Although direct evidence remains scarce, the endothelium may have altered antithrombotic and anticoagulant properties, and favors a prothrombotic, procoagulant, and antifibrinolytic slant [44]. Impaired endothelial vasomotor control likely involves a decrease in nitric oxide bioavailability [45]. In some cases, symptoms may improve with treatment with statins [46], estrogen [47], or ACE inhibitors [48], interventions that may improve endothelial vasodilator function, supporting endothelial involvement in this clinical scenario. The mechanisms that underlie these abnormalities in coronary microvascular function comprise a fruitful field for future research.

Reperfusion Injury

As previously detailed in this chapter, the endothelium actively participates in chronic and acute events that relate closely to acute coronary syndromes. Among these events are the formation, progression, and regulation of determinants of atherosclerotic plaque stability, superficial erosion, coronary artery spasm, and coronary microcirculatory disorders.

The endothelium also participates in events that follow: reperfusion injury after myocardial ischemia. The possibility that restoration of blood flow in a previously ischemic region could damage viable cells at the time of reperfusion was suggested in the 1970s [49–51] and confirmed in cardiac tissue by Braunwald and Kloner a few years later [52]. Since then, the pathophysiology of ischemia-reperfusion injury has undergone intensive study and several events involved in the phenomenon have become more clearly defined. Myocardial reperfusion occurs commonly in contemporary cardiology, after therapeutic reperfusion by percutaneous intervention and fibrinolysis, coronary artery bypass grafting, and transplantation of organs, rendering reperfusion injury particulary important.

Myocardial reperfusion produces an immediate tissue lesion that involves two interrelated phases [53]. The first corresponds to an early stage triggered by the endothelium, while the second is amplified later by neutrophils. The primary factor for reperfusion injury appears to be the generation of reactive oxygen species by the vascular intima [54]. During myocardial ischemia, adenosine triphosphate (ATP) is degraded to hypoxanthine, while the enzyme xanthine dehydrogenase converts to its oxidizing form, xanthine oxidase, on the endothelial cell surface. Upon reperfusion, xanthine oxidase catalyzes the conversion of hypoxanthine to uric acid, generating superoxide anion (O_2^-) as a coproduct. Thus, the xanthine oxidase pathway, among other mechanisms, can generate copious amounts of reactive oxygen species during myocardial reperfusion [55]. These highly toxic products harm the endothelium, amplifying existing endothelial dysfunction caused by traditional risk factors [56]. Activation of the endothelium allows for the expression of adhesion molecules, which recruit blood elements, especially circulating polymorphonuclear leucocytes that adhere to the endothelial surface and migrate into the intima, where they can generate further reactive oxygen species and chemotactic substances that perpetuate reperfusion injury [54]. Furthermore, the endothelium also interacts with circulating platelets, which in association with leukocytes form aggregates that can obstruct the microsculature, yielding a phenomenon known as no-reflow [57]. Thus, the endothelium plays a predominant role in reperfusion injury, both by the primary formation of reactive oxygen species and by their interaction with blood cells, which ultimately amplify the initial damage. Although many studies have aimed to reduce reperfusion injury, using interventions effective in animals, so far none have consistently shown a beneficial clinical effect [58,59].

Recovery of Perturbed Endothelial Function

In addition to fulfilling an important role in ACS and triggering reperfusion injury, it is quite likely that endothelial cells also contribute critically to vessel recovery after ischemic events. The endothelium can modify its behavior very quickly and therefore a positive endothelial action probably contributes to recovery from vascular disorders. Two important factors, biomechanical stimuli and the balance between pro- and antiinflammatory mediators, can counteract the noxious stimuli that endothelial cells encounter during ACS or reperfusion injury, and probably contribute to the restoration of vascular homeostasis. Laminar flow elicits antiinflammatory and antithrombotic activity in endothelial cells, fostering the restoration of normal endothelial function. Moreover, the recovery of endothelial homeostasis associates with a biochemical balance wherein the production of antiinflammatory cytokines and nitric oxide predominates over the generation of reactive oxygen species and proinflammatory cytokines. These measures may help restore the basal anticoagulant and fibrinolytic properties of endothelial cells.

In addition to inducing the expression of transcription factors that regulate healthy endothelial function such as KLF-2 (Krüppel-like factor 2), which augments nitric oxide production and has antiinflammatory effects that may mute atherosclerotic plaque evolution [60], shear stress also elevates nitric oxide production by endothelial cells. High shear stress causes calcium entry into endothelial cells, favoring dissociation of eNOS and caveolin, a protein forms caveolae, shifting eNOS to the cytosol where the enzyme becomes active [61]. Nitric oxide formed in the endothelium rapidly diffuses into the adjacent middle muscular layer, and stimulates relaxation of vascular smooth muscle cells [62]. Apart from diffusing into the vascular wall, nitric oxide enters the blood vessel lumen, acting on circulating platelets and leukocytes, limiting the interaction of these blood elements with the arterial intima. Nitric oxide may not only limit thrombus formation, but also may favor disaggregation of platelet clumps [63].

A healthy arterial endothelium has low levels of surface expression of adhesion molecules. In normal arteries, leukocytes and blood platelets do not adhere tightly to endothelium. Inflammatory activation can limit nitric oxide production and augment adhesion molecule expression on the luminal cell surface [64]. Restoration of normal endothelial function can contribute to the resolution of inflammatory and thrombotic complications of atherosclerosis and other vascular disorders.

Thus, improving endothelial function can become a target of therapy, and may explain some of the beneficial actions of treatments that improve cardiovascular outcomes and lessen the risk of first or recurrent ACS. The use of statins after ACS events provides an example.

The MIRACL trial (Myocardial Ischemia reduction with aggressive cholesterol lowering) evaluated the effects of early use of statins in ACS [65]. This randomized, double-blind trial compared early treatment with high doses of atorvastatin (80 mg/day) to placebo after ACS. After 16 weeks of follow-up, the atorvastatin group had a reduction in the primary endpoint, defined as total mortality, fatal myocardial infarction, nonfatal heart attack, and recurrent ischemia requiring hospitalization. An analysis of GRACE (Global Registry of Acute Coronary Events) examined the records of nearly 20,000 patients with ACS divided into four groups: those that started statins during hospitalization; patients that did not start statins during hospitalization; patients that suspended statin use during hospitalization; and patients that continued to use statins during hospitalization [66]. The study showed beneficial effects on mortality or complications in the groups that initiated or maintained the use of statins in comparison to groups that had suspended or never used the drug, supporting the hypothesis that statin therapy can modulate the pathophysiology involved in ACS early on. The RECIFE study (Reduction of cholesterol in ischemia and function of the endothelium) demonstrated early recovery of endothelium-dependent vasodilation of the brachial artery in patients with unstable angina and myocardial infarction without ST elevation treated with pravastatin (40 mg/day) [67]. These clinical studies, in combination with experimental studies [68] that support the beneficial effect of statins on endothelial cells, provide a theoretical basis for the clinical use of statins as adjunctive therapy in the acute treatment of ACS. The benefits of this treatment most likely result not only from a reduction in LDL levels (an aspect not discussed here), but also from improvement in endothelial function.

CONCLUSIONS

At the beginning of the millennium the participation of the endothelium in ACS furnished an attractive hypothesis, but was not actionable from a clinical perspective. A number of clinical and experimental studies now support the endothelium as a target of therapy in relation to prevention and treatment of myocardial infarction both in the acute phase and the longer term. New evidence supports a key role for endothelial involvement in superficial erosion, a cause of a substantial proportion of ACS. Research over the past three decades has helped us to understand atherosclerotic disease and improve morbidity and cardiovascular mortality with the implementation of effective treatments and measures to prevent onset, progression, and complications of the disease. Due to the adoption of these effective therapies, in particular the widespread use of statins, lesions now contain less lipid and fewer inflammatory cells than in the prestatin era [69]. Superficial erosion of the endothelial monolayer may account for some for some of the residual risk that persists despite current standard of care. Decades of study have led to understanding of the mechanisms of plaque rupture, and has informed us regarding how current therapies can combat this mode of ACS pathogeneis. Now we need to understand the mechanisms of plaque erosion, and treatments and management strategies that may address the distinct pathways that lead to this type of plaque complication [70].

The endothelium also participates in ACS related to spasm of the epicardial coronary arteries and microvascular coronary dysfunction. Considering the diverse ways in which the endothelium participates in ACS, this cell type assumes even more importance as therapeutic target. This challenge should spur further research into the mechanisms of endothelial dysfunction and ways to speed its resolution in the prevention and treatment of ACS.

References

[1] Suwaidi JA, Hamasaki S, Higano ST, et al. Long-term follow-up of patients with mild coronary artery disease and endothelial dysfunction. Circulation 2000;101(9):948–54.

[2] Crea F, Libby P. Acute coronary syndromes. Circulation 2017;136 (12):1155–66.

[3] Crea F, Camici PG, Bairey Merz CN. Coronary microvascular dysfunction: an update. Eur Heart J 2014;35(17):1101–11.

[4] Libby P. Mechanisms of acute coronary syndromes and their implications for therapy. N Engl J Med 2013;368(21):2004–13.

[5] Bentzon JF, Otsuka F, Virmani R, et al. Mechanisms of plaque formation and rupture. Circ Res 2014;114(12):1852–66.

[6] Libby P, Pasterkamp G. Requiem for the "vulnerable plaque" Eur Heart J 2015;36(43):2984–7.

[7] van Lammeren GW, den Ruijter HM, Vrijenhoek JE, et al. Time-dependent changes in atherosclerotic plaque composition in patients undergoing carotid surgery. Circulation 2014;129 (22):2269–76.

[8] Burke AP, Farb A, Malcom GT, et al. Effect of risk factors on the mechanism of acute thrombosis and sudden coronary death in women. Circulation 1998;97(21):2110–6.

[9] Finn AV, Nakano M, Narula J, et al. Concept of vulnerable/unstable plaque. Arterioscler Thromb Vasc Biol 2010;30(7):1282–92.

[10] Kolodgie FD, Burke AP, Wight TN, et al. The accumulation of specific types of proteoglycans in eroded plaques: a role in coronary thrombosis in the absence of rupture. Curr Opin Lipidol 2004; 15(5):575–82.

[11] Libby P, Ganz P, Schoen FJ, et al. The vascular biology of the acute coronary syndromes. In: Topol EJ, editor. Acute coronary syndromes. 2nd ed. New York: Marcel Dekker Inc.; 2000. p. 33–57.

[12] Rajavashisth TB, Liao JK, Galis ZS, et al. Inflammatory cytokines and oxidized low density lipoproteins increase endothelial cell expression of membrane type 1-matrix metalloproteinase. J Biol Chem 1999;274(17):11924–9.

[13] Dollery CM, Libby R. Atherosclerosis and proteinase activation. Cardiovasc Res 2006;69(3):625–35.

[14] Turner AW, Nikpay M, Silva A, et al. Functional interaction between COL4A1/COL4A2 and SMAD3 risk loci for coronary artery disease. Atherosclerosis 2015;242(2):543–52.

[15] Daugherty A, Dunn JL, Rateri DL, et al. Myeloperoxidase, a catalyst for lipoprotein oxidation, is expressed in human atherosclerotic lesions. J Clin Invest 1994;94(1):437–44.

[16] Sugiyama S, Okada Y, Sukhova GK, et al. Macrophage myeloperoxidase regulation by granulocyte macrophage colony-stimulating factor in human atherosclerosis and implications in acute coronary syndromes. Am J Pathol 2001;158(3):879–91.

[17] Vaisar T, Shao B, Green PS, et al. Myeloperoxidase and inflammatory proteins: pathways for generating dysfunctional high-density lipoprotein in humans. Curr Atheroscler Rep 2007; 9(5):417–24.

[18] Sugiyama S, Kugiyama K, Aikawa M, et al. Hypochlorous acid, a macrophage product, induces endothelial apoptosis and tissue factor expression: involvement of myeloperoxidase-mediated oxidant in plaque erosion and thrombogenesis. Arterioscler Thromb Vasc Biol 2004;24(7):1309–14.

[19] Ferrante G, Nakano M, Prati F, et al. High levels of systemic myeloperoxidase are associated with coronary plaque erosion in patients with acute coronary syndromes: a clinicopathological study. Circulation 2010;122(24):2505–13.

[20] Tricot O, Mallat Z, Heymes C, et al. Relation between endothelial cell apoptosis and blood flow direction in human atherosclerotic plaques. Circulation 2000;101(21):2450–3.

[21] Mullick AE, Soldau K, Kiosses WB, et al. Increased endothelial expression of Toll-like receptor 2 at sites of disturbed blood flow exacerbates early atherogenic events. J Exp Med 2008;205 (2):373–83.

[22] Dunzendorfer S, Lee HK, Tobias PS. Flow-dependent regulation of endothelial toll-like receptor 2 expression through inhibition of SP1 activity. Circ Res 2004;95(7):684–91.

[23] Scheibner KA, Lutz MA, Boodoo S, et al. Hyaluronan fragments act as an endogenous danger signal by engaging TLR2. J Immunol 2006;177(2):1272–81.

[24] Quillard T, Araujo HA, Franck G, et al. TLR2 and neutrophils potentiate endothelial stress, apoptosis and detachment: implications for superficial erosion. Eur Heart J 2015;36(22):1394–404.

[25] Demers M, Wagner DD. Neutrophil extracellular traps: a new link to cancer-associated thrombosis and potential implications for tumor progression. Oncoimmunology 2013;2(2)e22946.

[26] Brinkmann V, Reichard U, Goosmann C, et al. Neutrophil extracellular traps kill bacteria. Science 2004;303(5663):1532–5.

[27] Badimon L, Vilahur G. Neutrophil extracellular traps: a new source of tissue factor in atherothrombosis. Eur Heart J 2015; 36(22):1364–6.

[28] Arbab-Zadeh A, Nakano M, Virmani R, et al. Acute coronary events. Circulation 2012;125(9):1147–56.

[29] Prinzmetal M, Kennamer R, Merliss R, et al. Angina pectoris. I. A variant form of angina pectoris; preliminary report. Am J Med 1959;27:375–88.

[30] Lanza GA, Careri G, Crea F. Mechanisms of coronary artery spasm. Circulation 2011;124(16):1774–82.

[31] Kaneda H, Taguchi J, Kuwada Y, et al. Coronary artery spasm and the polymorphisms of the endothelial nitric oxide synthase gene. Circ J 2006;70(4):409–13.

[32] Moriyama Y, Tsunoda R, Harada M, et al. Nitric oxide-mediated vasodilatation is decreased in forearm resistance vessels in patients with coronary spastic angina. Jpn Circ J 2001;65(2):81–6.

[33] Okumura K, Yasue H, Matsuyama K, et al. Diffuse disorder of coronary artery vasomotility in patients with coronary spastic angina. Hyperreactivity to the constrictor effects of acetylcholine and the dilator effects of nitroglycerin. J Am Coll Cardiol 1996;27 (1):45–52.

[34] Ito K, Akita H, Kanazawa K, et al. Systemic endothelial function is preserved in men with both active and inactive variant angina pectoris. Am J Cardiol 1999;84(11):1347–9. A8.

[35] Egashira K, Inou T, Yamada A, et al. Preserved endothelium-dependent vasodilation at the vasospastic site in patients with variant angina. J Clin Invest 1992;89(3):1047–52.

[36] Yasue H, Nakagawa H, Itoh T, et al. Coronary artery spasm—clinical features, diagnosis, pathogenesis, and treatment. J Cardiol 2008;51(1):2–17.

[37] Yasue H, Mizuno Y, Harada E, et al. Effects of a 3-hydroxy-3-methylglutaryl coenzyme A reductase inhibitor, fluvastatin, on coronary spasm after withdrawal of calcium-channel blockers. J Am Coll Cardiol 2008;51(18):742–8.

[38] Motoyama T, Kawano H, Kugiyama K, et al. Vitamin E administration improves impairment of endothelium-dependent vasodilation in patients with coronary spastic angina. J Am Coll Cardiol 1998;32(6):1672–9.

[39] Camici PG, Crea F. Coronary microvascular dysfunction. N Engl J Med 2007;356(8):830–40.

[40] Lanza GA, Crea F. Primary coronary microvascular dysfunction: clinical presentation, pathophysiology, and management. Circulation 2010;121(21):2317–25.

[41] Crea F, Lanza GA. Angina pectoris and normal coronary arteries: cardiac syndrome X. Heart 2004;90(4):457–63.

[42] Lanza GA. Cardiac syndrome X: a critical overview and future perspectives. Heart 2007;93(2):159–66.

[43] Bairey Merz CN, Pepine CJ, Walsh MN, Fleg JL. Ischemia and No Obstructive Coronary Artery Disease (INOCA): developing evidence-based therapies and research agenda for the next decade. Circulation 2017;135(11):1075–92.

[44] Hurst T, Olson TH, Olson LE, et al. Cardiac syndrome X and endothelial dysfunction: new concepts in prognosis and treatment. Am J Med 2006;119(7):560–6.

[45] Kaski JC. Pathophysiology and management of patients with chest pain and normal coronary arteriograms (cardiac syndrome X). Circulation 2004;109(5):568–72.

[46] Kayikcioglu M, Payzin S, Yavuzgil O, et al. Benefits of statin treatment in cardiac syndrome-X1. Eur Heart J 2003;24(22):1999–2005.

[47] Rosano GM, Peters NS, Lefroy D, et al. 17-Beta-estradiol therapy lessens angina in postmenopausal women with syndrome X. J Am Coll Cardiol 1996;28(6):1500–5.

[48] Pizzi C, Manfrini O, Fontana F, et al. Angiotensin-converting enzyme inhibitors and 3-hydroxy-3-methylglutaryl coenzyme a reductase in cardiac syndrome X: role of superoxide dismutase activity. Circulation 2004;109(1):53–8.

[49] Hearse DJ. Reperfusion of the ischemic myocardium. J Mol Cell Cardiol 1977;9(8):605–16.

[50] Maroko PR, Libby P, Ginks WR, et al. Coronary artery reperfusion. I. Early effects on local myocardial function and the extent of myocardial necrosis. J Clin Invest 1972;51(10):2710–6.

[51] Braunwald E, Maroko PR, Libby P. Reduction of infarct size following coronary occlusion. Circ Res 1974;35(Suppl. III):192–201.

[52] Braunwald E, Kloner RA. Myocardial reperfusion: a double-edged sword? J Clin Invest 1985;76(5):1713–9.

[53] Lefer AM, Hayward R. The role of nitric oxide in ischemia-reperfusion. In: Loscalzo J, Vita JA, editors. Nitric oxide and the cardiovascular system. Totowa: Humana Press; 2000. p. 357–80.

[54] Zweier JL. Measurement of superoxide-derived free radicals in the reperfused heart. Evidence for a free radical mechanism of reperfusion injury. J Biol Chem 1988;263(3):1353–7.

[55] Zweier JL, Kuppusamy P, Lutty GA. Measurement of endothelial cell free radical generation: evidence for a central mechanism of free radical injury in postischemic tissues. Proc Natl Acad Sci U S A 1988;85(11):4046–50.

[56] Tsao PS, Aoki N, Lefer DJ, et al. Time course of endothelial dysfunction and myocardial injury during myocardial ischemia and reperfusion in the cat. Circulation 1990;82(4):1402–12.

[57] Davies MJ, Thomas AC, Knapman PA, et al. Intramyocardial platelet aggregation in patients with unstable angina suffering sudden ischemic cardiac death. Circulation 1986;73(3):418–27.

[58] Vander Heide RS, Steenbergen C. Cardioprotection and myocardial reperfusion: pitfalls to clinical application. Circ Res 2013;113 (4):464–77.

[59] Hausenloy DJ, Yellon DM. Myocardial ischemia-reperfusion injury: a neglected therapeutic target. J Clin Invest 2013;123(1):92–100.

[60] Parmar KM, Larman HB, Dai G, et al. Integration of flow-dependent endothelial phenotypes by Kruppel-like factor 2. J Clin Invest 2006;116(1):49–58.

[61] Michel T. Targeting and translocation of endothelial nitric oxide synthase. Braz J Med Biol Res 1999;32(11):1361–6.

[62] Arnold WP, Mittal CK, Katsuki S, et al. Nitric oxide activates guanylate cyclase and increases guanosine 3′:5′-cyclic monophosphate levels in various tissue preparations. Proc Natl Acad Sci U S A 1977;74(8):3203–7.

[63] Radomski MW, Palmer RM, Moncada S. Endogenous nitric oxide inhibits human platelet adhesion to vascular endothelium. Lancet 1987;2(8567):1057–8.

[64] Lefer DJ, Jones SP, Girod WG, et al. Leukocyte-endothelial cell interactions in nitric oxide synthase-deficient mice. Am J Phys 1999;276(6 Pt 2):H1943–50.

[65] Schwartz GG, Olsson AG, Ezekowitz MD, et al. Effects of atorvastatin on early recurrent ischemic events in acute coronary syndromes: the MIRACL study: a randomized controlled trial. JAMA 2001;285(13):1711–8.

[66] Spencer FA, Allegrone J, Goldberg RJ, et al. Association of statin therapy with outcomes of acute coronary syndromes: the GRACE study. Ann Intern Med 2004;140(11):857–66.

[67] Dupuis J, Tardif JC, Cernacek P, et al. Cholesterol reduction rapidly improves endothelial function after acute coronary syndromes. The RECIFE (reduction of cholesterol in ischemia and function of the endothelium) trial. Circulation 1999;99(25):3227–33.

[68] Satoh M, Takahashi Y, Tabuchi T, et al. Cellular and molecular mechanisms of statins: an update on pleiotropic effects. Clin Sci (Lond) 2015;129(2):93–105.

[69] Pasterkamp G, den Ruijter HM, Libby P. Temporal shifts in clinical presentations and underlying mechanisms of atherosclerotic disease. Nat Rev Cardiol 2017;14(1):21–9.

[70] Libby P. Superficial erosion and the precision management of acute coronary syndromes—not one size fits all. Eur Heart J 2017;38 (11):801–3.

Further Reading

[1] Bogaty P, Hackett D, Davies G, et al. Vasoreactivity of the culprit lesion in unstable angina. Circulation 1994;90(1):5–11.

[2] Chambers DE, Parks DA, Patterson G, et al. Xanthine oxidase as a source of free radical damage in myocardial ischemia. J Mol Cell Cardiol 1985;17(2):145–52.

32

Endothelium: A Coordinator of Acute and Chronic Inflammation

Marina Beltrami Moreira, Guillermo Garcia-Cardeña,
Marco Aurélio Lumertz Saffi, and Peter Libby

INTRODUCTION

The vascular endothelium—a multifaceted monolayer—furinishes the interface between the vessels and blood. It acts as a gateway for the passage of oxygen, nutrients, and inflammatory cells, maintains homeostasis, and mediates many responses to injury or infection. The healthy endothelium participates in many ways to maintain homeostasis. It provides basal vasodilation through production of nitric oxide—the endothelium-derived relaxation factor. Nitric oxide can also inhibit platelet aggregation, limit the proliferation of smooth muscle cells, and combat inflammation. In their basal state, endothelial cells have anticoagulant and fibrinolytic properties. This balance derives partially from molecule expression on the cell surface; these structures act to control adhesion, cellular proliferation, and vasomotion (Fig. 32.1). Together, these effects combat atherothrombosis [1,2].

Under homeostatic circumstances, the arterial endothelium does not attract or adhere to leukocytes that pass by it in the circulating blood. But, when the endothelial cell recognizes an injury signal, it can display crucial properties that protect against infections and injury—but which can also cause damage to the organism and foster diseases. Pathologic mediators that induce this switching of endothelial functions include pathogen-associated molecular patterns (PAMPs), danger-associated molecular patterns (DAMPs), and other small and large molecules that can mediate inflammation. The ability of endothelial cells to exhibit these functions in response to such stimuli yield a state frequently referred to as "endothelial activation." Activated endothelium exhibits fewer anticoagulant and profibrinolytic properties, and augments functions that promote thrombosis and inhibition of fibrinolysis. The normal oxidative balance, in which antioxidant agents prevail over prooxidant agents, becomes perturbed. Oxidative species can predominate, mainly due to reduced superoxide dismutase production by the endothelial cells. Excess oxygen reactive species can contribute to reduced nitric oxide bioavailability. Moreover, instead of resisting leukocyte recruitment, the activated endothelium expresses adhesion molecules that promote rolling and ultimately firm adhesion of several leukocyte types to its surface.

Endothelial cells can also produce chemokines and other chemoattractant substances that, along with those produced by underlying tissue cells, guide migration and accumulation of leukocytes, promoting inflammation of the involved tissues. The endothelium thus serves as the first line of defense against pathogen invasion. Once activated, it plays a key role in triggering a local inflammatory response; activated endothelium can promote vasoconstriction, oxidative stress, and thrombus formation and persistence. Several stimuli can unleash this endothelial activation. Examples of PAMPs include bacterial products such as Gram-negative endotoxin and Gram-positive wall components. Examples of DAMPs include products of dying or dead cells.

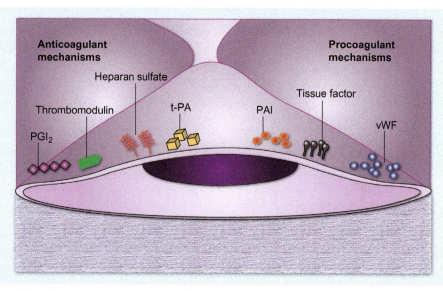

FIG. 32.1 Endothelial cells express molecules involved in anticoagulant and fibrinolytic mechanisms (left) and molecules involved in procoagulant and antifibrinolytic mechanisms (right). PGI₂, prostacyclin; t-PA, tissue plasminogen activator; PAI, plasminogen activator inhibitor; vWf, von Willebrand factor. Adapted from Bonow RO, Mann DL, Zipes DP, et al. Braunwald's heart disease: a textbook of cardiovascular medicine, vol I; 2011. p. 1949.

Counteracting the proinflammatory stimuli that endothelial cells encounter, two major factors promote and restore the endothelial cells' homeostatic function after perturbations: biomechanical stimuli and the proinflammatory and antiinflammatory/proresolving balance. In most physiologic circumstances, arterial endothelium experiences high-laminar shear stress levels: the force exerted by the blood flowing in parallel to the endothelial surface. In regions with blood flow disturbances (such as arterial branch points or flow dividers), endothelial cells lose the environmental cues conferred by high-laminar shear stress, which promotes the previously described homeostatic functions [3,4]. For further details on this topic, see Chapter 4.

At the molecular level, flow disturbances can activate nuclear factor kappa B (NF-κB), a proinflammatory transcriptional hub that regulates expression of multiple genes involved in this process [5]. This transcription factor may inhibit the expression of antiinflammatory factors such as the Krüppel-like factor 2 (KLF 2) and Krüppel-like factor 4 (KLF 4), and stimulate the expression of proatherogenic cellular receptors. In contrast, increased KLF 2 expression can stimulate nitric oxide production, inhibit IL-1 beta, and mitigate proinflammatory cell adhesion to human endothelial cells in culture [6]. In addition, in arterial regions with uniform laminar flow—atheroprotected regions—higher KLF 2 expression promotes antiinflammatory and

antithrombotic effects that may limit progression and complication of atherosclerotic lesions [7].

In addition to biochemical factors, the pro- and antiinflammatory cytokine balance critically determines endothelial cell properties. Today, we also recognize the importance of low-molecular weight mediators that promote active inflammation healing and homeostasis recovery once endothelial cells are activated. Such antiinflammatory and proresolving mediators prevent the inflammatory response's unrestrained expansion (Table 32.1).

TABLE 32.1 Endothelial "Proactivation" vs. "Proresolution" Factors

Activation promoters	Homeostasis promoters
PAMPs	*Mechanical factors*
Endotoxins	Shear stress, laminar flow
Lipopolysaccharides	*Biochemical factors*
Mycolic acid	Antiinflammatory cytokines: IL-10
Other microorganism cell wall components	Inflammation healing lipid mediators: resolvins, maresins
Exotoxins	
DAMPs	
Necrotic and pyroptotic cell content: ADP, IL-1α	

For abbreviations, please refer to the text.

Endothelial activation cell programs and their counter-regulatory mechanisms can be prompt and short-lived or sustained. Thus, the endothelium participates in both acute and chronic inflammatory conditions. We provide two examples commonly encountered in clinical practice to illustrate endothelial cell versatility as guardians of homeostasis and as key players in disease mechanisms. We shall first consider septic shock as an example of the acute scenario in which endothelial cells exert a major influence. To illustrate the role of endothelial cells in chronic disease, we will then consider their contribution to atherogenesis, atheroma progression, and thrombotic complications. For a detailed description of this process, please refer to Chapter 33.

ACUTE ACTIVATION OF ENDOTHELIAL CELLS IN SEPTIC SHOCK

Septic shock prevails as a frequent condition in contemporary medicine, particularly among the elderly, patients with comorbidities, patients with devices implanted in the cardiovascular system, and immunocompromised individuals. Regardless of the focus or microorganisms that cause the infection, PAMPs released in the blood circulation can activate endothelial cells. This activation promotes sepsis complications, leading to septic shock and in extreme cases multiple organ failure and death.

Endotoxins produced by Gram-negative bacteria are a classic example of PAMPs. Lipopolysaccharides and other constituents of the cellular wall of these microorganisms mediate their endotoxic effects: they stimulate endothelial cells through pattern-recognition receptors such as toll-like receptor 4 (TLR4) or toll-like receptor 2 (TLR2) to produce pro-inflammatory cytokines such as tumor necrosis factor (TNF), interleukin-1 (IL-1β), and interleukin-6 (IL-6), in addition to stimulating an antimicrobial response by way of NF-κB activation. These effects on activation of the innate immune system and TLR signaling not only contribute to atherogenesis, but also to atheromatous plaque destabilization [8,9]. Experiments in animals that received exogenous endotoxin administration demonstrated that the neutralization of these proinflammatory cytokines can attenuate deleterious effects and associated mortality. The most lethal manifestations of endotoxemia largely result from the downstream action of proinflammatory cytokines on the endothelium.

The complement system is part of the initial response against pathogens and can be activated by PAMPs. The complement system also participates in antibody-mediated immunity against both pathogens and autoantibodies. Complement and endothelium interact in a continuous manner. Under inflammatory stimuli, the endothelium can synthesize C3, C4, and factor B. Endothelial cells also express receptors for complement components, including C1q, C5a, and terminal complex C. The engagement of these receptors induces expression and/or relase of adhesion molecules, inflammatory cytokines, von Willebrand factor, tissue factor, etc. This response suggests that the complement system can directly influence endothelial functions and can produce cytotoxic damage to endothelial cells [10].

Imbalance in endogenous pro- and anticoagulant pathways promote thrombus formation in blood vessels. For example, increased expression of tissue factor (a powerful trigger of thrombosis) and decreased expression of tissue factor pathway inhibitor and protein C can promote thrombus formation. Concomitant increases in plasminogen activation inhibitor-1 (PAI-1) contributes to the stability of thrombi by inhibiting thrombolysis. Ischemic complications of septic shock may occur due to formation of microthrombi in target organs. Some examples of clinical manifestations of this dysregulated hemostasis include renal dysfunction due to glomerular or tubular damage, adrenal gland necrosis (Waterhouse-Friedrichsen syndrome, in particular in meningococcemia), and disseminated intravascular coagulation [11].

Endothelial activation also leads to leukocyte recruitment, in particular of acute inflammatory cells, notably neutrophils (Fig 32.2). Activated endothelial cells express adhesion molecules on their surface; polymorphonuclear leukocytes (PMN, neutrophils) bind to E-selectin, among other adhesion molecules. The local expression of chemokines such as IL-8 can spur the penetration of the bound inflammatory cells into tissues and/or cause leukocyte congestion in microvessels.

PMNs associated with the endothelial monolayer can generate neutrophil extracellular traps (NETs). Genetic material, chromatin, deiminated histones, and other proteins form a network that captures bacteria, more effectively exposing them to bactericidal substances such as reactive oxygen species

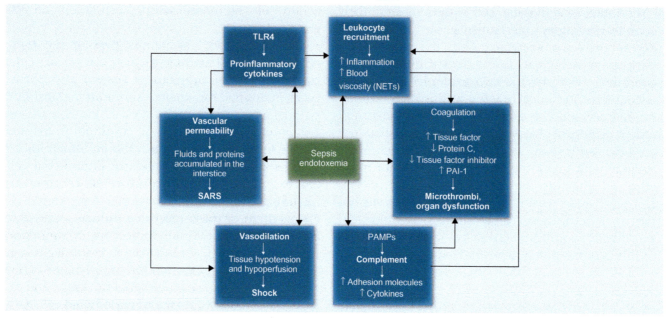

FIG. 32.2   Inflammatory mechanisms in endotoxemia and in sepsis. The endothelium takes part in the combat against infectious insults. However, the exacerbated response culminates in many clinical complications of severe sepsis. TLR4, toll like receptor 4; SARS, severe respiratory distress syndrome; NETs, neutrophil extracellular traps; PAI-1, plasminogen activator inhibitor-1; PAMPs, pathogen-associated molecular pattern.

produced by PMNs. In particular the DNA backbone of NETs binds myeloperoxidase released from PMN granules and provides a localized "solid state reactor" that produces the potent prooxidant hypochlorous acid (HOCl). NETs can further entrap leukocytes and platelets contributing to thrombosis. Histone concentration in the peripheral bloodstream has been associated with sepsis severity. in vitro observations suggest that NET production depends on PMN adhesion, highlighting the endothelium's importance in this process [12–14].

The activated endothelium, particularly when subjected to noxious produts of recruited leukocytes, develops impairment in the normal barrier function of these intimal lining cells. Excess endothelium permeability leads to fluid and protein extravasation to the interstitium or alveolar space. These phenomena promote the adult respiratory distress syndrome, and contribute to the hypoxemia that often accompanies hypotension and vasoplegia characteristic of septic shock. This clinical scenario illustrates how the dreaded effects of sepsis derive not only from the infectious agent, but also from altered functions of the micro- and macrovascular endothelium.

THE ENDOTHELIAL CELL IN A CHRONIC INFLAMMATORY STATE: ATHEROSCLEROSIS

In contrast with the acute inflammatory situation in sepsis described above, the consequences of chronic endothelial activation—as in atherosclerosis—develop over many decades. While in sepsis PMNs comprise the main leukocyte type recruited by endothelial cells, during atherogenesis inflammation mononuclear leukocytes (monocytes) prevail. The adhesion molecule most prominent in mediating monocyte recruitment by the endothelium is vascular cell adhesion molecule-1 (VCAM-1). Other adhesion molecules, such as fractalkine and integrins, also take part in monocyte recruitment during the development of atheromata. The stimuli that activate adhesion molecule expression in endothelial cells include products associated with lipoprotein oxidation, as well as proinflammatory cytokines produced in the arterial neointima. The production by endothelial cells of chemoattractant cytokines such as CCL2 cause the entry into the intima of bound monocytes, where they can mature into macrophages, imbibe lipids, and form foam cells. Chapter 36 contains further information on the interaction between the endothelium and immune system.

Monocytes and macrophages predominate in the innate immune system's response during atherogenesis. These cells cluster in the intima beneath the endothelial monolayer, and can eventually undergo apoptosis or oncosis. Together with their defective clearance, these mononuclear cells recruited by the endothelium typically form a "necrotic" lipid-rich core of the atheromatous lesion. Monocytes/macrophages contribute to atheroma plaque growth by stimulating chemotaxis of smooth muscle cells from the tunica media into the intima. These mesenchymal cells can form a fibrous cap beneath the endothelial monolayer that overlies the lesion's necrotic/lipid core. The plaque's volumetric growth, lipid core formation, and fibrous cap formation characterize advanced atherosclerotic lesions (Fig. 32.3).

Classic studies demonstrated endothelial barrier dysfunction at locations of the arterial tree subject to disturbed flow or low shear stress, such as at flow dividers and bifurcations where atherosclerotic lesions typically develop. In experimental atherosclerosis, these regions display extravasation of Evans Blue (a dye that binds to albumin), thus disclosing the endothelium's locally excessive

permeability. Such breakdown in the endothelial barrier function facilitates the accumulation of lipoproteins within the intima. Oxidized lipoproteins and crystallized cholesterol, among other substances that can accumulate in the arterial intima, can amplify the inflammatory response. Therefore, the dysfunctional endothelium provides one of the main triggers to atherogenesis and unleashes the cellular and molecular events that give rise to the lesions, promote their progression, and ultimate complications, as also discussed below and in Chapter 41.

In addition to responding to stimuli that augment adhesion molecule expression and chemokine production, endothelial cells can also participate in antiinflammatory cytokine expression, which contributes to control of inflammation. For example IL-10 can inhibit macrophage and T helper 1 cells (Th1) inflammatory functions, among others. These T helper 1 cells (Th1), when activated, secrete interferon-gamma, a proatherogenic stimulus. B cells can also exhibit pro- or antiatherogenic actions. Thus, while B2 cells promote experimental atherogenic effects, B1 cells secrete natural antibodies that, at least in mice, mitigate the atherosclerotic process. In addition to peptide mediators, several small lipid molecules actively promote resolution of the inflammatory process by disabling

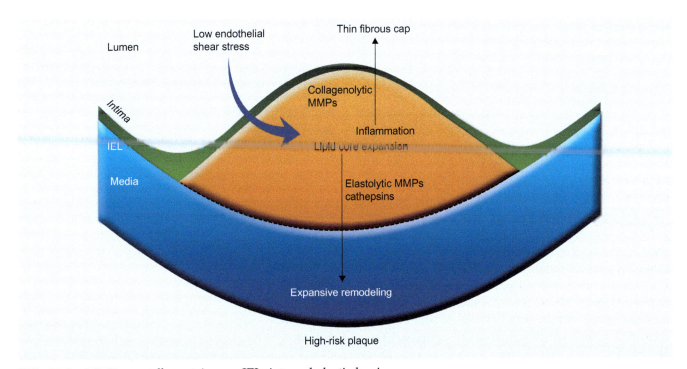

FIG. 32.3 MMPs, metalloproteinases; IEL, internal elastic lamina.

proinflammatory cell functions. Examples include lipoxins, resolvins, and protectins (Table 32.1). An imbalance between pro- and antiinflammatory pathways, the failure of resolution of inflammation, and accumulation of monocyte/macrophages in the arterial intima can all promote atherosclerotic lesion progression over time.

Endothelial cells participate pivotally in certain acute thrombotic complications of atherosclerosis. Although plaques progress insidiously over many years, the thrombi they provoke usually occur suddenly. Plaque rupture and the exposure of its necrotic core to blood provide one path to unleashing the coagulation cascade. The same molecular mechanisms that incite thrombotic complications in small caliber vessels during sepsis, as discussed above, also promote thrombus formation and accumulation in the larger caliber vessels that develop atherosclerosis.

Beyond plaque rupture-related events, endothelial cells seem to have particular importance in superficial erosion—a mechanism of acute coronary syndromes whose frequency we have argued has risen recently as a proportion of total coronary events [15]. TLR2 activation on the endothelial surface can elicit functions that promote local thrombosis in atherosclerotic plaque erosion conditions. PMNs and NETs likely participate in superficial erosion [16–19]. Thus, endothelial functions span from the alpha to the omega of atherosclerosis: pivotal in the initial steps of lesion formation, and often triggers to the thrombotic complications that cause the acute manifestations of this chronic disease.

CONCLUSIONS

Until a few decades ago, many viewed the endothelium as a passive structure with a semipermeable membrane function, acting as a passive filter between blood and tissue [20,21]. We now recognize the key role of endothelial cells in acute and chronic diseases, as depicted in this chapter. Endothelial dysfunctions provide targets for intervention. Several endothelial-leukocyte adhesion molecule inhibitors have undergone experimental evaluation in the context of sepsis, ischemic-reperfusion injury, and arterial intervention [22,23]. Such interventions may not apply to the prevention or treatment of chronic diseases that involve long-term endothelial dysfunction, such as atherosclerosis. Nonetheless,

the benefit of lifestyle changes that improve insulin sensitivity and limit pro-inflammatory stimuli (excess abdominal fat, insulin resistance, and high triglyceride rich lipoproteins concentrations, etc.) may result in part from the effects of these interventions on endothelial health. The multitudinous roles of vascular endothelial cells in health and disease offer fertile fields for further investigation and an attractive target for new therapies to maintain vascular homeostasis and combat cardiovascular diseases.

References

[1] Libby P, Okamoto Y, Rocha VZ, et al. Inflammation in atherosclerosis: transition from theory to practice. Circ J 2010;74(2):213–20.
[2] Tousoulis D, Kampoli AM, Tentolouris C, et al. The role of nitric oxide on endothelial function. Curr Vasc Pharmacol 2012;10(1):4–18.
[3] Gimbrone MA, García-Cardeña G. Vascular endothelium, hemodynamics, and the pathobiology of atherosclerosis. Cardiovasc Pathol. 2013 Jan-Feb 2013;22(1):9–15.
[4] Chatzizisis YS, Coskun AU, Jonas M, et al. Role of endothelial shear stress in the natural history of coronary atherosclerosis and vascular remodeling: molecular, cellular, and vascular behavior. J Am Coll Cardiol 2007;49(25):2379–93.
[5] Hajra L, Evans AI, Chen M, et al. The NF-KB signal transduction pathway in aortic endothelial cells is primed for activation in regions predisposed to atherosclerotic lesion formation. Proc Natl Acad Sci U S A 2000;97(16):9052–7.
[6] SenBanerjee S, Lin Z, Atkins GB, et al. KLF2 is a novel transcriptional regulator of endothelial proinflammatory activation. J Exp Med 2004;199(10):1305–15.
[7] Parmar KM, Larman HB, Dai G, et al. Integration of flow-dependent endothelial phenotypes by Kruppel-like factor 2. J Clin Invest 2006;116(1):49–58.
[8] Hovland A, Jonasson L, Garred P, et al. The complement system and toll-like receptors as integrated players in the pathophysiology of atherosclerosis. Atherosclerosis 2015;241(2):480–94.
[9] Curtiss LK, Tobias PS. Emerging role of toll-like receptors in atherosclerosis. J Lipid Res 2009;50(Suppl):S340–5.
[10] Fischetti F, Tedesco F. Cross-talk between the complement system and endothelial cells in physiologic conditions and in vascular diseases. Autoimmunity 2006;39(5):417–28.
[11] Levi M, Poll T. Coagulation in patients with severe sepsis. Semin Thromb Hemost 2015;41(1):9–15.
[12] Gupta AK, Joshi MB, Philippova M, et al. Activated endothelial cells induce neutrophil extracellular traps and are susceptible to NETosis-mediated cell death. FEBS Lett 2010;584(14):3193–7.
[13] Neeli I, Dwivedi N, Khan S, et al. Regulation of extracellular chromatin release from neutrophils. J Innate Immun 2009;1(3):194–201.
[14] Xu J, Zhang X, Pelayo R, et al. Extracellular histones are major mediators of death in sepsis. Nat Med 2009;15(11):1318–21.
[15] Pasterkamp G, et al. Temporal shifts in clinical presentation and underlying mechanisms of atherosclerotic disease. Nat Rev Cardiol 2017;14(1):21–9.
[16] Knight JS, Luo W, O'Dell AA, et al. Peptidylarginine deiminase inhibition reduces vascular damage and modulates innate immune responses in murine models of atherosclerosis. Circ Res 2014;114(6):947–56.
[17] Fuchs TA, Brill A, Duerschmied D, et al. Extracellular DNA traps promote thrombosis. Proc Natl Acad Sci U S A 2010;107(36):15880–5.

[18] Semeraro F, Ammollo CT, Morrissey JH, et al. Extracellular histones promote thrombin generation through platelet-dependent mechanisms: involvement of platelet TLR2 and TLR4. Blood 2011;118(7):1952–61.

[19] Quillard T, Franck G, Mawson T, et al. Mechanisms of erosion of atherosclerotic plaques. Curr Opin Lipidol 2017;28:434–41.

[20] Rajendran P, Rengarajan T, Thangavel J, et al. The vascular endothelium and human disease. Int J Biol Sci 2013;9:1057–69.

[21] Libby P, Aikawa M, Jain MK. Vascular endothelium and atherosclerosis. Handb Exp Pharmacol 2006;176(Pt 2):285–306.

[22] Bonow RO, Mann DL, Zipes DP, et al. Braunwald's heart disease: a textbook of cardiovascular medicine. vol. I. Grune & Stratton Inc.; 2011 p. 1949.

[23] Kumar V, Abbas AK, Fausto N, et al. Robbins and Cotran pathologic basis of disease. Professional Edition Expert Consult-Online Internet; 2009.

II. ENDOTHELIAL DYSFUNCTION AND CLINICAL SYNDROMES

33

Endothelium in Atherosclerosis: Plaque Formation and Its Complications

Protásio Lemos da Luz, Antonio Carlos Palandri Chagas, Paulo Magno Martins Dourado, and Francisco R.M. Laurindo

INTRODUCTION AND HISTORICAL PERSPECTIVE

Atherosclerosis refers to the slow process of plaque formation on artery walls that may result in acute complications, such as myocardial infarction, unstable angina, and cerebrovascular accident. It is the pathophysiological process that is behind the main causes of death in the Western world. The word atherosclerosis derives from the combination of the Greek term *athēra* (oatmeal) and *esclerose* (hardening or thickening); the word refers to the fat deposits and cellular remains in artery inner walls, which become progressively thicker due to the proliferative response of the vessel wall and are frequently calcified [1].

Typically, the atherosclerotic process begins on the first decades of life and progresses slowly, according to the presence and level of several risk factors, such as arterial hypertension, dyslipidemia, *diabetes mellitus*, and tobacco use [2–4]. Aging and genetic susceptibility play an essential role on atherosclerosis evolution. The process tends to be diffuse, affecting different arterial beds, and asymptomatic for several decades. Frequently there is compromise of the aorta, carotids, peripheral arteries, femoral arteries, smaller arteries in lower limbs, as well as mesenteric and penile arteries.

Over time, the pathophysiological understanding of atherosclerosis went through several stages. Initially, it was considered a passive process of simple deposit of cholesterol and cell debris on arterial walls [5–9]. In the 1960s and 1970s, the role of smooth muscle cells and their proliferation in the atherogenic process was emphasized. The last decades witnessed major developments regarding the pathophysiological mechanisms involved in atherogenesis. Today, it is understood as an active and dynamic process that involves multiple and complex interactions between different cell types and subtypes, mediated by numerous cytokines, based on proinflammatory changes in the vascular wall microenvironment. In summary, the disease is inflammatory/proliferative [10–12].

Changes in endothelial cells or endothelial dysfunction play a central role in atherosclerotic plaque formation, its progression, and complications. Therefore, the main goal of this chapter is to provide concise integrated review of the main factors that take part in atherosclerotic plaque formation, with specific focus on the endothelium. Detailed reviews of the roles of these factors are found in other chapters herein or in other revisions [4,13].

ATHEROGENESIS OVERVIEW

Essentially, the atherothrombotic process undergoes the following stages:

- damage to endothelial cells and consequent dysfunction;
- increased endothelial permeability to macromolecules (e.g., low density lipoprotein [LDL]);

Endothelium and Cardiovascular Diseases
https://doi.org/10.1016/B978-0-12-812348-5.00033-7

- retention and modification of cholesterol-rich particles in the subendothelium (e.g., LDL oxidation);
- recruitment of inflammatory cells, mainly monocytes and T-lymphocytes;
- differentiation of monocytes and macrophages and foam cells formation;
- foam cell necrosis and apoptosis and lipid core formation;
- smooth muscle cell migration and proliferation, with consequent production of collagen and formation of fibrous tissue; and
- plaque rupture or erosion and thrombi formation.

All these stages are regulated by mediators able to signal different processes, stimulating or inhibiting adhesion, proliferation, and differentiation of several cell types, as well as interfering with cell secreting function. The presence of cells and inflammatory mediators is a notable characteristic in the entire atherogenic process, from nascent plaque formation to thrombotic complications.

Participation of genetic alterations has been recognized in up to 50% of cases [14]. However, the characteristics and magnitude of such alterations are still unclear and will not be discussed herein.

ENDOTHELIAL DYSFUNCTION AS EARLY EVENT

Senescence, increased blood pressure, lipid and glycemic alterations, and tobacco use are some of the factors that affect the functions of endothelial cells. In particular, mechanical forces influence the functions of these cells and signaling of inflammatory pathways. Uniform shear stress reduces the expression of inflammatory proteins and recruitment of leukocytes to the vascular wall, while low fluctuation shear stress has opposite effects. Signaling of inflammatory pathways is also sensitive to pulse wave frequency and magnitude [15].

Endothelial dysfunction is recognized as an early event in atherosclerotic plaque formation [16,17]. Morphological alteration of endothelial cells and increased permeability to macromolecules allow passive penetration of LDL in the subendothelial space. This permeability is modulated by cytokines, and is increased, for example, by interferon gamma (IFN-γ) and tumor necrosis factor alpha (TNF-α) [18–20]. Interactions between LDL apolipoprotein B and extracellular matrix components, such as

proteoglycans, collaborate with particle retention in the subendothelium.

One of the main functions of the endothelium is that of semipermeable barrier that separates the blood and arterial walls. Removal of the endothelium by physical means, which causes deendothelization of the aorta in vivo, exposed the entire arterial wall to massive infiltration of lipids in rabbits subjected to hypercholesterolemic diet, as illustrated in Fig. 33.1 [21].

The atherosclerotic process begins with the penetration of apo B-rich circulating particles among endothelial cells, which are retained in the subendothelial space and suffer oxidative and enzymatic modifications [22,23]. These activate the endothelium, which starts to produce cytokines such as monocyte chemotactic protein-1 (MCP-1) that attract monocytes to the subendothelium. The interaction between monocyte adhesion molecules and endothelial cells favors monocyte adhesion to the endothelium, settlement, and, following, penetration in the subendothelial region. There, they differentiate into macrophages, and then into foam cells [24] (see below). Fig. 33.2 illustrates the fundamental points of normal endothelial functions and its alterations in atherothrombosis [14].

There is no consensus regarding the chronology of events that lead to plaque formation. On the one hand, Tabas et al. [17] proposed that subendothelial retention of apo-B lipoproteins is the process that initiates atherosclerosis; they called it the "atherogenesis response-to-retention model." On the other hand, other authors such as Libby et al. [25] and Ross et al. [10] suggest that everything begins with endothelial dysfunction, considering inflammation, endothelial dysfunction, and oxidation as main initiating factors. Differences aside, all these phenomena are essential components of the atherosclerotic process.

CYTOKINES, ADHESION MOLECULES, AND MONONUCLEAR CELLS RECRUITMENT

The vascular wall inflammatory process is regulated by several mediators [25,26]. The term "cytokine" refers to the diversified group of more than 100 low molecular weight proteins identified so far, combined in several classes, such as interleukins (IL), chemokines, colony stimulating factors

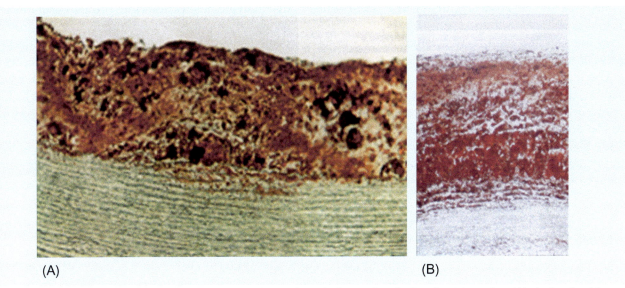

(A)   (B)

FIG. 33.1 Aorta histological section of rabbits subjected to cholesterol-rich diet for 14 weeks. (A) Animal with preserved endothelium and lipid infiltration restricted to the intimal layer; (B) Animal subjected to deendothelization before beginning the diet, with lipid deposits on the entire arterial wall [21].

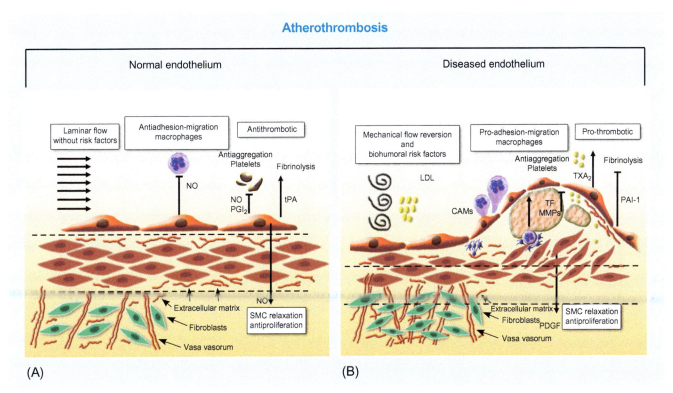

(A)   (B)

FIG. 33.2 Action of risk factors in the normal endothelium leading to endothelial dysfunction and development of atherothrombosis. (A) Normal endothelium under laminar flow conditions and without risk factors. Nitric oxide (NO) is involved in multifactorial pathways, which prevent monocyte adhesion, platelet aggregation, and smooth muscle cell proliferation. *PGI2*, prostacyclin 2; *SMC*, smooth muscle cells; *tPA*, tissue plasminogen activator. (B) Disease endothelium with nonlaminar flow, deposit of low density lipoproteins (LDL), expression of molecular adhesion cells (MAC), macrophage migration, tissue factor (TF), and expression of matrix metalloproteinases (MMP), causing smooth muscle cell proliferation and vasa vasorum neovascularization. *PDGF*, platelet-derived growth factor; *PAI-1*, plasminogen activator inhibitor-1; *TXA2*, thromboxane A2. Adapted from Fuster V, Moreno PR, Fayad ZA, et al. Atherothrombosis and high-risk plaque. J Am Coll Cardiol 2005;46:937–54.

(CSF), tumor necrosis factor (TNF), interferons (IFN), and transformation growth factors (TGF). They are secreted both by inflammatory and vascular wall cells [18–20].

The term "chemokine" refers to a large family of structurally related cytokines with chemoattractant properties [19]. They are divided in subgroups based on the position of N-terminal cysteine residues (CC, CXC, CX3C, XC). Expressed by vascular wall cells and emigrated leukocytes, they interact with receptors

that activate heterotrimeric G proteins and intracellular signaling pathways. Originally, chemokines were related to recruitment of leukocytes to inflammation sites; today, they are known to have properties that go beyond cell recruitment, interfering in vascular homeostasis and foam cell formation. Fig. 33.3 illustrates the several functions of chemokines in atherosclerosis [19].

In the context of endothelial dysfunction, activated endothelial cells express adhesion molecules, such as

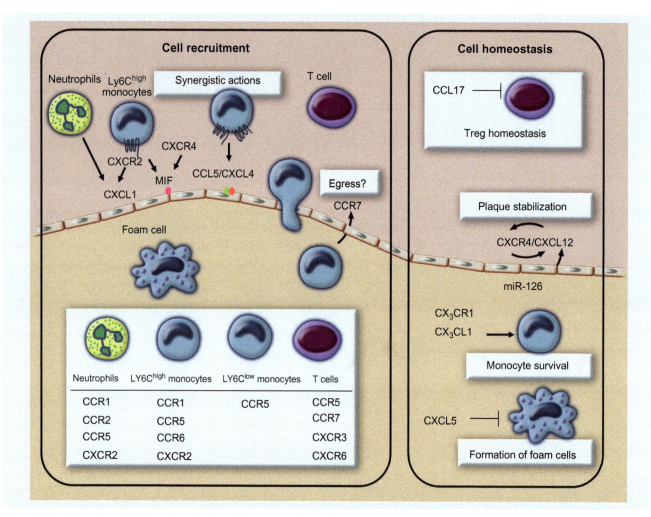

FIG. 33.3 Functional diversity of chemokines in atherosclerosis. Chemokines direct leukocytes towards inflammation sites, through chemokine receptors (see box on the left). Lysophosphatidic acid, a component of low density lipoproteins, induces endothelial CXC binder chemokine (CXCb) 1 to mobilize and recruit monocytes and neutrophils by means of chemokine receptors CXC (CXCR) 2. Under certain conditions, CCRT may mediate macrophage outflow. The migration inhibition factor (MIF) interacts directly with both CXCR2 and CXCR4. Synergistic interactions between CCL5 and CXCR4 in monocyte cell recruitment summarize the concept of functional chemokine interaction. In addition to recruitment, chemokines have homeostasis functions. CCL17 restricts regulating T cells (Tregs) to promote atherosclerosis. CXCL12 may be induced by microRNA (MIR) 126 in endothelial cells to stabilize atherosclerotic lesions. CX3CR1 provides essential survival signals to monocytes/macrophages and promotes protection against apoptosis. CXCL5 limits macrophage foam cells formation. Adapted from Zernecke A, Weber C. Chemokines in atherosclerosis: proceedings resumed. Arterioscler Thromb Vasc Biol 2014;34:742–50.

intercellular adhesion molecule-1 (ICAM-1), vascular cell adhesion molecule-1 (VCAM-1), and selectin, which interact with circulating leukocyte receptors, particularly monocytes and T-lymphocytes, recruiting them to the endothelium [27] (see Chapter 14).

Once adhered to the endothelial surface, mononuclear cells enter the intimal layer under the influence of mediators, for example, MCP-1. Accordingly, the absence of MCP-1 reduces atheroma plaque formation in experimental models [28]. Different types of monocytes also use different combinations of chemokines/chemokine receptors to infiltrate the intima [29–31]. In mice atherosclerosis models, 80% of the monocytes recruited are of the GR1 +LY6C^{hi} subtype, and the rest are of the GR1– LY6C^{low} subtype (Fig. 33.3).

GR1+LY6C^{hi} monocytes are related to differentiation in macrophages with proinflammatory profile (see below), and are elevated in the presence of hyperlipidemia. Chemokines induced by IFN-γ contribute to selective recruitment of lymphocytes to the nascent atherosclerotic plaque [19].

MONOCYTES AND ENDOTHELIUM

Monocytes are mononuclear cells characterized mainly by the expression of several surface receptors, representing the innate immune system, which play essential roles in atherosclerosis initiation, propagation, progression, and complications [29,31]. An important characteristic of monocytes is their diversity. Studies involving expression of different adhesion molecules identified two monocyte subpopulations based on surface receptors. Cells that produce CD14+/CD16—represent 80%– 90% of the "classic" circulating monocytes and express high levels of CCR2 (MCP-1 receptor) and low levels of CX-3CR1 (fractalkine receptor) [32–34]. When exposed to lipopolysaccharides (LPS)—important stimulators for the production and expression of inflammatory chemokines—they are weak producers of inflammatory cytokines. The monocytes that express CD14+/CD16+ receptors in their membrane represent 10%–20% of the "nonclassic" circulating monocytes, which are strong producers of CX3CR1, responsible for expressing most inflammatory cytokines and the largest contributor of atheroma expansion [29,30,32]. More recently, a third class has been described, characterized by CCr2 expression on the surface. During its patrolling function in circulation, monocytes are

activated by different factors and act on the damaged endothelium, participating in the atherosclerosis process in three distinctive moments [34]. First, during the long atherosclerotic plaque initiation and formation stage, probably accelerated by different factors such as tobacco use, hypertension, hyperglycemia and, more significantly, hyperlipidemia. During this stage, monocytes move towards the injured endothelium. The dysfunctional endothelium hyperexpresses MCP-1, VCAM-1, and ICAM-1 on its surface. After rolling and adhering to the endothelium, the monocytes cross the endothelial surface, determining the leukocyte extravasation phenomenon. In addition, neural factors that participate in monocyte recruitment have been described. Therefore, molecules of the netrin, semaphorin, and ephrin families [13,35,36] are expressed in endothelial cells and promote or protect from atherosclerosis by acting on arterial wall homeostasis. In the subendothelial space, monocytes proliferate and originate daughter cells or differentiate into macrophages by means of the macrophage colony stimulating factor (MCSF). Macrophages incorporate oxidized LDL through scavenger receptors A1, B1, CD36, as well as specific LDLox receptor LOX1 [12,37], thus forming foam cells. These undergo an apoptosis/necrosis process, which perpetuates the formation of new lipid-rich macrophages. Then, the monocytes participate in the acute inflammatory phase, which corresponds to clinical syndromes of unstable angina or myocardial infarction, during which the atherosclerotic plaque becomes unstable, with rupture and thrombus formation.

Monocytes also contribute to propagate thrombi; as they adhere to the extracellular matrix and induce matrix metalloproteinases, spreading out to the injured tissue, they promote expression of several cytokines such as TNF, IL-6, and IL-1, all potent inflammatory agents. Monocytes participate in tissue hypoxia lesion in the ischemia/perfusion cascade, especially through of reactive oxygen species (ROS) originated in mitochondria; ROS stimulates the expression of inflammatory molecules, as proposed by the "innate autoimmunity" model, largely represented by monocytes [34].

Later on, during the healing period, monocytes residing in myocardial tissue during the hypoxia phase, which occurs during acute coronary events, promotes myofibroblast accumulation, myocardial healing, and remodeling. Therefore, they also play a relevant role during the healing phase [33]. Fig. 33.4 illustrates monocyte participation in plaque formation.

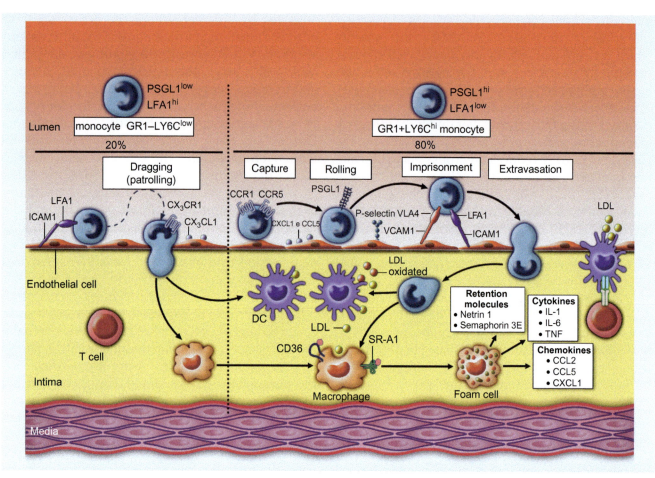

FIG. 33.4 Participation mechanism of monocytes in plaque formation. Hyperlipidemia increases the number of GR1+LY6C^{hi} monocytes, which constitute 80% of the monocytes recruited by atherosclerotic plaque of rats, with the remainder carried on by patrolling monocytes GR1–LY6C^{low}. These monocyte subgroups use different pairs of chemokine-chemokine receptors to infiltrate the intima, which is favored by endothelial adhesion molecules that include selectins, ICAM-1, and VCAM-1. In the intima, the monocytes recruited differentiate into macrophages or dendritic cells (DC) and interact with atherogenic lipoproteins. Macrophages absorb native and modified LDL through pathways mediated by the macropinocytosis or scavenger receptor (including the pathway of scavenger receptors A1 (SR-A1) and CD36), which results in foam cells formation that constitute the base of atherosclerotic plaque. These foam cells secrete proinflammatory cytokines, including IL-1, IL-6, and TNF, and chemokines (e.g., chemokine ligand CC-2 [CCL2], CCl5, and chemokine ligand CXC [CXCL1]), as well as macrophage retention factors (such as netrin 1 and semaphorin 3E), which enhance inflammatory response. *CX3CL1*, chemokine ligand CX3C; *CX3CR1*, CX3C chemokine receptor 1; *LFA1*, lymphocyte function associated with antigen 1; *PSGL1*, P-selectin glycoprotein ligand-1; *VLA4*, very late antigen 4. Adapted from Jaipersad AS, Lip GYH, Silverman S, et al. The role of monocytes in angiogenesis and atherosclerosis. J Am Coll Cardiol 2014;63:1–11.

MACROPHAGES

Once adhered to the endothelium, monocytes migrate to the subendothelial region, passing through endothelial cells, by interacting with monocyte receptor CCR2. MCP-1, a powerful chemotactic agent, plays an important role in monocyte migration. As seen before, in the intima the monocytes differentiate into macrophages or dendritic

cells, and form foam cells. Macrophages are heterogeneous cells, and their subpopulations are not entirely known [37–40] (Fig. 33.5).

Role of Different Macrophages

In principle, monocytes recruited to inflammation sites must contribute to resolve the process, removing cells from the location, performing apoptotic cell

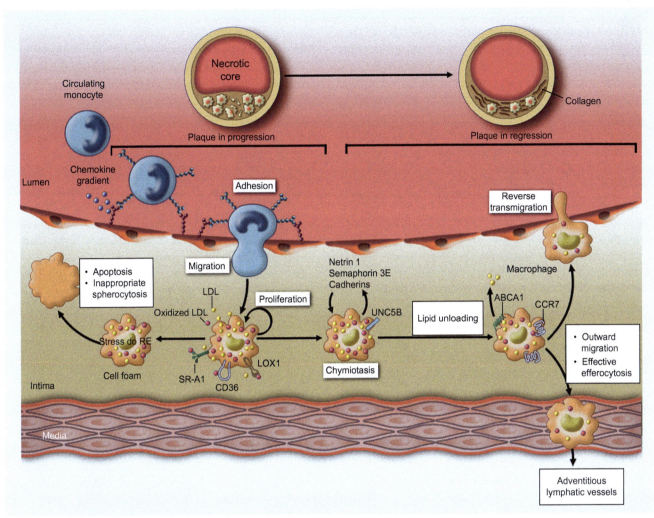

FIG. 33.5   Disregulated lipid metabolism in macrophages during plaque progression leads to macrophage retention and chronic inflammation. Macrophages overloaded with lipids express retention molecules (such as netrin and its receptor UNC5b, semaphorin 3E, and cadherins) that promote macrophage chemostasis. In this inflammatory medium, these macrophages sufferendoplasmic reticulum (ER) stress that, if prolonged, results in apoptosis. This cell death, together with defective spherocytosis, causes necrotic core formation. The mechanisms that promote lipid unloading from foam cells, including factors that hyperregulate the expression of ABCA1 in plaque macrophages, and cholesterol flow, revert the accumulation of foam cells. Plaque regression is characterized by hyperregulation of chemokine cc receptor 7 (CCR7) in cells of myeloid origin and reduced expression of retention factors. Therefore, there is evidence that factors that regulate macrophage migration contribute to macrophage migration from the plaque to the lumen or adventitious lymphatic vessels. Adapted from Moore KJ, Sheedy FJ, Fisher EA. Macrophages in atherosclerosis: a dynamic balance. Nature 2013;13:709–21.

spherocytosis, and limiting accumulation of new immune cells. These homeostatic mechanisms are impaired in atherogenesis, and the final result is continuous monocyte recruitment, reduced cell removal, reduced spherocytosis (renewal of dead cells), and permanent inflammatory process [41,42].

Furthermore, recent evidence indicates that the macrophage replacement time in mice atherosclerosis models is short, ~4 weeks, and replacement depends predominantly on local macrophage proliferation and not on monocyte influx. Such local macrophage proliferation involves the participation of scavenger receptor A (SR-A) [43].

Studies in vitro demonstrate that macrophages in atherosclerotic plaque do not behave uniformly, but may acquire different phenotypes. Therefore, macrophages can become large foam cells or small inflammatory cells. Such polarization and plasticity is influenced by the local microenvironment, including modified lipids, cytokines, and senescent erythrocytes. The diversity of macrophages in atherosclerotic plaque seems to be a dynamic phenomenon, and might influence the type of plaque, its development, progression, and possibly the risk of complications [43–45].

Classically, proinflammatory macrophages, M1, and antiinflammatory macrophages, M2, are recognized, but this differentiation seems to be more complex and such dichotomy is probably simplistic. Macrophage subtype M1 supposedly derives from LY6C^{high} monocytes and produces proinflammatory cytokines, such as IL-6, IL-12, and TNF-α. On the other hand, M2 macrophages seem to be derived from LY6C^{low} monocytes and secrete antiinflammatory cytokines, such as IL-10 and TGF-β, which could help inflammation resolution.

Cytokines derived from T-helper lymphocytes (Th-1), such as IFN-γ and IL-1β, favor macrophage differentiation into M1, while Th-2 cell cytokines, such as IL-4 and IL-3, are required to differentiate macrophages into M2 [46].

Monocytes differentiation into macrophages and respective proliferation are largely due to the action of M-CSF. Foam cells originate from macrophages or from SMC that migrate to the subendothelial space, as described before. On the other hand, fatty cells produce cytokines, which maintain the stimulus to attract leukocytes, promote macrophage replication, and increase the expression of removal receptors [40,42]. Once internalized, lipoproteins are hydrolyzed, and transformed into free cholesterol and fatty acids, and then eliminated by ABCA-1 receptors or by passive diffusion. The necrotic core observed in advanced lesions results from macrophage apoptosis, necrosis or defective phagocytosis (spherocytosis); dying cells release apoptotic bodies and microparticles, which form the potent procoagulant factor called tissue factor that is essential to form thrombi [12]. The balance between apoptosis/necrosis and spherocytosis is critical to stabilize lesions [44].

LDL MODIFICATIONS, FOAM CELLS FORMATION, AND PATHOPHYSIOLOGICAL IMPLICATIONS

Monocytes recruited to the intima may differentiate into macrophages or dendritic cells, according to cytokine signaling. Differentiation into macrophages is promoted by macrophage colony stimulating factors, for example, Ref. [45]. Most white cells found in atheroma are represented by the macrophage population. Macrophages express a series of standard recognition receptors, such as scavenger receptors, Toll-like (TLRs) receptors, and nucleotide-binding oligomerization domain (NOD)-like (NLR) receptors [47,48]. They include receptors that recognize apoptotic cells, which allows macrophages to exercise phagocytic action and contribute to atheroma necrotic core formation [48–51]. Mitochondrial dysfunction due to overload of free cholesterols in macrophages is one of the pathway that leads to apoptosis [48].

LDL that accumulates in the subendothelial space may undergoes modifications, such as oxidation, acetylation, and aggregation, and then becomes a foreign and antigenic element in the organism, triggering innate and adaptive immunological responses that are the essence of the atherogenesis inflammatory process. In particular, LDL retention and oxidative modification are crucial events for the onset of plaque formation.

LDL oxidation may occur due to the presence of oxidants derived activated endothelium, transition metals (e.g., bivalent iron cations), hemoglobin heme radicals, and enzymes such as lipoxygenases, myeloperoxidases, NADPH oxidases, and nitric oxide synthases [48,49,51].

In contrast to cell capture of native, unmodified LDL by the LDL receptor that is regulated by a negative feedback mechanism, oxidized LDL internalization in macrophages occurs without regulation, leading to great lipid accumulation inside the cells and formation of the so-called foam cells. In addition to the importance of oxidized LDL, minimally modified LDL induces genes that lead to proinflammatory modifications in macrophages [52,53].

The LDL retained and modified on the vascular wall is recognized and avidly internalized, mainly by macrophage scavenger receptors, such as SRA1, LOX1, SRB1, and CD36. Other mechanisms, such as pinocytosis, phagocytosis, and macropinocytosis, also contribute to lipoprotein influx, including native and modified LDL37 (Fig. 33.5).

Once inside the macrophage, lipids proceeding from lipoproteins are digested by lisosomes, resulting in discharge of free cholesterol that can be directed to the plasma membrane and suffer cell efflux or to the endoplasmic reticulum membrane, where it is esterified by acyl-cholesterol acyltransferase

(ACAT) and stored as lipid droplets in cytosol. This stored fat can also be directed to efflux. Excessive cholesterol in the cell also activates intracellular signaling pathways involved in free cholesterol transport to lipid poor apolipoprotein A-1 (APOA-I) forming nascent HDL, or to HDL particles that already have lipids, forming mature HDL [54]. Although most foam cells originate from mononuclear phagocytes, it is known that smooth muscle cells and endothelial cells also can accumulate lipids [17]. Recent evidence indicates that lipid accumulation inside cells may occur even in circulation: foam monocytes may be formed early in the blood of hypercholesterolemic mice and infiltrate the nascent atherosclerotic plaque.

Foam cells formation is also modulated by several cytokines. IFN-γ and TNF-α, which are proinflammatory cytokines, promote foam cells formation, while IL-1RA, IL-33, TGF-β1, and IL-10 inhibit it [55–57].

On the other hand, as cholesterol accumulation inside the cell increases, it promotes signaling via TLR4, which results in activation of nuclear factor κB (NF-κB) and production of proinflammatory cytokines. Oxidized LDL itself may induce TLR4 by means of a heterotrimeric complex consisting of CD36-TLR4-TLR6. Examples of proinflammatory cytokines secreted by foam ells include IL-1, IL-6, TNF, and chemokines (CCL2, CCL5, and CXCL1), besides macrophage retention factors (such as netrin 1 and semaphorin 3E) [57–61].

Recently, noncodifying RNAs (ncRNA)—that is, functional RNA nucleotides that do not suffer translation into protein, such as microRNAs (miRs) and long ncRNAs (lncRNAs)—have also been involved in scavenger receptor regulation, inflammatory cascade, and atherosclerosis [62].

The atherogenic actions of LDLox occur in part due to the specific LDLox lectin-like receptor, LOX-1, by which it is activated. LOX-1 is present in EC, macrophages, and SMC; this is a membrane protein type II, with lectin-like extracellular domain, which may be cleaved and, therefore, release its form, LOX-1. In normal conditions, it removes cell debris from apoptotic or injured cells. In the presence of several risk factors, such as hypertension, diabetes, and hypercholesterolemia, LOX-1 is abundantly found in blood vessels [57]. Angiotensin II and endothelin, two NO antagonists, mediate LOX-1 expression. Increased LOX-1 also elevates LDLox intake by cells, which, in turn, increases

LOX-1 synthesis; this reduces NO availability, aggravating endothelial dysfunction [16,22]. Increased levels of LOX-1 also induce apoptosis, increase expression of P-selectin, VCAM-1, and ICAM-1, trigger signaling pathway CD40/CD40L activity, increase ROS production, and modulate the action of metalloproteinases effects [21].

CONCEPT OF INFLAMMATION— A SYNOPSIS

In the last decades, the concept of inflammation obtained great relevance in atherosclerosis pathophysiology as widely discussed by Ross and Libby [10–12]. Bacteria such as *Chlamydia pneumoniae* was found in coronary lesions [63], in addition to signs of virus including nucleic acids. There is also a positive association between bacterial colonization and increased risk of CAD. However, studies with antibacterial agents did not demonstrate significant effects on cardiovascular events [64,65], suggesting that the infection itself is not the cause of atherosclerosis. The prevailing concept today is that the inflammatory process in atherosclerosis results from biochemical alterations due to the presence of inflammatory cells such as monocytes, macrophages, and lymphocytes. Inflammation is sterile. A large body of evidence corroborates this perspective. Lately, mechanisms involving "inflammasomes" have been described in degenerative diseases such as atherosclerosis, Alzheimer, and diabetes [56,66]. Inflammasomes constitute an intracellular signaling multimolecular complex that is the platform to caspase-1 activation and IL-1B and IL-18 maturation. Human atherosclerotic lesions exhibit cholesterol crystals inside macrophages and outside cells; for a long time, these crystals were considered inactive substances. Not anymore. In human macrophage culture, Rajamaki et al. [67] observed that these cells avidly absorb cholesterol crystals, storing them in the form of cholesterol esters. Cholesterol crystals induced secretion of mature IL-1B in macrophages and monocytes. This phenomenon depended on caspase-1, which indicates participation of the inflammasome pathway. By silencing the NLRP3 receptor, which is a crucial component in inflammasome, secretion of IL-1B was completely abolished, indicating that NLRP3 is the element responsible for the cholesterol crystal inflammatory phenomenon. Therefore, these findings establish an important connection between cholesterol metabolism and inflammation in atherosclerotic lesions. Sheedy et al. [66] also observed that CD36 coordinates inflammasome NLRP3 activation.

Kawana et al. [68] identified 83 genes involved in inflammasome NLRP3 activation in human monocytes by means of "genome wide gene expression profiling," clarifying the molecular bases of this inflammatory pathway.

Therefore, it is clear that several interleukins are involved in the inflammatory process. For example, interleukin-18 participates significantly in the atherosclerotic process [69–71]. It is expressed in atherosclerotic plaque, mainly in macrophages, and is an independent risk marker in patients with coronary disease. IL-18 induces recruitment of plaque inflammatory cells and stimulates ICAM-1 and VCAM-1 production. Blocking IL-8 has been suggested as therapeutic goal; therefore, it was noted that weight loss reduces its circulating levels [70]. In the Jupiter study in patients with low LDL and increased reactive protein C (CRP), Ridker et al. [72] demonstrated that rosuvastatin reduced cardiovascular events. Since it is not clear if this benefit is due to reduced LDL of reduced inflammation in itself, the authors planned the *Cantos* [73] study, whose therapeutic target is IL-1B. The *Cantos* study, which involves ~17,000 patients in 40 medical centers around the world, will investigate the effects of specific IL-1B blocking using monoclonal antibody Canakinumab over the incidence of infarction, strokes, and cardiovascular death in patients without evident dyslipidemia but with increased CRP. The population includes patients with stable coronary arterial disease (CAD) after myocardial infarction. Furthermore, the CIRT (The Cardiovascular Inflammation Reduction TR) [74] study will evaluate if low doses of methotrexate reduce the incidence of infarction, strokes, and cardiovascular death in patients with stable CAD and diabetes type II, or metabolic syndrome. Therefore, these studies shall offer a definitive answer regarding the clinical importance of inflammatory process in atherosclerosis evolution.

On the other hand, protease-activated receptors (PAR) are a family of membrane receptors bonded to protein G that mediate tissue lesions of several cellular responses, including inflammation and tissue reparation. Four PARs were identified and called PAR-1 to 4 [75–77]. Their levels are high in induced atherosclerotic and vascular lesions; they are present in EC, SMC, and platelets. The activation of PAR-1 and PAR-4 causes monocyte adhesion on the endothelium by releasing P-selectin and von Willebrand factor. PAR activation is also connected to liver production of IL-6, interleukin that promotes RPC synthesis. PAR activation is related to platelet activation as well, which is an important phenomenon in atherosclerosis evolution. Furthermore, activation of PAR-2 induced SMC migration in vitro, a phenomenon inhibited by antibody anti-PAR [76]. In general, PAR activation promotes inflammatory response in the intima, and therefore favor plaque development and progression. Peroxisomal proliferator-activated receptors (PPAR) modulate the initial stage of atherogenesis, because they regulate chemotaxis and adhesion of circulating cells to endothelial cells. PPAR-α and PPAR-γ activators inhibit the expression of endothelin-1 and MCP-1, modulate the proliferation of T-lymphocytes and immune response, and reduce the expression of VCAM-1 and ICAM-1. PPARs are nuclear receptors activated by fatty acids and derivatives, and participate in the regulation of plasmatic lipids, lipoproteins, insulin secretion, and inflammatory processes. In addition to the effects above, PPARs modulate platelet aggregation by reducing thromboxane A2 expression and the expression of monocyte and macrophage platelet activation receptors and tissue factor, therefore reducing vessel thrombogenic response Thus, the antiatherogenic effect of PPAR-γ is predominant and was documented in experimental atherosclerosis models. The antiatherogenic action of PPARs also results from inhibition of inflammatory genes and synthesis of cytokines such as TNF-α, IL-1β, IL-6, and IL-8 and metalloproteinases [75–77].

The presence of T-lymphocytes in the atherogenesis process, associated with macrophages, suggests that alongside the inflammatory reaction, there is a local immunological response that could be associated with the presence of a specific antigen. Other than representing large local lipid reserves, foam cells also are an important source of proinflammatory mediators (cytokines and chemokines) and producers of oxidant species, characterizing a significant role as amplifiers of the local inflammatory process called innate immunity [78,79].

ADAPTIVE IMMUNE RESPONSE

As mentioned above, the presence of modified LDL on the vascular wall may not only trigger the innate immune response to the foreign element in the microenvironment, but also triggers the organization of a more elaborate and precise immune response directed towards specific antigens, the so-called adaptive immune response. This reaction involves lymphocytes and contributes to the local inflammatory process. In fact, T- and B-lymphocytes are present in atherosclerotic plaque, although in smaller numbers than macrophages [80,81].

Cells that present antigens, such as macrophages and dendritic cells, engulf and process antigens, exposing them on its surface in association with major histocompatibility complexes (MHC). T-lymphocytes recognize these antigens and trigger the immune response characterized by the proliferation and activation of several types of lymphocytes, production of antibodies, and secretion of cytokines [78,82].

T naive CD4+ cells are able to differentiate in different subtypes of T helper cells (Th), defined according to the cytokine profile they produce. IL-12 and IL-18, produced by macrophages, are potent inducers of IFN-γ and promote the differentiation of T naive cell into T helper cells type 1 (Th-1), the most abundant type of lymphocyte in atherosclerotic plaque. In turn, these cells secrete IFN-γ, TNF-α, and IL-2, adopting proatherogenic behavior. On the other hand, regulating T cells (Tregs) CD4+, that may derive from T naive cells in peripheral or thymus lymphoid tissues, produce TGF-β and IL-10 and have antiatherogenic action. Th-2 cells, which secrete IL-4, IL-5, and IL-13, and Th-17 cells, which produce IL-17A/F, IL-22, and IL-23, still have a controversial role in atherosclerosis [78,79].

Cytotoxic T-lymphocytes, derived from antigens presented to T naive cells CD8+, promote death of the target cells that transport antigen and secrete IFN-γ, contributing for vascular inflammation and atherosclerotic lesion growth.

T-lymphocytes present in the lesion are activated by recognition of certain antigens, such as modified lipoproteins, beta-2-glycoprotein I-b, and infectious agents shown by cells that present antigens (macrophages and dendritic cells). Activated T cells may secrete cytokines able to modulate the entire atherogenesis process. T-helper cells CD4 are subdivided in two categories. Th-1 produces proinflammatory cytokines (interferon gamma, lymphotoxin, TNF-α), activating vascular wall cells and contributing with the destabilization and increased thrombogenicity process. On the other hand, Th-2 produces antiinflammatory cytokines, acting as local inflammation inhibitor [78,79]. Cytolytic T cells CD8 express cytotoxic factors that contribute to plaque progression and instabilization. It is easy to observe that humoral immunity may act as inflammatory or antiinflammatory factor depending on external stimuli and other circumstances [83,84]. Recently, the PK Shah's group tried to develop vaccines against atherosclerosis by antagonizing specific epitopes in the cells that participate in the atherosclerotic process [85,86]; these are ongoing, long-term projects whose final results are expected with great interest.

A caveat is needed here. The concepts mentioned above derive mainly from experiments in macrophage culture and in vivo studies in animal models. The extent to which they may apply to human beings is still quite obscure.

SMOOTH MUSCLE CELLS AND FIBROTIC PLAQUE FORMATION

The transition from nascent atheroma plaque to mature plaque involves two fundamental aspects: formation of the fibrous cap and constitution of the lipid core.

The essential component of the fibrous cap is collagen, mainly produced by smooth muscle cells. In turn, smooth muscle cells migrate from the medial layer to the intima and then proliferate, processes regulated by several growth factors produced by macrophages, endothelial cells, and T-lymphocytes. In the atheroma plaque, smooth muscle cells no longer have a contractile role and adopt a secreting phenotype. In addition to secreting collagen, smooth muscle cells are able to internalize oxidized LDL via scavenger receptors and become foam cells. In fact, treatment of smooth muscle cells with oxidized LDL induces the formation of typical foam cells [58,87,88]. The lipid core, many times called necrotic core, is formed from lipids and debris originated from the death of foam cells. The excessive free cholesterol is toxic for the cells and results in endoplasmic reticulum stress and cell death due to apoptosis and necrosis [89]. Furthermore, deficient removal of apoptotic cells contributes to lipid accumulation [52], which also suffers modulation by cytokines. TNF-α, for example, inhibits apoptotic cell clearance by macrophages.

PARTICIPATION OF THE RENIN-ANGIOTENSIN-ALDOSTERONE SYSTEM IN ATHEROSCLEROSIS

The renin-angiotensin-aldosterone system (RAAS) is a hormone complex that plays a major role in the regulation of blood pressure, salts, and fluids. Acting on the vascular system and on perivascular fat tissue, it is recognized as in important element involved in endothelial dysfunction that significantly contributes to arterial rigidity. Under the action of several risk factors, such as hypertension, tobacco use, and hypercholesterolemia, the vascular RAAS is stimulated by higher production of angiotensin II and activation of angiotensin receptor AT1, which increases vascular tissue oxidative stress and inflammation [90]. Angiotensin II induces endothelium ROS production, increases the expression of IL-6 and MCP-1, and hyperregulates VCAM-1. A-II is a very important NADPH oxidase agonist, resulting in superoxide production, NO inactivation, and therefore hypertension [91]; most of the A-II hypertensive effect is due to this mechanism. The intercommunication between angiotensin and aldosterone highlights the importance of mineralocorticoid receptors in modulating insulin resistance, reduced NO bioavailability, endothelial dysfunction, and arterial rigidity. Moreover, both innate and adaptive immunities are involved in local tissue activation by RAAS [56].

These modifications on the endothelium promote vessel inflammation, creating conditions for atherosclerosis development and arterial hypertension [54]. Experimental and clinical studies demonstrated that atherosclerosis evolution may be significantly reduced by blocking receptor AT1 with losartan [92] or angiotensin-converting enzyme inhibitors in men, as in the Hope [93] study, which included 9541 randomized patients to receive rampril or placebo; after 4.5 years of follow-up, there was significant reduction of the combined end-point of cardiovascular death, nonlethal infarction, and nonlethal stroke. The Europa

[94] study compared perindopril ($n=6110$) with placebo ($n=6108$) in patients with stable CAD, obtaining a reduction of 20% in risk in the combined end-point of cardiovascular death, infarction, and stroke. Therefore, experimental and clinical observations have demonstrated unquestionable participation of RAAS in the atherosclerotic process.

OXIDATIVE STRESS AND ATHEROSCLEROSIS

Oxidative stress is defined as imbalance between oxidative species production and cell antioxidant capacity, leading to altered cell signaling on redox pathways [95,96] and oxidative modifications of several cell components, such as lipids, proteins, and sugars, which may suffer partial or total inactivation or no inactivation at all. In practically all cell types, including vessels, a large amount of evidence clearly indicates modulation of different reactive mediators derived from oxygen in signaling cell proliferation, differentiation, migration, senescence, and apoptosis. In endothelial cells, excessive superoxide radical production directly reduces NO bioactivity, another gaseous free radical with vasculoprotective effects (inhibition of proliferation, protection against lipid peroxidation, inhibition of inflammation and thrombosis). Thus, endothelial dysfunction is essentially a redox signaling dysfunction.

At molecular level, several enzymatic systems contribute for ROS generation, in particular the NADPH oxidase family, whose specific function is dedicated to such production. In addition to NADPH oxidase, mitochondria, uncoupled NO synthases, xanthine oxidase, cytochrome P540, and cyclooxygenases, among others, may also contribute to this function [97,98].

Vascular NADPH oxidase is the main enzymatic complex that generates ROS with cell signaling functions on the vascular wall, both in endothelial cells and smooth muscle cells. The adventitial layer also shows great activity of this enzyme and is also an important ROS production site. Vascular NADPH oxidase expression and activity can be stimulated by several factors, including mechanical forces (cyclic vascular distention), growth factors (angiotensin II, platelet-derived growth factor), cytokines (TNF-α), autocoids (bradykinin), and coagulation factors (thrombin). Vascular NADPH oxidase activation represents the transduction of signals triggered by angiotensin receptor AT1 activation [91], involving proliferative kinases and general protein synthesis stimulation. The redox and metabolic role of mitochondria has been increasingly recognized. For example, hydrogen peroxide of mitochondrial origin generated by increased shear stress has vessel relaxing and hyperpolarizing activity on arterioles [97]. The role of mitochondria in atherosclerosis has been increasingly studied and may reveal important disease mechanisms and essential therapeutic targets.

Endothelial dysfunction genesis in atherosclerosis may also involve xanthine oxidase, which is expressed in small quantities in this cell, but may bind to it from the circulating pool [99,100]. In smokers, inhibition of xanthine oxidase with allopurinol corrects endothelial dysfunction [101].

On the other hand, currently there is great interest on the interaction between oxidative stress and the endoplasmic reticulum [102]. ER stress is a frequent condition that results from ER inability to perform its main function, which is to correctly fold and glycosylate recently synthesized proteins. ER stress generates a complex signaling network known as unfolded protein response (UPR), involved in the genesis of several diseases, including atherosclerosis and its risk factors, such as obesity and resistance to insulin [98]. UPR leads to higher oxidants production, and at the same time redox homeostasis disorders sustain UPR signaling [103]. During ER stress, the oxidative stress mechanisms involve several pathways, including NADPH oxidase Nox4, in addition to activation of Nox2 in final apoptosis phases. Importantly, ER-constituent proteins, such as protein disulfide isomerase (PDI), may be associate to and modulate the NADPH oxidase complex [104–106]. Furthermore, there are several functional and physical connections between the ER and mitochondria. Overall, ER is a new element to be considered to understand cellular redox signaling.

originates from the vessel wall itself and include proliferation messengers, migration, apoptosis, etc. Secondly, the inflammatory processes that characterize atherosclerosis are strongly associated with redox mechanisms. Thirdly, peroxidation of lipoprotein components is relevant in atheroma genesis—the so-called "atherogenesis oxidative theory" [107].

The redox effect of cell growth trophic processes in atherosclerosis is in accordance with the role of the redox pathways in cell proliferation. The interaction between inflammation and redox pathways is complex. Certainly, proinflammatory cytokines are able to trigger oxidative stress in several vascular wall cells. Moreover, redox pathways have direct effect on inflammatory cell activation. In fact, one of the first effects of oxidative stress and endothelial dysfunction is to expose adhesion molecules to inflammatory cells [107]. On the other hand, the fact that only part of the vessel oxidant production initiates inflammation shall not be minimized. For example, in contrast to Nox1 and Nox2 NADPH oxidases, Nox4 potentially has protective effects against vascular diseases [106].

Still, even in phagocyte, Nox2 activation and respiratory explosion activate several bactericide mechanisms, but at the same time limit inflammation or favor its resolution.

The atherogenesis oxidative theory, involving oxidative modifications of LDL, was one of the first logic links between lipids, cell proliferation, and plaque growth [107]. Today, this theory is solidly characterized and is well accepted. On the other hand, this does not mean that unspecific antioxidants are able to predictably change the course of the disease, as evidenced in clinical controlled studies. Vascular wall redox imbalance and lipoprotein oxidation are still strictly linked to mitochondrial dysfunction. Overall, vascular oxidative stress promotes and enhances endothelial dysfunction in atherosclerosis. Redox imbalance, inflammation, and endothelial dysfunction are strictly related to the pathophysiology of atherosclerosis and several vascular diseases.

INTERACTIONS BETWEEN ENDOTHELIAL DYSFUNCTION REDOX PATHWAYS, LIPOPROTEIN OXIDATION, AND ATHEROSCLEROSIS

The role of redox processes in atherosclerosis involves at least three important aspects. First, the biggest source of oxidants in the arterial wall

PLAQUE RUPTURE, EROSION, AND THROMBOSIS

The atheroma plaque is composed by cells, predominantly macrophages and lymphocytes, extracellular matrix (interstitial collagen types I and II and proteoglycans), smooth muscle cells, and lipid core. The composition percentage varies from plaque to plaque, and the same individual may have plaques of most different compositions [108].

In the Prospect study, Stone et al. [109] investigated 697 patients with acute coronary syndrome using intravascular ultrasound imaging and conventional angiography after coronary angioplasty. The objective was to assess cardiovascular events over 3 years after the intervention, that is, cardiovascular death, re-hospitalization, infarction, and cardiac arrest, and relate them both to culprit lesions and nonculprit lesions.

Cardiovascular events occurred in 20.4% of the patients, of which 12.9% were related to culprit lesions and 11.6% to nonculprit lesions. The latter showed mild stenosis, that is, $32.3 \pm 20.6\%$. However, these nonculprit lesions more frequently were associated with 70% or more plaque burden in comparison with those not associated with events, showed small lumen area, that is, 4.0 mm^2 or less, or showed thin—cap fibroatheroma (60–70 mm). Therefore, the authors concluded that postangioplasty events occurred equally in culprit and nonculprit lesions; the latter were frequently angiographically mild, had thin-cap atheroma, or large plaque burden; and, of course, combinations of these characteristics increased significantly the proportion of events, varying between 4.9% in thin-cap atheroma, 10.2% in thin-cap + lumen area ≤ 4 mm^2, 16.4% in thin-cap + 70%

plaque burden, and 18.2% in lesions that showed all three characteristics associated. One of the implications of the study is that postangioplasty treatment must focus on atherosclerosis as a whole process in order to avoid its progression and complications; treatment of culprit lesions alone is not enough.

On the other hand, Forrester et al. [110] described endothelial surface alterations with ulcerations in patients with acute coronary syndrome using intracoronary ultrasound. They noted intraluminal thrombus incorporated into the arterial wall, which contributes to lesion growth.

The most common atherosclerotic plaque instabilization mechanism is rupture, which allows close contact between circulating blood and the highly thrombogenic lipid core due to the presence of the tissue factor, a potent procoagulant expressed in macrophages and smooth muscle cells inside the plaques. This occurs in ~70% of the cases [111]. Thus, the coagulation cascade is triggered and the resulting thrombus may occlude the artery, impairs blood supply, causes ischemia and acute clinical syndromes (e.g., unstable angina or acute myocardial infarction) (Fig. 33.6). Intraplaque neovascularization due to adventitial vasa vasorum is also recognized as an independent predictive factor for bleeding and plaque rupture predictive factor [112]. On the other hand, as described by Glagov [113] and also observed by Motoyama [114], positive

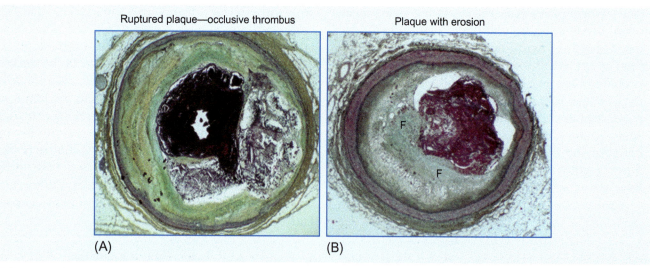

FIG. 33.6 (A) Section of the anterior descending artery of a patient that died due to acute myocardial infarction. Due to the atherosclerotic process, sub occlusive thrombus, thin-cap plaque rupture, large lipid core, and impairment of the entire arterial circumference are observed. (B) Plaque showing erosion thrombosis; plaque showing erosion is fibrous, has higher content of proteoglycans, and less inflammation that torn plaque, being more frequent in women. Contribution of Dr. Maria de Lourdes Higuchi. Pathology Department, InCor.

remodeling representing compensatory expansion is another characteristics of unstable plaques. In addition, microcalcification has also been observed in association with plaque instability and progression [115–117].

Thrombosis may occur in the atheromatous plaque due to erosion or, more rarely, in the presence of calcified nodules, which occurs more frequently in older patients with tortuous and highly calcified arteries. Plaque erosion as thrombosis substrate in patients that died from myocardial infarction, was studied by Arbustini et al. [118] in 298 necropsies of male patients and 109 females. Acute thrombi were found in 98% of the cases, and 25% of the cases showed plaque erosion as anatomic substrate, predominantly in women (37.4% in women versus 18.5% in men). Therefore, the authors concluded that erosion is an important thrombosis substrate in infarctions, especially in women. They also considered that even plaque without large lipid cores or thick-cap may originate thrombi, most probably due to endothelial dysfunction. It has also been noted that erosions are more common in female smokers with less than 50 years of age. This was observed by Burke et al. [111], who studied the coronaries of 51 cases of sudden death, comparing them with 15 cases of women that died due to trauma. The investigation showed that plaque erosion was associated with tobacco use by young female patients without significant plasmatic cholesterol increase. The mechanism is not well-known, but it is believed that local endothelial cell apoptosis may contribute to local predisposition to form obstructive thrombi. A detailed review regarding acute syndrome and the role of inflammation was recently presented by Libby et al. [12]. Increased production of IL-6 by hepatocytes is particularly significant; recent genetic studies strongly imply IL-6 in acute coronary syndrome pathogenesis [12]. Fig. 33.7 illustrates several mechanisms involving plaque rupture.

Together, the data obtained by several studies indicates that plaques that are more vulnerable to rupture are those with thin cap, few smooth muscle cells, pronounced lipid core, high-macrophage content, microcalcifications, and inflammatory infiltrates. For that matter, inflammatory mediator ligand CD40 [119,120] is a relevant activator of tissue factor expression in human macrophages, which significantly influences thrombi formation. On the other hand, plaque with little inflammation, few macrophages, small lipid core, considerable collagen, and thick cap are less vulnerable to instabilization, although not immune to it.

Therefore, in less frequent circumstances, plaque without the anatomic characteristics described above may also suffer acute destabilization, as is observed in cases or erosion, which demonstrates the importance of other factors (endothelial dysfunction, inflammation, and others). Furthermore, observations in coronario angiography of patients subjected to diagnostic examinations showed that plaque with "unstable" morphology is not necessarily associated with acute clinical syndromes. Since most of the anatomic studies regarding unstable plaque were conducted in patients that suffered acute coronary crisis (infarction, ACS), we are forced to recognize that our understanding regarding instabilization phenomenon is still incomplete.

Rupture reflects imbalance between factors that constitute the plaque and mechanical forces of the fibrous cap. Interstitial forms of collagen are responsible for plaque biomechanical resistance to rupture. Therefore, extracellular and pericellular matrix remodeling play a fundamental role in atherosclerotic plaque composition and propensity to rupture. This process is controlled by several proteases, in particular matrix metalloproteinases (MMP) and their tissue inhibitors (TIMP) produced by macrophages and other vascular cells [121]. MMPs, mostly formed as inactive proenzymes activated by proteolysis, degrade extracellular matrix components. In turn, MMP and TIMP expression and/or activity is regulated by cytokines. MMPs and TIMPS also modulate the production and secretion of cytokines and growth factors, mediating cell migration, proliferation and tissue remodeling. Fig. 33.8 illustrates some fundamental elements that constitute the plaque instabilization scenario, such as endothelial cell and smooth muscle cell apoptosis, inflammation, and extracellular and pericellular matrix disorder [122].

Different cytokines are related to plaque vulnerability. IFN-γ inhibits collagen synthesis by smooth muscle cells and, such as IL-18, GDF-15 (growth differentiation factor-15), and TWEAK (TNF-like weak inducer of apoptosis), is associated with plaque instabilization in experimental models. On the one hand, IFN-γ, TNF-α, and IL-1β promote macrophage

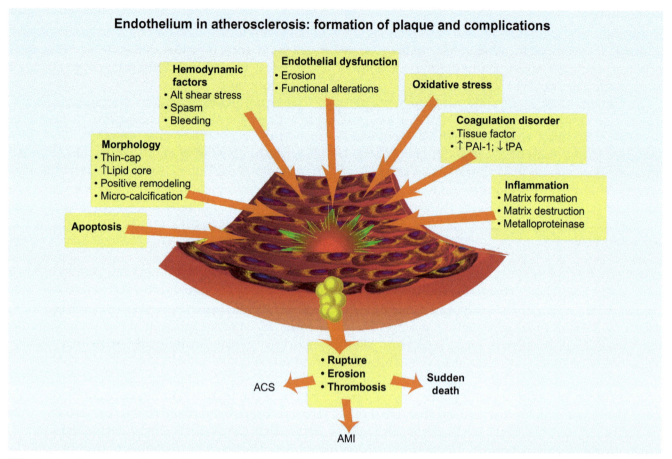

FIG. 33.7 Representation of the essential factors that participate in atherosclerotic plaque destabilization. Note that destabilization occurs mainly in plaque with special morphology (thin-cap, large lipid core, microcalcifications, positive remodeling). However, these anatomic characteristics per se are not sufficient to trigger destabilization. Other factors, that is, hemodynamic alterations, endothelial dysfunction, inflammatory process, and coagulation alterations, must also occur in order to form the thrombus and reduce blood flow. *ACS*, acute coronary syndrome; *AMI*, acute myocardial infarction; *PTA*, plasminogen tissue activator; *PAI*, plasminogen activator inhibitor; *MMP*, matrix metalloproteinases. Adapted from Da Luz PL, Laurindo FRM, Chagas ACP. Endotélio: doenças cardiovasculares. 1st ed. São Paulo: Atheneu; 2004.

apoptosis, increasing the lipid core, and, on the other hand, promote smooth muscle cells apoptosis, contributing to fibrous cap thinning. Conversely, cytokine TGF-β is associated with higher plaque stabilization [12].

In conclusion, the atherosclerosis development process involves risk factors, most of them related to lifestyle, genetic factors, interactions between inflammatory cells, lipid particles, endothelial cells, smooth muscle cells, vasa vasorum, and adventitious vessels. Atherosclerosis evolves over long periods and tends to be systemic. Essentially, this form of evolution causes slow plaque growth, leading to progressive of the coronary flow reduction, with the classic manifestation of *angina pectoris*.

On the other hand the atherosclerotic plaque instabilization process is very complex, involving anatomic, hemodynamic, and chemical factors. These factors seem to be in a dynamic balance state, in which alterations in one or more of them may collapse the entire plaque structure and lead to instabilization with consequent acute clinical manifestations, such as acute coronary syndrome or myocardial infarction.

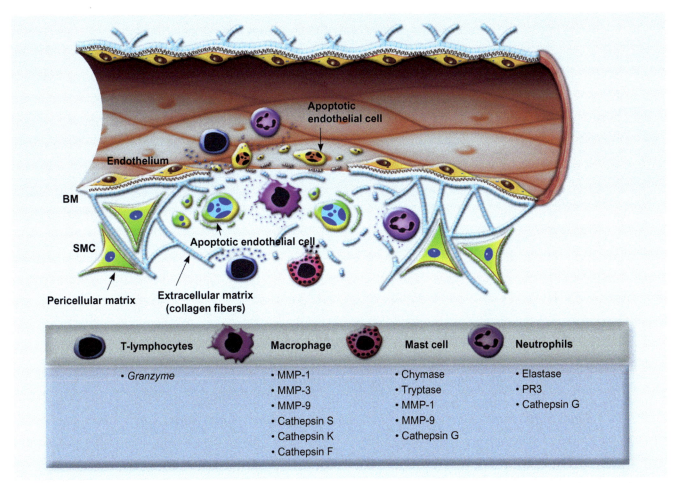

FIG. 33.8 Scheme representing the mechanisms that lead to endothelium rupture in acute coronary syndromes. Essentially, infiltration of the subendothelial space by macrophages, lymphocytes, mast cells, and neutrophils, and production of ROS, cytokines, and metalloproteinases, causes EC and SMC apoptosis and pericellular and extracellular matrix breakdown [122]. *MMP*, matrix metalloproteinases; *PR3*, antiproteinase antibody3; *SMC*, smooth muscle cells; *BM*, baseline membrane. Adapted from Lindstedt KA, Leskinen MJ, Kovanen PT. Proteolysis of the pericellular matrix: a novel element determining cell survival and death in the pathogenesis of plaque erosion and rupture. Arterioscler Thromb Vasc Biol 2004;24:1350–8.

References

[1] Faggiotto A, Ross R, Harker L. Studies of hypercholesterolemia in the nonhuman primate. I. Changes that lead to fatty streak formation. Arteriosclerosis 1984;4:323–40.

[2] Relationship of atherosclerosis in young men to serum lipoprotein cholesterol concentrations and smoking: a preliminary report from the Pathobiological Determinants of Atherosclerosis in Youth (PDAY) Research Group. JAMA 1990;264:3018–24.

[3] Berenson GS, Srinivasan SR, Bao W, et al. Association between multiple cardiovascular risk factors and atherosclerosis in children and young adults. N Engl J Med 1998;338:1650–6.

[4] Kannel WB, Dawber TR, Kagan A, et al. Factors of risk in the development of coronary heart disease—six year follow-up experience. The Framingham Study. Ann Intern Med 1961;55:33–50.

[5] Ignatowsky AI. Ueber die wirkung der tiershen einwesses auf der aorta. Virchows Arch Pathol Anat 1909;198:248.

[6] Stuckey NW. On the changes of rabbit aorta under the influence of rich animal food. São Petersburgo. Dissertação inaugural; 1910.

[7] Anichkov N, Chalatov S. Ueber expetimentelle cholesterin steatose-ihre bedeutung fur die enstehung einiger pathologischer proesen. Centrablatt Allgemeine Pathologie Pathologische Anatomie 1913;1:1.

[8] Dock W. Research in atherosclerosis—the first fifty years. Ann Intern Med 1958;49(3):699–705.

[9] Anichkov N. A history of experimentation on arterial atherosclerosis in animals. In: Blumental HT, editor. Cowdry's arteriosclerosis—a survey of the problem. 2nd ed. Springfield, IL: Charles C. Thomas; 1967.

[10] Ross R. Atherosclerosis—an inflammatory disease. N Engl J Med 1999;340:115–26.

[11] Libby P. Inflammation in atherosclerosis. Arterioscler Thromb Vasc Biol 2012;32:2045–51.

[12] Libby P, Tabas I, Fredman G, et al. Inflammation and its resolution as determinants of acute coronary syndromes. Circ Res 2014;114:1867–79.

[13] Van Gils JM, Derby MC, Fernandes LR, et al. The neuroimmune guidance cue netrin-1 promotes atherosclerosis by inhibiting the emigration of macrophages from plaques. Nat Immunol 2012;13:136–43.

[14] Roberts R. Genetics of coronary artery disease. Circ Res 2014;114:1890–903.

[15] Fernandes DC, Laurindo FRM, Araujo TL, et al. Forças hemo-dinâmicas no endotélio: da mecanotransdução às implicações no desenvolvimento de aterosclerose.

[16] Gimbrone Jr. MA, García-Cardeña G. Vascular endothelium, hemodynamics, and the pathobiology of atherosclerosis. Cardio-vasc Pathol 2013;22:9–15.

[17] Tabas I, Willians KJ, Borén J. Subendothelial lipoprotein retention as the initiating process in atherosclerosis—update and therapeutic implications. Circulation 2007;116:1832–44.

[18] Ait-Oufella H, Taleb S, Mallat Z, et al. Recent advances on the role of cytokines in atherosclerosis. Arterioscler Thromb Vasc Biol 2011;31:969–79.

[19] Zernecke A, Weber C. Chemokines in atherosclerosis: proceedings resumed. Arterioscler Thromb Vasc Biol 2014;34:742–50.

[20] Ramji DP, Davies TS. Cytokines in atherosclerosis: key players in all stages of disease and promising therapeutic targets. Cytokine Growth Factor Rev 2015;26:673–85.

[21] Da Luz PL, Uint L, Serrano Jr. CV, et al. Endotélio e aterosclerose. Rev Soc Cardiol Est de São Paulo 1996;2:160–70.

[22] Tsimikas S, Miller YI. Oxidative modification of lipoproteins: mechanisms, role in inflammation and potential clinical applications in cardiovascular disease. Curr Pharm Des 2011;5:27–37.

[23] Bae YS, Lee JH, Choi SH, et al. Macrophage generate oxygen species in response to minimally oxidized low density lipoprotein: toll-like receptor-4 and spleen tyrosine kinase dependent activation of NADPH oxidase 2. Circ Res 2009;104:210–8. 21p following 218.

[24] Fuster V, Moreno PR, Fayad ZA, et al. Atherothrombosis and high-risk plaque. J Am Coll Cardiol 2005;46:937–54.

[25] Libby P, Ridker PM. Inflammation atherothrombosis: from population biology and bench research to clinical practice. J Am Coll Cardiol 2006;48:A33–46.

[26] Ridker PM. Targeting inflammatory pathways for the treatment of cardiovascular disease. Eur Heart J 2014;35:540–3.

[27] Libby P, Ridker P, Hanson GK. Progress and challenges in translating the biology of atherosclerosis. Nature 2011;473:317–25.

[28] Gu L, Okada Y, Clinton SK, et al. Absence of monocyte chemoat-tractant protein-1 reduces atherosclerosis in low density lipopro-tein receptor-deficient mice. Mol Cell 1998;2:275–81.

[29] Gratchev A, Sobenin I, Orekhov A, et al. Monocytes as a diagnostic marker of cardiovascular diseases. Immunobiology 2012;217:476–82.

[30] Tapp LD, Shantsila E, Wrigley BJ, et al. The CD14++CD16+ mono-cyte subset and monocyte-platelet interactions in patients with ST-elevation myocardial infarction. J Thromb Haemost 2012;10:1231–41.

[31] Jaipersad AS, Lip GYH, Silverman S, et al. The role of monocytes in angiogenesis and atherosclerosis. J Am Coll Cardiol 2014;63:1–11.

[32] Landsman L, Bar-On L, Zernecke A, et al. CX3CR1 is required for monocyte homeostasis and atherogenesis by promoting cell sur-vival. Blood 2009;113:963–72.

[33] Ghattas A, Griffiths HR, Devitt A, et al. Monocytes in coronary artery disease and atherosclerosis. J Am Coll Cardiol 2013;62:1541–51.

[34] Hilgendorf I, Swirski FK, Robbins CS. Monocyte fate in atheroscle-rosis. Arterioscler Thromb Vasc Biol 2015;35:272–9.

[35] Wanschel A, Seibert T, Hewing B, et al. Neuroimmune guidance cue semaphoring 3E is expressed in atherosclerotic plaques and regulates macrophages retention. Arterioscler Thromb Vasc Biol 2013;33:886–93.

[36] Van Gils JM, Ramkhelawon B, Fernandes L, et al. Endothelial expression of guidance cues in vessel wall homeostasis dysregula-tion under proatherosclerotic conditions. Arterioscler Thromb Vasc Biol 2013;33:911–9.

[37] Moore KJ, Sheedy FJ, Fisher EA. Macrophages in atherosclerosis: a dynamic balance. Nature 2013;13:709–21.

[38] Robbins CS, Hilgendorf I, Weber GF, et al. Local proliferation dom-inates lesional macrophage accumulation in atherosclerosis. Nat Med 2013;19:1166–72.

[39] Wolfs IM, Donners MM, de Winther MP. Differentiation factors and cytokines in the atherosclerotic plaque micro-environment as a trigger for macrophage polarisation. Thromb Haemost 2011;106:763–71.

[40] Chinetti-Gbaguidi G, Colin S, Staels B. Macrophage subsets in ath-erosclerosis. Nat Rev Cardiol 2015;12:10–7.

[41] Bellingan GJ, Caldwell H, Howie SE. In vivo fate of the inflamma-tory macrophage during the resolution of inflammation: inflamma-tory macrophages do not die locally, but emigrate to the draining lymph nodes. J Immunol 1996;15(157):2577–85.

[42] Randolph GJ. Emigration of monocyte-derived cells to lymph nodes during resolution of inflammation and its failure in athero-sclerosis. Curr Opin Lipidol 2008;19:462–8.

[43] Moore KJ, Freeman MW. Scavenger receptors in atherosclerosis: beyond lipid uptake. Arterioscler Thromb Vasc Biol 2006;26:1702–11.

[44] Cardilo-Reis L, Gruber S, Schreier SM, et al. Interleukin-13 protects from atherosclerosis and modulates plaque composition by skew-ing the macrophage phenotype. EMBO Mol Med 2012;10:1072–86.

[45] Tabas I. Macrophage death and defective inflammation resolution in atherosclerosis. Nat Rev Immunol 2010;10:36–46.

[46] Tabas I. Consequences and therapeutic implications of macrophage apoptosis in atherosclerosis: the importance of lesion stage and phagocytic efficiency. Arterioscler Thromb Vasc Biol 2005;25:2255–64.

[47] Yao PM, Tabas I. Free cholesterol loading of macrophages induces apoptosis involving the fas pathway. J Biol Chem 2000;275:23807–13.

[48] Yao PM, Tabas I. Free cholesterol loading of macrophages is asso-ciated with widespread mitochondrial dysfunction and activation of the mitochondrial apoptosis pathway. J Biol Chem 2001;276:42468–76.

[49] Ball RY, Stowers EC, Burton JH, et al. Evidence that the death of macrophage foam cells contribute to the lipid core of atheroma. Atherosclerosis 1995;114:45–54.

[50] Thorp E, Subramanian M, Tabas I. The role of macrophages and dendritic cells in the clearance of apoptotic cells in advanced ath-erosclerosis. Eur J Immunol 2011;41:2515–8.

[51] Konior A, Schramm A, Czesnikiewicz-Guzik M, et al. NADPH oxi-dases in vascular pathology. Antioxid Redox Signal 2014;20:2794–814.

[52] Trogan E, Feig JE, Dogan S, et al. Gene expression changes in foam cells and the role of chemokine receptor CCR7 during atherosclero-sis regression in ApoE-deficient mice. Proc Natl Acad Sci U S A 2006;103:3781–6.

[53] Chaabane C, Coen M, Bochaton-Piallat ML. Smooth muscle cell phenotypic switch: implications for foam cell formation. Curr Opin Lipidol 2014;25:374–9.

[54] Feig JE, Rong JX, Shamir R, et al. HDL promotes rapid atheroscle-rosis, regression in mice and alters inflammatory properties of pla-que monocyte-derived cells. Proc Natl Acad Sci U S A 2011;108:7166–71.

[55] Rosenfeld ME, Ross R. Macrophage and smooth muscle cell prolif-eration in atherosclerotic lesions of WHHL and comparably hyper-cholesterolemic fat-fed rabbits. Arteriosclerosis 1990;10:680–7.

[56] Witztum JL, Lichtman AH. The influence of innate and adaptive immune responses on atherosclerosis. Annu Rev Pathol 2014;9:73–102.

[57] Steinberg D, Witztum JL. Oxidized low-density lipoprotein in atherosclerosis. Arterioscler Thromb Vasc Biol 2010;30:2311–6.

[58] Miller YI, Choi SH, Fang L, et al. Lipoprotein modification and macrophage uptake: role of pathologic cholesterol transport in atherogenesis. Subcell Biochem 2010;51:229–51.

[59] Higashimori M, Tatro JB, Moore KJ, et al. Role of toll-like receptor 4 in intimal foam accumulation in apolipoprotein deficient mice. Arterioscler Thromb Vasc Biol 2011;31:50–7.

[60] Hayashi C, Papadopoulos G, Gudino CV, et al. Protectin role for TLR4 signaling in atherosclerosis progression as revealed by infection with a common oral pathogen. J Immunol 2012;189:3681–8.

[61] Stewart CR, Stuart LM, Wilkinson K, et al. CD36 ligands promote sterile inflammation through assembly of a Toll-like receptor 4 and 6 heterodimer. Nat Immunol 2010;11:155–61.

[62] Dai Y, Condorelli G, Mehta JL. Scavenger receptors and non-coding RNAs: relevance in atherogenesis. Cardiovasc Res 2016;109(1):24–33.

[63] Tufano A, Di Capua M, Coppola A, et al. The infectious burden in atherothrombosis. Semin Thromb Hemost 2012;38:515–23.

[64] Grayston JT, Kronmal RA, Jackson LA. Azithromycin for the secondary prevention of coronary events. N Engl J Med 2005;352:1637–45.

[65] Tarbutton GL, Mitia AK. Is antibiotic treatment effective for coronary artery disease? J Appl Res 2007;7:39–49.

[66] Sheedy FJ, Grebe A, Rayner KJ. CD36 coordinates NLRP3 inflammasome activation by facilitating the intracellular nucleation from soluble to particulate ligands in sterile inflammation. Nat Immunol 2013;14:812–20.

[67] Rajamaki K, Lappalainen J, Oorni K, et al. Cholesterol crystals activate the NLRP3 inflammasome in human macrophages: a novel link between cholesterol metabolism and inflammation. PLoS One 2010;5:e11765.

[68] Kawana N, Yamamoto Y, Kino Y, et al. Molecular network of NLRP3 inflammasome activation responsive genes in a human monocyte cell line. J Clin Immunol 2014;1:1017.

[69] Mallat Z. Expression of interleukin-18 in human atherosclerotic plaques and relation to plaque instability. Circulation 2011;104:1598–603.

[70] Jefferis BJ, Papacosta O, Owen CG, et al. Interleukin-18 and coronary heart disease: prospective study and systematic review. Atherosclerosis 2011;217:227–33.

[71] Tiret L, Godefroy T, Lubos E, et al. Genetic analysis of the interleukin-18 system highlights the role of the interleukin-18 gene in cardiovascular disease. Circulation 2005;112:643–50.

[72] Ridker PM, Danielson E, Fonseca FA, et al. Reduction in C-reactive protein and LDL cholesterol and cardiovascular event rates after initiation of rosuvastatin: a prospective study of the JUPITER trial. Lancet 2009;373:1175–82.

[73] Ridker PM, Thuren T, Zalewski A, et al. Interleukin-1b inhibition and the prevention of recurrent cardiovascular events: rationale and design of the Canakinumab Anti-inflammatory Thrombosis Outcomes Study (CANTOS). Am Heart J 2011;162:597–605.

[74] Ridker PM. Testing the inflammatory hypothesis of atherothrombosis: scientific rationale for the cardiovascular inflammation reduction trial (CIRT). J Thromb Haemost 2009;7(Suppl. 1):332–9.

[75] Chinetti G, Fruchart JC, Staels B. Peroxisome proliferator-activated receptors (PPARs): nuclear receptors at the crossroads between lipid metabolism and inflammation. Inflamm Res 2000;49:497–505.

[76] Duez H, Fruchart JC, Staels B. PPARS in inflammation, atherosclerosis and thrombosis. J Cardiovasc Risk 2001;8:187–94.

[77] Moore KJ, Rosen ED, Fitzgerald ML, et al. The role of PPAR-gamma in macrophage differentiation and cholesterol uptake. Nat Med 2001;7:41–7.

[78] Song L, Leung C, Schindler C. Lymphocytes are important in early atherosclerosis. J Clin Investig 2001;108:251–9.

[79] Daugherty A. The effects of total lymphocyte deficiency on the extent of atherosclerosis in apolipoprotein E−/− mice. J Clin Investig 1997;100:1575–80.

[80] Shalhoub J, Falck-Hansen MA, Davies AH, et al. Innate immunity and monocyte-macrophage activation in atherosclerosis. J Inflamm (Lond) 2011;8:9.

[81] Lahoute C, Herbin O, Mallat Z, et al. Adaptive immunity in atherosclerosis: mechanisms and future therapeutic targets. Nat Rev Cardiol 2011;8:348–58.

[82] Johnson JL. Emerging regulators of vascular smooth muscle cell function in the development and progression of atherosclerosis. Cardiovasc Res 2014;103:452–60.

[83] Packard RR, Magantyo-Garcia E, Gotsman I, et al. CD11c dendritic cells maintain antigen processing, presentation capabilities, and CD4+T cell primary efficacy under hypercholesterolemic conditions associated with atherosclerosis. Circ Res 2008;103:965–73.

[84] Koltsova EK, Garcia Z, Chodaczek G, et al. Dynamic T cell-APC interactions sustain chronic inflammation in atherosclerosis. J Clin Investig 2012;122:3114–26.

[85] Nilsson J, Fredrikson GN, Björkbacka H, et al. Vaccines modulating lipoprotein autoimmunity as a possible future therapy for cardiovascular disease. J Intern Med 2009;266:221–31.

[86] Hansson GK, Nilsson J. Vaccination against atherosclerosis? Induction of atheroprotective immunity. Semin Immunopathol 2009;31:95–101.

[87] Steinberg D, Witztum JL. Oxidized low-density lipoprotein and atherosclerosis. Arterioscler Thromb Vasc Biol 2010;30:2311–6.

[88] Zhao L, Funk CD. Lipoxygenase pathways in atherogenesis. Trends Cardiovasc Med 2004;14:191–5.

[89] Goldstein JL, Ho YK, Basu SK, et al. Binding site on macrophages that mediates uptake and degradation of acetylated low density lipoprotein, producing massive cholesterol deposition. Proc Natl Acad Sci U S A 1979;76:333–7.

[90] Mueller C, Baudler S, Nickenig G, et al. Identification of a novel redox-sensitive gene, Id3, which mediates angiotensin II-induced cell growth. Circulation 2002;105:2423–8.

[91] Fernandes DC, Bonatto D, Laurindo FRM. The evolving concept of oxidative stress. In: Sauer H, Shah A, Laurindo FR, editors. Oxidative stress in clinical practice: cardiovascular diseases. New York: Springer; 2010. p. 1–41.

[92] Li HZN, Miao YDP. Losartan alleviates hyperuricemia-induced atherosclerosis in a rabbit model. Int J Clin Exp Pathol 2015;8 (9):10428–35.

[93] Sleight P. The HOPE study (heart outcomes prevention evaluation). J Renin-Angiotensin-Aldosterone Syst 2000;1:18–20.

[94] Scheen AJ, Legrand V. Clinical study of the month. The EUROPA study: cardiovascular protection with perindopril in patients with stable coronary heart disease. Rev Med Liege 2003;58:713–6.

[95] Winterbourn CC. Reconciling the chemistry and biology of reactive oxygen species. Nat Chem Biol 2008;4:278–86.

[96] Jones DP. Radical-free biology of oxidative stress. Am J Physiol Cell Physiol 2008;295:C849–68.

[97] Zhang DX, Gutterman DD. Mitochondrial reactive oxygen species-mediated signaling in endothelial cells. Am J Physiol Heart Circ Physiol 2007;292:H2023–31.

[98] Santos CX, Tanaka LY, Wosniak J, et al. Mechanisms and implications of reactive oxygen species generation during the unfolded protein response: roles of endoplasmic reticulum oxidoreductases, mitochondrial electron transport, and NADPH oxidase. Antioxid Redox Signal 2009;11:2409–27.

[99] White CR, Darley-Usmar V, Berrington WR, et al. Circulating plasma xanthine oxidase contributes to vascular dysfunction in hypercholesterolemic rabbits. Proc Natl Acad Sci U S A 1996;93:87445–9.

[100] Battelli MG, Polito L, Bolognesi A. Xanthine oxidoreductase in atherosclerosis pathogenesis: not only oxidative stress. Atherosclerosis 2014;237:562–7.

[101] Guthikonda S, Sinkey C, Barenz T, et al. Xanthine oxidase inhibition reverses endothelial dysfunction in smokers. Circulation 2003;107:416–21.

[102] Feng B, Yao PM, Li Y. The endoplasmic reticulum is the site of cholesterol-induced cytotoxicity in macrophages. Nat Cell Biol 2003;5:781–92.

[103] Eletto D, Chevet E, Argon Y, et al. Redox controls UPR to control redox. J Cell Sci 2014;27:3649–58.

[104] Laurindo FR, Pescatore LA, Fernandes Dde C. Protein disulfide isomerase in redox cell signaling and homeostasis. Free Radic Biol Med 2012;52:1954–69.

[105] Janiszewski M, Lopes LR, Carmo AO, et al. Regulation of NAD(P)H oxidase by associated protein disulfide isomerase in vascular smooth muscle cells. J Biol Chem 2005;280:40813–9.

[106] Fernandes DC, Manoel AH, Wosniak Jr. J, et al. Protein disulfide isomerase overexpression in vascular smooth muscle cells induces spontaneous preemptive NADPH oxidase activation and Nox1 mRNA expression: effects of nitrosothiol exposure. Arch Biochem Biophys 2009;484:197–204.

[107] Darley-Usmar V, Halliwell B. Blood radicals: reactive nitrogen species, reactive oxygen species, transition metal ions, and the vascular system. Pharm Res 1996;13:649–62.

[108] Falk E, Nakano M, Bentzon JF. Update on acute coronary syndromes: the pathologists' view. Eur Heart J 2013;34:719–28.

[109] Stone GW, Maehara A, Lansky AJ, et al. A prospective natural-history study of coronary atherosclerosis. N Engl J Med 2011;364:226–35.

[110] Forrester JS, Litvack F, Grundfest W, et al. A perspective of coronary disease seen through the arteries of living man. Circulation 1987;75:505–13.

[111] Burke AP, Farb A, Malcolm GT, et al. Coronary risk factors and plaque morphology in men with coronary disease who died suddenly. N Engl J Med 1997;336:1276–82.

[112] ten Kate GL, Sijbrands EJG, Valkema R, et al. Molecular imaging of inflammation and intraplaque vasa vasorum: a step forward to identification of vulnerable plaques? J Nucl Cardiol 2010;17:897–912.

[113] Glagov S, Weisenberg E, Zarins CK, et al. Compensatory enlargement of human atherosclerotic coronary arteries. N Engl J Med 1987;316:1371–5.

[114] Motoyama S, Sarai M, Harigaya H, et al. Computed tomographic angiography characteristics of atherosclerotic plaques subsequently resulting in acute coronary syndrome. J Am Coll Cardiol 2009;54:49–57.

[115] Katasoka Y, Wolski K, Uno K, et al. Spotty calcification as a marker of accelerated progression of coronary atherosclerosis: insights from serial intravascular ultrasound. J Am Coll Cardiol 2012;59:1592–7.

[116] Rambhia SH, Liang X, Xenos M. Microcalcifications increase coronary vulnerable plaque rupture potential: a patients-based micro-CT fluid—structure interaction study. Ann Biomed Eng 2012;40:1443–54.

[117] Ehara S, Kobayashi Y, Yoshiyama M, et al. Spotty calcification typifies the culprit plaque in patients with acute myocardial infarction: an intravascular ultrasound study. Circulation 2004;110:3424–9.

[118] Arbustini E, Dal Bello B, Morbini P, et al. Plaque erosion is a major substrate for coronary thrombosis in acute myocardial infarction. Heart 1999;82:269–72.

[119] Schönbeck U, Sukhova GK, Shimizu K, et al. Inhibition of CD40 signaling limits evolution of established atherosclerosis in mice. Proc Natl Acad Sci U S A 2000;97:7458–63.

[120] Mach F, Schönbeck U, Sukhova GK, et al. Reduction of atherosclerosis in mice by inhibition of CD40 signalling. Nature 1998;394:200–3.

[121] Siasos G, Tousoulis D, Kioufis S, et al. Inflammatory mechanisms in atherosclerosis: the impact of matrix metalloproteinases. Curr Top Med Chem 2012;12:1132–48.

[122] Lindstedt KA, Leskinen MJ, Kovanen PT. Proteolysis of the pericellular matrix: a novel element determining cell survival and death in the pathogenesis of plaque erosion and rupture. Arterioscler Thromb Vasc Biol 2004;24:1350–8.

Further Reading

[1] Da Luz PL, Laurindo FRM, Chagas ACP. Endotélio: doenças cardiovasculares. 1st ed. São Paulo: Atheneu; 2004.

34

Endothelial Function and Cardiovascular Risk Factors

Desidério Favarato and Protásio Lemos da Luz

INTRODUCTION

Cardiovascular diseases of atherosclerotic origin are the main causes of death in the Western world and in Brazil [1]. Atherosclerosis begins, is maintained and present complication due to endothelial dysfunction, which arise in the presence of risk factors. Endothelial dysfunction is the common trait of all cardiovascular disease risk factors.

The classic risk factors are: male gender, age, smoke use, diabetes, hypercholesterolemia, low HDL, and sedentary lifestyle. Over time, other factors were added, such as obesity, metabolic syndrome, depression, and emotional stress, besides familial predisposition, hyperhomocysteinemia, etc.

However, the INTERHEART study on risk factor prevalence in patients hospitalized with infarction diagnosis, 90% of the patients showed the classic risk factors [2].

NORMAL AND DYSFUNCTIONAL ENDOTHELIUM ACTIVITIES

The normal functioning endothelium presents vasodilation properties and antiplatelet, antiinflammatory, antiproliferative, and fibrinolysis actions. The substances produced in these circumstances are nitric oxide, prostacyclin, and endothelium-derived hyperpolarizing factor. The dysfunctional endothelium presents vasoconstriction, inflammation, platelet aggregation, and procoagulant and proliferative actions. The substances produced by these actions are angiotensin II, endothelin-1, oxygen-derived free radicals, and thromboxane A_2.

Therefore, the healthy endothelium presents the vasodilation phenotype with high levels of vasodilators, such as nitric oxide (NO) and prostacyclin (PGI2), and low levels of reactive oxygen species (ROS) and uric acid. Further, it has an anticoagulant phenotype with low levels of plasminogen activator inhibitor-1 (PAI-1), von Willebrand factor (vWF), and P-selectin. In addition, there are low levels of inflammation and soluble vascular cell adhesion molecules (sVCAM), soluble intracellular adhesion molecules (sICAM), E-selectin, C-reactive protein (CRP), alpha tumoral necrosis factor (TNF-α), and interleukin-6 (IL-6). Moreover, the endothelial progenitor cell population, which indicates the endothelial reparation capacity, is large, in contrast to the low levels of microparticles of endothelial origin and circulating endothelial cells that indicate stress or endothelial damage.

In endothelial dysfunction, the phenotype includes vasodilation impairment, increased oxidative stress, uric acid, peroxidized lipid radicals, and nitrotyrosine, low levels of NO, procoagulant and proinflammatory profile, with reduced vascular reparation capacity and increased number of endothelial microparticles (EMP) and circulating endothelial cells (CEC), 6-keto-prostaglandin F1-alpha, prostacyclin and ADMA (asymmetric dimethylarginine), NO_2 (nitrite), NO_3 (nitrate), and peroxynitrite (Fig. 34.1).

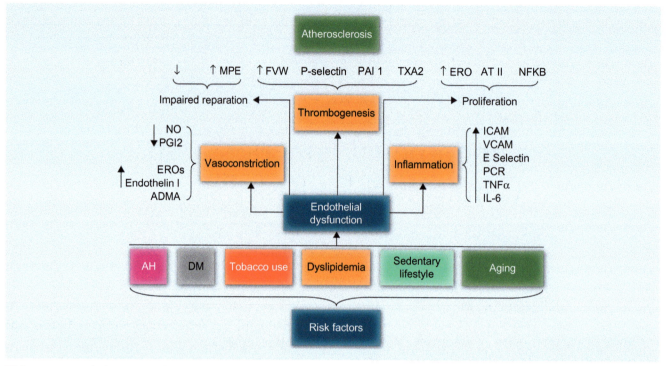

FIG. 34.1   Risk factors and endothelial dysfunction. *CEC*, circulating endothelial cells; *EMP*, endothelial microparticles; *NO*, nitric oxide; *PGI2*, prostacyclin; *ROS*, reactive oxygen species; *ADMA*, asymmetric dimethylarginine; *vWF*, von Willebrand factor; *P-selectin*, platelet selectin; *PAI-1*, plasminogen activator inhibitor-1; *TXA2*, thromboxane A2; *A-II*, angiotensin II; *NF-κB*, nuclear factor kappa B; *ICAM*, intercellular adhesion molecule; *VCAM*, vascular cell adhesion molecule; *E-selectin*, endothelial selectin; *CRP*, C-reactive protein; *TNF-α*, alpha tumoral necrosis factor; *IL6*, interleukin-6; *AH*, arterial hypertension.

ENDOTHELIAL ACTIVITY ASSESSMENT METHODS

There are several modes for endothelial activity assessment:

(a) *Noninvasive methods*: flow-mediated dilation (FMD); pulse wave analysis; venous occlusion plethysmography; peripheral arterial tonometry; Doppler laser flowmetry; magnetic resonance imaging; PET scan.

(b) *Invasive methods*: coronarography with intracoronary administration of vasoactive agents; intrabrachial administration of vasoactive agents.

(c) *Circulating markers of endothelial function*: asymmetric dimethylarginine (ADMA), soluble E-selectin, thrombomodulin, von Willebrand factor, circulating endothelial cells and endothelial progenitor cells, and circulating endothelial microparticles.

ENDOTHELIAL ACTIVITY AND PROGNOSIS

Several authors demonstrated that endothelial functional evaluation correlates with a patient's prognosis.

Therefore, by using coronary vasomotor response tests, Al Suwaidi et al. [3], evaluating individuals with coronary atherosclerosis without critical lesions, found that altered coronary vasomotion predicted higher occurrence of cardiac events; while Targonski et al. [4] established the correlation with increased occurrence of cerebrovascular events in the same type of patient.

In a comparative study between patients with coronary disease and lesion-free individuals, Halcox et al. [5] discovered that altered coronary vasomotricity was an independent predictor of cardiovascular events. The predictive value of peripheral circulation assessment was also proven in several subgroups studied. Venous occlusion plethysmography proved to be useful in distinguishing hypertensive patients at high risk of coronary events [6], patients with coronary diseases [7], postmenopausal women [8], and patients with acute coronary syndromes [9].

Altered flow-mediated vasodilation (FMD) correlates with cardiovascular events in populations with risk factors [10,11], hypertensive postmenopausal women [8,12], elderly people [13], patients with peripheral vascular disease [14], and patients with vascular disease in the preoperative period [15]. In coronary disease, the number of events increased in patients with infarction without ST-segment elevation [16], patients with stable coronary disease [17], and patients after stent implantation. The low amplitude capillary pulse and reactive hyperemia in peripheral arteries correlates with the presence of coronary disease and higher incidence of events [18,19]. In vasodilation due to intraarterial administration of acetylcholine in patients of coronary disease, the number of events in patients with below-average vasodilation doubled in respect to those with above-average vasodilation [7].

Table 34.1 shows the risk factors and some interventions that may improve endothelial dysfunction.

One of the most widely used methods is flow-mediated vasodilation in the brachial artery. In our Endothelium Laboratory, we have assessed the impact of risk factors on flow-mediated

TABLE 34.1 Factors That Influence Endothelial Function

Risk factors	Improve endothelial function
Aging	L-Arginine
Male gender	Antioxidants
Family history	Quit smoking
Hypercholesterolemia	Lowering cholesterol/statins
Tobacco	Converting enzyme inhibitors
Low HDL-cholesterol	Physical activity
Arterial hypertension	Mediterranean diet
Diabetes mellitus	
Obesity	
Diet rich in fats	

vasodilation; all risk factors reduced it, especially arterial hypertension and obesity (Fig. 34.2).

RISK FACTORS AND ENDOTHELIAL FUNCTION

Family History

Predisposition to coronary atherosclerotic disease (CAD) is multifactorial and polygenetic. In polygenetic diseases, single genes have little effect that is

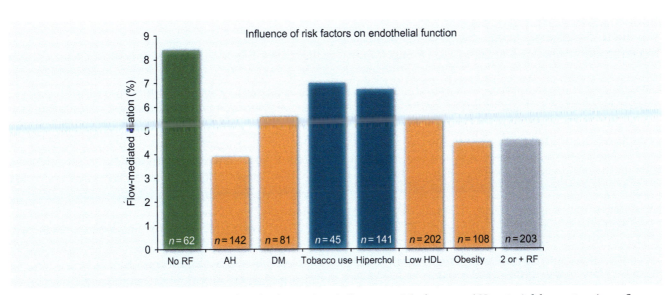

FIG. 34.2 Flow-mediated dilation, in brachial arteries, influences risk factors. *AH*, arterial hypertension. *Source*: Laboratório de Endotélio, Equipe de Aterosclerose InCor-HCFMUSP (InCor-HCFMUSP Endothelium Laboratory, Atherosclerosis Team).

difficult to document. Therefore, linkage analysis, in which a few hundred DNA markers are used, is not able to identify the genes that predispose to polygenetic diseases. Only the emergence of the GWAS (Gene-Wide Association Study) technique, which analyzes millions of DNA markers, allowed assessment of the contribution of several genes to define the final phenotypes in polygenetic diseases [20].

This technique, associated with quick genotyping platforms, allowed mapping CAD genetic variations [21–23]. The CARDIoGRAMplusC4D study found 36 genetic variants associated with CAD, and these variants were confirmed in populations different from those originally studied [24]. In the GWAS analysis, the genetic variants associated with CAD showed several common characteristics: frequent occurrence in population; each variant has little individual effects; less than half act through conventional risk factors, and effects on cholesterol prevail, followed by arterial hypertension, and ABO locus, increasing propensity to coronary thrombosis; most act by risk factors independent mechanisms; most SNPs (single nucleotide polymorphism) are located in noncoding protein regions; finally, the risk is proportional to the number of risk variants present.

It is important to emphasize that the genetic variants associated with increased coronary disease risk are not indicated in CAD prevention and treatment assessments, because most of their action mechanisms are still unknown; however, the GWAS techniques opened a wide research field for new therapeutic targets. Even the GWAS approach may not be sufficient to determine the heritability to which epistasis might be due, that is, gene-to-gene interactions. Gene interactions occur by means of networks instead of isolated units, and have synergistic effects beyond the simple sum of effects.

In the Framingham Offspring study, if the father or mother had premature coronary disease, the risk of their children to have DAC was increased three to five times (Fig. 34.3). However, in this study and in a study conducted in Scotland, risk factor frequency was higher in individuals with early DAC family history [25].

FMD in healthy youngsters descendants from patients who showed early coronary disease was 60% in respect to individuals without ancestors with coronary disease [26]. Schachinger et al. [27] observed that the baseline coronary flow of

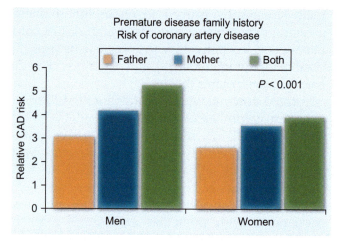

FIG. 34.3 Relative risk of heart ischemic disease due to early disease family history (Framingham Offspring Study). Created from data of Lloyd-Jones DM, Nam BH, D'Agostino Sr RB, et al. Parental cardiovascular disease as a risk factor for cardiovascular disease in middle-aged adults. A prospective study of parents and offspring. JAMA 2004;291:2204–11.

descendants of CAD patients was only 40% in comparison to nondescendants, and the maximum flow in response to administration of acetylcholine was 30% lower in children of patients CAD [26]. In the Northern Manhattan Family Study, by means of linkage analysis, the heritability of FMD was statistically significant amounting to 0.17 after adjustment for classic risk factors, their treatments, and brachial artery diameter [28]. In the Twins Heart Study, in a model with risk factor control, heritability was 39% [29].

Sedentarism

Prolonged rest reduces FMD in individuals with high-insulin levels, that is, insulin-resistant individuals, as well as healthy adults [30,31]. On the contrary, physical activity increases circulating endothelial progenitor cells, inhibits neointima formation, and enhances angiogenesis. It increases nitrite plasmatic levels and superoxide dismutase and decreases oxidative stress, besides superregulating NO-synthase transcription and decreasing NADPH oxidase activity. In contrast to sedentarism, regular physical activity is essential to maintain normal endothelium phenotype [32–34].

Sedentarism is followed by reduced number of circulating endothelial cells and increased endothelial microparticles, markers of elevated endothelial cell apoptosis, and both are correlated to reduced

FMD [35–38]. Conversely, exercise for 10 weeks improves endothelial dysfunction, and therefore Claskson et al. demonstrated an 80% increase in FMD youngsters. Although it increases 54% in young adults after only 1 week of training, it may return to the baseline values after 4 weeks of training interruption [36,39–41].

Aging

Aging is one of the main risk factors. Deaths due to ischemic heart disease in Brazil have risen from 10/100,000 between 30 and 34 years of age to 1200/100,000 at 80 years among men, and from 6/100,000 to 1000/100,000 among women. This is due to the increased risk factors prevalence such as hypertension and diabetes and to aging itself.

Aging is followed by FMD decline, with a reduction of 50% from youngsters to the elderly [8,42]. Between middle age and old age, the drop is ~25% [43–45] (Fig. 34.4). In postmenopause women, the risk of cardiovascular events increases four times for those in the lower third of FMD.

The mechanisms suggested for aging related dysfunction are vessel structural changes, but, mainly, the drop in nitric oxide (NO) availability due to NO-synthase isoform changes inherent to aging [46]. There is also a reduction in endothelium-derived hyperpolarizing factors, among them H_2O_2 [47]. Other mechanisms are those associated with endothelial cell senescence, such as splicing

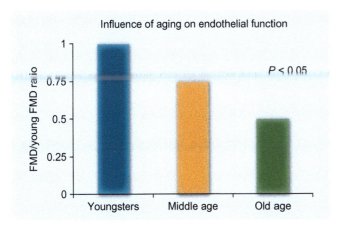

FIG. 34.4 FMD and aging. Adapted from Celermajer DS, Sorensen KE, Spiegelhaller DJ, et al. Aging is associated with endothelial dysfunction in healthy men years before the age-related decline in women. J Am Coll Cardiol 1994;24(2):471–6.

factor 1 (removal of introns and fusion of exons after RNA transcription) and serine-arginine (SRSF1), which lead to underregulation of telomerase catalytic units due to decreased endothelial reparation, together with reduction in circulating endothelial progenitor cells [48–51].

Regular physical activity preserves FMD, preventing the reduction that occurs due to the aging process [52–58]. A longitudinal study in adolescents demonstrated that FMD decreased between 13 and 17 years of age from 9.1% to 9% in sedentary individuals, but increased from 8.7% to 12% in those who incremental physical activity during this period [59].

The mechanism responsible for FMD augmentation is increased NO-synthase expression and nitric oxide bioavailability [41].

Gender

Men have higher incidence of atherosclerosis than women do. With aging the evolution of the lipid profile is different between genders; men show elevation in small and dense LDL atherogenic particles, and women show elevation in large and less atherogenic LDL particles. Regarding small HDL particles, the concentration in youth, and its ascension and fall in middle age and old age, are lower in women than in men [60]. The mechanisms of this distinct behavior would be women's hormonal protection in the reproductive phase.

The female sex hormones increase constitutive NO-synthase even more, by activating the estrogen alpha nuclear receptors [61–63], and also regulate prostacyclin synthase, which increases the vascular endothelial growth factor (VEGF) and inhibits smooth muscle cell apoptosis, migration, and proliferation [64]. Another action of female hormones is to inhibit LDL oxidation, thereby positively influencing endothelium-dependent relaxation [65–67].

Therefore, FMD, which is similar in men and young women in the first phase of the menstrual cycle, evolves differently in both sexes, with a drop from 40 years of age in men and 55 years of age in women [44,68].

FMD is higher in women than in men and, in our laboratory at InCor HC-FMUSP, the frequency of abnormal vasodilation was similar in both sexes: 36.5% in men and 35.5% in women. Women maintain FMD higher than men do from young age up to the eight decade of life [69] (Fig. 34.5).

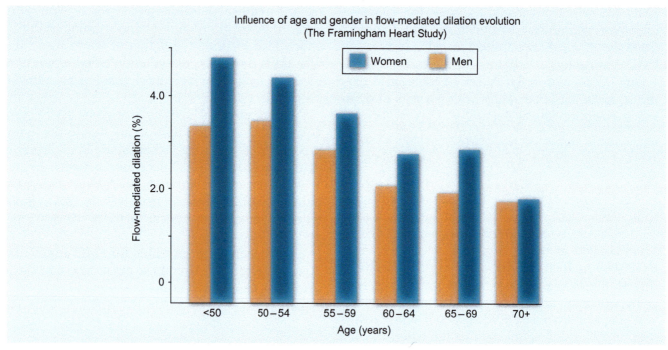

FIG. 34.5   Flow-mediated dilation influences gender [69]. Created from data extracted of Benjamin EJ, Larson MG, Keyes MJ, et al. Clinical correlates and heritability of flow-mediated dilation in the community. The FraminghamHeart Study. Circulation 2004;109:613–9.

Arterial Hypertension

Other than peripheral resistance, the endothelium influences other factors that may lead to hypertension, such as vascular rigidity and endothelial activation with vascular remodeling, and inflammatory response with expression of adhesion molecules and cytokines, such as the alpha tumoral necrosis factor (TNF-α).

Although vasoconstriction due to angiotensin II and aldosterone mineralocorticoid effects as well as endothelin production are all implicated in arterial hypertension genesis, an additional mechanism would be increased production of the oxygen-free radical, superoxide; angiotensin II increase NADPH oxidase activity with NO inactivation, generating peroxynitrite and also decreasing extracellular superoxide dismutase activity [70–73].

Thus, endothelial dysfunction is a usual finding in arterial hypertension, and is identical to aging. It may even precede it, because children of hypertensive individuals already show higher arterial rigidity in childhood and adolescence even with normal blood pressure [74–77]. The elasticity that is already reduced in prehypertensive individuals drops even more in individuals with established hypertension and increases the incidence of events three times (Fig. 34.6) [78–80].

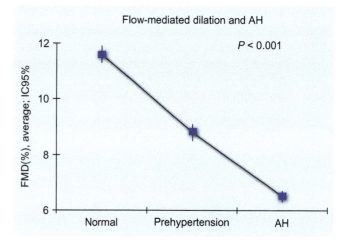

FIG. 34.6   Flow-mediated dilation and AH. Modified from Giannotti G, Doerris C, Mocharla PS, et al. Impaired endothelial repair capacity of early endothelial progenitor cells in prehypertension relation to endothelial dysfunction. Hypertension 2010;55:1389–97.

The brachial artery FMD of hypertensive individuals improves with the use of converting enzyme inhibitors, as they increase NO availability and reduce bradykinin degradation, which causes vasodilation by activating endothelium-derived hyperpolarizing factors [81–83].

Cigarette Smoking

Smoking depresses FMD in dose-dependent fashion, and this dysfunction is reversible by giving up smoking for ~1 year. FMD is one third to half in respect to nonsmokers (Fig. 34.7) [84–86].

The endothelial morphological alterations induced by smoking include irregular aspect, formation of vesicles, oxidation-mediated retraction, and collapse of the cytoskeleton tubulin system, and correlate to functional alterations such as reduced FMD due to reduced NO bioavailability, reduced prostacyclin production and increased expression of adhesion molecules. Smoking also increases endothelial permeability to LDL, platelet and macrophage adhesion, and leads to a procoagulant and inflammatory state.

Smoking increases LDL-cholesterol, reduces HDL-cholesterol, and causes qualitative lipid alterations. The oxygen-free radicals present on cigarettes itself and those induced in smokers lead to lipid peroxidation and biomolecule oxidation and inactivation, creating a medium that is favorable to atherosclerosis development and complications. Therefore, tobacco triggers cellular oxidative stress, which may be reverted by using antioxidants such as vitamin C [87]. The higher the redox imbalance, the higher the incidence of events [7]. The inflammatory state leads to activation of matrix metalloproteinases and reduction of its inhibitors (TIMP), promoting acute plaque complications [88–98].

Diabetes Mellitus

Diabetes systematically reduces FMD, as demonstrated by several authors (Fig. 34.8). Moreover, the reduction is proportional to the glucose level and diabetes duration [99,100]. The presence of diabetes elevates the risk of cardiovascular events two to four times [101,102]. Endothelial dysfunction, an early diabetes vascular disease marker, is a common independent predictor of cardiovascular events. It is also observed in conditions associated with diabetes type 2: obesity, sedentary lifestyle, and metabolic syndrome [103–106].

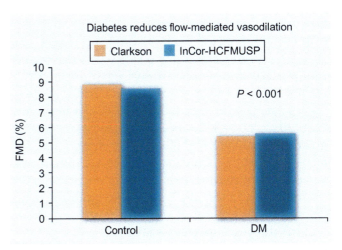

FIG. 34.8 Diabetes reduces flow-mediated vasodilation [99].

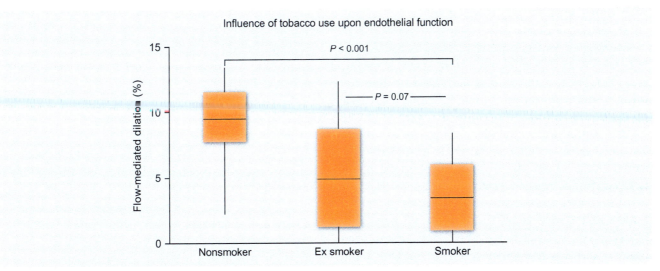

FIG. 34.7 Flow-mediated dilation and tobacco use. Created from data extracted from Celermajer DS, Sorensen KE, Georgakopoulos D, et al. Cigarette smoking is associated with dose-related and potentially reversible impairment of endothelium-depend dilation in healthy youngadults. Circulation 1993;88(Part I):2149–55; Celermajer DS, Sorensen KE, Gooch VM, et al. Non-invasive detection of endothelial dysfunction in children and adults at risk of atherosclerosis. Lancet 1992;340:1111–5; Zeiher AM, Schächinger V, Minners J. Long-term cigarette smoking impairs endothelium-dependent coronary arterial vasodilator function. Circulation 1995;92:1094–100.

The mechanisms that lead to endothelial dysfunction in diabetes start with hyperglycemia and increased free fatty acids. Leading to elevated cytosol redox potential due to increased concentration of NADH and glycerol-3-phosphate, which in turn stimulate electron transfer to the mitochondrial respiratory chain, with reduced Q cycle and complex III activity and decoupling of oxidative phosphorylation, lower ATP production, and increased superoxide radical production, which maintains this vicious cycle with lower constitutional NO-synthase activity and NO production [107–109].

There is evidence that a reciprocal relation between insulin resistance and endothelial dysfunction exists. Insulin resistance operates by means of cell signaling in the endothelium, fat tissue, and skeletal muscle. There is alterations in the insulin receptor substrate-1 (IRS-1) and the phosphatidylinositol 3-kinase Akt system, which besides decreasing production of constitutional NO-synthase and NO, also causes reduced translocation of the glucose-4 transporter (GLUT-4). Human studies have shown the relevance of these mechanisms. In healthy individuals, insulin administration stimulates vasodilation and increases blood flow to peripheral tissues, while this is blocked in diabetes and insulin-resistance cases [109–111].

Diabetic or obese patients show high-inflammatory markers levels, including reactive protein C (RPC), TNF-α, interleukin-6, and intercellular cell adhesion molecule-1 (ICAM-1), and these high levels are associated with increased cardiovascular risk [112–114].

The NF-κB nuclear transcription factor is a key regulator of endothelial activation, and is associated with insulin resistance. It is activated by free fatty acids, inflammatory cytokines, and advanced glycation end products receptor (RAGE) [113–115].

The signaling/metabolic pathway of protein kinase C beta (PKCB) is activated in diabetes, due to elevated diglycerides, and may explain the relation between inflammation, endothelial dysfunction, and insulin resistance, because PKCB inhibits the PI3 kinase Akt pathway, reducing NO-synthase phosphorylation, and activates NF-κB and generation of reactive oxygen species by NADPH oxidase, vascular adhesion molecules, proinflammatory cytokines, and growth factors [116–120].

The use of angiotensin II antagonists decreases oxidative stress and improves endothelial dysfunction in diabetic and hypertensive patients [121].

HYPERCHOLESTEROLEMIA

Since the classic Framingham study, it is clear that high-cholesterol levels are closely related to incidence of atherosclerotic disease and its complications. The increased risk occurs in all age groups, despite its attenuation with aging. However, the Seven Countries Study observed that the risk of events might vary according to the country in which the individuals live, with higher risk in countries in the north of Europe and United States, and lower risk in countries on the Mediterranean shoreline and Japan [122] (Fig. 34.9).

The particles mostly implicated in atherogenesis are LDL oxidized by reactive oxygen species. The small, dense, and oxidized LDL are found in atherosclerotic lesions, and their toxicity extends through the intima, releasing phospholipids that activate endothelial cells and produce inflammatory reactions [123].

Therefore, oxidized LDL induces adhesion molecules production in the endothelium, monocyte attraction molecules, have cytotoxic effects and increase activation of proinflammatory genes and growth factors in endothelial and vascular wall cells, all leading to endothelial dysfunction, which promotes platelet aggregation, metalloproteinases expression, thrombogenesis, and formation of foam cells. In turn, these processes increase oxidative stress and generation of reactive oxygen species, creating a vicious spiral. Several studies revealed that oxidized LDL inhibits FMD by inactivating NO, inactivating NO-synthase, or reducing arginine availability, which is NO-synthase substrate to generate NO. Therefore, elevated LDL is followed by reduced FMD [124,125].

Apart from the oxidative modifications, LDL also affects endothelial NO-synthase function due to elevated expression of caveolin-1 and stabilization of the heterocomplex of its bond with NO-synthase. Caveolin-1, the protein complex of the caveolar layer, and NO-synthase, are known to turn it inactive. And this effect is proportional to intracellular cholesterol increase, which modulates caveolin-1 gene transcription by means of the sterol regulatory element binding protein (SREBP). Thus NO bioavailability may decrease both by maintenance of its bond with caveolin-1 and by the action of asymmetric dimethyl L-arginine (ADMA) which blocks the active site of the enzyme. It occurs regardless

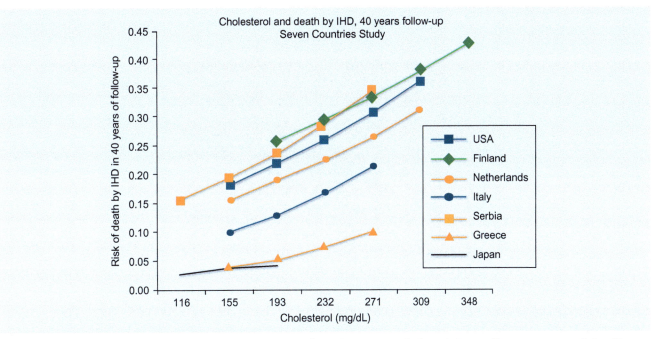

FIG. 34.9 Influence of cholesterol levels and country of residence upon ischemic heart disease, report of the 40-year follow-up of the Seven Countries Study. Adapted from Menotti A, Lanti M, Kromhout D, et al. Homogeneity in the relationship of serum cholesterol to coronary deaths across different cultures. 40-years follow-up of the Seven Countries Study. Eur J Cardiovasc Prev Rehabil 2008;15(6):719–25.

abundance NO synthase [126–130]. The ADMA release by endothelial cells increases due to the presence of native or oxidized LDL, possibly mediated by S-adenosilmetionine-dependent methyltransferase superregulation, with subsequent proteolysis of dimethyl-L-arginine. ADMA levels are higher in patients with coronary arterial disease and are markers of higher risk of cardiovascular events [131–134].

Contrary to LDL, HDL activates endothelial NO-synthase via SR-BI (scavenger receptor BI) through a process that requires apo-AI bond. This action, complementing its antioxidant and antiinflammatory activities, makes it a protector against atherosclerosis [134].

However, in certain conditions, such as diabetes and coronary disease, HDL might lose these beneficial activities by inactivation of paraoxanase-1 and other oxidative modifications. It becomes a blocker of NO-synthase by activating Lox1 (lectin-like oxidized LDL receptor) and PKC beta II (protein kinase C beta II) [134–136].

Cholesterol has direct actions upon potassium ion channels, among them inwardly rectifying potassium channels (Kir). Of these channels, cholesterol acts upon channels Kir2, Kir4, and Kir6.

Channel Kir2 is suppressed by elevated cholesterol in cell membrane and enhanced by its depletion. Fang et al. [137] demonstrated that hypercholesterolemia leads to endothelial cell membrane depolarization, impairing endothelium-mediated vasodilation. Higher cholesterol levels, even within the normal range, lead to reduced flow-mediated vasodilation and increased sympathetic tone [138]. The use of statins, causing decreased cholesterolemia, practically doubles flow-mediated dilation [139].

LOW HDL-CHOLESTEROL

HDL have multiple antiatherosclerotic, antioxidant, antiinflammatory, and vasodilation actions. Administration of human HDL improves FMD in individuals with hypercholesterolemia or diabetes. This action is immediate and due to increased NO-synthase expression and increased nitric oxide bioavailability. A more sustained response is ascribed to increased circulating endothelial progenitor cells and endothelium reparation, by its mobilization of bone marrow, and also reduced apoptosis

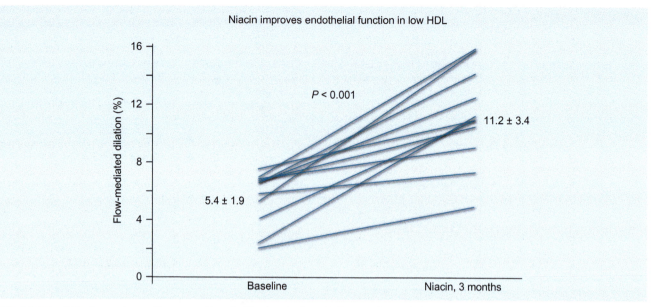

FIG. 34.10 Niacin restores flow-mediated dilation in low HDL. Adapted from Benjo AM, Maranhão RC, Coimbra SR, et al. Accumulation of chylomicron remnants and impaired vascular reactivity occur in subjects with isolated low HDL cholesterol: effects of niacin treatment. Atherosclerosis 2006;187(1):116–22.

of these cells due to increased NO bioavailability of in bone marrow and in the vascular endothelium.

HDL interacts with endothelial cells via apo-E with SR-BI (scavenger receptor BI), promotes increased intracellular calcium concentration, and activates the signaling Src (serine kinase) pathways and kinases PI3K, p38 MAPK, p42/44 MAPK, Rho, and Rac, which promote endothelial cell proliferation, migration, and endothelial NO-synthase expression, also reducing apoptosis by blocking the activation of caspases 3 and 9 [99,140–150].

Endothelial function improvement may result from increased HDL concentration or from the better functioning of these particles, as demonstrated in our laboratory by Benjo et al. [151], when endothelium-mediated vasodilation improved with the use of niacin, without changing the HDL-cholesterol concentration in individuals with low HDL (Fig. 34.10). Improvement of HDL functions, promoted by controlled aerobic exercises for a few months, without changing the plasmatic levels, was also demonstrated by Casella et al. [152] in our laboratory.

CONCLUSIONS

Endothelial dysfunction is a common denominator among risk factors, and is the link between these factors and atherosclerosis. Endothelial function

assessment might be used as a subclinical atherosclerosis marker and predictor of events in both asymptomatic individuals and those with manifested cardiovascular disease.

References

[1] Mozaffarian D, Benjamin EJ, Go AS, et al. Heart disease and stroke statistics—2015 update: a report from the American Heart Association. Circulation 2015;131:e99–e322.
[2] Yusuf S, Hawken S, Ounpuu S, et al. Effect of potentially modifiable risk factors associated with myocardial infarction in 52 countries (the INTERHEART study): case-control study. Lancet 2004;364:937–52.
[3] Al Suwaidi J, Hamasaki S, Higano ST, et al. Long-term follow-up of patients with mild coronary artery disease and endothelial dysfunction. Circulation 2000;101:948–54.
[4] Targonski PV, Bonetti PO, Pumper GM, et al. Coronary endothelial dysfunction is associated with an increased risk of cerebrovascular events. Circulation 2003;107:2805–9.
[5] Halcox JPJ, Schenke WH, Zalos G, et al. Prognostic value of coronary vascular endothelial dysfunction. Circulation 2002;106:653–8.
[6] Perticone F, Ceravolo R, Pujia A, et al. Prognostic significance of endothelial dysfunction in hypertensive patients. Circulation 2001;104:191–6.
[7] Heitzer T, Schiling T, Krohn K, et al. Endothelial dysfunction, oxidative stress, and risk of cardiovascular events in patients with coronary artery disease. Circulation 2001;104:2673–8.
[8] Rossi R, Nuzzo A, Origliani G, et al. Prognostic role of flow-mediated dilation and cardiac risk factors in post-menopausal women. J Am Coll Cardiol 2008;51(10):997–1002.
[9] Fichtlscherer S, Rosenberger G, Walter DH, et al. Elevated C-reactive protein levels and impaired endothelial vasoreactivity in patients with coronary artery disease. Circulation 2000;102:1000–6.

[10] Shimbo D, Grahame-Clarke C, Miyake Y, et al. The association between endothelial dysfunction and cardiovascular outcomes in a population-based multi-ethnic cohort. Atherosclerosis 2007;192(1):197–203.

[11] Shechter M, Shechter A, Koren-Morag N, et al. Usefulness of brachial artery flow-mediated dilation to predict long-term cardiovascular events in subjects without heart disease. Am J Cardiol 2014;113(1):162–7.

[12] Modena MG, Bonetti L, Coppi F, et al. Prognostic role of reversible endothelial dysfunction in hypertensive postmenopausal women. J Am Coll Cardiol 2002;40(3):505–10.

[13] Yeboah J, Folsom AR, Burke GL, et al. Predictive value of brachial flow-mediated dilation for incident cardiovascular events in a population-based study: the multiethnic study of atherosclerosis. Circulation 2009;120:502–9.

[14] Brevetti G, Silvestro A, Schiano V, et al. Endothelial dysfunction and cardiovascular risk prediction in peripheral arterial disease additive value of flow-mediated dilation to ankle-brachial pressure index. Circulation 2003;108:2093–8.

[15] Gokce N, Keaney JF, Hunter LM, et al. Predictive value of noninvasively determined endothelial dysfunction for long-term cardiovascular events in patients with peripheral vascular disease. J Am Coll Cardiol 2003;41(10):1769–75.

[16] Karatzis EN, Ikonomidis I, Vamyakou GD, et al. Long-term prognostic role of flow-mediated dilatation of the brachial artery after acute coronary syndromes without ST elevation. Am J Cardiol 2006;98(11):1424–8.

[17] Chan SY, Mancini BJ, Kuramoto L, et al. The prognostic importance of endothelial dysfunction and carotid atheroma burden in patients with coronary artery disease. J Am Coll Cardiol 2003;42(6):1037–43.

[18] Bonetti PO, Pumper GM, Higano ST, et al. Noninvasive identification of patients with early coronary atherosclerosis by assessment of digital reactive hyperemia. J Am Coll Cardiol 2004;44:2137–41.

[19] Rubinshtein R, Kuvin JT, Soffler M, et al. Assessment of endothelial function by non-invasive peripheral arterial tonometry predicts late cardiovascular adverse events. Eur Heart J 2010;31:1142–8.

[20] The International Hap Map Consortium. A haplotype map of the human genome. Nature 2005;437:1299–320.

[21] McPherson R, Pertsemlidis A, Kavaslar N. A common allele on chromosome 9 associated with coronary heart disease. Science 2007;316:1488–91.

[22] Helgadottir A, Thorleifsson G, Manolescu A, et al. A common variant on chromosome 9p21 affects the risk of myocardial infarction. Science 2007;316:1491–3.

[23] Preuss M, Konig IR, Thompson JR, et al. Design of the Coronary ARtery DIsease Genome-Wide Replication and Meta-Analysis (CARDIoGRAM) study: a genome-wide association meta-analysis involving more than 22,000 cases and 60,000 controls. Circ Cardiovasc Genet 2010;3:475–83.

[24] The CARDIoGRAMplusC4D Consortium. Coronary artery disease risk loci identified in over 190,000 individuals implicate lipid metabolism and inflammation as key causal pathways. Nat Genet 2013;45:25–33.

[25] Thompson HJ, Pell ACH, Anderson J, et al. Screening families of patients with premature coronary heart disease to identify avoidable cardiovascular risk: a cross-sectional study of family members and a general population comparison group. BMJ Res Notes 2010;3:132.

[26] Clarkson P, Celermajer DS, Powe AJ, et al. Endothelium-dependent dilatation is impaired in young healthy subjects with a family history of premature coronary disease. Circulation 1997; 96:3378–83.

[27] Schachinger V, Britten MB, Elsner M, et al. A positive family history of premature coronary artery disease is associated with impaired endothelium-dependent coronary flow regulation. Circulation 1999;100:1502–8.

[28] Suzuki K, Juo SHK, Rundek T, et al. Genetic contribution to brachial artery flow-mediated dilation: the Northern Manhattan Family Study. Atherosclerosis 2008;197(1):212–6.

[29] Zhao J, Cheema FA, Reddy U, et al. Heritability of flow-mediated dilation: a twin study. J Thromb Haemost 2007;5(12):2386–92.

[30] Sonne MP, Hojbjerre L, Alibegovic AC, et al. Endothelial function after 10 days of bed rest in individuals at risk for type 2 diabetes and cardiovascular disease. Exp Physiol 2011;96(10):1000–9.

[31] Nosova EV, Yen P, Chong KC, et al. Short-term physical inactivity impairs vascular function. J Surg Res 2014;190(2):672–82.

[32] Booth FW, Roberts CK. Linking performance and chronic disease risk: indices of physical performance are surrogates for health. Br J Sports Med 2008;42:950–2.

[33] Laughlin MH, Roseguini B. Mechanisms for exercise training-induced increases in skeletal muscle blood flow capacity: differences with interval sprint training versus aerobic endurance training. J Physiol Pharmacol 2008;59(Suppl. 7):71–88.

[34] Laughlin MH, Newcomer SC, Bender SB. Importance of hemodynamic forces as signals for exercise induced changes in endothelial cell phenotype. J Appl Physiol 2008;104:588–600.

[35] Edwards DG, Scholfield RS, Lennon SL, et al. Effect of exercise training on endothelial function in men with coronary artery disease. Am J Cardiol 2004;93(5):617–20.

[36] Hambrecht R, Wolf A, Gielen S, et al. Effect of exercise on coronary endothelial function in patients with coronary artery disease. N Engl J Med 2000;342:454–60.

[37] Laufs U, Wassman S, Czech T, et al. Physical inactivity increases oxidative stress, endothelial dysfunction, and atherosclerosis. Arterioscler Thromb Vasc Biol 2005;25:809–14.

[38] Boyle LJ, Credeur DP, Jenkins NT, et al. Impact of reduced daily physical activity on conduit artery flow-mediated dilation and circulating microparticles. J Appl Physiol 2013;115(10):1519–25.

[39] Clarkson P, Montgomery HE, Mullen MJ, et al. Exercise training enhances endothelial function in young health men. J Am Coll Cardiol 1999;33(5):1379–85.

[40] Tinken TM, Thijssen DHJ, Black MA, et al. Time course of change in vascular capacity in response to exercise training in humans. J Physiol 2008;20:5003–12.

[41] Hambrecht R, Adams V, Erbs S, et al. Regular physical activity improves endothelial function in patients with coronary artery disease by increasing phosphorylation of endothelial nitric oxide synthase. Circulation 2003;107:3152–8.

[42] DeVan AE, Seals DR. Vascular health in the ageing athlete. Exp Physiol 2012;97(3):305–10.

[43] Siasos G, Chrysohoou C, Tousoulis D, et al. The impact of physical activity on endothelial function in middle-aged and elderly subjects: the Ikaria study. Hell J Cardiol 2013;54:94–101.

[44] Celermajer DS, Sorensen KE, Spiegelhalter DJ, et al. Aging is associated with endothelial dysfunction in healthy men years before the age-related decline in women. J Am Coll Cardiol 1994;24(2):471–6.

[45] Wray DW, Nishiyama SK, Harris RA, et al. Acute reversal of endothelial dysfunction in elderly after antioxidant consumption. Hypertension 2012;59:818–24.

[46] Cau SBA, Carneiro FS, Tostes RC. Differential modulation of nitric oxide synthases in aging: therapeutic opportunities. Front Physiol 2012;3:1–11.

[47] Shimokawa H. Hydrogen peroxide as an endothelial-derived hyperpolarizating factor. Pflugers Arch—Eur J Physiol 2010;459:915–22.

[48] Blanco FJ, Bernabeu C. The splicing factor SRSF-1 as a marker for endothelial senescence. Front Physiol 2012;54:1–6.

[49] Voglauer R, Chang MW, Dampier B, et al. SNEV overexpression extends the life span of human endothelial cells. Exp Cell Res 2006;312:746–59.

II. ENDOTHELIAL DYSFUNCTION AND CLINICAL SYNDROMES

[50] Grillari J, Grillari-Voglauer R, Jansen-Durr P. Post-translational modification of cellular proteins by ubiquitin and ubiquitin-like molecules: role in cellular senescence and aging. Adv Exp Med Biol 2010;694:172–96.

[51] Schraml E, Voglauer R, Fortschegger K, et al. Haploinsufficiency of senescence evasion factor causes defects of hematopoietic stem cells functions. Stem Cells Dev 2008;17:355–66.

[52] Black MA, Cable NT, Thijssen DHJ, et al. Impact of age, sex, and exercise on brachial artery flow-mediated dilatation. Am J Physiol Heart Circ Physiol 2009;297:H1109–16.

[53] Moyna NM, Thompson PD. The effect of physical activity on endothelial function in man. Acta Physiol Scand 2004;180(2):113–23.

[54] Rinder MR, Spina RJ, Ehsani AA. Enhanced endothelium-dependent vasodilation in older endurance-trained men. J Appl Physiol 2000;88:761–6.

[55] Seals DR, DeSouza CA, Donato AJ, et al. Habitual exercise and arterial aging. J Appl Physiol 2008;105:1323–32.

[56] Seals DR, Walker AE, Pierce GL, et al. Habitual exercise and vascular ageing. J Physiol 2009;587(3):5541–9.

[57] Trott DW, Gunduz F, Laughlin MH, et al. Exercise training reverses age-related decrements in endothelium-dependent dilation ins skeletal muscle feed arteries. J Appl Physiol 2009;106:1925–34.

[58] DeSouza CA, Shapiro LF, Clevenger CM, et al. Regular aerobic exercise prevents and restores age-related declines in endothelium-dependent vasodilation in healthy men. Circulation 2000;102:1351–7.

[59] Pahkala K, Heinonen OJ, Simell O, et al. Association of physical activity with vascular endothelial function and intima-media thickness. A longitudinal study in adolescents. Circulation 2011;124:1956–63.

[60] Freedman DS, Otvos JO, Jeyarajah EJ, et al. Sex and age differences in lipoprotein subclasses measured by nuclear magnetic resonance spectroscopy: the Framingham Study. Clin Chem 2004;50(7):1189–200.

[61] Darblade B, Pendaries C, Krust A, et al. Estradiol alters nitric oxide production in the mouse aorta through the alfa, but not beta, estrogen receptor. Circ Res 2002;90:413–9.

[62] Evinger 3rd AJ, Levin ER. Requirements for estrogen receptor alpha membrane localization and function. Steroids 2005;70:361–3.

[63] Shearman AM, Cupples LA, Demissie S, et al. Association between estrogen receptor a gene variation and cardiovascular disease. JAMA 2003;290(17):2263–70.

[64] Weiner CP, Lizasoain I, Baylis SA, et al. Induction of calcium-dependent nitric oxide synthases by sex hormones. Proc Natl Acad Sci U S A 1994;91:5212–6.

[65] Chowienczyk, Weiner C, Baylis I, et al. Regulation of NO-synthase by sex hormones. Endothelium 1993;1:s1.

[66] Gilligan DM, Quyyumi AA, Cannon RO. Effects of physiological levels of estrogen on coronary vasomotor function in postmenopausal women. Circulation 1994;89:2545–51.

[67] Sack MN, Rader DJ, Cannon III RO. Oestrogen and inhibition of oxidation of low-density lipoproteins in postmenopausal women. Lancet 1994;343:269–70.

[68] Hashimoto M, Akishita M, Eto M, et al. Modulation of endothelium dependent flow-mediated dilatation of the brachial artery by sex and menstrual cycle. Circulation 1995;92:3431–5.

[69] Benjamin EJ, Larson MG, Keyes MJ, et al. Clinical correlates and heritability of flow-mediated dilation in the community. The Framingham Heart Study. Circulation 2004;109:613–9.

[70] Rajagopalan S, Kurz S, Munzel T, et al. Angiotensin II-mediated hypertension in the rat increases vascular superoxide production via membrane NADH/NADPH oxidase activation: contribution to alterations of vasomotor tone. J Clin Invest 1996;97:1916–23.

[71] Grunfeld S, Hamilton CA, Mesaros S, et al. Role of superoxide in the depressed nitric oxide production by the endothelium of genetically hypertensive rats. Hypertension 1995;26:854–7.

[72] Watson T, Goon PK, Lip GY. Endothelial progenitor cells, endothelial dysfunction, inflammation, and oxidative stress in hypertension. Antioxid Redox Signal 2008;10:1079–88.

[73] Lob HE, Vinh A, Li L, et al. Role of vascular extracellular superoxide dismutase in hypertension. Hypertension 2011;58:232–9.

[74] Quiroz R, Enserro DM, Xanthakis V, et al. Increased vascular stiffness in non-hypertensive offspring of hypertensive parents: the Framingham Heart Study. Circulation 2013;128:A15880.

[75] Panza JA, Quyyumi AA, Brush Jr JE, et al. Abnormal endothelium-dependent vascular relaxation in patients with essential hyper-tension. N Engl J Med 1990;323:22–7.

[76] Taddei S, Virdis A, Ghiadoni L, et al. The role of endothelium in human hypertension. Curr Opin Nephrol Hypertens 1998;7:203–9.

[77] John S, Schmieder RE. Impaired endothelial function in arterial hypertension and hypercholesterolemia: potential mechanisms and differences. J Hypertens 2000;18:363–74.

[78] Giannotti G, Doerris C, Mocharla PS, et al. Impaired endothelial repair capacity of early endothelial progenitor cells in prehypertension relation to endothelial dysfunction. Hypertension 2010;55:1389–97.

[79] Muisan MM, Massimo S, Paini A, et al. Prognostic role of flow-mediated dilatation of the brachial artery in hypertensive patients. J Hypertens 2008;26(8):1612–8.

[80] Weil BR, Stauffer BL, Greiner JJ, et al. Prehypertension is associated with impaired nitric oxide-mediated endothelium-dependent vasodilation in sedentary adults. Am J Hypertens 2011;24(9):976–81.

[81] Ghiadoni L, Magagna A, Versari D, et al. Different effect of antihypertensive drugs on conduit artery endothelial function. Hypertension 2003;41:1281–6.

[82] Buus NH, Jorgensen CG, Mulvany MJ, et al. Large and small artery endothelial function in patients with essential hypertension—effect of ACE inhibition and beta-blockade. Blood Press 2007;16:106–13.

[83] Ceconi C, Francolini G, Olivares A, et al. Angiotensin converting enzyme (ACE) inhibitors have different selectivity for bradykinin binding sites of human somatic ACE. Eur J Pharmacol 2007;577:168.

[84] Celermajer DS, Sorensen KE, Georgakopoulos D, et al. Cigarette smoking is associated with dose-related and potentially reversible impairment of endothelium-depend dilation in healthy young adults. Circulation 1993;88(Part I):2149–55.

[85] Celermajer DS, Sorensen KE, Gooch VM, et al. Non-invasive detection of endothelial dysfunction in children and adults at risk of atherosclerosis. Lancet 1992;340:1111–5.

[86] Zeiher AM, Schächinger V, Minners J. Long-term cigarette smoking impairs endothelium-dependent coronary arterial vasodilator function. Circulation 1995;92:1094–100.

[87] Heitzer T, Just H, Munzel T. Antioxidant vitamin C improves endothelial dysfunction in chronic smokers. Circulation 1996;94:6–9.

[88] Lavi S, Prasad A, Yang EH, et al. Smoking is associated with epicardial coronary endothelial dysfunction and elevated white blood cell count in patients with chest pain and early coronary artery disease. Circulation 2007;115:2621–7.

[89] Bernhard D, Csordas A, Henderson B, et al. Cigarette smoke metal-catalyzed protein oxidation leads to vascular endothelial cell contraction by depolymerization of microtubules. FASEB J 2005;19:1096–107.

[90] Pitrilo RM, Bull HA, Gulatis S, et al. Nicotine and cigarette smoking: effects on the ultrastructure of aortic endothelium. Int J Exp Pathol 1990;71:573–86.

[91] Pittilo RM, Woolf N. Cigarette smoking, endothelial cell injury and atherosclerosis. J Smok Rel Disord 1993;4:17–25.

[92] Garbin U, Pasini FA, Stranieri C, et al. Cigarette smoking blocks the protective expression of Nrf2/ARE pathway in peripheral

mononuclear cells of young heavy smokers favouring inflammation. PLoS One 2009;4:e8225.

[93] Morrow JD, Frei B, Longmire AW, et al. Increase in circulating products of lipid peroxidation (F2-isoprostanes) in smokers. Smoking as a cause of oxidative damage. N Engl J Med 1995;332:1198–203.

[94] Salonen JT, Ylä-Herttuala S, Yamamoto R, et al. Autoantibody against oxidised LDL and progression of carotid atherosclerosis. Lancet 1992;339:883–7.

[95] Yamaguchi Y, Haginaka J, Morimoto S, et al. Facilitated nitration and oxidation of LDL in cigarette smokers. Eur J Clin Investig 2005;35:186–93.

[96] Pilz H, Oguogho A, Chehne F, et al. Quitting cigarette smoking results in a fast improvement of in vivo oxidation injury (determined via plasma, serum and urinary isoprostane). Thromb Res 2000;99:209–21.

[97] Csordas A, Bernhard D. The biology behind the atherothrombotic effects of cigarette smoke. Nat Rev Cardiol 2013;10:219–30.

[98] Nordskog BK, Blixt AD, Morgan WT, et al. Matrix-degrading and pro-inflammatory changes in human vascular endothelial cells exposed to cigarette smoke condensate. Cardiovasc Toxicol 2003;3:101–17.

[99] Clarkson P, Celermajer DS, Donald AE, et al. Impaired vascular reactivity in insulin-dependent diabetes mellitus is related to disease duration and low density lipoprotein cholesterol levels. J Am Coll Cardiol 1996;28(3):573–9.

[100] Kawano H, Motoyama T, Hirashita O, et al. Hyperglycemia rapidly suppresses flow-mediated endothelium-dependent vasodilation of brachial artery. J Am Coll Cardiol 1999;34(1):146–54.

[101] Nitenberg A, Valensi P, Sachs R, et al. Prognostic value of epicardial coronary artery constriction to the cold pressor test in type 2 diabetic patients with angiographically normal coronary arteries and no other major coronary risk factors. Diabetes Care 2004;27(1):208–15.

[102] Hamilton SJ, Watts GF. Endothelial dysfunction in diabetes: pathogenesis, significance, and treatment. Rev Diabet Stud 2013;10(2–3):133–56.

[103] Benjamin EJ, Larson MG, Keyes MJ, et al. Clinical correlates and heritability of endothelial function in the community: the Framingham Heart Study. Circulation 2004;109:613–9.

[104] Lteif AA, Han K, Mather KJ. Obesity, insulin resistance, and the metabolic syndrome: determinants of endothelial dysfunction in whites and blacks. Circulation 2005;112:32–8.

[105] De Souza CA, Shapiro LF, Clevenger CM, et al. Regular aerobic exercise prevents and restores age related declines in endothelium-dependent vasodilation in healthy men. Circulation 2000;102:1351 7.

[106] Nitenberg A, Pham I, Antony I, et al. Cardiovascular outcome of patients with abnormal coronary vasomotion and normal coronary arteriography is worse in type 2 diabetes mellitus than in arterial hypertension: a 10 year follow-up study. Atherosclerosis 2005;183(1):113–20.

[107] Lind L, Berglund L, Larsson A, et al. Endothelial function in resistance and conduit arteries and 5-year risk of cardiovascular disease. Circulation 2011;123(14):1545–51.

[108] Woodman RJ, Chew GT, Watts GF. Mechanisms, significance and treatment of vascular dysfunction in type 2 diabetes mellitus: focus on lipid-regulating therapy. Drugs 2005;65(1):31–74.

[109] Kim J, Montagnani M, Koh KK, et al. Reciprocal relationships between insulin resistance and endothelial dysfunction: molecular and pathophysiological mechanisms. Circulation 2006;113(15):1888–904.

[110] Steinberg HO, Brechtel G, Johnson A, et al. Insulin-mediated skeletal muscle vasodilation is nitric oxide dependent. A novel action of insulin to increase nitric oxide release. J Clin Invest 1994;94:1172–9.

[111] Steinberg HO, Chaker H, Leaming R, et al. Obesity/insulin resistance is associated with endothelial dysfunction: implications for the syndrome of insulin resistance. J Clin Invest 1996; 97:2601–10.

[112] Lim SC, Caballero AE, Smakowski P, et al. Soluble intercellular adhesion molecule, vascular cell adhesion molecule, and impaired microvascular reactivity are early markers of vasculopathy in type 2 diabetic individuals without microalbuminuria. Diabetes Care 1999;22(11):1865–70.

[113] Festa A, D'Agostino Jr R, Howard G, et al. Chronic subclinical inflammation as part of the insulin resistance syndrome: the Insulin Resistance Atherosclerosis Study (IRAS). Circulation 2000;102:42–7.

[114] Dandona P, Weinstock R, Thusu K, et al. Tumor necrosis factor alpha in sera of obese patients: fall with weight loss. J Clin Endocrinol Metab 1998;83:2907–10.

[115] Bierhaus A, Schiekofer S, Schwaninger M, et al. Diabetes-associated sustained activation of the transcription factor nuclear factor-kappa B. Diabetes 2001;50:2792–808.

[116] Pierce GL, Lesniewski LA, Lawson BR, et al. Nuclear factor-{kappa}B activation contributes to vascular endothelial dysfunction via oxidative stress in overweight/obese middle-aged and older humans. Circulation 2009;119(9):1284–92.

[117] Yao D, Brownlee M. Hyperglycemia-induced reactive oxygen species increase expression of the receptor for advanced glycation end products (RAGE) and RAGE ligands. Diabetes 2010;59(1):249–55.

[118] Naruse K, Rask-Madsen C, Takahara N, et al. Activation of vascular protein kinase C-beta inhibits Akt-dependent endothelial nitric oxide synthase function in obesity-associated insulin resistance. Diabetes 2006;55:691–8.

[119] Geraldes P, King GL. Activation of protein kinase C isoforms and its impact on diabetic complications. Circ Res 2010;106 (8):1319–31.

[120] Giacco F, Brownlee M. Oxidative stress and diabetic complications. Circ Res 2010;107(9):1058–70.

[121] Flammer AJ, Hermann F, Wiesli P, et al. Effect of losartan, compared with atenolol, on endothelial function and oxidative stress in patients with type 2 diabetes and hypertension. J Hypertens 2007;25(4):785–91.

[122] Menotti A, Lanti M, Kromhout D, et al. Homogeneity in the relationship of serum cholesterol to coronary deaths across different cultures. 40-Years follow-up of the Seven Countries Study. Eur J Cardiovasc Prev Rehabil 2008;15(6):719–25.

[123] Leitinger N. Oxidized phospholipids as modulators of inflammation in atherosclerosis. Curr Opin Lipidol 2003;14:421–30.

[124] Liao JK, Shin WS, Lee WY, et al. Oxidized low-density lipoprotein decreases the expression of endothelial nitric oxide synthase. J Biol Chem 1995;270:319 24.

[125] Lind L. Flow-mediated vasodilation over five years in the general elderly population and its relation to cardiovascular risk factors. Atherosclerosis 2014;237(2):666–70.

[126] Michel JB, Feron O, Sase K, et al. Caveolin versus calmodulin. Counterbalancing allosteric modulators of endothelial nitric oxide synthase. J Biol Chem 1997;272:25907–12.

[127] Feron O, Dessy C, Moniotte S, et al. Hypercholesterolemia decreases nitric oxide production by promoting the interaction of caveolin and endothelial nitric oxide synthase. J Clin Invest 1999;103:897–905.

[128] Feron O, Dessy C, Desager JP, et al. Hydroxymethylglutarylcoenzyme a reductase inhibition promotes endothelial nitric oxide synthase activation through a decrease in caveolin abundance. Circulation 2001;103:113–8.

[129] Pelat M, Dessy C, Massion P, et al. Rosuvastatin decreases caveolin-1 and improves nitric oxide-dependent heart rate and blood pressure variability in apolipoprotein E−/− mice in vivo. Circulation 2003;107:2480–6.

[130] Boger RH, Sydow K, Borlak J, et al. LDL cholesterol upregulates synthesis of asymmetrical dimethylarginine in human endothelial cells: involvement of S-adenosylmethionine-dependent methyltransferases. Circ Res 2000;87:99–105.

[131a] Perticone F, Sciacqua A, Maio R, et al. Asymmetric dimethylarginine, L-arginine, and endothelial dysfunction in essential hyper-tension. J Am Coll Cardiol 2005;46:518–23.

[131] Melikian N, Wheatcroft SB, Ogah OS, et al. Asymmetric dimethylarginine and reduced nitric oxide bioavailability in young black African men. Hypertension 2007;49:873–7.

[132] Juonala M, Viikari JSA, Alfthan G, et al. Brachial artery flow mediated dilation and asymmetrical dimethylarginine in the cardiovascular risk in young Finns study. Circulation 2007;116:1367–73.

[133] Yuhanna IS, Zhu Y, Cox BE, et al. High-density lipoprotein binding to scavenger receptor-BI activates endothelial nitric oxide synthase. Nat Med 2001;7:853–7.

[134] Besler C, Heinrich K, Rohrer L, et al. Mechanisms underlying adverse effects of HDL on eNOS-activating pathways in patients with coronary artery disease. J Clin Invest 2011;121:2693–708.

[135] Xu S, Ogura S, Chen J, et al. LOX-1 in atherosclerosis: biological functions and pharmacological modifiers. Cell Mol Life Sci 2013;70(16):2859–72.

[136] Undurti A, Huang Y, Lupica JA, et al. Modification of high density lipoprotein by myeloperoxidase generates a pro-inflammatory particle. J Biol Chem 2009;284:30825–35.

[137] Fang Y, Mohler ER, Hsieh III E, et al. Hypercholesterolemia suppresses inwardly rectifying k+channels in aortic endothelium in vitro and in vivo. Circ Res 2006;98:1064–71.

[138] Steinberg HO, Bayazeed B, Hook G, et al. Endothelial dysfunction is associated with cholesterol levels in the high normal range in humans. Circulation 1997;96:3287–93.

[139] Yildz A, Cakar MA, Baskurt M, et al. The effects of atorvastatin therapy on endothelial function in patients with coronary artery disease. Cardiovasc Ultrasound 2007;5:51.

[140] Spieker LE, Sudano I, Hurlimann D, et al. High-density lipoprotein restores endothelial function in hypercholesterolemic men. Circulation 2002;105:1399–402.

[141] Nieuwdorp M, Vergeer M, Bisoendial RJ, et al. Reconstituted HDL infusion restores endothelial function in patients with type 2 diabetes mellitus. Diabetologia 2008;51:1081–4.

[142] Kaul S, Coin B, Hedayiti A, et al. Rapid reversal of endothelial dysfunction in hypercholesterolemic apolipoprotein E-null mice by recombinant apolipoprotein AI[Milano]-phospholipid complex. J Am Coll Cardiol 2004;44:1311–9.

[143] Van Ostrom O, Nieuwdorp M, Westerweel PE, et al. Reconstituted HDL increases circulating endothelial progenitor cells in patients with type 2 diabetes. Arterioscler Thromb Vasc Biol 2007;27:1864–5.

[144] Kuvin JT, Ramet ME, Patel AR, et al. A novel mechanism for the beneficial vascular effects of high-density lipoprotein cholesterol: enhanced vasorelaxation and increased endothelial nitric oxide synthase expression. Am Heart J 2002;144:165–72.

[145] Lupattelli G, Marchesi S, Lombardini R, et al. Mechanisms of high-density lipoprotein cholesterol effects on the endothelial function in hyperlipemia. Metabolism 2003;52:1191–5.

[146] Lundman P, Eriksson MJ, Stuhlinger M, et al. Mild-to-moderate hypertriglyceridemia in young men is associated with endothelial dysfunction and increased plasma concentrations of asymmetric dimethylarginine. J Am Coll Cardiol 2001;38:111–6.

[147] Sinkey CA, Chenard CA, Stumbo PJ, et al. Resistance vessel endothelial function in healthy humans during transient postprandial hypertriglyceridemia. Am J Cardiol 2000;85:381–5.

[148] Keogh JB, Grieger JA, Noakes M, et al. Flow-mediated dilatation is impaired by a high-saturated fat diet but not by a high-carbohydrate diet. Arterioscler Thromb Vasc Biol 2005;25:1274–9.

[149] Davis N, Katz S, Wylie-Rosett J. The effect of diet on endothelial function. Cardiol Rev 2007;15:62–6.

[150] Fisher EA, Feig JE, Hewing JE, et al. High density lipoprotein function, dysfunction, and reverse cholesterol transport. Arterioscler Thromb Vasc Biol 2012;32:2813–20.

[151] Benjo AM, Maranhão RC, Coimbra SR, et al. Accumulation of chylomicron remnants and impaired vascular reactivity occur in subjects with isolated low HDL cholesterol: effects of niacin treatment. Atherosclerosis 2006;187(1):116–22.

[152] Casella-Filho A, Chagas ACP, Maranhão RC, et al. Effect of exercise training on plasma levels and functional properties of high-density lipoprotein cholesterol in metabolic syndrome. Am J Cardiol 2011;107:1168–72.

Further Reading

[1] Patti G, Pasceri V, Melfi R, et al. Impaired flow-mediated dilation and risk of restenosis in patients undergoing coronary stent implantation. Circulation 2005;111:7005.

[2] Lloyd-Jones DM, Nam BH, D'Agostino Sr RB, et al. Parental cardiovascular disease as a risk factor for cardiovascular disease in middle-aged adults. A prospective study of parents and offspring. JAMA 2004;291:2204–11.

35

Sleep Disorders and Endothelial Dysfunction

Fernanda Fatureto Borges, Geraldo Lorenzi-Filho, and Luciano F. Drager

INTRODUCTION

Sleep consumes about a third of our lives. Its structure is classically divided into NREM (nonrapid eye movement) and REM (rapid eye movement) stages. NREM stage is subdivided into three phases. Each of them has a peculiar structure that can have a greater or lesser impact on the cardiovascular system. In general, it is well established that sleep modifies the autonomous nervous system [1,2] and modifies cardiovascular regulation with profound changes in blood pressure and heart rate [3].

There are currently about 80 sleep disorders listed in the International Code of Diseases (ICD). Within this group of diseases, special attention is given to obstructive sleep apnea (OSA). OSA consists of a respiratory disorder characterized by recurrent obstruction of the upper airways during sleep, leading to complete recurrent (apnea) or partial (hypopnea) respiratory pauses. These obstructions promote reduction of intrathoracic pressure, intermittent hypoxia, and frequent arousals [4].

OSA is a very common clinical condition, but still underdiagnosed, even in patients with cardiovascular diseases [5]. Recent epidemiologic data suggest that among adults from 30 to 70 years, approximately 13% of men and 6% of women present moderate to significant forms of OSA (apnea-hypopnea index > 15 events per hour of sleep) [6]. In the population of São Paulo, an epidemiological study showed that about a third of the adult population has some degree of OSA [7]; it is clearly emerging as a new cardiovascular risk factor [8]. When untreated, it is an independent risk factor for hypertension, myocardial ischemia, and stroke [9–11].

However, the association mechanisms between OSA and cardiovascular disease have not yet been fully elucidated. Recent research suggests that OSA directly promotes several unfavorable effects, such as persistent increase in sympathetic activity, inflammation, increased oxidative stress, increased insulin resistance, changes in lipid metabolism, and endothelial dysfunction [8]. With regard to the latter, there are several potential mechanisms through which OSA can affect endothelial integrity, including changes in vasomotor tone and repeated episodes of hypoxia/reoxygenation, causing oxidative stress and inflammatory activation and changing endothelial repair capacity [12–14]. Endothelial dysfunction may be an important link between OSA and the development of cardiovascular disease [15].

Another sleep disorder that deserves attention is sleep deprivation (SD); it has already been shown that this can trigger sympathetic activation [16,17] and systemic inflammation [18]. Chronic sleep disorders and SD are associated with increased incidence of metabolic and cardiovascular diseases in humans.

In this chapter, we shall review the impact of OSA and SD on endothelial function, also highlighting the effect of treatment of these sleep disorders on the endothelium.

ENDOTHELIAL FUNCTION IN OSA

Increasing evidence has shown that OSA contributes independently to endothelial dysfunction [19]. Endothelium is a dynamic tissue layer, which constitutes the source or target for multiple growth factors and vasoactive mediators involved in the systemic

Endothelium and Cardiovascular Diseases
https://doi.org/10.1016/B978-0-12-812348-5.00035-0

regulation of physical and biochemical properties of blood vessels, and vascular contractility and cell growth. Endothelial injury is an important initial event in atherogenesis, leading to thickening of the intima and formation of atherosclerotic plaques [20–24]. Repeated episodes of hypoxia/reoxygenation may affect endothelial function by altering vasomotor tone, promoting oxidative stress and inflammation (with direct reduction of NO production), hypercoagulable development and increased apoptosis (decreasing endothelial repair capacity).

Regulation of Vasomotor Tone in OSA

Observational and interventional studies have demonstrated a relationship between OSA and altered vasomotor tone. Kato et al. demonstrated that vasodilation in the forearm after infusion of acetylcholine (endothelium-dependent vasodilation) is reduced in patients with significant OSA compared to controls of the same age and body mass index (BMI) [25]. In the same manner, flow-mediated dilatation is reduced in healthy patients with OSA, indicating decrease NO bioavailability. The Sleep Heart Health Study demonstrated damaged flow-mediated dilatation of brachial artery in patients with OSA. Patients in this study maintained impaired vasodilatation even after adjusting for BMI to cardiovascular comorbidities. The association was stronger between brachial artery reactivity and the degree of hypoxemia than with apnea/hypopnea, suggesting an important role of the hypoxia/reoxygenation phenomena in reducing the bioavailability of NO and promoting endothelial dysfunction [14].

More recently, Jelic et al. demonstrated worse flow-mediated dilatation in patients with OSA compared with controls ($4.01 \pm 2.99\%$ vs. $9.52 \pm 2.79\%$; $P < 0.001$) [26]. Although these studies suggest reduction in NO bioavailability, other works show that there is even less amount of NO in plasma and endothelial cells from patients with OSA compared with controls and reduction of circulating NO levels in patients with untreated OSA [27,28]. Levels of asymmetric NG-dimethylarginine, an endogenous inhibitor in endothelial NO synthase, and soluble fraction of CD40 binding agent were elevated in patients with OSA regardless of the presence of other cardiovascular risk factors [29]. Expression of eNOS (the major source of endothelial NO baseline) and P-eNOS activated form eNOS are

reduced by 59% and 94%, respectively, in OSA patients compared to controls, while nitrotyrosine expression, a marker of oxidative stress, and COX-2, a marker of inflammation, was five times greater in patients with OSA than in controls (Fig. 35.1) [26].

On the other hand, evidence of increased production of vasoconstrictor substances such as angiotensin II and endothelin-1 (ET-1) in patients with OSA is still inconsistent. Plasma levels of aldosterone and angiotensin II in patients with OSA have been reported as elevated or similar to controls [30,31]. ET-1 is a potent vasoconstrictor peptide present in human vascular endothelial cells and has mitogenic properties [32]. A study demonstrated increased plasma levels of ET-1 and blood pressure in rats exposed to hypoxia/hypercapnia intermittent, as in OSA [33]. Although several studies have shown that patients with OSA have higher systemic levels of ET-1 to healthy persons [34–36], human studies evaluating ET-1 are still controversial (systemic levels are strongly affected by

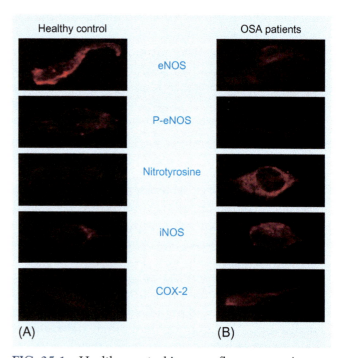

FIG. 35.1 Healthy control immunofluorescence images (A) and patients with OSA (B). The expression of eNOS, main source of endothelial NO, and P-eNOS, activated form of eNOS, in venous endothelial cells was lower in patients with OSA, while the expression of nitrotirosin (oxidative stress marker), iNOS and COX-2 were increased. Extracted from Jelic S, Padeletti M, Kawut SM, et al. Inflammation, oxidative stress, and repair capacity of the vascular endothelium in obstructive sleep apnea. Circulation 2008;117(17):2270–8.

cardiovascular comorbidities and do not correlate well with ET-1 tissue production) [12]. Thus, despite the increase of ET-1 playing a role in the development of hypertension in OSA, it has not yet been possible to establish association between OSA and ET-1 elevation, since most of the patients and controls in these groups surveyed had a history of hypertension and suggesting a possible association of endothelial dysfunction in both groups (and perhaps no significant difference in ET-1 level) [30,37]. Another study showed increase in ET-1 plasma precursors in nontreated patients with OSA [38]. However, plasma concentrations of ET-1 were at physiological levels. Finally, one study demonstrated elevation in ET-1 levels in severe to moderate OSA, but not mild OSA [31].

A study with coronary angiography and acetylcholine infusion, which showed association in the severity of OSA and inappropriate arterial vasoconstriction and endothelial dysfunction in the coronary arteries, also stands out [39].

Proinflammatory/Antiinflammatory Endothelial Homeostasis in OSA

Repetition of hypoxia/reoxygenation associated with OSA apnea and hypopnea increases production of inflammatory mediators, expression of adhesion molecules and production of reactive oxygen species [40]. Reactive oxygen species are highly reactive molecules that can damage cell tissue resulting in inflammation and endothelial activation. In healthy vasculature, its effect is balanced by antioxidant activity [19]. In experimental models, repeated episodes of hypoxia/reoxygenation may worsen endothelial function by directly reducing the production of NO in transcriptional and posttranscriptional levels and increase production of reactive oxygen species [41,42]. Increased levels of reactive oxygen species cause more oxidative stress, which in turn, reduces and destabilizes eNOS RNA messenger, while limiting the availability of cofactors required for NO production [43–47]. In addition, prolonged oxidative stress, as observed in OSA, reduced eNOS activity by suppressing its phosphorylation [48], superoxide production promoting the endothelial NO synthase pathway [49], and reducing NO bioavailability.

It was also observed that, in patients with OSA, circulating levels of ICAM-1 adhesion molecules, cell adhesion molecule to vascular endothelium-1 (VCAM-1), L-selectin and E-selectin are elevated

compared with healthy controls adjusted for age [50,51], suggesting endothelial activation. In general, accumulation and adhesion of circulating leukocytes to vascular endothelium leads to vascular inflammation and atherosclerosis progression [52,53]. Furthermore, monocyte expression of adhesion molecules CD15 and CD11c is increased in patients with OSA compared with controls adjusted for age and cardiovascular comorbidities [54]. Comparing patients with moderate/important OSA to patients with apnea-hypopnea index fewer than 10 events per hour, the lymphocyte production of interleukin-4 (procytokine inflammatory cytokine) is increased in the former group, while the production of interleukin-10 (potent antiinflammatory cytokine) is decreased [55]. Levels of inflammatory markers classically associated with atherosclerosis, including C-reactive protein with high sensitivity, interleukin-6, interleukin-8, tumor necrosis factor, serum amyloidosis A, cell adhesion molecules, chemoattractant protein-1 and selectins of monocyte, have been reported as high in OSA [56,57]. Levels of interleukin-6 proinflammatory cytokines and interleukin-8 are higher in patients with OSA compared to controls and relate to the severity of OSA [58]. Plasma levels of antioxidants such as glutathione peroxidase, g-glutamyltransferase, vitamins A, E, and B [12], pholate, and homocysteine are reduced in patients with OSA compared with controls with similar age and BMI.

Thus, repetitive hypoxia/reoxygenation observed in OSA adversely affects endothelial function by promoting oxidative stress and inflammation, as well as a reduction in NO availability. The proinflammatory/antiinflammatory balance is deregulated for vascular inflammation in patients with untreated OSA.

Coagulation Homeostasis in OSA

The factors secreted by normal endothelium—which decrease platelet aggregation, such as nitric oxide and prostacyclin, thrombomodulin, which promotes the generation of activated protein C and heparin, which acts as a cofactor for antithrombin III help maintaining normal blood flow [59]. Endothelial dysfunction can lead to homeostasis alterations resulting in an atherogenic and procoagulant state. Shifting coagulation homeostasis to a state of procoagulability contributes to the progression of atherosclerosis and promotes cardio-vascular events [60].

OSA has been associated with a state of hypercoa-gulability and excessive platelet activation [61]. XIIa, VIIa coagulation factors levels and thrombin-antithrombin complex are elevated in OSA [62]. Plasma levels of fibrinogen and type 1 plasminogen activator inhibitor are also increased [62–66]. How-ever, the role of OSA as a procoagulation indepen-dent stimulus remains uncertain. Coagulation has been evaluated in patients with OSA with comor-bidities that adversely affect coagulation [67], and increased levels of hypercoagulable markers such as thrombin-antithrombin III complex and D-dimer are more related to coexisting hypertension than OSA [68]. Furthermore, apnea-hypopnea index is not a significant predictor of plasminogen activa-tor inhibitor type 1 levels in the concomitant pres-ence of metabolic syndrome in patients with OSA [69]. Coexistent cardiovascular disease and hyper-tension appear to have a greater role in altering the coagulation in these patients.

Endothelial Repair Capacity in OSA

Endothelial dysfunction may result from direct damage to the endothelium or, alternatively, may be caused by reduced endothelial repair in response to damage. Endothelial progenitor cells (EPCs) are derived from the bone marrow and enter the sys-temic circulation to replace defective or prematurely damaged cells. Thus, EPCs are a marker of endothe-lial repair capacity. In general, reduced levels of EPC are associated with impaired endothelial function and increased cardiovascular risk [70–72].

There is evidence suggesting that OSA changes the regenerative capacity of the endothelium. For example, circulating progenitor cell levels are reduced in patients with OSA without other cardio-vascular diseases [26]. EPC reduced levels may exac-erbate endothelial dysfunction in patients with OSA because these cells are the major repositories of eNOS at the site of endothelial injury induced by ischemia/reperfusion [73]. Progenitor cellular cells are not only decreased in OSA, but it seems that the endothelial cells are damaged and apoptotic. Circulating apoptotic cell levels appear to be higher in obese patients with untreated OSA without other cardiovascular diseases, compared to nonobese age-adjusted controls, suggesting increased endothelial apoptosis [74]. The level of circulating apoptotic cells was positively related to index of apnea-hypopnea and dysfunction in endothelium-dependent vasodilation [74].

EFFECT OF TREATMENT FOR OSA IN THE ENDOTHELIAL FUNCTION

Interventional studies show that the improve-ment of endothelial dysfunction is dependent on the effective control of OSA. Thus, the therapeutic modality should be able to eliminate all or part of the respiratory sleep disorder, and good adherence to the treatment of OSA is necessary to produce sig-nificant and sustainable effects on the vasculature. Continuous positive airway pressure (CPAP) is a device that provides airflow through a nasal or oro-facial facemask acting as a pneumatic splint to main-tain airways patency during sleep. Observational studies show sustained improvement of endothelial dysfunction after 6 months of CPAP therapy [75,76]. The flow-mediated vasodilation in the forearm improves in 2 weeks of CPAP therapy in both nor-motensive and hypertensive patients with OSA [77]. Jelic et al. demonstrated a significant improve-ment in flow-mediated vasodilation in patients with OSA who adhered to CPAP therapy for more than 4 hours a day ($7.24 \pm 4.24\%$ vs. $3.71 \pm 3.44\%$; $P = 0.004$) [26]. Another study demonstrated that endothelium-dependent vasodilation increases, while the endothelium-independent vasodilation remains unchanged in healthy OSA patients after 3 months of CPAP therapy, suggesting an increase in the availability of NO [78].

A recent systematic review (Fig. 35.2) evaluated eight randomized clinical trials about the CPAP effect on endothelial function, showing that the ther-apeutic CPAP for 2 and 24 weeks leads to statisti-cally significant improvement in endothelial function assessed by flow-mediated dilation com-pared to no therapy or placebo (*sham* CPAP). The improvement of flow-mediated dilatation was esti-mated at 3.87% (95% CI: 1.93–5.8, $P < 0.001$), reinfor-cing that CPAP therapy may reduce overall cardiovascular risk by improving vascular function in OSA patients [79].

Circulatory levels of NO, which were reduced in patients with moderate to severe OSA compared with healthy subjects, can be normalized with CPAP two nights after initiation of therapy and remain ele-vated within 5 months [28]. In addition, there is a

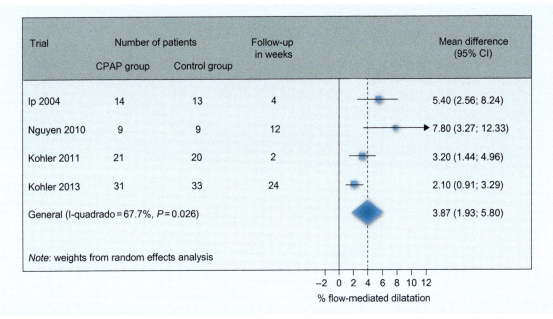

Trial	Number of patients		Follow-up in weeks		Mean difference (95% CI)
	CPAP group	Control group			
Ip 2004	14	13	4		5.40 (2.56; 8.24)
Nguyen 2010	9	9	12		7.80 (3.27; 12.33)
Kohler 2011	21	20	2		3.20 (1.44; 4.96)
Kohler 2013	31	33	24		2.10 (0.91; 3.29)
General (I-quadrado = 67.7%, P = 0.026)					3.87 (1.93; 5.80)

Note: weights from random effects analysis

−2 0 2 4 6 8 10 12
% flow-mediated dilatation

FIG. 35.2 Effect of continues positive airway pressure (CPAP) treatment in flow-mediated dilatation in patients with OSA. *Forest plot* showing differences in mean and 95% confidence interval of the treatment effect on flow-mediated dilatation. Adapted from Schwarz EI, Puhan MA, Schlatzer C, et al. Effect of cPAP therapy on endothelial function in obstructive sleep apnoea: a systematic review and meta-analysis. Respirology 2015;20:886–95.

reduction in systemic levels of ET-1 that were elevated in patients with OSA compared with healthy controls [36] and their plasma precursors with long-term therapy with CPAP [42], in addition to reduced plasma concentrations of asymmetric NG-dimer arginine (eNOS inhibitor) [80].

CPAP therapy reduces circulating levels of soluble adhesion molecules and tumor necrosis factor alpha and reduces the ability of monocyte adhesion in cultured endothelial cells, indicating reduced inflammation, leukocyte activation and endothelial leukocyte interaction [51,54,56,81]. This treatment also reduces the expression of adhesion molecules (CD15 and CD11c on monocyte), decreases reactive oxygen species production by monocytes CD11 and reduces monocyte adhesion to human endothelial cells in culture [50]. There is a reduction in serum levels of C-reactive protein, interleukin-6 and spontaneous of interleukin-6 production after 4 weeks of CPAP therapy. Plasma levels of inflammatory markers C-reactive protein, interleukin-6 and alpha tumor necrosis factor remain unchanged after withdrawal of CPAP for 7 days, despite the immediate return of OSA [82]. Treatment with CPAP is also associated with a decrease in the levels of fibrinogen and PAI-1 activity [66,83].

Long-term CPAP therapy improves endothelial repair capacity, evidenced by the increase in circulating levels of EPCs and reduced numbers of circulating apoptotic endothelial cells [26,74].

A recent randomized controlled trial showed that withdrawal of CPAP after 2 weeks (compared to maintenance) led to a return to sleepiness, increased morning blood pressure and recurrence in endothelial dysfunction, as well as recurrence of OSA [84].

Fig. 35.3 summarizes the possible mechanisms of association of OSA with endothelial dysfunction and effects of CPAP therapy.

Among the options for definitive surgical treatment, adenotonsillectomy reduces the level of the inflammatory markers C-reactive protein, interleukin-6, and ligand CD40, and increases levels of the antiinflammatory marker interleukin-10 in children with OSA [85,86]. Pulmonary artery pressure is reduced and the hyperemic response after occlusion of the brachial artery improved in these patients [87,88]. In adult patients, the uvulopalatopharyngoplasty decreases plasma levels of proinflammatory C-reactive protein mediators and tumor necrosis factor [89,90].

Mandibular support increase the transverse sectional area of the oropharynx and reduce the chance of airway collapse in OSA. Long-term therapy with these devices seem to have the potential to partially reverse endothelial dysfunction, although this does not completely eliminate the obstructive events of OSA, especially in its most advanced forms.

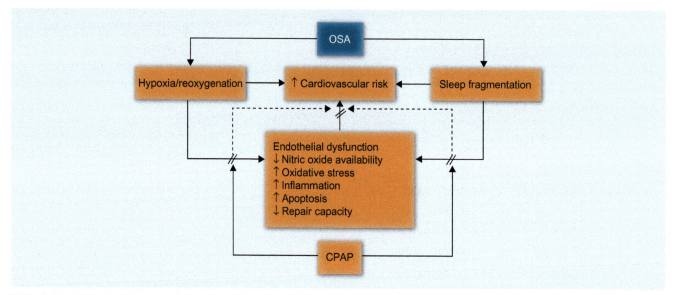

FIG. 35.3 Mechanisms associated with endothelial dysfunction in patients with OSA potentially contributing to increased cardiovascular risk. Repetitive hypoxia/reoxygenation and sleep fragmentation promote endothelial dysfunction by decreasing NO availability, increasing oxidative stress, inflammation and apoptosis, and reducing endothelial repair capacity. CPAP therapy improves these changes by reversing endothelial dysfunction and potentially reducing cardiovascular risk in OSA.

Vascular reactivity, assessed by peripheral arterial tonometry, and lipid peroxidation, an oxidative stress marker, were similar in patients with OSA and controls (adjusted for age, BMI, and cardiovascular comorbidities) after 1 year of suboptimal therapy with oral devices (apnea index residual hypopnea 19 per hour) [91].

SD and Endothelial Dysfunction

In recent decades, a progressive reduction in sleep duration has been observed, from approximately 9 hours per night in 1910 to approximately 7 hours per night at present [92]. There is growing evidence that chronic sleep restriction may be related not only to the change in cognitive function [93], but also to an increase in cardiovascular mortality and morbidity [94–97]. Endothelial dysfunction related to SD is a potential mechanism for this increased cardiovascular risk and may result in both acute deprivation and chronic restriction of sleep.

In healthy adults, total acute SD induces increased circulating levels of endothelial activation markers, such as ICAM-1 molecular adhesion molecules and E-selectin, and increased circulating levels of pro and anti-inflammatory markers, such as tumor necrosis factor alpha, interleukin-1β, interleukin-1 receptor antagonist, and interleukin-6

[98–100]. High levels of interleukin-6, ICAM-1, or E-selectin can be linked to the development of endothelial dysfunction and cardiovascular disease. It has already been shown that even short periods, such as a night of SD, may be associated with increased arterial stiffness in healthy adults [101]. Sauvet et al. evaluated 12 healthy men undergoing acute SD for 40 hours and found a reduction in vascular reactivity dependent and independent endothelial from the 29th hour of SD, with increase in plasma endothelial cell activation markers (E-selectin, interleukin-6, and cell adhesion molecule ICAM-1). This microvascular change appeared earlier than changes in systolic blood pressure, heart rate, and sympathetic activity (from the 32nd hour of SD) [102]. Garcia-Fernández et al. evaluated endothelial function of 15 cardiology residents after one day of common work and after 24 hours on duty. Despite the demonstration of a significant reduction of endothelium-dependent vasodilation after 24-hour on-call, the association of a second stress factor (on-call in a medical emergency setting) limits the causal association of SD with endothelial dysfunction in this study [103]. In a similar way, another study also evaluated the endothelial function of healthy doctors. Independent association of reduced hours of sleep (24-hour shift) with reduction of flow-mediated vasodilation was demonstrated, with no correlation of the endothelial alteration with the difficulty score of the duty (mental stress evaluation) or with coffee consumption. This data reinforces the concept of SD—and not mental stress—as a trigger for altered endothelial function [104].

Observational studies demonstrate alterations of endothelial function related to chronic sleep restriction. Compared to 50 adults with normal sleep duration (7–9 hours per night), 30 adults with chronic SD (<7 hours per night) had higher endothelium-1 mediated vasoconstrictor tone [105]. ET-1 is a potent vasoconstrictor peptide that is associated with endothelial vasomotor dysfunction, and increased cardiovascular risk. Another study has shown that chronic SD may be associated with a 50% decline in flow-mediated vasodilation [106].

The effect of partial sleep restriction on endothelial function has also been evaluated experimentally in humans. A Brazilian study evaluated 13 healthy subjects in a control period (7–9.5 hours of sleep per night) who underwent partial deprivation of sleep for five nights (3.5–5 hours of sleep). A significant increase in the modulation of cardiac and peripheral sympathetic activity after SD, associated with decreased endothelium-dependent venodilation, was demonstrated [107]. Calvin et al. subjected healthy adults to restriction of sleep for two thirds of their normal period and observed—after 8 days of restriction—significant worsening of flow-mediated vasodilation. The worsening of endothelial function in this study was similar to that previously reported in smokers and people with diabetes or coronary disease, highlighting the cardiovascular risk associated with SD [108].

Shearer et al. described a model comparing total deprivation with partial SD for 4 days. Elevation of soluble receptor levels of tumor necrosis factor-1 and interleukin-6 in the total deprivation group was demonstrated in relation to the partial deprivation group, reinforcing the role of SD as a potent inflammatory stimulus [109].

PERSPECTIVES

There is evidence that sleep disorders promote increased cardiovascular risk. The data presented show that OSA and SD are independently related to endothelial dysfunction and that identifying and treating these sleep disorders may have a beneficial effect on this important marker of cardiovascular risk. As subdiagnoses are still frequent, these and other evidences clearly indicate the need for strategies to improve recognition and treatment of sleep disorders. The detailed understanding of the mechanisms involved may favor the development of potential specific biomarkers of vascular injury in OSA and SD. Ultimately, these biomarkers may serve both to identify sleep disorders and to predict cardiovascular events in these patients.

References

[1] Smith RP, Veale D, Pepin JL, et al. Obstructive sleep apnoea and the autonomic nervous system. Sleep Med Rev 1998;2:69–92.

[2] Somers VK, Dyken ME, Mark AL, et al. Sympathetic-nerve activity during sleep in normal subjects. N Engl J Med 1993;328:303–7.

[3] Murali NS, Svatikova A, Somers VK. Cardiovascular physiology and sleep. Front Biosci 2003;8:s636–52.

[4] The Report of an American Academy of Sleep Medicine Task Force. Sleep-related breathing disorders in adults: recommendations for syndrome definition and measurement techniques in clinical research. Sleep 1999;22:667–89.

[5] Costa LE, Uchôa CH, Drager LF, et al. Potential underdiagnosis of obstructive sleep apnoea in the cardiology outpatient setting. Heart 2015;10(16):1288–92.

[6] Peppard PE, Young T, Barnet JH, et al. Increased prevalence of sleep-disordered breathing in adults. Am J Epidemiol 2013; 177(9):1006–14.

[7] Tufik S, Santos-Silva R, Taddei JA, et al. Obstructive sleep apnea syndrome in the Sao Paulo Epidemiologic Sleep Study. Sleep Med 2010;1(1):441–6.

[8] Drager LF, Togeiro SM, Polotsky VY, et al. Obstructive sleep apnea: a cardiometabolic risk in obesity and the metabolic syndrome. J Am Coll Cardiol 2013;62:569–76.

[9] Peppard PE, Young T, Palta M, et al. Prospective study of the association between sleep-disordered breathing and hypertension. N Engl J Med 2000;342:1378–84.

[10] Peker Y, Kraiczi H, Hedner J, et al. An independent association between obstructive sleep apnea and coronary artery disease. Eur Respir J 1999;13:179–84.

[11] Yaggi HK, Concato J, Kernan WN, et al. Obstructive sleep apnea as a risk factor for stroke and death. N Engl J Med 2005;353:2034–41.

[12] Atkeson A, Yeh SY, Malhotra A, et al. Endothelial function in obstructive sleep apnea. Prog Cardiovasc Dis 2009;51(5):351–63.

[13] Ip MS, Tse HF, Lam B, et al. Endothelial function in obstructive sleep apnea and response to treatment. Am J Respir Crit Care Med 2004;169(3):348–53.

[14] Nieto FJ, Herrington DM, Redline S, et al. Sleep apnea and markers of vascular endothelial function in a large community sample of older adults. Am J Respir Crit Care Med 2004;169 (3):354–60.

[15] Drager LF, Bortolotto LA, Lorenzi MC, et al. Early signs of atherosclerosis in obstructive sleep apnea. Am J Respir Crit Care Med 2005;172(5):613–8.

[16] Irwin M, Thompson J, Miller C, et al. Effects of sleep and sleep deprivation on catecholamine and interleukin-2 levels in humans: clinical implications. J Clin Endocrinol Metab 1999;84:1979–85.

[17] Irwin MR, Ziegler M. Sleep deprivation potentiates activation of cardiovascular and catecholamine responses in abstinent alcoholics. Hypertension 2005;45:252–7.

[18] Irwin MR, Wang M, Campomayor CO, et al. Sleep deprivation and activation of morning levels of cellular and genomic markers of inflammation. Arch Intern Med 2006;166:1756–62.

[19] Lui MM, Lam DC, Ip MS. Significance of endothelial dysfunctions in sleep related breathing disorder. Respirology 2013;18 (1):39–46.

[20] Ross R. Atherosclerosis: an inflammatory disease. N Engl J Med 1999;340(2):115–26.

[21] Shimokawa H. Primary endothelial dysfunction: atherosclerosis. J Mol Cell Cardiol 1999;31(1):23–7.

[22] Celermajer DS. Endothelial dysfunction: does it matter? Is it reversible? J Am Coll Cardiol 1997;30(2):325–33.

[23] Reddy KG, Nair RN, Sheehan HM, et al. Evidence that selective endothelial dysfunction may occur in the absence of angiographic or ultrasound atherosclerosis in patients with risk factors for atherosclerosis. J Am Coll Cardiol 1994;23(4):833–43.

[24] Celermajer DS, Sorensen KE, Gooch VM, et al. Non-invasive detection of endothelial dysfunction in children and adults at risk of atherosclerosis. Lancet 1992;340(8828):1111–5.

[25] Kato M, Roberts-Thomson P, Phillips BG, et al. Impairment of endothelium-dependent vasodilation of resistance vessels in patients with obstructive sleep apnea. Circulation 2000;102 (21):2607–10.

[26] Jelic S, Padeletti M, Kawut SM, et al. Inflammation, oxidative stress, and repair capacity of the vascular endothelium in obstructive sleep apnea. Circulation 2008;117(17):2270–8.

[27] Ip MS, Lam B, Chan LY, et al. Circulating nitric oxide is suppressed in obstructive sleep apnea and is reversed by nasal continuous positive airway pressure. Am J Respir Crit Care Med 2000;162(6): 2166–71.

[28] Schulz R, Schmidt D, Blum A, et al. Decreased plasma levels of nitric oxide derivatives in obstructive sleep apnea: response to CPAP therapy. Thorax 2000;55(12):1046–51.

[29] Barcelo A, de la Pena M, Ayllon O, et al. Increased plasma levels of asymmetric dimethylarginine and soluble CD40 ligand in patients with sleep apnea. Respiration 2009;77(1):85–90.

[30] Møller DS, Lind P, Strunge B, et al. Abnormal vasoactive hormones and 24-hour blood pressure in obstructive sleep apnea. Am J Hypertens 2003;16(4):274–80.

[31] Gjørup PH, Sadauskiene L, Wessels J, et al. Abnormally increased endothelin-1 in plasma during the night in obstructive sleep apnea: relation to blood pressure and severity of disease. Am J Hypertens 2007;20(1):44–52.

[32] Howard PG, Plumpton C, Davenport AP. Anatomical localization and pharmacological activity of mature endothelins and their precursors in human vascular tissue. J Hypertens 1992;10 (11):1379–86.

[33] Kanagy NL, Walker BR, Nelin LD. Role of endothelin in intermittent hypoxia-induced hypertension. Hypertension 2001;37(2 Pt 2):511–5.

[34] Zamarron-Sanz C, Ricoy-Galbaldon J, Gude-Sampedro F, et al. Plasma levels of vascular endothelial markers in obstructive sleep apnea. Arch Med Res 2006;37(4):552–5.

[35] Saarelainen S, Seppala E, Laasonen K, et al. Circulating endothelin-1 in obstructive sleep apnea. Endothelium 1997;5 (2):115–8.

[36] Phillips BG, Narkiewicz K, Pesek CA, et al. Effects of obstructive sleep apnea on endothelin-1 and blood pressure. J Hypertens 1999;17(1):61–6.

[37] Grimpen F, Kanne P, Schulz E, et al. Endothelin-1 plasma levels are not elevated in patients with obstructive sleep apnea. Eur Respir J 2000;15(2):320–5.

[38] Jordan W, Reinbacher A, Cohrs S, et al. Obstructive sleep apnea: plasma endothelin-1 precursor but not endothelin-1 levels are elevated and decline with nasal continuous positive airway pressure. Peptides 2005;26(9):1654–60.

[39] Kadohira T, Kobayashi Y, Iwata Y, et al. Coronary artery endothelial dysfunction associated with sleep apnea. Angiology 2011; 62(5): 397–400.

[40] Drager LF, Polotsky VY, Lorenzi-Filho G. Obstructive sleep apnea: an emerging risk factor for atherosclerosis. Chest 2011;140(2): 534–42.

[41] McQuillan LP, Leung GK, Marsden PA, et al. Hypoxia inhibits expression of eNOS via transcriptional and posttranscriptional mechanisms. Am J Physiol 1994;267(5 Pt 2):H1921–7.

[42] Liao JK, Zulueta JJ, Yu FS, et al. Regulation of bovine endothelial constitutive nitric oxide synthase by oxygen. J Clin Invest 1995;96 (6):2661–6.

[43] Takemoto M, Sun J, Hiroki J, et al. Rho-kinase mediates hypoxia-induced downregulation of endothelial nitric oxide synthase. Circulation 2002;106(1):57–62.

[44] Wang P, Zweier JL. Measurement of nitric oxide and peroxynitrite generation in the postischemic heart. Evidence for peroxynitrite-mediated reperfusion injury. J Biol Chem 1996;271(46):29223–30.

[45] Laursen JB, Somers M, Kurz S, et al. Endothelial regulation of vasomotion in ApoE deficient mice. Implications for interactions between peroxynitriteand tetrahydrobiopterin. Circulation 2001;103(9):1282–8.

[46] Kuzkaya N, Weissmann N, Harrison DG, et al. Interactions of peroxynitrite, tetrahydrobiopterin, ascorbic acid, and thiols: implications for uncoupling endothelial nitric-oxide synthase. J Biol Chem 2003;278(25):22546–54.

[47] Antoniades C, Shirodaria C, Warrick N, et al. 5-Methyltetrahydrofolate rapidly improves endothelial function and decreases superoxide production in human vessels: effects on vascular tetrahydrobiopterin availability and endothelial nitric oxide synthase coupling. Circulation 2006;114(11):1193–201.

[48] Tanaka T, Nakamura H, Yodoi J, et al. Redox regulation of the signaling pathways leading to eNOS phosphorylation. Free Radic Biol Med 2005;38(9):1231–42.

[49] Xia Y, Roman LJ, Masters BS, et al. Inducible nitric-oxide synthase generates superoxide from the reductase domain. J Biol Chem 1998;273(35):22635–9.

[50] Ohga E, Nagase T, Tomita T, et al. Increased levels of circulating ICAM-1, VCAM-1, and L-selectin in obstructive sleep apnea syndrome. J Appl Physiol 1999;87(1):10–4.

[51] Chin K, Nakamura T, Shimizu K, et al. Effects of nasal continuous positive airway pressure on soluble cell adhesion molecules in patients with obstructive sleep apnea syndrome. Am J Med 2000;109(7):562–7.

[52] Aird WC. Phenotypic heterogeneity of the endothelium: representative vascular beds. Circ Res 2007;100:174–90.

[53] Price DT, Loscalzo J. Cellular adhesion molecules and atherogenesis. Am J Med 1999;107(1):85–97.

[54] Dyugovskaya L, Lavie P, Lavie L. Increased adhesion molecules expression and production of reactive oxygen species in leukocytes of sleep apnea patients. Am J Respir Crit Care Med 2002;165(7):934–9.

[55] Dyugovskaya L, Lavie P, Lavie L. Lymphocyte activation as a possible measure of atherosclerotic risk in patients with sleep apnea. Ann N Y Acad Sci 2005;1051:340–50.

[56] Drager LF, Lopes HF, Maki-Nunes C, et al. The impact of obstructive sleep apnea on metabolic and inflammatory markers in consecutive patients with metabolic syndrome. PLoS One 2010;5: e12065.

[57] McNicholas WT. Obstructive sleep apnea and inflammation. Prog Cardiovasc Dis 2009;51(5):392–9.

[58] Minoguchi K, Yokoe T, Tazaki T, et al. Increased carotid intima-media thickness and serum inflammatory markers in obstructive sleep apnea. Am J Respir Crit Care Med 2005;172(5):625–30.

[59] Rosenberg RD, Aird WC. Vascular-bed-specific hemostasis and hypercoagulable states. N Engl J Med 1999;340(20):1555–64.

[60] Davies MJ. The contribution of thrombosis to the clinical expression of coronary atherosclerosis. Thromb Res 1996; 82(1):1–32.

[61] Bokinsky G, Miller M, Ault K, et al. Spontaneous platelet activation and aggregation during obstructive sleep apnea and its response to therapy with nasal continuous positive airway pressure. A preliminary investigation. Chest 1995;108(3):625–30.

[62] Robinson GV, Pepperell JC, Segal HC, et al. Circulating cardiovascular risk factors in obstructive sleep apnea. Thorax 2004;59(9): 777–82.

[63] Wessendorf TE, Thilmann AF, Wang YM, et al. Fibrinogen levels and obstructive sleep apnea in ischemic stroke. Am J Respir Crit Care Med 2000;162(6):2039–42.

[64] Nobili L, Schiavi G, Bozano E, et al. Morning increase of whole blood viscosity in obstructive sleep apnea syndrome. Clin Hemorheol Microcirc 2000;22(1):21–7.

[65] Rangemark C, Hedner JA, Carlson JT, et al. Platelet function and fibrinolytic activity in hypertensive and normotensive sleep apnea patients. Sleep 1995;18(3):188–94.

[66] von Kanel R, Loredo JS, Ancoli-Israel S, et al. Association between sleep apnea severity and blood coagulability: treatment effects of nasal continuous positive airway pressure. Sleep Breath 2006; 10(3):139–46.

[67] von Känel R, Dimsdale JE. Hemostatic alterations in patients with obstructive sleep apnea and the implications for cardiovascular disease. Chest 2003;124(5):1956–67.

[68] von Kanel R, Le DT, Nelesen RA, et al. The hypercoagulable state in sleep apnea is related to comorbid hypertension. J Hypertens 2001;19(8):1445–51.

[69] von Känel R, Loredo JS, Ancoli-Israel S, et al. Elevated plasminogen activator inhibitor 1 in sleep apnea and its relation to the metabolic syndrome: an investigation in 2 different study samples. Metabolism 2007;56(7):969–76.

[70] Hill JM, Zalos G, Halcox JP, et al. Circulating endothelial progenitor cells, vascular function, and cardiovascular risk. N Engl J Med 2003;348(7):593–600.

[71] Werner N, Kosiol S, Schiegl T, et al. Circulating endothelial progenitor cells and cardiovascular outcomes. N Engl J Med 2005;353(10): 999–1007.

[72] Urbich C, Dimmeler S. Endothelial progenitor cells: characterization and role in vascular biology. Circ Res 2004;95(4):343–53.

[73] Ii M, Nishimura H, Iwakura A, et al. Endothelial progenitor cells are rapidly recruited to myocardium and mediate protective effect of ischemic preconditioning via "imported" nitric oxide synthase activity. Circulation 2005;111(9):1114–20.

[74] El Solh AA, Akinnusi ME, Baddoura FH, et al. Endothelial cell apoptosis in obstructive sleep apnea. A link to endothelial dysfunction. Am J Respir Crit Care Med 2007;175(11):1186–91.

[75] Bayram NA, Ciftci B, Keles T, et al. Endothelial function in normotensive men with obstructive sleep apnea before and 6 months after CPAP treatment. Sleep 2009;32(10):1257–63.

[76] Duchna HW, Orth M, Schultze-Werninghaus G, et al. Long-term effects of nasal continuous positive airway pressure on vasodilatory endothelial function in obstructive sleep apnea syndrome. Sleep Breath 2005;9(3):97–103.

[77] Imadojemu VA, Gleeson K, Quraishi SA, et al. Impaired vasodilator responses in obstructive sleep apnea are improved with continuous positive airway pressure. Am J Respir Crit Care Med 2002; 165(7):950–3.

[78] Lattimore JL, Wilcox I, Skilton M, et al. Treatment of obstructive sleep apnoea leads to improved microvascular endothelial function in the systemic circulation. Thorax 2006;61(6):491–5.

[79] Schwarz EI, Puhan MA, Schlatzer C, et al. Effect of cPAP therapy on endothelial function in obstructive sleep apnoea: a systematic review and meta-analysis. Respirology 2015;20:886–95.

[80] Ohike Y, Kozaki K, Iijima K, et al. Amelioration of vascular endothelial dysfunction in obstructive sleep apnea syndrome by nasal continuous positive airway pressure—possible involvement of nitric oxide and asymmetric NG, NG-dimethylarginine. Circ J 2005;69:221–6.

[81] Ryan S, Taylor CT, McNicholas WT. Selective activation of inflammatory pathways by intermittent hypoxia in obstructive sleep apnea syndrome. Circulation 2005;112(17):2660–7.

[82] Phillips CL, Yang Q, Williams A, et al. The effect of short-term withdrawal from continuous positive airway pressure therapy on sympathetic activity and markers of vascular inflammation in subjects with obstructive sleep apnea. J Sleep Res 2007;16(2): 217–25.

[83] Chin K, Ohi M, Kita H, et al. Effects of NCPAP therapy on fibrinogen levels in obstructive sleep apnea syndrome. Am J Respir Crit Care Med 1996;153(6 Pt 1):1972–6.

[84] Kohler M, Stoewhas AC, Ayers L, et al. Effects of continuous positive airway pressure therapy withdrawal in patients with obstructive sleep apnea: a randomized controlled trial. Am J Respir Crit Care Med 2011;184(10):1192–9.

[85] Kheirandis-Gozal L, Capdevila OS, Tauman R, et al. Plasma C-reactive protein in nonobese children with obstructive sleep apnea before e after adenotonsillectomy. J Clin Sleep Med 2006;2:301–4.

[86] Gozal D, Serpero LD, Sana Capdevila O, et al. Systemic inflammation in non-obese children with obstructive sleep apnea. Sleep Med 2008;9:254–9.

[87] Yilmaz MD, Onrat E, Altuntas A, et al. The effects of tonsillectomy and adenoidectomy on pulmonary arterial pressure in children. Am J Otolaryngol 2005;26:18–21.

[88] Tezer MS, Karanfil A, Aktas D. Association between adenoidalnasopharyngeal ratio and right ventricular diastolic function in children with adenoid hypertrophy causing upper airway obstruction. Int J Pediatr Otorhinolaryngol 2005;69:1169–73.

[89] Kinoshita H, Shibano A, Sakoda T, et al. Uvulopalatopharyngoplasty decreases levels of C-reactive protein in patients with obstructive sleep apnea syndrome. Am Heart J 2006; 152:692.e1–5.

[90] Kataoka T, Enomoto F, Kim R, et al. The effect of surgical treatment of obstructive sleep apnea syndrome on the plasma TNF-alpha levels. Tohuku J Exp Med 2004;204:267–72.

[91] Itzhaki S, Dorchin H, Clark G, et al. The effects of 1-year treatment with a Herbst mandibular advancement splint on obstructive sleep apnea, oxidative stress, and endothelial function. Chest 2007;131:740–9.

[92] Askar V, Hirshkowitz M. Health effects of sleep deprivation. Clin Pulm Med 2003;10:47–61.

[93] Banks S, Dinges DF. Behavioral and physiological consequences of sleep restriction. J Clin Sleep Med 2007;3(5):519–28.

[94] Ayas NT, White DP, Manson JE, et al. A prospective study of sleep duration and coronary heart disease in women. Arch Intern Med 2003;163(2):205–509.

[95] Gangwisch JE, Heymsfleld SB, Boden-Albala B, et al. Short sleep duration as a risk factor for hypertension: analyses of the first National Health and Nutrition Examination Survey. Hypertension 2006;47(5):833–9.

[96] Gottlieb DJ, Punjabi NM, Newman AB, et al. Association of sleep time with diabetes mellitus and impaired glucose tolerance. Arch Intern Med 2005;165(8):863–7.

[97] Qureshi AI, Giles WH, Croft JB, et al. Habitual sleep patterns and risk for stroke and coronary heart disease: a 10-year follow-up from NHANES I. Neurology 1997;48(4):904–11.

[98] Dinges DF, Dougla SD, Zaugg L, et al. Leukocytosis and natural killer cell function parallel neurobehavioral fatigue induced by 64 hours of sleep deprivation. J Clin Invest 1994;93:1930–9.

II. ENDOTHELIAL DYSFUNCTION AND CLINICAL SYNDROMES

[99] Frey DJ, Fleshner M, Wright Jr KP. The effects of 40 hours of total sleep deprivation on inflammatory markers in healthy young adults. Brains Behav Immun 2007;21:1050–7.

[100] Vgontzas NA, Zoumakis E, Bixler EO, et al. Adverse effects of modest sleep restriction on sleepiness, performance, and inflammatory cytokines. J Clin Endocrinol Metab 2004;89:2119–26.

[101] Sunbul M, Kanar BG, Durmus E, et al. Acute sleep deprivation is associated with increased arterial stiffness in healthy young adults. Sleep Breath 2014;18:215–20.

[102] Sauvet F, Leftheriotis G, Gomez-Merino D, et al. Effect of acute sleep deprivation on vascular function in healthy subjects. J Appl Physiol 2010;108:68–75.

[103] Garcia-Fernández R, Pérez-Velasco JG, Milián AC, et al. Endothelial dysfunction in cardiologists after 24 hours on call. Rev Esp Cardiol 2002;55(11):1202–4.

[104] Amir O, Alroy S, Schliamser JE, et al. Brachial artery endothelial function in residents and fellows working night shifts. Am J Cardiol 2004;93:947–9.

[105] Weil BR, Mestek ML, Westby CM, et al. Short sleep duration is associated with enhanced endothelin-1 vasoconstrictor tone. Can J Physiol Pharmacol 2010;88:777–81.

[106] Takase B, Akima T, Uehata A, et al. Effect of chronic stress and sleep deprivation on both flow-mediated dilation in the brachial artery and the intracellular magnesium level in humans. Clin Cardiol 2004;27:223–7.

[107] Dettoni JL, Consolim-Colombo FM, Drager LF, et al. Cardiovascular effects of partial sleep deprivation in healthy volunteers. J Appl Physiol 2012;113:232–6.

[108] Calvin AD, Covassin N, Kremers WK, et al. Experimental sleep restriction causes endothelial dysfunction in healthy humans. J Am Heart Assoc 2014;3(6)e001143. https://doi.org/10.1161/JAHA.114.001143.

[109] Shearer WT, Reuben JM, Mullington JM, et al. Soluble TNF-alpha receptor 1 and IL-6 plasma levels in humans subjected to the sleep deprivation model of spaceflight. J Allergy Clin Immunol 2001;107:165–70.

36

Smoking and the Endothelium

Juan Carlos Yugar-Toledo, Rodrigo Modolo, and Heitor Moreno, Júnior

INTRODUCTION

History

The casual act of lighting a cigarette and sucking the smoke has its origin lost in time. It is practically impossible to determine how and when someone first had the idea of burning the dried tobacco leaves and sucking up their smoke. Long before Europeans arrived in America, smoking was a daily part of the native Americans' daily lives, and its function was much more related to their religious beliefs than to the pure and simple pleasure of tobacco consumption. In Brazil, at the time of the discovery, tobacco was part of the Indian rituals of all tribes that came in contact with the Portuguese. As it was later observed, and the reports of Pedro Álvares Cabral sailors confirmed, smoke obtained from the burning of the leaves was considered miraculous materialization of the breath of the shamans.

In Europe, it was Jean Nicot, the French ambassador to the Portuguese court, who introduced and disseminated tobacco. Nicot's initiative later caused the botanist De La Champ to baptize tobacco, scientifically, as *herbanicotiana*, giving the name of the ambassador to all kinds of plants to which tobacco belongs.

Only in the 1960s were the first epidemiological studies relating usual cigarette smoking to lung disease published [1,2], and later cardiovascular disease in smokers (active smoking) [3] and then, in nonsmokers (passive smoking) [4–6]. The relationship between smoking and increased mortality from cardiovascular disease was consistently and unequivocally demonstrated [3,7–9]; it is also known also that the risk for cardiovascular atherosclerotic disease

among smokers relates to the number of cigarettes burnt per day [10,11]. Smoking and addiction to nicotine were then seen as diseases that needed to be prevented, investigated, and treated [12].

According to the World Health Organization (WHO) [13,14], tobacco smoke is the main factor responsible for pollution indoors. On average, this polluted air contains three times more nicotine, three times more carbon monoxide, and up to 50 times more carcinogenic substances than the smoke that enters the smoker's mouth after passing through the cigarette filter.

In adult nonsmokers, there is a greater risk of cardiovascular disease caused by smoking, in proportion to the time of exposure to smoke, and the risk of myocardial infarction is 30% higher than in nonexposed nonsmokers [15]. Therefore, these findings reinforce government efforts to eradicate smoking in public places.

Epidemiology

Cigarette smoking—the most important form of tobacco use—is a serious public health problem in the world and specifically in Brazil [16]. According to the WHO, smoking is related, directly or indirectly, to the deaths of 6 million people worldwide each year. If effective measures to control smoking are not taken, by 2030 this number could reach 8 million deaths—80% of them in developing countries [17].

Recent estimates of tobacco consumption in the world indicate a relative reduction of prevalence estimated by smoking age groups from 1980 to 2012.

537

Among men, a reduction from 41.2% to 31.1% was observed, with an annual decline rate of 0.9%, and from 10.6% to 6.2% among women, with an annual decline rate of 1.7%. However, due to the growth of the population aged over 15 years, there was an increase in the absolute number of smokers in both sexes, from 721 million smokers in 1980 to 967 million in 2012 [18]. With regard to the number of cigarettes smoked per day a significant change was not observed with time, remaining at around 18 cigarettes a day.

In Brazil, estimated prevalence of smokers in the population over 15 years of age is 17.2% (24.6 million individuals). The prevalence among men is 21.6% (14.8 million individuals). Among women, prevalence is 13.1% (9.8 million inhabitants). The majority make daily use of tobacco products (15.1%), while the percentage of occasional smokers is only 2.1%. This pattern was observed in all regions of the country. The prevalence of use of industrialized cigarettes was 14.4%, while the prevalence of use of straw or hand-rolled cigarettes was 5.1%. The percentage of smokers of other tobacco products (cigars, pipes, cigarillos, Indian cigarettes, and hookah) is low: 0.8% on average [17,19].

Data on the prevalence of smoking in the United States, in individuals over 18 years old, showed a reduction from 20.9% in 2005 to 17.6% in 2013, with a greater reduction in the prevalence of smoking among men (20.5%) than women (15.3%). There was also a significant decrease in the number of inveterate smokers (>30 cigarettes/day) from 12.7% to 7.1% [16]. Smoking is responsible for approximately 45% of deaths in men younger than 65 years old and more than 20% of all deaths due to coronary disease in men over 65 years old. In addition, male smokers between 45 and 54 years old are almost three times more likely to die from myocardial infarction than nonsmokers from the same age group [20].

It is estimated that smoking is responsible for 40% of deaths due to coronary disease in women over 65 years of age. The risks of myocardial infarction, pulmonary embolism, and thrombophlebitis in young women who smoke and use oral contraceptives are ten times greater than those who do not smoke and use this method of birth control [21]. Once smoking is stopped, the risk of heart disease begins to decline. After 1 year, the risk is halved, and after 10 years, it is similar to that of individuals with no previous history of smoking [22,23].

Components of Cigarette Smoke

There are two ways to inhale cigarette smoke: (1) when the smoker sucks by absorbing toxic substances (primary stream) through the mouth; and (2) when smoke which is released freely from the burning tip of the cigarette or other tobacco derivative into ambient air endangers the health of nonsmokers (secondary current).

Cigarette smoke is a complex mixture of about 5000 different toxic substances, many of which are generated during the burning of the tobacco leaf, some in gas phase and others in particulate phase [24]. The gas phase represents approximately 60% of the smoke from the burning of tobacco, 99% of this phase is composed of nitrogen, oxygen, carbon dioxide, carbon monoxide, hydrogen, argon, and methane. The remaining 1% is represented by 43 other components. The aromatized hydrocarbons present in tobacco smoke considered carcinogenic contain from four to six condensed rings, the main representative of which is benzopyrene. Other toxic components less studied are: nitrosamines, radioactive substances, polonium 210 and carbon 14, DDT pesticides, benzene, heavy metals (lead and cadmium), nickel, hydrogen cyanide, ammonia, and formaldehyde.

The particulate phase contains nicotine and tar. The amount of tar from the smoke of a cigarette ranges from 3 to 40 mg according to the conditions of burning, condensation, cigarette size, filter presence, paper porosity, cigarette content, weight, and type of tobacco. Another determining factor in the composition of tobacco smoke is the burning temperature, which reaches 884°C during aspiration. It is estimated that the primary stream of tobacco smoke contains about 150 mg of metallic constituents, mainly potassium (90%), sodium (5%), arsenic (0.3%–1.4%), and traces of aluminum, calcium and copper. The inorganic components are mostly chlorides; however, beryl and chromium may be present in low quantities.

Nicotine is an alkaloid compound present in *Solanaceae* family of plants such as tobacco (*Nicotiana tabacum*) and coke plant (*Erythroxylum coca*). Its highly lipophilic composition allows it to be easily absorbed in the gastrointestinal tract, skin, and mucous membranes, and to cross the blood-brain barrier and fetoplacental barrier. Nicotine is primarily responsible for chemical dependence and cardiovascular disease related to smoking.

When aspirated it reaches the brain in 8 seconds, whereas if applied directly (intravenously) it would take 14 seconds. Between 50% and 90% of nicotine is absorbed during smoking, and it can be detected and quantified in plasma and urine for 24 hours, with a half-life of 120 minutes. Sixty percent of the substance is converted into cotinine by cytochrome P450 oxidation reactions. Cotinine, a metabolite of lower toxicity with nicotine-like psychoactive properties, is slowly cleared from the circulation (half-life of 15 hours) by predominantly hepatic elimination, and may be excreted by the kidneys, depending on the urinary pH [24–26].

Nicotine acts by binding to acetylcholine receptors on the cell membrane called nicotinic acetylcholine receptors (nAChR), which are pentameric proteins originating from a 370 kDa protein complex and form an extensive family of subunits (α1–α10, β1–β4, γ, δ, and ε) [27]. These receptors are widely distributed in the central and peripheral nervous system. Nicotine dependence occurs by increasing the expression of these nicotinic receptors in the central nervous system as a consequence of long-term exposure [28].

PATHOPHYSIOLOGICAL BASES

Mechanisms by Which Smoking Promotes Vascular Changes

Smoking is probably the most complex and least understood risk factor for cardiovascular diseases (CVD). This is because tobacco smoke contains more than 5000 different chemical substances—from atoms to particles in suspension, including those that generate reactive oxygen species (ROS) that cross the alveolar epithelium to reach blood circulation and promote cytotoxic lesions in the different tissue targets.

Interaction among cigarette smoke components, the endothelium, and the vascular wall is demonstrated by vascular morphological changes, such as those observed experimentally in cultures of endothelial cells and isolated organs, characterized by endothelial disruption, endothelial cell desquamation, apoptosis, cell repair, and cell proliferation [29]—a key process in maintaining vascular integrity [30].

Angiogenic properties of nicotine have also been observed involving the three relevant proangiogenic molecules: fibroblast growth factor (FGFb), platelet-derived growth factor (PDGF), and vascular

endothelial growth factor (VEGF). In addition, endothelial vascular growth factor receptors are expressed in the endothelial cells and in the blood vessels of individuals exposed to tobacco smoke.

Thus, the actions of nicotine on vascular endothelium are related to the functional expression of nAChR on the endothelial surface, as demonstrated experimentally [31]. The α7 nAChR-unit—through modifications of its configuration—is involved in mediating the antiproliferative effects of nicotine, including: modulation and neogenesis of vascular remodeling, migration, proliferation, and cell differentiation, that contribute to the development of the atherosclerotic disease and carcinogenesis [32].

Binding of nicotine to α7-nAChR through the endothelial cell promotes activation of tyrosine kinase Src and beta-arrestin, expression mediators of major angiogenic factors, such as transforming growth factor beta, fibroblasts growth factor (FGFb), PDGF, and VEGF. Both Src and VEGF stimulate the synthesis and release of extracellular matrix metalloproteinases (MMPs), which result in several pathological processes, including vascular wall alterations such as intimal hyperplasia, atherosclerosis progression, and restenosis. Additionally, a process of mobilization of progenitor cells from bone marrow and spleen is speculated via nicotinic stimulation at its nAChR receptor and incorporation of these cells into ischemic tissues. For these and other reasons discussed below, the interaction of the various components of cigarette smoke, especially nicotine, with the vascular endothelium has been linked to cytotoxic and harmful effects [33] (Fig. 36.1A and B).

Structural and Functional Changes of the Vascular Wall

Experimental evidence in humans and animals shows that smoking promotes functional changes (endothelial dysfunction (ED)) and structural changes in the wall of conductance arteries (large arteries), muscular arteries (medium caliber) and resistance arteries (microcirculation). These alterations include changes in compliance and stiffness of the vascular wall [34] with a significant reduction of the elastic properties of the vascular wall probably mediated by stimulation of the sympathetic nervous system, responsible for the effects on blood pressure, pulse pressure, and pulse wave profile amplification in the aorta with increased central

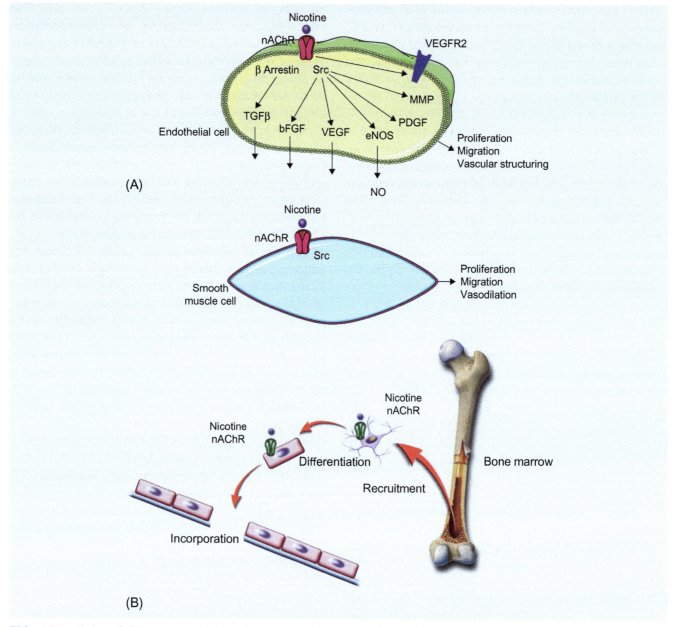

FIG. 36.1   (A) and (B) represent molecular mechanism by which smoking promotes vascular changes (acronyms, see text). Adapted from Costa F, Soares R. Nicotine: a pro-angiogenic factor. Life Sci 2009;84(23–24):785–90.

systolic pressure [35–38]. Contributing factors to this manifestation are ED, vascular remodeling by increasing the thickness of the intima and media layers of medium-sized arteries and microvascular bed, and reduction of lumen/wall thickness ratio [29,34,39–43].

Endothelial Dysfunction

The effects of tobacco smoke on endothelial mechanisms that control vascular tone have received considerable interest in in vivo and in vitro arteries and

in vivo veins [44]. ED is expressed as a modification of endothelium-dependent vasodilation in different vascular beds such as the coronary circulation [45–49], peripheral circulation (brachial artery, dorsal vein of the hand), and microcirculation [39,50].

Smoking promotes ED of multifactorial origin, as a consequence of: (1) reduction of NO bioavailability, associated with decreased expression of NO endothelial synthase (eNOS); (2) increase of oxidative stress with generation of ROS by activation of NADPH oxidase via C protein kinase; and (3) direct effect of oxidative substances present in tobacco smoke that causes plasma levels of antioxidant glutathione (GSH) to fall, with consequent increase in oxidative stress [51].

In addition, tobacco smoke promotes alterations in the concentration of vasoactive substances such as thromboxane A_2 (TXA$_2$), angiotensin II (A-II), endothelin-1 (ET-1), and prostacyclin (PGI). These substances participate in the alteration of the endothelium-dependent vasodilation, thrombogenesis, fibrinolisis reduction, platelet activation, expression of inflammatory factors, and vascular smooth muscle cell proliferation. This process occurs by stimulation of cytokine release from neutrophils, monocytes, T cells, and activated platelets by the effect of the particulate phase components in tobacco smoke, which contribute to the perpetuation of ED [52–61] (Fig. 36.2).

Coronary Vascular Reactivity

Changes in coronary vascular tone take place 5 minutes after the consumption of a cigarette and are characterized by a 7% decrease flow speed and increased coronary resistance by 21%, regardless of changes in heart rate and blood pressure. Coronary vasoconstriction induced by smoking is mediated by alpha-adrenergic stimulation, which causes immediate proximal and distal constriction of these arteries and increase of tonus in resistance vessels. This effect is attributed to the action of nicotine, which promotes local (norepinephrine) and systemic (epinephrine) release of catecholamines. There is also a reduction of prostacyclin levels (PGI), since nicotine decreases its synthesis in the endothelium of the coronary arteries without affecting TXA$_2$. The imbalance in the prostacyclin/thromboxane ratio leads to a predominance of the thromboxane effect on the endothelium of coronary arteries and to the basal and stimulated nitric oxide (NO) deficiency [49,62]. Coronary vasoreactivity compromised by tobacco smoke is accompanied by dysfunction of the coronary endothelial cells with subsequent release of prothrombotic factors such as the von Willebrand factor, tissue plasminogen

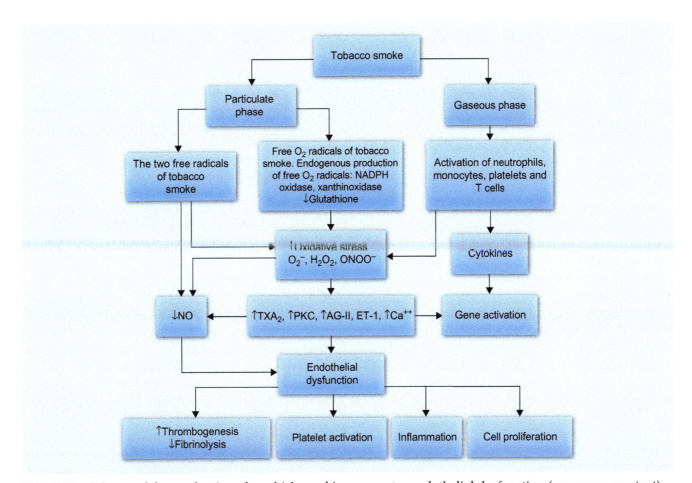

FIG. 36.2   Scheme of the mechanisms by which smoking promotes endothelial dysfunction (acronyms, see text).

activator factor, and tissue plasminogen activating factor inhibitors PAI-1 and PAI-2. These factors stimulate lipid peroxidation of LDL particles, attract inflammatory cells, stimulate smooth muscle cell proliferation, and promote the formation of atherosclerotic plaques and remodeling of coronary arteries [63,64].

Peripheral Vascular Reactivity

Endothelium-dependent vasodilator response can be assessed by intraarterial administration of muscarinic agonists such as acetylcholine and bradykinin (mediators of NO release). In conductance arteries, such as the brachial artery, using intravascular occlusion plethysmography with intraarterial injection of acetylcholine, a distinct reduction of the endothelium-dependent vasodilator response in smokers can be observed.

Another method to evaluate of brachial artery reactivity is flow-mediated vasodilatation (FMD) performed with high-resolution ultrasound, which also evaluates vascular function dependent on endothelium—which is compromised in active and passive smokers [65,66] (Fig. 36.3A).

Acute exposure of hand dorsal veins to nicotine infusion at plasma concentrations equivalent to those found in smokers is accompanied by a decrease in the vasodilator response to bradykinin (a NO and PGI release agonist). In these experiments, blocking the synthesis of prostacyclin with indomethacin suggests that the participation of prostacyclin is less important than that of NO in this condition [67,68].

Moreover, the dependent and independent vascular dysfunction of the endothelium of heavy smokers (>20 cigarettes per day) is normalized after 24 hours of smoking interruption, as demonstrated experimentally by Moreno and collaborators [44] (Fig. 36.3B).

Evaluation of the vasodilatory microcirculation response, by infusing nicotine into fragments of human skin properly prepared for this purpose, has demonstrated an accentuation of the vasoconstriction induced by norepinephrine [69]. Experimental studies of arteriolar circulation in oral mucous of hamsters show change of endothelium dependent vasodilation during nicotine infusion, which is prevented by infusion of superoxide dismutase (SOD), suggesting that the formation of oxygen-free radicals contributes to decrease the endothelium-dependent vascular response in the microcirculation [70].

Oxidative Stress and Lipid Peroxidation

Oxidative stress takes place as an organic response to conditions that facilitate oxidation, such as chronic inflammation, low intake of nutritional antioxidants, and smoking. Products derived from metabolism of arachidonic acid via noncyclooxigenase (COX), such as 8-iso-prostaglandin F α-I (IPGF α-I), produced during lipid peroxidation of arachidonic acid by free radicals, increased urinary output, reflecting increased oxidative stress in smokers [51,71].

The main ROS present in the gaseous phase of tobacco smoke are superoxide anions, hydrogen peroxide, hydroxyl, peroxynitrite and free radicals of organic compounds, which are highly reactive substances. Due to their short half-life, they are readily inactivated by antioxidants, such as catalase, GSH, and SOD, both cellular (SOD1—CuZnSOD cytoplasmic and SOD2—MgSOD mitochondrial), and extracellular (SOD3—CuZnSOD), also known as EC-SOD.

More stable substances such as alpha and beta-unsaturated aldehydes (acrolein and crotonaldehyde), unsaturated alpha and betaketone, and some saturated aldehydes—present in the gaseous phase of tobacco smoke—participate in the regulation of the enzymatic activity of nicotinamide adenine dinucleotide phosphate (NADPH) oxidase and xanthine oxidase, which catalyze the formation of superoxide anions increasing oxidative stress. Cytokines, such as tumor necrosis factor alpha (TNF-α), interleukins (IL-1α and IL-6) and vasoactive peptides, endothelin-1 and angiotensin II, also increase the expression and activity of NADPH oxidase promoting increases in production of ROS and reduction of plasma levels of GSH antioxidant with increase of lipid peroxidation products.

Thus, the most characteristic deleterious result of this disorder is the oxidation of the LDL-cholesterol molecules. In vitro oxidation studies by LDL-cholesterol incubation in nicotine and/or cotinine (main active metabolite of nicotine) and monitoring of lipid peroxidation markers (decrease hydroperoxide and increased thiobarbituric acid reactive substances—TBARS) confirm the hypothesis that the LDL of smokers are highly susceptible to oxidation [72,73].

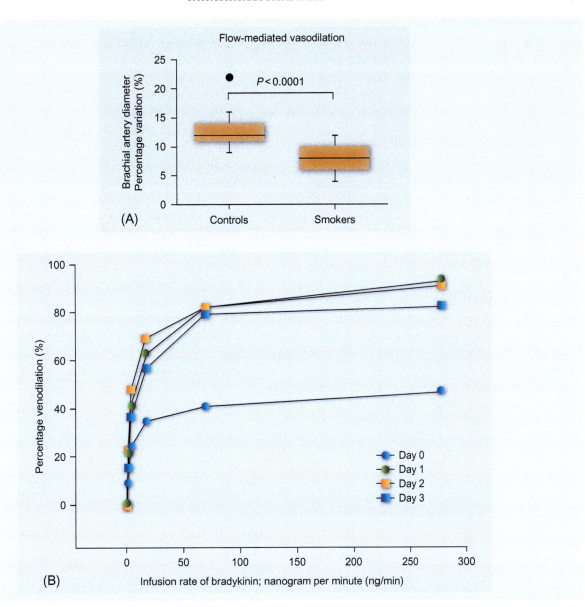

FIG. 36.3 (A) Endothelium-dependent vasodilation. Evaluation of flow-mediated vasodilation (FMD) performed with high-resolution ultrasound. Endothelium-dependent vascular function is compromised in active smokers compared to controls. $P < 0.0001$. (B) Reversibility of vascular impairment. Representative of the dose response curve with bradykinin (hand dorsal vein technique) in a smoker volunteer. The vasodilator response to bradykinin normalizes 24 hours after cessation of smoking, and is maintained for 48–72 hours. *$P < 0.05$ vs. day 0. Adapted from Moreno Jr H, Chalon S, Urae A, et al. Endothelial dysfunction in human hand veins is rapidly reversible after smoking cessation. Am J Physiol 1998;275:H1040–5.

Effects of Nicotine on Gene Expression, DNA Synthesis, Angiogenesis, Apoptosis, Migration, Survival, and Proliferation of Endothelial Cells

Nicotine promotes a bimodal DNA response in the endothelial cell. While experimental studies have shown that human endothelial cells incubated with low concentrations of nicotine respond with increased DNA synthesis, probably associated with cell proliferation, incubation with high concentrations of nicotine causes cytotoxicity.

Increased NADPH oxidase activity of the cellular and mitochondrial membranes, by direct effect and also modulated by phosphokinase C (PKC), is observed after exposure to components of tobacco smoke. The consequence of this exposure is increased peroxide anions (O_2^-), that at cytoplasmic level is accompanied by increased concentration of peroxynitrite (ONOO), which inactivates NO and promotes ED. In the mitochondria, the increase of ROS promotes mitochondrial oxidative stress, mitochondrial DNA damage, and mitochondrial dysfunction.

Excess peroxynitrite promotes activation of nuclear factor kappa-beta factor (NF-κβ) and the nuclear enzyme poly (ADP-ribose) polymerase (PARP-1). This activation contributes to a greater expression of NF-κβ, elevation of mRNA levels, translating nuclear DNA damage with alteration of gene expression of several systems involved in atherogenesis such as eNOS, ECA (vasomotor tonus modulators), vWF factor, t-PA, PAI-1 (mediators of thrombogenicity) and VCAM-1, promoting adherence of monocytes and T lymphocytes migrating into the subendothelial space. Excessive peroxynitrite inactives GSH and modulates the activity of mitogen-activated protein kinase (MAPK), which in combination with NF-κβ promote the atherogenic phenotype.

Excess hydrogen peroxide, derived from the inactivation of superoxide anions by superoxidedismutase, is involved in NF-κβ activation and also in DNA impairment, which in association with

previously mentioned alterations promote apoptosis, expression of adhesion molecules, activation of monocytes and finally expression of the atherogenic phenotype (Fig. 36.4).

The mechanism by which nicotine induces pathological angiogenesis involves the endothelial cholinergic pathway. It is known that the endothelial cell—which has acetylcholine receptors (ACh)—synthesizes ACh from acetyl-coenzyme A and choline via acetyltransferase from choline (ChAT). Choline reuptake is essential for ACh synthesis. The presence of a transporter with high affinity for choline has been demonstrated in endothelial cells and vascular smooth muscle. However, endothelial cells express acetylcholinesterase (AChe) and butyrylcholinesterase (BChe), which hydrolyze ACh restricting its action to a paracrine or autocrine effect. The importance of this effect lies in the possibility of maintaining the activity of α7-nAChR cholinergic receptors that are also activated by choline and thus promote angiogenic activity.

Therefore, even after cleavage of ACh, these receptors maintain signaling activity. Hence, activation of nAChR receptors by endogenous ACh or exogenous nicotine increases the permeability of

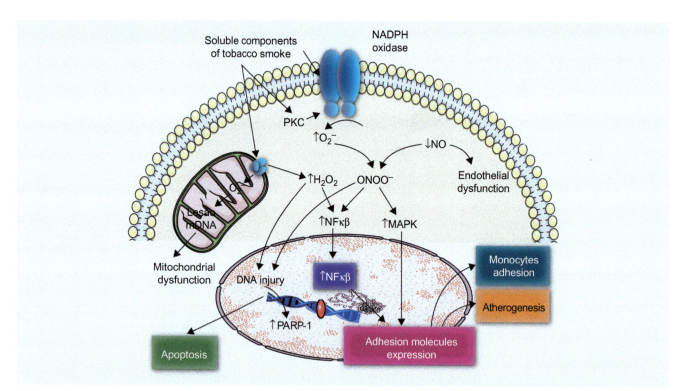

FIG. 36.4　Effects of nicotine on gene expression, mitochondrial dysfunction, apoptosis, expression of adhesion molecules and atherogenesis (acronyms, see text). Adapted from Csiszar A, Podlutsky A, Wolin MS, et al. Oxidative stress and accelerated vascular aging: implications for cigarette smoking. Front Biosci (Landmark Ed). 2009;14:3128–44.

calcium channels initially via activation of the α7-nAChR receptor and, subsequently, by opening of secondary channels such as the channels of potential receptors (trpC).

Increased cytoplasmic calcium stimulates action of phospholipase C (PLC γ) to form diacylglycerol (DAG) that activates PKC and inositol 3 phosphate (IP3), that in turn acts on the endoplasmic reticulum (ER) and releases calcium from intracellular reserves.

The high concentration of calcium activates the calcium-calmodulin complex that stimulates eNOs and increases the concentration of NO and citrulline, which is involved in cell migration. This increase in intracellular calcium via activation of the α7-nAChR receptor activates the kinases cascades. Activation of PKC induces NF-κβ stimulation involved in cell proliferation. Activation of PKA mobilizes the Raf-kinase cascade, followed by stimulation of MEK (from the MAPK cascade), and

subsequently the ERK 2-cascade which is specifically activated in response to nicotinic stimuli [74].

In addition, a prosurvival effect can also be observed after depletion of beta-arginine-1 by shRNA inhibition with an increase of Akt/PKB, which promotes phosphorylation and inactivation of apoptotic Bax and Bad proteins resulting in suppression of cell death. This inactivation is induced by the stimulation of PKC, PKA, MEK, and P13K (Fig. 36.5).

Changes in the Synthesis of Prostaglandins, Prostacyclin, and TXA₂

Smoking and the therapeutic use of nicotine increase the formation of TXA$_2$, a powerful vasoconstrictor and platelet proaggregating, product of arachidonic acid metabolism via cyclooxygenase, platelet, and lung macrophages [75,76].

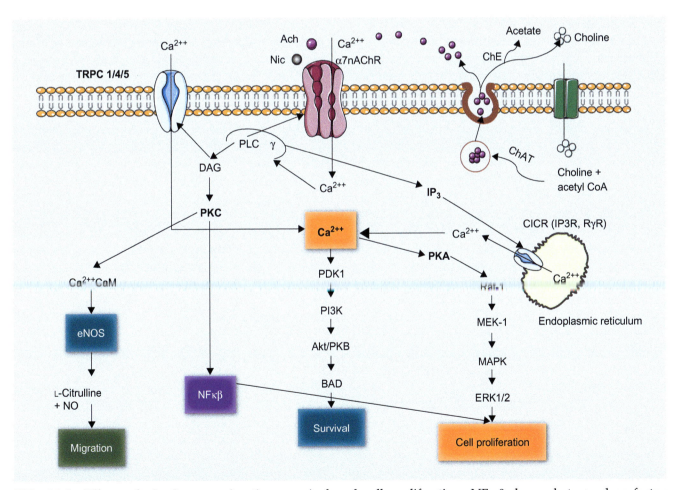

FIG. 36.5 Effects of nicotine on migration, survival and cell proliferation. NF-κβ: kappa-beta nuclear factor (acronyms, see text). Adapted from Lee J, Cooke JP. Nicotine and pathological angiogenesis. Life Sci 2012; 91(21–22):1058–64.

Nicotine inhibits the synthesis of prostacyclin, primary product of arachidonic acid metabolism of the endothelial cell, with vasodilator and antiplatelet platelet functions in isolated heart and blood flow in vitro.

However, human excretion of the metabolite 2–3 dinor-6-keto-prostaglandin is increased, suggesting that other sources (p. Ex.: platelets) and other factors may be involved in the synthesis of prostacyclin in smokers.

CURRENT STAGE OF KNOWLEDGE

Smoking and Atherosclerosis

Atherosclerosis associated with smoking is not necessarily an effect of nicotine, but probably the joint action of the various constituents of cigarette smoke. ROS from the gas phase of the smoke of tobacco contribute to the onset and progression of atherosclerosis.

Exposure of endothelial cells to tobacco smoke, which contains oxygen-free radicals, inactivates NO, which converted into peroxynitrite promotes protein nitrosylation, reduced activity of endothelial NO synthase (eNOS), increased expression of uncoupled eNOS, activation of NADPH oxidase (higher endogenous source of ROS), and increased oxidative stress. As a consequence of reduced NO bioavailability and increased oxidative stress, stimulation of the tumor necrosis factor NF-κB and immediate increase in the expression of adhesion molecules (ICAM, VCAM, etc.), selectin (E-selectin, P-selectin), inflammatory cytokine, recruiting platelets, macrophages, and lymphocytes, as well as activation and endothelial cell dysfunction and subsequent loss endothelial cells by apoptosis or necrosis occurs.

Reduction of NO derived endothelial cells promotes contraction of vascular smooth muscle cells (VSMC) of the medium layer of vascular wall (vasoconstriction).

As a result of NO reduction in endothelial cells, the following is observed: (1) expression of adhesion mole-cell receptors on the endothelial surface; and (2) activation of macrophages—that after joining the vessel wall, migrate into the subendothelial space, where they capture oxidized lipids by increased ROS production from tobacco smoke. This capture of oxidized lipids by macrophages is mediated by *scavenger* receptors, which induce formation of foam cells. With the subsequent death of these cells there is release of lipid content, contributing to the formation of atheromatous plaques. It is postulated also that tobacco smoke increases the proliferation and migration of VSMC (vascular smooth muscle cells) causing intimal-medial thickening in the vascular wall and subsequent formation of fatty streaks and atheromatous plaques.

Cell death by VSMC necrosis is another consequence of exposure to tobacco smoke triggered by an inflammatory process in progress, as well as release of intracellular proteolytic enzymes, which promote the cleavage of extracellular matrix proteins. The destruction of the extracellular matrix proteins is reinforced by the increased expression of MMPs and reduced expression of tissue inhibitors of MMPs—the TIMMPs [77] (Fig. 36.6).

Acute Cardiovascular Effects of Nicotine

Intravenous administration of nicotine in humans is clearly associated with increased blood pressure, peripheral vascular resistance and heart rate—resulting from the release of catecholamines in adrenergic terminals [78–80].

Heart rate variability (HRV) is an important marker of cardiovascular and autonomic activity. During acute exposure to nicotine, HRV is reduced and mediated by increased sympathetic activity; this effect can be abolished by beta-blockers. However, baroreceptors, when intact, act in contrast to this sympathetic stimulation by reducing adrenergic activity in the periphery. In young women greater impairment of baroreflex integrity can be observed, and thus higher sympathetic activity during acute exposure to the nicotine effect, which may explain at least part of the greater existing cardiovascular risk in smoker women.

Chronic Cardiovascular Effects of Nicotine

Chronic administration of nicotine causes tolerance (tachyphylaxis) and the acute hemodynamic effects of nicotine on the cardiovascular system stabilize or attenuate. Increased risk of ventricular and supraventricular arrhythmias, myocardial infarction, exacerbation of heart failure and sudden death are cardiovascular effects observed in smoking—a result of the activation of the sympathetic nervous system by nicotine [81–83].

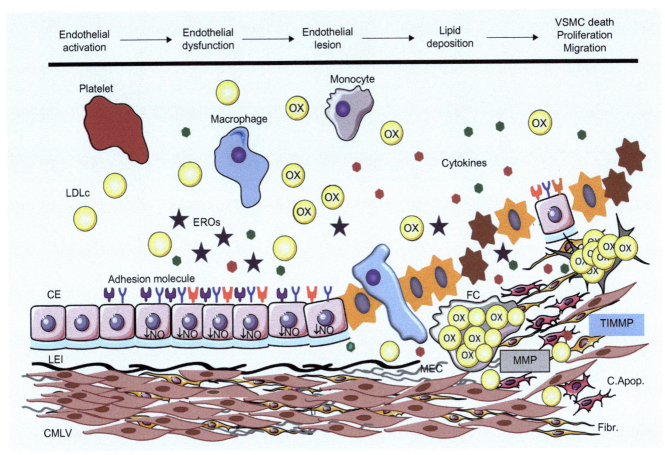

FIG. 36.6 Smoking and atherosclerosis. *VSMC*, vascular smooth muscle cell, *LDLc*, cholesterol fraction of low density; *EC*, endothelial cell, *IEL*, internal elastic layer, *NO*, nitric oxide; *ROS*, reactive oxygen species; *OX*, particles of oxidized LDLc; *MEC*, extracellular matrix; *FC*, foam cells, *MMP*, metalloproteinase; *TIMMP*, metalloproteinase tissue inhibitor; *C.Apop.*, apoptotic cell; *Fibr.*, fibroblast. Adapted from Jacob-Ferreira AL, Palei AC, Cau SB, et al. Evidence for the involvement of matrix metalloproteinases in the cardiovascular effects produced by nicotine. Eur J Pharmacol 2010;627:216–22.

Studies with heavy smokers (consuming more than 25 cigarettes per day) showed alterations in blood pressure circadian rhythm characterized by decreased night rest and morning elevation of blood pressure levels. This effect was reproduced experimentally in patients undergoing implantation of a transdermal nicotine patch after cessation of smoking for 24 hours [80].

Increased sympathetic activity in the central nervous system (CNS) takes place by two mechanisms. First, nicotine stimulates nicotinic receptors of the ventrolateral rostral region of the bulb promoting increased central sympathetic activity. Second, blockade of the inhibitory effect of CNS neurotransmitters mediated by NO [84]. The effect of exposure to tobacco smoke on the autonomic nervous system depends on integrity and baroreflex sensitivity that acts as pain inhibition of

sympathetic stimulation mediated by nicotine [84]. Several studies have shown that chronic tobacco use causes decreased the sensitivity of baroreceptors in smokers [85,86].

The tobacco smoke particulate phase stimulates C-type lung afferent fibers that by direct effect of nicotine or increased oxidative stress trigger increased sympathetic activation and induced neurogenic inflammation with perpetuation of sympathetic over activity cycle. Suppression of the baroreflex in smokers can be explained by three mechanisms: (1) nicotine induces neuroplasticity of the nucleus of the solitary tract, directly altering baroreflex responsiveness in smokers; (2) increased oxidative stress and decreased NO bioavailability; and (3) ED and increased arterial stiffness compromise the vascular wall response to pressure fluctuations in response to exposure to tobacco smoke (Fig. 36.7).

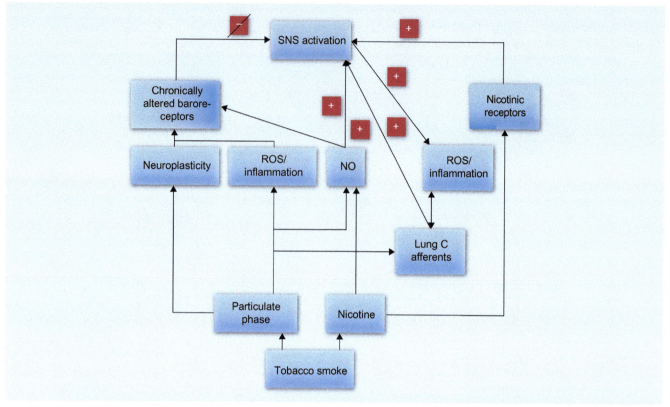

FIG. 36.7   Mechanisms of interaction between nicotine sympathetic nervous system and baroreflex. Adapted from Middlekauff HR, Park J, Moheimani RS. Adverse effects of cigarette and noncigarette smoke exposure on the autonomic nervous system: mechanisms and implications for cardiovascular risk. J Am Coll Cardiol 2014;64:1740–50.

Smoking Effects on Lipids

Through the release of catecholamines, nicotine induces lipolysis and release of free fatty acids, which are primarily captured by the liver, where they are transformed into VLDL-cholesterol. There is also increased lipoprotein lipase activity of skeletal and cardiac muscle and decreased activity of lipoprotein lipase adipocyte [87]. Smoking has an adverse effect on lipid metabolism, especially in heavy smokers (>25 cigarettes per day), such as increasing VLDL-cholesterol and triglycerides levels, and decreasing HDL-cholesterol (HDL mainly -2) and apolipoprotein A levels, actions that promote increased particle oxidized LDL cholesterol, with a greater atherogenic potential [88]. Changes in lipid profile can be reversed at least in part after 2 weeks of smoking cessation [89,90].

Oxidation studies in vitro by incubation of LDL-cholesterol with nicotine and/or cotinine and monitoring of lipid peroxidation markers confirmed the hypothesis that smokers LDL are highly susceptible to oxidation [91,92]. These oxidized LDL participate in progression and susceptibility to rupture of atherosclerotic plaque, since atheromatous plaques of smokers have higher lipid content in the extracellular space, suggesting a favorable effect of tobacco smoke on lipid composition and stability of the plaque. Important determinants of plaque instability are the presence of inflammatory cells, intraplaque hemorrhage, lipid core growth and highly thrombogenic extracellular matrix proteins.

Neutrophil Activation

There is an augmentation of circulating neutrophils associated with smoking, which decreases rapidly with smoking cessation [93,94]. The increase in circulating neutrophils appears to contribute to installation of acute coronary events, through the release of oxygen-free radicals, proteases, and leukotrienes. These mediators promote platelet aggregation, worsening ED and in turn aggravating the atherothrombotic process.

Increase in Thrombogenicity

Smoking causes a hypercoagulability state by: (1) increased circulating levels of procoagulant factors;

(2) reduction of anticlotting factors; (3) increased blood viscosity; (4) increment in circulating levels of fibrinogen; (5) increase in plasma levels of von Willebrand factor; and (6) the activation of factor VII and coagulation cascade and tissue factor derived macrophages.

Furthermore, there is a decrease due to reduced fibrinolysis inhibitor of the expression of tissue factor, decreased tissue plasminogen activator and increase of tissue plasminogen activator inhibitor that is positively correlated with the number of cigarettes burnt per day [95]. Association of these findings creates a high risk of atherothrombosis.

Carbon Monoxide Effects on Oxygen Transport

Smokers inhale carbon monoxide, and its circulating levels of carboxyhemoglobin may reach 5%–10% compared with 0.5%–2% in nonsmokers. Carbon monoxide binds to hemoglobin, reducing its concentration and the release of oxygen in the periphery, which increases blood viscosity.

Exposure of patients with coronary artery disease to elevated levels of carbon monoxide (6%), worsens of left ventricular function and increases the number and complexity of exercise-induced ventricular arrhythmias.

CLINICAL IMPLICATIONS AND PRACTICAL APPLICATIONS

Smoking and Vulnerable Plaques

The vulnerable plaques are more prone to rupture, thrombosis and clinical events such as myocardial infarction and stroke.

Initiation of plaque rupture is a complex process. The determining factors for plaque vulnerability are: composition of the lipid core, fibrous cap thickness <65 mm and recruitment of inflammatory cells (>25 macrophages per field). Smokers exhibit increased expression of inflammatory molecules within the plate, macrophage migration, increased synthesis and release of metalloproteinases which degrade extracellular matrix, phenomena of neovascularization, intraplaque hemorrhage, proliferation of smooth muscle cells, and inappropriate synthesis of collagen. In addition, sympathetic activation, increased blood pressure and greater mechanical stress in the transition zone of the plate contribute to destabilization and plaque rupture with subsequent

acute cardiovascular event [88]. A recent study with large population sample associated smoking or even the fact that someone has smoked in the past (even if they had already abandoned the addiction) with the presence of coronary subclinical atherosclerosis identified by tomography of coronaries.

However, the same study found that only active smokers who maintain the habit have stronger association with significant stenosis of coronary arteries, showing that for proper protection of coronary atherosclerosis, the ideal is not to stop smoking, but never to start the habit [96].

Smoking in Patients After Percutaneous Intervention and Myocardium Revascularization

Several epidemiological studies suggest that thrombosis mediated by smoking is the main triggering factor associated with myocardial infarction and sudden death. On the other hand, paradoxically the prognosis of myocardial infarction patients undergoing thrombolysis, is better in smokers than in nonsmokers. This is probably explained by the observation that infarcted smokers are younger, have fewer risk factors and coronary artery lesions that are less serious when compared to nonsmokers.

However, a recent meta-analysis of contemporary studies with follow-up greater than a year demonstrated that there is no convincing evidence to say that protection can be ascribed to smoking after interventional treatment of acute myocardial infarction. The differences observed in studies of previous decades have been attributed to methodological errors [97].

During the postinfarction evolution in patients subjected to myocardial revascularization by angioplasty or surgical treatment, cessation of smoking has shown higher rates of survival among smokers who quit smoking when compared to nonsmokers [97]. Perhaps this paradox occurs for the same reason described above.

Genetic Polymorphisms and Smoking

Despite the close relationship between smoking and death from cardiovascular disease, this association is seen only in half of smokers; the other half does not manifest CVD throughout life despite exposure to tobacco smoke for a long period. Certainly, genetic factors are involved in the different presentation of CVD in smokers.

The presence of single nucleotide polymorphisms (SNPs) of the eNOS were reported in active smokers and exsmokers, in particular T786C variants of the promoter region and G894T variant of exon 7, characterized by a conversion of guanine for thymine at position 894 of the gene and subsequent replacement of glutamine (common allele) by aspartate (rarer allele) at residue 298 of eNOS (Glu298Asp), which accompanies reduction of eNOS activity with consequent reduced bioavailability of NO [98,99].

The activity of cytochrome CYP450 plays an important role in the detoxification of many toxic compounds from tobacco smoke. The polymorphic variants of the CYP1A1 gene as MspI polymorphism or M1 (T6235C at the 3'), M2 polymorphism (A4889G in exon 7), M3 polymorphism (T5639C in the 3' position), and M4 polymorphism (C4887A in exon 7) are associated with severity of cardiovascular disease in smokers [100,101]. Smokers carriers of the C allele of MspI polymorphism are more likely to have triple vessel disease. In addition, GSTM1 polymorphism of detoxifying enzymes GSTT1 and participate in the development or progression of atherosclerotic disease [102]. Genes encoding the synthesis of fibrinogen, glycoprotein IIIa and the genes of factor XIII modify the phenotypic expression of smokers imparting increased cardiovascular risk by increased fibrinogen when the A 455 allele of the fibrinogen promoter gene (G-455a) is present. Likewise, the polymorphism of glycoprotein IIIa P1 (A2) is related to a higher risk of myocardial infarction with ST segment elevation. Factor XIII polymorphisms affect fibrin structure and architecture and fibrinolysis mechanisms. Polymorphisms of factor XIII affect the structure and architecture of fibrin and fibrinolysis mechanisms [103,104].

Lipid peroxidation plays a decisive role in the pathophysiology of atherosclerosis related to tobacco smoke. The enzyme associated with HDL cholesterol, paraoxonase (PON1) reduces the accumulation of lipid peroxides and hydrolyzes them in atherosclerotic plaques. Plasma levels and the antiatherogenic activity of PON1 are genetically modulated by two polymorphic variations: the first is a substitution of arginine by glutamine at position 192 (R/Q192) of the gene encoding PON1, while the second is a substitution of methionine by leucine at position 55 (M/L55). In addition to these polymorphisms, genetic variants of the PON2 and PON3 genes have been implicated in the increased cardiovascular risk associated with smoking [105,106].

The presence of the E4 allele of the polymorphism of apo-E is associated with higher plasma levels of small, dense LDL with more atherogenic potential. Smokers with this polymorphism are more prone to oxidation of LDL cholesterol particles and a higher chance of cardiovascular events [107]. Smokers who carry the polymorphism -344 TC CYP11B2 gene of aldosterone synthase have a higher risk of coronary events [108].

Knowledge of the interaction between smoking and the modulation of expression of genes encoding cellular, molecular, and biochemical functions of the structures exposed to tobacco smoke will allow better understanding the mechanisms involved in complex relation to assist the development of new therapeutic measures for effective control and prevention of cardiovascular events.

Clinical and Therapeutic Aspects

The relevance of cardiovascular involvement in active or passive smokers is based on robust epidemiological data. It should be added that tar smoking, even after declining between 1980 and 2010, has been gaining market share again for the broad dissemination of cigarettes without nicotine among children, adolescents, and youths.

From the pharmacological point of view, it is well known the means used by cigarette manufacturers to cause and maintain addiction: lower concentrations of nicotine in cigarettes leads to progressive need for greater number of cigarettes for the consumer to gain initial pleasure (tolerance mechanism). Therefore in terms of public health, there is no doubt that one must combat smoking, guiding smokers regarding the need of smoking cessation as soon as possible. In addition to smoking, other forms of tobacco, such as cigars and pipes, are also related to CVD.

Several commercial nicotine preparations (adhesive, nasal spray, chewing gum) are recommended to assist smokers to quit the habit, with about 32% success in 6 months. Without monitoring of specialized professionals, that number drops to half. Although the use of transdermal nicotine patches (21 mg) increases the blood pressure and heart rate frequently in nonsmokers and smokers, there are no changes in these variables in mild hypertension [80].

Bupropion, previously used as antidepressive, has been employed for the same purpose, with a therapeutic success of 30%. This drug can trigger convulsions in smokers with a history of epilepsy and in those cases it is contraindicated. Few small studies exist on the two therapies used simultaneously, and the results point to no more than 50% smoking cessation in 6 months.

Varenicline, a partial and selective agonist of the nAChR alpha4 beta2 (α4β2-nAChR) involved in enforcement mechanisms and dependence associated with nicotine, is a new drug developed specifically for smoking cessation. This drug partially stimulates nicotinic acetylcholine nicotinic receptors alpha4 beta2 (α4β2-nAChR), promotes the release of dopamine in the brain center of clearing and blocks the action of nicotine on these receptors. For this dual mechanism varenicline showed greater efficacy on the urgent need to smoke (crack) on withdrawal symptoms (depression, irritable accounting, frustration, anger, anxiety, and difficulty concentrating) and the effects of reward, satisfaction and reinforcement associated with smoking. The adverse effects associated with the use of varenicline are nausea, sleep disturbances, constipation, flatulence and vomiting. This drug is contraindicated during pregnancy.

According to studies comparing the effectiveness of the different methods discussed, varenicline showed to be superior to bupropion, chewing gum, and adhesive [109]. The antinicotine vaccine acts by stimulating the immune system to produce specific antibodies which bind with high affinity to nicotine in plasma and extracellular fluids. Initial clinical trials with antinicotine vaccine suggest that this method can improve smoking cessation rates. However, more studies are needed to confirm the efficacy of this new therapeutic method [110].

Therefore, we are still far from effective treatment to assist the definitive discontinuation of the habit of smoking.

References

[1] Blackburn H, Labarthe D. Stories from the evolution of guidelines for causal inference in epidemiologic associations: 1953–1965. Am J Epidemiol 2012;176:1071–7.
[2] Alberg AJ, Shopland DR, Cummings KM. The 2014 Surgeon General's report: commemorating the 50th Anniversary of the 1964 Report of the Advisory Committee to the US Surgeon General and updating the evidence on the health consequences of cigarette smoking. Am J Epidemiol 2014;179:403–12.
[3] Kannel WB, D'Agostino RB, Belanger AJ. Fibrinogen, cigarette smoking, and risk of cardiovascular disease: insights from the Framingham Study. Am Heart J 1987;113:1006–10.
[4] Glantz SA, Parmley WW. Passive smoking and heart disease. Epidemiology, physiology, and biochemistry. Circulation 1991;83:1–12.
[5] Glantz SA, Parmley WW. Passive smoking and heart disease. Mechanisms and risk. J Am Med Assoc 1995;273:1047–53.
[6] He J, Vupputuri S, Allen K, et al. Passive smoking and the risk of coronary heart disease—a meta-analysis of epidemiologic studies. N Engl J Med 1999;340:920–6.
[7] Doll R, Gray R, Hafner B, et al. Mortality in relation to smoking: 22 years' observations on female British doctors. Br Med J 1980;280:967–71.
[8] Peto R, Lopez AD, Boreham J, et al. Mortality from smoking worldwide. Br Med Bull 1996;52:12–21.
[9] Ezzati M, Lopez AD. Regional, disease specific patterns of smoking-attributable mortality in 2000. Tob Control 2004;13:388–95.
[10] Negri E, Franzosi MG, La Vecchia C, et al. Tar yield of cigarettes and risk of acute myocardial infarction. GISSI-EFRIM Investigators. BMJ 1993;306:1567–70.
[11] Parish S, Collins R, Peto R, et al. Cigarette smoking, tar yields, and non-fatal myocardial infarction: 14,000 cases and 32,000 controls in the United Kingdom. The International Studies of Infarct Survival (ISIS) Collaborators. BMJ 1995;311:471–7.
[12] Doll R, Peto R, Boreham J, et al. Mortality in relation to smoking: 50 years' observations on male British doctors. BMJ 2004;328:1519.
[13] WHO urges more countries to require large, graphic health warnings on tobacco packaging: the WHO report on the global tobacco epidemic, 2011 examines anti-tobacco mass-media campaigns. Cent Eur J Public Health 2011;19:133–51.
[14] Healton CG, et al. Butt really? The environmental impact of cigarettes. Tobacco Control 2011;20(Suppl 1):i1–i1.
[15] Barnoya J, Glantz SA. Cardiovascular effects of secondhand smoke: nearly as large as smoking. Circulation 2005;111:2684–98.
[16] Consumption of cigarettes and combustible tobacco—United States, 2000–2011. MMWR Morb Mortal Wkly Rep 2012;61:565–9.
[17] Giovino GA, Mirza SA, Samet JM, et al. Tobacco use in 3 billion individuals from 16 countries: an analysis of nationally representative cross-sectional household surveys. Lancet 2012;380:668–79.
[18] Ng M, Freeman MK, Fleming TD, et al. Smoking prevalence and cigarette consumption in 187 countries, 1980–2012. JAMA 2014;311:183–92.
[19] Health-care provider screening for tobacco smoking and advice to quit—17 countries, 2008–2011. MMWR Morb Mortal Wkly Rep 2013;62:920–7.
[20] Teo KK, Ounpuu S, Hawken S, et al. Tobacco use and risk of myocardial infarction in 52 countries in the INTERHEART study: a case-control study. Lancet 2006;368:647–58.
[21] Petitti DB. Clinical practice. Combination estrogen-progestin oral contraceptives. N Engl J Med 2003;349:1443–50.
[22] Huxley RR, Woodward M. Cigarette smoking as a risk factor for coronary heart disease in women compared with men: a systematic review and meta-analysis of prospective cohort studies. Lancet 2011;378:1297–305.
[23] Peters SA, Huxley RR, Woodward M. Smoking as a risk factor for stroke in women compared with men: a systematic review and meta-analysis of 81 cohorts, including 3,980,359 individuals and 42,401 strokes. Stroke 2013;44:2821–8.
[24] Fowles J, Dybing E. Application of toxicological risk assessment principles to the chemical constituents of cigarette smoke. Tob Control 2003;12:424–30.
[25] Borgerding MF, Milhous Jr LA, Hicks RD, et al. Cigarette smoke composition. Part 2. Method for determining major components in smoke of cigarettes that heat instead of burn tobacco. J Assoc Off Anal Chem 1990;73:610–5.
[26] Donner CF. Components of tobacco smoke. Ital Heart J 2001;2(Suppl. 1):22–4.
[27] Millar NS, Harkness PC. Assembly and trafficking of nicotinic acetylcholine receptors. Mol Membr Biol 2008;25:279–92 [Review].

[28] Barik J, Wonnacott S. Molecular and cellular mechanisms of action of nicotine in the CNS. Handb Exp Pharmacol 2009;192:173–207.

[29] Pittilo RM, Bull HA, Gulati S, et al. Nicotine and cigarette smoking: effects on the ultrastructure of aortic endothelium. Int J Exp Pathol 1990;71:573–86.

[30] Hsu PP, Li S, Li YS, et al. Effects of flow patterns on endothelial cell migration into a zone of mechanical denudation. Biochem Biophys Res Commun 2001;285:751–9.

[31] Macklin KD, Maus AD, Pereira EF, et al. Human vascular endothelial cells express functional nicotinic acetylcholine receptors. J Pharmacol Exp Ther 1998;287:435–9.

[32] Egleton RD, Brown KC, Dasgupta R. Nicotinic acetylcholine receptors in cancer: multiple roles in proliferation and inhibition of apoptosis. Trends Pharmacol Sci 2008;29:151–8.

[33] Hakki A, Friedman H, Pross S. Nicotine modulation of apoptosis in human coronary artery endothelial cells. Int Immunopharmacol 2002;2:1403–9.

[34] Esen AM, Barutcu I, Acar M, et al. Effect of smoking on endothelial function and wall thickness of brachial artery. Circ J 2004;68:1123–6.

[35] McVeigh GE, Morgan DJ, Finkelstein SM, et al. Vascular abnormalities associated with long-term cigarette smoking identified by arterial waveform analysis. Am J Med 1997;102:227–31.

[36] Ghiadoni L. Smoking and central blood pressure: a metabolic interaction? Am J Hypertens 2009;22(6):585.

[37] Santarelli MF, Landini L, Positano V. Can imaging techniques identify smoking-related cardiovascular disease? Curr Pharm Des 2010;16:2578–85.

[38] Takami T, Saito Y. Effects of smoking cessation on central blood pressure and arterial stiffness. Vasc Health Risk Manag 2011;7:633–8.

[39] Ijzerman RG, Serne EH, van Weissenbruch MM, et al. Cigarette smoking is associated with an acute impairment of microvascular function in humans. Clin Sci (Lond) 2003;104:247–52.

[40] van den Berkmortel FW, Wollersheim H, van Langen H, et al. Two years of smoking cessation does not reduce arterial wall thickness and stiffness. Neth J Med 2004;62:235–41.

[41] Leone A. Biochemical markers of cardiovascular damage from tobacco smoke. Curr Pharm Des 2005;11:2199–208.

[42] Rahman MM, Laher I. Structural and functional alteration of blood vessels caused by cigarette smoking: an overview of molecular mechanisms. Curr Vasc Pharmacol 2007;5:276–92.

[43] Lerant B, Christina S, Olah L, et al. The comparative analysis of arterial wall thickness and arterial wall stiffness in smoking and non-smoking university students. Ideggyogy Sz 2012;65:121–6.

[44] Moreno Jr H, Chalon S, Urae A, et al. Endothelial dysfunction in human hand veins is rapidly reversible after smoking cessation. Am J Physiol 1998;275:H1040–5.

[45] Vita JA, Treasure CB, Nabel EG, et al. Coronary vasomotor response to acetylcholine relates to risk factors for coronary artery disease. Circulation 1990;81:491–7.

[46] Campisi R, Czernin J, Schoder H, et al. Effects of long-term smoking on myocardial blood flow, coronary vasomotion, and vasodilator capacity. Circulation 1998;98:119–25.

[47] Ambrose JA, Barua RS. The pathophysiology of cigarette smoking and cardiovascular disease: an update. J Am Coll Cardiol 2004;43:1731–7.

[48] Morita K, Tsukamoto T, Naya M, et al. Smoking cessation normalizes coronary endothelial vasomotor response assessed with 15O-water and PET in healthy young smokers. J Nucl Med 2006;47:1914–20.

[49] Barua RS, Ambrose JA. Mechanisms of coronary thrombosis in cigarette smoke exposure. Arterioscler Thromb Vasc Biol 2013;33:1460–7.

[50] Fujii N, Reinke MC, Brunt VE, et al. Impaired acetylcholine-induced cutaneous vasodilation in young smokers: roles of nitric oxide and prostanoids. Am J Physiol Heart Circ Physiol 2013;304:H667–73.

[51] Seet RC, Lee CY, Loke WM, et al. Biomarkers of oxidative damage in cigarette smokers: which biomarkers might reflect acute versus chronic oxidative stress? Free Radic Biol Med 2011; 50:1787–93.

[52] McVeigh GE, Lemay L, Morgan D, et al. Effects of long-term cigarette smoking on endothelium-dependent responses in humans. Am J Cardiol 1996;78:668–72.

[53] Migliacci R, Gresele R. Smoking and impaired endothelium-dependent dilatation. N Engl J Med 1996;334:1674.

[54] Iida H, Iida M, Takenaka M, et al. Angiotensin II type 1 (AT1)-receptor blocker prevents impairment of endothelium-dependent cerebral vasodilation by acute cigarette smoking in rats. Life Sci 2006;78:1310–6.

[55] Varela-Carver A, Parker H, Kleinert C, et al. Adverse effects of cigarette smoke and induction of oxidative stress in cardiomyocytes and vascular endothelium. Curr Pharm Des 2010;16:2551–8.

[56] Grassi D, Desideri G, Ferri L, et al. Oxidative stress and endothelial dysfunction: say NO to cigarette smoking! Curr Pharm Des 2010;16:2539–50.

[57] Shih RH, Cheng SE, Hsiao LD, et al. Cigarette smoke extract upregulatesheme oxygenase-1 via PKC/NADPH oxidase/ROS/PDGFR/PI3K/Akt pathway in mouse brain endothelial cells. J Neuroinflammation 2011;8:104.

[58] Naya M, Morita K, Yoshinaga K, et al. Long-term smoking causes more advanced coronary endothelial dysfunction in middle-aged smokers compared to young smokers. Eur J Nucl Med Mol Imaging 2011;38:491–8.

[59] Barbieri SS, Zacchi E, Amadio P, et al. Cytokines present in smokers' serum interact with smoke components to enhance endothelial dysfunction. Cardiovasc Res 2011;90:475–83.

[60] Liao JK. Linking endothelial dysfunction with endothelial cell activation. J Clin Invest 2013;123:540–1.

[61] Messner B, Bernhard D. Smoking and cardiovascular disease: mechanisms of endothelial dysfunction and early atherogenesis. Arterioscler Thromb Vasc Biol 2014;34:509–15.

[62] Quillen JE, Rossen JD, Oskarsson HJ, et al. Acute effect of cigarette smoking on the coronary circulation: constriction of epicardial and resistance vessels. J Am Coll Cardiol 1993;22:642–7.

[63] Gaemperli O, Liga R, Bhamra-Ariza P, et al. Nicotine addiction and coronary artery disease: impact of cessation interventions. Curr Pharm Des 2010;16:2586–97.

[64] Hung MJ, Hu P, Hung MY. Coronary artery spasm: review and update. Int J Med Sci 2014;11:1161–71.

[65] Celermajer DS, Sorensen KE, Georgakopoulos D, et al. Cigarette smoking is associated with dose-related and potentially reversible impairment of endothelium-dependent dilation in healthy young adults. Circulation 1993;88:2149–55.

[66] Yugar-Toledo JC, Tanus-Santos JE, Sabha M, et al. Uncontrolled hypertension, uncompensated type II diabetes, and smoking have different patterns of vascular dysfunction. Chest 2004; 125:823–30.

[67] Sabha M, Tanus-Santos JE, Toledo JC, et al. Transdermal nicotine mimics the smoking-induced endothelial dysfunction. Clin Pharmacol Ther 2000;68:167–74.

[68] Chalon S, Moreno Jr H, Benowitz NL, et al. Nicotine impairs endothelium-dependent dilatation in human veins in vivo. Clin Pharmacol Ther 2000;67:391–7.

[69] Black CE, Huang N, Neligan PC, et al. Effect of nicotine on vasoconstrictor and vasodilator responses in human skin vasculature. Am J Physiol Regul Integr Comp Physiol 2001;281: R1097–104.

[70] Mayhan WG, Patel KP. Effect of nicotine on endothelium-dependent arteriolar dilatation in vivo. Am J Physiol 1997;272: H2337–42.

[71] Rangemark C, Benthin G, Granstrom EF, et al. Tobacco use and urinary excretion of thromboxane A2 and prostacyclin metabolites in women stratified by age. Circulation 1992;86:1495–500.

[72] Bernhard D, Wang XL. Smoking, oxidative stress and cardiovascular diseases—do anti-oxidative therapies fail? Curr Med Chem 2007;14:1703–12.

[73] Peluffo G, Calcerrada P, Piacenza L, et al. Superoxide-mediated inactivation of nitric oxide and peroxynitrite formation by tobacco smoke in vascular endothelium: studies in cultured cells and smokers. Am J Physiol Heart Circ Physiol 2009;296:H1781–92.

[74] Schaal C, Chellappan SP. Nicotine-mediated cell proliferation and tumor progression in smoking-related cancers. Mol Cancer Res 2014;12:14–23.

[75] Schmid P, Karanikas G, Kritz H, et al. Passive smoking and platelet thromboxane. Thromb Res 1996;81:451–60.

[76] Ahmadzadehfar H, Oguogho A, Efthimiou Y, et al. Passive cigarette smoking increases isoprostane formation. Life Sci 2006;78:894–7.

[77] Jacob-Ferreira AL, Palei AC, Cau SB, et al. Evidence for the involvement of matrix metalloproteinases in the cardiovascular effects produced by nicotine. Eur J Pharmacol 2010;627:216–22.

[78] Grassi G, Seravalle G, Calhoun DA, et al. Mechanisms responsible for sympathetic activation by cigarette smoking in humans. Circulation 1994;90:248–53.

[79] Narkiewicz K, van de Borne PJ, Hausberg M, et al. Cigarette smoking increases sympathetic outflow in humans. Circulation 1998;98:528–34.

[80] Tanus-Santos JE, Toledo JC, Cittadino M, et al. Cardiovascular effects of transdermal nicotine in mildly hypertensive smokers. Am J Hypertens 2001;14:610–4.

[81] Benowitz NL, Gourlay SG. Cardiovascular toxicity of nicotine: implications for nicotine replacement therapy. J Am Coll Cardiol 1997;29:1422–31.

[82] Hand S, Edwards S, Campbell IA, et al. Controlled trial of three weeks nicotine replacement treatment in hospital patients also given advice and support. Thorax 2002;57:715–8.

[83] Shinozaki N, Yuasa T, Takata S. Cigarette smoking augments sympathetic nerve activity in patients with coronary heart disease. Int Heart J 2008;49:261–72.

[84] Middlekauff HR, Park J, Moheimani RS. Adverse effects of cigarette and noncigarette smoke exposure on the autonomic nervous system: mechanisms and implications for cardiovascular risk. J Am Coll Cardiol 2014;64:1740–50.

[85] Benowitz NL, Hansson A, Jacob 3rd P. Cardiovascular effects of nasal and transdermal nicotine and cigarette smoking. Hypertension 2002;39:1107–12.

[86] Najem B, Houssiere A, Pathak A, et al. Acute cardiovascular and sympathetic effects of nicotine replacement therapy. Hypertension 2006;47:1162–7.

[87] Sztalryd C, Hamilton J, Horwitz BA, et al. Alterations of lipolysis and lipoprotein lipase in chronically nicotine-treated rats. Am J Physiol 1996;270:E215–23.

[88] Freeman DJ, Griffin BA, Murray E, et al. Smoking and plasma lipoproteins in man: effects on low density lipoprotein cholesterol levels and high density lipoprotein subfraction distribution. Eur J Clin Invest 1993;23:630–40.

[89] Glueck CJ, Heiss G, Morrison JA, et al. Alcohol intake, cigarette smoking and plasma lipids and lipoproteins in 12–19-year-old children. The Collaborative Lipid Research Clinics Prevalence Study. Circulation 1981;64:III48–56.

[90] Imaizumi T, Satoh K, Yoshida H, et al. Effect of cigarette smoking on the levels of platelet-activating factor-like lipid(s) in plasma lipoproteins. Atherosclerosis 1991;87:47–55.

[91] Puri BK, Treasaden IH, Cocchi M, et al. A comparison of oxidative stress in smokers and non-smokers: an in vivo human quantitative study of n-3 lipid peroxidation. BMC Psychiatry 2008;8(Suppl. 1):S4.

[92] Sliwinska-Mosson M, Mihulka E, Milnerowicz H. Assessment of lipid profile in non-smoking and smoking young health persons. Przegl Lek 2014;71:585–7.

[93] Petitti DB, Kipp H. The leukocyte count: associations with intensity of smoking and persistence of effect after quitting. Am J Epidemiol 1986;123:89–95.

[94] Jensen EJ, Pedersen B, Frederiksen R, et al. Prospective study on the effect of smoking and nicotine substitution on leucocyte blood counts and relation between blood leucocytes and lung function. Thorax 1998;53:784–9.

[95] Barua RS, Ambrose JA, Saha DC, et al. Smoking is associated with altered endothelial-derived fibrinolytic and antithrombotic factors: an in vitro demonstration. Circulation 2002;106:905–8.

[96] Yi M, Chun EJ, Lee MS, et al. Coronary CT angiography findings based on smoking status: do ex-smokers and never-smokers share a low probability of developing coronary atherosclerosis? Int J Cardiovasc Imaging 2015;31(Suppl. 2):169–76.

[97] Hammal F, Ezekowitz JA, Norris CM, et al. Smoking status and survival: impact on mortality of continuing to smoke one year after the angiographic diagnosis of coronary artery disease, a prospective cohort study. BMC Cardiovasc Disord 2014;14:133.

[98] Ciftci C, Melil S, Cebi Y, et al. Association of endothelial nitric oxide synthase promoter region (T-786C) gene polymorphism with acute coronary syndrome and coronary heart disease. Lipids Health Dis 2008;7:5.

[99] Ragia G, Nikolaidis E, Tavridou A, et al. Endothelial nitric oxide synthase gene polymorphisms-786T > C and 894G > T in coronary artery bypass graft surgery patients. Hum Genomics 2010;4:375–83.

[100] Zhang C, Guo L. Correlation of polymorphisms of adiponectin receptor 2 gene +33371Gln/Arg, cytochrome P4502E1 gene Rsa I and smoking with nonalcoholic fatty liver disease. Nan Fang Yi Ke Da Xue Xue Bao 2014;34:1481–7.

[101] Tang X, Guo S, Sun H, et al. Gene-gene interactions of CYP2A6 and MAOA polymorphisms on smoking behavior in Chinese male population. Pharmacogenet Genomics 2009;19:345–52.

[102] Grazuleviciene R, Nieuwenhuijsen MJ, Danileviciute A, et al. Gene-environment interaction: maternal smoking and contribution of GSTT1 and GSTM1 polymorphisms to infant birth-weight reduction in a Kaunas cohort study. J Epidemiol Community Health 2010;64:648.

[103] Barakat K, Kennon S, Hitman GA, et al. Interaction between smoking and the glycoprotein IIIa P1(A2) polymorphism in non-ST-elevation acute coronary syndromes. J Am Coll Cardiol 2001;38:1639–43.

[104] Humphries SE, Luong LA, Montgomery HE, et al. Gene-environment interaction in the determination of levels of plasma fibrinogen. Thromb Haemost 1999;82:818–25.

[105] Haj Mouhamed D, Ezzaher A, Mechri A, et al. Effect of cigarette smoking on paraoxonase 1 activity according to PON1 L55M and PON 1 Q192R gene polymorphisms. Environ Health Prev Med 2012;17:316–21.

[106] Han Y, Dorajoo R, Ke T, et al. Interaction effects between paraoxonase 1 variants and cigarette smoking on risk of coronary heart disease in a Singaporean Chinese population. Atherosclerosis 2015;240:40–5.

[107] Grammer TB, Hoffmann MM, Scharnagl H, et al. Smoking, apolipoprotein E genotypes, and mortality (the Ludwigshafen Risk and Cardiovascular Health study). Eur Heart J 2013; 34:1298–305.

[108] Jia EZ, Xu ZX, Guo CY, et al. Renin-angiotensin-aldosterone system gene polymorphisms and coronary artery disease: detection of gene-gene and gene-environment interactions. Cell Physiol Biochem 2012;29:443–52.

II. ENDOTHELIAL DYSFUNCTION AND CLINICAL SYNDROMES

[109] Eisenberg MJ, Filion KB, Yavin D, et al. Pharmaco therapies for smoking cessation: a meta-analysis of randomized controlled trials. Can Med Assoc J 2008;179:135–44.

[110] Pentel PR, LeSage MG. New directions in nicotine vaccine design and use. Adv Pharmacol 2014;69:553–80.

Further Reading

[1] Costa F, Soares R. Nicotine: a pro-angiogenic factor. Life Sci 2009; 84(23–24):785–90.

[2] Celermajer DS, Adams MR, Clarkson P, et al. Passive smoking and impaired endothelium-dependent arterial dilatation in healthy young adults. N Engl J Med 1996;334:150–4.

[3] Csiszar A, Podlutsky A, Wolin MS, et al. Oxidative stress and accelerated vascular aging: implications for cigarette smoking. Front Biosci (Landmark Ed) 2009;14:3128–44.

[4] Lee J, Cooke JP. Nicotine and pathological angiogenesis. Life Sci 2012;91(21–22):1058–64.

37

Endothelial Dysfunction and Coronary Microvascular Dysfunction in Women With Angina and Nonobstructive Coronaries

Jenna Maughan, Janet Wei, Erika Jones, C. Noel Bairey Merz, and Puja Mehta

INTRODUCTION

Chest pain evaluation remains a challenge in medical practice, comprising 7%–24% of the primary care population visits [1]. For women presenting for evaluation of suspected ischemic symptoms, a diagnosis of normal coronary arteries is five times more common in females than males [2]. The challenge to diagnose coronary disease in women is amplified with the presence of more atypical symptom burden, functional disability, and often a greater degree of comorbidity and clustering of cardiac risk factors in women when compared to men [3]. Prior studies demonstrate that the presence of obstructive coronary artery disease (CAD) can be predicted with less certainty in women than men [2], and that women are less likely than age matched men to have obstructive CAD [4].

Women with signs and symptoms of myocardial ischemia in the absence of obstructive coronary stenosis were previously labeled cardiac syndrome X (CSX); [5] however, this term should no longer be used. Angina due to coronary microvascular dysfunction (CMD) [6] is a common etiologic mechanism in women with signs and symptoms of ischemia. These conditions are increasingly investigated, but full elucidation of their pathogenesis remains lacking. This has resulted in a lack of consensus regarding diagnosis and treatment, and caused a considerable drain on health resources [7]. Understanding the pathogenesis and contributing factors

underlying chest pain and normal coronaries at angiography is of vital importance for appropriate management of ischemic heart disease (IHD).

In this chapter we shall overview and discuss the terminology and anatomy, diagnosis, and mechanistic pathways leading to adverse events in patients with signs and symptoms of ischemia with nonobstructive coronary arteries due to CMD.

CMD: Terminology and Anatomy

The term "cardiac syndrome X" [5] was popularized by Kemp in 1973 to describe patients with (1) angina-like chest pain, (2) ischemic changes in response to stress, and (3) angiographically normal coronary arteries. The terms "coronary microvascular dysfunction" and "microvascular angina" [6] (MVA) are a subgroup of cardiac syndrome X, accounting for at least half of these patients [8,9].

CMD [6] is further defined as: (1) the presence of myocardial ischemia; and (2) the presence of a coronary vascular dysfunction in either of the macro or microvascular bed (Fig. 37.1). These can result in subjectively manifested chest pain or objectively evidenced abnormal ischemic responses to stress with resultant hemodynamic changes [10,11], metabolic abnormalities [12], ECG change [13], myocardial perfusion [14], or regional wall motion [15] abnormalities. Patients with coronary spasm, left ventricular hypertrophy, or valvular heart disease are excluded from this diagnosis of CMD [6,16].

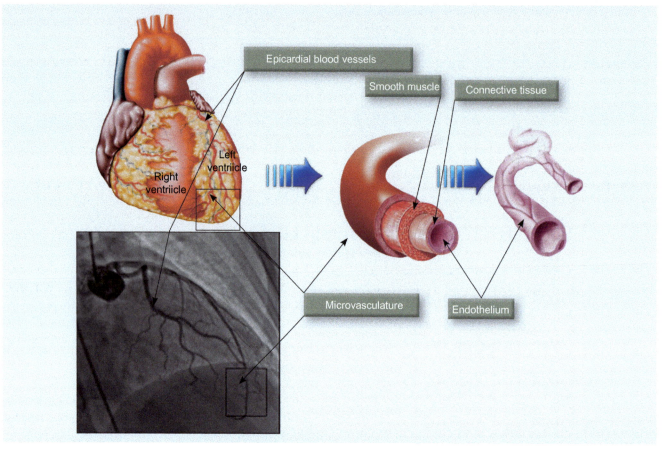

FIG. 37.1 Coronary microvasculature and endothelium.

CMD is classified as primary to be distinguished from CMD secondary to other specific disease settings [8].

Patients with symptoms and signs of myocardial ischemia in the absence of obstructive CAD are classically acknowledged as a female predominant disorder whereas nearly 70% are women [9]. In a large female cohort suspected to have myocardial ischemia and referred for clinically indicated coronary angiography, 41% of women studied showed nonsignificant epicardial coronary artery obstruction. Only 8% of the men studied showed the same nonsignificant epicardial coronary artery obstruction results [2]. The increased risk of abnormal results could be due to the fact that women with CMD are usually in their perimenopausal or menopausal stage of life with their onset of symptoms between 40 and 50 years [10].

Older epidemiologic studies based on intermediate-term follow-up studies that included women either exclusively [7,11,12], as a majority [13,14], or as a minority [15] suggested that intermediate-term survival was not adversely affected [13,16], and the coronary morbidity and mortality in these patients were similar to the overall population [15], while an increased adverse event risk was found mainly in patients with elevated risk factors [15]. However, these older data were based on heterogeneous enrollment and small numbers of patients, as well as nonuniform assessment of patients for factors like epicardial coronary spasm, endothelial dysfunction, and intravascular ultrasound (IVUS) characterization of coronary lesions [17]. Contemporary studies demonstrate an elevated adverse event risk [12,18,19], especially those incorporating in their methods endothelial function assessment and cohort longitudinal follow-up [14,20–22].

DIAGNOSIS

CMD is diagnosed invasively and noninvasively by its components including chest pain, endothelial dysfunction defined by reduced coronary blood flow (CBF) to acetylcholine, and myocardial ische-mia (Fig. 37.2). In a large cohort of women with chest pain and nonobstructive CAD by angiography, "persistent chest pain" was defined as chest pain (typical or atypical), that lasts over a year during follow-up, occurred in 45% of them and yet was associated with significantly more than twice the cardiovascular events, including MIs, strokes, congestive heart failure, and cardiovascular death compared to those without chest pain [12]. Subsequent "functional disability" secondary to chest pain was reported in half of women with nonobstructive CAD, the repeated angiography rate was 13.2%, and the repeated hospitalization after 1-year follow-up was 1.8-fold higher than those with 1-vessle disease [7]. Endothelial dysfunction predicts the later development of obstructive CAD [11]. In a recent report [20] from Women's Ischemia Syndrome Evaluation (WISE) study, 189 women were followed-up for a mean period of 5.4 years after having their basal coronary flow reserve (CFR) measured using intracoronary adenosine. In their study, lower CFR was associated with adverse CV outcome, whether women had or had no significant coronary obstruction. Also, CFR significantly improved the prediction of adverse CV outcome

over angiographic severity and other risk factors. Prior results from the same study have shown that in women suspected to have myocardial ischemia, abnormal response to intracoronary acetylcholine was an independent predictor of adverse CV events including hospitalization for worsening angina, MI, congestive heart failure, stroke, revascularization, and death [22]. The interrelation between the vessel wall structure versus its function has been suggested as critical in the prognosis of atherosclerotic disease, whereas endothelial dysfunction seems to modulate the impact of a given atheroma burden [23]. Thus, the worst expected prognosis occurs when severe grades of endothelial dysfunction are concomitant with high degrees of atheroma burden [23]. Follow-up results of 157 patients with mild coronary atherosclerosis associated with microvascular endothelial dysfunction revealed that cardiac events occurred only in those with severe degree of endothelial dysfunction while no adverse events were detected in individuals with normal or mild dysfunction [24] (Fig. 37.3).

- *Coronary endothelial dysfunction.* Coronary endothelial dysfunction is suggested to be at least one of the possible several mechanisms

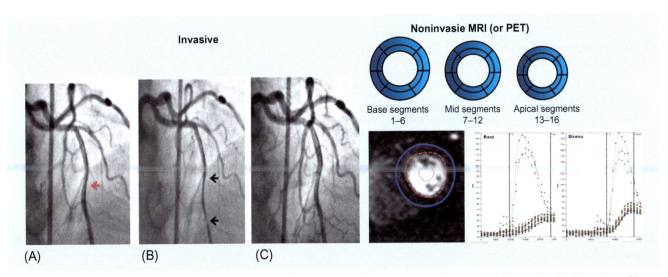

FIG. 37.2 Diagnosis of coronary microvascular dysfunction. *Symptoms*: persistent typical or atypical angina. *Signs*: abnormal stress testing or acute coronay syndrome with cardiac enzyme elevation. *Angiography*: nonobstructive coronary artery atherosclerosis. *Invasive*: (A) Baseline. (B) Vasoconstriction to ACH. (C) Postnitroglycerin. Doppler flow wire in the left anterior descending artery *(red arrow)* (A). In response to acetylcholine (ACH) infusion, there is abnormal coronary artery vasoconstriction *(black arrows)*, indicating endothelial dysfunction (B). This is resolved by intracoronary nitroglycerin (C). *Noninvasive MRI (or PET)*: MPRI contours were manually placed for endocardial and epicardial regions and time intensity curves were used to calculate the relative upslope of stress compared to rest first pass perfusion in 16 segments.

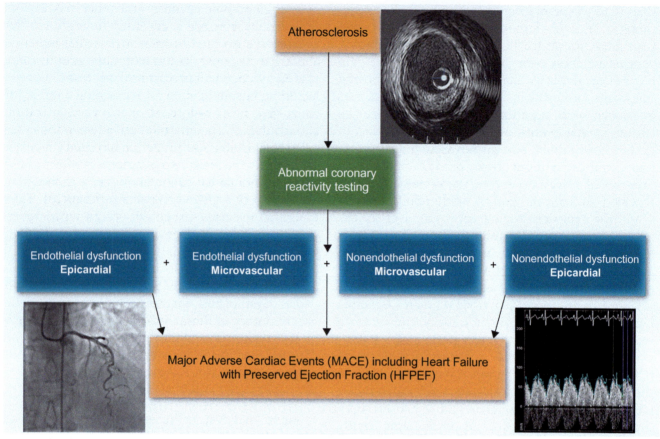

FIG. 37.3   Coronary microvascular dysfunction mechanistic pathways.

contributing to CMD [14,25,26]. In one study, a normal endothelial-dependent function was defined as an increase of more than 50% of CBF in response to acetylcholine administration, while an impaired coronary endothelium-independent function was defined as a flow velocity ratio to adenosine of ≤2.5 [14]. In large arteries, endothelial dysfunction is considered among the earliest changes associated with atherosclerosis before structural changes to the vessels are appreciated [21] and appears to predict the development of obstructive CAD [11]. The suggested underlying mechanisms for endothelial dysfunction are based upon the observations that endothelial vasodilator dysfunction correlates with failure of CBF to increase during dipyridamole [6,27] or acetylcholine [28] infusion, pacing [29], or cold exposure [30]. Recent evidence suggests that altered circulating endothelial progenitor cells, normally involved in the biological repair of vascular injury, constitutes an underlying

contributing factor to endothelial dysfunction encountered in many patients with CSX [31].

• *Nitric oxide-endothelin imbalance*: another proposed disrupted regulatory mechanism of microvascular circulation is an imbalance between the endothelial-derived nitric oxide (NO) (vasodilator) and endothelin-1 (ET-1) (vasoconstrictor). Reduced bioavailability of endogenous NO and increased plasma levels of ET-1 may be responsible for abnormal vasoreactivity in patients with signs and symptoms of ischemia and nonobstructive CAD. ET-1 was significantly higher and its baseline level showed abnormal coronary vascular response in these patients [32]. An abnormal response to ET-1 was also observed, even though its plasma concentration was not elevated [33]. Genetic abnormality underlying the altered endothelial production of NO has recently been suggested [34]. Findings from a patient-controlled genetic study of polymorphism of endothelial nitric oxide synthase gene (eNOS is the enzyme responsible for the synthesis of NO), displayed higher frequency of Intron 4aa genotype in controls compared to patients with signs and symptoms of ischemia and nonobstructive CAD, suggesting its protective effect.

- *Relation to CFR*: reduced CFR appears to be a common underlying factor noted in many of the studies exploring the pathogenesis of patients with chest pain and normal coronary angiograms [6,10,20,28,35]. Zeiher et al. examined patients with chest pain and normal or minimally affected coronary arteries, using single photon emission tomography (SPECT) imaging, and demonstrated that impaired endothelial-dependent regulation of CBF in the resistance arterioles was associated with stress-induced myocardial ischemia [30]. However, it was not clear whether this was due to an abnormal production or destruction of endothelium-derived relaxing factors like NO, an abnormality of endothelial cell membrane receptors, or a nonspecifically reduced sensitivity of vascular smooth muscle cell to relax [30]. A reduced CFR in response to intracoronary adenosine injection was also reported in 47% of women with chest pain and normal or minimal coronary artery irregularities, suggesting an endothelial-independent mechanism of microvascular dysfunction [8].

  Noninvasive diagnosis of impaired CFR using positron emission tomography (PET) imaging, suggests a heterogeneous distribution of the microvascular defects which is not confined to a single coronary artery distribution, the method which is commonly used during invasive CFR assessment [36]. These abnormalities in smooth muscle relaxation, as well as reduction in CFR in patients with no flow limiting stenosis on coronary angiography, supports the hypothesis that signs and symptoms was microvascular in origin [9,17,25,37].

- *Relation to myocardial ischemia.* Documenting the presence of myocellular ischemia as etiological for chest pain in individuals with normal coronary angiograms remains the focus of many current studies [6,30,36,38,39]. Testing of CMD is not well defined; however, noninvasive imaging has been used to determine whether ischemia is present or not and also to risk stratify patients with CMD [3,4]. The use of SPECT imaging revealed myocardial perfusion defects in response to

exercise in some of these patients [30]. An abnormal response of coronary resistance vasculature to acetylcholine during diagnostic coronary angiography is also suggested to identify patients likely to have myocardial perfusion defects in response to stress [30]. Using PET in women with chest pain and no coronary obstruction by angiogram, the adenosine-induced changes in myocardial perfusion reflected a heterogeneous pattern of microvascular dysfunction [36]. In a cohort of women with chest pain and no obstructive CAD studied by cardiac magnetic resonance imaging (MR), one fifth of them showed an abnormal decrease in myocardial high-energy phosphate (a metabolic marker of ischemia) during light handgrip exercise. The magnitude of the decrease was equal to or greater than that observed in patients with at least 70% epicardial stenosis of the left anterior descending artery.

Endothelial dysfunction in the absence of obstructive CAD may not consistently cause myocardial ischemia that can be detected noninvasively [14,25]. This can be explained by the fact that the commonly applied nuclear-based techniques for ischemia depends upon regional differences in perfusion and/or function that identified by normalizing radiotracer uptake across the myocardium. This will obfuscate detection of diffuse microvascular abnormalities. Further analyses demonstrated that even in apparently normal scans, the majority of these patients showed reduced thallium-201 uptake and washout in comparison to their controls [40]. Given the fact that traditional nuclear imaging techniques rely on detection of abnormalities that are compared to a normalized myocardium, diffuse CAD will appear as normal [3,25]. Recently, stress MR has been capable of defining epicardial as well as subendocardial hypoperfusion following administration of IV adenosine in women with signs and symptoms of ischemia but no obstructive CAD [39]. Adenosine may also induce global and regional left ventricular diastolic dysfunction as demonstrated by both radionuclide imaging and stress echocardiography in patients with CMD. In the same study, the long axis diastolic dysfunction detected by tissue-Doppler study of the mitral annular was also suggestive of subendocardial ischemia [41].

- *Slow flow phenomenon.* The "slow coronary flow phenomenon" (SCFP) is an angiographic finding that is characterized by angiographically normal coronaries with delayed opacification of the distal vasculature in the absence of significant epicardial coronary disease or stenosis [1,2]. This phenomenon is typically observed in patients who present with chest pain and undergo angiography to assess for acute coronary syndrome and is different from abnormal CFR where there is an abnormal capacity for coronary microvasculature to vasodilate in the face increase demand. As opposed to abnormal CFR, the slow flow phenomenon is associated with decreased coronary flow velocity at rest. Fineschi et al. demonstrated that these two mechanisms do not necessarily occur together. Their study demonstrated that CFR was within normal limits in the patients with coronary slow flow [2]. It is hypothesized that SCFP is caused by abnormally increased resting microvascular resistance, an effect that is associated with a reduced endothelial function [3,4].

FEMALE-SPECIFIC ISSUES

The Women's Ischemia Syndrome Evaluation (WISE) study [42] has contributed extensively to this issue of female-specific study of CMD. The contributions have lead towards not only better understanding of the pathophysiology and prognosis of IHD in women, but also have pointed toward future studies [43,43a].

Reports indicate that most of women are peri- or postmenopausal [2,6,8,16,27,28], suggesting a role of sex-hormones changes during this phase of women's cycle of life [3,20,27,28] and supporting the hypothesis that endogenous estrogen deficiency encountered in postmenopausal women is associated with reduction of many of its protective physiologic roles and regulating mechanisms to the endothelial and smooth muscle factors in the vessel wall [3,20,27]. Clustering of risk factors such as age, hypertension diabetes, obesity, and metabolic syndromes are frequent in postmenopausal women and sustain the hypothesis that clustering of risk factors may be responsible for increased atheroma burden and are associated with coronary macro and microvascular dysfunction in women [3,20,27]. However, the effect of traditional risk factors on

endothelial dysfunction is not consistent [8,29], suggesting other possible factors to be identified [20,27]. Other potential factors, including duration of exposure to risk factors versus merely their existence [29] or inflammation and inflammation-mediated autoimmune diseases commonly encountered in females and their relation to general vasculopathy, are all speculative and need to be further evaluated [3,27,33].

Recent experimental work raised the plausibility of genetic-differences which may exert its effect independent from gonadal function [44]. Their findings demonstrated the presence and maintenance of intrinsic sex-related differences in gene expression and cellular phenotyping by microvascular endothelial cells in a gonadal free environment. Furthermore, they concluded that intrinsic sex-cell likely contributes significantly to sexual dimorphism in cardiovascular function.

MECHANISTIC PATHWAYS LEADING TO ADVERSE EVENTS

Mechanisms and contributing factors [9,10,45] to CMD include altered regulation of coronary microcirculation through autonomic dysregulatory mechanisms and/or imbalance state between endothelial-derived vasodilator and vasoconstrictor factors, generalized vascular disorder, abnormal subendocardial perfusion, inflammation, hyperinsulinemia, enhanced sodium-hydrogen exchange, hormonal deficiency, abnormal pain perception, and lastly inherent pathogenetic pathways. Fig. 37.3 demonstrates the relation between nonobstructive atherosclerosis and dysfunction of the macro and/or microvasculature as mechanistic pathways to adverse cardiac events.

- *Insulin resistance or hyperinsulinemia*: some evidence has suggested that CMD might be related to insulin resistance or hyperinsulinemia [46], whereas interventions aiming at improving insulin sensitivity have been shown to improve endothelial function and decrease myocardial ischemia in patients with CMD [47]. However, *other* specific studies demonstrate that CMD per se is not associated with hyperinsulinemia or insulin resistance when other confounding factors are excluded [48].

- *Sodium-hydrogen exchange*: sodium-hydrogen exchange in red cells has been found enhanced threefold in older studies of CSX patients when compared to those with atherosclerosis or healthy subjects suggesting its potential role as a marker of coronary vascular dysfunction [49].
- *Hyperglycemia*: studying the role of chronic hyperglycemia in the pathogenesis of endothelial dysfunction reveals a significant reduction in both endothelial-dependent and endothelial-independent coronary vasodilator function in a high-risk group of women [50].
- *Inflammation*: high levels of C-reactive protein, as a marker of low grade chronic inflammation, were associated with increased frequency of ischemic episodes, detected by ambulatory ECG, regardless of whether chest pain was present or not [51]. It is well known that inflammatory processes are activated in the presence of oxidative stress. Some investigators described a high level of thioredoxine (known to be induced and released from cells by oxidative stress) in patients during coronary spasm [52].
- *Vascular and nonvascular smooth muscle abnormalities*: the hypothesis that CMD represents a more generalized abnormality of vascular and nonvascular smooth muscle function is supported by studies on forearm arterial function [37,42] and airway hyperresponsiveness frequently demonstrated in patients with MVA [43].

CONCLUSIONS

Women who have nonobstructive coronaries, but who also show symptoms of angina and other concerning symptoms are being classified as suspected CMD, which has major connections to the presence of endothelial dysfunction. Endothelial dysfunction is thought to be one of the several mechanisms that contribute to CMD. Compared to men, women are at higher risk of a reduction in protective physiologic mechanisms that can affect the endothelial muscles leading to the risk of developing CMD due to the rapid changes in hormone levels seen when a women enters perimenopause and menopause. However, endothelial dysfunction is not the only abnormality playing a role in the onset of these symptoms, leading to the opinion and belief that other mechanisms and syndromes also contribute

to the onset of alarming symptoms in these women displaying angina with nonobstructive coronaries.

References

[1] Kroenke K, Arrington ME, Mangelsdorff AD. The prevalence of symptoms in medical outpatients and the adequacy of therapy. Arch Intern Med 1990;150:1685–9.
[2] Sullivan AK, Holdright DR, Wright CA, et al. Chest pain in women: clinical, investigative, and prognostic features. BMJ 1994;308:883–6.
[3] Shaw LJ, Bairey Merz CN, Pepine CJ, et al. Insights from the NHLBI-sponsored Women's Ischemia Syndrome Evaluation (WISE) study: Part I: Gender differences in traditional and novel risk factors, symptom evaluation, and gender-optimized diagnostic strategies. J Am Coll Cardiol 2006;47:S4–S20.
[4] Mieres JH, Shaw LJ, Arai A, et al. Role of noninvasive testing in the clinical evaluation of women with suspected coronary artery disease: Consensus statement from the cardiac imaging committee, council on clinical cardiology, and the cardiovascular imaging and intervention committee, council on cardiovascular radiology and intervention, American Heart Association. Circulation 2005;111:682–96.
[5] Kemp Jr. HG. Left ventricular function in patients with the anginal syndrome and normal coronary arteriograms. Am J Cardiol 1973;32:375–6.
[6] Cannon 3rd. RO, Epstein SE. "Microvascular angina" as a cause of chest pain with angiographically normal coronary arteries. Am J Cardiol 1988;61:1338–43.
[7] Shaw LJ, Merz CN, Pepine CJ, et al. The economic burden of angina in women with suspected ischemic heart disease: results from the National Institutes of Health – National Heart, Lung, and Blood Institute – sponsored Women's Ischemia Syndrome Evaluation. Circulation 2006;114:894–904.
[8] Lanza GA, Crea F. Primary coronary microvascular dysfunction: clinical presentation, pathophysiology, and management. Circulation 2010;121:2317–25.
[9] Kaski JC. Pathophysiology and management of patients with chest pain and normal coronary arteriograms (cardiac syndrome X). Circulation 2004;109:568–72.
[10] Cannon 3rd. RO. Microvascular angina and the continuing dilemma of chest pain with normal coronary angiograms. J Am Coll Cardiol 2009;54:877–85.
[11] Bugiardini R, Manfrini O, Pizzi C, et al. Endothelial function predicts future development of coronary artery disease: a study of women with chest pain and normal coronary angiograms. Circulation 2004;109:2518–23.
[12] Johnson DD, Shaw LJ, Pepine CJ, et al. Persistent chest pain predicts cardiovascular events in women without obstructive coronary artery disease: results from the NIH-NHLBI-sponsored Women's Ischemia Syndrome Evaluation (WISE) study. Eur Heart J 2006;27:1408–15.
[13] Kaski JC, Rosano GM, Collins P, et al. Cardiac syndrome x: clinical characteristics and left ventricular function. Long-term follow-up study. J Am Coll Cardiol 1995;25:807–14.
[14] Suwaidi JA, Hamasaki S, Higano ST, et al. Long-term follow-up of patients with mild coronary artery disease and endothelial dysfunction. Circulation 2000;101:948–54.
[15] Lichtlen PR, Bargheer K, Wenzlaff P. Long-term prognosis of patients with angina-like chest pain and normal coronary angiographic findings. J Am Coll Cardiol 1995;25:1013–8.
[16] Schroeder C, Adams F, Boschmann M, et al. Phenotypical evidence for a gender difference in cardiac norepinephrine transporter function. Am J Physiol Regul Integr Comp Physiol 2004;286:R851–6.
[17] Bairey Merz CN, Shaw LJ, Reis SE, et al. Insights from the NHLBI-sponsored Women's Ischemia Syndrome Evaluation (WISE) study:

part II: gender differences in presentation, diagnosis, and outcome with regard to gender-based pathophysiology of atherosclerosis and macrovascular and microvascular coronary disease. J Am Coll Cardiol 2006;47(Suppl. 1):S21–9.

[18] Diver DJ, Bier JD, Ferreira PE, et al. Clinical and arteriographic characterization of patients with unstable angina without critical coronary arterial narrowing (from the TIMI-IIIA trial). Am J Cardiol 1994;74:531–7.

[19] Shaw LJ, Bugiardini R, Merz CN. Women and ischemic heart disease: evolving knowledge. J Am Coll Cardiol 2009;54:1561–75.

[20] Pepine CJ, Anderson RD, Sharaf BL, et al. Coronary microvascular reactivity to adenosine predicts adverse outcome in women evaluated for suspected ischemia: results from the National Heart, Lung and Blood Institute WISE (Women's Ischemia Syndrome Evaluation) study. J Am Coll Cardiol 2010;55:2825–32.

[21] Schächinger V, Britten MB, Zeiher AM. Prognostic impact of coronary vasodilator dysfunction on adverse long-term outcome of coronary heart disease. Circulation 2000;101:1899–906.

[22] von Mering GO, Arant CB, Wessel TR, et al. Abnormal coronary vasomotion as a prognostic indicator of cardiovascular events in women results from the National Heart, Lung, and Blood Institute-sponsored Women's Ischemia Syndrome Evaluation (WISE). Circulation 2004;109:722–5.

[23] Mancini GB. Vascular structure versus function: is endothelial dysfunction of independent prognostic importance or not? J Am Coll Cardiol 2004;43:624–8.

[24] Camici PG, Crea F. Coronary microvascular dysfunction. N Engl J Med 2007;356:830–40.

[25] Bugiardini R, Bairey Merz CN. Angina with "normal" coronary arteries: a changing philosophy. JAMA 2005;293:477–84.

[26] Quyyumi AA, Cannon 3rd. RO, Panza JA, et al. Endothelial dysfunction in patients with chest pain and normal coronary arteries. Circulation 1992;86(6):1864–71.

[27] Opherk D, Zebe H, Weihe E, et al. Reduced coronary dilatory capacity and ultrastructural changes of the myocardium in patients with angina pectoris but normal coronary arteriograms. Circulation 1981;63:817–25.

[28] Egashira K, Inou T, Hirooka Y, et al. Evidence of impaired endothelium-dependent coronary vasodilatation in patients with angina pectoris and normal coronary angiograms. N Engl J Med 1993;328:1659–64.

[29] Quyyumi AA, Cannon 3rd. RO, Panza JA, et al. Endothelial dysfunction in patients with chest pain and normal coronary arteries. Circulation 1992;86:1864–71.

[30] Zeiher AM, Krause T, Schachinger V, et al. Impaired endothelium-dependent vasodilation of coronary resistance vessels is associated with exercise-induced myocardial ischemia. Circulation 1995;91:2345–52.

[31] Huang PH, Chen YH, Chen YL, et al. Vascular endothelial function and circulating endothelial progenitor cells in patients with cardiac syndrome X. Heart 2007;93:1064–70.

[32] Cox ID, Botker HE, Bagger JP, et al. Elevated endothelin concentrations are associated with reduced coronary vasomotor responses in patients with chest pain and normal coronary arteriograms. J Am Coll Cardiol 1999;34:455–60.

[33] Newby DE, Flint LL, Fox KA, et al. Reduced responsiveness to endothelin-1 in peripheral resistance vessels of patients with syndrome X. J Am Coll Cardiol 1998;31:1585–90.

[34] Sinici I, Atalar E, Kepez A, et al. Intron 4 VNTR polymorphism of eNOS gene is protective for cardiac syndrome X. J Investig Med 2010;58:23–7.

[35] Zeiher AM, Drexler H, Wollschlager H, et al. Endothelial dysfunction of the coronary microvasculature is associated with coronary blood flow regulation in patients with early atherosclerosis. Circulation 1991;84:1984–92.

[36] Marroquin OC, Holubkov R, Edmundowicz D, et al. Heterogeneity of microvascular dysfunction in women with chest pain not attributable to coronary artery disease: implications for clinical practice. Am Heart J 2003;145:628–35.

[37] Pepine CJ, Kerensky RA, Lambert CR, et al. Some thoughts on vacular pathology of women with ischemic heart disease. J Am Coll Cardiol 2006;47(Suppl. 1):S30–5.

[38] Buchthal SD, den Hollander JA, Merz CN, et al. Abnormal myocardial phosphorus-31 nuclear magnetic resonance spectroscopy in women with chest pain but normal coronary angiograms. N Engl J Med 2000;342:829–35.

[39] Panting JR, Gatehouse PD, Yang GZ, et al. Abnormal subendocardial perfusion in cardiac syndrome X detected by cardiovascular magnetic resonance imaging. N Engl J Med 2002;346:1948–53.

[40] Rosano GM, Peters NS, Kaski JC, et al. Abnormal uptake and wash-out of thallium-201 in patients with syndrome X and normal-appearing scans. Am J Cardiol 1995;75:400–2.

[41] Vinereanu D, Fraser AG, Robinson M, et al. Adenosine provokes diastolic dysfunction in microvascular angina. Postgrad Med J 2002;78:40–2.

[42] Turiel M, Galassi AR, Glazier JJ, et al. Pain threshold and tolerance in women with syndrome X and women with stable angina pectoris. Am J Cardiol 1987;60:503–7.

[43] Cannon 3rd. RO, Peden DB, Berkebile C, et al. Airway hyperresponsiveness in patients with microvascular angina. Evidence for a diffuse disorder of smooth muscle responsiveness. Circulation 1990;82:2011–7.

[43a] Bairey Merz CN, Pepine CJ, Walsh MN, Fleg JL. Ischemia and no obstructive coronary artery disease (INOCA): developing evidence-based therapies and research agenda for the next decade. Circulation 2017;135(11):1075–92. https://doi.org/10.1161/CIRCULATIONAHA.

[44] Wang J, Bingaman S, Huxley VH. Intrinsic sex-specific differences in microvascular endothelial cell phosphodiesterases. Am J Physiol Heart Circ Physiol 2010;298:H1146–54.

[45] Merz CNB, Eteiba W, Pepine CJ, et al. Cardiac syndrome X: relation to microvascular angina and other conditions. Curr Cardiovasc Risk Rep 2007;1:167–75.

[46] Reaven GM. Role of insulin resistance in human disease (syndrome X): an expanded definition. Annu Rev Med 1993;44:121–31.

[47] Jadhav S, Ferrell W, Greer IA, et al. Effects of metformin on microvascular function and exercise tolerance in women with angina and normal coronary arteries: a randomized, double-blind, placebo-controlled study. J Am Coll Cardiol 2006;48:956–63.

[48] CavalloPerin P, Pacini G, Giunti S, et al. Microvascular angina (cardiological syndrome X) per se is not associated with hyperinsulinaemia or insulin resistance. Eur J Clin Invest 2000;30:481–6.

[49] Koren W, Koldanov R, Peleg E, et al. Enhanced red cell sodium-hydrogen exchange in microvascular angina. Eur Heart J 1997;18:1296–9.

[50] Di Carli MF, Janisse J, Grunberger G, et al. Role of chronic hyperglycemia in the pathogenesis of coronary microvascular dysfunction in diabetes. J Am Coll Cardiol 2003;41:1387–93.

[51] Cosin-Sales J, Pizzi C, Brown S, et al. C-reactive protein, clinical presentation, and ischemic activity in patients with chest pain and normal coronary angiograms. J Am Coll Cardiol 2003;41:1468–74.

[52] Miwa K, Fujita M, Sasayama S. Recent insights into the mechanisms, predisposing factors, and racial differences of coronary vasospasm. Heart Vessel 2005;20:1–7.

HEART FAILURE

38

Endothelial Alterations in Heart Failure—Mechanisms and Molecular Basis

Santiago A. Tobar, Daniel Umpierre, Michael Andrades, and Nadine Clausell

INTRODUCTION

Disturbances in the vascular endothelium's function have been widely studied in the context of heart failure. The multiple biological functions of the endothelium qualify it, at the same time, to serve as a target to different aggressors and cause different cardiovascular disorders with varied implications in heart failure progression. In this chapter we shall discuss the pathogenic mechanisms involved in endothelium dysfunction in heart failure and its potential as a therapeutic target.

PATHOGENESIS

The key element acting as a common route to several mechanisms involved in endothelial dysfunction in heart failure is nitric oxide (NO) and the imbalances in its metabolic pathways. The starting point are the alterations in NO-synthase production and regulation in several levels and secondarily to multiple signals. The heart, during heart failure, presents a well described alteration of its redox state with significant production of reactive oxygen species (ROS). In heart failure, the phenotypical vascular and cardiac alterations seem to result from an imbalance in the bioavailability of NO and oxidative stress [1]. In turn, this imbalance is caused by the neurohumoral activation involving the renin-angiotensin-aldosterone system, adrenergic activation, by production of inflammatory cytokines and by shear forces that modulate the gene expression leading to reduction in the availability of NO and increase in oxidative stress. The resulting endothelial dysfunction further increases cytokine production (activated endothelial cells are capable of producing inflammatory cytokines in the context of heart failure [2]), the negative regulation or decoupling of the endothelial NO-synthase (eNOS) and also the largest increase of oxidative stress [1,3,4]. Closing the cycle, there is pronounced endothelial dysfunction with reduced NO bioavailability, which in turn further exacerbates heart failure progression. The confluence of altered mechanisms converging to oxidative pattern alteration to some extent shows that the NO route and its imbalance have a key role in the heart failure syndrome, both in its central cardiac function and its peripheral vascular component. In addition to vascular tone regulation, other endothelial dysfunction characteristics are important in heart failure, namely: propensity to a pro-thrombotic state of the endothelium and the capability of attracting inflammatory cells to the endothelial surface or subendothelium, reflecting the loss of antithrombotic and antiinflammatory properties of the endothelium's normal physiology, respectively. This setting contributes to the clinical presentation with characteristic signs and symptoms, and also provides relevant prognosis indicators [5].

MOLECULAR BASES

NO is a free radical produced by endothelial cells. This radical has a very important role in the maintenance of endothelial functions. As described above, NO does not participate only as a vasodilator, but also reduces the adhesion of inflammatory cells and platelet aggregation, which ultimately lead to a decrease of thromboembolic events.

Nitric Oxide Synthesis

In the endothelium, NO is produced by the endothelial isoform of NO-synthase (eNOS). Despite the synthase function suggested in its nomenclature, eNOS is actually an enzyme system with oxidoreductase function, represented by its NADPH oxidase (reductase domain) and hemeoxidase (oxidase domain) domains, which participate in the oxidative deamination process removing an NH_2 moiety—of L-arginine, generating L-citrulline and NO as product (Fig. 38.1).

The correct performance of eNOS depends on a homodimer formation interaction between two identical eNOS subunits, only then the interaction of BH4 cofactor (tetrahydrobioprotein) and GMPc (guanosine monophosphate cyclic) and the Ca^{++}/Calmodulin protein and the L-arginine link will occur.

The eNOS enzymatic activity has 5 main steps, as follows: (1) NADPH oxidase domain captures electrons from NADPH; (2) electrons are transferred to the heme oxygenase domain using FAD, FMN and Ca++/Calmodulin; (3) oxygen linked to the heme oxygenase domain receives the electrons; (4) arginine binds to the heme oxygenase domain; and (5) with the aid of a BH4 cofactor the oxidative deamination of L-arginine occurs, generating L-citrulline and NO as products [6–8] (Fig. 38.2).

Understanding the mechanism of eNOS enzyme action is extremely important to comprehend the eNOS decoupling process and its association to the increase of ROS and to the induction of oxidant stress, as we shall see ahead.

Endothelial Protection: Nitric Oxide Function

eNOS is the endothelial isoform of NO-synthase. However, its production is not restricted to endothelium. eNOS may be detected in cardiomyocytes, platelets and certain brain neurons [9,10]. The NO produced by eNOS can control several cellular functions via protein nitrosylation such as: (1) guanylate cyclase activity [7]; and (2) control of mRNA transcription and translation through its link to the iron-responsive elements (IRP)—proteins that associate with mRNA and control its translation [11,12].

Guanylate cyclase activity's control by NO is the most studied and best described mechanism of action for it is through this action that NO controls vascular tonus and exerts its platelet antiaggregant action [7]. In this process, NO's interaction with guanylate cyclase enzyme promotes a raise in enzyme activity, culminating with an increase [13,14] of cGMP (guanonisemonophostate cyclic).

In smooth muscle cells, cGMP controls calcium release by the sarcoplasmatic reticulum and, as a result, promotes fiber relaxation [14]. In the platelet, cGMP will inhibit the thromboxane A2 receptor (TXA_2R) and prevent platelet activation and aggregation, reducing the likelihood of thromboeolic events [14] (Fig. 38.3). In addition, NO controls the expression of chemoattraction protein MCP-1 in the endothelium and reduces adhesion protein (CD11/CD18) expression in leukocytes. Thus, NO decreases activation and adhesion of inflammatory cells, thus preventing atherosclerosis [7,15,16].

Given all the above, there is no doubt that correct eNOS functioning is of extreme importance to

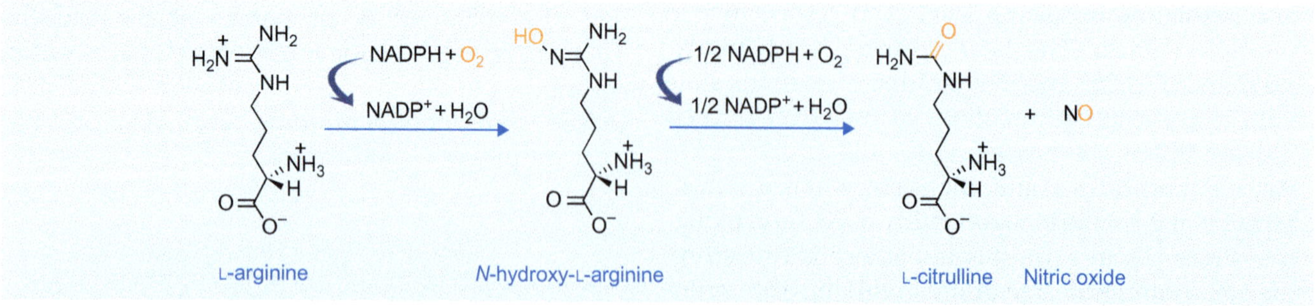

FIG. 38.1   Oxidative deamination of L-arginine occurs in two stages. In the first stage of oxireduction occurs the hydroxylation of L-arginine; in the second oxireduction stage occurs the oxidative deamination, which has L-citrulline and nitric oxide as its products (for acronyms, refer to the text). Adapted from Knowles RG, Moncada S. Nitric oxide synthases in mammals. Biochem J 1994;298(Pt 2):249–58.

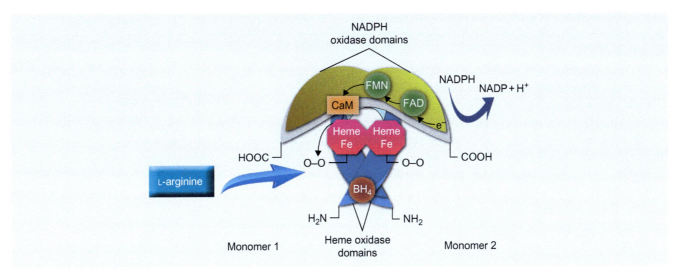

FIG. 38.2 eNOS structure and organization. The catalytic structure is formed by two identical eNOS monomers. Each monomer is formed by a NADPH oxidase domain and a hemeoxidase domain. The FAD and FMN cofactors and the Ca++/Calmodulin protein (CaM) are associated with the NADPH oxidase domain. In the hemeoxidase domain BH cofactor link and L-arginine link occurs. Electrons from NADPH are directed to the molecular oxygen linked to the heme cluster, where the reaction with L-arginine will take place. The complete reaction will occur in two oxidation cycles of the NADPH and the input of two molecular oxygen molecules (for acronyms, refer to the text). Adapted from Forstermann U, Sessa WC. Nitric oxide synthases: regulation and function. Eur Heart J 2012;33:829–37, 837a–d.

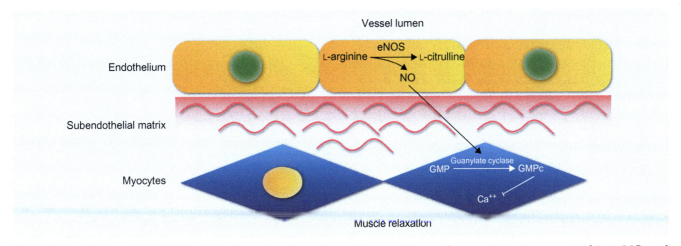

FIG. 38.3 Nitric oxide's (NO) role in the relaxation of vessels smooth muscle. L-arginine is converted into NO and L-citrulline by catalysis of the NO-synthase enzyme. NO spreads through the endothelium and reaches the muscle cells, promoting guanylate cyclase activation with GMPc synthesis, which controls calcium release by the sarcoplasmic reticulum and promotes smaller contraction (acronyms, refer to text).

vascular physiology, and that alterations in this enzyme's activity, due to alteration in the availability of its substrate and cofactors or by modification of its structure due to free radical, will cause the process called eNOS decoupling, which will lead to the disruption of vascular system normal functioning and, eventually, to the development of arterial hypertension.

OXIDATIVE STRESS AND eNOS ALTERATIONS

Free radicals are atoms or molecules that contain unpaired electrons in their molecular structures. This conformation makes radicals highly reactive with organic molecules (capable of losing electrons), such as proteins, lipids, and DNA. Such interactions can cause structural and conformational alterations in cellular macro-molecules and compromise their functions [17].

The production of free radicals in cells is constant and part of the chemical arsenal used to eliminate invading microorganisms and in intra and intercellular signaling processes. Sources that can generate these radicals are many, for example: enzymes of the mitochondria electron transport chain, NADPH oxidase, xanthine oxidase, and eNOS itself (Table 38.1). In all cases, the precursor radical is superoxide, formed by the partial reduction of molecular oxygen ($O_2 + 1e \rightarrow O_2^-$).

Superoxide, in spite of the "super" prefix, is a low reactive radical and its concentration is kept under control by action of the superoxide dismutase (SOD) enzyme, which chemically reduces superoxide to hydrogen peroxide (H_2O_2). H_2O_2 is an oxygen reactive species that may permeate through the cell membranes and reach cells and tissues at distant points from their origin. Under normal conditions, H_2O_2 levels are controlled by catalase enzymes (CAT) and glutathione peroxidase (GPx). CAT is a heme protein capable of reducing H_2O_2 in water and molecular oxygen. GPx in turn uses the reduction power of the tripeptide glutathione (GSH) to reduce H_2O_2 into water molecules [17] (Fig. 38.4).

Therefore, in normal conditions, most of the superoxide produced will be neutralized by the action of these enzymes (SOD, CAT and GPx). However, when there is an exacerbated production of the superoxide radical, there will be H_2O_2 accumulation that may react with metal ions, such as iron and copper, and generate the hydroxyl radical ($\cdot$OH). Hydroxyl radical is the free radical with greatest toxicity, due to its high reactivity (10^{-9} sec) and to a lack of enzymatic control of its levels [17].

Oxidative stress is a cellular condition in which production of free radicals exceeds the defense capacity and, as a result, damage to the cellular structures occur (lipids, proteins, and DNA) and

TABLE 38.1 Free Radicals and Reactive Species

Molecule	Symbol	Source in the organism
Superoxide	$O^{\bullet -}$	Mitochondria, NADPH oxidase, xanthine oxidase, and eNOS
Hydrogen peroxide	H_2O_2	From the action of the SOD enzyme
Hidroxil	$^{\bullet}OH$	Reaction of H_2O_2 with the reactive metallic ions (Fe^{2+} and Cu^+)
Nitric oxide	NO	NO-synthase
Peroxynitrite	^-ONOO	From the reaction of superoxide radical with nitric oxide

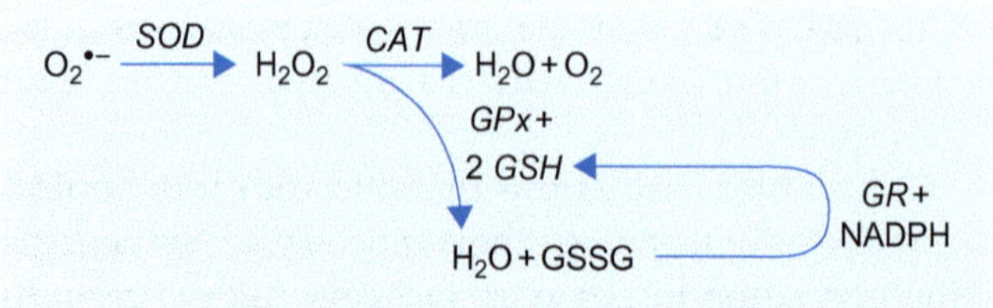

culminates in cell death and loss of viable tissue [17]. Endothelial dysfunction is characterized by the tissue's incapacity to generate significant amounts of NO, a condition commonly found in patients with cardiovascular disease or those exposed to cardiovascular risk factors (hypertension, hypercholesterolemia, *diabetes mellitus*, and smoking). Other common features in these patients are endothelial inflammation and increase in ROS production, which can trigger a state of oxidative stress [7,18].

In this context, the main cause of oxidative stress is the augmented expression of NAPDH oxidase in the endothelium, in the smooth muscle and adventitia of the vessels, especially when associated with high cholesterol levels, metabolic syndrome, smoking, and in response to angiotensin II (A-II) increase [7,18].

The first impact of increased superoxide production in the endothelium is reduction in NO availability due to the reaction of this radical with super-oxide, forming the peroxynitrite anion (—ONOO) (Fig. 38.5), which is a molecule with cytotoxic potential that can alter mitochondrial function, DNA structure, and induce apoptosis [6,18,19].

Moreover, the superoxide overproduction may lead to a process called eNOS decoupling. This process is characterized by the incapacity of coupling

$$O_2^{\bullet -} \xrightarrow{SOD} H_2O_2 \xrightarrow{CAT} H_2O + O_2$$

GPx+ 2 GSH / H_2O + GSSG / GR+ NADPH

FIG. 38.4 Ratio between production and consumption of free radicals by the enzymatic antioxidant system.

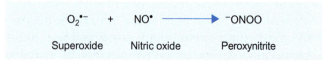

$$O_2^{\bullet -} + NO^{\bullet} \longrightarrow {}^-ONOO$$

Superoxide   Nitric oxide   Peroxynitrite

FIG. 38.5 Synthesis of peroxynitrite by reaction of superoxide with nitric oxide.

oxygen reduction (by the electron transfer from the eNOS' NADPH oxidase domain) for the oxidative deamination reaction in the hemeoxidase domain [7,20]. This dysfunction can occur by different mechanisms, which are described below.

Tetrahydrobiopterin (BH4) Oxidation and eNOS Decoupling

BH_4 acts as a cofactor in eNOS oxygenase motif, participating in the oxidative deamination of L-arginine at the enzyme's active site. Oxidation of BH_4 by superoxide radical ($O_2^{\cdot-}$) or by peroxynitrite (^-ONOO) produces $BH_3^{\cdot}$ radical, which is unable to bind to eNOS and cause the decoupling of the enzyme. Consequently, due to BH_4 absence, L-arginine is not oxidized, and as a result oxygen is only partially reduced at heme domain, leading to superoxide radical formation and contributing to the enhancement of oxidative stress [7,8,15]. The importance of BH_4 for accurate eNOS functioning and for preservation of endothelial health is unequivocal in patients with coronary diseases, in whom superoxide radical production increases while BH_4 levels decreases in vascular tissue followed by ineffectiveness of vasodilator agents [21].

Modification of Cysteine Residues and eNOS Decoupling

Despite the evidence regarding BH_4 oxidation and eNOS decoupling, they do not explain the whole mechanism for endothelial dysfunction. In fact, oxidative stress studies have showed that supplementation with BH_4 was unable to restore eNOS activity to basal levels [22]. This could be due to the higher levels of oxidized glutathione levels that are observed during a redox imbalance of oxidative stress [22]. Therefore, increased levels of GSSG may expose eNOS to glutathionylation—a reversible linkage of oxidized glutathione to cysteine residues [22]. In fact, S-glutathionylation occurs in two cysteine residues (Cys689 and Cys908) of eNOS resulting in reduction of the enzyme activity, depressing the NO formation rate [7,22], and increasing in more than five times the rate of $O_2^{\cdot-}$ formation [22]. On the other hand, reduction of oxidized glutathione by DTT—recycling GSH—in vitro, showed that only 80% of eNOS activity is reestablished, which suggest that oxidative stress may cause a permanent damage to the enzyme.

Therefore, even after endothelial cells restore its redox balance, the oxidative stress damage to eNOS may persist, leading to sustained endothelial dysfunction [22].

Physiologically, eNOS undergo another type of modification in cysteine residues. When in high NO concentrations, S-nytrosilation may occur—a linkage of NO to reduced cysteine—in Cys101 and Cys107 residues [23]. However, this oxidation seems to act as a mechanism for eNOS self-inhibition in order to avoid its prolonged action.

Oxidative Stress and Apoptosis

In the topics above, we discussed the role of oxidative stress in eNOS decoupling and its synergistic effect, which increases redox imbalance in the endothelium. However, it is worth noting that deleterious effects of free radicals affect other cellular structure, in addition to increasing the inflammatory process that may lead the cell to death. In acute insults of great magnitude, endothelial cells may perish by necrosis, which increases the inflammatory process and threating the neighboring cells, leading to greater tissue impairment. As per chronic insults, before cellular structures collapse, the apoptotic cell machinery—programmed cell death—takes action to prevent overflow of cell contents and spare neighbor cells. Even so, if the apoptotic process is sustained, there will be a great loss of endothelial cells, compromising their functions [24].

Oxidative stress can induce apoptosis in the endothelium by lipid peroxidation and by proinflammatory molecules formation, which may have a synergistic effect with signals of vasoactive peptides. In heart failure, the main circulating molecules capable of inducing endothelial cell apoptosis are TNF-α, IL-6, and angiotensin II, among others [25,26]. Pro-inflammatory cytokines, for example, trigger cell pathways that culminate in the activation of the transcription factor NF-κB, which maximizes the cellular inflammatory process and induces gene expression of adhesion molecules, such as E-selectin, VCAM-1, and ICAM-1. In addition, NF-κB activation is associated with the trigger the apoptotic process [27].

Apoptosis detection usually requires tools of molecular biology and isolated cells, which hampers its evaluation in patients. However, one alternative is an indirect observation, by means of detecting apoptotic microvesicles released into the blood stream and which bring along membrane markers that identify their endothelial origin. These microvesicles have already been identified in hypertension, coronary artery disease, acute myocardial infarction, obesity, and heart failure. In heart failure, the presence of these circulating microvesicles has a potential role in identifying patients with worse prognosis, suggesting the important role of endothelium in heart failure progression [24].

PATHOPHYSIOLOGIC IMPLICATIONS

- *In acute heart failure*: redox imbalance has an important role in acute heart failure pathophysiology. Ventricular function adjustment and vascular tonus that depends on NO suffer important alterations in the acute context. NO reduction in the vasculature induces vasoconstriction and decreased vascular compliance, both systemic and pulmonary, resulting in increased left and right systolic cardiac work. In addition, induces an increased production of endothelin-1, intensifying vasoconstriction, activating catecholamine release and decreasing sodium excretion by the kidney—all these elements contribute to greater amplitude of heart failure's clinical syndrome [28–30]. The excess ROS reacts with NO, breaking physiological signaling pathways and leading to production of toxic reactive molecules, such as peroxynitrite [31]. Therefore, assessing the degree of oxidative stress is important and can be measured in urine by dosage of isoprostane and in plasma by dosage of aminothiols. These products indicate an unfavorable deviation of the nitrous-redox axis and contribute to deleterious myocardium and vascular tonus effects in acute heart failure [32].

- *In chronic heart failure*: NO reduction with resulting endothelial dysfunction contributes to syndrome progression in many ways. In addition to exerting continued vasoconstriction in the systemic and pulmonary vascular bed, which increases the left and right ventricles afterload, the reduction in NO availability and worsening of the oxidative stress act decisively to promote alterations that will contribute to worse the adverse heart remodeling. This occurs through several deleterious mechanisms that include metalloproteinase activation, influencing cell

migration, myocardial hypertrophy, and destabilization of atherosclerotic plaques. Enhanced vascular tonus also happens in coronary circulation, compromising coronary flow, potentially causing/contributing to myocardial ischemia and reducing ventricular function. High levels of endothelin-1 in chronic heart failure lead to endothelial dysfunction, further increasing vascular resistance and promoting vascular and cardiomyocyte hypertrophy, and also inducing signaling pathways involved in the activation of extracellular matrix growth, with vascular and myocardial fibrosis - the main contributors to vascular and myocardial remodeling [33]. Reduced NO levels in heart failure also contribute to reduce the endothelial regenerative capacity in the myocardium, for they alter progenitor endothelial cells [34,35]. Inflammatory cytokines, known to be high in chronic heart failure, as the alpha tumor necrosis factor (TNF-α), promote negative regulation of NO synthesis by inhibiting eNOS, contributing to amplify endothelial dysfunction degree in addition to negatively impacting cardiac function—TNF-α is an agent that reduces cardiac inotropism [36,37].

- *In heart failure with systolic dysfunction*: the role of endothelial dysfunction in heart failure progression seems well established, with a clear reduction in NO production resulting from eNOS enzyme activity reduction, which culminates in systemic vascular tone alterations, causing increased afterload to the left ventricle. However, the trophic myocardial modifications also are influenced by increased production of ROS, promoting autophagic signaling in cardiomyocyte, apoptosis and tissue replacement due to interstitial fibrosis growth [20].

- *In heart failure with preserved ejection fraction*: the comprehension of heart failure with preserved ejection fraction pathogenesis has been subject to great controversies, and its limited extent explains the relative failure in achieving effective therapeutic strategies. Therapeutic tools for this condition are far from similar to what is available for heart failure with systolic dysfunction. Recently a new paradigm is being proposed, in which, ventricular remodeling in that condition is due to a multiplicity of mechanisms, namely: (a) high pro-inflammatory state; (b) this pro-inflammatory state induces production of ROS by endothelial cells, limiting NO bioavailability to adjacent cardiomyocytes; (c) NO reduction induces decreases the activity of protein kinase G in cardiomyocytes; (d) the reduction in protein kinase G activity removes/prevents the blocking of cardiomyocyte hypertrophy that will generate remodeling with concentric hypertrophy and increased myocardial stiffness; and (e) finally, the increase of ventricular stiffness and growth of collagen deposition would contribute to the clinical syndrome of heart failure with preserved ejection fraction [38]. In this context, endothelial dysfunction has secondary, although high relevant, participation due to reduction on NO bioavailability, which is caused primarily by the intense circulating inflammatory activity. Expression of adhesion molecules, e.g., VCAM-1 and E-selectin, on endothelial surface of cardiac microcirculation indicates endothelium dysfunction at that level. In turn, inflammatory activation creates a favorable environment for increased synthesis of ROS by endothelium vascular cells, via NADPH oxidase activation, yielding oxidative stress [38]. This may explain the high nitrosative/oxidative stress recently observed in the myocardium of heart failure with preserved ejection fraction [38].

THERAPEUTIC IMPLICATIONS INVOLVING PATHOPHYSIOLOGICAL ASPECTS OF ENDOTHELIAL DYSFUNCTION

The increased prooxidant state in the development of endothelial dysfunction is important because it impacts heart failure. Consequently, this prooxidant process is suitable as therapeutic target to minimize the consequences on vascular endothelium, both systemic and coronary. Accordingly, interventions that can reduce oxidative damage to the endothelium have potential benefit to patients.

Blocking of the renin-angiotensin-aldosterone is the most explored mechanism, since it is critical to inhibit several pathways that impact adverse remodeling of the left ventricle; one of these pathways, inhibit the generation of ROS and directly impacts the endothelial function and help to recovering it. There are several studies demonstrating the benefit of angiotensin-converting enzyme inhibitors (ACEi)

due to restoring the endothelium-dependent vasodilation that proves the improvement in endothelial dysfunction [20]. This effect takes place by inhibiting the enzyme kinase II, which leads to bradykinin formation and stimulates the releases of NO—as a result of endothelial Bradykinin B2 Receptor activation—, endothelium-derived hyperpolarizing factor, and prostacyclin. These actions contribute positively to the antiproliferative, antithrombotic and vasodilator effects of ACEi—all representing favorable modulation or recovery of the endothelial dysfunction and its multiple characteristics present in heart failure [4,20]. The use of hydralazine and nitrates is a known strategy that improves clinical outcomes such as total mortality and hospitalizations, in African-American patients, in addition to standard therapy. The proposed mechanism to support the physiopathology of this result is the beneficial effect on endothelial function, in which hydralazine act as a powerful antioxidant agent by inhibiting the formation of ROS through the suppressive action of nitroglycerin (GTN), those effects were demonstrated on vasculature, both in vitro and in vivo experiments [4,20]. Thus, hydralazine and nitrates cause a favorable balance in redox equilibrium and modulates beneficially the endothelial properties.

THE ROLE OF PHYSICAL EXERCISE IN IMPROVEMENT OF ENDOTHELIAL FUNCTION IN HEART FAILURE

Vasodilatory capacity contributes to control of vascular tone in situations of increased demand for blood flow, as occurs particularly in active areas (limbs in movement) during physical exercise. On the other hand, patients with heart failure present inadequate blood redistribution during effort [39], which is also expressed before thermal stress, demonstrating less circulatory adaptation in exposure to heat [40]. Such phenomena may be related to endothelial dysfunction prompted by endothelium-mediated dilatation in heart failure patients.

In this context, when compared with non-heart failure individuals, patients with heart failure present a marked reduction in endothelium-dependent dilatation in response to different doses of intra-arterial administration of a vasodilation agonist [41]. This may contribute to reduced circulatory capacity of patients with heart failure [42]. On the other hand, physical training induces endothelial function improvement in patients with cardiovascular disease, which occurs even in coronary arteries [43]. In heart failure, lower-limb exercise training over 6 months results in correction of endothelial function, as observed by the increase of blood flow in response to acetylcholine in the femoral artery [44]. These endothelial response modifications are positively associated with the gain in functional capacity ($r = .64$, $P < .005$), indicated by maximal oxygen consumption, which has prognostic value in heart failure [45].

Physical training apparently promotes systemic vascular function improvement. This has been reinforced not only by analyses that demonstrate reduction in total vascular resistance after unsupervised exercise [43], but also by studies designed specifically to test remote effects of exercise in heart failure. In this sense, a parallel a clinical trial in parallel with patients randomized into groups that exclusively trained lower limbs or had control intervention without exercise, showed that four weeks of training can induce systemic vascular improvement. In that study, endothelial function was assessed by intra-arterial acetylcholine infusion in different doses in the brachial artery, i.e., in a specific untrained site during the intervention (bicycle). In addition, our group has demonstrated that the remote effects (unexercised vasculature) seem also to occur after performing a single session of aerobic [46] or strength [47] exercise for patients with heart failure.

More recently, studies have shown that physical training seems to induce increments in quantity and migration capacity of endothelial progenitor cells [48–50]. These responses are induced even after short intervention periods such as 3 weeks of aerobic training. Acute responses show divergent results (increase or attenuation) in regard to changes in quantity and function of progenitor cells [51,52], which can be associated with the different exercise intensities. Yet, although young individuals may be less responsive to the effect of physical training on progenitor cells, it should be emphasized that increased quantity and migration capacity occurred after four weeks of intervention in patients with heart failure [50].

Therefore, regular exercise is a powerful adjuvant for endothelial improvement in heart failure. As briefly discussed above, this has been confirmed by vasomotor variables (vascular resistance, flow-mediated vasodilatation), as well as by the cellular study in response to acute or chronic physical exercise stimuli.

CONCLUSIONS

Endothelial dysfunction is an important part of heart failure pathophysiology, in which the altered redox state leads to an increase of ROS contributing to phenotypic alterations in the heart and peripheral areas of the disease, largely mediated by increased oxidative stress and of by modifications on NO bioavailability. This cycle, in turn, interacts with the neurohumoral and inflammatory activation present in heart failure, leading to a perpetuation of oxidative stress. The concept that matches these elements implies an understanding that endothelial dysfunction is significant in clinical manifestation of heart failure, and has prognostic and therapeutic implications.

References

[1] Marti CN, Gheorghiade M, Kalogeropoulos AP. Endothelial dysfunction, arterial stiffness, and heart failure. J Am Coll Cardiol 2012;60:1455–69.

[2] Voltan R, Zauli G, Rizzo P. In vitro endothelial cell proliferation assay reveals distinct levels of proangiogenic cytokines characterizing sera of healthy subjects and of patients with heart failure. Mediators Inflamm 2014;2014:257081.

[3] Bauersachs J, Bouloumie A, Fraccarollo D. Endothelial dysfunction in chronic myocardial infarction despite increased vascular endothelial nitric oxide synthase and soluble guanylate cyclase expression: role of enhanced vascular superoxide production. Circulation 1999;100:292–8.

[4] Munzel T, Harrison DG. Increased superoxide in heart failure: a biochemical baroreflex gone awry. Circulation 1999;100:216–8.

[5] Shantsila E, Wrigley BJ, Blann AD. A contemporary view on endothelial function in heart failure. Eur J Heart Fail 2012;14:873–81.

[6] Bec N, Gorren AFC, Mayer B. The role of tetrahydrobiopterin in the activation of oxygen by nitric-oxide synthase. J Inorg Bio-Chem 2000;81:207–11.

[7] Forstermann U, Sessa WC. Nitric oxide synthases: regulation and function. Eur Heart J 2012;33:829–37. 837a–d.

[8] Forstermann U, Munzel T. Endothelial nitric oxide synthase in vascular disease: from marvel to menace. Circulation 2006;1(13):1708–14.

[9] Forstermann U, Closs EI, Pollock JS. Nitric oxide synthase isozymes. Characterization, purification, molecular cloning, and functions. Hypertension 1994;23:1121–31.

[10] Wu KK. Regulation of endothelial nitric oxide synthase activity and gene expression. Ann N Y Acad Sci 2002;962:122–30.

[11] Pantopoulos K, Hentze MW. Nitric oxide signaling to iron-regulatory protein: direct control of ferritin mRNA translation and transferrin receptor mRNA stability in transfected fibroblasts. Proc Natl Acad Sci U S A 1995;92:1267–71.

[12] Liu XB, Hill P, Haile DJ. Role of the ferroportin iron-responsive element in iron and nitric oxide dependent gene regulation. Blood Cells Mol Dis 2002;29:315–26.

[13] Gudi T, Hong GK, Vaandrager AB. Nitric oxide and cGMP regulate gene expression in neuronal and glial cells by activating type II cGMP-dependent protein kinase. FASEB J 1999;13:2143–52.

[14] Wang GR, Zhu Y, Halushka PV. Mechanism of platelet inhibition by nitric oxide: in vivo phosphorylation of thromboxane receptor by cyclic GMP-dependent protein kinase. Proc Natl Acad Sci U S A 1998;95:4888–93.

[15] Kubes P, Suzuki M, Granger DN. Nitric oxide: an endogenous modulator of leukocyte adhesion. Proc Natl Acad Sci U S A 1991;88:4651–5.

[16] Khan BV, Harrison DG, Olbrych MT. Nitric oxide regulates vascular cell adhesion molecule 1 gene expression and redox-sensitive transcriptional events in human vascular endothelial cells. Proc Natl Acad Sci U S A 1996;93:9114–9.

[17] Halliwell B. Oxidative stress and neurodegeneration: where are we now? J Neurochem 2006;97:1634–58.

[18] Capettini LS, Montecucco F, Mach F. Role of renin-angiotensin system in inflammation, immunity and aging. Curr Pharm Des 2012;18:963–70.

[19] Gorren AC, Kungl AJ, Schmidt K. Electrochemistry of pterin cofactors and inhibitors of nitric oxide synthase. Nitric Oxide 2001;5:176–86.

[20] Munzel T, Gori T, Keaney Jr. JF. Pathophysiological role of oxidative stress in systolic and diastolic heart failure and its therapeutic implications. Eur Heart J 2015;36(38):2555–64.

[21] Antoniades C, Shirodaria C, Crabtree M. Altered plasma versus vascular biopterins in human atherosclerosis reveal relationships between endothelial nitric oxide synthase coupling, endothelial function, and inflammation. Circulation 2007;1(16):2851–9.

[22] Chen CA, Wang TY, Varadharaj S. S-glutathionylation uncouples eNOS and regulates its cellular and vascular function. Nature 2010;468:1115–8.

[23] Erwin PA, Lin AJ, Golan DE. Receptor-regulated dynamic S-nitrosylation of endothelial nitric-oxide synthase in vascular endothelial cells. J Biol Chem 2005;280:19888–94.

[24] Berezin A, Zulli A, Kerrigan S. Predictive role of circulating endothelial-derived microparticles in cardiovascular diseases. Clin Biochem 2015;48:562–8.

[25] Rossig L, Hoffmann J, Hugel B. Vitamin C inhibits endothelial cell apoptosis in congestive heart failure. Circulation 2001;104:2182–7.

[26] Rossig L, Haendeler J, Mallat Z. Congestive heart failure induces endothelial cell apoptosis: protective role of carvedilol. J Am Coll Cardiol 2000;36:2081–9.

[27] Sprague AH, Khalil RA. Inflammatory cytokines in vascular dysfunction and vascular disease. Biochem Pharmacol 2009;78:539–52.

[28] Bech JN, Nielsen CB, Ivarsen R. Dietary sodium affects systemic and renal hemodynamic response to NO inhibition in healthy humans. Am J Physiol 1998;274:F914–23.

[29] Sartori C, Allemann Y, Scherrer U. Pathogenesis of pulmonary edema: learning from high-altitude pulmonary edema. Respir Physiol Neurobiol 2007;159:338–49.

[30] Sartori C, Lepori M, Scherrer U. Interaction between nitric oxide and the cholinergic and sympathetic nervous system in cardiovascular control in humans. Pharmacol Ther 2005;106:209–20.

[31] Berry CE, Hare JM. Xanthine oxidoreductase and cardiovascular disease: molecular mechanisms and pathophysiological implications. J Physiol 2004;555:589–606.

[32] Kadiiska MB, Gladen BC, Baird DD. Biomarkers of oxidative stress study II: are oxidation products of lipids, proteins, and DNA markers of CCl4 poisoning? Free Radic Biol Med 2005;38:698–710.

[33] Massion PB, Feron O, Dessy C. Nitric oxide and cardiac function: ten years after, and continuing. Circ Res 2003;93:388–98.

[34] Bauersachs J, Widder JD. Endothelial dysfunction in heart failure. Pharmacol Rep 2008;60:119–26.

[35] Thum T, Fraccarollo D, Galuppo P. Bone marrow molecular alterations after myocardial infarction: impact on endothelial progenitor cells. Cardiovasc Res 2006;70:50–60.

[36] Agnoletti L, Curello S, Bachetti T. Serum from patients with severe heart failure downregulates eNOS and is proapoptotic: role of tumor necrosis factor-alpha. Circulation 1999;100:1983–91.

[37] Hermann C, Zeiher AM, Dimmeler S. Shear stress inhibits H_2O_2-induced apoptosis of human endothelial cells by modulation of the glutathione redox cycle and nitric oxide synthase. Arterioscler Thromb Vasc Biol 1997;17:3588–92.

[38] Paulus WJ, Tschope C. A novel paradigm for heart failure with preserved ejection fraction: comorbidities drive myocardial dysfunction and remodeling through coronary microvascular endothelial inflammation. J Am Coll Cardiol 2013;62:263–71.

[39] Chiappa GR, Roseguini BT, Vieira PJ. Inspiratory muscle training improves blood flow to resting and exercising limbs in patients with chronic heart failure. J Am Coll Cardiol 2008;51:1663–71.

[40] Green DJ, Maiorana AJ, Siong JH. Impaired skin blood flow response to environmental heating in chronic heart failure. Eur Heart J 2006;27:338–43.

[41] Kubo SH, Rector TS, Bank AJ. Endothelium-dependent vasodilation is attenuated in patients with heart failure. Circulation 1991;84:1589–96.

[42] Dall'Ago P, Chiappa GR, Guths H. Inspiratory muscle training in patients with heart failure and inspiratory muscle weakness: a randomized trial. J Am Coll Cardiol 2006;47:757–63.

[43] Hambrecht R, Gielen S, Linke A. Effects of exercise training on left ventricular function and peripheral resistance in patients with chronic heart failure: a randomized trial. JAMA 2000;283:3095–101.

[44] Hambrecht R, Fiehn E, Weigl C. Regular physical exercise corrects endothelial dysfunction and improves exercise capacity in patients with chronic heart failure. Circulation 1998;98:2709–15.

[45] Ribeiro JP, Stein R, Chiappa GR. Beyond peak oxygen uptake: new prognostic markers from gas exchange exercise tests in chronic heart failure. J Cardiopulm Rehabil 2006;26:63–71.

[46] Umpierre D, Stein R, Vieira PJ. Blunted vascular responses but preserved endothelial vasodilation after submaximal exercise in chronic heart failure. Eur J Cardiovasc Prev Rehabil 2009;16:53–9.

[47] Guindani G, Umpierre D, Grigoletti SS. Blunted local but preserved remote vascular responses after resistance exercise in chronic heart failure. Eur J Prev Cardiol 2012;19:972–82.

[48] Erbs S, Hollriegel R, Linke A. Exercise training in patients with advanced chronic heart failure (NYHA IIIb) promotes restoration of peripheral vasomotor function, induction of endogenous regeneration, and improvement of left ventricular function. Circ Heart Fail 2010;3:486–94.

[49] Gatta L, Armani A, Iellamo F. Effects of a short-term exercise training on serum factors involved in ventricular remodelling in chronic heart failure patients. Int J Cardiol 2012;155:409–13.

[50] Sandri M, Viehmann M, Adams V. Chronic heart failure and aging—effects of exercise training on endothelial function and mechanisms of endothelial regeneration: Results from the Leipzig Exercise Intervention in Chronic heart failure and Aging (LEICA) study. Eur J Prev Cardiol 2016;23(4):349–58.

[51] Van Craenenbroeck EM, Beckers PJ, Possemiers NM. Exercise acutely reverses dysfunction of circulating angiogenic cells in chronic heart failure. Eur Heart J 2010;31:1924–34.

[52] Van Craenenbroeck EM, Bruyndonckx L, Van Berckelaer C. The effect of acute exercise on endothelial progenitor cells is attenuated in chronic heart failure. Eur J Appl Physiol 2011;111:2375–9.

Further Reading

Knowles RG, Moncada S. Nitric oxide synthases in mammals. Biochem J 1994;298(Pt 2):249–58.

39

Heart Failure: Influence of Drug Interventions on Vessels

Fernando Bacal, Iáscara Wozniak de Campos, and José Leudo Xavier, Júnior

INTRODUCTION

For a long time, the only function credited to the vascular endothelium was that of a simple physiological barrier, which separated the blood from neighboring tissues. However, this began to change in the early 1980s with the works published by Furchgot and Zawadski, who demonstrated the role played by the endothelium in controlling vascular tonus, by means of acetylcholine [1]. Since then, other important functions of the blood vessel inner layer have been revealed: participation in local coagulation and fibrinolysis, and activation and adhesion of leukocytes and platelets, among others. Currently, it is considered a real endocrine organ, consisting of a total area of 1000 m^2, occupying a strategic position regarding vasomotor control: between the blood and vessel smooth muscle cells [2].

Endothelial dysfunction plays a role in the physiopathology of several relevant nosological entities, such as heart failure, arterial hypertension, *diabetes mellitus*, renal failure, and myocardial infarction [3]. The first stands out among them, which affects 23 million people around the world, with 2 million new cases diagnosed annually, and that, with aging population, has increased both in world prevalence and incidence, making it a serious public health issue [4]. In myocardial dysfunction, it is known that, as an adaptive response to reduced systolic volume due to initial cardiac muscle injury, there is increased activation of the adrenergic system and the renin-angiotensin-aldosterone system. This results in increased circulating levels of not only epinephrine, norepinephrine, and angiotensin-II, but

also other vasoconstriction factors, such as endothelin-1, vasopressin, neuropeptide Y, and thromboxane A2 (TXA2). The intensive resulting vasoconstriction affects the splenic territory, muscles, bone, skin, and kidneys, in order to maintain appropriate blood flow to the brain and heart. The high concentrations of vascular constriction factor also stimulate higher production of vasodilation agents, with the release of substances such as nitric oxide, bradykinin, PGI2, and PGE2. This vasodilation response is mainly important in physical strain situations, allowing compensatory arterial dilation, required in such circumstances [5]. However, heart failure evolution results in constant shear stress on the endothelium, systemic inflammatory reaction, and oxidative stress due to oxygen-free radicals, generating progressive endothelial dysfunction, which impairs any vasodilation response. The consequence of such inability is the well-known physical activity intolerance, typical of such heart diseases. The high concentrations of catecholamines, angiotensin-II, aldosterone, and endothelin-1 also interfere in cardiac remodeling, since these vasoconstriction agents are involved in myocyte apoptosis induction, increased myocardium inflammatory reaction, and fibroblast proliferation stimulation, leading to grater cardiac damage, lower systolic volume, and higher stimulation of catecholamine and renin-angiotensin-aldosterone action, generating a vicious cycle [6]. Therefore, it is only logical to conclude that vasodilation drugs improve heart failure symptoms, with possible effect of reducing the mortality of such a disease, which has such an important impact on the worldwide community.

Such vasodilation may be achieved by blocking vasoconstriction factors, such as, for example, angiotensin-II, in the case of angiotensin converting enzyme inhibitors (ACEI) or receptor AT1 blockers, or by means of direct vasodilation, as is the case of hydralazine, nitrates, serelaxin, and milrinone, among others [7].

ENDOTHELIUM AND VASOMOTOR CONTROL

The vascular smooth muscle tone results from the complex interaction between vasoconstriction and vasodilation factors. Just like in a scale that tilts to one side or the other, the predominance of dilation over constriction, or vice-versa, occurs according to the momentary needs of each tissue, or as an adaptive response to specific organic situations. Substances with systemic actions, produced remotely, sympathetic and parasympathetic innervations, and the endothelium itself, by releasing local action factors, known as endothelium-derived factors, are the main regulating agents of such balance [8].

Vasoconstriction Factors

Norepinephrine, a vasoactive amine released by sympathetic system nerve endings, has potent vasoconstriction action on arterioles and veins. It acts on adrenergic alpha 1 receptors, with important participation in blood pressure fine tuning and adaptive mechanisms triggered by organic perfusion deficit, resulting from low output in patients with heart failure [9]. Epinephrine is a less potent vasoconstrictor that norepinephrine, exerting even low vasodilation in specific vessels, as seen in coronaries, during increased cardiac work. The vasodilation action of catecholamines is due to its bond to adrenergic beta 2 receptors [10].

Neuropeptide Y is a vasoconstrictor peptide released together with norepinephrine in sympathetic nerve endings, acting on receptor Y1 located in the postsynaptic membrane of peripheral vessels. It maximizes the effect of other vasoconstrictors, such as adrenergic alpha and angiotensin-II agonists, and inhibits the acetylcholine released from parasympathetic nerve endings to the heart [11].

The increased serum concentration of decapeptide angiotensin-II results predominantly from the activation of the renin-angiotensin-aldosterone system. This is another example of a powerful vasoconstrictor substance that acts mainly by binding to AT1 receptors, which is the main subtype in vasculature. Stimulated by the low flow in glomerulus afferent arterioles, renin is released by the juxtaglomerular apparatus, converting the angiotenisinogen into angiotensin-I. The latter, in turn, is converted into angiotensin-II by means of the angiotensin converting enzyme (ACE). Just like adrenergic neurotransmitters, angiotensin-II plays an essential role in pressure regulation and in heart failure adaptive mechanisms. It also has a neuromodulator function, increasing the release of norepinephrine through the presynaptic membrane [12].

Vasopressin, also known as antidiuretic hormone, is produced in the hypothalamus and stored in the posterior pituitary. Besides promoting higher water retention in the nephron-collecting duct, it acts directly on the arterioles, generating vasoconstriction even more potent that angiotensin-II [13].

The vasoconstriction factors related to the endothelium are mainly endothelin-1 and thromboxane A2. The latter is produced by the action of an enzyme that exists in platelets, thromboxane synthase, over prostaglandins G2 and H2. After activation, it is released by the platelets by means of a process that has an intricate relationship with the endothelium. Its vasoconstriction function is exerted by increasing the concentration of intracellular calcium and inhibiting cyclical adenosine monophosphate (cyclic AMP) [14].

Endothelial damage, either by external mechanical trauma or even by the hyperflow generated, for example, by hypertension (shear stress), is the main stimulus to release endothelin, a peptide with 21 amino acids and 3 different types: endothelins 1, 2, and 3. Only endothelin-1 is produced by endothelial cells. Other secretion stimuli are hypoxia, catecholamines, and angiotensin-II. Endothelin-1 is produced as preendothelin and is converted into the active form by means of the endothelin converting enzyme (ECE), which may act on two types of receptors, ETA and ETB, to cause vasoconstriction. In several pathological situations, such as heart failure, myocardial infarction, essential hypertension, pulmonary hypertension, and renal failure, the levels of endothelin-1 in circulation are very high [15].

As we can see, all of these vasoconstriction agents act on different vascular smooth muscle cell membrane receptors. However, what they have in common is the fact that the transduction of the bonds with receptors in the final mechanical stimulus is by means of the second messenger system: the phosphatidylinositol system which, ultimately, leads to increased intracellular Ca^{2+} and vascular muscle contraction [16]. Fig. 39.1 shows the schematic representation of the balance between vasoconstriction and vasodilation factors.

Vasodilation Factors

Acetylcholine is the neurotransmitter released in most autonomous parasympathetic nervous system nerve endings. Physiologically, the parasympathetic system counterbalances the actions of the sympathetic system. Therefore, acetylcholine dilates blood vessels. Such dilation does not occur due to the direct action on smooth muscles, but instead by inhibiting the secretion of noradrenaline in sympathetic nerve endings and stimulating nitric oxide synthesis, which is a potent endothelium-derived vasodilation agent (see below) [17]. It acts on two types of receptors: actinic and muscarinic. Muscarinic receptors prevail in blood vessels: types M2 in the endothelium and M3 in vascular smooth muscle. It is important to emphasize that the action of acetylcholine prevails on M2 receptors, resulting in vasodilation. However, in the case of endothelial dysfunction, the acetylcholine binds predominantly to smooth muscle M3 receptors, which results in vasoconstriction. This is explained by the fact that M3 receptors are coupled to the same system of second messengers than the vasoconstriction factor receptors [18].

Bradykinin, a polypeptide with strong arteriole vasodilation action and that also significantly increases capillary permeability, comes from the cleavage of alpha 2 globulin by proteolytic enzymes, present in plasma and bodily fluids, with emphasis on kallikrein. Its synthesis is stimulated mainly by tissue inflammation [18]. Similarly to bradykinin, histamine also causes vigorous vasodilation and increased capillary porosity. It is synthesized mainly inside mast cells and basophils, and the release stimulus occurs in tissue injury or inflammation situations, and also in allergic reactions [19]. Adrenomedullin is a peptide with 52 amino acids with vasodilation properties. Its serum concentration is proportionally elevated according to the heart failure severity. Apparently, this peptide compensates the intense vasoconstriction pertinent to this condition [20].

Nitric oxide is, by far, the most important endothelium-derived relaxation factor. In 1980, Furchgot and Zawadski accidentally discovered that acetylcholine only exerted its vasodilation function in the presence of the endothelium. Later, it was demonstrated that this effect was mediated by a labile substance, a free radical, until, in 1987, Palmer identified it: it was nitric oxide. It was considered the molecule of the year in 1992 [21]. This is a lipophilic gas synthesized from L-arginine by means of nitric oxide synthase (NOS). This enzyme has three subtypes, with subtype II predominating in endothelial cells. The half-life of nitric oxide is approximately 5 s, being quickly inactivated by oxygen-free radicals, which explains why the action of this vasodilation factor is predominantly local. It diffuses in the endothelium up to the adjacent vascular smooth

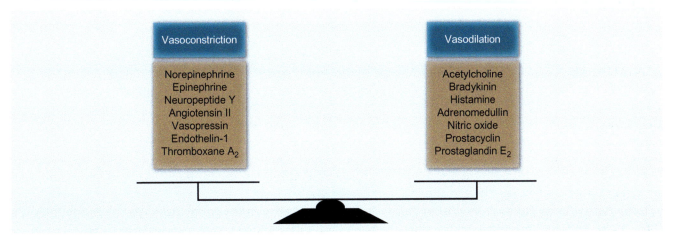

FIG. 39.1 Schematic representation of the balance between vasoconstriction and vasodilation factors.

muscle, where it stimulates the guanylate cyclase enzyme, leading to cyclic guanosine monophosphate (GPMc) synthesis that, finally, results in smooth muscle relaxation and vasodilation. Nitric oxide also acts as vasodilation mediator by means of other substances, such as acetylcholine, bradykinin, histamine, ATP, ADP, serotonin, and thrombin, among others [22] (Fig. 39.2).

In normal conditions, the vascular contracture baseline status is dictated by endothelial nitric oxide, released constantly to maintain moderated vasodilation, which is confirmed by experiments that demonstrate immediate vascular contraction after removing the endothelium. In the case of increased vessel blood flow, and consequent increased shear stress, there is a stimulus to produce more nitric oxide, which results in higher vasodilation. Vasoconstriction, on the other hand, results from the interaction between ceasing nitric oxide production and activation of vasoconstriction stimuli [23]. Endothelial dysfunction and consequent deficient vasodilation are present in pathological conditions such as hypertension, diabetes, atherosclerosis, and chronic renal failure, and are decisive in the physiopathology and prognosis of these diseases [24].

Prostaglandins are biologically active molecules produced by practically all body tissues, exerting different functions in each one of them. They received this name because they were primarily identified in human seminal fluid, in 1934, by Goldblatt and Von Euler [25]. Prostaglandins are the product of the action of an enzyme, cicloxigenase, on arachidonic acid, which comes from cell plasma membrane phospholipids due to the action of another enzyme, phospholipase A2 (PLA2). They are named according to the molecular structure, from A to I, and according to the number of double carbon-carbon bonds, from 1 to 3. The prostaglandin produced by the vascular endothelium is prostaglandin I2 (PGI2), also known as prostacyclin, which has vasodilation action. It acts in paracrine way on vascular smooth muscle, increasing

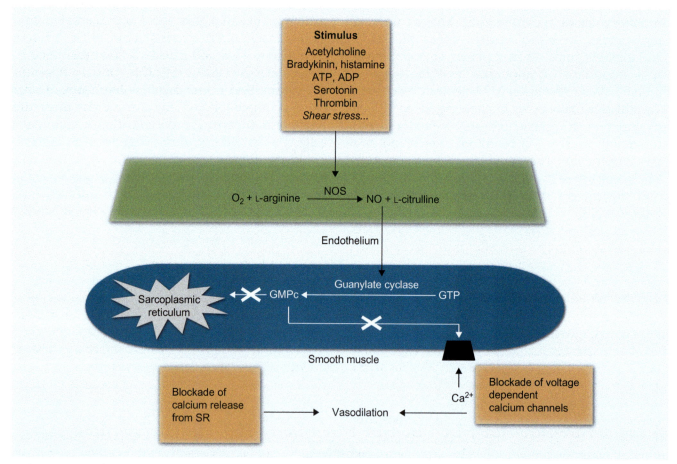

FIG. 39.2 Endothelial nitric oxide production, synthesis, and action stimulation in underlying vascular smooth muscle. *SR*, sarcoplasmic reticulum.

cyclic AMP levels in the intracellular environment, which, ultimately, results in vasodilation. As mentioned before, thromboxane A2 (TXA2), prostaglandin released by platelets, have antagonistic action in respect to prostacyclin, promoting vasoconstriction [26].

DRUG INTERVENTION IN ACUTE HEART FAILURE

Vasodilation Agents and Acute Heart Failure

The importance of vasodilation agents in heart failure treatment and prognosis began to be observed in the mid-1970s. However, the first large study came only in 1982. The Veterans Administration Cooperative Study, one of the rare large works to assess acute heart failure (AHF) and to study nitroprusside, whose vasodilation action is due probably to metabolization of the drug in nitric oxide. Although this is not an inotropic drug, by reducing postload by means of arteriole dilation, this nitrate ends up increasing cardiac output. It is also a potent venodilator, reducing the preload and therefore the ventricular filling pressures. This study was able to demonstrate benefits regarding mortality, in subgroup analysis, when the referred nitrate was initiated 9 h after the infarction (14.4% vs 22.3%, $P = .04$) [27]. In a more recent work conducted by the Cleveland Clinic, 78 patients with cardiac index ≤ 2.0 L/min/m^2, average blood pressure (ABP) ≥ 60 mmHg, and high filling pressure (PVC ≥ 8 mmHg and PCP ≥ 18 mmHg) were evaluated by means of a case-control retrospective study. Both groups showed improved hemodynamic parameters; however, mortality was reduced, for all causes, in favor of the group that used nitroprusside ($P = .005$), even when only patients with ABP < 85 mmHg ($P = .0001$) were assessed [28].

Another important AHF study, even more recent than the latter, was RELAX-AHF (serelaxin, recombinant human relaxin-2, for treatment of AHF: a randomized, placebo-controlled trial, 2013). This work used serelaxin, which is a recombinant form of relaxin-2. This peptide is one of the seven components of the relaxin family, which are hormones with vasodilation properties that occur naturally in human beings, discovered in 1929, and one of the main substance responsible for pregnancy hemodynamic alterations: increased cardiac output, reduced

systemic vascular resistance, and higher vascular compliance. Therefore, it was suggested that these hemodynamic adaptations might be beneficial in acute decompensated heart failure situations. In experimental works that preceded RELAX-AHF, serelaxin showed the capacity of reducing cardiomyocyte apoptosis, reducing myocardial injury due to inflammatory cells and free radicals, modulating cardiac interstice collagen and fibrosis proliferation, and reducing the serum concentration of markers related to myocardial injury. So, under the perspective or more than hemodynamic effects, with symptomatic improvement, but also in the physiopathology itself at heart failure cellular level, a study regarding serelaxin in myocardial dysfunction was initiated. In fact, the results in comparison to the placebo improved, both for vascular mortality ($P = .028$) and mortality due to any cause ($P = .02$), both in 180 days; nevertheless, these were secondary outcomes. The primary outcomes were dyspnea improvement, assessed by means of a visual analogue scale from the first to the fifth day ($P = .007$), and proportion of patients with moderate to important improvement of dyspnea in 6, 12, and 24 h, by means of the Likert scale ($P = .70$). As we can see, although the results suggest the possible benefit, RELAX-AHF was not powerful enough to assess mortality effectively, and so another larger study would be required to elucidate appropriately the role of serelaxin in AHF [29–31].

Pharmacology of Vasodilators in Acute Heart Failure

The endovenous nitrates used in AHF are nitroglycerin and sodium nitroprusside.

(a) Nitroglycerin is a coronary vasodilator, whose mechanism of action is due to the conversion of nitroglycerin in nitric oxide, which stimulates production of GMPs, leading to vascular smooth muscle relaxation, decreasing the preload, and generating systemic venodilation. Regardless of effect predominance at venous level, nitroglycerin acts quickly and produces dilation, both at arterial and venous levels, with small doses (30–40 µg/min) inducing venodilation and higher doses (250 µg/min) causing arteriole dilation. The biggest benefit in AHF is relieving the pulmonary congestion caused by vasodilation and increased coronary flow.

Headache, nausea, and dizziness are common side effects [32,33].

(b) Sodium nitroprusside is also a potent vasodilator, with short half-life that, even when administered in relatively low doses, decreases resistance to left ventricular ejection (afterload) and higher ventricular filling pressure (preload). The drug accumulates in vascular muscle cells, in which it decreases muscle tone due its own actuation or by forming active nitrite, therefore decreasing the oxygen consumption needs of the myocardium. In therapeutic doses, the substance is completely metabolized in a few minutes, and the reduced pre and after loads together with the lower oxygen consumption by the myocardium improve the biventricular systolic performance, increasing cardiac output in decompensated heart failure. The common dose is 0.5–10 µg/kg/min. The most common side effect is arterial hypotension, which may lead to hypoperfusion and renal function impairment. Sudden suspension of the medication is not recommended due to the rebound effect that may occur; gradual removal is suggested using oral vasodilators. With high doses and in the case of extended periods of use, especially in patients with renal and/or hepatic dysfunction, there is a risk of intoxication by thiocyanate and cyanide, and therefore thiocyanate must be monitored daily [28,33,34].

Inotropes and Acute Heart Failure

The main role of inotropic drugs is to treat AHF in which the patient shows signs of low output. The most used inotropes are dobutamine, milrinone, and levosimedan. Dobutamine is the most widely used inotropic in the world. It is a nonselective agonist of adrenergic beta 1 and beta 2 receptors, with variable action on alpha 1 receptors. At low doses, its action prevails over beta receptors, leading to positive inotropism and chronotropism, besides vasodilation. This results in increased cardiac output and reduced postload. At higher doses, the alpha 1 receptors are also stimulated, resulting in arterial and venous vasoconstriction. The recommended initial dose is 2–3 mcg/kg/min, which must be titrated according to the patient's clinical response. In general, the maximum dose is 15 mcg/kg/min, and may reach 20 mcg/kg/min, in patients with history of beta-blocker use.

Dobutamine shall not be suspended suddenly, but weaned according to the patient's clinical improvement in response to appropriate vasodilation. Weaning can be made in 2 mcg/kg/min steps [35].

In 2002, OPTIME-CHF (short Term Intravenous Milrinone for Acute Exacerbation of Chronic Heart Failure) assessed endovenousmilrinone in patients with exacerbated AHF. Milrinone is an inodilator of the bipyridine group, that is, it is simultaneously a positive inotropic and causes arterial dilation, with studies evidencing increased systolic volume and reduced ABP and pulmonary capillary pressure. In contrast to dobutamine, which is a pure inotropic, milrinone does not increase heart rate not myocardial consumption. It acts by inhibiting phosphodiesterase type III, leading to lower cyclic AMP degradation, both in cardiomyocytes and vascular smooth muscle. In the study in question, there were no differences between the groups in respect to the primary and secondary outcomes, with the group that used milrinone showing higher incidence of hypotension, fibrillation, and new atrial flutter. However, further analysis suggested that the subgroup of patients whose heart failure etiology was not ischemic might have neutral or positive effects from treatment with milrinone. It is important to emphasize that patients with signs of low output or peripheral hypoperfusion were excluded from the study [36].

Levosimedan acts by increasing affinity of troponin C by intracellular calcium in cardiomyocytes, and by opening ATP-dependent potassium channels in vascular smooth muscle, resulting, ultimately, in positive inotropism and vasodilation. The main study regarding levosimedan was Survive, from 2007 (levosimedan vs dobutamine for patients with acute decompensated heart failure). This study was randomized, controlled, and double-blind, and compared levosimedan with dobutamine in terms of mortality. There was no difference in primary outcome (mortality by any cause during the 180 days followed by the study) between these two inotropes ($P=.40$). It was not different with the secondary outcomes: mortality by any cause in the first month ($P=.29$), cardiovascular mortality during the 180 days followed by the study ($P=.33$), number of days alive and away from the hospital during the 180 days followed by the study ($P=.30$), and patient dyspnea assessment change ($P=NS$). The only outcome that showed difference between branches was reduced BNP (brain natriuretic peptide), which favored the group that used levosimedan ($P<.001$) [37]. A metanalysis study conducted later, including 19 very heterogeneous trials in terms of size and quantity, did not show any difference in mortality between levosimedan and placebo, but revealed increased mortality in patients treated with dobutamine when compared to levosimedan [38].

Pharmacology of Inotropes in Acute Heart Failure

(a) Dobutamine is a synthetic catecholamine, formulated as a racemic mix with alpha-agonist (-) and beta-agonist (+) isomers, with predominant effects in beta-adrenergic receptors, promoting increased cardiac output and reduced ventricular filling pressure, in the dose-dependent model. It also results in decreased central venous pressure and pulmonary capillary pressure due to improved heart performance; however, the pulmonary vascular resistance is altered. Up to 15 mg/kg/min, it promotes myocardial contraction without significant heart rate increase. Doses higher than 30 mg/kg/min may cause the emergence of ventricular arrhythmia and increased blood pressure. It is important to emphasize that dobutamine may promote reduced systemic vascular resistance due to its interaction with vascular beta-adrenergic receptors [39,40].

(b) Levosimedan is a moderately lipophilic molecule with double action mechanism: sensitization of myofilaments to calcium, causing therefore positive inotropic effects, besides coronary and peripheral vasodilation, by opening ATP-sensitive potassium channels. All this occurs without effectively increasing myocardial oxygen consumption. With a short elimination half-life of approximately 1 h, its active metabolite (OR 1896) remains in action for 14–18 days after 24 h of infusion. This residual effect has proven to be useful, because it provides a more stable profile to the patient until other therapeutic strategies are executed, many times unfeasible in patients in functional class IV or even III of the New York Heart Association (NYHA) rating [41–43].

(c) Milrinone is a positive inotropic agent and vasodilator, with little chronotropic activity, which also improves diastolic relaxation of the left ventricle. It is different from digitalis glycosides, catecholamines, or ACEI, both in structure and mode of action. In appropriate concentrations, to produce inotropic and vasodilation effects, milrinone is a selective inhibitor of cyclic AMP phosphodiesterase III in cardiac and vascular muscles. This inhibiting action is consistent with increased levels of ionized intracellular calcium and myocardium contractile force, mediated by cyclic AMP, as well as contractile protein phosphorylation and vascular muscle relaxation, also dependent on the cyclic AMP. Milrinone produces light increased conduction of the AV node, but without other significant electro-physiological effects. Clinical studies conducted in patients with congestive heart failure demonstrated that Primacor IV produces immediate improvement in hemodynamic indexes of congestive heart failure, including cardiac output, pulmonary capillary pressure, and vascular resistance, without clinically significant effects on heart rate or myocardium oxygen consumption, according to the dose as plasma levels. Both the inotropic and vasodilation effects are observed in milrinone plasma concentration in the range of 100–300 ng/mL. The dosage indicated is 50 µg/kg, which must be administered as loading dose, with the maintenance dose between 0.375 and 0.750 µg/kg/min, without exceeding daily doses of 1.13 mg/kg/day. Hemodynamic improvement occurs without significant increased oxygen consumption by the myocardium. The drug is metabolized hepatically and the medication is eliminated through the kidney, and must be adjusted according to the creatinine clearance. It may show side effects such as arrhythmia and headache, especially when associated with many drugs [33,36,44].

Table 39.1 shows the list of main studies regarding vasodilators and inotropic drugs in patients with acute decompensated heart failure.

DRUG INTERVENTION IN CHRONIC HEART FAILURE

Still in the 1980s, with the proliferation of studies regarding vasodilators in chronic heart failure, leveraged by the first results in AHF, a new study

TABLE 39.1   Main Studies Regarding Vasodilation and Inotropic Agents

Trial in acute IC	Year	Objective	Drawing	Number of patients	Results
VA Cooperative Study	1982	With nitroprusside × Without nitroprusside	Randomized controlled double-blind	812 (405 × 407)	14.4×22.3, $P = .04$
OPTIME-CHF	2002	Milrinone × Placebo	Randomized controlled double-blind	949 (477 × 472)	12.3×12.5, $P = .71$ Milrinone had more hypotension and atrial fibrillation and flutter
RELAX-AHF	2004	Serelaxin × Placebo	Randomized controlled double-blind	1161 (581 × 580)	Analogue scale, $P = .007$ Likert scale, $P = .70$ CV death, $P = .028$ death in general, $P = .02$
SURVIVE	2007	Levosimedan × Dobutamine	Randomized controlled double-blind	1327 (664 × 663)	General mortality $P = .40$ CV mortality, $P = .29$ days alive and away from the hospital, $P = .30$

From José Leudo Xavier Júnior.

was initiated: V-HeFT (Effect of Vasodilator Therapy on Mortality in Chronic Congestive Heart Failure: Results of a Veterans Administration Cooperative Study, 1986). Considered a landmark, this was the first randomized and double-blind trial to have real power to assess the mortality caused by heart failure in patients with chronic disease, with ejection fraction (EF) < 45%, and reduced tolerance to exercises. A total of 642 patients were included, randomized in three branches: hydralazine associated with isosorbidedinitrate, prazosin (an alpha-blocker), and placebo. The results showed that there was no difference between prazosin and placebo in any aspects. The patients in the hydralazine and nitrate group showed improved in tolerance to exercises and improved EF, and the relative risk of mortality was reduced by 22% [45]. In the following year, a Scandinavian group launched Consensus (Effects of Enalapril on Mortality in Severe Heart Failure: Results of the Cooperative North Scandinavian Enalapril Survival Study), in which patients with advanced heart failure, most of them of functional class (FC) IV of the New York Heart Association (NYHA), were randomized to use enalapril or placebo. There was significant reduction in the relative risk for all causes of death in 40%, during the first six months, with absolute risk reduction of 18% and NNT of 6 ($P = .002$). Enalapril is part of the ACE inhibitor group, whose final action is reducing the production of angiotensin-II, resulting therefore in vasodilation [46].

At the time, the paradigm had already changed, and heart failure treatment was no longer based only on digoxin and diuretics, with widespread use of vasodilators. However, the benefits of enalapril had only been demonstrated in advanced cases. Therefore, there was a prospect for a study that included patients with light and moderate heart failure, which were, and still are, the predominant severity profile. However, ethically speaking and in light of the new evidence, a work on vasodilators with placebo branch was no longer allowed. Then, in 1991, the V-HeFT II study emerged, to compare enalapril vs hydralazine and nitrate, in patients with the same profile of the first V-HeFT. The trial concluded that the hydralazine and nitrate branch obtained the same reduction or mortality as the previous study, but the reduction in mortality of the enalapril branch was even higher, with 18% mortality reduction in respect to the other group. This result reinforced the use of vasodilators in heart failure treatment, and, as the drugs apparently has independent beneficial effects, the combined use of enalapril and hydralazine and nitrate would be a possible strategy to increase survival and improve the symptoms [47]. Years later, the A-HeFT (African-American Heart Failure Trial Investigators, Combination of isossorbidedinitrate and hydralazine in blacks with heart failure, 2004) study would demonstrate the beneficial use of a combination of hydralazine and nitrate in black patients [48].

DRUG INTERVENTION IN CHRONIC HEART FAILURE

Specifically regarding ACEI, after the Consensus study, some other following studies confirmed this drug group as mandatory in the treatment. In 1991, the SOLVD (Studies of Left Ventricular Dysfunction. Effect of enalapril on survival in patients with reduced left ventricular EFs and congestive heart failure) trial demonstrated that all NYHA functional classes, and not only FC IV, benefited from enalapril, with risk reduction of 16% (5%–26%, $P < .0036$). The SAVE (Effects of captopril on mortality and morbidity in patients with left ventricular dysfunction after myocardial infarction, results of the Survival and Ventricular Enlargement Trial, 1992) study, which used captopril and not enalapril, as has been done until then, was the first study to demonstrate the impact on mortality with the introduction of an ACEI in patients with ventricular dysfunction, after acute myocardial infarction (AMI), but without signs or symptoms of failure (reduction of death risk by 19%, $P = .019$) [49].

Knowing that angiotensin-II can be synthesized in other ways different that ACE, the hypothesis suggested was that sequential blocking of the renin-angiotensin-aldosterone system could generate more benefits in terms of mortality in heart failure. Therefore, a new era of studies began, with Angiotensin-II AT1 Receptor Blockers (ARBs). In 2001, the time at which the association of certain beta-blockers (namely: Carvedilol, Metoprolol, and Bisoprolol) with ACEI was already consolidated with the foundation stone in heart failure treatment, another study was introduced: Val-HeFT (A randomized trial of the angiotensin-receptor blocker Valsartan in Chronic Heart Failure). In this study, valsartan was added to the treatment of heart failure with beta-blockers (35% of use in the population studied) and ACEI (93% of use in the population studied), compared to placebo. Regarding morbidity, some benefits were demonstrated (improved EF: 4.0% × 3.2%, $P = .001$; improved NYHA FC: 23.1% × 20.7%, $P < .001$, and improved life quality score, $P = .005$), benefit not shown regarding mortality (19.7% × 19.4% $P = 0.8$). The analysis of the subgroup also showed increased number of deaths in the group that used valsartan, when this drug was associated with patients that in fact were using beta-blockers and ACEI [50]. The Charm Added Trial (Effects of candesartan in patients with chronic heart failure and reduced left-ventricular systolic function taking angiotensin-converting-enzyme inhibitors, 2003) also added angiotensin-II receptor

beta-blockers ARB to heart failure treatment with ACEI, confirming the reduced morbidity of the previous study. However, differently from the Val-HeFT study, there was no increase in mortality when candesartan was initiated in patients taking the ACE inhibitor and beta-blockers [51]. Another study of the same group, and conducted in the same year, Charm-Alternative (Effects of candesartan in patients with chronic heart failure and reduced left-ventricular systolic function intolerant to angiotensin-convertin-enzyme inhibitors), randomized, for candesartan or placebo, 2028 patients that for some reason were intolerant to ACEI. It demonstrated reduction of 23% in relative risk to compound outcome of cardiovascular mortality and hospital admission due to heart failure (0.77, 0.67–0.89, $P = .0004$), with NNT of 14 similar to the results seen for enalapril in the SOLVD study [52].

Another drug that stands out due to its vasodilation properties, mainly in the pulmonary territory, is sildenafil. Increased pulmonary vascular resistance, with resulting pulmonary hypertension, is present in 68%–78% of the patients with severe myocardial dysfunction. Once present, these hemodynamic parameters become extremely important in heart failure prognosis. Sildenafil is a selective inhibitor of phosphodiesterase type V, an enzyme that degrades GMPc. The higher availability of GMPc in vascular smooth muscle cells results in vasodilation. Small studies have demonstrated that sildenafil is effective in reducing vascular resistance, blood pressure, and capillary pressure in the lugs, improving ventilatory efficiency, oxygen consumption recovery kinetics, and also improves tolerance to exercises [53–55]. In 2005, the Sildenafil Citrate Therapy for Pulmonary Arterial Hypertension double-blind placebo-controlled study evaluated 278 patients, randomized for placebo or sildenafil (20, 40, or 80 mg), 3×/day for 4 weeks. The study showed significant reduction in pulmonary blood pressure with all doses of sildenafil, as well as increased functional capacity demonstrated by the 6-min walking test in the sildenafil group. This study did not assess mortality [56]. Studies conducted with patients with severe heart failure showed improved cardiac performance by means of reduced heart rate, increased cardiac index, and reduced left ventricle postload by means of reduced systemic vascular resistance [57,58].

However, large randomized clinical trials are required, with patients with cardiac dysfunction, to elucidate the appropriate use of sildenafil, especially in associated with drugs that are already well-established in heart failure treatment.

The RELAX (Effect of phosphodiesterase-5 inhibition on exercise capacity and clinical status in heart failure with preserved EF, 2013) study failed to demonstrate the real benefit in patients with heart failure with normal EF. The role of phosphodiesterase inhibitors in systolic heart failure is still undetermined [59].

PARADIGM-HF (Angiotensin-neprelysin inhibition vs enalapril in heart failure, 2014) is the most recent study with vasodilators in heart failure. It compared enalapril to LCZ-696, which consists of a drug based on a molecule resulting from the fusion of valsartan with the neprilysin inhibitor called sacubitril. Neprilysin is an endopeptidase that degrades some vasodilation substances, such as bradykinin and adrenomedullin. Therefore, the proposal is to achieve even more intense vasodilation by blocking the renin-angiotensin-aldosterone system and increasing availability of organic vasodilators. After 25 years of ACEI hegemony, another drug overcame this class in terms of survival in heart failure. The study was interrupted before the allotted time. Regarding cardiovascular mortality, the risk was reduced by 0.80, $P < .001$. For hospital, admission due to heart failure worsening, risk reduction was 0.79, $P < .001$ (Table 39.2) [60].

TABLE 39.2   List of the Main Studies on Vasodilators in Chronic Heart Failure

Trial in acute IC	Year	Objective	Drawing	Number of patients	Results
V-HeFT	1986	Hydralazine/ nitrate × prazocin × placebo	Randomized controlled double-blind	642 (186 × 183 × 273)	General mortality 38.7% × 49.7% × 44% RRR 12%, RAR 5.3%
CONSENSUS	1987	Enalapril × Placebo	Randomized controlled double-blind	253 (127 × 126)	General mortality 16% × 44%, $P = .002$ RR 40%, RAR 18% NNT 6, in 6 months
V-HeFT II	1991	Enalapril × Hidralazine/ Nitrate	Randomized controlled double-blind	804 (403 × 401)	General mortality 18% × 25%, $P = .016$
SOLVD	1991	Enalapril × Placebo	Randomized controlled double-blind	2569 (1285 × 1284)	General mortality 35.2% × 39.7% RRR 16%, $P < .003$
SAVE	1992	Captopril × Placebo	Randomized controlled double-blind	2231 (1115 × 1116)	General mortality 20% × 25%, $P = .019$ VC mortality 17% × 20% $P = .014$
Val-HeFT	2001	Valsartan × Enalapril	Randomized controlled double-blind	5010 (2511 × 2499)	General mortality 19.7% × 19.4%, $P = .8$ mortality/ morbidity 28.8% × 32.1% RRR 0.87, $P = .009$
CHARM-add	2003	Candesartan × Placebo	Randomized controlled double-blind	2548 (1276 × 1272)	CV death or hospitalization due to HF 37.9% × 42.3%, $P = .011$ RRR 15%, RAR 4.4%, NNT23
CHARM-alternatie			Randomized controlled double-blind	2028 (1013 × 1015)	VC death or hospitalization due to HF 33% × 40%, $P = .0004$ RRR 23%, RAR 7%, NNT 14

TABLE 39.2 List of the Main Studies on Vasodilators in Chronic Heart Failure—cont'd

Trial in acute IC	Year	Objective	Drawing	Number of patients	Results
A-HeFT	2004	Hidralazine/ Nitrate × Placebo	Randomized controlled double-blind	1050 (518 × 532)	General death or hospitalization or drop QV, $P = .01$
RELAX	2013	Sildenafil × Placebo	Randomized controlled double-blind	216 (113 × 103)	Without VO2 change in 24 h, $P = .90$ mortality, $P = .25$ hospitalization, $P = .89$
PARADIGM-HF	2014	LCZ-696 × Enalapril	Randomized controlled double-blind	8442 (4187 × 4212)	CV mortality 13.3% × 16.5%, RR 0.80, $P < .001$ hospitalization due to IC 12.8% × 15.6%, RR 0.79, $P < .001$

From José Leudo Xavier Júnior.

Pharmacology of Vasodilators in Chronic Heart Failure

Action of ACEI—Renin-Angiotensin Receptor Blockers (ARBs)

ACEIs are drugs very similar among themselves. All examples tested were effective to treat systolic heart failure and can be used daily (Table 39.3). It is important to emphasize that the biggest benefit of its use occurs at high doses, and whenever possible, the highest doses tolerated by the patient should be used.

However, due to the risk of hypotension, the treatment must always begin with low doses that are increased progressively, according to the patient's tolerance, up to the target dose [61].

ACEIs increase the activity of renin, which is a protease with catalytic activity. Renin cleaves the

TABLE 39.3 IECA Used in Heart Failure and Their Usual Doses

Drugs	Minimum dose/day	Maximum dose/day
Captopril	6.25 mg 3 ×/day	50 mg 3 ×/day
Enalapril	2.5 mg 2 ×/day	20 mg 2 ×/day
Lisinopril	2.5–5 mg 1 ×/day	40 mg 1 ×/day
Ramipril	1.25–2.5 mg 1 ×/ day	10 mg 1 ×/day
Perindopril	2 mg 1 ×/day	16 mg 1 ×/day

From José Leudo Xavier Júnior.

angiotenisinogen, generating angiotensin-I, which is a precursor of the active product of the renin-angiotensin system, and when it comes in contact with the ACE, located on the vascular endothelium surface, it is converted in effector peptide angiotensin-II. Renin is formed in the kidney, from its inactive precursor, prorenin, and participates in the first stage of activation of the renin-angiotensin system, cleaving the Leu10—Val11 bond of the angiotenisinogen. Renin initiates and determines the speed of the entire enzymatic cascade of the renin-angiotensin system. The emergence of ACEI and, later ARBs, allowed confirming the important role of SRA blocking in heart failure. This intratissue formation of angiotensin-II may be essential in vascular disease progression, and moreover these alternative angiotensin-II formation pathways can be hyperactivated in pathological situations, such as heart pressure or volume overload, leading to systolic heart failure [62].

The ACEIs increase renin plasmatic activity (renin enzymatic activity level: angiotensin-II formation speed) and prorenin levels. The ACEIs inhibit the tissue and plasmatic ACE, and the level of angiotensin I increases significantly with the use of ACEI, while the level of angiotensin-II decreases. The renin-angiotensin system is very complex, generating products such as angiotensin-III, angiotensin-IV, and angiotensin-1–7, which seem to have important actions on vessels, acting on vasodilation. The ACEIs increase angiotensin-1–7, enabling the formation of angiotensin-II by means of pathways that are

TABLE 39.4 BRA Used in Heart Failure and Their Usual Doses

Drugs	Minimum dose/day	Maximum dose/day
Losartan	25 mg 1×/day	50–100 mg 1×/day
Candersartan	4–8 mg 1×/day	32 mg 1×/day
Valsartan	40 mg 1×/day	320 mg 2×/day

From José Leudo Xavier Júnior.

different from ACE and, with the great accumulation of angiotensin-I, outweigh ACEI blocking, which are competitive inhibitors. Furthermore, the ACEIs increase bradykinin that, via B2 receptors, release nitric oxide and prostacyclin to increase vessel vasodilation [49,62].

The clinical effect of the ARBs is very similar to the ACEIs. Its action mechanism is related to blocking angiotensin-II ATI receptors, leading to the same hemodynamic and neurohormonal effects of the ACEIs. The benefits of use and application are similar. They also show antiproliferative activity, with little effect on chronotropism and inotropism. They do not interfere with bradykinin degradation, reducing the incidence of cough. They are an interesting option for patients that show side effects, such as incoercible cough with the use of ACEI. However, they show incidence of hypotension, hyperkalaemia, and worsen renal function similarly as found with the use of ACEI. They are also contraindicated during pregnancy. As this is a homogeneous class of medications, any ARB can be used, always focusing on the highest dose tolerated by the patient [51,61] (Table 39.4).

Action of Nitrates × Hydralazine

In studies regarding mortality, benefits of the association between hydralazine and nitrate were demonstrated [63]. Vasodilation is due to the direct action on arteriole smooth muscle, probably involved with the release of intravesicularcatecholamines inside neurons. Hydralazine reduces systemic vascular resistance and increases cardiac output, with slight reduction of atrial pressures and discretely increased heart rate. It may increase renal blood flow, improving renal function due to secondary improvement of cardiac output. The common dose is 25 mg, 3–4 times/day. Its side effects include

vascular headache, swelling, redness, nausea, and vomiting, avoided with gradual increase of the doses and, frequently, disappearing with treatment continuation. At high doses (300 mg/day), a syndrome similar to lupus may develop; this disappears when the drug is suppressed [48].

Nitrates mainly promote reduction of the preload. Any long-term nitrate may be used in these situations. In patients with heart failure, with high filling and capillary pulmonary pressures, nitrates reduce atrial pressures and relieve the congestive symptoms. In addition, the pulmonary vasodilation and dilation effects on systemic arterioles, although discrete, are sufficient to cause slight cardiac output elevation, as long as the ventricular filling pressures are maintained at appropriate levels. In patients with persistent dyspnea despite ACEIs use, nitrates are an important therapeutic option in association with hydralazine. The nitrates induce vasodilation by regenerating the free nitric oxide radical or an akin, S-nitrosothiol. The effects on smooth muscles are due to the reduced concentration of calcium in cytosol and low phosphorylation of myosin light-chain, which leads to vasodilation. Low doses of isossorbidedinitrate (30 mg 3×/day) preferably dilate the venous system. Arterial vasodilation is typically associated with higher doses [61,64].

In acute and chronic congestive heart failure, both forms of the isosorbide dinitrate can be used. The choice can be based mainly on action duration, and not on response intensity, since this is the biggest difference observed in the presentation forms. In order to obtain the maximum therapeutic effect, it is important that the doses are individualized according to the needs of each patient, clinical response, and hemodynamic alterations. The treatment must begin with nitrate using the lower effective dose: isosorbide dinitrate: 40 mg 3–4×/day; isosorbidemononitrate: 20–40 mg 2–3×/day. In heart failure, the nitrate dosage strategy must prevent the onset of the tolerance phenomenon, by means of intermittent administration—that is, allowing a few hours of the day free from the action of the medication.

CONCLUSIONS

Several studies demonstrate that, for patients with heart failure, vasodilation improves symptoms and increases survival. Therefore, vasodilation should be pursued regardless of the strategy used, both by blocking vasoconstrictor factors and/or by means of direct vasodilation, in order to reach the maximum drug dosage tolerated by the patient.

References

[1] Furchgott RF, Zawadzki JV. The obligatory role endothelial cells in the relaxation of arterial smooth muscle by acetylcholine. Nature 1980;188:373–6.

[2] Nachman RL, Jafffe EA. Endothelial cell culture: beginnings of modern vascular biology. J Clin Invest 2004;114:1037.

[3] Cines DB, Pollak ES, Buck CA. Endothelial cells in physiology and in the pathophysiology of vascular disorder. Blood 1998;91:3527.

[4] Levy D, Kenchaiah S, Larson SD, et al. Long-term trends in the incidence of and survival with heart failure. N Engl J Med 2002;347:1397–402.

[5] Opie LH. Heart failure and neurohumoral responses. In: Opie LH, editor. The heart: physiology, from cell to circulation. Philadelphia, PA: Lippincott Williams & Wilkins; 1998. p. 475–511.

[6] Mann DL, Bristow MR. Mechanisms and models in heart failure: the biomechanical model and beyond. Circulation 2005;111(21):2837–49.

[7] Francis GS, Cohn JN, Johnson G. Plasma norepinephrine, plasma renin activity, and congestive heart failure: relations to survival and the effects of therapy in V-HeFT II. Circulation 1993;87:V140–8.

[8] Berne RM, Levy MN. The peripheral circulation and its control. In: Berne RM, Levy MN, editors. Physiology. St Louis, MO: Mosby; 1998. p. 442–57.

[9] Hoffman BB, Lefkowitz RJ, Taylor P. Neurotransmission: the autonomic and somatic motor nervous systems. In: Hardman JG, Limbird LL, Molinoff PB, editors. Goodman & Gilman's the pharmacological basis of therapeutics. New York: McGraw-Hill; 2012. p. 105–39.

[10] Guyton AC, Hall JE. Humoral and local tissue blood flow control. In: Guyton AC, Hall JE, editors. Textbook of medical physiology. Rio de Janeiro: Saunders; 2012. p. 201–11.

[11] Feng QP, Hedner T, Anderson B. Cardiac neuropeptide Y and noradrenaline balance in patients with congestive heart failure. Br Heart J 1994;71:261.

[12] Matsusaka T, Ichikawa I. Biological functions of angiotensin and its receptors. Annu Rev Physiol 1997;59:395–412.

[13] Tang WH, Bhavnani S, Francis GS. Vasopressin receptor antagonists in the management of acute heart failure. Expert Pin Investig Drugs 2003;14:593.

[14] Mombouli JV, Vanhoutte PM. Kinins and endothelial control of vascular smooth muscle. Annu Rev Pharmacol Toxicol 1995;35:679–705.

[15] Masaki T. The discovery of endothelins. Cardiovasc Res 1998;39(3):530–3.

[16] Vahoutte PM, Mombouli JV. Vascular endothelium: vasoactive mediators. Prog Cardiovasc Dis 1996;39:229–38.

[17] Person PB. Modulation of cardiovascular control mechanisms and their interaction. Physiol Rev 1996;76:193–244.

[18] Guyton AC, Hall JE. Nervous regulation of circulation. In: Guyton AC, Hall JE, editors. Textbook of medical physiology. Rio de Janeiro: Saunders; 2012. p. 213–23.

[19] Champion HC, Skaf MW, Hare JM. Role of nitric oxide in the pathophysiology of heart failure. Heart Fail Rev 2003;8:35–46.

[20] Klip IT, Voors AA, Anker SD. Prognostic value of mid-regional pro-adrenomedullin in patients with heart failure after an acute myocardial infarction. Heart 2011;97:892–8.

[21] Koschland Jr DE. The molecule of year. Science 1992;258:1861.

[22] Ignarro LJ. Biosynthesis and metabolism of endothelium-derived nitric oxide. Ann Rev Pharmacol Toxicol 1990;30:535–60.

[23] Davies MG, Fulton GJ, Hagen PO. Clinical biology of nitric oxide. Br J Surg 1995;82:1598–610.

[24] Wennmalm A. Endothelial nitric oxide and cardiovascular disease. J Int Med 1994;235:317–27.

[25] Needleman P, Turk J, Jackschik BA. Arachidonic acid metabolism. Ann Rev Biochem 1986;69:102.

[26] Dusting GJ, Moncada S, Vane JR. Prostaglandins, their intermediates and precursors: cardiovascular action and regulatory roles in normal and abnormal circulatory systems. Prog Cardiovasc Dis 1980;21:405–30.

[27] Cohn JN, Franciosa JA, Francis GS, et al. Effect of short-term infusion of sodium nitroprusside on mortality rate in acute myocardial infarction complicated by left ventricular failure: results of a Veterans Administration Cooperative study. N Engl J Med 1982;306(19):1129–35.

[28] Mullens W, Abrahams Z, Francis GS, et al. Sodium nitroprusside for advanced low-output heart failure. J Am Coll Cardiol 2008;52(3):200–7.

[29] Teerlink JR, Cotter G, Davison BA, et al. Serelaxin, recombinant human relaxin-2, for treatment of acute heart failure (RELAXAHF): a ramdomized, placeb-controlled trial. Lancet 2013;381:29–39.

[30] Teichman S, Dschietzig T, Unemori E, et al. Serelaxin demonstrated favorable hemodynamic effects in a pilot study in patients with chronic heart failure. Ann NY Acad Sci 2009;1160:387–92.

[31] Perna AM, Masini E, Nistri S, et al. Serelaxin reduces markers of myocardial damage in an in vivo porcine model of ischemia/reperfusion. FASEB J 2005;19:1525–7.

[32] Steinhorn BS, Loscalzo J, Michel T. Nitroglycerin and nitric oxide: a rondo of themes in cardiovascular therapeutics. N Engl J Med 2015;373(3):277–80.

[33] Mangini S, Pires PV, Braga FG. Decompensated heart failure. Einstein 2013;11(3):383–91.

[34] Elkayam U, Janmohamed M, Habib M, et al. Vasodilators in the management of acute heart failure. Crit Care Med 2008;36(1 Suppl):S95–S105.

[35] Bonow RO, Mann DL, Zipes DP. Diagnosis and management of acute heart failure syndromes. In: Braunwald's heart disease: a textbook of cardiovascular medicine. Amsterdam: Elsevier; 2012. p. 517–42.

[36] Cuff MS, Califf RM, Adams KF, et al. Short-term intravenous milrinone for acute exacerbation of chronic heart failure. J Am Med Assoc 2002;287(12):1541–7.

[37] Mebazaa A, Nieminen MS, Packer M, et al. Levosimendan vs dobutamine for patients with acute decompensated heart failure. J Am Med Assoc 2007;297(17):1883–91.

[38] Delaney A, Bradford C, McCaffrey J, et al. Levosimendan for the treatment of acute severe heart failure: a meta-analysis of randomized controlled trials. Int J Cardiol 2010;138:281–9.

[39] Stevenson LW. Clinical use of inotropic therapy for heart failure: looking backward or forward? Part I: Inotropic infusions during hospitalization. Circulation 2003;108:367–72.

[40] Fonseca J. Drogasvasoativas—Usoracional. Rio de Janeiro: Rev SOCERJ; 2001.

[41] Follath F, Cleland JG, Just H, et al. Efficacy and safety of intravenous levosimendan compared with dobutamine in severe low-output heart failure the LIDO study): a randomized double blind trial. Lancet 2002;360(9328):196–202.

[42] Nieminem MS, Akkila J, Hasenfuss G, et al. Hemodynamic and neurohormonal effects of continuous infusion of levosimendan in patients with congestive heart failure. J Am Coll Cardiol 2000;36(6):1903–12.

[43] Gerk AMR, Fonseca AG, Andrade ARV. Levosimedan: a new alternative for managing cardiac insufficiency in intensive care. Rio de Janeiro: Rev SOCERJ; 2005.

[44] Felker GM, Benza RL, Chandler AB. Heart failure etiology and response to milrinone in decompensated heart failure: results from the OPTIME-CHF study. J Am Coll Cardiol 2003;41(6):997–1003.

[45] Cohn JN, Archibald DG, Ziesche S, et al. Effects of vasodilator therapy on mortality in chronic congestive heart failure. Results of a Veterans Administration Cooperative Study. N Engl J Med 1986;314(24):1547–52.

[46] The CONSENSUS Trial Study Group. Effects of enalapril on mortality in severe congestive heart failure. Results of the Cooperative North Scandinavian Enalapril Survival Study. N Engl J Med 1987;316(23):1429–35.

[47] Cohn JN, Johnson G, Ziesche S, et al. A comparison of enalapril with hydralazine-isosorbide dinitrate in treatment of chronic congestive heart failure. N Engl J Med 1991;325(5):303–10.

[48] Taylor AL, Ziesche S, Yancy C, et al. African-American Heart Failure Trial Investigators. Combination of isosorbide dinitrate and hydralazine in blacks with heart failure. N Engl J Med 2004;351 (20):2049–57.

[49] The SOLVD Investigators. Effects of enalapril on survival in patients with reduced left ventricular ejection fractions and congestive heart failure. N Engl J Med 1991;325(5):293–302.

[50] Cohn JN, Tognoni G. A randomized trial of the angiotensin-receptor blocker valsartan in chronic heart failure. N Engl J Med 2001;345:1667–75.

[51] McMurray JJ, Ostergren J, Swedberg K, et al. Effects of candesartan in patients with chronic heart failure and reduced left-ventricular systolic function taking angiotensin-converting-enzyme inhibitors: the CHARM-Added Trial. Lancet 2003;362:767–71.

[52] Granger CB, McMurray JJ, Yusuf S, et al. Effects of candesartan in patients with chronic heart failure and reduced left-ventricular systolic function intolerant to angiotensin-converting-enzyme inhibitors: the CHARM-Alternative Trial. Lancet 2003;362:772–6.

[53] Guazzi M, Vicenzi M, Arena R. PDE-5 inhibition with sildenafil improves left ventricular diastolic function, cardiac geometry, and clinical status in patients with stable systolic heart failure: results of a 1-year, prospective, randomized, placebo controlled study. Cir Heart Fail 2011;4(1):8–17.

[54] Lewis GD, Shah R, Shazad K. Sildenafil improves exercise capacity and quality of life in patients with systolic heart failure and secondary pulmonary hypertension. Circulation 2011;124(2):164–74.

[55] Behling A, Rohde LE, Colombo FC. Effects of 5phosphodiesterase four-week long inhibition with sildenafil in patients with chronic heart failure: a double-blind, placebo-controlled clinical trial. J Card Faill 2008;14:189–97.

[56] Galiè N, Ghofrani HA, Torbicki A, et al. Sildenafil citrate therapy for pulmonary arterial hypertension. N Engl J Med 2005;353:2148–57.

[57] Freitas Jr AF, Bacal F, Oliveira Jr JL. Sildenafil vs. Nitroprussiato de Sódio Durante Teste de Reatividade Pulmonar Pré-Transplante Cardíaco. Arq Bras Cardiol 2012;99(3):848–56.

[58] Hirata K, Adji A, Vlachopoulos C, et al. Effect of sildenafil on cardiac performance in patients with heart failure. Am J Cardiol 2005;96(10):1436–40.

[59] Redfield MM, Chen HH, Borlaug BA. Effect of phosphosdiesterase-5 inhibition on exercise capacity and clinical status in heart failure with preserved ejection fraction. J Am Med Assoc 2013;309(12):1268–77.

[60] McMurray JJ, Packer M, Desai AS, et al. Angiotensin-neprelysin inhibition vs enalapril in heart failure. N Eng J Med 2014;371 (11):993–1004.

[61] Diretriz Brasileira de Insuficiência Cardíaca Crônica—Arquivos Brasileiros de Cardiologia vol. 93 no. 1 supl. 1 São Paulo 2009 [Internet]. Available at: https://doi.org/10.1590/S0066-782X2009002000001 [accessed 27.06.16].

[62] Feitosa GS, Carvalho EM. Sistemarenina-angiotensina e insuficiênciacardíaca: o uso dos antagonistas do receptor da angiotensina I. Rev Bras Hipertens 2000;7(3).

[63] Pfeffer MA, Braunwald E, Moyle LA, et al. Effect of captopril on mortality and morbidity in patients with left ventricular dysfunction after myocardial infarction. N Engl J Med 1992;327(10):669–77.

[64] Santos IS, Bittencourt MS. Insuficiênciacardíaca. Heart failure. São Paulo: Rev Med; 2008. p. 224–31.

Further Reading

[1] Sharma JN, Sharma J. Cardiovascular properties of the Kallikrein-Kinin system. Curr Med Res Opin 2002;18:10–7.

PERCUTANEOUS CORONARY INTERVENTIONS AND CARDIAC SURGERY

40

Postpercutaneous Interventions: Endothelial Repair

Julio Flavio Marchini, Vinicius Esteves, and Pedro A. Lemos

INTRODUCTION

Evolution of interventional cardiology over the years has brought unquestionable clinical benefits to patients undergoing invasive procedures. Rechanneling of occluded arteries and reestablishment or normal myocardial perfusion in vessels with severe atherosclerotic lesions brings, in addition to symptoms relief, improvement in survival.

However, despite all these benefits, percutaneous coronary intervention is directly associated with mechanical endothelial injury. Vascular trauma can result in excessive neointimal hyperplasia (NIH) formation and intra-stent restenosis or even acute stent thrombosis, usually related to severe ischemic events. Therefore, correct understanding of endothelial dysfunction process, as well as its regeneration, caused by the devices used in interventional procedures, is of great relevance both for prevention and for the treatment of adverse cardiac events.

Several histopathological studies showed the presence of monocytes and macrophages in atherosclerotic plaque in all stages of atherosclerotic process [1] (Fig. 40.1A). Atherosclerotic plaque may be composed of a varied amount of extracellular accumulation of lipids (lipid nucleus), calcified nuclei, hematoma and thrombosis areas, and necrotic nuclei. Permeating the plaque are macrophages and foam cells, smooth muscle cells (SMC), lymphocytes and mast cells. Capillaries develop on plaque margins. Plaque is delimited by a fibrous layer of smaller or larger thickness, composed mainly of collagen.

As a consequence of insufflation at high balloons pressures and apposition of rigid stents structures against the vessel wall, there is local vascular lesion with loss of endothelial layer, compression and rupture of atherosclerotic plaque, and lacerations that extend from the internal elastic membrane to the outer layer [2,3]. Platelets and fibrin deposit on nonendothelized surfaces (Fig. 40.1B). There is a local and systemic triggering of inflammatory reactions, leading to cellular response with recruitment of monocytes and macrophages, neutrophils and lymphocytes to arterial wall (Fig. 40.1C and D). The release of cytokines and growth factors stimulates migration and proliferation of muscle cells and fibroblasts, triggering NIH and intra-stent restenosis (Fig. 40.1E).

NEOINTIMA AND REESTENOSIS

The granulation phase or cell proliferation phase is characterized by the release of platelet, leukocyte and smooth muscle cell growth factors. Activated platelets express P-selectin and glycoprotein (GP) Ibα. These factors act on SMC themselves, stimulating proliferation and migration of media to intima on days after the injury. Platelet P-selectin binds to P-selectin GP ligand receptors present in circulating leukocytes and a rolling process begins. During rolling, stimulated by cytokines, leukocytes progress to adhesion by binding of Mac-1 (CD11b/CD18) to GPIbα and fibrinogen-GPIIb/IIIa and other receptors. After adhesion, the leukocytes migrate into atheroma due to tropism from cytokines and growth factors mentioned above. Resulting neointima consists of SMC, macrophages and extracellular matrix

591

composed of hyaluronium, fibronectin, osteopontin, and vitronectin. In the medium to long term, there is a remodeling of extracellular matrix with degradation and new synthesis of matrix proteins and impoverishment of cells present in the neointima (Fig. 40.1F).

In addition to NIH growth that may lead to restenosis, the vascular lumen may also "re-narrow" due to a neoatherosclerosis process. Histologically, it is similar to the initial atheromatosis process, but occurs late after NIH [4]. It is characterized by a cluster of macrophages covered by a lipids layer.

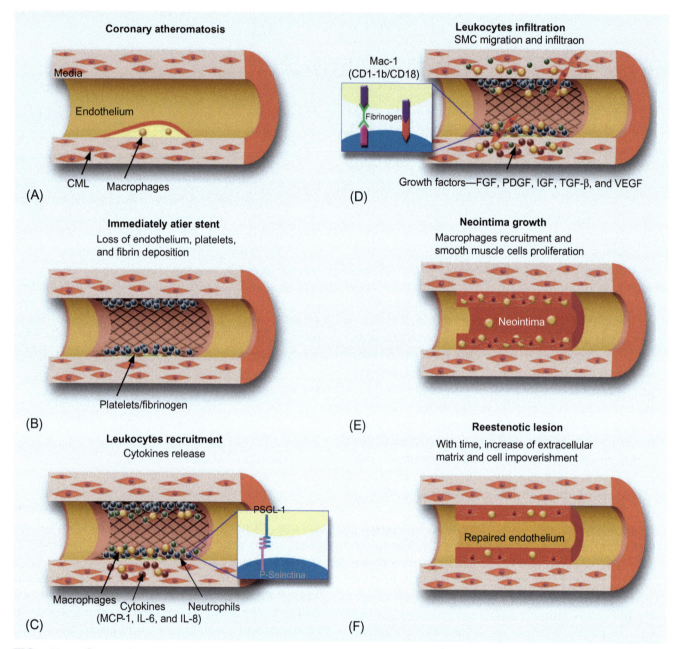

FIG. 40.1 Successive events in restenosis development. (A) Preintervention atherosclerotic coronary lesion. (B) Immediate result after stent implantation with loss of endothelium and platelets and fibrin deposition. (C) and (D) Leukocyte recruitment and infiltration, SMC proliferation and migration in the days following the lesion. (E) Neointimal thickening in weeks after lesion with continued proliferation of CML and monocyte recruitment. (F) In the long term, neointima converts from a cells-rich plaque to an extracellular matrix-rich plate and poor in cells. *SMC*, smooth muscle cells. Adapted from Welt FG, Rogers C. Inflammation and restenosis in the stent era. Arterioscler Thromb Vasc Biol 2002;22:1769–76.

These may be covered by necrotic or calcified tissue. This accumulation of macrophages may progress to formation of fibroatheromas and may be observed in the lumen or in deeper neointima layers. Neoatherosclerotic plaque necrotic nucleus contains a cellular debris and occasionally presents hemorrhagic processes with fibrin deposit, in generally with origin of luminal face fissures and ruptures. Furthermore, additional infiltration of macrophages in neointima results in the formation of thin cap atheroma, which often causes breakage of plates located within the stent. Calcifications can also be observed in neointimal, especially in patients undergoing stent implantation with longer evolution.

Below we describe the effects of devices intervention on coronary endothelium, from the use of balloons to application of the bioabsorbable housing.

REPAIR PROCESS

Balloon Angioplasty

After lesion process of vascular layers by percutaneous coronary intervention, repair process depends fundamentally on the device used. Balloon angioplasty provides modest acute light gain and still suffers from significant negative remodeling (Fig. 40.2A). However, NIH is relatively small, about 0.3–0.4 mm (Fig. 40.2B) [2,5]. The damage is not limited to the balloon contact area, it also extends a few millimeters in both directions. These coronary segments may also have negative remodeling and NIH.

The restenosis mechanism involves the activation of endothelial cells, which begin to express cell adhesion molecules like E-selectin and VCAM-17 [6] (vascular cell adhesion molecule). Cytokines, such as MCP-1 (monocyte chemoattractant protein-1) and IL-8, begin to be expressed even hours after intervention. A neutrophilic infiltrate occurs, which starts 30 min later, with a peak at 6 h. The more important this infiltrate, the greater the proliferation of SMC [8]. Neutrophils concentrate mainly in adventitia, which promote collagen synthesis and tissue contraction [7]. The CCR2 receptor blockade (MCP-1) does not alter remodeling process in balloon angioplasty, but a blockade of β2 integrin of CD18 beta subunit, related to neutrophil recruitment, reduces NIH with balloon [9].

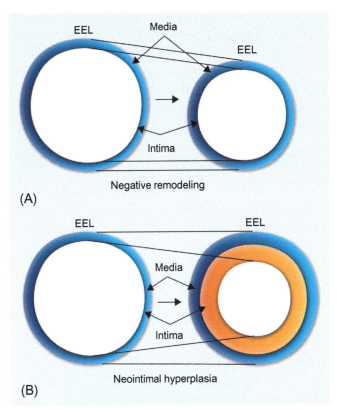

FIG. 40.2 Processes that contribute to the reduction of luminal area following angioplasty. (A) Negative remodeling [6] with reduction of area bounded by external elastic lamina. (B) Neointimal hyperplasia, which is an increase of atherosclerotic plaque area added to medium area. Remodeling is defined by the ratio of the area delimited by external elastic lamina (EEL) on the lesion and by EEL area in adjacent healthy tissue. Neointimal hyperplasia is defined by transverse area of plaque plus medias delimited externally by EEL.

The endothelium is reestablished by contiguity and by implantation of progenitor endothelial cells. Local signals such as SDF-1 (stromal cell-derived factor-1) derived from activated platelets recruit endothelial progenitor cells from the bone marrow to vascular lesion sites [10,11]. In general, the endothelium reestablishes in less than 30 days. This repair process may not be completely benign and, depending on local conditions, SDF-1 may also contribute to local inflammation. SDF-1 may contribute to monocyte recruitment. In addition, it has also been observed that progenitor endothelial cells have the ability to differentiate into SMC in the neointima, contributing with exaggerated reparatory response and restenosis.

ANGIOPLASTY WITH CONVENTIONAL STENT

Through its metallic framework, a stent prevents elastic recoil observed with balloon use. Thus, it achieves a higher gain than that obtained after simple balloon angioplasty. In addition, less negative remodeling occurs. However, NIH is much more exuberant than that observed with balloon angioplasty. Consequently, late loss of stent is worse than with the balloon [12], reaching 0.9–1 mm [13]. However, the balance between initial acute gain and late loss is even favorable to use a stent, compared to balloon use.

Several factors influence NIH:lesion extension, coronary diameter, presence of calcification and bifurcations, lesion expansion after intervention, presence of residual stenosis, and poor apposition of stent struts. There is an important recruitment of monocytes and macrophages in neointima [14], which can be detected at least within 14 days [15]. Typically, NIH takes place in 6–12 months from stent implantation [16]. Unlike intervention with a balloon, early blocking of CCR2 receptor interferes with monocytes recruitment and reduces NIH [9]. Inflammatory cells contribute directly with NIH by mass effect [17], generation of reactive oxygen species [18], secretion of growth and chemotactic factors [19], and production of enzymes such as metalloproteinases and cathepsin [20]. Enzymatic modulation nitrosilation is capable of reducing the NIH and increase endothelialization [21].

ANGIOPLASTY WITH DRUG-ELUTING STENT

A drug-eluting stent can significantly reduce NIH. Two families of drugs were successful for this role:sirolimus-like agents and paclitaxel. The first binds to the 12 binding protein, FK506, and the formed heterodimer binds to mTOR (mechanistic target of rapamycin) preventing its activation. MTOR is from PI3K-related kinase family that participates in critical steps in cell cycle [22]. The result is the interruption of the cell cycle between G1 and S [23]. It affects SMC, monocytes, and macrophages, reducing inflammation, migration of SMC and collagen synthesis and also reducing MCP-1 and IL6 cytokines expression [24,25]. The other drug is paclitaxel, which binds to beta-tubulin, promoting formation and stabilization of microtubules. Cell cycle stagnates between G2 and M phases. Like sirolimus, it interferes with migration and proliferation of CML, it also acts on leukocytes reducing inflammation [26]. The pharmacological stents were very successful in reducing the NIH, reaching 0.1 values to 0.3 mm luminal late loss [27]. In addition to inhibiting SMC and leukocytes, they also inhibit endothelial cells and endothelial progenitor cells [28]. Drug-eluting stents may not cover the endothelial layer by prolonged periods exceeding 40 months with stay exposed stems, and this is correlated with late and very late stent thrombosis [29,30].

NIH can take place even with drug-eluting stents. It is associated with suboptimal stent implantation, as underexpansion, fracture, plaque prolapse, or geographic miss [31]. It further increases the risk of restenosis and presence of diabetes, and treatment involves venous graft, restenotic lesions, and bifurcations [32].

Drug-eluting stents consist not only of medication, but also polymer, which allows controlled drug release. Polymer has been associated with hypersensitivity reactions, late and very late stent thrombosis, development of neoatherosclerosis, and reduction of endothelium-mediated vasodilation [33–37].

Again, atherosclerosis occurs late and is known as neoatherosclerosis. It is a process distinct from NIH with histology compatible with development of a new atherosclerotic plaque and, when associated with an acute event, has vulnerable plaque characteristics [4].

Several changes have been implemented in a new generation of stents to mitigate adverse effects seen in the first generation. Stent struts have become thinner, and the polymer layer is biodegradable in some cases. Another innovation is the application of polymer only on the abluminal side, i.e., the stent face in contact with vessel wall. To illustrate the evolution, one can compare first-generation stent cypher strut, which is 140 µm thick and 13.7 µm polymer layer, with a Xience V stent strut, which is 81 µm thick and 7.8 µm polymer layer.

In fact, the new generation of drug-eluting stents, compared to first-generation stents, maintained the same performance in reducing the NIH and also showed reduction of noncoated struts [38], decrease in fibrin deposition, lowest score of inflammation, less eosinophilic infiltration and giant cells, and lower hypersensitivity reaction. Associated with a lower prevalence of uncoated struts, less inflammation, and lower fibrin deposition, a lower frequency of late and very late thrombosis was observed [38–40]. However, there was no great difference in the incidence of neoatherosclerosis between different stents generations (cobalt chromium-everolimuseluting stent: 29%, sirolimuse-luting stent: 35%, and paclitaxel eluting stent: 19%).

ANGIOPLASTY WITH BIOABSORBABLE SCAFFOLDS

The most recent technology for percutaneous coronary intervention consists of bioabsorbable Scaffolds, formed by polymers of poly-L-lactic acid and polyglycolic acid, among others. Within a year, there is a loss of structural strength of the device and within 2 years dissolution of the stent itself takes place until complete resorption after 3 years.

In the porcine model, within a month from implant, there is no difference with Xience V implant [41]. From 12 months, one begins to observe positive remodeling, i.e., an increase in the luminal area, which is most important until 18 months and then in a slower pace. Initially, luminal area with Xience V is higher, but with positive remodeling the group with bioabsorbable stent reaches and even surpasses luminal area of the group with drug-eluting stents. Lesion and fibrin deposition scores were similar in both groups, but there was a greater presence of inflammatory cells in the absorb group for up to 36 months. Peak inflammation occurs at 18 months with the intensity of the inflammatory response proportional to stent biodegradation [42]. Two years after implantation, restoration of endothelial function, vasomotion, and late light gain can be observed. Studies are under way to test whether bioabsorbables will reduce the presence of neoatherosclerosis [43].

CONCLUSIONS

Interventional cardiology has undergone numerous changes in recent decades, and advances allowed a significant reduction in the incidence of unfavorable clinical outcomes to patients. For many years, in order to reduce complications resulting from percutaneous coronary interventions, it focused first on occlusion and acute vascular thrombosis, then on suppression of cell proliferation and NIH. Then, with progressive improvement of the implantable devices, gradual reduction of late and very late thrombosis was achieved. Intravascular image assessment methods, such as ultrasound and especially the optical coherence tomography, allow accurate assessment of lesion and detailed feedback from the interventional procedure. This advance can be of great importance for a better understanding of NIH and neoatherosclerosis development, and consequently in preventing stents failure mechanisms. New studies and strategies such as use of stents with bioresorbable vascular platforms are a reality and can envision treatments that will restore an artery to a healthy condition with its vasomotion preserved without device presence.

References

[1] Stary HC, Chandler AB, Dinsmore RE, et al. A definition of advanced types of atherosclerotic lesions and a histological classification of atherosclerosis. A report from the Committee on Vascular Lesions of the Council on Arteriosclerosis, American Heart Association. Circulation 1995;92:1355–74.

[2] Welt FG, Rogers C. Inflammation and restenosis in the stent era. Arterioscler Thromb Vasc Biol 2002;22:1769–76.

[3] Costa MA, Simon DI. Molecular basis of restenosis and drug-eluting stents. Circulation 2005;111:2257–73.

[4] Park SJ, Kang SJ, Virmani R, et al. In-stent neoatherosclerosis: a final common pathway of late stent failure. J Am Coll Cardiol 2012;59:2051–7.

[5] Schwartz RS, Topol EJ, Serruys PW, et al. Artery size, neointima, and remodeling: time for some standards. J Am Coll Cardiol 1998;32:2087–94.

[6] Mintz GS, Popma JJ, Pichard AD, et al. Arterial remodeling after coronary angioplasty: a serial intravascular ultrasound study. Circulation 1996;94:35–43.

[7] Okamoto E, Couse T, De Leon H, et al. Perivascular inflammation after balloon angioplasty of porcine coronary arteries. Circulation 2001;104:2228–35.

[8] Welt FG, Edelman ER, Simon DI, et al. Neutrophil, not macrophage, infiltration precedes neointimal thickening in balloon-injured arteries. Arterioscler Thromb Vasc Biol 2000;20:2553–8.

[9] Horvath C, Welt FG, Nedelman M, et al. Targeting CCR2 or CD18 inhibits experimental in-stent restenosis in primates: inhibitory potential depends on type of injury and leukocytes targeted. Circ Res 2002;90:488–94.

[10] Asahara T, Murohara T, Sullivan A, et al. Isolation of putative progenitor endothelial cells for angiogenesis. Science 1997;275:964–7.

[11] Chatterjee M, Gawaz M. Platelet-derived CXCL12 (SDF-1alpha): basic mechanisms and clinical implications. J Thromb Haemost 2013;11:1954–67.

[12] Fischman DL, Leon MB, Baim DS, et al. A randomized comparison of coronary-stent placement and balloon angioplasty in the treatment of coronary artery disease. Stent Restenosis Study Investigators. N Engl J Med 1994;331:496–501.

[13] Morice MC, Serruys PW, Sousa JE, et al. A randomized comparison of a sirolimus-eluting stent with a standard stent for coronary revascularization. N Engl J Med 2002;346:1773–80.

[14] Rogers C, Welt FG, Karnovsky MJ, et al. Monocyte recruitment and neointimal hyperplasia in rabbits. Coupled inhibitory effects of heparin. Arterioscler Thromb Vasc Biol 1996;16:1312–8.

[15] Paolini JF, Kjelsberg MA, Edelman ER, et al. Sustained expression of chemokines monocyte chemoattractant protein-1 and interleukin-8 after stent—but not balloon-induced arterial injury. J Am Coll Cardiol 2000;35:15.

[16] Kimura T, Yokoi H, Nakagawa Y, et al. Three-year follow-up after implantation of metallic coronary-artery stents. N Engl J Med 1996;334:561–6.

[17] Moreno PR, Bernardi VH, Lopez-Cuellar J, et al. Macrophage infiltration predicts restenosis after coronary intervention in patients with unstable angina. Circulation 1996;94:3098–102.

[18] Chen Z, Keaney Jr. JF, Schulz E, et al. Decreased neointimal formation in Nox2-deficient mice reveals a direct role for NADPH oxidase in the response to arterial injury. Proc Natl Acad Sci USA 2004;101:13014–9.

[19] Assoian RK, Fleurdelys BE, Stevenson HC, et al. Expression and secretion of type beta transforming growth factor by activated human macrophages. Proc Natl Acad Sci U S A 1987;84:6020–4.

[20] Sukhova GK, Shi GP, Simon DI, et al. Expression of the elastolytic-cathepsins S and K in human atheroma and regulation of their production in smooth muscle cells. J Clin Invest 1998;102:576–83.

[21] Manica A, Marchini JF, Travers R, et al. S-nitrosoglutathione reductase (GSNOR) modulates reendothelialization and vascular repair. Circulation 2011;124:A15820.

[22] Schmelzle T, Hall MN. TOR, a central controller of cell growth. Cell 2000;103:253–62.

[23] Marx SO, Jayaraman T, Go LO, et al. Rapamycin-FKBP inhibits cell cycle regulators of proliferation in vascular smooth muscle cells. Circ Res 1995;76:412–7.

[24] Suzuki T, Kopia G, Hayashi S, et al. Stent-based delivery of sirolimus reduces neointimal formation in a porcine coronary model. Circulation 2001;104:1188–93.

[25] Poon M, Marx SO, Gallo R, et al. Rapamycin inhibits vascular smooth muscle cell migration. J Clin Invest 1996;98:2277–83.

[26] Zhou X, Li J, Kucik DF. The microtubule cytoskeleton participates in control of beta2 integrin avidity. J Biol Chem 2001;276:44762–9.

[27] Saito S, Nakamura S, Fujii K, et al. Mid-term results of everolimus-eluting stent in a Japanese population compared with a US randomized cohort: SPIRIT III Japan Registry with harmonization by doing. J Invasive Cardiol 2012;24:444–50.

[28] Liu HT, Li F, Wang WY, et al. Rapamycin inhibits re-endothelialization after percutaneous coronary intervention by impeding the proliferation and migration of endothelial cells and inducing apoptosis of endothelial progenitor cells. Tex Heart Inst J 2010;37:194–201.

[29] Joner M, Finn AV, Farb A, et al. Pathology of drug-eluting stents in humans: delayed healing and late thrombotic risk. J Am Coll Cardiol 2006;48:193–202.

[30] Nakazawa G, Finn AV, Joner M, et al. Delayed arterial healing and increased late stent thrombosis at culprit sites after drug-eluting stent placement for acute myocardial infarction patients: an autopsy study. Circulation 2008;118:1138–45.

[31] Castagna MT, Mintz GS, Leiboff BO, et al. The contribution of "mechanical" problems to in-stent restenosis: an intravascular ultrasonographic analysis of 1090 consecutive in-stent restenosis lesions. Am Heart J 2001;142:970–4.

[32] Marchini JF, Manica A, Croce K. Stent thrombosis: understanding and managing a critical problem. Curr Treat Options Cardiovasc Med 2012;14:91–107.

[33] Virmani R, Guagliumi G, Farb A, et al. Localized hypersensitivity and late coronary thrombosis secondary to a sirolimus-eluting stent: should we be cautious? Circulation 2004;109:701–5.

[34] Maekawa K, Kawamoto K, Fuke S, et al. Images in cardiovascular medicine. Severe endothelial dysfunction after sirolimus-eluting stent implantation. Circulation 2006;113:e850–1.

[35] Togni M, Windecker S, Cocchia R, et al. Sirolimus-eluting stents associated with paradoxic coronary vasoconstriction. J Am Coll Cardiol 2005;46:231–6.

[36] Hofma SH, van der Giessen WJ, van Dalen BM, et al. Indication of long-term endothelial dysfunction after sirolimus-eluting stent implantation. Eur Heart J 2006;27:166–70.

[37] Nakazawa G, Finn AV, Vorpahl M, et al. Coronary responses and differential mechanisms of late stent thrombosis attributed to first-generation sirolimus- and paclitaxel-eluting stents. J Am Coll Cardiol 2011;57:390–8.

[38] Otsuka F, Vorpahl M, Nakano M, et al. Pathology of second-generation everolimus-eluting stents versus first-generation sirolimus-and paclitaxel-eluting stents in humans. Circulation 2014;129:211–23.

[39] Palmerini T, Biondi-Zoccai G, Della Riva D, et al. Stent thrombosis with drug-eluting and bare-metal stents: evidence from a comprehensive network meta-analysis. Lancet 2012;379:1393–402.

[40] Raber L, Magro M, Stefanini GG, et al. Very late coronary stent thrombosis of a newer-generation everolimus-eluting stent compared with early-generation drug-eluting stents: a prospective cohort study. Circulation 2012;125:1110–21.

[41] Otsuka F, Pacheco E, Perkins LE, et al. Long-term safety of an everolimus-eluting bioresorbable vascular scaffold and the cobalt-chromium XIENCE V stent in a porcine coronary artery model. Circ Cardiovasc Interv 2014;7:330–42.

[42] Shive MS, Anderson JM. Biodegradation and biocompatibility of PLA and PLGA microspheres. Adv Drug Deliv Rev 1997;28:5–24.

[43] Shibuya M, Cheng Y, Wang Q, et al. TCT-657 Effect of the absorb bioresorbable vascular scaffold (BVS) on features of neoatherosclerosis in familial hypercholesterolemic swine at 1-year follow-up as assessed by in vivo imaging. J Am Coll Cardiol 2014;64:.

41

Stents and the Endothelium

*J. Ribamar Costa, Jr., Daniel Chamie, J. Eduardo Sousa, and
Alexandre Abizaid*

INTRODUCTION

Since the first transluminal coronary angioplasty performed by Gruentzig 30 years ago [1], there have been major technological advances, expansion on the knowledge of the disease and vascular response to percutaneous treatment of atherosclerotic coronary disease, contributing to evolution of devices and to the expansion of its indications. Percutaneous coronary intervention with stent implantation is currently the main form of myocardial revascularization employed.

Recent estimates indicate the annual performance of more than 1 million coronary interventions with stents in the United States. Percutaneous intervention with the use of these metallic endoprostheses results in a variable degree of aggression to the endothelium and underlying vascular layers, causing different responses in every treated segment. This chapter addresses the changes processed in the segment covered with stents, from the vascular response considered physiologic to mechanisms of restenosis and endoprosthesis thrombosis, especially those that release drugs, a keenly discussed subject at the present time and which, at least in part, can be explained by the coronary response to this type of intervention.

POSTSTENT ACUTE ENDOTHELIAL RESPONSE: CELLULAR AND MOLECULAR MECHANISMS

Vascular damage produced by the mechanical action of the stent and the vessel wall's response to it trigger thrombotic coating and acute and chronic parietal inflammation reactions. Subsequent release of cytokines and growth factors induce activation of smooth muscle cells, which migrate and proliferate in the subintimal portion of the vessel wall and produce an extracellular matrix.

The long-term effect of migration and proliferation of smooth muscle cells and production of matrix results in development of neointimal hyperplasia, which, in association with reendothelialization, constitutes the damage repairing response. Specific studies regarding the vascular response process in humans is limited by the impossibility of directly examining tissue in situ, except in situations where atherectomy is used, a procedure that is rarely used in clinical practice. Thus, animal models are required, and much of what is known about the pathophysiology and prevention of restenosis comes from these experiments [2], although today it is clear that the healing time to interventions varies among different animal species (e.g., rabbit and pig) and also between different sites arteries (e.g., illac and coronary artery in humans) [3], in addition to the devices used (e.g., nonpharmacological and pharmacological metallic stents).

Thrombus Formation

Studies in animal models suggest that the earliest events in restenosis process are platelet activation and aggregation in response to endothelial injury caused by balloon and/or the stent. The process of thrombus generation is generally limited to the first 3 days after the intervention (Figs. 41.1 and 41.2A), although some investigators have demonstrated mural thrombus throughout the first month after PCI [4].

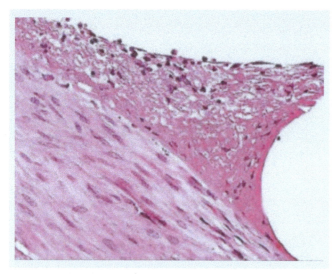

FIG. 41.1 Porcine coronary artery 4 days after implantation of nondrug eluting stents. Accumulation of mural thrombus along the prosthesis struts and acute inflammatory infiltrate is already evident.

Local thrombus formation and platelet activation result in the release of potent mitogens and vasoactive agents, including: platelet-derived growth factor, thrombin, and thromboxane A2. Thrombin induces production and secretion of the platelet-derived growth factor (among other growth factors) by smooth muscle cells, which is why it is believed that thrombin may lead indirectly to neointimal hyperplasia.

The generation of thrombin in vivo is accompanied by platelet activation and release of growth factors, such as serotonin. In vitro, thrombin and serotonin act synergistically inducing proliferation of smooth muscle cells [5]. It is postulated that thrombus-bound thrombin can potentiate the mitogenic effect of serotonin and maintain smooth muscle cells in a proliferative state for prolonged periods.

This thrombotic phase of vascular repair after damage caused by PCI is extended both by application of endovascular radiation (brachytherapy) and by local drug therapies (DES), which delay endothelization [6] and may be responsible for thrombosis at the treated site, if adequate care is not taken to prevent this complication.

Inflammation

Acute and chronic inflammatory cells participate in the vascular repair process after PCI. In experimental models, neutrophils infiltrate the vascular wall at the damaged place within 24 hours [7] (Fig. 41.2B). This acute phase of inflammation is followed by monocyte adhesion and infiltration. The extent of acute inflammatory cell infiltratation is dependent on the arterial substrate and the degree of damage caused by the intervention. Farb et al. observed that the presence of a lipid nucleus and deep arterial damage is significantly more associated with an increased acute infiltrate of inflammatory cells than the absence of vascular damage or an underlying fibrocellular plaque alone [4]. Chronic inflammatory cells also contribute to the release of cytokines, such as interleukin-1 (IL-1) and tumor necrosis factor alpha (TNF-α) and other paracrine substances, which stimulate migration and proliferation of smooth muscle cells [8]. Furthermore, intrastent restenosis is accompanied by infiltration of leukocytes, histiocytes, and giant cells around the stent struts, which contribute further to the proliferation of smooth muscle cells. The ubiquitous presence of these chronic inflammatory cells in human coronary arteries after stent implantation demonstrates the importance of the host response to the foreign body.

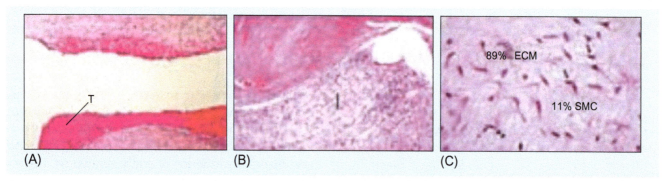

FIG. 41.2 Histological sections of a human coronary artery, after implantation of uncoated stent. Evolutionary phases of the vascular repair process after the damage caused by the instruments. (A) Thrombus formation (T); (B) inflammatory (I); (C) production of extracellular matrix (ECM). *SMC*, smooth muscle cells.

Proliferation of Smooth Muscle Cells

The growth of smooth muscle cells is an important component of the pathophysiology of virtually all forms of vascular disease, including atherosclerosis, arterial hypertension, and vasculopathy after cardiac transplantation. Increasingly, interventional therapies (percutaneous and surgical) have been indicated for treatment of vascular occlusive disease. However, mechanical damage associated with these procedures induces migration of smooth muscle cells from the media to the intima, where they proliferate, transforming themselves into secretory cells and synthesizing an extracellular matrix (Fig. 41.2C), which contributes to the growth of lesions.

Activated smooth muscle cells are able to release various growth factors including platelet-derived growth factor, fibroblast growth factor, angiotensin II, and chemotactic agents that participate in the proliferative process. In addition, the loss of protective factors produced by the endothelium, i.e., nitric oxide and prostacyclin, may accelerate the migration and proliferation of smooth muscle cells [9].

Mechanical trauma induced by devices is an important in trigger of proliferative response [3]. The type of damage in experimental animals, i.e., normal arteries that have receive stent, differs considerably from what takes place in human arteries with atherosclerosis. In normal arteries of pigs, for example, oversizing of the stent regarding the vessel (ratio stent-artery > 10%) can induce proliferative neointimal injury resulting from direct trauma of the stent struts in the middle layer (compression or laceration secondary to barotrauma). In contrast, in humans, about two-thirds of the struts of stents are in direct contact with the atherosclerotic plaque and not to the media; thus, media compression by the stent struts only takes place in one third of cases. Moreover, deep vascular damage also seems to be an important neointimal formation factor after stents in humans.

Extracellular Matrix

Nowadays it is known that the extracellular matrix has not only an inert and structural stabilizing role, but also contributes to vascular cell response to various mechanical and biochemical stimuli. Many of the extracellular components transmit their signals through specialized cell surface receptors, called integrins, and control a variety of cellular functions, such as migration and proliferation in response to mitogens.

This current and more accurate understanding of restenosis vascular biology, with identification of molecular events responsible for cytokine-mediated proliferation of smooth muscle cells and growth factors, allowed a more rational selection of potential therapeutic candidates and their association to a local drug treatment, having as a platform the coronary stent itself. This would inhibit neointimal hyperplasia by the antiproliferative action of a powerful drug.

INTRASTENT RESTENOSIS

Restenosis is defined angiographically as coronary obstruction >50% at the site previously treated and is the vascular response to the device damage caused during revascularization. Intrastent restenosis is usually manifested by gradual recurrence of symptoms of myocardial ischemia within the first 6–8 months after the intervention.

Recurrence of symptoms very early, namely <1 month after stent implantation suggests imperfect results (bad prosthetic expansion, dissection on the edges of the stents, final TIMI III flow) and incomplete revascularization or, when later (i.e., >1 year), progression of coronary disease [10].

About 10% of cases of restenosis after stainless steel stent are manifested as acute coronary syndrome with or without ST segment elevation, and up to 26% of patients have the manifestation unstable angina requiring hospitalization [11], denoting a potentially severe vascular response which requires effective prevention.

The incidence of restenosis ranges from 10% to 30% in patients treated with nondrug eluting stents, depending on clinical, angiographic factors, and genetic predisposition [12]. While significant reductions in the rates of restenosis occurred after introduction of drug-eluting stents [13], it can still occur in certain specific subgroups.

As described earlier, coronary restenosis is not, as opposed to initial propositions, an example of accelerated atherosclerosis, but arises from a temporal and pathophysiologically distinct process [14]. The vascular response following coronary intervention constitutes a sequence of complex events, depending not only on clinical aspects, but also on the device used (balloons, stents), on the technique (prosthesis release pressure, full coverage of the

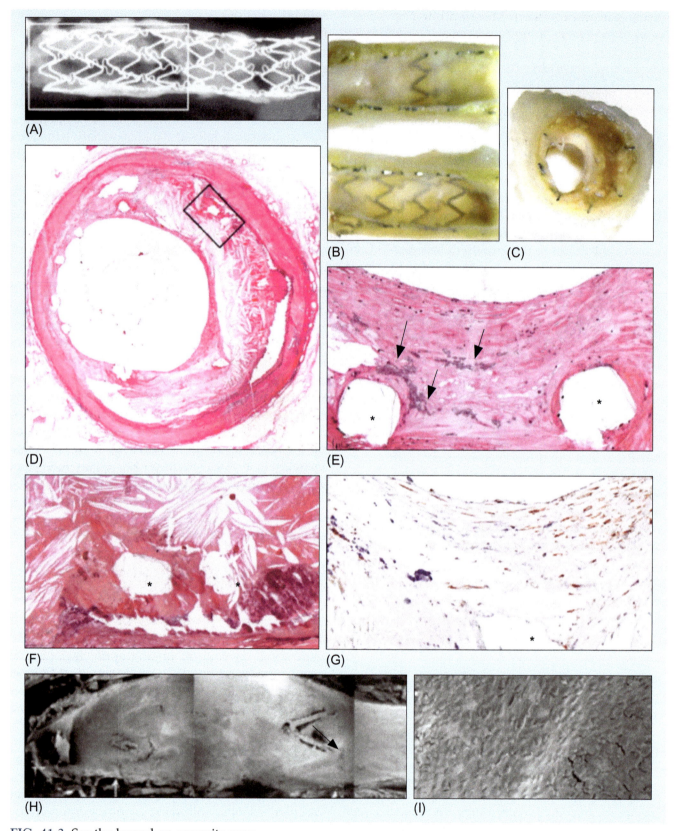

FIG. 41.3 See the legend on opposite page.

lesion, perfect expansion of the stent), and characteristics of the lesions (fibrotic, calcified, elastic) [15]. In a didactic way, it is attributed to three processes responsible for reparative neointimal hyperplasia: (1) acute and chronic changes in the vessel geometry, i.e., acute elastic retraction and cicatricial and late negative remodeling; (2) migration and proliferation of smooth muscle cell; and (3) excessive production of extracellular matrix [16].

Elastic Retraction of Vessel (or Acute Negative Remodeling)

This phenomenon was more frequently observed in balloon angioplasty, especially in calcified lesions and ostial location. It depends on the elastic properties of the vessel wall in the treated segment and can be defined as the difference between the maximum inflating diameter of the balloon and the vessel diameter after insufflation. It is of early occurrence (i.e., within the first 24 hours after the procedure) and should be differentiated from late negative cicatricial remodeling.

Chronic Negative Remodeling

Histopathological studies of vascular fragments obtained at necropsies demonstrate a predominantly fibromuscular response in sites previously treated by balloon angioplasty, due to the migration and proliferation of smooth muscle cells [17]. This mechanism constitutes the main determinant process of restenosis after balloon angioplasty [18], having been virtually abolished with the introduction of metal stents.

Migration of Smooth Muscle Cells and Excessive Production of Extracellular Matrix; Neointimal Hyperplasia

Studies with intracoronary ultrasound brought an unquestionable contribution to the understanding of restenosis after PCI, by demonstrating that, unlike angioplasty balloon, stents virtually eliminate late elastic retraction and negative remodeling; thus restenosis after stent implantation is due exclusively to the formation of neointimal tissue [18].

More recently, with the advent of drug-eluting stents, formation of this neointimal tissue was significantly reduced (Fig. 41.3), resulting in almost total abolition of intrastentrestenosis, which until then was the major limitation of percutaneous procedures.

STENT THROMBOSIS

Stents thrombosis has been a constant concern of the interventional cardiologist, since the introduction of these devices for the treatment of coronary disease [19,20]. In the initial experience with stents, thrombosis rates of up to 24% after 6 months were observed. As a result of thrombosis, anticoagulants were prescribed in association with antiplatelet agents, and hospitalization periods were prolonged, which made clinical applicability of these instruments difficult [21]. Furthermore, clinical consequences to those suffering from stent thrombosis are usually severe, with incidence of myocardial infarction of 60%–70% of cases and short-term mortality of 20%–25% [22].

FIG. 41.3 Morphological appearance of sirolimus-eluting stent 4 years after implantation in human coronary (anatomopathological evaluation). (A) Radiography reveals good expansion of the endoprosthesis in all its extension; the prominent region represents the proximal portion of the stent, cut longitudinally for evaluation by electron microscopy (H and I). The remainder of the endoprosthesis was embedded in resin and cut for sectional review (D). (B) At macroscopic evaluation, one can notice that the vascular light is evident, with a thin neointimal cicatricial layer; the struts are visible throughout the circumference of the vessel. At macroscopy most struts lying next to the vascular lumen and some closely penetrate the adjacent atherosclerotic plaque (C and D). (E) Note that all struts are enclosed in neointimal tissue, with minimal inflammation around. The *arrow* indicates small peri-strut calcification. (F) Note the presence of some cholesterol agglomerates, representing incipient neo-atherosclerotic response. (G) Smooth muscle cells are observed around the struts (*). (H and I) at electron microscopy, it can be observed almost complete neointimal coverage of the stent struts, except for a single strut, with a small side branch (*arrow*). (I) It is observed the presence of mature connective tissue. Modified from Sousa JE, Costa MA, Farb A, et al. Images in cardiovascular medicine. Vascular healing 4 years after the implantation of sirolimus-eluting stent in humans: a histopathological examination. Circulation 2004;110(1):e5–6.

With the evolution of physiopathological knowledge and technical improvement, stent thrombosis was reduced to <1.5% within the first year after coronary intervention, with both nonpharmacological and pharmacological stents. In fact, with postdilatation with high pressures, adequate expansion and complete apposition of the prosthesis rods to the vascular wall was ensured, and interventional pharmacotherapy could be reduced to only two antiplatelet drugs [23].

In the first few years after the introduction of drug-eluting stents, new questions emerged about the higher incidence of thrombosis [24]. Although drug-eluting stents determine significant reduction of restenosis and revascularization of the target vessel, the occurrence of stent thrombosis was not reduced. In clinical studies that determined the superior efficacy of drug-eluting stents, death and infarction rates were similar to those found with nonpharmacological stents after 6–9 months [13].

Other registries and meta-analyses also suggest that rates of very late thrombosis of drug-eluting stents (>360 days), previously rare with nonpharmacological stents, have been identified in a number of cases, in excess of 5000 patients treated when comparing drug-eluting stents vs. uncoated stents [25,26].

It is true that records include patients of higher complexity and situations not originally investigated in clinical trials (populations of "real world"), which may explain the higher rates of the phenomenon. The cause of stent thrombosis is multifactorial, relating not only with clinical and anatomical characteristics of patients and lesions treated, as well as with the technical aspects of the procedure. Biological properties of the drug-eluting stents have been identified as additional factors, namely: induction of tissue thrombogenic factors by carrier polymers and adjuvant drugs; delayed endothelialization of the treated segment; hypersensitivity reactions with positive vessel remodeling and malapposition of the struts; and endothelial dysfunction in adjacent stent segments [27].

Delayed Endothelialization and Endothelial Dysfunction

Preclinical studies conducted in rabbits and pigs treated with first-generation drug-eluting stents, sirolimus-eluting stents (Cypher, Johnson & Johnson) and paclitaxel-eluting (Taxus, Boston Scientific Corporation) showed that both systems may lead to delayed endothelial regeneration with persistence of fibrin on the treated vascular surface [28,29,3]. It is important to mention the fact that, when regenerated, the endothelium formed after use of the stents can frequently become dysfunctional, as demonstrated by the use of acetylcholine in experimental models and changes in response to physical exercise in humans [30–32].

The safety of first-generation drug-eluting stents was questioned mainly by suboptimal biocompatibility of the polymer and late endothelialization of the stent, something imperfect, and may, in rare cases, result in late or very late thrombosis. Both Cypher and Taxus stents use durable thick polymers to transport and release their antiproliferative agents.

In vivo studies using angioscopy, in patients treated with drug-eluting stents, confirmed the findings of preclinical models mentioned. Awata et al. analyzing with serial angioscopy (3, 10, and 21 months) 17 patients with Cypher stents and 11 nondrug eluting stents and found a significant lack of complete endothelialization among patients who received sirolimus-eluting stents, being this difference maintained over the months [33].

Furthermore, the presence of yellow plaques, a marker of instability during angioscopy, was more often observed after use of the Cypher stent. Another frequent response observed in these preclinical models is local inflammation characterized by eosinophilia and occurrence of granulomatosis secondary to the presence of durable polymer used for carrying the drug of these stents [34].

The clinical significance of the findings described above is still controversial, and may have a role in the genesis of late (>30 days) and very late (>360 days) thrombosis.

Because of these observations, a second generation of drug-eluting stents emerged and is currently predominant in clinical practice, represented by stents with new, more biocompatible polymers (Xience prime, Promus element, Rezolute integrity) or even bioabsorbable polymers (Bioametrix, Nobori), which significantly increased the safety profile of these devices.

Incomplete Apposition of the Stent Struts

Incomplete apposition of the stent strut scan be observed both after the use of adjunctive intracoronary brachytherapy and after stent implantation (nondrug eluting and drug-eluting), being detected only with the use of intracoronary ultrasound [35].

This abnormal finding is defined as the separation of at least one stent strut from the intimal surface of the vessel wall, with evidences, by intravascular ultrasound, of blood flow behind the strut(s) in a segment in which the presence of the secondary branch is not observed [36]. It can be classified into three types, according to serial ultrasound monitoring: (1) resolved, when present immediately after stent implantation but absent in evolution; (2) persistent, when present after implant and in evolutionary follow-up; and (3) late or acquired, when the change is observed only in the evolutionary phase, being absent in the evaluation performed immediately after the stent implantation. This classification allows the determination of the mechanism involved in this vascular process: while the persistent incomplete apposition is due to technical or mechanical factors, i.e., disproportion between the size of the implanted stent and the vessel; lower expansion of the prosthesis, especially on its borders, presence of calcium, resolved and late malapposition arise from biological factors. In the resolved malapposition, the growth of atherosclerotic or neointima tissue, behind the strut, leads to filling of the previously observed space.

Although not completely understood, the pathophysiology of late incomplete apposition would include: (1) regression of atherosclerotic plaque behind the stent; (2) evolutionary dissolution of thrombus, present at the time of stent implantation; and, mainly, (3) positive remodeling of the vessel, determined by acute or chronic vascular damage. The concomitant observations of positive remodeling and histopathological evidence of late hypersensitivity reactions, after implantation of drug-eluting stents, corroborate such a hypothesis [37]. Based on the fact that the release of drugs by the carrier polymer remains for a finite period and that the half-life of these drugs in the tissue is a few days, the possibility that the polymer induces such vascular response cannot be dismissed.

Late incomplete apposition after nondrug eluting stents occurs in 4%–5% of cases, usually found in prosthesis edges [38]. In a trial of 881 patients, survival free of major cardiac events (death from cardiac origin, myocardial infarction, and target lesion revascularization) after three years of follow-up was greater than 98% and was similar in groups with and without evidence of late incomplete apposition [39]. Increased incidence has been observed after drug-eluting stents. In a subanalysis of the Sirius trial with 80 patients, 8.7% of the cases presented late incomplete apposition [40]. In a registry with 557 patients treated with paclitaxel- and sirolimus-eluting stents, incidence of late incomplete apposition was 12% [41].

Late incomplete apposition has been identified as one of the factors responsible for the occurrence of stent thrombosis, since blood flow whirls between the stent struts and the vascular wall creates a hemodynamic niche for thrombus formation, which, associated with delayed endothelialization, may result in acute vessel occlusion. Vascular dilatation resulting from positive remodeling of the vessel could also be used to determine the aneurysm formation in the long-term, with the potential for rupture of the vascular wall. In a retrospective study that included 13 patients with very late thrombosis (>1 year) of drug-eluting stents were evaluated with intracoronary ultrasound; the results were compared to 144 patients without thrombosis. Incomplete apposition of the struts was the most prevalent finding in the group with thrombosis (77% vs. 12%; $P < 0.001$) [42]. In another study with 195 patients who underwent ultrasound after implantation and in the 6–8 months follow-up, 10 patients (5.1%) had late incomplete apposition, 2 of whom suffered stent thrombosis [43]. Thus, we can conclude that what is common to these experiences is the propensity to thrombosis to correlate with the degree of positive remodeling.

It is important to note that the phenomenon of late, or acquired, bad apposition rarely has been reported after the use of new generations of drug-eluting stents, which is attributed to better biocompatibility of their components, in particular the new polymers developed to carry antiproliferative drugs.

HYPERSENSITIVITY REACTIONS

Local hypersensitivity reactions have been demonstrated in pathological studies and, when related to stent presence, may result from interaction with the polymer, the drug or the metal platform. Hypersensitivity to some metals, such as nickel and chrome, is reported in the literature, with an incidence of up to 10% after implantation of nondrug-eluting stents and is identified as predisposed to restenosis and not to stent thrombosis [44]. However, delayed local hypersensitivity processes due to the presence of the polymer could result in excessive inflammation, destruction of the medium arterial layer, incomplete apposition of the stent, and aneurysm formation with late thrombosis. In an autopsy record of 40 patients treated with 68 drug-eluting stents for more than 30 days, 14 cases of late thrombosis were identified: in addition to the presence of thrombus, local inflammatory reaction was observed with delayed endothelialization, persistent deposition of fibrin and infiltration of eosinophils and lymphocytes [45].

Systemic hypersensitivity manifestations—such as fever, rash, myalgia, and arthralgia—are most commonly related to drugs used concomitantly and not exclusively to stents: of 5783 cases submitted to the Food and Drug Administration (FDA) in 2003–04, 17 cases were probably or demonstrably caused by stents. Four autopsies revealed intrastent eosinophilic inflammation, absence of endothelialization and thrombosis [46]. In 2067 patients included in the American Registry e-Cypher, 39 (1.9%) had hypersensitivity reactions. Restenosis, thrombosis, or major cardiac events were not observed.

BIOABSORBABLE VASCULAR SUPPORTS (BIORESORBABLE STENTS)

In the last decade, the idea of a transient vascular support, that for a certain period, prevented excessive intimal repairing hyperplasia and, at the same time, prevented the negative remodeling, and was then reabsorbed causing the endothelium to return to its normal functional condition, has gained prominence within the percutaneous approach of coronary disease.

The bioabsorbable vascular supports, also commonly called bioabsorbable stents, would be the devices that could meet these requirements, keeping the obvious potential advantages over current technology of drug-eluting stents: (1) reduction of late and very late stent thrombosis—as the release of drugs and the presence of foreign structures are temporary, remaining only until healing of the vessel (6–24 months), no material (potentially causing

thrombosis) persists in the long term; (2) removal by bioabsorption of rigid components of the vascular wall (fibrosis, calcium etc.), along with the stent—that causes restoration of arterial vasomotricity with adaptive shear stress, late luminal enlargement and positive late remodeling; in addition, it can also reduce the problems of ostium entrapment of the lateral branches, as noted in the permanent metal structures of current stents; (3) improvement in future global atherosclerosis treatment options—the treatment of complex multivessel disease, for example, often results in the long use of multiple drug-eluting stents; in such cases, a new revascularization procedure, through percutaneous or surgical revascularization method, is potentially challenging due to the metal brackets formed by the permanent metal stents implanted previously; and (4) use in the intervention in cardiopediatrics for treatment of stenosis of the pulmonary arteries, aortic coarctation, etc.—once the implanted device will be absorbed, it will not hinder the growth of the child, who may receive several devices during their lifetime, according to the dimensions of the treated vessel.

Efforts to create bioabsorbable stents began about 20 years ago. The first experimental studies, using nonbiodegradable polyethylene terephthalate intertwined mesh stents, in a porcine animal model, were published in 1996 [47]. In the following years, Van der Giessen, Lincoff, and Yamakawa, working separately, were able to prove, in experimental models, the concept of bioabsorption [48,49].

From the current generation of bioresorbable stents, the most commonly used polymer is poly-L-lactic acid (PLLA). This material is already largely employed in clinical practice, being used in resorbable surgical thread, soft tissue implants, orthopedic implants and dialysis filters. Among all clinical programs, without doubt ABSORB (Abbott Vascular) is the one in the most advanced stage, and has already demonstrated, in selected populations and in medium term, highly encouraging results, even having commercial use released in Europe, in some Middle East countries and recently also in Brazil. This bioresorbable stent has 150 mm thick struts directly joined by straight bridges. The ends of the stent have two radiopaque platinum adjacent markers.

The ABSORB device scaffold is composed of PLLA. The coating is made from poly-D,L-lactide

(PDLLA), a random copolymer of D-lactic acid and L-lactic acid with lesser crystallinity than the scaffold. This coating contains the antiproliferative drug everolimus and also promotes the control of its release. The PLLA and PDLLA polymers are completely bioreabsorbable. During the reabsorption, the long chains of PLLA and PDLLA are progressively shortened as the ester bonds between the repeating units of lactide are hydrolyzed, and small particles of less than 2 μm diameter are engulfed by macrophages. Finally, the PLLA and PDLLA polymers are degraded to lactic acid which is metabolized in the Krebs cycle. In a porcine coronary artery model, it was observed reduction in molecular weight over time, and 30% at 12 months, rising to 60% at 18 months and 100% at 24 months postimplant.

This device was evaluated in 101 patients in cohort B of the ABSORB clinical trial. The main findings included low late luminal loss of the ABSORB (0.19 mm), resulting in minimal clinical restenosis (2.4%), and from the safety point of view, there was complete endothelialization of the stent struts in more than 98% of cases, with no thrombosis in up to 4 years of follow-up. At the same time, with different imaging modalities (ultrasound, virtual histology, palpography, and optical coherence tomography), it was documented for the first time in vivo in the complete process of bioresorption of these stents, with restoration of coronary vasomotricity and recovery of vascular remodeling capacity, thus confirming the preclinical findings [50,51].

More recently, the trial AB-SORB II was presented; it is the first randomized comparison between a contemporary, drug-eluting metal stent (Xience with everolimus elution) and the bioresorbable stent ABSORB. This study had as its main investigator Patrick Serruys from the Netherlands, and included 501 patients with up to two coronary lesions of moderate complexity, randomized (2:1) to receive the bioreabsorbable or metal stent. The study was carried out in several centers in Europe, Oceania, and Canada, and has as its primary endpoint the comparison of the change in vascular motricity and between the smallest luminal diameter in the treated segment according to angiographic evaluation (new catheterization) after 3 years of the devices implantation.

Initially, secondary outcomes of the study were presented, including a combination of death, MI, and repeat revascularization due to ischemia, and a relapse of angina at the end of 12 months of follow-up [52]. As main results, similar adverse clinical events in both groups (5% in the ABSORB group vs. 3% Xience group, $P = 0.35$) are highlighted but with lower rates of recurrence of angina in the group treated with the absorbable stent (22% vs. 30%, $P = 0.04$). The thrombosis rate of the devices was relatively low (<1% in both groups) and similar between the two devices.

Although the finding of reduced angina should still be viewed with caution, the rationale for its occurrence is related to smaller changes in the dynamic of coronary flow (sheer stress) caused by a highly flexible polymeric device and with good vascular conformation compared to rigid metal stents. Although not a definitive study, especially when only the results of secondary outcomes are known in a relatively short follow-up period, the ABSORB II findings are nonetheless encouraging, especially in view of the excellent metal stent used as a comparator in this analysis.

Figs. 41.4 and 41.5 illustrate a case of patient treated in Dante Pazzanese Institute of Cardiology with bioabsorbable PLLA stent. This patient was included in a multicenter protocol of the institution and subjected to evaluation with coronary angiography and intravascular imaging methods (OCT) in different periods of his evolution (6, 18, and 36 months). At 36 months, traces of the stent were no longer noticed in the patient's coronary circulation; he underwent an endothelial response stimulation test, which demonstrated restoration of the vasoconstriction and dilation capacity in the segment previously treated with the bioabsorbable endoprosthesis.

CONCLUSIONS

The coronary response to percutaneous interventions represents a multifaceted process for which not all the variables involved are completely known and understood. Normal endothelial repair and excessive neointimal proliferation represent opposite ends of a process that begins with damage to the endothelium by the various instruments used by interventional cardiologists. Only a broader understanding of the mechanisms involved in this process will allow us to increase the success of percutaneous

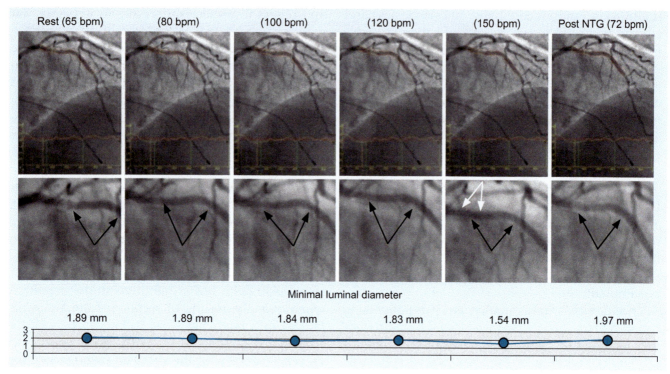

FIG. 41.4 Patient treated with bioabsorbable PLLA stent in Dante Pazzanese Institute. At 36 months after percutaneous coronary intervention (PCI), the patient underwent a new coronary angiography, which confirmed the patency of the vessel lumen and, through stimulation with temporary pacemaker, it was demonstrated, during peak elevation heart rate (150 bpm), the restoration of the vasodilation capacity in the segment previously treated with stent. This type of response, although representing the coronary physiological manifestation before chronotropic stimulus, is not usually seen in segments treated with metal stents, which, due to their permanent structure, limit the late vascular expansion capacity.

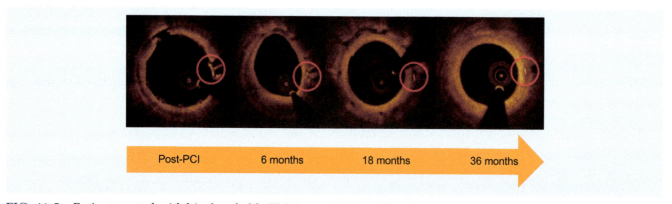

FIG. 41.5 Patient treated with bioabsorbable PLLA stent in Dante Pazzanese Institute and submitted to cinecoronariography in different periods (6, 18, and 36 months) after percutaneous coronary intervention (PCI), as part of the study protocol *First-in-man* (FIM). in vivo EVALUATION with optical coherence tomography (OCT) showing the various stages of healing and absorption of the stent graft. Noteworthy is the fact that at 36 months there is no longer any trace of the stent previously implanted in the coronary artery.

interventions, also ensuring greater safety of the procedure.

Finally, it seems fair to say that we are seeing the beginning of a new era in interventional cardiology, with the incorporation of bioabsorbable stents to clinical practice. If this new technology confirms the potential shown in preclinical evaluations and in the first studies in humans, soon, instead of discussing modalities of myocardial revascularization, we shall be discussing the so-called vascular restoration therapy, perhaps the most ambitious promise of interventional cardiology at the moment.

References

[1] Gruentzig AR, Senning A, Siegenthaler WE. Nonoperative dilatation of coronary-artery stenosis: percutaneous transluminal coronary angioplasty. N Engl J Med 1979;301:61–8.

[2] Schwartz RS, Huber KC, Murphy JG, et al. Restenosis and the proportional neointimal response to coronary artery injury; results in a porcine model. J Am Coll Cardiol 1991;19:267–74.

[3] Finn AV, Nakazawa G, Joner M, et al. Vascular responses to drug eluting stents: importance of delayed healing. Arterioscler Thromb Vasc Biol 2007;27(7):1500–10.

[4] Farb A, Sangiorgi G, Carter AJ, et al. Pathology of acute and chronic coronary stenting in humans. Circulation 1999;99:44–52.

[5] Pakala R, Benedict C. Synergy between thrombin and serotonin in inducing vascular smooth muscle cell proliferation. J Lab Clin Med 1999;134:659–67.

[6] Cheneau E, John MC, Fournadjiev J, et al. Time course of stent endothelialization after intravascular radiationtherapy in rabbit iliac arteries. Circulation 2003;107(16):2153–8.

[7] Tanguay JF, Hammoud T, Geoffroy P, et al. Chronic platelet and neutrophil adhesion: a causal role for neointimal hyperplasia in in-stent restenosis. J Endovasc Ther 2003;10(5):968–77.

[8] Libby P, Ordovas JM, Birinyi LK, et al. Inducible interleukin-1 gene expression in human vascular smooth muscle cells. J Clin Invest 1986;78:1432–8.

[9] Costa MA, Simon DI. Molecular basis of restenosis and drug-eluting stents. Circulation 2005;111:2257–73.

[10] Chan AW, Moliterno DJ. Restenosis: the clinical issue. In: Topol EJ, editor. Textbook in interventional cardiology. Amsterdam: Elsevier Science; 2003. p. 415.

[11] Chen MS, John JM, Chew DP, et al. Bare metal stent restenosis is not a benign clinical entity. Am Heart J 2006;151:1260–4.

[12] Kastrati A, Schomig A. Predictive factors of restenosis after coronary stent placement. J Am Coll Cardiol 1997;30:1428–36.

[13] Stone GW, Moses JW, Ellis SG, et al. Safety and efficacy of sirolimus- and paclitaxel-eluting coronary stents. N Engl J Med 2007;356:998–1008.

[14] Schwartz RS, Holmes Jr. DR, Topol EJ. The restenosis paradigm revisited: an alternative proposal for cellular mechanisms. J Am Coll Cardiol 1992;20:1284–93.

[15] Mercado N, Boersma E, Wijns W, et al. Clinical and quantitative coronary angiographic predictors of coronary restenosis: a comparative analysis from the balloon-to-stent era. J Am Coll Cardiol 2001;38:645–52.

[16] Welt FG, Rogers C. Inflammation and restenosis in the stent era. Arterioscler Thromb Vasc Biol 2002;22:1769–76.

[17] Schwartz RS, Topol EJ, Serruys PW, et al. Artery size, neointima, and remodeling: time for some standards. J Am Coll Cardiol 1998;32:2087–94.

[18] Mintz GS, Popma JJ, Pichard AD, et al. Arterial remodeling after coronary angioplasty: a serial intravascular ultrasound study. Circulation 1996;94:35–43.

[19] Honda Y, Fitzgerald P. Stent thrombosis: an issue revisited in a changing world. Circulation 2003;108:2–5.

[20] Wang F, Stouffer GA, Waxman S, et al. Late coronary stent thrombosis: early vs. late stent thrombosis in the stent era. Catheter Cardiovasc Interv 2002;55:142–7.

[21] Serruys PW, Strauss BH, Beatt KJ, et al. Angiographic follow-up after placement of a self-expanding coronary-artery stent. N Engl J Med 1991;324:13–7.

[22] Cutlip DE, Baim DS, Ho KK, et al. Stent thrombosis in the modern era: a pooled analysis of multicenter coronary stent clinical trials. Circulation 2001;103:1967–71.

[23] Colombo A, Hall P, Nakamura S, et al. Intracoronary stenting without anticoagulation accomplished with intravascular ultrasound guidance. Circulation 1995;91:1676–88.

[24] McFadden EP, Stabile E, Regar E, et al. Late thrombosis in drug-eluting stents after discontinuation of antiplatelet therapy. Lancet 2004;364:1519–21.

[25] Pfisterer M, Brunner-La Rocca H-P, Buser PT. Late clinical events after clopidogrel discontinuation may limit benefit of drug-eluting stents: an observational study of drug-eluting versus bare-metal stents. J Am Coll Cardiol 2006;48:2592–5.

[26] Nordman A, Briel M, Bucher HC. Mortality in randomized controlled trials comparing drug-eluting vs bare metal stents in coronary artery disease: a meta-analysis. Eur Heart J 2006;27:2784–814.

[27] Jaffe R, Strauss BH. Late and very late thrombosis of drug-eluting stents: evolving concepts and perspectives. J Am Coll Cardiol 2007;50:119–27.

[28] Parry TJ, Brosius R, Thyagarajan R, et al. Drug-eluting stents: sirolimus and paclitaxel differentially affect cultured cells and injured arteries. Eur J Pharmacol 2005;524(1–3):19–29.

[29] Axel DI, Kunert W, Göggelmann C, et al. Paclitaxel inhibits arterial smooth muscle cell proliferation and migration in vitro and in vivo using local drug delivery. Circulation 1997;96(2):636–45.

[30] Togni M, Räber L, Cocchia R, et al. Local vascular dysfunction after coronary paclitaxel-eluting stent implantation. Int J Cardiol 2007;120(2):212–20.

[31] Togni M, Windecker S, Cocchia R, et al. Sirolimus eluting stents associated with paradoxic coronary vasoconstriction. J Am Coll Cardiol 2005;46(2):231–6.

[32] Hofma SH, van der Giessen WJ, van Dalen BM, et al. Indication of long-term endothelial dysfunction after sirolimus-eluting stent implantation. Eur Heart J 2006;27:166–70.

[33] Awata M, Kotani J, Uematsu M, et al. Serial angioscopic evidence of incomplete neointimal coverage after sirolimus-eluting stent implantation, comparison with bare-metal stents. Circulation 2007;116(8):910–6.

[34] Joner M, Finn AV, Farb A, et al. Pathology of drug-eluting stents in humans: delayed healing and late thrombotic risk. J Am Coll Cardiol 2006;48(1):193–202.

[35] Mintz GS, Weissman NJ. Intravascular ultrasound in the drug-eluting stent era. J Am Coll Cardiol 2006;48:421–9.

[36] Mintz GS, Nissen SE, Anderson WD, et al. American College of Cardiology Clinical Expert Consensus Document on Standards for Acquisition, Measurement and Reporting of Intravascular Ultrasound Studies (IVUS). A report of the American College of Cardiology. J Am Coll Cardiol 2001;37:1478–92.

[37] Virmani R, Guagliumi G, Farb A, et al. Localized hypersensitivity and late coronary thrombosis secondary to a sirolimus-eluting stent: should we be cautious? Circulation 2004;109:701–5.

[38] Shah VM, Mintz GS, Apple S, et al. Background incidence of late malapposition after bare-metal stent implantation. Circulation 2002;106:1753–5.

[39] Hong MK, Mintz GS, Lee CW, et al. Incidence, mechanism, predictors, and long-term prognosis of late ISA after BMS implantation. Circulation 2004;109:881–6.

[40] Ako J, Morino Y, Honda Y, et al. Late incomplete stent apposition after sirolimus-eluting stent implantation: a serial intravascular ultrasound analysis. J Am Coll Cardiol 2005;46:1002–5.

[41] Hong MK, Mintz GS, Lee CW, et al. Late stent malapposition after drug-eluting stent implantation: an intravascular ultrasound analysis with long-term follow-up. Circulation 2006;113:414–9.

[42] Cook S, Wenaweser P, Togni M, et al. Incomplete stent apposition and very late thrombosis after drug-eluting stent implantation. Circulation 2007;115:2426–34.

[43] Siqueira DA, Abizaid AA, Costa J de R. Late incomplete apposition after drug-eluting stent implantation: incidence and potential for adverse clinical outcomes. Eur Heart J 2007;28:1304–9.

[44] Koster R, Vieluf D, Kiehn M, et al. Nickel and molybdenum contact allergies in patients with coronary in-stent restenosis. Lancet 2000;356:1895–7.

[45] Nebeker JR, Virmani R, Bennett CL. Hypersensitivity cases associated with drug-eluting coronary stents: a review of available cases from the Research on Adverse Drug Events and Reports (RADAR) Project. J Am Coll Cardiol 2006;47:175–81.

[46] FDS. Information for physicians on Sub-Acute Thromboses (SAT) and hypersensitivity reactions with use of the cordis CYPHER coronary stent. FDA Public Health Web Notification. Rockville, MD: Food and Drug Administration; 20032.

[47] van der Giessen WJ, Lincoff AM, Schwartz RS, et al. Marked inflammatory sequelae to implantation of biodegradable and non-biodegradable polymers in porcine coronary arteries. Circulation 1996;94:1690–7.

[48] Lincoff AM, Furst JG, Ellis SG, et al. Sustained local delivery of dexamethasone by a novel intravascular eluting stent to prevent restenosis in the porcine coronary injury model. J Am Coll Cardiol 1997;29:808–16.

[49] Yamawaki T, Shimokawa H, Kozai T, et al. Intramural delivery of a specific tyrosine kinase inhibitor with biodegradable stent suppresses the restenotic changes of the coronary artery in pigs in vivo. J Am Coll Cardiol 1998;32:780–6.

[50] Serruys PW, Onuma Y, Ormiston JA, et al. Evaluation of the second generation of a bioresorbable everolimus drug-eluting vascular scaffold for treatment of de novo coronary artery stenosis: six-month clinical and imaging outcomes. Circulation 2010;122:2301–12.

[51] Ormiston JA, Serruys PW, Onuma Y, et al. First serial assessment at 6 months and 2 years of the second generation of absorb everolimus-eluting bioresorbable vascular scaffold: a multi-imaging modality study. Circ Cardiovasc Interv 2012;5:620–32.

[52] Serruys PW, Chevalier B, Dudek D, et al. A bioresorbable everolimus-eluting scaffold versus a metallic everolimus-eluting stent for ischaemic heart disease caused by de-novo native coronary artery lesions (ABSORB II): an interim 1-year analysis of clinical and procedural secondary outcomes from a randomised controlled trial. Lancet 2015;385(9962):43–54.

Further Reading

[1] Sousa JE, Costa MA, Farb A, et al. Images in cardiovascular medicine. Vascular healing 4 years after the implantation of sirolimus-eluting stent in humans: a histopathological examination. Circulation 2004;110(1):e5–6.

42

Vascular Disease of the Transplanted Heart: Physiopathology and Therapeutic Options

Alfredo Inácio Fiorelli, Noedir Antonio Groppo Stolf, and Raul Cavalcante Maranhão

INTRODUCTION

Although the first experimental heart transplants were carried out at the beginning of the last century, the great historical achievement was only performed in humans successfully in 1967, by Barnard [1]. The method was introduced in Latin America in the following year by Zerbini, at Hospital das Clínicas, Universidade de São Paulo, also successfully [2]. However, this surgery was only routinely incorporated in clinical practice in the 1980s, mainly due to better rejection control. This was possible thanks, among other factors, to the introduction of cyclosporin in the immunosuppressive regimen and to endomyocardial biopsy, which allowed the identification of histological alterations resulting from rejection before graft dysfunction. Thus, cardiac transplantation has been consolidated as the most durable way of restoring the hemodynamic patterns of patients with irreversible myocardial failure.

The most recent International Registry documents more than 120,000 heart transplants performed worldwide since 1982 and with significant improvement in survival, reaching 81% in the first year and 69% in the fifth year after the operation [3]. Average survival is 13 years for those who overcome the first year after transplantation [4]. In the immediate follow-up, the main causes of death are acute dysfunction of the graft and rejection. In the late phase, when the graft is already adapted to the new vascular territory, vascular disease of the graft, neoplasias and kidney failure become the most common causes of death [3–5].

In the relationship between graft versus host immune phenomena deserve special attention, since the receptor recognition of the alloantigens of the donor heart as foreign proteins is the main reason for triggering the rejection mechanisms [6]. This process begins at the first exposure of the alloantigens of the graft to the immune elements of the receptor— that is, immediately after the reperfusion of the newly transplanted heart [7–10].

The purpose of this chapter is to analyze the physiopathology of vascular graft in the transplanted heart, with an emphasis on the endothelium, as well as exploring future treatment options.

TYPES OF REJECTION

Rejection is a phenomenon of immune origin that occurs due to the interaction between the donated heart and the recipient, has a dynamic and interactive character; from the didactic point of view, it is usually divided into apparently distinct entities [6,9,11]. However, what is observed in practice is the coexistence of different forms of aggression with greater predominance of one of them. Therefore, rejection in cardiac transplantation has been classified as follows.

Acute Cellular Rejection

It usually takes place around 4–6 days after transplantation and will accompany the patient throughout his/her life, being more severe and frequent in the first 3–6 months postoperatively. In acute cellular rejection, the antigen-presenting cells, directly or indirectly, carry the immune message of the graft to the T lymphocyte, a phenomenon known as allorecognition. In this process, T lymphocyte membrane is bombarded by multiple immune stimuli, which activate different effectors, especially calcineurin, which through interleukin-2, promotes the clonal expansion of T lymphocytes, leading to the production of different cell clones and enzymes [4,5,7,8,12,13]. The episodes of acute rejection are usually identified through routine endomyocardial biopsies, which guide the modulation of immunosuppressants [14–19]. The main elements that participate in the rejection phenomenon are as follows.

1. *T Helper lymphocytes (CD4—T-helper lymphocytes)*: identify antigens on the membrane of cells undergoing phagocytosis through macrophages and, thereby, activate specific immunity of the body.
2. *T Cytotoxic lymphocytes (CD8—T-killer lymphocytes)*: have the ability to induce lysis of target cells, in this case the graft cells.
3. *B lymphocytes*: are responsible for humoral immunity due to production of antibodies against foreign antigens and may give rise to plasma cells or to memory cells.
4. *Natural cytotoxic cells (natural killer cells)*: are granular lymphocytes that destroy target cells by adherence, similar to T-killer lymphocytes (CD8).
5. *Proliferation or rapamycin target enzyme signals (mTOR—Target of Rapamycin)*: regulates the messenger RNA transcript, acting in the growth, proliferation, motility, survival and lymphocyte protein synthesis

Antibody-Mediated Rejection

This is another modality of immune response, which usually has a more severe course because there are already preformed circulating antibodies in the receptor against the HLA system alloantigens (*Human Leukocyte Antigens*) of the graft. This is an extremely catastrophic situation, leading to acute dysfunction of the transplanted organ and the immunosuppressors have no immediate effect, since the antibodies are already formed and circulating [6,11,14,15,20].

The time of installation of this rejection modality is variable, and may occur within a few hours or immediately after reperfusion of the transplanted heart. The severity and precocity of aggression are linked, among other factors, to the concentration and affinity of the circulating antibodies to the endothelium of the graft, which represent the first interface with the immune elements of the receptor. The immune response triggers activation of different cascade systems, leading to progressive obstruction of the graft coronary arteries and their dysfunction. A therapeutic alternative is the use of plasmapheresis to try to clear the circulating antibodies to the maximum, associated with aggressive immunosuppression in order to inhibit new antibodies formation [20–22].

As a precautionary measure to prevent this serious event, previous knowledge of the potential reactivity of the recipient to a panel of lymphocytes and the prospective knowledge of the lymphocytes crossmatch may be useful. Accordingly, it is possible to allocate the donated hearts in a more rational and safe way to the most suitable recipients [20–23].

VASCULAR DISEASE OF THE GRAFT

Vascular disease of the graft represents a type of chronic rejection that leads to progressive obstruction of the coronary arteries. Endothelial dysfunction precedes the development of vasculopathy, which is characterized by its rapid onset and diffuse intimal proliferation, culminating in the development of stenosis and occlusion of small vessels [24–26]. This disease was first described by Alexis Carrel, at the University of Chicago, in 1910, in a canine carotid artery allograft. Late obstruction of the coronary arteries was initially described by Lower and collaborators in 1968 in transplanted hearts of dogs and confirmed later in clinical trials carried out by Thomson et al. in 1969 [6,27].

This immune phenomenon has received different denominations due to its multifaceted characteristics, such as posttransplant atherosclerosis, chronic rejection, transplanted heart vasculopathy, accelerated atherosclerosis, graft vasculopathy and others [8,9]. However, the designation vascular disease of the graft began to receive greater acceptance because it expresses more adequately this immunological phenomenon that is common to other solid organ transplanted [14,15,28,29].

Vascular disease of the graft affects epicardial vessels and microcirculation, which makes it difficult to identify the disease by coronary angiography and intravascular ultrasound, and may be suspected from abnormal coronary perfusion studies, physiological studies or by the presence of restrictive diastolic filling patterns. Recently, the International Society for Heart Lung Transplantation proposed a new nomenclature to define vascular disease of the graft based on angiography and also included myocardial function, assessed both by imaging and by the physiological behavior of the heart (Table 42.1) [14,15,28].

Incidence

The importance of this entity can be observed in the comparative analysis of the survival curves presented by the Heart Transplant Registry of the International Society for Heart and Lung Transplantation, where patients who develop this vasculopathy have a higher mortality rate regarding those who do not [3].

The incidence of the disease varies according to the form and how the diagnostic investigation is performed. This incidence increases when the investigation is accompanied by anatomopathological examination at necropsy. Vascular disease of the graft affects both sexes and patients of different ages, whose initial manifestations may occur in the first months after transplantation [3,30–32].

Vascular disease of the graft is responsible for 17% of deaths and can be detected already from the first year after transplantation, reaching, in the third year, 42% by coronary angiography and 75% by intravascular ultrasonography [24,30,33]. In a previous study at InCor-HCFMUSP and controlled trials with annual coronary angiography for 5 years, the annual incidence of coronary obstruction was 13.6%, 15.0%, 21.1%, 25.0%, and 44.4%, respectively [30,34]. This data is in agreement with those of the International Registry: 8% in the first year, 30% in the fifth year, and 50% in the tenth year after transplantation (Fig. 42.1) [3].

Physiopathology

Vascular disease of the graft is the main late complication after transplantation and is characterized by the persistent immune aggression to the coronary endothelium, limiting long-term survival of patients and the graft itself [24–26]. The endothelium of the coronary arteries represents the first barrier in the presentation of graft antigens to the immune system of the host. Previous lesions of the endothelium and nonimmunological factors have also been considered as coadjuvants in the genesis of the disease [7,35–37].

The endothelium plays a critical role in maintaining vascular tone by detecting stimuli and releasing vasoactive substances that promote vascular contraction and relaxation [38–41]. When this balance is broken, there is predisposition to vasoconstriction, leukocyte adhesion, platelet activation, mitogenesis, prooxidation, thrombosis, defective coagulation, vascular inflammation, and, in the heart transplantation receptor, predisposition to the vascular disease of the graft [26,42–44].

TABLE 42.1 Graft Vascular Disease Nomenclature according to ISHLT [28]

Degree	Severity	Definition
0	Nonsignificant	No detectable angiographic lesion
1	Discrete	Injury of the main trunk <50% or main vessel <70% or other branches with <70% and diffuse lesion in the absence of graft dysfunction
2	Moderate	Single vessel injury <70% or two-branches injury <70% in two systems Trunk injury <50% in the absence of graft dysfunction
3	Severe	Trunk lesion >50% or >two principal vessels >70% or injury >70% in three systems or degree 1 or 2 with graft dysfunction or with physiological restriction of the graft

Graft dysfunction—ejection fraction by echocardiography <45%.
Physiological restraint of the graft—by echocardiography: speed >2, reduction of isovolumetric relaxation time (<60 m) and shortened deceleration time (<150 ms); by hemodynamics: right atrial pressure >12 mmHg, capillary pressure >25 mmHg and cardiac rate <2 L/min/m^2.
ISHLT, *International Society for Heart Lung Transplantation.*

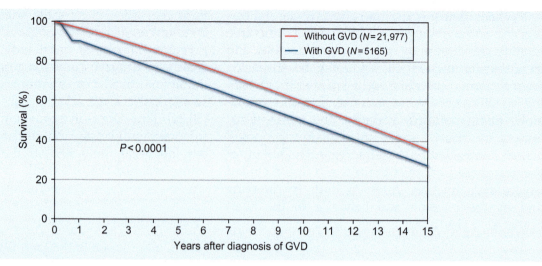

FIG. 42.1 Survival curves after heart transplantation, where it may be noted the negative interference of graft vascular disease (GVD) [3].

Endothelial dysfunction and intimal hyperplasia culminate with vascular remodeling in response to endothelial injury related to transplantation. This phenomenon is potentiated by inflammatory cytokines, growth factors, and chemotactic factors produced by activated endothelial cells. Nitric oxide plays an important role as a mediator in vascular relaxation [45–47].

The immune inflammatory response leads to endothelial dysfunction, migration, and proliferation of smooth muscle cells. Apoptosis and factors produced by the endothelium, as well as proliferation of smooth muscle cells, lead to fibrosis. Migration of these muscle cells together with fibrosis causes concentric intimal thickening of the coronary arteries, which is the main characteristic of vascular disease of the graft [10,39].

The intimate mechanisms involved in the physiopathology of vasculopathy are still uncertain and represent a great challenge for researchers in controlling this complication that is specific to solid organ transplants [24,26,42]. The immune characteristics of the donor and recipient are of particular interest in this context and represent the major factors in the evolution of the disease. It is a multifactorial disease that affects the allograft slowly and progressively [3]. Chronic obstruction of the coronary arteries may lead to slow and progressive myocardial necrosis, which can cause heart failure.

Although acute graft failure after transplantation has diminished in the last two decades, the same cannot be said in relation to its late evolution. In this scenario, vascular disease of the graft appears as the main complication after the first year of transplantation and lacks specific and effective therapy [3,48].

Vascular disease of the graft is a form of accelerated coronary vasculopathy in which the inflammatory response has not yet been fully elucidated; however, its main participant is the immune system. The most likely entry gate is endothelial injury secondary to immune bombardment, which allows direct attack of the subintimal layer and stimulates myointimal proliferation in the artery wall. The inflammatory process extends to the entire arterial wall and, occasionally, to the veins, sparing only the recipient's native vessels [24,35].

In the initial phase, endothelitis characterized by alignment of lymphocytes and macrophages under the vascular endothelium can be observed. The intima becomes thicker due to fibromuscular proliferation and an increased extracellular matrix. At this stage, the internal elastic lamina is still intact and the involvement is limited to the proximal arteries [24,36]. Subsequently, thickening progresses diffusely through coronary vasculature, with appearance of fibroadipose tissue plaques and gradual calcium deposit with the formation of future isolated atheromatous plaques [24,37]. The first intimal changes can already be seen from the sixth month after transplantation [37].

In the late phase, it can be observed that thickening of the intima is diffuse, hyperplasia and concentric fibrosis. A detailed study of the coronary arteries has shown incorporation of lipids and focal plaques of atheromas interspersed with disseminated arteritis.

Thickening of the arteries occurs by the infiltration of mononuclear inflammatory cells in response to alloimmune stimuli or by infection, and in the latter situation, participation of cytomegalovirus deserves special attention. As the disease progresses the medial layer of the arteries may be totally or partially replaced by fibrous tissue. Only vessels with little or no muscle layer can be spared [33,40]. Fig. 42.2 schematically shows the temporal evolution of allograft vasculopathy considering the main mechanisms involved in each of these phases and their manifestations of the disease [48–52].

In the first few months after transplantation, coronary endothelial function and fractional flow reserve display both change, even in asymptomatic patients. Abnormal coronary vasoconstrictor response to acetylcholine precedes the appearance of histological changes with intimal thickening [18]. Abnormal segments of the coronary arteries usually develop intimal thickening earlier than the other [12,53–56]. Early alterations in coronary endothelial function, before the identification of vascular disease of the graft, suggests that the rapid decline in endothelial function is a predictor of poor prognosis. Endothelium-independent abnormal coronary flow reserve, as assessed by dipyridamole, is associated with left ventricular dysfunction during exercise. On the other hand, endothelial dysfunction of the epicardial arteries, evaluated by the microcirculation resistance index, after the second year of transplantation, suggesting that the involvement of the microcirculation by vascular disease of the graft occurs later [57–60].

In vascular disease of the graft, concentric fibromuscular hyperplasia of the intima is one of the characteristics that differentiates it from the lesions produced by atherosclerotic disease. In the latter, lesions tend to be better defined in the proximal segments of the epicardial arteries and take decades to develop [24,57–61].

The morphology of vascular disease of the graft includes three types of lesions: fibromuscular hyperplasia of the intima, lipid deposit, and vasculitis. The most characteristic lesion observed most often is concentric fibromuscular intimal hyperplasia (Fig. 42.3).

It has been erroneously stated that vascular disease of the graft affects the coronary arteries in a centripetal fashion, that is, from vessels of smaller caliber to larger ones. What is actually observed is diffuse vascular involvement, including in the segment of the aorta corresponding to the graft. The aggression to the endothelium is diffuse and without great preference for the size of the vessels and this inflammatory process leads to the earlier obstruction of the microcirculation due to its smaller vessel diameter [25,57–60]. In vascular disease of the graft, intimal hyperplasia and cell proliferation of vascular smooth muscle lead to thickening of epicardial vessel walls, of intramyocardial arteries (50–20 μm), arterioles (20–10 μm) and of the capillaries (<10 μm), that is, the impairment is diffuse [33,57–62].

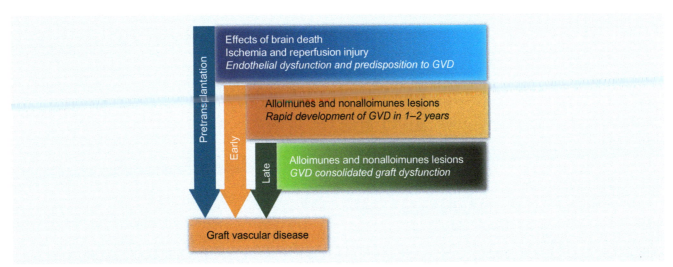

FIG. 42.2   Time evolution of graft vascular disease (GVD) emphasizing in the initial phase the installation of endothelial dysfunction in the graft due to donor characteristics; subsequently, the action of immunological and nonimmunological risk factors that begin in the early stage of transplantation; and, in late phase, the detection of vascular disease of the graft with graft dysfunction.

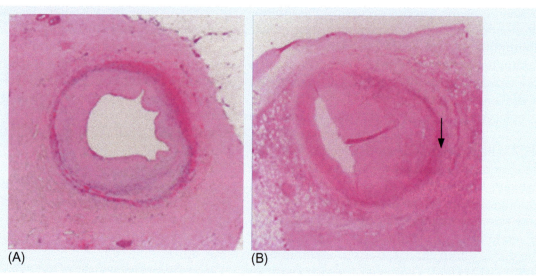

FIG. 42.3 Myocardial microphotography showing coronary arteries with different degrees of graft vascular disease. (A) Usual pattern with concentric thickening of the coronary artery wall due to intimal fibromuscular hyperplasia is noted. (B) Presence of significant involvement of the artery by graft vascular disease *(arrow)* with marked reduction of vessel lumen (hematoxylin-eosin, increase of 100×). *Source*: Alfredo Inácio Fiorelli.

Vascular disease of the graft does not spare even the youngest patients, such as children. The process of evolution of vascular changes follows the same pattern with poor clinical manifestation and the production of nonspecific inflammatory markers [12,29,44,53,63].

Fig. 42.4 shows schematically the major factors involved in the development of vascular disease of the graft, and it is worth highlighting the fundamental role played by endothelial function.

Risk Factors

Immunological and nonimmunological factors involved in the development of vascular disease of the graft do not act alone; in general, they compete in association with the predominance of one of them [24,35,52]. The participation of acute rejection is controversial. However, there are more specific factors that may potentiate the development of the entity, linked to immunosuppression, to the donor and to the recipient [16,17,19]. Among the risk factors stand out those that compromise the integrity of the endothelium [3,9,10,24,29,34,35,45,64,65] (Fig. 42.5), classified as follows:

1. *Regarding the immunological pathway*: hypersensitivity of the recipient with high lymphocytes panel, degree of HLA incompatibility, presence of donor-specific antibodies, presence of non-HLA antibodies, immune response for antigens (myosin and vimentin), rejection episodes, complications of immunosuppressive regimens, corticosteroids, calcineurin inhibitors, polytransfusion, and retransplantation.
2. *Regarding the donor*: etiology of brain death, old age, female gender, prior atherosclerotic disease, shock time and its clinical features.
3. *Regarding preservation of the graft*: hemodynamic conditions at cardiectomy, intravenous catecholamine use, ischemia time, ischemia and reperfusion injuries.
4. *Regarding the recipient*: age, sex, cytomegalovirus infection, *diabetes mellitus*, hypertension, nephropathy, dyslipidemia, smoking, obesity, low testosterone levels, hyperhomocysteinemia and cardiogenic shock before transplantation. Among them are hyperlipidemia and *diabetes mellitus*, with incidence between 50% and 80%.

Clinical Presentation and Diagnosis

Patients remain asymptomatic for many years, since this vascular disease usually develops silently and may manifest with graft failure, ventricular arrhythmias or sudden death. Thus, early clinical diagnosis of vascular disease of the graft is limited by graft denervation and, therefore, occurs in absence of angina [24,55,56].

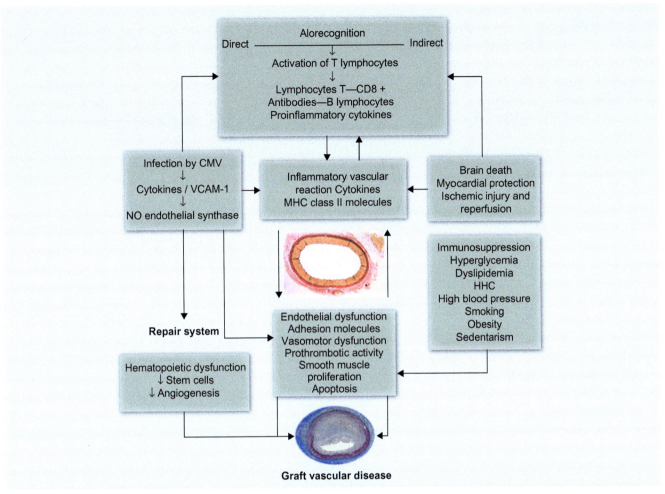

FIG. 42.4 Graft vascular disease. The first step in trigger of graft vascular disease is allorecognition, which occurs after reperfusion of the graft, aggravated by postanoxic endothelial dysfunction. Major cytokines involved in the rejection process are: interleukin-2 (IL-2), interferon-gamma (IFN-γ) and tumor necrosis factor alpha (TNF-a). IL-2 induces proliferation and differentiation of T lymphocytes; IFN-γ activates the macrophages, and TNF-α, alone, is cytotoxic to the transplanted heart. In addition, TNF-α increases the expression of MHC class I molecules, whereas IFN-γ increases MHC expression of both classes I and II. In general, these cytokines can lead to chronic rejection of the graft. IFN-γ and TNF-α induce production of vascular cell adhesion molecule 1 (VCAM-1), promoting monocyte adhesion and passage through the endothelium and, consequently, vascular graft disease. Explosive encephalic death promotes greater release of cytokines, adhesion molecules and increases the expression of class I and II antigens of the MHC system, promoting exacerbated inflammatory reaction in the heart of the potential donor and leading to endothelial dysfunction. *HHC*, human histocompatibility complex.

The dobutamine stress echocardiogram stands out among the noninvasive methods and has provided good results, and is included in analytical protocols because of its high negative predictive value [34,66–68]. The examination requires special attention in candidates who cannot undergo angiography. It presents reasonable sensitivity (80%) and specificity (88%) for vascular disease of the graft, providing predictive value of future events [34,68].

Other noninvasive test includes coronary angiography with a sensitivity of 70% and a specificity of 92% in relation to intravascular ultrasound [63,69–76]. Coronary CT angiography has gained more acceptance, since it allows quantification of calcium in the artery wall. Promising results have been found in the correlation of coronary vasa vasorum volume with plaque evolution in graft vasculopathy when analyzed by intracoronary ultrasonography [77]. This method is limited to vessels between 1 and 1.5 mm in diameter, so that those with the smallest size cannot be reliably detected. Magnetic resonance did not demonstrate sufficient sensitivity to

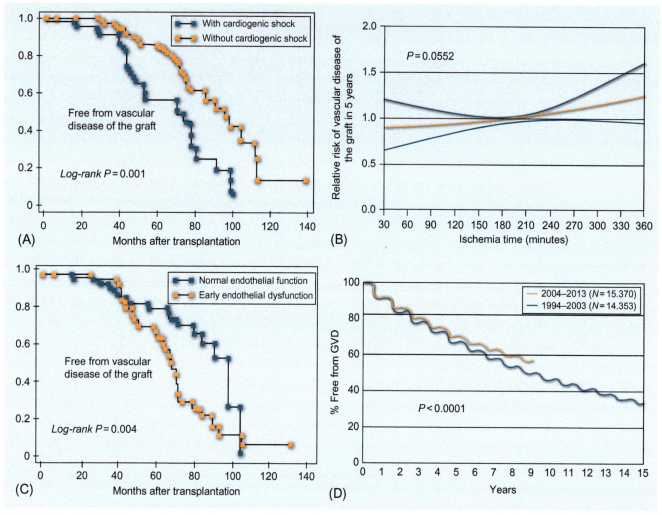

FIG. 42.5 (A) Risk factors for development of the graft vascular disease. Shock state at the time of transplantation [64]. (B) Cold ischemia time promotes endothelial damage and increases the risk of disease [3]. (C) Development of early endothelial dysfunction predisposes to the development of vascular disease of the graft [64]. (D) Incidence of the disease has declined in recent decades due to the preventive measures and changes of immunosuppression. Adapted from Lund LH, Edwards LB, Kucheryavaya AY, et al. The registry of the International Society for Heart and Lung Transplantation: thirty-second Official Adult Heart Transplantation Report—2015; focus theme: early graft failure. J Heart Lung Transplant 2015;34(10):1244–54.

detect vascular disease of the graft; however, there have been attempts to attribute a negative predictive value to the reduction of perfusion reserve [78,79].

Coronary angiography, associated with intracoronary ultrasound, is currently the gold standard in assessing graft impairment [33,73,80]. Nevertheless, early detection of coronary injury by any imaging method is difficult. It is not uncommon to detect severe vascular disease of the graft at necropsy and coronary angiography was considered normal in a recent past [42]. Assessment of coronary flow reserve in conjunction with coronary angiography may be useful for the detection of coronary disease affecting small vessels [17,33,59,60].

Table 42.2 presents a summary of some studies that analyze the endothelial functions in patients undergoing heart transplantation [18,22,26,39, 42,44,59,81,82].

Different protocols have recommended coronary angiographic evaluation after the first year of transplantation; it has also been recommended in elderly donors or with suspected coronary disease prior to acceptance of the organ or, in special situations, in the first few months after transplantation. Serial coronary angiography provides meaningful information about the evolutive prognostic of vascular disease of the graft [15,17,19,80].

In 1997, Barbir et al. showed that luminal narrowing below 25%–50% is a significant predictor of

TABLE 42.2 Parameters That Analyze Endothelial Function and Graft Vascular Disease (GVD) After Heart Transplantation [22]

Method	Result
Coronary flow reserve (CFR)	1. CRF $\leq$ 2.6 predicts risk of major adverse cardiac events in heart transplant and is superior to angiography standard in risk stratification for GVD. 2. Patients using dipyridamole with CFR <2.5 have a significant decline in ejection fraction, compared to individuals with CFR >2.5. 3. Abnormal response to acetylcholine precedes development of GVD and death.
Flow-mediated dilation (FMD)	1. FMD is reduced in young recipients of heart transplants. 2. FMD is reduced after heart transplantation.
Dilatation of the brachial artery by ultrasound	1. There is a progressive decrease in endothelium-dependent vasodilatation up to 12 months after heart transplantation when compared to controls.
Pulse wave at the radial artery	1. The elasticity of small arteries is significantly reduced in heart transplant recipients regarding control and is associated with the presence of GVD.

survival free of cardiac events in heart transplant recipients [71,72]. However, this method has limitations due to lack of sensitivity in the case of vascular disease of the graft [5,33,80,83].

Thus, there are scarce and little effective measures that can be adopted to prevent the advance of myointimal proliferation [15,26,32,81,82,84,85]. Table 42.3 presents the recommendations for diagnosis, prevention, and treatment of graft vascular disease, recommended by the II Directive of the Brazilian Society of Cardiology for Heart Transplant [17].

Laboratory Markers

It is well established that both cellular and humoral responses against HLA are related to the development of chronic rejection [12,13]. However, not only immune-mediated processes but also non immunology related factors play a role in the pathogenesis of the disease (Fig. 42.4). Endothelial damage mediated by both pathways, immune and nonimmune, gained greater importance on investigations into the development of this entity. Obviously, it would be extremely advantageous to predict which patients would develop graft vascular disease. Investigators have been pursuing the utility of blood and tissue markers that could predict the graft vascular disease [86–89].

Recent studies have analyzed different mediators that take part in this complex process [90–93]. Tables 42.4 and 42.5 present a summarized description of these biomarkers and mediators [86–89,92,93]. The ideal marker with higher specificity and sensitivity has not yet been determined.

Maybe in the future, some of these (or others) may become diagnostic markers of the inflammatory reaction of vascular disease of the graft, allowing early identification of the lesion and the best guidance in therapy [94,88].

Starling et al. published in 2016 the results of a large multicenter and observational study, in a cohort of 200 adult recipients of heart transplantation followed during the first year after transplantation [86]. The primary objective of this well designed study was analysis of mortality, graft loss or retransplantation, proven acute rejection by endomyocardial biopsy and graft vascular disease, confirmed by intracoronary ultrasound. These data were correlated with serial determinations of anti-HLA antibodies, auto-antibodies, angiogenic proteins, alloreactivity, and patterns of gene expression in peripheral blood. A significant correlation with older donor grafts and the presence of anti-HLA receptor antibodies was noted. Recipients that had negative serology for cytomegalovirus, regardless of donor standard, had a greater number of episodes of acute rejection and increases in plasma C-factor of the vascular endothelium, combined with the decrease in endothelin-1 and associated with the higher incidence of graft vascular disease. Other biomarkers showed no relationship to the study outcomes. The results of this multicenter study show the challenges to be faced in the search for biomarkers to identify the most susceptible receptors to death and graft vascular disease, observed by other authors [38,49,86,90].

TABLE 42.3   Recommendations for Diagnosis, Graft Vascular Disease (GVD) Prevention and Treatment—II Directive of the Brazilian Society of Cardiology for Heart Transplant [23]

Diagnose	Indication	Evidence level
Class I	Annual dobutamine stress echocardiography to identify patients at higher risk of cardiovascular events after cardiac transplantation.	C
Class IIa	Annual coronary angiography (after first year) for diagnosis and prognosis of patients undergoing heart transplantation.	C
	Annual intravascular ultrasound for diagnosis and prognosis of patients undergoing heart transplantation.	C
Class IIb	Conventional ergometric test/myocardial scintigraphy/angiotomography for diagnosis of ischemia after cardiac transplantation.	C
PREVENTION		
Class I	Guidance of patients for weight reduction, control of hypertension, diabetes and physical activity.	C
	Statins should be used early in all patients, regardless of cholesterol levels, in the absence of contraindications and with monitoring of liver and muscle enzymes.	A
	Diltiazem early as first-line drug for control of systemic arterial hypertension (SAH) and the prevention of GVD.	A
Class IIa	ACE inhibitors to prevent GVD, associated or not with diltiazem.	B
	Acetylsalicylic acid (ASA) for diabetic patients after heart transplantation to prevent GVD and cardiovascular events.	C
TREATMENT		
Class I	Angioplasty with stent placement for proximal lesions > 70% and ischemia documentation.	C
	Surgical revascularization for triple-vessel disease patients, with a favorable distal bed and documented ischemia.	C
Class IIa	Retransplantation for patients with multivessel diffuse GVD with distal bed involvement and significant ventricular dysfunction, not amenable to percutaneous or surgical treatment.	C

SAH, *systemic arterial hypertension; GVD, Graft vascular disease; ACE, angiotensin converting enzyme inhibitors.*

Treatment

More recently, researches have turned to the development of immunosuppressants that have associated antiproliferative action in view of the importance that the graft vascular disease in the evolution of transplantation. It is hoped that it will be possible to control the acute cellular rejection and reduce artery intimal hyperplasia [84,90,95–97].

It has been observed that the progression of graft vascular disease is slower in patients receiving mycophenolate mofetil instead of azathioprine. The combination of cyclosporine and mycophenolate mofetil has been associated with a reduction of 35% in 3-year mortality or loss of transplantation compared to patients treated with ciclosporin and azatioprin [97]. Also, a significantly lower progression of thickening of the intima was observed, indicating that prevention of vascular disease of the graft is greater with mycophenolate mofetil [97,98].

Mycophenolate mofetil reduces adhesion of leukocytes to the endothelium and inhibits smooth muscle cell proliferation. Inhibitors of proliferation sirolimus and everolimus, have the potential to reduce the incidence of microvasculopathy by intimal proliferation and consequently graft vascular disease. Everolimus is an analog of sirolimus and has similar results in reducing the severity and incidence of graft vascular disease [97].

TABLE 42.4 Biomarkers and Graft Vasculopathy [50,53,58]

Biomarker	Action
C-reactive protein	Inflammation marker (>3 mg/L)
Triglycerides/HDL	Resistance to insulin (>3)
BNP	May increase in cellular rejection (≥250 pg/mL)
Von Willenbrand factor	Endothelial dysfunction
Circulating apoptotic endothelial cell and endothelial microparticles	Lesion to endothelial cells
Vascular endothelial growth factor (VEGF) C, VEGF-A and platelet factor	Angiogenesis
High-sensitivity troponin	Cardiomyocyte injury

In this scenario, the actual clinical benefits of antiproliferative mycophenolate mofetil, sirolimus, everolimus, and the embedding of the therapeutic arsenal have been investigated. On the other hand, preventive measures that can slow graft vascular disease progression still represent the cornerstone of treatment of this disease [97,98].

Surgical revascularization of the myocardium or angioplasty with drug-eluting stents are tools used in some cases with serious injuries and have few benefits, since the disease affects the coronary arteries diffusely (Fig. 42.6) [26,99,10]. Retransplantation should also be considered in cases of graft failure, provided that it has no contraindications [3,31,10,100]. The International Registration accounts for about 2700 (2.3%) retransplantations conducted by 2013, being the vascular disease of the graft responsible for the largest contingent [3].

Experimental Investigation

More investigations are needed to identify fully the double interaction of both donor and recipient cells. Some studies have used animal models only or created alternative experimental models. The objectives aimed extrapolation to heart transplantation in humans and get better control of the disease. In recent years, experimental studies increased in number and diversity focusing on potential biomarkers involved.

At our institution, an experimental research line was developed aimed at this purpose [26–29]. The Laboratory of Metabolism and Lipids developed a LDL-cholesterol-based nanoparticle, called LDE, which has the property of carrying drugs. Therefore, LDE-paclitaxel and LDE-methotrexate particles were made under the same principles employed in producing other drugs carried in LDE [101–105]. These drugs were chosen to be of widely used in humans, including in transplants, and have antiinflammatory and antiproliferative properties. The aim is to find a more selective action with less toxicity and increased pharmacological action.

In the experimental model, animal chosen was the rabbit because of its inability to metabolize cholesterol and therefore its propensity to the development of atherosclerotic plaques. Surgical procedure consisted in heterotopic transplantation of the donor heart in the cervical region of the receptor. The main advantages of this model are based on the following items:

1. The model in rabbits has been considered more similar to humans.
2. The lesion develops tubular shape similar to what occurs in humans, unlike mice, which is segmented.
3. Short arteritis development time, between 5 and 6 weeks unlike mice, which require 12–24 weeks.

In this model, graft vascular disease is characterized by the presence of fibrous intimal hyperplasia exacerbated by fatty infiltration. In the genesis of the process a synergism of two events is observed: the immune inflammatory reaction on the coronaries of the graft due to rejection and induced hypercholesterolemia [101,103,104]

These investigations have been studying the interference of drugs associated with nanoparticles in the development of allograft vasculopathy. For that, the following parameters have been studied: histological analysis of the myocardium through hematoxylin-eosin and by immunohistochemistry; morphometric evaluation of the coronary arteries; gene expression of different cell receptors, inflammatory mediators and metalloproteinase analyzed by real-time PCR (Table 42.5).

Results obtained so far indicate that the experimental model shows histological changes in the coronary arteries similar to those observed in vascular disease of the graft in humans; both drugs

TABLE 42.5 Cell Receptors, Inflammatory Mediators and Matrix Metalloproteinases Analyzed by Real-Time PCR [29,38,46,49,86]

Acronym	Description	Activity
CELL RECEPTORS		
CD36	Thrombospondin receptor	Antiangiogenic. Participates in repair of lesions, migration and cellular proliferation
LDRL	Low density lipoprotein receptor	Carrier of captors of low-density lipoproteins
LRP-1	Low density lipoprotein receptor-related protein 1	Participates in the cell endocytic function, lipoproteins metabolism, proteases degradation and lysosome activation.
INFLAMMATORY MEDIATORS		
1L 1β	1β Interleukin	Proinflammatory. Origin of IL-1 by action of caspase-1. Induces the expression of E-selectin, VCAM-1 and ICAM-1.
IL-10	Interleukin-10	Inhibits proinflammatory cytokines, such as interferon gamma, IL-2, IL-3 and TNF-α. Inhibits macrophages and activates B lymphocytes.
IL-18	Interleukin-18	Proinflammatory. Stimulates macrophages, IL-12 and interferon-gamma.
TNF-α	Tumor necrosis factor-α	Migration of interstitial dendritic cells to lymphoid organs, activating T cells effectors.
VCAM-1	Vascular cell adhesion molecule-1	Adhesion of lymphocytes proteins, monocytes (vascular cell adhesion molecule-1), eosinophils and basophils to vascular endothelium.
MCP-1	Monocyte chemotactic protein-1	Diapedesis of monocytes, memory T cells and dendritic cells. Protective effect on cardiac restoration.
METALOPROTEINASES		
MMP-9	Matrix metalloproteinase 9	Facilitates invasion of leukocytes in tissues.
MMP-12	Matrix metalloproteinase 12	Participates in the breakdown of extracellular matrix. Increases in graft vascular disease.

ICAM-1, *intercellular adhesion molecule 1*; PCR, *polymerase chain reaction*.

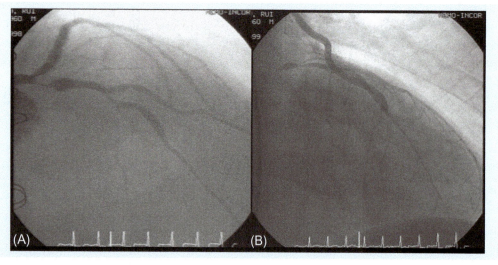

FIG. 42.6 (A) Coronary angiography performed 4 years after heart transplantation. Note the diffuse involvement in left—system injuries in different stages of evolution. (B) Coronary angiography control performed on a patient in the sixth year of a heart transplant who underwent coronary artery bypass graft anastomosis of the left internal thoracic artery to the anterior interventricular coronary artery. Note the severe diffuse involvement of the coronary arteries, which is a characteristic of allovascular disease of the graft and represents a challenge for the indication of surgical treatment. *Source*: Alfredo Inácio Fiorelli.

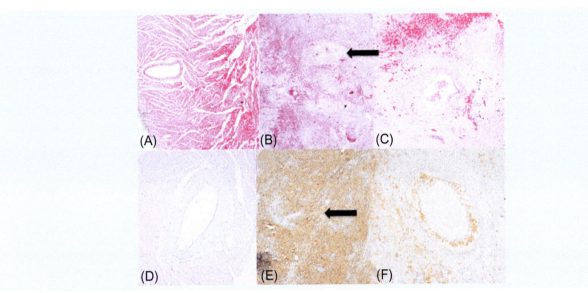

FIG. 42.7 Representative cut of native heart (control group) and transplanted. (A) Control group, there is preservation of myocardial tissue architecture without significant inflammatory reaction and well-defined artery part. (B) Note destruction of myocardial tissue with the presence of intense inflammatory reaction and fatty infiltration in the transplanted heart. The artery of the *arrow* has lost its normal structure and is fully occluded with imprecise limits. (C) Cut transplanted heart, where it can be seen that the myocardium suffered fatty infiltration and inflammatory reaction. However, changes were attenuated by methotrexate as the inflammatory reaction is less intense compared to the control group. It is also noted that the intimal thickening of the artery wall occurred, but still maintaining the lumen of the vessel open and the presence of clots remains in its postmortem inside (hematoxylin-eosin increase of 100×). (D) In the control group the presence of rare macrophages, indicative of little inflammatory reaction was observed. Coronary artery displays integrity of the wall and is permeable. (E) Myocardial transplanted heart, showing dense population of macrophages (stained brown) due to the exacerbated immune reaction, occupying most of the area analyzed. It is noted with great difficulty the presence of a totally occluded artery by intimal hyperplasia, mingling with adjacent tissue *(arrow)*. (F) heart Myocardial transplanted with increased population of macrophages due to immune reaction, but low compared to the control group. The artery is open wall thickened and displays the presence of macrophages, although its structure is better preserved (anti-RAM-11 antibody; increase of 100×). *Source*: Alfredo Inácio Fiorelli.

were effective in blocking vasculitis induced hypercholesterolemia in the rabbit transplanted heart, and the most intense effect of paclitaxel with a more intensive block than the methotrexate (Figs. 42.7–42.9). The same has been observed in significant mitigation in relative gene expression of different inflammatory factors [102–104].

Based on these results, clinical trials have been started with the use of methotrexate LDE-nanoparticles in patients with graft vascular disease, with the initial aim of evaluating possible blocking of the disease already installed with this therapeutic modality [103–105]. These studies are still in progress.

CONCLUSIONS

Vascular disease of the graft is a complex immune phenomenon whose pathogenesis is not fully clarified. Most likely the disease involves mechanisms that go beyond the interaction graft and host. The immune system of the receptor surely plays the most important role, since the immune activation of the receptor precipitates allograft immune responses, which ultimately leads to vascular damage. Immune cells derived from the donor and the transplanted heart are capable of increasing the immune response of the receptor, but apparently they are not capable of inducing independently vascular disease of the graft. Most likely, there is some kind of synergistic interaction between recipient and donor cells, which accelerates the pathogenesis of the disease.

In heart transplant endothelial function suffers constant attack by the immune response to the graft, resulting in injury that is installed in the coronary arteries. The myointimal proliferation that develops chronic progressive form is the principal limiting factor in a long-term transplant. Early clinical

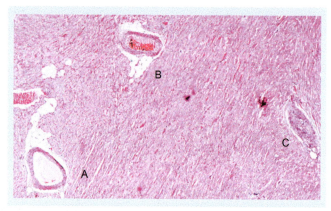

FIG. 42.8   Representative section of transplanted heart of the rabbit treated with LDE-methotrexate. The myocardium shows inflammatory reaction with intense fatty infiltration; however, it is attenuated by the action of methotrexate. In this section, it can be observed the presence of three coronary arteries in different stages of graft vascular disease, within partially conserved myocardium. Fatty infiltration and inflammatory reaction were partially blocked. (A) Patent coronary artery with intimal thickening. (B) Graft vascular disease is clearly evident with greater intimal thickening, and stenosis partially occluding the lumen of the vessel and postmortem clot. (C) Artery with the wall almost completely destroyed and obstruction by intimal hyperplasia (hematoxylin-eosin, $100\times$ magnification). *Source*: Alfredo Inácio Fiorelli.

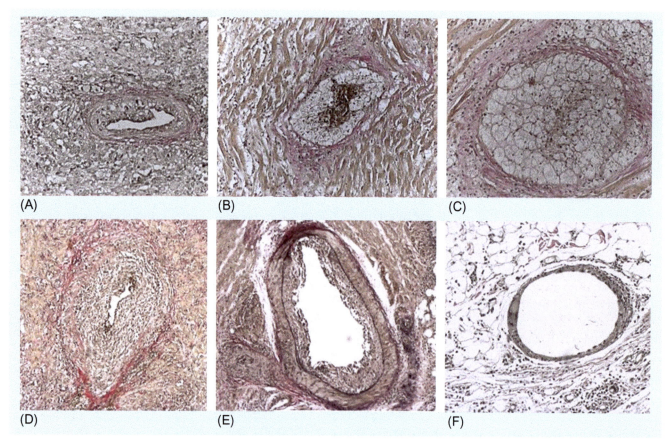

FIG. 42.9   Representative histological sections of the coronary arteries of transplanted hearts. In (A, B, and C) the coronary arteries of animals in the control group are shown exhibiting intense inflammatory reaction with intimal hyperplasia and causing severe obstruction of the vessel. In (D, E, and F), blocking of hyperplasia in coronary produced by the LDE-paclitaxel nanoparticles is observed. Note that the inflammatory response was less pronounced compared to control group. Method: Verhoeff Van Gieson (magnification $\times100$). *Source*: Alfredo Inácio Fiorelli.

diagnosis is difficult due to denervation of the heart, which limits its clinical expression, such as *angina pectoris*. However, microvascular disease has been widely demonstrated by histopathological findings in retransplantation and necropsy. The allograft vasculopathy is fertile ground for different research endeavors since current therapeutic approaches are ineffective. Advances in immunosuppression have aimed more at the control of cellular rejection and less at myointimal proliferation.

References

[1] Barnard CN. The operation. A human cardiac transplantation: an interim report of the successful operation performed at Groote Schuur Hospital Cape Town. S Afr Med J 1968;22:584–96.

[2] Zerbini EJ, Décourt LV. Experience on three cases of human heart transplantation. In: Symposium Mondial Deuxiemé Level Heart transplantation, Annals of the 2nd world symposium, Quebec; 1969. p. 179–82.

[3] Lund LH, Edwards LB, Kucheryavaya AY, et al. The registry of the International Society for Heart and Lung Transplantation: thirty-second Official Adult Heart Transplantation Report—2015; focus theme: early graft failure. J Heart Lung Transplant 2015; 34(10):1244–54.

[4] Zhu D, Cai J. Cardiac transplantation in the United States from 1988 to 2010: an analysis of OPTN/UNOS registry. Clin Transpl 2011;29–38.

[5] Rickenbacher PR, Pinto FJ, Chenzbraun A, et al. Incidence and severity of transplant coronary artery disease early and up to 15 years after transplantation as detected by intravascular ultrasound. J Am Coll Cardiol 1995;25:171–7.

[6] Lower RR, Kontos HA, Kosek JC, et al. Experiences in heart transplantation. Technic, physiology and rejection. Am J Cardiol 1968;22:766–71.

[7] Jukes JP, Jones ND. Immunology in the Clinic Review Series; focus on host responses: invariant natural killer T cell activation following transplantation. Clin ExpImmunol 2012;167:32–9.

[8] Diujvestijn AM, Derhaag JG, Van Breda Vriesman PJ. Complement activation by anti-endothelial cell antibodies in MHC-mismatched and MHC-matched heart allograft rejection: anti-MHC-, but not anti non-MHC alloantibodies are effective in complement activation. Transpl Int 2000;13:363–71.

[9] Velez M, Johnson MR. Management of allosensitized cardiac transplant candidates. Transplant Rev (Orlando) 2009;23(4):235–47.

[10] Jansen MA, Otten HG, de Weger RA, et al. Immunological and fibrotic mechanisms in cardiac allograft vasculopathy. Transplantation 2015;99(12):2467–75.

[11] Burke MM. Late cardiac allograft failure, cardiac allograft vasculopathy, and antibody-mediated rejection: untangling some knots? Am J Transplant 2016;16(1):9–10.

[12] Tran A, Fixler D, Huang R, et al. Donor-specific HLA alloantibodies: impact on cardiac allograft vasculopathy, rejection, and survival after pediatric heart transplantation. J Heart Lung Transplant 2016;35(1):87–91.

[13] Tambur AR, Pamboukian SV, Costanzo MR, et al. The presence of HLA-directed antibodies after heart transplantation is associated with poor allograft outcome. Transplantation 2005;80:1019–25.

[14] Winters GL, Marboe CC, Billingham ME. The International Society for Heart and Lung Transplantation grading system for heart transplant biopsy specimens: clarification and commentary. J Heart Lung Transplant 1998;17(8):754–60.

[15] Billingham ME, Cary NR, Hammond ME, et al. A working formulation for the standardization of nomenclature in the diagnosis of heart and lung rejection: Heart Rejection Study Group. The International Society for Heart Transplantation. J Heart Transplant 1990;9(6):587–93.

[16] Fiorelli AI, Coelho GB, Santos RH, et al. Successful endomyocardial biopsy guided by transthoracic two-dimensional echocardiography. Transplant Proc 2011;43(1):225–8.

[17] Bacal F, Souza-Neto JD, Fiorelli AI, et al. II Diretriz Brasileira deTransplante Cardíaco. Arq Bras Cardiol 2009;94(1 Suppl. 1): e16–73.

[18] Fish RD, Nabel EG, Selwyn AP, et al. Responses of coronary arteries of cardiac transplant patients to acetylcholine. J Clin Invest 1988;81:21–31.

[19] Fiorelli AI, Benvenuti L, Aielo V, et al. Comparative analysis of the complications of 5347 endomyocardial biopsies applied to patients after heart transplantation and with cardiomyopathies: a single-center study. Transplant Proc 2012;44(8):2473–8.

[20] Gonzalez-Stawinski GV, Cook DJ, Chui J, et al. A comparative analysis between survivors and nonsurvivors with antibody mediated cardiac allograft rejection. J Surg Res 2007;142:233–8.

[21] Loupy A, Toquet C, Rouvier P, et al. Late failing heart allografts: pathology of cardiac allograft vasculopathy and association with antibody-mediated rejection. Am J Transplant 2016;16(1):111–20.

[22] Colvin MM, Cook JL, Chang P, et al. Antibody-mediated rejection in cardiac transplantation: emerging knowledge in diagnosis and management: a scientific statement from the American Heart Association. Circulation 2015;131(18):1608–39.

[23] Yamani MH, Taylor DO, Rodriguez ER, et al. Transplant vasculopathy is associated with increased AlloMap gene expression score. J Heart Lung Transplant 2007;26:403–6.

[24] Weis M, Von Scheidt W. Coronary artery disease in the transplanted heart. Annu Rev Med 2000;5(1):81–100.

[25] Segura AM, Buja LM. Cardiac allograft vasculopathy: a complex multifactorial sequela of heart transplantation. Tex Heart Inst J 2013;40(4):400–2.

[26] Avery RK. Cardiac-allograft vasculopathy. N Engl J Med 2003;349:829–30.

[27] Thomson JG. Production of severe atheroma in a transplanted human heart. Lancet 1969;22:1088–92.

[28] Mehra MR, Crespo-Leiro MG, Dipchand A, et al. International Society for Heart and Lung Transplantation working formulation of a standardized nomenclature for cardiac allograft vasculopathy—2010. J Heart Lung Transplant 2010;29(7):717–27.

[29] Fenton M, Simmonds J, Shah V, et al. Inflammatory cytokines, endothelial function and chronic allograft vasculopathy in children: an investigation of the donor and recipient vasculature after heart transplantation. Am J Transplant 2016;16(5):1559–68.

[30] Fiorelli AI, Stolf NAG, Graziosi P, et al. Incidencia de coronariopatia após o transplante cardíaco ortotópico. Rev Bras Cir Cardiovasc 1994;9:69–80.

[31] Meyer DM, Rogers JG, Edwards LB, et al. The future direction of the adult heart allocation system in the United States. Am J Transplant 2015;15(1):44–54.

[32] Andrew J, Macdonald P. Latest developments in heart transplantation: a review. Clin Ther 2015;37(10):2234–41.

[33] Kobashigawa JA, Tobis JM, Starling RC, et al. Multicenter intravascular ultrasound validation study among heart transplant recipients: outcomes after five years. J Am Coll Cardiol 2005;45:1532–7.

[34] Bacal F, Veiga VC, Fiorelli AI, et al. Analysis of the risk factors for allograft vasculopathy in asymptomatic patients after cardiac transplantation. Arq Bras Cardiol 2000;75:421–8.

[35] Khan UA, Williams SG, Fildes JE, et al. The physiopathology of chronic graft failure in the cardiac transplant patient. Am J Transplant 2009;9(10):2211–6.

[36] Fishbein A. Predicting the development of cardiac allograft vasculopathy. Cardiovasc Pathol 2014;23(5):253–60.

[37] Labarrere CA, Nelson DR, Faulk WP. Myocardial fibrin deposits in first month after transplantation predict subsequent coronary artery disease and graft failure in cardiac allograft recipients. Am J Med 1998;105:207–13.

[38] Singh N, Heggermont W, Fieuws S, et al. Endothelium-enriched microRNAs as diagnostic biomarkers for cardiac allograft vasculopathy. J Heart Lung Transplant 2015;34(11):1376–84.

[39] Colvin-Adams M, Harcourt N, Duprez D. Endothelial dysfunction and cardiac allograft vasculopathy. J Cardiovasc Transl Res 2013; 6(2):263–77.

[40] Valantine HA. Cardiac allograft vasculopathy: central role of endothelial injury leading to transplant "atheroma" Transplantation 2003;76:891–9.

[41] Hirohata A, Nakamura M, Waseda K, et al. Changes in coronary anatomy and physiology after heart transplantation. Am J Cardiol 2007;99:1603–7.

[42] Hollenberg SM, Klein LW, Parrillo JE, et al. Coronary endothelial dysfunction after heart transplantation predicts allograft vasculopathy and cardiac death. Circulation 2001;104:3091–6.

[43] Davis SF, Yeung AC, Meredith IT, et al. Early endothelial dysfunction predicts the development of transplant coronary artery disease at 1 year posttransplant. Circulation 1996;93:457–62.

[44] Asante-Korang A, Amankwah EK, Lopez-Cepero M, et al. Outcomes in highly sensitized pediatric heart transplant patients using current management strategies. J Heart Lung Transplant 2015; 34(2):175–81.

[45] Kim MS, Kang SJ, Lee CW, et al. Prevalence of coronary atherosclerosis in asymptomatic healthy subjects: an intravascular ultrasound study of donor hearts. J Atheroscler Thromb 2013; 20(5):465–71.

[46] Lattmann T, Hein M, Horber S, et al. Activation of proinflammatory and anti-inflammatory cytokines in host organs during chronic allograft rejection: role of endothelin receptor signaling. Am J Transplant 2005;5:1042–9.

[47] Singh N, Van Craeyveld E, Tjwa M, et al. Circulating apoptotic endothelial cells and apoptotic endothelial microparticles independently predict the presence of cardiac allograft vasculopathy. J Am Coll Cardiol 2012;60(4):324–31.

[48] Chantranuwat C, Blakey JD, Kobashigawa JA, et al. Sudden, unexpected death in cardiac transplant recipients: an autopsy study. J Heart Lung Transplant 2004;23(6):683–9.

[49] Verma S, Buchanan MR, Anderson TJ. Endothelial function testing as a biomarker of vascular disease. Circulation 2003;108:2054–9.

[50] Caforio ALP, Tona F, Fortina AB, et al. Immune and nonimmune predictors of cardiac allograft vasculopathy onset and severity: multivariate risk factor analysis and role of immunosuppression. Am J Transplant 2004;4:962–70.

[51] Pratschke J, Neuhaus P, Tullius S. What can be learned from brain-death models? Transpl Int 2005;18:15–21.

[52] Johansson I, Andersson R, Friman V, et al. Cytomegalovirus infection and disease reduce 10-year cardiac allograft vasculopathy-free survival in heart transplant recipients. BMC Infect Dis 2015;15 (1):582.

[53] Kindel SJ, Law YM, Chin C, et al. Improved detection of cardiac allograft vasculopathy: a multi-institutional analysis of functional parameters in pediatric heart transplant recipients. J Am Coll Cardiol 2015;66(5):547–57.

[54] Lima ML, Fiorelli AI, Vassallo DV, et al. Comparative experimental study of myocardial protection with crystalloid solutions for heart transplantation. Rev Bras Cir Cardiovasc 2012;27(1):110–6.

[55] Payne GA, Hage FG, Acharya D. Transplant allograft vasculopathy: role of multimodality imaging in surveillance and diagnosis. J Nucl Cardiol 2016;23:713–27.

[56] Hansson GK. Inflammation, atherosclerosis, and coronary artery disease. N Engl J Med 2005;352:1685–95.

[57] Mannam VK, Lewis RE, Cruse JM. The fate of renal allografts hinges on responses of the microvascular endothelium. Exp Mol Pathol 2013;94(2):398–411.

[58] Hollenberg SM, Klein LW, Parrillo JE, et al. Changes in coronary endothelial function predicts progression of allograft vasculopathy after heart transplantation. J Heart Lung Transplant 2004;23:265–71.

[59] Weis M, Hartmann A, Olbrich HG, et al. Prognostic significance of coronary flow reserve on left ventricular ejection fraction in cardiac transplant recipients. Transplantation 1998;65(1):103–8.

[60] Barbir M, Lazem F, Banner N, et al. The prognostic significance of non-invasive cardiac tests in heart transplant recipients. Eur Heart J 1997;18:692–6.

[61] Rahmani M, Cruz R, Granville D, et al. Allograft vasculopathy versus atherosclerosis. Circ Res 2006;99:801–15.

[62] Costanzo MR, Naftel DC, Pritzker MR, et al. Heart transplant coronary artery disease detected by coronary angiography: a multi-institutional study of preoperative donor and recipient risk factors. Cardiac Transplant Research Database. J Heart Lung Transplant 1998;17:744–53.

[63] Tomai F, De Luca L, Petrolini A, et al. Optical coherence tomography for characterization of cardiac allograft vasculopathy in late survivors of pediatric heart transplantation. J Heart Lung Transplant 2016;35(1):74–9.

[64] Lopez-Fernandez S, Manito-Lorite N, Gómez-Hospital JA, et al. Cardiogenic shock and coronary endothelial dysfunction predict cardiac allograft vasculopathy after heart transplantation. Clin Transplant 2014;28(12):1393–401.

[65] Dasari TW, Saucedo JF, Krim S, et al. Clinical characteristics and in hospital outcomes of heart transplant recipients with allograft vasculopathy undergoing percutaneous coronary intervention: insights from the National Cardiovascular Data Registry. Am Heart J 2015;170(6):1086–91.

[66] Spes CH, Klauss V, Rieber J, et al. Functional and morphological findings in heart transplant recipients with a normal coronary angiogram: an analysis by dobutamine stress echocardiography, intracoronary Doppler and intravascular ultrasound. J Heart Lung Transplant 1999;18:391–8.

[67] Eroglu E, D'hooge J, Sutherland GR, et al. Quantitative dobutamine stress echocardiography for the early detection of cardiac allograft vasculopathy in heart transplant recipients. Heart 2008;94:e3.

[68] Bacal F, Abuhab A, Mangini S, et al. Dobutamine stress echocardiography in heart transplant recipients' evaluation: the role of reinnervation. Transplant Proc 2010;42(2):539–41.

[69] Gregory SA, Ferencik M, Achenbach S, et al. Comparison of sixty-four-slice multidetector computed tomographic coronary angiography to coronary angiography with intravascular ultrasound for the detection of transplant vasculopathy. Am J Cardiol 2006;98:877–84.

[70] Sigurdsson G, Carrascosa P, Yamani MH, et al. Detection of transplant coronary artery disease using multidetector computed tomography with adaptative multisegment reconstruction. J Am Coll Cardiol 2006;48:772–8.

[71] Fearon WF, Shah M, Ng M, et al. Predictive value of the index of microcirculatory resistance in patients with ST-segment elevation myocardial infarction. J Am Coll Cardiol 2008;51:560–5.

[72] Barbir M, Lazem F, Bowker T, et al. Determinants of transplant-related coronary calcium detected by ultrafast computed tomography scanning. Am J Cardiol 1997;79:1606–9.

[73] JangI K, Bouma BE, Kang DH, et al. Visualization of coronary atherosclerotic plaques in patients using optical coherence tomography: comparison with intravascular ultrasound. J Am Coll Cardiol 2002;39:604–9.

[74] Garrido IP, García-Lara J, Pinar E, et al. Optical coherence tomography and highly sensitivity troponin T for evaluating cardiac allograft vasculopathy. Am J Cardiol 2012;110:655–61.

[75] Mirelis JG, García-Pavía P, Cavero MA, et al. Magnetic resonance for noninvasive detection of microcirculatory disease associated with allograft vasculopathy: intracoronary measurement validation. Rev Esp Cardiol 2015;68(7):571–8.

[76] Beitzke D, Berger-Kulemann V, Schöpf V, et al. Dual-source cardiac computed tomography angiography (CCTA) in the follow-up of cardiac transplant: comparison of image quality and radiation dose using three different imaging protocols. Eur Radiol 2015;25(8):2310–7.

[77] Sato T, Seguchi O, Ishibashi-Ueda H, et al. Risk stratification for cardiac allograft vasculopathy in heart transplant recipients—annual intravascular ultrasound evaluation. Circ J 2016;80(2):395–403.

[78] Muehling OM, Wilke NM, Panse P, et al. Reduced myocardial perfusion reserve and transmural perfusion gradient in heart transplant arteriopathy assessed by magnetic resonance imaging. J Am Coll Cardiol 2003;42:1054–60.

[79] Colvin-Adams M, Petros S, Raveendran G, et al. Qualitative perfusion cardiac magnetic resonance imaging lacks sensitivity in detecting cardiac allograft vasculopathy. Cardiol Res 2011;2:282–7.

[80] Nogueira LG, Santos RH, Ianni BM, et al. Myocardial chemokine expression and intensity of myocarditis in Chagas cardiomyopathy are controlled by polymorphisms in CXCL9 and CXCL10. PLoS Negl Trop Dis 2012;6(10):e1867.

[81] Gao SZ, Hunt SA, Schroeder JS, et al. Does rapidity of development of transplant coronary artery disease portend a worse prognosis? J Heart Lung Transplant 1994;13(6):1119–24.

[82] Furchgott R, Zawadzki J. The obligatory role of endothelial cells in the relaxation of arterial smooth muscle by acetylcholine. Nature 1980;288:373–6.

[83] Tuzcu EM, Kapadia SR, Sachar R, et al. Intravascular ultrasound evidence of angiographically silent progression in coronary atherosclerosis predicts long-term morbidity and mortality after cardiac transplantation. J Am Coll Cardiol 2005;45:1538–42.

[84] O'Neill BJ, Pflugfelder PW, Singh NR, et al. Frequency of angiographic detection and quantitative assessment of coronary arterial disease one and three years after cardiac transplantation. Am J Cardiol 1989;63:1221–6.

[85] Crespo-Leiro MG, Marzoa-Rivas R, Barge-Caballero E, et al. Prevention and treatment of coronary artery vasculopathy. Curr Opin Organ Transplant 2012;17:546–50.

[86] Starling RC, Stehlik J, Baran DA, et al. Multicenter analysis of immune biomarkers and heart transplant outcomes: results of the clinical trials in organ transplantation-05 study. Am J Transplant 2016;16(1):121–36.

[87] Pham MX, Teuteberg JJ, Kfoury AG, et al. Gene-expression profiling for rejection surveillance after cardiac transplantation. N Engl J Med 2010;362:1890–900.

[88] Crespo-Leiro MG, Stypmann J, Schulz U, et al. Performance of gene-expression profiling test score variability to predict future clinical events in heart transplant recipients. BMC Cardiovasc Disord 2015;15:120.

[89] Crespo-Leiro MG, Stypmann J, Schulz U, et al. Clinical usefulness of gene-expression profile to rule out acute rejection after heart transplantation: CARGO II. Eur Heart J 2016;37:2591–601.

[90] Nogueira LG, Santos RH, Fiorelli AI, et al. Myocardial gene expression ofT-bet, GATA-3, Ror-gt, FoxP3, and hallmark cytokines in chronic Chagas disease cardiomyopathy: an essentially unopposed TH1-type response. Mediators Inflamm 2014;2014:914326.

[91] Otton J, Hayward C, Macdonald P. Gene-expression profiling after cardiac transplantation. N Engl J Med 2010;363(14):1374.

[92] Schneeberger S, Amberger A, Mandl J, et al. Cold ischemia contributes to the development of chronic rejection and mitochondrial injury after cardiac transplantation. Transpl Int 2010;23(12):1282–92.

[93] Sambiase NV, Higuchi ML, Nuovo G, et al. CMV and transplant-related coronary atherosclerosis: an immunohistochemical, in situ hybridization, and polymerase chain reaction in situ study. Mod Pathol 2000;13(2):173–9.

[94] Colvin-Adams M, Agnihotri A. Cardiac allograft vasculopathy: current knowledge and future direction. Clin Transplant 2011;25(2):175–84.

[95] Kilic A, Allen JG, Weiss ES. Validation of the United States-derived Index for Mortality Prediction After Cardiac Transplantation (IMPACT) using international registry data. J Heart Lung Transplant 2013;32(5):492–8.

[96] Kindel SJ, Pahl E. Current therapies for cardiac allograft vasculopathy in children. Congenit Heart Dis 2012;7(4):324–35.

[97] Watanabe T, Seguchi O, Nishimura K, et al. Suppressive effects of conversion from mycophenolate mofetil to everolimus for the development of cardiac allograft vasculopathy in maintenance of heart transplant recipients. Int J Cardiol 2016;203:307–14.

[98] Vecchiati A, Tellatin S, Angelini A, et al. Coronary microvasculopathy in heart transplantation: consequences and therapeutic implications. World J Transplant 2014;4(2):93–101.

[99] Schnetzler B, Drobinski G, Dorent R, et al. The role of percutaneous transluminal coronary angioplasty in heart transplant recipients. J Heart Lung Transplant 2000;19:557–65.

[100] Goldraich LA, Stehlik J, Kucheryavaya AY, et al. Retransplant and medical therapy for cardiac allograft vasculopathy: International Society for Heart and Lung Transplantation registry analysis. Am J Transplant 2016;16:301–9.

[101] Maranhão RC, Garicochea B, Silva EL, et al. Increased plasma removal of microemulsions resembling the lipid phase of low-density lipoproteins (LDL) in patients with acute myeloid leukemia: a possible new strategy for the treatment of the disease. Braz J Med Biol Res 1992;25:1003–7.

[102] Maranhão RC, Tavares ER, Padoveze AF, et al. Paclitaxel associated with cholesterol-rich nanoemulsions promotes atherosclerosis regression in the rabbit. Atherosclerosis 2008;197:959–66.

[103] Lourenço-Filho DD, Maranhão RC, Méndez-Contreras CA, et al. An artificial nanoemulsion carrying paclitaxel decreases the transplant heart vascular disease: a study in a rabbit graft model. J Thorac Cardiovasc Surg 2011;141:1522–8.

[104] Fiorelli AI. Ação da nanopartícula LDE-Metotrexato no desenvolvimento da doença vascular do enxerto em coração transplantado de coelho. Tese apresentada par obtenção do Título de Professor Livre Docente pela Faculdade de Medina da USP; 2013.

[105] Bacal F, Veiga VC, Fiorelli AI, et al. Treatment of persistent rejection with methotrexate in stable patients submitted to heart transplantation. Arq Bras Cardiol 2000;74:141–8.

DIABETES AND ERECTILE DYSFUNCTION

43

Erectile Dysfunction and the Endothelium

Fabiola Zakia Mónica and Gilberto De Nucci

INTRODUCTION

The erectile cycle is composed of four phases: flaccidity, tumescence, erection and detumescence. The erection process involves the participation of synergic and simultaneous events, namely: dilation of arteries and penile arteries with increased blood flow, relaxation of smooth muscles and entrapment of blood entering the expanding sinusoids, compression of the subalbugal venular plexus between the tunica albuginea and peripheral sinusoids, reducing venous blood efflux, maximal stretching of the tunica, occlusion of the emissary veins between the inner and outer longitudinal circular layers of the tunica albuginea, and an even greater decrease in venous efflux, increase in oxygen pressure and intracavernous pressure, which leads the penis to its erect position (complete erection phase) and an even greater increase in pressure with contraction of the ischiochondromatic (rigid phase) musculature. The events described involve the release of contractile and relaxing substances from both nerve fibers and the endothelial layer, which covers the smooth muscles of the sinusoids (Fig. 43.1).

Erectile dysfunction is defined as the inability to get or keep the erection during sexual activity. It is known that obesity, *diabetes mellitus*, hypercholesterolemia, hypertension, coronary artery disease, and advanced age increases the incidence. Loss of endothelial functional integrity and, subsequently, endothelial dysfunction contribute to development of erectile dysfunction. This chapter aims to address the role of the endothelium in erectile physiology as well as erectile dysfunction, especially in situations in which there is a risk factor associated with vascular diseases.

CONTRACTILE AND RELAXING SUBSTANCES DERIVED FROM ENDOTHELIUM AND NERVOUS FIBERS

Contractile Substances

In the absence of sexual stimulation, the penis is kept in a state of flaccidity by the action, mainly, of basal sympathetic tone. Sympathetic tone triggers the release of noradrenaline, which activates the alpha 1 adrenergic receptors present in the smooth musculature, leading to the tonic contraction of arteries and veins that irrigate the cavernous body, as well as the smooth musculature of the cavernous sinusoids. This muscle contraction results in a resistance against the influx of arterial blood, keeping the penis flaccid [1].

In addition to noradrenaline, other substances such as angiotensin-II, endothelin-1 and histamine, through the activation of H1 receptors, also lead to contraction of the cavernous smooth musculature, arteries and veins. In human cavernous corpus endothelial cell culture, the genetic and protein expression of endothelin-1 in the sinusoids and in the smooth muscle was observed [2]. In isolated corpus cavernosum from rat the presence of endothelin-1 and its receptor was observed on endothelial WTP layer [3], while in human corpus cavernosum the ETA and ETB receptors are expressed on smooth muscle [4]. Activation of these receptors by endothelin-1 leads to a slow and lasting contraction of cavernous smooth muscles, arteries and veins [5]. Although in vitro studies showed that endothelin-1 contracts smooth muscles, its physiological importance in the process of detumescence needs to be better exploited.

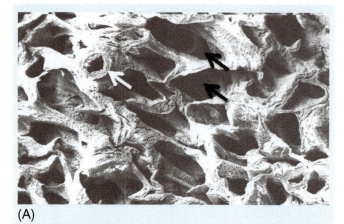

(A)

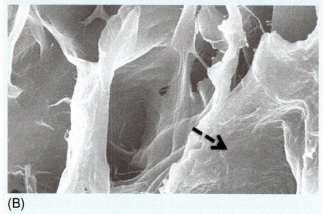

(B)

FIG. 43.1 Images obtained by microscopy of human isolated corpus cavernosum scan. (A) *Black and white arrows* show the sinusoids and artery, respectively. (B) Amplification of the endothelial layer overlying the inner portion of sinusoids (*dashed arrow*).

In the isolated corpus cavernosum of rats and dogs, expression of angiotensin-II was seen in the endothelial layer of the arteries and cavernous smooth muscles. Intracavernosal injection of angiotensin-II reduced intracavernous pressure in an anesthetized dog, whereas losartan, nonselective antagonist for the AT-1 and AT-2 receptors, increased this pressure [6]. Despite the in vitro findings, it is not known whether the angiotensin-II is a significant agent that regulates the tone of the cavernous smooth muscle. In a study of 124 patients with moderate and severe erectile dysfunction, one arm received losartan or tadalafil while the other received losartan and tadalafil for 12 weeks. Erectile function was evaluated by the International Erectile Function Index (IIFE). The combination of losartan and tadalafil showed more significant improvements in IIFE than when given separately in patients with erectile dysfunction [7]. However, more studies need to be performed with angiotensin-II antagonists to evaluate the

effectiveness of this class in the treatment of erectile dysfunction. There is no clinical evidence that this pharmacological class adversely affects erectile function, as it does with other antihypertensive agents.

The incidence of sexual impotence is three to four times higher in hypertensive individuals treated with antihypertensives. There is evidence that diuretics, central sympatholytic drugs and beta-blockers have a negative impact on sexual function when compared to calcium channel antagonists, angiotensin-II receptors, and angiotensin-converting enzyme inhibitors [8–10]. A study conducted by the Treatment of Mild Hypertension Study (TOMHS) evaluated the incidence of erectile dysfunction in hypertensive men treated with different antihypertensive agents. After 24 months of treatment, patients who used chlorthalidone had an increased incidence of erectile problems (15.7%) compared to the placebo group (4.9%), whereas the other agents did not show significant differences in incidence of sexual dysfunction (acebutolol—7.9%, amlodipine—6.7%, enalapril—6.5% and doxazosin—2.8%) [11]. According to the data obtained after two years of treatment, the incidence of disorders in erectile function was 10.1% in the placebo group and 22.6% in the group using bendrofluazide (Medical Research Council, 1981, http://www.thelancet.com/journals/lancet/article/PIIS0140-6736(81)90936-3/abstract). In addition, other studies have correlated reduced libido and sexual function with frequent use of diuretics such as hydrochlorothiazide and chlorthalidone [8].

Relaxing Substances

As opposed to contraction, the major second messengers involved in smooth muscle relaxation are cyclic adenine monophosphate (cAMP) and cyclic guanosine monophosphate (cGMP). These molecules activated cAMP- and cGMP-dependent kinases, which, in turn, phosphorylate proteins and ion channels primarily resulting in the opening of potassium channels and hyperpolarization, in addition to intracellular calcium sequestration by the endoplasmic reticulum and inhibition of voltage-dependent calcium channels, resulting in lower cytoplasmic concentrations of calcium. Thus, substances that inhibit the degradation of cGMP or cAMP lead to relaxation of the cavernous and arterial smooth muscle, and therefore to penile erection [12].

The most studied relaxation pathway regarding peripheral mechanisms of erection is that of nitric oxide. NO is produced when L-arginine substrate is transformed into L-citrulline in a reaction catalyzed by endothelial nitric oxide synthase enzyme (eNOS) and neuronal NOS (nNOS) in the presence of oxygen and calmodulin cofactor, tetrahydrobiopterin (BH$_4$) NADPH, heme, flavin-adenine dinucleotide Deo (FAD), flavin mononucleotide (FMN) and calcium (Ca^{2+}). Nitric oxide released diffuses into the cavernous smooth muscles and converts the soluble guanine triphosphate (GTP) enzyme into cGMP [13,14]. In the penis, cGMP is hydrolyzed to its inactive metabolite 5'-GMP by the action, mainly, of phosphodiesterase-5 enzyme (PDE5) (Fig. 43.2).

In 1992, Rajfer et al. [15] demonstrated that electrical stimulation induced frequency-dependent relaxation in isolated human corpus cavernosum, being this effect significantly reduced in the presence of NOS and GCs inhibitors. Since then, several studies have been published showing the importance of nitric oxide in the physiology of erection and that in erectile dysfunction its effect is significantly reduced because of its lower bioavailability.

In addition to nitric oxide, it has been seen that other substances also lead to the relaxation of the human corpus cavernosum, such as prostaglandin E1 (PGE1), vasoactive intestinal peptide (VIP) [16], histamine, through the activation of H2 [17] receptors, and more recently, the gas hydrogen sulfide

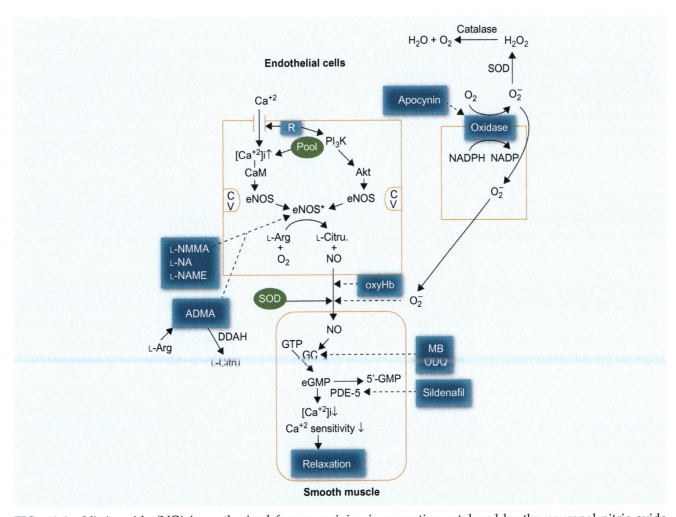

FIG. 43.2 Nitric oxide (NO) is synthesized from L-arginine in a reaction catalyzed by the neuronal nitric oxide synthase (nNOS) or endothelial (eNOS) enzyme. This reaction involves the participation of calcium ions (Ca^{2+}), NO diffuses into the smooth muscle of sinusoids and stimulates soluble guanylate cyclase (GCs), which converts guanosine triphosphate (GTP) to cyclic guanosine monophosphate (cGMP). CGMP is degraded to 5-GMP by the action of phosphodiesterase type 5 (PDE5). Substances that inhibit the activity of NOS (L-NMMA, L-NA, L-NAME) or GCs (MB and ODQ) are used only as pharmacological tools. Adapted from Toda N. Age-related changes in endothelial function and blood flow regulation. Pharmacol Ther 2012;133:159–76.

(H₂S) [18] generated from the catabolism of sulfonated amino acids. However, if the last three mediators have any physiological relevance in erection and erectile dysfunction, more studies need to be performed.

ROLE OF ENDOTHELIUM IN ERECTILE FUNCTION AND DYSFUNCTION: FOCUS ON THE NO-GCS-PDE5 PATHWAY

Penile erection is triggered by neural signals from the spinal column, increasing nNOS activity and therefore blood flow to the cavernous tissue [19]. The eNOS enzyme, in turn, is activated by the action of agonist (acetylcholine, bradykinin, ADP, endothelin-1) or by shear force caused by increased blood flow and the expansion of the sinusoidal spaces of the corpus cavernosum. This expansion causes activation of the enzyme phosphatidylinositol 3 kinase (PI3K) and AKT protein kinase leading to phosphorylation of Ser1177/1179 eNOS, resulting in an increased release of nitric oxide [20] and, therefore, in the maintenance of erection (Fig. 43.3). In isolated cavernous body of the mouse, relaxation induced by electrical stimulation or the intracavernous addition of papaverine increased the phosphorylation of the phosphatidylinositol 3 kinase/AKT and eNOS signaling pathway. Furthermore, erection induced by papaverine was significantly reduced in the presence of PI3K/AKT pathway inhibitors, as well as *knockout* animals for eNOS gene [21], suggesting that this pathway is significant in maintaining erection.

In vascular diseases, alterations in the functional integrity of the endothelium contribute to the lower responsiveness of this layer to hemodynamic stimuli and paracrine and autocrine factors, a condition denominated endothelial dysfunction. The term is commonly used to signify a decrease in the relaxation of the smooth muscle dependent on the endothelium mainly due to the reduction of nitric oxide bioavailability, which may be due to the less expression and/or eNOS activity; absence of substrates or cofactors essential for eNOS activation; decoupling of eNOS and degradation of nitric oxide by the action of species reactive to oxygen and nitrogen.

Several studies have shown that the incidence of erectile dysfunction is higher in patients with cardiovascular disease. In both situations, endothelial dysfunction is present, leading to atherosclerosis, and therefore, significantly altering the penile and

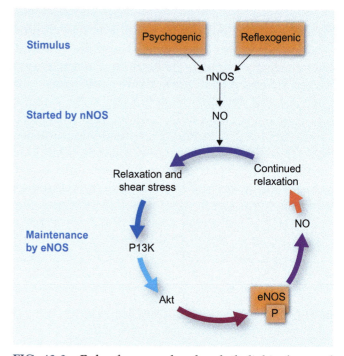

FIG. 43.3 Role of neuronal and endothelial isoforms of NO synthases (nNOS and eNOS) in the initiation and maintenance of penile erection [21]. Phosphatidylinositol 3 kinase (PI3K) enzyme; protein kinase AKT. Adapted from Hurt KJ, Musicki B, Palese MA, et al. Akt-dependent phosphorylation of endothelial nitric-oxide synthase mediates penile erection. Proc Natl Acad Sci U S A 2002;99:4061–6.

coronary circulations [22]. Since the diameter of the penile artery is smaller when compared to the coronary arteries, the same level of endothelial dysfunction may cause a more significant reduction in penile blood flow compared to the coronary circulation. Therefore, from a clinical point of view, the erectile dysfunction can be used as an early marker of a possible vascular disease [22,23] (Fig. 43.4).

Several preclinical trials have shown that oxidative stress, particularly nitric oxide reaction with superoxide anion (O_2^-) (Fig. 43.2), is one of the main factors that lead to endothelial dysfunction in vascular diseases such as diabetes, hypertension, hypercholesterolemia, advanced age, and erectile dysfunction. Most studies investigating the role of superoxide anion in endothelial and erectile dysfunctions were performed in animal models and observed a reduction in the relaxation induced by acetylcholine or by electrical stimulation that releases nitric oxide from nitrergic nervous fibers [24–26], increase in oxygen reactive species and a significant improvement in endothelium-dependent

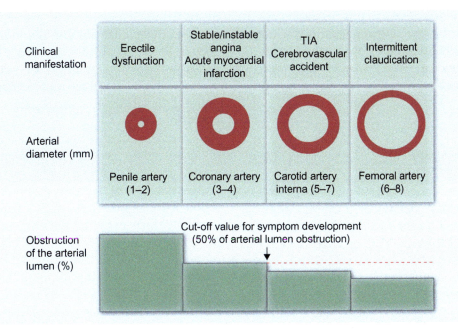

FIG. 43.4 Hypothesis "artery" diameter. Since the diameter of the penile arteries (1–2 mm) is smaller compared to coronary arteries (3–4 mm), carotid (5–7 mm), and femoral (6–8 mm), narrowing of the arteries due to the atherosclerosis process, will produce early clinical manifestations (erectile dysfunction). Transient ischemic attack (TIA). Adapted from Watts GF, Chew KK, Stuckey BG. The erectile-endothelial dysfunction nexus: new opportunities for cardiovascular risk prevention. Nat Clin Pract Cardiovasc Med 2007;4:263–73; Gandaglia G, Briganti A, Jackson G, et al. A systematic review of the association between erectile dysfunction and cardiovascular disease. Eur Urol 2014;65:968–78; Montorsi P, Ravagnani PM, Galli S, et al. The artery size hypothesis: a macrovascular link between erectile dysfunction and coronary artery disease. Am J Cardiol 2005;96:19M–23M.

and independent relaxation in the presence of the antioxidant apocynin [26]. To date, there is no clinical trial that evaluated whether prolonged use of antioxidants would have any efficacy in erectile dysfunction.

Another factor that has increasingly been studied in endothelial dysfunction and causes erectile dysfunction is the augmentation in expression and activity of the endogenous inhibitor of nitricoxidesynthase (NOS), called asymmetric di-methylarginine (ADMA). In the rabbit ischemic cavernous body, relaxations against acetylcholine or electrical stimulation were significantly reduced, as were NOS activity and cGMP levels, while higher levels of the ADMA inhibitor were observed [27]. In men with vasculogenic erectile dysfunction, ADMA enzyme levels were significantly higher compared to men with nonvasculogenic erectile dysfunction [28,29]. A recent study showed that *knockout* animals for cyclooxygenase-2 ($COX-2^{-/-}$) had higher plasma levels of ADMA compared to the control group. Relaxation induced by acetylcholine but not the donor of sodium nitroprusside nitric oxide was lower in

aorta of $COX-2^{-/-}$ animals as compared to control group. Administration of celecoxib (200 mg twice daily) for 7 days in healthy volunteers increased plasma ADMA levels. The same effect was observed with nonselective antiinflammatory, naproxen (500 mg, $2 \times$/day) [30].

PDE5 inhibitors such as sildenafil, tadalafil, vardenafil, and lodenafil carbonate [31,32] are drugs of first choice for the treatment of erectile dysfunction. In isolated cavernous cavernosum from humans, these inhibitors produce concentration-dependent relaxation, significantly reduced in the presence of the NOS inhibitor. Furthermore, acetylcholine-induced relaxation and electrical stimulation were significantly increased in the presence of these inhibitors. These data suggest that endogenous and exogenous nitric oxide contribute to the relaxation induced by PDE5 inhibitors [31]. In severe cases of endothelial dysfunction, the efficacy of these drugs may be reduced, and therefore, there is a number of patients with erectile dysfunction not responsive to the therapy with PDE5 inhibitors, and even at the highest recommended dose, no improvement is observed [33]. In cases of unresponsiveness to PDE5 inhibitors, the use of intracavernous injection of papaverine, phentolamine, or prostaglandin E2 becomes the second option. However, many patients do not react to these substances and the penile prosthesis is considered as an alternative.

ANDROGENS AND ERECTILE FUNCTION

Androgens are recognized as powerful modulators of male sexual behavior, although there is controversy regarding their role in the maintenance of penis responsiveness [34–36]. It is notorious, however, that episodes of nocturnal penile tumescence do not occur in men with hypogonadism [37]. In animal models of erectile function, surgical castration or antiandrogenic therapy reduce the magnitude of the responses induced by in vivo ganglion stimulation, while treatment with androgens reverses this effect [35,38–40]. Reduction of erectile dysfunction caused by castration is associated with alterations that lead to reduced NOS expression and activity in rat corpus cavernosum [41,42]. In a model of hypogonadotropic hypogonadism in rabbits, the genetic and protein expression of PDE5 and relaxation induced by sildenafil was significantly reduced compared to the control group, while testosterone replacement restored these levels, suggesting that testosterone regulates PDE5 positively [43]. In patients with hypogonadism and unresponsive to PDE5 inhibitors it has been observed that testosterone replacement improved erectile function [44]. In another study in men with erectile dysfunction and hypogonadism, intramuscular injection of testosterone (250 mg) for three weeks improved the symptoms of late-onset hypogonadism (loss of libido, poor energy and irritability) as well as erectile function (from 9 to 13.1) according to the IIFE [45].

POSSIBLE THERAPEUTIC ROLE OF SUBSTANCES THAT ACT DIRECTLY IN SOLUBLE GUANILATO CYCLASE FOR THE TREATMENT OF ERETRIC DYSFUNCTION

With the discovery that the NO-cGMP system is the most relevant and effective mechanism of penile erection, several substances have been identified with therapeutic potential for treatment of erectile dysfunction. These agents are divided into two groups: those that inhibit degradation of cGMP, such as PDE-5 inhibitors, and those that raise cGMP levels by direct activation of soluble guanylate cyclase, which in turn are represented by nitric oxide donors, including nitroglycerin and sodium nitroprusside.

Two classes of substances that directly act on soluble guanylate cyclase have been developed by pharmaceutical company Bayer: the stimulators and activators of soluble guanylate cyclase and, unlike nitrovasodilators, do not cause pharmacological tolerance and act independently of nitric oxide [46]. These substances have not been approved by regulatory agencies for the treatment of erectile dysfunction. Regarding the first class, there are the YC-1, BAY 41-2272, BAY 41-8543 and BAY 63-2521 (riociguat). The latter was recently approved by the regulatory agencies for the treatment of pulmonary hypertension. YC-1 is the precursor of this class, but less potent and selective in relation to BAY 41-2272 BAY 41-8543 [47,48] and BAY 63-2521 [49].

Removal of heme grouping or its oxidation (Fe^{3+}) decreases the effectiveness of nitric oxide. Endogenous changes of the redox state of the enzyme can be induced by oxygen and nitrogen reactive species, such as peroxynitrite ($ONOO^-$) and the free radical superoxide (O_2^-), which are generated in oxidative stress conditions [50]. This change in redox state compromises the NO-GCs-GMPc signaling, causing the enzyme to be unresponsive to endogenous and exogenous nitric oxide [51]. In this regard, unlike the stimulators of soluble guanylate cyclase, activators, BAY 58-2667 (cinaciquat), HMR 1766 (ataciguat) and BAY 60-2770 bind to soluble guanylate cyclase when the iron is in the ferrous state (Fe^{2+}), ferric (Fe^{3+}) or even absent [51] (Fig. 43.5).

In the isolated corpus cavernosum of rabbits and humans [52,53] or mice [54,55], it was seen that the stimulator BAY 41-2272 produced concentration-dependent relaxation, which was significantly decreased in the presence of soluble guanylate cyclase inhibitors and NOS, suggesting that the efficacy of the stimulators is reduced when the iron is in oxidized state and/or when bioavailability of the nitric oxide is reduced. Regarding the activators, BAY 60-2770 produced concentration-dependent relaxation in isolated rabbit cavernous body, and this response is potentiated in the presence of the soluble guanylate cyclase inhibitor, mainly due to the higher

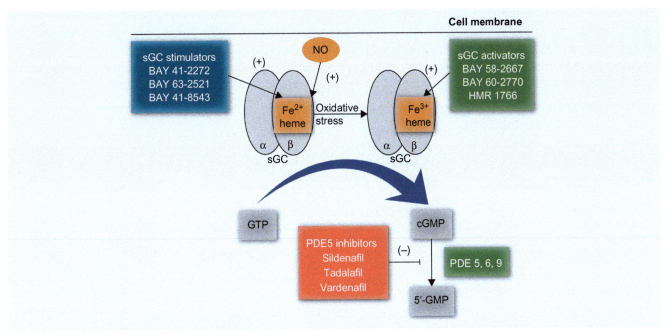

FIG. 43.5 Mechanism of action of the stimulators and activators of soluble guanylate cyclase (sGC). The potency of sGC stimulators is greater in the presence of NO and when iron is in the reduced form (Fe^{2+}), while sGC activators act in a more powerful way when iron is in its oxidized form (Fe^{3+}). Absence of nitric oxide does not interfere with pharmacological parameters of sGC activators. Adapted from Monica FZ, Murad F, Bian K. Modulating cGMP levels as therapeutic drug targets in cardiovascular and non-cardiovascular diseases. OA Biochem 2014;2(1):3–12.

levels of cGMP [56]. Intracavernosal injection of BAY 60-2770 (350 ng/kg) in rats increased intracavernosal pressure, with this effect being even stronger in the presence of soluble guanylate cyclase inhibitor [57].

Based on preclinical findings, it is possible to suppose that soluble guanylate cyclase stimulators and activators are a therapeutic alternative for the treatment of erectile dysfunction. Furthermore, it may be said that soluble guanylate cyclase activators have certain advantages over guanylate cyclase stimulators and PDE5 inhibitors, since the pharmacological parameters do not change in situations of lower nitric oxide bioavailability and are even more potent when the soluble guanylate cyclase is oxidized.

CONCLUSIONS

Endothelial dysfunction is one of the main causes leading to vasculogenic erectile dysfunction. Erectile dysfunction is a significant predictor of cardiovascular diseases, whose mechanisms involved are mainly related to alterations in the NO-sGC-PDE5 pathway, either due to the lower bioavailability of nitric oxide and/or lower production of the second cGMP messenger. Thus, substances that improve

endothelial dysfunction would be of extreme value in treating erectile dysfunction and, consequently, slowing the progression of cardiovascular diseases. In addition, substances that act independently of nitric oxide may be a therapeutic alternative for cases that are irresponsive to PDE5 inhibitors.

References

[1] Andersson K, Stief C. Penile erection and cardiac risk: pathophysiologic and pharmacologic mechanisms. Am J Cardiol 2000;86: 23F–26F.

[2] Saenz de Tejada I, Carson MP, de las Morenas A, et al. Endothelin: localization, synthesis, activity, and receptor types in human penile corpus cavernosum. Am J Physiol 1991;261:H1078–85.

[3] Bell CR, Sullivan ME, Dashwood MR, et al. The density and distribution of endothelin 1 and endothelin receptor subtypes in normal and diabetic rat corpus cavernosum. Br J Urol 1995;76(2):203–7.

[4] Christ GJ, Lerner SE, Kim DC, et al. Endothelin-1 as a putative modulator of erectile dysfunction: I. Characteristics of contraction of isolated corporal tissue strips. J Urol 1995;153:1998–2003.

[5] Holmquist F, Kirkeby HJ, Larsson B, et al. Functional effects, binding sites and immunolocalization of endothelin-1 in isolated penile tissues from man and rabbit. J Pharmacol Exp Ther 1992;261: 795–802.

[6] Kifor I, Williams GH, Vickers MA, et al. Tissue angiotensin II as a modulator of erectile function. I. Angiotensin peptide content, secretion and effects in the corpus cavernosum. J Urol 1997;157: 1920–5.

[7] Chen Y, Cui S, Lin H, et al. Losartan improves erectile dysfunction in diabetic patients: a clinical trial. Int J Impot Res 2012;24:217–20.

[8] Düsing R. Angiotensin II-receptor blocker dosages: how high should we go? Int J Clin Pract 2006;60:179–83.

[9] Düsing R. Overcoming barriers to effective blood pressure control in patients with hypertension. Curr Med Res Opin 2006;22:1545–53.

[10] Papatsoris AG, Korantzopoulos PG. Hypertension, antihypertensive therapy, and erectile dysfunction. Angiology 2006;57:47–52.

[11] Grimm Jr. RH, Grandits GA, Prineas RJ, et al. Long-term effects on sexual function of five antihypertensive drugs and nutritional hygienic treatment in hypertensive men and women. Treatment of Mild Hypertension Study (TOMHS). Hypertension 1997;29:8–14.

[12] Dean RC, Lue TF. Physiology of penile erection and pathophysiology of erectile dysfunction. Urol Clin North Am 2005;32:379–95.

[13] Katsuki S, Arnold W, Mittal C, et al. Stimulation of guanylate cyclase by sodium nitroprusside, nitroglycerin and nitric oxide in various tissue preparations and comparison to the effects of sodium azide and hydroxylamine. J Cyclic Nucleotide Res 1977;3:23–35.

[14] Murad F, Mittal CK, Arnold WP, et al. Guanylate cyclase: activation by azide, nitro compounds, nitric oxide, and hydroxyl radical and inhibition by hemoglobin and myoglobin. Adv Cyclic Nucleotide Res 1978;9:145–58.

[15] Rajfer J, Aronson WJ, Bush PA, et al. Nitric oxide as a mediator of relaxation of the corpus cavernosum in response to nonadrenergic, noncholinergic neurotransmission. N Engl J Med 1992;326:90–4.

[16] Kirkeby HJ, Fahrenkrug J, Holmquist F, et al. Vasoactive intestinal polypeptide (VIP) and peptide histidine methionine (PHM) in human penile corpus cavernosum tissue and circumflex veins: localization and in vitro effects. Eur J Clin Invest 1992;22:24–30.

[17] Cará AM, Lopes-Martins RA, Antunes E, et al. The role of histamine in human penile erection. Br J Urol 1995;75:220–4.

[18] d'Emmanuele di Villa Bianca R, Sorrentino R, Maffia P, et al. Hydrogen sulfide as a mediator of human corpus cavernosum smooth-muscle relaxation. Proc Natl Acad Sci U S A. 2009;106:4513–18.

[19] Burnett AL. Novel nitric oxide signaling mechanisms regulate the erectile response. Int J Impot Res 2004;16:S15–9.

[20] Dimmeler S, Fleming I, Fisslthaler B, et al. Activation of nitric oxide synthase in endothelial cells by Akt-dependent phosphorylation. Nature 1999;399:601–5.

[21] Hurt KJ, Musicki B, Palese MA, et al. Akt-dependent phosphorylation of endothelial nitric-oxide synthase mediates penile erection. Proc Natl Acad Sci U S A 2002;99:4061–6.

[22] Watts GF, Chew KK, Stuckey BG. The erectile-endothelial dysfunction nexus: new opportunities for cardiovascular risk prevention. Nat Clin Pract Cardiovasc Med 2007;4:263–73.

[23] Gandaglia G, Briganti A, Jackson G, et al. A systematic review of the association between erectile dysfunction and cardiovascular disease. Eur Urol 2014;65:968–78.

[24] Ahn TY, Gómez-Coronado D, Martínez V, et al. Enhanced contractility of rabbit corpus cavernosum smooth muscle by oxidized low density lipoproteins. Int J Impot Res 1999;11:9–14.

[25] Rubbo H, Trostchansky A, Botti H, et al. Interactions of nitric oxide and peroxynitrite with low-density lipoprotein. Biol Chem 2002;383:547–52.

[26] Silva FH, Mónica FZ, Báu FR, et al. Superoxide anion production by NADPH oxidase plays a major role in erectile dysfunction in middle-aged rats: prevention by antioxidant therapy. J Sex Med 2013;10:960–71.

[27] Masuda H, Tsujii T, Okuno T, et al. Accumulated endogenous NOS inhibitors, decreased NOS activity, and impaired cavernosal relaxation with ischemia. Am J Physiol Regul Integr Comp Physiol 2002;282:R1730–8.

[28] Ioakeimidis N, Vlachopoulos C, Rokkas K, et al. Relationship of asymmetric dimethylarginine with penile Doppler ultrasound parameters in men with vasculogenic erectile dysfunction. Eur Urol 2011;59:948–55.

[29] Paroni R, Barassi A, Ciociola F, et al. Asymmetric dimethylarginine (ADMA), symmetric dimethylarginine (SDMA) and L-arginine in patients with arteriogenic and non-arteriogenic erectile dysfunction. Int J Androl 2012;35:660–7.

[30] Ahmetaj-Shala B, Kirkby NS, Knowles R, et al. Evidence that links loss of cyclooxygenase-2 with increased asymmetric dimethylarginine: novel explanation of cardiovascular side effects associated with anti-inflammatory drugs. Circulation 2015;131:633–42.

[31] Toque HA, Teixeira CE, Lorenzetti R, et al. Pharmacological characterization of a novel phosphodiesterase type 5 (PDE5) inhibitor lodenafil carbonate on human and rabbit corpus cavernosum. Eur J Pharmacol 2008;591:189–95.

[32] Mendes GD, dos Santos Filho HO, dos Santos Pereira A, et al. A phase I clinical trial of lodenafil carbonate, a new phosphodiesterase type 5 (PDE5) inhibitor, in healthy male volunteers. Int J Clin Pharmacol Ther 2012;50:896–906.

[33] Eardley I. Optimisation of PDE5 inhibitor therapy in men with erectile dysfunction: converting "non-responders" into "responders". Eur Urol 2006;50:31–3.

[34] Arver S, Dobs AS, Meikle AW, et al. Improvement of sexual function in testosterone deficient men treated for 1 year with a permeation enhanced testosterone transdermal system. J Urol 1996;155:1604–8.

[35] Mills TM, Lewis RW. The role of androgens in the erectile response: a 1999 perspective. Mol Urol 1999;3:75–86.

[36] Carruthers M. The diagnosis of late life hypogonadism. Aging Male 2008;11:45–6.

[37] Granata AR, Rochira V, Lerchl A, et al. Relationship between sleep-related erections and testosterone levels in men. J Androl 1997;18:522–7.

[38] Bivalacqua TJ, Rajasekaran M, Champion HC, et al. The influence of castration on pharmacologically induced penile erection in the cat. J Androl 1998;19:551–7.

[39] Marin R, Escrig A, Abreu P, et al. Androgen-dependent nitric oxide release in rat penis correlates with levels of constitutive nitric oxide synthase isoenzymes. Biol Reprod 1999;61:1012–6.

[40] Traish AM, Park K, Dhir V, et al. Effects of castration and androgen replacement on erectile function in a rabbit model. Endocrinology 1999;140:1861–8.

[41] Lugg JA, González-Cadavid NF, Rajfer J. The role of nitric oxide in erectile function. J Androl 1995;16(1):2–4.

[42] Baba K, Yajima M, Carrier S, et al. Effect of testosterone on the number of NADPH diaphorase-stained nerve fibers in the rat corpu cavernosum and dorsal nerve. Urology 2000;56:533–8.

[43] Zhang XH, Morelli A, Luconi M, et al. Testosterone regulates PDE5 expression and in vivo responsiveness to tadalafil in rat corpus cavernosum. Eur Urol 2005;47:409–16.

[44] Yassin AA, Saad F. Testosterone and erectile dysfunction. J Androl 2008;29:593–604.

[45] Heidari R, Sajadi H, Pourmand A, et al. Can testosterone level be a good predictor of late-onset hypogonadism? Andrologia 2015;47:433–7.

[46] Stasch JP, Schmidt P, Alonso-Alija C, et al. NO- and haem-independent activation of soluble guanylyl cyclase: molecular basis and cardiovascular implications of a new pharmacological principle. Br J Pharmacol 2002;136:773–83.

[47] Stasch JP, Becker EM, Alonso-Alija C, et al. NO-independent regulatory site on soluble guanylate cyclase. Nature 2001;410:212–5.

[48] Stasch JP, Dembowsky K, Perzborn E, et al. Cardiovascular actions of a novel NO-independent guanylyl cyclase stimulator, BAY 41-8543: in vivo studies. Br J Pharmacol 2002;135:344–55.

[49] Schermuly RT, Stasch JP, Pullamsetti SS, et al. Expression and function of soluble guanylate cyclase in pulmonary arterial hypertension. Eur Respir J 2008;32:881–91.

[50] Zou MH, Hou XY, Shi CM, et al. Modulation by peroxynitrite of Akt- and AMP-activated kinase-dependent Ser1179 phosphorylation of endothelial nitric oxide synthase. J Biol Chem 2002;277:32552–7.

[51] Stasch JP, Schmidt PM, Nedvetsky PI, et al. Targeting the heme-oxidized nitric oxide receptor for selective vasodilatation of diseased blood vessels. J Clin Invest 2006;116:2552–61.

[52] Kalsi JS, Rees RW, Hobbs AJ, et al. BAY41-2272, a novel nitric oxide independent soluble guanylate cyclase activator, relaxes human and rabbit corpus cavernosum in vitro. J Urol 2003;169:761–6.

[53] Baracat JS, Teixeira CE, Okuyama CE, et al. Relaxing effects induced by the soluble guanylyl cyclase stimulator BAY 41-2272 in human and rabbit corpus cavernosum. Eur J Pharmacol 2003;477:163–9.

[54] Teixeira CE, Priviero FB, Webb RC. Effects of 5-cyclopropyl-2-[1-(2-fluoro-benzyl)-1H-pyrazolo[3,4-b]pyridine-3-yl]pyrimi-din-4-ylamine (BAY 41–2272) on smooth muscle tone, soluble guanylyl cyclase activity, and NADPH oxidase activity/expression in corpus cavernosum from wild-type, neuronal, and endothelial nitric-oxide synthase null mice. J Pharmacol Exp Ther 2007;322:1093–102.

[55] Nimmegeers S, Sips P, Buys E, et al. Role of the soluble guanylyl cyclase alpha1-subunit in mice corpus cavernosum smooth muscle relaxation. Int J Impot Res 2008;20:278–84.

[56] Estancial CS, Rodrigues RL, De Nucci G, et al. Pharmacological characterisation of the relaxation induced by the soluble guanylate cyclase activator, BAY 60-2770 in rabbit corpus cavernosum. BJU Int 2015;116(4):657–64.

[57] Lasker GF, Pankey EA, Frink TJ, et al. The sGC activator BAY 60-2770 has potent erectile activity in the rat. Am J Physiol Heart Circ Physiol 2013;304:H1670–9.

Further Reading

[1] Toda N. Age-related changes in endothelial function and blood flow regulation. Pharmacol Ther 2012;133:159–76.

[2] Montorsi P, Ravagnani PM, Galli S, et al. The artery size hypothesis: a macrovascular link between erectile dysfunction and coronary artery disease. Am J Cardiol 2005;96:19M–23M.

[3] Monica FZ, Murad F, Bian K. Modulating cGMP levels as therapeutic drug targets in cardiovascular and non-cardiovascular diseases. OA Biochem 2014;2(1):3–12.

44

Obesity, Diabetes, and Endothelium: Molecular Interactions

Mario J.A. Saad

INTRODUCTION

Obesity increases the risk of developing metabolic and cardiovascular diseases. Normal endothelial function is essential for preservation of homeostasis, maintaining blood flow and vascular tone, preventing cardiovascular diseases. Endothelial dysfunction, with reduced release of relaxing or vasodilatory factors and increased production of vasoconstricting mediators, is the first step that leads to regional blood flow alterations, resulting in functional and histological damage in organs and tissues that will evolve to the atherosclerotic cardiovascular disease. It is widely recognized that the increased risk for cardiovascular disease in obese individuals is associated with endothelial dysfunction. In these individuals, dilatation of the endothelium-dependent brachial artery is inversely associated with the mass of visceral adipose tissue [1]. In normal volunteers, modest weight gains (~4 kg) are accompanied by endothelial dysfunction, even in the absence of changes in blood pressure. Next, we shall discuss molecular mechanisms that contribute to endothelial dysfunction in the obese, and later in diabetes. It is crucial to emphasize that in diabetes, particularly in type 2 *diabetes mellitus*, there is a superposition of the molecular mechanisms described in obesity with hyperglycemia-induced damage.

MECHANISMS RESPONSIBLE FOR ENDOTHELIAL DISORDER IN THE OBESE

Insulin Resistance

Insulin resistance is defined as a subnormal biological response to a given hormone concentration. Increasingly prevalent in our society, it accompanies various clinical situations, such as obesity, type 2 *diabetes mellitus*, hypertension, infectious processes, some endocrine diseases, and polycystic ovary syndrome [2–5]. This hormone resistance has attracted great attention, both in clinical research and in basic, due to its association with cardiovascular disease [6,7]. However, molecular mechanisms linking insulin resistance, endothelial dysfunction, blood pressure, and development and progression of atherosclerosis are not fully understood, despite the progress made in recent decades [6,8].

Most often, the term "insulin resistance" is used with reference to glycemic control, reflecting an inadequate insulin effect on glucose homeostasis. However, insulin has pleiotropic actions, modulating various cellular functions, such as lipid and protein metabolism, ion and amino acid transport, proliferation and cell cycle, cell differentiation, apoptosis, and nitric oxide synthesis [2,9,10]. Thus, when considering insulin resistance situations, we should not take into account only the glucose metabolism, but the whole range of metabolic actions, growth and vascular insulin. We must also consider that insulin resistance may affect these functions in a heterogeneous way [6]. However, it should be noted that most insulin resistance situations, detected in clinical practice or in the laboratory, are accompanied by insulin resistance in the endothelium.

In most situations of insulin sensitivity reduction, we observed compensatory hyperinsulinemia. We understand today that the appearance of insulin resistance is not uniform in relation to the target tissues of this hormone, nor in relation to the intracellular pathways activated by insulin. Thus, at the same time that insulin resistance may negatively affect certain metabolic functions of this hormone, compensatory hyperinsulinemia may intensify others.

Insulin Signaling of in the Endothelium

The insulin receptor belongs to the family of membrane receptors that have intrinsic tyrosine kinase capacity. It is composed of two extracellular alpha subunits and two transmembrane beta subunits, bound by disulfide bonds [9]. Insulin binds to the receptor subunit α, provoking a conformational change in the subunit b, which leads to its autophosphorylation in tyrosine and activates a tyrosine kinase capacity. Once activated, the insulin receptor is able to phosphorylate several intracellular substrates, including insulin receptor substrate (IRS-1–4) and Shc protein (Src homology collagen) [2,11,12]. These proteins, once phosphorylated, recruits and activates several intracellular effectors with many different cellular functions [9]. In the endothelium, the ERK pathway is primarily involved in the control of growth, mitogenesis and production of endothelin, while activation of PI-3 kinase by IRS-1 is preferably linked to the metabolic

actions of insulin and nitric oxide production [6,13] (Fig. 44.1).

Recently, it was demonstrated that the regulation of Insulin signaling may be associated with a balance between the positive modulation, which occurs the IRS-1 tyrosine phosphorylation, and the negative modulation, which occurs by serine phosphorylation of the same IRS-1 [14,15]. It is well established that insulin and IGF-1 induces tyrosine phosphorylation of IRS-1, as agents known to lead to insulin resistance, such as TNF-α, free fatty acids, cell stress and hyperinsulinemia, induce activation of serine/threonine kinases that phosphorylate IRS-1 in serine, inhibiting its function [14,15]. The inhibitory phosphorylation of IRS-1, in serine, may be a unifying molecular mechanism of many insulin resistance-triggering factors. Insulin resistance inducers in the endothelium in obesity can be hormones, adipocytokines, or lipids, originating from adipose tissue, which act independently or through the innate immune system, and use cell signals emanating from oxidative stress and stress of endoplasmic reticulum. Regardless of the mechanism, the effect of insulin resistance on the vessel of obese animals is extremely deleterious for the endothelial function, since there is a reduction in the activity of the IRS/PI3K/Akt/eNOS pathway, with a

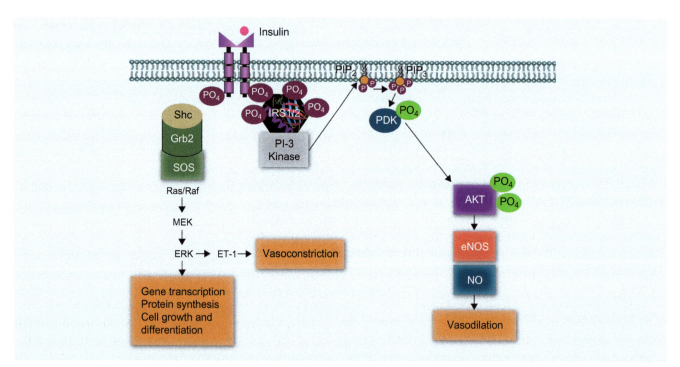

FIG. 44.1 Endothelium insulin molecular signaling, highlighting the IRS/PI3K/Akt/eNOS and ERK pathway (acronyms, see text).

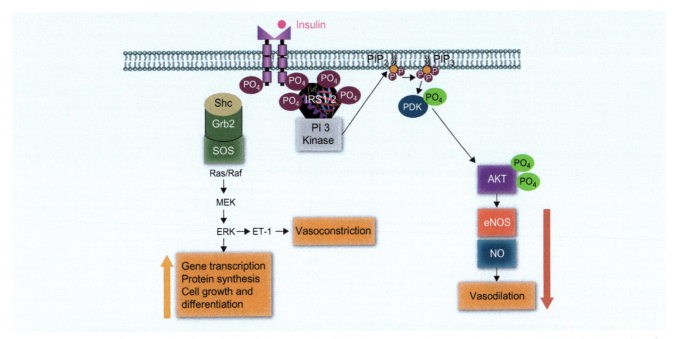

FIG. 44.2 Molecular signaling in endothelial insulin, highlighting IRS pathway activity/PI3K/Akt/eNOS and pathway activity increased ERK (acronyms, see text).

reduction of nitric oxide production, and in parallel, increase in activity of ERK pathway, with mitogenic stimulation and increased production of endothelin [15] (Fig. 44.2).

OBESITY, RESISTANCE TO INSULIN AND ANGIOTENSIN

Several clinical trials have shown that blocking of the renin-angiotensin system, either with converting enzyme inhibitors or with blockers or receiver of angiotensin II AT1, reduce the incidence of type 2 *diabetes mellitus* in high-risk patients. A meta-analysis, involving data of more than 33,100 patients, showed a reduction in the incidence of type 2 *diabetes mellitus* compared to other antihypertensive regimens [16]. Obesity is related to hypertrophy of adipose tissue, active endocrine organ that secretes several hormones, which may lead to development of insulin resistance, including angiotensin II and other cytokines [17]. Angiotensinogen secreted by adipose tissue plays an important role in development of adipose tissue itself, as well as arterial hypertension associated with obesity [18]. Obese women, when compared to those who are nonobese, show high plasma levels of converting enzyme, angiotensinogen, renin, and aldosterone, and increased angiotensinogen gene expression in

adipose tissue. An average loss of 5 kg in the group of obese women causes significant decreases in the activation of the renin-angiotensin system, reflected by the falls in the plasma concentrations of the mentioned mediators [19].

In recent decades, several researchers, including ourselves, have sought to understand how insulin signaling and angiotensin systems correlate, and have made some advances, showing that this interaction occurs at different levels of cell signaling, by means of different cellular proteins. This interaction modulates several cellular functions of insulin cells in a diverse way, affecting different cellular functions.

MOLECULAR EVENTS RELATED TO INTERACTION OF INSULIN VERSUS ANGIOTENSIN

Angiotensin and the IRS-1/PI-3 Kinase Pathway

Several studies have shown that angiotensin-II, the main effector peptide of the renin-angiotensin system, plays a key role both in the development of hypertension and in insulin resistance [8]. Agents that inhibit the action of angiotensin II, such as inhibitors of the enzyme conversion of angiotensin blockers and their AT1 receptor, leading to a reduction in blood pressure and also an increase in insulin sensitivity in hypertensive patients previously resistant to insulin [20,21].

Taking into account these clinical findings and the epidemiological association between insulin resistance, hypertension, and cardiovascular disease, several studies investigated the interaction of the initial signaling stages of these two hormones. Studies carried out in vivo and in cell culture showed that similarly to insulin, angiotensin-II is able to stimulate the phosphorylation of IRS-1 tyrosine and in the IRS-2 [22–24]. These phosphorylations are induced rapidly and secondary to activation of JAK-2, member of the *Janus kinases* family. Signaling through JAK family was initially described for cytokines family receptors, including interleukins and interferon receptors, but subsequent studies showed that JAK/STAT cascade can be activated also by receptors associated with G protein [25]. In fact, studies carried out in vivo and in cell culture have shown that angiotensin II is able rapidly to induce the tyrosine phosphorylation of JAK-2, activating its catalytic capacity and inducing its association and coimmunoprecipitation with IRS-1 and IRS-2 [23,24]. This phenomenon happens in parallel with the tyrosine phosphorylation of both IRS-1 and IRS-2 [23]. Furthermore, this rapid tyrosine phosphorylation of JAK2 and association with the IRS-1 and IRS-2 suggests the formation of a large signaling complex with proteins such AT1 receptor after stimulation with angiotensin-II [23–27].

Whenever IRS-1/2 tyrosine phosphorylation takes place, binding of these proteins to the p85

regulatory subunit of PI-3 kinase occurs. However, in an antagonistic way to that observed after insulin stimulation, PI-3 kinase activity associated with IRS-1 and IRS-2 is decreased after the angiotensin-II stimulus in a dose-dependent manner [23]. Studies carried out in vivo in the cardiac muscle of rats showed that angiotensin-II is capable of stimulating the tyrosine phosphorylation of IRS-1 and IRS-2 and their respective associations with PI-3 kinase, but inhibiting the catalytic activity of that enzyme [23,24]. AT1 receptor blockers are able to prevent this phenomenon [28]. In conclusion, studies both in vivo and in cell culture showed that angiotensin II is able to inhibit insulin signaling through PI-3 chi xylanase pathway, possibly by activating its AT1 receptor.

Reduction of tyrosine phosphorylation and increases of serine phosphorylation of the beta subunit of the insulin receptor, as well as IRS-1, were proposed as molecular mechanisms of insulin resistance [24,29,30]. In this context, angiotensin II is capable of inducing phosphorylation on serine of three components of insulin signaling pathway: the insulin receptor itself, IRS-1 and p85 regulatory subunit of PI-3 kinase [24] (Fig. 44.3). Studies on rat aortic smooth muscle cell culture showed that angiotensin II induced serine phosphorylation at the two receptors. This serine phosphorylation of IRS-1 decreased its ability to bind to the beta subunit of the insulin receptor. In addition, the use of serine/threonine phosphatase inhibitors mimicked the effects of angiotensin-II on IRS-1 and PI-3 kinase

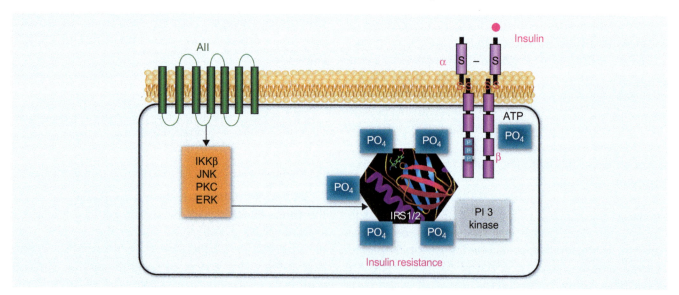

FIG. 44.3 Serine kinases activated by AII and able to phosphorylate IR and IRS-1 in serine and induce insulin resistance (acronyms, see text).

[24]. It was recently demonstrated that angiotensin II is able to block the vasodilator effects of insulin through the serine phosphorylation of IRS-1 on Ser[312] and Ser[616] residues, via activation of JNK and ERK1/2 serine kinase, which leads to negative modulation of the IRS-1/PI-3 kinase/Akt/eNOS, decreasing production of insulin-induced nitric oxide [31].

The Role of SOCS-3 in the Insulin/Angiotensin Interaction

As discussed previously, activation of the angiotensin-II AT1 receptor leads to activation of the intracellular JAK/STAT pathway. This route, in addition to activating transcription of several different intracellular effectors, also induces expression of the family of cytokine signaling suppressor proteins (SOCS), including the SOCS-3 which has a negative feedback effect on this pathway, and possibly, characterizing the most enduring form of signaling inhibition through the JAK/STAT pathway [32].

Both insulin and angiotensin-II are able to induce the expression of proteins of the SOCS family into animal tissues in vivo and in cell cultures [33–37]. After induced, SOCS-3 binds to Tyr[960] residue of the insulin receptor and reduces its capacity to phosphorylate in tyrosine the STAT-5b, IRS-1 and IRS-2 [33,35,38]. Furthermore, SOCS-3 is able to bind to IRS-1/2, causing proteosomic degradation through a ubiquitination-dependent mechanism [39] (Fig. 44.4).

The SOCS-3 interaction induced by angiotensin-II, with insulin signaling pathways proteins, has molecular and functional implications. At the molecular level, this interaction prevents the IRS-1 and IRS-2 phosphorylation in thyrosine and phosphorylation in serine and activation of AKT [36]. Furthermore, induction of SOCS-3 by angiotensin-II prevents activation of JAK-2/STAT-5b pathway by insulin [36]. At a functional level, increased expression of SOCS-3 induced by angiotensin-II prevents intracellular GLUT4 translocation to the surface of heart cells membrane and isolated heart cells [36]. Therefore, SOCS-3 represents a distal interface in insulin and angiotensin-II signaling systems. On one hand, this may represent protection of the insulin target organs against a constant growth stimulus, on the other induction of SOCS-3 expression may prevent an efficient transmission of the insulin signal through the metabolic pathway, making it difficult the acquisition of energy, and in the endothelium, preventing insulin-induced vasodilatation.

The Role of MAP Kinase in Insulin/Angiotensin Interaction

ERK1/2 is another distal point of interaction between Insulin and angiotensin-II signaling pathways. The insulin signal can activate ERK1/2 by two different molecular mechanisms. Phosphorylation and activation of the insulin receptor trigger its interaction and tyrosine phosphorylation of a protein called SHC. Once phosphorylated, SHC recruits

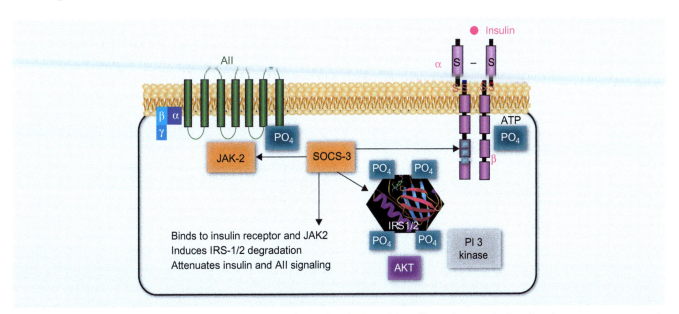

FIG. 44.4 Activation of SOCS-3 protein, cross-talk insulin/IIA and insulin resistance induction (acronyms, see text).

another protein called GRB2 and induces the activation of the RAS-RAF-MEK-ERK pathway [40,41]. Activated insulin receptor also phosphorylates in tyrosine the IRS-1 and IRS-2 substrates, which can interact directly with GRB2 then activate the RAS-RAF-MEK-ERK pathway [42].

Angiotensin-II can also activate ERK1/2 in two distinct ways. The first is through its binding to the AT1 receptor, which activates the Gq protein, leading to increased cytoplasmic Ca^{++} content and subsequent activation of the EGF receptor. Once activated, the EGF receptor recruits SHC/GRB2 and then activates the RAS-RAF-MEK-ERK pathway [43,44]. The second mechanism also depends on activation of the Gq protein by AT1 and increase of the cytoplasmic Ca^{++} content, which causes activation of PKC that activates the RAF1, which can activate MEK and ERK1/2, regardless of the RAS activation [45,46].

In endothelial vascular cells, insulin stimulates nitric oxide production through activation of the IRS-1/PI-3 kinase/AKT pathway, which leads to phosphorylation and activation of eNOS [47,48]. Pretreatment of these cells with angiotensin II leads to activation of ERK 1/2 and JNK by a molecular mechanism dependent on the AT1 activation. These serine kinases, once activated, promote serine phosphorylation of IRS-1, preventing activation of eNOS mediated by insulin [31]. Therefore, in this tissue, angiotensin-II is capable of inducing insulin resistance through the IRS-1 pathway by a mechanism dependent on JNK and ERK 1/2 (Fig. 44.4).

According to this finding, the simultaneous treatment of guinea pigs with insulin and angiotensin leads to a blockade of insulin signaling via IRS-1/PI-3 kinase/AKT pathway, at the same time as it leads to an increase in signaling through the ERK pathway [27]. This mechanism is constitutively active in an animal model of metabolic syndrome and seems to have a prominent role in insulin resistance-related myocardial hypertrophy [27].

Angiotensin Induces Insulin Resistance in the Vessel and Endothelial Dysfunction

Thus far, we have shown that angiotensin-II is capable of negatively modulating transmission of insulin signal by blocking the IRS/PI-3 kinase/AKT pathway, and that conversion enzyme inhibitors can reverse these effects in insulin resistance situations. It is established that this same pathway is

capable of inducing of nitric oxide production and, therefore, vasodilation, in endothelial cells, in which AKT, after activated by the action of PI-3 kinase, is able to phosphorylate eNOS and cause its activation [5]. These stimulatory effects of insulin on eNOS and nitric oxide may be important in prevention of endothelial dysfunction and early stages of atherosclerosis induced by LDL oxidized by smoking and other factors [5]. Angiotensin-II, having inhibitory effect on this signaling pathway, can decrease nitric oxide production and at the same time it stimulates of oxygen-free radicals production, increasing oxidative stress and increasing the degradation of nitric oxide. Therefore, angiotensin-II simultaneously has a deleterious effect on insulin-mediated glucose metabolism in skeletal and cardiac muscle; it may also affect insulin signaling in blood vessels, contributing to endothelial dysfunction [5].

Effects of Conversion Enzyme Inhibitors and Increases in Bradykinin Levels on Metabolic and Cellular Insulin Actions

In clinical practice, beneficial effects of converting enzyme and ATI inhibitors in patients with type 2 *diabetes mellitus* have been known for a long time [49–51]. Hyperactivity of the renin-angiotensin system impairs insulin intracellular signaling and contributes to insulin resistance observed in essential hypertension [8]. This hyperactivity of the renin-angiotensin system modulates insulin signaling, decreasing muscle uptake of insulin-induced glucose [51,52].

At the molecular level, captopril improves insulin sensitivity by increasing the tyrosine phosphorylation of IRS-1 and its binding and activation of PI-3 kinase after insulin stimulation [28]. This data suggests that the improved sensitivity to insulin induced by treatment with captopril is related to its action in early stages of insulin signal transmission, which shows the importance of the interaction between insulin and angiotensin-II.

It should be noted that the conversion enzyme inhibition effects go beyond merely decreasing angiotensin production, a powerful vasoconstrictor, but also include the reduction of bradykinin degradation, an effective vasodilator. In fact, the conversion enzyme has more affinity for bradykinin than for angiotensin-I, and may be considered as a kinase rather than an angiotensininase [53]. Previous studies show that bradykinin can improve insulin sensitivity in regard to glucose metabolism [54], and some authors relate this improvement to brady-kinin ability to enhance the capillary blood flow, improving distribution of insulin and glucose for the muscle tissue [49].

Bradykinin causes vasodilation, since it induces release of nitric oxide by endothelial cells and, as a consequence, increases the transcapillary glucose transport [54]. Therefore, bradykinin is responsible for some of the beneficial effects of conversion enzyme inhibitors on the insulin transmission signal, since it is able to increase tyrosine phosphorylation of both IR and IRS-1 [28].

Inflammatory Pathways and Insulin Resistance in the Endothelium

Although low-grade chronic inflammation is evident in the adipose tissue of obese individuals, the mechanisms by which inflammatory cytokines and fatty acids mediate insulin resistance are not yet fully understood [2,4,55–58]. Inhibition of insulin intracellular signaling pathways is a primary mechanism by which inflammatory signals may lead to insulin resistance in different tissues, including the endothelium.

Treatment of hepatocytes and adipocytes, as well as endothelial cells with TNF-α, IL-1, IL-6, or high doses of free fatty acids, reduces insulin-stimulated autophosphorylation of IR by 20%–50%. In addition, other experiments with different cell types in culture have demonstrated that TNF-α enhances the phosphorylation of IRS-1 and IRS-2 on serine residues [55]. These modifications are enough to reduce the IRS-1 and IRS-2 ability to interact with IR, as well as block the subsequent events of insulin signaling, resulting in insulin resistance states. Inhibitory phosphorylation of IRS-1 on serine has also been observed in obesity and other insulin resistance situations. Genetic and biochemical studies indicated the JNK (c-Jun N-terminal kinase) pathway as one of the responsible for this inhibitory phosphorylation of IRS-1, and documented activation of the JNK pathway by TNF-α. Subsequent studies have shown that activation of JNK results in phosphorylation of IRS-1 in serine residue 307, and IRS-1 phosphorylation in that residue was sufficient to mediate TNF-α-induced insulin resistance in some models [2,55,56].

In animal models of diet-induced obesity and insulin resistance, there is increased activation of JNK in the liver, muscle and vessels. Obese mice that do not express JNK isoform 1 (JNK-1) have protection against insulin resistance associated with obesity. In addition, animals with JNK-1 deficiency have reduced adiposity, suggesting that this kinase is involved in the regulation of obesity and diabetes. Pharmacological inhibition of JNK also avoids the phosphorylation of IRS-1 in serine 307 stimulated by TNF-α and restores IRS-1 phosphorylation in insulin-stimulated tyrosine, increasing sensitivity to insulin. Therefore, as JNK can be activated by TNF-α and other cytokines, in addition to fatty acids, and these are elevated in the insulin resistance syndrome, JNK seems to be a key element for converging different signals that reduce the action of insulin. The JNK pathway is also an important route of inflammatory response, and thus can establish a connection between stress/inflammation and metabolic regulation, as well as being a pathway with therapeutic potential for diabetes and atherosclerosis [55].

Recently, it has also been reported that mTOR (mammalian target of rapamycin) can phosphorylate IRS-1 in serine in the presence of TNF-α. Suppression of serines/threonine-phosphatases or the activation of protein tyrosine phosphatases (PTPases) may also be important in insulin resistance caused by TNF-α [2,4,56].

In addition to the JNK pathway, another TNF-α-activated inflammatory pathway has received much attention due to its potential to establish connections between inflammatory response and insulin resistance: the -IKK-NFκB pathway. In cell culture, blocking the activity of this pathway may prevent the onset of insulin resistance induced by TNF-α. In animals with genetically or diet-induced obesity, blocking the IKKβ activity via administration of high doses of salicylates or mutation in an IKKβ allele results in improved insulin sensitization. More recently, another IKK isoform, IKKε, has been highlighted. Knockout animals for IKKε are protected from insulin resistance [55–57].

IKKε and IKKβ [53] may interfere with insulin signaling in at least two ways: first, they can phosphorylate IRS-1 directly on serine residues; second, they may indirectly activate nuclear transcription factor κB (NF-κB), a transcription factor that, among other targets, may stimulate production of various inflammatory mediators, including TNF-α and IL-6. Activation of these kinases in obesity, especially IKKS and JNK, highlights the overlap of metabolic and inflammatory pathways: these are the same kinases that are activated in the innate immune response by TLR (Toll-like receptor) in response to LPS, peptidoglycan, double-stranded RNA, and other microbial products [57,59–62]. TLRs play a critical role in the innate immune response in mammals. Studies in animals with mutation or genetic inactivation of TLR4 are protected against insulin resistance in liver and muscle and animals with mutation of this receptor are protected against diet-induced obesity. These results suggest that TLR4 is a cross-talk modulator between the metabolic and inflammatory pathways [57,59,60] (Fig. 44.5).

Inducible nitric oxide synthase (iNOS) and SOCS (suppressors of cytokine signaling) proteins, whose genes are targets of IKK and JNK pathways, are also implied in insulin resistance promoted by TNF-α. iNOS expression is stimulated by TNF-α and is elevated in obesity; mice with mutations in the iNOS gene develop less resistance to insulin associated with obesity than their intact iNOS gene controls [63,64]. Expression of various SOCS isoforms, especially of SOCS-3, increases in TNF-α in the presence of obesity and can induce insulin resistance, probably due to increased degradation of IRS-1 mediated by proteosomes [32–35].

Another insulin resistance mechanism has been described: S-nitrosation of insulin receptor, of IRS-1 and AKT. Nitric oxide produced by iNOS can induce insulin resistance in muscle by means of a mechanism that involves S-nitrosation of IR, IRS-1 and Akt in vitro and in animal models of obesity and insulin resistance [63,64].

ENDOPLASMIC RETICULUM STRESS AND ENDOTHELIAL DYSFUNCTION

The endoplasmic reticulum is an organelle that integrates and regulates the synthesis, folding and transport of at least one third of the proteins of a cell. Protein folding is essential for survival and cellular functions, and numerous processes help the cell to preserve homeostasis [65–68].

Thus, some physiological or pathological conditions challenge this endoplasmic reticulum homeostasis, especially its protein assembly capacity. The organelle adapts to the new situation using an adaptive response system known as unfolded protein response—UPR (unfolded protein response). Such conditions that can activate the UPR include increased protein synthesis, presence of mutant proteins, inhibition of protein glycosylation, unbalanced levels of calcium, energy and glucose deprivation, high-fat diet, hypoxia, pathogens or components or toxins associated with pathogens [65,66]. This adaptive response helps the cell to adapt to endoplasmic reticulum stress. Three signaling pathways are activated in UPR: PERK-eIF2α, IRE1α and activation of ATF6α transcription factor (Fig. 44.6). Thus, upon external insults or excessive protein synthesis, inappropriate folding of proteins in organelle takes place, and this accumulation of malformed proteins is known as endoplasmic reticulum stress. In response to this stress, there is UPR activation to restore homeostasis, and if the situation is not resolved, cell death may occur.

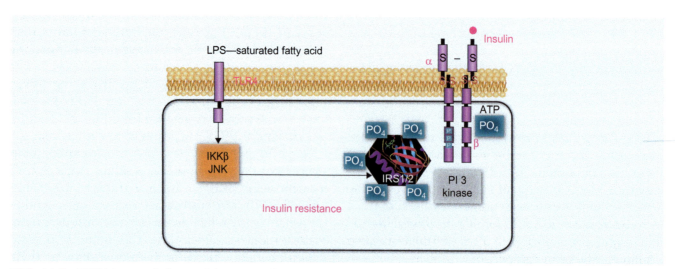

FIG. 44.5   TLR4 is a modulator of the *cross-talk* between metabolic and inflammatory pathways, and their activation (by saturated LPS or AG) can induce insulin resistance through activation of serine kinases (acronyms, see text).

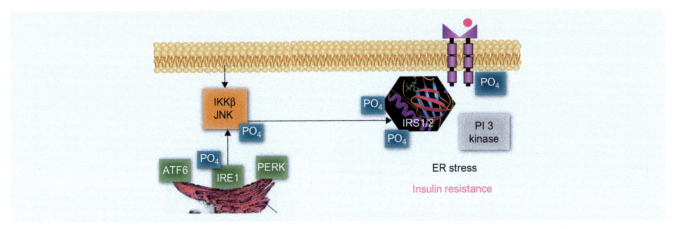

FIG. 44.6   Activation of endoplasmic reticulum stress and insulin resistance (acronyms, see text).

UPR, by means of the three activated pathways, attempts to reduce protein synthesis and simultaneously degrade malformed proteins. It is noteworthy that in the activation of one of these pathways, the IRE-1α, subsequently, takes place activation of JNK protein, which is certainly the serine kinase most studied as an inducer of IRS-1 serine phosphorylation with consequent insulin resistance (Fig. 44.6).

Obesity and DM2 overload the functional capacity of the endoplasmic reticulum, thus inducing stress. The endoplasmic reticulum stress and inflammatory pathways can be connected in various ways. The activation of the IRE1α induces activation of JNK and AP-1, increasing the expression of several proinflammatory genes.

In addition, IRE1α and PERK activate IKKβ-NFκB pathway, leading to the inflammatory response [55–57].

MECHANISMS RESPONSIBLE FOR ENDOTHELIAL DYSFUNCTION IN DIABETES

Progression of vascular complications of diabetes depends on the severity and duration of hyperglycemia. One of the key steps in the induction of endothelial dysfunction induced by hyperglycemia is oxidative stress originating both in the cytoplasm and in the mitochondria [69–79].

Endothelial cells are capable of producing reactive oxygen species (ROS) from a variety of enzyme sources. Mechanisms contributing to the increase of oxidative stress in endothelium, in diabetes, are shown in Figs. 44.7 and 44.8, and include: elevation of hyperglycemia-induced ROS both from source of cytosolic and mitochondrial; change on glycolytic flow to alternative metabolic pathways, induced by hyperglycemia; blocking of pentoses pathways (PPP) induced by hyperglycemia; formation of glycation products and advanced glycation end products (AGE); and activation of protein kinase C (PKC) [70,71]. These mechanisms have wide demonstration in vitro in the presence of hyperglycemia, but still need better studies in vivo.

HYPERGLYCEMIA INDUCES INCREASES IN ROS PRODUCTION

Increased Cytosolic ROS Production

The increase of ROS production, induced by hyperglycemia, may be the result of autoxidation of glucose or activation of NADPH oxidase [76,77]. The increase of ROS derived from NADPH oxidase and xanthine can induce eNOS uncoupling (eNOS ceases to produce nitric oxide and citrulline and instead produces ROS) in addition to the fact that superoxide anion can react directly with nitric oxide to form peroxynitrite thus contributing to reduce it. The uncoupling of eNOS results in increased oxidative stress due to superoxide generation in place of nitric oxide, creating a vicious circle. Furthermore, nitric oxide reduction contributes significantly to endothelial dysfunction. Another mechanism by which there can be eNOS decoupling is the reduced availability of arginine, nitric oxide precursor and tetrahydrobiopterin cofactor (BH4), O-glycosylation via increased HBP, as well as an increase of AGE.

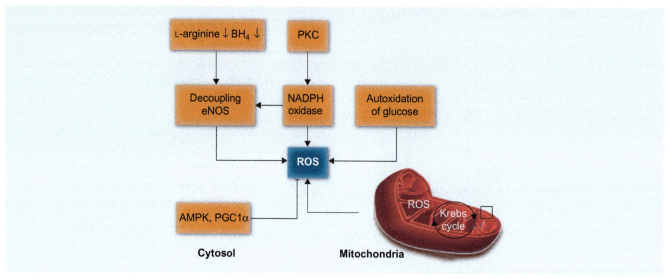

FIG. 44.7   Increased ROS production induced by hyperglycemia with mitochondrial and cytosolic components (acronyms, see text).

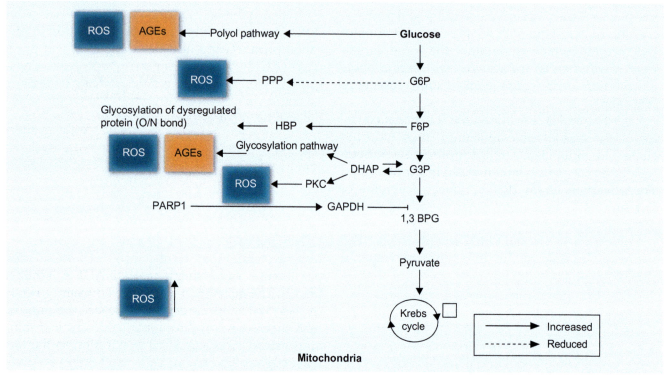

FIG. 44.8   Changes in endothelial metabolic pathways induced by hyperglycemia and ROS production (acronyms, see text).

Increased Mitochondrial ROS Production

Hyperglycemia induces mitochondriopathy, evinced by increased biogenesis and autophagy, change in mitochondrial function and increased fragmentation of this organelle.

Endothelial cell produces its energy primarily from anaerobic glycolysis (and not from oxidative phosphorylation in the mitochondria), and endothelial mitochondria are essential for the calcium homeostasis and ROS generation, and are also cell death sensors and initiators. As a result of mitochondrial dysfunction, induced by hyperglycemia, there is an increase in ROS production (the electron transport chain) and calcium overload, worsening oxidative stress and endothelial dysfunction [71].

It is important to mention that the AMPK protein and PGC1α transcription factor can participate in the regulation of ROS production increases by hyperglycemia. Nitric oxide, through elevation of PGC1α levels, protects against ROS damage by regulation of mitochondrial antioxidant defense. However, this point is controversial, and data indicate that hyperglycemia can raise PGC1α levels.

On the other hand, it is well established that AMPK activation reduces generation of hyperglycemia-induced ROS, since it limits NADPH oxidase activity, increases the expression of mitochondrial antioxidant enzymes and increases mitochondrial biogenesis. Thus, AMPK activation reduces endothelial dysfunction induced by hyperglycemia [71].

Change of Glycolytic Flow to Alternative Pathway Induced by Hyperglycemia

Increased production of ROS and reactive nitrogen species (RNS) activate the poly-ADP-ribose polymerase (PARP1), an enzyme for DNA repair, as a consequence of oxidative DNA damage and subsequent inhibition of glyceraldehyde 3-phosphate dehydrogenase (GAPDH) by ribosylation. Inhibition of GAPDH induces accumulation of glycolytic intermediates and subsequent activation of three metabolic pathways: the polyols pathways, the hexosamine biosynthetic pathway (HBP), and the glycation pathway.

GLYCOLYTIC FLOW CHANGE TO ALTERNATIVE PATHWAY INDUCED BY HYPERGLYCEMIA

Flow Increase by the Polyols Pathway

Aldose reductase is the first polyol pathway enzyme. It is an enzyme that has a low affinity for glucose (high K_m), and, therefore, under normal glucose levels, glucose metabolization through this pathway is greatly reduced. However, in a hyperglycemic environment, increases in intracellular glucose results in greater enzymatic conversion of glucose in sorbitol, NADPH-consuming reaction.

In this polyol pathway, sorbitol is oxidized to fructose by the sorbitol dehydrogenase enzyme, with transformation of NAD+ into NADH. This pathway, for the same level of hyperglycemia, has variable activity depending on the tissue analyzed,

suggesting that the contribution of the polyol pathway to the complications of diabetes is tissue-specific. Its activation may contribute to the onset of chronic complications, leading to increased oxidative stress, because the NA-DPH consumed on the pathway will no longer regenerate reduced glutathione (GSH), which is an important defense mechanism against ROS. Furthermore, by reducing the GAPDH activity, the glyceraldehyde 3-phosphate levels increase, intensifying the DAG and AGE precursors synthesis, with subsequent activation of PKC [70].

CHANGE IN GLYCOLYTIC FLOW TO ALTERNATIVE PATHWAY INDUCED BY HYPERGLYCEMIA

Increased Flow Through the BPH

Excess glucose enters into the BPH pathway through fructose-6-phosphate, which is converted to glucosamine-6-phosphate and then into N-acetylglucosamine uridine 5'-phosphate (UDP-GlcNAc). Under normal conditions, UDP-GlcNAc is important for glycosylation, but in hyperglycemia this glycosylation is deregulated and results in alteration of expression of certain proteins, such as eNOS [71].

Increased Flow by the Glycation Pathway and Increased Production of Advanced Glycation End Products

At first, it was thought that the AGE originated from the nonenzymatic reaction between glucose and extracellular proteins. However, the rate of AGE formation from glucose is much slower than from glucose dicarbonyl precursors generated intracellularly, and today it is believed that the intracellular hyperglycemia is the primary event initiating the formation of AGE in intra and extracellular means. The AGE can be produced in the following ways: autoxidation of glyoxal glucose; decomposition of Amadori products in 3-deoxiglucosona; and fragmentation of glyceraldehyde 3-phosphate and dihydroxyacetone phosphate in methylglyoxal. These intracellular dicarbonyls react with amine groups of intra- and extra-cellular proteins to form AGEs, which can be found in retinal vessels and glomeruli of diabetic patients.

Production of intracellular AGE precursors induces target cell injury by three different mechanisms: intracellular proteins modified by AGE have changed functions; extracellular matrix components modified by AGE precursors interact abnormally with other components and with matrix receptors (integrins) on cells; and plasma proteins modified by AGE precursors bind to AGE receptors on endothelial cells, mesangial, and macrophages, inducing the production of ROS. Binding to AGE (RAGE) receptors and production of ROS activate NF-κB, causing pathological changes in gene expression.

Regarding the first two, some matrix proteins and components modified by AGE are: proteins involved in endocytosis of macromolecules, some peptidases, collagen types I and IV. This can modify the normal intracellular degradation process of proteins, and also change functional properties of matrix molecules, changing the elasticity of vessels and endothelial function.

Regarding the third, the cell membrane proteins that bind to AGE are: OST-48, 80K-H, galectin-3, and RAGE. Such modified bonds induce cellular modifications maintained, including increased expression of cytokines and growth factors (IL-1, TNF-α, TGF-β, PDGF) by macrophages and mesangial cells, and increased expression of procoagulant and pro inflammatory (thrombomodulin, tissue factor, VCAM-1) by endothelial cells.

A randomized, double-blind, placebo-controlled trial showed that the use of an AGE inhibitor (aminoguanidine) in *diabetes mellitus* type 1 with nephropathy has reduced proteinuria and progression of nephropathy and retinopathy.

Experimental trials have shown that blockade of RAGE suppresses the development of macrovascular disease in mice as well as development of nephropathy and periodontal disease.

BLOCKING OF PPP INDUCED BY HYPERGLYCEMIA

Glucose 6-phosphate, fructose 6-phosphate and glyceraldehyde 3-phosphate are glycolytic intermediates generated outside the PPP. In hyperglycemic conditions, PPP may have a protective role, reducing the flow in the three aforementioned ways, reducing damage and also by increasing NADPH and GSH, which are protectors against oxidative stress. However, hyperglycemia reduces the flow by PPP pathway, increasing oxidative stress and reducing nitric oxide production [71].

ACTIVATION OF PKC INDUCED BY HYPERGLYCEMIA

Protein kinase C is a broad expression enzyme, which participates in various cell signaling pathways. Its activity is increased in the vascular tissue of diabetic patients, including large vessels, retina and glomerulus. Out of the ten PKC isoforms, alpha, beta, and delta are the most commonly implicated in vascular complications of diabetes. In animals that do not express any of these PKC isoforms, there is protection of nephropathy, retinopathy or atherosclerosis. Treatment with specific inhibitors of PKC-beta isoform reduces the activity of this enzyme in the retina and glomerulus of diabetic animals, and, consequently, normalizes the average time of circulation in the retina and the glomerular filtration rate, and partially corrects microalbuminuria and glomerular expansion [70,71].

The beta and delta isoforms can be activated by the second diacylglycerol (DAG) messenger. Intracellular hyperglycemia increases DAG content in microvascular cells of the retina and glomerulus of diabetic animals, once it again increases synthesis of this lipid from the dihydroxyacetone-phosphate glycolytic intermediate through reduction of glycerol 3-phosphate. PKC can also be activated by ROS.

In experimental models of diabetes, activation of PKC isoforms appears to mediate changes in blood flow in the retina and in the kidney by reducing nitric oxide production—consequence of the larger decoupling of eNOS—and by increased activity of endothelin-1. Other effects arising from the PKC activation include of NADPH-dependent membranes oxidase and the nuclear transcription factor NF-κB (increasing apoptosis in retinal pericytes), increased TGF-β expression, fibronectin, collagen IV, and also enhanced expression of activator 1 of the plasminogen inhibitor (PAI-1) and VEGF.

Initial studies in humans show that inhibitors of PKCβ induce improvement of retinopathy and nephropathy, but phase III clinical trials are awaited. However, in order to obtain significant effects on prevention of nephropathy and retinopathy, it would be important the development of multiple isoform inhibitors such as alpha, beta, and delta.

CONCLUSIONS

The impact of hyperglycemia and, potentially, glycemic variations in endothelial cell function seems to be the connection between diabetes and accelerated atherosclerosis. Increased glucose metabolism produces elevated levels of ROS, with clear consequences on endothelial dysfunction. In endothelial cells, monocytes and platelets also are exposed to blood glucose fluctuations in diabetes and have increased ROS production, with effect in inflammatory and thrombotic phenomena [79]. Thus, the impact of hyperglycemia on endothelial cells, monocytes, and platelets with increased ROS production probably represents the causal connection between diabetes and atherosclerosis.

References

[1] Toda N, Okamura T. Obesity impairs vasodilatation and blood flow increase mediated by endothelial nitric oxide: an overview. J Clin Pharmacol 2013;53(12):1228–39.

[2] Samuel VT, Shulman GI. The pathogenesis of insulin resistance: integrating signaling pathways and substrate flux. J Clin Invest 2016;126(1):12–22.

[3] Kahn BB, Flier JS. Obesity and insulin resistance. J Clin Invest 2000;106:473–81.

[4] Lackey DE, Olefsky JM. Regulation of metabolism by the innate immune system. Nat Rev Endocrinol 2016;12(1):15–28.

[5] Reaven G. The metabolic syndrome or the insulin resistance syndrome? Different names, different concepts, and different goals. Endocrinol Metab Clin North Am 2004;33:283–303.

[6] Wang CC, Goalstone ML, Draznin B. Molecular mechanisms of insulin resistance that impact cardiovascular biology. Diabetes 2004;53:2735–40.

[7] Reaven G, Abbasi F, McLaughlin T. Obesity, insulin resistance, and cardiovascular disease. Recent Prog Horm Res 2004;59:207–23.

[8] Natali A, Ferrannini E. Hypertension, insulin resistance, and the metabolic syndrome. Endocrinol Metab Clin North Am 2004;33:417–29.

[9] Shulman GI. Cellular mechanisms of insulin resistance in humans. Am J Cardiol 1999;84:3J–10J.

[10] Saltiel AR, Kahn CR. Insulin signalling and the regulation of glucose and lipid metabolism. Nature 2001;414:799–806.

[11] Saad MJ, Carvalho CR, Thirone AC, et al. Insulin induces tyrosine phosphorylation of JAK2 in insulin-sensitive tissues of the intact rat. J Biol Chem 1996;271:22100–4.

[12] Velloso LA, Carvalho CR, Rojas FA, et al. Insulin signalling in heart involves insulin receptor substrates-1 and -2, activation of phosphatidylinositol 3-kinase and the JAK 2- growth related pathway. Cardiovasc Res 1998;40(1):96–102.

[13] Van Gaal LF, Mertens IL, De Block CE. Mechanisms linking obesity with cardiovascular disease. Nature 2006;444:875–80.

[14] Hotamisligil GS, Peraldi P, Budavari A, et al. IRS-1-mediated inhibition of insulin receptor tyrosine kinase activity in TNF-alpha- and obesity-induced insulin resistance. Science 1996;271:665–8.

[15] Zecchin HG, Bezerra RM, Carvalheira JB, et al. Insulin signalling pathways in aorta and muscle from two animal models of insulin resistance—the obese middle-aged and the spontaneously hypertensive rats. Diabetologia 2003;46:479–91.

[16] Scheen AJ. VALUE: analysis of results. Lancet 2004;364:932–3 [Author reply 935].

[17] Knowler WC, Barrett-Connor E, Fowler SE, et al. Reduction in the incidence of type 2 diabetes with lifestyle intervention or metformin. N Engl J Med 2002;346:393–403.

[18] Massiera F, Bloch-Faure M, Ceiler D, et al. Adipose angiotensinogen is involved in adipose tissue growth and blood pressure regulation. FASEB J 2001;15:2727–9.

[19] Engeli S, Bohnke J, Gorzelniak K, et al. Weight loss and the renin-angiotensin-aldosterone system. Hypertension 2005;45:356–62.

[20] Feldman R. ACE inhibitors versus AT1 blockers in the treatment of hypertension and syndrome X. Can J Cardiol 2000;16(Suppl. E):41E–44E.

[21] Scheen AJ. Prevention of type 2 diabetes mellitus through inhibition of the renin-angiotensin system. Drugs 2004;64:2537–65.

[22] Saad MJ, Velloso LA, Carvalho CR. Angiotensin II induces tyrosine phosphorylation of insulin receptor substrate 1 and its association with phosphatidylinositol 3-kinase in rat heart. Biochem J 1995;310 (Pt 3):741–4.

[23] Velloso LA, Folli F, Sun XJ, et al. Cross-talk between the insulin and angiotensin signaling systems. Proc Natl Acad Sci U S A 1996;93: 12490–5.

[24] Folli F, Kahn CR, Hansen H, et al. Angiotensin II inhibits insulin signaling in aortic smooth muscle cells at multiple levels. A potential role for serine phosphorylation in insulin/angiotensin II crosstalk. J Clin Invest 1997;100:2158–69.

[25] Marrero MB, Schieffer B, Paxton WG, et al. Direct stimulation of Jak/STAT pathway by the angiotensin II AT1 receptor. Nature 1995;375:247–50.

[26] Venema RC, Venema VJ, Eaton DC, et al. Angiotensin II-induced tyrosine phosphorylation of signal transducers and activators of transcription 1 is regulated by Janus-activated kinase 2 and Fyn kinases and mitogen-activated protein kinase phosphatase 1. J Biol Chem 1998;273:30795–800.

[27] Carvalheira JB, Calegari VC, Zecchin HG, et al. The cross-talk between angiotensin and insulin differentially affects phosphatidylinositol 3-kinase- and mitogen-activated protein kinase-mediated signaling in rat heart: implications for insulin resistance. Endocrinology 2003;144:5604–14.

[28] Carvalho CR, Thirone AC, Gontijo JA, et al. Effect of captopril, losartan, and bradykinin on early steps of insulin action. Diabetes 1997;46:1950–7.

[29] Tanti JF, Gremeaux T, van Obberghen E, et al. Serine/threonine phosphorylation of insulin receptor substrate 1 modulates insulin receptor signaling. J Biol Chem 1994;269:6051–7.

[30] Mothe I, Van Obberghen E. Phosphorylation of insulin receptor substrate-1 on multiple serine residues, 612, 632, 662, and 731, modulates insulin action. J Biol Chem 1996;271:11222–7.

[31] Andreozzi F, Laratta E, Sciacqua A, et al. Angiotensin II impairs the insulin signaling pathway promoting production of nitric oxide by inducing phosphorylation of insulin receptor substrate-1 on Ser312 and Ser616 in human umbilical vein endothelial cells. Circ Res 2004;94:1211–8.

[32] Krebs DL, Hilton DJ. SOCS: physiological suppressors of cytokine signaling. J Cell Sci 2000;113(Pt 16):2813–9.

[33] Emanuelli B, Peraldi P, Filloux C, et al. SOCS-3 is an insulin-induced negative regulator of insulin signaling. J Biol Chem 2000;275:15985–91.

[34] Sadowski CL, Choi TS, Le M, et al. Insulin induction of SOCS-2 and SOCS-3 mRNA expression in C2C12 skeletal muscle cells is mediated by Stat5*. J Biol Chem 2001;276:20703–10.

[35] Emanuelli B, Peraldi P, Filloux C, et al. SOCS-3 inhibits insulin signaling and is up-regulated in response to tumor necrosis factor-alpha in the adipose tissue of obese mice. J Biol Chem 2001;276:47944–9.

[36] Calegari VC, Bezerra RM, Torsoni MA, et al. Suppressor of cytokine signaling 3 is induced by angiotensin II in heart and isolated cardiomyocytes, and participates in desensitization. Endocrinology 2003;144:4586–96.

[37] Torsoni MA, Carvalheira JB, Calegari VC, et al. Angiotensin II (AngII) induces the expression of suppressor of cytokine signaling

(SOCS)-3 in rat hypothalamus—a mechanism for desensitization of AngII signaling. J Endocrinol 2004;181:117–28.

[38] Ueki K, Kondo T, Kahn CR. Suppressor of cytokine signaling 1 (SOCS-1) and SOCS-3 cause insulin resistance through inhibition of tyrosine phosphorylation of insulin receptor substrate proteins by discrete mechanisms. Mol Cell Biol 2004;24:5434–46.

[39] Rui L, Yuan M, Frantz D, et al. SOCS-1 and SOCS-3 block insulin signaling by ubiquitin-mediated degradation of IRS1 and IRS2. J Biol Chem 2002;277:42394–8.

[40] Giorgetti S, Pelicci PG, Pelicci G, et al. Involvement of src-homology/collagen (SHC) proteins in signaling through the insulin receptor and the insulin-like-growth-factor-I-receptor. Eur J Biochem 1994;223:195–202.

[41] Holt KH, Kasson BG, Pessin JE. Insulin stimulation of a MEK dependent but ERK-independent SOS protein kinase. Mol Cell Biol 1996;16:577–83.

[42] Sarbassov DD, Peterson CA. Insulin receptor substrate-1 and phosphatidylinositol 3-kinase regulate extracellular signal-regulated kinase-dependent and -independent signaling pathways during myogenic differentiation. Mol Endocrinol 1998;12:1870–8.

[43] Eguchi S, Iwasaki H, Ueno H, et al. Intracellular signaling of angiotensin II-induced p70 S6 kinase phosphorylation at Ser (411) in vascular smooth muscle cells. Possible requirement of epidermal growth factor receptor, Ras, extracellular signal-regulated kinase, and Akt. J Biol Chem 1999;274:36843–51.

[44] Werry TD, Sexton PM, Christopoulos A. "Ins and outs" of seven-transmembrane receptor signalling to ERK. Trends Endocrinol Metab 2005;16:26–33.

[45] Hunyady L, Turu G. The role of the AT1 angiotensin receptor in cardiac hypertrophy: angiotensin II receptor or stretch sensor? Trends Endocrinol Metab 2004;15:405–8.

[46] Zou Y, Komuro I, Yamazaki T, et al. Protein kinase C, but not tyrosine kinases or Ras, plays a critical role in angiotensin II-induced activation of Raf-1 kinase and extracellular signal-regulated protein kinases in cardiac myocytes. J Biol Chem 1996;271:33592–7.

[47] Zeng G, Nystrom FH, Ravichandran LV, et al. Roles for insulin receptor, PI3-kinase, and Akt in insulin-signaling pathways related to production of nitric oxide in human vascular endothelial cells. Circulation 2000;101:1539–45.

[48] Zecchin HG, Priviero FB, Souza CT, et al. Defective insulin and acetylcholine induction of endothelial cell-nitric oxide synthase through insulin receptor substrate/Akt signaling pathway in aorta of obese rats. Diabetes 2007;56(4):1014–24.

[49] Jauch KW, Hartl W, Guenther B, et al. Captopril enhances insulin responsiveness of forearm muscle tissue in non-insulin-dependent diabetes mellitus. Eur J Clin Invest 1987;17:448–54.

[50] Moan A, Risanger T, Eide I, et al. The effect of angiotensin II receptor blockade on insulin sensitivity and sympathetic nervous system activity in primary hypertension. Blood Press 1994;3:185–8.

[51] Kurtz TW, Pravenec M. Antidiabetic mechanisms of angiotensin-converting enzyme inhibitors and angiotensin II receptor antagonists: beyond the renin-angiotensin system. J Hypertens 2004;22:2253–61.

[52] Fukuda N, Satoh C, Hu WY, et al. Endogenous angiotensin II suppresses insulin signaling in vascular smooth muscle cells from spontaneously hypertensive rats. J Hypertens 2001;19:1651–8.

[53] Weissmann L, Quaresma PG, Santos AC, et al. IKKε is key to induction of insulin resistance in the hypothalamus, and its inhibition reverses obesity. Diabetes 2014;63(10):3334–45.

[54] Damas J, Garbacki N, Lefebvre PJ. The kallikrein-kinin system, angiotensin converting enzyme inhibitors and insulin sensitivity. Diabetes Metab Res Rev 2004;20:288–97.

[55] Gregor MF, Hotamisligil GS. Inflammatory mechanisms in obesity. Annu Rev Immunol 2011;29:415–45.

[56] Glass CK, Olefsky JM. Inflammation and lipid signaling in the etiology of insulin resistance. Cell Metab 2012;15(5):635–45.

[57] Velloso LA, Folli F, Saad MJ. TLR4 at the crossroads of nutrients, gut microbiota, and metabolic inflammation. Endocr Rev 2015;36(3):245–71.

[58] Jin C, Henao-Mejia J, Flavell RA. Innate immune receptors: key regulators of metabolic disease progression. Cell Metab 2013;17(6):873–82.

[59] Tsukumo DM, Carvalho-Filho MA, Carvalheira JB, et al. Loss-of-function mutation in Toll-like receptor 4 prevents diet-induced obesity and insulin resistance. Diabetes 2007;56(8):1986–9.

[60] Caricilli AM, Picardi PK, de Abreu LL, et al. Gut microbiota is a key modulator of insulin resistance in TLR 2 knockout mice. PLoS Biol 2011;9(12):e1001212.

[61] Carvalho BM, Oliveira AG, Ueno M, et al. Modulation of double-stranded RNA-activated protein kinase in insulin sensitive tissues of obese humans. Obesity (Silver Spring) 2013;21(12):2452–7.

[62] Carvalho-Filho MA, Carvalho BM, Oliveira AG, et al. Double-stranded RNA-activated protein kinase is a key modulator of insulin sensitivity in physiological conditions and in obesity in mice. Endocrinology 2012;153(11):5261–74.

[63] Carvalho-Filho MA, Ueno M, Carvalheira JB, et al. Targeted disruption of iNOS prevents LPS-induced S-nitrosation of IRbeta/IRS-1 and Akt and insulin resistance in muscle of mice. Am J Physiol Endocrinol Metab 2006;291(3):E476–82.

[64] Carvalho-Filho MA, Ueno M, Hirabara SM, et al. S-nitrosation of the insulin receptor, insulin receptor substrate 1, and protein kinase B/Akt: a novel mechanism of insulin resistance. Diabetes 2005;54(4):959–67.

[65] Hummasti S, Hotamisligil GS. Endoplasmic reticulum stress and inflammation in obesity and diabetes. Circ Res 2010;107(5):579–91.

[66] Hotamisligil GS. Endoplasmic reticulum stress and the inflammatory basis of metabolic disease. Cell 2010;140(6):900–17.

[67] Wang M, Kaufman RJ. Protein misfolding in the endoplasmic reticulum as a conduit to human disease. Nature 2016;529(7586):326–35.

[68] Lenna S, Han R, Trojanowska M. Endoplasmic reticulum stress and endothelial dysfunction. IUBMB Life 2014;66(8):530–7.

[69] Shaw A, Doherty MK, Mutch NJ, et al. Endothelial cell oxidative stress in diabetes: a key driver of cardiovascular complications? Biochem Soc Trans 2014;42(4):928–33.

[70] Brownlee M. The pathobiology of diabetic complications: a unifying mechanism. Diabetes 2005;54:1615–25.

[71] de Zeeuw P, Wong BW, Carmeliet P. Metabolic adaptations in diabetic endothelial cells. Circ J 2015;79(5):934–41.

[72] Monnier L, Mas E, Ginet C, et al. Activation of oxidative stress by acute glucose fluctuations compared with sustained chronic hyperglycemia in patients with type 2 diabetes. JAMA 2006;295:1681–7.

[73] Quagliaro L, Piconi L, Assaloni R, et al. Intermittent high glucose enhances apoptosis related to oxidative stress in human umbilical vein endothelial cell. The role of protein kinase C and NAD(P)H-oxidase activation. Diabetes 2003;52:2795–804.

[74] El-Osta A, Brasacchio D, Yao DC, et al. Transient high glucose causes persistent epigenetic changes and altered gene expression during subsequent normoglycemia. J Exp Med 2008;205:2409–17.

[75] Quijano C, Castro L, Peluffo G, et al. Enhanced mitochondrial superoxide in hyperglycemic endothelial cells: direct measurements and formation of hydrogen peroxide and peroxynitrite. Am J Physiol Heart Circ Physiol 2007;293:H3404–14.

[76] Brandes RP, Kreuzer J. Vascular NADPH oxidases: molecular mechanisms of activation. Cardiovasc Res 2005;65:16–27.

[77] Dikalov S. Cross talk between mitochondria and NADPH oxidases. Free Radic Biol Med 2011;51:1289–301.

[78] Sweet IR, Gilbert M, Maloney E, et al. Endothelial inflammation induced by excess glucose is associated with cytosolic glucose 6-phosphate but not increased mitochondrial respiration. Diabetologia 2009;52:921–31.

[79] Fink BD, Herlein JA, O'Malley Y, et al. Endothelial cell and platelet bioenergetics: effect of glucose and nutrient composition. PLoS One 2012;7(6):e39430.

NON-CARDIOVASCULAR DISEASES AND ENDOTHELIUM

CHAPTER

45

Endothelial Mechanisms in Preeclampsia

Soubhi Kahhale, Rossana Pulcineli Vieira Francisco, and Marcelo Zugaib

INTRODUCTION

Hypertensive syndromes are the most frequent complications in pregnancy and constitute the first cause of maternal death in Brazil, especially in severe forms, such as eclampsia and HELLP syndrome (hemolysis, elevated liver enzymes, and thrombocytopenia) (Fig. 45.1) [1].

Preeclampsia may be responsible for complications such as acute renal failure, acute pulmonary edema, hepatic failure, intracerebral hemorrhage, coagulopathy, placental abruption, and fetal distress. Hypertensive disease determines greater perinatal death, also causing substantial number of affected newborns, when they survive the perinatal hypoxia damage [2].

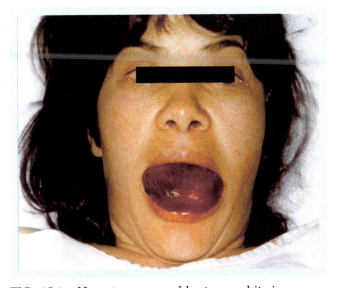

FIG. 45.1 Hematoma caused by tongue bite in a pregnant woman with eclampsia and HELLP syndrome.

Hypertensive syndromes comprise two entities of completely different etiology. One is the gestational hypertensive disease (GHD) or preeclampsia, which reverts after delivery. The other is chronic hypertension that coincides with gestation. Eventually, preeclampsia may be present in a chronic hypertensive pregnant woman, a condition called superimposed preeclampsia.

Preeclampsia is defined as the development of hypertension, with proteinuria and/or hand or face edema. It occurs after the 20th week of pregnancy, or before that period, in the gestational trophoblastic disease. Preeclampsia is predominantly a primiparous pathology. Eclampsia is characterized by the appearance of generalized seizures, without other established causes, in pregnant women with preeclampsia. Among the severity criteria, HELLP syndrome is considered a clinical entity that occurs in preeclampsia and eclampsia, characterized by a set of signs and symptoms associated with microangiopathic hemolysis, thrombocytopenia, and alterations in liver function. The term HELLP was initially used by Louis Weinstein in 1982 and was based on the initials of the words hemolisys, elevated liver functions tests and low platelet counts, i.e., hemolysis, elevation of liver enzymes, and thrombocytopenia (Fig. 45.2).

Etiology of preeclampsia is still unknown. Numerous theories and factors have been suggested to explain its cause, but most have not been confirmed. Currently, immunological genetic aspects and failure in placental invasion are unanimously accepted. There is increased reactivity and vascular permeability and activation of coagulation with damage mainly to the vascular endothelium, kidneys, central nervous system, liver, and placenta;

Endothelium and Cardiovascular Diseases
https://doi.org/10.1016/B978-0-12-812348-5.00045-3

655

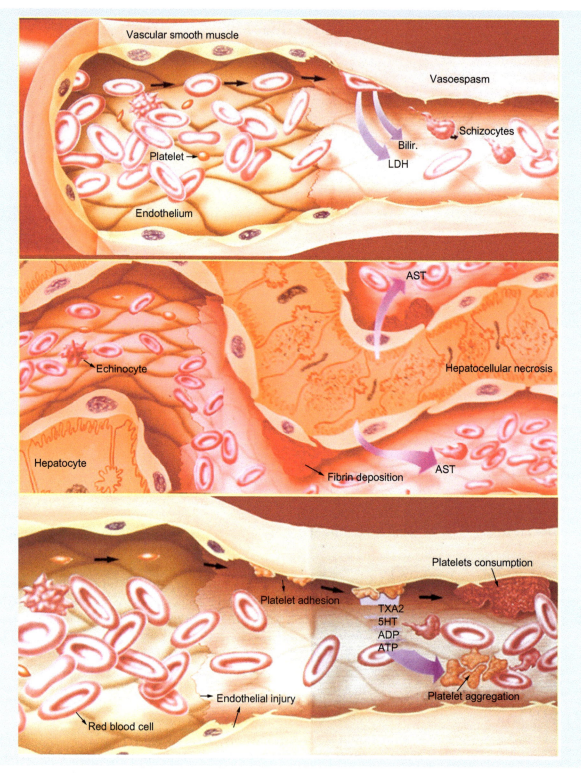

FIG. 45.2   Pathophysiology of hemolysis, elevation of liver enzymes, and thrombocytopenia.

as a result, patients may have involvement of multiple organs with different degrees of severity. Many pathophysiological abnormalities found in this pathology are similar to those of acute atherosclerosis (Fig. 45.3).

It is not possible to prevent preeclampsia due to a lack of knowledge of its etiology, and low-dose aspirin may be used for this purpose (100 mg/day). Once the disease has been diagnosed, treatment consists of prevention of maternal-fetal complications,

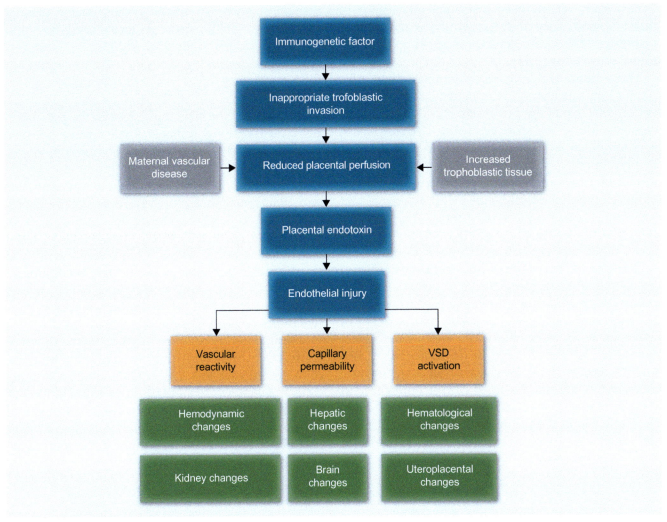

FIG. 45.3 Pathophysiology of preeclampsia. Adapted from Kahhale S, Zugaib M. Síndromes Hipertensivas na Gravidez. São Paulo: Atheneu; 1995.

such as premature placental abruption, stroke, acute pulmonary edema, renal insufficiency, and worsening of clinical conditions that predispose to severe preeclampsia, HELLP syndrome and eclampsia. As regards as the fetus, premature labor and respiratory discomfort of the newborn must be prevented.

Magnesium sulfate is the drug of choice for the control of eclamptic seizures. The best treatment for preeclampsia continues to be correct prenatal care, diagnosis, and early clinical treatment (low-salt diet, sedation and antihypertensive drugs such as methyldopa, pindolol, and amlodipine), adequate time for hospitalization and delivery the fetus and the placenta, which is the definitive treatment.

Three endothelial functions are particularly relevant for the study of preeclampsia: maintenance of vascular integrity by controlling its permeability,

modulation of vascular wall tone and prevention of disseminated intravascular coagulation [3,4].

EVIDENCE OF ENDOTHELIAL DAMAGE IN PREECLAMPSIA

Endothelial cells, when injured, not only lose the ability to carry on normal functions, but may also express new functions, producing vasoconstricting substances, such as increased plasma levels of endothelin, procoagulants such as activating factor XII and tissue factor and mitogens. These new functions of the endothelium are appropriate responses to the endothelial lesion secondary to the section of a vessel, as a result of which the patient is able to reduce and stop a hemorrhage.

However, when these functions take place in intact vessels, they may determine inappropriate responses and severe consequences, such as those found in preeclampsia.

There are several indications demonstrating morphological, functional, and biochemical evidence of endothelial injury in preeclampsia. Morphological changes found in kidneys, spiral arteries, liver, and umbilical arteries suggest endothelial damage in preeclampsia. The most consistent abnormality found in pregnant women with preeclampsia is renal glomerular endotheliosis [5], which is not present in other forms of arterial hypertension (Fig. 45.4). Swelling of the glomerular endothelium is one of the first indications that the endothelial cell will be affected in patients with preeclampsia. Direct functional evidence of increased endothelial cell permeability is the increased rate of elimination of Evans blue from the intravascular compartment of gestation with preeclampsia, as well as decreased expression of the endothelial oxidonitric synthase (eNOS) enzyme [6,7]. Biochemical findings include increased thromboxane (TXA2)/prostacyclin (PGI2) ratio and elevations in plasma of substances normally found within the membranes of endothelial cells or so-called markers of endothelial activation, such as: von Willebrand factor, cellular fibronectin (cFN), growth factor activity, and plasma endothelin (Fig. 45.4) [1,8].

Consequences in Eicosanoids

Biochemical studies suggest that in the preclinical period a functional imbalance takes place between vasodilator and vasoconstrictor eicosanoid products. PGI2 and TXA2 are the most important eicosanoids that regulate the interaction between vascular wall and platelets with biologically opposite functions. TXA2 is a potent vasoconstrictor and inducer of platelet aggregation. PGI2 has an opposite effect; it is a potent vasodilator and antiplatelet agent. PGI2 is reduced in pregnant women with preeclampsia while TXA2 released by active platelets is increased [8]. It is believed that preeclampsia is a state of relative deficiency of PGI2 and dominance of TXA2.

The imbalance between vasodilatory and vasoconstrictive prostaglandins, with a predominance of the latter, such as TXA2 and angiotensin-II, especially in the uteroplacental circulation and in the kidneys, may be the main factor in the development of preeclampsia and explains many of its clinical manifestations. The absence of normal stimulation of the

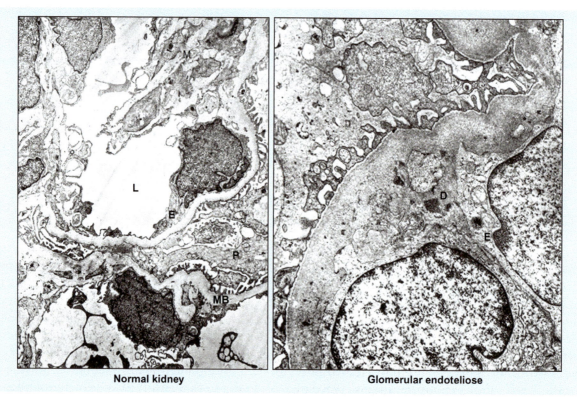

Normal kidney   **Glomerular endoteliose**

FIG. 45.4 Electron microscopy of normal kidney and with glomerular endotheliosis [5]. *L*, lumen; *E*, endothelium; *BM*, basal membrane; *D*, electro-dense deposits in the subendothelial space.

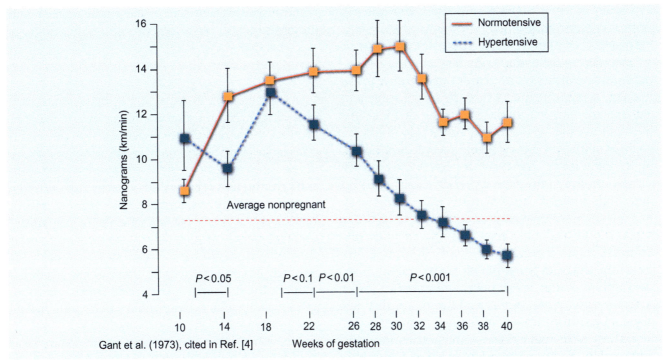

FIG. 45.5 Comparison of the mean dose of angiotensin-II required to provoke pressure response in 120 primigravidae who remained normal and 72 primigravidae who developed preeclampsia. Adapted from Wang Y, Gu Y, Zhang Y. Evidence of endothelial dysfunction in pre-eclampsia: decreased endothelial nitric oxide synthase expression is associated with increased cell permeability in endothelial cells from pre-eclampsia. Am J Obstet Gynecol 2004;190(3):817–24.

renin-angiotensin-aldosterone in spite of significant hypovolemia and increased vascular sensitivity for angiotensin-II and norepinephrine may be explained by a simple mechanism, i.e., endothelial lesion with imbalance between PGI2/TXA2. This increment in vascular reactivity to angiotensin-II can be identified 12 weeks before the clinical onset of the disease (Fig. 45.5). In addition, the increase in the TXA2/PGI2 rate may be the cause of selective platelet destruction. The main function of PGI2 production is as a protective mechanism, especially in the periods of ischemia and hypoxia.

Kidney Consequences

Alterations in renal function and morphology are found in preeclampsia. These changes also are involved in the maintenance of hypertension. The classic lesion, glomerular endotheliosis, characterized by swelling of the glomerular endothelium, suggests endothelial injury. The intrarenal production of PGI2 and PGE2 in preeclampsia is decreased. This deficiency in the production of vasodilatory substances may be the cause of decreased renal

plasma flow, glomerular filtration rate, urate clearance, and proteinuria development. Deficiency of intrarenal vasodilators may also result in an unrestrained vascular action of intrarenal angiotensin-II, activating the renin-angiotensin-aldosterone system, causing failure in the ability of the kidney to excrete sodium and consequently increasing vascular tone and the blood pressure. Proteinuria and generalized edema are clinical signs that indicate increased endothelial cell permeability.

ACTIVATION OF THE COAGULATION SYSTEM

Platelet activation takes place in preeclampsia. Platelets will adhere and release serotonin and TXA2; with this, more circulating platelets will be recruited. Coagulation will be activated, locally generating thrombin, contributing to platelet aggregation and inducing fibrin formation to stabilize the platelet thrombus, which may occlude the blood flow to the maternal cotyledon, leading to placental infarction. Recent research shows that activation of nitric oxide is important in fetus placentary regulation [9].

The effect of platelet-derived serotonin on vascular smooth muscle depends on the integrity of the endothelial layer. In vessels with intact endothelium, serotonin interacts with its S1 receptors, causing an increase in the synthesis of PGI2 and EDRF. However, in the damaged vessels, serotonin interacts with S2 subendothelial receptors, amplifying the platelet aggregation process. Since platelets are the major suppliers of circulating serotonin, increased platelet aggregation in preeclampsia may be the cause of high serotonin levels found in the blood and placentas of patients with preeclampsia. The increased level of circulating free serotonin derived from platelets can cause direct contraction at the serotonergic receptors located in the vascular smooth muscle and amplify the vasoconstricting action of other neurohumoral mediators, in particular catecholamines and angiotensin-II. The involvement of serotonin in the pathophysiology of preeclampsia is shown by the therapeutic effects of S2 Serotonin antagonists blockers, such as ketanserin. This substance is effective both in reducing blood pressure and in platelet aggregation of preeclampsia.

Bleeding time may be increased. The degree of this extension is not correlated with the degree of thrombocytopenia. Activation of the coagulation system in preeclampsia can be evidenced by increasing the rate between antigen-related factor VIII and factor VIIIc, and a progressive increase in thrombin and antithrombin III complex. These findings and decreasing protein C levels in preeclampsia may cause a trend towards thromboembolism.

In normotensive pregnant women, fibrinolysis is reduced but returns to normal 1 hour after delivery. This decrease mediated by tPA activity takes place earlier in patients with preeclampsia. The imbalance between tPA and PAI activity contributes to deposition of fibrin in uteroplacental vessels and renal microcirculation. Decreased fibrinolytic activity in patients with preeclampsia has been attributed to elevated plasma levels of both PAI1 and PAI2. The level of PAI1 rises after the 20th week of pregnancy and is secreted by endothelial cells, fibroblasts, platelets, and hepatocytes. These levels increase more precociously and intensely in patients with preeclampsia. PAI2 (placental type) correlates significantly with placental and newborn weight, and is reduced in pregnant women with unfavorable perinatal outcomes.

Endothelin

Plasma endothelin levels are normal or decreased in normal pregnant women. Recently, studies have shown an increment in endothelin levels in preeclampsia [1]. The increase in endothelin could be the cause of treatment-resistant severe hypertension and multiple organ failure occurring in some patients with preeclampsia severe and HELLP syndrome. Increased levels of endothelin could be a consequence of extensive endothelial damage. However, since the lung inactivates most of the circulating endothelin, further studies are needed to support the hypothesis that endothelin is involved in pulmonary and multiple organ insufficiency found in severe forms of preeclampsia.

Immunological Aspects

When the vascular endothelium is damaged, it can produce antigens that make these important cells immunological targets. In a significant proportion of patients with severe preeclampsia, vascular endothelial anticell antibodies have been found [10]. These antibodies, endothelial anticellular and immunocomplexes against endothelial fragments, can alter PGI2 secretion, increase platelet adhesion, activate the cascade of complement, and break this monolayer. In addition, the mitogenic activity of these patients' blood is increased before delivery. It has been hypothesized that this increase in mitogenic activity and growth factor are results of the release of direct mitogens from endothelial cells and platelets activated by the injured endothelium. The differences between mitogenic indices in hypertensive and normotensive pregnant women disappear 6 weeks after delivery, when clinical and biochemical evidence of preeclampsia is no longer detected.

Cause of Endothelial Injury

To date, the cause(s) of endothelial injury is unknown and controversial. Poorly perfused trophoblast is believed to produce substances toxic to endothelial cells, eventually leading to clinical preeclampsia. The exact nature of this substance remains unknown. Roberts and Lain believe it to be an anticell endothelial antibody, lipid peroxidation, or platelet-derived mitogenic factor [8].

Zeeman and Dekker believe that the formation of oxygen-free radicals and increased peroxidation can establish a link between the hypothetical immunological mechanisms, trophoblastic endovascular and endothelial injury, which occur in preeclampsia. Endothelial injury would be due to the activation of neutrophils and free radicals that can cause lipid peroxidation or activation of macrophages and lymphocytes, probably T [11].

SERUM CYTOTOXIC FACTOR

Different investigators have demonstrated that aortas of rabbits perfused with serum from patients with preeclampsia presented greater reactivity to angiotensin-II and noradrenaline, compatible with a decrease in the vasodilator function of the endothelium. Other authors have demonstrated that the serum of patients with preeclampsia placed in the culture of human umbilical endothelial cells provoked a greater increase in the release of chromo 51, marker of human endothelial lesion in vitro, when compared to the serum of the same pregnant woman after delivery or other gestation without the disease [8].

Lipid peroxidation, causing vasoconstriction and inhibition of PGI2 synthesis, is also increased in the serum of patients with preeclampsia. One possible mechanism of endothelial damage, recently suggested, is the endothelial anticell antibody. Research has shown that patients with preeclampsia have higher values of autoantibodies directed at endothelial cells [10].

Hence, there is growing evidence that endothelial damage is an early characteristic of preeclampsia and endothelial cell dysfunction may explain many of the pathophysiological changes in the disease.

FREE RADICALS

A free radical can be defined as any freely diffuse small molecule, which has one or more unpaired electrons in its outermost orbit. Free radicals are highly reactive, with a half-life of microseconds produced during normal physiological processes, but they may be increased in ischemia, immune

reactions and under the influence of exogenous stimuli, such as ionizing radiations, cigarettes and others. Neutrophil activation, which takes place with the immune response and prolonged periods of ischemia, leads to a 2- to 20-fold increase in oxygen consumption and, consequently, the production of free radicals as superoxide anions and hydrogen peroxide. Another mechanism for the production of free radicals by neutrophils derives from the metabolism of arachidonic acid.

Much attention has been given to the superoxide anion, which is formed when oxygen turns into simple electron. The superoxide anion alters the PGI2/TXA2 balance in favor of TXA2, increasing platelet aggregation and vasoconstriction. It may be cytotoxic to endothelial cells by the oxidative conversion of unsaturated fatty acids present in the membranes of lipid peroxides, which, in turn, damage the endothelium and are thrombogenic when interacting with the plasma coagulation system. Thus, increased lipid peroxidation may be involved in endothelial injury. In addition, superoxide anion and lipid peroxides directly induce vascular smooth muscle contraction and may cause vasoconstriction.

During normal pregnancy, the activity of free radicals increases. This may be the result of increased cell turnover or decreased free radical scavenging antioxidant mechanisms. The activity of lipid peroxidation is directly related to the time of gestation and decreases after delivery. In addition, augmentation in lipid peroxides in the normotensive gestation may be involved with the increased incidence of thromboembolic complications. On the other hand, a pregnant woman with preeclampsia presents alterations in the balance between oxidants and antioxidants due to inadequate placental perfusion [12,13].

In preeclampsia, the postulated inappropriate trophoblastic invasion and ischemic involvement may result in increased free radical formation by the activation of neutrophils, macrophages, and T cells. Involvement of free oxygen radicals in the pathophysiology of preeclampsia has been demonstrated by the hepatocellular deposition of the lipofuscin pigment in the liver of patients with preeclampsia. In addition, these pigments derived from lipid peroxidation are present in high concentrations in placentas of patients with preeclampsia. The placental lipid peroxidation in this disease provides an additional origin of high circulating peroxidation products. In patients with preeclampsia particularly severe, high levels of oxygen-free radicals are detected before initiation clinical manifestations and correlates with maternal blood pressure levels. The increase in lipid peroxidation is found particularly in high density lipoprotein. It is possible that lipid peroxides produced by cell membrane are transferred to the fraction of high-density lipoprotein and circulate in the blood, resulting in disseminated endothelial damage and, consequently, preeclampsia. In addition, oxygen-free radicals contribute to the deficiency of PGI2 and inactivation of EDRF [8].

ENDOTHELIUM RESEARCH METHODS

Investigations of vascular endothelium have progressed from histology and morphology to physiology, pharmacology, pathophysiology, and clinical setting. Thus, endothelial studies can vary from cell and molecular aspects, using methodologies such as tissue culture and molecular biology, to clinical investigations in humans, with invasive and noninvasive methods of evaluating vasoreactivity dependent on the endothelium (intravascular angiography and ultrasound, high-resolution two-dimensional ultrasonography, and plethysmography) and determination of plasma substances that indicate activation and endothelial injury, including physiological preparations in experimental animals and isolated organs. Each of these methods of investigation is adequate to elucidate a specific type of scientific questioning, from the molecular intimacy of the intra- and intercellular signaling mechanisms, through both the homeostasis mechanisms of circulation and coagulation, as well as the pathophysiology of inflammation and vascular response to the lesion, to clinical implications for diagnosis, prognosis, and treatment of a particular pathology [3].

The possibility of studying vessels obtained from preeclampsia has allowed us to know in more detail the vascular reactivity to several vasoactive substances. This way, the endothelium can be better studied, since it plays a fundamental role in the pathophysiology of preeclampsia. The main vessels studied are obtained from the omentum and the subcutaneous cellular tissue of patients with preeclampsia, to normal pregnant and non-pregnant women.

Studies evaluating EDRF dependent reactivity in vessels of patients with preeclampsia showed decreased release of EDRF. Even so, the number of studies that analyzed the vessel isolated in vitro is small, due to the difficulty of obtaining specimens. The studies concluded that pregnant women with preeclampsia suffer a deficiency of EDRF in relation to normal pregnant women, contributing to the increase of blood pressure in this disease. In addition, rhythmic wave contractions also were observed in vessels from pregnant women with preeclampsia called vasomotricity, which may contribute to the great oscillation of the blood pressure levels found in these patients [14–16]. Takiuti et al. also found less endothelium-dependent relaxation in the aorta of stressed pregnant rats and greater vasoreactivity [17], suggesting that stress may participate in preeclampsia pathophysiology [18].

Endothelial dysfunction can be evaluated by the flow-mediated dilatation of the brachial artery which is reduced, a fact observed both in women who presented preeclampsia in a previous pregnancy [19] and in those who subsequently developed the disease [20,21].

CONCLUSION AND PHYSIOPATHOLOGICAL MODEL

Preeclampsia occurs in the presence of placental tissue; it is a multifactorial entity influenced by environmental, immunological and genetic factors of pregnant women. Inadequate trophoblast invasion (Fig. 45.6) and poorly perfused trophoblast produce toxic substances that damage the endothelium, leading to preeclampsia. In its pure form, results from poor placental perfusion due to an inappropriate trophoblastic invasion, the etiology of which is still unknown. In its secondary form due to other causes, this poor perfusion is a consequence of vasculitis and maternal atherosclerosis, multiple and molar gestation. As a result of endothelial injury, integrity of the vascular system, production of endogenous vasodilators and maintenance of anticoagulation are compromised. There is an increase in vascular reactivity and activation of coagulation cascade.

The success of physiological placentation depends on the regulation of angiogenic factors (PLGF) and antiangiogenic factors (sFlt-1). The most recent studies associate the decrease in PLGF and the increase of sFlt-1, as well as the increase in the sFlt-1/PLGF ratio with the prediction, diagnosis, and prognosis [22] of pregnant women with preeclampsia (Fig. 45.7).

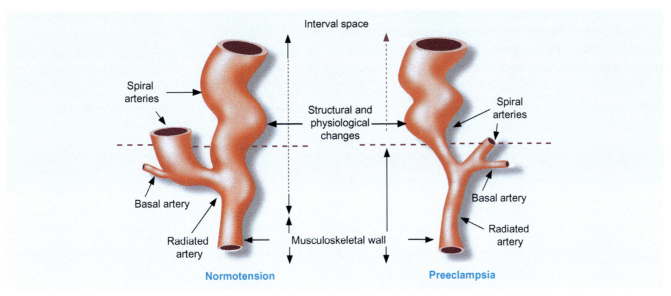

FIG. 45.6 Inappropriate trophoblastic invasion in preeclampsia. Adapted from Kahhale S, Zugaib M. Síndromes Hipertensivas na Gravidez. São Paulo: Atheneu; 1995.

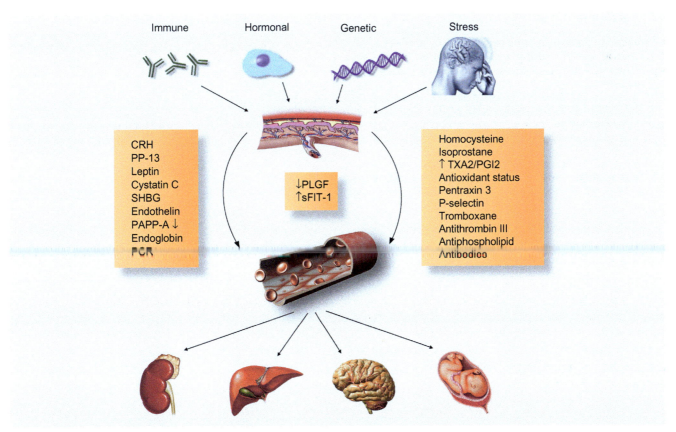

FIG. 45.7 Pathophysiology of preeclampsia. Angiogenic (PLGF) and antiangiogenic factors (sFlT-1) (acronyms, see text). Adapted from Verlohren S, Herraiz I, Lapaire O. The sFlt-1/PLGF ratio in different types of hypertensive pregnancy disorders and its prognostic potential in preeclamptic patients. Am J Obstet Gynecol 2012;206:58,e1–8.

The next few years will provide answers to the various questions about the pathophysiology of this great enigma, which is preeclampsia and its possible correlation with the endothelium.

References

[1] Baksu B, Davas I, Baksu A. Plasma nitric oxide, endothelin-1 and urinary nitric oxide and cyclic guanosine monophosphate levels in hypertensive pregnant women. Int J Gynaecol Obstet 2005;90 (2):112–7.

[2] Vega C, Kahhale S, Zugaib M. Maternal mortality due to arterial hypertension in São Paulo city (1995–1999). Clinics 2007;62:679–84.

[3] Luz PL, Laurindo FRM, Chagas ACP. Endotélio e doenças cardiovasculares. Atheneu: São Paulo; 2003.

[4] Kahhale S, Zugaib M. Síndromes Hipertensivas na Gravidez. Atheneu: São Paulo; 1995.

[5] Barros ACSD, Saldanha LB, Paula FJ. Correlação entre os diagnósticos clínico e histopatológico renal na doença hipertensiva específica da gestação. Rev Ginecol Obstet 1990;1(1):47–54.

[6] Wang Y, Gu Y, Zhang Y. Evidence of endothelial dysfunction in pre-eclampsia: decreased endothelial nitric oxide synthase expression is associated with increased cell permeability in endothelial cells from pre-eclampsia. Am J Obstet Gynecol 2004;190(3):817–24.

[7] Buhimschi IA, Saade GR, Chwalisz K. The nitric oxide pathway in pre-eclampsia: pathophysiological implications. Hum Reprod Update 1998;4(1):25–42.

[8] Roberts JM, Lain KY. Recent insights into the pathogenesis of pre-eclampsia. Placenta 2002;23(5):359–72.

[9] Karteris E, Vatish M, Hillhouse EW. Pre-eclampsia is associated with impaired regulation of the placental nitric oxide-cyclic guanosine monophosphate pathway by corticotropin-releasing hormone (CRH) and CRH-related peptides. J Clin Endocrinol Metab 2005;90 (6):3680–7.

[10] Rappaport VJ, Hirata G, Yap HK. Anti-vascular endothelial cell antibodies in severe pre-eclampsia. Am J Obstet Gynecol 1990;162(1):138–46.

[11] Zeeman GG, Dekker GA. Pathogenesis of pre-eclampsia: a hypothesis. Clin Obstet Gynecol 1992;35(2):317–37.

[12] Bisseling TM, Maria Roes E, Raijmakers MT. N-acetylcysteine restores nitric oxide-mediated effects in the fetoplacental circulation of preeclamptic patients. Am J Obstet Gynecol 2004;191(1):328–33.

[13] Many A, Hubel CA, Fisher SJ. Invasive cytotrophoblasts manifest evidence of oxidative stress in pre-eclampsia. Am J Pathol 2000;156 (1):321–31.

[14] Ashworth JR, Warren AY, Baker PN. Loss of endothelium-dependent relaxation in myometrial resistance arteries in pre-eclampsia. Br J Obstet Gynaecol 1997;104(10):1152–8.

[15] Cockell AP, Poston L. Flow-mediated vasodilatation is enhanced in normal pregnancy but reduced in pre-eclampsia. Hypertension 1997;30(2 Pt 1):247–51.

[16] Pascoal IF, Lindheimer MD, Nalbantian-Brandt C. Pre-eclampsia selectively impairs endothelium-dependent relaxation and leads to oscillatory activity in small omental arteries. J Clin Invest 1998;101(2):464–70.

[17] Takiuti NH, Kahhale S, Zugaib M. Stress in pregnancy: a new Wistar rat model for human pre-eclampsia. Am J Obstet Gynecol 2002;186(3):544–50.

[18] Takiuti NH, Kahhale S, Zugaib M. Stress-related pre-eclampsia: an evolutionary maladaptation in exaggerated stress during pregnancy? Med Hypotheses 2003;60(3):328–31.

[19] Chambers JC, Fusi L, Malik IS. Association of maternal endothelial dysfunction with pre-eclampsia. JAMA 2001;285(12):1607–12.

[20] Savvidou MD, Hingorani AD, Tsikas D. Endothelial dysfunction and raised plasma concentrations of asymmetric dimethylarginine in pregnant women who subsequently develop pre-eclampsia. Lancet 2003;361(9368):1511–7.

[21] Serrano NC, Casas JP, Diaz LA. Endothelial NO synthase genotype and risk of pre-eclampsia: a multicenter case-control study. Hypertension 2004;44(5):702–7.

[22] Verlohren S, Herraiz I, Lapaire O. The sFlt-1/PLGF ratio in different types of hypertensive pregnancy disorders and its prognostic potential in preeclamptic patients. Am J Obstet Gynecol 2012;206 (58):e1–8.

Further Reading

[1] Ghosh SK, Raheja S, Tuli A. Is serum placental growth factor more effective as a biomarker in predicting early onset pre-eclampsia in early second trimester than in first trimester of pregnancy? Arch Gynecol Obstet 2012;287:865–73.

46

Endothelium and Nitric Oxide: Interactions in Cancer Evolution

Roberta Eller Borges, Wagner Luiz Batista, Elaine Guadelupe Rodrigues, and Hugo Pequeno Monteiro

Signaling pathways are essential for cell growth, proliferation, and division. Cells of a multicellular organism respond to a specific set of extracellular signals produced by other cells. These signals act in various combinations regulating cellular behavior. Cell signaling requires, in addition to the extracellular signaling molecules, a complementary set of receptor proteins in each cell, allowing it to respond characteristically to a given extra or intracellular signal. In addition, for this whole process to occur—i.e., migration, survival, differentiation, and proliferation—recognition and association of ligands with their receptors located in specific domains of the plasma membrane are necessary. Among these receptors, we highlight the tyrosine kinase (RTK), receptors, and the G protein coupled receptors (GPCR). RTKs undergo transphosphorylation after association with their specific ligands by initiating cell signaling cascades that will mobilize monomeric G proteins (low molecular weight GTPases), adapter proteins and other protein kinases [1]. GPCRs transmit the signal of the ligand into the cell via interactions with heterotrimeric G proteins [2]. Individual or combined action of both types of receptors can activate or inhibit many signal transduction pathways, altering the cell phenotype. Cell signaling processes are highly regulated and the

loss of such regulation may result in cellular transformation and consequently carcinogenesis.

Cancer is the name given to a series of 100 diseases characterized by uncontrolled growth and cell division, the product of uncontrolled activation of several cellular signaling processes, not respecting tissue boundaries and spreading through different sites. Tumor cells differ from the normal in many respects, including loss of differentiation, changes in genetic material, disordered growth associated with rejections to inhibitory signals of cell growth in reference to the space occupied, increasing invasiveness, and morphological changes [3]. Malignant process induces biochemical and morphological changes in tumor cells, by activation of oncogenes and/or deletion of tumor suppressor genes, giving the malignized cell properties such as the ability to make neighbor endothelial cells create new blood vessels for its own supply (tumor angiogenesis), separate them from the tumor tissue for subsequent migration and infiltration into normal neighboring tissues, inducing accelerated uncontrolled cell division, as well as promoting the appearance of secondary tumors or metastases at sites far from the primary tumor [4–6].

Participation of signaling proteins during the processes of angiogenesis and metastasis in tumors is

☆ The Scientific production of the Cell Signaling Laboratory of the Center for Cellular and Molecular Therapy of UNIFESP cited in this chapter was the product of the work of students, collaborators, and the indispensable financial support of the funding institutions: Fundação de Amparo à Pesquisa do Estado de São Paulo (FAPESP), Conselho Nacional de Pesquisa (CNPq), Coordenação de Aperfeiçoamento de Pessoal de Nível Superior (CAPES).

already well documented [7]. We mention some examples, such as the phosphatidylinositol 3-kinase/protein kinase (PI3K), which promotes cell migration through the activation of another protein kinase, the focal adhesion kinase (FAK) protein [8]. Akt (also known as protein kinase B or PKB) performs a critical role in regulation of cellular process as in cell growth, cell survival, and tumor progression [9–11]. PI3K can also regulate migration and cell polarity through independent mechanisms, in which low molecular weight Rho GTPases, Rac1e, and Cdc42 activated are associated with nucleotide GTP and promote cell movement, triggering cell signaling pathways, which operate from the ends of the cells [12,13].

The Ras protein family plays a critical role in control of cell growth and differentiation, being a fundamental component of signaling pathways associated with cellular proliferation, survival, and migration [14]. Ras activation initiates a complex network of signal transduction pathways, including PI3K/Akt pathway, involved in cell survival signaling [15]; the Rac/Rho pathway acts on cytoskeletal remodeling and cellular mobility [16]; and the Rac/JNK and Rac/P38 pathways, are both involved in response to stress, inhibition of growth, and induction of apoptosis [17].

Rho GTPases integrates the Ras super family and control a wide variety of essential signaling pathways in all eukaryotic cells. In mammalian cells, proteins from the Ras and Rho family are of particular interest, since the signaling pathways can be altered with each change in the intra- and intercellular environment [18]. The GTPases occur in two states: the inactive conformation (associated with GDP) and active (associated with GTP) [19]. Guanine nucleotide exchange factors catalyze the release of GDP, followed by binding to GTP (its intracellular concentration is higher than GDP). Finally, the GTPase activity, complete the cycle and the GTPases return to the inactive state, associated with GDP [13,20,21]. In the active state, associated with GTP, Ras and Rho GTPases recruit effector protein signaling pathways related to proliferation, survival, and migration (Fig. 46.1).

Members of the Rho GTPases family, including Rac1, Cdc42, and Rho, participate in the regulation of several processes relevant to cell migration, including cell-substrate adhesion, cell-cell adhesion, protein secretion, vesicle trafficking and transcription [22]. Rho GTPases also participate in the

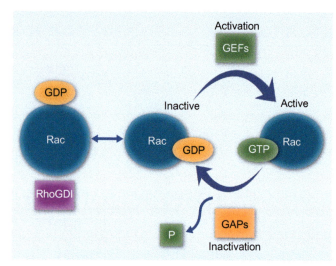

FIG. 46.1   Schematic representation of regulation of the cyclic state of activation of the GTPase family proteins. They are inactive when connected to GDP and active when connected to GTP. Regulation of this mechanism, occurring through a GDP-GTP cycle, is controlled by the opposing activities of the GEFs, which catalyze the exchange of GDP by GTP, and the GAPs (which stimulate the hydrolysis of GTP to GDP) (see plan). GDP dissociation inhibitors (GDIs), which prevent the GDP-GTP exchange, also regulate Rho proteins. GTPases interact with several effector proteins, influencing activity and/or location of these effectors.

regulation of cell polarity. Rac1 specifically participates in the formation of the NADPH oxidase enzyme complex and consequently the generation of reactive oxygen species (ROS) [18,19,23].

Different classes of transmembrane receptors, such as RTKs, activate Rho GTPases and when these receptors bind to polypeptide growth factors such as platelet-derived growth factor (PDGF) and vascular endothelial growth factor (VEGF), they transmit these signals to their effector proteins. In these signaling pathways, in addition to the low molecular weight GTPases, kinases and adapter proteins and protein phosphatases participate [24], as well as transcription factors that lead to the expression of genes necessary for the morphological changes that accompany cell migration. However, under certain conditions, they have been implicated in invasion and in some aspects of metastasis [7,25] (Fig. 46.2).

The importance of GTPase Rac1 expression in tumors was first demonstrated in breast cancer [26]. Recent observations documented an overexpression of several members of the Rho GTPases family in many different types of human tumors [27].

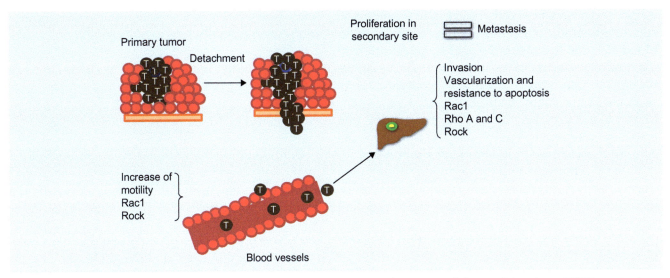

FIG. 46.2 Schematic representation of metastasis development mechanism and the participation of GTPases in these processes.

Angiogenesis is a complex process and involves different phases. It is initiated by the induction of endothelial cells proliferation, with subsequent migration and formation of capillary structures, the initial form of blood vessels [28,29]. Angiogenesis is regulated by signaling cascades initiated by RTKs stimulated by their respective ligands. Polypeptide growth factors such as the basic fibroblast growth factor (FGFb) and VEGF are potent stimulators of proliferation and migration of endothelial precursor cells. The later phase involves redifferentiation of endothelial cells [29–31].

In addition to polypeptide growth factors mentioned above, Ras and Rac1 GTPases, calcium-regulated proteins, Ras-ERK1/2 MAP kinase cascade signaling proteins, PKC and PI3K protein kinases, phospholipase-C and -D, and nitric oxide (NO) participate in migration process of angiogenesis [19,32,33].

Cellular motility occurs due to recruitment of a large numbers of proteins: integrins, GTPases, protein kinases and structural, constituents of the focal adhesion complex [12], growth factors, and NO. Motility involves two central features: transduction of signals activated in response to stimuli that can induce motility and cellular components that mediate mechanical aspects of motility or migration [34]. Along the cytoskeletal changes in adhesion and extracellular matrix, cell migration and invasion are constitutive processes of angiogenesis and characteristic of tumor cell progression [8].

Angiogenesis promotes tumor neovascularization, which triggers a double effect on tumor growth, since, in addition to supplying newly formed endothelial cells with nutrients and oxygen, these same cells stimulate growth of adjacent malignant cells by secreting peptides such as PDGF and interleukin-1, and serve as a pathway for dissemination of tumor cells to reach other sites by metastasis [29,35].

An important connective element between the angiogenesis and tumor metastasis processes is NO; in the following sections, we shall discuss pertinent evidence.

THE ENDOTHELIUM AND NITRIC OXIDE

The vascular system is a complex network of vessels connecting the heart with various organs and tissues to maintain homeostasis in response to physiological and pathological needs. Endothelial cells form the inner layer of the arterial and venous wall and play an important role in the development and remodeling of blood and lymphatic vessels, transmigration of leukocytes, angiogenesis, maintenance of vascular tone, blood flow, coagulation, exchange of nutrients, and the development of organs. It functions as a semipermeable barrier that regulates the exchange of nutrients and oxygen between blood cells and underlying tissues [36–38]. The endothelium controls not only small and large molecules traffic, and even whole cells, but also local expansion and contraction, in response either to changes in blood flow or to vasoactive agents, in addition to maintaining vascular wall structure [39]. It is structurally simple, but functionally complex; it is responsible for many processes, including regulation of blood pressure, inflammation, and immune response; endothelial cells are as important as smooth muscle cells in inflammatory processes.

Physical, chemical, and/or hormonal stimuli produce a variety of substances in endothelial cells, including the endothelium-derived relaxation factor: NO. This free radical with signaling properties plays a fundamental role in the regulation of vascular tone and homeostasis [40]. The family of enzymes known as NO synthase (NOS) is responsible for NO biosynthesis in mammalian cells. Three distinct isoforms of NOS are well characterized: neuronal NOS (nNOS or NOS I), inducible NOS (iNOS or NOS II) and endothelial NOS (eNOS or NOS III). NOS are dimeric enzymes and catalyze the production of NO from the O_2 and L-arginine, producing NO and L-citrulline. And this NOS enzymatic activity depends on the cofactors (Ca^{2+}/calmodulin, BH4, FMN, FAD, NADPH, heme) undergoing dimerization [41]. The endothelial enzyme NOS (eNOS), initially characterized in the vascular endothelium, is constitutively expressed and has its activity regulated by posttranslational modifications, specifically phosphorylation in Ser1177 residue, which promotes the activity of the enzyme [42].

A number of contractile factors are also produced by the endothelium. The main factors are prostaglandin H2, thromboxane A2, angiotensin II, endothelin-1, and superoxide anion radical (O_2^-), a product of one-electron reduction of oxygen and one of the ROS. In addition, the platelet-activating factor and the factor hyperpolarization derived from endothelium are also important contracting factors [36].

Under physiological conditions, there is an equilibrium in the release of these factors. However, in several pathological conditions, this balance is altered with a consequent attenuation of the vasodilator effects of the endothelium. This apparent decrease in vascular relaxation dependent on endothelial factors is called endothelial dysfunction. During endothelial dysfunction, development of vascular diseases such as hypertension [37], atherosclerosis [43], aneurysm formation [44], vascular disease associated with diabetes [45], and tumor angiogenesis (which is essential in the development of tumor metastases) may occur [46,47].

NITRIC OXIDE AS A CONNECTING ELEMENT BETWEEN ANGIOGENESIS AND TUMOR DEVELOPMENT

Bradykinin (BK) is a vasoactive peptide produced by the kallikrein-kinin system. Its broad spectrum of action is mediated by GPCR receptors classified as kinin type 1 receptors (B1R) and kinin type 2 receptors (B2R) [48]. BK receptors have been associated with tumorigenesis and angiogenesis processes. Clinical trials and animal experiments showed that the expression of the B1 receptor levels are elevated in colon tumors [49], while the levels of B2 receptor expression are elevated on gliomas [50], in gastric tumors, hepatic, duodenal, prostate, and lung tumors [51]. Furthermore, the involvement of BK as a positive mediator of angiogenesis is well documented [52], as well as NO participation in the process [53].

Studies conducted by our research group using human umbilical cord vein endothelial cells (HUVEC) and rabbit aortic endothelial cells (RAEC) evidenced the central involvement of NO in angiogenesis process stimulated by BK. BK binding to B2 receptor on HUVEC cells promoted eNOS activation and endogenous NO production. The actions of NO, derived from BK stimulated HUVEC cells, promoted phosphorylation of specific tyrosine residues and S-nitrosylation of specific EGFR cysteine residue(s), stimulating the activity of this receptor. NO also promoted S-nitrosylation of the preserved cysteine residue essential for phosphatase activity of the enzyme SHP-1 [54] and the cysteine residue (cysteine118) located in the guanine nucleotides binding domain of the GTPase Ras [55]. The subsequent activation of ERK1/2 MAP kinases and VEGF expression induction of this cascade of events highlights the importance of NO in the process [54].

In another recent study, we showed that in RAEC endothelial cells, the S-nitrosylation of the GTPase Ras leads to the recruitment of another GTPase, Rac1 and PI3K, resulting in stimulation of migration of these cells [56] (Fig. 46.3).

An increasing number of observations have emphasized the importance of NO-mediated posttranslational (nitrosylation/nitration) modifications in signaling proteins associated with the regulation

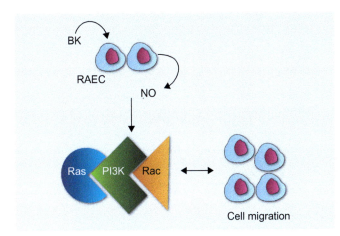

FIG. 46.3 Schematic representation of the signaling pathway induced by nitric oxide and associated migration of endothelial cells from rabbit aorta (RAEC).

of various signaling cascades [57,58]. Among the signaling proteins that can be modified by NO, whose stimulated activity correlates with cellular transformation, there are the GTPase Ras and Src protein-tyrosine kinase [55,59]. In both cases, it has been demonstrated that S-nitrosylation promoted by NO in cysteine residues of both proteins is associated with tumor progress [60,61].

NO may promote or inhibit tumor progression and metastasis in different tumors, but this depends on the activity and location of NOS isoforms, concentration and time of NO exposure, as well as the sensitivity of the target-cell [62–64]. In melanomas, NO derived from eNOS activation plays an important role in recruitment and induction of migration of tumor cells precursors, which will form blood capillaries in the tumor angiogenesis processes, in B16F10 murine melanoma model [65].

Studies involving B2R receptor antagonists have shown that these drugs inhibit angiogenesis and the growth of Walker 256 carcinoma implanted in experimental animals [66]. Studies like these evidenced the connections between angiogenesis processes and tumor progression.

In our most recent study [67], we showed that NO production, by stimulation with BK, promotes cell migration and invasion of melanomas with high-metastatic potential. These events are directly related to NO production by mediating Ras activation, which promoted the recruitment of the PI3K protein and activation of Rac1, a mechanism analogous to that described above for the RAEC endothelial cells [56]. Interestingly, murine melanomas with preferential growth to the primary site

have a high-basal level of NO production, and the presence of BK promoted inhibition in the cell migration and invasion process.

Murine melanoma cells showed another common feature with endothelial cells in terms of interactions between two GTPases: Ras and Rac1. The inhibition of Ras activity resulted in total inhibition of cell migration and invasion process in the primary site melanoma line. The activity of Rac1 was regarded as essential for the migration process of the two strains. However, different to what happens in endothelial cells, activation of the Ras/PI3K/Rac1 signaling pathway in the line of primary murine melanoma occurs by means of the action of O_2^-, an ROS. Two signaling proteins located downstream, with respect to Ras and Rac1, have their activities regulated by ROS. These proteins are the ERK1/2 MAP kinases and the protein kinase PAK, respectively. ERK1/2 MAP kinases are Ras effectors while PAK is an effector of Rac that mediates cytoskeletal remodeling; it is responsible for cell migration and also acts on angiogenesis [68,69]. ROS are also products of cellular oxidative metabolism and participate as mediators and modulators of signaling pathways that control biological processes, such as cell proliferation, differentiation, and migration [58].

The activation of Ras/PI3K/Rac1 signaling pathway, mediated by NO, occurs in the metastatic lineage. These processes illustrate the importance of NO reactive species and O_2^- during tumor development.

Our observations, combined with that of other researchers, led us to suggest the occurrence of two mechanism of regulation of Ras/PI3K/Rac1 signaling pathway. In primary melanoma cells, it is possible that a direct regulation of Rac1 by the action of O_2^- occurs. This assumption is supported by the fact that O_2^- promoted acceleration of GDP nucleotide dissociation in Rac1, while NO almost did not change the basal levels of this dissociation [70]. In metastatic melanoma cells, pathway activation occurs by Ras S-nitrosylation in the same way to that which we have described for endothelial cells.

Increasing evidence favors the assertion that, under pathological conditions, NO production is not altered, but its bioavailability instead is, due to oxidative inactivation, resulting from excessive production of O_2^- in the vascular wall [71,72]. This same type of NO bioavailability regulation should be operative in tumor cells, with cells derived from primary tumors with high-endogenous levels of O_2^- and other ROS [73] and the cells derived from metastatic tumors with high-endogenous levels of NO [74].

This set of observations suggests that NO and, by extension, the endothelium and ROS, play important roles in cell migration associated angiogenesis and the stage of tumor cell development. A better understanding of these phenomena will help in the comprehension of mechanisms involved in cell invasion and migration during angiogenesis and cancer development.

References

[1] Lemmon MA, Schlessinger J. Cell signaling by receptor tyrosine kinases. Cell 2010;141:1117–34.

[2] Neves SR, Ram PT, Iyengar R. G proteins pathways. Science 2002;296:1636–9.

[3] Croce CM. Oncogenes and cancer. N Engl J Med 2008;358:502–11.

[4] Fidler IJ. The pathogenesis of cancer metastasis: the "seed and soil" hypothesis revisited. Nat Rev Cancer 2003;3:1–6.

[5] Freitas ZF, Rodrigues EG, Oliveira V, et al. Melanoma heterogeneity: differential, invasive, metastatic properties and profiles of cathepsin B, D and L activities in subclones of the B16F10-NEX2 cell line. Melanoma Res 2004;14:333–44.

[6] Chiang AC, Massagué J. Molecular basis of metastasis. N Engl J Med 2008;359:2814–23.

[7] Weinberg RA. The biology of cancer. New York: Garland Science Taylor and Francis Group; 2007. p. 844.

[8] Kallergi G, Agelaki S, Markomanolaki H, et al. Activation of FAK/PI3K/Rac1 signaling controls actin reorganization and inhibits cell motility in human cancer cells. Cell Physiol Biochem 2007;20:977–86.

[9] Krasilnikov MA. Phosphatidylinositol-3 kinase dependent pathways: the role in control of cell growth, survival, and malignant transformation. Biochemistry (Mosc) 2000;65:59–67.

[10] Morales-Ruiz M, Fulton D, Sowa G, et al. Vascular endothelial growth factor-stimulated actin reorganization and migration of endothelial cells is regulated via the serine/threonine kinase Akt. Circ Res 2000;86:892–6.

[11] Carnero A, Blanco-Aparicio C, Renner O, et al. The PTEN/PI3K/AKT signalling pathway in cancer, therapeutic implications. Curr Cancer Drug Targets 2008;8:187–98.

[12] Ridley AJ, Schwartz MA, Burridge K, et al. Cell migration: integrating signals from front to back. Science 2003;302:1704–9.

[13] Huntterlocher A. Cell polarization mechanisms during directed cell migration. Nature 2005;7:336–7.

[14] Ahearn IM, Haigis K, Bar-Sagi D, et al. Regulating the regulator: post-translational modification of RAS. Nat Rev Mol Cell Biol 2011;13:39–51.

[15] Kauffmann-Zeh A, Rodriguez-Viciana P, Ulrich E, et al. Suppression of c-Myc induced apoptosis by Ras signaling through PI(3)K and PKB. Nature 1997;385:544–8.

[16] Lamarche N, Tapon N, Stowers L, et al. Rac and Cdc42 induce actin polymerization and G1 cell cycle progression independently of p65PAK and JNK/SAPK MAP kinase cascade. Cell 1996;87:519–29.

[17] Xia Z, Dickens J, Raingeaud J, et al. Opposing effects of ERK and JNK-p38 MAP kinases on apoptosis. Science 1995;270:1326–31.

[18] Etienne-Manneville S, Hall A. Rho GTPases in cell biology. Nature 2002;420:629–35.

[19] Hall A. Rho GTPases and the actin cytoskeleton. Science 1998;279:509–14.

[20] Takai Y, Kaibuchi K, Kikuchi A, et al. Small GTP binding proteins. Int Rev Cytol 1992;133:187–230.

[21] Bokoch GM. Assay of CDC42, Rac1 and Rho GTPase activation by affinity methods. Methods Enzymol 2003;345:349–59.

[22] Ridley AJ. Rho family proteins: coordinating cell responses. Trends Cell Biol 2001;11:471–7.

[23] Bokoch GM, Knaus UG. NADPH oxidases: not just for leukocytes anymore! Trends Biochem Sci 2003;28:502–8.

[24] Chrzanowska-Wodnicka M, Burridge K. Rho-stimulated contractility drives the formation of stress fibers and focal adhesions. J Cell Biol 1996;133:1403–15.

[25] Schmitz AA, Govek EE, Böttner B, et al. Rho GTPases: signaling, migration and invasion. Exp Cell Res 2000;261:1–12.

[26] Fritz G, Brachetti C, Bahlmann F, et al. Rho GTPases in human breast tumours: expression and mutation analyses and correlation with clinical parameters. Br J Cancer 2002;87:635–44.

[27] Gómez del Pulgar T, Benitah SA, Valerón PF, et al. Rho GTPase expression in tumourigenesis: evidence for a significant link. Bioessays 2005;27:602–13.

[28] Folkman J, Shing Y. Angiogenesis. J Biol Chem 1992;267:10931–4.

[29] Carmeliet P. Angiogenesis in health and disease. Nat Med 2003;9:653–60.

[30] Zhang Z, Nie F, Chen X, et al. Upregulated periostin promotes angiogenesis in keloids through activation of the ERK 1/2 and focal adhesion kinase pathways, as well as the upregulated expression of VEGF and angiopoietin 1. Mol Med Rep 2015;11:857–64.

[31] Rivas-Fuentes S, Salgado-Aguayo A, Pertuz Belloso S, et al. Role of chemokines in non-small cell lung cancer: angiogenesis and inflammation. J Cancer 2015;6(10):938–52.

[32] Ridley AJ. Rho GTPases and migration. J Cell Sci 2001;114:2713–22.

[33] Oliveira CJ, Curcio MF, Moraes MS, et al. The low molecular weight S-nitrosothiol, S-nitroso-N-acetylpenicillamine, promotes cell cycle progression in rabbit aortic endothelial cells. Nitric Oxide 2008;18:241–55.

[34] Goligorsky MS, Budzikowski AS, Tsukahara H, et al. Co-operation between endothelin and nitric oxide in promoting endothelial cell migration and angiogenesis. Clin Exp Pharmacol Physiol 1999;26:269–71.

[35] Brasileiro F. Disturbios do crescimento e diferenciação celular. Patologia geral. 2nd ed. Rio de Janeiro: Guanabara; 1998. p. 148–92.

[36] Flavahan NA. Atherosclerosis or lipoprotein-induced endothelial dysfunction. Potential mechanisms underlying reduction in EDRF/nitric oxide activity. Circulation 1992;85:1927–38.

[37] Moncada S, Higgs EA. Nitric oxide and the vascular endothelium. Handb Exp Pharmacol 2006;176:213–54.

[38] Marinković G, Heemskerk N, van Buul JD, et al. The ins and outs of small GTPase Rac1 in the vasculature. J Pharmacol Exp Ther 2015;354:91–102.

[39] Carvalho MHC, Nigro D, Lemos VS, et al. Hipertensão arterial: o endotélio e suasmúltiplasfunções! Rev Bras Hipertens 2001;8:76–88.

[40] Furchgott RF, Zawadzki JV. The obligatory role of endothelial cells in the relaxation of arterial smooth muscle by acetylcholine. Nature 1980;288:373–6.

[41] Nathan C, Xie QW. Regulation of biosynthesis of nitric oxide. J Biol Chem 1994;269:13725–8.

[42] Fulton D, Gratton JP, Sessa WC. Post-translational control of endothelial nitric oxide synthase: why isn't calcium/calmodulin enough? J Pharmacol Exp Ther 2001;299:818–24.

[43] Libby P. Atherosclerosis: disease biology affecting the coronary vasculature. Am J Cardiol 2006;98:3Q–9Q.

[44] Rateri DL, Moorleghen JJ, Balakrishnan A, et al. Endothelial cell-specific deficiency of Ang II type 1a receptors attenuates AngII-induced ascending aortic aneurysms in LDL receptor−/− mice. Circ Res 2011;108:574–81.

[45] Kim JA, Montagnani M, Koh KK, et al. Reciprocal relationships between insulin resistance and endothelial dysfunction: molecular and pathophysiological mechanisms. Circulation 2006;113:1888–904.

[46] Rüegg C, Mariotti A. Vascular integrins: pleiotropic adhesion and signaling molecules in vascular homeostasis and angiogenesis. Cell Mol Life Sci 2003;60:1135–57.

[47] Kowanetz M, Ferrara N. Vascular endothelial growth factor signaling pathways: therapeutic perspective. Clin Cancer Res 2006;12:5018–22.

[48] Leeb-Lundberg LM, Marceau F, Muller-Esterl W, et al. International union of pharmacology. XLV. Classification of the kinin receptor family: from molecular mechanisms to pathophysiological consequences. Pharmacol Rev 2005;57:27–77.

[49] Zelawski W, Machnik G, Nowaczyk G, et al. Expression and localisation of kinin receptors in colorectal polyps. Int Immunopharmacol 2006;6:997–1002.

[50] Zhao Y, Xue Y, Liu Y, et al. Study of correlation between expression of bradykinin B2 receptor and pathological grade in human gliomas. Br J Neurosurg 2005;19:322–6.

[51] Wu J, Akaike T, Hayashida K, et al. Identification of bradykinin receptors in clinical cancer specimens and murine tumor tissues. Int J Cancer 2002;98:29–35.

[52] Colman RW. Regulation of angiogenesis by the kallikrein-kinin system. Curr Pharm Des 2006;12:2560–99.

[53] Sessa WC. Molecular control of blood flow and angiogenesis: role of nitric oxide. J Thromb Haemost 2009;7(Suppl. 1):35–7.

[54] Moraes MS, Costa PE, Batista WL, et al. Endothelium-derived nitric oxide (NO) activates the NO-epidermal growth factor receptor-mediated signaling pathway in bradykinin-stimulated angiogenesis. Arch Biochem Biophys 2014;558:14–27.

[55] Lander HM, Ogiste JS, Pearce SF, et al. Nitric oxide-stimulated guanine nucleotide exchange on p21ras. J Biol Chem 1995;270:7017–20.

[56] Eller-Borges R, Batista WL, da Costa PE, et al. Ras, Rac1, and phosphatidylinositol-3-kinase (PI3K) signaling in nitric oxide induced endothelial cell migration. Nitric Oxide 2015;47:40–51.

[57] Hess D, Matsumoto A, Kim S, et al. Protein S-nitrosylation: purview and parameters. Nat Rev Mol Cell Biol 2005;6:150–66.

[58] Monteiro HP, Arai RJ, Travassos LR. Protein tyrosine phosphorylation and protein tyrosine nitration in redox signaling. Antioxid Redox Signal 2008;10:843–89.

[59] Curcio MF, Batista WL, Linares E, et al. Regulatory effects of nitric oxide on Src kinase, FAK, p130Cas, and receptor protein tyrosine phosphatase alpha (PTP-Alfa): a role for the cellular redox environment. Antioxid Redox Signal 2010;13:109–25.

[60] Lim KH, Ancrile BB, Kashatus DF, et al. Tumour maintenance is mediated by eNOS. Nature 2008;452:646–9.

[61] Rahman MA, Senga T, Ito S, et al. Nitrosylation at cysteine 498 of c-Src tyrosine kinase regulates nitric oxide-mediated cell invasion. J Biol Chem 2010;285:3806–14.

[62] Fukumura D, Kashiwagi S, Jain RK. The role of nitric oxide in tumour progression. Nat Rev Mol Cell Biol 2006;7:521–34.

[63] Oliveira GA, Rosa H, Reis AKCA, et al. A role for nitric oxide and for nitric oxide synthases in tumor biology. For Immunopathol Dis Therap 2012;3:169–82.

[64] Monteiro HP, Costa PE, Reis AKCA, et al. Nitric oxide: protein phosphorylation and S-nitrosylation in cancer. Biomed J 2015;38:380–8.

[65] Kashiwagi S, Izumi Y, Gohongi T, et al. NO mediates mural cell recruitment and vessel morphogenesis in murine melanomas and tissue-engineered blood vessels. J Clin Invest 2005;115:1816–27.

[66] Ikeda Y, Hayashi I, Kamoshita E, et al. Host stromal bradykinin B2 receptor signaling facilitates tumor-associated angiogenesis and tumor growth. Cancer Res 2004;64:5178–85.

[67] Eller-Borges R, Monteiro H. Estudodopapel do oxidonítriconaativacao da GTPase Rac-1 emcélulasendoteliais e de melanoma murino. Tese de Doutorado em Ciências Biológicas, UNIFESP, Brasil. Ano de Obtenção; 2015.

[68] Radisky DC, Levy DD, Liu H, et al. Rac1b and reactive oxygen species mediate MMP-3induced EMT and genomic instability. Nature 2005;436:123–7.

[69] Ferraro D, Corso S, Fasano E, et al. Pro-metastatic signaling by c-Met through RAC-1 and reactive oxygen species (ROS). Oncogene 2006;25:3689–98.

[70] Heo J, Campbell SL. Mechanism of redox-mediated guanine nucleotide exchange on redox-active Rho GTPases. J Biol Chem 2005;280:31003–10.

[71] Kojda G, Harrison D. Interactions between NO and reactive oxygen species: pathophysiological importance in atherosclerosis, hypertension, diabetes and heart failure. Cardiovasc Res 1999;43:562–71.

[72] Pacher P, Beckman JS, Liaudet L. Nitrico oxide and peroxynitrite in health and disease. Physiol Rev 2007;87:315–424.

[73] Hussain SP, Hofseth LJ, Harris CC. Radical causes of cancer. Nat Rev Cancer 2003;3:276–85.

[74] Rinaldi R, Ogata FT, Salo T, et al. A mestatatic cell line permanently silenced for iNOS (SW620-I12) resembles the primary tumor in many important phenotypes: the importance of nitric oxide in the progression of human colon carcinoma. FEBS J 2013;280:386.

II. ENDOTHELIAL DYSFUNCTION AND CLINICAL SYNDROMES

47

Endothelial Function in Skin Microcirculation

Cristina Pires Camargo and Rolf Gemperli

INTRODUCTION

Skin is the largest organ of the human body and performs several functions: protection against external agents (i.e., microorganism, ultraviolet irradiation, traumas), thermoregulation, excretion of minerals, drugs, water among others. Due to the thermoregulatory action, skin presents high-vessel density. In addition, these vessels are distributed in plexuses that are located near the external surface of the skin, allowing the study of systemic endothelial function [1–3]. Cutaneous microcirculation is characterized by two plexuses: dermal and subdermal, that allows several pieces of equipment such as laser flowmetry [4] and oxymeter devices [5] measures the systemic endothelial function. This chapter addresses the anatomy of skin microvascularization, the examinations performed for analysis of the endothelial function, and clinical conditions that can interfere in the endothelial function evaluation.

ANATOMY OF SKIN MICROVASCULARIZATION

Skin microvascularization is organized in two plexuses:

- *Dermal*: located 1–1.5 mm below the skin surface; and
- *Subdermal*: located in the dermal-subdermal junction.

The subdermal plexus is organized by perforating vessels originating from adjacent musculature and subadjacent adipose tissue. From this plexus, communicating vessels migrate into the dermal level

forming the dermal plexus [1–3] (Fig. 47.1). These two plexuses are connected by means of ascending arterioles and descending venules.

The caliber of dermal arterioles varies between 17 and 26 µm and its main function is to promote the skin's vascular resistance.

The arteriolar structure of the plexus is organized as follows:

- endothelial cells;
- smooth muscle cells; and
- elastic fibers.

Muscle cells and elastic fibers are surrounded by the basement membrane, which depending on the vessel caliber can be distributed throughout the vascular extension either continuously or discontinuously (when the vessel caliber is less than 12 µm) [2,3]. Endothelial cells of the arterioles rest over two layers of smooth muscle cells. The muscle cells are organized in two layers: in inner layer, the cells are configured longitudinally; in external layer, the cells are configured in a spiral format. This spatial configuration is critical to control blood flow and vascular resistance [2,3] (Figs. 47.2 and 47.3).

As the diameter of the vessels decreases, certain structures are no longer identified individually, for example, when reaching a diameter of 15 µm the elastic fibers and muscle cells of the vessel wall are no longer continuous. At this level, muscle cells appear in a single layer, emitting myo filaments that surround the endothelial cells. This type of organization suggests the formation of precapillary sphincters, with vasomotor function [4].

From 12 µm down these vessels are classified as capillaries. Arterial capillaries have external diameters between 10 and 12 µm and inner diameter of

673

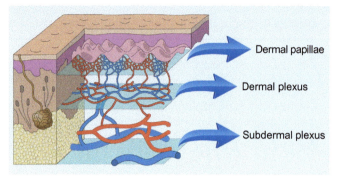

FIG. 47.1 Schematic drawing of dermal and subdermal plexuses. In highlight, capillary network that communicates these two plexuses.

4–6 μm. The basal layer is homogeneous and presents pericytes, which emit filaments in the basement membrane.

These pericytes, since they contain contractile protein microfilaments, probably working as a flow control structure [6]. At this level the vessel walls are discontinuous, allowing the exchange of substances between capillaries and extracellular matrix. From this network, venous capillaries are formed that coalesce to form the postcapillary venules [3].

The dermal papillae have an important role in thermoregulation; the vascular components of this region are 1–2 mm from the surface of the skin. The dermal plexus emits terminal arterioles, arterial and venous capillaries, and postcapillary venules. Each dermal papilla receives a single capillary loop originating from the terminal arteriole and is directed to a postcapillary venule. This structural organization increases the skin irrigation area [7].

BLOOD FLOW REGULATION

Capillaries are responsible for the exchange of gasses, nutrients and cellular excreta; for this exchange to be effective, regulation of blood flow must be performed uninterruptedly and rapidly. This regulation is determined by several factors, such as:

(a) myogenic response of arterioles, capillaries and microcirculation venules;
(b) flow vasodilatation due to wall shear stress; and
(c) neural and metabolic control.

Changes in vessel diameter can take place due to alterations in the microenvironment. They are caused by modifications in the intraluminal pressure of the microcirculation and are independent from systemic arterial pressure. The endothelium plays an important role in releasing nitric oxide, which acts as a vasodilator of this microenvironment. Other vasodilation pathways of the muscles arterioles are prostacyclin (PGI2), epoxy-ethyatrienoic acids (TSEs) and hyperpolarizing factor derived from the endothelium [4].

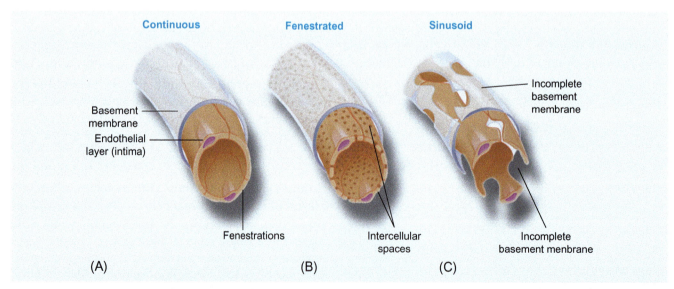

FIG. 47.2 Schematic representation of a cross section of the dermis arteriole. (A) Arterioles with the continuous wall; all components of the muscular wall are present. (B) Arteriole in the capillary transition. There are fenestrations in the wall for the exchange of gases and nutrients. (C) Sinusoid, at that level some of the vessel wall components are absent, allowing exchanges in the microcirculation.

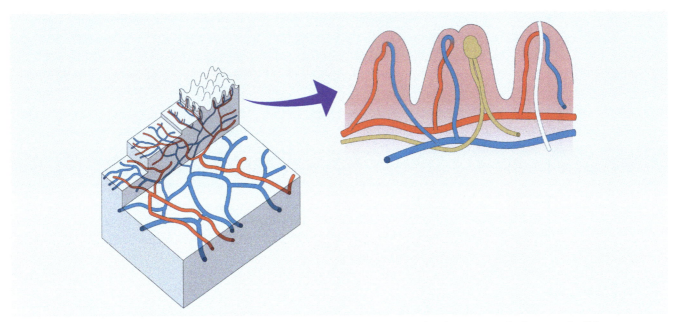

FIG. 47.3 Schematic representation of the cutaneous microcirculation. The deep arterial plexuses *(red)* intercommunicate with superficial vessels and finally with terminal arterioles, which initiate the afferent branch of the capillary loops. These loops project themselves into the dermal papilla in a perpendicular orientation to the skin surface. The efferent branch of the capillary loop joins the superficial subpapillary venous plexus *(blue,* PVS) which communicates with deeper venous plexuses. Only the capillaries on the dermal papillae and the subpapillary venous plexus are visualized at the Periungueal capillaroscopy. Nervous terminal *(yellow).*

The interaction between vessel walls and blood components generates a response from the endothelium, which secretes a greater or smaller amount of nitric oxide, causing vasodilation of this system. As an example, changes in the hematocrit lead to the alteration of plasma stress forces and cells against the vessels wall, causing a change in their caliber [8,9].

The last two factors that interfere with vasodilatation and vasoconstriction of the vessels are neural and metabolic factors. The first factor acts via the sympathetic nervous system, promoting vasoconstriction by releasing epinephrine and vasodilation by releasing substance P and calcitonin gene related peptide (GCRP) [10]. Some metabolic conditions. For example, diabetes [11], autoimmune diseases [4], or vitamin D deficiency; [12] causes changes in endothelial function and rheological properties.

Endothelial Function

As described previously, the skin is one of the windows that allow analysis of systemic endothelial function, once it is located superficially, encompasses a large body surface and presents a linear correlation between the time of ischemia and the time of hyperemia. The study of the time of ischemia and hyperemia shows the functional condition of the vessels [4].

Endothelial function in the skin can be measured through several techniques; such as the following.

Videocapillaroscopy

Capillary density and recruitment can be assessed by means of a magnifying glass. Videocapillaroscopy in the region of the nailfold, where capilaries are parallel to the surface, allows evaluation of autoimmune diseases, for example, systemic sclerosis. Other techniques, such as polarization spectral imaging and sidestream darkfield imaging, also allow evaluation of microvascularization [13]. Also, in patients with infection, analysis of sublingual microvasculature is a factor that can predict risk of severe sepsis [4].

Laser Flowmetry

Basically, most of the lasers devices measure the blood cells flow and density of red cells that pass by a certain area. Depending on the equipment, blood flow does not provide enough information of endothelial function assessment. To solve this

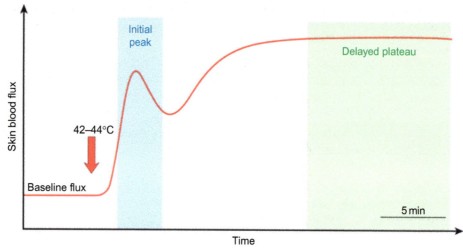

FIG. 47.4 Post occlusion reactive hyperemia test chart. The cross-sectional area corresponds to the time of hyperemia.

limitation, procedures that cause vascular reactivity are associated with this technique [14]. Thus, local stimulation can be performed to test vascular reactivity. The most commonly used techniques are: arterial occlusion (mechanical stimulation), pharmacological stimulation, and thermal stimulation [4].

Stimulation by Occlusion Pressure

Arterial occlusion causes reactive postocclusion hyperemia. For its accomplishment a pressure apparatus is attached to the upper limb or thigh and insufflation is carried out until the operator observes absence of the downstream flow. This occlusion time can be from 1 to 15 min; usually about 5 min is sufficient at the end of which the compression is released. Immediately after analysis of the stream should started by evaluating absolute peak, the peak flow relative to the initial and the area under the curve until vascularization reaches the baseline values [4,15] (Fig. 47.4).

Pressure Stimulation

Pressure induced stimulation is useful to avoid pressure ulcers, as it occurs in patients immobilized for long periods of time. In healthy tissue, nonnociceptive stimulation leads to hyperemia due to prolonged vasodilatation at the site of pressure stimulus [16].

Thermal Stimulation

The most widely used protocol in thermal stimulation is the temperature elevation between 42°C and 44°C, since this temperature does not cause tissue injury or pain. This type of stimulation gives rise to a biphasic response of blood flow (Fig. 47.5).

For this test to be effective, the observation time should be at least 30 min and may extend up to 45 min [17].

Stimulation by Iontophoresis

Iontophoresis uses acetylcholine and sodium nitroprusside to analyze the effects of endothelium-

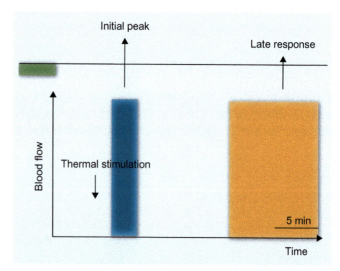

FIG. 47.5 Test chart of reactive hyperemia by temperature stimulation. Biphasic curve. The cross-sectional areas correspond to the time of hyperemia.

dependent and endothelium-independent reactive hyperemia, respectively.

The underlying mechanism by which this technique induces acetylcholine release from the skin microvasculature is not entirely clear, but one hypothesis is the interaction with cyclooxygenase (COX) receptors [4,14]. Topical anesthesia applied prior to this procedure partially neutralizes the effect of reactive hyperemia caused by iontophoresis.

Electric Stimulation

A light electric current can also be used to study skin microvasculature and endothelial function. The mechanism of vasodilation is probably caused by C-fibers and probably, according to preclinical studies, mediated by prostanoid (PGI2) [4].

Pharmacological Stimulation

For this type of evaluation, local drugs are used, with no systemic effect; the dose is controlled by the investigator. These drugs can be used intradermally. In the specific case of the skin, the most frequent substances employed to study microcirculation by microdialysis are interleukin-6 (IL-6), albumin, and neuropeptides [18].

It is a minimally invasive technique requiring previous anesthesia for insertion of the transcutaneous fibers. This method allows control of the dose and rate of infusion; however, the anesthetic can interfere with the results [15,18].

All of these techniques have advantages and limitations, which are summarized in Table 47.1.

All patients undergoing these tests should abstain from coffee, cigarettes, and teas containing caffeine. The environment must have a standardized temperature (24°C) and be quiet, since stress, temperature and certain substances as described above can interfere in blood flow in the skin.

CHANGES IN THE ENDOTHELIAL FUNCTION DUE TO CONDITIONS OF THE SKIN ITSELF

Aging

Aging leads to decreased vascular reactivation of the skin [19]. Elderly individuals have flatter dermal papillae with decreased vascularization; the extracellular matrix is diminished, influencing the regeneration and apoptosis of the skin angiogenesis [20]; All these factors lead to change in the assessment of endothelial function (Table 47.1).

TABLE 47.1 Techniques Used for Vascular Reactivity Analysis With Indications and Limitations of Each Method

Technique	Indications/characteristics	Limitations
Arterial occlusion	Simple	Interference by the use of oral contraceptives and menstrual cycle. Increased reaction in the presence of risk factors for cardiovascular disease, decrease reaction with statins and antihypertensive drugs
Pressure stimulation	Simple	Decreased reaction in the elderly and diabetic patients. This equipment is available exclusively for research
Thermal stimulation	Several techniques under standardization	Lower response in the elderly. Interferences in hypertension, polycystic ovary and use of statins
Pharmacological stimulation	Need for specific equipment	Topical anesthesia may interfere with the results. Minimally invasive technique (insertion of fibers)
Electrical stimulation	Although not invasive, iontophoresis should have control in the patient himself; however, if the control solution has ions the result can be changed	Iontophoresis is altered with the use of topical anesthetic and metformin in diabetic women with angina

Autoimmune Diseases of Collagen

Recent studies showed the influence of autoimmune diseases in the pathogenesis of vascular dysfunction. The effects of acute inflammation on the vascular system are critical to healing infections. However, a chronic inflammatory condition causes irreversible and deleterious changes to the organism. This type of reaction leads to a slight increase in peripheral resistance and consequently to altered responses to pressure and ischemia, perpetuating the inflammation state and originating a cycle that affects in the vascular components [21].

For example, in the Raynaud's phenomenon, it is showed an alteration when the stimulation of the vascular reactivity is thermal, mainly at low temperatures, by the pathophysiology effect of the disease. Currently, this test is performed in patients with this phenomenon to evaluate the of the treatment efficacy [4]. Another clinical condition is systemic scleroderm, in addition to systemic vascular abnormalities, affects vessels of great caliber of the lung, and in the medium and long term can cause cardiovascular disease due to pulmonary hypertension [22].

Metabolic Diseases

Metabolic diseases can interfere in the vascular structure, for this reason, the assessment of endothelial function changes can predict systemic diseases [23].

Moreover, the endothelial function analysis can also evaluate the effect of the medical treatment in the microvasculature system, to assess the therapeutic effectiveness. Oxygen-free radicals are responsible for vascular dysfunction, among other harmful effects to the body. Several studies have shown that vitamins and antioxidants supplementation improve endothelial function, such as vitamins A, C, E, and D [24]. Vitamin A acts on lipid metabolism and consequently in arteriosclerosis. In addition, there are evidences that retinoids and their derivatives interfere with the release of nitrous oxide and in the endothelium function [25]. Currently, several studies suggest that one of the extra-bone tissue effects of vitamin D is the effect on vascular function. The probable pathway of cardiovascular protection is the antiinflammatory effect of this vitamin in its active form 1.25 hydroxyvitamin D [26].

In conclusion, endothelial function in the skin allows the study of systemic microcirculation, as well as diseases restricted to this organ. In addition, it can be used for the control of therapies for skin disease that compromise microvascularization.

References

[1] Farage M, Miller K, Maibach HI. Textbook of aging skin. Berlin: Springer; 2010. p.13–8.
[2] Braveman IM. The cutaneous microcirculation. J Investig Dermatol Symp Proc 2000;5:3–9.
[3] Braveman IM. The cutaneous microcirculation: ultrastructure and microanatomical organization. Microcirculation 1997;4:329–40.
[4] Roustitand M, Cracowski JL. Assessment of endothelial and neurovascular function in human skin microcirculation. Trends Pharmacol Sci 2013;34:373–84.
[5] Thorn CE, Kyte H, Slaff DW, et al. An association between vasomotion and oxygen extraction. Am J Physiol Heart Circ Physiol 2011;301:H442–9.
[6] Joyce NC, Haire MF, Palade GE. Contractile proteins in pericytes. II. Immunocytochemical evidence for the presence of two isomyosins in graded concentrations. J Cell Biol 1985;100:1387–95.
[7] Yen A, Braveman IM. Ultrastructure of the human dermal microcirculation: the horizontal plexus of the papillary dermis. J Invest Dermatol 1976;66:131–42.
[8] Tsai AG, Friesenecker B, McCarthy M, et al. Plasma viscosity regulates capillary perfusion during extreme hemodilution in hamster skinfold model. Am J Physiol 1998;275:H2170–80.
[9] Tsai AG, Acero C, Nance PR, et al. Elevated plasma viscosity in extreme hemodilution increases perivascular nitric oxide concentration and microvascular perfusion. Am J Physiol Heart Circ Physiol 2005;288:H1730–9.
[10] Hodges GJ, Del Pozzi AT. Noninvasive examination of endothelial, sympathetic, and myogenic contributions to regional differences in the humancutaneous microcirculation. Microvasc Res 2014;93:870–91.
[11] Stirban A. Microvascular dysfunction in the context of diabetic neuropathy. Curr Diab Rep 2014;14:541–50.
[12] Abdi-Ali A, Nicholl DD, Hemmelgarn BR, et al. 25-Hydroxyvitamin D status, arterial stiffness and the renin-angiotensin system in healthy humans. Clin Exp Hypertens 2014; 36:386–91.
[13] Levy BI, Schiffrin EL, Mourad JJ, et al. Impaired tissue perfusion: a pathology common to hypertension, obesity, and diabetes mellitus. Circulation 2008;118:968–76.
[14] Tew GA, Klonizakis M, Moss J, et al. Reproducibility of cutaneous thermal hyperaemia assessed by laser Doppler flowmetry in young and older adults. Microvasc Res 2011;81:177–82.
[15] Cracowski JL, Minson CT, Salvat-Melis M, et al. Methodological issues in the assessment of skin microvascular endothelial function in humans. Trends Pharmacol Sci 2006;27:503–8.
[16] Fromy B, Abraham P, Bouvert C, et al. Early decrease of skin blood flow in response to locally applied pressure in diabetic subjects. Diabetes 2002;51:1214–7.
[17] Brunt VE, Minson CT. Cutaneous thermal hyperemia: more than skin deep. J Appl Physiol 2011;111:5–7.
[18] Clough GF. Microdialysis of large molecules. AAPS J 2005;26:E686–92.
[19] Thijssen DH, Carter SE, Green DJ. Arterial structure and function in vascular ageing:"are you as old as your arteries"? J Physiol 2016;594(8):2275–84.
[20] Chang E, Yang J, Nagavarapu U, et al. Aging and survival of cutaneous microvasculature. J Invest Dermatol 2002;118:752–8.
[21] McCarthy CG, Goulopoulou S, Wenceslau CF, et al. Toll-like receptors and damage-associated molecular patterns: novel links

between inflammation and hypertension. Am J Physiol Heart Circ Physiol 2014;306:H184–96.

[22] Ghiadoni L, Mosca M, Tani C, et al. Clinical and methodological aspects of endothelial function in patients with systemic autoimmune diseases. Clin Exp Rheumatol 2008;26:680–7.

[23] Kraemer-Aguiar LG, Laflor CM, Bouskela E. Skin microcirculatory dysfunction is already present in normoglycemic subjects with metabolic syndrome. Metabolism 2008;57:1740–6.

[24] Ozkanlar S, Akcay F. Antioxidant vitamins in atherosclerosis—animal experiments and clinical studies. Adv Clin Exp Med 2012;21:115–23.

[25] Rhee EJ, Nallamshetty S, Plutzky J. Retinoid metabolism and its effects on the vasculature. Biochim Biophys Acta 1821;2012: 230–40.

[26] Mozoz I, Marginean O. Links between vitamin D deficiency and cardiovascular diseases. Biomed Res Int 2015;2015:109275.

TREATMENT OF ENDOTHELIAL DYSFUNCTION

48

Clinical Endothelial Dysfunction: Prognosis and Therapeutic Target

Elisa Alberton Haas, Marcelo Nishiyama, and Protásio Lemos da Luz

INTRODUCTION

The endothelium is an important regulator of vascular homeostasis. Endothelial dysfunction is a pathological condition characterized mainly by imbalance among substances with vasodilator, antimitogenic and antithrombotic properties and substances with vasoconstricting, prothrombotic and proliferative characteristics [1]. Among the most important vasodilator molecules, particularly in the muscular arteries, is nitric oxide (NO), which also inhibits other key events in the development of atherosclerosis, such as adhesion and platelet aggregation, leukocyte adhesion and migration, and smooth muscle cell proliferation. In general, loss of NO bioavailability indicates a broadly dysfunctional phenotype in many endothelial properties [2].

Prostacyclin acts synergistically with NO to inhibit platelet aggregation [3]. Bradykinin stimulates the release of NO, prostacyclin and endothelium-derived hyperpolarizing factor, another vasodilator, which contributes to the inhibition of platelet aggregation [4]. Bradykinin also stimulates the production of activated tissue plasminogen activator (t-PA) and thus may play an important role in fibrinolysis. The endothelium also produces vasoconstrictive substances, such as endothelin and angiotensin II. Angiotensin II not only acts as a vasoconstrictor, but is also prooxidant [5] and stimulates the production of endothelin. Endothelin and angiotensin II promote smooth muscle cell proliferation and, therefore, contribute to the formation of atherosclerotic plaques. [4] Activated macrophages and vascular smooth muscle cells, cell components characteristic of atherosclerotic plaques,

produce large amounts of endothelin [6]. Damage to the endothelium disturbs the balance between vasoconstriction and vasodilation, and initiates a series of processes that promote or exacerbate atherosclerosis; these include increased endothelial permeability, platelet aggregation, leukocyte adhesion, and generation of cytokines [7]. The decrease in the production or activity of NO, manifested as impaired vasodilation, may be an early sign of atherosclerosis.

Thus, evaluation of endothelium vasodilatory properties resulting from NO and other molecules may provide more extensive information on endothelial integrity and function. Most, if not all, cardiovascular risk factors are associated with endothelial dysfunction [8], and correction of risk factors leads to improved vascular function. Endothelial dysfunction has been detected in the coronary vasculature, in various peripheral arteries, and in veins; therefore, it can be deemed a systemic condition. It should be noted that the process of atherosclerosis begins early in life and that endothelial dysfunction contributes to atherogenesis and precedes the development of vascular morphological changes [9].

Prognostic Value

The literature widely documents that endothelial dysfunction is associated with almost all conditions predisposing to atherosclerosis and cardiovascular disease. For example, endothelial dysfunction has been observed in: patients with hypertension [10]; patients with a family history of early atherosclerotic disease [11]; smokers [12] and passive smokers [13];

patients with dyslipidemia [14]; elderly [15]; patients with *diabetes mellitus* [16]; obese individuals [17,18]; patients with hyperhomocysteinemia; and individuals with inflammatory or infectious diseases [19,20]. Furthermore, the effects of endothelium in cardiovascular risk can be seen in children aged 7 years old [21–23].

Table 48.1 shows evaluation of endothelial dysfunction in various clinical settings and its impact on cardiovascular events.

The fact that endothelial dysfunction is a systemic condition may explain why the peripheral, microvascular, and macrovascular endothelial functions correlate with the endothelial function in coronary arteries [37].

There is good evidence that endothelial dysfunction is significantly associated with the cardiovascular risk load and can be considered a "thermometer" of the total risk load: "the risk factor of risk factors."

TABLE 48.1 Endothelial Dysfunction and Impact on Cardiovascular Events

Presence of endothelial dysfunction × prognosis

Trial	No. of patients	Diagnose/population	Follow-up (average ± SD)	Cardiovascular events[a] or general mortality
Perticone et al. [10]	225	Hypertension	31.5 months	2.084 (RR)
Modena et al. [24]	400	Postmenopause women	67 months	3.5 vs 0.51 ($p < 0.0001$)
Rossi et al. [25]	2264	Postmenopause women	45 months	4.42 (RR)
Hirsch et al. [26]	268	Healthy individuals, with no apparent CVD	45 months	14.1% vs. 0.7% ($p = 0.007$)
Shechter et al. [27]	465	Healthy individuals, with no apparent CVD	32 months	2.7 (OR)
Yeboah et al. [28]	3026	Individuals with no CVD—MESA trial	5 years	0.61 $p < 0.02$ (RR)
Fichtlscherer et al. [29]	198	ACS	47.7 months	0.54 $p < 0.02$ (RR)
Shechter et al. [30]	82	Ischemic HF	14 months	2.04 (HR)
Heitzer et al. [31]	289	HF	4.8 years	Independent predictor for events 0.97 (HR) $P = 0.001$
Kübrich et al. [32]	185	Post heart transplant	25 months	1.97 (RR)
Schächinger et al. [33]	147	Patients undergoing routine diagnostic catheterization or PTCA	7.7 years	Independent predictor for events
Halcox et al. [34]	308	Patients undergoing cineangiocoronariography	46 months	Survival free from significant events curve ($p = 0.037$)
Schindler et al. [35]	130	Patients undergoing cineangiocoronariography with normal results	45 months	0.95 (RR) $P = 0.040$
Gokce et al. [36]	187	Patients undergoing vascular surgery	30 days	3.7 (OR) $P = 0.007$

[a] *Cardiovascular events: acute myocardial infarction; myocardial revascularization or coronary angioplasty; stroke; cardiovascular death.*
SD, *standard deviation;* RR, *relative risk;* CV, *cardiovascular;* CVD, *cardiovascular disease;* ED, *endothelial dysfunction;* OR, *odds ratio;* HR, *hazard ratio;* ACS, *acute coronary syndrome;* HF, *heart failure;* PTCA, *percutaneous transluminal coronary angioplasty.*

Prognostic Value in Asymptomatic Patients

The value of endothelial function in the primary prevention scenario is of great interest. In a study with 268 healthy subjects without heart disease and with low clinical risk who were followed for 45 months, the brachial artery flow-mediated dilatation (FMD) was an independent predictor of adverse cardiovascular events, with a significantly higher number of events in patients with low FMD compared to those with normal FMD (14.1% versus 0.7%, $p = 0.007$) [26]. In another trial with 435 patients without cardiovascular disease who were followed for 32 months, the altered FMD was significantly associated with adverse cardiovascular events compared to patients with normal FMD: 11.8% versus 4.7% $p = 0.007$, odds ratio (OR) of 2.78 ($p = 0.003$) [27]. In the Cardiovascular Healthy Study, the relationship between endothelial function and subsequent cardiovascular events was assessed in 2792 apparently healthy individuals older than 72 years. For more than 5 years of follow-up, the event-free survival was significantly higher in patients with normal endothelial function, even after adjustment for traditional risk factors, hazard ratio (HR) of 0.91 ($P = 0.02$) [38]. Similarly, in the Mesa (Multi-Ethnic Study of Atherosclerosis) trial with 3026 Caucasian, black, Latin-American, and Chinese patients without cardiovascular disease who were followed for 5 years, FMD predicted future cardiovascular events even after adjusting for the Framingham risk score (HR 0.84, $p = 0.04$). In addition, FMD allowed for the better classification of cardiovascular risk in combination with the Framingham score, as compared to FMD or the Framingham score alone [28].

In these studies, adjustments for traditional risk factors have weakened the correlation of endothelial function and outcomes. This finding is not surprising, since endothelial dysfunction is a key biologic mechanism by which cardiovascular risk factors exert influence on atherosclerosis and adverse events [2].

Prognostic Value in Hypertension

Endothelial dysfunction is closely related to systemic arterial hypertension [39–41]. There possibly is a bidirectional causal relationship between them. Recent data imply an increase in oxidative stress and vascular inflammation as essential in hypertension pathogenesis [42,43]. These are also central features in endothelial dysfunction, which, when reduced, revert endothelial dysfunction [44]. The question of the prognostic value of endothelial dysfunction in hypertension has been studied in several studies, as discussed below.

Recent publications suggest that hypertension is associated with increased production of oxygen reactive species (ORS) of mitochondria and NADPH oxidase (nicotinamide adenine dinucleotide phosphate) in the vascular endothelium [45–47].

Isolated rat carotids that were exposed to an increase in intraluminal pressure showed a reduction of endothelium-dependent vasodilation with acetylcholine, and an increase in both production of superoxide and NADPH oxidase activity [46]. Another trial showed coordinated expression of NADPH oxidase ORS and mitochondria in rats [47].

The relevance of adipose tissue in the regulation of metabolism and inflammation through the production of inflammatory and antiinflammatory adipokines has been increasingly recognized [48]. However, adipose inflammation is not restricted to obesity and insulin resistance: recent studies have evaluated the effect of perivascular adipose tissue in vascular homeostasis in hypertension. In hypertensive rats, the adipose tissue inserted into segments of the thoracic aorta did not suppress vasoconstriction induced by phenylephrine, in contrast with the adipose tissue of normotensive animals [49]. Similarly, obese and hypertensive rats with perivascular inflammation show damaged endothelial function when compared to normotensive animals [50]. This data suggest that perivascular adipose tissue and inflammation influence the local and systemic regulation of vascular homeostasis.

On the other hand, recent data suggests relevance of innate and adaptive immune responses in the regulation of endothelial function in hypertension. Activation of the complement pathway in innate immunity may negatively impact vascular endothelial function in hypertension [51], whereas increased expression of interleukin-10 (which is antiinflammatory) in the adaptive immune response attenuates the adverse effects of angiotensin II on the endothelial function associated with hypertension [52]. Circulating endothelial progenitor cells (EPCs), derived from pluripotent myeloid stem cells, which originate from mature mononuclear cells, also play an important role in the maintenance of the endothelium homeostasis through its regeneration and repair mechanism. A reduction of EPCs levels circulating in men correlates with impaired vascular endothelial function [53], whereas an infusion of EPCs help to reverse endothelial dysfunction in atherosclerosis-prone mice.

The regeneration and repair capacity of EPCs in humans newly diagnosed as prehypertensive or hypertensive patients is reduced in relation to healthy controls [54]. Furthermore, the level of circulating EPCs is negatively influenced by fragments of C3a activated complement in hypertensive humans [51]. Impairment of vascular endothelial function and vascular damage related to C-reactive protein depends on the presence of C3 [55]. Taken together, this data suggests that vascular endothelial dysfunction associated with hypertension is related to local vascular inflammation and systemic inflammation. Excess oxidative stress and inflammation result in endothelium-dependent vasomotor dysfunction. Endothelial dysfunction may subsequently worsen hypertension (Fig. 48.1) [56].

The prognostic value of endothelial dysfunction was evaluated in 225 hypertensive patients, men and women between 35 and 54 years who were followed for 31 months. A relative risk of 2.084 was found for major cardiovascular events, i.e., cerebral, cardiac, and peripheral, even when adjusting

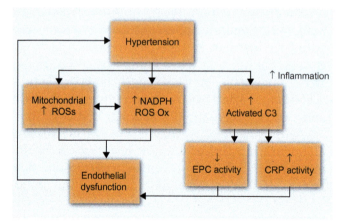

FIG. 48.1 Potential mechanism of endothelial dysfunction associated with hypertension. Mitochondrial ROS, reactive oxygen species of mitochondrial origin; NADPH ROS Ox, NADPH reactive oxygen species (*nicotinamide adenine dinucleotide phosphate*) oxidase in the vascular endothelium; activated C3, activated complement 3; EPC, endothelial progenitor cells; CRP, C-reactive protein. Adapted from Dharmashankar K, Widlansky ME. NIH public access. Curr Hypertens Rep 2011;12:448–55.

for other risk factors, including 24-h blood pressure values [10]. In the Modena et al. trial [24], 400 postmenopausal women were included with mild to moderate hypertension and altered FMD who were followed for 67 months. In the group of 150 patients who continued to present with endothelial dysfunction, even after treatment, the occurrence of cardiovascular events was significantly higher (3.5 versus 0.51 events/100 person/year; $p < 0.0001$) compared to the group who had reversal of endothelial dysfunction with treatment. This finding indicates that the reversal of endothelial dysfunction beneficially influences clinical evolution, at least in this population.

Insulin Resistance, Diabetes, and Endothelial Dysfunction

Endothelial dysfunction is associated with *diabetes mellitus* (DM) and insulin resistance in clinical and experimental trials [57].

Cross-sectional clinical trials show reduction of endothelium-dependent vasodilation in coronary and peripheral arteries of patients with *diabetes mellitus* type 1 [58,59] and type 2 [16,60]. Endothelial dysfunction is also observed in conditions associated with DM2, including obesity [61,62], sedentary lifestyle, and metabolic syndrome [63,64]. In addition to compromising vasodilator function, DM is also associated with increased levels of circulating adhesion molecules and endothelium-derived biomarkers, as plasminogen activator inhibitor-1 (PAI-1) [65], which is reflected in a pro-inflammatory and prothrombotic endothelial phenotype.

The importance of inflammation for the pathogenesis of atherosclerosis is well established [7]. Under physiological conditions, NO prevents the adhesion of leukocytes and the endothelium remains in a state of antiinflammatory rest [44]. In the presence of risk factors, the endothelium can be activated to express adhesion molecules such as vascular cell adhesion molecule-1 (VCAM-1) and intercellular adhesion molecule-1 (ICAM-1), necessary for the adhesion of leukocytes to the endothelial surface [66]. The activated endothelium also expresses chemotactic factors, such as monocyte chemoattractant protein-1 (MCP-1) and other proinflammatory cytokines such as macrophage colony-stimulating factor (MCSF) and beta tumoral necrosis factor (TNF-β) [66]. Endothelial expression of these factors contributes to the development of inflammation in the arterial wall and promotes atherogenesis [67]. In addition to regulating inflammation of the vessel wall, the vascular endothelium produces a number of other molecules that affect blood fluidity and thrombosis [44]. Endothelial production of pro-thrombotic molecules, such as plasminogen activator inhibitor-1, thromboxane, tissue factor, and von Willebrand factor (vWF), is counteracted by the production of antithrombotic molecules, such as NO, heparins, prostacyclin, t-PA, and thrombomodulin. Risk factors, including *diabetes mellitus,* are associated with modifications in this balance state towards a prothrombotic and antifibrinolytic state.

Endothelial dysfunction may precede the development of DM. Healthy nondiabetic subjects who have a first-degree relative with type 2 DM show reduced endothelium-dependent vasodilation as well as increased markers of endothelial cell activation [68–70]. In addition to these cross-sectional trials, prospective studies showed that blood markers of endothelial activation, such as vWF, E-selectin, and CD49, predict the incidence of type 2 diabetes after adjusting for other risk factors, including body mass index, physical activity level, lipids, family history of *diabetes mellitus* and glucose tolerance [65]. Likewise, reduced FMD and polymorphisms of NO synthase (eNOS) are multivariate predictors of the incidence of type 2 DM [71,72].

The occurrence of endothelial dysfunction before the development of type 2 DM suggests that there are common pathophysiological mechanisms, increasing the possibility of a certain causal link between insulin resistance and endothelial dysfunction.

Prognostic Value In Coronary Artery Disease

Although endothelial dysfunction plays an important role in the pathogenesis of atherothrombotic diseases, the question of its prognostic value in coronary disease (CAD) deserves special analysis. Early evidence emerged in patients with nonobstructive CAD, in whom the incidence of cardiovascular and cerebrovascular events in the presence of decreased coronary vascular function was significantly elevated. In 503 patients subjected to coronary angiography, followed for 9 years, the 305 patients who had coronary endothelial dysfunction had a greater chance of cerebrovascular events (OR 4.32) [88]. Similarly, peripheral endothelial

dysfunction evaluated with FMD and venous occlusion plethysmography predicted CV events in patients with stable CAD [89] and in patients after acute coronary syndrome [29]. In 281 patients, men and postmenopausal women who were followed for 4.5 years after elective coronary angiography, endothelial dysfunction assessed by FMD and plethysmography at the time of examination was an independent predictor of major cardiovascular events with OR of 0.9 ($p < 0.01$) [89].

In the established CAD scenario, patients with endothelial dysfunction have higher rates of adverse cardiovascular events compared to those with normal endothelial function [90]; in addition, reduced FMD has proven to be an independent predictor of intra-stent stenosis after single-vessel coronary interventions [91].

Prognostic Value in Acute Myocardial Infarction

There are several prognostic indexes for acute myocardial infarction (AMI); here, the endothelial dysfunction will be specifically evaluated [92–95]. For example, the no-reflow phenomenon at angiography strongly predicts mortality in 5 years, independent of infarct size, in patients with acute myocardial infarction with elevation in the ST segment. In 1046 patients followed after angioplasty, 410 had no-reflow, and the mortality rate was significantly higher in this group, with HR of 1.66; ($P = 0.004$) [96]. Interestingly, no-reflow may be reversible in some cases, which is associated with better prognosis [97].

Endothelial dysfunction in peripheral circulation and in the coronary arteries is not only a cardiovascular risk marker but also contributes to the progression of atherosclerosis [98] and cardiovascular events. It is interesting to note that atherosclerotic epicardial segments that show more endothelial dysfunction are those with characteristics of vulnerable atherosclerotic plaques [99]. They are characterized by loss of NO activity, increased endothelin-1 activity [100], and greater propensity to progress to obstructive disease [101].

It should be emphasized that microvascular dysfunction may contribute to poor regulation of myocardial perfusion, reducing the capacity to increase perfusion in response to exercise or mental stress, a circumstance that may lead to myocardial ischemia [102]. In the context of AMI, microvascular endothelial dysfunction is an important mediator of the event, not just a consequence [103]. This is probably due to reduction in coronary blood flow by shear stress alterations at the epicardial level and by endothelial function impairment, facilitating the formation of thrombi. Diabetes and the accumulation of risk factors in metabolic syndrome, for example, have significant deleterious effects on myocardial perfusion and on infarct size in patients with AMI [104–107].

In addition, patients with alterations in preprocedure microvascular function changes in angioplasty are more likely to have postprocedure microvascular dysfunction, as well as injuries related to the procedure and diminished outcomes. In the Debate II study [108], 379 patients were submitted to pre and postprocedure coronary flow reserve assessment by Doppler; a good preprocedure flow predicted best result in stent implantation (OR 1.97, $p < 0.05$). In contrast, low flow reserve was a strong predictor of major cardiovascular events in 30 days (OR 4.71, $p = 0.034$). Thus, preexisting endothelial microvascular dysfunction leads to greater vulnerability to myocardial injury, highlighting the relevance of dysfunctional microcirculation and its damages.

Prognostic Value in Ischemic Heart Failure and After Cardiac Transplantation

In heart failure, there are several prognostic indices, both hemodynamic and plasmatic. In this context, endothelial function has been considered.

Thus, in 82 patients of whom 75 were men that were followed for 14 months with ischemic heart failure functional class IV of NYHA and mean ejection fraction of 22%, endothelial dysfunction evaluated by brachial FMD was significantly associated with a composite endpoint of global mortality, AMI and hospitalizations for decompensated heart failure. Twenty-two events (53.6%) were observed in the group with endothelial dysfunction against 8 events (19.5%) in the group without dysfunction with $p < 0.01$ [30].

In patients with vascular disease of the graft, normal endothelial function is associated with decreased progression of coronary intimal thickening, and epicardial endothelial dysfunction independently predicts outcome. In 73 patients, 64 of them being men who were followed after heart transplantation, 14 presented with cardiovascular

disease of the graft or cardiac death; in these, endothelial dysfunction was significantly more prevalent: epicardial endothelial dysfunction (constriction of 11.1%±2.9 versus expansion of 1.7%±2.2, $p = 0.01$) and microvascular endothelial dysfunction (flow increase of 75%±20 vs. 149%±16, $p = 0.03$) compared to patients who did not have major cardiac outcomes [109]. Another trial evaluated 185 patients of both sexes after 60 months since cardiac transplantation. Seventy three percent of them were free of events. Epicardial endothelial dysfunction was an independent predictor for events in multivariate analysis (RR 1.97; $p = 0.028$) [32].

Prognostic Value in Postmenopausal Women

The prognostic value of endothelial dysfunction in postmenopausal women has been analyzed in different circumstances. In the Italian study of Rossi et al. [25], 2264 postmenopausal women were subjected to FMD evaluation and followed for 45 months. There were 90 major cardiovascular events in the period. The tertile of patients with the lower FMD had RR for events of 4.42 and accounted for 56.6% of total events. Even when added to other conventional cardiovascular risk factors, i.e., age, smoking, hypercholesterolemia, diabetes, and hypertension, FMD significantly contributed to the prediction model of cardiovascular events (chi-square: 10.22; $P < 0.0001$).

On the other hand, reversal of endothelial dysfunction in postmenopausal women is associated with improved prognosis. In the follow-up of 400 postmenopausal women with mild to moderate hypertension for 67 months, FMD was evaluated in the initial period and after 6 months of ideal blood pressure control with antihypertensive therapy. After 6 months of treatment, FMD had not changed ($\leq 10\%$ regarding baseline) in 37.5% of the 400 women (group 1), while it improved significantly ($\geq 10\%$ regarding baseline) in the remaining 62.5% (group 2). During follow-up, there were 3.50 events per 100 person-years in group 1 and 0.51 per 100 person-years in group 2 ($p < 0.0001$). The study shows, therefore, that a significant improvement in endothelial function can be achieved after 6 months of antihypertensive therapy, clearly identifying patients who have a more favorable prognosis [24].

ENDOTHELIUM AS A THERAPEUTIC TARGET

In this section, we shall analyze whether reversal of endothelial dysfunction can be achieved through the use of medications and changes in lifestyle.

Statin therapy significantly improves peripheral and coronary vascular function. Early evidence of this arose from two controlled trials conducted in 1995, in which cholesterol reduction improved endothelial function [110,111]. There is now convincing evidence that treatment with statins has a beneficial effect on coronary and peripheral endothelial function [112], in addition to the lipid-lowering effect, probably due to its antiinflammatory and antioxidant properties, as well as restoration of vascular NO bioavailability [113].

Oxidized LDL (oxLDL) plays an important role in endothelial dysfunction, promoting inflammation and vascular oxidation, which leads to vascular endothelial activation. Circulating levels and tissue immunoreactivity in endothelin-1 (ET-1), a powerful vasoconstrictor and mitogenic substance, are elevated in patients with advanced atherosclerosis and acute coronary syndrome [100]. Exposure to oxLDL increases the production and release of ET-1 [114]. Increased ET-1 in combination with platelet-derived growth factor promotes proliferation of vascular smooth muscle cells in the neointima of atherosclerotic lesions [115].

The mechanisms by which LDL inhibits NO activity of endothelial cells include downregulation of the endothelial NO synthase (eNOS) expression [116], a decrease in receptor-mediated NO release [117] and inactivation of NO by anion superoxide [118]. Furthermore, LDL facilitates the development of atherosclerosis, increasing the adhesion of monocytes to endothelial cells in vitro [119], a process that can be mediated by increased expression of adhesion molecules, such as intercellular adhesion molecule-1 (Fig. 48.2) [120].

Hypercholesterolemia impairs endothelial function and blocks the conversion of 3-hydroxy-3-methylglutaryl coenzyme A (HMG-CoA) to

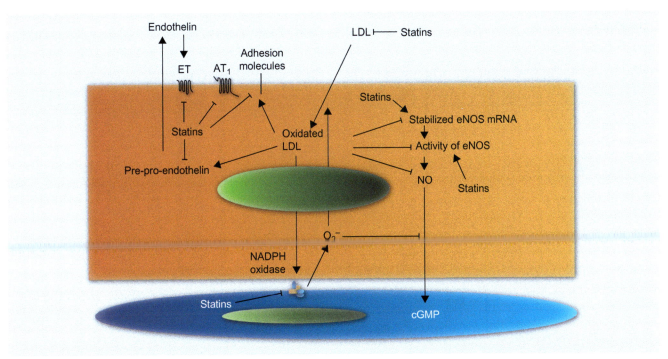

FIG. 48.2 Statins mechanisms of action on the endothelial function: oxidized LDL impairs endothelial function and leads to endothelial cell activation. Statins neutralize these effects by reducing circulating LDL levels and, in part, by direct action on endothelial cells, leading to increased eNOS activity, reduced expression of vasoconstricting agents, and decreased reactive oxygen species production. LDL, low-density lipoprotein; Ox LDL, oxidized LDL; CGMP, cyclic guanosine monophosphate; ET, endothelin; AT1, angiotensin receptor subtype 1; ENOS, endothelial nitric oxide synthase; MRNA, messenger RNA. Adapted from Wolfrum S. Endothelium-dependent effects of statins. Arterioscler Thromb Vasc Biol 2003;23:729–36.

mevalonate, thus limiting a key step in cholesterol biosynthesis. However, restoration of endothelial function occurs before significant decreases in cholesterol serum levels [121], suggesting other effects on endothelial function in addition to cholesterol reduction.

Inhibition of HMG-CoA reductase by statins not only reduces cholesterol production but also inhibits the formation of various isoprenoid intermediates [122]. Protein isoprenylation allows a covalent bond, subcellular localization, and intracellular traffic of several membrane associated proteins. Their inhibition, such as geranylgeranyl transferases or Rho, leads to increased eNOS expression [123]. Statins also regulate eNOS expression by prolonging the half-life of eNOS mRNA, but not its transcription [123]. Hypoxia, oxLDL and cytokines such as tumor necrosis factor (TNF-α) decrease eNOS expression, reducing the stability of eNOS mRNA. The ability of statins to prolong the half-life of eNOS may turn them into effective agents against conditions that negatively regulate eNOS expression. In addition, statins can activate the Akt protein kinase [124]. The serine-threonine kinase Akt pathway is an important regulator of many cellular processes including cell metabolism and apoptosis. Activation of Akt by statins inhibits apoptosis and increases NO production in cultured endothelial cells.

Another mechanism by which statins may improve endothelial function is by their TES antioxidant effects—for example, reducing free radicals generation induced by angiotensin II through inhibition of NAPH oxidase and downregulation of expression of the AT1 receptor [125].

An initial step in atherogenesis involves adhesion of monocytes to the endothelium and penetration into the subendothelial spaces. Statins can reduce the number of inflammatory cells in atherosclerotic plaques; therefore, they have antiinflammatory properties. The mechanisms involve inhibition of adhesion molecules and cytokines such as interleukin-6, leptin, resistin [126]. In a study with 30 men after SCA and 26 controls, administration of atorvastatin 40 mg for three months improved endothelial function: it was able to significantly improve FMD and reduce CRP levels [79]. In another evaluation after ACS, 87 patients without previous use of statins were randomized to receive 5 or 40 mg of rosuvastatin. FMD was assessed at baseline and after 6 and 12 months. There was a significant reduction in LDL in both groups ($p = 0.001$),

but FMD increased significantly only in the high-dose rosuvastatin group [76].

Atorvastatin induced, in 26 patients with heart failure, improved endothelial function assessed by FMD through the mobilization of endothelial progenitor cells, a possible new mechanism for endothelial protection of statins in heart failure [78]. It is important to observe that two doses of atorvastatin 10 and 40 mg were tested; improvement in endothelial dysfunction was significantly greater at the 40 mg dose ($p = 0.001$).

Among statins, pravastatin has been shown to play an important role in the reversal of endothelial dysfunction in experimental models and in some patients with preeclampsia, and may be a valuable candidate for the treatment of this condition [77,127].

In antihypertensive therapy with angiotensin converting enzyme (ACE) inhibitors, angiotensin receptor blockers, calcium channel blockers (CCB) and certain beta-blockers, particularly in the group containing the nebivolol molecule, the actions on the endothelium were analyzed. ACE inhibitors improved endothelial function by inhibiting the angiotensin converting enzyme and reducing Angiotensin II. In addition, ACE inhibitors promote the stabilization of bradykinin, which induces NO and prostacyclin release and reduces oxygen-free radicals production by means of NADPH vascular oxidase, which is stimulated by angiotensin II [128]. In the Trend trial [129], compared to placebo at six months, quinapril treatment produced improvement in endothelial dysfunction in normotensive patients with coronary artery disease. These benefits probably occur due to the attenuation of the vasoconstricting effects ascribed to the generation of superoxide by angiotensin II and to the increase in the release of cellular endothelial NO secondary to the reduction of bradykinin breakdown. A recent meta-analysis has shown that ACE inhibitors improve endothelial function in patients with endothelial dysfunction caused by various conditions and are superior to calcium channel blockers and beta-blockers. There was no significant difference between the effects of ACE inhibitors and angiotensin receptor blockers (ARBs) on peripheral endothelial function [83].

On the other hand, studies with ARBs have shown a positive effect on endothelial function, which endorses the important role of angiotensin II in the development of atherosclerosis [84]. In the

Island trial [84], ARBs improved endothelial function and reduced inflammatory markers, which implies an important role in the pathogenesis of atherosclerosis.

As for beta-blockers, there was a historic development. Unlike the first- and second-generation beta-blockers, the third generation, such as carvedilol [130] and nebivolol [131], has favorable effects on endothelial function. Both drugs stimulate β3 receptors, which activate eNOS, have antioxidant effects and increase NO release [130–132]. In a recent randomized study, it was observed that, in comparison with metoprolol, carvedilol significantly improved endothelial function in patients with hypertension and type 2 *diabetes mellitus* when administered for 5 months, in addition to its customary antihypertensive drugs [85]. As for calcium channel blockers (CCB), they reduce calcium entry into voltage-dependent L-type channels in vascular muscle cells, promoting peripheral and coronary arterial vasodilatation. In addition, some CCBs activate endothelial NO-synthetase or have antioxidant properties, thus increasing NO availability [133]. The Encore-1 [134] and Encore 2 [87] studies showed long acting nifedipine improves coronary endothelial function in patients with stable CAD, an effect that persisted even after drug discontinuation.

In a recent meta-analysis involving 150 patients, treatment of Obstructive Sleep Apnea Syndrome (OSAS) with positive ventilation by CPAP (continuous positive airway pressure) showed important improvement of endothelial function [82]. Compared to the control group, patients who received treatment with CPAP showed significant increase in FMD of 3.87% (HF of 95%: 1.93–5.80, $p < 0.001$). To rate the magnitude of this FMD improvement, other well-established treatments of endothelial dysfunction have not achieved the same increase, as statins (increased 1.5% FMD) [112], ACE inhibitors (1.3%) [83], and exercise (2.1%) [135].

Among antidiabetic drugs, empaglifozine deserves mention. In the recent trial EMPA-REG [73] with 7020 diabetic patients who were followed by 3 years, this sodium-glucose cotransporter 2 (SGLT-2) in the renal tubule, associated with standard treatment, reduced rates of individual outcomes and combinations of hospitalizations for heart failure or cardiovascular death in type 2 diabetic patients with or without heart failure. In this study, the drug produced a significant improvement in endothelial function. A longer follow-up of these patients is necessary, but it is possible that the improvement in endothelial function is associated with some of the cardiovascular benefits found in this cohort.

Data from clinical and bench studies indicate that metformin exerts a direct effect on the endothelium and thus provides protection against the development of hyperglycemia-induced vascular disease. In a study with 42 type 1 diabetic patients, metformin improved endothelial function and oxidative stress, independent of weight loss or glycemic control [136]. Furthermore, metformin has emerged as a promising drug in the management of preeclampsia. In placentas from women with preeclampsia, metformin reduced endothelial dysfunction, vasodilation in the arteries of the omentum, and angiogenesis [75]. It was noted in this study that metformin reduced the levels of two anti-angiogenic substances that induce endothelial dysfunction: soluble fms-like tyrosine kinase 1 (sFlt-1) and soluble endoglin (sEND). This decrease occurred possibly through inhibition of the mitochondrial electron transport chain. The activity of the mitochondrial electron transport chain was increased in placentas tissue of premature babies by preeclampsia. Thus, metformin appears to have potential to prevent or treat preeclampsia.

In the scenario of erectile dysfunction, a recent clinical trial with 54 diabetic men, tested for 6 months, vardenafil caused a significant improvement in endothelial function—measured by FMD and IL6, with $p = 0.04$ and 0.19, respectively. In addition, there was improvement in erectile dysfunction index and testosterone levels [81]. Table 48.2 illustrates several drugs influence on endothelial function.

Reduction in Salt Intake

Substantial reductions in morbidity, mortality and health care costs can be achieved with modest and achievable restriction of salt in the diet [137]. Interestingly, the antihypertensive effect of sodium restricted diet are relatively modest (about 5 mm Hg reduction in systolic arterial pressure and about 2.7 mm Hg reduction of arterial diastolic blood pressure) according to a recent meta-analysis of studies with sodium restriction to reduce blood pressure [138].

There is evidence that reduction in salt intake may reduce morbidity and cardiovascular mortality, reversing endothelial dysfunction induced by high

TABLE 48.2   Pharmacological Influence on Endothelial Dysfunction

Drug influence on endothelial dysfunction

Drug	Population	No.	Endothelial function	Clinical outcome
Empaglifozin [73]	T2DM	7020	↑↑↑	↓CV death, hospitalization for HF, all-cause mortality
Metformin [74]	T2DM	44	↑↑	–
Metformin [75]	Preeclampsia	In vitro and ex vixo	↑	–
Rosuvastatin [76]	ACS	87	↑↑	–
Pravastatin [77]	Preeclampsia	3	↑	–
Atorvastatin [78]	Ischemic HF	26	↑↑	–
Atorvastatin [79]	ACS	56	↑↑	–
Fenofibrate [80]	T2DM	193	↑ in 4 weeks, = in 2 years	–
Vardenafil [81]	ED	54	↑↑	↑ of erectile function
CPAP [82]	OSA	150	↑↑↑	–
ACEI [83]	Meta-analysis: Several clinical conditions	1129	↑↑	–
Irbesatana [84]	Metabolic syndrome	58	↑↑	↓ of inflammatory markers
Carvediol [85]	Hypertension and T2DM	34	↑↑	–
Nebivolol [86]	Hypertension	12	↑↑↑	–
Nifedipino [87]	ACS	454	↑↑	–

T2DM, type 2 *diabetes mellitus*; *HF*, heart failure; *CV*, cardiovascular; *ACS*, acute coronary syndrome; *HF*, heart failure; *ED*, erectile dysfunction; *CPAP*, *continuous positive airway pressure*; *OSA*, obstructive sleep apnea; *ACEI*, angiotensin converting enzyme inhibitors.

salt intake. In animal models, the antihypertensive effect of salt loading seems mechanically linked to increased oxidative stress and reduction of NO bioavailability [139,140]. In addition, the hypertensive effect of sodium load is exacerbated by NO inhibition [141].

Overall, epidemiological and mechanistic data suggests two possible explanations for the beneficial effects of sodium restriction, despite its modest general antihypertensive effects: (1) the direct sodium restriction improves endothelial function in men with hypertension; and (2) the interventions that enhance NO bioavailability and reduce systemic oxidative stress can also counteract the effects of endothelial dysfunction induced by a high salt diet and reduce overall cardiovascular risk. These hypotheses were the subject of two recent studies

in humans [142,143]. In a cross-sectional study with hypertensive and prehypertensive phase I patients, those with self-reported sodium intake lower than 100 mmol/day had significantly higher brachial FMD than those with an intake of between 100 and 200 mmol/day; there was a strong negative correlation of FMD with self-reported sodium intake [142]. In another trial, 147 hypertensive subjects were randomized to receive one, three or six servings of fresh fruits and vegetables per day for 12 weeks, after a pretest period of four weeks with less than the one serving per day. Systolic blood pressure decreased 5.4 mm Hg in groups receiving three and six portions, although the difference was not statistically significant. Endothelium-dependent brachial blood flow increased significantly, i.e., 6.2% for each increase of a portion of fresh fruits and vegetables [143]. The sodium content in the diet was not reported, but it is likely that those randomized to more servings of fruits and vegetables also have had a concomitant reduction of salt intake because of caloric food replacement which tend to have higher salt content. Part of the effect can also be secondary to the high polyphenol content in the fruits and vegetables consumed, which may improve NO bioavailability by mechanisms in addition to the antioxidant activity (see Chapter 25) [144].

Lifestyle Changes

Many interventions are beneficial to micro- or macrovascular endothelial function, increasing NO bioavailability, such as: physical exercise; [145–147] weight reduction [148,149], including bariatric surgery; [150,151] and dietary interventions with foods rich in polyphenols, especially fruits, tea, cocoa [152], and red wine [153]. Experimental studies indicate that the most beneficial effects of wine consumption are attributed to flavonoid found in red wine, in grape juice and various fruits and vegetables. The mechanisms include antiplatelet actions, increased high-density lipoprotein, oxidation, reduction of endothelin-1, and increased expression of endothelial NO-synthase. These results lead to the concept that moderate consumption of red wine, in the absence of contraindications, may be beneficial for patients who are at risk of atherosclerotic cardiovascular events. There is also the possibility that resveratol and other polyphenols protect against deterioration of cognitive function [154]. In addition, a diet high in fruits and vegetables that contain flavonoids can be even more beneficial [155,156]. More information on diets and endothelium can be obtained in Chapters 20 and 25.

An important change of lifestyle with impact on endothelial function is smoking cessation, which clearly demonstrates a favorable effect on coronary epicardial endothelial function [157].

Responders *Versus* Nonresponders

Endothelial function measurements can differentiate responders from nonresponders to treatment [158]. In secondary prevention, it was observed that patients who do not respond to drug interventions with improvement in endothelial function are at increased risk for future events. This preliminary data suggests that individual therapy guided by measures of endothelial function is feasible, but more studies are needed to evaluate whether treatments guided by endothelial function effectively improve results.

By using the Framingham risk score, we can deal with patients in the category of high or low risk. However, many patients have intermediate risk, in which the recommendations are less clear. Thus, the reclassification of patients with intermediate risk according to their endothelial function appears to be feasible and reasonable, although more studies are needed in this area.

In conclusion, pathophysiology of vascular dysfunction is complex and probably changes in different vascular beds, as well as micro- and macro-circulation. The fundamental objective of the assessment of endothelial function is to identify patients who will develop future events. The fact that some do not respond properly to treatment that usually improves endothelial function also has prognostic value. Many interventions preserve endothelial function, including the use of medications as well as other factors related to a healthy lifestyle. Advances in both assessment techniques and interventions that can preserve endothelial function can be expected in the near future.

References

[1] Virdis A, Ghiadoni L, Taddei S. Human endothelial dysfunction: EDCFs. Pflugers Arch 2010;459:1015–23.

[2] Flammer AJ, Anderson T, Celermajer DS, et al. The assessment of endothelial function: from research into clinical practice. Circulation 2012;126:753–67.

[3] Lüscher TF, Barton M. Biology of the endothelium. Clin Cardiol 1997;20:II-3–II-10.

[4] Drexler H. Factors involved in the maintenance of endothelial function. Am J Cardiol 1998;82:3S–4S.

[5] Sowers JR. Hypertension, angiotensin II, and oxidative stress. N Engl J Med 2002;346:1999–2001.

[6] Kinlay S, Behrendt D, Wainstein M, et al. Role of endothelin-1 in the active constriction of human atherosclerotic coronary arteries. Circulation 2001;104(10):1114–8.

[7] Ross R. Atherosclerosis—an inflammatory disease. N Engl J Med 1990;340:115–26.

[8] Anderson TJ, Gerhard MD, Meredith IT, et al. Systemic nature of endothelial dysfunction in atherosclerosis. Am J Cardiol 1995;75: 71B-4B.

[9] Juonala M, Viikari JS, Laitinen T, et al. Interrelations between brachial endothelial function and carotid intima-media thickness in young adults: the cardiovascular risk in young Finns study. Circulation 2004;110:2918–23.

[10] Perticone F, Ceravolo R, Pujia A, et al. Prognostic significance of endothelial dysfunction in hypertensive patients. Circulation 2001;104:191–6.

[11] Clarkson P, Celermajer DS, Powe AJ, et al. Endothelium-dependent dilatation is impaired in young healthy subjects with a family history of premature coronary disease. Circulation 1997;96:3378–83.

[12] Zeiher AM, Schächinger V, Minners J. Long-term cigarette smoking impairs endothelium-dependent coronary arterial vasodilator function. Circulation 1995;92:1094–100.

[13] Celermajer DS, Adams MR, Clarkson P, et al. Passive smoking and impaired endothelium-dependent arterial dilatation in healthy young adults. N Engl J Med 1996;334:150–4.

[14] Mäkimattila S, Virkamäki A, Groop PH, et al. Chronic hyperglycemia impairs endothelial function and insulin sensitivity via different mechanisms in insulin-dependent diabetes mellitus. Circulation 1996;94:1276–82.

[15] Linder L, Kiowski W, Bühler FR, et al. Indirect evidence for release of endothelium-derived relaxing factor in human forearm circulation in vivo. Blunted response in essential hypertension. Circulation 1990;81:1762–7.

[16] Steinberg HO, Chaker H, Leaming R, et al. Obesity/insulin resistance is associated with endothelial dysfunction. Implications for the syndrome of insulin resistance. J Clin Invest 1996;97:2601–10.

[17] Al Suwaidi J, Higano ST, Holmes DR, et al. Obesity is independently associated with coronary endothelial dysfunction in patients with normal or mildly diseased coronary arteries. J Am Coll Cardiol 2001;37:1523–8.

[18] Apovian CM, Bigornia S, Mott M, et al. Adipose macrophage infiltration is associated with insulin resistance and vascular endothelial dysfunction in obese subjects. Arterioscler Thromb Vasc Biol 2008;28:1654–9.

[19] Parchure N, Zouridakis EG, Kaski JC. Effect of azithromycin treatment on endothelial function in patients with coronary artery disease and evidence of Chlamydia pneumoniae infection. Circulation 2002;105:1298–303.

[20] Hürlimann D, Forster A, Noll G, et al. Anti-tumor necrosis factor-alpha treatment improves endothelial function in patients with rheumatoid arthritis. Circulation 2002;106:2184–7.

[21] de Jongh S, Lilien MR, Roodt J, et al. Early statin therapy restores endothelial function in children with familial hypercholesterolemia. J Am Coll Cardiol 2002;40:2117–21.

[22] Charakida M, Donald AE, Terese M, et al. Endothelial dysfunction in childhood infection. Circulation 2005;111:1660–5.

[23] Sorensen KE, Celermajer DS, Georgakopoulos D, et al. Impairment of endothelium-dependent dilation is an early event in children with familial hypercholesterolemia and is related to the lipoprotein(a) level. J Clin Invest 1994;93:50–5.

[24] Modena MG, Bonetti L, Coppi F, et al. Prognostic role of reversible endothelial dysfunction in hypertensive postmenopausal women. J Am Coll Cardiol 2002;40(3):505–10.

[25] Rossi R, Nuzzo A, Origliani G, et al. Prognostic role of flow-mediated dilation and cardiac risk factors in post-menopausal women. J Am Coll Cardiol 2008;51:997–1002.

[26] Hirsch L, Shechter A, Feinberg MS, et al. The impact of early compared to late morning hours on brachial endothelial function and long-term cardiovascular events in healthy subjects with no apparent coronary heart disease. Int J Cardiol 2011;151:342–7.

[27] Shechter M, Isaachar A, Marai I, et al. Long-term association of brachial artery flow-mediated vasodilation and cardiovascular events in middle-aged subjects with no apparent heart disease. Int J Cardiol 2009;134:52–8.

[28] Yeboah J, Delaney JA, Nance R, et al. Predictive value of brachial flow-mediated dilation for incident cardiovascular events in a population-based study: the multi-ethnic study of atherosclerosis. Circulation 2009;120:502–9.

[29] Fichtlscherer S, Breuer S, Zeiher AM. Prognostic value of systemic endothelial dysfunction in patients with acute coronary syndromes: further evidence for the existence of the "vulnerable" patient. Circulation 2004;1(10):1926–32.

[30] Shechter M, Matetzky S, Arad M, et al. Vascular endothelial function predicts mortality risk in patients with advanced ischaemic chronic heart failure. Eur J Heart Fail 2009;11:588–93.

[31] Heitzer T, Baldus S, von Kodolitsch Y, et al. Systemic endothelial dysfunction as an early predictor of adverse outcome in heart failure. Arterioscler Thromb Vasc Biol 2005;25:1174–9.

[32] Kübrich M, Petrakopoulou P, Kofler S, et al. Impact of coronary endothelial dysfunction on adverse long-term outcome after heart transplantation. Transplantation 2008;85:1580–7.

[33] Schächinger V, Britten MB, Zeiher AM. Prognostic impact of coronary vasodilator dysfunction on adverse long-term outcome of coronary heart disease. Circulation 2000;101:1899–906.

[34] Halcox JPJ, Schenke WH, Zalos G, et al. Prognostic value of coronary vascular endothelial dysfunction. Circulation 2002;106:653–8.

[35] Schindler TH, Horning B, Buser PT, et al. Prognostic value of abnormal vasoreactivity of epicardial coronary arteries to sympathetic stimulation in patients with normal coronary angiograms. Arterioscler Thromb Vasc Biol 2003;23:495–501.

[36] Gokce N, Keaney Jr JF, Hunter LM, et al. Risk stratification for postoperative cardiovascular events via noninvasive assessment of endothelial function: a prospective study. Circulation 2002;105:1567–72.

[37] Anderson TJ, Uehata A, Gerhard MD, et al. Close relation of endothelial function in the human coronary and peripheral circulations. J Am Coll Cardiol 1995;26:1235–41.

[38] Yeboah J, Crouse JR, Hsu FC, et al. Brachial flow-mediated dilation predicts incident cardiovascular events in older adults: the Cardiovascular Health Study. Circulation 2007;1(15):2390–7.

[39] Treasure CB, Manoukian SV, Klein JL, et al. Epicardial coronary artery responses to acetylcholine are impaired in hypertensive patients. Circ Res 1992;71:776–81.

[40] Panza JA, Casino PR, Kilcoyne CM, et al. Role of endothelium-derived nitric oxide in the abnormal endothelium-dependent vascular relaxation of patients with essential hypertension. Circulation 1993;87:1468–74.

[41] Panza JA, Quyyumi AA, Brush JE, et al. Abnormal endothelium-dependent vascular relaxation in patients with essential hypertension. N Engl J Med 1990;323:22–7.

[42] Harrison DG, Gongora MC. Oxidative stress and hypertension. Med Clin North Am 2009;93:621–35.

[43] Kizhakekuttu TJ, Widlansky ME. Natural antioxidants and hypertension: promise and challenges. Cardiovasc Ther 2010;28: e20–32.

[44] Widlansky ME, Gokce N, Keaney JF, et al. The clinical implications of endothelial dysfunction. J Am Coll Cardiol 2003;42:1149–60.

[45] Widder JD, Fraccarollo D, Galuppo P, et al. Attenuation of angiotensin II-induced vascular dysfunction and hypertension by overexpression of Thioredoxin 2. Hypertension 2009;54:338–44.

[46] Vecchione C, Carnevale D, Di Pardo A, et al. Pressure-induced vascular oxidative stress is mediated through activation of integrin-linked kinase 1/betaPIX/Rac-1 pathway. Hypertension 2009;54:1028–34.

[47] Doughan AK, Harrison DG, Dikalov SI. Molecular mechanisms of angiotensin II-mediated mitochondrial dysfunction: linking mitochondrial oxidative damage and vascular endothelial dysfunction. Circ Res 2008;102:488–96.

[48] Shoelson SE, Lee J, Goldfine AB. Inflammation and insulin resistance. J Clin Invest 2006;116:1793–801.

[49] Zeng ZH, Zhang ZH, Luo BH, et al. The functional changes of the perivascular adipose tissue in spontaneously hypertensive rats and the effects of atorvastatin therapy. Clin Exp Hypertens 2009;31:355–63.

[50] Marchesi C, Ebrahimian T, Angulo O, et al. Endothelial nitric oxide synthase uncoupling and perivascular adipose oxidative stress and inflammation contribute to vascular dysfunction in

a rodent model of metabolic syndrome. Hypertension 2009;54:1384–92.

[51] Magen E, Feldman A, Cohen Z, et al. Potential link between C3a, C3b and endothelial progenitor cells in resistant hypertension. Am J Med Sci 2010;339:415–9.

[52] Didion SP, Kinzenbaw DA, Schrader LI, et al. Endogenous interleukin-10 inhibits angiotensin II-induced vascular dysfunction. Hypertension 2009;54:619–24.

[53] Hill JM, Zalos G, Halcox JP, et al. Circulating endothelial progenitor cells, vascular function, and cardiovascular risk. N Engl J Med 2003;348:593–600.

[54] Giannotti G, Doerries C, Mocharla PS, et al. Impaired endothelial repair capacity of early endothelial progenitor cells in prehypertension: relation to endothelial dysfunction. Hypertension 2010;55:1389–97.

[55] Hage FG, Oparil S, Xing D, et al. C-reactive protein-mediated vascular injury requires complement. Arterioscler Thromb Vasc Biol 2010;30:1189–95.

[56] Dharmashankar K, Widlansky ME. NIH public access. Curr Hypertens Rep 2011;12:448–55.

[57] Creager MA, Lüscher TF, Cosentino F, et al. Diabetes and vascular disease: pathophysiology, clinical consequences, and medical therapy: part I. Circulation 2003;108:1527–32.

[58] Johnstone MT, Creager SJ, Scales KM, et al. Impaired endothelium-dependent vasodilation in patients with insulin-dependent diabetes mellitus. Circulation 1993;88:2510–6.

[59] Nicolls MR, Haskins K, Flores SC. Oxidant stress, immune dysregulation, and vascular function in type I diabetes. Antioxid Redox Signal 2007;9:879–89.

[60] McVeigh GE, Brennan GM, Johnston GD, et al. Impaired endothelium-dependent and independent vasodilation in patients with type 2 (non-insulin-dependent) diabetes mellitus. Diabetologia 1992;35:771–6.

[61] Benjamin EJ, Larson MG, Keyes MJ, et al. Clinical correlates and heritability of flow-mediated dilation in the community: the Framingham Heart Study. Circulation 2004;109:613–9.

[62] Hamdy O, Ledbury S, Mulloly C, et al. Lifestyle modification improves endothelial function in obese subjects with the insulin resistance syndrome. Diabetes Care 2003;26:2119–25.

[63] Lteif AA, Han K, Mather KJ. Obesity, insulin resistance, and the metabolic syndrome: determinants of endothelial dysfunction in whites and blacks. Circulation 2005;112:32–8.

[64] Hamburg NM, Larson MG, Vita JA, et al. Metabolic syndrome, insulin resistance, and brachial artery vasodilator function in Framingham Offspring participants without clinical evidence of cardiovascular disease. Am J Cardiol 2008;101:82–8.

[65] Meigs JB, Hu FB, Rifai N, et al. Biomarkers of endothelial dysfunction and risk of type 2 diabetes mellitus. JAMA 2004;291:1978–86.

[66] Libby P, Ridker PM, Maseri A. Inflammation and atherosclerosis. Circulation 2002;105:1135–43.

[67] Li H, Cybulsky MI, Gimbrone MA, et al. An atherogenic diet rapidly induces VCAM-1, a cytokine-regulatable mononuclear leukocyte adhesion molecule, in rabbit aortic endothelium. Arterioscler Thromb 1993;13:197–204.

[68] Balletshofer BM, Rittig K, Enderle MD, et al. Endothelial dysfunction is detectable in young normotensive first-degree relatives of subjects with type 2 diabetes in association with insulin resistance. Circulation 2000;101:1780–4.

[69] Caballero AE, Arora S, Saouaf R, et al. Microvascular and macrovascular reactivity is reduced in subjects at risk for type 2 diabetes. Diabetes 1999;48:1856–62.

[70] Tesauro M, Rizza S, Iantorno M, et al. Vascular, metabolic, and inflammatory abnormalities in normoglycemic offspring of patients with type 2 diabetes mellitus. Metabolism 2007;56:413–9.

[71] Rossi R, Cioni E, Nuzzo A, et al. Endothelial-dependent vasodilation and incidence of type 2 diabetes in a population of healthy postmenopausal women. Diabetes Care 2005;28:702–7.

[72] Monti LD, Barlassina C, Citterio L, et al. Endothelial nitric oxide synthase polymorphisms are associated with type 2 diabetes and the insulin resistance syndrome. Diabetes 2003;52:1270–5.

[73] Zinman B, Wanner C, Lachin JM, et al. Empagliflozin, cardiovascular outcomes, and mortality in type 2 diabetes. N Engl J Med 2015;373:2117–28.

[74] Mather KJ, Verma S, Anderson TJ. Improved endothelial function with metformin in type 2 diabetes mellitus. J Am Coll Cardiol 2001;37:1344–50.

[75] Brownfoot FC, Hastie R, Hannan NJ, et al. Metformin as a prevention and treatment for preeclampsia: effects on soluble fms-like tyrosine kinase 1 and soluble endoglin secretion and endothelial dysfunction. Am J Obstet Gynecol 2016;214:356. e1–356, e15.

[76] Egede R, Jensen LO, Hansen HS, et al. Effect of intensive lipid-lowering treatment compared to moderate lipid-lowering treatment with rosuvastatin on endothelial function in high risk patients. Int J Cardiol 2012;158:376–9.

[77] Brownfoot FC, Tong S, Hannan NJ, et al. Effects of pravastatin on human placenta, endothelium, and women with severe preeclampsia. Hypertension 2015;66:687–97.

[78] Oikonomou E, Siasos G, Zaromitidou M, et al. Atorvastatin treatment improves endothelial function through endothelial progenitor cells mobilization in ischemic heart failure patients. Atherosclerosis 2015;238:159–64.

[79] Altun I, Oz F, Arkaya SC, et al. Effect of statins on endothelial function in patients with acute coronary syndrome: a prospective study using adhesion molecules and flow-mediated dilatation. J Clin Med Res 2014;6:354–61.

[80] Harmer JA, Keech AC, Veillard AS, et al. Fenofibrate effects on arterial endothelial function in adults with type 2 diabetes mellitus: A FIELD substudy, for the FIELD Vascular Study Investigators. Atherosclerosis 2015;242:295–302.

[81] Santi D, Granata AR, Guidi A, et al. Six months of daily treatment with vardenafil improves parameters of endothelial inflammation and of hypogonadism in male patients with type 2 diabetes and erectile dysfunction: a randomized, double-blind, prospective trial. Eur J Endocrinol 2016;174:513–22.

[82] Schwarz EI, Puhan MA, Schlatzer C, et al. Effect of CPAP therapy on endothelial function in obstructive sleep apnoea: a systematic review and meta-analysis. Respirology 2015;20:889–95.

[83] Shahin Y, Khan JA, Samuel N, et al. Angiotensin converting enzyme inhibitors effect on endothelial dysfunction: a meta-analysis of randomised controlled trials. Atherosclerosis 2011;216:7–16.

[84] Sola S, Mir MQ, Cheema FA, et al. Irbesartan and lipoic acid improve endothelial function and reduce markers of inflammation in the metabolic syndrome: results of the Irbesartan and Lipoic Acid in Endothelial Dysfunction (ISLAND) study. Circulation 2005;111:343–8.

[85] Bank AJ, Kelly AS, Thelen AM, et al. Effects of carvedilol versus metoprolol on endothelial function and oxidative stress in patients with type 2 diabetes mellitus. Am J Hypertens 2007;20:777–83.

[86] Tzemos N, Lim PO, MacDonald TM. Nebivolol reverses endothelial dysfunction in essential hypertension: a randomized, double-blind, crossover study. Circulation 2001;104:511–4.

[87] Lüscher TF, Pieper M, Tendera M, et al. A randomized placebo-controlled study on the effect of nifedipine on coronary endothelial function and plaque formation in patients with coronary artery disease: the ENCORE II study. Eur Heart J 2009;30: 1590–7.

[88] Targonski PV, Bonetti PO, Pumper GM, et al. Coronary endothelial dysfunction is associated with an increased risk of cerebrovascular events. Circulation 2003;107:2805–9.

[89] Heitzer T, Schlinzig T, Krohn K, et al. Endothelial dysfunction, oxidative stress, and risk of cardiovascular events in patients with coronary artery disease. Circulation 2001;104:2673–8.

[90] Lerman A, Zeiher AM. Endothelial function: cardiac events. Circulation 2005;111:363–8.

[91] Patti G, Pasceri V, Melfi R, et al. Impaired flow-mediated dilation and risk of restenosis in patients undergoing coronary stent implantation. Circulation 2005;111:70–5.

[92] Morishima I, Sone T, Okumura K, et al. Angiographic no-reflow phenomenon as a predictor of adverse long-term outcome in patients treated with percutaneous transluminal coronary angioplasty for first acute myocardial infarction. J Am Coll Cardiol 2000;36:1202–9.

[93] Sorajja P, Gersh BJ, Costantini C, et al. Combined prognostic utility of ST-segment recovery and myocardial blush after primary percutaneous coronary intervention in acute myocardial infarction. Eur Heart J 2005;26:667–74.

[94] Wu KC, Zerhouni EA, Judd RM, et al. Prognostic significance of microvascular obstruction by magnetic resonance imaging in patients with acute myocardial infarction. Circulation 1998;97:765–72.

[95] Prasad A, Stone GW, Aymong E, et al. Impact of ST-segment resolution after primary angioplasty on outcomes after myocardial infarction in elderly patients: an analysis from the CADILLAC trial. Am Heart J 2004;147:669–75.

[96] Ndrepepa G, Tiroch K, Fusaro M, et al. 5-Year prognostic value of no-reflow phenomenon after percutaneous coronary intervention in patients with acute myocardial infarction. J Am Coll Cardiol 2010;55:2383–9.

[97] Galiuto L, Lombardo A, Maseri A, et al. Temporal evolution and functional outcome of no reflow: sustained and spontaneously reversible patterns following successful coronary recanalisation. Heart 2003;89:731–7.

[98] Halcox JP, Donald AE, Ellins E, et al. Endothelial function predicts progression of carotid intima-media thickness. Circulation 2009; 1(19):1005–12.

[99] Lavi S, Bae JH, Rihal CS, et al. Segmental coronary endothelial dysfunction in patients with minimal atherosclerosis is associated with necrotic core plaques. Heart 2009;95:1525–30.

[100] Lerman A, Edwards BS, Hallett JW, et al. Circulating and tissue endothelin immunoreactivity in advanced atherosclerosis. N Engl J Med 1991;325:997–1001.

[101] Stone GW, Maehara A, Lansky AJ, et al. A prospective natural-history study of coronary atherosclerosis. N Engl J Med 2011;364:226–35.

[102] Hasdai D, Gibbons RJ, Holmes DR, et al. Coronary endothelial dysfunction in humans is associated with myocardial perfusion defects. Circulation 1997;96:3390–5.

[103] Lerman A, Holmes DR, Herrmann J, et al. Microcirculatory dysfunction in ST-elevation myocardial infarction: cause, consequence, or both? Eur Heart J 2007;28:788–97.

[104] Angeja BG, de Lemos J, Murphy SA, et al. Impact of diabetes mellitus on epicardial and microvascular flow after fibrinolytic therapy. Am Heart J 2002;144:649–56.

[105] Kurisu S, Inoue I, Kawagoe T, et al. Diabetes mellitus is associated with insufficient microvascular reperfusion following revascularization for anterior acute myocardial infarction. Intern Med 2003;42:554–9.

[106] Celik T, Turhan H, Kursaklioglu H, et al. Impact of metabolic syndrome on myocardial perfusion grade after primary percutaneous coronary intervention in patients with acute ST elevation myocardial infarction. Coron Artery Dis 2006;17:339–43.

[107] Clavijo LC, Pinto TL, Kuchulakanti PK, et al. Metabolic syndrome in patients with acute myocardial infarction is associated with increased infarct size and in-hospital complications. Cardiovasc Revasc Med 2006;7:7–11.

[108] Albertal M, Voskuil M, Piek JJ, et al. Coronary flow velocity reserve after percutaneous interventions is predictive of periprocedural outcome. Circulation 2002;105:1573–8.

[109] Hollenberg SM, Klein LW, Parrillo JE, et al. Coronary endothelial dysfunction after heart transplantation predicts allograft vasculopathy and cardiac death. Circulation 2001;104:3091–6.

[110] Anderson TJ, Meredith IT, Yeung AC, et al. The effect of cholesterol-lowering and antioxidant therapy on endothelium-dependent coronary vasomotion. N Engl J Med 1995;332:488–93.

[111] Treasure CB, Klein JL, Weintraub WS, et al. Beneficial effects of cholesterol-lowering therapy on the coronary endothelium in patients with coronary artery disease. N Engl J Med 1995;332:481–7.

[112] Reriani MK, Dunlay SM, Gupta B, et al. Effects of statins on coronary and peripheral endothelial function in humans: a systematic review and meta-analysis of randomized controlled trials. Eur J Cardiovasc Prev Rehabil 2011;18:704–16.

[113] Bonetti PO, Lerman LO, Napoli C, et al. Statin effects beyond lipid lowering—are they clinically relevant? Eur Heart J 2003;24: 225–48.

[114] Martin-Nizard F, Houssaini HS, Lestavel-Delattre S, et al. Modified low density lipoproteins activate human macrophages to secrete immunoreactive endothelin. FEBS Lett 1991;293:127–30.

[115] Weissberg PL, Witchell C, Davenport AP, et al. The endothelin peptides ET-1, ET-2, ET-3 and sarafotoxin S6b are co-mitogenic with platelet-derived growth factor for vascular smooth muscle cells. Atherosclerosis 1990;85:257–62.

[116] Liao JK, Shin WS, Lee WY, et al. Oxidized low-density lipoprotein decreases the expression of endothelial nitric oxide synthase. J Biol Chem 1995;270:319–24.

[117] Liao JK. Inhibition of Gi proteins by low density lipoprotein attenuates bradykinin-stimulated release of endothelial-derived nitric oxide. J Biol Chem 1994;269:12987–92.

[118] Ohara Y, Peterson TE, Harrison DG. Hypercholesterolemia increases endothelial superoxide anion production. J Clin Invest 1993;91:2546–51.

[119] Alderson LM, Endemann G, Lindsey S, et al. LDL enhances monocyte adhesion to endothelial cells in vitro. Am J Pathol 1986;123:334–42.

[120] Smalley DM, Lin JH, Curtis ML, et al. Native LDL increases endothelial cell adhesiveness by inducing intercellular adhesion molecule-1. Arterioscler Thromb Vasc Biol 1996;16:585–90.

[121] O'Driscoll G, Green D, Taylor RR. Simvastatin, an HMG-coenzyme A reductase inhibitor, improves endothelial function within 1 month. Circulation 1997;95:1126–31.

[122] Goldstein JL, Brown MS. Regulation of the mevalonate pathway. Nature 1990;343:425–30.

[123] Laufs U, Liao JK. Post-transcriptional regulation of endothelial nitric oxide synthase mRNA stability by Rho GTPase. J Biol Chem 1998;273:24266–71.

[124] Kureishi Y, Luo Z, Shiojima I, et al. The HMG-CoA reductase inhibitor simvastatin activates the protein kinase Akt and promotes angiogenesis in normocholesterolemic animals. Nat Med 2000;6:1004–10.

[125] Wassmann S, Laufs U, Bäumer AT, et al. Inhibition of geranylgeranylation reduces angiotensin II-mediated free radical production in vascular smooth muscle cells: involvement of angiotensin AT1 receptor expression and Rac1 GTPase. Mol Pharmacol 2001;59:646–54.

[126] Ferreira Grosso A, de Oliveira SF, Higuchi Mde L, et al. Synergistic anti-inflammatory effect: simvastatin and pioglitazone reduce inflammatory markers of plasma and epicardial adipose tissue of coronary patients with metabolic syndrome. Diabetol Metab Syndr 2014;6:1–8.

[127] Costantine MM, Cleary K, Hebert MF, et al. Safety and pharmacokinetics of pravastatin used for the prevention of preeclampsia in high-risk pregnant women: a pilot randomized controlled trial. Am J Obstet Gynecol 2016;214:720. e1–720, e17.

[128] Rajagopalan S, Harrison DG. Reversing endothelial dysfunction with ACE inhibitors. A new trend. Circulation 1996;94:240–3.

[129] Mancini GB, Henry GC, Macaya C, et al. Angiotensin-converting enzyme inhibition with quinapril improves endothelial vasomotor dysfunction in patients with coronary artery disease. The TREND (Trial on Reversing ENdothelial Dysfunction) Study. Circulation 1996;94:258–65.

[130] Kalinowski L, Dobrucki LW, Szczepanska-Konkel M, et al. Third-generation beta-blockers stimulate nitric oxide release from endothelial cells through ATP efflux: a novel mechanism for antihypertensive action. Circulation 2003;107:2747–52.

[131] Toblli JE, DiGennaro F, Giani JF, et al. Nebivolol: impact on cardiac and endothelial function and clinical utility. Vasc Health Risk Manag 2012;8:151–60.

[132] Khan MU, Zhao W, Zhao T, et al. Nebivolol: a multifaceted antioxidant and cardioprotectant in hypertensive heart disease. J Cardiovasc Pharmacol 2013;62:445–51.

[133] Tang EHC, Vanhoutte PM. Endothelial dysfunction: a strategic target in the treatment of hypertension? Pflugers Arch 2010; 459:995–1004.

[134] ENCORE Investigators, Effect of nifedipine and cerivastatin on coronary endothelial function in patients with coronary artery disease: the ENCORE I Study (Evaluation of Nifedipine and Cerivastatin On Recovery of coronary Endothelial function). Circulation 2003;107:422–8.

[135] Brockow T, Conradi E, Ebenbichler G, et al. The role of mild systemic heat and physical activity on endothelial function in patients with increased cardiovascular risk: results from a systematic review. Forsch Komplementärmed 2006;18:24–30.

[136] Pitocco D, Zaccardi F, Tarzia P, et al. Metformin improves endothelial function in type 1 diabetic subjects: a pilot, placebo-controlled randomized study. Diabetes Obes Metab 2013;15:427–31.

[137] Bibbins-Domingo K, Chertow GM, Coxson PG, et al. Projected effect of dietary salt reductions on future cardiovascular disease. N Engl J Med 2010;362:590–9.

[138] He FJ, MacGregor GA. Effect of longer-term modest salt reduction on blood pressure. Cochrane Database Syst Rev 2004;3:CD004937.

[139] Kopkan L, Majid DSA. Enhanced superoxide activity modulates renal function in NO-deficient hypertensive rats. Hypertension 2006;47:568–72.

[140] Kopkan L, Majid DS. Superoxide contributes to development of salt sensitivity and hypertension induced by nitric oxide deficiency. Hypertension 2005;46:1026–31.

[141] Majid DSA, Kopkan L. Nitric oxide and superoxide interactions in the kidney and their implication in the development of salt-sensitive hypertension. Clin Exp Pharmacol Physiol 2007;34:946–52.

[142] Jablonski KL, Gates PE, Pierce GL, et al. Low dietary sodium intake is associated with enhanced vascular endothelial function in middle-aged and older adults with elevated systolic blood pressure. Ther Adv Cardiovasc Dis 2009;3:347–56.

[143] McCall DO, Mc Gartland CP, McKinley MC, et al. Dietary intake of fruits and vegetables improves microvascular function in hypertensive subjects in a dose-dependent manner. Circulation 2009;119:2153–60.

[144] Widlansky ME, Duffy SJ, Hamburg NM, et al. Effects of black tea consumption on plasma catechins and markers of oxidative stress and inflammation in patients with coronary artery disease. Free Radic Biol Med 2005;38:499–506.

[145] Hambrecht R, Fiehn E, Weigl C, et al. Regular physical exercise corrects endothelial dysfunction and improves exercise capacity in patients with chronic heart failure. Circulation 1998;98: 2709–15.

[146] Clarkson P, Montgomery HE, Mullen MJ, et al. Exercise training enhances endothelial function in young men. J Am Coll Cardiol 1999;33:1379–85.

[147] Hambrecht R, Wolf A, Gielen S, et al. Effect of exercise on coronary endothelial function in patients with coronary artery disease. N Engl J Med 2000;342:454–60.

[148] Dod HS, Bhardwaj R, Sajja V, et al. Effect of intensive lifestyle changes on endothelial function and on inflammatory markers of atherosclerosis. Am J Cardiol 2010;105:362–7.

[149] Meyer AA, Kundt G, Lenschow U, et al. Improvement of early vascular changes and cardiovascular risk factors in obese children after a six-month exercise program. J Am Coll Cardiol 2006;48:1865–70.

[150] Sturm W, Tschoner A, Engl J, et al. Effect of bariatric surgery on both functional and structural measures of premature atherosclerosis. Eur Heart J 2009;30:2038–43.

[151] Gokce N, Vita JA, McDonnell M, et al. Effect of medical and surgical weight loss on endothelial vasomotor function in obese patients. Am J Cardiol 2005;95:266–8.

[152] Sudano I, Spieker LE, Hermann F, et al. Protection of endothelial function: targets for nutritional and pharmacological interventions. J Cardiovasc Pharmacol 2006;47(Suppl 2):S136–50. discussion S172–6.

[153] da Luz P, Nishiyama M, Chagas AC. Drugs and lifestyle for the treatment and prevention of coronary artery disease—comparative analysis of the scientific basis. Braz J Med Biol Res 2011;44:973–91.

[154] Da Luz PL, Fialdini RC, Nishiyama M. Red wine and vascular aging: implications for dementia and cognitive decline. In: - Preedy VR, editor. Diet and nutrition in dementia and cognitive decline. Washington, DC: Academy Press; 2015. p. 943.

[155] Da Luz PL, Coimbra SR. Wine, alcohol and atherosclerosis: clinical evidences and mechanisms. Braz J Med Biol Res 2004;37:1275–95.

[156] Coimbra SR, Lage SH, Brandizzi L, et al. The action of red wine and purple grape juice on vascular reactivity is independent of plasma lipids in hypercholesterolemic patients. Braz J Med Biol Res 2005;38:1339–47.

[157] Lavi S, Prasad A, Yang EH, et al. Smoking is associated with epicardial coronary endothelial dysfunction and elevated white blood cell count in patients with chest pain and early coronary artery disease. Circulation 2007;115:2621–7.

[158] Ganz P, Hsue PY. Individualized approach to the management of coronary heart disease: identifying the nonresponders before it is too late. J Am Coll Cardiol 2009;53:331–3.

Further Reading

[1] Wolfrum S. Endothelium-dependent effects of statins. Arterioscler Thromb Vasc Biol 2003;23:729–36.

49

Physical Exercise and the Endothelium

Carlos Eduardo Negrão, Ana Cristina Andrade, Maria Janieire de Nazaré Nunes Alves, and Allan Robson Kluser Sales

INTRODUCTION

Physical exercise can be defined as a behavior that intensifies the organic functioning to meet energy needs, especially those related to skeletal musculature. It triggers integrated responses involving different systems, including the cardiovascular system, with elevated cardiac output and decreased peripheral vascular resistance in the muscle vascular bed [1,2]. This response remains for hours after the activity—a phenomenon known as acute effect of exercise. The sum of these responses, induced over weeks or months, can lead to chronic adaptations of the cardiovascular system, which favors tissue perfusion, with important implications for both healthy individuals and patients with cardiovascular disease.

This chapter presents the mechanisms involved in blood flow control during exercise, with an emphasis on endothelial function, and the effects of practice for months and years, as health promotion and treatment of cardiovascular diseases.

VASCULAR BLOOD FLOW REGULATORY MECHANISMS DURING THE EXERCISE

Vascular blood flow is controlled by extrinsic and intrinsic mechanisms that act on resistance vessels, located in the small arteries and arterioles [2,3]. The extrinsic mechanism is constituted by the central nervous system, responsible for modulating contractile state of vascular smooth muscle via constrictor sympathetic fibers; the endocrine system, which releases hormone action on vessel specific receptors, and shear stress in the vascular wall.

Resistance vessels are richly innervated by noradrenergic, preganglionic sympathetic fibers that release noradrenaline [3,4]. The action of this neurotransmitter in alpha-2-adrenergic receptors triggers vasoconstriction. The extrinsic control of blood vessels also involves stimulation of alpha-2-adrenergic receptors due to adrenaline released by the medulla of the adrenal gland [3,4]. When stimulated, these receptors cause relaxation of vascular smooth muscle, which results in increased local blood flow. The shear stress may cause the release of substances either as a vasodilator or vasoconstrictor. The increase in perfusion pressure causes hyperpolarization of the cell membrane and increase in nitric oxide production [3–5]. On the other hand, a reduced or swirling action of shear stress (oscillatory) increases the release of vasoconstrictive substances such as endothelin-1 and inflammatory cytokines [6–8].

Myogenic reflex control is one of the most important intrinsic mechanisms of blood flow control. It consists of a self-regulating process that controls blood flow under conditions of sudden changes in perfusion pressure [2,3]. This response is very important because it prevents an excessive increase in local blood flow by a sudden rise in blood pressure and inversely, blood flow reduction due to drop in blood pressure. Local metabolic control, in turn, is dependent on changes in the concentration of local metabolites, such as potassium, adenosine, adenine nucleotides (ATP, ADP, AMP), and hydrogen ions. Increases in concentrations of these metabolites results in vascular relaxation and vasodilatation.

Blood flow control is even more complex during exercise. The increase in blood flow caused by exercise intensifies shear stress or shear force on the surface of the endothelial cells [8], resulting in an increase in nitric oxide production. This process begins with an increase in calcium concentration in the endothelial cell, which acts as a cofactor of the enzyme nitric oxide synthase (NOS) [9,10]. This enzyme converts L-arginine into nitric oxide and citrulline [9,10]. Nitric oxide diffuses into the adjacent vascular smooth muscle, activating guanylatecyclase enzyme, that catalyzes the conversion of guanosine triphosphate (GTP) into cyclic monophosphate guanosine (cGMP). These responses lead to vascular smooth muscle relaxation and vasodilation (Fig. 49.1).

It is important to highlight that the laminar shear stress (unidirectional) caused by exercise differs much from oscillatory shear stress (bidirectional), typical of pathological conditions such as arterial hypertension [11]. While exercise shear stress increases nitric oxide production and has an antiatherogenic role, the pathological shear stress is associated with increased production of inflammatory substances, oxidative stress, and vascular damage [6–8].

Modulation of nitric oxide in muscle vasodilator response during exercise can be demonstrated by the administration of NG-monomethyl L-arginine (L-NMMA). This NOS blocker, infused into the brachial artery, significantly reduces the muscle vasodilator response [3,12]. This reduction in flow is even clearer when the inhibition of nitric oxide production is associated with the blockade of prostaglandin production [2,13]. There is also evidence that adrenaline modulates muscle vasodilator response during exercise. This hormone acts on beta-2-adrenergic receptors, causing intense vasodilation. Intraarterial infusion of propranolol to antagonize adrenaline action significantly restricts muscle vasodilator response during exercise [4,14].

Metabolites produced in the skeletal muscles also contribute to the control of muscle flow at the moment of the activity. However, this mechanism depends on the intensity and duration of the exercise. Decreased oxygen partial pressure, elevated

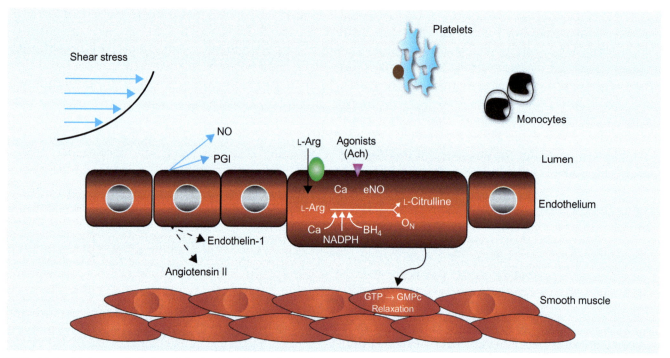

FIG. 49.1　Release of nitric oxide derived from the endothelium. The shear stress increases the concentration of endothelial Ca^{2+}, catalyzing the conversion of L-arginine to citrulline and nitric oxide. Nitric oxide diffuses into adjacent vascular smooth muscle, activating the guanylatecyclase enzyme that catalyzes the conversion of GTP to cGMP, whose result is vascular smooth muscle relaxation. *NO*, nitric oxide; *PGI*, prostacyclin. Adapted from Palmer RM, Ashton DS, Moncada S. Vascular endothelial cells synthesize nitric oxide from L-arginine. Nature 1988;333(6174):664–6.

partial pressure of carbon dioxide and, especially, increased hydrogen concentration, provoke vascular relaxation, favoring increased flow in the muscle territories involved in the movement [2,5]. In the first seconds, muscle vasodilation depends mainly on the increase in extracellular potassium intracellular adenosine concentrations [2,5].

Vasoconstrictor mechanisms also contribute to blood flow control during exercise. They assist in blood pressure regulation and in the increase of blood flow to the skeletal muscles. At the beginning of the exercise, sympathetic activation in the heart and peripheral vessels is triggered by cortical irradiation in the rostroventrolateral region of the medulla (vasomotor center) and by augmentation in skeletal muscle tone. Muscle tonus stimulates mechanoreceptors, which, through an afferent pathway, project their signals into the vasomotor center, further increasing sympathetic nervous activity [15,16]. A second control mechanism of skeletal muscle involves metabolic acidosis. Chemosensitive terminals, known as metaboreceptors, are stimulated and, by an afferent pathway, project in the vasomotor center, further enhancing efferent reflex sympathetic nerve activation [15–17]. This response is critical to blood pressure elevation and even cardiac output during exercise. Taken together, these reflex responses, modulated by skeletal muscle, are known as an ergoreflex [1,15–17]. Sympathetic nerve activation also is important in redistribution of cardiac output. It acts in resistance vessels, in territories not directly involved in exercise, to redirect the cardiac output to the skeletal muscle that is involved in the exercise [1,17,18]. Sympathetic activation also takes place in muscle blood vessels. This response may be demonstrated by increased contralateral forearm flow during exercise with intraarterial infusion of phentolamine (an alpha 1-adrenergic antagonist). However, sympathetic system-mediated vasoconstriction is overcome by the vasodilator mechanisms associated with physical exercise. This mechanism is known as functional sympatholysis [5,19].

Nonadrenergic vasoconstricting mechanisms also contribute to blood flow control in skeletal muscle during exercise. Among them, neurotransmitters ATP and neuropeptide Y (NPY), angiotensin-II, and endothelin-1 are of paramount significance. The neurotransmitters ATP and NPY are released together with noradrenaline at the sympathetic nerve terminals [5,20,21]. Animal studies show that ATP release in interstices by the sympathetic nerve fibers, causes vasoconstriction in vascular smooth muscle via purinergic P2X receptors. In contrast, blockade of P2X receptors during exercise causes vasodilation in skeletal muscle, regardless of alpha-adrenergic stimulation [5]. In relation to NPY, the infusion of an agonist of the receptor Y1 during exercise increases muscle vasoconstriction [5]. This response occurs by a mechanism independent of alpha-adrenergic receptors [5].

Angiotensin-II is a potent vasoconstrictor of the renin-angiotensin system [22]. Its vasoconstrictor action takes place via AT1 receptors in vascular smooth muscle [22–24]. It has an important role in regulating blood flow in muscle contraction. The reduction in flow mediated by AT1 receptors is similar to that observed in the stimulation of alpha-1-adrenergic receptors, albeit by different mechanism [24]. Adrenergic stimulation through alpha-1-adrenergic receptors causes reduction in vessel diameter [24]. Stimulation of angiotensin-II through AT1 receptors causes a reduction in vascular blood velocity [24]. In addition, experimental animal studies show that the vasoconstrictor effect via stimulation of alpha-1-receptors is less evident in precapillary arteriolar level [25], while angiotensin-II is more potent in resistance vessels [26]. These findings suggest a heterogenic distribution of AT1 and alpha-1-receptors in artery beds, with implications in vascular responses.

Endothelin-1 is recognized as one of the most powerful endogen vasoconstrictors [27]. This peptide is synthesized by endothelial cells in response to various physiological stimuli such as shear stress, mechanical stress, hypoxia, and reduced blood pH [5,13]. It acts on ETA and ETB receptors, located at the vascular smooth muscle, causing severe vasoconstriction. Recent studies, both in animals and humans, show that the increase in circulating endothelin-1 concentration depends on exercise intensity [27–29]. The role of endothelin-1 in vascular flow control during exercise was elegantly demonstrated by the blockade of the ETA receptor through BQ-123 (ETA receptor antagonist) [28]. Blocking of this receptor caused increased flow and vascular conductance of the leg during lower limb exercise (knee extension). It was also observed that these responses are related to exercise intensity [28].

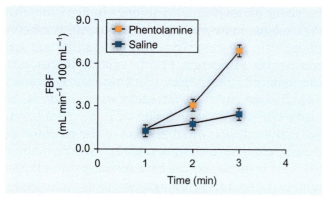

FIG. 49.2 Blood flow in contralateral forearm (FBF) during exercise, with intraarterial administration of phentolamine (alpha-1-adrenergic receptor antagonist) in patients with heart failure. Note that phentolamine significantly increases muscle blood flow. Adapted from Alves MJ, Rondon MU, Santos AC, et al. Sympathetic nerve activity restrains reflex vasodilatation in heart failure. Clin Auton Res 2007;17(6):364–9.

VASCULAR BLOOD FLOW REGULATORY MECHANISMS DURING EXERCISE IN PATHOLOGICAL CONDITIONS

Vascular behavior can be profoundly altered under pathological conditions. One of the most striking examples is heart failure. Patients with this syndrome present intense peripheral vasoconstriction, attributed to neurohumoral exacerbation. This behavior is demonstrated by the significant increase in forearm vasodilatation after infusion of phentolamine in the brachial artery in patients with heart failure [30] (Fig. 49.2).

No less important is the role of endothelium in the vasoconstrictor state in heart failure. Decreased muscle vasodilation in response to acetylcholine infusion and absence of altered muscle blood flow (MBF) after infusion of L-NMMA into the brachial artery are evidence of endothelial dysfunction in patients with heart failure [31]. Endothelial dysfunction has implications for certain physiological behaviors. Patients with heart failure have decreased muscle vasodilator response during exercise and mental stress [31–33]. This vascular behavior was also demonstrated in a recent study on response to hypoxia. Infusion of L-NMMA into the brachial artery virtually eliminates the vasodilatory response to hypoxia in healthy individuals, while in patients with heart failure, the infusion of this drug does not alter the vascular response to hypoxia (Fig. 49.3) [34].

In the heart, chronic stimulation of beta-adrenergic receptors, increases of postinflammatory factors (tumor necrosis factor, interleukin-1 and 6), and the rise in induced nitric oxide synthase expression (iNOS) causes profound changes in microcirculation

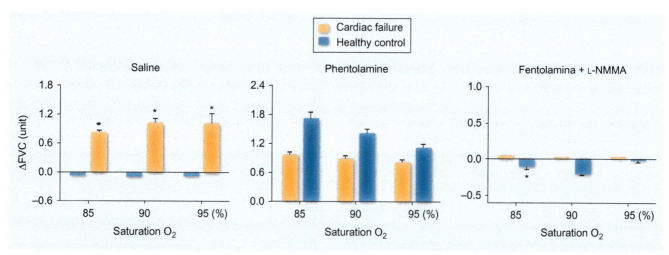

FIG. 49.3 Forearm vascular conductance (FVC) during hypoxia with infusion of saline control, phentolamine (alpha-1-adrenergic receptor antagonist) and phentolamine + L-NMMA (alpha-1-adrenergic receptor antagonist plus nitric-oxide synthase enzyme inhibitor), in patients with heart failure and healthy individuals. Note that phentolamine + L-NMMA reduced FVC in healthy individuals, but not in patients with heart failure. Comparison with a heart failure group, $P < 0.05$. Adapted from Nazare Nunes Alves MJ, dos Santos MR, Nobre TS, et al. Mechanisms of blunted muscle vasodilation during peripheral chemoreceptor stimulation in heart failure patients. Hypertension 2012;60(3):669–76.

and contractile function [35]. These responses lead to a reduction in the number and sensitivity of cardiac beta 1 and beta 2 receptors, aggravating cardiac dysfunction [35]. Furthermore, the exacerbated chronic stimulation of beta-adrenergic receptors on blood vessel increases membrane depolarization in presynaptic terminals and expression of calmodulin and intracellular calcium, which contributes further to the vasoconstrictor state in cardiac failure [36].

Vascular changes in heart failure have clinical implications. Decreased muscle flow limits oxygen supply and energy substrates, contributing to effort intolerance in patients suffering from this syndrome [32,37,38].

EFFECTS OF PHYSICAL TRAINING IN ENDOTHELIAL FUNCTION

Regular physical exercise causes important adaptations in vascular structure (remodeling and angiogenesis) and function. It is also known that the degree of these adaptations may depend on the modality, intensity and duration of physical exercise, and the presence of risk factors for cardiovascular diseases or even cardiovascular disease itself, as will be discussed below.

In Healthy Conditions

Previous studies have shown that a physical training program, performed in cycle ergometer (70% of peak oxygen consumption) for 8 weeks, increases concentrations of nitrite and nitrate (stable end products of nitric oxide degradation) and decreases endothelin-1 concentration in the circulation, which shows an improvement in endothelial function and in vasoconstrictor state, respectively [39]. The authors also observed that plasma levels of nitrite and nitrate and endothelin-1 had returned to pretraining values after 4 weeks of completion of the exercise program, confirming that changes in vascular function were due to physical training. Goto et al. showed that, unlike low- and high-intensity training, moderate-intensity training causes an increase in acetylcholine-mediated vasodilation (endothelium-dependent) [40]. These results suggest that moderate exercise causes more benefit to the endothelial function. More recently,

Spence et al. found that resistance or aerobic training, performed over a period of 6 months at moderate intensity, results in an improvement in function and structure (reduction in wall thickness and increase in vessel light) of conductance arteries [41]. These results are consistent with angiography studies, in which it was shown that athletes had resistance arteries with a greater cross-sectional area than healthy control subjects [42]. Together, these findings suggest that the improvement in architecture and endothelial function of blood vessels depends on the type, time, and intensity of training.

Regarding endothelium-independent function, studies show that administration of sodium nitroprusside by iontophoresis provokes a similar vasodilatory response in trained and sedentary individuals of the same age [36,43]. Similar results were achieved with nitroglycerin [43,44]. These findings show that physical training does not alter vascular function by an endothelium-independent pathway (directly on the vascular smooth muscle).

In the Presence of Risk Factors and Diseases

Studies developed in the last decades have shown that endothelial alterations predict the risk of atherosclerotic disease [6,45]. Thus, improvement in endothelial function produced by physical training should be understood as a strategy to prevent cardiovascular disease [46–49]. The benefits of exercise are explained by changes that occur directly on the vascular wall, especially at the endothelium level, as shown previously, and by changes in cardiovascular risk factors. A meta-analysis suggests that changes in cardiovascular risk factors account for 59% of the improvement in vascular function. The remaining [50] may be attributed to direct effects of shear stress upon the vascular wall, independent of other factors [46,51–53]. This phenomenon is known as vascular conditioning [46,52]. Next, the effects of exercise on endothelial function in the presence of cardiovascular risk factors will be discussed.

Dyslipidemia

Kraus et al. observed changes in the LDL fractions of sedentary men with hypercholesterolemia who underwent 8 months of physical training at varying intensities and volumes (quantity). The high-intensity group showed a decrease in LDL and small LDL particle concentrations, while moderately trained group had an increase in the mean LDL particle size. The influence of training model is even more evident in the HDL particles. That is, the greater the intensity and volume of training, the more significant is the increase in HDL, reaching 9.8% of the initial value [54]. These results were confirmed later in a meta-analysis [55]. Other researchers suggest that the most striking effect of exercise training on lipid profile is related to the LDL kinetics. Trained individuals present increased LDL inflow and outflow from the circulation when compared to untrained individuals [56]. Furthermore, physical training increases the time the LDL remains in a reduced form in the circulation, which means that it takes longer to reach the oxidized form (more atherogenic) [57]. In relation to the relationship of vascular function and lipid profile, studies show that the improvement in vascular response to acetylcholine in trained individuals is not necessarily dependent on changes in the lipid profile [58].

High-Blood Pressure

Hypertension is a disease that causes profound changes in vascular function [59]. The muscle vasodilator responses to intraarterial infusion of acetylcholine is decreased in hypertensive patients and is associated with cardiovascular events [60]. In addition, hypertension alters the muscle vasodilator response during exercise [61] and mental stress [62].

Regular aerobic exercise causes a significant reduction in blood pressure in hypertensive patients [63–66]. Although the mechanisms involved in blood pressure reduction after physical training are not yet fully known, there is consensus that changes achieved in blood vessels are important. The increase in shear force exerted by the blood in the vascular wall during repeated bouts of exercise leads to an increase in nitric oxide release [67–69]. This response contributes to the reduction in blood pressure in hypertensive patients. A recent study shows that 3 months of aerobic physical training, at an intensity of 50% of peak oxygen consumption, performed three times a week, increases the vascular response to intraarterial infusion of acetylcholine

[60]. There are also extrinsic adjustments to the vessel. Exercise reduces sympathetic nervous activity [70] and circulating levels of angiotensin II [71], which decreases peripheral vascular resistance.

Regular practice of resistance exercises also causes blood pressure to drop [72–75]. However, the effect of this type of exercise is smaller than with aerobic exercise [72,73,76]. Furthermore, it is important to emphasize that the increase in blood pressure during high intensity resistance exercise is very intense [77,78]; it is therefore not recommended for patients with cardiovascular disease or elderly. Resistance exercise should be performed at moderate intensity and in a way that does not exceed 10–15 repetitions. Observations with intraarterial measurements show that even moderate resistance exercises can cause a very intense increase in blood pressure if the number of repetitions is high [77–79].

Diabetes

Epidemiological evidences show that reduction in cardiovascular events in patients with risk factors, particularly type II diabetes, is associated with modification in lifestyle [80–84]. A study cohort followed up for 14 years showed that regular physical exercise reduces mortality from all causes [84], in addition to preventing or delaying diabetes development [82,83].

Regarding control of diabetes, studies show that exercise causes reduction in blood glucose and fasting insulin concentrations, leading to a decrease in oral hypoglycemic and insulin doses [81–83]. As far as the effects of physical exercise in vascular function in diabetic patients, there is evidence that an aerobic exercise program combined with resistance exercise causes significant improvement in endothelium-dependent function [85,86]. The vascular response to intraarterial infusion of acetylcholine is significantly increased in diabetic patients undergoing physical training [87]. Physical training also has effects on vascular remodeling. It increases the lumen and reduces the intima-media thickness of peripheral conduction arteries [86], which means a protective effect of atherosclerotic disease that may particularly benefit diabetic patients.

Recent studies suggest that the benefits achieved in the diabetic patient may vary with the intensity of exercise. Glycemic control with high-intensity intermittent training is more significant than continuous moderate training [88–90]. Gibala et al. have shown that high-intensity intermittent training improves fasting blood glucose and insulin sensitivity (measured by oral glucose tolerance test or homeostasis model assessment (HOMA)) [88–90]. Wisloff et al. found that the improvement in endothelium-dependent vasodilation achieved intermittent training in patients with metabolic syndrome is higher than that with moderate training [91]. Although intermittent exercise is still a reason for much questioning regarding safety, recent statistical data shows that an activity corresponding to 175,820 hours of intermittent exercise did not cause any serious event in patients with heart failure and coronary artery disease [77].

Coronary Disease

Physical training is an important nonpharmacological approach in the treatment of patients with coronary artery disease. A classic study has shown that physical training for 4 weeks attenuates (59%) the paradoxical vasoconstricting response coronary arteries to the administration of acetylcholine and increases (29%) the coronary flow reserve in response to intracoronary infusion of adenosine, in patients with coronary artery disease [92]. Studies also show that physical training reduces neointima formation, increases luminal circumference, and causes angiogenesis [93]. This last parameter is associated with the growth factor derived from the endothelium and fibroblasts (VEGF and FGF, respectively). Increased mobilization of endothelial progenitor cells and expression of endothelial nitric oxide synthase (eNOS) has also been described in trained patients [93].

Other important results of physical training are related to the progression of arteriosclerotic disease [94]. Patients with coronary artery disease undergoing physical training associated with a low-fat diet for 1 year, as opposed to the control group that received only diet, showed no progression of coronary disease [59].

The effects of exercise training in coronary vessels are related to energy expenditure and improvement in physical capacity. There is a close link between increases in the diameter of coronary arteries with stenosis and the amount of energy spent per week in physical training [92]. The improvement in angina threshold has high correlation with increases in functional capacity [15,25,26], and the increment in collateral circulation is related to the increase in peak oxygen consumption and workload [15,25].

Cardiac Failure

One of the most striking effects of a physical training program in patients with heart failure is attenuation in peripheral vasoconstriction. This adaptation is due to two factors: reduced neurohumoral and increased endothelial function. Investigations from our group showed that physical training decreases muscle sympathetic nerve activity in patients with heart failure, regardless of sex, age, and presence of a sleep disorder [95–97]. Regarding endothelial function, physical training improves eNOS expression and the vasodilator response to acetylcholine [98]. Decreases in the of VCAM-1 and ICAM-1 [38,99], angiotensin-II [71], and endothelin-1 [100] have also been documented in patients with heart failure. Physical training also induces changes in vascular structure. It decreases wall thickness and increases the vascular diameter, thus decreasing the wall/lumen ratio [101].

The increase in peripheral blood flow in the heart failure achieved by physical training is very important and resembles that caused by medications. The improvement in peripheral circulation contributes to the reduction of pro-inflammatory and oxidative stress [99]. Together, these responses improved skeletal myopathy, whose clinical outcomes are the increase in physical capacity and quality of life.

Final Considerations

Increased muscle blood flow during exercise is an endothelial-dependent response that may be influenced by neurohumoral control, especially in the presence of cardiovascular diseases. Physical training improves endothelium-dependent vascular function. In pathological conditions, this treatment model based on exercise contributes to correction of vascular alterations. It improves endothelial function and reduces neurohumoral activity and levels of angiotensin II, which results in increases in peripheral perfusion, decreased cardiac work and reversal of muscle skeletal myopathy. These responses are critical to cardiovascular health (Fig. 49.4).

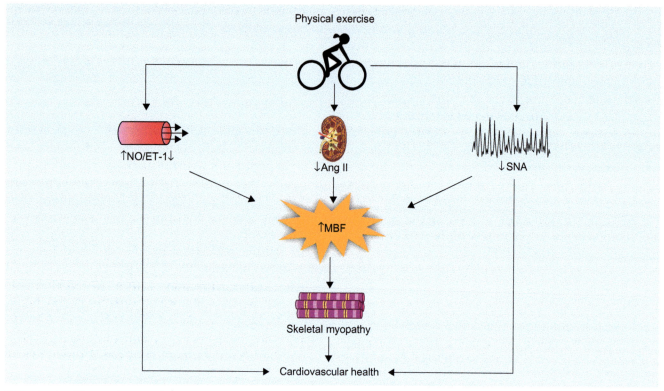

FIG. 49.4 Physical exercise causes a reduction in sympathetic nerve activity (SNA) and improves endothelium-dependent function by increasing nitric oxide bioavailability and decreasing endothelin-1. Together, these effects induce an increase in muscle blood flow (MBF) and an improvement in skeletal myopathy, resulting in improved cardiovascular health of patients with cardiovascular diseases (acronyms, see text).

References

[1] Rowell LB. Ideas about control of skeletal and cardiac muscle blood flow (1876–2003): cycles of revision and new vision. J Appl Physiol (1985) 2004;97(1):384–92.

[2] Joyner MJ, Casey DP. Regulation of increased blood flow (hyperemia) to muscles during exercise: a hierarchy of competing physiological needs. Physiol Rev 2015;95(2):549–601.

[3] Green DJ, O'Driscoll G, Blanksby BA, et al. Control of skeletal muscle blood flow during dynamic exercise: contribution of endothelium-derived nitric oxide. Sports Med 1996;21(2):119–46.

[4] Joyner MJ, Halliwell JR. Sympathetic vasodilation in human limbs. J Physiol 2000;526(Pt 3):471–80.

[5] Holwerda SW, Restaino RM, Fadel PJ. Adrenergic and non-adrenergic control of active skeletal muscle blood flow: implications for blood pressure regulation during exercise. Auton Neurosci 2015;188:24–31.

[6] Chatzizisis YS, Coskun AU, Jonas M, et al. Role of endothelial shear stress in the natural history of coronary atherosclerosis and vascular remodeling: molecular, cellular, and vascular behavior. J Am Coll Cardiol 2007;49(25):2379–93.

[7] Newcomer SC, Thijssen DH, Green DJ. Effects of exercise on endothelium and endothelium/smooth muscle cross talk: role of exercise-induced hemodynamics. J Appl Physiol (1985) 2011;111(1):311–20.

[8] Laughlin MH, Newcomer SC, Bender SB. Importance of hemodynamic forces as signals for exercise-induced changes in endothelial cell phenotype. J Appl Physiol (1985) 2008;104(3):588–600.

[9] Palmer RM, Ashton DS, Moncada S. Vascular endothelial cells synthesize nitric oxide from L-arginine. Nature 1988;333(6174):664–6.

[10] Hemmens B, Mayer B. Enzymology of nitric oxide synthases. Methods Mol Biol 1998;100:1–32.

[11] Paravicini TM, Touyz RM. Redox signaling in hypertension. Cardiovasc Res 2006;71(2):247–58.

[12] Maxwell AJ, Schauble E, Bernstein D, et al. Limb blood flow during exercise is dependent on nitric oxide. Circulation 1998;98(4):369–74.

[13] Joyner MJ, Casey DP. Muscle blood flow, hypoxia, and hypoperfusion. J Appl Physiol (1985) 2014;116(7):852–7.

[14] Dyke CK, Proctor DN, Dietz NM, et al. Role of nitric oxide in exercise hyperaemia during prolonged rhythmic handgripping in humans. J Physiol 1995;488(Pt 1):259–65.

[15] Antunes-Correa LM, Nobre TS, Groehs RV, et al. Molecular basis for the improvement in muscle metaboreflex and mechanoreflex control in exercise-trained humans with chronic heart failure. Am J Physiol Heart Circ Physiol 2014;307(11):H1655–66.

[16] Negrao CE, Middlekauff HR, Gomes-Santos IL, et al. Effects of exercise training on neurovascular control and skeletal myopathy in systolic heart failure. Am J Physiol Heart Circ Physiol 2015;308(8):H792–802.

[17] Rowell LB. Neural control of muscle blood flow: importance during dynamic exercise. Clin Exp Pharmacol Physiol 1997;24(2):117–25.

[18] Fisher JP, Young CN, Fadel PJ. Autonomic adjustments to exercise in humans. Compr Physiol 2015;5(2):475–512.

[19] Saltin B, Mortensen SP. Inefficient functional sympatholysis is an overlooked cause of malperfusion in contracting skeletal muscle. J Physiol 2012;590(Pt 24):6269–75.

[20] Buckwalter JB, Hamann JJ, Clifford PS. Vasoconstriction in active skeletal muscles: a potential role for P2X purinergic receptors? J Appl Physiol (1985) 2003;95(3):953–9.

[21] Buckwalter JB, Hamann JJ, Kluess HA, et al. Vasoconstriction in exercising skeletal muscles: a potential role for neuropeptide Y? Am J Physiol Heart Circ Physiol 2004;287(1):H144–9.

[22] Saris JJ, van Dijk MA, Kroon I, et al. Functional importance of angiotensin-converting enzyme-dependent in situ angiotensin II generation in the human forearm. Hypertension 2000;35 (3):764–8.

[23] Zucker IH, Schultz HD, Patel KP, et al. Modulation of angiotensin II signaling following exercise training in heart failure. Am J Physiol Heart Circ Physiol 2015;308(8):H781–91.

[24] Brothers RM, Haslund ML, Wray DW, et al. Exercise-induced inhibition of angiotensin II vasoconstriction in human thigh muscle. J Physiol 2006;577(Pt 2):727–37.

[25] Faber JE. In situ analysis of alpha-adrenoceptors on arteriolar and venular smooth muscle in rat skeletal muscle microcirculation. Circ Res 1988;62(1):37–50.

[26] Vicaut E, Montalescot G, Hou X, et al. Arteriolar vasoconstriction and tachyphylaxis with intraarterial angiotensin II. Microvasc Res 1989;37(1):28–41.

[27] Wray DW, Nishiyama SK, Donato AJ, et al. Endothelin-1-mediated vasoconstriction at rest and during dynamic exercise in healthy humans. Am J Physiol Heart Circ Physiol 2007;293(4): H2550–6.

[28] Barrett-O'Keefe Z, Ives SJ, Trinity JD, et al. Taming the "sleeping giant": the role of endothelin-1 in the regulation of skeletal muscle blood flow and arterial blood pressure during exercise. Am J Physiol Heart Circ Physiol 2013;304(1):H162–9.

[29] Glaus TM, Grenacher B, Koch D, et al. High altitude training of dogs results in elevated erythropoietin and endothelin-1 serum levels. Comp Biochem Physiol A Mol Integr Physiol 2004;138 (3): 355–61.

[30] Alves MJ, Rondon MU, Santos AC, et al. Sympathetic nerve activity restrains reflex vasodilatation in heart failure. Clin Auton Res 2007;17(6):364–9.

[31] Katz SD, Yuen J, Bijou R, et al. Training improves endothelium-dependent vasodilation in resistance vessels of patients with heart failure. J Appl Physiol (1985) 1997;82(5):1488–92.

[32] Calbet JA, Lundby C. Skeletal muscle vasodilatation during maximal exercise in health and disease. J Physiol 2012;590 (Pt 24):6285–96.

[33] Middlekauff HR, Nguyen AH, Negrao CE, et al. Impact of acute mental stress on sympathetic nerve activity and regional blood flow in advanced heart failure: implications for "triggering" adverse cardiac events. Circulation 1997;96(6):1835–42.

[34] Nazare Nunes Alves MJ, dos Santos MR, Nobre TS, et al. Mechanisms of blunted muscle vasodilation during peripheral chemoreceptor stimulation in heart failure patients. Hypertension 2012; 60(3):669–76.

[35] Murray DR, Prabhu SD, Chandrasekar B. Chronic beta-adrenergic stimulation induces myocardial proinflammatory cytokine expression. Circulation 2000;101(20):2338–41.

[36] Creager MA, Quigg RJ, Ren CJ, et al. Limb vascular responsiveness to beta-adrenergic receptor stimulation in patients with congestive heart failure. Circulation 1991;83(6):1873–9.

[37] Esposito F, Mathieu-Costello O, Shabetai R, et al. Limited maximal exercise capacity in patients with chronic heart failure: partitioning the contributors. J Am Coll Cardiol 2010;55(18):1945–54.

[38] Negrao CE, Middlekauff HR. Adaptations in autonomic function during exercise training in heart failure. Heart Fail Rev 2008; 13(1):51–60.

[39] Maeda S, Miyauchi T, Kakiyama T, et al. Effects of exercise training of 8 weeks and detraining on plasma levels of endothelium-derived factors, endothelin-1 and nitric oxide, in healthy young humans. Life Sci 2001;69(9):1005–16.

[40] Goto C, Higashi Y, Kimura M, et al. Effect of different intensities of exercise on endothelium-dependent vasodilation in humans: role of endothelium-dependent nitric oxide and oxidative stress. Circulation 2003;108(5):530–5.

[41] Spence AL, Carter HH, Naylor LH, et al. A prospective randomized longitudinal study involving 6 months of endurance or resistance exercise. Conduit artery adaptation in humans. J Physiol 2013;591(Pt 5):1265–75.

[42] Currens JH, White PD. Half a century of running. Clinical, physiologic and autopsy findings in the case of Clarence DeMar ("Mr. Marathon"). N Engl J Med 1961;265:988–93.

[43] Green DJ, Spence A, Halliwill JR, et al. Exercise and vascular adaptation in asymptomatic humans. Exp Physiol 2011;96 (2):57–70.

[44] Birk GK, Dawson EA, Atkinson C, et al. Brachial artery adaptation to lower limb exercise training: role of shear stress. J Appl Physiol (1985) 2012;112(10):1653–8.

[45] Huveneers S, Daemen MJ, Hordijk PL. Between Rho(k) and a hard place: the relation between vessel wall stiffness, endothelial contractility, and cardiovascular disease. Circ Res 2015;116 (5):895–908.

[46] Joyner MJ, Green DJ. Exercise protects the cardiovascular system: effects beyond traditional risk factors. J Physiol 2009;587(Pt 23):5551–8.

[47] Malloy MJ. Effects of exercise on coronary atherosclerotic lesions. J Am Coll Cardiol 1993;22(2):478–9.

[48] Blair SN, Morris JN. Healthy hearts and the universal benefits of being physically active: physical activity and health. Ann Epidemiol 2009;19(4):253–6.

[49] Lavie CJ, Arena R, Swift DL, et al. Exercise and the cardiovascular system: clinical science and cardiovascular outcomes. Circ Res 2015;117(2):207–19.

[50] Mora S, Cook N, Buring JE, et al. Physical activity and reduced risk of cardiovascular events: potential mediating mechanisms. Circulation 2007;116(19):2110–8.

[51] Green DJ, Eijsvogels T, Bouts YM, et al. Exercise training and artery function in humans: nonresponse and its relationship to cardiovascular risk factors. J Appl Physiol (1985) 2014;117 (4):345–52.

[52] Green DJ, O'Driscoll G, Joyner MJ, et al. Exercise and cardiovascular risk reduction: time to update the rationale for exercise? J Appl Physiol (1985) 2008;105(2):766–8.

[53] Green DJ, Walsh JH, Maiorana A, et al. Exercise-induced improvement in endothelial dysfunction is not mediated by changes in CV risk factors: pooled analysis of diverse patient populations. Am J Physiol Heart Circ Physiol 2003;285(6): H2679–87.

[54] Kraus WE, Houmard JA, Duscha BD, et al. Effects of the amount and intensity of exercise on plasma lipoproteins. N Engl J Med 2002;347(19):1483–92 [Epub 2002/11/08].

[55] Kodama S, Tanaka S, Saito K, et al. Effect of aerobic exercise training on serum levels of high-density lipoprotein cholesterol: a meta-analysis. Arch Intern Med 2007;167(10):999–1008.

[56] Vinagre CG, Ficker ES, Finazzo C, et al. Enhanced removal from the plasma of LDL-like nanoemulsioncholesteryl ester in trained men compared with sedentary healthy men. J Appl Physiol (1985) 2007;103(4):1166–71.

[57] Ribeiro IC, Iborra RT, Neves MQ, et al. HDL atheroprotection by aerobic exercise training in type 2 diabetes mellitus. Med Sci Sports Exerc 2008;40(5):779–86.

[58] Walsh JH, Yong G, Cheetham C, et al. Effects of exercise training on conduit and resistance vessel function in treated and

untreated hypercholesterolaemic subjects. Eur Heart J 2003;24 (18):1681–9.

[59] Schuler G, Hambrecht R, Schlierf G, et al. Regular physical exercise and low-fat diet. Effects on progression of coronary artery disease. Circulation 1992;86(1):1–11.

[60] Higashi Y, Sasaki S, Kurisu S, et al. Regular aerobic exercise augments endothelium-dependent vascular relaxation in normotensive as well as hypertensive subjects: role of endothelium-derived nitric oxide. Circulation 1999;100(11):1194–202.

[61] Paniagua OA, Bryant MB, Panza JA. Role of endothelial nitric oxide in shear stress-induced vasodilation of human microvasculature: diminished activity in hypertensive and hypercholesterolemic patients. Circulation 2001;103(13):1752–8.

[62] Cardillo C, Kilcoyne CM, Cannon 3rd RO, et al. Impairment of the nitric oxide-mediated vasodilator response to mental stress in hypertensive but not in hypercholesterolemic patients. J Am Coll Cardiol 1998;32(5):1207–13.

[63] Hagberg JM, Montain SJ, Martin 3rd WH, et al. Effect of exercise training in 60- to 69-year-old persons with essential hypertension. Am J Cardiol 1989;64(5):348–53.

[64] Perticone F, Ceravolo R, Pujia A, et al. Prognostic significance of endothelial dysfunction in hypertensive patients. Circulation 2001;104(2):191–6.

[65] Sales AR, Silva BM, Neves FJ, et al. Diet and exercise training reduce blood pressure and improve autonomic modulation in women with prehypertension. Eur J Appl Physiol 2012;112 (9):3369–78.

[66] Whelton SP, Chin A, Xin X, et al. Effect of aerobic exercise on blood pressure: a meta-analysis of randomized, controlled trials. Ann Intern Med 2002;136(7):493–503.

[67] Green DJ, Bilsborough W, Naylor LH, et al. Comparison of forearm blood flow responses to incremental handgrip and cycle ergometer exercise: relative contribution of nitric oxide. J Physiol 2005;562(Pt 2):617–28.

[68] Green DJ, Maiorana A, O'Driscoll G, et al. Effect of exercise training on endothelium-derived nitric oxide function in humans. J Physiol 2004;561(Pt 1):1–25.

[69] Halliwill JR, Buck TM, Lacewell AN, et al. Postexercise hypotension and sustained postexercise vasodilatation: what happens after we exercise? Exp Physiol 2013;98(1):7–18.

[70] Laterza MC, de Matos LD, Trombetta IC, et al. Exercise training restores baroreflex sensitivity in never-treated hypertensive patients. Hypertension 2007;49(6):1298–306.

[71] Gomes-Santos IL, Fernandes T, Couto GK, et al. Effects of exercise training on circulating and skeletal muscle renin-angiotensin system in chronic heart failure rats. PLoS One 2014;9(5)e98012.

[72] Pescatello LS, Franklin BA, Fagard R, et al. American College of Sports Medicine position stand. Exercise and hypertension. Med Sci Sports Exerc 2004;36(3):533–53.

[73] Keese F, Farinatti P, Pescatello L, et al. A comparison of the immediate effects of resistance, aerobic, and concurrent exercise on postexercise hypotension. J Strength Cond Res 2011;25 (5):1429–36.

[74] Moraes MR, Bacurau RF, Simoes HG, et al. Effect of 12 weeks of resistance exercise on post-exercise hypotension in stage 1 hypertensive individuals. J Hum Hypertens 2012;26(9):533–9.

[75] Tibana RA, de Sousa NM, da Cunha Nascimento D, et al. Correlation between acute and chronic 24-hour blood pressure response to resistance training in adult women. Int J Sports Med 2015;36(1): 82–9.

[76] Gomes Anunciacao P, DoederleinPolito M. A review on post-exercise hypotension in hypertensive individuals. Arq Bras Cardiol 2011;96(5):e100–9.

[77] Rognmo O, Moholdt T, Bakken H, et al. Cardiovascular risk of high- versus moderate-intensity aerobic exercise in coronary heart disease patients. Circulation 2012;126(12):1436–40.

[78] MacDougall JD, Tuxen D, Sale DG, et al. Arterial blood pressure response to heavy resistance exercise. J Appl Physiol (1985) 1985;58(3):785–90.

[79] Gomides RS, Costa LA, Souza DR, et al. Atenolol blunts blood pressure increase during dynamic resistance exercise in hypertensives. Br J Clin Pharmacol 2010;70(5):664–73.

[80] Knowler WC, Barrett-Connor E, Fowler SE, et al. Reduction in the incidence of type 2 diabetes with lifestyle intervention or metformin. N Engl J Med 2002;346(6):393–403.

[81] Schreuder TH, Maessen MF, Tack CJ, et al. Life-long physical activity restores metabolic and cardiovascular function in type 2 diabetes. Eur J Appl Physiol 2014;114(3):619–27.

[82] Schuler G, Adams V, Goto Y. Role of exercise in the prevention of cardiovascular disease: results, mechanisms, and new perspectives. Eur Heart J 2013;34(24):1790–9.

[83] Sieverdes JC, Sui X, Lee DC, et al. Physical activity, cardiorespiratory fitness and the incidence of type 2 diabetes in a prospective study of men. Br J Sports Med 2010;44(4):238–44.

[84] Tanasescu M, Leitzmann MF, Rimm EB, et al. Physical activity in relation to cardiovascular disease and total mortality among men with type 2 diabetes. Circulation 2003;107(19):2435–9.

[85] Schreuder TH, Green DJ, Nyakayiru J, et al. Time-course of vascular adaptations during 8 weeks of exercise training in subjects with type 2 diabetes and middle-aged controls. Eur J Appl Physiol 2015;115(1):187–96.

[86] Schreuder TH, Van Den Munckhof I, Poelkens F, et al. Combined aerobic and resistance exercise training decreases peripheral but not central artery wall thickness in subjects with type 2 diabetes. Eur J Appl Physiol 2015;115(2):317–26.

[87] Maiorana A, O'Driscoll G, Cheetham C, et al. The effect of combined aerobic and resistance exercise training on vascular function in type 2 diabetes. J Am Coll Cardiol 2001;38(3):860–6.

[88] Gibala MJ, Little JP, Macdonald MJ, et al. Physiological adaptations to low-volume, high-intensity interval training in health and disease. J Physiol 2012;590(Pt 5):1077–84.

[89] Gillen JB, Little JP, Punthakee Z, et al. Acute high-intensity interval exercise reduces the postprandial glucose response and prevalence of hyperglycaemia in patients with type 2 diabetes. Diabetes Obes Metab 2012;14(6):575–7.

[90] Little JP, Gillen JB, Percival ME, et al. Low-volume high-intensity interval training reduces hyperglycemia and increases muscle mitochondrial capacity in patients with type 2 diabetes. J Appl Physiol (1985) 2011;111(6):1554–60.

[91] Tjonna AE, Lee SJ, Rognmo O, et al. Aerobic interval training versus continuous moderate exercise as a treatment for the metabolic syndrome: a pilot study. Circulation 2008;118(4):346–54.

[92] Hambrecht R, Wolf A, Gielen S, et al. Effect of exercise on coronary endothelial function in patients with coronary artery disease. N Engl J Med 2000;342(7):454–60.

[93] Laufs U, Werner N, Link A, et al. Physical training increases endothelial progenitor cells, inhibits neointima formation, and enhances angiogenesis. Circulation 2004;109(2):220–6.

[94] Niebauer J, Hambrecht R, Velich T, et al. Attenuated progression of coronary artery disease after 6 years of multifactorial risk intervention: role of physical exercise. Circulation 1997;96(8):2534–41.

[95] Antunes-Correa LM, Kanamura BY, Melo RC, et al. Exercise training improves neurovascular control and functional capacity in heart failure patients regardless of age. Eur J Prev Cardiol 2012;19(4):822–9.

[96] Antunes-Correa LM, Melo RC, Nobre TS, et al. Impact of gender on benefits of exercise training on sympathetic nerve activity and muscle blood flow in heart failure. Eur J Heart Fail 2010;12(1): 58–65.

[97] Ueno LM, Drager LF, Rodrigues AC, et al. Effects of exercise training in patients with chronic heart failure and sleep apnea. Sleep 2009;32(5):637–47.

[98] Hambrecht R, Fiehn E, Weigl C, et al. Regular physical exercise corrects endothelial dysfunction and improves exercise capacity in patients with chronic heart failure. Circulation 1998;98(24): 2709–15.

[99] Brum PC, Bacurau AV, Medeiros A, et al. Aerobic exercise training in heart failure: impact on sympathetic hyperactivity and cardiac and skeletal muscle function. Braz J Med Biol Res 2011;44 (9): 827–35.

[100] Thijssen DH, Ellenkamp R, Kooijman M, et al. A causal role for endothelin-1 in the vascular adaptation to skeletal muscle deconditioning in spinal cord injury. Arterioscler Thromb Vasc Biol 2007;27(2):325–31.

[101] Maiorana AJ, Naylor LH, Exterkate A, et al. The impact of exercise training on conduit artery wall thickness and remodeling in chronic heart failure patients. Hypertension 2011;57(1):56–62.

50

Endothelium and Immunological Alterations in Atherosclerosis

Prediman K. Shah

A large amount of experimental and clinical data support the potential role of immune-mediated proinflammatory mechanisms in atherogenesis [1–4]. Cells of both the innate and adaptive immune systems, such as macrophages, dendritic cells (DCs), B and T lymphocytes, and mast cells, are present in atherosclerotic plaques in experimental animals as well as in humans [1–4]. Other components of the immune system such as immune-inflammatory mediators, pathogen-and-danger-associated molecular pattern molecules, immunoglobulins, cytokines, chemokines, and complement proteins are all present to varying degrees in the atherosclerotic plaque [1–4]. There is also experimental evidence suggesting that atheroprone segments of the aorta contain a larger number of antigen presenting dendritic cells especially in the adventitia, even before hyperlipidemia or atherosclerosis is induced, raising the possibility that there is proinflammatory immune priming of atherosusceptible segments of normal aorta [5].

THE POTENTIAL ROLE OF INNATE IMMUNITY IN ATHEROGENESIS

Innate immunity consists of a nonspecific rapid response to neutralize pathogens and other danger signals. Innate immune response is orchestrated by dendritic cells and macrophages, which act as first responders, sampling the host environment to detect molecular signals of damage or danger, such as oxidatively modified low-density lipoprotein (oxLDL) [4]. These molecular patterns of damage or danger act as molecular insults to the vascular wall by activating toll-like receptors which act as pattern-recognition receptors, leading to the activation of proinflammatory genes with the eventual release of inflammatory cytokines that characterize the acute innate immune response. Disruption of genes involved in this innate immune response, such as MyD88 or TLR-4, reduce atherosclerosis, plaque inflammation, and circulating inflammatory cytokines in mice, independent of circulating cholesterol levels [6].

Furthermore, the proatherogenic roles for cells of the innate immune system are demonstrated by the deletion of genes involved in their differentiation and proliferation. Thus, severely hypercholesterolemic mice deficient in the macrophage colony-stimulating factor gene, which is essential for macrophage survival and proliferation, have markedly reduced atherosclerosis [7]. A body of evidence also suggests phenotypic diversity in the mononuclear cells involved in innate immune responses. Experimental studies in mice have identified two broad subtypes of monocytes: the proinflammatory monocytes (M1 subtype) bearing certain specific surface markers and antiinflammatory (M2 subtype) also bearing specific surface markers [8–10]. Although this simplified view of monocyte-macrophages helps investigators to better understand the phenotypic heterogeneity of monocyte-macrophages in vitro, in reality, in vivo phenotypic changes are likely to be much more complex and not quite so discrete [8,10,11]. Bone

711

marrow transplantation experiments as well as cell depletion experiments have also shown that natural killer (NK) cells, yet another component of innate immunity, also contribute atherogenesis [12,13]. The proatherogenic function of NK cells appears to be dependent on mediators such as perforin and granzyme B [13].

Mast cells also accumulate in atherosclerotic lesions and hypercholesterolemic mice lacking mast cells have reduced inflammation and atherosclerotic lesion formation [14–16]. Interestingly, two of these studies reported lower circulating cholesterol levels in mast cell-deficient mice, suggesting a link between innate mast cell signaling and cholesterol homeostasis [14–16].

Dendritic cells are the most efficient antigen presenting cell and help bridge the innate immune system with the adaptive immune response. Resident intimal DCs rapidly ingest lipid in hypercholesterolemic conditions, whereas depleting DCs using CD11c-specific diphtheria toxin receptor (DTR) transgenic mice results in reduced early lesion formation in LDLR null mice supporting a proatherogenic role for conventional DCs [17]. These results are supported by other experiments showing genetic deletion of CD11c reduces atherosclerosis in apoE null mice [18]. Murine and human atherosclerotic lesions contain plasmacytoid dendritic cells (pDCs). Exposure to oxLDL enables pDCs to elicit antigen-specific T cell responses. Depleting pDCs using antimouse plasmacytoid dendritic cell antigen-1 antibody in apoE null mice reduced atherosclerosis in the aortic sinus and was associated with reduced plaque inflammation and global suppression of T cell activation [19]. The atherosclerosis-enhancing role of pDCs was mediated by activation of immune responses to autoantigens through protein deoxyribonucleic acid (DNA) complexes [20].

NEUTROPHILS, NETS, AND ATHEROSCLEROSIS

The potential role of neutrophils in human and experimental atherosclerosis was largely ignored for many years, but is now being increasingly recognized [21]. Several experimental studies have recently highlighted the potential role of neutrophils and neutrophil derived structures, neutrophil extracellular traps (NETS), as mediators of inflammatory responses in atherosclerosis through activation/

priming of NLRP3 inflammasome [22]. Further studies are needed to evaluate fully the potential role of neutrophils as mediators of atherosclerosis and as a potential target for therapy.

THE ROLE OF ADAPTIVE IMMUNITY IN ATHEROSCLEROSIS

In contrast to the rapid, nonspecific nature of the acute innate immune response, the adaptive immune response is slower, more specific and develops over time through stochastic rearrangement during immunoblast development, generating a wide variety of T and B cell receptors that recognize specific antigens [1,3]. Innate immune cells, such as DCs and macrophages, present antigens in the context of the major histocompatibility complex and costimulatory molecules for recognition by T cells. T cell activation occurs upon presentation of the antigen in the setting of an inflammatory state, resulting in clonal proliferation. CD8+ T cell clonal proliferation involves increased cytokine production and cytotoxic function targeted against cells presenting the specific antigen. CD4+ T cell activation also results in cytokine production, which, in turn, skews subsequent B cell activation to produce specific immunoglobulins.

Prior work with B cells suggested a protective role against atherosclerosis in hypercholesterolemic mice [23]. A series of elegant studies by the Witztum group showed that natural antibodies of the immunoglobulin (Ig) M isotype are reactive with the phosphorylcholine head group present in oxidized low-density lipoprotein (LDL), apoptotic cells, and the cell wall of Pneumococcus. These natural antibodies of the IgM isotype attenuated atherosclerosis [24–26], supporting the role of molecular mimicry in atherogenesis. These IgM antibodies are produced by self-renewing B1 cells, lending support to the protective role of this subtype of B cells in atherogenesis [27]. In contrast to these studies, B cell depletion using antiCD20 antibody reduced atherosclerosis [28], suggesting that a more intricate balance of B cell subtypes is likely involved. Studies targeting the B cell activating factor pathway in murine atherosclerosis support this concept [29,30]. B cell activating factor deletion resulted in reduced B2 cells with a preserved B1 cell population, and was associated with reduced atherosclerosis. Thus, cumulative evidence suggests that there is cell subtype specificity in the role of B cells in atherosclerosis, with B1 cells having an atheroprotective effect, whereas B2 cells have a proatherogenic effect.

Human atherosclerotic plaques contain macrophages and DCs, both of which can function as antigen-presenting cells (APCs), as well as T cells that express markers of activation [31]. The evidence of APC-T cell interaction suggests antigen-specific immune activation through immunologic synapses in the plaque [32]. CD4+ T cells were initially reported to have a generalized proatherogenic role. This was supported by HLA-DR restricted oxLDL activation of CD4+ T cells cloned from atherosclerotic plaques [33]. Adoptive transfer of CD4+ T cells into immunodeficient hypercholesterolemic mice aggravated atherosclerosis [34]. The proatherogenic role of CD4+ T cells is, in part, due to the exaggerated proinflammatory effect of the Th1 cytokine response, particularly interferon-γ [35,36]. Increased atherosclerosis after adoptive transfer of CD4+ T cells reactive to LDL supported the postulated role of LDL as an autoantigen for CD4+ T cell responses in atherosclerosis [37]. Other reports suggested that certain Th2 cytokine responses, such as IL-10, are protective against atherosclerosis [38,39]. However, not all Th2 cytokine responses are atheroprotective. Deficiency of IL-4, another prototypical Th2 cytokine, in bone marrow-derived cells achieved by a bone marrow transplant strategy in LDLR null mice resulted in reduced atherosclerosis in specific sites of the vascular tree, suggesting a proatherogenic role of IL-4 [40]. ApoE null mice genetically deficient in IL-4 also developed less atherosclerosis when compared with apoE null mice with normal IL-4 genotype, further supporting a proatherogenic role of IL-4 [41]. In addition to CD4+ Th2 cells, other types of immune secrete Th2 cytokines. The heterogeneous involvement of these many different types of cells, and their possible interaction with Th1 response, makes a unified interpretation of the role of Th2 response in atherogenesis difficult [42]. Another subtype of CD4+ T cells that has gained attention in atherosclerosis is CD4+ regulatory T (Treg) cells [43,44]. The CD4+Treg cells are of two subtypes: natural T-regs or adaptive Tregs. Natural Tregs develop in the thymus with high CD25 expression and specificity against self-antigens, whereas adaptive Tregs develop from mature T-cell populations under antigenic stimulation with variable levels of CD25 expression and specificity against tissue and foreign antigens [45].

However, CD25 is not exclusively a marker for Tregs. A more reliable Treg marker is the transcription factor, FoxP3, which distinguishes CD4+CD25+Treg cells from their CD4+CD25+T effector cell counterparts. Depletion of CD+CD25+Treg cells by deletion of CD80/CD86, CD28 or with CD25 neutralizing antibody results in increased atherosclerosis [44,46]. Selective deletion of FoxP3+ cells using either a DTR transgenic approach or by vaccination against FoxP3 resulted in increased atherosclerosis in hypercholesterolemic mice [47,48]. Thus, experimental evidence from preclinical studies supports the notion that CD4+Tregs have atheroprotective properties. Clinical studies also showed that the number of CD4+CD25+Tregs in peripheral blood from patients with acute coronary syndrome was reduced and their ability to suppress responder CD4+ T cell proliferation was compromised compared with cells from patients with stable angina and subjects with normal coronary arteries [49,50]. Similar reductions of FoxP3 expression in peripheral CD4+CD25+Tregs and decreased frequency of CD4+CD25+FoxP3+ T cells in patients with aortic aneurysm have also been reported [51]. Another prospective study revealed that low levels of baseline CD4+FoxP3+Treg cells were associated with an increased risk of developing acute coronary syndrome 11–14 years later, suggesting that patients with low baseline levels of CD+FoxP3+ T cells have the propensity to develop a higher burden of atherosclerotic coronary disease [52].

Investigators are also keenly interested in the natural killer subtype of T cells (NKT), due to their ability to recognize lipid antigens presented in a CD1-restricted manner. CD1 molecules are present in atherosclerotic plaques and colocalize in areas of the arterial walls containing T cells, indicating potential interaction between CD1+ cells and T cells [53]. Genetic deletion of CD1d in hypercholesterolemic mice reduced lesion size and activation of NKT cells using a synthetic glycolipid, α galactosylceramide, increased atherosclerosis [54–56]. However, not all NKT cells are atherogenic. ApoE null mice rendered NKT cell-deficient by day-3 neonatal thymectomy developed smaller atherosclerotic lesions. Adoptive transfer of CD4+, but not double negative NKT cells, into thymectomized apoE null mice promoted atherosclerotic lesion formation. These data suggest differential roles of NKT cell subtypes in atherogenesis [57].

CD8+ T cells were initially sidelined, partly due to the lack of observed effects on atherogenesis in mice with gene deletions that severely reduced or eliminated CD8+ T cells [58,59]. However, subsequent studies reported that CD8+ T cell activation occurred earlier than CD4+ T cell response in the setting of hypercholesterolemia [60]. A hypercholesterolemic milieu was reported to enhance T cell activation and function in both CD4+ and CD8+ cell subtypes [61]. A study that used depleting antibodies in hypercholesterolemic mice showed a pathogenic role for CD8+ T cells in atherosclerosis, supporting the role of CD8+ T cells in atherogenesis [62]. In a recent prospective clinical study, a higher fraction of CD8+ T cells correlated with characteristics of insulin resistance such as a high-waist-hip ratio and high fasting plasma glucose, insulin, and triglyceride levels. Patients with the two highest tertiles of CD8+ T cells displayed a trend toward increased incidence of coronary events during follow-up for 11–14 years [63]. Additional work with CD8+ T cells in atherogenesis showed complexity in the subtypes involved. CD8+CD25+ T cells adoptively transferred into hypercholesterolemic mice reduced atherosclerosis associated with immunomodulatory functions [64]. Thus, similar to CD4+ T cells, regulatory subtypes of CD8+ T cells may also function to downregulate immune responses in atherosclerotic diseases. This function of CD8+ T cells is in line with the historic reference to this cell type as suppressor T cells. Further investigations are still needed for a clearer understanding of the role of CD8+ T cell subtypes in atherogenesis.

CLINICAL IMPLICATIONS

Immune dysregulation plays an important role in mediating inflammation in atherosclerosis which in turn contributes to initiation and progression of atherosclerosis. Therefore modulation of the immune system with either: (a) activation/enhancement of atheroprotective immunity, perhaps through upragulation of antigen specific T-regs; and/or (b) suppression of atheropromoting immune responses may provide a novel approach to preventing or stabilizing atherosclerosis. Several experimental studies, including those from our laboratories, have shown that immune modulation using apo B-100 related antigens in vaccine formulations hold

promise in this regard and further development and refinements of this idea are likely to pay off [65].

CONCLUSION

As summarized in the preceding text, numerous components of the innate and adaptive immune systems participate in modulating atherogenesis, with intricate interactions among these components that can lead to either worsening or amelioration of atherosclerosis. Given a major role for the adaptive immune system in atherosclerosis, its modulation with vaccination provides an opportunity for a potential atherosclerosis treatment strategy. Several candidate antigens for vaccination are under investigation in our laboratories and those of other investigators, with LDL and its protein component, apoB-100, at the forefront.

References

[1] Chyu KY, Nilsson J, Shah PK. Immune mechanisms in atherosclerosis and potential for an atherosclerosis vaccine. Discov Med 2011;11:403–12.
[2] Libby P, Lichtman AH, Hansson GK. Immune effector mechanisms implicated in atherosclerosis: from mice to humans. Immunity 2013;38:1092–104.
[3] Andersson J, Libby P, Hansson GK. Adaptive immunity and atherosclerosis. Clin Immunol 2010;134:33–46.
[4] Lundberg AM, Hansson GK. Innate immune signals in atherosclerosis. Clin Immunol 2010;134:5–24.
[5] Cybulsky MI, Jongstra-Bilen J. Resident intimal dendritic cells and the initiation of atherosclerosis. Curr Opin Lipidol 2010;21:397–403.
[6] Michelsen KS, Wong MH, Shah PK, et al. Lack of Toll-like receptor 4 or myeloid differentiation factor 88 reduces atherosclerosis and alters plaque phenotype in mice deficient in apolipoprotein E. Proc Natl Acad Sci U S A 2004;101:10679–84.
[7] Rajavashisth T, Qiao JH, Tripathi S, et al. Heterozygous osteopetrotic (op) mutation reduces atherosclerosis in LDL receptor-deficient mice. J Clin Invest 1998;101:2702–10.
[8] Leitinger N, Schulman IG. Phenotypic polarization of macrophages in atherosclerosis. Arterioscler Thromb Vasc Biol 2013;33:1120–6.
[9] Galkina E, Ley K. Leukocyte influx in atherosclerosis. Curr Drug Targets 2007;8:1239–48.
[10] Moore KJ, Sheedy FJ, Fisher EA. Macrophages in atherosclerosis: a dynamic balance. Nat Rev Immunol 2013;13:709–21.
[11] Williams HJ, Fisher EA, Greaves DR. Macrophage differentiation and function in atherosclerosis: opportunities for therapeutic intervention? J Innate Immun 2012;4:498–508.
[12] Whitman SC, Rateri DL, Szilvassy SJ, et al. Depletion of natural killer cell function decreases atherosclerosis in low-density lipoprotein receptor null mice. Arterioscler Thromb Vasc Biol 2004;24:1049–54.
[13] Selathurai A, Deswaerte V, Kanellakis P, et al. Natural killer (NK) cells augment atherosclerosis by cytotoxic-dependent mechanisms. Cardiovasc Res 2014;102:128–37.
[14] Smith DD, Tan X, Raveendran VV, et al. Mast cell deficiency attenuates progression of atherosclerosis and hepatic steatosis in

apolipoprotein E-null mice. Am J Physiol Heart Circ Physiol 2012;302:H2612–21.

[15] Sun J, Sukhova GK, Wolters PJ, et al. Mast cells promote atherosclerosis by releasing proinflammatory cytokines. Nat Med 2007;13:719–24.

[16] Heikkila HM, Trosien J, Metso J, et al. Mast cells promote atherosclerosis by inducing both an atherogenic lipid profile and vascular inflammation. J Cell Biochem 2010;109:615–23.

[17] Paulson KE, Zhu SN, Chen M, et al. Resident intimal dendritic cells accumulate lipid and contribute to the initiation of atherosclerosis. Circ Res 2010;106:383–90.

[18] Wu H, Gower RM, Wang H, et al. Functional role of CD11c+monocytes in atherogenesis associated with hypercholesterolemia. Circulation 2009;119:2708–17.

[19] MacRitchie N, Grassia G, Sabir SR, et al. Plasmacytoid dendritic cells play a key role in promoting atherosclerosis in apolipoprotein E-deficient mice. Arterioscler Thromb Vasc Biol 2012;32:2569–79.

[20] Doring Y, Manthey HD, Drechsler M, et al. Auto-antigenic protein-DNA complexes stimulate plasmacytoid dendritic cells to promote atherosclerosis. Circulation 2012;125:1673–83.

[21] Pende A, Artom N, Bertolotto M, et al. Role of neutrophils in atherogenesis: an update. Eur J Clin Invest 2015; https://doi.org/10.111/eci.12566 [Epub ahead of print, Review].

[22] Leavy O. Inflammation: NETing a one-two-punch. Nat Rev Immunol 2015;15(9):526–7. https://doi.org/10.1038/nri3898 [Epub 2015 July 31].

[23] Caligiuri G, Nicoletti A, Poirier B, et al. Protective immunity against atherosclerosis carried by B cells of hypercholesterolemic mice. J Clin Invest 2002;109:745–53.

[24] Binder CJ, Horkko S, Dewan A, et al. Pneumococcal vaccination decreases atherosclerotic lesion formation: molecular mimicry between *Streptococcus pneumoniae* and oxidized LDL. Nat Med 2003;9:736–43.

[25] Shaw PX, Horkko S, Chang MK, et al. Natural antibodies with the T15 idiotype may act in atherosclerosis, apoptotic clearance, and protective immunity. J Clin Invest 2000;105:1731–40.

[26] Faria-Neto JR, Chyu KY, Li X, et al. Passive immunization with monoclonal IgM antibodies against phosphorylcholine reduces accelerated vein graft atherosclerosis in apolipoprotein E-null mice. Atherosclerosis 2006;189:83–90.

[27] Kyaw T, Tay C, Krishnamurthi S, et al. B1a B lymphocytes are atheroprotective by secreting natural IgM that increases IgM deposits and reduces necrotic cores in atherosclerotic lesions. Circ Res 2011;109:830–40.

[28] Ait-Oufella H, Herbin O, Bouaziz JD, et al. B cell depletion reduces the development of atherosclerosis in mice. J Exp Med 2010;207:1579–87.

[29] Kyaw T, Cui P, Tay C, et al. BAFF receptor mAb treatment ameliorates development and progression of atherosclerosis in hyperlipidemic ApoE(−/−) mice. PLoS One 2013;8:e60430.

[30] Sage AP, Tsiantoulas D, Baker L, et al. BAFF receptor deficiency reduces the development of atherosclerosis in mice—brief report. Arterioscler Thromb Vasc Biol 2012;32:1573–6.

[31] Hansson GK, Jonasson L. The discovery of cellular immunity in the atherosclerotic plaque. Arterioscler Thromb Vasc Biol 2009;29:1714–7.

[32] Koltsova EK, Garcia Z, Chodaczek G, et al. Dynamic T cell-APC interactions sustain chronic inflammation in atherosclerosis. J Clin Invest 2012;122:3114–26.

[33] Stemme S, Faber B, Holm J, et al. T lymphocytes from human atherosclerotic plaques recognize oxidized low density lipoprotein. Proc Natl Acad Sci U S A 1995;92:3893–7.

[34] Zhou X, Nicoletti A, Elhage R, et al. Transfer of CD4(+) T cells aggravates atherosclerosis in immunodeficient apolipoprotein E knockout mice. Circulation 2000;102:2919–22.

[35] Whitman SC, Ravisankar P, Elam H, et al. Exogenous interferon-gamma enhances atherosclerosis in apolipoprotein E−/− mice. Am J Pathol 2000;157:1819–24.

[36] Gupta S, Pablo AM, Xc Jiang, et al. IFN-gamma potentiates atherosclerosis in ApoE knock-out mice. J Clin Invest 1997;99:2752–61.

[37] Zhou X, Robertson AK, Hjerpe C, et al. Adoptive transfer of CD4+ T cells reactive to modified low-density lipoprotein aggravates atherosclerosis. Arterioscler Thromb Vasc Biol 2006;26:864–70.

[38] Pinderski OL, Hedrick CC, Olvera T, et al. Interleukin-10 blocks atherosclerotic events in vitro and in vivo. Arterioscler Thromb Vasc Biol 1999;19:2847–53.

[39] Mallat Z, Besnard S, Duriez M, et al. Protective role of interleukin-10 in atherosclerosis. Circ Res 1999;85:e17–24.

[40] King VL, Szilvassy SJ, Daugherty A. Interleukin-4 deficiency decreases atherosclerotic lesion formation in a site-specific manner in female LDL receptor−/− mice. Arterioscler Thromb Vasc Biol 2002;22:456–61.

[41] Davenport P, Tipping PG. The role of interleukin-4 and interleukin-12 in the progression of atherosclerosis in apolipoprotein E-deficient mice. Am J Pathol 2003;163:1117–25.

[42] Ait-Oufella H, Sage AP, Mallat Z, et al. Adaptive (T and B cells) immunity and control by dendritic cells in atherosclerosis. Circ Res 2014;114:1640–60.

[43] Taleb S, Tedgui A, Mallat Z. Regulatory T-cell immunity and its relevance to atherosclerosis. J Intern Med 2008;263:489–99.

[44] Ait-Oufella H, Salomon BL, Potteaux S, et al. Natural regulatory T cells control the development of atherosclerosis in mice. Nat Med 2006;12:178–80.

[45] Bluestone JA, Abbas AK. Natural versus adaptive regulatory T cells. Nat Rev Immunol 2003;3:253–7.

[46] Gotsman I, Grabie N, Gupta R, et al. Impaired regulatory T-cell response and enhanced atherosclerosis in the absence of inducible costimulatory molecule. Circulation 2006;114:2047–55.

[47] van Es T, van Puijvelde GH, Foks AC, et al. Vaccination against Foxp3(+) regulatory T cells aggravates atherosclerosis. Atherosclerosis 2009;209:74–80.

[48] Klingenberg R, Gerdes N, Badeau RM, et al. Depletion of FOXP3+ regulatory T cells promotes hypercholesterolemia and atherosclerosis. J Clin Invest 2013;123:1323–34.

[49] Cheng X, Yu X, Ding YJ, et al. The Th17/Treg imbalance in patients with acute coronary syndrome. Clin Immunol 2008;127:89–97.

[50] Mor A, Luboshits G, Planer D, et al. Altered status of CD4(+) CD25(+) regulatory T cells in patients with acute coronary syndromes. Eur Heart J 2006;27:2530–7.

[51] Yin M, Zhang J, Wang Y, et al. Deficient CD4+CD25+ T regulatory cell function in patients with abdominal aortic aneurysms. Arterioscler Thromb Vasc Biol 2010;30:1825–31.

[52] Wigren M, Bjorkbacka H, Andersson L, et al. Low levels of circulating CD4+FoxP3+ T cells are associated with an increased risk for development of myocardial infarction but not for stroke. Arterioscler Thromb Vasc Biol 2012;32:2000–4.

[53] Melian A, Geng YJ, Sukhova GK, et al. CD1 expression in human atherosclerosis. A potential mechanism for T cell activation by foam cells. Am J Pathol 1999;155:775–86.

[54] Nakai Y, Iwabuchi K, Fujii S, et al. Natural killer T cells accelerate atherogenesis in mice. Blood 2004;104:2051–9.

[55] Major AS, Wilson MT, McCaleb JL, et al. Quantitative and qualitative differences in proatherogenic NKT cells in apolipoprotein E-deficient mice. Arterioscler Thromb Vasc Biol 2004;24:2351–7.

[56] Tupin E, Nicoletti A, Elhage R, et al. CD1d-dependent activation of NKT cells aggravates atherosclerosis. J Exp Med 2004;199:417–22.

[57] To K, Agrotis A, Besra G, et al. NKT cell subsets mediate differential proatherogenic effects in ApoE−/− mice. Arterioscler Thromb Vasc Biol 2009;29:671–7.

[58] Elhage R, Gourdy P, Brouchet L, et al. Deleting TCR alpha beta+ or CD4+ T lymphocytes leads to opposite effects on site-specific

atherosclerosis in female apolipoprotein E-deficient mice. Am J Pathol 2004;165:2013–8.

[59] Kolbus D, Ljungcrantz I, Soderberg I, et al. TAP1-deficiency does not alter atherosclerosis development in Apoe−/− mice. PLoS One 2012;7:e33932.

[60] Kolbus D, Ramos OH, Berg KE, et al. CD8+ T cell activation predominate early immune responses to hypercholesterolemia in Apoe(/) mice. BMC Immunol 2010;11:58.

[61] Chyu KY, Lio WM, Dimayuga PC, et al. Cholesterol lowering modulates T cell function in vivo and in vitro. PLoS One 2014;9: e92095.

[62] Kyaw T, Winship A, Tay C, et al. Cytotoxic and proinflammatory CD8+ T lymphocytes promote development of vulnerable atherosclerotic plaques in apoE-deficient mice. Circulation 2013;127:1028–39.

[63] Kolbus D, Ljungcrantz I, Andersson L, et al. Association between CD8(+) T-cell subsets and cardiovascular disease. J Intern Med 2013;274:41–51.

[64] Zhou J, Dimayuga PC, Zhao X, et al. CD8(+)CD25(+) T cells reduce atherosclerosis in apoE(−/−) mice. Biochem Biophys Res Commun 2014;443:864–70.

[65] Shah PK, Chyu KY, Dimayuga PC, et al. Vaccine for atherosclerosis. J Am Coll Cardiol 2014;64:2779–91.

Index

Note: Page numbers followed by *f* indicate figures, and *t* indicate tables.

mechanosensors, 87, 88*t*, 89*f*
MicroRNAs and, 175–177
nicotine effects on, 543–545
precursors, 18–19
research methods, 662
senescence in human atherosclerosis, 239*f*
in septic shock, acute activation of, 487–488
under shear stress, physiological responses of, 88–92
Endothelial coronary MESH, 16–17
Endothelial cyclooxygenase, 423–424
Endothelial cytoskeleton reorganization, 189
Endothelial damage, 576
Endothelial disease treatment, 165–166
Endothelial dysfunction (ED), 7–8, 108, 117, 141–142, 153, 177, 218, 231, 239–240, 366–367, 404, 421, 427, 429–430, 462–463, 468, 473–474, 478, 490, 513, 575–576, 610, 612, 632, 639. *See also* Prognostic value, endothelial dysfunction
in acute coronary syndromes, 473–478
angiotensin induces insulin resistance in vessel and, 644
arterial hypertension, prognostic value of, 434
asymmetric di-methylarginine, 633
atherosclerosis, 494
biomarkers, 231–232, 232*f*
in cardiac diseases, 271–273
cardiovascular risk factors, 699–700, 704
delayed endothelialization and, 602
endoplasmic reticulum stress and, 646–647
in general population, 269–271
genesis, 505
impact on cardiovascular events, 684*t*, 699
induced by inflammation in hypertension, 198
insulin resistance, diabetes, 703
interrelation between arterial stiffening and, 241
LOX-1 role in, 287–288*t*
mechanisms of, 274–276
oxidative stress, 632–633
pharmacological influence on, 692*t*, 703
potential mechanism of, 700*f*, 702
redox pathways and atherosclerosis, 505
risk factors and, 514*f*
sleep deprivation and, 532–533
smoking, 540–541
SRA on, 206–209
vascular alterations associated with, 422*t*
Endothelial factors, coagulation activation mediated by, 148–149
Endothelial function. *See also* Exercise
after disturbances, recovery of, 480–481
arterial bed measurements, 367–368
biomarkers of, 424–425
cardiac failure, 705
coronary disease, 705
diabetes, 704–705
dyslipidemia, 703–704
final considerations, 705
genes involved in, 159–165
in healthy conditions, 703
high-blood pressure, 704

in humans, evaluation of, 217–218
integrity, 86–87, 218
neurohumoral control, 705
nitric oxide synthase, 700, 700*f*
nuts, 368, 370–373
olive oil, 368–370
red wine, 368, 373–374 (*see also* Red wine)
risk factors and
aging, 517
arterial hypertension, 518
cigarette smoking, 519
diabetes mellitus, 519–520
family history, 515–516
gender, 517
sedentarism, 516–517
risk factors and diseases, 703
seafood, 368, 374–378
in skin, 675, 677–678
statins mechanisms of, 702*f*
whole dietary pattern and, 366–367
Endothelial glycocalyx, 37–38
Endothelial hyperpolarizing factors (EDHF), 71, 107–111, 456
Endothelial injury, 527–528
Endothelial intercellular adhesion molecule (ICAM)-1, 365
Endothelial isoforms of NO synthases, 632*f*
Endothelialization, 602
Endothelial junctions in postcapillary venules, 40
Endothelial lesions, 426
Endothelial microparticles (EMP), 225, 229–230
Endothelial monolayer, 481
Endothelial nitric oxide synthase (eNOS), 106, 120, 159–160, 176, 292, 344–345, 405–406, 431, 478, 668, 689–690, 705
activation component, 102–103
decoupling, 569
heart failure, 567–570
nitric oxide function, 566–567
nitric oxide synthesis, 566
structure and organization, 567*f*
Endothelial nitric oxide synthase (eNOS), 658
Endothelial platelet adhesion, 148
Endothelial progenitor cells (EPCs), 56–58, 56*f*, 224–225, 407, 530, 702
Endothelial protection, nitric oxide function, 566–567
Endothelial protein C receptor (EPCR), 150
Endothelial vascular cells, 644
Endothelial/vascular pathophysiology model, vascular repair to injury as, 139–141
Endothelin (ET), 73–74, 118–119, 424, 501, 660 system, 165
Endothelin-1 (ET-1), 118, 120*f*, 276, 404, 431, 440, 457, 460, 528–529, 576, 689
role in atheromatosis, 461*f*
vasoconstrictor, 701
Endothelin converting enzyme (ECE), 118, 120*f*
Endothelin receptor antagonists, 8
Endothelium, 85–87, 149–151
activities
assessment methods, 514
normal and dysfunctional, 513
and prognosis, 514–515

anticoagulant properties of, 149–150
and arterial rigidity in hypertension, 433–434
in blood, 224–225
cellular signaling in, 31–35
physiopathology, 611
in pulmonary vascular disease, 439–440, 442–443
redox processes in shear stress induced signal transduction in, 92
shear stress, 85–87
treatment of, 434–436
Endothelium antiapoptosis, high density lipoproteins, 303
Endothelium-dependent arterial dilation, 310, 311*f*
Endothelium-dependent contracting factors (EDCF), 121, 123
Endothelium-dependent hyperpolarization, 108
Endothelium-dependent vasodilating drugs, 221
Endothelium-dependent vasodilation, 57, 97, 217, 238*f*, 239, 248–249, 543*f*
associated with aging, impairment of, 237
mediators, 110–111
Endothelium-dependent vasodilator response, 542
Endothelium-dependent vasorelaxation dysfunction, 103
Endothelium-derived contracting factors, 69–70, 117–123
Adenosine UridineTetraphosphate (Up4A), 122–123
Angiotensin II, 119–120
cyclooxygenase, products of, 117–118
Endothelins, 118–119
Reactive Oxygen Species (ROS), 120–122
Endothelium-derived hyperpolarizing factor (EDHF), 71, 136
Endothelium-derived NO, 194
Endothelium-derived relaxation factor (EDRF), 217
sex hormones on, 69–70
Endothelium function, treatment options for, 276–279
Endothelium independent vasodilation, 237
Endothelium-mediated vasodilation, 241
Endothelium reparation function, high density lipoproteins, 303
Endotoxins, 487
Endovascular elastase (EVE), 440–441
eNOS. *See* Endothelial nitric oxide synthase (eNOS)
Enzymatic superoxide radical sources in vessels, 132–136
EPCs. *See* Endothelial progenitor cells (EPCs)
Epicardial coronary artery disease, 478
Epicardium, 13–16
in coronary genesis, role of, 15–16
Epicardium markers, 17
Epidermal growth factor (EGF) domain, 190
Epigallocatechin, 407
Epigenetics, 171
Epinephrine, 576
EPO. *See* Erythropoietin (EPO)
Erectile cycle, phases of, 629